1983年9月至1986年2月，李曙光在美国麻省理工学院地球与行星科学系 Stan Hart 教授的同位素实验室访问，测定了大别山第一个榴辉岩 Sm-Nd 矿物等时线年龄和祝家铺辉石岩 Sr、Nd 同位素组成，以及华北克拉通新太古代浅变质五台群和中级变质登封群全岩 Sm-Nd、Rb-Sr 年龄。左上图是李曙光在做 Rb、Sr、Sm、Nd 离子交换分离实验；左下图是李曙光在做同位素质谱测定。右下图是唯一一张李曙光与 Hart 教授在实验室的合影。

1991年李曙光、孙卫东与西北大学教师在东秦岭野外

2002年李曙光、刘贻灿、洪吉安陪同张国伟院士考察大别山地质

1986年李曙光（右一）、刘德良（左一）与许靖华（瑞士）（右二）、辛格（土耳其）（右三）、Okay（土耳其）（左三）、罗杰斯（美国）（左二）等外国科学家一起考察大别山地质

右上图，1991年李曙光（左一）与中国科学院地质研究所从柏林（右二）、翟明国（右三）等组队在大别山双河做万分之一地质填图。

右下图，1992年从柏林（右一）、张儒媛（右二）、江博明（左二）、李曙光（左一）坐在苏北青龙山榴辉岩露头上合影。

左图，1993年李曙光在苏北青龙山榴辉岩露头上工作。

右上图，1997年李曙光在西安参加全国同位素地质大会期间与涂光炽先生在陕北延安参观时合影。

右下图，1997年李曙光在西安全国同位素地质大会作大会报告。

左下图，1997年李曙光与丁悌平在陕北黄河壶口瀑布合影。

1998年李曙光在国际同位素地质大会作大会报告

1998年李曙光在国际同位素地质大会招待会上与储雪蕾交谈

1998年李曙光（右）与 E. Jagouz（左）在国际同位素地质大会上合影

2001年李曙光为研究生开了新课——"造山带野外工作方法"，通过穿越大别山纵剖面认识超高压变质岩和大别造山带结构。

右上图为一天野外观察后，在现场进行总结和讨论。

右下图和左下图是两个不同年级的研究生在上该课时的合影。

2001年李曙光与研究生李秋立（左一）和黄方（右一）在大别山天堂寨

2002年研究生李秋立（中），李王晔（左一）和本科生王永刚（右一）在大别山双河

右上图，1998年12月在旧金山美国地球物理学会(AGU)会议与在美留学中国科学技术大学学生李洁、王李平合影。

右下图，1998年12月参观斯坦福大学的SHRIMP离子探针实验室。

左图，1999年11月至2000年1月在香港大学访问。

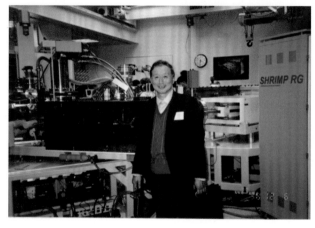

2005年李曙光（右二）获何梁何利基金科学与技术进步奖

2010年12月"变质同位素年代学及华北与华南陆块碰撞过程"获国家自然科学奖二等奖，李曙光代表团队参加国家科学技术奖励大会

安徽省科学技术奖
证　书

为表彰安徽省科学技术奖获得者，特颁发此证书。

项目名称：变质同位素年代学及华北-华南陆块碰撞过程

奖励等级：一等奖

获 奖 者：李曙光

2009年1月4日

证书号：2008-1-R1

国家自然科学奖
证　书

为表彰国家自然科学奖获得者，特颁发此证书。

项目名称：变质同位素年代学及华北与华南陆块碰撞过程

奖励等级：二等

获 奖 者：李曙光(中国科学技术大学)

证书号：2010-Z-104-2-04-R01

李曙光院士论文选集

（卷一）

变质同位素年代学及华北-华南陆块碰撞过程

李曙光　刘贻灿　何永胜　等　编

科学出版社
北京

内 容 简 介

本卷有选择地汇集了李曙光研究团队 1987~2018 年在变质同位素年代学和秦岭-大别-苏鲁造山带的各种变质岩（绿片岩、蓝片岩、斜长角闪岩、麻粒岩、榴辉岩和花岗片麻岩等高压-超高压变质岩）的同位素年代学和地球化学方面的主要学术论文。本卷汇集的论文涉及影响变质岩同位素定年的各种因素，应用同位素年代学和地球化学揭示华北和华南陆块的拼合过程，测定其碰撞时代，探讨陆块碰撞时发生的陆壳深俯冲导致的超高压变质岩的折返过程与机制和流体活动导致的微量元素（Nb/Ta）分异等许多重要科学问题。这些论文有很高的引用率，在国际超高压变质岩年代学与地球化学研究领域有较大影响，并在 2008 年获安徽省科学技术奖一等奖，在 2010 年获国家自然科学奖二等奖。

本文集所收集论文源自多种学术期刊，各源刊格式标准可能不统一，本着尊重历史、忠于原文的精神，所用物理量单位、符号、图例、参考文献等尽量保留了原文风貌。

本书可供地质学领域科技工作者和相关领域教师、研究生与本科生参考。

图书在版编目(CIP)数据

李曙光院士论文选集. 卷一，变质同位素年代学及华北-华南陆块碰撞过程/李曙光等编. —北京：科学出版社，2022.4
ISBN 978-7-03-068589-6

Ⅰ. ①李… Ⅱ. ①李… Ⅲ. ①地球物理学-文集 ②同位素年代学-文集 ③碰撞构造-文集 Ⅳ. ①P3-53 ②P597-53 ③P542.2-53

中国版本图书馆 CIP 数据核字(2021) 第 064201 号

责任编辑：王 运 韩 鹏／责任校对：张小霞
责任印制：肖 兴／封面设计：北京图阅盛世

科学出版社出版
北京东黄城根北街16号
邮政编码：100717
http://www.sciencep.com

中国科学院印刷厂 印刷
科学出版社发行 各地新华书店经销

*

2022年4月第 一 版　开本：889×1194　1/16
2022年4月第一次印刷　印张：39 1/4　插页：6
字数：1 300 000

定价：398.00 元
(如有印装质量问题，我社负责调换)

《李曙光院士论文选集》（卷一）
编委会名单

李曙光　刘贻灿　何永胜

王水炯　朱建明　刘金高

刘盛遨　安诗超　孙卫东

李王晔　李秋立　肖益林

沈　骥　侯振辉　黄　方

前　言

本文集有选择地汇集了李曙光研究团队 1987~2018 年在变质同位素年代学和秦岭–大别–苏鲁造山带的各种变质岩（绿片岩、蓝片岩、斜长角闪岩、麻粒岩、榴辉岩和花岗片麻岩等高压–超高压变质岩）的同位素年代学和地球化学方面的主要学术论文。这些论文涉及研究影响变质岩同位素定年的各种因素，以及应用同位素年代学和地球化学揭示华北与华南陆块的拼合过程，测定其碰撞时代，探讨陆块碰撞时发生的陆壳深俯冲导致的超高压变质岩的折返过程与机制和流体活动导致的微量元素（Nb/Ta）分异等许多重要科学问题。这些论文有很高的引用率，在国际超高压变质岩年代学与地球化学研究领域有较大影响，并在 2008 年获安徽省科学技术奖一等奖，在 2010 年获国家自然科学奖二等奖。

精确测定地壳运动中发生的变质作用时代可恢复地壳运动的温压变化历史，对大陆动力学研究非常重要。1984 年在挪威西部和西阿尔卑斯以及 1989 年在大别山发现了含柯石英超高压变质岩，表明比重轻的大陆地壳可以俯冲到至少 90 km 深的地幔中去，然后折返出露地表。这一大陆动力学的重要发现立即在全世界掀起了超高压变质岩的研究热潮。超高压变质岩的同位素年代学和地球化学研究是其中一个重要研究领域。其重要科学意义涉及 3 个方面：①超高压变质岩是在大陆碰撞时陆壳深俯冲过程中形成的，测定超高压变质岩的同位素变质年龄是确定大陆碰撞时代的最好途径；②超高压变质岩是如何从大于 100 km 的深度快速折返到地表已成为大陆深俯冲研究的著名科学问题，应用同位素年代学精确测定超高压变质岩抬升和折返过程的冷却史是观测其折返过程的最有效手段；③超高压变质岩的地球化学研究对揭示陆壳俯冲和变质过程中的流体活动及其地球化学效应，如导致微量元素（如 Nb/Ta）分异和壳幔相互作用，有重要意义。为此，早在 1983~1986 年李曙光在美国麻省理工学院访问时，就用 Sm-Nd 同位素方法测定了大别山第一个榴辉岩的矿物等时线年龄，给出华北和华南陆块在三叠纪早期碰撞的结论。回国后，1987 年李曙光在国家自然科学基金的资助下率先开展了大别–苏鲁超高压变质岩的同位素年代学和地球化学研究。1989 年李曙光等在《中国科学》和《科学通报》杂志上先后发表了三个大别山榴辉岩和石榴辉石岩的 Sm-Nd 同位素年龄，它们一致证明这些榴辉岩形成于三叠纪早期，并据此提出华北和华南陆块在三叠纪早期碰撞，拼合成中国中–东部大陆。同年，美国斯坦福大学刘忠光团队在 *Geology* 报道大别山榴辉岩中发现了柯石英；安徽省地质科学研究所徐树桐团队与土耳其 Okay 等合作在《欧洲矿物学报》报道了在大别山榴辉岩中发现柯石英。1992 年，徐树桐团队与土耳其 Okay 等合作进一步在大别山超高压变质岩中发现了金刚石，并发表在美国 *Science* 杂志上。这些工作一致证明华北和华南陆块在三叠纪碰撞时也发生了大陆地壳的深俯冲，大别–苏鲁造山带是世界上出露面积最大的超高压变质带，并成为国际研究超高压变质和大陆深俯冲的热点地区。此后，在国家自然科学基金、973 计划和中国科学院知识创新项目等持续资助下，李曙光团队在该领域持续研究了 30 多年。秦岭与大别–苏鲁造山带同是华北–华南陆块碰撞的产物，属同一造山带，为了全面认识华北–华南陆块的碰撞过程，该研究领域扩展到东秦岭与陆块汇聚，碰撞过程相关岩石的同位素年代学和地球化学研究。在这些研究中，变质作用的精确定年是关键，其成功与否在于我们对变质过程中同位素地球化学行为是否有正确的了解和不同变质阶段产物的准确取样和分析。为获得精确的变质年龄和正确理解其地质意义，迫切要求我们对超高压变质作用中同位素体系和定年方法做详细的研究。为此，变质同位素年代学的基础理论和定年方法研究也成为该研究的核心内容之一。本文集汇集的 61 篇论文涉及以下 4 个方面的贡献。

1. 变质同位素年代学的重要发现与贡献

（1）**榴辉岩中多硅白云母的过剩氩的发现**。此前，国际上长期认为白云母不含过剩氩，因而是最适合做 Ar 定年的矿物。然而，在大别山和阿尔卑斯等造山带的高压变质岩定年中都存在多硅白云母 $^{40}Ar/^{39}Ar$ 年龄显著老于 Sm-Nd 或锆石 U-Pb 年龄的矛盾，从而成为国际上存在普遍争议的一个问题。李曙光等应用 Sm-Nd，Rb-Sr 和 $^{40}Ar/^{39}Ar$ 这 3 种不同的同位素体系对同一块含多硅白云母榴辉岩进行定年从而发现并证实了榴辉岩中多硅白云母含大量过剩 Ar，指出它是导致大别山和阿尔卑斯山榴辉岩氩同位素年龄与其他同位素年龄矛盾的主要原因。这不仅解决了这些地区长期存在的年代学争议，而且对正确进行超高压变质岩定年具有普遍性意义。该成果结论在国际一流地球化学杂志 Chemical Geology (Li S G et al., 1993) 上首先报道，并于 1994 年在 Chemical Geology（Li S G et al., 1994）详细全文发表，从而在国际上掀起研究多硅白云母过剩 Ar 的高潮。

（2）**证明了榴辉岩中高压变质矿物与退变质矿物存在 Nd-Sr 同位素不平衡并指出对同位素定年结果的影响**。Sm-Nd 和 Rb-Sr 矿物等时线定年是榴辉岩定年常用的方法，然而文献中报道的同一地区榴辉岩的 Sm-Nd 或 Rb-Sr 年龄有较大差异。参与定年的变质矿物应该在变质时达到 Nd-Sr 同位素平衡是实现 Sm-Nd 和 Rb-Sr 矿物等时线精确定年的基本条件。李曙光等通过详细分析榴辉岩的所有变质矿物证明了超高压变质岩的高压矿物之间同位素可以达到平衡，但它的退变质过程是开放体系，在高压矿物与退变质矿物之间存在 Sr、Nd 同位素不平衡。这一发现揭示了含有退变质矿物的超高压变质矿物及全岩样品参与定年是导致文献中 Sr、Nd 同位素等时线年龄误差增大的主要原因，为正确进行高压变质矿物的定年提供了理论基础。依据这一原则，我们获得的大别山榴辉岩峰期变质年龄为 226±3Ma，已被近年来高精度离子探针锆石 U-Pb 年龄多次证实。该成果是 1998 年第 9 届国际同位素地质大会特邀报告的重要组成部分，并相继发表在 International Geology Review（Li S G et al., 1999）和 Geochimica et Cosmochimica Acta（Li S G et al., 2000）上。

（3）**观察到在低级变质过程中火山岩的全岩 Sm-Nd 同位素体系可以被重置**。此前人们已注意到低级变质作用中稀土元素的活动性，但是它对 Sm-Nd 同位素定年的影响尚不清楚。在太古宙低级变质火山岩的 Sm-Nd、Rb-Sr 年代学研究中发现其全岩 Sm-Nd 同位素体系可以被重置，可以给出与 Rb-Sr 等时线年龄一样的变质年龄。这一发现对正确解释变质岩 Sm-Nd 年龄具有重要意义，如它是测定的南秦岭勉略带蛇绿岩三叠纪 Sm-Nd 年龄被解释为变质时代的重要理论依据。此外，它也成为近年来人们质疑太古宙变质岩 Nd 同位素示踪结果的依据。该成果发表在 Precambrian Research（Li S G et al., 1990）上。

（4）**首次实现榴辉岩中金红石的 U-Pb 精确定年**。榴辉岩中金红石的 U-Pb 封闭温度适中，是测定超高压变质岩冷却 T-t 曲线的关键矿物，然而由于金红石中 U、Pb 含量低，其精确 U-Pb 定年一直是同位素定年中的难题。李秋立在其博士学位论文研究工作中，首次精确测定了大别山含柯石英榴辉岩中金红石的 U-Pb 年龄，并建立了用绿辉石 Pb 同位素组成扣除金红石普通 Pb 的定年方法，从而精确测定了榴辉岩金红石的 U-Pb 年龄，为测定超高压变质岩冷却曲线提供了关键数据。该成果发表在 Chemical Geology（Li Q L et al., 2003）上。

2. 华北与华南陆块的碰撞时代及其拼合过程的同位素年代学证据

华北和华南陆块的碰撞时代与拼合过程是中国大陆动力学研究的关键课题。其中华北和华南陆块的碰撞时代曾长期存在争议，严重影响了我们对中国东部岩石圈演化的理解。陆–陆碰撞可导致陆壳深俯冲和超高压变质岩的形成。测定超高压变质岩的同位素变质年龄是测定陆–陆碰撞时代的最好途径。华北和华南陆块碰撞形成的秦岭–大别–苏鲁造山带有世界上出露面积最大的超高压变质带，从而为进行这一研究提供了条件。此外，陆–陆碰撞过程是一个包括碰撞前洋壳俯冲消减，多岛洋的微陆块拼合和洋盆的最终闭合、陆–陆碰撞与陆壳俯冲、俯冲板片断离和造山带去根与垮塌的复杂过程。

李曙光团队应用变质同位素年代学在解决华北和华南陆块的碰撞时代与拼合过程及秦岭–大别造山带演化这一中国大陆动力学关键研究课题中取得如下重要成果：

(1) **华北与华南陆块的碰撞时代**。最早测定出大别山超高压岩石的 Sm-Nd 年龄为三叠纪，并最早通过 Sr、Nd 同位素示踪证明该带含柯石英榴辉岩是陆壳俯冲成因，从而获得华北与华南陆块在三叠纪最终碰撞的结论。这些成果发表在《中国科学（B 辑）》（李曙光等，1989）、《科学通报》（李曙光等，1989，1992）和 *Chemical Geology*（Li S G et al., 1993）上。

(2) **大陆碰撞前秦岭洋消减及古岛弧的时代**。系统测定了北秦岭丹凤群及二郎坪群古生代岛弧火山岩和侵入体的侵位及变质年龄为 400Ma 左右；查明南秦岭勉–略构造带古洋壳削减发生在石炭纪，及古洋盆闭合时代为早三叠纪。这些工作为秦岭造山带两条地缝合线的厘定及秦岭多岛洋闭合历史和多陆块拼合模型的建立提供了重要依据。研究成果发表在《中国科学》（李曙光等，1989）、《中国科学（D 辑）》（李曙光等，2003）、《地质学报》（李曙光等，1993）、*The Journal of Geology*（Li S G and Sun W D，1996）、*Journal of Asian Earth Sciences*（Sun W D et al., 2002）上。

为解决大别山陆壳俯冲前是否存在洋壳俯冲问题，首次在大别山西北部定远组发现与古生代洋壳俯冲有关的岩浆弧，并测定了其形成时代，可与东秦岭丹凤群对比。将锆石微量元素、包裹体矿物成分和高精度离子探针 U-Pb 定年结合起来系统研究了其南侧的浒湾低温榴辉岩年代学，证明浒湾带低温榴辉岩的高压变质时代为石炭纪，变质岩的原岩为古生代俯冲洋壳；而更南侧的红安地体低温榴辉岩为三叠纪陆壳俯冲成因。该工作解决了国内该类榴辉岩形成时代的争议，发现了大陆碰撞前的洋壳俯冲成因榴辉岩，从而找到了从洋壳俯冲过渡到陆壳俯冲的连接证据。该成果发表在 *Physics and Chemistry of the Earth*（A）（Li S G et al., 2001）和 *Journal of Metamorphic Geology*（Sun W D et al., 2002）上。

(3) **与俯冲板块断离有关的同碰撞花岗岩时代**。系统测定了与俯冲板块断离有关的南秦岭同碰撞花岗岩的锆石 U-Pb 年龄，证明它们的形成时代（225~205 Ma）与大别山超高压变质岩第一次快速冷却时代（226±3~219±7 Ma）一致，从而确定了华南俯冲板块断离的时代，并证明超高压岩石第一次快速抬升与此有关。该成果发表在 *The Journal of Geology*（Sun W D et al., 2002）上。

(4) **碰撞后山根去根时代**。对大别山罗田穹隆的惠兰山基性麻粒岩的年代学和冷却史研究，证明它在早白垩世受上涌地幔的热作用而发生麻粒岩相变质作用，并与罗田穹隆一起发生快速抬升。这一研究揭示的山体快速抬升时代与大别山碰撞后大规模岩浆作用时代的耦合关系，证明早白垩世是大别山山根拆离、垮塌发生的时代。该成果发表在《中国科学（D 辑）》（侯振辉等，2005）。

3. 超高压变质岩多岩板、多阶段差异折返模型的建立及其同位素年代学和地球化学证据

超高压变质岩是如何从大于 100 km 的深度快速折返到地表已成为大陆深俯冲研究的著名科学问题。研究这一问题有助于理解大陆地壳深俯冲与碰撞过程。国外流行观点是俯冲陆壳整体一次快速折返。然而，这种观点很难解释大别–苏鲁超高压变质带由多个不同岩片组成的情况。由于大别–苏鲁超高压变质带是世界上出露面积最大的超高压变质带，因而它可能携带了更为丰富的大陆深俯冲过程信息，为进行这方面研究提供了有利条件。

李曙光团队通过系统工作提出俯冲陆壳多层解耦及多岩片三阶段折返模型并提供了相关证据。这是该团队对超高压变质岩折返机制建立的核心创新理论，它的主要发现和关键证据是：

（1）**大别山超高压榴辉岩及其围岩的冷却史**。首次测定出大别山超高压榴辉岩及其围岩具有相同的二次快速冷却 T-t 曲线，它揭示了超高压变质岩经历了两次快速抬升。据此提出超高压岩石多阶段快速抬升的折返机制，并为证明超高压岩石与围岩是原位关系提供了重要的年代学制约；该成果应邀在1998年第9届国际同位素地质大会上做特邀报告，并发表在地球化学权威学术刊物 *Geochimica et Cosmochimica Acta*（Li S G et al., 2000）上（在2000年 *Geochimica et Cosmochimica Acta* 所刊论文中引用率排名第二），从而为解决国际上长期存在的超高压变质岩与围岩关系的"外来说（Foreign）"与"原地说（In-situ）"的争论提供了直接的同位素年代学证据，有助于大陆碰撞动力学过程的正确理解。

（2）**北大别的大地构造属性和演化过程**。由于1997年以前人们未能在大别山北大别带发现超高压变质岩，因而北大别带的大地构造属性和演化过程一直存在争论并极大影响人们对大别山造山带的形成与演化过程以及超高压变质岩折返机制的正确理解。李曙光团队率先开展北大别高级变质岩的详细年代学、地球化学和岩石学等方面的研究，系统、精确测定了北大别榴辉岩及片麻岩的超高压变质时代为三叠纪，证明北大别带也是华南俯冲陆壳的一部分，而且还论证了北大别带与南大别超高压带在折返过程、峰期变质时代和退变质历史等方面均存在明显差异，说明这两个不同超高压岩片有不同的俯冲和折返历史。这为大别山超高压变质岩多岩板折返机制的建立提供了关键的年代学、岩石学和地球化学方面证据。相关成果发表在 *Journal of Asian Earth Sciences*（Liu Y C et al., 2005）、《科学通报》（刘贻灿和李曙光，2008）、*Journal of Metamorphic Geology*（Liu Y C et al., 2007）和 *Lithos*（Liu Y C et al., 2007）上。

（3）**超高压变质岩的 U-Pb 同位素地球化学——俯冲陆壳多岩片拆分、解耦的证据**。对大别山南大别带和北大别带的 Pb 同位素地球化学填图揭示出，已折返到地表的南大别超高压变质岩具有上地壳特征，而北大别片麻岩具有下地壳特征，并据此提出在俯冲过程中上、下陆壳之间发生拆分、逆冲，导致深俯冲上地壳逆冲上升的折返模型。该文发表在《中国科学（D辑）》（李曙光等，2001; Shen J et al., 2014）。对苏北中国大陆科学深钻 100~2000 m 岩心的 Pb 同位素系统测量发现超高压变质带在垂向上也是由多个岩片叠置而成，并首次指出地壳内的古断层流体活动通道形成的地壳薄弱带是导致俯冲陆壳拆离成薄板的重要因素，支持了上述多岩板解耦理论。这方面成果发表在 *Tectonophysics*（Li S G et al., 2009）、*Geochimica et Cosmochimica Acta*（Shen J et al., 2014）上。

综合上述成果，提出了大别山超高压变质岩的俯冲陆壳内多层解耦和多岩板三阶段差异折返模型。该模型揭示了大陆地壳由于岩石组成和力学性质的不均一性导致的与海洋板块显著不同的俯冲过程。该模型相继发表在《中国科学（D辑）》（李曙光等，2001）、《岩石学报》（刘贻灿和李曙光，2005；李曙光等，2005）和 *Lithos*（Liu Y C et al., 2007）上。

4. **陆壳俯冲过程中的流体和 Nb/Ta 地球化学分异**

发现华南陆壳俯冲与折返过程中均存在流体活动，并首次发现了陆壳俯冲过程中流体活动可导致 Nb/Ta 分异的证据，为揭示壳幔分异过程的 Nb/Ta 之谜提供了重要依据。相关成果论文发表在 *Geochimica et Cosmochimica Acta*（Xiao Y L et al., 2006）上。

为了方便阅读或浏览，该论文选集的每篇论文开头都撰写了该文的"亮点介绍"。由于研究工作的逐步深入，对某些问题的认识也在深化，故少量已经发现错误而放弃或修改的早期论文的某些观点或结论也在"亮点介绍"中指出。

目 录

前言

第一部分　变质同位素年代学

Rb-Sr and Sm-Nd isotopic dating of an Early Precambrian spilite-keratophyre sequence in the Wutaishan area, North China: preliminary evidence for Nd-isotopic homogenization in the mafic and felsic lavas during low-grade metamorphism ··· 3

河南中部登封群全岩 Sm-Nd 同位素年龄及其构造意义 ··· 13

Excess argon in phengite from eclogite: evidence from dating of eclogite minerals by Sm-Nd, Rb-Sr and $^{40}Ar/^{39}Ar$ methods ··· 16

大别山-苏鲁地体超高压变质年代学——Ⅰ. Sm-Nd 同位素体系 ·· 22

大别山-苏鲁地体超高压变质年代学——Ⅱ. 锆石 U-Pb 同位素体系 ······································· 30

构造剪切与变质角闪石中过剩 Ar 的引入：北秦岭丹凤群斜长角闪岩的 Sm-Nd、Rb-Sr 和 $^{40}Ar/^{39}Ar$ 测年证据 ··· 36

Sm-Nd, Rb-Sr and $^{40}Ar/^{39}Ar$ isotopic systematics of the ultrahigh-pressure metamorphic rocks in the Dabie-Sulu Belt, Central China: a retrospective view ·· 42

大别山北部榴辉岩和英云闪长质片麻岩的锆石 U-Pb 年龄及多期变质增生 ··························· 53

青岛仰口榴辉岩的 Nd 同位素不平衡及二次多硅白云母 Rb-Sr 年龄 ······································ 58

Effects of retrograde-zoning of garnet on Sm-Nd isotopic dating of eclogite and oxygen isotopic disequilibrium between eclogitic minerals ··· 63

A high precision U-Pb age of metamorphic rutile in coesite-bearing eclogite from the Dabie Mountains in Central China: a new constraint on the cooling history ·· 72

大别山金河桥榴辉岩矿物 O-Nd-Pb 同位素体系及其对扩散速率的制约 ······························· 83

大别造山带超高压变质岩和镁铁质岩浆岩锆石 U-Pb 年代学的 TIMS 和 SIMS 法定年结果比较 ············ 90

青龙山榴辉岩高压变质新生锆石 SHRIMP U-Pb 定年、微量元素及矿物包裹体研究 ············· 97

超高压变质岩的放射性同位素体系及年代学方法 ·· 105

变质岩同位素年代学：Rb-Sr 和 Sm-Nd 体系 ·· 121

Modification of the Sm-Nd isotopic system in garnet induced by retrogressive fluids ····················· 131

第二部分　华北和华南陆块碰撞时代及拼合过程的同位素年代学和地球化学研究

中国华北、华南陆块碰撞时代的钐-钕同位素年龄证据 ·· 151

大别山北翼大别群中 C 型榴辉岩的 Sm-Nd 同位素年龄及其构造意义 ································ 157

大别山南麓含柯石英榴辉岩的 Sm-Nd 同位素年龄 ··· 160

胶南榴辉岩的形成时代及成因——Sr、Nd 同位素地球化学及年代学证据 ··························· 163

中国中部蓝片岩的形成时代 ··· 167

青岛榴辉岩及胶南群片麻岩的锆石 U-Pb 年龄——胶南群中晋宁期岩浆事件的证据 ············· 172

华北与扬子陆块的碰撞时代及过程 ·· 176

Collision of the North China and Yangtse Blocks and formation of coesite-bearing eclogites: timing and processes ··· 178

北秦岭拉垃庙苏长辉长岩的痕量元素和 Sr，Nd 同位素地球化学 ·································· 198

蛇绿岩生成构造环境的 Ba-Th-Nb-La 判别图 ··· 208

ε_{Nd}-La/Nb、Ba/Nb、Nb/Th 图对地幔不均一性研究的意义——岛弧火山岩分类及 EMII 端元的分解 ········ 215

北秦岭黑河丹凤群岛弧火山岩建造的发现及其构造意义 ··· 222

南秦岭勉略构造带黑沟峡变质火山岩的年代学和地球化学——古生代洋盆及其闭合时代的证据 ····· 229

A middle Silurian-early Devonian magmatic arc in the Qinling Mountains of Central China: a discussion ··········· 236

北秦岭西峡二郎坪群枕状熔岩中一个岩枕的年代学和地球化学研究 ································· 239

Sm-Nd and Rb-Sr ages and geochemistry of volcanics from the Dingyuan Formation in Dabie Mountains, Central China: evidence to the Paleozoic magmatic arc ·· 247

Accretional history of the North and South China Blocks, and the microcontinents between them: implications for dispersion of Gondwanaland ·· 254

大别山北部榴辉岩的地球化学特征和 Sr、Nd 同位素组成及其大地构造意义 ······················· 258

Geochemical and geochronological constraints on the suture location between the North and South China Blocks in the Dabie orogen, Central China ·· 266

Carboniferous and Triassic eclogites in the western Dabie Mountains, east-central China: evidence for protracted convergence of the North and South China Blocks ··· 289

Mid-paleozoic collision in the North Qinling: Sm-Nd, Rb-Sr and $^{40}Ar/^{39}Ar$ ages and their tectonic implications ········ 309

Timing of synorogenic granitoids in the South Qinling, Central China: constraints on the evolution of the Qinling-Dabie orogenic belt ·· 318

南秦岭勉略构造带三岔子古岩浆弧的地球化学特征及形成时代 ····································· 330

大别造山带惠兰山镁铁质麻粒岩 Sm-Nd 和锆石 SHRIMP U-Pb 年代学及锆石微量元素地球化学 ······ 340

西秦岭关子镇蛇绿混杂岩的地球化学和锆石 SHRIMP U-Pb 年龄 ·································· 349

青海东昆南构造带苦海辉长岩和德尔尼闪长岩的锆石 SHRIMP U-Pb 年龄及痕量元素地球化学——对"祁-柴-昆"晚新元古代-早奥陶世多岛洋南界的制约 ·· 358

第三部分　超高压变质岩的冷却史、退变质 *P-T-t* 轨迹与多阶段差异折返机制

大别山石马地区榴辉岩 *P-T-t* 轨迹及其构造意义 ·· 369

P-T-t path for coesite-bearing peridotite-eclogite assciation in the Bixiling, Dabie Mountains ··················· 378

Sm-Nd and Rb-Sr isotopic chronology and cooling history of ultrahigh-pressure metamorphic rocks and their country rocks at Shuanghe in the Dabie Mountains, Central China ·· 381

大别山双河超高压变质岩及北部片麻岩的 U-Pb 同位素组成——对超高压岩石折返机制的制约 ······ 403

Geochemical constraints of the eclogite and granulite facies metamorphism as recognized in the Raobazhai complex from North Dabie Shan, China ·· 410

大别山北部镁铁-超镁铁质岩带中榴辉岩的分布与变质温压条件 ······ 430

大别山北部榴辉岩的大地构造属性及冷却史 ······ 443

Geochemistry and geochronology of eclogites from the northern Dabie Mountains, Central China ······ 448

大别山超高压变质岩的冷却史及折返机制 ······ 464

大别山下地壳岩石及其深俯冲 ······ 472

Zircon SHRIMP U-Pb dating for gneisses in northern Dabie high T/P metamorphic zone, Central China: implications for decoupling within subducted continental crust ······ 480

Ultrahigh-pressure eclogite transformed from mafic granulite in the Dabie orogen, east-central China ······ 497

俯冲陆壳内部的拆离和超高压岩石的多板片差异折返：以大别-苏鲁造山带为例 ······ 513

Common Pb of UHP metamorphic rocks from the CCSD project (100–5000 m) suggesting decoupling between the slices within subducting continental crust and multiple thin slab exhumation ······ 525

A granulite record of multistage metamorphism and REE behavior in the Dabie orogen: constraints from zircon and rock-forming minerals ······ 540

Common Pb isotope mapping of UHP metamorphic zones in Dabie orogen, Central China: implication for Pb isotopic structure of subducted continental crust ······ 568

第四部分　大陆俯冲过程中的流体及Nb-Ta分异

大陆俯冲过程中的流体 ······ 593

Making continental crust through slab melting: constraints from niobium-tantalum fractionation in UHP metamorphic rutile ······ 599

致谢 ······ 616

第一部分　变质同位素年代学

Rb-Sr and Sm-Nd isotopic dating of an Early Precambrian spilite-keratophyre sequence in the Wutaishan area, North China: preliminary evidence for Nd-isotopic homogenization in the mafic and felsic lavas during low-grade metamorphism*

Shuguang Li[1], S.R. Hart[2] and Tieshan Wu[3]

1. Department of Earth and Space Sciences, University of Science and Technology of China, Hefei 230026, Anhui, China
2. Center of Geochemistry, Department of Earth, Atmosphere and Planetary Sciences, Massachusetts Institute of Technology, Cambridge, MA 02139, U.S.A.
3. Regional Geological Surveying Team of Shanxi Province, Yuci 030600, Shanxi, China

亮点介绍：对前寒武纪五台群细碧岩-角斑岩系进行了全岩 Rb-Sr 和 Sm-Nd 等时线定年，发现：使用整个火山岩系的样品进行 Sm-Nd 等时线拟合会给出偏老的"假等时线"年龄；石英角斑岩和细碧岩样品分别拟合给出一致的 Sm-Nd 等时线年龄（1977 ± 385 Ma 和 1981 ± 178 Ma），而且这两个年龄与 Rb-Sr 等时线年龄以及区域绿片岩相变质作用的时代一致，表明在低级变质过程中细碧岩和石英角斑岩能够分别达到 Nd 同位素组成的局部均一化，但不能使整个细碧岩-角斑岩系 Nd 同位素组成均一化。细碧岩和石英角斑岩的初始 ε_{Nd} 值分别为 2.6 ± 3.5 和 3.0 ± 6.1，表明它们可能来自有相同亏损历史的地幔源区。

Abstract The spilite-keratophyre sequence in the Wutaishan area, North China has been dated by zircon from the quartz-keratophyre, and a volcanic crystallization age of 2522 Ma (Liu Dunyi et al., 1985) was obtained. This volcanic rock sequence experienced a low-grade metamorphism of greenschist facies during 1800–2000 Ma. However, combined Sm-Nd whole rock data from the volcanic rock sequence, if interpreted in terms of an isochron, gives a younger age of 2250 ± 182 Ma. If separately calculated, Sm-Nd isochrons for quartz-keratophyre and spilite yield two consistently younger ages of 1977 ± 385 Ma and 1981 ± 178 Ma. These Sm-Nd isochron ages are consistent with a Rb-Sr age of 1871 ± 98 Ma, and with the time of regional greenschist facies metamorphism. Consequently, we interpret the 2250 Ma alignment on the Sm-Nd diagram as an errorchron. The two Sm-Nd metamorphic ages suggest that partial homogenization of Nd isotopic composition in water-rich spilite-keratophyric meta-volcanics can be achieved during low-grade metamorphism as a result of REE mobility.

Using the ε_{Nd} values at the time of metamorphism, the average $^{147}Sm/^{144}Nd$ ratios of the spilite and the quartz-keratophyre, initial ε_{Nd} values of 2.6 ± 3.5 and 3.0 ± 6.1, respectively are estimated at the time of volcanic eruption (2522 Ma).

Similar positive initial ε_{Nd} values for the spilites and quartz-keratophyres suggest that they might have been derived from the same depleted mantle source, or that the quartz-keratophyres are derived from a short-lived mafic source.

1 Introduction

To successfully apply the Sm-Nd isotopic dating method to define eruption ages of Precambrian metavolcanic rock sequences, two basic assumptions need to be satisfied, i.e., (1) samples must have the same initial $^{143}Nd/^{144}Nd$ ratios and (2) rare earth elements (REE) are immobile during metamorphism. However, these two conditions are not always satisfied. Several investigations have demonstrated that volcanic rocks erupted from a single volcano may

---

* 本文发表在：Precambrian Research, 1990, 47: 191-203

show significant isotopic variation as a result of the mantle source heterogeneity and/or crustal contamination (Arculus and Johnson, 1981; Staudigel et al., 1984; Zindler et al., 1984). Furthermore, mobility of REE may be significant during low-grade metamorphism and hydrothermal alteration of basic lavas (Hellman and Henderson, 1977; Hellman et al., 1979; Wood et al., 1976; Nystrom, 1984; Windrim et al., 1984; Vocke et al., 1987).

Several studies have already shown that the Sm-Nd whole rock dating method does not always yield accurate ages for Archaean greenstone belts (Hegner et al., 1984; Cattel et al., 1984; Chauvel et al., 1985). It is also known that under some circumstances Sm-Nd whole-rock isochrons can be reset during high-grade tectonothermal events (McCulloch and Black, 1984; Black and McCulloch, 1987). In this paper, we present Nd and Sr isotopic data for an early Precambrian spilite-keratophyre sequence in the Wutaishan area, Shanxi Province, China to show the effect of the REE mobility during low-grade metamorphism on the Sm-Nd systematics. This is accomplished by comparing results of Sm-Nd, Rb-Sr dates with U-Pb zircon ages.

2 Geological setting and samples

Precambrian metamorphic rock complexes are widely distributed in the Wutaishan area which is a part of North China Craton (Fig. 1). The metamorphic complexes can be broadly subdivided into three stratigraphic units, namely the Fuping, Wutai and Hutuo Groups from bottom to top. All of the three groups have been dated by U-Pb zircon method (Liu et al., 1985) (Fig. 2).

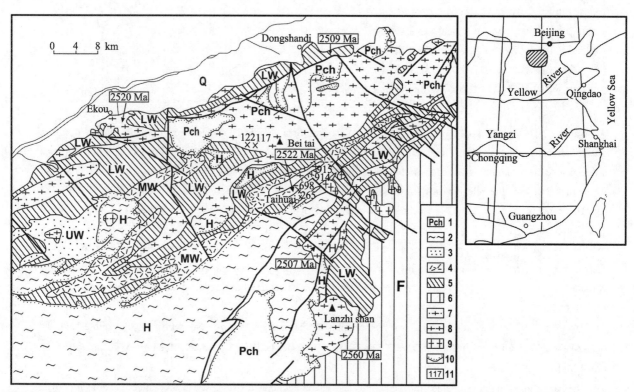

Fig. 1 Simplified sketch map of the Wutaishan area showing Precambrian rock units and sample localities. 1 = Paleozoic strata. 2 = Hutuo Group. 3 = Upper Wutai Group. 4 = Middle Wutai Group. 5 = Lower Wutai Group. 6 = Fuping Group. 7 = K-granite. 8 = Na-granite. 9 = Mesozoic granite. 10 = unconformity. 11 = sample no.

The Fuping Group consists predominantly of mafic granulites, paragneisses, quartzites and amphibolites, which have experienced high to medium grade metamorphism. A minimum age of 2560 ± 9 Ma for the Fuping Group is defined by a zircon age of the Lanzhishan Granite which intrudes the Fuping Group (Liu et al., 1985). A maximum age of 2800 Ma is defined by a zircon date from granulite and amphibolite-grade paragneisses in the Fuping Group (Liu et al., 1985).

The basal part of the Wutai Group unconformably overlies the Fuping Group. From its composition and sedimentological features, the Wutai Group was subdivided into three subgroups separated by two unconformities

(Sun and Lu, 1985). However, recent mapping by the Regional Geological Survey team of Shanxi Province has revised these subdivisions (Table 1).

Liu et al. (1985) have dated zircons from a quartz-keratophyre unit in the Hongmenyan Formation by U-Pb techniques. They obtained a concordia upper intercept age of 2522 ± 17 Ma which was interpreted as a depositional age of the Wutai Group. In contrast to this interpretation, Wang and Bai (1986) and Wu Tieshan (coauthor of this paper) consider the "quartz-keratophyre lava" from which the zircons were extracted as an intrusion (Na-rich granite), and the age of 2522 Ma as a crystallization age of that granite.

According to isotopic ages of the Wutaishan area, the Wutai Group has experienced at least two metamorphic events (Liu et al., 1985).

The first metamorphic event (Wutai movement) took place around 2500 Ma which is defined by a zircon age (2508 ± 2 Ma) derived from the regional metamorphic amphibolitic schist and paragneisses in the Lower Wutai Subgroup (Liu et al., 1985) and a zircon age (2520 ± 30 Ma) derived from the Ekou granite which intrudes the Lower Wutai Subgroup (Liu et al., 1985).

The second metamorphic event (Luliang movement) took place during 1800–2000 Ma, following deposition of the Hutuo Group. This metamorphism is observed both in the Wutai and the overlying Hutuo Groups and its deformation affects the Wutai Group, resulting in parallel regional structures of both the Wutai and Hutuo Groups (Li et al., 1986). K-Ar ages of 1600–2000 Ma for samples from these groups reflect this metamorphic event. Most of the K-Ar ages for the Wutai Group concentrate around 1850 Ma. Whole rock Rb-Sr isochron ages of 1767 Ma and 1851 Ma for the Hutuo Group also record this metamorphic event (Sun and Lu, 1985), which locally caused remelting of older granite (age = 2520 Ma ±). Some of these anatectic granites intrude the Hutuo Group at 1800 Ma. This metamorphic event has also caused Pb loss from zircon and/or recrystallization in the amphibolite- and granulite-facies terrains in the Fuping Group at 1800–2000 Ma (Liu et al., 1985).

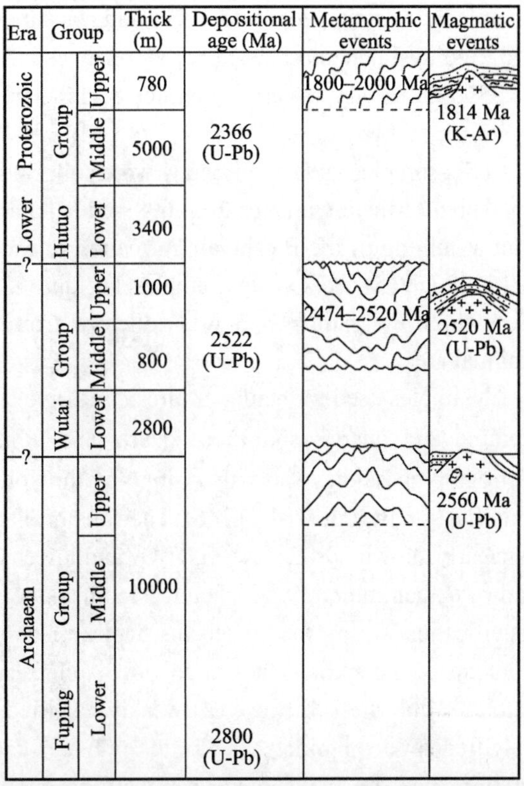

Fig. 2 Schematic cross-section showing the major stratigraphic units and ages of the major geological events of Wutaishan area. The boundary between Archaean and Proterozoic in this area is still controversial.

Table 1 Subdivisions of the Wutai Group

Strata unit		Thickness (m)	Lithological character		Metamorphic facies	
Subgroup	Formation					
Upper (Gaofan) Subgroup		1000	Phylite, metasiltstone and quartzite		greenschist facies	
Middle (Taihai) Subgroup	Hongmenyan Formation	800	meta-spilite-keratophyre sequence		greenschist facies	
	Luzuitou Formation	0–90	sericite-quartzite and feldspathic quartzite			
Lower (Shizui) Subgroup	Baizhiyan and Wenxi Formation	1000	(Baizhiyan Fm.) meta-spilite-keratophyre sequence intercalated with BIF	(Wenxi Fm.) meta-tholeiite and meta-dacite intercalated with BIF	greenschist facies ←→ lateral metamorphic facies change	amphibolite facies
	Zhuangwang Formation	600	biotite-gneiss intercalated with amphibolite		amphibolite facies	
	Jingangku Formation	1000	Biotite-gneiss intercalated with amphibolite and BIF			
	Banyukou Formation	200–600	biotite-gneiss, biotite-quartz schist, tremolite-marble and quartzite			

The Hutuo Group unconformably overlies the Wutai Group. It consists predominantly of clastic, volcanic and chemical (quartzite, dolomite) rocks and has experienced low-grade greenschist-facies metamorphism. Zircons from the Liudingsi metamorphosed basic volcanics in the middle part of Hutuo Group yield a U-Pb zircon concordia age of 2366 ± 103 Ma, which is interpreted as the eruption age of the mafic volcanics of the Hutuo Group by Wu et al. (1986).

Six samples used in this study were collected from three localities in Taihuai and north of Taihuai area (see Fig. 1). Three of them (samples 265, 698 and 914) are quartz-keratophyres and were taken from a metafelsic volcanic unit at the top of the Baizhiyan Formation of the Lower Wutai Subgroup in the Taihuai area. Three other samples (117, 122, 915) are spilites. Samples 117 and 122 from an enclave in a granite at Beitai belong to the Baizhiyan Formation and sample 915 was collected from the Hongmenyan Formation of the Middle Wutai Subgroup in Taihuai area.

Samples used for dating are fresh, massive and without vesicular textures. They mainly contain chlorite, albite, apidote and quartz (spilite) of quartz and albite (quartz-keratophyre). Samples 806-117 also contain calcite. Petrographic studies show that albite of the spilites are often replaced by epidote or calcite, which was termed calcification by Liu et al. (1986). Their major element compositions are given in Table 2. The effect of spilitization is clearly shown by their Na_2O, H_2O^+ and lower CaO contents than metatholeiites (Figs. 3 and 4). Liu et al. (1986) indicate that mineral and chemical compositions of spilite in the Wutai Group are heterogeneous caused by calcification during the greenschist-facies metamorphism. According to mineral assemblages and Na_2O and CaO contents of the spilites, they subdivided spilites in the Wutai Group into three types (see Fig. 4): (1) albite spilites (albite + chlorite + quartz), (2) weakly calcified spilites (chlorite + albite + epidotite + quartz), and (3) strongly calcified spilites (epidote + chlorite + albite + calcite + quartz). Our samples are included in the strongly calcified group (Fig. 4), which suggests that the mobility of Na_2O and CaO was significant during the regional greenschist-facies metamorphism.

Table 2 Major elements compositions of rocks from Wutaishan area, North China[a] (wt.%)

Sample No.	806-915	806-117	806-122	806-914	806-698	806-265
Rock type	spilite	spilite	spilite	quartz-keratophyre	quartz-keratophyre	quartz-keratophyre
SiO_2	52.06	47.04	51.28	73.72	71.64	73.80
TiO_2	0.65	0.53	0.86	0.08	0.15	0.10
Al_2O_3	17.07	14.40	14.62	14.84	15.12	15.03
Fe_2O_3	1.80	1.81	3.34	0.66	0.76	0.44
FeO	6.51	7.03	7.68	0.55	1.01	0.60
MnO	0.13	0.16	0.14	0.02	0.03	0.01
MgO	6.78	8.22	6.86	0.44	0.70	0.48
CaO	7.66	7.88	7.88	0.82	2.34	1.00
Na_2O	3.25	3.72	3.28	5.40	5.39	5.62
K_2O	0.64	0.15	0.32	1.90	1.32	1.63
P_2O_5	0.08	0.04	0.10	0.08	0.08	0.07
H_2O^+	2.34	4.30	2.08	0.88	0.60	0.74
CO_2	0.10	4.04	0.10	0.10	0.10	0.10
Total	99.07	99.32	98.54	99.49	99.24	99.62
CaO/Na_2O	2.36	2.12	2.40	0.15	0.43	0.18

[a] The data were obtained by gravimetric analysis (SiO_2, H_2O^+), volumetric analysis (Al_2O_3, Fe_2O_3, FeO), atomic absorption spectrum (CaO, MgO, Na_2O, K_2O), photoelectric colorimeter (P_2O_5, TiO_2, MnO) and nonaqueous titration (CO_2) from the analytical center of Geological Bureau, Anhui Province.

3 Sample preparation and analytical techniques

Rock samples were wrapped in cotton cloth and plastic (to avoid metal contamination) and broken into approximately 1–2 cm size species. Approximately 100 g of material were selected, avoiding altered portions and attempting to obtain a fresh and representative sample. These pieces were ultrasonically cleaned in H_2O, then crushed to a powder (–200 mesh) using a tungstencarbide shatterbox. The shatterbox was carefully cleaned by H_2O before crushing a new sample.

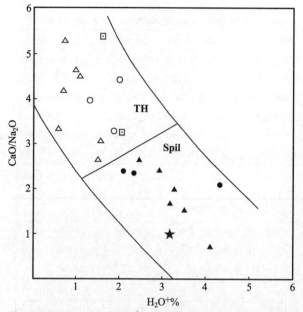 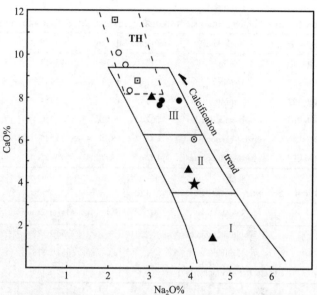

Fig. 3 CaO/Na_2O vs. H_2O^+% diagram for metatholeiite (TH) and spilite. It shows a negative correlation between CaO/Na_2O ratios and H_2O^+ contents of these rocks. Field of TH and field of spilite are separate. The samples (No. 806-915, 117, 122) shown by ● are plotted in field of spilite. ★ = average composition of spilite in the Wutai Group (Liu et al., 1986). ▲ = average compositions of spilite in other areas (Xia and Huang, 1979; Liu et al., 1986). ⊡ = average compositions of Archaean tholeiites (DAT and EAT) (Condie, 1981). △ = Dengfeng Group (China)-metatholeiite (S. Li, unpublished data). ○ = Wutai Group-metatholeiite (Liu et al., 1986).

Fig. 4 CaO% vs. Na_2O% diagram for metatholeiite (TH) and spilite in the Wutai Group. I = field of albitization spilite. II = field of weak calcification spilite. III = filed of strong calcification spilite. The field I, II, III are on the basis of the data from Liu et al. (1986). ▲ = average compositions of the above three type of spilites (Liu et al., 1986). Other symbols and data sources are the same as in Fig. 3.

The Sm, Nd, Rb, Sr concentrations and Nd, Sr isotopic data were obtained at M.I.T. using techniques described by Zindler et al. (1979). $^{87}Sr/^{86}Sr$ values are reported relative to a value of 0.70800 for the E & A standard, and $^{143}Nd/^{144}Nd$ values are reported relative to a value of 0.512600 for the BCR-1 standard. ε_{Nd}^i values were calculated based on present day reference values for CHUR: $(^{143}Nd/^{144}Nd)_{CHUR} = 0.51264$ and $(^{147}Sm/^{144}Nd)_{CHUR} = 0.1967$ (Jacobsen and Wasserburg, 1984). Isochron ages (2σ errors) were calculated with a YORK 2 regression (York, 1969).

4 Results

The Nd, Sr isotopic compositions with 2σ errors and $^{147}Sm/^{144}Nd$, $^{87}Rb/^{86}Sr$ ratios are reported in Table 3, and the isochron results are given in Table 4. The Sm-Nd and Rb-Sr isochrons are illustrated in Figs. 5 and 6.

As shown in Fig. 5, the $^{147}Sm/^{144}Nd$ ratios for the spilite and quartz-keratophyre in the Wutai Group are positively correlated with $^{143}Nd/^{144}Nd$. If these rock units were genetically related and had crystallized over a short period of time from magmas with the same initial $^{143}Nd/^{144}Nd$ ratios, they would define an isochron age of 2250 ± 182 Ma with an initial $^{143}Nd/^{144}Nd$ of 0.50975 ± 17 (2σ) ($\varepsilon_{Nd}^i = 0.4 \pm 3.3$). This age is significantly lower than the depositional and first metamorphic ages and higher than the second metamorphic age observed in the Wutai Group (see above). When data for three spilitic samples and three quartz-keratophyre samples were calculated separately, two younger ages of 1981 ± 178 (2σ) Ma with an initial $^{143}Nd/^{144}Nd$ of 0.51003 ± 18 ($\varepsilon_{Nd}^i = -0.9 \pm 3.5$) for the spilite and 1977 ± 385 (2σ) Ma with an initial ^{143}Nd/^{144}Nd of 0.50996 ± 31 (ε_{Nd}^{i} = −2.4 ± 6.1) for the quartz-keratophyre are obtained (Table 4). These younger ages are not only consistent with each other but are also identical with the second metamorphic ages of 1800–2000 Ma observed in the Wutai Group.

Table 3 Sm, Nd, Rb, Sr concentrations (ppm) and Nd, Sr isotopic ratios (± 2σ error) of rocks from Wutaishan area, North China

Sample	Rock type	Sm(ppm)	Nd(ppm)	^{147}Sm/^{144}Nd	^{143}Nd/^{144}Nd	Rb(ppm)	Sr(ppm)	^{87}Rb/^{86}Sr	^{87}Sr/^{86}Sr
806-117	spilite	1.37	4.94	0.1679	0.512223±16	2.97	165.9	0.0521	0.707465±28
806-915	spilite	2.31	9.60	0.1453	0.511916±18	1.96	322.7	0.0173	0.706347±28
806-122	spilite	2.75	12.27	0.1354	0.511808±23	5.15	237.7	0.0627	0.707496±27
806-265	quartz-keratophyre	4.32	20.70	0.1260	0.511593±19	39.43	215.9	0.5292	0.717100±26
806-914	quartz-keratophyre	3.85	18.70	0.1245	0.511579±20	42.95	165.1	0.7544	0.726067±38
806-698	quartz-keratophyre	4.22	21.85	0.1167	0.511474±17	33.70	428.2	0.2280	0.712622±29

Table 4 Sm-Nd whole-rock isochron ages

Samples used	Age (Ma) (± 2σ error)	(^{143}Nd/^{144}Nd)$_i$	ε_{Nd}^{i}
whole six samples	2250±182	0.50975±17	0.4±3.3
spilite (3)	1981±178	0.51003±18	−0.9±3.5
quartz-keratophyre (3)	1977±385	0.50996±31	−2.4±6.1

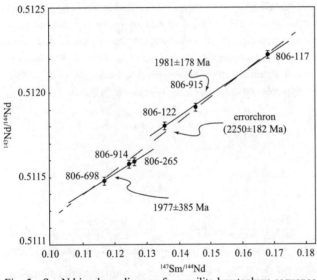 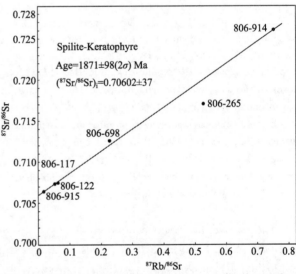

Fig. 5 Sm-Nd isochron diagram for a spilite-keratophyre sequence in the Wutai Group. Two solid lines for the spilite and quartz-keratophyre give metamorphic ages. The line of dashes for all of data points is an errorchron.

Fig. 6 Rb-Sr isochron diagram for a spilite-keratophyre sequence in the Wutai Group. Two-solid line is for all samples except 806-265.

Figure 6 shows that except for sample 806-265, the data points yield a Rb-Sr isochron age of 1871 ± 98 (2σ) Ma with an initial ^{87}Sr/^{86}Sr of 0.70602 ± 37. This Rb-Sr age is in good agreement with an unpublished Rb-Sr isochron age (~1850 ± 20 Ma) of Chung Fu-Dao on spilite-keratophyre samples from the same area (provided by S.-S. Sun). It is also consistent with the whole rock Rb-Sr isochron age of 1851 Ma for the Hutuo Group which is interpreted as a metamorphic age (Sun and Lu, 1985).

5 Discussion

5.1 Homogenization of Nd isotopic composition during the low grade metamorphic at 1980 Ma

Sm-Nd isotope systematics of spilite-keratophyre sequence in the Wutai Group discussed above clearly indicate that these two elements were mobile, and Sm-Nd whole-rock isochrons for spilite-keratophyre volcanics were reset during low-grade metamorphism. Due to the limited number of samples (6) used in this study (collected from 3 localities 3–10 km apart), it is not possible to evaluate the detailed mobility behavior of these elements during metamorphism. A tentative interpretation is given in Fig. 7. This model suggests that the effect of the isotopic homogenization caused by metamorphism on whole rock isotopic dating not only depends on the scale length of the isotopic homogenization, but also on the scale length of the heterogeneities of the parent/daughter ratios (Sm/Nd or Rb/Sr) in volcanic rocks. If the scale length of the isotopic homogenization caused by metamorphism is similar to the scale length of the heterogeneity for Sm/Nd (or Rb/Sr) ratios in a volcanic sequence, the effect of isotopic homogenization on whole rock isotopic dating will be significant, no matter how great the distances between samples are.

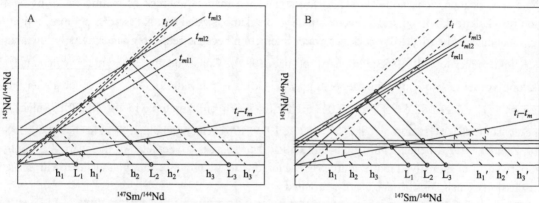

Fig. 7 Effects of Nd isotope homogenization caused by metamorphism on whole rock Sm-Nd dating for the two cases. A. L_1, L_2, L_3 are samples from three localities, which are significant distances ($h \times$ km) apart. They have different Sm/Nd ratios and the same initial Nd isotopic ratios. The h_1–h_1', h_2–h_2' and h_3–h_3' are the ranges of $^{147}Sm/^{144}Nd$ ratios for the samples from the three localities respectively. A shows that the Sm/Nd ratio of the rocks is relatively homogeneous on a small ($m - n \times 10m$) scale and heterogeneous on a large scale. If the scale length of Nd isotope homogenization caused by metamorphism is smaller ($m - n \times 10m$), the average Nd isotope compositions of the samples from different localities will not be influenced by the metamorphism so that an isochron with a depositional age (t_i) within small error limits (dashed lines) will be formed and cannot yield a metamorphic age. The samples from the same locality may give a metamorphic age; however, this age may not be reliable because of the small range of $^{147}Sm/^{144}Nd$ ratios. B. Samples from the same locality have a large range of $^{147}Sm/^{144}Nd$ ratios, such as h_1–h_1', h_2–h_2' or h_3–h_3', and the average Sm/Nd ratios (L_1, L_2 and L_3) of samples from different localities are similar. This suggests that the Sm/Nd ratios of the rocks are heterogeneous on a small scale and are homogeneous on a large scale. In this case, the samples from different localities can give a metamorphic age within error limits (t_{ml3} and t_{ml1}), which results from Nd isotope homogenization on a smaller scale during metamorphism, and cannot give any meaningful depositional age because of the large error limits (dashed lines) or small range of $^{147}Sm/^{144}Nd$ ratios.

However, many original ages of Archaean greenstone belts have been obtained by Sm-Nd isotopic dating, such as in the Onverwacht, Pilbara and Abitibi greenstone belts which have experienced low grade metamorphism (Hamilton et al., 1979, 1981; Zindler, 1982; Basu et al., 1984). The question arises as to why the low grade metamorphism did not influence the Nd isotopic compositions of those greenstone belts. One explanation is that the spilite-keratophyric rocks are the products of hydrothermal seawater alteration of volcanic rocks (Shaw et al., 1977; Seyfried et al., 1978; Mengel et al., 1987; Xia and Xia, 1987). Thus, they contain more water and complexing agents such as CO_2, Cl (see Table 2), which can yield higher fluid/rock ratios and lower solid-fluid bulk partition coefficients (D) for REE than those in general greenstone during metamorphism (Windrim et al., 1984). Since diffusion of an element through a fluid phase is more effective than volume diffusion in a solid, the REE should be more mobile in spilite-keratophyric rocks than in other basalts during low grade metamorphism.

5.2 Initial isotopic compositions of the spilite and the quartz-keratophyre

Accepting the 2560–2522 Ma as an eruption age and the 1800–2000 Ma dates as metamorphic ages, the 2250 ± 182 Ma alignment in the Sm-Nd diagram is a errorchron which may be caused by a range of initial $^{143}Nd/^{144}Nd$ ratios of the samples or by the homogenization of Nd isotopic composition in limited space as mentioned above. The following model calculation suggests that the errorchron may be caused only by the partial homogenization of Nd isotopic composition in space. If initial $^{143}Nd/^{144}Nd$ ratios of the spilite and the quartz-keratophyre at 2522 Ma were identical, the age calculated based on the two average $^{143}Nd/^{144}Nd$ and $^{147}Sm/^{144}Nd$ ratios for spilites and keratophyres at the present time should be larger than 2250 Ma and close to 2500 Ma (see Fig. 8). Table 5 shows that the calculation based on these limited number of 6 samples ($t = 2424$ Ma) supports this expectation.

Making this assumption, we used the ε_{Nd} values at the time of metamorphism and the average $^{147}Sm/^{144}Nd$ of the spilite and the quartz-keratophyre to calculate the average initial $^{143}Nd/^{144}Nd$ ratios of 0.50950 ± 18 ($\varepsilon_{Nd}^i = 2.6 \pm 3.5$) and 0.50952 ± 31 ($\varepsilon_{Nd}^i = 3.0 \pm 6.1$) for the spilite and the quartz-keratophyre, respectively, at the eruption age of 2522 Ma (Table 6).

Since Sr isotope compositions of the spilite-keratophyre have been modified by seawater interaction and metamorphic mobilization, it is not possible to investigate the magma sources of spilitic rocks by Sr isotopic composition of whole rocks. In contrast, several studies have demonstrated that Nd isotope composition of oceanic basalts are not significantly affected by rock-sea water interaction because sea water has extremely low abundances of REE (Klinkhammer et al., 1983; Michard et al., 1983). Consequently, ε_{Nd}^i values calculated for the spilite-keratophyre at the time of their deposition (2.5 Ga) are relevant to their source character (± crustal contamination). ε_{Nd}^i values of 2.6 ± 3.5 and 3.0 ± 6.1 for the spilite and keratophyre are similar to values for other Archaean greenstone rocks (Hamilton et al., 1977, 1978, 1979, 1981; Huang et al., 1986; Fletcher et al., 1984; Basu et al., 1984; Zindler, 1982; Li et al., 1988), and suggest that their sources are depleted or the quartz-keratophyre may have been derived from short-lived mafic source. However, due to a large uncertainty associated with these values, detailed discussion of their significance is not warranted.

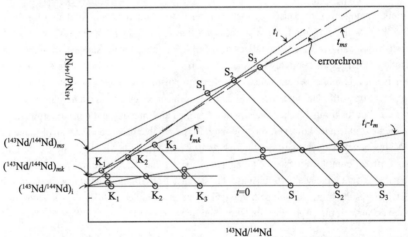

Fig. 8 Evolution diagrams of the Sm-Nd system in a metaspilite-keratophyre sequence for a case in which all samples have the same initial $^{143}Nd/^{144}Nd$ value. The samples K_1, K_2, K_3, are quartz-keratophyre, and $(^{147}Sm/^{144}Nd)_{K2} = [(^{147}Sm/^{144}Nd)_{K1} + (^{147}Sm/^{144}Nd)_{K3}]/2$. The samples S_1, S_2, S_3, are spilite, and the $(^{147}Sm/^{144}Nd)_{S2}$ value is also equal to the average $^{147}Sm/^{144}Nd$ value of the spilite unit. Nd in the samples is homogenized in the spilite and quartz-keratophyre units separately by low-grade metamorphism. Consequently, two parallel isochrons (t_{ms}, t_{mk}) with $(^{143}Nd/^{144}Nd)_{ms}$ and $(^{143}Nd/^{144}Nd)_{mk}$ for the spilite and quartz-keratophyre respectively yield the same metamorphic ages $t_{ms} = t_{mk} = t_m$. The solid line (t_i) which passes through points $S_{2'}$ and $K_{2'}$ gives an eruption age since the Nd isotopic compositions of samples S_2 and K_2 were not changed during the low-grade metamorphism. The dashed line based on all of the data is errorchron, and the slope of which is lower than that of t_i and higher than that of t_{ms} and t_{mk}.

Table 5 Age calculation based on the average $^{143}Nd/^{144}Nd$ and $^{147}Sm/^{144}Nd$ ratios

Rock unit	Average $(^{143}Nd/^{144}Nd)_{now}$	Average $^{147}Sm/^{144}Nd$	T^a (Ma)
spilite	0.511982	0.1495	2424
quartz-keratophyre	0.511549	0.1224	

Table 6 Average initial $^{143}Nd/^{144}Nd$ ratios at 2522 Ma

Rock unit	Average $(^{143}Nd/^{144}Nd)_m$	Average $^{147}Sm/^{144}Nd$	Time range^a T (Ma)	$(^{143}Nd/^{144}Nd)_i$ (at 2522 Ma)	ε_{Nd}^i (at 2522 Ma)
spilite	0.51003 ± 18	0.1495	541	0.50950 ± 18	2.6 ± 3.5
quartz-keratophyre	0.50996 ± 31	0.1224	545	0.50952 ± 31	3.0 ± 6.1

^{a} The "time range" is for the first stage evolutionary history.

6 Conclusions

(1) The homogenization of Nd isotopic composition in spilite-keratophyric volcanic rocks can be achieved in a limited rock unit during low-grade metamorphism because of REE mobility. The Nd isotopic compositions were homogenized in the spilite and quartz-keratophyre units separately during the metamorphism to greenschist facies in the Wutaishan area, and two consistent metamorphic ages of 1981 ± 178 Ma and 1977 ± 385 Ma were obtained from the samples of those rock units. However, we cannot obtain any meaningful ages from the samples of the whole rock sequence. Consequently, this problem should be considered when we apply Sm-Nd dating method to metaspilite-keratophyric rocks.

(2) The effect of isotopic homogenization caused by metamorphism on dating not only depends on the scale length of the isotopic homogenization but also depends on the scale length of the heterogeneity for Sm/Nd (or Rb/Sr) ratios in the volcanic rocks. If the scale length of the isotopic homogenization is similar to the scale length of the heterogeneity for Sm/Nd (or Rb/Sr) ratios in the volcanic rocks, the effect of isotopic homogenization on dating will be significant, no matter how large the distances between samples are.

(3) The Nd isotopic compositions of original rocks or parental magma are not obviously affected by spilitization, consequently, we can use Nd isotopes as tracers to investigate the magma sources of spilite-keratophyric volcanic rocks. Like Precambrian greenstone belts, the positive initial ε_{Nd} values of 2.6 ± 3.5 and 3.0 ± 6.1 of the Precambrian spilite and quartz-keratophyre of the Wutai Group suggest that the sources of these rocks were depleted in LREE for a long time prior to their eruptions and had similar Sm/Nd ratios. It is possible that the quartz-keratophyre may either be derived from the same source as the spilite since they have similar initial ε_{Nd} values, or be derived from a short-lived mafic source which was derived from a mantle source with a similar Sm/Nd ratio to the source of the spilite.

Acknowledgements

S. Li thanks Academia Sinica and S.R. Hart for financial support during laboratory work at MIT. Zhang Zonqing, Yang Jiedong and Wang Yinxi helped in isochron calculations. We also thank Dr. S.-S. Sun and Dr. M. Wilks for constructive criticism and suggestion during preparation of this paper.

References

Arculus, R.J. and Johnson, R.W., 1981. Island-arc magma sources: a geochemical assessment of the roles of slab-derived components and crustal contamination. Geochem. J., 15: 109-133.
Basu, A.R., Goodwin, A.M. and Tatsumoto, M., 1984. Sm-Nd study of Archaean alkalic rocks from the Superior Province of the Canadian Shield. Earth Planet. Sci. Lett., 70: 40-46.
Black, L.P. and McCulloch, M.T., 1987. Evidence for isotopic equilibration of Sm-Nd whole-rock systems in early Archaean crust of Enderby Land, Antarctica. Earth Planet. Sci. Lett., 82: 15-24.
Cattel, A., Krogh, T.E. and Arndt, N.T., 1984. Conflicting Sm-Nd whole rock and U-Pb zircon ages for Archaean lavas from Newton Township, Abitibi belt, Ontario. Earth Planet. Sci. Lett., 70: 280-290.
Chauvel, C., Dupre, B. and Jenner, G.A., 1985. The Sm-Nd age of Kambalda volcanics is 500 Ma to old. Earth Planet. Sci. Lett., 74: 315-324.
Condie, K.C., 1981. Archaean Greenstone Belts. Elsevier, Amsterdam, p. 97.
Fletcher, T.R., Rosman, K.J.R., Williams, I.R., Hickman, A.H. and Baxter, J.L., 1984. Sm-Nd geochronology of greenstone belts in the Yilgarn Block, Western Australia. Precambrian Res., 26: 333-361.
Hamilton, P.J., O'Nions, R.K. and Evensen, N.M., 1977. Sm-Nd dating of Archaean basic and ultrabasic volcanics. Earth Planet. Sci. Lett., 36: 263-268.
Hamilton, P.J., O'Nions, R.K., Evensen, N.M., Bridgwater, D. and Allaart, J.H., 1978. Sm-Nd isotopic investigations of Isua supracrustals and implications for mantle evolution. Nature, 272: 41-43.
Hamilton, P.J., Evensen, N.M., O'Nions, R.K., Smith, H.S. and Erlank, A.J., 1979. Sm-Nd dating of Onverwacht Group volcanics, southern Africa. Nature, 279: 298-300.
Hamilton, P.J., Evensen, N.M., O'Nions, R.K., Glikson, A.Y. and Hickman, A.H., 1981. Sm-Nd dating of the North Star basalt,

Warrawoona group, Pilbara Block. Spec. Publ. Geol. Soc. Aust., 7: 187-192.

Hegner, E., Kröner, A. and Hofmann, A.W., 1984. Age and isotope geochemistry of the Archaean Pongola and Usushwana suites in Swaziland, Southern Africa: a case for crustal contamination of mantle-derived magma. Earth Planet. Sci. Lett., 70: 267-279.

Hellman, P.L. and Henderson, P., 1977. Are rare earth elements mobile during spilitization? Nature, 267: 38-40.

Hellman, P.L., Smith, R.E. and Henderson, P., 1979. The mobility of the rare earth elements: evidence and implications from selected terrains affected by burial metamorphism. Contrib. Mineral. Petrol., 71: 23-44.

Huang Xuan, Bi Ziwei and DePaolo, D.J., 1986. Sm-Nd isotope study of early Archaean rocks, Qianan, Hebei Province, China. Geochim. Cosmochim. Acta, 50: 625-631.

Jacobsen, S.B. and Wasserburg, G.J., 1984. Sm-Nd isotopic evolution of chondrites and achondrites II. Earth Planet. Sci. Lett., 67: 137-150.

Klinkhammer, G., Elderfield, H. and Hudson, A., 1983. Rare earth elements in seawater near hydrothermal vents. Nature, 305: 185-188.

Li, S., Hart, S.R., Guo, A. and Zhang, G., 1987. Sm-Nd whole-rock isochron age of the Dengfeng Group in Central Henan and tectonic implications. Kexue Tongbao (Bull. Sci.), 33 (20): 1714-1717.

Li, S.X., He, G.P., Zhang, L.J. and Liu, B.R., 1986. Evolution of metamorphism. In: Bei Jin (Editor), The Early Precambrian Geology of Wutaishan. Tianjin Sci. Technol. Press, Tianjin, pp. 242-247 (in Chinese).

Liu, B.R., Li, S.G. and He, G.P., 1986. Rocks in the Wutai Group. In: Bei Jin (Editor), The Early Precambrian Geology of Wutaishan. Tianjin Sci. Technol. Press, Tianjin, pp. 106-150 (in Chinese).

Liu, D., Page, R.W., Compston, W. and Wu, J., 1985. U-Pb zircon geochronology of Late Archaean metamorphic rock in the Taihangshan-Wutaishan area, North China. Precambrian Res., 27: 85-109.

McCulloch, M.T. and Black, L.P., 1984. Sm-Nd isotopic systematics of Enderby Land granulites and evidence for the redistribution of Sm and Nd during metamorphism. Earth Planet. Sci. Lett., 71: 46-58.

Mengel, K., Borsuk, A.M., Gurbanov, A.G., Wedepohl, K.H., Baumann, A. and Hoefs, J., 1987. Origin of spilitic rocks from the southern slope of the Greater Caucasus. Lithos, 20: 115-133.

Michard, A., Albarede, F., Michard, G., Minster, J.F. and Charlous, J.L., 1983. Rare-earth elements and uranium in high-temperature solutions from East Pacific Rise hydrothermal vent field (13° N). Nature, 303: 795-797.

Nystrom, J.O., 1984. Rare earth element mobility in vesicular lava during low-grade metamorphism. Contrib. Mineral. Petrol., 88: 328-331.

Seyfried, W.E., Mottl, M.J. and Bischoff, J.L., 1978. Seawater/basalt ratio effects on the chemistry and mineralogy of spilites from the ocean floor. Nature, 275: 211-213.

Shaw, D.M., Vatin-Pergnon, N. and Muysson, J.R., 1977. Lithium in spilites. Geochim. Cosmochim. Acta, 41: 1601-1607.

Staudigel, H., Zindler, A., Hart, S.R., Leslic, T., Chen, C.Y. and Clague, D., 1984. The isotope systematics of a juvenile intraplate volcano: Pb, and Sr isotope ratios of basalts from Loihi seamount, Hawaii. Earth Planet. Sci. Lett., 69: 13-29.

Sun, D. and Lu, S., 1985. A subdivision of the Precambrian of China. Precambrian Res., 28: 137-162.

Vocke, R.O., Hanson, G.N. and Grunenfelder, M., 1987. Rare earth element mobility in the Roffna Gneiss, Switzerland. Contrib. Mineral. Petrol., 95: 145-154.

Wang R.J. and Bai J., 1986. Ages of the Wutai Group. In: Bai J.(Editor), The Early Precambrian Geology of Wutaishan. Tianjin Sci. Technol. Press, Tianjin, pp. 364-370 (in Chinese).

Windrim, D.P., McCulloch, M.T., Chappell, B.W. and Cameron, W.E., 1984. Nd isotopic systematics and chemistry of Central Australian sapphirine granulites: in example of rare earth element mobility. Earth Planet. Sci. Lett., 70: 27-39.

Wood, D.A., Gibson, I.L. and Thompson, R.N., 1976. Element mobility during zeolite facies metamorphism of the tertiary basalts of eastern Iceland. Contrib. Mineral. Petrol., 55: 241-254.

Wu, J., Liu, D. and Jin, L., 1986. The zircon U-Pb age of metamorphosed basic volcanic lavas from the Hutuo Group in Wutai mountain area, Shanxi Province. Geol. Rev., 32: 178-184 (in Chinese).

Xia, L. and Xia, Z., 1987. Some problems on the spilite-keratophyric volcanic rocks. Bull. Xian Inst. Geol. Min. Res., Chin. Acad. Geol. Sci., 19: 1-30 (in Chinese).

Xia, L. and Huang, Y., 1979. Genesis of some spilite-keratophyric suites in Northwestern China as illustrated by the structural state of soda plagioclases. Geochimica, 1: 13-27 (in Chinese).

York, D., 1969. Least-squares fitting of a straight line with correlated errors. Earth Planet. Sci. Lett., 5: 320-324.

Zindler, A., 1982. Nd and Sr isotopic studies of komatiites and related rocks. In: N.T. Arndt and E.G. Nisbet (Editors), Komatiites. Allen and Unwin, pp. 399-420.

Zindler, A., Hart, S.R., Frey, F.A. and Jacobsen, S.P., 1979. Nd and Sr isotope ratios and rare earth elements abundances in Reykjanes Peninsula basalts: Evidence for mantle heterogeneity beneath Iceland. Earth Planet. Sci. Lett., 45: 249-262.

Zindler, A., Staudigel, H. and Batiza, R., 1984. Isotope and trace element geochemistry of young Pacific seamounts: implications for the scale of upper mantle heterogeneity. Earth Planet. Sci. Lett., 70: 175-195.

河南中部登封群全岩 Sm-Nd 同位素年龄及其构造意义[*]

李曙光[1]，S. R. Hart[2]，郭安林[3]，张国伟[3]

1. 中国科学技术大学地球与空间科学系，合肥 230026
2. 美国麻省理工学院地球、大气与行星科学系，剑桥 02139
3. 西北大学地质系，西安 710069

> **亮点介绍**：全岩 Sm-Nd 等时线定年给出河南登封群下部角闪岩相变质火山岩的生成年龄为 2509.2 ± 16.1 Ma；说明角闪岩相中级变质作用不能使该火山岩系的 Nd 同位素组成均一化重置。其 Nd 同位素特征(ε_{Nd}^i = +2.16 ± 0.39)揭示出 25 亿年以前华北克拉通南缘存在亏损地幔。

河南登封群角闪岩相的变质火山-沉积岩系被认为是中朝地台南缘一个东西延伸的绿岩带[1]。在南面，它与由太华群和霍邱群组成的高级变质区相邻；在北面，隔黄河与太行山的赞皇群相望[2]。很显然，搞清登封绿岩带及其两侧相邻高级变质区的生成时代关系对认识该绿岩带发育的构造背景及中朝地台西缘的地壳演化历史无疑是十分重要的。

然而，除了一些反映变质时代的全岩 Rb-Sr 等时年龄外[3]，登封群还没有一个能真正反映其生成时代的年龄。为此，我们对登封群中下部变质火山岩作了全岩 Sm-Nd 等时年龄测定，以求得其生成年龄，并据此对该绿岩带生成环境和物质来源做进一步讨论。

1 测试样品地质特征

登封群划分为三个组。下部郭家窑组以角闪片岩、斜长角闪岩、变粒岩、浅粒岩为主，含磁铁石英岩。中部金家门组及上部老羊沟组以云母石英片岩、云母片岩及绢云母石英片岩为主[1]。样品采自登封君召，属登封群郭家窑组。从剖面底部至顶部，逐层均匀采取样品。样品主要为斜长角闪岩(原岩为拉斑玄武岩)和少量浅粒岩(原岩为中酸性火山岩)[4]。岩石变质相为角闪岩相。样品新鲜。

2 样品加工及测试方法

为避免样品的污染，岩块用棉布及软塑料薄板包裹后，用铁锤击成碎块。选内部完全新鲜的碎块，在去离子水内用超声波洗净，烘干，并放在碳化钨摇震盒内粉碎。每粉碎一个样品后，摇震盒要用去离子水仔细洗净。

样品的 Sm、Nd 含量及 Nd 同位素比值分析是在麻省理工学院地球、大气和行星科学系同位素实验室进行的。分析流程已有文献描叙[5]。质谱分异校正以 $^{146}Nd/^{144}Nd$ = 0.72190 为基准。文中给出的 Nd 同位素比值是相对于标准样(BCR-1) $^{143}Nd/^{144}Nd$ = 0.512600 的。ε_{Nd}^i 计算所采用的参考值为 $(^{143}Nd/^{144}Nd)^0_{CHUR}$ = 0.51264 和 $(^{147}Sm/^{144}Nd)_{CHUR}$ = 0.1967[6]。等时线拟合采用 YOURK 1 计算[7]，并给出 1σ 误差。λ 值采用 6.54 × $10^{-12}a^{-1}$。

3 测试结果及讨论

样品的 Sm、Nd 含量和 $^{143}Nd/^{144}Nd$ 值及其 2σ 误差列于表 1。年龄计算结果列于表 2。等时线示于图 1。

---

[*] 本文发表在：科学通报，1987，32(22)：1728-1731
对应的英文版论文为：Li Shuguang, Hart S.R., Guo Anlin, et al., 1988. Sm-Nd whole-rock isochron age of the Dengfeng Group in Central Henan and its tectonic implication. Kexue Tongbao (Science Bulletin), 33(20): 1714-1717

表 1 登封群变火山岩 Sm、Nd 含量及 Nd 同位素比值

样品	岩类	Sm/ppm	Nd/ppm	$^{147}Sm/^{144}Nd$	$^{143}Nd/^{144}Nd$
JG-11	斜长角闪岩	1.96	5.80	0.20436	0.512899 ± 26
JG-5 (1)	斜长角闪岩	2.21	6.65	0.20125	0.512814 ± 19
JG-5 (2)	斜长角闪岩	2.22	6.69	0.20091	0.512824 ± 20
平均		2.22	6.67	0.20108	0.512819 ± 14*
JG-1	斜长角闪岩	2.24	6.75	0.20025	0.512803 ± 17
JG-3	斜长角闪岩	2.60	7.92	0.19851	0.512776 ± 18
MT-2	斜长角闪岩	2.45	10.91	0.13545	0.511754 ± 16
MT-5	斜长角闪岩	2.40	10.84	0.13381	0.511719 ± 20
JG-4-2	浅粒岩	1.66	8.82	0.11363	0.511376 ± 18
JG-4-1	浅粒岩	1.70	9.39	0.10946	0.511301 ± 18

* 平均误差计算据 $\sigma^2 = \dfrac{1}{\Sigma(1/\sigma^2)}$。

表 2 全岩 Sm-Nd 等时年龄计算结果

参与计算的样品	年龄/Ma	$(^{143}Nd/^{144}Nd)_i$	ε_{Nd}^i
6 个斜长角闪岩样品	2486.3 ± 23.9	0.50953 ± 3	+2.15 ± 0.60
全部 8 个火山岩样品	2509.2 ± 16.1	0.50950 ± 2	+2.16 ± 0.39

表 1 表明数据测量精度是很高的。JG-5 的两个不同时间分析的平行样结果表明实验的相对误差也很小。

表 2 表明等时线的拟合程度也很高。全部 8 个样品拟合的等时线给出一个很精确的年龄 2509.2 ± 16.1 Ma，和初始 Nd 同位素比值 $(^{143}Nd/^{144}Nd)_i = 0.50950 \pm 2$，$\varepsilon_{Nd}^i = +2.16 \pm 0.39$。长英质变火山岩常常可能并不与玄武质火山岩同源或混有碎屑沉积，从而具有与玄武质岩石不同的初始同位素比值。将这种岩石与玄武质岩石一起拟合等时线有可能给出虚假的年龄[8]。为此，我们去掉两个长英质火山岩样品，仅用 6 个玄武岩样品计算，同样也得到一个较好的年龄 2486.3 ± 23.9 Ma 和初始 $(^{143}Nd/^{144}Nd)_i = 0.50953 \pm 3$。这一年龄值与全部 8 个样品的年龄值在误差范围内是一致的。说明我们所选取的长英质火山岩样品可能与玄武质样品是同源的，或者至少它们的初始同位素比值是十分接近的。因此，2509.2 ± 16.1 Ma 这一年龄值是可信的。

该年龄值与最近 Kröner 等人做的该群上部长英质火山岩中锆石 U-Pb 一致线上交点年龄完全一致[16]。此外，陈好寿等人曾报道了许昌登封群中部武庄组(相当登封的金家门组) 4 个样品的全岩 Rb-Sr 等时年龄为 2562 ± 170 Ma。它在其误差范围内与我们的 Sm-Nd 年龄一致。此外，该年龄是采用老的 ^{87}Rb 衰变常数 $\lambda = 1.39 \times 10^{-11} a^{-1}$ 计算的。如果采用新校正的 ^{87}Rb 衰变常数 $\lambda = 1.42 \times 10^{-11} a^{-1}$[9] 计算，则该 Rb-Sr 等时年龄应为 2506 ± 165 Ma，它与我们的 Sm-Nd 等时年龄非常一致。这说明该 Rb-Sr 年龄可代表生成时代。上述讨论表明，登封群下、中、上各组的形成年龄完全一致。说明登封群是在一个较短时间范围内完成了各组沉积的全过程，也说明在太古宙，地幔对流速度较现代快，从而裂谷的发生及闭合也较迅速[10]。

如果将这一年龄与相邻高级变质区下部地体年龄进行比较，我们可发现登封群绿岩的发育晚于邻区高级

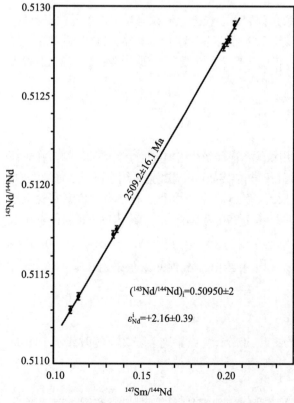

图 1 河南登封群变火山岩系 Sm-Nd 全岩等时线图

区的长英质地壳。如登封群以南的太华群,其底部长英质片麻岩的锆石 U-Pb 年龄为 2620 Ma[3]。太华群的东延部分,安徽霍邱群吴集组,全岩 Rb-Sr 等时年龄为 2744 ± 64 Ma[11]。登封群的北邻,豫北赞皇群未见同位素年龄报道。但与赞皇群相邻,层位相当,岩性相似,同处太行山区的河北阜平群底部,年龄为 27 亿~28 亿年[12]。这些资料表明,在登封绿岩带形成之前,华北地台南缘已有一片相当 27 亿年年龄的硅铝质地壳。在 25 亿年前,地台南缘发生了一东西向裂谷,登封绿岩带就是在这一硅铝壳裂谷环境上发育的。

登封群变火山岩的初始 Nd 同位素比值与陨石值相比是亏损的(ε_{Nd}^i = +2.16 ± 0.39)。这说明其岩浆是来自亏损地幔,并较少受硅铝壳混染的影响。这与世界上其他地区绿岩带变火山岩 Nd 同位素特征相同[9,13-15]。它从 Nd 同位素特征上再次证明登封群变火山岩系具有绿岩带特征。在该绿岩发育时,裂谷张开较大,内部已具洋壳性质。然而,陈好寿等人对登封群变质火山岩系给出了一个非常高的初始 $^{87}Sr/^{86}Sr$ 值(0.7124 ± 0.0007)。我们很难用海水蚀变或变质作用解释如此高的初始 Sr 同位素比值。由于他们的样品采自许昌地区,无论在空间上还是层位上均不能与我们的样品相比较。因此,这种高初始 $^{87}Sr/^{86}Sr$ 值仅反映了在许昌地区武庄组沉积时,已有相当多的古老陆源物质参与了火山沉积过程。这可能暗示着许昌地区在当时更接近盆地的边缘,而登封君召地区则相对接近盆地的中心。

4 结论

(1)登封绿岩带的生成年龄为 2509.2 ± 16.1 Ma。

(2)在 25 亿年以前,在华北克拉通南缘之下,存在亏损地幔。其 Nd 同位素特征是初始$(^{143}Nd/^{144}Nd)_i$ = 0.50950 ± 2,ε_{Nd}^i = +2.16 ± 0.39。该亏损地幔可能是登封太古宙绿岩的岩浆源。

(3)登封绿岩带是发育在一个较老的(27 亿年左右)硅铝质地壳的裂谷环境中。

(4)登封绿岩带与其他绿岩带类似,它不同层位的生成年龄极为接近,说明当时裂谷的开启、闭合相当迅速。

致谢

李曙光非常感谢中国科学院和 S. R. Hart 教授的支持,使他能有机会访问麻省理工学院,并完成该 Sm-Nd 同位素定年的实验。

参考文献

[1] Zhang Guo-wei et al., Precamb. Rea., 27(1985), 7-35.
[2] 中国地质科学院,中华人民共和国地质图集,1973.
[3] 陈好寿等,中国地质科学院院报,1(1980), 88-102.
[4] 张国伟等,西北大学学报前寒武纪地质专辑,1982, 1-10.
[5] Zindler, A., Hart, S. R. et al., Earth Planet. Sci. Lett., 45(1979), 249-262.
[6] Jacobesen, S. B. and Wasserburg, G. J., Earth Planet. Sci. Lett., 67(1984), 137-150.
[7] Faure, G., Principles of Isotope Geology, John Wiley & Sons, New York, 1979.
[8] Claoue-Long, J. C., Thirlwall, M. F. & Nesbitt, R. W., Nature, 307(1984), 697-701.
[9] Zindler, A., in Komatiite (Eds. Arndt, N. T. and Nisbet, E. G.), George Allen & Unwin, 1982, 399-420.
[10] Jahn, B. M., Glikson, A. Y., Peucat, J. J., Hikman, A. H., Geochim. Cosmochim. Acta, 45(1981), 1633-1652.
[11] 安徽省地质局地质科学研究所,安徽省前寒武纪变质铁矿地质特征与找矿方向的研究,1981, 79-87.
[12] 刘敦一,R. W. 佩吉等,中国地质科学院院报, 8(1984), 57-83.
[13] Hamilton, P. J., O'Nions, R. K. and Evensen, N. M., Earth Planet. Sci. Lett., 36(1977), 263-268.
[14] Hamilton, P. J., Evensen, N. M., O'Nions, R. K., Smith, H. S. and Erlank, A. J., Nature, 279(1979), 298-300.
[15] Jahn, B. M., Gruau, G. and Glikson, A. Y., Contrib. Mineral. Petrol., 80(1982), 25-40.
[16] Kröner, A., Composton, W., Zhang, G.W. and Guo, A.L., Age and tectonic setting of Late Archean greenstone-gneiss terrain in Henan Province, China, as revealed by single-grain zircon dating. Geology, 1988, 16(3), 211-215.

Excess argon in phengite from eclogite: evidence from dating of eclogite minerals by Sm-Nd, Rb-Sr and $^{40}Ar/^{39}Ar$ methods*

Shuguang Li[1], Songshan Wang[2], Yizhi Chen[3], Deliang Liu[1], Ji Qiu[2], Hongxing Zhou[3] and Zhimin Zhang[4]

1. Department of Earth and Space Sciences, University of Science and Technology of China, Hefei 230026, Anhui, China
2. Institute of Geology, Academia Sinica, Beijing 100029, China
3. No. 6 Geological Team of Jiangsu Province, Donghai 222300, Jiangsu, China
4. Regional Geological Survey Team of Shandong Province, Jiaozhou 266300, Shandong, China

亮点介绍：通过对同一块标本分选出来的榴辉岩矿物的 Sm-Nd、Rb-Sr 和 $^{40}Ar/^{39}Ar$ 多同位体系联合定年，发现多硅白云母 $^{40}Ar/^{39}Ar$ 年龄（943-877 Ma）明显老于石榴子石-绿辉石-多硅白云母的 Sm-Nd（228-226 Ma）和 Rb-Sr（223-219 Ma）等时线年龄，从而证明榴辉岩中的多硅白云母普遍存在过剩 Ar，其 $^{40}Ar/^{39}Ar$ 年龄不能反映真实的超高压变质年龄。该研究解决了存在于大别-苏鲁和阿尔卑斯造山带榴辉岩 $^{40}Ar/^{39}Ar$ 年龄老于其他同位素年龄产生的争议，并带动了高压和超高压变质岩 Ar 同位素年代学的过剩 Ar 研究高潮。

Abstract The existence of excess Ar in phengites from coesite-bearing ultrahigh-pressure eclogites of the Dabie Mountains and Su-Lu terrain, China, has been confirmed by dating of minerals using Sm-Nd, Rb-Sr and $^{40}Ar/^{39}Ar$ methods. $^{40}Ar/^{39}Ar$ ages of phengites for two samples studied are much older than the Sm-Nd and Rb-Sr mineral isochron ages: sample QL-1 has Rb-Sr age of 219.5±0.5 Ma, Sm-Nd age of 226.3±4.5 Ma and phengite $^{40}Ar/^{39}Ar$ age of 877.5±15.9 Ma; the corresponding ages for sample ZB-4 are: 223.9±0.9 Ma (Rb-Sr), 228.4±6.0 Ma (Sm-Nd) and 943.3±12.4 Ma ($^{40}Ar/^{39}Ar$, phengite). An age of ~ 225 Ma based on the Sm-Nd and Rb-Sr mineral isochron ages is consistent with the time inferred from paleomagnetic and geological data for collision of the North China and Yangtse Blocks and formation of these eclogites.

We call for greater caution in $^{40}Ar/^{39}Ar$ dating of phengites from other occurrences of ultrahigh-pressure environments.

1 Introduction

In general, phengite is an ideal mineral for K-Ar or $^{40}Ar/^{39}Ar$ dating because of its high K content and good retention properties for radiogenic argon (McDougall and Harrison, 1988). However, contradiction between $^{40}Ar/^{39}Ar$ ages of phengite and Sm-Nd mineral isochron or zircon U-Pb ages for some high-pressure metamorphic rocks has been observed (Tilton et al., 1991; Li et al., 1993). In the case of an ultrahigh-pressure collision zone in the Dabie mountains, China (Fig. 1), Mattauer et al. (1991) suggest that it was formed during a Proterozoic Yangtze orogenic event. Their conclusion is based on $^{40}Ar/^{39}Ar$ ages of 500−700 Ma obtained from amphibole, biotite and phengite of coesite-bearing eclogite. In contrast, Li et al. (1989a, b, 1992, 1993) proposed an early Mesozoic collision event between the North China and Yangtse Blocks on the basis of Sm-Nd mineral isochron ages (208−244 Ma). An obstacle to resolving this controversial issue is that all of the $^{40}Ar/^{39}Ar$ (or K-Ar) ages (200−1100 Ma) and Sm-Nd ages for eclogites in the Dabie Mountains and the Su-Lu terrain were obtained from different samples and not be compared directly. To clarify this paradoxical issue, we have carried out a comprehensive study by applying Sm-Nd, Rb-Sr and $^{40}Ar/^{39}Ar$ dating methods on the same eclogite samples. This approach surprisingly has not been attempted by previous workers either for the Dabie Mountains or for the Alps of Central Europe.

---

* 本文发表在：Chemical Geology, 1994, 112: 343-350

2 Samples and analyses

The east-west striking Qinling-Dabie orogenic belt is a collision zone between NCB and YB (Fig. 1). Its eastern extension, the Su-Lu terrain, has been displaced ~500 km to the north by the Tanlu fault (J. Xu et al., 1987). Various types of eclogites including coesite-bearing and diamond-bearing ones formed under ultrahigh-pressure conditions exist in the Dabie Mountains and Su-Lu terrain (Okay et al., 1989; X. Wang et al., 1989; S. Xu et al., 1992). Our two samples, QL-1 and ZB-4, were collected from Qinglongshan and Zhubian, respectively, in the Su-Lu terrain (Fig. 1). At these localities, larger eclogite bodies are en-cased in the Proterozoic leucocratic gneiss with tectonic contact. Eclogites in Qinglongshan and Zhubian are strongly foliated. Sample QL-1 contains garnet + omphacite + phengite + rutile + kyanite + quartz. Coesite pseudo-morphs have been recognized in eclogite from Qinglongshan (Zhang et al., 1991). Sample ZB-4 is more retrogressive than QL-1 and contains garnet + omphacite + phengite + amphibole + rutile. Phengite in both eclogite samples have been developed along foliations. The structural relationship of these minerals suggests that phengite was formed later than garnet, omphacite and rutile, and perhaps during the exhumation of the eclogite (Okay and Sengör, 1992; Li et al., 1993).

Pure mineral separates of garnet, omphacite, phengite for sample QL-1, and garnet and phengite for sample ZB-4 were collected using a magnetic separator followed by hand-picking under a binocular microscope. Sm-Nd and Rb-Sr data (Table 1) were obtained in the isotope laboratory at the Modern Analysis Center of Nanjing University using standard techniques, which were similar to those described by Cohen et al. (1988). The corresponding mineral isochrons are shown in Fig. 2. $^{40}Ar/^{39}Ar$ analyses of phengite were conducted at the Institute of Geology, Academia Sinica, following the procedure described by S. Wang et al. (1987). The $^{40}Ar/^{39}Ar$ step experimental data and apparent ages for the phengites are given in Table 2 and the age spectra are shown in Fig. 3.

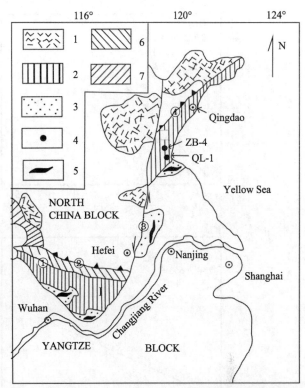

Fig.1 Sketch map of geology and tectonics of the Dabie Mountains (I) and Su-Lu terrain (II) (①=Tongbo-Mozitan fault;②=Xinyang-Shucheng fault;③=Tanlu fault;④=Wulian-Rongcheng fault. Other symbols are: 1=Archean complexes, 2=Dabie and Jiaonan Groups containing eclogites, 3= Susong, Zhangbaling and Haizhou Groups containing blueschist; 4=location of the eclogite sample; 5 = blueschist; 6= Xinyang and Fuziling Groups (lower or middle Paleozoic); 7=North Qinling terrane).

Table 1 Sm-Nd and Rb-Sr isotope data of eclogite from Subei-Jiaonan Rise

Sample No.	Rock or mineral	Sm /ppm	Nd /ppm	$\frac{^{147}Sm}{^{144}Nd}$	$\frac{^{143}Nd}{^{144}Nd} \pm 2\sigma$	Rb /ppm	Sr /ppm	$\frac{^{87}Rb}{^{86}Sr}$	$\frac{^{87}Rb}{^{86}Sr} \pm 2\sigma$
QL-1$_w$	whole rock	7.260	34.03	0.1290	0.511865±0.000011	7.915	545.6	0.04134	0.706556±0.000011
QL-1$_{ph}$	phengite	0.2627	1.025	0.1550	0.511897±0.000013	131.6	50.87	7.501	0.729821±0.000012
QL-1$_{omp}$	omphacite	0.8962	2.708	0.2002	0.511977±0.000015	0.6472	46.94	0.03998	0.706470±0.000019
QL-1$_{gar}$	garnet	0.2355	0.4284	0.3326	0.512165±0.000009				
ZB-4$_w$	whole rock	7.453	31.48	0.1432	0.511652±0.000013	9.465	658.3	0.04297	0.712901±0.000016
ZB-4$_{gar}$	phengite	0.4075	0.8642	0.2851	0.511877±0.000015	147.8	43.62	9.826	0.743998±0.000018
ZB-4$_{gar}$	garnet	2.411	3.271	0.4460	0.512106±0.000011	0.08213	11.08	0.02149	0.712675±0.000022

All mineral separates for Sm-Nd and Rb-Sr dating were washed with 1:1 HCl at 100 ℃ for 1 h before dissolution to remove any surface contamination and then they were ultrasonically cleaned in H_2O. Total procedure blanks for Nd and Sr are 60 pg and 2 ng, respectively. Sample weights are 200–300 mg in these analyses. Mass fractionations were normalized to $^{146}Nd/^{144}Nd=0.72190$. The $^{143}Nd/^{144}Nd$ ratio of BCR-1 standard and the $^{87}Sr/^{86}Sr$ ratio of NBS 987 standard measured at Nanjing University during the course of this work are 0.512633 and 0.71022, respectively.

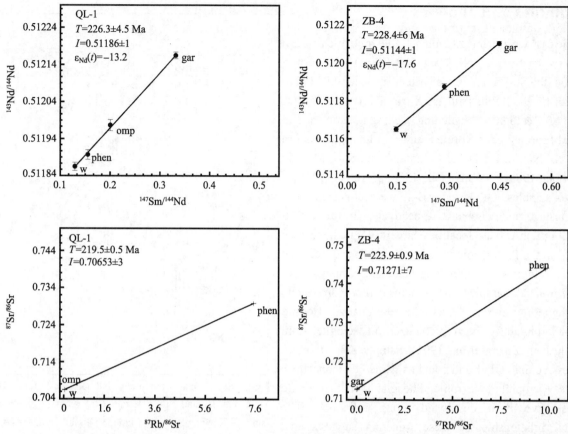

Fig. 2 Sm-Nd and Rb-Sr isochron diagrams for sample QL-1 and ZB-4. Isochron ages were calculated with York II weighted regression and the decay constant $\lambda_{Sm} = 6.54 \times 10^{-12}$ a^{-1}. $\varepsilon_{Nd}(t)$ was calculated based on present-day reference values for chondritic uniform reservoir (CHUR): $(^{143}Nd/^{144}Nd)_{CHUR} = 0.51264$ and $(^{147}Sm/^{144}Nd) = 0.1967$. Ages are given with 2σ error.

Table 2 Analytical data of $^{40}Ar/^{39}Ar$ step experiment and apparent ages for phengite samples from the Su-Lu terrain

Sample No.	T /°C	$\left(\dfrac{^{40}Ar}{^{39}Ar}\right)_m$	$\left(\dfrac{^{36}Ar}{^{39}Ar}\right)_m$	$\left(\dfrac{^{37}Ar}{^{39}Ar}\right)_m$	^{39}Ar /(10^{-12}mol)	^{39}Ar (% of total)	$\dfrac{^{40}Ar^*}{^{39}Ar}$	Apparent age ± 1σ /Ma
QL-1	400	50.36	0.05423	1.229	0.3874	4.28	34.53±0.50	557.2±7.9
	700	62.53	0.03937	0.3396	0.7653	8.46	50.98±0.62	772.2±8.7
	850	60.11	0.02125	0.6696	1.121	12.38	53.93±0.60	808.3±8.2
	1,100	60.25	0.006221	0.02361	4.282	47.32	58.41±0.60	861.6±8.0
	1,250	62.65	0.005120	0.09929	2.329	25.74	61.15±0.62	893.4±8.2
	1,500	153.50	0.2419	0.7044	0.1639	1.82	82.45±1.53	1,123±17
ZB-4	400	46.33	0.04936	1.070	0.2213	0.77	31.91±0.46	520.5±7.4
	700	53.49	0.01951	0.2750	0.4310	1.50	47.78±0.53	732.4±7.6
	800	66.99	0.05243	0.2299	0.7219	2.51	51.59±0.67	779.7±9.3
	1,000	67.64	0.003357	0.01703	22.92	79.69	66.65±0.68	955.7±8.5
	1,200	66.77	0.007904	0.02324	4.079	14.20	64.44±0.67	930.9±8.5
	1,450	130.40	0.1775	0.5515	0.3868	1.32	78.28±1.30	1,080±15

m=mixing ratio.
*Radiogenic ^{40}Ar.

3 Results and discussions

Fig. 2 shows that the data points for the whole rock and mineral separates, including phengite, fit a straight line on the Sm-Nd isochron diagrams, even though the phengite apparently was formed later than garnet and omphacite. This good isochron fit suggests that the age of phengite is very close to the age of the garnet and omphacite, and they have similar initial $^{143}Nd/^{144}Nd$ ratios. The Sm-Nd ages of 226.2 ± 4.5 and 228.4 ± 6 Ma defined by the two isochrons in Fig. 2 are within the range of other Sm-Nd ages (208–244 Ma) for eclogite from this belt (Li et al., 1989a, b, 1992, 1993). Very low initial ε_{Nd}-values of –13 and –17 suggest that the protoliths of these eclogites are ancient crustal rocks.

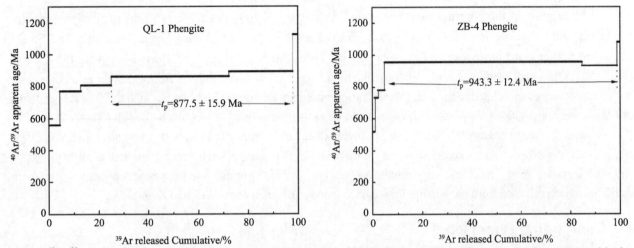

Fig. 3 $^{40}Ar/^{39}Ar$ age spectra of phengite from eclogite sample QL-1 and ZB-4. The following factors and constants were used in the calculations: $(^{40}Ar/^{39}Ar)_K=7.15\times10^{-3}$, $(^{36}Ar/^{37}Ar)_{ca}=2.64\times10^{-4}$, $(^{39}Ar/^{37}Ar)_{ca}=6.87\times10^{-4}$, $\lambda_{40K}=5.543\times10^{-10}a^{-1}$ and $^{40}K/K=1.167\times10^{-4}$ (atom/atom) (S. Wang et al., 1987). Analytical uncertainties (1σ) are represented by vertical width of bars. Sample ZB-4 gives a "plateau" age of 943.3 + 12.4 (1σ) Ma. Sample QL-1 gives an incremental apparent age spectrum. Most of its ^{39}Ar (73.06%) is released at high temperature and defines a "plateau" age of 877.5 ± 15.9 (1σ) Ma.

Since the $^{87}Rb/^{86}Sr$ ratios of garnet, omphacite and whole rock of these two eclogites are close to zero, the slopes of the Rb-Sr isochrons in Fig. 2 are mainly defined by data points of phengite. Rb-Sr ages of 219.5 ± 0.5 and 223.9 ± 0.9 Ma given by these isochrons controlled by phengites are consistent with but probably slightly lower than Sm-Nd mineral isochron ages.

In contrast to the Sm-Nd and Rb-Sr ages, phengites yield much higher $^{40}Ar/^{39}Ar$ "plateau" ages of 877.5 ± 15.9 and 943.3 ± 12.4 Ma. These "plateau" age spectra cover >73% and > 93% of the total ^{39}Ar release for samples QL-1 and ZB-4, respectively. Plateau age spectra with no U-shape and no indication of excess ^{40}Ar may lead some authors to think that the $^{40}Ar/^{39}Ar$ ages of phengite from eclogite have a certain geological significance. However, it is now well documented that biotites containing excess ^{40}Ar can exhibit flat release patterns (Pankhurst et al., 1973; Roddick et al., 1980; Foland, 1983; McDougall and Harrison, 1988), and the isochron analysis gives no indication of excess ^{40}Ar (Roddick et al., 1980; Foland, 1983). A recent study demonstrated that structural breakdown of the hornblende during heating in the vacuum system can yield plateau age spectra, regardless of any original Ar spatial gradients in the mineral (Lee et al., 1991). In this context, excess ^{40}Ar in the phengite cannot be recognized by the step heating technique. Independent evidence may be required to establish whether a plateau date from the phengite is truly a geologically significant age.

There are several other lines of arguments which suggest that the above $^{40}Ar/^{39}Ar$ ages derived from phengites do not have any geological significance:

(1) The high $^{40}Ar/^{39}Ar$ ages contradict the observation that all phengites appear to occur in the foliation later than garnet and omphacite.

(2) The Ar closure temperature of 350 ℃ for phengite is lower than that of Nd and Sr for garnet and phengite under the same conditions (Dodson, 1973; Vance and O'Nions, 1990). If eclogites developed in this belt were formed in the Precambrian and Sm-Nd ages were reset in the Triassic by retrograde metamorphism and recrystallization of garnet and omphacite during their early uplift time, it is inconceivable that the $^{40}Ar/^{39}Ar$ age would not be reset by the same process. The possibility of increased Ar closure temperature for phengite under ultrahigh pressure can be eliminated by examining the P-T-t path of eclogite from the Dabie Mountains. Li et al.

(1993) show that metamorphic pressures decreased dramatically and temperature lowered very slowly in the early stage of uplift. When eclogite rose to a level corresponding to amphibolite facies, the metamorphic temperatures were still higher than 400 ℃ (i.e. above Ar closure temperature of phengite) while the pressures were lowered to 4–7 kbar (Li et al., 1993). Some experiments have suggested that pressure has little influence upon Ar diffusion coefficient for $P<10$ kbar (Giletti and Tullis, 1977); therefore, the phengite should have been reset at this P-T condition and yielded a younger age.

(3) In contrast to the phengite from eclogite, $^{40}Ar/^{39}Ar$ (or K-Ar) ages (204–245 Ma) of phengite, biotite, amphibole and riebeckite from felsic rocks and glaucophane schist in the Qinling-Dabie Mountains and Su-Lu terrain are similar to the Sm-Nd ages of eclogite (208–244 Ma) (Mattauer et al., 1985,1991; Sang et al., 1987; Cong et al., 1991; Chen et al., 1992; Li et al., 1993). If the high $^{40}Ar/^{39}Ar$ ages of phengites from eclogites reflect the time for high-pressure event related to the Proterozoic Yangtse orogeny as suggest by Mattauer et al. (1991), it is difficult to explain why phengites from gneiss, mica schist and blueschist developed in this belt do not give such old ages.

Assuming phengite ages of 223 and 226 Ma (average of Sm-Nd and Rb-Sr ages) for samples QL-1 and ZB-4, respectively, the excess ^{40}Ar contents estimated for them are 2.7×10^{-9} and 5.1×10^{-9} mol·g^{-1}, which amount to 78.5% and 80.7% of their total ^{40}Ar. These large amounts of excess ^{40}Ar incorporated into the younger phengite of eclogite could have been released from the hosting Proterozoic leucocratic gneiss that is rich in K and ^{40}Ar.

4 Concluding remarks

For the first time, the existence of large amounts of excess Ar in phengites was demonstrated by using different dating methods on the same sample. We would suggest that $^{40}Ar/^{39}Ar$ or K-Ar method in many occasions may not be suitable for dating of phengite from orogenic eclogite because of its potential excess argon problem.

Similar to the Dabie Mountains, contradiction between $^{40}Ar/^{39}Ar$ ages (100 Ma) of phengites and Sm-Nd mineral isochron ages (40 Ma) for coesite-bearing rocks has also been reported in the Dora Maira massif, Western Alps (Monié and Chopin, 1991; Tilton et al., 1991). Scaillet et al. (1992) have suggested that localized introduction of an extraneous excess ^{40}Ar may be responsible for the high $^{40}Ar/^{39}Ar$ apparent ages ($\geqslant130$ Ma) of phengites from the Dora Maira massif. However, they do not discuss the contradiction between $^{40}Ar/^{39}Ar$ ages (100 Ma) and Sm-Nd ages (40 Ma). In our opinion, in addition to the possibility of different P-T-t history for different tectonic blocks (studied by different dating methods), the excess argon problem in phengite from eclogites may be more common than previously thought.

Acknowledgements

We thank S.-S. Sun, James K.M. Lee, J.C. Roddick and Ching-Hua Lo for their helpful discussion and/or review and English corrections of this manuscript. This work was supported by the National Natural Science Foundation of China.

References

Chen, W., Harrison, T.M., Heizler, M.T., Liu, R., Ma, B. and Li, J., 1992. The cooling history of melange zone in north Jiangsu-south Shandong region: evidence from multiple diffusion domain ^{40}Ar-^{39}Ar thermal geochronology. Acta Petrol.Sin., 8: 1-17 (in Chinese, with English abstract).

Cohen, A.S., O'Nions, R.K., Siegenthaler, R. and Griffin, W. L., 1988. Chronology of the pressure-temperature history recorded by a granulite terrain. Contrib. Mineral. Petrol., 98: 303-311.

Cong, B., Zhang, R., Li, S., Wang, S. and Chen, Y.C., 1991. Isotopic geochronology of Subei-Jiaonan-Jiaodong high pressure metamorphic belt, China. Annu. Rep. Lab. Lithostr. Tectonic Evolution, Inst. Geol. Acad. Sin., Press of Sciences and Technology of China, Beijing, pp. 68-72 (in Chinese).

Dodson, M.H., 1973. Closure temperature in cooling geochronological and petrological systems. Contrib. Mineral. Petral., 40: 259-274.

Foland, K.A., 1983. $^{40}Ar/^{39}Ar$ incremental heating plateaus for biotites with excess argon. Isot. Geosci., 1: 3-21.

Giletti, B.J. and Tullis, J., 1977. Studies in diffusion, IV. Pressure dependence of Ar diffusion in phlogopite mica. Earth Planet. Sci. Lett.,

35: 180-183.

Lee, J.K.W., Onstott, T.C., Cashman, K.V., Cumbest, R.J. and Johnson, D., 1991. Incremental heating of hornblende in vacuo: Implications for $^{40}Ar/^{39}Ar$ geochronology and the interpretation of thermal histories. Geology, 19: 872-876.

Li, S., Ge, N., Liu, D., Zhang, Z., Ye, X., Zheng, S. and Peng, C., 1989a. The Sm-Nd isotopic age of C-type eclogite from the Dabie Group in the Northern Dabie Mountains and its tectonic implications. Chin. Sci. Bull., 34: 1625-1628.

Li, S., Hart, S.R., Zheng, S., Liu, D., Zhang, G. and Guo, A., 1989b. Timing of collision between the North and South China Blocks -- the Sm-Nd isotopic age evidence. Sci. China (Ser. B), 32: 1391-1400.

Li, S., Liu, D., Chen, Y., Ge, N., Zhang, Z. and Ye, X., 1992. Sm-Nd isotope age of coesite-bearing from the Southern Dabie Moutains. Chin. Sci. Bull., 37: 1638-1641.

Li, S., Xiao, Y., Zhang, Z., Chen, Y., Sun, S.S., Cong, B., Liu, D., Ge, N., Hart, S.R. and Zhang, R., 1993. Collision of the North China and Yangzi Blocks and formation of coecite-bearing ecologites: Timing and processes. Chem. Geol., 109: 89-111.

Mattauer, M., Matte, P., Malavieille, J., Tapponnier, P., Maluski, H., Xu, Z., Lu, Y. and Tang, Y., 1985. Tectonics of the Qinling Belt: build-up and evolution of eastern Asia. Nature (London), 317:496-500.

Mattauer, M., Matte, P., Maluski, H., Xu, Z., Zhang, Q. and Wang Y., 1991. Paleozoic and Triassic plate boundary between North and South China —New structural and radiometric data on the Dabie-Shan (Eastern China). C.R. Acad. Sci. Paris, Sér. II, 312: 1227-1233 (in French, with English abstract).

McDougall, I. and Harrison, T.M., 1988. Geochronology and Thermo-chronology by the $^{40}Ar/^{39}Ar$ Method. Oxford University Press, Oxford, 110 pp.

Monié, P. and Chopin, C., 1991. $^{40}Ar/^{39}Ar$ dating in coesite-bearing and associated units of the Dora Maira massif, western Alps. Eur. J. Mineral., 3: 239-262.

Okay, A.I. and Sengör, A.M.C., 1992. Evidence for intracontinental thrust-related exhumation of the ultra-high-pressure rocks in China. Geology, 20: 411-414.

Okay, A.I., Xu, S. and Sengör, A.M.C., 1989. Coesite from the Dabie Shan eclogites, Central China. Eur. J. Mineral., 1: 595-598.

Pankhurst, R.J., Moorbath, S., Rex, D.C. and Turner, G., 1973. Mineral age patterns in ca. 3700 m.y. old rocks from West Greenland. Earth Planet. Sci. Lett., 20: 157-170.

Roddick, J.C., Cliff, R.A. and Rex, D.C., 1980. The evolution of excess argon in Alpine biotites—A $^{40}Ar/^{39}Ar$ analysis. Earth Planet. Sci. Lett., 48: 185-208.

Sang, B., Chen, Y. and Shao, G., 1987. Rb-Sr isotopic age of Susong Group on the Southern root of the Dabie Mountains and its tectonic implications. Region. Geol. China, No. 4, pp. 364-370 (in Chinese).

Scaillet, S., Feraud, G., Ballèvre, M. and Amouric, M., 1992. Mg/Fe and [(Mg,Fe)Si-Al$_2$] compositional control on argon behaviour in high-pressure wite micas: a $^{40}Ar/^{39}Ar$ continuous laser-probe study from the Dora-Maira nappe of the internal Western Alps, Italy. Geochim. Cosmochim. Acta, 56: 2851-2872.

Tilton, G.R., Schreyer, W. and Schertl, H.P., 1991. Pb-Sr-Nd isotopic behaviour of deeply subducted crustal rocks from the Dora Maira Massif, Western Alps, Italy, II. What is the age of the ultrahigh-pressure metamorphism? Contrib. Mineral. Petrol., 108: 22-23.

Vance, D. and O'Nions, R.K., 1990. Isotopic chronometry of zoned garnets: growth kinetics and metamorphic histories. Earth Planet. Sci. Lett., 97: 227-240.

Wang, S., Hu, S., Zhai, M., Song, H. and Qiu, J., 1987. An application of the $^{40}Ar/^{39}Ar$ dating technique to the formation time of Qingyuan granite-greenstone terrain in NE China. Acta Petrol. Sin., No. 4, pp. 55-62 (in Chinese).

Wang, X., Liou, J.G. and Mao, H.K., 1989. Coesite-bearing eclogite from the Dabie Mountains in central China. Geology, 17: 1085-1088.

Xu, J., Liu, D. and Li, X., 1987. Integration between the Southern and Northen Blocks, Eastern China in Mesozoic. Proc. Geol. Symp. on Mesozoic and Cenozoic, Geological Publishing House, Beijing, 99-112 (in Chinese).

Xu, S., Okay, A.I., Ji, S., Sengör, A.M.C., Liu, Y. and Jiang, L., 1992. Diamond from the Dabie Shan Metamorphic rocks and its implication for tectonic setting. Science, 256: 80-92

Zhang, R., Cong, B., Hirajima, T. and Banno, S., 1990. Coesite eclogite in Su-Lu region, Eastern China. Eos (Trans. Am. Geophys. Union), 71: 1708 (abstract).

Zhang, R., Hirajima, T., Banno, S., Cong, B. and Li, J., 1991. Discovery of coecite-bearing eclogite from Donghai in the Jiangsu Province, China, and its geological implications. Annu. Rep. Lab. Lithosstr. Tectonic Evolution, Inst. Geol. Acad. Sin., Press of Sciences and Technology of China, Beijing, pp. 57-60 (in Chinese).

大别山-苏鲁地体超高压变质年代学
——I. Sm-Nd 同位素体系*

李曙光[1]，E. Jagoutz[2]，肖益林[1]，葛宁洁[1]，陈移之[1]

1. 中国科学技术大学地球和空间科学系，合肥 230026
2. 德国马普化学所宇宙化学研究室，德国

> **亮点介绍**：本文系统讨论了影响榴辉岩 Sm-Nd 同位素定年的各种因素，指出岩石后期蚀变及与深部流体的相互作用是影响 Sm-Nd 定年的最主要因素。大别山含柯石英榴辉岩的可靠 Sm-Nd 同位素年龄集中在 221~232 Ma，大别山碧溪岭榴辉岩 Sm-Nd 同位素年龄系统偏年轻的原因是高压变质时矿物之间的 Nd 同位素未达到平衡。但该文将碧溪岭榴辉岩高压变质矿物未达到平衡的原因归结为其原岩类型是辉长岩，矿物颗粒粗且缺少变质流体的结论不正确。我们后续的深入研究证明是退变质流体局部改造了石榴石和绿辉石的 Sm-Nd 体系所致（An et al., 2018）。

摘要 原岩是辉长岩的含柯石英榴辉岩，其超高压变质作用将在"干"条件下进行，矿物具有较高的封闭温度，并能观察到 Nd 同位素不平衡现象。若榴辉岩原岩是变玄武岩，在超高压变质时，角闪石将分解并释放出水，因而矿物 Nd 同位素封闭温度较低，并可达到 Nd 同位素平衡。岩石后期蚀变及与深部流体的相互作用是影响 Sm-Nd 定年的最主要因素。大别山含柯石英榴辉岩的可靠 Sm-Nd 同位素年龄集中在 221~232 Ma，虽然数值略低，但接近榴辉岩的峰期变质时代。苏家河低温榴辉岩有可能是古生代洋壳俯冲成因。

关键词 大别山；苏鲁地体；超高压变质岩；Sm-Nd 同位素定年

华中大别山与苏鲁地体超高压变质岩是华北与扬子两大陆块碰撞的产物，准确测定其同位素年龄有重要构造意义。自 1989 年以来，大多数已发表的大别山-苏鲁地体超高压变质岩同位素年龄，证明该超高压变质作用发生在三叠纪[1-11]。但是，大别山是否存在多期超高压变质作用仍是一个有争议的问题[12-15]。这一分歧可能主要是由于我们对超高压变质条件下同位素体系认识不足，与部分方法的适用性及样品质量有关。现在尽管对超高压变质的许多影响因素还不很清楚，但对某些同位素体系的认识已有较大进步。例如，对蓝片岩的多期变质过程中 Sr 的均一化尺度进行了动力学分析，为正确解释 Rb-Sr 全岩等时线年龄与单矿物年龄的差异提供了理论依据[16]；榴辉岩多硅白云母中大量过剩 Ar 的发现[9]，解决了 $^{40}Ar/^{39}Ar$ 或 K-Ar 年龄老于 Sm-Nd 年龄的矛盾，然而对超高压岩石的 Sm-Nd 体系还有许多问题值得深入探索。如一些作者在某些榴辉岩和石榴橄榄岩中观察到在石榴石和单斜辉石之间存在同位素不平衡现象[17,18]，这一问题在大别山超高压变质岩中是否也可能存在呢？近年来，一些作者强调榴辉岩的 Sm-Nd 矿物等时线年龄不能指示超高压变质时代，而仅是退变质改造年龄[15]。这样，一个尖锐的问题被提出来了，在榴辉岩退变质过程中，其 Sm-Nd 体系能够被重新平衡吗？榴辉岩的 Sm-Nd 矿物等时线年龄的真实地质意义是什么？

为回答上述问题，本文将深入探讨在超高压变质条件下的 Sm-Nd 同位素体系、可能存在的问题及对大别山超高压变质岩定年的影响，并对已发表的大别山-苏鲁地体超高压变质岩 Sm-Nd 同位素年龄（见表 1）的可靠性、影响因素及其意义进行讨论。

---

* 本文发表在：中国科学（D 辑），1996, 26: 249-257
对应的英文版论文为：Li Shuguang, Jagoutz E., Xiao Yilin, et al., 1996. Chronology of ultrahigh-pressure metamorphism in the Dabie Mountains and Su-Lu terrane: I Sm-Nd isotope system. Sciences in China (Ser. D), 39: 597-609

表 1 大别山-苏鲁地体高压变质岩 Sm-Nd 同位素年龄

构造单元	样号	地点	岩石类型	等时线矿物组合	年龄 (Ma±2σ)	$\varepsilon_{Nd}(t)$	参考文献
苏家河混杂岩带	Hy23-1	熊店	低温榴辉岩	3gar+2omp	544±14[a]	+2.1	[13]
				3gar+w	422±67	+5.1	[19]
北大别地体	R-4	饶拔寨	石榴辉石岩	gar+cpx+w	244±11	−3.0	[1]
	G-1	高坝岩	石榴辉石岩	3gar	224±20	−4.7	[2]
南大别超高压变质带	ND	石马	柯石英榴辉岩	2gar+omp	221±5	−6.2	[3]
	T16-1	石马	石榴黑云片麻岩	3gar+w	229±3	−7.1	[6]
	HT-4	双河	柯石英榴辉岩	gar+omp+rut	226.2±2.9	−11.9	本文
				gar+w	234.6±3.2[a]	−13	本文
	HS4-4	双河	柯石英榴辉岩	gar+omp+ph+w	225±2	−4.6	[20]
	HT-12	双河	退变质榴辉岩	gar+w	242±3.6[a]	10.9	本文
				am+sph	197±44	−10.1	本文
	DX36A-1	碧溪岭	柯石英榴辉岩	gar+rut+omp+ky	210.1±1.5[a]	−3	本文
				gar+rut+omp+ky	239±5.8[a]	−2.7	本文
	93DB-1	碧溪岭	石榴石橄榄岩	gar+cpx+w	219±7[a]	0.7	本文
	TB87	碧溪岭	柯石英榴辉岩	gar+omp+w	243.2±0.3[a]	15.9	[12]
	DX67-1	碧溪岭	柯石英榴辉岩	gar+omp+w	230.1±8.3	+4.6	[14]
	2300	五庙	退变质榴辉岩	gar+w	246.1±8[a]	−19.2	[8]
	M105-4	蜜蜂尖	退变质榴辉岩	3gar+3w	481±25[a]	−10.3	[12]
苏鲁超高压变质带	E87N9-1	日照的哑口	柯石英榴辉岩	gar2+2omp	221±6	−13.1	[5]
	E87M1	日照的马家沟	柯石英榴辉岩	gar1+2omp	280±14[a]		[5]
				gar1+omp1	211±83[a]	−14	[5]
				gar1+omp2	208±33[a]		[5]
	E87Y-3	荣成的迟家店	柯石英榴辉岩	gar+omp	232	−5	[4]
	E87Y-4	荣成的迟家店	柯石英榴辉岩	gar+omp	294	+5.5	[4]
	QL-1	青龙山	柯石英榴辉岩	gar+omp+ph+w	226.3±4.5	−13.2	[9]
	ZB-4	莒南的朱边	柯石英榴辉岩	gar+ph+w	228.4±6	17.6	[9]
	EcpV-8	荣成的腾家集	柯石英榴辉岩	2gar+cpx+w	128±4[a]	−13	[11]
	Ec31-2	诸城桃林	柯石英榴辉岩	gar+cpx+w	222±6	−19.4	[11]
红安群低温榴辉岩带	DJ9-7	高桥	低温榴辉岩	gar+omp+w	719±48[a]	+3.5	[13]

a) 年龄值受某种因素影响而偏高或偏低(详见正文).

应当说明, 大别山高压变质岩种类繁多, 有含柯石英榴辉岩及其他超高压变质岩、低温榴辉岩、石榴辉石岩和蓝片岩等. 它们在空间上各自形成特征的变质带, 在成因及形成时代上也可能存在某些不同[19]. 因此, 在讨论其同位素年龄时, 区分其所处构造带位置及岩性是很重要的. 图1显示了各采样点位置.

1 超高压条件下 Nd 同位素不平衡的可能性及其影响

Jagoutz 和 Thoin 等, 在一些榴辉岩及石榴橄榄岩中观察到在石榴石和单斜辉石之间存在 Nd 同位素不平衡现象[17,18], 他们指出在大多数情况下, 石榴石-单斜辉石年龄比真实年龄年轻. 不过这些 Nd 同位素不平衡现象都是在粗粒、低温榴辉岩及金伯利岩中的榴辉岩包体中观察到的. 在超高压变质条件下, 上述 Nd 同位素不平衡现象是否存在及影响程度还不清楚. 对此, 我们以大别山超高压变质岩为例做了较深入研究, 该研究选择了两类榴辉岩进行精确的 Sm-Nd 同位素定年. 第一类以碧溪岭含柯石英榴辉岩为代表, 为粗粒结

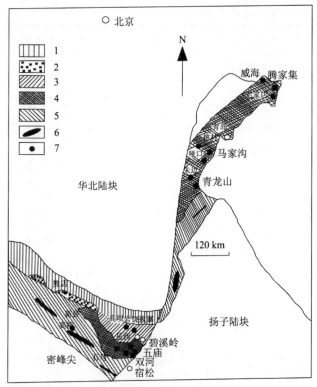

图1 大别山及苏鲁地体高压及超高压岩石 Sm-Nd 同位素年龄样品位置示意图

1-北淮阳角闪岩-绿片岩相变质带; 2-苏家河构造混杂岩带; 3-北大别高温变质带; 4-南大别及苏鲁超高压变质带; 5-红安群、宿松群、低温榴辉岩及蓝片岩带; 6-蓝片岩; 7-样品位置

构,条带状构造,原岩为一巨大的($0.7~km^2$)超镁铁-镁铁质岩深成侵入体。其中辉长岩变质为榴辉岩,橄榄岩变质为石榴橄榄岩。第二类以双河含柯石英榴辉岩为代表,为均匀细粒结构,与超高压变泥质岩、大理岩互层,原岩是玄武岩。为了揭示可能存在的 Nd 同位素不平衡现象,每个样品都尽可能多地选出高压变质矿物进行测定。

碧溪岭榴辉岩测定的高压矿物有石榴石 + 角闪石 + 绿辉石 + 金红石 + 蓝晶石 + 多硅白云母,它们均早于后成合晶阶段,属榴辉岩相矿物,但多硅白云母受后期蚀变影响而部分水云母化,其 Sm-Nd 同位素体系可能受到影响而不参与等时线年龄计算。图 2(a) 显示这些矿物数据点的线性排列不好。在所有可能的拟合中,石榴石 + 金红石 + 绿辉石 + 蓝晶石拟合给出最低的年龄值 210.1 ± 1.5 Ma;而角闪石 + 金红石 + 绿辉石 + 蓝晶石拟合给出最高年龄值 239 ± 5.8 Ma。该岩体石榴橄榄岩的石榴石 + 单斜辉石 + 全岩给出 219 ± 7 Ma (图 2(b))。这些年龄值都在德国马普化学所测定,不存在实验室间的误差。同一样品高压矿测定值的离散性,石榴石 + 其他矿物给出的较低年龄值,说明碧溪岭榴辉岩在高压变质过程中,石榴石与其他矿物的 Nd 同位素未达到完全平衡。但是,它也表明该岩体 Nd 同位素不平衡现象的影响并不严重,其年龄比真实 Sm-Nd 年龄(见下面) 仅低 10~20 Ma。碧溪岭榴辉岩之所以存在同位素不平衡现象,可能与它的粗粒结构、原岩缺少含水矿物因而变质时缺少流体有关(详见第 4 节)。但由于它很高的峰期变质温度 (841~930 ℃)[20],使得这一不平衡现象并不严重。

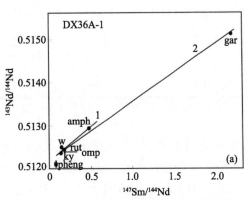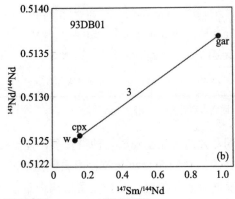

图2 碧溪岭含柯石英榴辉岩(a)和石榴橄榄岩(b)Sm-Nd 等时线图

1-$\varepsilon_{Nd}(t) = -2.7$, $t = 239 \pm 5.8$ Ma; 2-$\varepsilon_{Nd}(t) = -3$, $t = 210.1 \pm 1.5$ Ma; 3-$\varepsilon_{Nd}(t) = -0.7$, $t = 219 \pm 24$ Ma。gar-石榴石, amph-角闪石, rut-金红石, omp-绿辉石, ky-蓝晶石, pheng-多硅白云母, w-全岩, cpx-单斜辉石

双河含柯石英榴辉岩(HT-4),测定了 3 个高压矿物 (石榴石 + 绿辉石 + 金红石)及全岩,这 3 个矿物点的线性较好,并给出 226.2 ± 2.9 Ma 的 Sm-Nd 矿物等时线年龄(图 3(a))。双河另一含柯石英榴辉岩 (HS4-4)的石榴石 + 绿辉石 + 全岩等时线给出与 HT-4 几乎一致的年龄值 225 ± 2 Ma (图 3(b))[21]。3 个以上单矿物点的良好线性关系及不同样品测定年龄值的一致性,说明双河榴辉岩的同位素在超高压变质过程中达到了平衡。这可能与其细粒结构,尤其与它的原岩(变玄武岩)含有较多含水矿物(角闪石)有关。

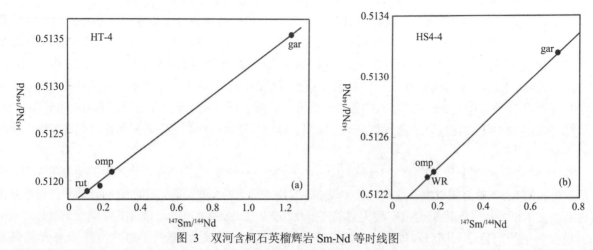

图 3 双河含柯石英榴辉岩 Sm-Nd 等时线图

(a)德国马普化学所测定。$\varepsilon_{Nd}(t) = -11.9$，$\varepsilon_{Nd}(t) = 226.2 \pm 2.9$ Ma；(b)南京大学测定。$\varepsilon_{Nd}(t) = -4.6$，$225.2 \pm 1.4$ Ma。gar-石榴石，omp-绿辉石，rut-金红石，WR-全岩

上述关于 Nd 同位素平衡或不平衡现象的观测，对评价大别山及苏鲁地体其他 Sm-Nd 年龄值有很大意义。如山东日照马家沟榴辉岩(E87M1)给出的两个较低的石榴石 + 绿辉石年龄 211 ± 83 Ma 和 208 ± 3 Ma 很可能是 Nd 同位素不平衡所致(表 1)。其他大多数榴辉岩和石榴辉石岩 Sm-Nd 同位素年龄均未表现出 Nd 同位素不平衡现象，尤其高坝岩石榴辉石岩(G-l)的等时线是由 3 个石榴石样品确定的[2]，这样就排除了石榴石与单斜辉石同位素不平衡的可能性。苏北青龙山含柯石英榴辉岩(QL-l)的等时线是由 3 个矿物确定的，同样也证明它们 Nd 同位素达到了平衡，其年龄值(226.3 ± 4.5 Ma) 与双河榴辉岩几乎一致[9]。

尽管大别山含柯石英榴辉岩的 Nd 同位素可以达到或接近平衡，但是在一个手标本尺度范围内 (6~8 cm)，观察到了 Nd 同位素的不平衡[6]，因此，榴辉岩的 Sm-Nd 全岩等时线定年是不可靠的。

2 石榴石 Nd 同位素封闭温度及榴辉岩 Sm-Nd 年龄的意义

石榴石 Sm-Nd 体系的封闭温度是一个有争议的问题。Hamphries 等根据 Sm 的扩散系数计算了冷却速率为 10 ℃/Ma，直径为 1 mm 的镁铝榴石的封闭温度为 500 ℃ 和钙铝榴石的封闭温度为 700 ℃[21]，Mezger 等通过比较同一地质单元的石榴石 Sm-Nd 年龄和其他已知封闭温度矿物的 U-Pb 年龄，获得了在缓慢冷却条件下(2~4 ℃/Ma)石榴石 Sm-Nd 体系封闭的温度为 600 ± 30 ℃[22]。

然而也有其他一些证据表明，石榴石中 Sm-Nd 体系的扩散远比上述实验结果要慢。Cohen 等在研究挪威反应边岩(麻粒岩相)时发现，在这些样品中石榴石的 Sm-Nd 与主要元素一样具有环带结构，它表明 REE 在石榴石中的扩散不比主要元素快[23]。根据已计算的石榴石的冷却速率，石榴石 Fe/Mg 的封闭温度估计为 760 ℃[24]。Jagogutg 在研究金伯利岩榴辉岩包体的 Sm-Nd 体系后，也获得石榴石的 Nd 封闭温度在干条件下可高达 850 ℃ 的结论[17]。

本文认为上述有关石榴石 Sm-Nd 封闭温度的差异主要是由于作者所用的方法不同，因而 Sm-Nd 扩散条件不同造成的。根据扩散系数计算封闭温度是在矿物表面的零浓度假设基础上进行的[25]，对于榴辉岩只有粒间存在相互连通的孔隙流体时才能满足这一假设。因此，根据扩散系数计算的封闭温度对应的是存在粒间流体的"湿"变质条件，而根据包体及反应边岩观察估计的较高封闭温度对应的是"干"变质条件。

榴辉岩相变质时的粒间流体主要来自原岩中含水矿物。因此，榴辉岩变质的"干""湿"条件主要取决于它的原岩性质。如果榴辉岩原岩是辉长岩，如碧溪岭榴辉岩，它基本无含水矿物。因此，它们在榴辉岩相变质时缺乏粒间流体，矿物的 Sm-Nd 封闭温度高。如果榴辉岩原岩是玄武岩，如双河榴辉岩，在榴辉岩相变质前它们已先变质为斜长角闪岩。在榴辉岩相变质时，其主要造岩矿物角闪石转变为单斜辉石并释放出水。特别是一些含钠闪石，它们要在超高压条件下变得不稳定而分解成硬玉及滑石[26]。因此，这类榴辉岩在变质峰期以前的高压变质作用是在"湿"条件下完成的，其矿物 Sm-Nd 封闭温度较低。

碧溪岭榴辉岩的峰期变质温度高达 841~930 ℃，已超过石榴石"干"条件下估计的最高封闭温度(850 ℃)[17]。但它仍表现有石榴石与其他高压矿物的同位素不平衡现象。这可能是因为对榴辉岩这种双矿物岩石，其同位素平衡取决于封闭温度最高的矿物。根据 Sneeringer 等测定的透辉石高压条件下(P = 2 GPa) Sm 的扩散系数可计算出在冷却速率为 10 ℃/Ma 和 1 mm 直径条件下，单斜辉石封闭温度在 800 ℃左右(a 轴方向)[27]，它高于同样条件下 Sm 在石榴石中的封闭温度[21]。可以预料在干条件下，单斜辉石的 Sm-Nd 封闭温度可能会接近 1000 ℃。这样，单斜辉石在形成后即已封闭，即使石榴石还没有封闭，已无法实现完全的同位素平衡。

双河榴辉岩的峰期变质温度为 733~882 ℃[28]，平均达 810 ℃左右，已高于在"湿"变质条件下石榴石及单斜辉石的 Sm-Nd 封闭温度。这是它能实现 Nd 同位素平衡的主要原因。由于它的峰期变质温度高出辉石封闭温度不很多，因此其 Sm-Nd 矿物等时线年龄(226.2 ± 2.9 Ma)很可能已接近其峰期变质时代。双河榴辉岩根据共生的粗颗粒石榴石和辉石边部测定值计算的变质温度仅为 655~760 ℃，它低于单斜辉石的 Sm-Nd 封闭温度。显然，榴辉岩的 Sm-Nd 体系不可能在这一低变质温度条件下重新平衡。因此，双河榴辉岩的 226 Ma 年龄指示的不是榴辉岩抬升过程中的"重结晶"时代。

3 退变质作用对 Sm-Nd 定年的影响

超高压岩石的退变质温度小于 700 ℃。因此，即使有流体介入，退变质作用也不可能使石榴石和绿辉石的 Nd 同位素重新平衡。但是，退变质过程中形成的各种含水矿物的 Sm-Nd 体系可以被退变质作用重置。为避免退变质对 Sm-Nd 定年的影响，样品质量至关重要。用于测试的单矿物样品必须是在双目镜下逐颗挑选完全新鲜，没有任何蚀变的颗粒。在大多数情况下，辉石比石榴石更容易发生退变质和蚀变，由于来自浅部长英质地壳的流体具有较低的 ^{143}Nd/^{144}Nd 值，蚀变的单斜辉石将给出偏低的 Nd 同位素比值，从而加大石榴石与它连线的斜率而给出较高的年龄。同理，由于全岩样品无法剔除其中的退变质或蚀变产物，石榴石 + 全岩"等时线"也要给出偏高的年龄值。例如图 3(a) 的 HT-4，它的全岩数据点却落在了矿物等时线下面。它与石榴石的连线给出 234.6 ± 3.2 Ma 的年龄，它高于由 3 个矿物确定的等时线年龄(226.2 ± 2.9 Ma)。图 4 显示的是来自双河大理岩中强退变榴辉岩包体(HT-12)的 Sm-Nd 等时线图。该样品的绿辉石已全部退变成角闪石，金红石退变成榍石和钛铁矿。该样品与 HT-4 产于同一高压岩片中，应有相同的变质时代，但它的石榴石 + 全岩"等时线"给出了较高的年龄值(242 ± 3.6 Ma)。退变质矿物、角闪石和榍石不在石榴石与全岩的连线上，它们给出较低的年龄值(197 ± 44 Ma)。这说明在退变质时，该样品除石榴石外，其他矿物的 Sm-Nd 体系由于蚀变开放了。据此，我们可以判定下面几个年龄值都是由于退变质作用使其年龄偏高。

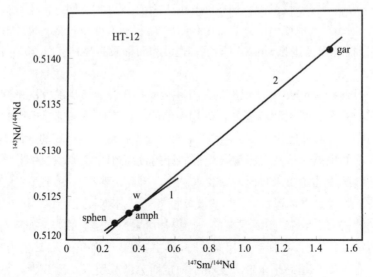

图 4 双河强退变质榴辉岩 Sm-Nd 同位素等时线图

1-t = 197 ± 4.4 Ma, 2-t = 242 ± 3.6 Ma。gar-石榴石, w-全岩, sphen-榍石, amph-角闪石

Okay 等发表了一个石榴石 + 全岩等时线年龄 (246 ± 8 Ma) (表1)[8]。据他们的描述，该样品为五庙大理岩中-强退变质榴辉岩，矿物组合中只有角闪石，而无单斜辉石。

最近报道的湖北英山蜜蜂尖含柯石英榴辉岩的 3 个石榴石 + 3 个全岩样品的"等时线"年龄为 481 ± 25 Ma (表1)[12]，并作为存在古生代超高压变质的证据，但是该样品也是强退变质榴辉岩，辉石已全部退变为后成合晶及角闪石(据张泽明谈话及出示的薄片)。

此外，另一个碧溪岭柯石英榴辉岩年龄(243.2 ± 0.3 Ma)[12]也可能因辉石及全岩含退变质产物的偏高，因为其 $\varepsilon_{Nd}(t) = -15.9$，大大低于他人的+4.6～-3 值(表1)。

上述讨论说明，除了非常新鲜，未退变的榴辉岩，石榴石 + 全岩的 Sm-Nd 定年数据要慎用。尤其对强退变质榴辉岩，它往往给出错误的高年龄值。

4 深部流体对榴辉岩 Sm-Nd 体系改造的可能性

在榴辉岩 Sm-Nd 同位素定年中有时还可观察到另外一种现象，即石榴石具有异常低的 $^{147}Sm/^{144}Nd$ 值(低于球粒陨石值 0.1967)，从而给出异常高的年龄值。例如从柏林等报道的荣成迟家店 E87Y-4 样给出 294 Ma 的老年龄[4]，该样的石榴石具有反常的 REE 模型，其 $^{147}Sm/^{144}Nd$ 值 < 0.18。与该样产于同一超镁铁岩体的 E87Y-3 样的石榴石，就具有正常的 HREE 富集模型 ($^{147}Sm/^{144}Nd > 0.64$)，并给出正常的年龄值 232 Ma[4]。从柏林指出这两个样可能有不同的演化历史。我们认为最大的可能是该岩体在榴辉岩变质作用以后，局部 (E87Y-4) 受到深部流体作用而显著降低了石榴石的 Sm/Nd。

类似情况也见于简平等报道的红安群高桥低温榴辉岩 Sm-Nd 等时线年龄(719 ± 48 Ma)[13]，该年龄值明显不合理，它大大高于红安群斜长角闪岩的角闪石 K-Ar 年龄(225 Ma)[12]。角闪石的 Ar 封闭温度高达 500 ℃，与红安群低温榴辉岩的变质温度相同。因此，红安群变质岩的 Sm-Nd 和 K-Ar 年龄应该有类似的值。高桥榴辉岩的石榴石也具有异常低的 $^{147}Sm/^{144}Nd$ 值(0.1799)[13]。此外，石榴石-单斜辉石温度计给出该样品的变质温度仅 330 ℃[13]，大大低于 Zhou 等测定的红安群低温榴辉岩变质温度 (500 ℃)[29]。高桥榴辉岩的上述异常年龄和异常变质温度，使我们有理由怀疑其 Sm/Nd、Mg/Fe 体系均被后期深部流体改造了。

上述石榴石与深部流体相互作用降低其 Sm/Nd 的假设，与 REE 在水溶液中的地球化学行为相符。已有研究证明，REE 在水溶液中是以络合物形式存在，其中 HREE 络合物的稳定性大大高于 LREE 的稳定性[30]。因此，在深部流体与石榴石相互作用时，较多的 HREE 优先进入流体，从而使石榴石的 Sm/Nd 下降。如果这种深部水-岩作用未充分进行，则有可能只显著改变了矿物的 Sm/Nd 而未使 Nd 同位素均一化，这样就不可能给出有意义的年龄，反之则可给出水-岩作用时代，如山东荣成滕家集榴辉岩 128 ± 4 Ma 年龄可能反映的正是燕山期一次深部流体与榴辉岩相互作用的时代。当然，这种深部流体与榴辉岩相互作用使石榴石 Sm/Nd 下降的假说还有许多细节不清楚，值得今后作为一个重要的课题去深入研究。

从表 1 删掉那些由于上述因素影响而偏低或偏高的年龄，我们可以看到，大别山及苏鲁地体超高压变质岩 Sm-Nd 的年龄集中在 221～232 Ma 之间。这可能便是大别山及苏鲁地体超高压变质岩 Sm-Nd 年龄的真实值。它们的中间值 (226.5 Ma)与双河、青龙山 3 个矿物精确 Sm-Nd 年龄一致，可成为大别山超高压变质岩 Sm-Nd 年龄的最佳估计值，这与最近 Marayand 等报道的大别山高压片麻岩锆石变质增生部分的离子探针 U-Pb 年龄为 238～216 Ma (中间值 227 Ma) 一致[10]，也与 Ames 等报道的 4 个超高压变质锆石的 $^{207}Pb/^{206}Pb$ 年龄平均值(226.1 Ma)一致(将在另一篇有关超高压变质岩锆石同位素体系论文中详述)。因此，上述 Sm-Nd 年龄的最佳估计值应指示大别山超高压变质峰期变质时代。

5 不同构造单元高压岩石 Sm-Nd 同位素年龄的差异及其意义

已报道了北大别地体的两个石榴辉石岩的 Sm-Nd 同位素年龄[1,2]，其中比较精确的是饶钹寨超镁铁岩体中的石榴辉石岩年龄(244 ± 11 Ma) (表1)，它老于南大别地体含柯石英榴辉岩年龄(221～226 Ma)。最近研究表明，北大别地体不是扬子俯冲陆壳的一部分，它极可能是位于华北与扬子陆块之间的微古陆[19]。在陆-陆碰撞时,扬子陆块向该微型陆块及华北陆块下俯冲,北大别地体作为仰冲壳楔而抬升。因此,北大别地体深部的混杂岩块要较南大别超高压岩块更早抬升、冷却,从而给出较老的 Sm-Nd 年龄。饶钹寨的 244 Ma 指的不是北大别地体仰冲的时代,它更接近两大陆块开始碰撞的时代。

苏家河低温榴辉岩仅有简平等发表的一个年龄数据(表 1)[13],Li 等认为这一年龄值显著偏高了[19],并指出该低温榴辉岩变质时代不应老于(422 ± 6.7 Ma),其 $\varepsilon_{Nd}(t) > +5.1$,这与南大别地体含柯石英榴辉岩非常低的 ε_{Nd} 值形成明显对照(表 1)。因此,Li 等提出苏家河低温榴辉岩是与美国西海岸弗朗西斯科低温榴辉岩的洋壳俯冲成因的榴辉岩类似。它是在古生代,扬子与华北陆块碰撞前,古秦岭洋向华北陆块下俯冲形成,并在 400 Ma 弧-陆碰撞时卷入到苏家河构造混杂岩中的[19]。

致谢

感谢张泽明介绍了英山含柯石英榴辉岩定年样品的情况,并出示了有关岩石薄片及熊店榴辉岩薄片。

参考文献

[1] 李曙光, Hart S R, 郑双根等. 中国华北、华南陆块碰撞时代的 Sm-Nd 同位素年龄证据. 中国科学, B 辑, 1989, (3): 312~319.
[2] 李曙光, 葛宁洁, 刘德良等. 大别山北翼大别群中 C 型榴辉岩的 Sm-Nd 同位素年龄及其构造意义. 科学通报, 1989, 34(7): 522~525.
[3] 李曙光, 刘德良, 陈移之等. 大别山南麓含柯石英榴辉岩的 Sm-Nd 同位素年龄. 科学通报. 1992, 37(4): 397~300.
[4] 从柏林, 张儒瑗, 李曙光等. 中国苏北-胶东南高压变质带的同位素地质年代学初探. 见: 中国科学院地质研究所岩石圈构造演化开放实验室年报(1989-1990). 北京: 中国科学技术出版社, 1991, 68~72.
[5] 陈移之, 李曙光, 从柏林等. 胶南榴辉岩的形成时代及成因——Sm-Nd 同位素地球化学及年代学证据. 科学通报, 1992, 37(23): 2167~2172.
[6] Li S, Xiao Y, Liou D et al. Collision of the North China and Yangtse Blocks and formation of coesite-bearing eclogites: timing and processes. Chemical Geology, 1993, 109: 89~111.
[7] Ames L, Tilton G R, Zhou G. Timing of collision of the Sino-korean and Yangtse cratons: U-Pb zircon dating of coesite-bearing eclogites. Geology, 1993. 21: 339~342.
[8] Okay A I, Sengor A M C. Tectonics of an ultrahigh-pressure metamorphic terrane: the Dabie Shan/Tongbai Shan Orogen, China. Tectonics, 1993,12:1320~1334.
[9] Li S, Wang W, Chen Y et al. Excess argon in phengite from eclogite: evidence from dating of eclogite minerals by Sm-Nd, Rb-Sr and $^{40}Ar/^{39}Ar$ methods. Chemical Geology, 1994, 112: 343~350.
[10] Maruyama S, Liou J G, Zhang R. Tectonic evolution of the ultrahigh-pressure (UHP) and high-pressure (HP) metamorphic belts from Central China. The Island Arc, 1994, 3: 112~121.
[11] 韩宗珠, 赵广涛, 胡受奚等. 中国东部榴辉岩的岩石学特征及其 P-T-t 轨迹研究. 南京大学学报(自然科学版), 1994, 30: 86~97.
[12] 杨巍然, 王林森. 韩郁菁等. 大别蓝片岩-榴辉岩年代学研究. 见: 陈好寿主编. 同位素地球化学研究. 杭州: 浙江大学出版社, 1994, 175~186.
[13] 简平, 叶伯丹, 李志昌等. 大别山带榴辉岩同位素年代学 P-T-t 轨迹及其构造意义. 见: 陈好寿主编. 同位素地球化学研究. 杭州: 浙江大学出版社, 1994, 205~213.
[14] 刘若新. 樊棋诚, 李惠民. 大别山碧溪岭石榴橄榄岩-榴辉岩体的原岩性质及同位素年代学的启示. 岩石学报, 1995, 11(3): 243~256.
[15] 曹荣龙, 朱寿华. 大别山含柯石英榴辉岩区——一个晚太古代的超高压变质带. 地质学报, 1995, 69 (3): 232~242.
[16] 李曙光, 刘德良, 陈移之等. 中国中部蓝片岩的形成时代. 地质科学, 1993, 28(1): 21~27.
[17] Jagoutz E. Nd and Sr systematics in an eclogite xenolith from Tanzania: evidence for frozen mineral equilibria in the continental lithosphere. Geochim Cosmochim Ada, 1988, 52: 1285~1293.
[18] Thoni M, Jagoutz E. Some new aspects of dating eclogites in orogenic belts: Sm-Nd, Rb-Sr and Pb-Pb isotopic results from the Austroalpine Saualpe and koralpe type-locality. Geochim Cosrpochim Acta, 1992, 56: 347~368.
[19] Li S, Jagoutz E, Zhang Z et al. Structure of High-pressure metamorphic belt in the Dabie Mountains and its tectonic implications. Chinese Sci Bull, 1995, 40(Supplement): 138~140.
[20] Xiao Y, Li S, Jagoutz E et al. P-T-t path for coesite-bearing peridotite-eclogite association in the Bixiling, Dabie Mountains. Chinese Sci, Bull, 1995, 40 (Supplement): 156~158.

[21] Hamphries F J, Cliff R A. Sm-Nd dating and cooling history of Scourian granulites, Sutherland. Nature, 1982, 295: 515~517.
[22] Mezger K, Essene E J, Halliday A N. Closure temperatures of the Sm-Nd system in metamorphic garnets. Earth Planet Sci Lett, 1992, 113: 397~409.
[23] Cohen A S, O'Nions R K, Siegenthaler R et al. Chronology of the pressure-temperature history recorded by a granulite terrane. Contrib Mineral Petrol, 1988, 98: 303~311.
[24] Vance D, O'Nions R K. Isotopic chronometry of zoned garnets: growth kinetics and metamorphic histories. Earth Planet Sci Lett, 1990, 97: 227~240.
[25] Dodson M H. Closure temperature in cooling geochronological and petrological systems. Contrib Mineral Petrol, 1973, 40: 259~274.
[26] Zhang R Y, Liou J G. Coesite-bearing eclogite in Henan Province, central China: detailed petrography, glaucophane stability and PT-path. Eur J Mineral, 1994, 6: 217~233.
[27] Sneeringer M, Hart S R, Shimizu N. Strontium and samarium diffusion in diopside. Geochimica et Cosmochimica Acta, 1984, 48(8): 1589~1608.
[28] Cong B, Zhai M, Carswell D A et al. Petrogenesis of ultralhigh-pressure rocks and their country rocks at Shuanghe in Dabieshan, central China. Eur J Mineral, 1995, 7: 119~138.
[29] Zhou G, Liu Y J, Eide E A et al. High-pressure/low-temperature metamorphism in northern Hubei province, central China. J Metamorphic Geol, 1993, 11: 561~574.
[30] Humphirs S E. Rare earth element geochemistry. In: Henderson P, ed. Elsevier Amsterdam-Okford-New York-Tokyo: 1984, 317~342.

大别山-苏鲁地体超高压变质年代学
——II. 锆石 U-Pb 同位素体系*

李曙光[1]，李惠民[2]，陈移之[1]，肖益林[1]，刘德良[1]

1. 中国科学技术大学地球和空间科学系，合肥 230026
2. 地质矿产部天津地质矿产研究所，天津 300170

> **亮点介绍：** 报道了一组新的大别山超高压变质岩及其围岩的锆石 U-Pb 年龄数据，并首次给出大别山超高压变质岩峰期变质时代为(228 ± 2)Ma。

摘要 超高压变质锆石在高压变质及退变质期间可发生少量 Pb 的连续扩散丢失，它不会显著破坏其 $^{207}Pb/^{203}U$ 和 $^{206}Pb/^{235}U$ 年龄的一致性，却使它们小于 $^{207}Pb/^{206}Pb$ 年龄。不含继承组分且普通 Pb 含量低的变质锆石 $^{207}Pb/^{206}Pb$ 年龄，可能更接近超高压变质岩的峰期变质时代。报道了一组新的大别山超高压变质岩及其围岩的锆石 U-Pb 年龄数据，并给出峰期超高压变质时代为(228 ± 2) Ma 的结论。

关键词 大别山；苏鲁地体；超高压变质岩；锆石 U-Pb 定年

自 1993 年以来，用锆石 U-Pb 法测定大别山超高压变质岩变质年代受到地质学家的格外重视[1-2]，近两年来，各种方法的不同锆石年龄值相继报道出来。据此，不同作者对大别山超高压变质时代得出完全不同的结论：计有太古代[3]、元古代[4]、古生代加里东期[5-6]和三叠纪[1-2,7]。对同一高压变质带，甚至对同一榴辉岩(如碧溪岭)获得如此多样的年龄值和完全不同的解释，说明了用锆石 U-Pb 法测定超高压变质岩变质时代的复杂性。这一复杂性主要是由于超高压变质岩中锆石类型及其成因的多样性，以及与锆石 U-Pb 体系在变质过程中的复杂行为有关。本文对这两个问题进行了讨论，还报道了一组新的大别山超高压变质岩及其片麻岩围岩的锆石 U-Pb 年龄，并根据本文及已发表的超高压变质锆石 U-Pb 年龄，估计了大别山超高压变质的峰期时代。

1 超高压变质对继承锆石的影响及变质锆石特征

前人的研究已证明在高压及超高压变质条件下，原岩中的锆石都可以继承下来，并且只遭受了部分 Pb 丢失[8-9]。即使在北哈萨克斯坦 Kokchetav 含金刚石超高压片麻岩(其峰值变质温度高达 900~1000 ℃，压力大于 4 GPa)中，仍然发现了继承锆石的存在并能部分保持其原有 Pb 同位素信息[9]。

由于继承锆石的存在，超高压变质岩中锆石的成因是很复杂的。继承锆石不可能给出指示高压变质时代的 U-Pb 一致年龄。已有研究指出，如果继承锆石只在超高压变质时经过 1 次 Pb 丢失，则其不一致线下交点年龄可以指示超高压变质时代，上交点指示锆石形成时代；如果它们经历了 2 次 Pb 丢失，则其下交点无地质意义并介于 2 次丢失事件之间[10]。

所有已发表并由不同方法验证过的可靠高压和超高压变质 U-Pb 年龄都是变质锆石给出的[1-9,11]。在这些文献中，所有的变质锆石都具有类似的特征：晶体呈浑圆形、卵形，多晶面清晰可见，一些晶面上有锆石增生体，透明、强玻璃或金刚光泽，地区不同锆石颜色不同，有浅褐、浅红、无色。一些高压变质锆石中可见

---

* 本文发表在：中国科学（D 辑），1997, 27(3)：200-206

有高压矿物包裹体,如微粒金刚石、柯石英、石榴石、绿辉石、白云母、金红石等[2,9]。这些高压矿物包裹体证明这些圆形锆石不是碎屑锆石,而是在高压变质条件下结晶的。有的变质锆石还可见到有老的锆石核。此外,在常规 U-Pb 年龄谐和图上,高压变质锆石的数据点都落在一致线上或不一致线下交点附近(若含有老锆石残留组分),而碎屑锆石(也呈圆形)及含有明显老锆石核的变质锆石因 Pb 丢失则会显著偏离一致线,这也是区分变质锆石与碎屑锆石的一个重要依据。

上述讨论表明,为了获得超高压变质年龄,我们必须正确鉴别超高压变质锆石并对其进行测定。由于超高压变质岩中锆石成因的复杂性,锆石年龄同样也需要与其他定年方法的测试结果相互验证才能获得正确结论[8,9]。

2 大别山超高压变质岩及其围岩的锆石 U-Pb 定年

根据上述原则,本文对大别山双河和石马地区 2 个超高压片麻岩及其围岩(花岗片麻岩)的锆石进行了年代学研究。这些样品均测定了 Sm-Nd 矿物等时线年龄,以便进行直接对比。

2.1 样品及地质背景

样品 92HT-1 为超高压变质的石榴石绿帘云母片麻岩,采自潜山县双河村附近(图1),它与含柯石英榴辉岩,含榴辉岩团块大理岩及硬玉石英岩互层并组成一超高压岩片。该榴辉岩 Sm-Nd 矿物等时线年龄为(226.2 ± 2.9) Ma[12]。样品 92HT-1 是从一 40 kg 大样中分离出来的锆石。除变质锆石外,大多数为不同程度磨圆的碎屑锆石,无色、光泽度低。这表明该片麻岩为副片麻岩。变质锆石为浑圆状、浅红色、强玻璃光泽。

样品 92T-16 为石榴黑云母片麻岩,采自太湖县石马乡(图1)。该片麻岩中有榴辉岩、大理岩团块分布。Wang 等在该片麻岩的石榴石中发现有柯石英假象包体[13],表明该片麻岩为超高压片麻岩。其 Sm-Nd 矿物等时线年龄为(229 ± 3) Ma[14]。从一 20 kg 大样中选出的锆石多为变质锆石,少量为圆形具蚀坑的继承锆石。图 2 为该样品的高压变质锆石照片,它们呈浑圆状,小晶面清晰并有明显增生体,有些锆石可见有老锆石核。

样品 92SH-1 为含石榴石黑云母花岗片麻岩。它与样品 92HT-1 采自同一地区,是该超高压岩片的围岩,为一低压角闪岩相变质的花岗片麻岩。该岩石的 Sm-Nd 矿物等时线年龄为(213.3 ± 4.6) Ma (李曙光等未发表数据)。从 40 kg 大样中选出的锆石主要为短柱状(长度比为 1:2~3)和长柱状(长度比为 1:5~6)两种,未见圆形重结晶变质锆石。

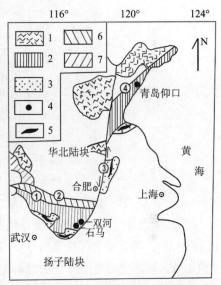

图 1 大别山-苏鲁地体地质略图
1 示华北太古界;2 示大别山-苏鲁地体变质杂岩;3 示红安群、松宿群、张八群、海州群;4 示锆石 U-Pb 年龄取样点;5 示蓝片岩;6 示北淮阳地体;7 示北秦岭地体;① 示桐坡-磨子潭断裂;② 示信阳-舒城断裂;③ 示郯庐断裂;④ 示五莲-荣城断裂

样品 Y90-11 为云母斜长片麻岩,采自青岛仰口湾 (图1),该样品详情已有前文描述并由中国地质科学院地质研究所用常规锆石 U-Pb 法测定获得不一致线下交点年龄(202 ± 13) Ma 和上交点年龄(871 ± 46) Ma[2],本文对该样用单颗粒锆石法进行重复测定。

2.2 测试方法与结果

锆石 U-Pb 同位素分析由天津地质矿产研究所同位素实验室测试,所测定的锆石颗粒特征见表 1。分析流程已有专文介绍[15]。本次实验所用为 ^{208}Pb-^{235}U 混合稀释剂,数据处理及普通铅扣除使用美国地质调查所 Ludwin 编写的 PBDAT(1898.8 版)和 ISOPLOT 软件进行,测定结果及 2σ 误差列于表中,对应的 U-Pb 年龄谐和图见图 3。

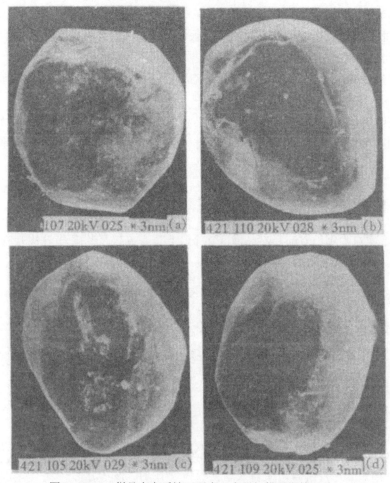

图 2 92T-16 样品中变质锆石形态（电子扫描显微镜照片）

它们均呈浑圆形。(a)、(b)可见两端锥形及增生晶面；(c)可见内部有柱状继承锆石核；(d)下端可见不平坦的增生小晶体

92HT-1 的 4 号颗粒为高压变质锆石，它的 t_{206}, t_{207}, $t_{207/206}$ 表面年龄在误差范围内一致，并落到一致线上（图3(a)）。因此可取较精确的 t_{206} 年龄，(238.4 ± 1.3) Ma 作为谐和年龄值，它表明该锆石不含有老锆石的残留组成，基本上是超高压变质时新生成锆石，该年龄指示双河地区超高压变质时代为三叠纪，但它略高于同一高压岩片的榴辉岩 Sm-Nd 年龄 (226.2 ± 2.9) Ma。92HT-1 的 1，2，3 号虽然形态各异，但都给出了古老的表面年龄值，且 3 种表面年龄不一致，并服从 $t_{206} < t_{207} < t_{207/206}$ 的规律。这说明它们有明显的 Pb 丢失。根据其 $^{207}Pb/^{206}Pb$ 年龄，它们为早元古代残留锆石。

92T-16 的一号颗粒为高压变质锆石。它的 t_{206} 和 t_{207} 年龄在误差范围内一致（表1），并落到一致线上（图3(b)）。其较精确的 t_{206} 表面年龄为(236.2 ± 2.4) Ma 与 92HT-1 高压变质锆石的 $^{206}Pb/^{238}U$ 年龄一致。然而它的 t_{207} 年龄较老（表1），显示了该锆石可能含有少量老锆石残留组分。2 号颗粒为古老的残留锆石，$t_{207/206}$ 表面年龄高达 1266 Ma，并远离一致线。它含有较多普通 Pb，其年龄值仅作参考。

92SH-1 的短柱状和长柱状锆石各测得 2 个数据。其中短柱状锆石的 $t_{207/206}$ 均显著高于 t_{206} 和 t_{207}，显示了大量 Pb 丢失的特征。它们连线的上交点在 1222 Ma，下交点在 129 Ma 左右（图 3(c)）。由于测点少及较大测量误差，下交点年龄有较大误差，但它仍很接近该样品的 Sm-Nd 矿物等时线年龄(123.3 ± 4.6) Ma。一个长柱状锆石（3 号)给出了异常低的 $t_{207/206}$ (133.3 Ma)（表1）。这可能是由于该锆石捕获了较多 ^{206}Pb 所致，因而该数据不予采用。另一个长柱状锆石给出接近一致的 t_{206}, t_{207} 和 t_{207}/t_{206}（表1）。该一致年龄显示该锆石形成于古生代，并在印支期变质事件中有轻微 Pb 丢失。将代表变质时代的上述不一致线下交点 (219 Ma) 与 4 号长柱状锆石点连线获得上交点为 400 Ma（图 3(c)）。上述分析表明，长柱状锆石可能是该花岗片麻岩原岩的岩浆锆石，其原岩可能为一加里东期花岗岩；而短柱状锆石为捕虏晶，它来自一更古老的围岩。应当指出，该样的 2，3，4 号锆石由于普通 Pb 含量较高，其结果只能视为参考年龄。

Y90-11 的短柱状和长柱状锆石各测定了一个颗粒。它们 $t_{207/206}$ 年龄均显著高于 t_{270} 和 t_{206} 年龄 (表1)。这表明这两个岩浆岩均不同程度发生了明显的 Pb 丢失。两个数据点连线给出下交点年龄 220 Ma 和上交点年龄 625 Ma (图3(d))。该下交点年龄也指示了苏-鲁地体的三叠纪超高压变质时代。

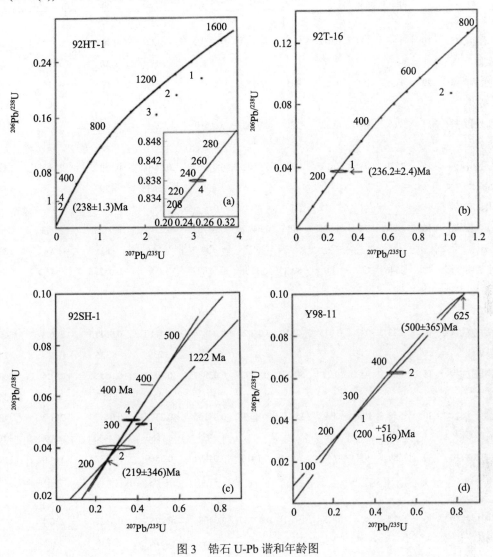

图 3 锆石 U-Pb 谐和年龄图

3 讨论: 超高压变质锆石的连续扩散 Pb 丢失及其峰期变质时代

如前述新报道的大别山锆石 U-Pb 定年结果仍然指示大别山超高压变质时代发生在三叠纪。这与 Ames 等和 Maruyama 等的锆石定年结论基本一致[1,16]。然而就具体年龄值而言，已报道的较精确锆石 U-Pb 一致年龄值也存在一些差异。如 Ames 等发表的 4 个超高压变质锆石 U-Pb 一致年龄(209~215 Ma)略低于本文及刘晓春等报道的 238 Ma 和 234 Ma[7] (表 2)。造成这种误差的一个原因可能是由于所用稀释剂不同，因为其他实验室普遍使用 ^{205}Pb 稀释剂。另一个原因可能是样品的差异。此外，Ames 等的 U-Pb 年龄明显低于大别山榴辉岩 Sm-Nd 年龄(212~223 Ma)[12]，这很难用实验室差异来解释，它很可能与超高压变质锆石 U-Pb 同位素的行为有关。

超高压变质岩冷却史研究表明，大别山超高压变质岩在 180~160 Ma 之间有一个快速冷却历史，可以从 500 ℃ 迅速冷却到 300 ℃ 以下。因此，超高压变质期间新生成的锆石在 180 Ma 以前，可以在较高温度下(>500 ℃)有一段连续扩散 Pb 丢失的历史。这种扩散 Pb 丢失可以在 160 Ma 以后因超高压变质岩迅速冷却至 300℃ 以下而基本停止。由于大别山超高压变质锆石很年轻，其 Pb 扩散丢失的 $^{207}Pb/^{238}U$-$^{206}Pb/^{235}U$ 变化轨迹非常接近一致线。因此，不含继承组分的新生成高压变质锆石仍给出接近一致的 t_{206} 和 t_{207} 年龄。

表 1　大别山及苏-鲁地体超高压片麻岩及其围岩中单颗粒锆石 U-Pb 同位素测定结果 [a]

样号	锆石特征	质量/μg	浓度 U/μg·g⁻¹	浓度 Pb/μg·g⁻¹	普通 Pb 含量/ng	$^{206}Pb/^{204}Pb$	$^{208}Pb/^{206}Pb$	同位素原子 $^{206}Pb/^{238}U\pm2\sigma$	$^{207}Pb/^{235}U\pm2\sigma$	$^{207}Pb/^{206}Pb\pm2\sigma$	表面年龄值/Ma $^{206}Pb/^{238}U\pm2\sigma$	$^{207}Pb/^{235}U\pm2\sigma$	$^{207}Pb/^{206}Pb\pm2\sigma$
92HT-1	大别山，双河												
1	橘红色透明柱状晶体	10	886	197	0.012	5140	0.1011	0.2112±8	3.134±17	0.1076±4	1235±5	1441±8	1760±7
2	浅紫红色透明圆形晶体	20	626	186	0.55	119	0.06687	0.1890±7	2.676±30	0.1027±10	1116±40	1322±15	1673±16
3	黄棕色透明方柱状晶体	15	968	180	0.22	330	0.02527	0.1610±5	2.255±21	0.1016±9	962±3	1198±11	1653±15
4	浅红色透明浑圆状晶体	15	2530	104	0.13	383	0.01028	0.03767±20	0.2661±163	0.05123±297	238.4±1.3	239.6±14.7	251.2±14.6
92T-16	大别山，石马												
1	浅红色透明浑圆状晶体	10	583	24	0.021	409	0.04906	0.03733±38	0.2741±444	0.05326±815	236.2±2.4	246.0±39.8	340.0±52.0
2	浅红色半透明浑圆状晶体	10	2646	435	1.0	89.3	0.1166	0.08827±31	1.008±11	0.08286±82	545±2	708±8	1266±13
92SH-1	大别山，双河												
1	无色透明短柱状晶体	10	233	14	0.015	284	0.08926	0.04874±32	0.4110±239	0.06115±335	306.8±2	349.6±20.3	644.4±35.3
2	无色透明短柱状晶体	15	405	48	0.27	50.27	0.1202	0.03977±71	0.3033±913	0.05531±1571	251.4±4.5	269.0±36.6	424.9±120.7
3	无色透明长柱状晶体	30	558	108	1.4	35.45	0.07194	0.04135±93	0.2776±1160	0.04870±1933	261.2±5.9	248.8±104	133.3±52.9
4	无色透明长柱状晶体	30	586	61	0.48	77.5	0.1410	0.04991±32	0.3669±342	0.05331±470	314.0±2.0	317.3±29.6	342.0±30.2
Y90-11	苏-鲁地体，青岛												
1	无色透明短柱状晶体	30	774	42	0.18	191.1	0.07757	0.04085±19	0.2952±130	0.05276±216	256.4±1.2	262.6±11.6	318.4±13.0
2	淡黄色长柱状透明晶体	20	385	37	0.11	155.4	0.1840	0.06199±36	0.4894±406	0.05726±447	387.7±2.2	404.5±33.5	501.5±39.1

a) $^{206}Pb/^{207}Pb$ 已对空白（Pb=50 pg，U=2 pg）及稀释剂作了校正，其他同位素比值中的铅同位素均为放射成因铅同位素，锆石质量是根据颗粒大小估计的质量，仅供参考

表 2　大别山超高压变质锆石 4 组 U-Pb 年龄对比表

样品	年龄/Ma, ±σ $^{206}Pb/^{204}Pb$	$^{206}Pb/^{238}U$	$^{207}Pb/^{235}U$	$^{207}Pb/^{206}Pb$	数据来源
92HT-1	383	238.4±1.3	239.6±14.7	251.2±14.6	本文
DB-91-28					
>149 μm	14223	209.7±1.3	211.1±1.3	226.7±2.0	[1]
74~149 μm	11821	209.0±1.3	210.6±1.3	228.5±3.9	[1]
平均		209.4±0.9[a]	210.9±0.9[a]	227.6±1.8[a]	
DB-91-34					
>149 μm	358	212.3±2.1	213.2±2.5	223.2±15.0	1)
74~149 μm	238	215.2±2.2	216.1±2.7	225.9±17.9	1)

a) 平均误差式：$\sigma^2=1/[\Sigma(1/\sigma)^2]$

由于 Pb 丢失, t_{206} 和 t_{207} 要小于 $t_{207/206}$ 年龄。这就是这种经历了连续扩散 Pb 丢失的超高压变质锆石的同位素特征。表 2 列出了符合这一特征的超高压变质锆石数据, 根据这一超高压变质锆石连续扩散 Pb 丢失模型, 表 2 中的那些普通 Pb 含量低 ($^{206}Pb/^{204}Pb$ 高) 的较精确 $^{207}Pb/^{206}Pb$ 年龄能更好地代表这些超高压变质锆石的平均结晶时代。如前所述, 变质锆石可以在整个榴辉岩相变质阶段连续生长, 其平均结晶时代应接近峰期变质时代。按此标准, 表 2 中 DB-91-28 样的 2 个测定值的平均 $^{207}Pb/^{206}Pb$ 年龄 (227.6±1.8) Ma 可指示大别山超高压变质岩的峰期变质时代, 它与我们测定的较精确的大别山双河及青龙山含柯石英榴辉岩的 3 种矿物 Sm-Nd 等时线年龄[(226.2+2.9) Ma 和 (226.3+4.5) Ma]非常一致[12], 也与 Maruyama 等报道的离子探针锆石 U-Pb 年龄的中间值(227 Ma)一致[16]。

参考文献

[1] Ames L, Tilton G R, Zhou G. Timing of collision of the Sino-korean and Yangtse cratons: U-Pb zircon dating of coesite-bearing eclogites. Geology, 1993, 21: 339~324.

[2] 李曙光, 张志敏, 张巧大, 等. 青岛榴辉岩及胶南群片麻岩的锆石 U-Pb 年龄——胶南群中晋宁期岩浆事件的证据. 科学通报. 1993, 38(19): 1773~1777.

[3] 曹荣龙, 朱寿华. 大别山含柯石英榴辉岩区——一个晚太古代的超高压变质带. 地质学报, 1995, 69(3): 232~242.

[4] 刘若新, 樊祺诚, 李惠民, 等. 大别山碧溪岭石榴橄榄石-榴辉岩体的原岩性质及同位素年代学的启示. 岩石学报, 1995, 11(3): 243~256.

[5] 杨巍然, 王林森, 韩郁菁, 等. 大别蓝片岩-榴辉岩年代学研究. 见: 陈好寿主编. 同位素地球化学研究. 杭州: 浙江大学出版社, 1994, 175~186.

[6] You Z, Han Y, Zhang Z, et al. Paleozoic dates from the Dabie Mountains, central China: implications for ultrahigh-pressure metamorphism in the Qinling Orogenic Belt. EOS, 1995, 46 (46): 712.

[7] 刘晓春, 李惠民, 左义成. 大别山超高压石榴多硅白云母片岩中锆石 U-Pb 年龄. 矿物岩石地球化学通报, 1996, 15(l): 10~13.

[8] Gebauer D, Grunenfelder M. U-Pb zircon and Rb-Sr mineral dating of eclogites and their country rocks. Example: Munchberg gneiss massif, northeast Bavaria. Earth Planet Sci Lett, 1979, 42: 35~44.

[9] Claoue-Long J C, Sobolev N V, Shatsky V S. Zircon response to diamond-pressure metamorphism in the Kokchetav massif, USSR. Geology, 1991, 19: 710~713.

[10] 李曙光, 宋明春, 李淳, 等. 胶东海洋所斜长角闪岩的锆石 U-Pb 年龄——多期变质作用对锆石不一致线年龄影响的实例. 地球学报, 1994, (1~2): 37~42.

[11] Peucat J J, Vidal Ph, Godard G, et al. Precambrian U-Pb zircon ages in eclogites and garnet Pyroxenites from South Britany(France): an old oceanic crust on the West European Hercynian belt? Earth Planet Sci Lett, 1982, 60: 70~78.

[12] 李曙光, Jagoutz E, 肖益林, 等. 大别山-苏鲁地体超高压变质年代学——I. Sm-Nd 同位素体系. 中国科学, D 辑, 1996. 26(3): 249~257.

[13] Wang X, Liou J G. Regional ultrahigh-pressure metamorphic terrane on central China: evidence from coesite-bearing eclogite, marble, and metapelite. Geology, 1991, 19: 933~936.

[14] Li S, Xiao Y, Liou D, et al. Collision of the north China and Yangtse blocks and formation of coesite-bearing eclogites: timing and processes. Chemical Geology, 1993, 109: 89~111.

[15] 陆松年, 李惠民. 蓟县长城系大红峪组火山岩的单颗粒锆石 U-Pb 法准确定年. 中国地质科学院院报, 1991, 22: 137~146.

[16] Maruyama S, Liou J G, Zhang R. Tectonic evolution of the ultralhigh-pressure (UHP) and high-pressure (HP) metamorphic belts from cental China. The Island Arc, 1994, 3: 112~121.

构造剪切与变质角闪石中过剩 Ar 的引入：北秦岭丹凤群斜长角闪岩的 Sm-Nd、Rb-Sr 和 $^{40}Ar/^{39}Ar$ 测年证据*

孙卫东[1]，李曙光[1]，陈 文[2]，孙 勇[3]，张国伟[3]

1. 中国科学技术大学地球和空间科学系，第三世界科学院地球科学和天文学高级研究中心，合肥 230026
2. 中国地质科学院地质研究所，北京 100000
3. 中国地质科学院地质研究所，北京 100000

> **亮点介绍**：本文对北秦岭丹凤群斜长角闪岩样品进行了 Sm-Nd、Rb-Sr 和 $^{40}Ar/^{39}Ar$ 测年，取得了丹凤群火山岩的变质时代，证明构造剪切作用对变质角闪石中过剩氩的引入有显著影响。

摘要 本文对两个北秦岭丹凤群斜长角闪岩样品进行了 Sm-Nd、Rb-Sr 和 $^{40}Ar/^{39}Ar$ 测年，发现其中受构造改造轻微的黑河丹凤群斜长角闪岩样品的角闪石 $^{40}Ar/^{39}Ar$ 年龄与其 Rb-Sr、Sm-Nd 矿物等时线年龄接近，而遭受强烈构造剪切作用的蒲峪丹凤群斜长角闪(片)岩的角闪石 $^{40}Ar/^{39}Ar$ 年龄则明显高于其 Rb-Sr、Sm-Nd 矿物等时线年龄，指示构造剪切作用对变质角闪石中过剩氩的引入有显著影响。在利用 $^{40}Ar/^{39}Ar$ 进行造山带年代学研究时这一问题应引起注意。

关键词 构造剪切；过剩 Ar；丹凤群；北秦岭

1 引言

由于角闪石、白云母、黑云母等矿物的封闭温度已精确测定，因此它们的 $^{40}Ar/^{39}Ar$ 法被广泛用来研究造山带变质岩的热演化历史。但是在高级变质岩内，这些矿物中都曾发现过有过剩 Ar 的存在[1,2]。因此，弄清上述变质矿物引入过剩 Ar 的条件和机制，对在造山带热演化史研究中正确运用 $^{40}Ar/^{39}Ar$ 定年方法是十分重要的。近年来的研究已认识到变质流体可能是引入过剩 Ar 的重要因素。在碰撞造山带的强剪切构造带内，岩石强烈剪切变形，为变质流体的活动提供了条件，在流体活动的同时是否将过剩 Ar 引入到变质矿物中是一个值得注意的问题。本文通过对陕西省商县蒲峪丹凤群的强烈剪切、片理化斜长角闪片岩和陕西省周至县黑河丹凤群无明显片理化的斜长角闪岩的对比研究，观察到了前者角闪石中过剩 Ar 对年龄有明显的干扰，而后者无过剩 Ar 存在的明显迹象。上述现象表明构造剪切作用对变质角闪石中过剩 Ar 的引入有明显的影响。

2 地质背景及样品

丹凤群为一套沿商丹断裂北缘出露的绿片岩相至低角闪岩相变质岩，主要由镁铁质及安山质岩石组成，伴生成熟度较低的碎屑岩[3]，表明北秦岭岩浆弧的存在，并在秦岭造山带构造演化中占有重要的地位，受到较为广泛的重视[3-10]。研究丹凤群的抬升历史对于揭示秦岭造山带形成演化历史具有重要意义。

黑河丹凤群变质岩位于周至县黑河(图 1)。实验所采用的样品 QH-3 采自周洋公路边虎豹河口丹凤群变质火山岩中的辉绿岩脉，属低角闪岩相，无明显片理化，有绿帘石化现象：主要矿物有斜长石、角闪石、石

---

* 本文发表在：高校地质学报，1996, 2(4)：382-389

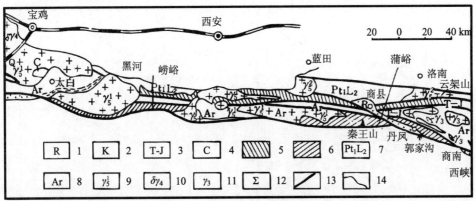

图 1 北秦岭丹凤群地质简图

1.新生界；2.白垩系；3.上三叠统—侏罗系；4.石炭系；5.二郎坪蛇绿岩；6.丹凤蛇绿岩；7.宽坪群；8.秦岭群；9.印支期花岗岩；10.海西期花岗闪长岩；11.加里东期花岗岩；12.松树沟超基性岩；13.断裂；14.地质界线

英及少量绿帘石；轻稀土富集。Nb、Ti、P 负异常等特点表明属于岛弧火山岩[10]。

蒲峪丹凤群位于商县县城东 12 km 的蒲峪沟内(图 1)。实验所采用的样品采自蒲峪冠庄-构造剪切带内，与张成立等研究的蒲峪丹凤群 REE 平坦型玄武岩采样层位接近[11]。岩石属绿帘角闪岩相、中粒结构，主要由角闪石、斜长石、石英及绿帘石组成，并已强片理化。该样品未做详细的地球化学工作，根据表1，球粒陨石均一化钐钕比值$(Sm/Nd)_N$ 为 0.88，呈现轻稀土微富集的特点。

表 1 北秦岭蒲峪、黑河丹凤群斜长角闪岩的 Sm、Nd、Rb、Sr 含量及 Nd、Sr 同位素组成

样号	名称	Rb (10^{-6})	Sr (10^{-6})	$^{87}Rb/^{86}Sr$	$^{87}Sr/^{86}Sr$	Sm (10^{-6})	Nd (10^{-6})	$^{147}Sm/^{144}Nd$	$^{143}Nd/^{144}Nd$
QSP-10	全岩	18.92	229.1	0.2354	0.710578±26	1.587	5.529	0.17362	0.512809±18
	角闪石	133.5	49.28	1.052	0.715342±18	1.219	3.591	0.20543	0.512890±14
	斜长石	20.54	475.6	0.1231	0.709912±22	0.3834	1.407	0.16482	0.512775±22
QH-3	全岩	27.14	452.3	0.1711	0.709338±16	4.339	18.84	0.13927	0.512697±19
	角闪石	35.47	505.4	0.2001	0.709509±13	3.878	12.28	0.19106	0.512837±14

注：实验由南京大学现代分析中心完成。

3 实验结果

大量研究表明，即使含有大量过剩 Ar 的角闪石、云母样品，也往往可以给出很均一的年龄分布图，因此，年龄谱的形状已经不是判断过剩 Ar 的必要条件[1,2]。许多过剩 Ar 的发现都是通过多种定年方法的对比测试发现的[1]。本文采用对同一样品的 Sm-Nd、Rb-Sr 和 $^{40}Ar/^{39}Ar$ 定年结果进行相互比较的方法来观察是否有过剩 Ar 的存在。

3.1 单矿物 Sm-Nd、Rb-Sr 定年

从样品中挑选出纯净的角闪石、斜长石单矿物样，外加全岩样，进行 Rb-Sr、Sm-Nd 同位素分析。实验由南京大学现代分析中心完成，结果列于表1，Sm-Nd、Rb-Sr 单矿物等时线示于图2。样品 QSP-10 年龄值 (图2c, d)为：Sm-Nd, 414 ± 38(2σ)Ma；Rb-Sr, 410.2 ± 1.9(2σ)Ma。ε_{Nd}= +4.5, $(^{87}Sr/^{86}Sr)_i$= 0.711944。由于只有两点，所以这两个年龄值的可靠程度较低，但它们彼此一致性很好，且均与 QSP-10 的 Sm-Nd、Rb-Sr 年龄非常一致，因此仍可以认为它们是可信的。其 ε_{Nd}= +4.2, $(^{87}Sr/^{86}Sr)_i$= 0.708329。

上述单矿物等时线年龄值与北秦岭构造带内不同的年代学方法所获得的大量 400 Ma 左右的年龄值[12-16]一致，说明蒲峪、黑河丹凤群卷入了在北秦岭普遍发生的加里东期变质事件。同时上述单矿物等时线年龄这种与区域变质年龄的一致性也反过来为它们的可靠性提供了佐证。

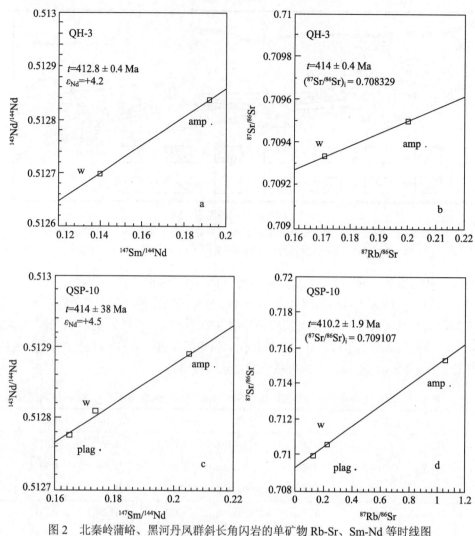

图 2 北秦岭蒲峪、黑河丹凤群斜长角闪岩的单矿物 Rb-Sr、Sm-Nd 等时线图
w. 全岩；amp. 角闪石；plag. 斜长石

3.2 角闪石 $^{40}Ar/^{39}Ar$ 测年

对样品中的角闪石单矿物进行 $^{40}Ar/^{39}Ar$ 同位素测定。实验在中国质科学院地质所完成。J 值的误差约为 ±0.8%。采用常数 $\lambda^{40}K=5.543\times10^{-10}a^{-1}$ 来计算年龄结果，并按 1σ 给出年龄误差。角闪石 Ar 同位素阶段升温结果列入表 2a，b，年龄谱示于图 3。测定的钾、钙干扰同位素校正因子结果为：$(^{39}Ar/^{37}Ar)_{Ca}=8.06\times10^{-4}$；$(^{40}Ar/^{39}Ar)_K=4.78\times10^{-3}$；$(^{36}Ar/^{37}Ar)_{Ca}=2.4\times10^{-4}$。

表 2a 北秦岭黑河丹凤群斜长角闪岩角闪石 $^{40}Ar/^{39}Ar$ 阶段升温数据表（重量= 80 mg，J= 0.01835）

温度 (℃)	$(^{40}Ar/^{39}Ar)_m$	$(^{36}Ar/^{39}Ar)_m$	$(^{37}Ar/^{39}Ar)_m$	$^{40}Ar^*/^{39}Ar$	^{39}Ar (mol) ×10^{-14}	视年龄（Ma）	^{39}Ar 累积百分数（%）
620	136.9200	0.0800	3.2226	113.7992	1.50	2034.20 ± 254.220	0.4
780	23.2871	0.0473	5.6095	9.7540	6.13	297.00 ± 47.80	2.2
965	20.8004	0.0356	1.4229	10.3769	15.43	314.40 ± 29.60	6.7
1055	17.8553	0.0200	4.6071	12.3000	16.46	367.10 ± 10.00	11.6
1175	18.2758	0.0176	17.5921	14.5244	69.90	426.20 ± 6.70	32.1
1215	16.9258	0.0129	18.0374	14.5914	122.23	428.00 ± 6.40	67.9
1295	18.9164	0.0199	16.2709	14.3733	80.40	422.30 ± 7.50	91.5
1400	21.5898	0.0286	17.3259	14.5001	28.72	427.70 ± 11.80	100

*表示放射性成因 ^{40}Ar。

表 2b　北秦岭萍峪丹凤群斜长角闪岩角闪石 $^{40}Ar/^{39}Ar$ 阶段升温数据表（重量= 131mg，J= 0.01851）

温度 (℃)	$(^{40}Ar/^{39}Ar)_m$	$(^{36}Ar/^{39}Ar)_m$	$(^{37}Ar/^{39}Ar)_m$	$^{40}Ar^*/^{39}Ar$	^{39}Ar (mol) $\times 10^{-14}$	视年龄（Ma）	^{39}Ar 累积百分数（%）
600	112.9085	0.2539	6.9553	38.5721	8.31	972.00 ± 79.00	0.9
785	59.4455	0.0817	3.4487	35.6290	17.13	913.80 ± 28.00	2.9
870	27.9282	0.0391	1.4061	16.4947	14.08	480.70 ± 27.60	4.6
965	44.4386	0.0401	1.8397	32.7542	16.44	855.00 ± 10.00	6.5
1055	19.5820	0.0244	2.5284	12.5776	41.02	377.80 ± 8.70	11.3
1175	18.6848	0.0125	13.3842	16.1179	183.66	471.00 ± 4.50	32.8
1215	15.2045	0.0087	12.9870	13.6434	212.08	406.20 ± 4.60	57.7
1275	18.7326	0.0104	13.3922	16.7904	247.56	488.22 ± 4.48	88.66
1400	19.7830	0.0135	12.8957	16.8760	110.31	490.40 ± 8.33	100.00

*表示放射性成因 ^{40}Ar。

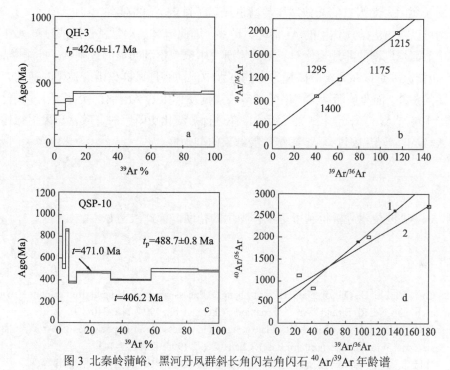

图 3　北秦岭蒲峪、黑河丹凤群斜长角闪岩角闪石 $^{40}Ar/^{39}Ar$ 年龄谱

a. 样品 QH-3 角闪石 $^{40}Ar/^{39}Ar$ 年龄谱，t_p 代表坪年龄；b. 样品 QH-3 角闪石高温阶段等时线图；c. 样品 QSP-10 角闪石 $^{40}Ar/^{39}Ar$ 年龄谱，t_p 代表坪年龄；d. 样品 QSP-10 角闪石 $^{40}Ar/^{39}Ar$ 相关图，其中，实方框代表 1275 ℃和 1400 ℃两个数据点（拟合直线 1），空心方框代表其他高温阶段数据点（拟合直线 2）

由图 3a 可见，黑河斜长角闪岩角闪石的 $^{40}Ar/^{39}Ar$ 高温阶段年龄谱谱型较好，1175 ℃，1215 ℃，1295 ℃，1400 ℃四个高温阶段给出的坪年龄为 426.0 ± 1.7 Ma。角闪石高温阶段在 $^{40}Ar/^{39}Ar$ 相关图上相关性很好(图 3b)，在 Y 轴上的截距为 290.9 ± 1.7，接近大气氩的 $^{40}Ar/^{36}Ar$ 值(295.5)；上述四个高温阶段给出的等时线年龄为 427.7 Ma，与坪年龄非常一致，均无过剩氩存在的迹象。坪年龄和等时线年龄均与其 Rb-Sr、Sm-Nd 单矿物等时线年龄在误差范围内一致，上述 Rb-Sr、Sm-Nd、$^{40}Ar/^{39}Ar$ 三种定年结果的一致性进一步证明黑河斜长角闪岩的角闪石中没有过剩氩存在。

由图 3c，蒲峪斜长角闪片岩中角闪石的 $^{40}Ar/^{39}Ar$ 高温阶段释出 Ar 的数据波动较大，年龄谱型复杂，1215 ℃的低年龄(占 ^{39}Ar 析出量的 25%)使整个谱型呈马鞍状，反映了放射成因 Ar 在该角闪石中分布的不均一性。其最后两个高温坪的年龄为 488.7 ± 0.8 Ma (占 ^{39}Ar 析出量的 42.3%)，以各阶段 ^{39}Ar 为权重加权计算出的全年龄为 478.23 Ma，均高于其 Rb-Sr、Sm-Nd 单矿物等时线年龄，显示该角闪石样品中存在过剩 Ar 且对 $^{40}Ar/^{39}Ar$ 年龄有明显的干扰。

通常放射成因 Ar 在矿物中分布的不均一性是受后期扰动或过剩 Ar 引入造成的。1215 ℃的低年龄为 406.2 Ma，与其 Rb-Sr、Sm-Nd 矿物等时线年龄在误差范围内一致，由此判断该低年龄值不是放射成因 Ar 丢失的结果。因此推断 QSP-10 马鞍型年龄谱可能是过剩 Ar 引入的结果。此外，该样品各加热阶段视年龄相差较大，且多数视年龄值偏离正确年龄值，因此它们拟合的"等时线"无意义。关于马鞍型谱形成的确切机制则需要用激光探针对矿物中 Ar 的分布做深入研究。

4 讨论和结论

（1）黑河与蒲峪斜长角闪岩的 Sm-Nd、Rb-Sr 年龄很一致，说明这两个地区的丹凤群岛弧火山岩系具有相同或相似的冷却历史，因此它们可代表北秦岭丹凤群火山岩的变质时代。

（2）蒲峪丹凤群斜长角闪片岩的角闪石 $^{40}Ar/^{39}Ar$ 定年结果高于同一样品的 Sm-Nd、Rb-Sr 角闪石、斜长石单矿物等时线年龄，也高于丹凤群其他 $^{40}Ar/^{39}Ar$ 年龄。没有过剩 Ar 的存在很难解释上述年龄差异。

此外，蒲峪斜长角闪岩角闪石的马鞍型年龄谱也显示了过剩 Ar 的存在。

（3）流体活动可以将古老地质体中的 Ar 活化、运移，因此流体活动常常是造成过剩 Ar 的直接原因[17]。由于笔者对北秦岭黑河丹凤群的非片理化斜长角闪岩研究中未观察到明显的过剩 Ar，推测样品 QSP-10 中出现过剩 Ar 的一个可能原因是：在北秦岭碰撞造山的过程中，蒲峪丹凤群火山岩被强烈剪切、片理化，为流体活动提供了必要的条件。而流体活动又将秦岭群等古老地质体中的 Ar 活化、运移至丹凤群变质角闪岩中，产生过剩 Ar，从而影响 $^{40}Ar/^{39}Ar$ 定年结果。因此，在进行热演化史研究时，应避免采用强构造剪切的斜长角闪岩做 $^{40}Ar/^{39}Ar$ 定年，以免因过剩 Ar 影响年龄结果的准确性。

致谢

样品 Rb-Sr、Sm-Nd 同位素分析得到了王银喜先生的帮助，谨此致谢！

参考文献

[1] Li, S., Wang, S., Chen, Y., Liu, D., Qiu, J., Zhou, H., and Zhang, Z. Excess argon in phengite from eclogite: evidence from dating of eclogite minerals by Sm-Nd, Rb-Sr and $^{40}Ar/^{39}Ar$ methods. Chem. Geol., 1994, 112: 340-350.

[2] Ruffet. G., Feraud, G., Balevre, M., Kirnast J., R., Plateau ages and excess argon in phengites: $^{40}Ar/^{39}Ar$ laser probe study of Alpine micas (Sesia Zone. Western Alps northern Italy). Chem. Geol., 1995, 121: 327-343.

[3] 孙勇, 于在平, 张国伟. 东秦岭蛇绿岩的地球化学. 见《秦岭造山带的形成及其演化（张国伟等著）》. 西安：西北大学出版社，1988，65-74.

[4] 张国伟, 于在平, 孙勇, 程顺有, 薛峰, 张成立. 秦岭商丹断裂边界地质体基本特征及其演化. 见《秦岭造山带的形成及其演化（张国伟等著）》. 西安：西北大学出版社，1988，29-47.

[5] 于在平, 孙勇, 张国伟. 商丹地区秦岭缝合带弧前沉积楔形体初探. 见《秦岭造山带的形成及其演化》. 西安：西北大学出版社，1988，48-64.

[6] 于在平, 孙勇, 张国伟. 秦岭商丹沉积岩系基本地质特征. 见《秦岭造山带学术讨论会论文选集》. 西安：西北大学出版社，1991，78-88.

[7] 许志琴, 牛宝贵, 刘志刚, 王水敏等. 秦岭-大别"碰撞-陆内"型复合山链的构造体制及陆内板块动力学机制. 见《秦岭造山带学术讨论会论文选集》. 西安：西北大学出版社，1991，139-147.

[8] 张国伟, 周鼎武, 于在平, 郭安林, 程顺有, 李桃红, 张成立, 薛锋. 秦岭岩石圈组成、结构与演化. 见《秦岭造山带学术讨论会论文选集》. 西安：西北大学出版社，1991，121-138.

[9] 李曙光, 张宗清, 张国伟. 北秦岭拉垃庙苏长辉长岩的痕量元素地球化学. 地质学报，1993，67（4）：310-322.

[10] 孙卫东, 李曙光, 肖益林, 孙勇, 张国伟. 北秦岭丹凤群岛弧火山岩建造的发现及其构造意义. 大地构造与成矿学，1995，19（3）：227-236.

[11] 张成立, 周鼎武, 韩松. 陕西商州地区丹凤变质火山岩的地球化学特征. 地质科学，1994，29：384-392.

[12] 李曙光, S. R. Hart, 邱双根, 韩安林, 刘德良, 张国伟. 中国华北、华南板块碰撞时代的钐钕同位素年龄证据. 中国科学 B 辑，1989，38(3)：312-319.

[13] 刘志刚, 富云莲, 牛宝贵, 任纪舜. 大别山北坡苏家河群及原信阳群龟山组变基性杂岩 $^{40}Ar/^{39}Ar$ 测年及其地质意义. 科

学通报，1993，38（13）：1214-1218.
[14] 张宗清，刘敦一，富国民. 北秦岭变质地层秦岭、宽坪、陶湾群同位素年代学研究. 北京：地质出版社，1993.
[15] Lerch., M. Xue, F., Kroner, A., Zhang, G. W. and Todl, W. A Middle Silurian Early Devonian magmatic arc in the Qinling Mountains of central China. J. Geology, 1995, 103：437-449.
[16] Li., S. and Sun, W. A Middle Silurian Early Devonian magmatic arc in QinIing Mountains of central China：a discussion. J. Geology, 1996, 104：501-503.
[17] Renne, P. R. Excess ^{40}Ar in biotite and hornblende from the Noril'sk 1 intrusion, Siberia: implications for the age of the Siberian Traps. Earth Planet. Sci. Lett., 1995, 131: 165-174.

Sm-Nd, Rb-Sr and ^{40}Ar/^{39}Ar isotopic systematics of the ultrahigh-pressure metamorphic rocks in the Dabie-Sulu Belt, Central China: a retrospective view*

Shuguang Li[1], Emil Jagoutz[2], Ching-Hua Lo[3], Yizhi Chen[1], Qiuli Li[1] and Yilin Xiao[1]

1. Department of Earth and Space Sciences, University of Science and Technology of China, Hefei 230026, Anhui, China
2. Department for Cosmochemistry, Max-Planck-Institut für Chemie, Mainz D55020, Germany
3. Department of Geology, Taiwan University, Taipei, Taiwan, China

> 亮点介绍：系统综述了针对大别-苏鲁超高压变质岩石所进行的 Sm-Nd、Rb-Sr 和 Ar-Ar 体系定年结果；指出退变质作用对 Sm-Nd 和 Rb-Sr 同位素体系的改造；提出过剩 Ar 可能的判别标志；首次报道了超高压岩石的两次快速冷却历史。

Abstract Because of a complicated metamorphic history, the isotopic systematics of the ultrahigh-pressure (UHP) metamorphic rocks in the Dabie-Sulu belt, east China, appear to be rather different from what were expected. Depending on the degree of retrograde metamorphism and on the retentivity of isotopes, the radiogenic isotopic systematics in the UHP metamorphic rocks yielded a wide range of radiometric ages. Some of these ages are geologically meaningful, but others may not be. In some fine-grained UHP metamorphic rocks, Sm/Nd isotopic systematics appear to be in equilibrium among the UHP phases, showing the best estimate for the age of peak metamorphism at 226 ± 3 Ma. On the other hand, retrograde overprinting often makes the interpretation of isotopic data more difficult. It is common to find that the Sm/Nd and Rb/Sr isotopic systematics among the UHP phases and retrograde phases are not in equilibrium. Regression of isotopic data involving both UHP and retrograde minerals in isotopic correlation diagrams often yields geologically meaningless ages. Although ^{40}Ar-^{39}Ar dating of UHP metamorphic rocks has been reported not to be very helpful in establishing the thermal history because of the presence of excess argon, a good correlation between excess argon and rock type in the Dabie-Sulu belt would provide a criterion in identifying the possible sources of excess argon. By taking all the possible effects into consideration, a T-t path with two rapid cooling stages for UHP metamorphic rocks from the Dabie-Sulu belt can be postulated. An initial rapid cooling stage in the period from 226 to 219 Ma may have resulted from rapid exhumation of UHP metamorphic rocks immediately after the peak metamorphism. The second rapid cooling stage, from 450 ℃ to 300 ℃, may have been caused by the exhumation of the entire terrane, including UHP metamorphic units and their host gneisses, during the period from 180 to 167 Ma.

1 Introduction

One of the most important achievements in the study of continental geodynamics during the past 15 years is the discovery of ultrahigh-pressure (UHP) minerals such as coesite and diamond in eclogite and metapelite within orogenic belts (Chopin, 1984; Smith, 1984; Okay et al., 1989; Wang et al., 1989; Sobolev and Shatsky, 1990; Xu et al., 1992). It suggests that the continental crust can be subducted to mantle depths during continent-continent collision. Yet scientists are still pondering the causes of the exhumation of the UHP metamorphic rocks from mantle depths. How were UHP minerals preserved under high-pressure and high-temperature conditions during the long history of subduction and exhumation? Answers to these questions are necessarily constrained by precise thermo-chronological data. Unfortunately, previous thermo-chronological results for the Dabie-Sulu belt appear to

---

* 本文发表在: International Geology Review, 1999, 41(12): 1114-1124

be quite controversial, mainly because of the complicated metamorphic history.

The E-W-trending Qinling-Dabie orogenic belt was formed through the collision between the North China Block and South China Block. Geologically, the belt is truncated and separated by two major elements—the Nayang basin and the Tanlu fault. The belt therefore can be subdivided into three sections—the Qinling terrane (west), the Dabie terrane (middle), and the Sulu terrane (east) (Fig. 1). It is generally believed that the Sulu terrane is the eastern extension of the Dabie terrane and was displaced ~500 km by the left-lateral movement of the Tanlu fault. UHP rocks occur widely in the Dabie-Sulu terranes, and it has been proposed that the Dabie-Sulu UHP belt may be the largest UHP terrane on the Earth's surface. In addition to the occurrence of UHP rocks in the Dabie-Sulu terrane, a blueschist belt also crops out along the northern margin of the South China block.

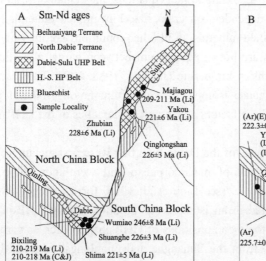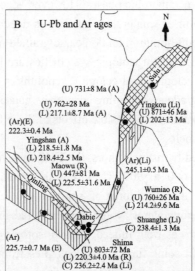

Fig. 1 A. Sketch map of the Dabie-Sulu UHP belt, with published Sm/Nd ages for coesite-bearing eclogites. Data sources: C & J = Chavagnac and Jahn, 1996; Li = Li, Xiao et al., 1993; Li et al., 1994, 1996. B. Sketch map with published U/Pb zircon ages for the UHP metamorphic rocks and $^{40}Ar/^{39}Ar$ ages of blueschists. Data sources: A = Ames et al., 1996; E = Eide et al., 1994; L = Li, Xiao et al., 1993; Li et al., 1997; R = Rowley et al., 1997. Symbols: U = U/Pb upper intercept age; L = lower intercept age; Ar = $^{40}Ar/^{39}Ar$ age; C = concordant age.

Since the first Sm/Nd age of coesite-bearing eclogite from the Dabie terrane was published in 1992 (Li et al., 1992), there have been quite a few Sm/Nd mineral isochron and U-Pb discordia ages published for the coesite-bearing eclogites and UHP gneisses in the Dabie-Sulu terrane. The published Sm/Nd ages—in the range from 209 to 246 Ma (Fig.1A) (Li et al., 1992, 1993, 1994, 1996, 1999; Okay and Sengör, 1993; Chavagnac and Jahn, 1996) —are in general agreement with the published U/Pb lower intercept ages, in the range from 202 to 238 Ma (Fig.1B) (Ames et al., 1993, 1996; Li, Chen et al., 1993; Li et al., 1997; Rowley et al., 1997; Hacker et al., 1998). This agreement indicates that the ultrahigh-pressure metamorphism in the Dabie-Sulu belt is Triassic in age. Furthermore, previous $^{40}Ar/^{39}Ar$ dating of blueschists found in the northern margin of the South China block showed ages of 222 to 245 Ma (Fig.1B). This may further confirm that the suturing of the North China and the South China blocks may indeed have occurred in the Triassic (Li, Xiao et al., 1993; Eide et al., 1994). The reported U/Pb upper-intercept ages, however, were much older, at 600 to 800 Ma, which would suggest that the protoliths of most of the UHP rocks in the Dabie-Sulu belt may have been late Proterozoic in age (Fig. 1B) (Ames et al., 1993, 1996; Li, Chen et al., 1993; Li et al., 1997; Rowley et al., 1997; Hacker et al., 1998).

Although all the reported Sm/Nd and U/Pb ages seem to be consistent in suggesting a Triassic age for the time of a UHP event for the Dabie-Sulu belt, it should be noted that the age range shown by these isotopic systematics is much larger than the analytical errors, and also larger than what would be expected if a simple cooling history is assumed. Isotopic equilibrium among the UHP phases at the peak of metamorphism and the possible effects of retrograde metamorphic overprinting are the two most likely mechanisms responsible for such a dispersion of isotopic ages. Without further clarification of this age dispersion, it may be difficult to employ those geochronological data in precisely constraining the age of the UHP event and in revealing the cooling history of the

UHP rocks in the Dabie-Sulu belt. In order to resolve this age dispersion, the present study carefully examines the Sm/Nd, Rb/Sr, and K/Ar isotopic systematics in a variety of rock types in the Dabie-Sulu belt. It is hoped that the present results will shed some light on the thermal evolution of the UHP metamorphic rocks and assist in revealing the tectonic evolution of the Dabie-Sulu belt.

2 Isotopic systematics in UHP metamorphic rocks

2.1 Sm/Nd systematics

There is general consensus that high metamorphic temperatures can induce a rapid exchange rate for isotopic species among metamorphic phases. However, according to rate equations, the degree of reaction (equilibration) not only is a function of temperature but also is dependent on grain size. Finegrained minerals therefore more easily reach isotopic equilibrium than do coarse-grained minerals, if all other conditions are the same. Indeed, it has been reported that the Sm/Nd isotopic systematics are not in equilibrium between garnet and clinopyroxene for some coarse-grained eclogites and eclogitic xenoliths in kimberlite (Jagoutz, 1988, 1994; Thöni and Jagoutz, 1992). In addition, retrograde overprinting also may cause isotopic disequilibrium among mineral phases. It would be interesting to examine the possible effects of these factors on isotopic systematics of the UHP metamorphic rocks in the Dabie-Sulu belt.

As reported by many authors, the UHP rocks of the Dabie-Sulu belt were first metamorphosed at temperature conditions of ~600 to 900 ℃ during the peak of metamorphism and were then subjected to a retrograde metamorphism at temperatures of ~480 to 800 ℃ (see Liou et al, 1995, for a review). Presumably, these high metamorphic temperatures would have resulted in an isotopic equilibration among UHP phases and among the retrograde phases as well. Indeed, Sm/Nd isotopic compositions of UHP phases (such as garnet, omphacite, rutile, and phengite) in some fine-grained eclogites from the Shuanghe and Qinglongshan areas define isochrons with reasonable goodness of fit in the range from 0.35 to 0.886 (Figs. 2A and 2B) (Li et al., 1994, 1999). The Sm/Nd isochron ages inferred from these isochron plots are not only in agreement with each other (226.2 ± 2.9 Ma and 226.3 ± 4.5 Ma), but also agree well with reported U/Pb zircon ages of 225.5 +3/−6 Ma and 225 ± 4 Ma for UHP metamorphic rocks in the Dabie terrane (Rowley et al., 1997; Hacker et al., 1998). This concordance may therefore suggest that the well-defined Sm/Nd isochron ages could be considered as the best estimate for the age of peak metamorphism in the Dabie-Sulu belt. However, it also was determined that the Sm/Nd isotopic compositions of mineral phases in retrograded UHP metamorphic rocks in the Dabie-Sulu belt usually do not form a good linear array in the isochron plot. As shown in Figure 2C, the Sm/Nd isotopic compositions of the retrograde phases (i.e., biotite and epidote in a UHP gneiss from the Shuanghe area) are located below the linear array defined by garnet and phengite in the same rock sample. It is not surprising to find a lower ^{143}Nd/^{144}Nd value by retrograde phases, because retrograde metamorphism occurred much later than peak metamorphism. However, it is interesting to find that the linear array defined by garnet and phengite indicates an isochron age of 226.5 ± 2.8 Ma, which is in perfect agreement with the Sm/Nd isochron and the U/Pb zircon ages mentioned above. These would demonstrate that the Sm/Nd isotopic systematics of UHP phases may have remained closed and scarcely modified by later retrograde metamorphism, although the temperature conditions of retrograde metamorphism found in the UHP rocks of the Dabie-Sulu belt are as high as ~480 to 800 ℃. This clearly demonstrates that the Sm/Nd isotopic compositions of UHP phases are not in equilibrium with those of retrograde phases in the same sample.

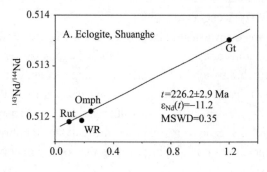
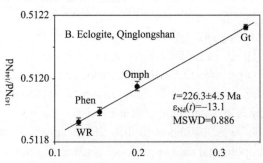

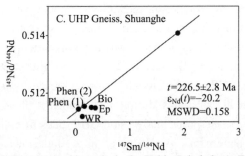

Fig. 2 Sm/Nd isochron plots for UHP metamorphic rocks in the Dabie-Sulu belt. A. Eclogite from Shuanghe in the Dabie terrane. B. Eclogite from Qinglongshan in the Sulu terrane. C. UHP gneiss from Shuanghe in the Dabie terrane. Data sources: Li et al. (1994, 1999). MSWD indicates the goodness of fit (least-squares regression). Abbreviations: Bio = biotite; Ep = epidote; Omph = omphacite; Phen = phengite; Rut = rutile; WR = whole-rock.

Accordingly, any linear array defined by the UHP phases and retrograde phases/whole-rock should be considered as an "errorchron" or a "mixing line" instead of an "isochron". As a result, depending on the Sm/Nd compositions of the retrograde phases, such an errorchron would give an abnormal age either older or younger than the true one. As shown in Figure 2C, the UHP gneiss from the Shuanghe area contains retrograde epidote and biotite; both minerals are low in Sm content. The linear array defined by garnet and retrograde phases would yield an abnormally old age. A similar case can be found in a retrograde eclogite sample from the Wuhe area. The omphacite in this sample has been replaced by amphibole, which results in an abnormally old garnet-whole-rock Sm/Nd isochron age (246 ± 8 Ma) (Okay and Sengör, 1993). On the other hand, if the retrograde phase is rich in Sm, a younger Sm/Nd errorchron age would be expected. As an example, the coarse-grained eclogites and garnet-peridotites from Bixiling in the Dabie terrane are often found to contain second-generation garnets (pink colored) that exhibit higher $^{143}Nd/^{144}Nd$ values than do the earlier-generation garnets (light pink in color). Regression of the data for the retrograde garnet and other UHP phases yields a younger age (210 ± 1.5 Ma) with respect to that for peak metamorphism (i.e., sample DX36A-1, discussed in Li et al., 1996). It is suspected that similar young Sm/Nd isochron ages (210 to 218 Ma) obtained by Chavagnac and Jahn (1996) for the eclogite and garnet-peridotite samples from the same area might have resulted from such an unequilibrated Sm/Nd isotopic system among the UHP and retrograde phases.

In summary, although the temperature conditions of retrograde metamorphism in the UHP rocks in the Dabie-Sulu belt may have been as high as ~480 to 800 ℃, the Sm/Nd isotopic systematics of UHP phases may have remained closed and may not have been re-equilibrated with that of retrograde phases during retrograde metamorphism. Depending on the Sm/Nd compositions of retrograde phases, the Sm/Nd isochron plot for the retrograded UHP rocks may yield abnormal ages that are either older or younger than the true age. Some abnormally young and old Sm/Nd isochron ages reported by previous studies may have resulted from such nonequilibrated isotopic systems in retrograded UHP rocks. After careful examination, highly consistent Sm/Nd isochron ages of ~ 226 Ma can be obtained for fine-grained and unretrograded samples and are considered to be the best estimate for the age of ultrahigh-pressure metamorphism in the Dabie-Sulu belt.

2.2 $^{40}Ar/^{39}Ar$ systematics

In general, the $^{40}Ar/^{39}Ar$ technique is widely accepted as one of the most powerful dating tools in revealing the thermal history of metamorphic rocks (McDougall and Harrison, 1988). However, it may be difficult to obtain reasonable $^{40}Ar/^{39}Ar$ ages for the UHP metamorphic rocks because in many cases the $^{40}Ar/^{39}Ar$ ages are much older than the Sm/Nd and U/Pb ages in the same sample (Mattauer et al., 1991; Monié and Chopin, 1991; Tilton et al., 1991; Li et al., 1992; Ames et al., 1993; Li, Xiao et al., 1993). One of the major drawbacks of $^{40}Ar/^{39}Ar$ dating for UHP metamorphic rocks is the presence of excess argon (Li et al., 1994; Arnaud and Kelly, 1995; Hacker and Wang, 1995; Ruffet et al., 1995; Scaillet, 1996). Yet scientists are still pondering whether the excess argon was trapped or inherited. How did the excess argon reside in UHP phases? And from where did the excess argon come? Many studies have shown that almost all minerals with excess argon in the UHP rocks yield perfect flat profiles for age spectra, indicating a homogeneous distribution of argon throughout the mineral grain (Li et al., 1994; Ruffet et al., 1995; Scaillet, 1996). This would suggest that the excess argon probably was incorporated into the minerals before and/or during the closure of the isotopic system and probably was residing either in structural defects (dislocations and vacancies) or in vacant sites (Ruffet et al., 1995; Scaillet, 1996, 1998). Ruffet et al. (1995) suspected that the excess argon residing in phengite in the UHP terrane probably was conveyed by a fluid phase enriched in ^{40}Ar through the degassing of adjacent minerals or rocks during UHP metamorphism. This postulation has been confirmed by the results of in-situ ^{40}Ar/^{39}Ar laserprobe analyses on UHP rocks from the Dora-Maira nappe (Scaillet 1996). On the basis of the distribution pattern of ^{40}Ar/^{39}Ar ages in a single mica grain, Scaillet (1996) further suggested that the incorporation of excess argon in micas may have been controlled by very low lattice and grain-boundary diffusion under dry conditions.

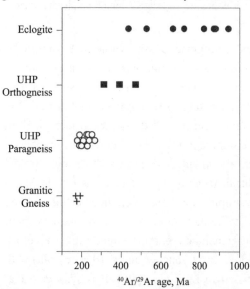

Fig. 3 Diagram showing the correlation between the ^{40}Ar/^{39}Ar apparent ages of micas from UHP metamorphic rocks and their country rocks in the Shuanghe area (cited from Li and Lo, 1999).

In order to further reveal the source of excess argon and the scope of effects from excess argon, a systematic ^{40}Ar/^{39}Ar dating survey was deployed by Li and Lo (1999) on UHP metamorphic rocks in the Shuanghe area. The results showed apparent ^{40}Ar/^{39}Ar ages of 431 to 943 Ma for micas in eclogites, 309 to 471 Ma in interlayering UHP orthogneisses, 187 to 266 Ma in UHP paragneisses, and 171 to 173 Ma in granitic country rocks (Fig. 3). The ^{40}Ar/^{39}Ar ages for micas in eclogites and UHP gneisses are much older than the Sm/Nd and U/Pb ages mentioned earlier, indicating that these abnormally old ^{40}Ar/^{39}Ar ages may have resulted from the presence of excess argon. This clearly indicates that ^{40}Ar/^{39}Ar dating may not be helpful in elucidating the thermal history of UHP rocks in the Dabie-Sulu belt. However, it is interesting to find that micas from the granitic country rocks at the same outcrop yield ^{40}Ar/^{39}Ar ages of 170 ± 1 Ma and 177 ± 1 Ma (Li and Lo, 1999), which are consistent with their Rb/Sr ages of 171 ± 2 Ma and 173 ± 3 Ma (Li et al., 1999). Since these ages are reasonable in view of the cooling history of the Dabie UHP rocks, as discussed below, they demonstrate that excess argon may not be present in the country rocks. On the other hand, substantial amounts of excess argon were residing in the UHP rocks and were not expelled from the rock body during retrograde metamorphism.

The difference of ^{40}Ar/^{39}Ar ages among the interlayering UHP eclogites and gneisses from the same outcrop also suggests that the UHP metamorphic rocks failed to equilibrate Ar isotopes with each other on a centimeter to meter scale. Certainly, there may be a number of mechanisms—for example, the density of schistosity, which is considered to be a potential conduit for the release of excess argon—controlling the amount of excess argon residing in the rock body. The fact that excess argon was widely distributed in the UHP rocks but was not present in the country rocks may suggest that excess ^{40}Ar has been derived internally from the UHP rocks, through the degassing of early-generation minerals during the UHP and/or retrograde metamorphism. However, the excess argon may not have been expelled from the rock body before the closure of the isotopic system. Thus, it is concluded that although ^{40}Ar/^{39}Ar dating of UHP rocks may not be helpful in elucidating the thermal history because of the presence of excess argon, useful information for regional tectonics can still be obtained from the ^{40}Ar/^{39}Ar dating of the country rocks.

2.3 Rb/Sr systematics

Similar to the case involving the Sm/Nd isotopic system, Rb/Sr isotopic dating has been carried out through the isochron-plot technique, which assumes that all minerals have remained in isotopic equilibrium since they formed. However, as discussed earlier, many phases in the UHP rocks are not in isotopic equilibrium with each other because of retrograde metamorphism. In such a case, careful examination of petrographic characteristics and isotopic systematics is required. If UHP metamorphic rocks experienced significant retrograde overprinting, the earlier-generation UHP minerals (e.g., garnet and phengite) might have remained isotopically closed while the later-generation minerals (e.g., biotite and epidote) grew during the retrograde metamorphism (Fig. 4). As a result of the time lag, these two sets of minerals would have grown in two different $^{87}Sr/^{86}Sr$ isotopic systems, and the latergeneration phases would have a higher initial $^{87}Sr/^{86}Sr$ value. As time passed, these two Rb/Sr isotopic systems would evolve independently (shown as the solid and dashed [Bio + WR and 450 ℃] lines in Fig. 4, respectively). In this case, regressions of data for these two separate isotopic systems will yield two different isochron ages that may closely represent the cooling ages after metamorphic events. However, any linear regression involving two generation phases (shown as the dotted line [Phen + WR] in Fig. 4) should be considered a "mixing line" or an "errorchron" rather than as an "isochron". The age implied by the errorchron regression is geologically meaningless. It could be either older or younger than the two ages of metamorphic events, depending on the isotopic compositions of the retrograde phases.

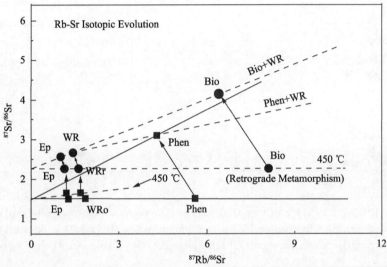

Fig. 4 Schematic diagram showing the evolution of Rb/Sr isotopic systematics in the UHP and retrograde phases (from Li et al., 1999). Abbreviations: WRo = whole-rock before retrograde metamorphism; WRr = whole-rock during the retrograde metamorphism; WR = whole-rock after retrograde metamorphism; Bio = biotite; Ep = epidote; Phen = phengite.

In the Dabie-Sulu belt, some eclogitic rocks are not significantly altered by the retrograde metamorphism and still exhibit UHP mineral assemblages, such as garnet-omphacite-phengite. Their Rb/Sr isochron plots yield consistent ages of ~214 to 224 Ma (Fig. 5) (Li et al., 1994; Chavagnac and Jahn, 1996), which are reasonable for rocks cooling through the closure temperature of phengite Rb/Sr isotopic systematics (~500 ℃). On the other hand, some UHP rocks have been significantly overprinted by retrograde effects and have been reported to exhibit much younger Rb/Sr isochron ages (169 to 181 Ma) (Li et al., 1994, 1999; Chavagnac and Jahn, 1996). These younger Rb/Sr isochron ages obviously result from retrograde metamorphism because the Rb/Sr ratios of retrograde biotites are much higher than those of other minerals in the same rock. If this is the case, the reported ages could approximate the cooling ages of the UHP rocks after the retrograde metamorphic even.

In order to further clarify this problem, the Rb/Sr isotopic compositions of the UHP phases, as well as those of the retrograde phases, from a UHP gneiss (HT-1-1) from the Shuanghe area are shown in Fig. 6 as an example. It is clear that the data points for UHP phases (i.e., garnet and phengite), retrograde phase (i.e., biotite), and whole-rock cannot define a statistically meaningful isochron. Alternatively, the linear array defined by garnet and phengite yields an isochron age of 218.5 ± 3.3 Ma, which is in perfect agreement with that for the unretrograded UHP rocks shown in Fig. 5. On the other hand, the data for the retrograde phase (biotite) and whole-rock form another isochron with an age of 173.9 ±1.7 Ma, which agrees with those reported for the retrograded eclogites in the Dabie-Sulu belt. Since biotite is high in Rb/Sr ratio, this isochron age could be considered to indicate the time the rock cooled through the closure temperature of biotite, which is ~300 ℃. However, if we consider the mixing line formed by phengite and the whole-rock, a geologically meaningless errorchron age of 130.1 ± 4.1 Ma would be calculated.

This demonstrates that during retrograde metamorphism, the Rb/Sr isotopic systematics of the UHP phases may not have been significantly altered, and the isotopic systematics of the retrograde phases may have formed an isotopic system independent from that defined by the UHP phases.

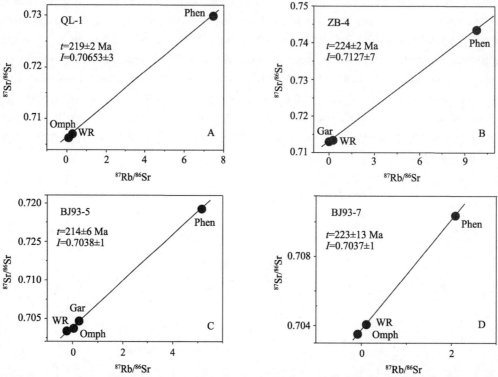

Fig. 5　Rb/Sr isochrons for unretrograded eclogites from Qinglongshan (QL-1) (A) and Zhubian (ZB-4) (B) in the Sulu terrane (from Li et al., 1994). C and D. Rb/Sr isochrons for unretrograded eclogites from Bixiling (BJ93-5, 7) in the Dabie terrane (from Chavagnac and Jahn, 1996). Abbreviations: Ep = epidote; Gar = garnet; Phen = phengite; WR = whole-rock; Omph = omphacite

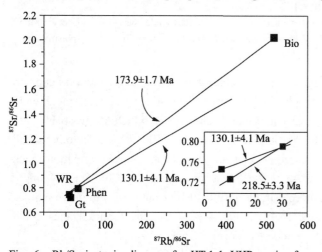

Fig. 6　Rb/Sr isotopic diagram for HT-1-1 UHP gneiss from Shuanghe (from Li et al., 1999). See Figs. 2, 4, and 5 for abbreviations.

In addition to the minerals discussed above, it is also interesting to see how the Rb/Sr isotopic systematics of amphibole evolved during the retrograde metamorphism, because amphibole was found to be abundant in retrograded UHP rocks in the Dabie-Sulu belt; amphibole also is one of the minerals with a high Rb/Sr ratio, and hence is suitable for Rb/Sr dating. Unfortunately, amphibole from the Dabie-Sulu belt has seldom been dated by the Rb/Sr method. The only case was reported by Li et al. (1999) for a retrograded eclogite collected from the Shuanghe area. In this sample, a well-defined isochron formed by retrograde phases—including amphibole, titanite, and whole-rock—yielded an age of 167 ± 5 Ma (Li et al., 1999). This age appears to be in perfect agreement with the published Rb/Sr isochron ages (169 to 181 Ma) defined by retrograde phases in the Dabie-Sulu belt (Li et al., 1994, 1999; Chavagnac and Jahn, 1996). This age could be considered as the time when the sample cooled through the closure temperature of the Rb/Sr isotope systematics in amphibole, because the Rb/Sr ratio of amphibole is much higher than that of titanite and the whole-rock. Unfortunately, the closure temperature of the Rb/Sr isotopic systematics in amphibole has not yet been well established. However, the fair agreement between the Rb/Sr isochron age defined by amphibole + titanite + whole-rock and the $^{40}Ar/^{39}Ar$ age for amphibole in the same sample may suggest that the closure temperature for the Rb/Sr system in amphibole might be equal to or slightly lower than the Ar closure temperature of amphibole (i.e., ~450 ± 50 ℃) (Li et al., 1999). Thus, the aforementioned Rb/Sr ages of 167 ± 5 Ma would indicate the time of the retrograded UHP rock cooling through a temperature of ~450 ± 50 ℃.

3 Tectonic implications

3.1 Timing of ultrahigh-pressure metamorphism

As discussed above, because of the complicated metamorphic history, any interpretation of geochronological data for the UHP rocks must be cautious and should take such factors as closed/open system, closure temperature, and isotopic equilibration of minerals into consideration. In the Dabie-Sulu belt, the retrograde metamorphism has induced the formation of retrograde phases in UHP rocks. The Sm/Nd and Rb/Sr isotopic systematics of these retrograde phases equilibrated with each other and remained isotopically closed during the subsequent cooling. On the other hand, Sm/Nd and Rb/Sr isotopic systematics of the UHP phases seem to have not been significantly affected by the retrograde metamorphism and have preserved a closed system during the retrograde metamorphism. It is therefore possible to obtain an age constraint for the UHP metamorphism from these immobile isotopic systematics (Sm/Nd and Rb/Sr) in the UHP phases with high closure temperatures. As demonstrated above, the Sm/Nd isotopic systematics of UHP phases (i.e., garnet-phengite-omphacite-rutile) in three UHP rocks in the Dabie-Sulu belt yield well-defined isochrons with a narrow age range of 226.2 to 226.5 Ma (Fig. 2). These Sm/Nd isochron ages also agree well with published U/Pb zircon ages for the UHP rocks in the same belt. Thus, it is suggested that the best age estimate for the peak ultrahigh-pressure metamorphism in the Dabie-Sulu belt might be 226 ± 3 Ma.

3.2 T-t path for UHP rocks in the Dabie-Sulu belt

As demonstrated above, the Rb/Sr isotopic systematics of the UHP phases in the Dabie-Sulu belt seem to have not been significantly altered by retrograde metamorphism, and the Rb/Sr isochron ages would effectively record the time of subsequent cooling through closure temperatures. Considering the Rb/Sr isochron ages with their corresponding closure temperatures, a T-t path for UHP rocks from the Dabie-Sulu belt can be obtained, as shown in Fig. 7. This T-t path shows that after peak metamorphism, the UHP rocks in the Dabie-Sulu belt experienced rapid cooling to ~500 ℃ at a rate of ~40 ℃/Ma. The UHP rocks then went through a nearly isothermal stage from ~219 Ma to ~175 Ma. Starting from ~175 Ma, the rocks cooled rapidly again to ~300 ℃ at a rate of >50 ℃/Ma. The first rapid cooling would probably reflect the initial exhumation of the UHP rocks from mantle depths to a middle crustal level characterized by pressures of ~6 to 8 kbar. It is interesting to note that the cooling path of the UHP rocks below ~500 ℃ is similar to that recorded in the country rocks after 220 Ma, as reported by many thermochronological results based on Rb/Sr and Ar/Ar dating studies (Chen et al., 1992, 1995; Li et al., 1995; Chavagnac and Jahn, 1998). This suggests that both the UHP rocks and their country rocks may have exhumed together from the middle-crustal level to the upper-crustal level. Change in the cooling rate during exhumation, shown as a two-stage rapid cooling curve in Fig. 7, is not uncommon in UHP terranes (Michard et al., 1993; Gebauer et al., 1997; Grasemann et al., 1998). It could simply reflect the change of thermal gradient during the exhumation of UHP rocks from depths at a constant rate in a compression regime (Chemenda et al., 1995; Ernst

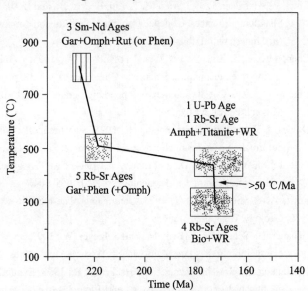

Fig. 7 T-t path of the UHP rocks in the Dabie-Sulu belt. Data sources: Li et al. (1994, 1999), Chavagnac and Jahn (1996), Li, H. (pers. commun.). Abbreviations: Amph = amphibole; Bio = biotite; Gar = garnet; Phen = phengite; Omph = omphacite; Rut = rutile; WR = whole-rock.

and Liou, 1995; Grasemann et al., 1998), or it may indeed indicate a thermal upwelling resulting from the breakoff of a subducted slab (Davies and von Blanckenburg, 1995; Jahn et al., 1999). Because the early rapid cooling record was only found in UHP rocks in the Dabie-Sulu belt, it is plausible that the two rapid cooling stages were caused by two different tectonic mechanisms. The earlier rapid cooling may reflect the exhumation of UHP rocks from mantle depths to the middle-crustal level by compression tectonics during the subduction and/or collision, whereas the later rapid cooling may reflect the exhumation of the entire subducted continental crust from middlecrustal levels to the surface via extension tectonics as a result of the breakoff of a subducted slab.

Acknowledgments

This study was financially supported by the National Natural Science Foundation of China (Nos. 49132040, 49573190, 49794012) and partly supported by Max-Planck-Institut für Chemie, Germany. The authors thank Dr. Tzen-Fu Yui for helpful discussion and comments.

References

Ames, L., Tilton, G. R., and Zhou, G., 1993, Timing of collision of the Sino-Korean and Yangtze cratons: U-Pb zircon dating of coesite-bearing eclogites: Geology, v. 21, p. 339-342.

Ames, L., Zhou, G., and Xiong, B., 1996, Geochronology and isotopic character of ultrahigh-pressure metamorphism with implications for collision of the Sino-Korean and Yangtze cratons, central China: Tectonics, v. 15, no. 2, p. 472-489.

Arnaud, N. O., and Kelly, S., 1995, Evidence for excess Ar during high pressure metamorphism in the DoraMaira (western Alps, Italy), using an ultraviolet laser ablation microprobe $^{40}Ar/^{39}Ar$ technique: Contrib. Mineral. Petrol., v. 121, p. 1-11.

Chavagnac, V., and Jahn, B., 1996, Coesite-bearing eclogites from the Bixiling complex, Dabie Mountains, China: Sm/Nd ages, geochemical characteristics, and tectonic implications: Chem. Geol., v. 133, p. 29-51.

Chavagnac, V., and Jahn, B., 1998, Geochronological evidence for the in situ tectonic relationship in the Dabie UHP metamorphic terrane, central China. Mineral. Magazine, v.62A, p.312-313.

Chemenda, A. I., Mattauer, M., Bokun, A. N., 1996, Continental subduction and a mechanism for exhumation of high-pressure metamorphic rocks : new modeling and field data from Oman. Earth Planet. Sci. Lett. v.143, p.173-182.

Chemenda, A. I., Mattauer, M., Malavieille, J., and Bokun, A. N., 1995, A mechanism for syncollisional rock exhumation and associated normal faulting: results from physical modeling: Earth Planet. Sci. Lett., v. 132, p. 225-232.

Chen, W., Harrison, T., Heizerler, M. T., Liu, R., Ma, B., and Li, J., 1992, The cooling histories of the mélange zone of N. Jiangsu and S. Shandong provinces: evidence from multiple diffusion domain $^{40}Ar/^{39}Ar$ thermal geochronology: Acta Petrol. Sinica, v. 8, p. 1-17 (in Chinese with English abstract).

Chen, J., Xie, Z., Liu, S., Li, X, and Foland, K. A., 1995, Cooling age of the Dabie orogen, China, determined by $^{40}Ar/^{39}Ar$ and fission-track techniques: Science in China (ser. B), v. 38, p. 749-757.

Chopin, C., 1984, Coesite and pure pyrope in high-grade blueschists of the Western Alps: a first record and some consequences. Contrib. Mineral. Petrol., v.86, p.107-118.

Davies, J. H., and von Blanckenburg, F., 1995, Slab breakoff : a model of lithosphere detachment and its test in the magmatism and deformation of collisional orogens: Earth Planet. Sci. Lett., v. 129, p. 85-102.

Eide, L., McWilliams, M. O., and Liou, J. G., 1994, $^{40}Ar/^{39}Ar$ geochronologic constraints on the exhumation of HP-UHP metamorphic rocks in east-central China: Geology, v. 22, p. 601-604.

Ernst, W. G., and Liou, J. G., 1995, Contrasting plate-tectonic styles of the Qinling-Dabie-Sulu and Franciscan metamorphic belts: Geology, v. 23, p. 353-356.

Gebauer, D., Schertl, H. R, Brix, M., and Schreyer, W., 1997, 35 Ma old ultrahigh-pressure metamorphism and evidence for very rapid exhumation in the Dora Maira Massif, Western Alps: Lithos, v. 41, p. 5-24.

Grasemann, B., Ratschbacher, L., and Hacker, B. R., 1998, Exhumation of ultrahigh-pressure rocks: Thermal boundary conditions and cooling history, in Hacker, B. R., and Liou, J. G., eds., When continents collide: geodynamics and geochemistry of ultrahigh pressure rocks: Dordrecht, Kluwer Acad. Publ., p. 117-139.

Hacker, B. R., and Wang, Q., 1995 , Ar/Ar geochronology of ultrahigh-pressure metamorphism in central China: Tectonics, v. 14, p. 994-1006.

Hacker, B. R., Ratschbacher, L., Webb, L., Ireland, T., Walker, D., and Dong, S., 1998, U/Pb zircon ages constrain the architecture of the ultrahigh-pressure Qinling-Dabie Orogen, China: Earth Planet. Sci. Lett., v. 161, p. 215-230.

Jahn, B., Wu, F., Lo, C. H., and Tsai, C. H., 1999, Maficultramafic intrusions of the Northern Dabie Complex, central China: geochemical and isotopic evidence for post-collisional crust-mantle interaction: Chem. Geol., v. 157, p. 119-146.

Jagoutz, E., 1988, Nd and Sr systematics in an eclogite xenolith from Tanzania: Evidence for frozen mineral equilibrium in the continental lithosphere. Geochim. Cosmochim. Acta, v.52, p.1285-1293.

Jagoutz, E., 1994, Isotopic systematics of metamorphic rocks. [abs.], U. S. Geol. Surv. Circ. 1107, Abs. for ICOG-8, Berkeley, p. 156.

Li, S., Chen, Y., Ge, N., Liu, D., Zhang, Z., Zhang, Q., and Zhao, D., 1993, U-Pb zircon ages of eclogite and gneiss from the Qingdao area—evidence for the later Proterozoic magmatism in the Qiaonan Group: Chinese Sci. Bull., v. 38, no. 19, p. 1773-1777.

Li, Q., Chen, W., Ma, B., Wang, Q., and Sun, M., 1995, Thermal evolution history after collision of North China plate with Yangtze plate: Seismol. Geol., v. 17, p. 193-203.

Li, S., Jagoutz, E., Chen, Y., and Li, Q., 2000, Sm-Nd and Rb-Sr isotopic chronology and cooling history of ultrahigh pressure metamorphic rocks and their country rocks at Shuanghe in the Dabie Mountains, central China. Geochim. Cosmochim. Acta. Submitted, v. 64, p. 1077-1093

Li, S., Jagoutz, E., Xiao, Y., Ge, N., and Chen, Y., 1996, Chronology of ultrahigh-pressure metamorphism in the Dabie mountains and Su-Lu terrane: I. Sm-Nd isotope system: Science in China (ser. D), v. 39, p. 597-609.

Li, S., Li, H., Chen, Y., Xiao, Y., and Liu, D., 1997, Chronology of ultrahigh-pressure metamorphism in the Dabie mountains and Su-Lu terrane: II. U-Pb isotope system of zircons: Science in China (ser. D), v. 40, p. 200-206.

Li, S., Liu, D., Chen, Y., and Ge, N., 1992, The Sm/Nd isotopic age of coesite-bearing eclogite from the Southern Dabie Mountains: Chinese Sci. Bull., v. 37, p. 1638-1641.

Li, S., and Lo, C. H., 1999, Inherited origin of excess argon in micas from UHPM rocks in the Dabie-Sulu belt: Chem. Geol., submitted.

Li, S., Wang, S., Chen, Y., Zhou, H., Zhang, Z., Liu, D., and Qiu, J., 1994, Excess argon in phengite from eclogite: evidence from dating of eclogite minerals by SmNd, Rb-Sr and 40Ar/39Ar methods: Chem. Geol., v. 112, p. 343-350.

Li, S., Xiao, Y., Liu, D., Chen, Y., Ge, N., Zhang, Z., Sun, S-S., Cong, B., Zhang, R., Hart, S. R., and Wang, S., 1993, Collision of the North China and Yangtze blocks and formation of coesite-bearing eclogites: timing and processes: Chem. Geol., v. 109, p. 89-111.

Liou, J. G., Wang, Q., Zhai, M., Zhang, R. Y., and Cong, B., 1995, Ultrahigh-P metamorphic rocks and their associated lithologies from the Dabie Mountains, Central China: a field trip guide to the 3rd International Eclogite Field Symposium: Chinese Sci. Bull., v. 40 (suppl.), p. 1-71.

Mattauer, M., Matte, P., Maluski, H., Xu, Z., Zhang, Q., and Wang, Y., 1991, Paleozoic and Triassic plate boundary between North and South China—New structural and radiometric data on the Dabie-Shan (Eastern China): Comptes Rendus Acad. Sci. Paris, v. 312, p. 1227-1233.

McDougall, I., and Harrison, T. M., 1988, Geochronology and thermochronology by the $^{40}Ar/^{39}Ar$ method: Oxford, UK, Oxford University Press, p. 110.

Michard, A., Chopin, C., and Henry, C., 1993, Compression versus extension in the exhumation of the Dora-Maira coesite-bearing unit, Western Alps, Italy: Tectonophys., v. 221, p. 173-193.

Monié, P., and Chopin, C, 1991, 40Ar/39Ar dating in coesite-bearing and associated units of the Dora Maira massif, Western Alps: Eur. Jour. Mineral., v. 3, p. 239-269.

Okay, A. I., and Sengör, A. M. C., 1993, Tectonics of an ultrahigh-pressure metamorphic terrane: the Dabie Shan/Tongbai Shan orogen, China: Tectonics, v. 12, p. 1320-1334.

Okay, A. I., Xu, S., and Sengör, A. M. C., 1989, Coesite from the Dabie Shan eclogites, central China: Eur. Jour. Mineral., v. 1, p. 595-598.

Rowley, D. B., Xue, F., Tucker, R. D., Peng, Z. X., Baker, J., and Davis, A., 1997, Ages of ultrahigh pressure metamorphism and protolith orthogneisses from the eastern Dabie Shan: U/Pb zircon geochronology: Earth Planet. Sci. Lett., v. 151, p. 191-203.

Ruffet, G., Feraud, G., Balevre, M., and Kienast, J. R., 1995, Plateau ages and excess argon in phengites: An $^{40}Ar/^{39}Ar$ laser probe study of Alpine micas (Sesia Zone, Western Alps, northern Italy): Chem. Geol., v. 121, p. 327-343.

Scailete, S., 1996, Excess ^{40}Ar transport scale and mechanism in high-pressure phengites: a case study from an eclogitized metabasite of the Dora-Maira mappe, western Alps. Geoch. Cosmoch. Acta, v.60, p.1075-1090.

Scailete, S., 1998, K-Ar ($^{40}Ar/^{39}Ar$) geochronology of ultrahigh pressure rocks, in Hacker, B. R., and Liou, J. G., eds., When continents collide: geodynamics and geochemistry of ultrahigh-pressure rocks: Dordrecht, Kluwer Acad. Publ., p. 161-202.

Smith, D. C., 1984, Coesite in clinopyroxene in the Caledonides and its implications for geodynamics: Nature, v. 310, p. 641-644.

Sobolev, N. V., and Shatsky, V. S., 1990, Diamond inclusions in garnets from metamorphic rocks: Nature, v. 343, p. 742-746.

Thöni, M., and Jagoutz, E., 1992, Some new aspects of dating eclogites in orogenic belts: Sm-Nd, Rb-Sr and Pb-Pb isotopic results from the Austroalpine, Savalpine, and Koralpe type-locality: Geochim. et Cosmochim. Acta, v. 56, p. 347-368.

Tilton, G. R., Schreyer, W., and Schertl, H. P, 1991, Pb-Sr-Nd isotopic behavior of deeply subducted crustal rocks from the Dora Maira

Massif, Western Alps, Italy, II. What is the age of the ultrahigh-pressure metamorphism? Contrib. Mineral. Petrol., v. 108, p. 22-23.

Wang, X., Liou, J. G., and Mao, H. J., 1989, Coesite-bearing eclogites from the Dabie Mountains in central China: Geology, v. 17, p. 1085-1088.

Xu, S., Okay, A. I., Ji, S., Sengör, A. M. C., Su, W., Liu, Y., and Jiang, L., 1992, Diamond from the Dabie Shan metamorphic rocks and its implication for tectonic setting: Science, v. 256, p. 80-82.

大别山北部榴辉岩和英云闪长质片麻岩的锆石 U-Pb 年龄及多期变质增生*

刘贻灿[1,2]，李曙光[2]，徐树桐[1]，李惠民[3]，
江来利[1]，陈冠宝[1]，吴维平[1]，苏 文[1]

1. 安徽省地质科学研究所，合肥 230001
2. 中国科学技术大学地球和空间科学系，合肥 230026
3. 国土资源部天津地质矿产研究所，天津 300170

> **亮点介绍：** 本文最早证明北大别榴辉岩和片麻岩的峰期变质时代为 226～230 Ma 以及原岩时代为新元古代，为解决北大别的大地构造属性和大别山深俯冲地壳岩石折返机制的建立提供了关键的年代学制约。

摘要 大别山北部榴辉岩及英云闪长质片麻岩的锆石 U-Pb 年龄分析表明：北部榴辉岩相峰期变质时代为 226～230 Ma 左右；北部塔儿河一带英云闪长质片麻岩经历过印支期变质事件；大别山北部与南部超高压岩石中一致的 (226～230 Ma) 高压或超高压变质年龄表明，北部镁铁-超镁铁质岩带中部分岩石也曾作为扬子俯冲陆壳的一部分，在印支期发生过高压或超高压变质作用；本区锆石发生过两期变质增生事件，一是印支期高压或超高压变质，另一期是燕山期热变质事件；榴辉岩及英云闪长质片麻岩的原岩形成时代为晚元古代；锆石 U-Pb 年龄可用多期变质增生模型来解释。

关键词 锆石 U-Pb 年龄；榴辉岩相峰期变质时代；榴辉岩；英云闪长质片麻岩；大别山北部

1 引言

近年来，大别山北部（指五河-水吼剪切带以北）地区获得了辉石岩、辉长岩和英云闪长质片麻岩的部分锆石 U-Pb 年龄[1-5]，其中相当一部分样品取自燕山期岩体。片麻岩一般给出大体一致的 700～800 Ma 上交点年龄和 130～140 Ma 左右的下交点年龄。对这些 U-Pb 年龄的解释存在较大分歧。一种观点认为，片麻岩的 130～140 Ma 左右的下交点年龄代表原岩形成年龄，而上交点代表继承锆石或老的锆石捕虏晶[2]；另一种观点则认为，片麻岩的 700～800 Ma 上交点年龄代表原岩形成年龄，但燕山期下交点 130～140 Ma 左右年龄指示燕山期受热变质影响事件[3,4]。这主要是由研究区片麻岩组成复杂、岩石类型多样所致。为此，需要对它们分别进行定年并给出合理解释。目前，作者等在大别山北部地区发现了两种产状榴辉岩[6,7]，并对其 Sm-Nd 同位素年龄进行了初步研究（石榴子石+绿辉石+全岩等时线年龄为 210 Ma 左右）①，但由于这些榴辉岩在榴辉岩相变质时温度较高，峰期变质之后又经历了一个高温、高压麻粒岩相退变质阶段②而造成 Sm-Nd 同位素体系封闭推迟，从而给出冷却年龄（210 Ma 左右）。因而，榴辉岩相峰期变质的精确时代还没有真正确定。此外，研究区片麻岩是否经历过印支期变质事件也尚无定论。针对这些问题，作者对大别山北部榴辉岩及英云闪长质片麻岩的锆石 U-Pb 年龄进行了详细测定，首次确定了研究区榴辉岩相峰期变质时代为 226～230 Ma 左右，并发现了塔儿河一带英云闪长质片麻岩也曾经历过印支期变质事件；同时，证明它们的原岩形成时代为晚元古代并经受过燕山期热变质事件影响。它们较复杂的锆石 U-Pb 数据用锆石 2 次变质增生模型可以获得较好的解释。

---

* 本文发表在：高校地质学报，2000, 6(3): 417-423
① 刘贻灿等，2000，大别山北部榴辉岩的 Sm-Nd 同位素年龄及其地质意义，待刊
② 刘贻灿等，2000，大别山北部镁铁-超镁铁质岩带中的榴辉岩及其分布与变质 P-T 条件，待刊

2 样品及地质背景

本项研究共采集了 2 个样品用于锆石 U-Pb 年龄分析，一个为榴辉岩（样号为 99104-2），另一个是英云闪长质片麻岩（样号为 98423）。它们均产于大别山北部镁铁-超镁铁质岩带中（图 1）。榴辉岩采集于饶钹寨橄榄岩中（呈构造透镜体产于面理化橄榄岩中），英云闪长质片麻岩采集于塔儿河水电站旁。榴辉岩已部分退变，它的主要矿物组合为石榴子石、绿辉石和金红石，含少量石英、斜方辉石、尖晶石、斜长石、角闪石等，经受过榴辉岩相高压或超高压变质阶段[7,8]及随后的麻粒岩相和角闪岩相退变质阶段[6,7]；英云闪长质片麻岩的主要矿物组合为斜长石+石英+角闪石+黑云母（棕色、富钛）+单斜辉石+斜方辉石±石榴子石，经过麻粒岩相及角闪岩相退变质作用[9,10]。

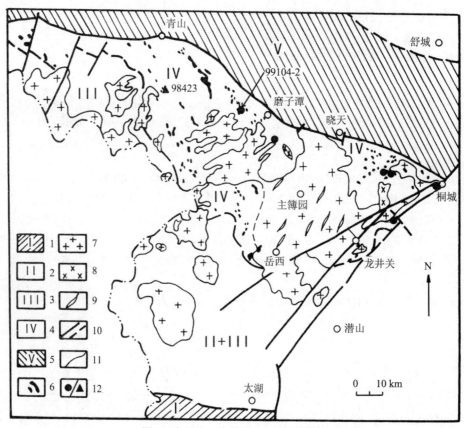

图 1 大别山（安徽部分）地质简图

1. 扬子板块俯冲盖层（宿松杂岩）；2. 超高压变质带；3. 扬子板块俯冲基底（大别杂岩）；4. 镁铁-超镁铁质岩带（可能包括部分大别杂岩）；5. 变质复理石推覆体；6. 镁铁-超镁铁质岩；7. 花岗岩；8. 辉长岩；9. 正长斑岩脉；10. 断层；11. 地质界线；12. 榴辉岩产地/采样点及样号

3 实验方法

在本文研究中，采用单颗粒锆石同位素稀释法进行测定。样品的碎样及锆石的初选工作由安徽省地质实验测试中心完成，锆石的精选、化学处理和锆石 U-Pb 同位素分析在天津地质矿产研究所同位素实验室完成，分析流程已有专文介绍[11]。本次实验使用 ^{205}Pb-^{235}U 混合稀释剂，质谱测定用 VG-354 热离子质谱。数据处理使用美国地调所 Ludwig[12]编写的 PBDAT 和 ISOPLOT 软件。结果见表 1。

4 结果与讨论

99104-2 的 1、2、3 号颗粒的 t_{206}、t_{207}、$t_{207/206}$ 表面年龄一致，并落在一致线上（图 2A），其 ^{206}Pb/^{238}U 表面年龄平均值为 230±6 Ma，与饶钹寨橄榄岩中石榴辉石岩的 Sm-Nd 年龄（244±11 Ma）在误差范围内一致[13]。这说明该年龄指示的是榴辉岩的高压或超高压变质时代。同时显示该锆石不含或含有很少的老锆石的残留，基本上是高压或超高压变质时新生成锆石（类似于南部超高压带）[14,15]。该年龄指示大别山北部榴辉岩高压或超高压变质时代为三叠纪。99104-2 的 4、5 号颗粒给出了较古老的 $t_{207/206}$ 表面年龄值和落在一致线右下方，且 $t_{206} < t_{207} < t_{207/206}$，这说明它们有明显的古老锆石继承核，并在变质过程中增生（尽管该数据也可用锆石丢失放射成因 Pb 来解释，但是，Hacker 等[1]和陈道公等[15]研究认为，大别山高级变质岩中存在变质增生锆石即一个老锆石核外围有新的增生边）。99104-2 的 6 号颗粒的极年轻的 t_{206}、t_{207} 表面年龄（162～165 Ma）和 215 Ma 的 $t_{207/206}$ 表面年龄，表明该锆石颗粒可能是在三叠纪高压或超高压变质锆石基础之上又经历了退变质增生，它与 4、5 号颗粒构成一很好的不一致线，它们和一致线的上、下交点分别得到 796±113 Ma 和 151±28 Ma，其 MSWD 值为 0.00076。由 1、2、3 和 4 号颗粒构成的不一致线，得到上、下交点年龄分别为 957±143 Ma 和 225±14 Ma（MSWD 值为 0.063）。这些数据表明，99104-2 样（榴辉岩）的原岩形成时代为元古代晚期（由于数据点靠近下交点，上交点年龄可能不准确，但仍能指示该样品的原岩形成时代为元古代晚期），其所含锆石在印支期高压或超高压变质及燕山期退变质过程中有过两次增生。应当指出的是，下交点年龄 151±28 Ma 可能是这两次增生的混合结果，真实的燕山期锆石增生事件年龄可能要比 151 Ma 年轻。

表 1 大别山北部榴辉岩及英云闪长质片麻岩中锆石 U-Pb 同位素组成

样品及点号	重量 (μg)	U (μg/g)	Pb (μg/g)	普通 Pb (ng)	同位素原子比率 $^{206}Pb/^{204}Pb$	$^{208}Pb/^{206}Pb$	$^{206}Pb/^{238}U$	$^{207}Pb/^{235}U$	$^{207}Pb/^{206}Pb$	表面年龄 (Ma) $^{206}Pb/^{238}U$	$^{207}Pb/^{235}U$	$^{207}Pb/^{206}Pb$
榴辉岩（99104-2）												
1 浅紫红色短柱（粒）状晶体	10	237	11	0.020	289	0.1846	0.0365 (29)	0.2560 (310)	0.0509 (42)	231	231	236
2 浅黄色透明浑圆状晶体	20	290	16	0.088	170	0.2829	0.0363 (11)	0.2570 (120)	0.0513 (17)	230	232	252
3 浅黄色透明不规则粒状晶体	20	142	9	0.076	105	0.1729	0.0362 (21)	0.2540 (240)	0.0509 (35)	230	230	234
4 浅紫红-无色浑圆（等轴）状晶体	20	152	12	0.020	575	0.3482	0.0596 (23)	0.5030 (270)	0.0613 (20)	373	414	649
5 浅黄色透明长柱状晶体	20	184	11	0.090	93	0.3292	0.0298 (16)	0.2190 (190)	0.0532 (34)	190	201	336
6 浅紫红色柱状晶体	10	404	17	0.076	116	0.1544	0.0254 (15)	0.1770 (170)	0.0504 (36)	162	165	215
英云闪长质片麻岩(98423)												
1 无色透明短柱状晶体	20	671	17	0.003	7418	0.2442	0.0228 (3)	0.1552 (33)	0.0494 (72)	145	147	167
2 浅紫红色透明短柱状晶体	15	671	21	0.077	192	0.2237	0.0229 (71)	0.1543 (78)	0.0489 (179)	146	146	144
3 无色透明长柱状晶体	15	674	23	0.021	868	0.2640	0.0282 (4)	0.2021 (45)	0.0520 (8)	179	187	285
4 浅黄色透明长柱状晶体	25	565	19	0.035	761	0.1795	0.0293 (4)	0.2122 (43)	0.0525 (7)	186	195	309
5 无色透明长柱状晶体	25	458	19	0.006	3861	0.3813	0.0326 (4)	0.2464 (44)	0.0548 (7)	207	224	402
6 无色透明短柱状晶体	10	731	123	0.015	1134	4.330	0.0356 (10)	0.2511 (106)	0.0511 (149)	226	228	247
7 浅紫红色透明短圆柱状晶体	15	366	22	0.130	112	0.1691	0.0356 (11)	0.3490 (160)	0.0711 (20)	226	304	959

注：锆石 U-Pb 同位素分析在国土资源部天津地质矿产研究所同位素实验室完成。$^{206}Pb/^{204}Pb$ 已对实验空白（Pb=0.05 ng, U=0.002 ng）及稀释剂作了校正。其他比率中的铅同位素均为放射成因铅同位素，括号内的数字为（2σ）绝对误差，例如：0.0356（11）表示 0.0356±0.0011（2σ）。

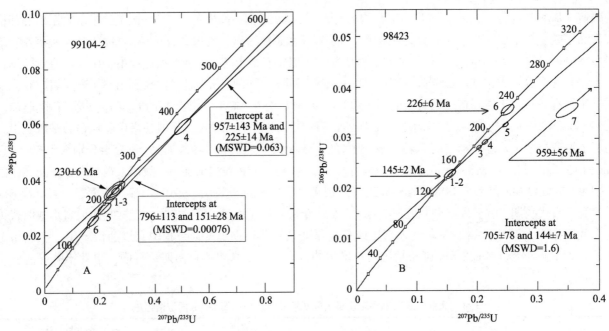

图 2 锆石 U-Pb 年龄曲线谐和图
A. 榴辉岩（样号为 99104-2）；B. 英云闪长质片麻岩（样号为 98423）

从表 1 及图 2 中可以看出，98423 样类似于 99104-2，锆石也明显表现为两期变质增生事件。即：一是印支期高压或超高压变质新生成的锆石及在老锆石基础之上发生变质增生的混合锆石；二是燕山期又发生一次退变质增生。98423 的 6 号颗粒为高压变质锆石，它的 t_{206}、t_{207}、$t_{207/206}$ 表面年龄一致，并落在一致线上（图 2B），取 $^{206}Pb/^{238}U$ 表面年龄值 226±6 Ma 作为谐和年龄值，与 99104-2 一致，代表印支期变质时新生成锆石。1、2 号颗粒的 t_{206}、t_{207}、$t_{207/206}$ 表面年龄一致，并落在一致线上，取 $^{206}Pb/^{238}U$ 表面年龄平均值 145±2 Ma 作为谐和年龄，代表燕山期退变质新生成锆石。由 1～5 号颗粒构成不一致线，它们与一致线的上、下交点分别为 705±78 Ma 和 144±7 Ma，这与 Hacker 等[1]结果一致。由于其下交点指示退变质锆石增生年龄，则上交点应代表原岩时代（但数据点靠近下交点，上交点可能不太准确）。7 号颗粒未能落在 1～5 号颗粒确定的不一致线上，其 $t_{207/206}$ 给出了更古老的表面年龄（959±56 Ma），它可能为古老锆石捕房晶；但 7 号颗粒的 t_{206}=226 Ma，说明该锆石颗粒可能在老锆石基础之上于印支期发生过变质增生。因此，该样品（英云闪长质片麻岩）也明显经历过印支期变质及强烈受燕山期热事件影响（少数锆石除外）。

以上分析表明：

（1）北部榴辉岩相的峰期变质时代为 226～230 Ma 左右。塔儿河一带英云闪长质片麻岩也曾经历过印支期变质事件。至于英云闪长质片麻岩是否经历过高压或超高压变质作用，目前还没有变质岩石学方面的证据（有待于进一步研究）。

（2）大别山北部 226～230 Ma 高压或超高压变质年龄，与南部超高压岩石中获得的年龄[1,14,15]一致。它表明，大别山北部镁铁-超镁铁质岩带中部分岩石也曾作为扬子俯冲陆壳的一部分，在印支期发生过高压或超高压变质作用。

（3）大别山北部榴辉岩及英云闪长质片麻岩的原岩均形成于晚元古代。

（4）大别山北部榴辉岩及英云闪长质片麻岩中锆石明显表现为两期变质增生事件：印支期高压或超高压变质锆石和燕山期退变质增生锆石。单纯的这两种变质锆石 U-Pb 数据应落在一致线上的 230 Ma 及 145 Ma 处。它们与原岩岩浆结晶锆石（位于上交点处）的 U-Pb 数据在谐和图上占据一三角形的三个顶点上。如果一颗锆石由这三种锆石成分组成，则会脱离一致线并落在这三角形内，它类似于锆石多次 Pb 丢失的情况[16]。本文数据表明，大别山北部高级变质岩均存在高压变质锆石为主的颗粒和由 2 次变质增生锆石或加上古老岩浆锆石核组成的混合锆石颗粒。这一锆石多期变质增生模型有助于合理解释高级变质岩的复杂锆石 U-Pb 数据。

致谢

本文的撰写得到安徽省地质矿产局资助,甚为感谢!

参考文献

[1] Hacker B K, Ratschbacher L, Webb L, et al. U/Pb zircon ages contain the architecture of the ultrahigh-pressure Qinling-Dabie orogen, China[J]. EPSL, 1998, 161: 215-230.

[2] Xue F, Rowley D B, Turker R D, et al. U-Pb ages of granitoid rocks in the northern Dabie complex, eastern Dabieshan, China[J]. Geol., 1997, 105: 744-753.

[3] Xie Z, Chen J, Zhou T, et al. U-Pb zircon ages of the rocks in northern Dabie terrain, China[J]. Scientia Geologica Sinica, 1998, 7 (4): 501-511.

[4] 李曙光,洪吉安,李惠民,等. 大别山辉石岩-辉长岩体的锆石 U-Pb 年龄及其地质意义[J]. 高校地质学报,1999,5(3): 351-355.

[5] 葛宁洁,侯振辉,李惠民,等. 大别山造山带岳西沙村镁铁-超镁铁岩体的锆石 U-Pb 年龄[J]. 科学通报,1999,44(19): 2110-2114.

[6] Xu S, Liu Y, Su W, et al. Discovery of the eclogite and its petrography in the Northern Dabie Mountain[J]. Chinese Science Bulletin, 2000, 45 (3): 273-278.

[7] 刘贻灿,徐树桐,江来利,等. 大别山北部镁铁-超镁铁质岩带的研究新进展[J]. 安徽地质,1999,9(4): 262-267.

[8] Tsai Chin-Ho, Liou J. G. Eclogite-facies relics and inferred ultrahigh-pressure metamorphism in the North Dabie Complex, central-eastern China[J]. American Mineralogist, 2000, 85: 1-8.

[9] 徐树桐,刘贻灿,江来利,等. 大别山的构造格局和演化[M]. 北京:科学出版社,1994,1-175.

[10] 刘贻灿,徐树桐,江来利,等. 大别山北部中酸性片麻岩的岩石地球化学特征及其古大地构造意义[J]. 大地构造与成矿学,1999,23(3): 222-229.

[11] 陆松年,李惠民. 蓟县长城系大红峪组火山岩的单颗粒锆石 U-Pb 法准确定年[J]. 中国地质科学院院报,1991,22:137-146.

[12] Ludwig K R. Isoplot—A plotting and regression program for radiogenic-isotope data[A]. USGS Open-file Report, Version 2.92. 1997, 91-445.

[13] 李曙光,Hart S R,郑双根,等. 中国华北、华南陆陆碰撞时代的钐-钕同位素年龄证据[J]. 中国科学(B 辑),1989,(3): 312-319.

[14] 李曙光,李惠民,陈移之,等. 大别山-苏鲁地体超高压变质年代学——II. 锆石 U-Pb 同位素体系[J]. 中国科学(D 辑),1997,27(3): 200-206.

[15] 陈道公,Isachen C,支霞臣,等. 安徽潜山片麻岩锆石 U-Pb 年龄[J]. 科学通报,2000,45(2): 2110-2114.

[16] 李曙光,陈移之,宋明春,等. 胶东海阳所斜长角闪岩的锆石 U-Pb 年龄——多期变质作用对锆石不一致线年龄影响的实例[J]. 地球学报,1994,(1~2): 37-42.

青岛仰口榴辉岩的Nd同位素不平衡及二次多硅白云母Rb-Sr年龄*

李曙光[1]，孙卫东[1]，张宗清[2]，李秋立[1]

1. 中国科学技术大学地球及空间科学系化学地球动力学研究实验室，合肥 230026
2. 中国地质科学院地质研究所，北京 100037

> **亮点介绍**：苏鲁超高压带青岛仰口含柯石英榴辉岩含两期白云母，并以沿构造面理发育的二次多硅白云母为主；证明由于退变质作用而引起石榴子石与绿辉石以及二次多硅白云母之间均存在Nd、Sr同位素不平衡，给出不合理的矿物等时线年龄。然而二次多硅白云母与退变质绿辉石组合给出的Rb-Sr年龄[(183 ± 4) Ma]可能更接近该云母的形成时代。

摘要 退变质的青岛仰口含柯石英榴辉岩含两期白云母，并以沿构造面理发育的二次多硅白云母为主。由于退变质作用，石榴石与退变质绿辉石以及二次多硅白云母之间均存在Nd、Sr同位素不平衡。因此多硅白云母+石榴石或含石榴石全岩给出的Rb-Sr年龄[(193 ± 4)~(195 ± 4) Ma]高估了该白云母的形成时代。二次多硅白云母与退变质绿辉石组合给出的较年轻Rb-Sr年龄[(183 ± 4) Ma]可能更接近该云母的形成时代。该年龄指示的是榴辉岩经历的退变质事件的时代而非超高压榴辉岩快速冷却至500 ℃（白云母Rb-Sr封闭温度）的年龄，这对文献中发表的类似白云母Rb-Sr年龄的解释有重要意义。

关键词 榴辉岩；Nd同位素不平衡；Rb-Sr定年；苏鲁超高压变质带

测定超高压变质岩具有不同封闭温度矿物的冷却年龄，并构筑其冷却T-t轨迹，是研究超高压变质岩快速抬升并出露地表的构造机制的重要途径。由于超高压变质岩中的多硅白云母普遍含过剩Ar[1~9]，因此Rb-Sr同位素定年就成为测定超高压变质岩多硅白云母冷却年龄的重要方法。然而，大别山及苏鲁带的超高压变质岩多硅白云母Rb-Sr定年显现出较为复杂的情况。一方面，一些榴辉岩中的多硅白云母给出年龄范围为224~214 Ma[平均(219 ± 5) Ma]的Rb-Sr年龄[1,10,11]；另一方面，一些花岗片麻岩(超高压岩片的围岩)的多硅白云母给出较年轻的Rb-Sr年龄(198~194 Ma)[12]。这些年龄均已被解释为超高压变质岩的冷却时代[10~12]。然而，多硅白云母Rb-Sr年龄的这一差异究竟是反映了榴辉岩与花岗片麻岩的不同冷却历史，还是反映了这些多硅白云母的不同成因，仍是一个待解决的问题。查清这一问题显然对研究超高压变质岩及其围岩(花岗片麻岩)的冷却史及相互关系有重要意义。本文通过对苏鲁超高压变质带青岛仰口榴辉岩的Sr、Nd同位素分析，证明了多硅白云母+全岩(195 ± 3.8) Ma的Rb-Sr年龄可能更接近退变质白云母的形成时代，而非该榴辉岩冷却到500 ℃时的时代。

1 地质概况及样品

青岛仰口镇榴辉岩出露于崂山脚下海滨，位于苏鲁超高压变质带北缘。对其地质产状及岩石学已有详细研究[13~15]。该超高压岩片主要以玄武-安山质榴辉岩及少量长石石英岩和蛇纹岩构成，并包裹在云母斜长片

---

* 本文发表在：科学通报，2000，45(20)：2223-2226
对应的英文版论文为：Li Shuguang, Sun Weidong, Zhang Zongqing et al., 2001. Nd isotope disequilibrium between minerals and Rb-Sr age of the secondary phengite in eclogite from the Yangkou area, Qindao, eastern China. Chinese Science Bulletin, 46(3): 252-255

麻岩中[13]。在大别-苏鲁超高压变质带中,仰口是第一个发现了粒间柯石英及从含石榴石变辉长岩到榴辉岩的过渡性岩石的地方[15]。

本文测试的样品(Y-12)为榴辉岩。地质略图及采样位置见文献[13]。与石榴石斜长角闪岩相比较,该榴辉岩挤压面理很发育。榴辉岩主要由石榴石(30%~40%)、绿辉石(20%~50%)和多硅白云母(约10%)组成,并含少量石英、金红石、楣石、锆石、磷灰石、绿帘石和绿泥石。石榴石和绿辉石颗粒细小,显现出在超高压条件下经历的重结晶作用较弱。该榴辉岩退变质较严重,其中绿辉石多被退变质后成合晶所取代。白云母分两期: 早期云母细小,无定向排列,蚀变严重,与石榴石、绿辉石共生;晚期白云母粗大,沿片理分布。电子探针分析表明该粗粒二次白云母为多硅白云母,Si 原子数为 3.43~3.56[13]。

从该榴辉岩样品中选取石榴石、粗粒多硅白云母和退变质绿辉石以及全岩做 Sr、Nd 同位素分析。其中石榴石和多硅白云母为 100%纯净、无蚀变和退变质颗粒;然而由于绿辉石多已退变质,很难挑选到完全透明的新鲜颗粒,因此只能挑选出颜色(淡绿色)均一、半透明的退变质绿辉石颗粒(主要由细粒后成合晶组成)以观察退变质作用的影响。

2 分析方法

所有单矿物在熔样前均用稀盐酸(2 mol/L)加热淋洗,以除去矿物表面的污染物。全岩样也分出一份做盐酸淋洗,以便与未淋洗全岩对比。Sr、Nd 同位素分析在中国地质科学院地质研究所同位素实验室完成。实验流程见文献[16, 17]。实验中,空白本底为 Nd = 40 pg, Sm = 20 pg, Sr = 1 ng, Rb = 1 ng。Nd 同位素质谱分异校正相对于 $^{146}Nd/^{144}Nd = 0.72190$。本实验测定 Nd 同位素标准样 BCR-1 为 $^{143}Nd/^{144}Nd = 0.512612$, Sr 同位素标准样 NBS987 为 $^{87}Sr/^{86}Sr = 0.71030$。年龄值用 ISOPLOT 软件计算(Ludwig, 1994)。在年龄计算中, $^{147}Sm/^{144}Nd$ 和 $^{87}Rb/^{86}Sr$ 的误差分别给定 0.2%和 2%,而 $^{143}Nd/^{144}Nd$ 和 $^{87}Sr/^{86}Sr$ 误差采用质谱测量的 2σ 误差。Nd、Sr 同位素测定结果列于表 1,并分别示于图 1 (a)和(b)。

表 1 青岛仰口榴辉岩(Y-12)Sm-Nd 和 Rb-Sr 同位素组成 a)

样号	样品描述	Sm/μg·g^{-1}	Nd/μg·g^{-1}	$^{147}Sm/^{144}Nd$	$^{143}Nd/^{144}Nd \pm 2\sigma$	Rb/μg·g^{-1}	Sr/μg·g^{-1}	$^{87}Rb/^{86}Sr$	$^{87}Sr/^{86}Sr \pm 2\sigma$
Y-12 w	全岩	6.907	39.43	0.1059	0.511706 ± 6	49.89	520.2	0.2777	0.71192 ± 6
Y-12 w$_{HCl}$	酸洗的全岩	3.055	15.19	0.1215	0.511476 ± 13	34.14	410.1	0.2410	0.71171 ± 4
Y-12 aom	蚀变绿辉石	1.272	6.450	0.1192	0.511720 ± 6	7.469	4102	5.272×10^{-3}	0.71406 ± 7
Y-12 phn	多硅白云母	7.525	13.62	0.3340	0.511880 ± 14	406.4	71.65	16.42	0.75672 ± 7
Y-12 gar	石榴石	1.745	2.132	0.4947	0.512552 ± 9	0.9624	255.5	0.01091	0.71164 ± 3

a) 不同样品组合的 Sm-Nd 年龄: gar + w = (332.4 ± 4.2) Ma, gar + w(HCl)= (440 ± 6.4) Ma, gar + aom = (338.4 ± 4.4) Ma, gar + phn = (638 ± 16) Ma; 不同样品组合的 Rb-Sr 年龄: phn + w = (195.1 ± 3.8) Ma, phn + w(HCl) = (195.6 ± 3.8) Ma, phn + aom = (182.7 ± 3.6) Ma, phn + gar = (193.2 ± 3.7) Ma.

3 结果与讨论

图 1(a)显示该组样品的 Sm-Nd 同位素数据是散点分布,不能构成等时线。将石榴石与其他样品点两两组合计算得出的年龄均大于 330 Ma [见表 1,图 1(a)],远高于大别-苏鲁高压变质岩的峰期变质时代(230~221 Ma)[1, 2, 10, 11, 17~21]。已有研究证明石榴石+全岩给出偏老的 Sm-Nd 年龄主要是全岩含有退变质产物所致[2, 10]。同理,该样品石榴石+退变质绿辉石给出老的 Sm-Nd 年龄也是该退变质绿辉石含大量退变质产物的结果。它说明在退变质过程中,全岩 Sm-Nd 体系开放,一些新形成的退变质矿物具有较低的 $^{143}Nd/^{144}Nd$ 值(因而应当具有较高的 $^{87}Sr/^{86}Sr$ 值),且与石榴石等高压矿物不平衡[10]。在一般的超高压变质岩中,石榴石与多硅白云母是共生的超高压变质矿物并能形成良好的 Sm-Nd 等时线和给出有意义的年龄[1, 10]。然而图 1(a)显示,仰口榴辉岩的粗粒二次多硅白云母与石榴石 Nd 同位素不平衡,并且二者连线给出了异常高的年龄[(638 ± 16) Ma]。这进一步证明了该粗粒二次多硅白云母是在退变质阶段形成的。根据矿物结构关系,它可能是在榴辉岩构造

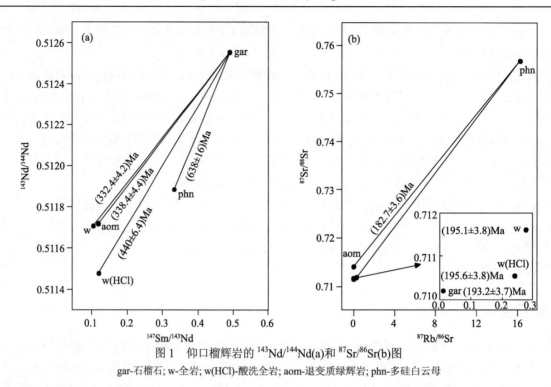

图 1 仰口榴辉岩的 $^{143}Nd/^{144}Nd$(a) 和 $^{87}Sr/^{86}Sr$(b) 图
gar-石榴石; w-全岩; w(HCl)-酸洗全岩; aom-退变质绿辉岩; phn-多硅白云母

抬升并被强烈面理化时形成的[13]。表 1 和图 1 还显示,全岩粉末样品的盐酸淋洗可使某种富稀土矿物溶解,造成 Sm、Nd(尤其是 Nd)的大量丢失,且该矿物具有较高的 $^{143}Nd/^{144}Nd$ 值。因此,酸洗全岩粉末对 Sm-Nd 同位素定年是不可取的。

图 1(b) 显示,该组样品的 Sr 同位素也存在不平衡。退变质绿辉石较石榴石和全岩具有较高的 $^{87}Sr/^{86}Sr$,这显然是由它所含高比例退变质产物造成的。尽管全岩含有相当数量的多硅白云母及退变质绿辉石,石榴石及全岩之间的较小 $^{87}Sr/^{86}Sr$ 变化范围说明全岩可能含有其他具有低 $^{87}Sr/^{86}Sr$ 的组分。多硅白云母与没有酸洗或酸洗过的全岩连线分别给出 (195.1 ± 3.8) 和 (195.6 ± 3.8) Ma(见表 1,图 1(b))。盐酸淋洗并没有明显改变全岩的 Rb/Sr 值和 $^{87}Sr/^{86}Sr$ 组成。多硅白云母+石榴石也给出类似的 Rb-Sr 年龄 ((193.2 ± 3.7) Ma)。前已述及,该多硅白云母与石榴石的 Sm-Nd 同位素体系是不平衡的。考虑到石榴石的 Sr 同位素封闭温度 (750 ℃) 略高于 Sm-Nd 封闭温度 (700 ℃)[22],可以期望该多硅白云母与石榴石的 Sr 同位素体系也是不平衡的。因此,严格来说,这一组由多硅白云母+石榴石或含石榴石全岩所给出的 196~193 Ma 的 Rb-Sr 年龄是无意义的。由于石榴石较退变质矿物 $^{87}Sr/^{86}Sr$ 偏低,因此,该组年龄偏老。退变质绿辉石与多硅白云母的连线给出了较年轻的 Rb-Sr 年龄 (182.7 ± 3.6) Ma,由于二次多硅白云母也是退变质成因,这一 Rb-Sr 年龄比较接近该多硅白云母的真实年龄。但是由于白云母很高的 Rb/Sr 值,这两组 Rb/Sr 年龄相差不大。

令人感兴趣的是尽管仰口榴辉岩二次多硅白云母+全岩 Rb-Sr 年龄 ((195 ± 3.5) Ma) 无地质意义,但它与花岗片麻岩的多硅质 (phengitic) 白云母+全岩 Rb-Sr 年龄 (198~194 Ma) 一致[12]。Carswell 等人[23]根据详细的岩石学研究指出,花岗片麻岩的多硅质白云母 (Si = 3.3) 也是退变质成因的。由于大别山超高压变质带的花岗片麻岩均含有高压变质矿物的残留物 (如石榴石、斜黝帘石等)[23],因此这些年龄值在 195 Ma 左右的白云母 Rb-Sr 年龄也可能高估了退变质多硅白云母的形成年龄。即使考虑到花岗片麻岩中残留的高压变质矿物较榴辉岩少,这些年龄有可能接近该退变质多硅白云母的形成时代,但它指示的是退变质事件年龄,而非超高压变质岩在峰期变质以后快速冷却到 500 ℃时的年龄。

4 结论

大别-苏鲁超高压变质岩的多硅白云母 (或多硅质白云母) 并非都是在峰期变质阶段形成,它们也可以在构造抬升过程中形成。因此不同成因的多硅白云母的 Rb-Sr 年龄含义是不同的。对此,在年代学工作中要特别小心。超高压阶段形成的多硅白云母与石榴石保持 Nd 同位素平衡，而退变质阶段形成的二次多硅白云母与石榴石 Nd 同位素不平衡，这是判别多硅白云母成因的一个有用的标志。本文工作表明，将文献中所发表的 (195 ± 4) Ma 左右的白云母+全岩 Rb-Sr 年龄用来指示超高压岩石快速冷却到 500 ℃时的冷却时代是危险的，它们可能指示的是一次退变质事件年龄。

致谢

本工作为国家重点基础研究发展规划(批准号：G1999075503)和国家自然科学基金(批准号：49794042)资助项目。

参考文献

[1] Li S, Wang S, Chen Y et al. Excess argon in phengite from eclogite: evidence from dating of eclogite minerals by Sm-Nd, Rb-Sr and ^{40}Ar/^{39}Ar methods. Chem. Geol., 1994, 112 : 343-350

[2] Li S, Jagoutz E, Lo C H et al. Sm-Nd, Rb-Sr and ^{40}Ar-^{39}Ar isotopic systematics of the ultrahigh pressure metamorphic rocks in the Dabie-Sulu Belt, central China: a retrospective view. Inter. Geol. Rev., 1999, 41：1114-1124

[3] Arnaud N O and Kelly S. Evidence for excess Ar during high-pressure metamorphism in the Dora-Maira (Western Alps, Italy), using a ultra-violet laser ablation microprobe ^{40}Ar/^{39}Ar technique. Contrib. Mineral. Petrol., 1995, 121: 1-11

[4] Hacker B R and Wang Q. Ar/Ar geochronology of ultrahigh pressure metamorphism in central China. Tectonics, 1995, 14: 994-1006

[5] Ruffet G, Feraud G, Balevre M et al. Plateau ages and excess argon in phengites : an ^{40}Ar/^{39}Ar laser probe study of Alpine micas (Sesia zone, Western Alps, northern Italy). Chem. Geol. (sot. Geosci. Section), 1995, 121: 327-343

[6] Ruffet G., Gruan G, Ballevre M et al. Rb-Sr and Ar/Ar laser probe dating of high-pressure phengites from the Sesia zone (Western Alps) : underscoring of excess argon and new age constraints on the high-pressure metamorphism. Chem. Geol., 1997, 141: 1-58

[7] Scaillet S. Excess Ar transport scale and mechanism in high-pressure phengite : a case study from an eclogitized metabasite of the Dora-Maira mappe, Western Alps. Geochim. Cosmochim. Acta, 1996, 60: 1075-1090

[8] Scaillet S. K-Ar (^{40}Ar/^{39}Ar) geochronology of ultrahigh pressure rocks. In B R Hacker and J G Liou, eds. When continents collide: geodynamics and geochemistry of ultrahigh-pressure rocks. Boston, Kluwer Academic Publishers, 1998: 161-201

[9] Boundy T M, Hall C M, Li G et al. Fine-scale isotopic heterogeneities and fluids in the deep crust : a ^{40}Ar/^{39}Ar laser ablation and TEM study of muscovites from a granite-eclogite transition zone. Earth Planet. Sci. Lett., 1997, 148: 223-242

[10] Li S, Jagoutz E, Chen Y et al. Sm-Nd and Rb-Sr isotopic chronology and cooling history of ultrahigh-pressure metamorphic rocks and their country rocks at Shuanghe in the Dabie Mountains, central China. Geochim. Cosmochim. Acta, 2000, 64(6): 1077-1093

[11] Chavagnac V and Jahn B M. Coesite-bearing eclogites from the Bixiling complex Dabie Mountains, China: Sm-Nd ages, Geochemical characteristics and tectonic implications. Chem. Geol., 1996, 133: 29-51

[12] Chavagnac V and Jahn B M. Geochronological evidence for the in situ tectonic relationship in the Dabie UHP metamorphic terrane, central China. Mineral Magazine, 1998, 62A: 312-313

[13] 李曙光，孙卫东，葛宁洁等. 青岛榴辉岩相蛇绿混杂岩的岩石学证据及退变质 P-T 轨迹. 岩石学报，1992，8(4)：351-362

[14] 叶凯，平岛崇男，石渡明，等. 青岛仰口榴辉岩中粒间柯石英的发现及其意义. 科学通报，1996，41(5)：1047-1048

[15] Zhang R and Liou J G. Partial transformation of gabbro to coesite-bearing eclogite from Yangkou, the Sulu terrane, eastern China. J. Metamorphic Geol., 1997,15: 183-202

[16] 李曙光，葛宁洁，刘德良等. 大别山北翼大别群中 C 型榴辉岩的 Sm-Nd 同位素年龄及其构造意义. 科学通报，1989，(7)：522-525

[17] Li S, Xiao Y, Liu D et al. Collision of the North China and Yangtze Blocks and formation of coesite-bearing eclogites : timing and processes. Chem. Geol., 1993, 109 :89-111

[18] 李曙光，刘德良，陈移之等. 大别山南麓含柯石英榴辉岩的 Sm-Nd 同位素年龄. 科学通报，1992，(4)：346-349

[19] Ames L, Zhou G and Xiong B. Geochronology and isotopic character of ultrahigh-pressure metamorphism with implications for collision of the Sino-korean and Yangtze cratons, central China. Tectonics, 1996, 15(2) : 472-489

[20] Rowley D B, Xue F, Tucker R D, et al. Ages of ultrahigh pressure metamorphism and protolith orthogneisses from the eastern

Dabie Shan : U/Pb zircon geochronology. Earth Planet Sci. Lett., 1997, 151: 191-203

[21] Hacker B R, Ratschbacher L, Webb L, et al. U/Pb zircon ages constrain the architecture of the ultrahigh-pressure Qinling-Dabie orogen, China. Earth Planet. Sci. Lett., 1998, 161: 215-230

[22] Burton K W, Kohn M. J, Cohen A S, et al. The relative diffusion of Pb, Nd, Sr and O in garnet. Earth Planet. Sci. Lett., 1995, 133:199-211

[23] Carswell D A, Wilson R N, Zhai M. Metamorphic evolution, mineral chemistry and thermobarometry of schists and orthogneisses hosting ultra-high pressure eclogites in the Dabieshan of central China. Lithos, 2000, 52: 121-155

Effects of retrograde-zoning of garnet on Sm-Nd isotopic dating of eclogite and oxygen isotopic disequilibrium between eclogitic minerals*

Li Shuguang[1], Liu Yican[1,2], Xu Shutong[2] and Zheng Yongfei[1]

1. School of Earth and Space Sciences, University of Science and Technology of China, Hefei 230026, Anhui, China
2. Anhui Institute of Geology, Hefei 230001, Anhui, China

亮点介绍：本文根据北大别饶钹寨面理化橄榄岩中的强变形榴辉岩的石榴子石成分、结合已发表的榴辉岩锆石 U-Pb 和矿物 Sm-Nd 同位素年龄及氧同位素分析，揭示了榴辉岩中石榴子石的退变质环带也可以造成石榴子石和绿辉石之间显著的 Nd 同位素和氧同位素不平衡现象，并给出异常的矿物 Sm-Nd "等时线" 年龄。

Abstract The compositional zoning of the garnet in a strongly deformed eclogite from Raobazhai foliated peridotite has been recognized. The CaO concentrations of the garnet are decreased from the core to the rim, while its MnO concentrations are increased, suggesting the retrograde origin of such CaO-MnO zoning. The tie line of garnet + omphacite from this eclogite gives a Sm-Nd age of (187 ± 5) Ma, which is less significant than the Sm-Nd ages of (221 ± 5) – (228 ± 3) Ma and (210 ± 6) – (214 ± 6) Ma for ultrahigh-pressure eclogites in the southern Dabie zone and in the northern Dabie zone, respectively. This younger Sm-Nd age could result from the ^{143}Nd/^{144}Nd ratio decrease of the retrograde zone in the garnet. The δ^{18}O values of the garnet and omphacite show that their fractionation values are less than the equilibrium fractionation value between the garnet and omphacite at 500–900 ℃ which suggests an oxygen isotopic disequilibrium between them.

Keywords eclogite; Sm-Nd isotopic dating; oxygen isotope; zonation of garnet; Dabie Mountains

Garnet is an important mineral for Sm-Nd isotopic dating of eclogite and other high-grade metamorphic rocks. If the analyzed garnet and other coexisting metamorphic minerals are pure and fresh without alteration grain, in general, we can obtain a meaningful Sm-Nd age. Geologists have not yet realized that whether the compositional zoning of garnet can influence the Sm-Nd isotopic dating or not, so that we rarely conduct a study on the compositional zoning of garnet in an Sm-Nd dating research. In recent years, we observed that the modification of the garnet caused by the retrograde metamorphism may seriously affect the Sm-Nd dating of eclogite[1] and result in Nd and O isotopic disequilibrium between the garnet and omphacite[2]. However, it is still a remained question that whether the retrograde-zoning of garnet can result in the observable Nd and O isotopic disequilibrium or not. In order to clarify this question, we have studied the mineral chemistry, Sm-Nd chronology and oxygen isotopic compositions of the garnet and omphacite in a strongly deformed eclogite from Raobazhai foliated peridotite in the northern Dabie zone (NDZ) and discussed the questions concerning effects of retrograde-zoning of the garnet on Sm-Nd isotopic dating of the eclogite and oxygen isotope fractionation between the garnet and omphacite.

1 Geological background and sample description

The eclogite sample studied was collected from Raobazhai foliated peridotite[3], which is located on the north margin of the NDZ, south of the Xiaotian-Mozitan fault and tectonically emplaced into the banded gneiss. Country rocks near the contact zone are strongly mylonitized[4]. Foliations are well developed in the peridotite and a

---

* 本文发表在 Science in China (Ser. B), 2002, 45: 28-37

mylonite zone crosscuts the massif[4]. One garnet-pyroxenite sample and one eclogite sample from the peridotite have been dated by the Sm-Nd and U-Pb method, respectively, which yield a Sm-Nd mineral (garnet + diopside + whole rock) isochron age of (244 ± 11) Ma[4] and metamorphic zircon U-Pb age of (230 ± 6) Ma[5]. These ages and geochemical studies[6] suggest that the eclogite-bearing foliated peridotites could be scraped off the overlying lithospheric mantle by erosion of the subducted continental crust of South China Block and were subducted with the continental crust into depth[7]. Thus, the Raobazhai peridotite also experienced the Triassic high- or ultrahigh-pressure metamorphism[3-5]. The eclogite sample (99104-2) studied was strongly deformed and foliated, and was firstly retrograded at the granulite-facies and then retrograded to the amphibolite-facies[3, 8-10].

Mineral assemblages of the eclogite sample (99104-2) include the garnet, omphacite, rutile, quartz, diopside, hyperthene, spinel, amphibole and plagioclase. Among them, the garnet, omphacite and rutile are major high-pressure metamorphic minerals, and others are retrograde minerals. Omphacite in this sample can be occurred both in the matrix and as inclusion in the garnet. The omphacite grains (e.g. end member of jadeite is Jd = 20) with the quartz exsolution usually have a diopside rim (e.g. end member of jadeite is Jd = 2.4) (see Table 1) formed by the retrograde metamorphism. The symplectites and corona structures in this eclogite are well developed. The omphacite grains are usually replaced at margin by the diopside, while the diopside grains show the characteristic replacement at margin with a symplectitic intergrowth of the amphibole and albite. The corona structures around the relic garnet are characterized by a double symplectites: the fine-grain inner symplectitic intergrowth of the amphibole + plagioclase, and the very fine-grain outer symplectitic intergrowth of the hypersthene + diopside + spinel + plagioclase. The primary petrological studies[3, 8-10] identified that at least three generations of the mineral assemblages can be discerned: (i) eclogite-facies assemblage, which is represented by the omphacite + garnet + rutile, and the quartz exsolution in the omphacite and exsolutions of the rutile, clinopyroxene and apatite in garnet indicate that the eclogite experienced ultrahigh-pressure metamorphism with the peak metamorphic pressure of ⩾ 2.5 GPa and 713−874 ℃[3, 9, 10]; (ii) retrograde high-pressure granulite-facies assemblage, which is represented by the fine-grain symplectite of the garnet + diopside + hypersthene + ilmenite ± corundum + spinel + plagioclase formed by the retrograde interaction between the garnet and omphacite under the condition of P = 1.10−1.37 GPa and T = 817−909 ℃; (iii) retrograde amphibolite-facies assemblage, which is represented by the replacement at margins of the garnet and diopside by an assemblage intergrowth of the amphibole + plagioclase + magnetite formed under the condition of P = 0.5−0.6 GPa and T = 500−600 ℃.

The garnet, omphacite and quartz from the studied sample 99104-2 were separated for Sm-Nd and O isotopic analysis. In addition, the omphacite separates used for Sm-Nd isotopic dating in a polished thin-section have been analyzed by the electron microprobe (see Table 2). The electron microprobe analysis results (20 grains) show that all the analyzed grains are omphacite (see Fig. 1) with the Na_2O concentrations of 3.5% − 4.5% and the end member of jadeite (Jd) ranging 15 − 31. No diopside grain has been observed.

Table 1 Electron probe analyses of a zoned clinopyroxene grain from eclogite (99104-2) at Raobazhai in northern Dabie Mountains

Composition	Sample 99104-2			
	core	core-mantle	mantle	rim
SiO_2	54.82	54.23	53.5	53.09
TiO_2	0.14	0.09	0.19	0.46
Al_2O_3	4.02	4.18	3.67	1.34
FeO	6.39	8.34	13.32	9.3
Fe_2O_3	1.8	0.7	0	0.03
MnO	0.05	0.07	0.15	0.28
MgO	10.64	10.71	12.26	12.67
CaO	19.07	18.76	16.17	22.94
Na_2O	3.19	2.85	1.56	0.34
K_2O	0.08	0.05	0.12	0.25
Total	100.2	99.98	100.94	100.7

Continued

Composition	Sample 99104-2			
	core	core-mantle	mantle	rim
O	6	6	6	6
Si	2.012	2	1.981	1.976
Al	0.174	0.182	0.16	0.058
Fe^{3+}	0.024	0.018	0	0.001
Ti	0.004	0.002	0.005	0.013
Fe^{2+}	0.222	0.259	0.413	0.289
Mg	0.582	0.589	0.677	0.703
Mn	0.002	0.002	0.005	0.009
Ca	0.75	0.741	0.642	0.915
Na	0.227	0.204	0.112	0.025
K	0.004	0.002	0.006	0.012
WEF	77.4	79.6	88.6	97.5
Jd	19.8	18.5	11.4	2.4
Ae	2.8	1.9	0	0.1

Compositional zoning has been observed in the part of garnets in this eclogite. A compositional profile of one garnet grain ($r = 400$ μm) (Fig. 2) (the original data are listed in Table 3) shows that the CaO concentrations in the core are higher than that at the rim. From the rim to core, the CaO concentrations are increased from 7.60%, through 9.61%, 10.81%, 9.70%, and 10.33%, to 10.22 %, or the end members of the gross are increased from 10.6, through 15.5, 18.9, 24.7, and 23.7, to 29.8. In contrast, the MnO concentration at the rim is higher than that in the core. From the rim to core, the MnO concentrations are decreased from 0.78%, through 0.54%, 0.43%, 0.23%, and 0.36%, to 0.33%. It is, in general, accepted that the garnet with higher CaO content is assumed to have formed at the higher-pressure condition[11]. Hence, the pressure condition of the core of the garnet formed at is relatively high compared with that of the rim, which reflects a pressure decrease during its exhumation process following peak metamorphism[3,5,9,10]. Therefore, the CaO and MnO compositional zonation of the garnet was caused by the retrograde metamorphism, which may affect the metamorphic temperature calculation and Sm-Nd isotopic dating (see details below).

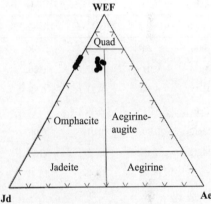

Fig. 1 WEF-Jd-Ae diagrams of the omphacites used for isotope analyses from the eclogite (99104-2) at Raobazhai in northern Dabie Mountains.

Table 2 Electron probe analyses (%) of the omphacites used for Sm-Nd dating from eclogite (99104-2) at Raobazhai in northern Dabie Mountains

Sample	99104-2-1	99104-2-2	99104-2-3	99104-2-4	99104-2-5	99104-2-6	99104-2-7	99104-2-8	99104-2-9	99104-2-10
SiO_2	52.63	53.12	53.43	53.07	52.32	53.40	53.76	53.47	52.69	52.84
TiO_2	0.05	0.06	0.05	0.07	0.09	0.07	0.09	0.07	0.04	0.04
Al_2O_3	6.92	6.14	6.57	6.77	7.06	6.27	7.23	6.21	7.26	6.35
FeO	5.08	5.68	5.24	5.49	5.16	4.95	4.90	5.57	5.45	5.12
MnO	0.07	0.04	0.01	0.03	0.04	0.10	0.09	0.06	0.05	0.00
MgO	11.68	11.94	11.30	11.89	11.48	11.28	11.03	12.27	11.27	11.54
CaO	19.43	19.07	18.44	18.78	18.35	18.72	18.38	19.40	18.83	18.99
Na_2O	4.09	3.89	4.30	4.11	4.52	4.12	4.48	3.74	4.38	4.09
K_2O	0.01	0.00	0.02	0.00	0.00	0.02	0.00	0.00	0.01	0.03
Total	99.96	99.94	99.36	100.25	99.08	98.96	99.96	100.79	99.98	99.00

Continued

Sample	99104-2-1	99104-2-2	99104-2-3	99104-2-4	99104-2-5	99104-2-6	99104-2-7	99104-2-8	99104-2-9	99104-2-10
O						6				
Si	1.891	1.914	1.933	1.903	1.892	1.942	1.931	1.911	1.893	1.918
Al^{IV}	0.109	0.086	0.067	0.097	0.108	0.058	0.069	0.089	0.107	0.082
Al^{VI}	0.184	0.175	0.213	0.189	0.192	0.210	0.237	0.173	0.200	0.189
Fe^{3+}	0.000	0.000	0.153	0.000	0.000	0.133	0.138	0.000	0.000	0.000
Ti	0.001	0.002	0.001	0.002	0.002	0.002	0.002	0.002	0.001	0.001
Fe^{2+}	0.153	0.171	0.006	0.165	0.156	0.018	0.010	0.167	0.164	0.155
Mg	0.626	0.642	0.609	0.636	0.619	0.612	0.591	0.654	0.603	0.624
Mn	0.002	0.001	0.000	0.001	0.001	0.003	0.003	0.002	0.002	0.000
Ca	0.748	0.736	0.715	0.721	0.711	0.729	0.707	0.743	0.725	0.738
Na	0.285	0.272	0.302	0.286	0.317	0.291	0.312	0.259	0.305	0.288
K	0.000	0.000	0.001	0.000	0.000	0.001	0.000	0.000	0.000	0.001
WEF	72.8	74.0	68.8	72.7	70.1	70.1	67.7	75.1	71.0	72.5
JD	27.2	26.0	18.2	27.3	29.9	18.3	20.4	24.9	29.0	27.5
AE	0.0	0.0	13.0	0.0	0.0	11.6	11.9	0.0	0.0	0.0

Sample	99104-2-11	99104-2-12	99104-2-13	99104-2-14	99104-2-15	99104-2-16	99104-2-17	99104-2-18	99104-2-19	99104-2-20
SiO_2	53.28	53.32	53.40	53.02	53.01	52.90	53.14	52.84	51.97	54.05
TiO_2	0.07	0.07	0.06	0.07	0.03	0.05	0.07	0.11	0.03	0.08
Al_2O_3	6.58	6.21	6.89	7.33	6.33	6.33	6.43	6.71	6.67	7.30
FeO	5.35	5.29	5.33	5.12	6.08	6.08	5.03	5.54	5.34	4.93
MnO	0.00	0.01	0.05	0.03	0.02	0.05	0.04	0.06	0.00	0.03
MgO	12.04	11.79	11.20	10.90	11.22	11.28	11.26	11.57	12.52	11.08
CaO	18.90	19.49	18.49	18.17	19.17	19.40	19.26	19.37	19.08	18.49
Na_2O	3.55	3.80	4.51	4.53	4.09	3.86	3.79	3.89	3.78	4.51
K_2O	0.01	0.00	0.01	0.00	0.00	0.01	0.02	0.02	0.00	0.01
Total	99.82	99.98	99.94	99.17	99.95	99.96	99.04	100.11	99.39	100.48
O						6				
Si	1.925	1.922	1.918	1.919	1.914	1.912	1.936	1.903	1.877	1.931
Al^{IV}	0.075	0.078	0.082	0.081	0.086	0.088	0.064	0.097	0.123	0.069
Al^{VI}	0.205	0.186	0.209	0.231	0.183	0.182	0.211	0.187	0.160	0.238
Fe^{3+}	0.112	0.153	0.000	0.000	0.000	0.172	0.116	0.000	0.000	0.138
Ti	0.002	0.002	0.002	0.002	0.001	0.001	0.002	0.003	0.001	0.002
Fe^{2+}	0.049	0.007	0.160	0.155	0.184	0.012	0.037	0.167	0.161	0.010
Mg	0.649	0.634	0.600	0.588	0.604	0.608	0.611	0.621	0.674	0.590
Mn	0.000	0.000	0.002	0.001	0.001	0.002	0.001	0.002	0.000	0.001
Ca	0.732	0.753	0.712	0.705	0.741	0.751	0.752	0.747	0.738	0.708
Na	0.249	0.266	0.314	0.318	0.286	0.271	0.268	0.272	0.265	0.312
K	0.000	0.000	0.000	0.000	0.000	0.000	0.001	0.001	0.000	0.000
WEF	74.2	72.4	70.1	69.5	72.8	71.7	72.3	73.9	74.8	67.7
JD	16.7	15.2	29.9	30.5	27.2	14.5	17.8	26.1	25.2	20.5
AE	9.1	12.4	0.0	0.0	0.0	13.7	9.8	0.0	0.0	11.8

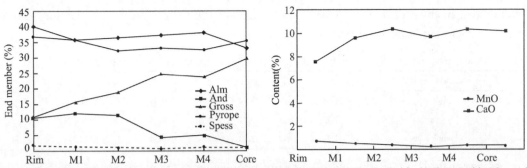

Fig. 2 Compositional diagram of a zoned garnet grain from the eclogite (99104-2) at Raobazhai in northern Dabie Mountains.

Table 3 Electron probe analyses (%) of a zoned garnet grain from the eclogite (99104-2) at Raobazhai in northern Dabie Mountains

Sample No.	99104-2-1	99104-2-2	99104-2-3	99104-2-4	99104-2-5	99104-2-6
SiO_2	38.52	38.49	39.46	37.86	39.03	38.75
TiO_2	0.10	0.08	0.13	0.05	0.10	0.02
Al_2O_3	22.53	22.40	19.27	23.30	22.83	23.46
FeO	18.32	17.25	17.48	18.88	17.74	17.76
Fe_2O_3	3.62	3.97	3.87	1.34	1.65	0.24
MnO	0.78	0.54	0.43	0.23	0.36	0.33
MgO	9.45	8.93	8.37	8.03	8.47	8.51
CaO	7.60	9.61	10.86	9.70	10.33	10.22
Na_2O	0.09	0.02	0.12	0.03	0.07	0.00
Total	100.92	101.29	99.99	99.42	100.58	99.29
O			12			
Si	2.902	2.893	3.021	2.891	2.940	2.943
Al^{IV}	0.098	0.107	0.000	0.109	0.060	0.057
Al^{VI}	1.901	1.875	1.738	1.986	1.965	2.041
Fe^{3+}	0.205	0.224	0.223	0.077	0.094	0.014
Ti	0.001	0.005	0.007	0.003	0.005	0.001
Fe^{2+}	1.155	1.084	1.119	1.206	1.118	1.128
Mg	1.062	1.000	0.955	0.914	0.952	0.964
Mn	0.050	0.034	0.028	0.015	0.023	0.021
Ca	0.613	0.774	0.891	0.793	0.833	0.832
Na	0.013	0.003	0.019	0.005	0.010	0.000
Alm	40.1	35.7	36.4	37.3	38.2	33.2
And	10.7	12.0	11.4	4.2	4.8	0.8
Gross	10.6	15.5	18.9	24.7	23.7	29.8
Prope	36.9	35.6	32.4	33.3	32.5	35.4
Spess	1.7	1.2	0.9	0.5	0.8	0.8

Sample Nos. 99104-2-1, 99104-2-6 represent analytical points from the rim to core in a garnet grain.

2 Analytical methods

Chemical separation of Sm and Nd was performed via a routine two-column (AG50W×8(H+) and HDEHP) iron exchange technique at the Isotope Laboratory, Institute of Geology, Chinese Academy of Geological Sciences (CAGS). Zhang et al.[12] have described the analytical procedure in detail. Sm and Nd isotopic ratios were measured on a Finnigan MAT-262 mass spectrometer in the Chemical Geodynamics Laboratory, University of Science and Technology of China (USTC). Mass fractionation during Nd isotopic analysis was corrected against $^{146}Nd/^{144}Nd = 0.72190$. During the period of data acquisition, the La Jolla standard gives $^{143}Nd/^{144}Nd = 0.511849 \pm 9$. Sm-Nd blanks for the whole chemical procedures are ~50 pg. The age was calculated using the ISOPLOT software provided by

Ludwig[13] with the input errors (2σ) of 0.2% for $^{147}Sm/^{144}Nd$ and 0.005% for $^{143}Nd/^{144}Nd$. The Sm-Nd isotopic data of the Raobazhai eclogite are listed in Table 4.

Table 4 Whole-rock and mineral Sm and Nd data of the eclogite from Raobazhai in northern Dabie Mountains

Sample	Rock or mineral	$Sm/mg·g^{-1}$	$Nd/mg·g^{-1}$	$^{147}Sm/^{144}Nd$	$^{143}Nd/^{144}Nd$	±2σ
99104-2Wr	whole-rock	1.96	3.84	0.3087	0.512916	11
99104-2Gt	garnet	1.72	0.71	1.4654	0.51438	11
99104-2Omp	omphacite	1.76	3.73	0.2854	0.51294	10

Oxygen isotope analysis was carried out by the laser fluorination technique of Sharp[14] with a 25 W MIR-10 CO_2 laser in the Chemical Geodynamics Laboratory, USTC. O_2 extracted from 2 mg minerals heated by the laser in BrF_5 atmosphere was directly transferred to the Finnigan Delta + mass spectrometer for isotope measurement. Oxygen isotope analytical results were given by $\delta^{18}O$ against SMOW standard with the errors ±0.1‰ (for the whole analytical procedures)[15]. The O isotope data of the Raobazhai eclogite are listed in Table 5.

Table 5 Oxygen isotope compositions of minerals from the eclogite at Raobazhai in northern Dabie Mountains

Mineral	Quartz	Omphacite	Garnet
$\delta^{18}O(‰)$	4.40	4.25	3.87
	4.28	4.18	3.74

3 Results and discussion

3.1 Effects of retrograde-zoning in garnet on Sm-Nd isotope dating of eclogite

Figure 3 shows that the tie line of the garnet and omphacite gives an age of (187 ± 5) Ma, which is significantly lower than the U-Pb zircon age of (230 ± 6) Ma for the same eclogite sample[5] and the Sm-Nd age of (210 ± 6) Ma[16] for another eclogite in NDZ. It is also lower than the Sm-Nd ages of (221 ± 5) – (226 ± 3) Ma for the coesite-bearing eclogite in the southern Dabie zone (SDZ)[17-19], but similar to the Rb-Sr age of 180–170 Ma for the biotite from UHPM rocks in SDZ[19]. Two possible interpretations for this younger age could be proposed that (i) it indicates a retrograde metamorphic time of 180–190 Ma, at which Nd isotopic reequilibrium between the garnet and omphacite was reached during the retrograde metamorphism; (ii) the younger age was resulted from the Nd isotopic disequilibrium between the garnet and omphacite, thus is geological meaningless. We separately discuss these two possible interpretations below.

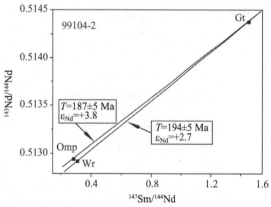

Fig.3 Sm-Nd mineral isochrons of the eclogite (99104-2) at Raobazhai in northern Dabie Mountains.

The viability of the first interpretation hinges on the likelihood of a recrystallization of both the garnet and omphacite during the retrograde metamorphism, which resulted in the Nd isotopic rehomogenization. Obviously, the P-T conditions for the recrystallization and stabilization of the garnet and omphacite must be equal to those of eclogite-facies. As mentioned above, the eclogites including the Raobazhai eclogite in NDZ experienced the retrograde metamorphism of granulite-facies after the eclogite-facies metamorphism, and then followed by the amphibolite-facies retrograde metamorphism. The Sm-Nd geochronological studies have documented that the granulite-facies retrograde metamorphism was terminated at 210 Ma[16], while the Rb-Sr isochron defined by the amphibole + whole rock of the eclogite from NDZ yields an age of (172 ± 3) Ma, indicating a retrograde metamorphic time of amphibolite-facies[20]. These geochronological results suggest that the recrystallization with eclogite-facies can only occur before 210 Ma, while 180 Ma is the retrograde time of the amphibolite-facies for the eclogite in NDZ. Obviously, it is impossible that the recrystallization of both the garnet and omphacite occurred under the *P-T* conditions of amphibolite-facies. Therefore, the first interpretation to explain the garnet + omphacite Sm-Nd age of (187 ± 5) Ma as a retrograde metamorphic time can be precluded.

We prefer the second interpretation based on the following two reasons: (i) The retrograde compositional zoning of the garnet suggests that the retrograde rim and the high-pressure core of the garnet are chemical disequilibrium, hence it is reasonable to believe that the retrograde rim of the garnet and omphacite formed in the high-pressure metamorphic stage may also be chemical disequilibrium because the Sm-Nd diffusion in omphacite is slower than that in garnet[19]. (ii) Fig. 3 shows that the whole-rock (containing retrograde minerals) data point is plotted below the tie line of garnet and omphacite in the Sm-Nd isochron diagram. It is suggested that the $^{143}Nd/^{144}Nd$ ratio of whole rock was reduced by the retrograde metamorphic minerals. For the same reason, the $^{143}Nd/^{144}Nd$ ratio of the retrograde rim in garnet could be lower than those of the high-pressure core in garnet, so that the average $^{143}Nd/^{144}Nd$ value of the whole garnet grain becomes lower. Therefore, the gradient of the tie line between the garnet and omphacite becomes smaller, which is correspondence to a younger age. Obviously, this younger age is geological meaningless. The hypothesis that the $^{143}Nd/^{144}Nd$ ratio of the retrograde rim of garnet was lowered by the retrograded metamorphism is supported by the relative old Sm-Nd age of the Raobazhai garnet-pyroxenite. The Sm-Nd age of (244 ± 11) Ma for the Raobazhai garnet-pyroxenite (R-4) reported by Li et al. [18] was given by a tie line of the garnet and diopside. This Sm-Nd age of the garnet + diopside is significantly older than the Sm-Nd age (210 ± 6) Ma of the garnet + omphacite[16] for the eclogite developed in NDZ. As mentioned above, the diopside in the Raobazhai eclogite was retrograded from the omphacite and, in general, formed a retrograde rim around the relic omphacite, which is textural-similar to the retrograded rim of the garnet. Therefore, the relative old Sm-Nd age (244 ± 11) Ma of the garnet + diopside could be resulted from the lower $^{143}Nd/^{144}Nd$ ratio of diopside than that of omphacite, which results in a higher gradient of the tie line of the garnet and diopside. Obviously, this interpretation requires that the retrograde metamorphic fluid that caused the retrograde compositional zonations of the garnet and omphacite must have a relative low $^{143}Nd/^{144}Nd$ ratio.

3.2 Effects of retrograde-zoning of garnet on oxygen isotopic equilibrium between eclogitic minerals

The oxygen isotope compositions of the garnet, omphacite and quartz for the eclogite (99104-2) from the Raobazhai are listed in Table 5. In terms of the oxygen isotope equilibrium fractionation values under the metamorphic temperature of eclogite-facies, the smaller $\delta^{18}O$ fractionation values between the minerals from the studied sample suggest oxygen isotope disequilibrium fractionation. The oxygen isotope equilibrium fractionation values for the quartz-omphacite and quartz-garnet pairs are 3.24‰–1.57‰ and 4.95‰ –2.38‰ under the conditions of 500 – 900 ℃, respectively[21,22]. However, the measured oxygen isotope fractionations for quartz-omphacite and quartz-garnet pairs are 0.03‰ – 0.22‰ and 0.41‰ – 0.66‰, respectively, which are significantly less than the known equilibrium fractionation values and suggest that their oxygen isotope systems are not at equilibrium. Although the ^{18}O-rich sequence of the measured garnet and omphacite has not been overturn relative to that of the equilibrium fractionation between the two minerals under the high temperature conditions, the measured fractionations of 0.31‰ – 0.51‰ for omphacite-garnet pairs are significantly lower than the equilibrium fractionation values of 1.76‰ – 0.83‰ for omphacite-garnet pairs under the conditions of 500 – 900 ℃[21,22]. The smaller measured fractionation values for omphacite-garnet pairs are correspondence to unreasonably higher temperatures, suggesting the oxygen isotopic disequilibrium between the minerals. In another words, the primary oxygen isotope equilibrium between the garnet and omphacite in studied sample was destroyed, and the reequilibrium of the oxygen isotope exchange between them was not achieved under the conditions of the retrograde metamorphism. Therefore, this work again suggests that the state of the oxygen equilibrium between cogenetic minerals can provide a critical test for the validity of the Sm-Nd mineral chronometer. In terms of the principle mentioned above, i.e. comparing the measured oxygen isotopic fractionation values of omphacite-garnet pairs with the equilibrium fractionation under the eclogite-facies conditions of 500 – 900 °C, we can examine the two Sm-Nd ages reported for other eclogites from NDZ[16]. The $\delta^{18}O$ values for the omphacite and garnet from the eclogite (sample 98702) at Huangweihe are 4.06‰ and 2.89‰, respectively, and their fractionation is 1.77‰, which is correspondence to the equilibrium fractionation value. Therefore, their Sm-Nd ages of (210 ± 6) – (214 ± 6) Ma for sample 98702 should be meaningful, indicating a retrograde metamorphic time of granulite-facies[16]. However, the $\delta^{18}O$ values for the omphacite and garnet from the eclogite (sample 9801) at Maohuayan[16] are 4.11‰ and 3.58‰, respectively, and their fractionation is 0.53‰, which is less than the known range for equilibrium fractionation values, suggesting that the oxygen isotope fractionation between the omphacite and garnet was not at the equilibrium. For this reason, we infer that the Nd isotope systems of the garnet and omphacite were not at the equilibrium and the Sm-Nd age of 208 Ma may be lower than the real age. Considering that the eclogite at Maohuayan is located near the Tan-Lu Fault, it could be dragged by the Tan-Lu Fault from SDZ. In addition, no evidence for the granulite-facies retrograde metamorphism has been found from the sample 9801[23]. Therefore, its real age should be similar to that of the eclogite in SDZ (226 Ma)[17].

4 Conclusions

The garnet in the Raobazhai eclogite (99104-2) displays a retrograded Ca-Mn compositional zoning, which suggests relatively fast diffusions of Ca and Mn acting in the garnet. Though the retrograde zoning of the garnet cannot strongly modify the oxygen isotopic composition of whole garnet grain enough to overturn the ^{18}O-rich sequence of the garnet and omphacite, but it can significantly lower the oxygen isotopic fractionation values between the eclogitic minerals and seriously influence the Sm-Nd age of the garnet + omphacite. Because the Sm-Nd diffusion coefficients of the garnet and omphacite are similar to those of oxygen, the oxygen isotope disequilibrium between the garnet with retrograde zoning and omphacite in the eclogite can indicate that the Sm-Nd isotopic system between the minerals has not been re-equilibrium under the retrograde metamorphic conditions. Consequently, the Sm-Nd age of (187 ± 5) Ma defined by the tie line of the garnet and omphacite from the Raobazhai eclogite does not indicate its retrograde metamorphic time with any geological significance. We call for caution to carefully check whether the garnet has retrograde zoning or not in the Sm-Nd isotopic chronological studies of eclogites.

Acknowledgements

The authors would like to thank Prof. Zhang Zongqing, Drs. Wang Jinhui and Tang Suohan for the technical assistance in the chemical separation of Sm-Nd at the Institute of Geology, CAGS, and Prof. Peng Zicheng, Drs. Zhang Zhaofeng and Huo Jianfeng for the technical assistance in the mass-spectrometer analysis at USTC. Liu Yican thanks Tang Suohan and Huang Fang for the age calculation and Prof. Wang Rucheng for the electron microprobe analysis in the Department of Earth Sciences at Nanjing University. This study was supported by the National Natural Science Foundation of China (Grant Nos. 40033010 and 49794041), the Major State Research Program of China (Grant No. G1999075503) and Anhui Bureau of Geology and Mineral Resources.

References

[1] Li, S., Jagoutz, E., Xiao, Y. et al., Chronology of ultrahigh-pressure metamorphism in the Dabie Mountains and Su-Lu terrane, I. Sm-Nd isotope system, Science in China, Series D, 1996, 39(6): 597-609.

[2] Zheng, Y. F., Wang, Z. R., Li, S. et al., Oxygen isotope equilibrium between eclogite minerals and its constrains on mineral chronometer, Geochim. Cosmochim. Acta, 2002, 66: 625-634.

[3] Liu, Y., Xu, S., Li, S. et al., Distribution and metamorphic P-T condition of the eclogites from the mafic-ultramafic belt in the northern part of the Dabie Mountains, Acta Geologica Sinica (in Chinese), 2001, 75(3): 385-395.

[4] Li, S., Hart, S., Zhang, S. et al., Timing of collision between the North and South China Blocks: Sm-Nd isotopic age evidence, Science in China, Series B, 1989, 32(11): 1391-1400.

[5] Liu, Y., Li, S., Xu, S. et al., U-Pb zircon ages of the eclogite and tonalitic gneiss from the northern Dabie Mountains, China and multi-overgrowths of metamorphic zircons, Geological Journal of China Universities (in Chinese), 2000, 6(3): 417-423.

[6] Zhang, Q., Ma, B., Liu, R. et al., A remnant of continental lithospheric mantle above subduction zone: geochemical constraints on ultramafic rock from Raobazhai area, Anhui Province, Science in China, Series B, 1995, 38(12): 1522-1529.

[7] Li, S., Huang, F., Nie, Y. et al., Geochemical and geochronological constrains on the suture location between the North and South China Blocks in the Dabie orogen, Central China, Physics and Chemistry of the Earth (A), 2001, 26: 655-672.

[8] Liu, Y. C., Xu S. T., Li S. G. et al., Eclogites from the northern Dabie Mountains, eastern China: geochemical characteristics,

Sr-Nd isotopic compositions and tectonic implications, Science in China, Series D, 2000, 43 (Supp.): 178-188.

[9] Xu, S., Liu, Y., Wu, W. et al., Eclogite in the northern Dabie Mountains: petrography and tectonic implications, in Academic Papers for the 80th Anniversary of the Geological Society of China (ed. Chen, Y.) (in Chinese), Beijing: Geological Publishing House, 2002, 92-104.

[10] Liu, Y., Xu, S., Li, S. et al., Tectonic setting and cooling history of the eclogites from northern Dabie Mountains, Earth Science—Journal of China University of Geosciences (in Chinese), 2003, 28(1): 1116.

[11] Carswell, D. A., Wilson, R. N., Zhai, M., Metamorphic evolution, mineral chemistry and thermobarometry of schists and orthogneisses hosting ultra-high pressure eclogites in the Dabieshan of central China, Lithos, 2000, 52: 121-155.

[12] Zhang, Z., Liu, D., Fu, G., Isotope chronology of metamorphic stratigraphy in Northern Qinling (in Chinese), Beijing: Geological Publishing House, 1994, 5-7.

[13] Ludwig, K. R., Isoplot A plotting and regression program for radiogenic-isotope data, USGS Open-Files Report, Version 2.92, 1997, 91-445.

[14] Sharp, Z. D., A laser-based microanalytical method for the in situ determination of oxygen isotope ratios of silicates and oxides, Geochim. Cosmochim. Acta, 1990, 54: 1353-1357.

[15] Gong, B., Zheng, Y. F., Zhao, Z. et al., CO_2-laser fluorination technique of oxygen isotope compositions in silicalite and metal oxides, Bulletin of Mineralogy, Petrology and Geochemistry (in Chinese), 2001, 20: 428-430.

[16] Liu, Y., Li, S., Xu, S. et al., Sm-Nd dating of eclogites from northern Dabie Mountains and its constrains on the timing of granulite-facies retrogression, Geochimica (in Chinese), 2001, 30(1): 79-87.

[17] Li, S., Liu, D., Chen, Y. et al., The Sm-Nd isotopic age of coesite-bearing eclogite from the Southern Dabie Mountains, Chinese Science Bulletin, 1992, 37(19): 1638-1641.

[18] Li, S., Xiao, Y., Liu, D. et al., Collision of the North China and Yangtze Blocks and formation of coesite-bearing eclogites: Timing and processes, Chem. Geol., 1993, 109: 89-111.

[19] Li, S., Jagoutz, E., Chen, Y. et al., Sm-Nd and Rb-Sr isotope chronology of ultrahigh-pressure metamorphic rocks and their country rocks at Shuanghe in the Dabie Mountains, Central China, Geochim. Cosmochim. Acta, 2000, 64 (6): 1077-1093.

[20] Liu, Y., Xu, S., Li, S. et al., Metamorphic characteristics and Rb-Sr isotopic age of garnet-bearing amphibolite at Lu tushipu in the northern Dabie Mountains, Geology of Anhui (in Chinese), 2000, 10(3): 194-198.

[21] Zheng, Y. F., Fu, B., Li, Y. L. et al., Oxygen and hydrogen isotope geochemistry of ultrahigh pressure eclogites from the Dabie Mountains and the Sulu terrane, Earth Planet. Sci. Lett., 1998, 155: 113-129.

[22] Zheng, Y. F., On calculations of oxygen isotope fractionation in minerals, Episodes, 1999, 22: 99-106.

[23] Liu, Y. C., Petrology, geochemistry and isotopic chronology of the eclogites from the northern Dabie Mountains, Ph. D. Thesis, Hefei: University of Science and Technology of China, 2000, 1-69.

A high precision U-Pb age of metamorphic rutile in coesite-bearing eclogite from the Dabie Mountains in Central China: a new constraint on the cooling history*

Qiuli Li[1], Shuguang Li[1], Yongfei Zheng[1], Huimin Li[2], Hans Joachim Massonne[3] and Qingchen Wang[4]

1. School of Earth and Space Sciences, University of Science and Technology of China, Hefei 230026, Anhui, China
2. Tianjin Institute of Geology and Mineral Resources, Tianjin 300170, China
3. Institut für Mineralogie & Kristallchemie, Universität Stuttgart, Azenbergstr 18, 70174 Stuttgart, Germany
4. Institute of Geology and Geophysics, Chinese Academy of Sciences, Beijing 100029, China

> **亮点介绍**：首次通过两种数据处理方法，等时线法和用绿辉石 Pb 同位素组成扣除普通 Pb 法，获得榴辉岩中低 Pb 含量金红石的高精度 U-Pb 等时线年龄（218±2.5 Ma）和谐和年龄（218±1.2 Ma）；根据该榴辉岩冷却速率计算得出金红石 U-Pb 体系封闭温度为 460 ℃左右，该年龄揭示了该榴辉岩折返到中地壳水平的时代。

Abstract This paper first reports a high precision U-Pb age of 218±1.2 Ma for rutile in coesite-bearing eclogite from Jinheqiao in the Dabie Mountains, east-central China. This work shows that the U-Pb mineral (rutile+omphacite) isochron age of 218±2.5 Ma and conventional rutile U-Pb concordia age of 218±1.2 Ma obtained by common Pb correction based on the Pb isotopic composition of omphacite in the same eclogite sample are consistent, proving that the omphacite with low U/Pb ratio (μ=2.8) can be used for common Pb correction in U-Pb dating of rutile. Oxygen isotope analysis of rutile aliquots gave the consistent $\delta^{18}O$ values of −6.1‰±0.1‰, demonstrating oxygen isotope homogenization in the rutile of different grains as inclusion in garnet and grain in matrix. Oxygen isotope thermometry yields temperatures of 695±35 ℃ and 460±15 ℃ for quartz-garnet and quartz-rutile pairs, respectively. These oxygen isotopic observations suggest that the diffusion of oxygen in rutile as inclusion in garnet is not controlled by garnet. According to field-based thermochronological studies of rutile, an estimate of the T_c of about 460 ℃ for U-Pb system in rutile under rapid cooling conditions (~20 ℃/Ma) was advised. Based on this U-Pb age as well as the reported chronological data with their corresponding metamorphic and/or closure temperature, an improved T-t path has been constructed. The T-t path confirms that the UHPM rocks in South Dabie experienced a rapid cooling following the peak metamorphism before 220 Ma and a long isothermal stage from 213−180 Ma around 425 ℃.

Keywords rutile; coesite-bearing eclogite; U-Pb dating; cooling history; closure temperature

1 Introduction

The Dabie-Sulu ultrahigh-pressure metamorphic (UHPM) belt in east-central China is known to be one of the largest UHP terranes in the world. Occurrences of coesite or diamond in eclogites from the Dabie Mountains indicate that these UHPM rocks have been subducted to depths of more than 100 km and then exhumed rapidly to the earth's surface (Wang et al., 1989; Okay et al., 1989; Xu et al., 1992). How the UHPM rocks were exhumed from the depth of >100 km to the surface has been an interesting scientific question. On the basis of determination of cooling ages of minerals with different closure temperatures (T_c) in UHPM rocks, construction of cooling T-t paths is the most direct means to provide constraints on tectonic processes of exhumation of UHPM rocks. Although

---

* 本文发表在：Chemical Geology, 2003, 200: 255-265

many isotopic dating results of UHPM rocks from the Dabie Mountains have been reported (Li et al., 1993,1994, 1997,1999,2000; Ames et al., 1996; Chavagnac and Jahn, 1996,2001; Hacker et al., 1998; Rowley et al., 1997; Ayers et al., 2002), their peak metamorphic age and some cooling ages are still in debate. For example, based on U-Pb zircon dating, Ames et al.(1996) and Rowley et al.(1997) suggested that the peak metamorphic age could be 219 Ma, however, other scientists suggest 226-228 Ma and 235-245 Ma, respectively, for it based on their Sm-Nd and U-Pb zircon dating (Li et al., 2000; Hacker et al., 1998; Ayers et al., 2002). On the other hand, because of the presence of excess Ar in phengite from UHPM rocks (Li et al., 1994), the Rb-Sr phengite dating method has been chosen to determine the 500 ℃ cooling time of the UHPM rocks. Unfortunately, the reported phengite Rb-Sr ages show a considerable variation (194-224 Ma, Li et al., 1994, 1999, 2000, 2001; Chavagnac and Jahn, 1996; Chavagnac et al., 2001).

Determination of U-(Th)-Pb ages of metamorphic monazite, rutile and titanite is another way to obtain cooling ages. Ayers et al. (2002), recently, reported a group of Th-Pb ages of monazite from UHPM rocks in the Dabie Mountains. Rutile is another interesting mineral formed under UHPM conditions. It has high U/Pb ratios and has been shown to yield precise U-Pb ages (Mezger et al., 1989a, 1991; Richards et al., 1988; Corfu and muir, 1989; Schandl et al., 1990; Wong et al., 1991; Davis et al., 1994; Corfu and Easton, 1995; Davis, 1997; Cox et al., 1998; Christofel et al., 1999; Connelly et al., 2000; Norcross et al., 2000; Bibikova et al., 2001; Cox et al., 2002; Hirdes and Davis, 2002; Schmitz and Bowring, 2003). According to the estimate of T_c (400-500 ℃) for the U-Pb system in rutile estimated by Mezger et al. (1989a), rutile U-Pb ages may be used to determine cooling time of UHPM rocks near 500 ℃ or slightly lower temperature. However, Cherniak (2000) suggested that the T_c for the U-Pb system in rutile should be about 600 ℃ on the basis of experimental measure. Until now there is no report of precise U-Pb ages in rutile from eclogites because rutile, especially in eclogite, contains a relatively large proportion of common Pb. Here, we report the first precise rutile U-Pb age for the coesite-bearing eclogite from the Dabie Mountains. Associated with rutile O isotopic analysis, this age will contribute to understanding of the closure temperature of U-Pb in rutile and cooling history of UHPM rocks in the Dabie Mountains.

2 Geological setting and sample description

The Qinling-Dabie orogenic belt is a collision zone of the North China Block and Yangtze Block. The Dabie Mountains represent the eastern section of this belt and can be subdivided into four metamorphic zones shown in Fig. 1 (Li et al., 2001b). Coesite and/or diamond-bearing eclogites are restricted to the South Dabie UHPM zone. In this work, we first analyzed many rutile samples from several eclogite localities (e.g. Bixiling, Shuanghe, Jinheqiao) in the South Dabie UHPM zone where detailed studies of geochronology and stable isotope geochemistry were accomplished (e.g., Li et al., 1993, 2000; Chavagnac and Jahn, 1996; Zheng et al., 1998, 1999; Li YL et al., 2001; Ayers et al., 2002). The analytical results show that most of the rutiles in the eclogites from the Dabie Mountains have low $^{206}Pb/^{204}Pb$ ratios (<24). Only rutile from the Jinheqiao eclogite has a relatively higher $^{206}Pb/^{204}Pb$ ratio (>30), which allows us to obtain high precision rutile U-Pb ages. Therefore, only the analytical results of rutile from

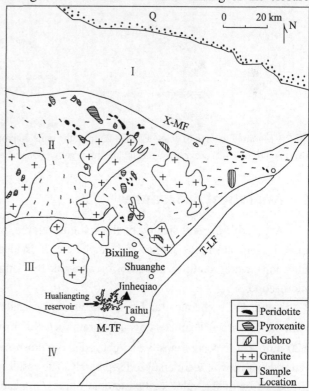

Fig. 1 Sketch map of Geology in the Dabie Mountains and sample location. I: Beihuaiyang zone; II: North Dabie HPM zone; III: South Dabie UHPM zone; IV: Susong HPM zone. X-M F: Xiaotian-Mozitan Fault; T-L F: Tancheng-Lujiang Fault; M-T F: Mamiao -Taihu Fault.

the Jinheqiao eclogite are reported here. The sample of Jinheqiao eclogite (99JHQ-1, WGS84 coordinates: N30°30.01′, E116°15.96′) was collected from a riverbed close to the northeastern shore of the Hualiangting reservoir in Taihu county, Anhui province. Tectonically it is located in the South Dabie UHPM zone (see Fig.1).

Eclogite samples from Jinheqiao show metamorphic banding with thickness in the mm to cm range. The bands consist of mm-size large grains mainly consisting of grass-green omphacite, red garnet and white quartz, so that the color banding is clearly discernable with the naked eye. Minor and accessory phases that were probably stable at the peak *P-T* conditions are kyanite, rutile, coesite, apatite and zircon (Fig. 2). Occurrences of coesite indicate that this eclogite has experienced eclogite facies metamorphism with temperature estimated to be above 650 ℃ and pressure > 2.7 GPa (Cong et al., 1995; Carswell et al., 2000). The rutile content is slightly above 1 vol% from point counting on a representative thin-section. Rutile appears mainly as inclusions in garnet and omphacite, indicating that rutile formed during prograde and/or peak eclogite-facies metamorphism. Larger grains occur also in the matrix. Various colors and shapes were observed for this phase, which was stable during the entire eclogitic evolution.

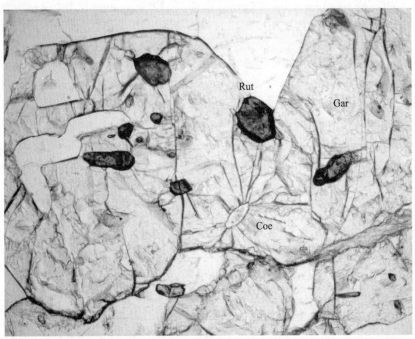

Fig. 2 Microphotograph of eclogite DS98-29 from the same outcrop as eclogite 99JHQ-1 at Jinheqiao in the Dabie Mountains. Coesite (Coe) and rutile (Rut) occur mostly as inclusions in garnet (Gar). Field of view is 1.0 mm horizontally.

3 Analytical procedures

After crushing and sieving of the eclogite sample, magnetic and heavy liquid separations were performed in order to separate rutile, garnet and omphacite. Individual mineral grains without alteration features and visible inclusions were carefully selected for analysis by handpicking from the purified separates under a binocular microscope.

Rutile U-Pb chemical and mass-spectrometric analyses were completed at Tianjin Institute of Geology and Mineral Resources. Diameters of rutile grains selected for analyses are limited to 0.16–0.24 mm with an average of 0.20 mm. Rutile occurs mostly as light red and reddish-brown to dark-brown grains. The two color types, light red and reddish-brown, were analyzed separately. The results showed no systematic difference in the concentrations of U and Pb or of the value of Pb isotopic ratios. Omphacite and garnet grains for analysis are transparent with a homogenous color. A mixed ^{235}U-^{205}Pb tracer was used during the analyses. The analytical procedures were described elsewhere (Mezger et al., 1989a, b). The differences to the previous methods are that H_3PO_4 was used instead of H_2SO_4 during dissolution and U was purified using H_2O instead of HNO_3. Rutiles were analyzed along with repeated analyses of blanks. The total procedural blanks for Pb and U are 0.20–0.24 ng and 0.01–0.05 ng, respectively. The Pb isotope compositions for the blank were basically the same as the average for the three-month period during this work. Data were processed with PBDAT (Ludwig, 1998) and ISOPLOT software (Ludwig, 2000). The results are shown in Table 1.

Table 1 U-Pb analytical results for minerals in eclogite from Jinheqiao

	Sample	Concentration		Isotopic ratios ($\pm 2\sigma$)					Apparent age (Ma)[c] ($\pm 2\sigma$)	
		U (ppm)	Pb (ppm)	$^{206}Pb^{(a)}/^{204}Pb$	$^{206}Pb^{(b)}/^{204}Pb$	$^{238}U^{(b)}/^{204}Pb$	$^{206}Pb^{(c)}/^{238}U$	$^{207}Pb^{(c)}/^{235}U$	$^{206}Pb/^{238}U$	$^{207}Pb/^{235}U$
1	99JHQ-1 Rutile	1.75	0.16	43.4	54.1 (1.5)	1074 (4)	0.0345 (5)	0.2554 (162)	218.4 ±3.2	231 ±14.7
2	99JHQ-1 Rutile	1.91	0.12	55.7	81.3 (4.3)	1853 (6)	0.0347 (5)	0.2390 (106)	219.6 ±3.4	218 ±9.7
3	99JHQ-1 Rutile	2.25	0.12	73.1	111.4 (6.5)	2745 (9)	0.0344 (4)	0.2382 (70)	217.9 ±2.5	217 ±6.4
4	99JHQ-1 Rutile	1.76	0.29	30.3	32.7 (0.3)	458 (2)	0.0342 (5)	0.2543 (319)	217.0 ±2.9	230 ±28.9
5	99JHQ-1 Rutile	2.05	0.19	43.8	51.3 (1.0)	993 (3)	0.0345 (4)	0.2496 (163)	218.5 ±2.4	226 ±14.8
6	99JHQ-1 Omph	0.07	1.50	17.0	17.0 (0.01)	2.8 (0.1)	—	—	—	—
7	99JHQ-1 Garnet	0.21	0.20	17.6	17.6 (0.2)	63 (1)	—	—	—	—

(a) Measured ratio.
(b) Corrected for mass fractionation, spike and blank.
(c) Corrected for mass fractionation, spike, blank and common Pb based on the Pb isotopic composition of omphacite.

Oxygen isotope analysis of minerals was carried out by laser fluorination technique of Sharp (1990) with a 25 W MIR-10 CO_2 laser in Laboratory for Chemical Geodynamics, University of Science and Technology of China at Hefei (Zheng et al., 2002). O_2 was directly transferred to the Delta+ mass spectrometer for the measurement of $^{18}O/^{16}O$ and $^{17}O/^{16}O$ ratios (Rumble et al., 1997). Two reference minerals were used: $\delta^{18}O$ = 5.8‰ for UWG-2 garnet (Valley et al., 1995), and $\delta^{18}O$ = 5.2‰ for SCO-1 olivine (Eiler et al., 1995). Reproducibility for repeat measurements of each standard on a given day was better than ±0.1‰ (1σ) for $\delta^{18}O$. Results are listed in Table 2.

Table 2 Oxygen isotope compositions of minerals in eclogite from Jinheqiao

Sample	$\delta^{18}O$ (‰)	$\delta^{18}O$ (‰) (average)	Minerals pair	$\Delta^{18}O$ (‰)	T^*(°C)
Quartz	1.06, 1.16	1.1			
Garnet	−2.23, −2.31	−2.3	Quartz-Garnet	3.4	695±35
Rutile	−6.01, −6.06, −6.21, −6.15, −6.02, −6.13, −5.98, −6.07, −6.18, −6.26, −5.99, −6.05	−6.1	Quartz-Rutile	7.2	460±15

*Calculated in terms of the calibration of Zheng (1991, 1993).

4 Results

4.1 Rutile U-Pb analysis

As shown in Table 1, the $^{206}Pb/^{204}Pb$ ratios of five fractions of rutile range from 30 to 73. Due to relatively low $^{206}Pb/^{204}Pb$ ratios, common Pb correction is very important in age calculation based on the isotope compositions of rutile in a single analysis. However, the diversity in $^{206}Pb/^{204}Pb$ ratios of five fractions of rutile offers a chance to construct an isochron, which can avoid the problem of common Pb correction. Fig. 3 shows that the data points of five fractions of rutile and omphacite define an isochron yielding an age of 218±2.5 Ma. The low MSWD of 0.1 suggests Pb isotopic equilibrium between rutile and omphacite. The isotopic data of garnet, however, deviates from this isochron slightly (see Fig. 3). It could be the result of difficulties in garnet dissolution that may cause analytical errors.

Since the isochron method requires more analyses of the coexisting minerals with large U/Pb fractionation than the conventional U-Pb dating method, it is necessary to find appropriate minerals, the Pb isotope composition of which can be used for common Pb correction. The compositions of K-feldspar and/or whole rocks are often used for common Pb corrections. However, there is no feldspar in eclogite. In addition, caution has been advised when we consider the use of the isotope composition of whole rock, the average of all minerals including retrograde products, in chronology (Li et al., 1996, 1999, 2000).

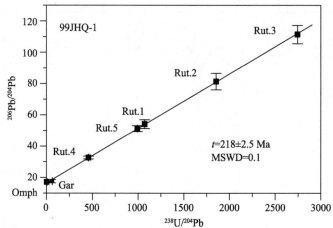

Fig. 3 ^{238}U-^{206}Pb isochron diagram for rutile from eclogite at Jinheqiao. Rut = rutile; Omph=omphacite; Gar=garnet.

As shown in Table 1, the omphacite in eclogite has very low U/Pb ratio (μ=2.8). Based on the results of analyses on the chemical diffusivity of Pb in clinopyroxene, the activation energy for Pb diffusion in clinopyroxene is relatively high and Pb diffusivity is thus fairly low (Cherniak, 1998). Consequently, the Pb isotope composition of clinopyroxene can hardly be influenced by later thermal events. Removing the data of omphacite, the isochron of five rutiles gives the initial ^{206}Pb/^{204}Pb ratio of 16.8±0.3, which is consistent with the ^{206}Pb/^{204}Pb ratio (17.0) of omphacite within analytical uncertainty. So, we suggest that the Pb isotope composition of omphacite can be used for the common Pb correction for the U-Pb analyses of rutile.

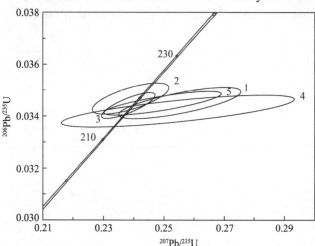

Fig.4 U-Pb age concordia diagram for rutile from eclogite at Jinheqiao.

Using Pb isotope composition of omphacite for common Pb correction, the calculated ages for all rutile fractions are shown in Table 1 and Fig. 4. Fig. 4 shows that five replicated analyses of rutile in Jinheqiao eclogite give precise and consistent ^{206}Pb/^{238}U ages. The ^{207}Pb/^{235}U ages with larger errors are consistent with ^{206}Pb/^{238}U ages within the uncertainty limits. Among the five-rutile analyses, No. 2 and No. 3 analyses with relatively large proportion of radiogenic Pb (^{206}Pb/^{204}Pb>55) have consistent ^{206}Pb/^{238}U ages and ^{207}Pb/^{235}U ages, while ^{207}Pb/^{235}U ages of the other three analyses (^{206}Pb/^{204}Pb<45) are slightly higher than ^{206}Pb/^{238}U ages. Additionally, they have relatively larger errors. This is due to the larger percentage correction on ^{207}Pb due to blank and common Pb for the samples with lower U/Pb ratios. Obviously, the ^{206}Pb/^{238}U ages are more credible than ^{207}Pb/^{235}U ages under this condition. The weighted average of five ^{206}Pb/^{238}U ages is 218.2±1.2 Ma. The perfect consistence between the isochron age and the average rutile ^{206}Pb/^{238}U age demonstrates that in this case the Pb isotope composition of omphacite can be used for common Pb correction in U-Pb dating of rutile.

4.2 Oxygen isotopes

An important premise in isochron dating is the presence of the same initial Pb ratios when the isochron minerals formed. In order to evaluate the validity of the rutile U-Pb isochron obtained in this study, oxygen isotope analysis was carried out for rutile aliquots leftover from the U-Pb dating as well as for coexisting quartz and garnet. As shown in Fig. 5, there are almost the same $\delta^{18}O$ values of $-6.1‰±0.1‰$ for different rutile grains from the same specimen within the analytical uncertainties, suggesting attainment of oxygen isotope homogenization in rutile within the specimen scale during the eclogite-facies metamorphism and preservation during amphibolite-facies retrograde metamorphism.

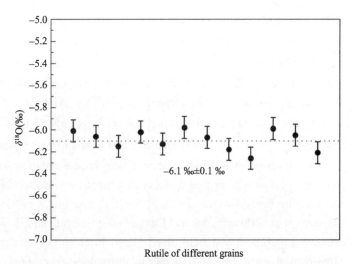

Fig. 5 Oxygen isotope composition of rutile of different grains from the same eclogite specimen for U-Pb dating

According to the oxygen isotope composition of coexisting quartz and other minerals in the eclogite sample, oxygen isotope geothermometry of quartz-mineral pairs is made using the fractionation curves of Zheng (1991, 1993). The temperature defined by quartz-garnet pair is 695±35 ℃, consistent with the peak UHP metamorphic temperatures of coesite-bearing eclogite from the Dabie Mountains (Cong et al., 1995; Carswell et al., 2000), suggesting the temperature of eclogite facies metamorphism. Quartz-rutile pair gives the temperature of 460±15 ℃, indicating equilibrium temperature of oxygen isotope exchange between quartz and rutile during eclogite cooling (Giletti, 1986; Zheng and Fu, 1998).

5 Discussion

5.1 Closure temperature

The Rb-Sr phengite dating method has been chosen to determine the 500 ℃ cooling time of the UHPM rocks. Li et al. (2000, 2001a) have detailedly discussed the problems that cause the considerable variation of Rb-Sr ages of phengite in UHPM rocks from the Dabie-Sulu belt. They suggested that only UHPM phengite + UHPM minerals (such as garnet, omphacite) Rb-Sr isochrons can define the meaningful cooling time corresponding to 500 ℃. Carswell et al. (2000) observed that the high Si UHPM phengite (Si>3.4) could be transformed into low Si phengitic muscovite (Si<3.4) developing in the matrix schistosity during amphibolite facies metamorphism. The Sr and Nd isotopic disequilibrium between the secondary phengitic muscovite and UHPM minerals (garnet or omphacite) or between UHPM phengite and whole rock sample containing retrograde minerals have been observed, which, in general, results of younger Rb-Sr ages (210 to 190 Ma) (Li et al., 2000, 2001a; Chavagnac et al., 2001). In this regard, we prefer to take the average Rb-Sr age of those isochrons defined by phengite + UHP minerals as the best estimate of cooling age at 500 ℃. So far, only five phengite + UHPM minerals of Rb-Sr isochron ages have been reported (Li et al., 1994, 2000; Chavagnac and Jahn, 1996), which yield an average of 220±5 Ma in the Dabie-Sulu UHP belt. This age is perfectly consistent with the phengite + garnet Rb-Sr isochron age of 219±6.6 Ma from UHPM rocks at Shuanghe in the South Dabie Mountains (Li et al., 1999, 2000).

Since the rutile occurs mainly as inclusion in garnet and omphacite, indicating that rutile formed before or during UHP metamorphic stage under high T condition (T>690 ℃)(Carswell et al., 2000), the U-Pb age of rutile can't be interpreted as the growth time during exhumation. Compared with the Rb-Sr cooling age of 220±5 Ma, the rutile U-Pb age of 218±1.2 Ma should correspond to the cooling time of around or slightly lower than 500 ℃ during cooling of eclogite, i.e., the closure temperature of Pb diffusion in rutile from this coesite-bearing eclogite could be similar or slightly lower than 500 ℃. This is consistent with the conclusion of Mezger (1989a) who estimated 420±30 ℃ for slow cooling rate of 1.5 ℃/Ma at grain radii of 0.09 to 0.25 mm based on field study. However, recent experiments by Cherniak (2000) on diffusion coefficients of lead in rutile suggest that Pb closure temperatures in rutile are around 600 ℃ for rutile grains of ~0.1 mm under dry condition or higher than 600 ℃

under wet condition. Based on this closure temperature, the U-Pb rutile age of 218±1.2 Ma could indicate the cooling time around 600 ℃ or close to the peak metamorphic temperature. This interpretation is contradictory with the Rb-Sr cooling age of 220±5 Ma at 500 ℃ for phengite. Similar contradictions between the experimental results and field-based studies also have been observed by many authors. According to Mezger et al. (1991), the U-Pb ages for rutile are similar or slightly younger than $^{40}Ar/^{39}Ar$ ages for hornblendes typically correspond to T_{cs} in the range of 450–550 ℃ (Harrison, 1981) and obviously younger than titanite U-Pb ages corresponding to T_{cs} in the range of 550–650 ℃ (Cherniak, 1993; Heaman and Parrish, 1991; Scott and St. Onge, 1995). A survey of the literatures reveals that an overwhelming number of field-based studies on rutile from slowly cooled terrains have demonstrated younger U-Pb ages compared to coexisting titanite and/or hornblende Ar-Ar ages (Mezger et al., 1989a, 1991; Corfu and Easton, 1995; Cox et al., 1998,2002; Christofel et al., 1999; Connelly et al., 2000; Norcross et al., 2000; Bibikova et al., 2001; Hirdes and Davis, 2002; Schmitz and Bowring, 2003). Causes of the substantial disagreement between experiment and field-based observation remain unclear. Further works are required before applying the experimental diffusion data to thermochronological studies on natural samples. Until the causes of why the experimental Pb diffusion data contradicts field estimates of Pb diffusivity in rutile are resolved, we prefer the field-based closure temperature estimates.

In view of our rock sample as coesite-bearing eclogite, the preservation of coesite requires a rapid uplift and cooling of eclogite following peak metamorphism. The T-t path studies show that the UHPM rocks from the Dabie Mountains experienced a rapid cooling with cooling rate about 40 ℃/Ma following the peak metamorphism (Chavagnac and Jahn, 1996; Li et al., 2000). Considering the effect of cooling rate on closure temperature, based on the Dodson equation (Dodson, 1973) for T_c and activation energy (40–70 kcal/mol) for Pb diffusion in rutile (estimated by Mezger, 1989a) as well as the radii (~0.1 mm) of rutile grains analyzed in this work, we estimate the T_c of 460±50 ℃ for the U-Pb system in rutile under a rapid cooling condition of ~20 ℃/Ma (T_c and cooling rate are calculated based on the function of cooling rate vs. T_c deduced from Dodson's equation and an assumption of simple cooling from 220 Ma at 500 ℃ to 218 Ma at the closure temperature of Pb diffusion in rutile). Therefore, 218±1.2 Ma should be the cooling age of the coesite-bearing eclogite in the Dabie Mountains at 460±50 ℃, which is concordant not only with the Rb-Sr cooling age of 220±5 Ma at 500 ℃ for phengite from UHPM rocks, but also with the oxygen isotope temperature of 460±15 ℃ for the quartz-rutile pair (Table 2). These concordances may imply the similarity in rates of O and Pb diffusion in rutile under UHPM conditions. Therefore, the oxygen isotope homogenization for the rutile of different grains (Fig. 5) lends support to the homogenization of initial Pb isotopes and thus the validity of the rutile U-Pb isochron dating (Fig. 3).

One may argue that the closure temperature of rutile enclosed within garnet is different from the closure temperature of rutile in rock matrix, as the diffusivity of Pb in garnet then controls closure of the rutile. In this study, Pb isotope homogenization in rutile from the inclusions and matrix has been proved by the oxygen isotope homogenization (Fig. 5) and the perfect U-Pb isochron as defined by five fractions of rutile (Fig. 3). Moreover, as mentioned above, the quartz-rutile pair gave the temperature of 460±15 ℃, which is significantly lower than the quartz-garnet pair temperature of 695±35 ℃. It is suggested that though most of the rutiles occur as inclusion in garnet, the diffusion of oxygen in rutile is not controlled by garnet. In addition, according to Mezger et al. (1989a, b), the rutile used for U-Pb analysis is commonly included in other phases, particularly garnet and sillimanite. They also concluded that the U-Pb rutile ages correspond to closure temperatures at around 420 ℃ though garnet has a high Pb closure temperature of excess 800 ℃ (Mezger et al., 1989b). The diffusion channels of rutile as inclusions in garnet remain unclear, but isotope homogenization in different rutile grains is observed. Of course, this issue calls for further works.

5.2 Cooling history

Based on the U-Pb rutile age in this paper as well as other new progress of geochronology and petrology of the UHPM rocks in the Dabie Mountains, the T-t path reported previously by Li et al. (2000) can be improved. Carswell et al. (2000) have evaluated the temperature estimates of UHPM rocks in the Dabie Mountains and indicate that the P_{max} mineral assemblage probably did not form at the peak temperatures attained in the subdction/exhumation metamorphic cycle. They suggest a "best" P-T path for the UHPM rocks in South Dabie, which is through P_{max} condition at 690–715 ℃ and 36 kbar to T_{max} at 710–755 ℃ and 18 kbar prior to their more extensive late-stage re-equilibration and re-crystallization at 400–450 ℃ and 6 kbar, corresponding to retrograde amphibolite facies metamorphism. Based on these new P-T estimates, we can improve our understanding of metamorphic temperatures corresponding to some Sm-Nd isochron ages for UHPM rocks at Shuanghe (Li et al., 2000). Li et al.(2000) suggested that the Sm-Nd mineral isochron age of 226±3 Ma may indicate the average age of recrystallization time for garnet, omphacite and rutile, which may be close to the peak metamorphic time. If P_{max} condition and T_{max} condition for UHPM rocks do not overlap on one point as indicated by Carswell et al. (2000) and recrystallization of minerals was much faster at high-T conditions, the Sm-Nd age of 226±3 Ma seems more likely to correspond to T_{max} condition (710–755 ℃). On the other hand, the Sm-Nd isochron (with age of 213.3±4.8 Ma) defined by garnet + biotite + epidote from granitic gneiss at Shuanghe suggests the Nd isotopic re-equilibration within garnet, biotite and epidote during retrograde amphibolite facies metamorphism (Li et al., 2000). Therefore, the age of 213.3±4.8 Ma may indicate the re-equilibration and recrystallization time at 425±25 ℃ and 6 kbar.

Recently, Ayers et al. (2002) reported Th-Pb monazite core age of 223±1 Ma and monazite rim age of 209±3 Ma for the Shuanghe jadeite quartzite. Because the T_{max} (710 to 755 ℃) of the UHPM rocks from South Dabie is higher than the Pb T_c of 650 to 700°C for monazite (Mezger et al., 1991; Smith and Giletti, 1997), the monazite core age should indicate the cooling time at T = 675±25 ℃, while the rim age of 209±3 Ma is corresponding to recrystallization-growth of monazite occurring during retrograde amphibolite facies metamorphism (Ayers et al., 2002). Actually, the re-crystallization time in the amphibolite facies metamorphic stage could last to 180 Ma based on Sm-Nd and Rb-Sr isochron ages defined by the amphibolite-facies mineral assemblages (Li et al., 2000).

On the basis of above chronological data with corresponding closure or metamorphic temperatures, an improved cooling T-t path in the range of temperature from 755 ℃ to 400 ℃ for UHPM rocks in South Dabie can be obtained. As illustrated in Fig. 6, this T-t path shows that UHPM rocks in South Dabie experienced a rapid cooling following the peak metamorphism with a cooling rate of about 55 ℃/Ma before 220 Ma at ~500 ℃. There was a nearly isothermal stage from 213 Ma to 180 Ma (Li et al., 2000) between 450 and 400 ℃ to offer enough time for extensive re-crystallization-growth of minerals during retrograde amphibolite facies metamorphism. The interval in a range of 500–450 ℃ is an important transitional stage from rapid cooling to relative isothermal stage. The rutile U-Pb ages reported in this paper provided a new constraint on this stage.

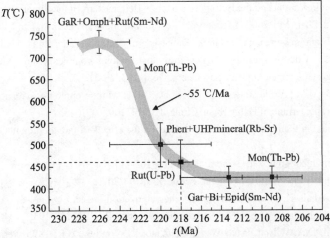

Fig. 6 T-t path of UHPM rocks in the South Dabie Mountains. Gar=garnet, Omph=omphacite, Rut=rutile, Phen=phengite, Mon=monazite, Bi=biotite, Epid=epidote. Data sources: Li et al., 2000; Ayers et al., 2002; this article. (Corresponding metamorphic or closure temperatures see text).

6 Conclusion

Common Pb correction is a key problem for U-Pb dating of metamorphic minerals such as rutile and garnet because of the small proportion of radiogenic Pb in these minerals. This work shows that the rutile + omphacite isochron age and conventional rutile U-Pb age obtained by common Pb correction based on the composition of omphacite are consistent. In this case, the omphacite in the same eclogite sample has low enough U/Pb ratio (μ=2.8) to be used for common Pb correction in U-Pb dating of rutile.

This work first reports a high precision U-Pb age (218±1.2 Ma) of rutile in eclogite from the South Dabie UHPM zone. The validity of the rutile U-Pb isochron is confirmed by the homogeneity of oxygen isotope composition in rutile of different grains. Based on field-based studies on rutile U-Pb system and phengite Rb-Sr datings in South Dabie, an estimate of the T_c of about 460 ℃ for U-Pb system in rutile (grain size ~0.1 mm) under rapid cooling condition (~20 ℃/Ma) was advised. Combined the rutile U-Pb age and other reported chronological data with their corresponding metamorphic and/or closure temperature, an improved T-t path has been constructed. The T-t path confirms that the UHPM rocks in South Dabie experienced a rapid cooling following the peak metamorphism before 220 Ma and a long isothermal stage from ~213 to 180 Ma around 425 ℃. This study indicates that rutile U-Pb dating could be a powerful method for constructing the cooling history of UHPM rocks, especially contributing to fix on the important transitional stage of the T-t path between 500 and 450 ℃.

Acknowledgements

This work is a part of the PhD thesis of Li Qiuli. We would like to thank Sun Weidong for his help during field trips, Zhang Xun for her help in the lab and Shen-su Sun, K. Mezger, J.C. Ayers, Xu Ronghua and an anonymous reviewer for their constructive comments that helped to improve the manuscript. This work was supported by the State Key Basic Research Development Program (Grant No. G1999075503) and the Natural Science Foundation of China (Grant Nos. 49973020 and 40033010).

References

Ames, L., Zhou, G.Z., Xiong, B.C., 1996. Geochronology and isotopic character of ultrahigh-pressure metamorphism with implications for collision of the Sino-Korean and Yangtze cratons, central China. Tectonics 15(2), 472-489.

Ayers, J.C., Dunkle, S., Gao, S., Miller, C.F., 2002. Constraints on timing of peak and retrograde metamorphism in the Dabie Shan Ultrahigh-Pressure Metamorphic Belt, east-central China, using U-Th-Pb dating of zircon and monazite. Chem. Geol. 186, 315-331.

Bibikova, E., Skiold, T., Bogdanova, S., Gorbatschev, R., Slabunov, A., 2001. Titanite-rutile thermochronometry across the boundary between the Archean Craton in Karelia and the Belomorian Mobile Belt, eastern Baltic Shield. Precambrian Res. 105, 315-330.

Carswell. M.D.A., Wilson, R.N., Zhai, M., 2000. Metamorphic evolution, mineral chemistry and thermobarometry of schists and orthogneisses hosting ultra-high pressure eclogites in the Dabieshan of central China. Lithos, 52: 121-155.

Chavagnac, V., Jahn, B.M., 1996. Coesite-bearing eclogites from the Bixiling complex Dabie Mountains, China: Sm-Nd ages, Geochemical characteristics and tectonic implications. Chem. Geol. 133, 29-51.

Chavagnac, V., Jahn, B.M., Villa, I.M., Whitehouse, M.J., Liu, D., 2001. Mutichronologic evidence for an in situ origin of the ultrahigh-pressure metamorphic terrane of Dabie Shan. China. J. Geol. 109, 633-646.

Cherniak, D.J., 1993. Lead diffusion in titanite and preliminary results on the effects of radiation damage on Pb transport. Chem. Geol., 110, 177-194.

Cherniak, D.J., 1998. Pb Diffusion in clinopyroxene. Chem. Geol. 115, 105-117.

Cherniak, D.J., 2000. Pb diffusion in rutile. Contrib. Mineral. Petrol. 139, 198-207.

Christofell, C.A., Connelly, J.N., Ahall, K.L., 1999. Timing and characterization of recurrent pre-Sveconorwegian metamorphism and deformation in the Varberg-Halmstad region of SW Sweden. Precambrian Res. 98, 173-195.

Cong, B.L., Zhai, M.G, Carswell, D.A. Wilson, R.N., Wang, Q., Zhao, Z., Windley, B.F., 1995. Petrogeneisis of ultrahigh-pressure rocks and their country rocks at Shuanghe in Dabieshan, central China. Eur. J. Mineral 7, 119-138.

Connelly, J.N., Van Goo, J.A.M., Mengel, F.C., 2000. Temporal evolution of a deeply eroded orogen: the Nagssugtoqidian orogen, West Greenland. Can. J. Earth Sci. 37, 1121-1142.

Corfu, F., Muir, T.L., 1989. The Hemlo-Heron bay greenstone belt and Hemlo Au-Mo deposit, superior Province, Ontario, Canada. 2. Timing of metamorphism. Alteration and Au mineralization from titanite,rutile and monazite U-Pb geochronology. Chem. Geol. Iso. Geosci. 79, 201-223.

Corfu, F., Easton, R.M., 1995. U-Pb geochronology of the Mazinaw terrane, an imbricate segment of the Central Metasedimentary Belt, Grenville Province, Ontario. Can. J. Earth Sci. 32, 959-976.

Cox, R.A., Dunning, G.R., Indares, A., 1998. Petrology and U-Pb geochronology of mafic, high-pressure, metamorphic coronites from the Tshenukutish domain, eastern Grenville Province. Precambrian Res. 90, 59-83.

Cox, R.A., Indares, A., Dunning, G.R., 2002. Temperature-time paths in the high-P Manicouagan Imbricate zone, eastern Grenville Province: evidence for two metamorphic events. Precambrian Res. 117, 225-250.

Davis, W. J., 1997. U-Pb zircon and rutile ages from granulite xenoliths in the slave province: evidence for mafic magmatism in the lower crust coincident with Proterozoic dike swarms. Geology 25, 343-346.

Davis, D.W., Schandl, E.S., Wasteneys, H.A., 1994. U-Pb dating of minerals in alteration halos of Superior Province massive sulfide deposits: syngenesis versus metamorphism. Contrib. Mineral. Petrol. 115, 427-437.

Dodson, M. H., 1973. Closure temperature in cooling geochronological and petrological systems. Contrib. Mineral. Petrol. 40, 259-274.

Eiler, J.M., Farley, K.A., Valley, J.W., Stolper, E.M., Hauri, E., Craig, H., 1995. Oxygen isotope evidence against bulk recycled sediment in the source of Pitcairn island lavas. Nature 377, 138-141.

Giletti, B.J. 1986. Diffusion effects on oxygen isotope temperatures of slowly cooled igneous and metamorphic rocks. Earth Planet. Sci. Lett. 77, 218-228.

Hacker, B.R., Ratschbacher, L., Webb, L., Ireland, T., Walker, D., Dong, S., 1998. U/Pb zircon ages constrain the architecture of the ultrahigh-pressure Qinling-Dabie Orogen, China. Earth Planet. Sci. Lett. 161, 215-230.

Heaman L., Parrish R., 1991. U-Pb geochronology of accessory minerals. In: Heaman L., Ludden J.N. (eds) Applications of radiogenic isotope systems to problems in geology. Short Course Handbookla 19, Mineralogical Association of Canada, Toronto, 69-102.

Hirdes, W., Davis, D.W., 2002. U-Pb zircon and rutile metamorphic ages of Dahomeyan garnet-hornblende gneiss in southeastern Ghana, West Africa. Journal of African Earth Sciences, 35, 445-449.

Li, S., Xiao, Y., Liou, D. Chen, Y., Ge, N., Zhang, Z., Sun, S.S., Cong, B., Zhang, R., Hart, S., Wang, S., 1993. Collision of the North China and Yangtse Blocks and formation of coesite-bearing eclogites: Timing and process. Chem. Geol. 109, 89-111.

Li, S., Wang, S., Chen, Y., Liu, D., Qiu, J., Zhou, H., Zhang, Z., 1994. Excess argon in phengite from eclogite: evidence from dating of eclogite minerals by Sm-Nd, Rb-Sr and $^{40}Ar/^{39}Ar$ methods. Chem. Geol. 112, 343-350.

Li, S., Jagoutz, E., Xiao, Y., Liu, D., 1996. Chronology of ultrahigh-pressure metamorphism in the Dabie Mountains and Su-Lu terrane: I. Sm-Nd isotopic system. Sci. in China(Ser. D) 39, 597-609.

Li, S., Li, H., Chen, Y., Xiao, Y., Liu, D., 1997. Chronology of ultrahigh-pressure metamorphism in the Dabie mountains and Su-Lu terrane: II. U-Pb isotopic system of zircons. Sci. in China (Ser. D) 27(3), 200-206.(in Chinese).

Li, S., Jagoutz, E., Lo, C. H., Li, Q., Chen, Y., 1999. Sm-Nd, Rb-Sr and ^{40}Ar-^{39}Ar isotopic systematics of the ultrahigh pressure metamorphic rocks in the Dabie-Sulu Belt, central China. A retrospective view. Inter. Geol. Rev. 41, 1114-1124.

Li, S., Jagoutz, E., Chen, Y., Li, Q., 2000. Sm-Nd and Rb-Sr isotopic chronology and cooling history of ultrahigh pressure metamorphic rocks and their country rocks at Shuanghe in the Dabie Mountains, Central China. Geochim. Cosmochim. Acta. 64(6), 1077-1093.

Li, S., Zhang, Z., Chen, Y., Li, Q., 2001a. Nd isotopic disequibrium minerals and Rb-Sr age of the secondary phengite in eclogite from the Yangkou area, Qingdao, eastern China. Chinese Sci. Bull. 46(3), 252-255.

Li, S., Huang, F., Nie, Y., Han, W., Long, G., Li, H., Zhang, S., Zhang, Z., 2001b. Geochemical and geochronological constraints on the suture location between the North and South China Blocks in the Dabie orogen, central China. Physics and Chemistry of the Earth(A) 26, 655-672.

Li, Y.L., Zheng, Y. F., Fu, B. et al., 2001. Oxygen isotope composition of quartz-vein in ultrahigh-pressure eclogite from Dabieshan and implications for transport of high-pressure metamorphic fluid. Phys. Chem. Earth (A) 26, 695-704.

Ludwig, K.R., 1998. On the treatment of concordant uranium-lead ages. Geochim. Cosmochim. Acta. 62, 655-676.

Ludwig, K.R., 2000. Users Manual for Isoplot/Ex: A Geochronoligical Toolkit for Microsoft excel Berkeley Geochronology center Special Publication, Berkeley, CA, USA, p.53.

Mezger, K., Hanson, G. N., Bohlen, S. R., 1989a. High-precision U-Pb ages of metamorphic rutiles: application to the cooling history of high-grade terranes. Earth Planet. Sci. Lett. 96, 106-118.

Mezger, K., Hanson, G.N., Bohlen, S.R., 1989b. U-Pb systematics of garnet: dating the growth of garnet in the Late Archean Pikwitonei granulite domain at Cauhon and Natawahunan Lakes, Manitoba, Canada. Contrib. Mineral. Petrol. 101, 136-148.

Mezger, K., Rawnsley, C., Bohlen, S.R., Hanson, G.N., 1991. U-Pb garnet, sphene, monazite, and rutile ages: implications for the duration of high-grade metamorphism and cooling histories Adirondack Mountains, New York. J. Geol. 99, 415-428.

Norcross, C., Davis, D.W., Spooner, E.T.C., Rust, A., 2000. U-Pb and Pb-Pb age constraints on Paleoproterozoic magmatism, deformation and gold mineralization in the Omai area, Guyana Shield. Precambrian Res. 102, 69-86.

Okay, A.L., Xu, S., Sengör, A.M.C., 1989. Coesite from the Dabie Shan eclogite, Central China. Eur J Mineral 1, 595-598.

Rowley, D.B., Xue, F., Tucker, R.D., Peng, Z.X., Baker, J., Davis, A., 1997. Ages of ultrahigh pressure metamorphism and protolith orthogneisses from the eastern Dabie Shan: U/Pb Zircon geochronology. Earth Planet. Sci. Lett. 151, 191-203.

Richards, J. P., Krogh, T.E., Spooner, E.T.C., 1988. Fluid inclusions characteristics and U-Pb rutile age of late hydrothermal alteration veining at the Musoshi stratiform copper deposit, central African copper belt. Econ. Geol. 83, 118-139.

Rumble, D., Farquhar, J., Young, E.D., Christensen, C.P., 1997. In situ oxygen isotope analysis with an excimer laser using F_2 and BrF_5 reagents and O_2 gas as analyte. Geochim. Cosmochim. Acta 61, 4229-4234.

Schandl, E.S., Davis, D.W., Krogh, T.E., 1990. Are the alteration halos of massive sulfide deposits syngenetic? Evidence from U-Pb dating of hydrothermal rutile at the Kidd volcanic center, Abitibi subprovince, Canada. Geology 18, 505-508.

Schmitz, M.D., Bowring, S.A., 2003. Constraints on the thermal evolution of continental lithosphere from U-Pb accessory mineral thermochronometry of lower crustal xenoliths, southern Africa. Contrib. Mineral. Petrol. 144, 592-618.

Scott D.J., St-Onge M.R., 1995. Constraints on Pb closure temperature in titanite based on rocks from the Ungava Orogen, Canada: implications for U-Pb geochronology and *P-T-t* path determinations. Geology, 23, 1123-1126.

Sharp, Z.D., 1990. A laser-based microanalytical method for the in situ determination of oxygen isotope ratios of silicates and oxides. Geochim. Cosmochim. Acta 54, 1353-1357.

Smith, H.A., Giletti, B.J., 1997. Lead diffusion in monazite. Geochim. Cosmochim. Acta 61, 1047-1055.

Valley, J.W., Kitchen, N., Kohn, M.J., Niendorf, C.R., Spicuzza, M.J., 1995. UWG-2, a garnet standard for oxygen isotope ratio: strategies for high precision and accuracy with laser heating. Geochim. Cosmochim. Acta 59, 5223-5231.

Wang, X.M., Liou, J.G., Mao, H.K., 1989. Coesite-bearing eclogites from the Dabie Mountains in central China. Geology 17, 1085-1088.

Wong, L., Davis, D.W., Krogh, T.E., Robert, F., 1991. U-Pb zircon and rutile chronology of Archean greenstone formation and gold mineralization in the Val d'Or region, Quebec. Earth Planet Sci. Lett. 104, 325-336.

Xu, S., Okay, A.L., Ji, S., Sengor, A.M.C., Sun, W., Liu, Y., Jiang, L., 1992. Diamond from the Dabie Shan metamorphic rocks and its implication for tectonic setting. Science 256, 80-82.

Zheng, Y.F., 1991. Calculation of oxygen isotope fractionation in metal oxides. Geochim. Cosmochim. Acta 55, 2299-2307

Zheng, Y.F., 1993. Calculation of oxygen isotope fractionation in anhydrous silicate minerals. Geochim. Cosmochim. Acta 57, 1079-1091

Zheng, Y.F., Fu, B., 1998. Estimation of oxygen diffusivity from anion porosity in minerals. Geochem. Jour. 32, 71-89

Zheng, Y.F., Fu, B., Li, Y., Xiao, Y., Li, S., 1998. Oxygen and hydrogen isotope geochemistry of ultrahigh pressure eclogites from the Dabie Mountains and the Sulu terrane. Earth Planet. Sci. Lett. 155, 113-129

Zheng, Y.F., Fu, B., Xiao, Y., Li, Y., Gong, B., 1999. Hydrogen and oxygen isotope evidence for fluid–rock interactions in the stages of pre- and post-UHP metamorphism in the Dabie Mountains. Lithos 46, 677-693

Zheng, Y.F., Wang, Z.R., Li, S., Zhao, Z., 2002. Oxygen isotope equilibrium between eclogite minerals and its constraints on mineral Sm-Nd chronometer. Geochim. Cosmochim. Acta 66, 625-634.

大别山金河桥榴辉岩矿物 O-Nd-Pb 同位素体系及其对扩散速率的制约[*]

李秋立，李曙光，郑永飞，龚 冰，洪吉安

中国科学技术大学地球和空间科学学院，合肥，230026

> **亮点介绍**：对大别山金河桥榴辉岩的主要矿物进行 Sm-Nd 体系、氧同位素分析对比研究；石榴石与绿辉石之间处于氧同位素不平衡状态，指示 Sm-Nd 体系未达到平衡，故石榴石+绿辉石 Sm-Nd "等时线"给出了较低年龄 210±3 Ma。金红石中氧扩散速率与 Pb 相近，较 Nd 显著快。

摘要 对大别山太湖金河桥超高压榴辉岩作了矿物 Sm-Nd 内部等时线定年研究和激光氧同位素分析。石榴石+绿辉石 Sm-Nd 等时线给出了较低年龄 210±3 Ma，石榴石+金红石 Sm-Nd 等时线给出了较高年龄 237±4 Ma。岩相学观察发现，绿辉石具有角闪石退变质边，氧同位素分析表明，石榴石与绿辉石之间的氧同位素体系处于不平衡状态，据此，石榴石+绿辉石 Sm-Nd 同位素体系因退变质作用导致 Nd 同位素不平衡而给出不合理偏低年龄。较老的石榴石+金红石 Sm-Nd 年龄有可能指示了榴辉岩相前期阶段的时代，且在温度变质峰期没有使它们之间的 Nd 同位素再次均一化，它指示 Nd 在金红石中的扩散速率较慢，可能与石榴石相当。矿物对氧同位素测温得到，石英-石榴石对温度为 695±35 ℃，石英-金红石对为 460±15 ℃，与根据金红石 U-Pb 内部等时线估计的 Pb 扩散封闭温度 470±50 ℃ 一致。对比表明，O 在石榴石中的扩散速率与 Nd 相当或略低，而 O 和 Pb 在金红石中的扩散速率相近，且均比 Nd 快。

关键词 榴辉岩；金红石；O 同位素；Nd 同位素；Pb 同位素；扩散；温度

Sm-Nd 矿物等时线法是测定含石榴石高级变质岩变质年龄最常用的方法之一。在榴辉岩年代学研究中，通常采用石榴石+绿辉石(+金红石) 等高压矿物 Sm-Nd 等时线法测定高压-超高压变质时代(Thoeni and Jagoutz, 1992; Li et al., 1993, 1994, 1996, 2000; Chavagnac and Jahn, 1996)。对于矿物内部等时线年龄测定来说，其地质年代学的有效性取决于所分析各矿物之间是否达到并保存了初始同位素平衡。三个以上不同矿物组成的等时线，矿物之间的同位素平衡通常可根据数据拟合相关系数来判断。但对两矿物等时线定年，则要借助其他途径来给出判据。已知在没有重结晶的情况下，元素或同位素在矿物之间的迁移或交换主要是以元素扩散和/或离子交换机制来进行的，同一矿物中不同元素或同位素的扩散速率有快有慢。因此，如果可以给出一定条件下多种同位素在同一矿物中的扩散快慢顺序，则根据某种元素同位素体系的平衡与否可以对其他元素同位素体系的平衡关系做出判断。例如，Zheng 等(2002)通过榴辉岩矿物的氧同位素分析和 Sm-Nd 等时线定年研究得到，石榴石和绿辉石中的氧扩散速率较其 Sm-Nd 扩散速率相似或略高，因此石榴石和绿辉石之间的氧同位素平衡与否可指示此两矿物的 Sm-Nd 同位素是否达到了平衡，从而可用来判别此两矿物 Sm-Nd 等时线定年结果的有效性。

金红石作为榴辉岩中一种常见的高压矿物，可用于 Sm-Nd 等时线定年。金红石中 Nd 的扩散尚没有较好的实验研究过，但是 Li 等(2000)对大别山双河地区榴辉岩中三种高压矿物(石榴石+绿辉石+金红石)的 Sm-Nd 同位素分析得到线性非常好的等时线年龄 226±3 Ma (MSWD = 0.35)，说明金红石与石榴石和绿辉石

---

[*] 本文发表在：高校地质学报，2003，9(2): 218-226

之间在 Nd 同位素上是平衡的,因此金红石与石榴石和绿辉石具有类似的 Sm-Nd 扩散封闭温度(干燥条件下,>750 ℃)(Li et al., 1996, 2000)。金红石又是一良好的 U-Pb 定年矿物,李秋立等(2001)获得了大别山金河桥榴辉岩的金红石 U-Pb 等时线年龄为 218.2±1.2 Ma。依据 Mezger 等(1989)的经验参数估算的 Pb 扩散封闭温度约为 470±50 ℃。显然 Sm-Nd 和 U-Pb 在金红石中表现出不同的扩散速率。Zheng 等(1998)对双河地区榴辉岩中金红石的氧同位素分析得到,石英-金红石矿物对多给出较低的氧同位素温度(约 500 ℃),指示金红石中 O 比 Sm-Nd 有较快的扩散速率,可能与 U-Pb 类似。然而,上述进行氧同位素分析的金红石样品与 Sm-Nd 和 U-Pb 等时线定年所采用的金红石均来自不同产地的榴辉岩样品,因此有关推测不能作为金红石中 O、Nd、Pb 之间扩散速率比较的依据。为了更好地揭示自然界金红石中 O 与 Nd 和 Pb 之间的相对扩散速率,本文对已做过金红石 U-Pb 定年的金河桥榴辉岩样品进行 O 和 Sm-Nd 同位素分析,结果对金红石中 O、Nd、Pb 之间的相对扩散速率提供了制约。

1 地质背景与测试方法

大别山是华北与扬子陆块的碰撞造山带。大别山东段(商-麻断裂以东)自北向南可以划分为四个基本构造单元(Li et al., 1998)(见图 1)。含柯石英榴辉岩仅分布于南大别超高压带,它是扬子陆壳俯冲最深的岩片。本文分析的榴辉岩样品(99JHQ-1, N30°30.01′, E116°15.96′)采自太湖县花凉亭水库东北缘金河桥,位于南大别超高压变质带南缘(图 1),与红安-宿松低温榴辉岩带相邻。初步岩石学研究表明,金河桥含柯石英榴辉岩的高压变质温度 600~700 ℃(据王清晨,2003,私人通信)略低于南大别其他地区高压榴辉岩的变质温度 700~800 ℃(Cong et al., 1995;Carswell et al., 2000)。

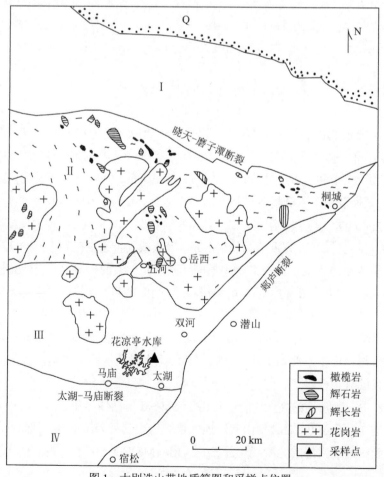

图 1 大别造山带地质简图和采样点位置

Ⅰ-北淮阳低级变质带;Ⅱ-北大别高温高压带;Ⅲ-南大别超高压变质带;Ⅳ-宿松高压变质带

所选金河桥榴辉岩为粗粒结构,条带状构造,红色石榴石、草绿色绿辉石及白色石英在毫米至厘米级尺度上互为条带,副矿物为金红石、多硅白云母、磷灰石、蓝晶石、柯石英、锆石。柯石英的产出指示该榴辉岩曾经历过超高压变质作用。岩相学及单矿物的显微镜下检查表明,该样品的绿辉石颗粒不同程度上具有淡黄绿色角闪石退变质边,在单矿物挑选中很难将其与内部绿辉石分离开,因此绿辉石中可能带有退变质成分。金红石主要以包裹体形式产出于石榴石和绿辉石中,粒间也见较大颗粒。对该样品中的金红石已进行过 U-Pb 同位素分析,获得了 218.2±1.2 Ma 的等时线年龄(Li et al., 2002)。岩石样品经破碎、淘洗、重液分离、电磁选后,分离出金红石、绿辉石、石榴石、多硅白云母和石英,再经双目镜下手工挑纯后做进一步的分析。

矿物氧同位素分析在中国科学技术大学化学动力学研究实验室完成。化学提取采用激光氟化技术(Sharp, 1990),反应采用 MIR-10 型 25W 的 CO_2 激光器进行加热,并将所得到 O_2 直接在 Delta+气体质谱仪上进行同位素比值测定(Zheng et al., 2002)。结果以 $\delta^{18}O$ 标记报道(相对于 SMOW),全流程分析误差小于±0.1‰。两个参照标准分别为:UWG-2 石榴石,$\delta^{18}O$=5.8‰(Valley et al., 1995); SCO-1 橄榄石,$\delta^{18}O$=5.2‰(Eiler et al., 1995)。为了保证数据的可靠性,对所有样品均进行了重复测定,结果 $\delta^{18}O$ 值误差均小于±0.1‰(1σ)。为了判断手标本尺度不同金红石颗粒的氧同位素均一化程度,对不同颗粒金红石进行了 12 次分析,所有结果列于表1。

表 1 大别山金河桥榴辉岩(99JHQ-1)中矿物氧同位素组成及测温结果

样品	测定 $\delta^{18}O$ (‰)	平均 $\delta^{18}O$ (‰)	测温矿物对	$\Delta^{18}O$ (‰)	温度*(℃)
石英	1.06, 1.16	1.1			
多硅白云母	−1.53, −1.64	−1.6	石英—多硅白云母	2.7	535±35
绿辉石	−2.33, −2.49	−2.4	石英—绿辉石	3.5	460±20
石榴石	−2.23, −2.31	−2.3	石英—石榴石	3.4	695±35
金红石	−6.01, −6.06, −6.21, −6.15, −6.02, −6.13, −5.98, −6.07, −6.18, −6.26, −5.99, −6.05	−6.1	石英—金红石	7.2	460±15

*根据理论分馏系数(Zheng, 1991, 1993a, 1993b)计算。

Sm-Nd 同位素分析实验室化学工作在中国地质科学院地质所进行,质谱测定在中国科学技术大学 MAT-262 型热离子质谱计上完成。化学处理采用常规同位素稀释法,样品溶解后,分出部分溶液加入 ^{149}Sm、^{146}Nd 混合稀释剂作含量分析。在阳离子交换柱(20 cm×φ1 cm,AG50W×8(H^+),200~400 目)分离出稀土元素,以二(2-乙基己基)正磷酸(HDEHP)法分离 Sm 和 Nd,最后用 MAT-262 热离子发射固体质谱计测定同位素比值。Sm-Nd 同位素分析结果列于表 2,年龄计算采用美国地质调查所 Ludwig 编写的 ISOPLOT 软件。

表 2 大别山金河桥榴辉岩(99JHQ-1)Sm-Nd 同位素组成及计算结果

矿物	Sm(μg/g)	Nd(μg/g)	^{147}Sm/^{144}Nd	^{143}Nd/^{144}Nd	年龄计算矿物对	年龄(Ma)	$\varepsilon_{Nd}(t)$
金红石	0.199	0.278	0.4328	0.512461±20			
石榴石	0.531	0.189	1.6994	0.514431±25	金红石+石榴石	237±4	−10.6
绿辉石	0.290	0.492	0.3563	0.512587±10	绿辉石+石榴石	210±3	−5.2

2 结果

对金红石样品的 12 次氧同位素分析表明,不同颗粒的 $\delta^{18}O$ 值基本落在分析误差 6.1‰±0.1‰ 的范围内(表1和图2),说明该榴辉岩样品中的金红石在手标本尺度上达到了不同颗粒之间的氧同位素均一化。由表1可见,绿辉石与石榴石之间的氧同位素分馏为 −0.1‰,明显不同于高温热力学平衡条件下的正值(400~900 ℃条件下为 0.4‰~0.8‰;Zheng, 1993a),指示它们之间的 ^{18}O 富集顺序发生了倒转,即绿辉石与石榴石之间出现氧同位素分馏不平衡。

根据理论计算的矿物对氧同位素分馏系数(Zheng, 1991, 1993a, 1993b),对不同的石英-矿物对进行了氧同位素温度计算,结果列于表1并标绘在等温线图3上。石英-石榴石对温度为 695±35 ℃,它与大别山含柯石英榴辉岩的变质温度范围(650~850 ℃)一致(Cong et al., 1995; 张泽明等, 1999; Carswell et al., 2000), 因此它记录了榴辉岩相变质温度。然而该温度比大别山其他地区含柯石英榴辉岩的峰期变质温度(715~850 ℃)略低。石英-多硅白云母对温度为535±35 ℃, 石英-金红石对为460±15 ℃, 它们分别代表了石英与多硅白云母和金红石之间在榴辉岩形成后简单冷却过程中的氧同位素退化交换再平衡温度(Giletti, 1986)。石英-绿辉石对温度为460±20 ℃(它较正常榴辉岩中石英-绿辉石对温度偏低, Zheng et al., 2002)。

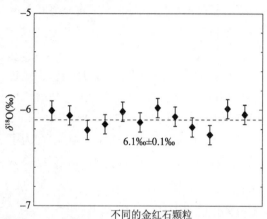

图2 金河桥榴辉岩中不同颗粒金红石的氧同位素组成

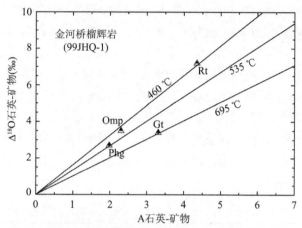

图3 大别山金河桥榴辉岩矿物氧同位素分馏等温线图解
(A 为分馏方程参数, 引自 Zheng, 1999)
Rt-金红石; Gt-石榴石; Omp-绿辉石; Phg-多硅白云母

在金河桥榴辉岩矿物 Sm-Nd 等时线图解中, 石榴石、绿辉石和金红石三高压矿物不能构成一条 Sm-Nd 等时线(图4), 表明三矿物之间存在 Nd 同位素不平衡。计算得到石榴石与金红石的连线年龄为 237±4 Ma, 石榴石与绿辉石的连线年龄为 210±3 Ma(图4)。这两个年龄的含义见讨论部分。

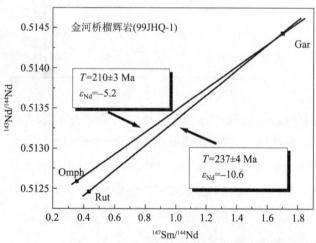

图4 大别山金河桥榴辉岩矿物 Sm-Nd 等时线图解

3 讨论

3.1 石榴石-绿辉石氧同位素不平衡的原因和年代学效应

导致绿辉石与石榴石之间氧同位素不平衡分馏的原因可能有二: ① 这两个矿物不是榴辉岩相条件下同时形成的。显然这一解释与该含柯石英榴辉岩的变质成因相矛盾, 不能成立; ② 角闪岩相退变质作用破坏了这两个矿物之间的原始氧同位素平衡。如前所述, 该样品的绿辉石普遍具有淡黄绿色角闪石蚀变边, 在单矿物挑选中很难将其与内部绿辉石分离开; 石英-绿辉石对氧同位素温度为 460 ℃, 对应于角闪岩相退变质条件(Cong et al., 1995), 可能指示绿辉石受到了角闪岩相退变质作用的影响, 导致它与石榴石之间在 O 和 Sm-Nd 同位素上的不平衡。因此, 本文倾向于第二种解释。

在榴辉岩矿物 Sm-Nd 等时线定年工作中(Thoeni and Jagoutz, 1992; Li et al., 1996), 已经发现在条带状榴辉岩中常常会出现绿辉石与石榴石之间的 Sm-Nd 同位素不平衡。Zheng 等(2002)对胶南榴辉岩的研究发现, O 在石榴石和绿辉石中的扩散速率较 Sm-Nd 略高或相似, 因此石榴石与绿辉石之间的氧同位素平衡状态可以用来判断这两个矿物之间的 Sm-Nd 同位素平衡状态。本文金河桥榴辉岩中石榴石与绿辉石之间的氧同位素分馏处于不平衡状态(表1), 它们之间的 Sm-Nd 同位素也可能同样处于不平衡状态。石榴石+绿辉石 Sm-Nd 等时线给出 210±3 Ma 的年龄, 显著低于大别山其他含柯石英榴辉岩的石榴石+绿辉石 Sm-Nd 等时线年龄 220~230 Ma(Li et al., 1993, 1994, 1996, 2000)。很明显, 这种 Sm-Nd 同位素的不平衡同样是由于角闪岩相退变质作用造成的，对应的石榴石与绿辉石 Sm-Nd 连线给出较低的年龄 210±3 Ma (图 4)，该年龄不具地质意义。

3.2 金红石中 Nd 和 O 之间的扩散比较

如前已述，虽然金红石中 Nd 的扩散尚没有较好的实验研究，但是 Li 等(2000)对大别山双河地区榴辉岩中三种高压矿物(石榴石＋绿辉石＋金红石)的 Sm-Nd 同位素分析得到线性非常好的 Sm-Nd 等时线年龄(226±3 Ma, MSWD = 0.35)，这证明金红石与石榴石和绿辉石之间在超高压变质条件下 Nd 同位素能够达到平衡，指示金红石的 Nd 扩散封闭温度与石榴石相近(>600 ℃)(Li et al., 1996, 2000)。本工作中石榴石与金红石 Sm-Nd 连线年龄为 237±4 Ma，大于大别山其他含柯石英榴辉岩的石榴石＋绿辉石＋金红石 Sm-Nd 等时线年龄（220~230 Ma）(Li et al., 1993, 1994, 1996, 2000)。这种 Sm-Nd 年龄偏老现象不可能是角闪岩相退变质作用的影响，因为退变质作用造成的结果是使得绿辉石与石榴石连线年龄偏低(210 Ma)，如果金红石同样受到了同期退变质作用的影响，则其结果应该是金红石与石榴石 Sm-Nd 连线年龄也会偏低，这与事实不符。考虑到金河桥榴辉岩位于大别山超高压变质带南部冷热榴辉岩接触带，其峰期变质温度较低，这种年龄偏老现象最可能是重结晶作用没有使矿物间达到 Nd 同位素再平衡而导致的结果。超高压变质前的矿物，如斜长石和角闪石，在高压变质时的分解反应有可能使新生成的高压矿物的 Nd 同位素均一化。这一榴辉岩相变质反应在前进变质阶段达到榴辉岩相时就可以完成，无须等到峰期变质阶段，如果在榴辉岩相温度峰期没有使金红石和石榴石之间的 Nd 同位素再次均一化，则矿物等时线年龄应该代表了榴辉岩相前进变质阶段的时代。此外，该年龄与双河硬玉石英岩的锆石 U-Pb 年龄(238 Ma)一致（Ayers et al., 2002)，而该变质锆石主体可能是副变质岩中古老蜕晶化碎屑锆石（1.8 Ga）在超高压变质条件下的重结晶锆石，其重结晶作用在 T>600 ℃ 时可迅速发生（Mezger et al., 1989），故作者认为该 U-Pb 年龄可能指示的是前进变质温度达到 600~650 ℃ 时的时代。因此，该金红石＋石榴石 Sm-Nd 连线给出榴辉岩相前进变质阶段时代（237±4 Ma)。如果这一解释是正确的，它表明 Nd 同位素在金红石中有较慢的扩散速率和较高的扩散封闭温度，从而使它在峰期变质温度接近 700 ℃ 时不能与石榴石实现再平衡。

石英–石榴石对的氧同位素温度为 695±35 ℃，石英–金红石对的氧同位素温度为 460±15 ℃，两者相差约 235 ℃。已知金红石可以退变质为榍石或钛铁矿，而对大别山其他地区榍石和钛铁矿的氧同位素分析表明，它们均具有较高的氧同位素封闭温度(Zheng et al., 1998)，因而即使有退变质现象发生，榍石和钛铁矿会继承金红石的氧同位素而不会对其产生大的影响。由于本文研究样品鲜见金红石退变质为榍石或钛铁矿的现象，以及不同颗粒金红石的氧同位素均一性(图 2)，所以作者认为这一温度差异主要反映了石榴石与金红石之间在氧扩散封闭温度上的差异，即氧在石榴石中的扩散速率与 Nd 相当或略高，而氧在金红石中比在石榴石中有较快的扩散速率，后一个观察与已知的经验估计(Zheng et al., 1998)一致。前面已述，Nd 同位素在金红石中的扩散速率较慢，金红石的 Sm-Nd 同位素可能具有与石榴石 Sm-Nd 类似的高封闭温度。如果金红石的 Sm-Nd 扩散速率与石榴石相当，则金红石中 O 的扩散速率比 Nd 快。已知在热力学平衡条件下金红石相对于石榴石和绿辉石显著亏损 ^{18}O(Zheng, 1999)，尽管金红石与石榴石在 O 扩散速率上的差异不会使它们之间的氧同位素富集顺序发生颠倒，但是矿物对氧同位素温度会出现明显差异(图 3)。因此，在应用金红石与石榴石之间的 O 同位素平衡分馏与否来判断 Sm-Nd 等时线年龄的有效性时，需要注意金红石在 O 和 Nd 扩散速率上的显著差别。

3.3 金红石中 Pb 与 O 之间的扩散比较

对金红石 Pb 的扩散系数或封闭温度存在不同认识。Mezger 等(1989)通过冷却年龄比较指出，金红石 Pb 的封闭温度在 400~500 ℃。而 Cherniak(2000)通过干燥条件下的实验研究认为，金红石的 Pb 封闭温度应大于 600 ℃。同一榴辉岩样品中的金红石 U-Pb 内部等时线年龄为 218.2±1.2 Ma(李秋立等, 2001)，与金红石＋石榴石的 Sm-Nd 等时线年龄 237±4 Ma 相差近 20 Ma，如果 218 Ma 代表的是 600 ℃ 的时代，则暗示了榴辉岩在 600 ℃ 以上的环境下滞留了近 20 Ma，这与本文氧同位素分析结果不符，因而该年龄应该代表了更低温度的时代。这对根据Mezger等(1989)经验估计的Pb在金红石中的扩散封闭温度(470±50 ℃)是一个有力的支持。Mezger等(1989)进行经验估计的岩石是缓慢冷却的麻粒岩,而本文金河桥榴辉岩为快速抬升的超高压变质岩。显然,Cherniak(2000)干燥条件下的实验数据在应用到这两类高级变质岩时都遇到了困难,其原因何在是一个值得深入研究的课题。

本文工作的氧同位素分析表明,不同颗粒金红石之间同样达到了氧同位素均一化(图2),石英-金红石对氧同位素温度为460±15 ℃(图3)。通常来说,石英-矿物对氧同位素温度属于直接测定值,在岩石经历简单冷却的条件下它通常是对后一个矿物中O扩散封闭温度的近似(Giletti, 1986; Eiler et al., 1993; Jenkin et al., 1994)。从这个角度来看,本文石英-金红石对氧同位素温度460±15 ℃可以近似代表榴辉岩冷却过程中O在金红石中的扩散封闭温度。它与金红石中Pb扩散封闭温度470±50 ℃之间的一致性不是偶然的,说明在金河桥榴辉岩的冷却过程中金红石的O与Pb的扩散速率具有相似性。

根据上述讨论,可见O和Pb在金河桥榴辉岩金红石中的扩散速率具有一定的可比性。因此,不同颗粒金红石样品具有相对一致的氧同位素组成(图2)表明,该榴辉岩样品中的金红石在手标本尺度上同样达到了Pb同位素均一化。这对先前不同颗粒金红石U-Pb内部等时线定年结果218.2±1.2 Ma(Li et al., 2002)的有效性是一个有力支持,因为等时线测定的一个重要前提是矿物同位素体系达到初始同位素平衡并且未受到后期地质事件的扰动(Zheng, 1989)。此外,金河桥榴辉岩金红石中O扩散速率与Pb扩散速率相当的事实说明,Moore等(1998)实验测定的金红石中O扩散系数对天然样品具有一定的应用性。

4 结论

为了获得一较大温度范围的冷却曲线,人们不得不对同一组样品应用不同同位素体系进行定年。因此,不同同位素体系的封闭温度能否相互比较就是一个非常重要的问题。近年来随着实验条件的不断改善,实验室测定的矿物扩散系数也越来越多,但是与大量的地球化学应用需求相比仍然存在不足。同时,对扩散系数的实验测定仍然存在一定困难,特别是氧逸度和水压力的影响难以准确测定。前人对麻粒岩相石榴石的U-Pb、Sm-Nd和Rb-Sr定年发现,Pb与Nd和Sr在石榴石中的扩散速率存在明显差异(Burton et al., 1995)。本文通过对榴辉岩金红石不同同位素体系的年代学和氧同位素研究,比较了自然界金红石中O、Nd和Pb的相对扩散速率,这对O-Nd-Pb三种同位素体系封闭温度的相互标定是一种重要约束。本文对大别山榴辉岩的研究表明,自然界金红石中O和Pb的扩散速率较Nd要快,而O和Pb具有相似的扩散速率。因此,对金红石及其共生矿物的氧同位素分析,不仅能够用来判断U-Pb和Sm-Nd同位素体系的平衡与否,而且可以用来确定这些放射成因同位素在同一矿物中的相对扩散速率。

致谢

野外工作得到王清晨研究员的帮助,质谱测定工作得到彭子诚教授的指导,余红承担了大部分单矿物的挑选工作,论文初稿承蒙张有学、俞震甫、储雪蕾教授审阅并提出宝贵意见,在此一并表示感谢。

参考文献

Ayers, J.C., Dunkle, S., Gao, S. et al. 2002. Constraints on timing of peak and retrograde metamorphism in the Dabie Shan Ultrahigh-Pressure Metamorphic Belt, east-central China, using U-Th-Pb dating of zircon and monazite. Chem. Geol., 186: 315-331.

Burton KW, Kohn MJ, Cohen AS et al. 1995. The relative diffusion of Pb, Nd, Sr and O in garnet. Earth Planet. Sci. Lett., 133: 199-211.

Carswell DA, Wilson RN, Zhai M. 2000. Metamorphic evolution, mineral chemistry and thermobarometry of schists and orthogneisses hosting ultra-high pressure ecolgites in the Dabieshan of central China. Lithos, 52: 121-155.

Chavagnac V and Jahn BM. 1996. Coesite-bearing eclogites from the Bixiling complex, Dabie Mountains, China: Sm-Nd ages, geochemical characteristics and tectonic implications. Chem. Geol., 133: 29-51.

Cherniak DJ. 2000. Pb diffusion in rutile. Contrib. Mineral. Petrol., 139: 198-207.

Cong BL, Zhai MG, Carswell DA et al. 1995. Petrogenesis of ultrahigh-pressure rocks and their country rocks at Shuanghe in Dabieshan, central China. Eur. J. Mineral, 7: 119-138.

Eiler JM, Valley JW and Baumgartner LP. 1993. A new look at stable isotope thermometry. Geochim. Cosmochim. Acta, 57: 2571-2583.

Eiler JM, Farley KA, Valley JW et al. 1995. Oxygen isotope evidence against bulk recycled sediment in the source of Pitcairn island lavas. Nature, 377: 138-141.

Giletti BJ. 1986. Diffusion effects on oxygen isotope temperatures of slowly cooled igneous and metamorphic rocks. Earth Planet. Sci. Lett., 77: 218-228.

Jenkin GRT, Farrow CM, Fallick AE et al. 1994. Oxygen isotope exchange and closure temperatures in cooling rocks. Jour. Metamor. Geol., 12: 221-235.

Li QL, Li SG, Zhou HY et al. 2002. Rutile U-Pb age for the ultrahigh-pressure eclogite from the Dabie Mountains, central China: evidence for rapid cooling. Chinese Science Bulletin, 47(1): 62-65.

Li SG, Xiao YL, Liu DL et al. 1993. Collision of the North China and Yangtse Blocks and formation of coesite-bearing eclogites: timing and processes. Chem. Geol., 109: 89-111.

Li SG, Wang S, Chen YZ et al. 1994. Excess argon in phengite from eclogite: evidence from dating of eclogite minerals by Sm-Nd, Rb-Sr and $^{40}Ar/^{39}Ar$ methods. Chem. Geol., 112: 343-350.

Li SG, Jagoutz E, Xiao YL et al. 1996. Chronology of ultrahigh-pressure metamorphism in the Dabie Mountains and Su-Lu terrane: I. Sm-Nd isotopic system. Sciences in China (D), 39: 597-609.

Li SG, Jagoutz E, Zhang ZQ et al. 1998. Geochemical and geochronological constraints on the tectonic outline of the Dabie Mountains, central China: a continent-microcontinent-continent collision model. Continental Dynamics, 3:14-31.

Li SG, Jagoutz E, Chen YZ et al..2000. Sm-Nd and Rb-Sr isotopic chronology and cooling history of ultrahigh pressure metamorphic rocks and their country rocks at Shuanghe in the Dabie Mountains, central China. Geochim. Cosmochim. Acta., 64: 1077-1093.

Mezger K, Hanson GN and Bohlen SR. 1989. High-precision U-Pb ages of metamorphic rutiles: application to the cooling history of high-grade terranes. Earth Planet. Sci. Lett., 96:106-118.

Sharp Z. 1990. A laser-based microanalytical method for the in situ determination of oxygen isotope ratios of silicates and oxides. Geochim. Cosmochim. Acta, 54: 1353-1357.

Thoeni M and Jagoutz E. 1992. Some new aspects of dating eclogites in orogenic belts: Sm-Nd, Rb-Sr and Pb-Pb isotopic results from the Austroalpine Saualpe and Koralpe type-locality. Geochim. Cosmochim. Acta, 56: 347-368.

Valley JW, Kitchen N, Kohn MJ et al. 1995. UWG-2, a garnet standard for oxygen isotope ratio: strategies for high precision and accuracy with laser heating. Geochim. Cosmochim. Acta, 59: 5223-5231.

Zhang ZM, Xu ZQ and Xu HF. 1999. Metamorphism of the eclogite from the ZK703 drillhole in Donghai, South Sulu(Jiangsu-Shandong) ultrahigh-pressure metamorphic belt, eastern China. Acta Geological Sinica, 73 (4):331-333.

Zheng YF. 1989. Influence of the nature of the initial Rb-Sr system on isochrom validity. Chem. Geol., 80: 1-16.

Zheng YF. 1991. Calculation of oxygen isotope fractionation in metal oxides. Geochim. Cosmochim. Acta, 55: 2299-2307.

Zheng YF. 1993a. Calculation of oxygen isotope fractionation in anhydrous silicate minerals. Geochim. Cosmochim. Acta, 57: 1079-1091.

Zheng YF. 1993b. Calculation of oxygen isotope fractionation in hydroxyl-bearing minerals. Earth Planet. Sci. Lett., 120: 247-263.

Zheng YF, Fu B. 1998. Estimation of oxygen diffusivity from anion porosity in minerals. Geochem. Jour., 32: 71-89.

Zheng YF, Fu B, Li YL et al. 1998. Oxygen and hydrogen isotope geochemistry of ultrahigh pressure eclogites from the Dabie Mountains and the Sulu terrane. Earth Planet. Sci. Lett., 155: 113-129.

Zheng YF. 1999. On calculation of oxygen isotope fractionation in minerals. Episodes, 22: 99-106.

Zheng YF, Fu B, Xiao YL et al. 1999. Hydrogen and oxygen isotope evidence for fluid-rock interactions in the stages of pre- and post-UHP metamorphism in the Dabie Mountains. Lithos, 46: 677-693.

Zheng YF, Wang ZR, Li SG et al. 2002. Oxygen isotope equilibrium between eclogite minerals and its constraints on mineral Sm-Nd chronometer., Geochim. Cosmochim. Acta., 66(4): 625-634.

李秋立，李曙光，周红英等. 2001. 超高压榴辉岩中金红石 U-Pb 年龄：快速冷却的证据. 科学通报，46(19)：1655-1658.

张泽明，许志琴，徐慧芬. 1999. 南苏鲁超高压变质带东海 ZK703 钻孔榴辉岩的变质作用. 地质学报，73(4)：321-333.

大别造山带超高压变质岩和镁铁质岩浆岩锆石 U-Pb 年代学的 TIMS 和 SIMS 法定年结果比较*

侯振辉，李曙光

中国科学技术大学地球和空间科学学院，合肥 230026

> **亮点介绍**：本文对大别造山带已发表的 ID-TIMS 和 SIMS 锆石 U-Pb 年龄（至 2002 年）进行了统计分析，对大别造山带三叠纪碰撞前后的两次重要地质事件发生的精确年代给出了比较明晰的制约，大别山超高压变质的峰期变质时代应为 223~231 Ma，碰撞后镁铁-超镁铁岩浆侵入事件发生在 123~130 Ma。

摘要 大别山已报道的锆石 U-Pb 年龄数据的综合对比表明大别山超高压变质岩的变质锆石的 ID-TIMS 年龄变化范围为 212~238 Ma（平均值为 225.8 Ma），SHRIMP 单个样品年龄平均值变化范围 219~231 Ma（平均值为 224.2 Ma），二者吻合得很好。已报道的 Cameca IMS 1270 锆石 U-Pb 年龄显示较大变化。瑞典及美国有关实验室用 Cameca IMS 1270 得到的年龄值与 ID-TIMS 法给出的结果一致，但法国 CNRS-CRPG 离子探针国家实验室用 Cameca IMS 1270 获得的两个年龄值（248±16 Ma，254±38 Ma）的中心值显著高于其他 ID-TIMS 和 SHRIMP 的测定结果。对大别山镁铁-超镁铁岩岩浆锆石 U-Pb 年龄的统计结果表明，用 ID-TIMS 法得到的年龄均集中在 123~130 Ma，SHRIMP 年龄变化范围为 125~129 Ma，二者仍吻合得很好。而在法国 CNRS-CRPG 离子探针国家实验室得到的年龄平均值为 144.5 Ma，也显著高于 ID-TIMS 和 SHRIMP 法的定年结果。因此已报道的法国 CNRS-CRPG 实验室用 Cameca IMS 1270 得到的大别山 U-Pb 锆石年龄结果可能存在偏高的系统偏差，其年龄数据不适于用来讨论地质事件中的精细年代学问题。此外，锆石 SHRIMP 单点分析年龄值具有较大的随机误差，在一个符合正态（高斯）分布的数据群中，只有平均值才有意义，而那些位于正态分布边缘的少量偏离平均值的单点分析年龄值本身并不具有特殊的地质意义。那种将数据群中的最高值赋予特殊地质意义的做法是违背数理统计原理的。

关键词 锆石 U-Pb 定年；超高压变质岩；镁铁-超镁铁岩侵入体；大别造山带

1 引言

大别造山带的锆石 U-Pb 年代学工作已经开展了十多年，且已经积累了相当多的数据。最常用的锆石 U-Pb 定年方法是 TIMS（Thermal Ionization Mass Spectrometry）和 SIMS（Secondary Ionization Mass Spectrometry）法，其中前者常用 ID-TIMS（Isotope Dilution - TIMS）方法，后者一般用 SHRIMP（Sensitive High Resolution Ion MicroProbe）和 Cameca IMS 1270 型离子探针来定年。这三种方法测定的锆石 U-Pb 年龄对大别造山带三叠纪碰撞前后的年代学框架给出了比较明晰的制约（Ames et al., 1993, 1996；李曙光等，1997; Rowley et al., 1997; Hacker et al., 1998; Maruyama et al., 1998; Li Shuguang et al., 2000；程裕淇等，2000；陈道公等，2000, 2002；刘贻灿等，2000；侯振辉，2000；谢智等，2001；吴元保等，2001; Chavagnac et al., 2001; Ayers et al., 2002）。例如，它们均一致证明大别山超高压变质作用发生在三叠纪（见表 1）。但是，近年来依据这些数据来讨论有关地质事件发生的精确年代方面还存在许多分歧。如 Ames et al. (1996) 和 Rowley et al. (1997) 根据他们报道的 ID-TIMS 下交点年龄认为峰期变质发生在 218 Ma；Hacker et al. (1998) 根据他们所测的两组 SHRIMP 年龄，认为其中的最高值 245 Ma 应代表峰期变质时代；Li et al. (2000) 建议他们所测的

---

* 本文发表在：岩石学报，2003, 19(3): 490-496

精确的 Sm-Nd 年龄 226±3 Ma，及已报道的变质锆石 SHRIMP 年龄的平均值（224～231 Ma）代表了峰期变质时代（Hacker et al., 1998; Maruyama et al., 1998）；Chavagnac et al.(2001)根据他们的 Sm-Nd 和锆石 Cameca IMS 1270 年龄数据也指出大别山超高压变质岩的峰期变质时代为约 230 Ma；陈道公等(2002) 依据他们用 Cameca IMS 1270 获得的锆石 U-Pb 下交点年龄，认为峰期变质时代上限应为 248 Ma；Ayers et al. (2002)根据他们的 Cameca IMS 1270 锆石 U-Pb 年龄认为超高压变质的峰期变质时代应为 230～237 Ma。造成这些分歧的原因固然有对锆石在超高压变质过程中的重结晶作用和增生历史认识上的分歧，而如何正确理解 SIMS 单点分析值的意义及不同测年方法和不同实验室所给出的年龄数据的可信度及可比性，也是一个值得重视的问题。

本文尝试通过对这些年龄结果进行综合比较，结合 ID-TIMS、SHRIMP 和 Cameca IMS 1270 定年方法的原理和特点，对这三种方法所测年龄结果的可比性及准确性做出一个客观的讨论和评价。

2 常用锆石 U-Pb 定年方法的原理及特点

在 Krogh 建立了超低本底化学流程以及以后的不断改进的基础上（Krogh, 1973, 1982），锆石的同位素稀释热电离质谱（ID-TIMS）法已成为目前最为成熟和普及的定年分析方法。由于样品量相对较多（10 μg 级），因此分析精度很高（0.1%），对那些地质历史简单的锆石进行测定往往可以得到精确的年龄值。例如对于单纯的岩浆成因锆石，它可以给出精确的谐和年龄；对于一次 Pb 丢失的岩浆锆石，或符合两端元混合模式的锆石（如岩浆锆石核+变质增生边）在谐和曲线图上也可给出有意义的上、下交点年龄。而对于复杂成因的锆石，如锆石经历了两次以上 Pb 丢失（李曙光等，1994），两次以上变质增生（刘贻灿等，2000），或锆石有两种以上的继承组分，则往往难以给出有价值的信息，并使测定年龄结果的解释有很大不确定性。

二次离子探针质谱（SIMS）法是通过高能一次离子轰击样品靶产生的二次离子对样品进行微区原位分析，从而得到样品的元素或同位素组成。因此离子探针分析的最大优点是能对复杂锆石中具有明确单一成因的部位进行微区分析，从而获得意义明确的年龄值。为体现这一优点，它要求一次离子束微区分析点的直径要小于复杂锆石中相同成因部位（如变质增生边）的宽度，同时还要能获得足够大的信号强度。显然这二者是矛盾的。为解决这一矛盾，一方面要提高离子探针的灵敏度，灵敏度越高的仪器，可允许用半径更细小的离子束进行微区原位分析；另一方面它必须对具有相同成因的锆石部位进行多次分析，获得一组具有正态分布的谐和年龄，并求平均值以获得高精度的年龄结果。从统计理论上说，那些位于正态分布边缘的少量偏离平均值的单点分析年龄值本身并不具有特殊的地质意义，而仅反映了分析过程中的随机误差。

目前，SHRIMP 和 Cameca IMS 1270 是地质年代学领域两种常用的 SIMS 仪器。SHRIMP 开发较早（第一台出现于 1980 年）(Compston, 1996; 宋彪等，2002)，和其他离子探针最大的不同之处在于它在高质量分辨率下仍有很高的灵敏度。因此它可以用较小的离子束直径（~25 μm）对锆石进行分析，从而获得具有单一成因意义的谐和年龄。近年来，许多实验室也应用 Cameca IMS 1270 进行锆石 U-Pb 定年工作。一些实验室（如美国加州大学洛杉矶分校）可将二次离子束直径调试至 20 μm 进行分析(Ayers et al., 2002)，但也有一些实验室（如 Stockholm 的瑞典自然历史博物馆实验室，法国国家科研中心的岩石及地球化学中心（CNRS-CRPG）离子探针国家实验室）需要用较大的离子束直径（约 50 μm）进行分析，以获得足够大的信号强度（Chavagnac et al., 2001；吴元保等，2001；陈道公等，2002）。这样，如果锆石的变质增生边宽度小于分析的离子束直径，那么就不可能获得具有单一成因信息的谐和年龄值，离子探针的优越性就不能得到体现。

3 大别山锆石 U-Pb 定年年龄结果比较

3.1 大别造山带超高压变质岩

大别山超高压变质岩的抬升、出露历史是大陆深俯冲研究的重要课题之一，精确确定超高压变质作用发生的峰期变质时代及其退变质时代对判断大陆深俯冲的发生和结束时代以及超高压变质岩折返历史具有重要的意义。但是，大别山超高压变质作用的峰期变质时代一直就是一个有争议的话题（Ames et al., 1996; Rowley et al., 1997; Hacker et al., 1998; Maruyama et al., 1998; Li Shuguang et al., 2000; Chavagnac et al., 2001; 陈道公等, 2002; Ayers et al., 2002）。表1列出了近年来大别山超高压变质时代的锆石 U-Pb 年龄测定结果。表中显示, 11 个 ID-TIMS 锆石变质年龄测定结果变化范围为 212～238 Ma, 平均值为 225.8 Ma。除了 3 个年龄低于 219 Ma, 2 个年龄高于 235 Ma 之外, 大多数年龄数据结果在 220～230 Ma 之间。表1还显示, 变质锆石的 SHRIMP 单点年龄测定值变化范围较大, 为 209～250 Ma, 但每个样品的统计平均值（误差<±5 Ma）变化范围为 219～231 Ma（平均年龄为 224.2 Ma）, 它与大多数 ID-TIMS 测定的变质年龄变化范围（220～230 Ma）一致。因此, ID-TIMS 和 SHRIMP 的统计平均定年结果吻合得很好。然而已报道的用 Cameca IMS 1270 得到的高压变质年龄除两个较精确的多点分析平均谐和年龄（230±4 Ma 和 238±3 Ma）外, 大多数是具有较大误差的不一致线下交点年龄：233±21 Ma（Chavagnac et al., 2001）, 236±32 Ma（Ayers et al., 2002）, 248±16 Ma（陈道公等, 2002）, 254±38 Ma（吴元保等, 2001）。由于它们较大的年龄误差, 我们可以说在这样大的误差范围之内, 它们与 ID-TIMS 和 SHRIMP 的测定结果一致。如果抛开这样大的误差不谈, 仅从测定结果的中心值来看, 陈道公等（2002）和吴元保等（2001）报道的在法国岩石及地球化学中心（CNRS-CRPG）离子探针国家实验室用 Cameca IMS 1270 得到的两个年龄数据（248±16 Ma 和 254±38 Ma）显著高于其他 ID-TIMS 和 SHRIMP 的测定结果（见表1, 图1）。这种差异以及 SHRIMP 单点分析年龄值的较大变化范围已经被一些作者赋予特殊地质意义。

例如, Hacker et al. (1998)将符合高斯分布的 SHRIMP 单点分析年龄值变化范围中的高值（245 Ma）视为峰期变质时代。首先这种简单地将锆石 U-Pb 年龄中的高值视为峰期变质时代的解释是缺乏理论依据的。一方面它忽视了在前进变质过程中当温度>600 ℃时, 蜕晶化锆石可以迅速重结晶, 以及该重结晶作用并不一定能使原放射成因 Pb 完全排除出来的事实（Mezger et al., 1997.）; 另一方面, 变质锆石 SHRIMP 单点分析年龄值具有较大的随机误差, 在一个符合正态（高斯）分布的数据群中, 只有平均值才有意义。那种将该数据群中的最高值（245 Ma）视为峰期变质时代, 如前述, 是违背数理统计原理的。其次, 由于未对这些锆石分析点做相应的微量元素分析, 因此至少从成因上缺乏将高值视为峰期变质时代的证据。考虑到 SHRIMP 法给出的均是变质锆石边的谐和年龄, 这些变质锆石既包括前进变质阶段形成的重结晶锆石, 也有高压变质和（或）退变质过程中增生的锆石。在未能很好区分这些不同成因变质锆石的情况下, 这些年龄测定结果的平均值（223～231 Ma）应当最接近峰期变质时代。表1显示双河硬玉石英岩或超高压副片麻岩中锆石 TIMS 年龄（238.4± 1.3 Ma）（李曙光等, 1997）和 SIMS 年龄（238±3 Ma）（Ayers et al., 2002）相对于正片麻岩或榴辉岩的年龄偏高。这可能是由于它们的原岩是沉积岩, 含有较多古老（早元古代）碎屑锆石（Ayers et al., 2002）, 因此它们的蜕晶化程度高于那些原岩年龄仅为 700～800 Ma 岩浆锆石。这样在双河超高压副变质岩的变质锆石中, 重结晶的蜕晶化锆石所占比例较高。由于蜕晶化锆石在重结晶过程中可能未将变质前积累的放射成因 Pb 排除干净从而给出较老的年龄值（Mezger et al., 1997）, 因此 238Ma 应该指示的是峰期变质时代的上限。

再如, 陈道公等（2002）将 Cameca IMS 1270 给出的 248 Ma 的年龄值解释为超高压变质作用的峰期变质时代上限。虽然这一解释仅对峰期变质时代给出了较宽松的约束, 但该年龄值较 ID-TIMS 和 SHRIMP 年龄值系统偏高的现象不能不引起我们的注意, 因为已有其他研究者在不同实验室得到了与 TIMS 法定年一致的 SIMS 年龄结果。例如, Ayers et al. (2002) 报道了用 Cameca IMS 1270 获得的毛屋榴辉岩的锆石 U-Pb 谐和年龄为 230±4 Ma, 与相应的 ID-TIMS 结果（225.4+3/−6 Ma）（Rowley et al.,

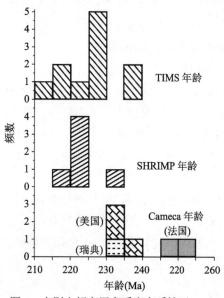

图1 大别山超高压变质岩变质锆石 U-Pb 年龄分布统计直方图

1997）在误差范围内一致。Ayers et al. (2002) 还报道了双河硬玉石英岩中锆石边部的 SIMS 谐和年龄为 238±3 Ma，也与相应的 TIMS 结果（238.4±1.3 Ma）（李曙光等，1997）一致（见表1）。Bingen et al. (2001)报道了挪威西部榴辉岩 Cameca IMS 1270 法锆石 U-Pb 年龄为 455±29 Ma（8 点平均），也与同一样品的 ID-TIMS 年龄（456±7 Ma）一致。因此已报道的系统偏高的 Cameca 离子探针年龄值并不是仪器本身的问题，而更可能是反映了该实验室实验测定或数据处理本身的问题。应当指出，由于超高压变质岩的快速抬升和冷却，仅当年龄误差小于±5 Ma 时才对约束其冷却史有意义，误差大于±10 Ma 的低精度年龄不适宜用来讨论超高压变质及冷却过程中的精细年代学问题。

表1 大别山超高压变质岩变质锆石 U-Pb 年龄（Ma）

地点	岩性	年龄值（Ma）		分析方法	数据来源
塔儿河水电站	英云闪长质片麻岩	226±6	(1 点)	ID-TIMS 谐和年龄	刘贻灿等，2000
饶钹寨	榴辉岩	230±6	(3 点平均)	ID-TIMS 谐和年龄	刘贻灿等，2000
石竹河	英云闪长质片麻岩	229±18		ID-TIMS 下交点	谢智等，2001
燕子河	英云闪长质片麻岩	227±10		ID-TIMS 下交点	侯振辉，2000
潜山三祖寺	英云闪长质片麻岩	221.9±6.8		ID-TIMS 下交点	陈道公等，2000
太湖石马乡	超高压副片麻岩 92T16	236.2±2.4	(1 点)	ID-TIMS $^{206}Pb/^{238}U$ 年龄	李曙光等，1997
五庙	榴辉岩	212±11		ID-TIMS 下交点	Ames et al., 1993
五庙	英云闪长质片麻岩（93-006）	222.6±5	(23 点平均)	SHRIMP 谐和年龄（变质增生边）	Maruyama et al., 1998
英山	榴辉岩 DB91-34	218.4±2.5		ID-TIMS 下交点	Ames et al., 1996
英山附近	副片麻岩 DS158	224±4	(10 点平均)	SHRIMP 谐和年龄	Hacker et al., 1998
岳西东南	正片麻岩 DS107	231±2	(15 点平均)	SHRIMP 谐和年龄（变质增生边）	Hacker et al., 1998
石马	片麻岩	218.5±1.7		ID-TIMS 下交点	Rowley et al., 1997
毛屋	榴辉岩	225.5 +3/−6		ID-TIMS 下交点	Rowley et al., 1997
		230±4	(25 点平均)	Cameca 谐和年龄（变质增生边）	Ayers et al., 2002
碧溪岭	深色榴辉岩	223±3	(12 点平均)	SHRIMP 谐和年龄（变质增生边）	程裕淇等，2000
	花岗片麻岩	219±3	(14 点平均)		
	石榴橄榄岩	254±38		Cameca 下交点	吴元保等，2001
潜山双河	榴辉岩（SH3）	248±16		Cameca 下交点	陈道公等，2002
	超高压片麻岩	233±21		Cameca 下交点	Chavagnac et al., 2001
	硬玉石英岩（Sh-JQ） (核)	236±32		Cameca 下交点	Ayers et al., 2002
	硬玉石英岩（Sh-JQ） (边)	238±3	(6 点平均)	Cameca 谐和年龄（变质增生边）	
	超高压副片麻岩 92HT1	238.4±1.3	(1 点)	ID-TIMS 谐和年龄	李曙光等，1997

3.2 北大别带镁铁-超镁铁岩侵入体的锆石 U-Pb 年龄

北大别燕山期碰撞后岩浆岩主要为镁铁-超镁铁岩侵入体、闪长-花岗闪长岩以及中酸性花岗岩。其中镁铁-超镁铁岩侵入体已经做了大量的锆石 U-Pb 年代学工作（Hacker et al., 1998；李曙光等，1999；王江海等，2002；Xue et al., 1997；葛宁洁等，1999；陈道公等，2001）。它们的锆石基本为成因简单的岩浆结晶锆石。对此类锆石的 ID-TIMS 法定年应该可以给出更精确的年龄值。从表2 中可以看出，用 ID-TIMS 法得到的这些岩体的侵入时代比较一致，均集中在 120~130 Ma。而 Hacker et al. (1998)对磨子潭辉长岩中锆石进行了 SHRIMP 分析，其 $^{206}Pb/^{238}U$ 年龄变化范围为 123.5~131.5 Ma，平均值为 129±2 Ma。这些结果与 ID-TIMS 法得到的年龄结果是一致的，这说明 SHRIMP 法在锆石定年方面具有与 ID-TIMS 法可比较的准确度。而陈道公等（2001）在法国 CNRS-CRPG 实验室用 Cameca IMS 1270 获得的道士冲辉石岩锆石 U-Pb 年龄变化范围为 134~159 Ma，平均值为 144.5±6.2 Ma，它显著高于 ID-TIMS 和 SHRIMP 法的定年结果。上述三种定年方法得到的 $^{206}Pb/^{238}U$ 年龄的统计直方图（见图2）表明 ID-TIMS 和 SHRIMP 法定年得到的年龄结果是非常一致的,而法国 CNRS-CRPG 实验室用 Cameca IMS 1270 离子探针得到的年龄结果明显偏高,而且数据点也比较分散。这一比较再次显示了该实验室已报道的 Cameca IMS 1270 离子探针年龄系统偏高的事实。因此大别造山带北部镁铁-超镁铁岩的形成时代应为 123~130 Ma,而并非 145 Ma。这与同时期的中酸性花岗岩的锆石 U-Pb 年龄是一致的(李石等,1991)。

4 结论

通过上述年龄数据分析,我们认为在地质年代学定年中,用 ID-TIMS 和 SHRIMP 方法得到的年龄结果比较一致,精度也比较好,可信度高。不同实验室用 Cameca IMS 1270 离子探针得到的锆石 U-Pb 年龄测定结果差异较大。Stockholm 的瑞典自然历史博物馆和美国加州大学洛杉矶分校的实验室测定的锆石 U-Pb 年龄结果与 ID-TIMS 一致,但除了后者报道了谐和的锆石 U-Pb 年龄数据外,目前已报道的大别山超高压变质岩锆石 Cameca IMS 1270 离子探针 U-Pb 年龄数据结果多为不一致线下交点年龄,且年龄误差较大(>±16 Ma),这些年龄数据不适合用来讨论精细地质年代学问题。需要指出的是,法国岩石及地球化学中心(CNRS-CRPG)离子探针国家实验室已报道的 Cameca IMS 1270 离子探针年龄普遍比 ID-TIMS 和 SHRIMP

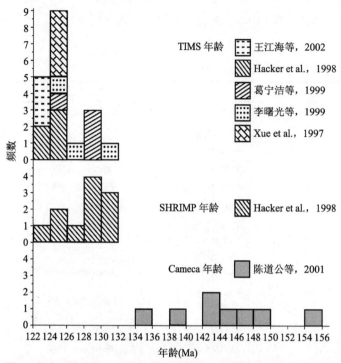

图 2 北大别带镁铁-超镁铁侵入体锆石 U-Pb 年龄分布统计直方图

年龄偏高,反映了该实验室的 SIMS 年龄数据存在着偏高的系统偏差。

表 2 北大别带镁铁-超镁铁侵入体的锆石 U-Pb 年龄(Ma)

地 点	岩 性	年 龄 值		分 析 方 法	数 据 来 源
祝家铺	伟晶状辉长岩脉	130.2±1.4	(1 点)		
小河口	辉石岩	125.3±0.8	(1 点)		李曙光等,1999
	辉石闪长岩	127±5.5	(1 点)		
椒 岩	辉长岩	124±16	ID-TIMS 下交点	ID-TIMS 谐和年龄	李曙光等,1999
湖北黄冈漆柱山	辉长岩(JM14)	122.9±0.6	(3 点平均)	ID-TIMS 谐和年龄	王江海等,2002
磨子潭地区	辉长岩(DS58)	129±2	(11 点平均)	SHRIMP 谐和年龄	Hacker et al.,1998
		125±2	(5 点平均)	ID-TIMS 谐和年龄	
岳西沙村	辉长岩	128.1±2.0	(4 点平均)	ID-TIMS 谐和年龄	葛宁洁等,1999
道士冲	辉石岩 DSCDB8	144.5±0.9	(7 点平均)	Cameca 谐和年龄	陈道公等,2001

由于 SIMS 单点分析年龄值误差较大,应对具有相同成因的锆石部位进行多次分析,获得一组具有正态分布的谐和年龄,并求平均值以获得高精度的年龄结果。依据统计学原理,那些位于正态分布边缘的少量偏离平均值的单点分析年龄值本身并不具有特殊的地质意义。

大别山造山带 ID-TIMS 和 SHRIMP 锆石年龄数据的统计平均值显示,大别山超高压变质的峰期变质时代应为 223~231 Ma,碰撞后镁铁-超镁铁岩浆侵入事件发生在 123~130 Ma。

致谢

成文过程中郑永飞教授提出许多宝贵意见,在此表示衷心感谢。

参考文献

Ames L, Tilton G R, Zhou G Z. 1993. Timing of collision of the Sino-Korean and Yangtze Cratons: U-Pb Zircon dating of coesite-bearing eclogites. Geology, 21: 339−342

Ames L, Zhou G Z, Xiong B C. 1996. Geochronology and isotopic character of ultrahigh pressure metamorphism with implications for collision of the Sino-Korean and Yangtze cratons, central China. Tectonics, 15: 472−489

Ayers J C, Dunkle S, Gao Shan, Miller C F. 2002. Constraints on timing of peak and retrograde metamorphism in the Dabie Shan ultrahigh-pressure metamorphic belt, east-central China, using U-Th-Pb dating of zircon and monazite. Chemical Geology, 186: 315−331

Bingen B, Austrheim H, Whitehouse M. 2001. Ilmenite as a source for zirconium during high-grade metamorphism? Textural evidence from the Caledonides of western Norway and implications for zircon geochronology. Journal of Petrology, 42(2): 355−375

Chavagnac V, Jahn B M, Villa I M, Whitehouse M J, Liu Dunyi. 2001. Multichronometric evidence for an in situ origin of the ultrahigh-pressure metamorphic terrane of Dabieshan, China. The Jour of Geol, 109: 633−646.

Chen Daogong, Deloule E, Xia Qunke, Wu Yuanbao, Cheng Hao. 2002. Metamorphic zircon from Shuanghe ultrahigh pressure eclogite, Dabie Mountains: ion microprobe and internal micro-structure study. Acta Petrologica Sinica, 18(3): 369−377 (in Chinese with English abstract)

Chen Daogong, Isachsen C, Zhi Xiachen, Zhou Taixi, Cheng Hao, Xia Qunke. 2000. Zircon U-Pb ages of gneiss from Qianshan of Anhui Province. Chin Sci Bull, 45(8): 764−767.

Chen Daogong, Wang Xiang, E. Deloule, Li Binxian, Xia Qunke, Cheng Hao, Wu Yuanbao. 2001. Zircon SIMS ages and chemical compositions from Northern Dabie Terrain: its implication for pyroxenite genesis . Chin Sci Bull, 46(12): 1047−1050

Cheng Yuqi, Liu Dunyi, Williams I S, Jian Ping, Zhuang Yuxun, Gao Tianshan. 2000. SHRIMP U-Pb dating of zircons of a dark-coloured eclogite and a garnet-bearing gneissic-granitic rock from Bixiling, eastern Dabie mountains. Acta Geologica Sinica, 74(3): 193−205 (in Chinese with English abstract).

Compston W. 1996. SHRIMP: origins, impact and continuing evolution. Journal of Royal Society of Western Australia, 79: 109−117

Ge Ningjie, Hou Zhenhui, Li Huimin, Chen Jiangfeng, Liu Bin, Ruan Jun, Qin Liping. 2000. Zircon U-Pb ages of Shacun gabbro body, Yuexi, Dabie orogen, and its geological implications. Chin Sci Bull, 45(1): 74−79

Hacker B R, Ratschbacher L, Webb L, Ireland T, Walker D, Dong Shuwen. 1998. U/Pb zircon ages constrain the architecture of the ultrahigh-pressure Qinling-Dabie Orogen, China. Earth Planet Sci Lett, 161: 215−230

Hou Zhenhui. 2000. Geochemistry and chronology of granulite and TTG gneiss from the northern Dabie mountains. M. A. thesis, University of Science and Technology of China (in Chinese with English abstract)

Krogh T E. 1973. A low contamination method for hydrothermal decomposition of zircon and extraction of U and Pb for isotopic age determinations. Geochim Cosmochim Acta, 37: 485−494

Krogh T E. 1982. Improved accuracy of U-Pb zircon ages by the creation of more concordant systems using an air abrasion technique. Geochim Cosmochim Acta, 46: 637−649

Li Shi, Wang Tong. 1991. Geochemistry of granitoids from Tongbai-Dabie mountains. Wuhan: China University of Geosciences Press, 208

Li Shuguang, Chen Yizhi, Song Mingchun, Zhang Zhimin, Yang Chun, Zhao Dunmin. 1994. Zircon U-Pb ages of amphibolite from the Haiyangsuo area, eastern Shandong province—An example for influence of multimetamorphism to lower and upper intercepts of zircon discordia line at the concordia curve. Acta Geoscientia Sinica, (1-2): 37−42 (in Chinese with English abstract)

Li Shuguang, Hong Ji'an, Li Huimin, Jiang Laili. 1999. Zircon U-Pb ages of pyroxenite-gabbro intrusions in Dabie Mountains and their geological implications. Geological Journal of China Universities, 5(3): 351−355 (in Chinese with English abstract)

Li Shuguang, Jagoutz E, Chen Yizhi , Li Qiuli. 2000. Sm-Nd and Rb-Sr isotopic chronology and cooling history of ultrahigh pressure metamorphic rocks and their country rocks at Shuanghe in the Dabie Mountains, Central China. Geochim Cosmochim Acta, 64 (6): 1077−1093

Li Shuguang, Li Huimin, Chen Yizhi, Xiao Yilin, Liu Deliang. 1997. Chronology of ultra-high pressure metamorphism in the Dabie mountains and Su-Lu terrane: II. U-Pb isotope system of zircon. Science in China (Series D), 27(3): 200−206 (in Chinese)

Liu Yican, Li Shuguang, Xu Shutong, Li Huimin, Jiang Laili, Chen Guanbao, Wu Weiping, Su Wen. 2000. U-Pb zircon ages of the eclogite and tonalitic gneiss from the northern Dabie mountains, China and multi-overgrowths of metamorphic zircons.

Geological Journal of China Universities, 6(3): 417−423 (in Chinese with English abstract)

Maruyama S, Tabata H, Nutman A P, Morikawa T, Liou J G. 1998. Shrimp U-Pb geochronology of ultrahigh-pressure metamorphic rocks of the Dabie Mountains, central China. Continental Dynamics, 3 (1~2): 72−85

Mezger K, Krogstad E J. 1997. Interpretation of discordant U-Pb zircon ages: An evaluation. J metamorphic Geol, 15: 127−140

Rowley D B, Xue F, Tucker R D, Peng Z X, Baker J, Davis A. 1997. Ages of ultrahigh pressure metamorphic and protolith orthgenisses from the eastern Dabie Shan: U/Pb zircon geochronology. Earth Planet Sci Lett, 151: 191−203

Song Biao, Zhang Yuhai, Liu Dunyi. 2002. Introduction to the naissance of SHRIMP and it contribution to isotope geology. Journal of Chinese Mass Spectrometry Society, 23(1): 58−62 (in Chinese with English abstract)

Wang Jianghai, Deng Xixian. 2002. Emplacement age for the mafic-ultramafic plutons in the northern Dabie Mts. (Hubei): zircon U-Pb, Sm-Nd and ^{40}Ar-^{39}Ar dating. Science in China (Series D), 45(1): 1−12

Wu Yuanbao, Chen Daogong, Cheng Hao, Deloule E, Xia Qunke. 2001. SIMS U-Pb dating and CL images of zircons from garnet peridotite of Bixiling. Geological Journal of China Universities, 7(3): 356−362 (in Chinese with English abstract)

Xie Zhi, Chen Jiangfeng, Zhang Xun, Gao Tianshan, Dai Shengqian, Zhou Taixi, Li Huimin. 2001. Zircon U-Pb dating of gneiss from Shizhuhe in Northern Dabie and its geologic implications. Acta Petrologica Sinica, 17(1): 139−144 (in Chinese with English abstract)

Xue F, Rowley D, Tucker R D, Peng Z C. 1997. U-Pb zircon ages of granitoid rocks in the North Dabie complex, Eastern Dabie Shan, China. J of Geology, 105: 744−753

陈道公, Deloule E, 夏群科, 吴元保, 程昊. 2002. 大别山双河超高压榴辉岩中变质锆石: 离子探针和微区结构研究. 岩石学报, 18(3): 369−377

陈道公, Isachsen C, 支霞臣, 周泰禧, 程昊, 夏群科. 2000. 安徽潜山片麻岩锆石 U-Pb 年龄. 科学通报, 45(2): 214−217

陈道公, 汪相, Deloule E, 李彬贤, 夏群科, 程昊, 吴元保. 2001. 北大别辉石岩成因: 锆石微区年龄和化学组成. 科学通报, 46(7): 586−590

程裕淇, 刘敦一, Williams I S, 简平, 庄育勋, 高天山. 2000. 大别山碧溪岭深色榴辉岩和片麻状花岗质岩石 SHRIMP 分析. 地质学报, 74(3): 193−205

葛宁洁, 侯振辉, 李惠民, 陈江峰, 刘斌, 阮俊, 秦礼萍. 1999. 大别造山带岳西沙村镁铁-超镁铁岩体的锆石 U-Pb 年龄. 科学通报, 42(19): 2110−2114

侯振辉. 2000. 大别造山带北部麻粒岩和 TTG 片麻岩的地球化学和年代学研究(中国科学技术大学硕士学位论文), 15−24

李石, 王彤. 1991. 桐柏山-大别山花岗岩类地球化学. 武汉: 中国地质大学出版社, 208

李曙光, 陈移之, 宋明春, 张志敏, 杨淳, 赵敦敏. 1994. 胶东海洋所斜长角闪岩的锆石 U-Pb 年龄——多期变质作用对锆石不一致线的影响. 地球学报, (1-2): 37−42

李曙光, 洪吉安, 李惠民, 江来利. 1999. 大别山辉石-辉长岩体的锆石 U-Pb 年龄及其地质意义. 高校地质学报, 5(3): 351−355

李曙光, 李惠民, 陈移之, 肖益林, 刘德良等. 1997, 大别-苏鲁地体超高压变质年代学——II. 锆石 U-Pb 同位素体系. 中国科学(D), 27(3): 200−206.

刘贻灿, 李曙光, 徐树桐, 李惠民, 江来利, 陈冠宝, 吴维平, 苏文. 2000. 大别山北部榴辉岩和英云闪长质片麻岩锆石 U-Pb 年龄及多期变质增生. 高校地质学报, 6(3): 417−423

宋彪, 张玉海, 刘敦一. 2002. 微量原位分析仪器 SHRIMP 的产生与锆石同位素地质年代学. 质谱学报, 23(1): 58−62

王江海, 邓尚贤. 2002. 湖北北大别镁铁-超镁铁质侵入体的时代: 锆石 U-Pb, Sm-Nd 和 ^{40}Ar-^{39}Ar 定年结果. 中国科学(D辑), 32(1): 1−9

吴元保, 陈道公, 程昊, Deloule E, 夏群科. 2001. 碧溪岭岩体中石榴橄榄岩的锆石显微结构及离子探针定年. 高校地质学报, 7(3): 356−362

谢智, 陈江峰, 张巽, 高天山, 戴圣潜, 周泰禧, 李惠民. 2001. 大别造山带北部石竹河片麻岩的锆石 U-Pb 年龄及其地质意义. 岩石学报, 17(1): 139−144.

青龙山榴辉岩高压变质新生锆石 SHRIMP U-Pb 定年、微量元素及矿物包裹体研究[*]

李秋立[1,2]，李曙光[1]，侯振辉[1]，洪吉安[1]，杨蔚[1]

1. 中国科学技术大学地球和空间科学学院，合肥 230026
2. 中国科学院地质与地球物理研究所，北京 100029

> **亮点介绍**：通过锆石阴极发光图像、Th/U 值、矿物包裹体及稀土元素配分模式综合证明青龙山榴辉岩中锆石的变质成因；通过 SHRIMP 微区定年获得锆石 U-Pb 年龄为 227.4±3.5 Ma；该年龄代表了青龙山榴辉岩的峰期变质时代，同时证实了三平衡矿物 Sm-Nd 等时线年龄的正确性。

摘要 对青龙山一榴辉岩的锆石成因及年龄进行了综合研究，阴极发光(CL)显微图像显示该锆石无继承锆石核，其矿物包裹体成分及微量元素特征表明它是高压变质新生锆石，并获得该锆石的高精度离子探针（SHRIMP）平均 U-Pb 年龄为 227.4±3.5 Ma。该年龄给出了青龙山榴辉岩峰期变质时代的最佳估计。

关键词 SHRIMP U-Pb 定年；微量元素；包裹体；峰期变质时代

对高级变质岩中的变质锆石进行 U-Pb 定年以获得变质年龄是人们常用的方法。早期人们通过变质成因锆石浑圆的外形及阴极发光图像缺少岩浆锆石韵律环带特征来识别变质锆石[1-6]，后来人们又进一步发现低 Th/U 也是变质锆石的主要特征[7-9]，然而对符合上述特征的变质锆石进行定年却给出了较大的年龄范围，如大别-苏鲁超高压带，变质锆石 U-Pb 年龄的范围可达 205～245 Ma[6,8,10-17]。以后人们逐渐认识到，变质锆石可以有多种成因[18-26]。大别-苏鲁超高压带内锆石微区包体矿物性质及相应阴极发光图像研究表明，单颗锆石有可能具有多圈层的生长带，自核心向边缘一般为：原岩岩浆锆石核，变质重结晶锆石，超高压变质增生锆石，退变质增生锆石[27-29]。值得注意的是，单凭 CL 图像很难确切区分高压变质新生锆石和变质重结晶锆石区域，因为它们都不具有岩浆锆石特有的韵律环带结构，又大都处在岩浆锆石核外围，并被最外层退变质锆石边所包裹[26,30,31]。因而，在不能很好区分上述不同成因变质锆石的情况下，所获得的变质锆石 U-Pb 年龄仍然是一种混合年龄，不具确切地质意义。

要获得确切的超高压峰期变质时代，应选取超高压变质时期的新生锆石进行测定。但这一方法的重要前提条件是我们能正确地区分、判断不同锆石微区的成因，以及该微区必须具有足够的宽度。锆石所含矿物包裹体的种类大多可以帮助判别含包体微区的生长条件，近年来许多作者根据锆石包裹体矿物的性质区分锆石不同成因区域并进行 SHRIMP 分析[13,15-17,30,32,33]。通过矿物包体的种类可以识别出超高压岩石中的锆石一般具有原岩残留锆石、高压增生锆石、退变质时期增生锆石。值得注意的是在某些具有韵律环带特征的继承岩浆锆石核中也发现有柯石英包体[29]，虽然这种柯石英包体是沿裂纹挤进去的还是由其他机制形成的尚不清楚，但说明仅凭 CL 图像和包裹矿物成分还不足以确切地判定超高压变质成因锆石。此外，考虑到 CL 图像不能很好区分变质重结晶锆石和高压变质新生锆石，这样如果不能提供其他方面的判据（如微量元素），就有可能带来所测区域为继承锆石或重结晶锆石的危险。

---

[*] 本文发表在：科学通报, 2004, 49（22）：2329-2334
对应的英文版论文为：Li Qiuli, Li Shuguang, Hou Zhenhui, et al., 2005. A combined study of SHRIMP U-Pb dating, trace element and mineral inclusions on high-pressure metamorphic overgrowth zircon in eclogite from Qinglongshan in the Sulu terrane. Chinese Science Bulletin, 50: 1-7

因为变质新生锆石与变质重结晶锆石不一样,尤其若重结晶锆石未能将原蜕晶化锆石所含放射成因 Pb 排除干净的话,会给出偏高的年龄值[18]。一些作者通过锆石微区微量元素的特征来研究所测微区的成因属性,取得了较大的成功[14,26,30,33-35]。例如,Hermann et al.[30]对北哈萨克斯坦超高压岩石中的锆石的研究发现,具有类似 CL 特征的区域可能具有不同的成因,其稀土元素特征明显不同。Hoskin and Black[26]研究了锆石固态变质重结晶域的微量元素特征,指出变质过程对锆石的稀土元素影响不明显。Rubatto[34]将锆石的稀土元素特征与其生长条件对应起来。显然,如果结合以上两种判别方法,确切区分所测锆石区域的成因属性,则可给出有明确地质含义的锆石年龄。Sun et al.[33]综合应用锆石包体矿物成分及锆石微量元素的判据,对大别山浒湾剪切带古生代洋壳俯冲成因榴辉岩的锆石成因进行了研究,并获得了很好的结果。最近刘敦一和简平[17]对双河硬玉石英岩中锆石进行了类似研究,提出了对超高压岩石峰期变质时代为 243±1 Ma 的新观点。然而对于原岩为沉积岩的硬玉石英岩,与榴辉岩相比,在相同的温压地质条件下,其所发生的变质反应不同,因而其锆石增生历史应该有一定的差别。

本工作在青龙山—榴辉岩中找到了不含岩浆锆石核的纯变质锆石样品。通过综合应用阴极发光显微图像分析、矿物包裹体研究及微量元素测定,论证了所测锆石为高压变质阶段新生成锆石,其 SHRIMP U-Pb 定年结果的平均值应该是榴辉岩峰期变质阶段时代的最佳估计。

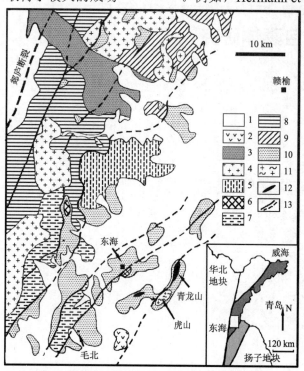

图 1 苏鲁东海地区地质简图 (据刘福来等[36])

1—第四系;2—第三纪玄武岩;3—白垩纪盆地沉积;4—造山期后未变质花岗岩;5—含霓石和角闪石的二长花岗质片麻岩;6—角闪黑云斜长花岗质片麻岩;7—含石榴子石和黑云母的斜长(二长)花岗质片麻岩;8—含黑云母二长花岗质片麻岩;9—钾长花岗质片麻岩;10—表壳岩系;11—含黄铁矿绿帘黑云二长花岗质片麻岩;12—榴辉岩和超基性岩;13—剪切带或断层

1 地质背景及样品

苏鲁高压-超高压变质带是大别山地体被郯庐断裂平移的东延部分,江苏东海县青龙山位于苏鲁超高压带的西南部(图 1)。青龙山榴辉岩因多硅白云母过剩氩的发现[37]和极低的 $\delta^{18}O$ 值的发现[38,39]而闻名。其地质概况可参见相关文献[36-41]。Li et al.[37]对青龙山榴辉岩进行了高压矿物 Sm-Nd 同位素定年及多硅白云母 Rb-Sr 分析,分别给出 226.3±4.5 Ma(Sm-Nd)和 219±0.5 Ma(Rb-Sr)年龄,刘福来等[15]测定了东海地区片麻岩中含柯石英包体的变质锆石幔部平均年龄为 229±4 Ma,含石英等低压矿物包体的变质锆石退变边平均年龄为 211±4 Ma,其中含柯石英锆石微区年龄值与南大别超高压变质岩的定年结果一致[42-44]。青龙山超高压变质带温压条件估计为 700-890 ℃,压力>28 kbar[41]。本文样品(02QL-2)为青龙山含柯石英榴辉岩,样品片理化发育,主要矿物组合为石榴石+绿辉石+多硅白云母+金红石+蓝晶石+石英。

2 分析方法

在天津地质矿产研究所从大约 20 kg 的榴辉岩中经常规分选程序分离出锆石。锆石的 U-Th-Pb 分析在北京离子探针中心的 SHRIMP II 型离子探针仪器上完成,标准测定流程见文献[45]。将锆石颗粒与标样(RSES)一起置于环氧树脂样品座中,然后磨至一半大小,使锆石内部暴露,用于进行透射光、反射光及阴极发光(CL)显微图像分析,选择没有裂隙及包裹体的颗粒或视域进行 SHRIMP 分析。应用 RSES 参考锆石 TEM (417 Ma)

进行元素间的分馏校正。应用 SL13(年龄，572 Ma；U 含量，238 μg/g)标定样品的 U、Th 和 Pb 含量。普通 Pb 校正采用实测 ^{204}Pb 值，使用 Stacey and Kramer 的两阶段模式进行扣除。因为是年轻锆石，所报道年龄数据为 ^{206}Pb/^{238}U 年龄。各种同位素比值及年龄误差均为 1σ。数据处理采用 Ludwig SQUID1.0 及 ISOPLOT 程序[46,47]。

锆石包裹体研究在北京大学地球与空间科学学院通过显微拉曼光谱完成。该仪器为英国 Renishaw 公司 RM-1000 型激光拉曼光谱仪，工作条件为 514 nm 激光器，发射功率 20 mW，样品接收功率 4.5 mW，扫描时间 10 s。

锆石的微区原位微量元素分析在西北大学地质学系大陆动力学实验室的 LA-ICPMS 上完成，详细分析流程参见有关文献[48]。实验时 ArF 激光束工作波长为 193 nm，束斑直径为 40 μm，频率为 10 Hz，激光束能量为 170 mJ。以锆石的 SiO_2 含量作为内部标准，以 NIST612 为外部标准，数据处理采用 Glitter 程序。

3 分析结果

所测青龙山榴辉岩(02QL-2)中锆石为无色透明浑圆状，粒径 50～80 μm，CL 显微图像（图 2）显示无核边结构及韵律环带特征。该锆石富含矿物包裹体，利用电子探针定性及激光拉曼光谱分析了大多数锆石的矿物包裹体成分，鉴别出为金红石、石榴石、绿辉石等高压矿物(图 3)，未见低压矿物包裹体。金红石标识峰为 611～615 cm^{-1}，石榴石标识峰为 904～915 cm^{-1}，绿辉石为 677～688 cm^{-1}。在部分锆石年龄测试点位置进行了稀土元素和其他微量元素测定，结果见表 1，稀土元素配分模型见图 4。表 2 列出了 SHRIMP U-Pb 分析的 12 个点数据，除 9 号分析点外，所有锆石分析点数据在谐和图上成群分布，^{206}Pb/^{238}U 年龄值属同一母体，其加权平均值为 227.4±3.5 Ma（图 5），9 号锆石分析点数据明显与其他锆石不同，Th/U=0.29，年龄为 300 Ma，它可能是未完全重置的变质重结晶锆石。

表 1 青龙山榴辉岩（02QL-2）锆石微量元素数据(μg/g)

分析点编号	1	2	5	6	7
La	0.01	0.02	0.03	0.03	0.04
Ce	0.83	1.46	1.08	0.83	1.17
Pr	0.04	0.03	0.04	0.05	0.03
Nd	0.18	0.24	0.24	0.23	0.31
Sm	0.19	0.13	0.12	0.20	0.15
Eu	0.08	0.10	0.13	0.06	0.12
Gd	0.43	0.86	0.72	0.54	0.75
Tb	0.16	0.30	0.28	0.16	0.18
Dy	2.62	2.50	2.57	1.97	2.81
Ho	0.88	0.87	0.74	0.68	0.69
Er	2.99	2.78	2.43	2.45	2.87
Tm	0.59	0.45	0.37	0.38	0.45
Yb	5.66	3.84	3.10	3.64	3.75
Lu	1.07	0.53	0.55	0.50	0.63
ΣREE	15.73	14.13	12.39	11.72	13.93
Ca	1153	1317	1089	1217	1232
Sc	336	361	362	359	369
Ti	49	55	46	91	157
V	9.06	5.63	4.6	5.13	5.18
Rb	0.23	0.27	0.24	0.31	0.48
Sr	0.49	0.3	0.42	0.36	0.33
Y	29	28	24	22	29

续表

分析点编号	1	2	5	6	7
Nb	0.29	0.31	0.17	0.37	0.46
Ta	0.04	0.06	0.03	0.04	0.06
Cs	0.02	0.04	0.03	0.05	0.04
Ba	0.29	0.33	0.88	2.3	0.37
Hf	8410	9213	8921	9063	8448
P	78	79	82	76	84
Pb	0.31	0.37	0.30	0.41	0.34
年龄	224 Ma	227 Ma	218 Ma	235 Ma	232 Ma

表 2　青龙山榴辉岩(02QL-2)锆石 SHRIMP 分析数据 a)

编号	$^{206}Pb_c$/%	U $\mu g \cdot g^{-1}$	Th $\mu g \cdot g^{-1}$	Th/U	$^{206}Pb^*$ $\mu g \cdot g^{-1}$	$^{207}Pb^*/^{206}Pb^*$ (%)	$^{206}Pb^*/^{238}U$ (%)	$^{207}Pb^*/^{235}U$ (%)	年龄/Ma $^{206}Pb/^{238}U$
1	1.35	105	3	0.03	3.23	0.0488 (12)	0.0354 (1.5)	0.238 (12)	224.3 ± 3.4
2	1.11	106	5	0.05	3.30	0.0483 (11)	0.0358 (1.5)	0.239 (11)	226.8 ± 3.3
3	2.23	67	2	0.03	2.10	0.0597 (13)	0.0357 (1.6)	0.294 (13)	226.3 ± 4.1
4	3.12	82	2	0.03	2.63	0.0390 (17)	0.0363 (1.7)	0.195 (17)	229.8 ± 3.9
5	0.83	68	3	0.04	2.01	0.0535 (9.0)	0.0344 (1.7)	0.254 (9)	218.1 ± 3.6
6	0.70	92	3	0.03	2.96	0.0502 (6.4)	0.0371 (1.4)	0.257 (6.5)	235.0 ± 3.3
7	0.50	93	4	0.05	2.95	0.0653 (8.4)	0.0366 (1.6)	0.330 (8.6)	231.8 ± 3.5
8	0.00	86	5	0.06	2.66	0.0631 (7.0)	0.0362 (1.4)	0.315 (7)	229.0 ± 3.4
9	0.62	148	42	0.29	6.07	0.0529 (6.5)	0.0476 (1.3)	0.347 (6.6)	299.8 ± 3.7
10	2.38	63	4	0.07	1.94	0.0400 (28)	0.0349 (2.1)	0.191 (28)	221.2 ± 4.6
11	2.97	89	3	0.04	2.75	0.0475 (17)	0.0348 (2.0)	0.228 (17)	220.5 ± 4.3
12	1.27	141	4	0.03	4.47	0.0462 (9.9)	0.0366 (1.4)	0.233 (10)	231.5 ± 3.1

a) 误差为 1σ, Pb_c 和 Pb^* 分别为普通和放射成因 Pb, 标准的误差是 0.26%。普通 Pb 校正采用实际测量 ^{204}Pb 值。

4 讨论

单从本工作的青龙山榴辉岩锆石 CL 图像（图 2）来看，内部结构均一，不具核边结构及韵律环带特征，它或者是单纯的变质新生锆石，或者是完全变质重结晶锆石，确切判别成因还需要其他判据。

锆石的包裹体矿物主要为石榴石、金红石、绿辉石等矿物，此外，薄片观察发现，锆石主要作为高压矿物（石榴石、绿辉石等）包裹体形式存在。这些细粒矿物包裹体可能是榴辉岩相早期阶段和峰期超高压阶段的锆石结晶生长过程中捕获而成。

表 1 可见，所测五颗锆石的微量元素含量均一，稀土元素球粒陨石标准化配分模型一致(图 4)，说明这些锆石生长于相同的环境。Yb_n/Dy_n 值介于 1.8～3.2 之间，与阿尔卑斯榴辉岩中锆石的稀土元素配分模型（图 4）进行比较，该榴辉岩中锆石重稀土组分配分模型与阿尔卑斯榴辉岩中变质锆石一致，均较平坦，显示了与富重稀土的石榴石同期结晶的特点，弱或无 Eu 异常说明没有长石的存在，证明锆石的生成环境不是岩浆条件[33,34]。锆石的低 Nb、Ta 含量（分别为 0.17～0.46 μg/g、0.03～0.06 μg/g）证明它是与富 Nb、Ta 的金红石同期结晶的[14]。

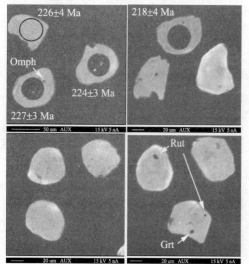

图 2　青龙山榴辉岩(02QL-2)典型锆石 CL 图像
锆石中黑圆坑为 LA-ICPMS 熔蚀坑，同时为 SHRIMP 分析点位

SHRIMP U-Pb 分析结果见表 2。9 号测试点年龄明显老于其他分析点，而在 CL 图像上该锆石与其他 11 个测试锆石颗粒的 CL 图像没有区别，都是均一无核为特征，但其高 Th/U 值(0.29)指示它与其他高压变质新生锆石(<0.07)有明显差别，它可能是未完全重结晶的原岩残留锆石。这进一步说明 CL 图像不能帮助判别区分变质重结晶锆石和变质新生锆石。其他 11 个分析点的 $^{206}Pb/^{238}U$ 年龄为 218~235 Ma，在一致曲线图中数据点成群分布，其加权平均值为 227.4±3.5 Ma（MSWD = 2.1，图 5）。锆石较低的 U 含量（<150 μg/g）暗示丢失放射成因 Pb 的可能性很小，因而加权平均年龄值年龄反映的是锆石平均生成时代。

据青龙山榴辉岩中锆石的包裹体及微量元素特征综合判定，该锆石与石榴石、金红石等高压矿物同期结晶。青龙山地区超高压岩石峰期变质温度可达 800 ℃，石榴石一般不能保留成分生长环带。但对超高压榴辉岩中大颗粒石榴石的详细矿物学研究表明，石榴石中心部位含有低压矿物包体成分，而边缘则包裹有金红石、绿辉石等高压矿物[49]。由于矿物的重结晶作用主要取决于温度，则超高压榴辉岩中石榴石应主要生长于温度峰期前的进变质过程中，因而与石榴石同期生长锆石的随机测定年龄值的统计平均可能高估了峰变质时代，但考虑到超高压岩石在温度峰期后的快速冷却[44]，它应非常接近峰期变质时代。超高压变质岩 Zr 的地球化学研究表明榴辉岩相前进变质

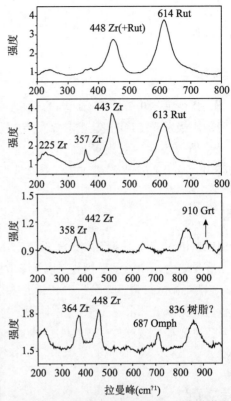

图 3 青龙山榴辉岩(02QL-2)锆石包裹体激光拉曼光谱图

Rut = 金红石; Grt = 石榴石; Zr = 锆石; Omph=绿辉

反应及压力升高导致的石榴子石 Zr 含量下降可释放 Zr 从而提供超高压变质锆石增生所需 Zr 的来源[50]，而降压过程中仅金红石退变为榍石的角闪岩相退变质反应可释放 Zr[1)]。因此，榴辉岩中超高压变质增生锆石很可能主要是在峰期前和峰期时形成的，其加权平均年龄值最接近超高压变质岩石的峰期变质时代。本文青龙山榴辉岩中与石榴石同期生长锆石的 SHRIMP 平均测定值为 227.4±3.5 Ma，与刘福来等报道的东海地区超高压片麻岩中含柯石英包裹体的锆石幔部年龄(229 ± 4 Ma)一致[15]，最近刘福来和许志琴[16]对超高压岩石中含柯石英及流体包裹体的变质增生锆石域定年结果为 233.7±4.3 Ma，也获得了与上述年龄一致的结果。采自同一岩体榴辉岩样品的三个高压矿物(石榴石+绿辉石+金红石)Sm-Nd 等时线年龄（226.3 ± 4.5 Ma）[37]与

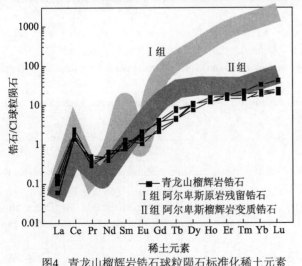

图 4 青龙山榴辉岩锆石球粒陨石标准化稀土元素配分模型（I, II 组数据取自文献[34]）

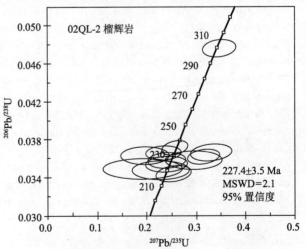

图 5 青龙山榴辉岩(02QL-2)锆石U-Pb年龄谱和图

---

1) 侯振辉，2004，中国科学技术大学博士论文

以上锆石年龄一致，它支持将达到高压变质平衡的三矿物 Sm-Nd 等时线年龄(226 Ma)视为大别-苏鲁榴辉岩超高压峰期变质时代的论点[43,44]。

5 结论

超高压变质岩中变质锆石成因的复杂性是人们对锆石 U-Pb 定年结果的解释产生较大争议的主要原因。综合应用阴极发光(CL)图像、锆石矿物包裹体研究及微量元素测定，可帮助我们判定变质锆石的成因。本工作的综合研究表明，所测青龙山榴辉岩锆石是在榴辉岩相变质阶段新生成的锆石，它的高精度离子探针（SHRIMP）定年结果平均值 227.4±3.5 Ma 较好地反映了青龙山榴辉岩的峰期变质时代。

致谢

SHRIMP 定年工作得到刘敦一教授工作组的悉心帮助，陶华老师担任了锆石的制靶工作，简平、宋彪研究员在锆石分析中给予了指导，张立飞教授、刘景波研究员、任景秋等对拉曼光谱工作给予了指导，激光 ICP-MS 锆石微区微量元素分析得到袁洪林的帮助，德国马普化学所 J. Huth 博士对锆石的 CL 图像拍摄提供了条件，本文承蒙郑永飞、简平和刘福来研究员审阅并提出宝贵修改意见，在此一并表示感谢。本工作得到国家重点基础研究发展规划项目(G1999075503)、国家自然科学基金(40173014)和中国博士后科学基金资助项目资助。

参考文献

[1] Gebauer D, Graunenfelder M. U-Pb zircon and Rb-Sr mineral dating of eclogites and their country rocks example: Munchberg gneiss massif, Northeast Bavaria. Earth Planet Sci Lett, 1979, 42: 35~44
[2] Peucat J J, Vidal P H, Godard G, et al. Precambrian U-Pb zircon ages in eclogites and garnet pyroxenites from South Brittany (France) : an old oceanic crust in the west European Hercynian belt? Earth Planet Sci Lett, 1982, 60: 70~78
[3] Pidgeon R T. Recrystallization of oscillatory-zoned zircon: some geochronological and petrological implications. Contributions to Mineralogy and Petrology, 1992, 110: 463~472
[4] Gebauer D, Schertl H P, Brix M, et al. 35 Ma old ultrahigh-pressure metamorphism and evidence for very rapid exhumation in the Dora Maira Massif, Western Alps. Lithos, 1997, 41: 5~24
[5] 李曙光, 李惠民, 陈移之, 等. 大别-苏鲁地体超高压变质年代学——II. 锆石 U-Pb 同位素体系. 中国科学(D), 1997, 27(3): 200~206
[6] Hacker B R, Ratschbacher L, Webb L, et al. U/Pb zircon ages constrain the architecture of the ultrahigh-pressure Qinling-Dabie Orogen, China. Earth Planet Sci Lett, 1998, 161: 215~230
[7] Vavara G, Schmid R, Gebauer D. Internal morphology, habit and U-Th-Pb microanalysis of amphibole to granulite facies zircon: geochronology of the Ivren Zones (Southern Alps). Contributions to Mineralogy and Petrology, 1999, 134: 380~404
[8] Rowley D B, Xue F, Tucker R D, et al. Ages of ultrahigh pressure metamorphic and protolith orthgenisses from the eastern Dabie Shan: U/Pb zircon geochronology. Earth Planet Sci Lett, 1997, 151: 191~203
[9] Rubatto D, Gebauer G, Compagnoni R. Dating of eclogite-facies zircons: the age of Alpine metamorphism in the Sesia-Lanzo Zone (Western Alps). Earth Planet Sci Lett, 1999, 167: 141~158
[10] Ames L, Tilton G R, Zhou G Z. Timing of collision of the Sino-Korean and Yangtze Cratons: U-Pb Zircon dating of coesite-bearing eclogites. Geology, 1993, 21: 339~342
[11] Ames L, Zhou G Z, Xiong B C. Geochronology and isotopic character of ultrahigh pressure metamorphism with implications for collision of the Sino-Korean and Yangtze cratons, central China. Tectonics, 1996, 15: 472~489
[12] Ayers J C, Dunkle S, Gao S, et al. Constraints on timing of peak and retrograde metamorphism in the Dabie Shan ultrahigh-pressure metamorphic belt, east-central China, using U-Th-Pb dating of zircon and monazite. Chem Geol, 2002, 186: 315~331
[13] Yang J S, Wooden J L, Wu C L, et al. SHRIMP U-Pb dating of coesite-bearing zircon from the ultrahigh-pressure metamorphic rocks, Sulu terrane, east China. Journal of Metamorphic Geology, 2003, 21: 551~560
[14] 吴元保, 陈道公, 夏群科, 等. 大别山黄镇榴辉岩锆石的微区微量元素分析：榴辉岩相变质锆石的微量元素特征. 科学通报, 2002, 47(11): 861~863

[15] 刘福来, 许志琴, 宋彪. 苏鲁地体超高压和退变质时代的厘定：来自片麻岩锆石微区 SHRIMP U-Pb 定年的证据. 地质学报, 2003, 77(2): 229~237

[16] 刘福来, 许志琴. 南苏鲁超高压岩石含柯石英锆石中的流体包裹体. 科学通报, 2004, 49(2): 181~189

[17] 刘敦一, 简平. 大别山双河硬玉石英岩的超高压变质和退变质事件——SHRIMP 测年的证据. 地质学报, 2004, 78(2): 211~217

[18] Mezger K, Krogstad E J. Interpretation of discordant U-Pb zircon ages: an evaluation. J Metamorphic Geol, 1997, 15: 127~140.

[19] Roberts M P and Finger F. Do U-Pb zircon ages from granulites reflect peak metamorphic conditions? Geology, 1997, 25: 319~322

[20] Fraser G, Eillis D, Eggins S M. Zirconium abundance in granulite-facies minerals, with implications for zircon geochronology in high-grade rocks. Geology, 1997, 25: 607~610

[21] Pan Y. Zircon- and monazite-forming metamorphic reactions at Manitouwadge, Ontario. Canadian Mineralogist, 1997, 35: 105~118

[22] Williams I S, Buick I S, Cartwright I. An extended episode of early Mesoproterozoic fluid in the Reynolds Range, central Australia. Journal of Metamorphic Geology, 1996, 14: 29~47

[23] Black L P, Williams I S, Compston W. Four zircon ages from one rock: the history of a 3930 Ma-old granulite from Mount Sones, Enderby Land, Antarctica. Contributions to Mineralogy and Petrology, 1986, 94: 427~437

[24] Friend C R L and Kinny P D. New evidence for protolith ages of Lewisian granulites, northwest Scotland. Geology, 1995, 23: 1027~1030

[25] Bowring S A and Williams I S. Priscoan (4.00–4.03 Ga) orthogneisses from northwest Canada. Contributions to Mineralogy and Petrology, 1999, 134: 3~16

[26] Hoskin P W O and Black L P. Metamorphic zircon formation by solid-state recrystallization of protolith igneous zircon. Journal of Metamorphic Geology, 2000, 18: 423~439

[27] Liu F L, Xu Z Q, Katayama I, et al. Mineral inclusions in zircons of para- and orthogneiss from pre-pilot drillhole CCSD-PP1, Chinese Continental Scientific Drilling Project. Lithos, 2001, 59: 199~215

[28] Ye K, Yao Y P, Cong B, et al. Areal extent of ultra-high pressure metamorphism in the Sulu terrane of east China: evidence from coesite inclusions in zircon from country rock granitic gneiss. Lithos, 2000, 52: 157~164

[29] Liu J B, Ye K, Maruyama S, et al. Mineral inclusions in zircon from gneisses in the ultrahigh pressure zone of the Dabie Mountains, China. J Geology, 2001, 109: 523~535

[30] Hermann J, Rubatto D, Korsakov A, et al. Multiple zircon growth during fast exhumation of diamondiferous deeply subducted continental crust (Kokchetav Massif, Kazakhstan). Contributions to Mineralogy and Petrology, 2001, 141: 66~82

[31] 陈道公, Deloule E, 程昊等. 大别－苏鲁变质岩锆石微区氧同位素特征初探：离子探针原位分析. 科学通报, 2003, 48(16): 1732~1739

[32] Katayama I, Maruyama S, Parkinson C D, et al. Ion micro-probe U-Pb zircon geochronology of peak and retrograde stages of ultrahigh-pressure metamorphic rocks from the Kokchetav massif, northern Kazakhstan. Earth and Planetary Science Letters, 2001, 188: 185~198

[33] Sun W D, Williams I, Li S G. Carboniferous and Triassic eclogites in the western Dabie Mountains, east-central China: evidence for protracted convergence of the North and South China Blocks. Journal of Metamorphic Geology, 2002, 20: 873~886

[34] Rubatto D. Zircon trace element geochemistry: partitioning with garnet and the link between U-Pb ages and metamorphism. Chemical Geology, 2002, 184: 123~138.

[35] Katayama I, Muko A, Lizuka T, et al. Dating of zircon from Ti-clinohumite-bearing garnet peridotite: implication for timing of mantle metasomatism. Geology, 2003, 31: 713~716

[36] 刘福来, 张泽明, 许志琴. 2003. 苏鲁地体超高压矿物的三维空间分布. 地质学报, 77: 69～84

[37] Li S G, Wang S S, Chen Y Z, et al. Excess argon in phengite from eclogite: evidence from dating of eclogite minerals by Sm-Nd, Rb-Sr and $^{40}Ar/^{39}Ar$ methods. Chem Geol, 1994. 112: 343~350

[38] Yui T F, Rumble D, Chen C H, et al. Stable isotope characteristics of eclogites from the ultra-high-pressure metamorphic terrain, east-central China. Chem. Geol, 1997, 137: 135~147

[39] Zheng Y F, Fu B, Li Y L, et al. Oxygen and hydrogen isotope geochemistry of ultrahigh pressure eclogites from the Dabie Mountains and the Sulu terrane. Earth Planet. Sci. Lett., 1998, 155: 113~129

[40] Rumble D and Yui T F. The Qinglongshan oxygen and hydrogen isotope anomaly near Donghai in Jiangsu province, China. Geochim Cosmochim Acta, 1998, 62: 3307~3321

[41] Zhang R Y, Hirajima T, Banno S, et al. Petrology of ultrahigh-pressure rocks from the southern Su-Lu region, eastern China. J Metamor Geol, 1995, 13: 659~675

[42] Li S G, Xiao Y L, Liou D, et al. Collision of the North China and Yangtze blocks and formation of coesite-eclogites: timing and

processes. Chem Geol, 1993, 109: 89~111

[43] Li S G, Jagoutz E, Chen Y Z, et al. Sm-Nd, Rb-Sr and ^{40}Ar-^{39}Ar isotopic systematics of the ultrahigh pressure metamorphic rocks in the Dabie-Sulu belt, Central China: a retrospective view. International Geol Review, 1999, 41(12): 1114~1124

[44] Li S G, Jagoutz E, Chen Y Z, et al. Sm-Nd and Rb-Sr isotopic chronology and cooling history of ultrahigh pressure metamorphic rocks and their country rocks at Shuanghe in the Dabie Mountains, Central China. Geochim Cosmochim Acta, 2000, 64 (6): 1077~1093

[45] Compston W, Williams I S, Kirschvink J L, et al. Zircon U-Pb ages for the Early Cambrian time-scale. J Geol Soc Lond, 1992, 149: 171~184

[46] Ludwig K R. Squid 1.02: A User Manual. Berkeley Geochronological Center Special Publication, 2:19

[47] Ludwig K R. Users Manual for Isoplot/Ex: A Geochronoligical Toolkit for Microsoft excel Berkeley Geochronology Center Special Publication, Berkeley, CA, USAA, 2000, 53

[48] Gao S, Liu X M, Yuan H L, et al. Determination of forty-two major and trace elements in USGS and NIST SRM glasses by laser ablation-inductively coupled plasma-mass spectrometry. The Journal of Geostandards and Geoanalysis, 2002, 26(2): 181~196

[49] Enami M, Nagasaki A, Prograde *P-T* path of kyanite eclogites from Junan in the Sulu ultrahigh-pressure province, eastern China. The Island Arc, 1999, 8: 459~474

[50] Donohue C L, Manning C E, Essene E J. The pressure and temperature dependence of Zr and Ti substitution in almandine. Abstract of GSA annual meeting, 2001

超高压变质岩的放射性同位素体系及年代学方法*

李曙光，侯振辉，李秋立

中国科学院壳幔物质与环境重点实验室，中国科学技术大学地球与空间科学学院，合肥 230026

> **亮点介绍：** 在超高压变质及退变质过程中，只有变质矿物同位素体系达到平衡才能给出精确有意义的 Sm-Nd、Rb-Sr 等时线年龄。高压变质、退变质增生锆石以及完全变质重结晶锆石 U-Pb 体系才能给出准确的变质时代。深刻理解这些同位素体系的地球化学行为对获得超高压变质岩准确并有明确意义的年龄值是非常重要的。

摘要　深刻理解同位素在超高压变质及退变质过程中的地球化学行为对获得超高压变质岩准确并有明确意义的年龄值是非常重要的。对 Sm-Nd、Rb-Sr 同位素体系，只有变质矿物同位素体系达到平衡才能给出精确有意义的等时线年龄。研究表明，与副变质岩互层的细粒榴辉岩的高压变质矿物之间，或者强退变质岩石的退变质矿物之间，其 Nd、Sr 同位素可以达到平衡；然而高压变质矿物与退变质矿物之间 Nd、Sr 同位素不平衡。由于全岩样品总是含有数量不等的退变质矿物，因此石榴石＋全岩 Sm-Nd 法或多硅白云母＋全岩 Rb-Sr 法将有可能给出无地质意义的年龄。通常低温榴辉岩的高压变质矿物之间存在 Nd 同位素不平衡。超高压变质岩多硅白云母所含过剩 Ar 主要源于榴辉岩原岩中角闪石在变质分解时释放出来的放射成因 Ar。因此，不含榴辉岩的花岗片麻岩多硅白云母基本不含过剩 Ar。对变质锆石成因的准确判断是正确理解锆石 U-Pb 年龄意义的关键。本文对不同成因锆石的判别标志及年龄意义做了总结，并指出将阴极发光图形、锆石痕量元素组成及矿物包裹体鉴定相结合是进行锆石成因鉴定的有效方法。高压变质或退变质增生锆石组成单一，是理想变质定年对象。然而变质重结晶锆石域常是重结晶锆石和继承晶质锆石的混合区，因而给出混合年龄。只有完全变质重结晶锆石才能给出准确变质时代。

关键词　同位素年代学；超高压变质；Sm-Nd 体系；Rb-Sr 体系；Ar-Ar 体系；锆石 U-Pb 定年

1 引言

在大陆深俯冲研究中超高压变质岩的年代学研究一直占有重要地位。精确测定超高压变质岩的峰期变质时代及其冷却年龄或退变质时代是判断大陆深俯冲的发生和结束时代，以及超高压变质岩折返历史的最直接手段。由于超高压变质岩的抬升和冷却相当快，仅当年龄误差小于±5 Ma 时才能约束其冷却史，这对同位素定年的精度提出了更高的要求。然而，超高压变质岩复杂的变质历史使得这一工作变得十分困难，并导致了部分测定结果的不一致或者年龄解释上的分歧。这主要是因为我们对超高压变质过程中各同位素体系的特殊行为缺乏了解所致。因此，查清与同位素定年有关的主要同位素体系在超高压变质及退变质过程中的行为对获得准确的有明确意义的年龄值是非常重要的。Sm-Nd 和 Rb-Sr 变质矿物等时线定年，锆石 U-Pb 定年和富 K 变质矿物的 $^{40}Ar/^{39}Ar$ 定年是超高压变质岩年代学研究中最常用的几种方法。本文将根据近年来的工作分别对它们在超高压变质及退变质条件下的同位素体系进行讨论。

2 Sm-Nd 同位素体系

Sm-Nd 矿物等时线法是测定榴辉岩及其他含石榴子石高级变质岩变质年龄的最有效方法之一。因为重稀土元素在石榴子石中比轻稀土元素有较高的分配系数，因而石榴石有很高的 Sm/Nd 值，而其他高压变质

---

* 本文发表在：岩石学报，2005，21：1229-1242

矿物，如绿辉石、金红石、多硅白云母等有较低的 Sm/Nd 值，从而使我们可以获得一条由石榴子石及其他高压变质矿物确定的高精度 Sm-Nd 等时线。但该等时线仅在它们之间在变质过程中达到 Nd 同位素平衡，且后来一直保持 Sm-Nd 体系封闭的条件下才是有意义的。

等时线年龄代表它们的 Sm-Nd 同位素体系封闭的时代。该年龄的地质意义取决于它们的 Sm-Nd 体系封闭温度与各阶段变质温度的对比关系，以及使该组矿物 Nd 同位素组成达到均一化的机制与变质过程的关系。因此，对超高压变质岩的 Sm-Nd 同位素体系，我们最关心的是如下 4 方面问题：

(1) 在高压或超高压变质条件下，一组共生的高压变质矿物的 Nd 同位素组成能否达到平衡或者均一化？

(2) 在退变质过程中，高压变质矿物的 Sm-Nd 体系能否保持封闭？如果不能保持封闭，它们可否在退变质条件下重新平衡？

(3) 在退变质阶段形成的一组退变质矿物可否达到 Nd 同位素平衡？它们与高压变质矿物之间是否存在 Nd 同位素不平衡及对定年的影响？

(4) 重要变质矿物的封闭温度。

这些问题引起了同位素年代学家的高度兴趣和深入研究。

2.1 超高压变质条件下的 Sm-Nd 同位素体系

2.1.1 低温榴辉岩的 Nd 同位素不平衡问题

Jagoutz（1988，1994）和 Thöni and Jagoutz（1992）最早在金伯利岩中的榴辉岩包体及某些阿尔卑斯低温榴辉岩中观察到石榴石和单斜辉石之间的 Nd 同位素不平衡，并指出这一不平衡往往导致石榴石＋单斜辉石 Sm-Nd 年龄偏低。导致年龄偏低的原因可以由图 1 说明。当辉长岩变质为榴辉岩时，榴辉岩中的绿辉石主要继承的辉长岩中单斜辉石的 Nd 同位素组成，因而它有较高的初始 $^{143}Nd/^{144}Nd$ 值；榴辉岩中的石榴子石主要继承了富 Al 斜长石的 Nd 同位素组成，因此它的初始 $^{143}Nd/^{144}Nd$ 较低。如果初始形成的绿辉石和石榴石在高压变质阶段不能通过扩散使其 Nd 同位素组成达到平衡，则其初始绿辉石与石榴石连线具有负斜率（见图 1），从而导致该连线在经过时间演化后所具有的斜率低于真正等时线斜率（其初始斜率为 0）并给出较年轻年龄（Jagoutz，1994）。已经发现的石榴石和单斜辉石之间存在 Nd 同位素不平衡现象的这些榴辉岩均不含柯石英，原岩为辉长岩的粗粒榴辉岩。由于变质温度较低或者由于辉长岩无含水矿物，从而导致高压变质阶段缺少粒间水，增加了矿物间同位素扩散平衡的困难。此外，它们较粗的颗粒直径提高了其封闭温度，也可能是导致 Nd 同位素不平衡的原因。由于这一问题的存在，低温榴辉岩的 Sm-Nd 同位素定年往往是不成功的，我们对大别山低温榴辉岩的 Sm-Nd 同位素定年均发现了这种同位素不平衡问题。早期文献发表的部分低温榴辉岩 Sm-Nd 年龄（如熊店，红安高桥）（简平等，1994）均被后来的锆石离子探针 U-Pb 定年证明是错误的（简平等，2000）。

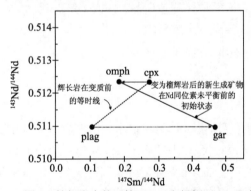

图 1 榴辉岩中的石榴石(gar)和绿辉石(omph)与原岩（辉长岩）中的单斜辉石(cpx)和斜长石(plag)的 Nd 同位素继承关系。在高压变质时，如果新生成的石榴石和绿辉石未达到 Nd 同位素平衡，它们的初始连线具有负斜率

2.1.2 超高压变质矿物的 Nd 同位素体系

含柯石英榴辉岩在超高压变质条件下（$P>27$ kbar，$T=700\sim900$ ℃）是否也存在石榴子石与绿辉石之间的 Nd 同位素不平衡？这对超高压变质岩的 Sm-Nd 同位素定年是一个至关重要的问题。为了判断高压变质矿物是否达到 Nd 同位素平衡，至少要分析 3 个高压变质矿物。Li et al.（1994，2000）报道了三个大别-苏鲁超高压变质岩石的三矿物 Sm-Nd 同位素等时线：大别山双河榴辉岩的石榴石＋绿辉石＋金红石等时线（$t=226.2\pm2.9$ Ma，MSWD=0.35）；大别山双河超高压片麻岩的石榴石＋2 个多硅白云母等时线（$t=226.5\pm2.8$ Ma，

MSWD=0.158）；苏北青龙山榴辉岩的石榴石＋绿辉石＋多硅白云母＋全岩等时线（t=226.3±4.5 Ma，MSWD=0.886）（见图 2）。三个矿物之间良好的线性关系（MSWD<1）以及它们年龄的高度一致性说明这些超高压变质矿物之间 Nd 同位素达到了平衡。这一类含柯石英榴辉岩及超高压片麻岩均为矿物均匀分布的细粒岩石。它们能在超高压变质条件下达到 Nd 同位素平衡，除因它们矿物颗粒较细外，其原岩性质可能也是重要因素。在上述定年样品中，超高压片麻岩的原岩为沉积岩，榴辉岩为与之互层的基性火山岩或斜长角闪岩。因此，在发生榴辉岩相变质时，角闪石分解可提供粒间水，从而有利于矿物间的 Nd 同位素扩散平衡。此外，超高压变质岩较高的温压变质条件也是加快 Nd 同位素扩散平衡的重要因素之一。

2.1.3 石榴石、绿辉石 Sm-Nd 封闭温度、平衡机制及超高压变质岩 Sm-Nd 年龄的意义

为了能正确理解上述已被证明在超高压变质时 Nd 同位素达到平衡的 Sm-Nd 年龄究竟是代表了超高压变质岩的峰期变质时代还是冷却年龄，我们需要对石榴石和绿辉石的 Sm-Nd 封闭温度以及它们实现 Nd 同位素平衡的机制进行更详细的讨论。

石榴石的 Sm-Nd 封闭温度是一个尚未很好测定并存在争议的问题。Humphries and Cliff（1982）依据 Sm 在石榴石中的扩散系数，计算了石榴石粒径为 1 mm、冷却速率为 10 ℃/Ma 条件下镁铝榴石的 Sm-Nd 封闭温度为 500 ℃，而钙铝榴石的封闭温度为 700 ℃。通过将石榴子石＋全岩 Sm-Nd 等时线年龄与同一麻粒岩地体精确测定的 T-t 冷却曲线进行对比，Mezger 等（1992）指出，在缓慢冷却条件下（2~4 ℃/Ma）石榴子石的 Sm-Nd 封闭温度为 600±30 ℃。然而其他一些证据却表明石榴石的 Sm-Nd 体系应有更高的封闭温度。依据对金伯利岩中榴辉岩包体 Sm-Nd 体系的研究，Jagoutz（1988）提出在干条件下石榴石的 Sm-Nd 封闭温度大于 850 ℃。Hensen and Zhou（1995）也指出在干的基性麻粒岩中石榴子石的 Sm-Nd 封闭温度要大于 700~750 ℃。事实上，任一给定元素在石榴石中的封闭温度取决于许多因素，包括矿物颗粒的大小、有无粒间流体、主要元素组成、与石榴石共生的矿物性质、初始温度及冷却速率等（Dodson，1973；Burton et al., 1995；Gangaly et al., 1998）。因此，Li et al.（2000）指出石榴石似乎并不具有固定的封闭温度，而是有一个 500~850 ℃ 的变化范围。对一特定样品，要估计它的封闭温度值则需要考虑上述所有因素。

依据在高压条件下 Sm 在透辉石的扩散系数（Sneeringer et al., 1984），我们可计算出，在矿物颗粒为 1 mm 和冷却速率为 10~40 ℃/Ma 条件下，单斜辉石的 Sm-Nd 封闭温度高达 1040~1090 ℃，这一封闭温度显然大大高于石榴石可能有的封闭温度（500~850 ℃）。Burton 等（1995）也观察到 Sm-Nd 在单斜辉石中的扩散速率低于在石榴石中的扩散速率。由于榴辉岩基本上是二矿物岩石，如果单斜辉石已对 Sm-Nd 扩散封闭了，则石榴石就不可能再与之进行同位素交换。考虑到大别山含柯石英榴辉岩的平均峰期变质温度在 800 ℃左右，它低于单斜辉石的封闭温度，则石榴石与绿辉石之间要通过扩散达到 Nd 同位素平衡是很困难的。因此我们不得不考虑超高压变质岩中超高压矿物之间实现 Nd 同位素平衡的其他方式。

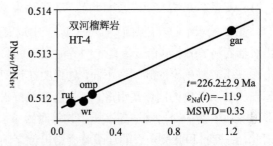
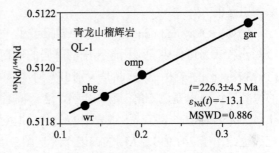
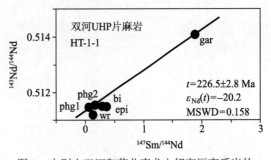

图 2　大别山双河和苏北青龙山超高压变质岩的 Sm-Nd 等时线图（据 Li et al., 1999）

gar-石榴石；omp-绿辉石；rut-金红石；phg-多硅白云母；epi-绿帘石；bi-黑云母；wr-全岩

Villa（1998）曾指出，在有流体参与的重结晶过程中同位素扩散速率较不存在重结晶作用的纯体积扩散要高出几个数量级。Li et al.（2000）在评述了各种可能的同位素平衡机制后也指出石榴石和绿辉石的重结

晶作用对它们之间的同位素平衡可能是一个重要因素。因此，上述 Sm-Nd 矿物等时线年龄（226±3 Ma）应指示重结晶作用的平均年龄。如果石榴石和绿辉石在整个高压和超高压变质阶段（包括榴辉岩相范围内前进变质和退变质阶段）都在通过重结晶连续增长，则重结晶的平均时代应当接近其峰期变质时代。

这一解释获得了 UHPM 岩石的其他同位素定年结果的支持。例如，Rowley et al.（1997）和 Ayers et al.（2002）报道的大别山毛屋榴辉岩的锆石 U-Pb 年龄分别为 225.5+2.9/−6.3 Ma 和 230±4 Ma。Hacker（1998）报道的大别山 UHPM 片麻岩锆石变质增生边的 SHRIMP 年龄的平均值也是 224±4 Ma 和 231±2 Ma。最近，刘福来等（2003）、李秋立等（2004）报道了苏北东海地区超高压片麻岩和榴辉岩高压变质增生锆石 SHRIMP 锆石 U-Pb 年龄分别为 229±4 Ma 和 227.4±3.5 Ma。这些高压变质锆石 U-Pb 年龄值均与上述 Sm-Nd 年龄一致。因此，上述 Sm-Nd 矿物等时线年龄 226±3 Ma 应当代表了大别山超高压变质岩的真实峰期变质时代。

2.2 超高压变质岩退变质过程中的 Sm-Nd 同位素体系

2.2.1 超高压矿物与退变质矿物的 Nd 同位素不平衡

图 2 显示了超高压片麻岩（HT-1-1）中退变质成因的黑云母和绿帘石以及含有退变质矿物的全岩均落在由石榴石＋2 个多硅白云母确定的等时线下面。与此类似，双河榴辉岩（HT-4）含退变质矿物的全岩也落在由超高压矿物确定的等时线下面。这说明在退变质过程中全岩体系开放了，新生成的退变质矿物具有不同于超高压变质矿物的同位素组成，并与仍然保持封闭的超高压变质矿物处于同位素不平衡状态。退变质矿物及含退变质矿物的全岩落在超高压矿物等时线下面说明，退变质流体引进的物质具有较低的 $^{143}Nd/^{144}Nd$ 值。因此，石榴石和此类全岩的连线就会给出偏老的年龄。在文献中一些偏老的石榴石＋全岩年龄可能都是这一原因引起的。

例如，Okay and Sengör（1993）曾报道了一个偏老的石榴石＋全岩的 Sm-Nd 年龄为 246±8 Ma，该值被一些作者认为可代表大别山超高压榴辉岩的峰期变质时代（Hacker et al., 1998）。然而，依据 Okay and Sengör（1993）自己的描述，该样品是大别山五河地区一个强烈退变质榴辉岩，它以团块状分布于大理岩中，所有的绿辉石均已退变质为角闪石。显然，这一偏老年龄是由该全岩含有大量退变质矿物造成的。类似地，Li et al.（2000）对于产于双河大理岩中的一个强退变质榴辉岩包体的 Sm-Nd 同位素分析表明，其石榴石＋全岩连线也给出了 241.9±3.2 Ma 的偏老年龄（图3）。此外，我们最早报道的饶钹寨石榴辉石岩的石榴石＋透辉石 Sm-Nd 等时线年龄（243.9±5.6 Ma）（Li et al., 1989），因现已证明该透辉石是绿辉石退变质产物（Xu et al., 2000）也是偏老的。

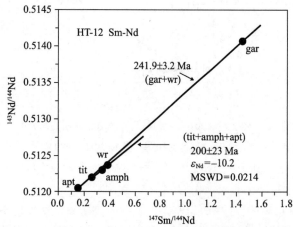

图 3 大别山双河退变质榴辉岩（HT-12）的 Sm-Nd 等时线图
（据 Li et al., 2000）
gar-石榴石；amph-角闪石；tit-榍石；apt-磷灰石；wr-全岩

退变质作用改变全岩 Nd 同位素组成的程度是因地而异的，它取决于榴辉岩围岩的性质及退变质流体的 Nd 同位素组成。上面的例子均是产于大理岩或橄榄岩中的榴辉岩。对于产于片麻岩中的一些退变质榴辉岩，其全岩 $^{143}Nd/^{144}Nd$ 值可下降很大，从而给出更加古老的石榴石＋全岩 Sm-Nd 年龄。例如，山东仰口片理化含柯石英榴辉岩（Y-12）是一强退变质榴辉岩，其绿辉石多已退变质为后成合晶，但仍然保持辉石假象。图 7 显示该样品的石榴石＋全岩给出了古生代 Sm-Nd 年龄（332.4±4.2 Ma），石榴石＋退变质绿辉石后成合晶也给出了 338±4.4 Ma 的 Sm-Nd 年龄。当用 HCl 洗去全岩中的可溶组成，石榴石＋全岩竟然给出了 440±6.4 Ma 的 Sm-Nd 年龄（李曙光等，2000）。显然这些年龄值毫无地质意义。

此外，被当做大别山存在古生代超高压变质作用证据的英山蜜蜂尖榴辉岩 Sm-Nd 年龄（481±25 Ma）（杨巍然等，1994）也是由石榴石＋强退变全岩给出的。该榴辉岩样品的绿辉石也全部退变质成为后成合晶。因此，应用石榴石+全岩来获取高压或者超高压变质年龄是十分危险的，它可能误导我们获得错误的结论。

2.2.2 同期退变质矿物的 Nd 同位素平衡

我们可否对同一退变质阶段形成的一组退变质矿物利用 Sm-Nd 法获得其退变质时代呢？这一定年成功的前提是该组退变质矿物的初始 Nd 同位素组成是均一的，亦即在它们形成或 Sm-Nd 同位素体系封闭时，其 Nd 同位素组成是达到平衡的。Li et al.（2000）在分析大别山双河强退变质榴辉岩（HT-12）时对这一问题做了较详细的检验。前已述及，样品 HT-12 是一强退变质榴辉岩，所有的退变质后成合晶均已在角闪岩相变质条件下重结晶为粒度较粗的角闪石、榍石（由金红石退变质形成）、斜长石、石英以及磷灰石。对退变质成因的角闪石、榍石和磷灰石做 Sm-Nd 同位素分析获得了线性很好的三矿物 Sm-Nd 等时线（MSWD=0.0214），并给出角闪岩相退变质重结晶时代为 200±23 Ma（图 3）。这说明经过退变重结晶，一组共生的角闪岩相退变质矿物的 Nd 同位素组成可以达到平衡。但如前所述，它们与高压矿物石榴石 Nd 同位素不平衡，这说明角闪岩相退变质重结晶作用对高压矿物的 Sm-Nd 体系不产生影响。

3 Rb-Sr 同位素体系

多硅白云母、角闪石、黑云母是超高压变质岩中常见的高压变质或退变质矿物，它们有较高的 Rb/Sr 值，因而很适合 Rb-Sr 同位素定年。多硅白云母的 Rb-Sr 封闭温度为 500 ℃（Cliff, 1993），而黑云母为 300 ℃（Dodson, 1973），因此它们的 Rb-Sr 年龄对揭示超高压变质岩冷却史有重要意义。尤其在发现超高压变质岩中这类矿物含有过剩 Ar 后（Li et al., 1994），Rb-Sr 法就成为超高压变质岩冷却史研究中一种广为应用的重要方法。然而对超高压变质岩而言，它们的 Rb-Sr 定年远比想象的要复杂，尤其是多硅白云母的 Rb-Sr 定年给出了较大的变化范围（224~194 Ma）（如 Chavagnac and Jahn, 1996）。为了获得准确的冷却年龄，我们有必要对这些矿物的 Rb-Sr 同位素体系在超高压及退变质条件下的行为做深入的了解和讨论。

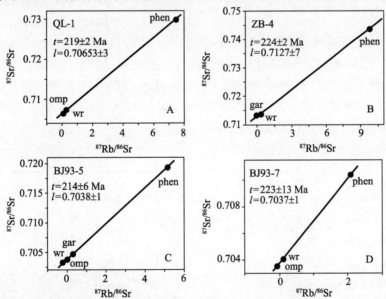

图 4 大别-苏鲁超高压变质带榴辉岩的多硅白云母（phen）+石榴石（gar）或绿辉石（omp）+全岩（wr）Rb-Sr 等时线图
A-苏北青龙山榴辉岩(Li et al., 1994)；B-胶南株边榴辉岩（Li et al., 1994）；C 和 D-大别山碧溪岭榴辉岩（Chavagnac and Jahn, 1996）

3.1 多硅白云母的 Rb-Sr 体系

3.1.1 多硅白云母与超高压变质矿物的 Sr 同位素平衡

在超高压变质岩中，多硅白云母是在榴辉岩阶段形成的高压变质矿物，它与石榴石、绿辉石为高压变质阶段的共生矿物。已知单斜辉石和石榴石 Sr 的封闭温度与 Sm-Nd 类似（Sneeringer et al., 1984; Burton et al., 1995）。因此，有理由假设在超高压及高压变质和退变质阶段，这些高压矿物的 Rb-Sr 同位素体系与 Sm-Nd 同位素类似，即在高压和超高压变质阶段，它们的 Sr 同位素组成可以因重结晶作用达到平衡，并在退变质

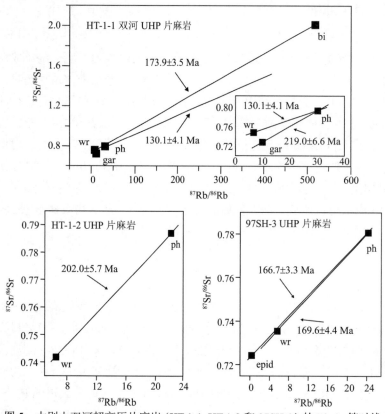

图 5 大别山双河超高压片麻岩（HT-1-1, HT-1-2 和 97SH-3）的 Rb-Sr 等时线图（据 Li et al., 2000）

gar-石榴石; ph-多硅白云母; bi-黑云母; epid-绿帘石; wr-全岩

阶段保持 Rb-Sr 同位素体系封闭。由于多硅白云母的 Sr 封闭温度（500 ℃）低于单斜辉石和石榴石的 Sr 封闭温度，对于缓慢冷却岩石来说，多硅白云母与石榴石或单斜辉石的连线将会给出较大的年龄误差，但考虑到超高压变质岩在 500 ℃以上高温阶段的快速冷却（≥ 40 ℃/Ma）(Li et al., 2000)，多硅白云母的封闭时代与石榴石、绿辉石的重结晶平均时代不会相差太多（≤ 5 Ma），加之单斜辉石和石榴石的 Rb/Sr 值很低，在如此短的时间范围内其 Sr 同位素组成不会有大的变化，因此多硅白云母与高压变质矿物（如石榴子石和绿辉石）还是可以落在同一等时线上。由于多硅白云母的 Rb/Sr 大大高于其他高压矿物的 Rb/Sr，该等时线斜率主要由多硅白云母控制，因而它所给出的年龄应该代表多硅白云母 Rb-Sr 体系封闭时的冷却年龄。这一点已被多硅白云母+高压变质矿物连线给出年龄的一致性验证。例如 Chavagnac and Jahn（1996）报道的大别山碧溪岭榴辉岩（BJ93-5，BJ93-7）的多硅白云母+石榴石（或+绿辉石）+全岩 Rb-Sr 等时线年龄分别为 214±6 Ma 和 223±13 Ma（见图 4），Li et al. (1994, 2000) 报道的苏北青龙山榴辉岩（QL-1）、胶南朱边榴辉岩（ZB-4）和大别山双河高压片麻岩（HT-1-1）的多硅白云母+石榴石（或+绿辉石+全岩）Rb-Sr 等时线年龄分别为 219±2 Ma，224±2 Ma 和 219.0±6.6 Ma（图 4 和图 5）。它们均在误差范围内一致，其平均年龄为 219±5 Ma（Li et al., 1999, 2000）。

3.1.2 多硅白云母与退变质矿物的 Sr 同位素不平衡

文献中也报道了大量的多硅白云母+全岩（或+退变质矿物）Rb-Sr 年龄，它们一般给出较上述多硅白云母+高压矿物 Rb-Sr 年龄年轻的年龄。如碧溪岭榴辉岩多硅白云母+全岩 Rb-Sr 年龄为 198±4 Ma～212±5 Ma（Chavagnac and Jahn, 1996），而碧溪岭高压片麻岩的多硅白云母+全岩+退变质矿物（绿帘石，磷灰石）仅给出 194±4 Ma～198±5 Ma，它们低于相同矿物的 Ar-Ar 年龄 205～212 Ma（Chavagnac et al., 2001）。Li et al.（2000）多次测定双河超高压片麻岩的多硅白云母+全岩 Rb-Sr 年龄获得较年轻且变化很大的结果，它们分别

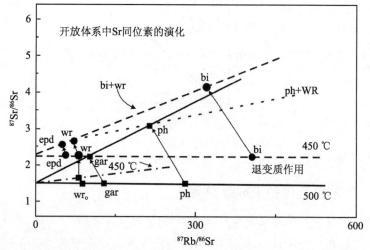

图 6 超高压变质矿物（多硅白云母 ph+石榴石 gar）和退变质矿物（绿帘石 epd+黑云母 bi）以及全岩（wr）在开放体系中的 Rb-Sr 同位素体系演化图。wr_0 为退变质前全岩初始 Sr 同位素组成；wr 为退变质全岩的 Sr 同位素组成（据 Li et al., 2000）

为 202±5.7 Ma, 169.6±4.4 Ma 和 130.1±4.1 Ma(图5)。其中样品 HT-1-1 的多硅白云母＋全岩Rb-Sr 年龄（130.1±4.1 Ma）甚至大大小于同一样品黑云母＋全岩的 Rb-Sr 年龄（173.9±3.5 Ma）。这是一个完全反常的结果，因为多硅白云母的Rb-Sr 封闭温度（500 ℃）高于黑云母 Rb-Sr 封闭温度（300 ℃）约 200 ℃，它理应要有比黑云母 Rb-Sr 年龄更老的年龄。对于这一反常情况，Li et al.（2000）指出这是由于全岩中含有大量的退变质矿物，在退变质过程中为开放体系，而且退变质矿物与多硅白云母处于 Sr 同位素不平衡的结果。

图 6 为超高压变质岩各种矿物 Rb-Sr 同位素的演化模式图，可清楚解释上述反常现象。在该图中，多硅白云母和石榴石的 Rb-Sr 体系在退变质过程中保持封闭，其 $^{87}Sr/^{86}Sr$ 值自500 ℃封闭以来持续增长至今。由它们所确定的等时线（图 4-7 中的黑实线）斜率可计算出多硅白云母 500 ℃ 的冷却年龄。然而退变质矿物黑云母、绿帘石以及含此类矿物的全岩（其 Sr 同位素组成主要由富含 Sr 的绿帘石确定）的 Sr 同位素组成则在退变质开放体系中被改造，从而具有较高的 $^{87}Sr/^{86}Sr$ 初始比值（见图 4-6 对应 450 ℃时的黑虚线）。因而它们形成不同于高压矿物的另一个 Rb-Sr 演化体系（如图 6 的黑虚线所示）。由退变质矿物黑云母＋绿帘石＋全岩所确定的等时线可正确地给出黑云母的 Rb-Sr 冷却年龄。然而图 6 显示，多硅白云母与全岩的连线的斜率有可能小于黑云母＋绿帘石＋全岩等时线的斜率，从而给出异常小的年龄。因此，我们不推荐用多硅白云母＋全岩（或＋退变质矿物）测定 Rb-Sr 年龄，而建议采用多硅白云母＋共生高压矿物测定 Rb-Sr 年龄。

3.1.3 晚期（或二次）多硅白云母与高压矿物的 Sr-Nd 同位素不平衡

值得注意的是多硅白云母并非都形成于超高压岩石的峰期变质阶段，一些沿晚期面理发育的多硅白云母显然是在超高压岩石构造抬升过程中形成的。已观察到，这种退变质阶段形成的晚期多硅白云母与超高压变质阶段形成的石榴石、绿帘石存在 Sr-Nd 同位素不平衡。图 7 显示了采自山东青岛仰口面理化含多硅白云母榴辉岩的 Sm-Nd、Rb-Sr 同位素测定结果。其中沿面理发育的晚期多硅白云母与石榴石的连线给出不合理的 Sm-Nd 老年龄（638±16 Ma）（图7a），证明它们之间的 Nd 同位素不平衡。该榴辉岩的绿辉石均已退变质成为具有辉石假象的后成合晶（aom）。它以及全岩（w）与石榴石的连线同样给出了不合理的 Sm-Nd 老年龄（338±4.4 Ma 和 332.4±

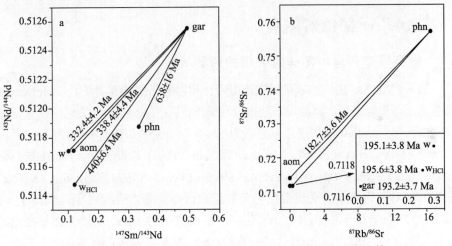

图 7　山东仰口榴辉岩的 Sm-Nd（a）和 Rb-Sr（b）等时线图（据李曙光等，2000）
gar-石榴石；w-全岩；w_{HCl}-酸性全岩；aom-退变质绿辉石；phn-多硅白云母

4.2 Ma）。这说明石榴石与退变质后成合晶之间也存在 Nd 同位素不平衡。由于石榴石的 Sr 封闭温度（750 ℃）略高于其 Nd 的封闭温度（700 ℃）（Burton et al.，1995），我们有理由相信该晚期多硅白云母与石榴石之间存在 Sr 同位素不平衡。因此，图 7b 中多硅白云母与石榴子石连线所给出的年龄 193.2±39 Ma 是无意义的。图 7b 还显示，该晚期多硅白云母与全岩（w）或酸洗过的全岩（w_{HCl}）样品连线也给出与石榴石连线类似的 Sm-Nd 年龄（195.1±3.8 Ma 和 195.6±3.8 Ma）。该全岩含大量石榴石，显然这些年龄也是没有意义的。由于石榴石比退变质产物有较低的 $^{87}Sr/^{86}Sr$ 值，因此 193 Ma 比多硅白云母的实际年龄要老。图 7b 显示多硅白云母与退变质后成合晶连线给出较年轻的 Rb-Sr 年龄（182.7±3.6 Ma）。由于它们都是退变质产物，该年龄有可能更接近多硅白云母的真实年龄。

3.2 退变质黑云母和角闪石的 Rb-Sr 同位素体系

在超高压变质岩中，黑云母是退变质矿物，因此，只有在退变质过程中全岩 Sr 同位素体系被重置，因而黑云母或角闪石与其他共生退变质矿物（如绿帘石）存在 Sr 同位素平衡时（如图 6 所示），我们才可以获

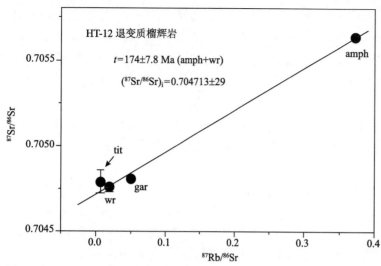

图 8　大别山双河退变质榴辉岩（HT-12）Rb-Sr 等时线图（据 Li et al., 2000）
gar-石榴石；amph-角闪石；tit-榍石；wr-全岩

得有意义的冷却年龄。显然，强退变质的岩石可以满足这一条件，尽管在退变岩石中石榴石等残留的高压矿物的 Rb-Sr 体系可能未被重置，但由于它含量少，又由于黑云母非常高的 Rb/Sr 值，这少量的残留高压矿物对黑云母＋全岩 Rb-Sr 年龄所产生的影响是较小的。大别山大量的黑云母＋退变质全岩的 Rb-Sr 年龄的一致性可证明这一点（见表 1）。表 1 显示对同一超高压变质岩露头区（如双河）的黑云母 Rb-Sr 年龄有约 10 Ma 的变化范围，这一年龄变化范围可能是全岩中高压矿物（如石榴石）含量不同造成的。如 HT-6 退变质较弱，含较多的高压矿物，因而使全岩的 $^{87}Sr/^{86}Sr$ 值较低，从而产生较高的黑云母＋全岩 Rb-Sr 年龄。因此，双河地区超高压变质岩黑云母 Rb-Sr 年龄的最佳估计值为 170～173 Ma。

退变质角闪石的 Rb-Sr 定年测定较少，Li et al. (2000) 报道了双河一强退变质榴辉岩（HT-12）的角闪石＋全岩 Rb-Sr 年龄为 174±7.8 Ma（图 8），它与该黑云母的 Rb-Sr 年龄是一致的。

4　^{40}Ar-^{39}Ar 同位素体系

4.1　超高压变质岩中多硅白云母的过剩 Ar

通常 $^{40}Ar/^{39}Ar$ 定年技术是研究造山带变质岩热历史的有力工具（McDougall and Harrison，1988）。然而对超高压变质岩来说，我们很难获得有意义的 $^{40}Ar/^{39}Ar$ 年龄。大量年代学研究结果表明，超高压变质岩的多硅白云母 $^{40}Ar/^{39}Ar$ 年龄老于该岩石的 Sm-Nd，Rb-Sr 和锆石 U-Pb 年龄，并因此引起关于超高压变质时代的争议（Mattauer et al., 1991；Monie and Chopin, 1991；李曙光等，1989a, b, 1992；陈移之等，1992；Li et al., 1993；Ames et al., 1993；Tilton et al., 1991）。Li et al.（1994）通过应用 Sm-Nd，Rb-Sr，$^{40}Ar/^{39}Ar$ 三种不同定年方法对苏鲁超高压变质带两个含多硅白云母榴辉岩进行定年，获得了一致的 Sm-Nd 和 Rb-Sr 年龄（228~219 Ma）和非常古老的 $^{40}Ar/^{39}Ar$ 年龄（877~943 Ma），并据此最早提出超高压榴辉岩的多硅白云母含大量过剩 Ar，不适合做 Ar 年代学研究的结论。此后，众多的研究证明，在大别山、西阿尔卑斯、挪威西部片麻岩区的超高压变质岩中，多硅白云母均含过剩 Ar（Arnaud and Kelly, 1995；Hacker and Wang, 1995；Ruffet

表 1　大别山超高压变质岩黑云母+全岩 Rb-Sr 年龄

地点	岩石	样品	年龄/Ma	数据来源
双河	UHP 片麻岩	HT-1-1	173.9±3.5	Li et al., 2000
	UHP 片麻岩	H-6	169.2±3.3	Li et al., 2000
	UHP 片麻岩	HT-6	181.2±3.5	Li et al., 2000
	花岗片麻岩	Sh-1	170.8±3.4	Li et al., 2000
	花岗片麻岩	Sh-2	173.2±3.4	Li et al., 2000
	花岗片麻岩	—	181	Jahn et al., 1994
碧溪岭	榴辉岩	BJ93-2	179±4	Chavagnac and Jahn, 1996
	花岗片麻岩	CF96-21a	170±4	Chavagnac et al., 2001
	花岗片麻岩	CF96-21d	187±4	Chavagnac et al., 2001
	花岗片麻岩	BJ95-08	180±4	Chavagnac et al., 2001

et al., 1995, 1997; Scaillet, 1996; Boundy et al., 1997; 王松山等, 1999; Giorgis et al., 2000)。尽管超高压变质岩中多硅白云母含大量过剩 Ar 已成为国际学术界的共识, 但有关过剩 Ar 成因的许多问题尚不很清楚。人们感兴趣的问题主要集中在三方面: ①过剩 Ar 在多硅白云母中的存在状态; ②过剩 Ar 的来源; ③在超高压地体中过剩 Ar 分布的不均匀性以及是否存在不含过剩 Ar 的岩石, 从而使我们仍然能获得有意义的 $^{40}Ar/^{39}Ar$ 年龄信息。这些方面的研究已获得一些初步进展。

4.2 多硅白云母过剩 Ar 的成因

许多研究显示尽管超高压变质岩的多硅白云母含大量过剩 Ar, 但多数都能给出很平坦的年龄谱, 它表明过剩 Ar 在矿物颗粒内部是均匀状态分布 (Li et al., 1994; Ruffet et al., 1995; Scaillet, 1996; Sherlock et al., 1999; Giorgis et al., 2000), 并说明过剩 Ar 是在矿物 Ar 同位素体系封闭以前进入到矿物中去, 且存在于 K 离子空位或充填于晶体缺陷中 (Ruffet et al., 1995; Scaillet, 1996, 1998; 王松山, 1999)。

关于过剩 Ar 的来源曾存在一些分歧, Ruffet et al. (1995) 提出过剩 Ar 可能来自邻近围岩及矿物在变质过程中通过去气作用释放的富 $^{40}Ar/^{39}Ar$ 流体; 而 Scaillet (1996) 认为过剩 Ar 来自榴辉岩原岩自身, 并依据矿物颗粒间 ^{40}Ar 分布的不均一性指出, 在超高压条件下含过剩 Ar 流体的局部循环尺度是非常有限的。为解决这一争议, Li et al. (1999) 注意到了过剩 Ar 在超高压岩石中分布的不均一性。图 9 显示了大别-苏鲁造山带各类超高压岩石中云母 $^{40}Ar/^{39}Ar$ 表面年龄具有规律性的变化。其中榴辉岩的云母的 $^{40}Ar/^{39}Ar$ 表面年龄最高, 因而过剩 Ar 含量最高; 其次为与榴辉岩呈薄互层的正片麻岩 (原岩为中酸性火山岩); 与榴辉岩互层的厚层超高压副片麻岩云母过剩 Ar 含量最低; 而作为超高压岩片的围岩, 花岗质片麻岩中的云母则不含过剩 Ar。这说明榴辉岩中多硅白云母大量的过剩 Ar 不是来自围岩, 而是从自身所在的原岩中继承的。此外, 图 9 还表明 $^{40}Ar/^{39}Ar$ 定年方法对超高压变质带的花岗质片麻岩围岩还是适用的。

在超高压变质带中, 榴辉岩、副片麻岩和花岗片麻岩均经历了超高压变质作用, 为什么它们的过剩 Ar 会有如此大的差异呢? 对此, 李曙光和侯振辉 (2001) 给出了一个解释, 即长英质花岗岩的主要含 K 矿物为钾长石、黑云母和白云母, 它们的 Ar 封闭温度低。白云母 Ar 封闭温度为 350 ℃, 黑云母仅有 300 ℃ (Dodson, 1973; McDougall and Harrison, 1998), 而钾长石虽具有多重扩散域及不确定的封闭温度, 但不同扩散域的封闭温度变化范围也在 350~150 ℃ (陈文寄, 李齐, 1999)。因此, 当陆壳俯冲达到蓝片岩相变质阶段时, 这些含 K 矿物的 Ar 同位素体系就开放了, 这些矿物所积累的放射成因 ^{40}Ar 会释放出来。这些释放出来的 ^{40}Ar 会被俯冲陆壳析出流体带走, 从而使得该花岗岩在经历超高压变质形成花岗片麻岩时其环境的 ^{40}Ar 分压很低。在这种高压环境下形成的多硅白云母不含过剩 Ar。与花岗质岩石相反, 榴辉岩的原岩主要为玄武质岩石。它们在变质为榴辉岩以前为斜长角闪岩, 其主要含 K 矿物为角闪石。它的 Ar 封闭温度在 500 ℃左右 (Harrison, 1981)。也就是说榴辉岩原岩中角闪石所积累的放射成因 ^{40}Ar 要在变质温度>500 ℃才开始释放出来。据对大别山红安地体低温榴辉岩的研究, 其变质温度为 500 ℃左右 (Zhou et al., 1993)。因此, 这些角闪石积累的放射成因 ^{40}Ar 是在开始发生榴辉岩变质时才释放出来, 而此时俯冲陆壳大规模脱水作用 (绿泥石分解) 早已停止, 少量基性岩中角闪石分解所产生的游离态水仅能在小范围环流 (李曙光和侯振辉, 2001; Ernst, 2000)。因此在榴辉岩变质阶段, 角闪石分解释放出来的 ^{40}Ar 将不会散失掉, 并会大大提高榴辉岩中的 ^{40}Ar 分压, 使得此时形成的多硅白云母能含有较高的过剩 Ar。那些与榴辉岩成薄互层的片麻岩尽管也是长英质岩石, 在超高压变质时自身没有角闪石释放 ^{40}Ar, 但因其与榴辉岩互层, 受榴辉岩层高 ^{40}Ar 分压的扩散及小尺度流体环流的影响, 其 ^{40}Ar 分压也会有不同程度的提高, 从而使其高压变质成因的多硅白云母也含有过剩 Ar, 但其含量将低于榴辉岩中的多硅白云母。

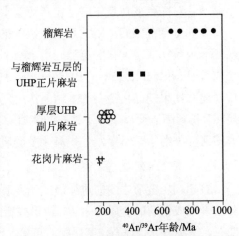

图 9 大别-苏鲁带各种不同类型超高压变质岩的云母 $^{40}Ar/^{39}Ar$ 年龄与岩性相关图(据 Li et al., 1999)

综上所述，控制不同类型的超高压变质岩中多硅白云母过剩 Ar 含量的主要因素是原岩的岩性，原岩中含 K 矿物的种类及 Ar 封闭温度，以及与榴辉岩的空间关系。

5 锆石 U-Pb 同位素体系

5.1 超高压变质岩中锆石的成因类型及可提供的年龄信息

超高压变质岩中的锆石按成因可分为以下三种类型。

5.1.1 继承锆石

它们是超高压变质岩原岩中所含的锆石。如果原岩是岩浆岩，这些继承锆石的主体是与岩浆同时结晶的岩浆成因锆石，此外还有少量从围岩中捕虏的锆石。如果原岩是沉积岩，则这些继承锆石为经过搬运、沉积的碎屑锆石。如果没有蜕晶化，该晶质锆石可有很高的 U-Pb 封闭温度。尽管至今实验研究尚不能提供可靠的 Pb 在锆石中的扩散参数，但是通过对各类岩石的锆石实际定年结果分析，我们可以获得这样一个信息：晶质锆石对 Pb 有很高的保存能力，即使在很高的温度条件下，也不可能使 Pb 完全通过扩散丢失。例如，许多花岗岩中都有继承锆石，花岗岩的岩浆温度可高达 700~900 ℃，然而经历了这一岩浆作用的继承锆石的 U-Pb 体系并未被重置，仍保留有它自己古老的年龄信息。再如北哈萨克斯坦的含金刚石超高压变质岩峰期变质温度高达 950~1000 ℃，峰期变质年龄为 530±7 Ma。它们当中的继承锆石仍保留了它们原岩的年龄（2000 Ma）信息，并落在 U-Pb 年龄谐和线上（Claoue-Long et al., 1991）。这说明结晶质锆石在高温条件下很少有 Pb 丢失。因此，如果超高压变质岩的原岩为火成岩，它的继承锆石的 U-Pb 定年可以获得其原岩年龄。

5.1.2 变质重结晶锆石

如果超高压变质岩从原岩中继承下来的锆石，由于长时间放射性衰变所造成的晶体损伤而发生蜕晶化，这种蜕晶化锆石在 $T>600$ ℃时可发生重结晶（Mezger and Krogstad, 1997）。在重结晶过程中，蜕晶化锆石中所保留的放射成因 Pb 因在锆石晶格中没有合适的位置可在重结晶作用过程中随孔隙流体一起排除。因此重结晶的蜕晶化锆石可重新开始 U-Pb 同位素计时。如果我们用 SHRIMP 对完全变质重结晶锆石域进行定年，我们可以获得其变质年龄。应当指出的是，如果一个继承锆石颗粒并未全部蜕晶化，蜕晶化锆石呈弥散状分布，则所分析的变质重结晶锆石域是变质重结晶锆石和继承晶质锆石的混合区，我们称它为部分变质重结晶锆石域。它给出的 U-Pb 年龄是混合年龄值，且会在 U-Pb 年龄谐和图上形成一条不一致线。它是由继承锆石和重结晶锆石确定的混合线。因此，上交点为继承锆石年龄，下交点为变质重结晶锆石年龄。如果继承锆石的某些部位 U 含量很高，因而蜕晶化很强，在这些部位已无残留继承晶质锆石，它可以形成完全变质重结晶域。对完全变质重结晶域定年可获得谐和的变质年龄。

变质重结晶锆石年龄的意义取决于锆石变质重结晶作用完成的时代。如果蜕晶化锆石的重结晶作用在峰期变质阶段完成，则它的年龄应指示峰期变质时代；如果蜕晶化锆石的重结晶作用在 $T=600~650℃$ 时迅速完成（Alines and Rossman, 1985; Farges, 1994），它的年龄应指示榴辉岩相前进变质时代，略老于峰期变质时代；然而如果蜕晶化锆石的重结晶作用延迟到石英榴辉岩阶段才完成，它的年龄也可能晚于峰期变质时代。显然，在超高压变质条件下锆石的变质重结晶作用速率对解释其年龄意义是非常重要的。变质重结晶锆石形成的前提是继承锆石要发生蜕晶化及变质温度要高于 600 ℃。

5.1.3 变质增生锆石

在超高压变质岩的前进变质或退变质反应中，原生矿物中的微量 Zr 会被释放出来，并增生到已有的锆石上去形成锆石的变质增生边，或形成新的变质锆石。在这种变质锆石中发现了许多高压矿物包裹体，如石榴石、绿辉石、金红石，甚至柯石英等。这足以证明这类锆石是在超高压变质过程中生长的。已有的研究表明榴辉岩的高压变质锆石增生所需的 Zr 的来源主要为原岩辉长岩中辉石和原生榍石的进变质分解反应以及因压力升高而从石榴石中释放出的 Zr，而其峰期高压变质矿物石榴石、绿辉石等几乎不含 Zr，这表明在峰期之后的退变质过程的高压阶段已没有足够的 Zr 源使锆石发生变质增生。因此高压变质锆石的增生主要发生在峰期之前的前进高压变质过程中,其 SHRIMP U-Pb 年龄值应略老于压力峰期时代(侯振辉,2003)。

高级变质岩各种变质矿物 Zr 的地球化学研究表明,变质锆石的增生也可以在退变质过程中发生。如 Fraser et al. (1997) 发现,麻粒岩的石榴石和角闪石是重要含 Zr 矿物(Zr 含量达 $n\times10$ ppm),在退变质过程中,它们分解生成的退变质矿物(斜长石、斜方辉石、黑云母、堇青石)的 Zr 含量均很低。考虑到 Zr 的地球化学的不活动性,他们认为石榴石和角闪石在退变质过程中释放的 Zr 会形成变质增生锆石。这种变质增生锆石的 U-Pb 年龄仅指示退变质分解反应发生的时代。

在通常情况下,由于锆石颗粒中多会有继承锆石核,而变质增生锆石总是表现为不规则的变质增生边形式。因此,用常规的热电离质谱(TIMS)方法所获得的单颗粒或多颗粒锆石数据多形成一条由继承锆石核和变质增生锆石边确定的混合线(不一致线),它与谐和线的上交点给出继承锆石的年龄,它与谐和线的下交点给出变质增生锆石的年龄。此外,也正因为超高压变质岩中锆石的这种变质增生边+继承核结构使得单颗粒锆石逐层蒸发技术不适用于变质岩定年。因为这种方法所获得高温段年龄基本指示的是古老继承核的年龄或它与变质锆石的混合年龄。

5.2 变质锆石与岩浆锆石的区分标志

上述讨论表明,变质成因锆石(包括变质重结晶锆石和变质增生锆石)和继承的岩浆锆石的年龄意义不同。因此,正确判断锆石的成因是能否对超高压变质岩锆石 U-Pb 年龄地质意义给出正确解释的关键。我们可以从以下四个方面特征进行区分。

5.2.1 晶形特征

岩浆锆石均有很好的柱状晶体,其晶棱锋锐,清晰;而变质锆石外形则是圆形,卵形,不规则状,一般延长度小。如果变质锆石是在一长柱状岩浆锆石基础上重结晶或增生的,它也可以有柱状,或长柱状外形,但其晶棱均已圆化,变得不那么锋锐、清晰(Gebauer and Graunenfelder, 1979; Peucat et al., 1982; 李曙光等, 1997)。

5.2.2 内部结构

锆石的阴极发光图像可清楚地观察锆石的内部结构。岩浆锆石有细密的韵律环带结构,它是岩浆成分呈韵律状振荡造成的。变质增生成因锆石无此类结构(Gebauer et al., 1997),它总是在继承锆石外围形成一个宽窄不匀的增生边(内部无韵律环带),有时沿继承锆石核的不同晶面方向形成扇状结构。变质重结晶锆石带常在古老继承锆石中不规则分布,多集中于含 U 高的部位(阴极发光暗)。如果它是蜕晶化的古老岩浆锆石重结晶形成,且重结晶部分蜕晶化很彻底,则重结晶锆石不再保留有原岩浆锆石环带的任何痕迹;如果重结晶部分蜕晶化不彻底,仍有一些未蜕晶化锆石残留,则重结晶锆石区仍可有隐约可见的环带残留,显得该区不如最外面的变质增生锆石边那么干净(Pidgeon, 1992)。

5.2.3 包裹体矿物

超高压变质岩的变质增生锆石可包含有石榴石、绿辉石,甚至柯石英等超高压矿物。它是证明这些变质增生锆石是在超高压变质阶段生长的最好证据。岩浆锆石核中或退变质增生边中的矿物包裹体则是非高压岩浆矿物,如斜长石、石英、磷灰石等。因此,锆石中矿物包裹体的性质可以作为判定锆石成因的一个主要判据(Liu et al., 2001)。需要指出的是,一些岩浆锆石核中也发现有柯石英(Liu et al., 2001),这可以有多种解释:如继承锆石原来的石英包裹体可在超高压变质时变为柯石英,或如果继承锆石有裂纹,石英可沿裂纹进入该锆石。但是,从未发现继承锆石含有其他高压矿物(如石榴石、绿辉石),因此,这类非相变成因的高压矿物包体更有指示意义。

5.2.4 痕量元素

Th/U 值:由于 Zr^{4+} 的离子半径小于 U^{4+} 和 Th^{4+},因此在蜕晶化锆石重结晶时 U^{4+} 和 Th^{4+} 被从蜕晶化锆石中排除出去而使 U、Th 含量降低。然而由于 Th^{4+} 的离子半径比 U^{4+} 大,其丢失量更多,因而在变质重结晶锆石中 Th 含量更低,从而导致较低的 Th/U 值。此外,由于变质流体中 Th 含量很低,从而使变质增生锆石的 Th/U 值也很低（<0.1）（Rowley et al., 1997; Rubatto, 2002），而岩浆锆石的 Th/U>0.1。需要指出的是麻粒岩的变质增生锆石有时会出现 Th/U>0.1 的例外。

Zr/Hf 值：已有数据表明，岩浆锆石一般具有非常低的 Hf 含量(Pan, 1997; Vavra et al., 1996)，变质重结晶作用不会对锆石 Hf 含量产生明显的影响，因此变质重结晶锆石和原岩浆锆石具有相近的高 Zr/Hf 值。对于变质增生锆石，锆石生长所需要的 Zr 主要来源于矿物的分解释放或变质流体的携带。我们的研究表明，原岩辉长岩中的辉石是锆石增生的主要 Zr 源（侯振辉，2003），辉石 Zr、Hf 的分配系数研究表明 $D_{Hf} > D_{Zr}$，$D_{Hf}/D_{Zr} \approx 1.7$（Johnson, 1998; Green et al., 2000），因此单斜辉石分解释放的 Zr、Hf 具有较低的 Zr/Hf（约 0.3～0.4）(Green et al., 2000)，显然以它们为物源形成的高压变质增生锆石应富 Hf 并具有较低的 Zr/Hf 值。在变质流体中 Zr 比 Hf 更容易形成稳定的配合物而使更多的 Hf 进入锆石晶格（Andrade et al., 2002）从而使其具有高的 Hf 含量。因此上述变质增生锆石均显示了低 Zr/Hf 值特征，这明显不同于岩浆锆石和变质重结晶锆石高 Zr/Hf 值特征。

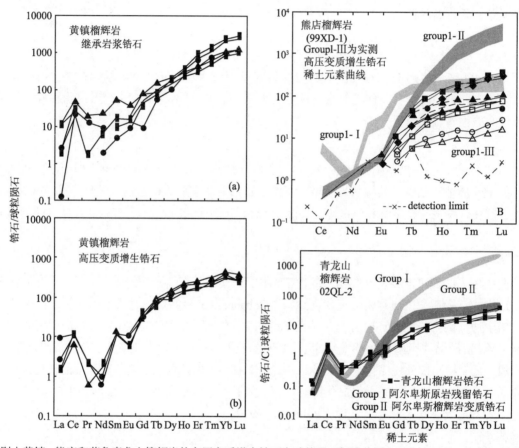

图 10 大别山黄镇、熊店和苏鲁青龙山榴辉岩的高压变质增生锆石在球粒陨石标准化图上具有扁平的 HREE 配分模型（Sun et al., 2002; 吴元保等, 2002; Li et al., 2005）

稀土元素：稀土元素丰度，尤其 HREE 富集程度是区分岩浆锆石、变质重结晶锆石和变质增生锆石的重要指标。岩浆锆石含有较高的 REE 含量和陡立的 HREE 富集模型。对于变质重结晶锆石，其轻稀土元素因较重稀土元素有较大的离子半径而更容易在变质重结晶过程中从锆石晶格中排除出来（Hoskin et al., 2000）。但 Hoskin et al.（2000）认为 Er 之后的重稀土元素（Er、Tm、Yb、Lu）的 3+离子半径仅比 Zr^{4+} 半径稍大（<16%），它们在八配位上可以取代 Zr^{4+}，因此重结晶作用对这些元素并没有明显的影响，从而其重稀土元素含量及分布特征与原岩岩浆锆石相比具有继承性。然而由于镧系收缩的影响 Er 之后的重稀土元素离子半径存在一定程度上的差异，据此可以推测重结晶过程中重稀土原子序数较低的元素（如 Dy）由于较大的离子半径，其离子被排出量高于原子序数较低的元素（如 Yb），从而导致变质重结晶锆石比原岩岩浆锆石具有更趋陡直的重稀土富集模型。在超高压变质条件下，因为与变质锆石增生同时还有富重稀土的石榴石

重结晶，因而使得变质锆石具有较低的重稀土富集程度（Rubatto, 2002）。例如，大别山黄镇、熊店及苏鲁青龙山榴辉岩锆石的 REE 分析显示变质增生边的 HREE 含量明显低于岩浆锆石核，呈扁平分布（图10）（吴元保等，2002; Sun et al, 2002; Li et al., 2005）。

退变质锆石增生时，由于没有石榴子石重结晶，锆石稀土元素曲线与岩浆锆石类似呈 HREE 陡立的富集模型，然而其稀土元素含量显著低于岩浆锆石和高压变质增生锆石。这可能是由退变质流体中的稀土元素含量较低造成的。此外，由于退变质流体中 Pb 含量较高，因此退变质增生锆石的普通 Pb 含量普遍较高。

Nb、Ta 及 Nb/Ta： 由于在超高压变质锆石增生时有富含 Nb、Ta 和高 Nb/Ta 值的金红石与之平衡，变质锆石的 Nb、Ta 含量及 Nb/Ta 值均低于岩浆锆石。如大别山黄镇榴辉岩变质锆石和岩浆锆石的 Nb、Ta 含量及 Nb/Ta 值有显著差别。其中变质锆石 Nb、Ta 含量及 Nb/Ta 值分别为 0.5~1.4 ppm，0.7~1.5 ppm 和 0.3~1.3。而岩浆锆石核的 Nb、Ta 含量及 Nb/Ta 值分别为 3.8~19.7 ppm，2.7~12.1 ppm 和 1.0~4.6（吴元保等，2002）。因此，锆石 Nb、Ta 含量及 Nb/Ta 值也是区分变质锆石和岩浆锆石的很好指标。

事实上，由于锆石成因的复杂性，单独使用上述任何一种判别指标都可能出现例外的特殊情况，从而影响不同成因锆石类型的准确区别。只有将锆石的晶形、内部 CL 特征、包裹体和痕量元素结合起来进行综合研究，才有可能对不同成因锆石做出正确的区分，并对其 U-Pb 年龄地质意义给出正确解释。例如，Sun et al.（2002），Li et al.（2005）将 CL 图像、矿物包裹体及锆石痕量元素特征相结合识别高压超高压变质增生锆石获得了很好的效果。

致谢

本工作受国家重点基础研究发展规划项目（G1999075503）、国家自然科学基金（批准号：40173014）和中国博士点（批准号：20010358026）科学基金资助。陈福坤审阅了此文并提供了修改建议，在此表示感谢。

参考文献

陈文寄，李齐. 1999. 碱性长石中氩的多重扩散域(MDD)模式与构造非平稳运动. 化学地球动力学，郑永飞主编，科学出版社，北京，358-388

陈移之，李曙光，从柏林，等. 1992. 胶南榴辉岩的形成时代及成因——Sr、Nd 同位素地球化学及年代学证据. 科学通报，37(23): 2169-2172

侯振辉. 2003. 大别-苏鲁造山带高级变质岩锆的地球化学、锆石微量元素特征及年代学效应. 博士学位论文，中国科学技术大学

简平，刘敦一，杨巍然，等. 2000. 大别山西部河南罗山熊店加里东期榴辉岩锆石特征及 SHRIMP 分析结果. 地质学报，74(3): 259-264

简平，叶伯丹，李志昌，等. 1994. 大别山造山带榴辉岩同位素年代学、PTt 轨迹及其构造意义. 同位素地球化学研究，陈好寿主编，浙江大学出版社，杭州，205-213

李秋立，李曙光，侯振辉，等. 2004. 青龙山榴辉岩高压变质新生锆石 SHRIMP U-Pb 定年、微量元素及矿物包裹体研究，科学通报，49(22): 2329-2334

李曙光，Hart S R，郑双根，等. 1989a. 中国华北、华南陆块碰撞时代的钐-钕同位素年龄证据. 中国科学(B 辑)，(3): 312-319

李曙光，葛宁洁，刘德良，等. 1989b. 大别山北翼大别群中 C 型榴辉岩的 Sm-Nd 同位素年龄及其构造意义. 科学通报，34(7): 522-525

李曙光，侯振辉. 2001. 大陆俯冲过程中的流体，地学前缘，8(3):123-129

李曙光，李惠民，陈移之，等. 1997. 大别山-苏鲁地体超高压变质年代学——II. 锆石 U-Pb 同位素体系. 中国科学（D 辑），27(3): 200-206

李曙光，刘德良，陈移之，等. 1992. 大别山南部含柯石英榴辉岩的 Sm-Nd 年龄. 科学通报，37: 346-349

李曙光，孙卫东，张宗清，等. 2000. 青岛仰口榴辉岩的 Nd 同位素不平衡及二次多硅白云母 Rb-Sr 年龄. 科学通报，45: 2223-2227

刘福来，许志琴，宋彪，2003. 苏鲁地体超高压和退变质时代的厘定：来自片麻岩锆石微区 SHRIMP U-Pb 定年的证据. 地质学报，77(2): 229-237

王松山，葛宁洁，桑海清，等. 1999. 多硅白云母过剩 Ar 成因及绿辉石 Ar-Ar 年龄谱意义：以南大别超高压榴辉岩为例. 科学通报，44(24): 2607-2613

吴元保，陈道公，夏群科，等. 2002. 大别山黄镇榴辉岩锆石微区微量元素分析：榴辉岩相变质锆石的微量元素特征. 科学通报，47(11): 859-863

杨巍然，王林森，韩郁箐，等. 1994. 大别蓝片岩-榴辉岩年代学研究. 同位素地球化学研究，陈好寿主编，杭州：浙江大学出版社.

Alines A D, Rossman G R. 1985. The high temperature behavior of trace hydrous components in silicate minerals. American Mineralogist, 70: 1169-1179

Ames L, Tilton G R, Zhou G. 1993. Timing of collision of the Sino-Korean and Yangtze cratons: U-Pb zircon dating of coesite-bearing eclogites. Geology, 21: 339-342

Andrade F R D D E, Möller P, Dulski P. 2002. Zr/Hf in carbonatites and alkaline rocks: new data and a re-evaluation. Revista Brasileira de Geociências, 32(3):361-370

Arnaud N O, Kelly S. 1995. Evidence for excess Ar during high pressure metamorphism in the Dora Maira (western Alps, Italy), using an ultraviolet laser ablation microprobe $^{40}Ar/^{39}Ar$ technique. Contrib Mineral Petrol, 121: 1-11

Ayers J C, Dunkle S, Gao S, et al. 2002. Constraints on timing of peak and retrograde metamorphism in the Dabie Shan Ultrahigh-pressure Metamorphic belt, east-central China, using U-Th-Pb dating of zircon and monazite. Chem. Geol., 186: 315-331

Boundy T M, Hall C M, Li G, et al. 1997. Fine-scale isotopic heterogeneities and fluids in the deep crust: a $^{40}Ar/^{39}Ar$ laser ablation and TEM study of muscovites from a granulite-eclogite transition zone. Earth Planet. Sci. Lett, 148: 223-242

Burton K W, Kohn M J, Cohen A S, et al. 1995. The relative diffusion of Pb, Nd, Sr and O in garnet. Earth Planet. Sci. Lett, 133: 199-211

Chavagnac V, Jahn B.M Villa I M, et al. 2001. Mutichronometric evidence from in situ origin of the ultrahigh-pressure metamorphic terrane of Dabieshan, China. Jour. of Geol, 109:633-646

Chavagnac V, Jahn B M. 1996. Coesite-bearing eclogites from the Bixiling complex, Dabie Mountains, China: Sm-Nd ages, geochemical characteristics and tectonic implications. Chemical Geology, 133: 29-51

Claoue-long J C, Sobolev N V, Shatsky V S, et al. 1991. Zircon response to diamond-pressure metamorphism in the Kokchetav massif, USSR. Geology, 19: 710-713

Cliff R A. 1993. Isotopic dating in metamorphic belts. J. ged. Soc. London. 142: 97-110

Dodson M H. 1973. Closure temperature in a cooling geochronological and petrological system. Contrib. Mineral. Petrol, 40: 259-274

Ernst W G. 2000. H_2O and ultrahigh-pressure subsolidus phase relations for mafic and ultramafic systems. In: Earnst W G and Liou J G, (eds.) Ultrahigh-pressure metamorphism and geodynamics in collision-trype orogenic belts: Bellwether Publishing Led, p.121-129

Farges F. 1994. The structure of zircon: a temperature-dependent EXAPS study. Physics and Geochemistry of Minerals, 20: 504-514

Fraser G, Davie E, Strphen E. 1997. Zirconium abundance in granulite-facies minerals with implication for zircon geochronology in high-grade rocks. Geology, 25(7): 607-610

Ganguly J, Tirrone M, Hervig R L. 1998. Diffusion kinetics of Samarium and Neodymium in garnet, and a method for determing cooling rates of rocks. Science, 281: 805-807

Gebauer D, Graunenfelder M. 1979. U-Pb zircon and Rb-Sr mineral dating of eclogites and their country rocks example: Munchberg gneiss massif, Northeast Bavaria. Earth Planet. Sci. Lett., 42: 35-44

Gebauer D, Schertl H P, Brix M, et al. 1997. 35Ma old ultrahigh-pressure metamorphism and evidence for very rapid exhumation in the Dora Maira Massif. Western Alps. Lithos, 41: 5-24

Giorgis D, Cosca M, Li S. 2000. Distribution and significance of extraneous argon in UHP eclogite (Sulu terrain, China): insight from in situ $^{40}Ar/^{39}Ar$ UV-laser ablation analysis. Earth Plant. Sci. Lett., 181: 605-615

Green T H, Blundy J D, Adam J, et al. SIMS determination of trace element partition coefficients between garnet, clinopyroxene and hydrous basaltic liquids at 2-7.5 GPa and 1080-1200 ℃. Lithos, 2000, 53: 165-187

Hacker B R, Wang Q. 1995. Ar/Ar geochronology of ultrahigh-pressure metamorphism in central china. Tectonics, 14: 994-1006

Hacker B R, Ratschbacher L, Webb L, et al. 1998. U/Pb zircon ages constrain the architecture of the ultrahigh-pressure Qinling-Dabie Orogen, China. Earth Planet. Sci. Lett, 161: 215-230

Harrison T M. 1981. Diffusion of ^{40}Ar in hornblende. Contrib. Mineral. Petrol, 78: 324-331

Hensen B J, Zhou B. 1995. Retention of isotopic memory in garnets partially broken down during an overprinting granulite-facies metamorphism: implications for the Sm-Nd closure temperature. Geology, 23:225-228

Hoskin P W O, Black L P. 2000. Metamorphic zircon formation by solid-state recrystallization of protolith igneous zircon. J Metamorphic Geol, 18: 423-439

Humphries F J, Cliff R A. 1982. Sm-Nd dating and cooling history of Scourian granulates, Sutherland. Nature, 295: 515-527

Jagoutz E. 1988. Nd and Sr systematics in an eclogite xenolith from Tanzania : Evidence for frozen mineral equilibrium in the continental lithosphere. Geochim. Cosmochim. Acta, 52,:1285-1293

Jagoutz E. 1994. Isotopic systematics of metamorphic rocks. Abstact of ICOG-8 at Berkeley, U.S.A. U. S. Geol. Survey Circular, 1107: 156

Jahn B M, Cornichet J, Henin O, et al. 1994. Geochemical and isotopic investigation of ultrahigh pressure (UHP) metamorphic terranes in China: Su-Lu and Dabie complexes, Abstracts of First workshop on UHP metamorphism and tectonics at Stanford, U. S. A. A, 71-74

Johnson K T M. 1998. Experimental determination of partition coefficients for rare earthand high-field-strength elements between clinopyroxene, garnet, and basaltic melt at high pressures. Contrib Mineral Petrol, 133: 60-68

Li S, Jagoutz E, Chen Y, et al. 2000. Sm-Nd and Rb-Sr isotopic chronology and cooling history of ultrahigh pressure metamorphic rocks and their country rocks at Shuanghe in the Dabie Mountains, Central China. Geochim. Cosmochim. Acta, 64(6): 1077-1093

Li S, Jagoutz E, Lo C H, et al. 1999. Sm-Nd, Rb-Sr and ^{40}Ar-^{39}Ar isotopic systematics of the ultrahigh pressure metamorphic rocks in the Dabie-Sulu Belt, central China: A retrospective view. Inter. Geol. Rev, 41: 1114-1124

Li S, Wang S, Chen Y, et al. 1994. Excess argon in phengite from eclogite: evidence from dating of eclogite minerals by Sm-Nd, Rb-Sr and ^{40}Ar/^{39}Ar methods. Chem. Geol, 112: 343-350

Liu J, Ye K, Maruyama S, et al. 2001. Mineral inclusions in zircon from gneisses in the ultrahigh-pressure zone of the Dabie Mountains, China. The J. of Geol, 109: 523-535

Mattauer M, Matte P, Maluski H, et al. 1991. Paleozoic and Triassic plate boundary between North and South China—New structural and radiometric data on the Dabie-Shan(Eastern China): Comptes Rendus Acad. Sci. Paris, 312: 1227-1233

McDougall I, Harrison T M. 1988. Geochronology and thermochronology by the ^{40}Ar/^{39}Ar method: Oxford, Uk, Oxford University Press, 110

Mezger K, Krogstad E J. 1997. Interpretation of discordant U-Pb zircon ages: an evaluation. J.Metamorphic Geol, 15: 127-140

Mezger K, Essene E J, Halliday A N. 1992. Closure temperatures of the Sm-Nd system in metamorphic garnet. Earth Planet Sci. Lett, 113: 397-409

Monié P, Chopin C. 1991. ^{40}Ar/^{39}Ar dating in coesite-bearing and associated units of the Dora Maira massif, Western Alps. Eur. Jour. Mineral, 3: 239-269

Okay A I, Sengör A M C. 1993. Tectonics of an ultrahigh-pressure metamorphic terrane: the Dabie Shan- Tongbai Shan orogen, China. Tectonics, 12: 1320-1334

Pan Y M. 1997. Zircon- and monazite-forming metamorphic reactions at Manitouwadge, Ontario. The Canadian Mineralogist, 35: 105~118

Peucat J J, Vidal P h, Godard G, et al. 1982. Precambrian U-Pb zircon ages in eclogites and garnet pyroxenites from South Brittany (France): an old oceanic crust in the west European Hercynian belt? Earth Planet. Sci. Lett, 60: 70-78

Pidgeon R T. 1992, Recrystallisation of oscillatory-zoned zircon: some geochronological and petrological implications. Contrib. Mineral. Petrol, 110: 463-472

Rowley D B, Xue F, Tucker R D, et al. 1997. Ages of ultrahigh pressure metamorphism and protolith orthogneisses from the eastern Dabie Shan: U/Pb Zircon geochronology. Earth Planet. Sci. Lett, 151: 191-203

Rubatto D. 2002. Zircon trace element geochemistry: partitioning with garnet and the link between U-Pb ages and metamorphism. Chemical Geology, 184: 123-138

Ruffet G, Feraud G, Balevre M, et al. 1995. Plateau ages and excess argon in phengites: an ^{40}Ar/^{39}Ar laser probe study of Alpine micas (Sesia Zone, Western Alps, northern Italy). Chem. Geol, 121: 327-343

Ruffet G, Gruau G, Féraud G, et al. 1997. Rb-Sr and ^{40}Ar-^{39}Ar laser probe dating of high-pressure phengites from the Sesia Zone (Western Alps): underscoring of excess argon and new age constraints on the high-pressure metamorphism. Chem. Geol, 141: 1-18

Scaillet S. 1996. Excess ^{40}Ar transport scale and mechanism in high-pressure phengites: case study from an eclogitized metabasite of the Dora Maira nappe, Western Alps. Geochim. Cosmochim. Acta, 60: 1075-1090

Scaillet S. 1998. K-Ar (^{40}Ar/^{39}Ar) geochronology of ultrahigh pressure rocks. In: Hacker B R and Liou J G (eds) When continents collide: Geodynamics and geochemistry of ultrahigh-pressure rocks: Dordrecht, Kluwer Acad. Publ, p. 161-202

Sherlock S, Arnaud N O. 1999. Flat plateau and impossible isochrones: Apparent ^{40}Ar/^{39}Ar geochronology in a high-pressure terrain. Geochim. Cosmochim. Acta, 63: 2835-2838

Sneeringer M, Hart S R, Shimizu N. 1984. Strontium and Samarium Diffusion in diopside. Geochim. Coesmochim. Acta, 48: 1589-1608

Thöni M, Jagoutz E. 1992. Some new aspects of dating eclogites in orogenic belts: Sm-Nd, Rb-Sr and Pb-Pb isotopic results from the Austroalpine Savalpine and Koralpe type - locality. Geochim. Cosmochim. Acta, 56: 347-368

Tilton G R, Schreyer W, Schertl H-P. 1991. Pb-Sr-Nd isotopic behavior of deeply subducted crustal rocks from the Dora Maira Massif,

Western Alps, Italy, II. What is the age of the ultrahigh-pressure metamorphism? Contrib. Mineral. Petrol, 108: 22-23

Vavra G, Gebauer D, Schmid R, et al. 1996. Multiple zircon growth and recrystallization during polyphase late Carboniferous to Triassic metamorphism in granulites of the Ivrea Zone (Southern Alps): an ion microprobe (SHRIMP) study. Contrib Mineral Petrol, 122: 337~358

Villa L M. 1998. Isotopic closure. Terra Nova, 10: 42-47

Xu S T, Liu Y C, Sun W, et al. 2000. Discovery of the eclogite and its petrography in the Northern Dabie Mountains, Chinese Science Bulletin, 45(3): 273-278.

Zhou G, Liu Y J, Eide E A, et al. 1993. High-pressure/Low-temperature metamorphism in northern Hubei Province, central China. J. Metamor. Geol, 11: 561-574

变质岩同位素年代学：Rb-Sr 和 Sm-Nd 体系*

李曙光[1,2]，安诗超[2]

1. 中国地质大学（北京）科学研究院地质过程与矿产资源国家重点实验室，北京 100083
2. 中国科学院壳幔物质与环境重点实验室；中国科学技术大学地球空间科学学院，合肥 230026

> **亮点介绍**：详细论述了变质同位素年代学若干基本概念和原理，特别指出超高压变质岩的 Rb-Sr 和 Sm-Nd 两个同位素定年体系在不同级别变质作用发生时，同位素体系重置或均一化尺度的差异性及其导致它们全岩或矿物等时线定年的适用范围；提出对超高压变质岩严格鉴别与分选不同阶段形成的石榴子石和同阶段共生单斜辉石做 Sm-Nd 和 Lu-Hf 同位素分析是获得准确变质时代的关键。

摘要 对变质岩经历的进变质和退变质作用定年并构建其 P-T-t 轨迹是观测地壳运动过程的重要途径。全岩等时线和矿物等时线是变质岩 Rb-Sr 和 Sm-Nd 定年的两个基本方法。在变质过程中同位素均一化尺度是影响全岩等时线定年的主要因素。在一般情况下，变质过程中 Rb-Sr 同位素体系的均一化尺度远大于 Sm-Nd 体系，从而 Rb-Sr 全岩等时线可以给出有意义的变质年龄，而 Sm-Nd 数据不能。然而，对于低级变质作用，因其较高级变质作用有更丰富的流体，其 Nd 同位素均一化尺度可能大，从而使得一些全岩 Sm-Nd 等时线给出和 Rb-Sr 年龄一致的有意义变质年龄。对于矿物等时线定年，在变质作用时矿物之间能否达到同位素平衡则是关键。已经证明超高压变质(UHPM)岩的退变质作用是开放体系，然而 UHPM 矿物的 Sr-Nd 同位素体系仍保持封闭。已观测到 UHPM 矿物和退变质矿物之间的 Sr-Nd 同位素不平衡，因此，高压矿物（如石榴石、多硅白云母）和退变质矿物或全岩的连线将会给出没有意义的偏老的 Sm-Nd 年龄和偏年轻的 Rb-Sr 年龄。由 3 个以上很好分开的矿物确定的等时线的良好线性，不同定年方法获得的年龄的一致性，以及确定等时线矿物之间的氧同位素平衡已被用来判定矿物间 Nd 同位素是否达到平衡。由于石榴石具有高 Sm/Nd 和 Lu/Hf 值，石榴石是榴辉岩或石榴辉石岩 Sm-Nd 或 Lu-Hf 定年最重要的矿物。然而由于石榴石非常宽的 P-T 稳定范围，石榴石可以在高级变质岩的前进变质和退变质作用中生长，从而具有复杂的环带结构。因此，如何从具有复杂结构的石榴石不同部位取样和分析，并判断其成因就成为榴辉岩或石榴辉石岩 Sm-Nd 或 Lu-Hf 矿物等时线定年的一个挑战。这需要今后做更进一步研究。

对变质岩经历的各变质阶段进行精确定年是了解地壳运动过程的温度-压力-时间变化轨迹的基本途径。Rb-Sr，Sm-Nd 等时线法是变质岩定年常用方法。如何获得有意义的准确年龄，和正确理解年龄含义是从事定年研究的关键问题。本文总结作者在该领域的工作经验，就有关问题进行了探讨。

1 变质同位素年代学的几个基本概念

1.1 变质相及 P-T-t 轨迹

地壳岩石的变质作用是随地壳的抬升、沉降和伴随的温度、压力变化而发生的，它是一个连续变化过程。在一定压力-温度（P-T）范围内的一组相对稳定的矿物组合，当 P-T 发生变化时，原矿物组合变得不稳定，而由一组新的稳定矿物组合取代，称为变质相变。这一过程可以在压力-温度图上用 P-T 演化路径（P-T path）来描述，如地壳俯冲成因榴辉岩的升温、升压 P-T 轨迹为绿片岩相-蓝片岩相或角闪岩相-榴辉岩相[1]。高级

---

* 本文发表在：地学前缘，2014，21(3)：246-255

变质相取代低级的变质相过程称为进变质过程；反之为退变质过程。变质同位素年代学任务是测定每一个变质相发生的时代（t），从而构建变质过程的 P-T-t 轨迹。

1.2 封闭温度

放射性同位素年龄是根据定年体系内积存的放射性衰变产生的子体 D 的量计算的，它假定在此年龄期间放射性衰变产生的子体 D 没有任何丢失。但是在高温时，一定年体系内放射性衰变产生的子体 D 会因高的扩散速率而逃逸，不在体系内积存，因而计时过程没有启动；在低温时，由于扩散速率近似为零，放射性衰变产生的子体在体系内完全累计下来，放射性同位素体系完整计时，D/P 与 t 呈指数线性关系。

$$D = P(e^{\lambda t} - 1)$$

在上述两个极端情况之间的温度条件下，会存在一过渡状态。将完整计时的指数线性关系向高温外推到 $D/P=0$，可获得时间 t_c 和对应温度 T_c，前者为计算所得年龄，后者为同位素"时钟"表面开始计时的温度，称为"封闭温度"（图1）。每个年龄都对应一个开始计时的封闭温度。

一个矿物的封闭温度不是一个确定值，而是一个范围。它取决于放射性母体的半衰期，所产生的子体在该矿物的活化能，矿物粒径大小和冷却速率。常用定年同位素体系和矿物的封闭温度如下：(1) K-Ar 体系。角闪石，550~400 ℃；黑云母，300 ℃[3,4]。(2) U-Pb 体系。锆石，>850 ℃；石榴石，750 ℃；独居石，650 ℃；榍石，600 ℃；金红石，470~420 ℃[5,6,7]。(3) Rb-Sr 体系。白云母，500 ℃；角闪石，450 ℃；黑云母. 300 ℃[2,8,9]。

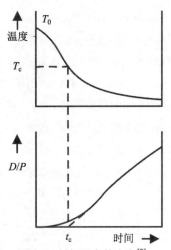

图 1 封闭温度的定义[2]

上图为同位素体系温度 T 随时间 t 冷却曲线；下图为体系内放射性衰变产生的子体 D 的累积量（以 D/P=子体/母体表示）随时间变化曲线。t_c 是计算所获得的封闭年龄，T_c 为对应该年龄的封闭温度

1.3 冷却曲线测定

构建完整的 P-T-t 轨迹需要与岩石学研究结合以测定前进变质和退变质经历的各变质相的变质温度和压力，由于早期变质相矿物组合多被后期变质相矿物组合交代而仅有少量残留，这为测定早期变质相年龄带来困难。但年代学研究可以根据不同矿物定年同位素体系的封闭温度差异给出可刻画退变质过程的 T-t 冷却曲线。

如果地质体是因为构造抬升而冷却，测定地质体的冷却曲线是变质同位素年代学提供的最直接观察地壳抬升过程的方法。构建 T-t 冷却曲线的样品需来自同一个连续地质体，以保证它们有相同冷却史。地质体的一个抬升过程可以在 10~20 Ma 完成，为此要求每个年龄的测量精度 <5 Ma。封闭温度的准确估计则要求根据测定样品的颗粒度和冷却史具体计算其封闭温度[2]。

1.4 等时线定年的基本原理和要求

众所周知，我们测定的岩石或矿物的同位素组成是由它们形成时的初始同位素组成加上后来放射性衰变积累的衰变子体构成的，也即：

$$^{87}\text{Sr}/^{86}\text{Sr} = \left(^{87}\text{Sr}/^{86}\text{Sr}\right)_0 + {}^{87}\text{Rb}/^{86}\text{Sr}(e^{\lambda_{\text{Rb}}t} - 1)$$

$$^{143}\text{Nd}/^{144}\text{Nd} = \left(^{143}\text{Nd}/^{144}\text{Nd}\right)_0 + {}^{147}\text{Sm}/^{144}\text{Nd}(e^{\lambda_{\text{Sm}}t} - 1)$$

据此，Rb-Sr，Sm-Nd 等时线法定年原理可简述为，一组具有相同初始同位素组成[$(^{87}\text{Sr}/^{86}\text{Sr})_0$ 或 $(^{143}\text{Nd}/^{144}\text{Nd})_0$]，且同时形成（封闭）的岩石矿物样品，在封闭条件下经过相同一段时间（t）演化，其同位素组成（$^{87}\text{Sr}/^{86}\text{Sr}$）或（$^{143}\text{Nd}/^{144}\text{Nd}$）与它们的母子体比值（$^{87}\text{Rb}/^{86}\text{Sr}$ 或 $^{147}\text{Sm}/^{144}\text{Nd}$）成正比，这组样品的数据在同位素组成对母子体比值二维图上（如 $^{87}\text{Sr}/^{86}\text{Sr}$-$^{87}\text{Rb}/^{86}\text{Sr}$ 图或 $^{143}\text{Nd}/^{144}\text{Nd}$-$^{147}\text{Sm}/^{144}\text{Nd}$ 图）呈线性排列，对其回归拟合的直线称为等时线，根据它的斜率（$e^{\lambda t}-1$）可计算年龄 t。因此，一个合格的等时线需满足下列 3 个条件：①一组定年样品同时形成或同位素体系同时封闭；②它们的初始同位素组成均一；③自它们形成后

同位素体系一直保持封闭。变质作用定年的基本问题始终是围绕是否满足这三个基本条件展开。

2 Rb-Sr、Sm-Nd 全岩等时线定年

2.1 全岩等时线定年采样原则

变质岩的 Rb-Sr、Sm-Nd 全岩等时线定年既可以获得原岩时代，也可获得变质时代。为了获得有意义年龄，正确采样是关键。

（1）同位素均一化尺度要大于采样范围，从而保证样品具有相同的初始同位素组成。为了获得原岩时代，测年对象应是同期喷出或侵位的岩浆岩，这是基于假定同一次岩浆活动的同位素组成是均匀的。因此，应在野外查明正变质岩的原岩期次和相互关系基础上采样。为了获得变质时代，则需要了解该变质作用可能导致的同位素均一化尺度，所有定年样品要在同位素均一化可达到的尺度内采取。通常，在变质过程中 Rb-Sr 较 Sm-Nd 活泼，前者在变质过程中全岩同位素体系往往在较大范围被重置，即 Sr 同位素组成被均一化而有了相同的初始同位素组成，因而全岩 Rb-Sr 等时线给出变质时代；而 Sm-Nd 体系因活动性差，其 Nd 同位素体系在变质过程中有可能被不同程度改造，但不能在全岩采样涉及的空间范围内被均一化或重置，因而无法获得等时线。对缺少流体的高级变质岩系，其 Sm-Nd 体系甚至不受变质作用影响而往往给出原岩时代。例如，Sun et al.[10]对东秦岭二郎坪蛇绿岩中一个小枕状熔岩进行了 Rb-Sr、Sm-Nd 全岩等时线定年。该枕状熔岩经历了绿片岩相变质。他们采取了一个 30 cm 长的未变形小岩枕，将其切割成小岩块样品，做全岩 Rb-Sr、Sm-Nd 定年。Rb-Sr 全岩数据有很好的线性关系，给出 406±22 Ma 的等时线变质年龄（图 2），证明该绿片岩相变质作用使其 Sr 同位素在此岩枕范围内达到均一化。然而，该组样品的 Sm-Nd 全岩数据离散（图 2），这说明该绿片岩相变质作用干扰了这组样品原岩的 Sm-Nd 同位素体系，但没能使 Nd 同位素在 30 cm 范围内达到均一化。

（2）为获得较精确的等时线年龄，样品的母体/子体值要有大的分异。为满足这一条件，一组定年全岩样品的组成最好有较大差异。显然，（2）与（1）的要求是矛盾的。我们只能在保证要求（1）的情况下，尽可能采集组成差异大的样品。例如，分离结晶作用形成的堆晶岩由于不同堆晶矿物比例不同可有较大成分差异，但初始同位素组成均一。再如喷发时代相近而成分差异较大的双峰式火山岩，其原岩同位素组成可能有差异；但它们的薄互层产状有可能在变质时由于变质流体作用达到相邻基性岩层和酸性岩层的同位素均一化。因此，薄互层变质双峰式火山岩的小范围采样可满足变质时代定年要求。

2.2 低级区域变质作用对全岩 Sm-Nd 体系的影响

尽管上文的实例表明蛇绿岩的低级变质未能导致一个 30 cm 的小岩枕的 Sm-Nd 同位素体系被重置，但是也有一些实例表明，在持续时间较长的区域低级变质条件下，其全岩 Sm-Nd 同位素体系可以被重置。

实例 1，Li et al.[11]报道了晚太古代五台群变质细碧角斑岩系的全岩 Rb-Sr、Sm-Nd 定年结果。该岩系经历了古元古代绿片岩相变质作用。对 3 个上部层位的石英角斑岩样品和 3 个中部层位的细碧岩样品做 Rb-Sr 和 Sm-Nd 全岩等时线定年，6 个样品获得一全岩 Rb-Sr 等时线年龄为 1871±98 Ma（图 3），与区域变质时代一致，说明变质时该火山岩系 Sr 同位素被均一化。然而 6 个样品的 Sm-Nd 数据不能给出一条等时线，但细碧岩和石英角斑岩各自给出了一致的（1981 Ma 和 1977 Ma）等时线年龄（图 3），且与 Rb-Sr 年龄在误差范围内一致。这说明在低级变质作用下 REE 活动了，在有限范围内使细碧岩和石英角斑岩各自 Nd 同位素被均一化。

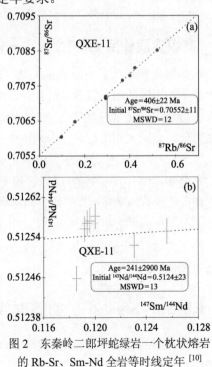

图 2 东秦岭二郎坪蛇绿岩一个枕状熔岩的 Rb-Sr、Sm-Nd 全岩等时线定年[10]

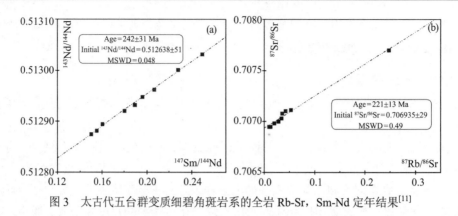

图 3 太古代五台群变质细碧角斑岩系的全岩 Rb-Sr, Sm-Nd 定年结果[11]

实例 2, 李曙光等[12]报道了南秦岭勉略带黑沟峡双峰式变质火山岩系定年[12]。黑沟峡双峰式绿片岩相变质火山岩位于南秦岭勉-略带地缝合线位置，已强烈变形，呈黑白条带状构造。该研究在一个露头上密集采集 9 个基性和酸性全岩样品，并同时获得了具良好线性的 Rb-Sr 和 Sm-Nd 等时线，分别给出 242±31 Ma (Sm-Nd) 和 221±13 Ma (Rb-Sr) 一致的三叠纪变质年龄（图 4）。这说明强变形的低级变质条带状双峰式火山岩系可在米级范围内达到 Nd 同位素均一化，小范围取样，也可获得有意义的变质 Sm-Nd 全岩等时线年龄。

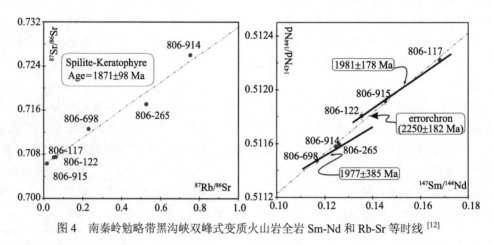

图 4 南秦岭勉略带黑沟峡双峰式变质火山岩全岩 Sm-Nd 和 Rb-Sr 等时线[12]

2.3 高级变质作用 Sr 同位素均一化尺度

在高级变质条件下，由于流体少，变质作用产生的同位素均一化尺度大大缩小，全岩 Sm-Nd 和 Rb-Sr 等时线法均不能给出高级变质时代。例如大别山蓝片岩曾被认为形成于新元古代，依据的就是红安群变质火山岩的全岩 Rb-Sr 等时线年龄（752 Ma）[13]。这组样品主要为绿片岩，含一个蓝片岩，因而被解释为高压变质时代。然而大别山蓝片岩中高压变质矿物多硅白云母的 Ar-Ar 年龄为 245.1 Ma[13]，与该带榴辉岩形成时代一致，它们都与三叠纪华南和华北陆块碰撞有关。造成这两个变质年龄差异的原因是红安群的蓝片岩相变质是叠加在绿片岩之上的。绿片岩相变质作用有较多流体活动，而由绿片岩相转变到蓝片岩相仅压力升高，温度相同，没有流体释放，导致这两种变质条件的 Sr 扩散尺度差异。Li et al.[13]计算了 Sr 的典型扩散距离 X ($X=(Dt)^{1/2}$, D 为扩散系数)。 计算表明，绿片岩相有粒间水时 Sr 的典型扩散距离为 560 m/Ma；而蓝片岩相干条件下 Sr 的典型扩散距离为 2～3 cm/Ma。故 752 Ma 全岩 Rb-Sr 年龄记录的是早期绿片岩相变质时代而非高压变质时代[13]。

3 Rb-Sr、Sm-Nd 矿物等时线定年

矿物等时线方法的样品是从同一个 cm 级尺度的样品中选取的，变质作用更容易在这样一个尺度范围内使同位素组成均一化；且测定矿物相及其封闭温度明确，因此是更被广泛采用的变质作用定年方法。它所涉

及的问题也被更深入地研究。

3.1 矿物等时线定年要求

根据等时线定年原理，变质作用的矿物等时线定年应满足以下 3 个要求：

（1）被测定的一组变质矿物应同时形成（或重结晶）（等时要求）——它们必须是同一变质阶段（相）的共生变质矿物。为此，必须在详细岩相学研究基础上选择同一变质阶段的共生矿物进行定年。根据封闭温度概念，这一组共生变质矿物还应当有相同的封闭温度才能符合"等时"要求，但这一要求很难满足。好在封闭温度的差异对低 Rb/Sr，或低 Sm/Nd 矿物影响不大，Rb-Sr、Sm-Nd 矿物等时线的斜率主要取决于高 Rb/Sr 或高 Sm/Nd 矿物，其年龄指示的是高 Rb/Sr 或高 Sm/Nd 矿物的封闭年龄。

（2）被测定变质矿物的初始同位素组成应相等——共生变质矿物之间要达到同位素平衡是必要条件。因此，为了检验年龄是否有意义，必须检查被测定矿物是否在变质时达到同位素平衡。

（3）被测定变质矿物同位素体系要保持封闭——要求被测定变质矿物不得有任何后期变化（如蚀变、退变质、风化等）。必须按严格标准精心挑选单矿物样品。

考虑到一组定年矿物样品的母体与子体比值需有较大差异才能获得较精确的等时线，Rb-Sr 体系多用于含云母（具有高 Rb/Sr）变质岩定年；Sm-Nd 体系多用于含石榴石（具有高 Sm/Nd）变质岩定年。

3.2 片麻岩的云母+全岩 Rb-Sr 定年问题

片麻岩常含有云母，很适用于 Rb-Sr 法定年。人们常误以为片麻岩的低 Rb/Sr 矿物或全岩对定年影响不大，故采用简单的云母+全岩 Rb-Sr "等时线"定年。由于全岩中含有不同期次的变质矿物，这可能导致定年严重错误。Li et al.[9] 详细研究了大别山双河超高压变质片麻岩的 Rb-Sr 体系。该片麻岩含有高压变质矿物多硅白云母（ph）和石榴石（gar）（少量），和角闪岩相退变质矿物黑云母（bi）和绿帘石（epid）（大量）。图 5 显示了对双河 3 个超高压片麻岩样品的这些矿物和全岩样品的分析结果，它表明这些定年结果既不确定，且相互矛盾。例如，同一岩片的片麻岩 ph+wr 分别给出 130.1±4.1 Ma（HT-1-1），202.0±5.7 Ma（HT-1-2）和 169.6±4.4 Ma（97SH-3），非常不一致。此外，多硅白云母（ph）和黑云母（bi）的 Rb-Sr 体系封闭温度分别是 T_c(ph)=500 ℃；T_c(bi)=300 ℃，多硅白云母的封闭温度显著高于黑云母，但同一样品 HT-1-1 的 ph+wr 年龄（130.1±4.1 Ma）反而常低于 bi+wr 年龄（173.9±3.5 Ma），这是反常不合理的。

造成这种不一致和反常的原因是多硅云母是榴辉岩相高压变质矿物，而片麻岩全岩含有大量角闪岩相退变质矿物（黑云母和绿帘石），高压变质矿物和退变质矿物是不同时代产物，且在退变质时全岩体系开放，其同位素组成被外来退变质流体改造，但残留高压变质矿物和退变质矿

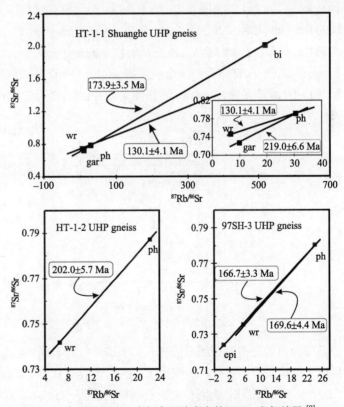

图 5　大别山双河 3 个超高压片麻岩的 Rb-Sr 定年结果[9]

物之间没有重新达到同位素平衡所致。因此，所有的 ph+wr 和 ph+epid（166.7±3.3 Ma）（97SH-3）年龄都是没有意义的[9]。石榴石（gar）也是高压变质矿物，它与多硅白云母同期形成，如果它们在高变质时达到同位素平衡，且没有被后来的退变质作用破坏，ph+gar 年龄 219.0±6.6 Ma（T_c=500 ℃）是有意义的[9]。这已被

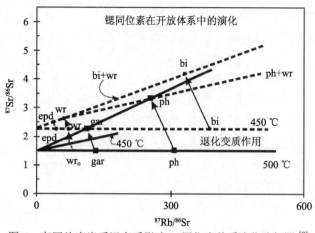

图 6 高压片麻岩受退变质影响 Sr 同位素体系演化路径图[9]

大别山榴辉岩金红石的 U-Pb 年龄 218.2±1.2 Ma（T_c=470 ℃）所证实[7]。苏鲁超高压变质带含多硅白云母榴辉岩的 ph+gar+omp 等高压矿物 Rb-Sr 等时线也给出了 219.5±0.5 Ma 和 223.9±0.9 Ma 类似的年龄[14]，也证明了苏鲁超高压岩片与大别山超高压岩片第一次折返抬升历史的一致性。由于全岩含大量退变质矿物，黑云母是退变质矿物，故 bi+wr 年龄（173.9±3.5 Ma）也是有意义的[9]。

这一高压变质岩退变质 Rb-Sr 开放体系的同位素演化关系可用图 6 更清楚说明。图中实线是高压矿物 ph 和 gar 的演化线，因被测定矿物未遭受退变质，保持封闭体系演化，故 ph+gar 等时线年龄有意义，指示 ph 冷却到 500 ℃时的年龄。图中虚线是该片麻岩遭受角闪岩相退变质后的演化线。在退变质时，体系开放，受外来退变质流体影响其 Sr 同位素比值升高，bi、epid 和 wr 具有相同和类似初始同位素组成，故 bi+wr 等时线年龄有意义，bi 冷却到 300 ℃时的年龄。在这种情况下，ph+wr 连线的斜率低于 bi+wr 连线，故给出更年轻的反常不合理年龄。

3.3 退变质矿物对 Sm-Nd 等时线定年的影响

退变质开放体系一般可导致高压变质岩的全岩 Nd 同位素比值下降，如果石榴石未受退变质改造，可导致 gar+wr 的 Sm-Nd 年龄偏老。双河 HT-4 榴辉岩和双河 HT-1-1 高压片麻岩的 3 高压变质矿物等时线均给出一致的 Sm-Nd 年龄：226.3±3.2 Ma 和 226.5±2.3 Ma（图 7）。证明高压变质矿物间的 Nd 同位素达到平衡。双河 HT-4 榴辉岩的 wr 落在高压变质矿物 gar+omp+rut 等时线之下（图 7）。双河高压片麻岩的退变质矿物 bi 和 epid，以及 wr 都落在高压变质矿物 gar+ph1+ph2 等时线之下（图 7），表明高压变质矿物与退变质矿物之间 Nd 同位素不平衡。gar+wr 给出偏老的 Sm-Nd 年龄 246.0±2.1Ma。

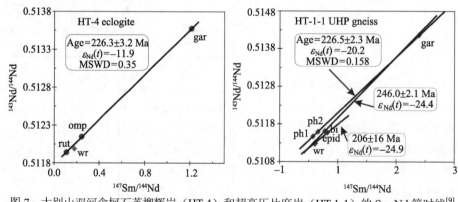

图 7 大别山双河含柯石英榴辉岩（HT-4）和超高压片麻岩（HT-1-1）的 Sm-Nd 等时线[9]

需要指出的是榴辉岩或石榴辉石岩中的单斜辉石是比较容易发生退变质变化的，它可以局部或沿解理面被后成合晶或角闪石置换。那些挑选不纯，带有退变质产物的单斜辉石颗粒可以严重干扰 Sm-Nd 等时线定年。Xie et al.[15]在对北大别百丈岩石榴辉石岩进行 Sm-Nd 定年时分别对纯净的单斜辉石单矿物（cpx-A），颗粒边部带有少量后成合晶的蚀变单斜辉石（cpx-B），和不纯的带有后成合晶、部分还带有其他矿物连生体的单斜辉石颗粒（cpx-C），连同石榴石（crt）、金红石（rut）和全岩（wr）分别进行了分析，结果 cpx-A+grt+rut+wr 呈良好线性关系，给出有意义的 219±11 Ma 年龄，但 cpx-B 和 cpx-C 均落在等时线下方，蚀变最严重的 cpx-C 偏离最远（图 8）。

3.4 高压变质条件下 Nd 同位素平衡问题

在变质过程中，原岩矿物分解，元素重新组合成新的变质矿物组合。在这一变质反应过程中，变质矿物的主要组分继承了原岩某矿物的，因而其同位素也主要继承了该原岩矿物的同位素组成。因此，新生成变质矿物需要在新的变质条件下彼此同位素交换平衡，才能具有近似一致的初始同位素组成。以辉长岩变质为榴辉岩为例，辉长岩的单斜辉石主体成分被榴辉岩的绿辉石继承，因此后者基本继承了前者 Nd 同位素组成；这一转变使绿辉石较普通辉石失去了部分 Mg、Fe，吸收了斜长石的 Na，且 Sm/Nd 下降（图9）。另外在榴辉岩变质作用中辉长岩的斜长石主体成分和 Nd 同位素组成基本被榴辉岩的石榴石继承，且 Sm-Nd 大大提高（图9）。变质新生成的石榴石（gar）和绿辉石（omp）只有通过再平衡才能使它们的初始 Nd 同位素组成相等（图9垂直箭头所示）。如果新生成的 gar 和 omp 的 Nd 同位素没有再平衡，它们之间连线具有初始负斜率（图9）。这样一组 gar+omp 矿物组合演化到今天的斜率将给出偏低的年龄。奥地利阿尔卑斯洋壳俯冲成因榴辉岩（600 ℃/14 kbar）观察到 gar 和 cpx 之间的 Nd 同位素不平衡[16]。gar+cpx 仅给出 53 Ma 年龄；而 2 个 gar + rut 给出 94 Ma，后者接近真实榴辉岩相变质年龄（图10）[16]。

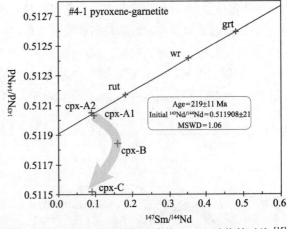

图 8 北大别百丈岩石榴辉石岩的 Sm-Nd 矿物等时线[15]

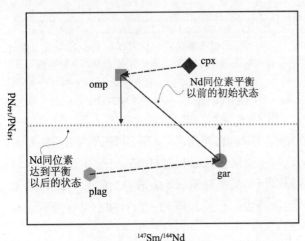

图 9 辉长岩榴辉岩相变质的矿物转化和 Nd 同位素继承关系图解（解释见正文）

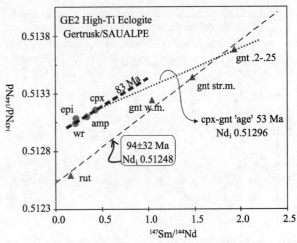

图 10 奥地利阿尔卑斯洋壳俯冲成因榴辉岩的 Sm-Nd 定年结果[16]

那么如何判断定年变质矿物是否达到 Nd 同位素平衡呢？理论上矿物颗粒之间同位素是否在可能的变质时间期间达到平衡可通过相关扩散系数和方程进行估算，但往往缺少必要参数。一个常用的简便方法是用三个以上共生变质矿物定年，且它们 Sm/Nd 彼此充分分开，如果平衡，应该有良好线性。此外，同一地区多个样品不同方法定年应有一致结果也是常用方法。如大别山双河和苏鲁青龙山榴辉岩及超高压片麻岩的三个以上超高压矿物 Sm-Nd 等时线均具有良好线性，并给出一致的 226±3 或±4.5 Ma 年龄[17]。

Zheng et al.[18] 指出可在 Sm-Nd 定年时同时测定变质矿物的氧同位素组成，如果不同矿物之间的氧同位素分馏达到平衡，可用来指示 Nd 同位素也达到平衡。如图 8 构成 cpx-A+grt+rut+wr 等时线的矿物之间氧同位素达到平衡，而偏离该等时线的 cpx-B 和 cpx-C 与它们氧同位素不平衡[15]。

3.5 退变质石榴石对榴辉岩 Sm-Nd、Lu-Hf 定年的影响

近年的研究发现超高压变质岩的退变质阶段可生长石榴石。由于石榴石的高 Sm/Nd 和 Lu/Hf 值，它是 Sm-Nd、Lu-Hf 矿物等时线定年的重要矿物，这一发现对这类定年的影响值得重视。

石榴石有很宽的结晶 P-T 范围，因而高级变质岩的石榴石可以有较宽的生长环带。不同部位的石榴石形成于不同时代，并有不同化学组成，这对依赖石榴石的 Sm-Nd 或 Lu-Hf 定年会带来不可忽略的影响。大别山超高压榴辉岩在折返过程所经历的退变质作用可相继有麻粒岩相和角闪岩相退变质作用。其中，麻粒岩相退变质期间石榴石可继续生长。Wang et al.[19] 报道了大别山北大别带退变质麻粒岩（0909QT-6）的年代学和岩石学研究结果。该岩石石榴石斑晶有残留核和生长边两部分。由核到外边，gar 的 HREE 连续降低（图 11），反映了在封闭条件下 gar 的连续生长。该样品的锆石根据矿物包体组合，微量元素和年龄可分为榴

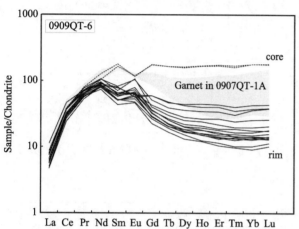

图 11 北大别带退变质麻粒岩（0909QT-6）石榴石斑晶不同部位的稀土元素组成[19]

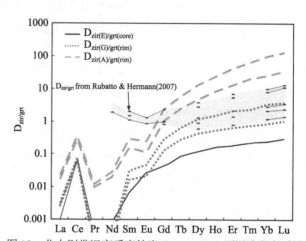

图 12 北大别带退变质麻粒岩（0909QT-6）不同阶段变质锆石与石榴石斑晶不同部位的稀土元素的表观分配系数（$D_{zir/grt}$）与实验测定结果（灰色区域）对比图[19]

图中 zir(E)-榴辉岩相结晶锆石；zir(G)-麻粒岩相结晶锆石；zir(A)-角闪岩相结晶锆石；grt(core)-核部石榴石；grt(rim)-边部石榴石

辉岩相结晶锆石、麻粒岩相结晶锆石和角闪岩相结晶锆石。分别计算它们与不同部位石榴石的 HREE 分配系数 D。图 12 显示只有麻粒岩相结晶锆石和 gar 边部的 HREE 分配系数 D 落在实验测定范围内，证明该斑晶边部石榴石是麻粒岩退变质阶段生长。

Cheng et al.[20] 报道了南大别低温榴辉岩退变质石榴石导致 Sm-Nd、Lu-Hf 年龄偏低的实例。该榴辉岩样品采自南大别带黄镇东，石榴石具有明显的环带结构。它明显具有 3 个环带；中心部分低 Ca-Mn，高 Mg，中间部分高 Ca-Mn，低 Mg，最外部又具有低 Ca-Mn，高 Mg 特征。Cheng et al.[20] 判定最外部的冠状边是退变质阶段生长石榴石。中心和中间的石榴石均含包裹物，仅最外部冠状边它不含包裹物，比较干净，可据此将其挑选出来。选择不含包裹物石榴石+omp+wr 获得 Lu-Hf 和 Sm-Nd 等时线年龄分别为 221±2.3 Ma 和 226±42 Ma（图 13）[20]。尽管该 Sm-Nd 年龄误差较大，但 Lu-Hf 年龄较精确，它们均显著小于黄镇榴辉岩 Sm-Nd 年龄 236.1±4.2 Ma 和锆石 U-Pb 年龄 242±3 Ma[21]，因此被解释为退变质流体活动时代[20]。

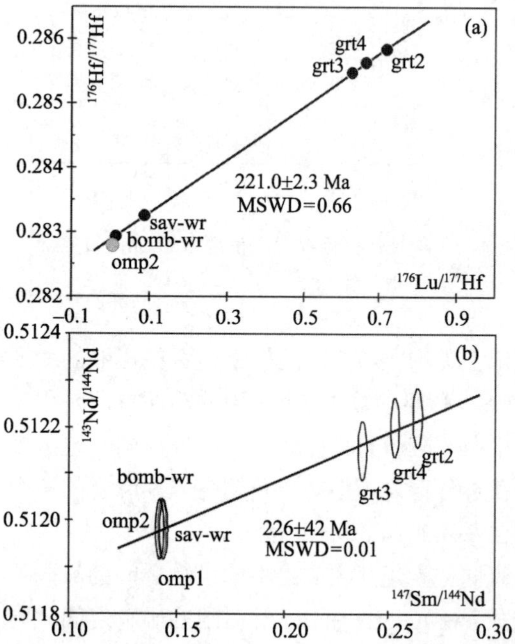

图 13 大别山黄镇东低温榴辉岩 Lu-Hf、Sm-Nd 定年[20]

图中 grt-garnet，石榴石；omp-omphacite，绿辉石；bomb-wr 表示用 bomb 溶解的全岩；sav-wr 表示 savillex 溶解的全岩

高级变质岩石榴石结构和成因的复杂性向它们的 Sm-Nd、Lu-Hf 变质年代学提出了具有挑战性的问题。这既涉及如何精确地对石榴石不同部位取样和分析的技术问题，又涉及如何准确地判定不同部位石榴石成因的科学问题。解决这些问题对解释现在仍存在的 Sm-Nd、Lu-Hf 定年结果与锆石 U-Pb 定年结果矛盾和年龄的准确度问题是关键。

致谢

吕逸文绘制本文部分图件，特致谢意。

参考文献

[1] Ernst, W.G., Maruyama, S. and Wallis, S. Buoyancy-driven, rapid exhumation of ultrahigh-pressure metamorphosed continental crust. Proc. Natl. Acad. Sci. USA, 1997, 94: 9532-9537.

[2] Dodson, M. I. Closure temperature in cooling geochronological and petrological systems. Contr. Mineral. and Petrol., 1973, 40: 259-274.

[3] Harrison, T.M. Diffusion of ^{40}Ar in hornblende. Contr. Mineral. Petrol., 1981, 78: 324-331.

[4] Harrison, T.M., Duncan, I. and McDougall, I. Diffusion of ^{40}Ar in biotite: temperature, pressure and compositional effects, Geochim. Cosmochim. Acta, 1985, 49: 2461-2468.

[5] Mezger, K., Hanson, G.N. and Bohlen, S.R. High-precision U-Pb ages of metamorphic rutile: application to the cooling history of high-grade terranes. Earth and Planetary Science Letters, 1989, 96: 106-118.

[6] Mezger, K., Hanson G.N. and Bohlen, S.R. U-Pb systematics of garnet: dating the growth of garnet in the Late Archean Pikwitonei granulite domain at Cauchon and Natawahunan Lakes, Manitoba, Canada. Contrib Mineral Petrol, 1989, 101:136-148.

[7] Li, Q.L., Li, S. G., Zheng, Y.F., et al. A high precision U-Pb age of metamorphic rutile in coesite-bearing eclogite from the Dabie Mountains in central China: a new constraint on the cooling history. Chem. Geol., 2003, 200: 255-265.

[8] Cliff, R.A. Isotopic dating in metamorphic belts. J. Ged. Soc. London, 1985, 142: 97-110.

[9] Li, S.G., Jagoutz, E., Chen, Y.Z., et al. Sm-Nd and Rb-Sr isotopic chronology and cooling history of ultrahigh pressure metamorphic rocks and their country rocks at Shuanghe in the Dabie Mountains, Central China. Geochimica et Cosmochimica Acta, 2000, 64 (6): 1077-1093.

[10] Sun, W.D., Li, S.G., Sun, Y., et al. Mid-paleozoic collision in the north Qinling: Sm-Nd, Rb-Sr and ^{40}Ar/^{39}Ar ages and their tectonic implications. Journal of Asian Earth Sciences, 2002, 21: 69-76.

[11] Li, S.G., Hart, S. R. and Wu, T.S., Rb-Sr and Sm-Nd isotope dating of an early Precambrian splite-keratophyre sequence for Nd-isotopic homogenization in the mafic and felsic lavas during low-grade metamorphism. Precambrian Research, 1990, 47: 191-203.

[12] 李曙光等. 南秦岭勉略构造带黑沟峡变质火山岩的年代学和地球化学—古生代洋盆及其闭合年代的证据. 中国科学（D辑），1996, 26: 223-230 (英文版：Li, S.G., et al., Chronology and geochemistry of metavolcanic rocks from Heigouxia Valley in the Mian-Lue tectonic zone, South Qinling: evidence for a Paleozoic oceanic basin and its close time. Sciences in China (ser. D), 1996, 39: 300-310.)

[13] Li, S.G., et al. Collision of the North China and Yangtz Blocks and formation of coesite-bearing eclogites: timing and processes. Chemical Geology, 1993, 109: 89-111.

[14] Li, S.G., Wang, S.S., Chen, Y.Z., et al. Excess argon in phengite from eclogite: evidence from dating of eclogite minerals by Sm-Nd, Rb-Sr and ^{40}Ar/^{39}Ar methods,Chemical Geology, 1994, 112, 343-350.

[15] Xie, Z., Zheng, Y.F., Jahn, B. M., et al. Sm-Nd and Rb-Sr dating of pyroxene-garnetite from North Dabie in east-central China: problem of isotope disequilibrium due to retrograde metamorphism. Chemical Geology, 2004, 206: 137-158.

[16] Thoni, M. and Jagoutz, E. Some new aspects of dating eclogites in orogenic belts: Sm-Nd, Rb-Sr and Pb-Pb isotopic results from the Austroalpine Savalpine and Koralpe-type locality. Geochim. Cosmochim. Acta, 1992, 56: 347-368.

[17] Li, S.G., Jagoutz, E., Lo, C.H., et al. Sm-Nd, Rb-Sr and ^{40}Ar-^{39}Ar isotopic systematics of the ultrahigh pressure metamorphic rocks in the Dabie-Sulu belt, Central China: a retrospective view. International Geol. Review, 1999, 41(12): 1114-1124.

[18] Zheng, Y.F., Wang, Z.R., Li, S.G. et al. Oxygen isotope equilibrium between eclogite minerals and its constraints on mineral Sm-Nd chronometer. Geochim. Cosmochim. Acta, 2002, 66(4): 625-634.

[19] Wang, S.J., Li, S.G., An, S.C., et al. A granulite record of multistage metamorphism and REE behavior in the Dabie orogen: constraints from zircon and rock-forming minerals. Lithos, 2012, 136-139: 109-125.

[20] Cheng, H., Nakamura, E., Zhou, Z.Y. Garnet Lu-Hf dating of retrograde fluid activity during ultrahigh-pressure metamorphic eclogites exhumation. Miner, Petrol., 2009, 95: 315-326.

[21] Li, X.P., Zheng, Y., Wu, Y., et al. Low-T eclogite in the Dabie terrane of China: petrological and isotopic constrains on fluid activity and radiometric dating. Contributions to Mineralogy and Petrology, 2004, 148: 443-470.

Modification of the Sm-Nd isotopic system in garnet induced by retrogressive fluids

Shi-Chao An[1], Shu-Guang Li[1,2] and Zhen Liu[1]

1. CAS Key Laboratory of Crust-Mantle Materials and Environments, School of Earth and Space Sciences, University of Science and Technology of China, Hefei 230026, Anhui, China
2. State Key Laboratory of Geological Processes and Mineral Resources, China University of Geosciences, Beijing 100083, China

> 亮点介绍：该文发现含柯石英超高压榴辉岩的高级退变质作用在不导致石榴石分解的情况下，也会改造石榴子石的 Sm-Nd 同位素体系，导致矿物 Sm-Nd 等时线年龄变年轻。该工作通过严格分选大别山碧溪岭榴辉岩超高压变质石榴石和被退变质流体改造的石榴石，获得了与大别山其他榴辉岩一致的超高压变质 Sm-Nd 矿物等时线年龄，证明了前人报道的碧溪岭榴辉岩偏年轻的 Sm-Nd 同位素年龄是未区分超高压变质和退变质石榴石造成的。

Abstract Garnet, as a major constitutive mineral of eclogite, is important for Sm-Nd dating of eclogite due to its high Sm/Nd ratio and its stability during retrogression. However, a comprehensive study of the petrography, mineral chemistry, garnet water content, and Sm-Nd isotopic composition of eclogites from the Bixiling massif, Central Dabie Zone (CDZ), reveals significant modification of the Sm-Nd isotopic system in garnet as a result of retrogression. This problem constitutes a challenge for Sm-Nd dating of the Bixiling eclogites, with the Sm-Nd isochron ages of 218±4 to 210±9 Ma reported in the literature younger than 226±3 Ma, which is the generally accepted peak metamorphic age of the CDZ. Petrographic analysis reveals heterogeneity in colour within single fractured garnet grains. There are light-pink garnet (Grt-P) and red garnet (Grt-R) types that possess distinct chemical compositions. Compared to Grt-P, Grt-R has higher Fe and andradite contents but lower Al and grossular contents. Grt-P also has lower water contents (15−35 ppm) than Grt-R (34−65 ppm), which, together with the spatial association between Grt-R and fractures, suggests that the colour change is related to fluid alteration. Grt-P is an ultra-high-pressure (UHP) mineral, and Grt-R is the product of the interaction between Grt-P and a fluid during retrogression. Moreover, Grt-R features lower Sm and Nd contents but higher Sm/Nd ratios than Grt-P.

The Sm-Nd isochrons defined by UHP minerals (Grt-P + Omp + Rt or Grt-P + Cpx + WR) from three eclogite samples yield consistent ages of 226.0±3.8 Ma, 225.0±3.9 Ma and 226.2±6.9 Ma, which are identical to the peak metamorphic age of 226±3 Ma for the CDZ. The retrogressed garnet (i.e., Grt-R), omphacite and rutile, together define a pseudoisochron with younger ages of 218.9±5.9 to 202.8±4.8 Ma, which are geologically meaningless. The increase in the Sm/Nd ratio with constant or lower $^{143}Nd/^{144}Nd$ ratios during the transformation of Grt-P to Grt-R was probably the cause of these younger ages.

Keywords Dabie zone; eclogite; garnet; REE mobility during retrogression; Sm-Nd geochronology

1 Introduction

Utilizing Sm-Nd mineral isochrons constitutes one of the most useful and efficient methods for determining the age of eclogite and other garnet-bearing high-grade metamorphic rocks. Generally, garnet, with high Sm/Nd ratios (up to 1−2) that are much higher than those of other ultra-high-pressure (UHP) minerals, such as omphacite (Omp), rutile (Rt), and phengite (Ph), is a key mineral for obtaining accurate isochrons. Thus, the formation history and compositional variation of garnet are important for Sm-Nd isochron dating. Several studies have explored garnet

---

* 本文发表在：Journal of Metamorphic Geology, 2018, 36: 1039-1048

zonation and its influence on Sm-Nd dating (e.g., Gatewood et al., 2015; Pollington and Baxter, 2010). However, interaction between retrogressive fluid and garnet and its influence on Sm-Nd dating have not been studied.

The Dabie-Sulu belt in east-central China is the world's largest UHP belt and thus has been the focus of Sm-Nd studies to determine the time scale of its orogenic development. Since Li et al. (1989, 1993) reported the first group of Sm-Nd ages (from 244±11 to 208±33 Ma) for coesite-bearing eclogite from the Dabie-Sulu belt, numerous other studies have obtained Triassic Sm-Nd ages for the coesite-bearing eclogite, ranging between 240 and 200 Ma (e.g., Chavagnac and Jahn, 1996; Chavagnac et al., 2001; Li et al., 1994, 2000, 2004; Xie et al., 2004; Zhao et al., 2006). Although the large Sm-Nd age variation for the coesite-bearing eclogite mentioned above could be partly attributed to the different subduction and exhumation histories of the different UHP metamorphic (UHPM) slices in the Dabie-Sulu UHP belt (Liu and Li, 2008), the Sm-Nd age variation within a given UHPM slice, such as the different Sm-Nd ages for the Bixiling eclogite (218±4 to 210±9 Ma) (Chavagnac and Jahn, 1996) and the Shuanghe UHPM rocks (226±3 Ma) (Li et al., 2000) from the Central Dabie Zone (CDZ) (Fig. 1a), cannot be explained by this reason. Therefore, it is necessary to confirm the validity of these isochrons and identify the influences responsible for the age spread. Two critical questions regarding the Sm-Nd isotopic systems of UHPM minerals are (1) is equilibrium attained under peak conditions and (2) can this equilibrium be disturbed during retrogression. Given full consideration of these influencing factors, Li et al. (2000) evaluated the peak metamorphic ages of UHPM rocks from Shuanghe in the CDZ, using the Grt + Omp + Rt and Grt+ 2 Ph Sm-Nd isochrons for different types of UHP rocks, including coesite-bearing eclogite and UHP gneiss, which yielded two identical ages of 226±3 Ma. Some other studies of SHRIMP U-Pb dating of metamorphic zircon with coesite inclusions from UHPM rocks from the CDZ have reported ages of 225±4 Ma (Hacker et al., 1998) and 227±9 Ma (Liu et al., 2005), which are identical to the Sm-Nd ages reported by Li et al. (2000). These combined data therefore constrain the best estimate of the peak UHP metamorphic age of the CDZ to 226±3 Ma.

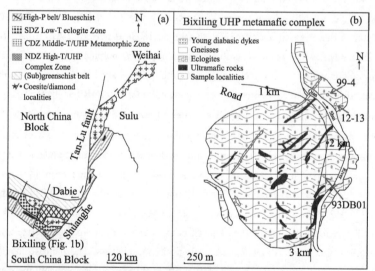

Fig. 1(a) Sketch map of the Dabie-Sulu orogenic belt and Tan-Lu fault in east-central China. (b) Simplified geological map of the Bixiling UHP complex. Maps are modified after Zhang et al. (1995). SDZ = South Dabie low-T eclogite Zone; CDZ = Central Dabie middle-T/UHP metamorphic Zone; NDZ = North Dabie high-T/UHP complex Zone.

Although it is now generally accepted that the peak metamorphism occurred at approximately 226±3 Ma in the CDZ (Li et al., 2000), the causes of other younger or older Sm-Nd ages remain unclear. Li et al. (1996) reported Sm-Nd ages ranging from 219 to 210 Ma for the Bixiling eclogites, similar to those reported by Chavanac and Jahn (1996), and suggested that the age variability could result from isotopic disequilibrium between the dated minerals. This disequilibrium was attributed to the coarse grains of the UHP minerals and the lack of hydrous minerals in their protoliths, which induced fluid-poor conditions during UHP metamorphism (Li et al., 1996).

In this study, petrographic analysis shows that the garnet from the Bixiling eclogites is variably fractured. Some fractured garnet pieces exhibit marked variation in colour between the core and the rim, and heavily fractured garnet pieces have different colour throughout. This feature indicates that some process might have altered the compositional attributes, including the isotopic composition, of the garnet. Thus, the aim of this work is to analyze these different colour variations to establish whether resetting induced by garnet-fluid interaction occurred and whether this setting could explain the variation in ages reported for these eclogites.

2 Geological setting and samples

The Dabie-Sulu orogenic belt is the result of Triassic continental collision between the North China Block (NCB) and the South China Block (SCB) (Fig. 1a) (Ames et al., 1993; Li et al., 1989, 1993). The Sulu orogen, as the eastern section of this orogenic belt, has been displaced 500 km to the north by the Tan-Lu fault (Xu et al., 1987; Ye et al., 2000). The occurrence of UHP minerals, such as micro-diamond and coesite, as inclusions in minerals in eclogites and gneisses indicates that the continental crust was subducted to mantle depths (Okay et al., 1989; Wang et al., 1989; Xu et al., 1992). Studies of the stable isotope composition and cooling history of the UHPM rocks revealed that the continental crust of the SCB experienced rapid subduction, a short mantle residence time, and rapid initial exhumation processes during continental collision (Li et al., 2000; Zheng et al., 2003; Zheng, 2009). The Dabie UHPM belt is usually subdivided into three UHPM slices with different subduction and exhumation histories, i.e., the North Dabie high-temperature/UHP complex zone (NDZ), the Central Dabie middle-temperature/UHP metamorphic zone (CDZ) and the South Dabie low-temperature eclogite zone (SDZ), which were progressively exhumed from the south to the north (Liu and Li, 2008) (Fig. 1a).

The Bixiling massif in the CDZ in the Dabie orogen is the largest UHP outcrop in the Dabie-Sulu belt, and is composed of a UHP meta-mafic and ultramafic complex occurring as a tectonic block within foliated UHP quartz-feldspathic gneisses (Fig. 1b). This complex mainly consists of banded eclogites with some elongated lenses of meta-ultramafic rocks (Zhang et al., 1995). Petrological and geochemical investigations suggest that the Bixiling eclogites had an igneous protolith with the gabbroic texture of cumulates (Chavagnac and Jahn, 1996; Xiao et al., 2000; Zhang et al., 1995).

Three eclogite samples, 93DB01, 99-4 and 12-13 were collected in the Bixiling massif for this study (Fig. 1b). A comprehensive investigation of the mineral chemistry, including the major and trace element compositions, water content and Sm-Nd isotopes, of sample 99-4 was conducted to establish whether compositional changes accompanied the colour variation in the garnet, thus potentially influencing the Sm-Nd dating of eclogite. In contrast, samples 12-13 and 93DB01 were selected for Sm-Nd dating only. The analytical methods used in this study are presented in Appendix S1. The major and trace element contents of garnet and clinopyroxene are listed in Tables S1 and S2, respectively. The water contents of the garnet, and the Sm and Nd concentrations and Nd isotope ratios of the Bixiling eclogites are listed in Tables 1 and 2, respectively.

3 Results

3.1 Petrography

The eclogites from Bixiling are coarse grained (~1–2 mm) and composed dominantly of garnet (55%) and clinopyroxene (35%–40%), with minor rutile and other accessory minerals (Fig. 2a). The mineral modes are estimated by the point counting method on three thin sections.

The petrographic analysis demonstrated that the most noteworthy feature of the sample is diversity in the colour of the garnet (Fig. 2a). The garnet grains are variably fractured and broken into pieces (Fig. 2b). The pieces typically have cores that are light pink, changing to red at the edges or where intensely fractured (Fig. 2c, d). Thus, the light-pink garnet and the red garnet are labelled Grt-P and Grt-R, respectively.

In contrast to garnet, the clinopyroxene is dominantly euhedral with no obvious colour difference in thin section (Fig. 2a). However, electron microprobe (EMP) analysis revealed that the clinopyroxene compositions vary from omphacite to aegirine-augite (see section 3.2). Although there is no evidence of a retrogressive event that has led to the breakdown of garnet + omphacite to lower grade hydrous assemblages such as amphibole + sodic plagioclase or symplectite, however, the evidence of the different garnet colour and the presence of altered clinopyroxenes (aegirine-augite) indicates some form of retrogression.

3.2 Mineral compositions

Compared to Grt-P, Grt-R has higher FeO_T and lower Al_2O_3 contents, i.e., FeO_T = 29.1±0.6 (wt %) and Al_2O_3 =19.9±0.6 in Grt-R (1SD, n=29) and FeO_T =27.5±1.1 and Al_2O_3=21.1±0.2 in Grt-P (1SD, n=31). Though the means 29.1±0.6 and 27.5±1.1 of FeO_T contents overlap, the t-Test shows that there are obvious differences between the two sets of data (see Appendix S2 for the detailed results of the t-Test). Correspondingly, the number of Fe^{3+} atoms per garnet formula unit is 0.15±0.03 in Grt-P and increases to 0.23±0.03 in Grt-R, although the number of Fe^{2+} atoms per garnet formula unit is similar in Grt-P and Grt-R (1.66±0.07 for Grt-P and 1.69±0.04 for Grt-R). Additionally, the number of Al atoms per garnet formula unit decreases from 1.96±0.02 in Grt-P to 1.86±0.05 in Grt-R (Table S1). Furthermore, although the number of Ca atoms per garnet formula unit overlaps in the range ~0.55–0.75 for the two garnet varieties, the Grt-P values extend to much higher values of ~0.80–0.90 (Fig. 3).

The andradite content increases significantly from 3.1±0.9 (mol %) in Grt-P to 7.9±2.3 in Grt-R, whereas the grossular content decreases from 18.6±3.9 in Grt-P to 12.4±2.1 in Grt-R (Table S1; Fig. 4b). The almandine and pyrope contents overlap in the ranges of ~54–58 and 19%–22%, respectively, in the two garnet types, but the Grt-P values extend to much lower values of ~50 and 17%, respectively (Figs. 4a, c).

The light rare earth element (LREE) values are lower than the detection limit of LA-ICP-MS and are thus not presented. Grt-P has a higher medium REE (MREE) and lower heavy REE (HREE) distribution pattern (Table S1; Fig. 5), which is typical of Ca-rich UHP garnet (e.g., Wang et al., 2012), while Grt-R has a lower MREE and higher HREE distribution pattern (Fig. 5).

Table 1 Water contents of Grt-P and Grt-R in sample 99-4, and related information.

Thickness (cm)	Grt-P			Grt-R		
	Spot	Area	Water (ppm)	Spot	Area	Water (ppm)
0.019	1	0.45	17	1	0.90	34
	2	0.69	26	2	1.48	56
	3	0.56	21	3	0.96	36
	4	0.39	15	4	1.60	61
	5	0.48	18	5	1.38	52
	6	0.45	17	6	1.38	52
	7	0.43	16	7	1.11	42
	8	0.56	21	8	1.38	52
	9	0.47	18	9	1.41	53
	10	0.48	18	10	0.90	34
	11	0.49	19	11	1.41	53
	12	0.48	18	12	1.50	57
	13	0.48	18	13	1.46	55
	14	0.47	18	14	1.48	56
	15	0.93	35	15	1.71	65
	16	0.58	22	16	1.38	52
	17	0.43	16	17	1.21	46
	18	0.58	22	18	1.44	55
	19	0.70	27	19	1.36	51
	20	0.56	21	20	1.23	47
	21	0.46	17	21	1.40	53

The clinopyroxene has a wide jadeite (Jd) range from 11.3% to 29.9% (Table S2) and exhibits a trend of increasing Fe from omphacite compositions to aegirine-augite compositions (Fig. 6). Although the clinopyroxene has no significant colour variations in thin section, its separates show obvious colour change from light green to dark green. The fresh clinopyroxene separates with a light green colour hand-picked for dating all have distinct omphacite compositions with a narrow Jd range of 26.9%–33.3% (Table S2; Fig. 6).

3.3 Garnet water contents

Grt-P has low water contents of 15–35 ppm, with a mean of 20±5 ppm (1SD, n=21), whereas Grt-R has high water contents of 34–65 ppm, with a mean of 51±8 ppm (1SD, n=21) (Table 1; Fig. 7). Consequently, the differences in the garnet water content between Grt-P and Grt-R are significant within error.

Table 2 Sm and Nd concentration and Nd isotope ratios of three eclogite samples from the Bixiling massif

Sample No.	Rock or mineral	Mass(mg)	Sm(ppm)	Nd(ppm)	Sm loaded(ng)	Nd loaded(ng)	$^{147}Sm/^{144}Nd$	$^{143}Nd/^{144}Nd\pm 2\sigma$
99-4	Grt-P	85.2	1.3895	0.6065	115	50	1.3855	0.514445±8
	Grt-R	130.7	1.3137	0.5270	166	67	1.5075	0.514496±7
	Omp	28.7	0.8803	1.7749	24	48	0.2995	0.512825±9
	Rt1	116.4	0.1197	0.3848	13	43	0.1881	0.512675±13
	Rt2	99.8	0.0923	0.4122	8.9	38	0.1354	0.512603±11
12-13	Grt-P	152.5	1.0818	0.4659	162	70	1.4042	0.514459±5
	Grt-R	148.6	1.0525	0.3861	155	56	1.6485	0.514694±6
	Omp	67.2	0.6711	1.3441	44	89	0.3019	0.512835±10
	Rt	195.1	0.1778	1.4199	33	268	0.0757	0.512503±14
93DB01	Grt-P	153.6	0.8670	0.6329	130	95	0.8280	0.513539±15
	Grt-R	177.2	0.8037	0.4332	140	73	1.1220	0.513825±17
	Grt-W	195.6	0.8166	0.5181	155	94	0.9526	0.513685±10
	Cpx	24.4	3.620	13.35	84	275	0.1639	0.512561±12
	WR	54.4	1.518	6.960	74	317	0.1318	0.512504±13

3.4 Sm-Nd isotopic data

In all samples, Grt-P, Grt-R and omphacite were chosen for isotopic dating. Notably, for sample 93DB01, besides Grt-P, Grt-R and omphacite, dating was performed on garnet grains that were not separated on the basis of colour (Grt-W) and on whole rock (WR) samples. In addition, two rutile separates (Rt1 and Rt2) were collected from sample 99-4, and only one was collected from sample 12-13.

For sample 99-4, the four UHPM minerals Grt-P + Omp + 2 Rt define a regression line that yields an age of 226.0±3.8 Ma, while Grt-R + Omp + 2 Rt defines a regression line that yields an age of 211.0 ±3.4 Ma (Fig. 8a). Similarly, for sample 12-13, the regression line defined by Grt-P + Omp + Rt yields an age of 225.0±3.9 Ma, and that by Grt-R+ Omp+ Rt yields an age of 212.2±3.2 Ma (Fig. 8b). In sample 93DB01, three mineral regression lines are obtained. The line defined by Grt-P + Omp + WR yields an age of 226.2±6.9 Ma, that by Grt-R + Omp + WR yields an age of 202.8±4.8 Ma, and that by Grt-W + Omp + WR yields an age of 218.9±5.9 Ma (Fig. 8c).

4 Discussion

The garnet in the Bixiling eclogites varies in colour, and Grt-P and Grt-R possess distinct geochemical characteristics. Two pertinent questions then arise: (1) what process caused this phenomenon, and (2) how does this process influence the garnet REE compositions and Sm-Nd dating of the Bixiling eclogites?

4.1 Genesis of Grt-P and Grt-R

Figs. 2c,d show that Grt-P is located mainly at the core of the garnet pieces, while Grt-R typically occurs at the rims of the garnet pieces or in intensely fractured areas. Thus, Grt-R appears to have originated from some retrogressive metamorphic process after garnet fracturing.

The garnet water content may provide direct evidence of the relation between Grt-R and an invasive fluid. In general, pressure is considered to be the main factor controlling garnet water content in UHPM rocks, with two main trends involving either a diffusional loss of hydrogen during early exhumation (e.g., Sheng, et al., 2007), or an increase in structural hydroxyl with increasing pressure (e.g., Katayama et al., 2006; Lu and Keppler, 1997).

Fig. 2 Photomicrographs of sample 99-4. (a) This sample is dominantly garnet and clinopyroxene with minor rutile. (b) The area bounded by the blue dashed line is an elongate garnet grain, which is broken into small pieces by fractures, and the diversity of colour in the garnet pieces is shown in this sample. The areas bounded by the red lines are enlarged in panel (c) and (d). (c) The area bounded by the solid circle shows a large garnet piece with Grt-P at the core and Grt-R at the rim. The area bounded by the dashed circle shows that Grt-R is also located in an intensely fractured area. (d) Another example similar to panel (c). Mineral abbreviations: Grt = garnet; Cpx = clinopyroxene; Rt = rutile; Ap = apatite.

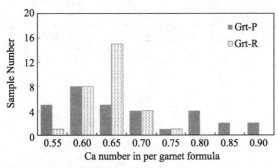

Fig. 3 Histograms of the Ca number in garnet formula in the Grt-P and Grt-R in sample 99-4.

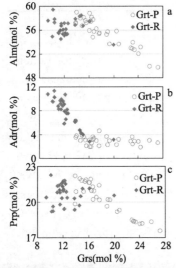

Fig.4 Endmember content plots of grossular (Grs) vs. (a) almandine (Alm); (b) andradite (Adr); and (c) pyrope (Prp) for Grt-P and Grt-R in sample 99-4.

In both cases, Grt-R, as a retrograded product, is predicted to have a lower water content than Grt-P. However, Grt-P exhibits low water contents of 15-35 ppm, while Grt-R has higher water contents of 34-65 ppm (Table 1). Thus, it is reasonable to conclude that the formation of Grt-R is related to mineral-fluid interaction, which resulted in the compositional differences between the two garnet types in the Bixiling eclogite. For clinopyroxenes, omphacite transformed to aegirine-augite (Fig. 6) can be resulted from two possible metamorphic reactions. (1) The interaction between Fe rich fluid and omphacite, similar to Grt-P transforming to Grt-R, as shown above. (2) Retrogression of eclogite under P-T conditions of granulite facies, leading decomposition of omphacite to augite + sodic plagioclase. The latter case can be excluded because the retrogressive assemblage (augite + sodic plagioclase) has not been observed in the samples of this study. Hence, it is more likely that Grt-R and aegirine-augite are both products of the interaction between UHP minerals and Fe rich fluid.

The retrogression assemblage of sample 99-4, i.e., Grt-R + aegirine-augite without symplectite or amphibole + sodic plagioclase, indicates the metamorphic grade was higher than amphibolite facies. Thus, this retrogression likely occurred during the initial stage of exhumation of the UHPM slice, during which the garnet and pyroxene did not transform into amphibole + plagioclase or symplectite.

The Grt-R could be formed by recrystallization from a retrogressive fluid through a dissolution and reprecipitation mechanism (e.g., Chen et al., 2015; Gatewood et al., 2015) or by element exchange reaction between the garnet and the fluid. The Cpx-fluid interaction resulted in an increase in Fe from the omphacite to the aegirine-augite without decomposition of the clinopyroxene, again suggesting that the interaction occurred under metamorphic conditions higher than amphibolite facies.

Notably, Grt-R possesses significantly more Fe^{3+} (0.23±0.03) than Grt-P (0.15±0.03) (number per formula unit) but similar Fe^{2+} values (Table S1), which means that the fluid was Fe-rich and oxidizing.

4.2 Modification of the garnet REE system

The general formula of anhydrous natural silicate garnet can be expressed as $\{X_3\}[Y_2](Z_3)\Phi_{12}$, where X = Na, Mg, Ca, Mn^{2+}, Fe^{2+}, or Y; Y = Mg, Al, Si, Sc, Ti, V, Cr, Mn^{3+}, Fe^{3+}, or Zr; Z = Al, Si, or Fe^{3+}; and Φ corresponds to the anion O or F (Locock, 2008). The REEs all partition exclusively into the garnet X site (Van et al. 1999), and thus variation in the major cations in the X site may influence the REE contents and distribution patterns of garnet. For the most common Mg^{2+} or Fe^{2+}-Ca^{2+} exchanges in natural silicate garnet, coupled REE changes may be theoretically predicted.

According to Shannon (1976), the effective radii of REE^{3+} for eightfold coordination, from La^{3+} to Lu^{3+}, decrease from 1.16 to 0.97 Å (1.11 Å for Nd^{3+} and 1.08 Å for Sm^{3+}), which is referred to as lanthanide contraction, and the effective radii of Ca^{2+}, Mg^{2+} and Fe^{2+} are 1.12, 0.89 and 0.92 Å, respectively. Theoretically, when Mg^{2+} or Fe^{2+} is replaced by Ca^{2+}, the MREE content could increase, while the HREE content could decrease (e.g., Chen et al., 2015; Wang et al., 2012).

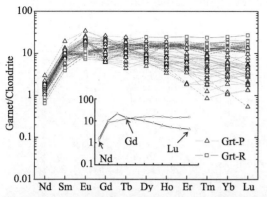
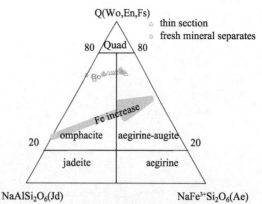

Fig. 5 Chondrite-normalized REE patterns of Grt-P and Grt-R in sample 99-4. The chondrite values are after Sun and McDonough (1989). Note that La, Ce, and Pr are absent because the LREE compositions in garnet are below the detection limit of LA-ICPMS. Two lines in the lower left inset denote the average values of chondrite-normalized REEs for Grt-P and Grt-R.

Fig. 6 Ternary diagram of the endmember contents of clino-pyroxene in sample 99-4. The green triangles denote pyroxene domains in thin section, and the red circles denote pyroxene separates selected for Sm-Nd dating.

In sample 99-4, the almandine and pyrope in Grt-R overlap with those of Grt-P in high values (Figs. 4a, c), the andradite increased from 3.1±0.9 (mol %) in Grt-P to 7.9±2.4 in Grt-R (Fig. 4b; Table S1) and the grossular decreased from 18.6±3.9 in Grt-P to 12.4± 2.2 in Grt-R (Fig. 4). Additionally, the Ca number decreased from Grt-P to Grt-R (Fig. 3), suggesting that Ca^{2+}-Fe^{2+} or Ca^{2+}-Mg^{2+} exchange in the X site was associated with the Al^{3+}-Fe^{3+} exchange in the Y site during the transformation from Grt-P to Grt-R. The contrasting REE patterns between Grt-P (enriched in MREEs and depleted in HREEs), and Grt-R (depleted in MREEs and flat HREEs) (inset Fig. 5) reflect REE modifications accompanying the Ca^{2+}-Fe^{2+} or Ca^{2+}-Mg^{2+} exchange during the garnet-fluid interaction. Specifically, the Nd to Gd concentrations decreased in Grt-R because their ionic radii are closer to that of Ca^{2+} than to that of Fe^{2+} or Mg^{2+}, while the Dy to Lu concentrations increased in Grt-R because their ionic radii are closer to that of Fe^{2+} or Mg^{2+} than that of Ca^{2+}. Notably, the Nd content in Grt-R decreased significantly more than the Sm content because, in comparison to Sm^{3+}, the radius of Nd^{3+} is larger and closer to that of Ca^{2+}, resulting in a higher Sm/Nd ratio in Grt-R than in Grt-P.

4.3 Influencing the Sm-Nd dating of the Bixiling eclogites

The three ages of 226.0±3.8 Ma, 225.0±3.9 Ma, and 226.2±6.9 Ma based on Grt-P + Omp + Rt or WR from the three samples in this study are identical to each other within error, which indicates that isotopic equilibrium was achieved among the UHP minerals. Moreover, these ages are in agreement with the generally accepted peak metamorphic time of the CDZ, i.e., 226±3 Ma (Li et al., 2000). Consequently, they are isochrons representing the peak metamorphic time of the Bixiling massif (Fig. 8). The Grt-R (or Grt-W) + Omp + Rt (or WR) ages (218.9±5.9 to 202.8±4.8 Ma) for the three samples, which are similar to the Sm-Nd ages (218±4 to 210±9 Ma) reported in Chavagnac and Jahn (1996), are younger than the above peak metamorphic ages and exhibit a wide range, indicating isotopic

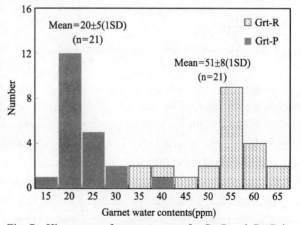

Fig. 7 Histograms of water contents for Grt-P and Grt-R in sample 99-4. The blue bar and red bar indicate Grt-P and Grt-R, respectively, which can be clearly distinguished from each other by colour during the FT-IR analysis.

disequilibrium between the UHP minerals (Omp, Rt) and retrograde garnet (i.e., Grt-R). Thus, these ages are geologically meaningless.

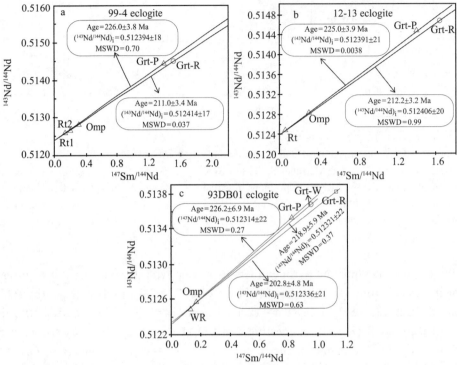

Fig. 8 Sm-Nd isochrons for three eclogites from the Bixiling massif: (a) sample 99-4; (b) sample 12-13; (c) sample 93DB01.

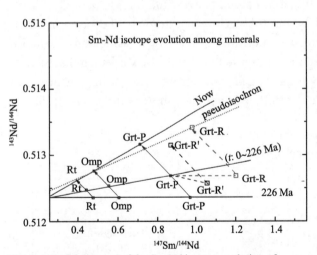

Fig. 9 Model diagram of the Sm-Nd isotope evolution of minerals in the Bixiling eclogites. See text for further details.

As mentioned above, the REE compositions of the garnet in the Bixiling eclogite have been modified by interaction between the minerals and a retrogressive fluid. The REE modification of garnet during retrogression can significantly influence the Sm-Nd isotopic dating of eclogite, as shown in the Sm-Nd isotopic evolution diagram (Fig. 9).

When eclogite formed during the UHPM period, the Nd isotopic system of UHPM minerals, such as garnet, omphacite and rutile, reached equilibrium (Li et al., 2000), and thus all UHPM minerals have identical $^{143}Nd/^{144}Nd$ ratios. The eclogite was then subject to retrogression during the subsequent exhumation of the UHPM slices. During exhumation, Grt-P was modified by a fluid, forming Grt-R. The Sm/Nd ratio of Gar-R increased, as mentioned above. However, the Nd isotopic composition can change in three possible ways: (1) The Nd isotopic composition remains constant because, as a heavy element with a high mass number, it exhibits little isotopic fractionation during the transformation of Grt-P to Grt-R. In this case, Grt-R is regarded to be the product of Grt-P by translation rightward in the Sm-Nd isotopic evolution diagram at time t (horizontal red dashed line in Fig. 9). The Sm-Nd system then closed again, and the reset isotopic clock resumed to the present day. Consequently, Grt-P + Omp + Rt would define an isochron corresponding to the timing of UHP metamorphism, while Grt-R + Omp + Rt would define a pseudoisochron with a lower slope, yielding a younger age. (2) The Nd isotope value of the fluid is lower than that of garnet, resulting in a decrease in the $^{143}Nd/^{144}Nd$ ratio in Grt-R relative to Grt-P. Grt-P then moves downward and rightward to form Grt-R' (blue sub-horizontal dashed line in Fig. 9), eventually yielding a younger age, similar to case (1). (3) The fluid has a higher Nd isotope value than the garnet. This case is obviously less likely, because garnet usually has the highest Sm/Nd and ^{143}Nd/^{144}Nd ratios in an eclogite system.

5 Conclusions

The eclogite from Bixiling underwent rock-fluid interaction during the initial stage of exhumation of the UHP slices. In addition, the Sm-Nd isotopic system of garnet in eclogite can be modified by iron-rich and oxidizing fluids. The increased Sm/Nd ratio in such modified garnet is probably responsible for Sm-Nd ages that postdate the actual eclogite peak metamorphism of the eclogite in the literature. Caution regarding compositional modification of garnet by mineral-fluid interaction during retrogression is thus suggested, and this factor should be taken into account to correctly apply the Sm-Nd method to date eclogite or other garnet-bearing high-grade metamorphic rocks.

Acknowledgement

This study was financially supported by the Natural Science Foundation of China (Nos. 40973016 and 41473036). Comments and suggestions on the manuscript, as well as help with the English from Editor Dr. D. Robinson, and Dr. Baxter, as well as another anonymous reviewer, are greatly appreciated. We thank Emil Jagoutz for the analysis of the mineral Sm-Nd isotopes of samples 93DB01 at MPI in Germany and Zhuyin Chu for the analysis of mineral Sm-Nd isotopes in samples 99-4 and 12-13 at IGG, CAS in China. We also express gratitude to Min Feng and Jiangling Xu for the EMP analysis of the mineral major element contents, Zhaochu Hu and Wenxiu Tang for the LA-ICPMS analysis of the mineral trace element contents, and Ji Shen for his assistance with field work.

References

Ames, L., Tilton, G. R., & Zhou, G. Z. (1993). Timing of collision of the Sino-Korean and Yangtze Cratons—U-Pb zircon eating of coesite-bearing eclogites. Geology, 21, 339-342.

Chavagnac, V. & Jahn, B.M. (1996). Coesite-bearing eclogites from the Bixiling Complex, Dabie Mountains, China: Sm-Nd ages, geochemical characteristics and tectonic implications. Chemical Geology, 133, 29-51.

Chavagnac, V., Jahn, B. M., Villa, I. M., Whitehouse, M. J., & Liu, D. Y. (2001). Multichronometric evidence for an in situ origin of the ultrahigh-pressure metamorphic terrane of Dabieshan, China. Journal of Geology, 109, 633-646.

Chen, Y. X., Zhou, K., Zheng, Y. F., Chen, R. X., & Hu, Z. C. (2015). Garnet geochemistry records the action of metamorphic fluids in ultrahigh-pressure dioritic gneiss from the Sulu orogen. Chemical Geology, 398, 46-60.

Gatewood, M. P., Dragovic, B., Stowell, H. H., Baxter, E. F., Hirsch, D. M., & Bloom, R., (2015). Evaluating chemical equilibrium in metamorphic rocks using major element and Sm-Nd isotopic age zoning in garnet, Townshend Dam, Vermont, USA. Chemical Geology, 401, 151-168.

Hacker, B. R., Ratschbacher, L., Webb, L., Ireland, T., Walker, D., & Dong, S. (1998). U/Pb zircon ages constrain the architecture of the ultrahigh-pressure Qinling-Dabie Orogen, China. Earth and Planetary Science Letters, 161, 215-230.

Katayama, I., Nakashima, S., & Yurimoto, H. (2006). Water content in natural eclogite and implication for water transport into the deep upper mantle. Lithos, 86, 245-259.

Li, S.G., Jagoutz, E., Chen, Y. Z., & Li, Q. (2000). Sm-Nd and Rb-Sr isotopic chronology and cooling history of ultrahigh pressure metamorphic rocks and their country rocks at Shuanghe in the Dabie Mountains, Central China. Geochimica et Cosmochimica Acta, 64, 1077-1093.

Li, S.G., Hart, S. R., Zheng, S. G., Liu, D. L., Zhang, G. W., & Guo, A. L. (1989). Timing of collision between the North and South-China Blocks—The Sm-Nd isotopic age evidence. Science in China (Series B), 32, 1393-1400.

Li, S.G., Jagoutz, E., Xiao, Y. L., Ge, N. J., & Chen, Y. Z. (1996). Geochronology of Dabie-Sulu UHP terrane—— I. Sm-Nd isotopic system. Science in China (series D), 26, 249-257.

Li, S.G., Wang, S. S., Chen, Y. Z., Liu, D. L., Qiu, J., Zhou, H. X., & Zhang, Z. M. (1994). Excess argon in phengite from eclogite: Evidence from dating of eclogite minerals by Sm-Nd, Rb-Sr and ^{40}Ar/^{39}Ar methods. Chemical Geology, 112, 343-350.

Li, S.G., Xiao, Y. L., Liou, D. L., Chen, Y. Z., Ge, N. J., Zhang, Z. Q., Sun, S.S., Cong, B., Zhang, R.Y., Hart, S.R., & Wang, S. S. (1993). Collision of the North China and Yangtse Blocks and formation of coesite—bearing eclogites—timing and processes.

Chemical Geology, 109, 89-111.

Li, X. P., Zheng, Y. F., Wu, Y. B., Chen, F., Gong, B., & Li, Y. L. (2004). Low-T eclogite in the Dabie terrane of China: petrological and isotopic constraints on fluid activity and radiometric dating. Contributions to Mineralogy and Petrology, 148, 443-470.

Liu, F., Liou, J. G., & Xu, Z. (2005). U-Pb SHRIMP ages recorded in the coesite-bearing zircon domains of paragneisses in the southwestern Sulu terrane, eastern China: New interpretation. American Mineralogist, 90, 790-800.

Liu, Y.C. and Li, S.G. (2008). Detachment within subducted continental crust and multi-slice successive exhumation of ultrahigh-pressure metamorphic rocks: Evidence from the Dabie-Sulu orogenic belt. Chinese Scicence Bulettin, 53, 3105-3119.

Liu, Y. S., Gao, S., Hu, Z. C., Gao, C. G., Zong, K. Q., & Wang, D. B. (2010). Continental and oceanic crust recycling-induced melt-peridotite interactions in the Trans-North China Orogen: U-Pb dating, Hf isotopes and trace elements in zircons from mantle xenoliths. Journal of Petrology, 51, 537-571.

Locock, A. J. (2008). An Excel spreadsheet to recast analyses of garnet into end-member components, and a synopsis of the crystal chemistry of natural silicate garnets. Computers & Geosciences, 34, 1769-1780.

Lu, R., & Keppler, H. (1997). Water solubility in pyrope to 100 kbar. Contributions to Mineralogy and Petrology, 129, 35-42.

Okay, A. I., Xu, S. T., & Sengör, A. M. C. (1989). Coesite from the Dabie Shan eclogites, Central China. European Journal of Mineralogy, 1, 595-598.

Pollington, A. D., & Baxter, E. F. (2010) High resolution Sm-Nd garnet geochronology reveals the uneven pace of tectonometamorphic processes. Earth and Planetary Science Letters, 293, 63-71.

Shannon, R. D. (1976). Revised effective ionic-radii and systematic studies of interatomic distances in halides and chalcogenides. Acta Crystallographica Section A, 32, 751-767.

Sheng, Y.M., Xia, Q.K., Dallai, L., Yang, X.Z., & Hao, Y. T. (2007). H_2O contents and D/H ratios of nominally anhydrous minerals from ultrahigh-pressure eclogites of the Dabie orogen, eastern China. Geochimica et Cosmochimica Acta, 71, 2079-2103.

Sun, S.S., & McDonough, W. F. (1989). Chemical and isotopic systematics of oceanic basalts: implications for mantle composition and processes. Geological Society, London, Special Publications, 42, 313-345.

Van, W., Blundy, J., & Wood, B. (1999). Crystal-chemical controls on trace element partitioning between garnet and anhydrous silicate melt. American Mineralogist, 84, 838-847.

Wang, S. J., Li, S.G., An, S. C., & Hou, Z. H. (2012). A granulite record of multistage metamorphism and REE behavior in the Dabie orogen: constraints from zircon and rock-forming minerals. Lithos, 136, 109-125.

Wang, X.M., Liou, J. G., & Mao, H. K. (1989). Coesite-bearing eclogite from the Dabie Mountains in Central China. Geology, 17, 1085-1088.

Xiao, Y. L., Hoefs, J., Van Den Kerkhof, A. M., Fiebig, J., & Zheng, Y. F. (2000). Fluid history of UHP metamorphism in Dabie Shan, China: a fluid inclusion and oxygen isotope study on the coesite-bearing eclogite from Bixiling. Contributions to Mineralogy and Petrology, 139, 1-16.

Xie, Z., Zheng, Y. F., Jahn, B. M., Ballevre, M., Chen, J. F., Gautier, P., Gao, T.S., Gong, B & Zhou, J. B. (2004). Sm-Nd and Rb-Sr dating of pyroxene-garnetite from North Dabie in east-central China: problem of isotope disequilibrium due to retrograde metamorphism. Chemical Geology, 206, 137-158.

Xu, J. W., Zhu, G., Tong, W. X., Cui, K. R., & Liu, Q. (1987). Formation and evolution of the Tancheng-Lujiang Wrench Fault System: a major shear system to the northwest of the Pacific-Ocean. Tectonophysics, 134, 273-310.

Xu, S. T., Wen, S., Liu, Y.C., Jiang, L.L., Jr, S.Y., Okay, A. I., & Sengör, A. M. C. (1992). Diamond from the Dabie-Shan metamorphic rocks and its implication for tectonic setting. Science, 256, 80-82.

Ye, K., Cong, B. L., & Ye, D. I. (2000). The possible subduction of continental material to depths greater than 200 km. Nature, 407, 734-736.

Zhang, R. Y., Liou, J. G., & Cong, B. L. (1995). Talc-bearing, magnesite-bearing and ti-clinohumite-bearing ultrahigh-pressure meta-mafic and ultramafic complex in the Dabie Mountains, China. Journal of Petrology, 36, 1011-1037.

Zhao, Z. F., Zheng, Y. F., Gao, T. S., Wu, Y. B., Chen, B., Chen, F. K., Wu, F. Y. (2006). Isotopic constraints on age and duration of fluid-assisted high-pressure eclogite-facies recrystallization during exhumation of deeply subducted continental crust in the Sulu orogen. Jounal of Metamorphic Geology, 24, 687-702.

Zheng, Y. F. (2009). Fluid regime in continental subduction zones: petrological insights from ultrahigh-pressure metamorphic rocks. Journal of the Geological Society, 166, 763-782.

Zheng, Y. F., Fu, B., Gong, B., & Li, L. (2003). Stable isotope geochemistry of ultrahigh pressure metamorphic rocks from the Dabie-Sulu orogen in China: implications for geodynamics and fluid regime. Earth Science Reviews, 62, 105-161.

Supporting Information

Table S1 Major (wt%) and rare earth element (ppm) compositions for Grt-P and Grt-R in sample 99-4.

	Grt-P														
No.	1	2	3	4	5	6	7	8	9	10	11	12	13	14	15
SiO_2	37.46	37.23	37.18	37.18	37.5	37.61	37.32	37.54	37.13	37.83	37.24	37.53	37.37	37.73	37.64
TiO_2	0	0.03	0	0.05	0.06	0.03	0.05	0	0.04	0	0	0	0.02	0.04	0.05
Al_2O_3	20.84	21.03	20.99	21.06	21.27	21.02	20.28	20.86	21.25	21.24	21.2	21.31	21.18	21.12	20.9
Cr_2O_3	0.03	0.01	0.03	0.03	0.01	0.03	0.02	0.03	0.02	0	0	0	0.02	0	0
FeO	28.57	28.83	28	28.07	27.84	28.3	29.1	29.15	27.5	28.02	26.35	25.91	24.92	28.08	26.07
MnO	0.32	0.39	0.4	0.37	0.32	0.43	0.42	0.33	0.44	0.32	0.25	0.39	0.36	0.35	0.36
MgO	5.33	5.68	5.34	5.21	5.6	5.62	5.05	5.34	5.17	5.54	4.64	4.65	4.49	5.37	4.7
CaO	6.69	6.25	7.18	7.24	6.56	6.33	7.22	6.37	7.99	6.42	9.58	9.35	10.66	6.8	9.53
NiO	0.03	0.05	0	0	0.06	0	0	0	0.03	0.06	0	0	0	0.01	0.02
Total	99.26	99.5	99.11	99.21	99.22	99.37	99.46	99.62	99.57	99.42	99.27	99.13	99.02	99.5	99.25
Si	2.96	2.93	2.94	2.93	2.95	2.96	2.95	2.96	2.92	2.98	2.93	2.96	2.94	2.97	2.96
Ti	0	0	0	0	0	0	0	0	0	0	0	0	0	0	0
Al	1.94	1.95	1.95	1.96	1.97	1.95	1.89	1.94	1.97	1.97	1.97	1.98	1.97	1.96	1.94
Cr	0	0	0	0	0	0	0	0	0	0	0	0	0	0	0
Fe^{3+}	0.14	0.18	0.18	0.16	0.11	0.12	0.21	0.15	0.2	0.08	0.17	0.11	0.15	0.1	0.13
Fe^{2+}	1.74	1.71	1.67	1.69	1.72	1.74	1.72	1.77	1.61	1.76	1.56	1.6	1.49	1.74	1.59
Mn	0.02	0.03	0.03	0.02	0.02	0.03	0.03	0.02	0.03	0.02	0.02	0.03	0.02	0.02	0.02
Mg	0.63	0.67	0.63	0.61	0.66	0.66	0.6	0.63	0.61	0.65	0.54	0.55	0.53	0.63	0.55
Ca	0.57	0.53	0.61	0.61	0.55	0.53	0.61	0.54	0.67	0.54	0.81	0.79	0.9	0.57	0.8
Uv	0.1	0	0.1	0.1	0	0.1	0.1	0.1	0.1	0	0	0	0.1	0	0
Sps	0.7	0.9	0.9	0.8	0.7	1	0.9	0.7	1	0.7	0.6	0.9	0.8	0.8	0.8
Prp	20.9	22.2	20.9	20.4	21.9	22	19.8	20.9	20.2	21.6	18.2	18.2	17.6	21	18.4
Alm	58.1	57.1	55.8	56.3	57.5	58	57.3	59	53.7	58.8	52.1	53.2	49.8	58.2	52.9
Grs	15.2	13.9	16.8	17.2	16.3	14.6	13.9	14	19.3	16.1	24.1	24.4	27.2	16.4	23
Adr	3.6	3.5	3.4	3	2	3	6.3	3.8	3	2	2.8	1.9	2.6	2.6	3.6
other	1.4	2.4	2.2	2.2	1.6	1.3	1.7	1.5	2.8	0.8	2.3	1.5	2	1.1	1.3
Nd	0.77	0.82	0.73	0.71	0.81	0.71	0.65	0.62	0.95	0.73	0.63	0.62	0.85	0.77	0.69
Sm	1.57	1.64	1.72	1.52	1.48	1.38	1.46	1.28	2.06	1.33	1.25	1.54	1.6	1.34	1.55
Eu	1.18	1.27	0.96	1.14	1.01	1.22	0.89	1.04	1.98	0.97	1.33	1.18	1.22	1.18	1.11
Gd	1.51	3.97	1.86	1.94	3.47	2.94	1.42	2.09	4.85	3.64	1.31	1.51	1.94	3.82	1.48
Tb	0.33	0.63	0.33	0.45	0.62	0.52	0.3	0.44	0.52	0.41	0.23	0.24	0.45	0.72	0.37
Dy	2.52	3.91	1.88	2.66	4.29	2.19	2.03	3.45	2.18	2.33	1.26	2.19	3.31	4.98	2.17
Ho	0.31	0.8	0.43	0.58	0.65	0.35	0.36	0.7	0.49	0.38	0.3	0.51	0.48	0.95	0.29
Er	1.04	1.61	0.8	1.18	1.65	0.76	1.03	2.12	1.22	0.86	0.58	1.19	1.43	1.44	1.08
Tm	0.11	0.23	0.1	0.2	0.18	0.05	0.14	0.25	0.19	0.08	0.09	0.24	0.18	0.17	0.15
Yb	0.75	1.03	0.97	0.81	0.94	0.48	0.84	2.03	0.77	0.71	0.39	1.37	0.91	0.79	1.04
Lu	0.17	0.11	0.14	0.13	0.09	0.05	0.11	0.35	0.08	0.07	0.04	0.2	0.09	0.11	0.14

Continued

	Grt-P															
No.	16	17	18	19	20	21	22	23	24	25	26	27	28	29	30	31
SiO_2	37.61	37.57	37.23	37.44	37.31	37.36	37.51	37.68	37.6	37.71	37.1	37.87	37.03	37.54	37.47	37.42
TiO_2	0.01	0	0.02	0.02	0	0	0.03	0	0.09	0.08	0.04	0	0.04	0	0	0
Al_2O_3	21.05	21.41	21.24	21.13	21.05	21.18	20.57	21.11	21.25	20.99	21.1	21.16	21.3	21.22	21.17	21.19
Cr_2O_3	0	0	0.04	0	0.02	0	0.01	0.01	0	0	0	0.01	0.05	0.01	0	0.04
FeO	28.44	27.79	28.19	26.31	26.93	26.65	27.4	27.9	25.29	28.29	28.21	28.37	27.74	26.44	27.16	28.22
MnO	0.3	0.36	0.46	0.3	0.45	0.33	0.37	0.33	0.31	0.44	0.36	0.37	0.31	0.34	0.38	0.44
MgO	5.18	5.2	5.58	4.68	4.93	4.73	5.14	5.06	4.72	5.61	5.45	5.6	5.47	4.7	4.89	5.56
CaO	7.09	7.61	6.33	9.07	8.54	9.14	7.93	8.07	10.45	6.8	7.25	6.64	7.08	9.41	8.06	6.87
NiO	0	0.01	0	0.08	0	0	0	0	0	0	0.06	0	0.01	0	0.02
Total	99.69	99.94	99.09	99.03	99.23	99.39	98.95	100.15	99.7	99.93	99.51	100.08	99.01	99.66	99.14	99.75
Si	2.96	2.94	2.94	2.96	2.94	2.94	2.97	2.94	2.94	2.95	2.92	2.96	2.92	2.94	2.96	2.93
Ti	0	0	0	0	0	0	0	0	0.01	0	0	0	0	0	0	0
Al	1.95	1.97	1.98	1.97	1.95	1.96	1.92	1.94	1.96	1.94	1.95	1.95	1.98	1.96	1.97	1.96
Cr	0	0	0	0	0	0	0	0	0	0	0	0	0	0	0	0
Fe^{3+}	0.14	0.15	0.14	0.12	0.16	0.16	0.15	0.17	0.15	0.15	0.21	0.13	0.17	0.15	0.12	0.17
Fe^{2+}	1.73	1.67	1.72	1.62	1.61	1.59	1.67	1.66	1.5	1.7	1.64	1.73	1.66	1.58	1.68	1.68
Mn	0.02	0.02	0.03	0.02	0.03	0.02	0.02	0.02	0.02	0.03	0.02	0.02	0.02	0.02	0.03	0.03
Mg	0.61	0.61	0.66	0.55	0.58	0.55	0.61	0.59	0.55	0.66	0.64	0.65	0.64	0.55	0.58	0.65
Ca	0.6	0.64	0.54	0.77	0.72	0.77	0.67	0.68	0.87	0.57	0.61	0.56	0.6	0.79	0.68	0.58
Uv	0	0	0.1	0	0.1	0	0	0	0	0	0	0.2	0	0	0.1
Sps	0.7	0.8	1	0.7	1	0.7	0.8	0.7	0.7	1	0.8	0.8	0.7	0.8	0.8	1
Prp	20.2	20.2	21.9	18.4	19.3	18.5	20.2	19.7	18.3	21.8	21.3	21.7	21.4	18.3	19.2	21.6
Alm	57.8	55.7	57.2	54	53.7	53	55.5	55.3	50	56.7	54.8	57.6	55.4	52.7	55.8	55.9
Grs	16.6	19	15.6	23.1	20.8	22.8	17.6	18.8	25.9	14.9	16.6	15.4	17.6	23.4	20.4	16
Adr	3.2	2.3	2.1	2.4	3.2	2.9	4.7	3.7	3	3.9	3.6	3.1	2.1	2.9	2.3	3.1
other	1.5	2	2.1	1.5	2	2.1	1.2	1.8	2.1	1.7	2.9	1.3	2.7	1.9	1.5	2.2
Nd	0.51	1.03	0.69	0.98	1.38	0.85	0.79	0.89	0.56	0.93	0.87	1.04	0.62	0.56	0.86	0.49
Sm	1.4	1.96	1.38	1.92	2.97	1.74	1.72	1.21	0.87	1.97	1.63	2	1.15	1.45	1.52	0.83
Eu	0.98	1.24	1.07	1.13	1.97	1.36	1.19	1.34	0.99	1.4	1.5	1.39	1.07	1.06	1.13	1.11
Gd	5.36	3.92	2.25	2.84	4.48	1.92	1.24	1.27	3.4	4.14	4.28	4.48	4.13	2.22	2.19	2.9
Tb	0.9	0.68	0.4	0.45	0.41	0.31	0.16	0.2	0.57	0.43	0.41	0.69	0.53	0.49	0.3	0.38
Dy	5.04	2.85	2.22	2.3	2.23	1.54	1.57	1.74	1.5	1.91	2.46	3.84	2.28	2.53	2.34	1.65
Ho	1.06	0.46	0.37	0.34	0.32	0.23	0.27	0.33	0.22	0.27	0.46	0.87	0.26	0.57	0.39	0.19
Er	2.67	0.88	0.7	0.73	0.58	0.58	0.67	1.03	0.3	0.32	1.22	1.67	0.57	1.3	1	0.3
Tm	0.3	0.12	0.1	0.11	0.07	0.05	0.08	0.12	0.05	0.02	0.17	0.22	0.03	0.11	0.1	0.06
Yb	1.54	0.46	0.49	0.6	0.55	0.39	0.53	0.99	0.19	0.2	0.78	1.12	0.32	1.03	0.84	0.18
Lu	0.14	0.07	0.09	0.09	0.05	0.08	0.09	0.24	0.03	0.01	0.08	0.13	0.05	0.12	0.11	0.03

Continued

								Grt-R							
No.	1	2	3	4	5	6	7	8	9	10	11	12	13	14	15
SiO_2	37.13	36.72	37.14	37.42	37.2	37.56	37.09	37.29	37.3	36.82	37.36	37.37	36.9	37.53	37.33
TiO_2	0.07	0.05	0.09	0.02	0.07	0.08	0.07	0.07	0.11	0.11	0.16	0	0.11	0.13	0.07
Al_2O_3	20.19	19.72	19.64	20.45	19.67	19.18	20.4	19.91	19.63	19.82	19.39	21.02	19.88	19.37	20.96
Cr_2O_3	0.04	0.02	0	0.06	0	0	0	0	0.04	0	0.02	0.01	0.04	0	0.07
FeO	28.84	29.5	29.31	29.34	28.98	29.79	29.4	29.06	28.66	29.78	28.81	28.12	29.42	29.98	26.6
MnO	0.44	0.38	0.41	0.32	0.43	0.46	0.36	0.38	0.35	0.38	0.39	0.39	0.41	0.39	0.28
MgO	5.54	5.33	5.68	5.21	5.33	5.18	5.48	5.17	5.48	5.29	5.15	5.39	5.43	4.87	5.25
CaO	6.78	7.36	7.05	7.18	7.53	7.19	6.3	7.19	7.43	7.13	7.73	6.71	7.02	7.09	8.36
NiO	0.05	0.02	0	0.02	0	0.02	0	0	0.01	0	0	0.03	0	0	0.02
Total	99.08	99.1	99.33	100	99.21	99.46	99.11	99.07	99.02	99.32	99	99.03	99.21	99.37	98.93
Si	2.94	2.91	2.94	2.94	2.95	2.98	2.94	2.96	2.96	2.92	2.97	2.95	2.92	2.98	2.95
Ti	0	0	0.01	0	0	0	0	0	0.01	0.01	0.01	0	0.01	0.01	0
Al	1.88	1.84	1.83	1.89	1.84	1.79	1.9	1.86	1.83	1.85	1.82	1.96	1.86	1.81	1.95
Cr	0	0	0	0	0	0	0	0	0	0	0	0	0	0	0
Fe^{3+}	0.22	0.32	0.29	0.22	0.27	0.25	0.21	0.21	0.24	0.3	0.23	0.13	0.28	0.21	0.15
Fe^{2+}	1.69	1.64	1.65	1.71	1.65	1.73	1.74	1.72	1.66	1.67	1.68	1.72	1.67	1.78	1.61
Mn	0.03	0.03	0.03	0.02	0.03	0.03	0.02	0.03	0.02	0.03	0.03	0.03	0.03	0.03	0.02
Mg	0.65	0.63	0.67	0.61	0.63	0.61	0.65	0.61	0.65	0.62	0.61	0.63	0.64	0.58	0.62
Ca	0.58	0.63	0.6	0.6	0.64	0.61	0.53	0.61	0.63	0.6	0.66	0.57	0.6	0.6	0.71
Uv	0.1	0.1	0	0.2	0	0	0	0.1	0	0.1	0.1	0	0.1	0	0.2
Sps	1	0.8	0.9	0.7	1	1	0.9	0.8	0.8	0.9	0.7	0.9	0.9	0.9	0.6
Prp	21.8	21	22.3	20.3	21	20.4	20.4	21.6	20.8	20.3	21.8	21.2	21.4	19.2	20.6
Alm	56.2	54.5	54.9	56.9	55.1	57.6	57.2	55.4	55.6	56.1	57.2	57.5	55.5	59.4	53.5
Grs	12.3	11.5	10.1	13.8	12	9.4	12.6	11.8	11.1	11.8	12	16.1	11.1	10.2	20
Adr	6.6	9.2	9.6	6.2	9.1	10.8	7.6	8.8	8.8	9.6	5.8	2.8	8.3	9.5	3.2
other	2.1	2.9	2.2	2	1.9	0.9	1.5	1.6	2.9	1.3	2.4	1.6	2.7	0.8	1.9
Nd	0.5	0.57	0.63	0.52	0.74	0.62	0.81	0.59	0.71	0.68	0.62	0.58	0.62	0.65	0.57
Sm	1.17	1.35	1.4	0.88	1.65	0.93	1.58	1.43	1.52	1.48	1.26	1.13	1.45	1.46	1.36
Eu	0.66	0.57	0.51	0.56	0.75	0.63	0.42	0.63	0.59	0.51	0.71	0.49	0.53	0.57	0.56
Gd	2.51	1.64	1.78	2.3	2.36	2.31	1.38	2.4	2.8	2.06	2.41	4.62	2.01	1.29	5.06
Tb	0.73	0.38	0.44	0.51	0.55	0.65	0.43	0.52	0.53	0.46	0.5	0.44	0.5	0.44	0.5
Dy	6.23	4.09	3.97	3.42	4.13	3.81	3.47	4.47	4.36	3.27	4.41	2.08	3.37	3.79	2.27
Ho	1.38	1	1.03	0.82	0.98	0.89	0.94	0.94	0.85	0.93	0.9	0.94	0.77	0.9	0.95
Er	4.11	2.72	3.25	2.46	2.43	2.46	2.71	2.39	2.62	2.61	2.16	2.11	2.09	2.74	2.63
Tm	0.61	0.43	0.4	0.45	0.34	0.36	0.45	0.4	0.28	0.29	0.35	0.32	0.33	0.42	0.39
Yb	4.23	2.74	3.54	2.59	2.18	2.44	3.17	2.49	2.3	2.34	2.02	1.86	2.36	3.16	2.45
Lu	0.69	0.47	0.5	0.51	0.37	0.41	0.45	0.38	0.29	0.37	0.3	0.41	0.37	0.42	0.44

Continued

	Grt-R													
No.	16	17	18	19	20	21	22	23	24	25	26	27	28	29
SiO_2	37.31	37.35	37.27	37.54	37.27	37.57	37.16	37.41	37.35	37.37	37.13	37.14	37.72	37.12
TiO_2	0.11	0.14	0.04	0.04	0.03	0.06	0.04	0.11	0.09	0.05	0.03	0.07	0.1	0.12
Al_2O_3	19.3	18.91	20.9	19.91	20.6	20.69	20.14	19.27	19.7	20.17	19.93	19.16	19.46	20.06
Cr_2O_3	0	0.02	0	0.01	0.02	0.03	0.06	0	0	0.02	0.03	0	0	0.02
FeO	29.18	29.51	28.66	29.52	29.15	28.68	29.21	29.82	28.86	29.09	29.25	29.11	28.68	29.32
MnO	0.39	0.36	0.38	0.38	0.37	0.32	0.37	0.32	0.38	0.44	0.32	0.35	0.39	0.4
MgO	5.01	4.85	5.4	5.32	5.02	5.23	4.9	4.97	5.43	5.52	5.32	4.91	5.45	5.38
CaO	7.66	7.8	6.82	7.08	6.89	6.84	7.2	7.42	7.51	7.19	7.4	8.12	7.56	7.24
NiO	0	0	0.04	0.01	0.03	0	0.01	0	0.07	0	0.02	0	0	0.02
Total	98.96	98.94	99.5	99.8	99.38	99.41	99.09	99.31	99.39	99.85	99.42	98.86	99.35	99.69
Si	2.97	2.98	2.94	2.96	2.95	2.96	2.95	2.97	2.95	2.94	2.93	2.96	2.98	2.93
Ti	0.01	0.01	0	0	0	0	0	0.01	0.01	0	0	0	0.01	0.01
Al	1.81	1.78	1.94	1.85	1.92	1.92	1.89	1.8	1.83	1.87	1.86	1.8	1.81	1.86
Cr	0	0	0	0	0	0	0	0	0	0	0	0	0	0
Fe^{3+}	0.24	0.25	0.19	0.23	0.18	0.14	0.2	0.24	0.25	0.25	0.27	0.27	0.22	0.27
Fe^{2+}	1.7	1.72	1.7	1.71	1.75	1.75	1.74	1.74	1.66	1.66	1.66	1.66	1.68	1.66
Mn	0.03	0.02	0.03	0.03	0.02	0.02	0.02	0.02	0.03	0.03	0.02	0.02	0.03	0.03
Mg	0.59	0.58	0.63	0.62	0.59	0.62	0.58	0.59	0.64	0.65	0.63	0.58	0.64	0.63
Ca	0.65	0.67	0.58	0.6	0.58	0.58	0.61	0.63	0.64	0.61	0.63	0.69	0.64	0.61
Uv	0	0.1	0	0	0.1	0.1	0.2	0	0	0.1	0.1	0	0	0.1
Sps	0.9	0.8	0.9	0.9	0.8	0.7	0.8	0.7	0.9	1	0.7	0.8	0.9	0.9
Prp	19.8	19.2	21.1	20.8	19.7	20.5	19.4	19.6	21.3	21.5	20.9	19.4	21.4	21.1
Alm	56.7	57.3	56.7	57.1	58.3	58.5	57.9	57.9	55.2	55.3	55.4	55.5	56	55.4
Grs	11.5	10.4	15	11.5	14.5	14.7	13.8	10.5	11.9	12.4	12.5	12.2	11.4	12.1
Adr	9.9	11.3	4.1	8.2	4.8	4.3	6.3	10.3	9	7.6	8.2	10.7	9.7	7.9
other	1.2	0.9	2.2	1.5	1.8	1.3	1.7	1.1	1.7	2.2	2.2	1.5	0.8	2.6
Nd	0.49	0.64	0.61	0.57	0.63	0.59	0.67	0.62	0.88	0.36	0.51	0.4	0.3	0.42
Sm	1.06	1.68	1.24	1.36	1.62	1.38	1.75	1.61	1.75	0.71	1.01	1.02	0.6	0.68
Eu	0.55	0.65	0.91	0.57	0.53	0.54	0.47	0.58	0.59	0.59	0.52	0.59	0.52	0.65
Gd	2.45	2.45	3.58	1.54	3.37	2.37	1.57	2.65	2.65	2.87	2.58	2.18	2.27	2.86
Tb	0.7	0.53	0.59	0.42	0.6	0.56	0.34	0.88	0.56	0.62	0.59	0.45	0.56	0.6
Dy	4.21	3.55	3.65	3.75	4.12	4.83	2.84	3.63	4.73	3.78	3.76	4.01	3.92	4.28
Ho	0.89	0.84	0.73	1.06	0.71	1.04	0.81	0.84	0.9	0.83	0.79	0.81	0.89	0.95
Er	2.61	2.48	2.16	2.99	2.21	2.79	2.24	2.32	2.51	2.19	2.37	2.43	2.74	2.85
Tm	0.35	0.32	0.25	0.46	0.32	0.32	0.35	0.34	0.33	0.34	0.28	0.31	0.43	0.32
Yb	2.23	2.38	1.92	3.05	1.84	2	1.98	2.44	2.21	2.34	2.41	1.91	2.33	2.3
Lu	0.41	0.28	0.27	0.36	0.31	0.27	0.49	0.29	0.36	0.27	0.3	0.3	0.34	0.34

Table S2 Major element (wt.%) and end-member (mol.%) contents for clinopyroxene in sample 99-4.

	Pyroxene													
	Thin section													
No.	1	2	3	4	5	6	7	8	9	10	11	12	13	14
SiO$_2$	54.34	55.27	55.25	54.03	54.79	53.65	53.92	55.02	53.87	54.91	54.27	55.57	54.49	55.68
TiO$_2$	0	0.01	0	0	0.01	0.09	0.07	0.02	0.02	0	0	0.01	0.07	0.05
Al$_2$O$_3$	2.99	4.53	4.65	3.07	3.82	2.99	3.19	4.44	3.07	4.13	2.87	5.44	3.07	5.34
FeO	11.05	9.23	8.92	10.74	9.98	10.78	10.82	9.65	10.78	9.82	11.13	8.15	11.03	8.14
MnO	0.01	0.02	0.02	0.05	0.01	0.06	0	0	0.03	0.02	0.01	0.02	0.03	0.03
MgO	10.09	9.78	9.85	10.16	9.97	10.07	9.9	9.88	10.25	9.93	10.16	9.56	9.98	9.6
CaO	16.9	16.02	16.16	16.8	16.53	16.65	16.78	16.35	17.07	16.44	16.78	15.41	16.92	15.59
K$_2$O	0.02	0.01	0.01	0	0	0	0	0.01	0.01	0.01	0.02	0	0.02	0.01
Na$_2$O	4.42	4.66	4.61	4.44	4.53	4.27	4.42	4.57	4.14	4.56	4.58	4.86	4.28	4.93
Total	99.81	99.53	99.47	99.29	99.64	98.56	99.09	99.94	99.23	99.82	99.81	99.02	99.88	99.37
Si	2	2.03	2.02	1.99	2.01	2	2	2.01	1.99	2.01	1.99	2.04	2	2.04
Ti	0	0	0	0	0	0	0	0	0	0	0	0	0	0
Al(T)	0	0	0	0.01	0	0	0	0	0.01	0	0.01	0	0	0
Al(M1)	0.13	0.2	0.2	0.13	0.17	0.13	0.13	0.19	0.13	0.18	0.12	0.24	0.13	0.23
Fe^{3+}(T)	0	0	0	0	0	0	0	0	0	0	0	0	0	0
Fe^{3+}(M1)	0.19	0.08	0.08	0.2	0.14	0.18	0.18	0.11	0.18	0.13	0.22	0.03	0.16	0.05
Fe^{2+}	0.15	0.2	0.2	0.13	0.17	0.16	0.15	0.18	0.16	0.17	0.12	0.22	0.18	0.2
Mn	0	0	0	0	0	0	0	0	0	0	0	0	0	0
Mg	0.55	0.53	0.54	0.56	0.55	0.56	0.55	0.54	0.57	0.54	0.56	0.52	0.55	0.52
Ca	0.66	0.63	0.63	0.66	0.65	0.66	0.67	0.64	0.68	0.64	0.66	0.61	0.67	0.61
K	0	0	0	0	0	0	0	0	0	0	0	0	0	0
Na	0.31	0.33	0.33	0.32	0.32	0.31	0.32	0.32	0.3	0.32	0.33	0.35	0.31	0.35
Jd	12.5	22.9	23.3	12.5	17.6	13	13.4	20.4	12.4	18.8	11.3	29.9	13.8	28.6
Ae	19.1	9.8	9	19.4	14.5	17.9	18.3	11.8	17.4	13.4	21.4	4.1	16.7	5.8
Quad	68.44	67.28	67.62	68.1	67.94	69.14	68.27	67.78	70.2	67.76	67.25	66.08	69.51	65.66

	Pyroxene											
	Thin section					Mineral separates						
	15	16	17	18	19	M1	M2	M3	M4	M5	M6	M7
SiO$_2$	53.79	55.37	54.68	54.44	55.35	56.17	55.71	55.55	55.64	54.96	55.4	55.31
TiO$_2$	0.03	0.01	0.01	0.06	0.08	0	0.05	0.03	0.05	0.07	0.06	0.11
Al$_2$O$_3$	2.94	4.73	3.72	2.8	5.88	5.3	5.37	5.39	5.31	5.58	5.82	5.94
FeO	10.87	9.09	10.35	10.73	8.4	7.8	7.8	8.4	8.48	8.51	8.34	8.47
MnO	0.05	0.01	0.02	0.03	0.04	0.03	0.04	0.02	0.02	0.02	0.06	0.02
MgO	10.07	9.67	9.98	10.09	9.21	9.61	9.57	9.52	9.63	9.52	9.3	9.12
CaO	16.98	15.86	16.61	16.96	15.04	15.43	15.58	15.42	15.6	15.4	15.15	14.93
K$_2$O	0.02	0.01	0.03	0.01	0.02	0.01	0.02	0.02	0	0.01	0.01	0.02
Na$_2$O	4.37	4.76	4.49	4.36	5.09	4.81	4.89	4.83	4.96	4.91	5.01	5.16
Total	99.12	99.51	99.89	99.47	99.11	99.16	99.04	99.16	99.7	98.96	99.15	99.07
Si	1.99	2.03	2	2.01	2.03	2.06	2.04	2.04	2.03	2.02	2.03	2.03
Ti	0	0	0	0	0	0	0	0	0	0	0	0

Continued

		Pyroxene									
	Thin section						Mineral separates				
15	16	17	18	19	M1	M2	M3	M4	M5	M6	M7
0.01	0	0	0	0	0	0	0	0	0	0	0
0.12	0.2	0.16	0.12	0.25	0.23	0.23	0.23	0.23	0.24	0.25	0.26
0	0	0	0	0	0	0	0	0	0	0	0
0.2	0.08	0.15	0.17	0.05	0	0.03	0.04	0.06	0.07	0.04	0.05
0.13	0.2	0.16	0.16	0.21	0.24	0.21	0.22	0.19	0.19	0.21	0.21
0	0	0	0	0	0	0	0	0	0	0	0
0.56	0.53	0.55	0.55	0.5	0.52	0.52	0.52	0.52	0.52	0.51	0.5
0.67	0.62	0.65	0.67	0.59	0.61	0.61	0.61	0.61	0.61	0.59	0.59
0	0	0	0	0	0	0	0	0	0	0	0
0.31	0.34	0.32	0.31	0.36	0.34	0.35	0.34	0.35	0.35	0.36	0.37
11.6	24.1	16.4	12.9	30	33.3	30.2	29.3	27	26.9	30	30.2
20	9.2	15.5	18.2	5.7	0	3.9	4.4	7.6	7.7	5.1	6
68.48	66.65	68.1	68.95	64.31	66.72	65.9	66.25	65.44	65.35	64.88	63.81

Appendix S1. Analytical methods.

1. Electron microprobe (EMP) analysis

Mineral compositions were determined using a Shimadzu Electron Probe Micro Analyzer (EMPA) 1600 at the CAS Key Laboratory of Crust-Mantle Materials and Environments, the University of Science and Technology of China (USTC), Hefei, with the following operating conditions: 15 kV accelerating voltage; 20 nA beam current and <5 mm beam diameter. Natural minerals and synthetic oxides were utilized as standards, and a program based on the ZAF procedure was used for data correction (Xia et al., 2013).

2. LA-ICP-MS analysis

Major and trace element analyses were performed by LA-ICP-MS at the State Key Laboratory of Geological Processes and Mineral Resources, China University of Geosciences, Wuhan. Detailed operating conditions for the laser ablation system, the ICP-MS instrument and data reduction are the same as given by Liu et al. (2008). Laser sampling was performed using a GeoLas 2005. An Agilent 7500a ICP-MS instrument was employed to acquire ion-signal intensities. Helium was applied as a carrier gas. Argon was used as the make-up gas, and mixed with the carrier gas via a T-connector prior to entering the ICP. Nitrogen was added into the central gas flow (Ar+He) of the Ar plasma to decrease the detection limit and improve precision (Hu et al., 2008). Each analysis incorporated a background acquisition of approximately 20-30 s (gas blank) followed by 50s data acquisition from the sample. The Agilent Chemstation was utilized for the acquisition of each individual analysis. Element contents were calibrated against multiple-reference materials (BCR-2G, BIR-1G, and BHVO-2G) without applying internal standardization (Liu et al., 2008). The preferred values of element concentrations for the USGS reference glasses are from the GeoReM database (http://georem.mpch-mainz.gwdg.de/). Off-line selection and integration of background and analyte signals, and time-drift correction and quantitative calibration were performed by ICPMSDataCal (Lin et al., 2016; Liu et al., 2008, 2010).

3. Garnet OH contents

The contents of structural hydroxyl (OH) in garnet were analyzed on thin sections by a Fourier Transform Infrared Spectroscopy (FTIR) at the Micro-FTIR Laboratory, School of Earth Science at Zhejiang University. Double-polished thin sections were prepared with a thickness of 0.1-0.2 mm. The thin sections were immersed in acetone for 24 h, followed by repeated cleaning with ethanol and distilled water. They were then heated in an oven at 110 ℃ for ~12 h to remove the surface absorbed water. Unpolarized spectra were obtained from 2500-4500 cm^{-1} on a Nicolet 5700 FTIR spectrometer coupled with a Continuum microscope, using a KBr beam splitter and a liquid nitrogen-cooled MCT-A detector. A total of 256 scans were acquired for each spectrum at a 4 cm^{-1} resolution, using an aperture size of 30 μm × 30 μm. Optically clean areas without inclusions and cracks were selected for measurements under a continuous dry air flush. The H$_2$O content of garnet (denoting hereafter the contents of structural hydroxyl in ppm H$_2$O unless specifically mentioned) was calculated by a modified form of the Beer–Lambert Law: $c=A / (I \times t \times \gamma)$, where c is the water contents (ppm H$_2$O); A is the integrated area (cm^{-2}) of absorption bands; I is the integral specific absorption coefficient (ppm^{-1} cm^{-2}); t is thickness (cm); and γ is the orientation factor(γ=1 for garnet; Paterson, 1982). The OH absorption bands were integrated between 3000 and 3800 cm^{-1} for garnet. The integral specific coefficient of 1.39/(ppm H$_2$O cm^2) was used to calculate the H2O contents in garnet (Bell et al., 1995). The thickness of the thin sections was measured by a digital micrometer and reported as the average of 10 measurements for each domain. The baseline corrections were carried out by hand at least three times for each spectrum, with an uncertainty of less than ±5%. The uncertainty of sample thickness measurement was small (<±3%), whereas that of the absorption coefficients was< ±10% for garnet crystals with different compositions (Bell et al., 1995). Consequently, the total uncertainty of calculated water in garnet was estimated to be less than ±10%–20%.

4. Sm-Nd isotope analysis

The eclogite sample was broken into small than 1 mm pieces in a hardened tool steel mortar. The pieces were washed, then ground in agate mortar to –60 and +80 mesh so that the Grt-P and Grt-R were separated. Hand-picking under a binocular microscope separated the three-types (Grt-P,Grt-R, and Grt-W) that did not contain inclusions or exhibit alteration features. The mineral fragments were checked for good transparencey and cleanliness for each surface and direction. For pyroxene, only omphacite with light green color was selected for dating analysis. In order to examine whether the mineral separates selected for dating were omphacite, a method to check their composition was used, because alteration of mineral can easily affect Sm-Nd dating (Xie et al., 2004). The pyroxene separates were mounted on an epoxy resin disc and polished to section the crystals in half, and then major compositions were determined by EMPA following the procedure of section 3.1.

Samarium-Neodymium (Sm-Nd) isotopic analyses of two samples(99-4 and 12-13) were performed using an IsoProbe-T thermal ionization mass spectrometer (TIMS) at the State Key Laboratory of Lithospheric Evolution, Institute of Geology and Geophysics, Chinese Academy of Sciences (IGGCAS). Samarium isotopes were measured using single Ta filaments. Neodymium isotopes were determined using single tungsten filaments with TaF$_5$ as an ionization activator. Detailed column chemistry and mass spectrometer operation conditions have been described previously (Chu et al., 2012). ^{143}Nd/^{144}Nd ratios were corrected for mass bias using ^{146}Nd/^{144}Nd = 0.7219. During the period of data collection, the measured value for the JNdi-1 Nd standard was 0.512117 ± 10 (2SD), corresponding to a LaJolla value of 0.511860±7 (Tanaka et al., 2000). Measured ^{149}Sm/^{147}Sm ratios were corrected for mass bias using ^{152}Sm/^{147}Sm =0.56081. Blanks for the entire chemical procedures were lower than 15 pg for Sm and 25 pg for Nd at IGGCAS (Chu et al., 2012).

Sm-Nd isotopic data of another sample (93DB01) were obtained at the Max Plank Institute für Chemie (MPI), Germany by Li (one of the authors), following procedures described elsewhere (Li et al., 2000). All mineral separates of sample 93DB01were leached by 2.5 N HCl at about 70°–80 ℃ for 1 h, and then for 15 min in cold 5% HF to remove any surface contamination before analyses. Blanks for all of the chemical procedures were 50 pg for Sm and Nd at MPI. ^{143}Nd/^{144}Nd ratios were also normalized against the value of ^{146}Nd/^{144}Nd = 0.7219. All Nd isotope ratios for sample 93DB01 reported in this article were relative to values of 0.511878 ±14 (2SD) for the LaJolla Nd standard. Because the Sm, Nd isotopic data of each eclogite sample were obtained at the same laboratory, and the Sm-Nd isotopic data were used for dating rather than to compare the initial Nd isotopic composition among eclogites, thus, no correction was made for isotopic data from different laboratories. Ages were calculated using the ISOPLOT software of Ludwig (1994), and given with 2σ error. Input errors used in age computations were ^{147}Sm/^{144}Nd = 0.2% and ^{143}Nd/^{144}Nd= 0.005%.

References

Bell, D. R., Ihinger, P. D., & Rossman, G. R. (1995). Quantitative analysis of trace OH in garnet and pyroxenes. American Mineralogist, 80, 465-474.

Chu, Z. Y., Guo, J. H., Yang, Y. H., Qi, L., Chen, L., Li, X. C., & Gao, J. F. (2012). Evaluation of sample dissolution method for Sm-Nd isotopic analysis of scheelite. Journal of Analytical Atomic Spectrometry, 27, 509-515.

Hu, Z. C., Gao, S., Liu, Y. S., Hu, S. H., Chen, H. H., & Yuan, H. L. (2008). Signal enhancement in laser ablation ICP-MS by addition of nitrogen in the central channel gas. Journal of Analytical Atomic Spectrometry, 23, 1093-1101.

Lin, J., Liu, Y., Yang, Y., & Hu, Z. (2016). Calibration and correction of LA-ICP-MS and LA-MC-ICP-MS analyses for element contents and isotopic ratios. Solid Earth Sciences 1, 5-27. doi.org/10.1016/ j.sesci.2016.04.002

Liu, Y. S., Gao, S., Hu, Z. C., Gao, C. G., Zong, K. Q., & Wang, D. B. (2010). Continental and oceanic crust recycling-induced melt-peridotite interactions in the Trans-North China Orogen: U-Pb dating, Hf Isotopes and trace elements in zircons from mantle xenoliths. Journal of Petrology, 51, 537-571.

Liu, Y. S., Hu, Z. C., Gao, S., Gunther, D., Xu, J., Gao, C. G. & Chen, H. H. (2008). In situ analysis of major and trace elements of anhydrous minerals by LA-ICP-MS without applying an internal standard. Chemical Geology, 257, 34-43.

Ludwig, K. R. (1994). Isoplot: a plotting and regression program for radiogenic-isoplot date, version 2.75. Open-File Report (U. S. Geological Survey) 91-445, 47.

Paterson, M. S. (1982). The determination of hydroxyl by infrared absorption in quartz, silicate glasses and similar materials. Bulletin de Mineralogie, 105, 20-29.

Tanaka, T., Togashi, S., Kamioka, H., Amakawa, H., Kagami, H., Hamamoto, T., … Dragusanu, C. (2000). JNdi-1: a neodymium isotopic reference in consistency with LaJolla neodymium. Chemical Geology, 168, 279-281.

Xia, Q. K., Hao, Y. T., Liu, S. C., Gu, X. Y., & Feng, M. (2013). Water contents of the Cenozoic lithospheric mantle beneath the western part of the North China Craton: Peridotite xenolith constraints. Gondwana Research, 23, 108-118.

Appendix S2. The t-Test results of FeO^T contents for Grt-P and Grt-R.

Sets of data:

Grt-P = {28.57 28.83   28.00   28.07   27.84   28.30   29.10   29.15   27.50   28.02
26.35   25.91   24.92   28.08   26.07   28.44   27.79   28.19   26.31   26.93   26.65
27.40   27.90   25.29   28.29   28.21   28.37   27.74   26.44   27.16   28.22}, n = 31;

Grt-R = {28.84   29.50   29.31   29.34   28.98   29.79   29.40   29.06   28.66   29.78
28.81   28.12   29.42   29.98   26.60   29.18   29.51   28.66   29.52   29.15   28.68
29.21   29.82   28.86   29.09   29.25   29.11   28.68   29.32}, n = 29.

Results:

F-Test Two-Sample for Variances ($\alpha = 0.05$)		
	Variable 1	Variable 2
Mean	27.54970968	29.09103448
Variance	1.175984546	0.40372382
Observations	31	29
df	30	28
F	2.912844096	
P(F<=f) one-tail	0.002812994	
F Critical one-tail	1.868709158	

$\alpha = 0.05 > P$ ($F \leq f$) one-tail, or $F > F$ Critical one-tail, indicates that two samples possess unequal variances. Then t-Test: Two-Sample Assuming Unequal Variances were performed, with a result below.

t-Test: Two-Sample Assuming Unequal Variances ($\alpha = 0.05$)		
	Variable 1	Variable 2
Mean	27.54970968	29.09103448
Variance	1.175984546	0.40372382
Observations	31	29
Hypothesized Mean Diff	0	
df	49	
t Stat	-6.768502322	
P(T<=t) one-tail	7.57581E-09	
t Critical one-tail	1.676550893	
P(T<=t) two-tail	1.51516E-08	
t Critical two-tail	2.009575237	

P ($T \leq t$) one-tail (7.57581E-09) or P ($T \leq t$) two-tail (1.51516E-08) $\ll \alpha = 0.05$, means obvious differences exist between these two sets of data.

第二部分 华北和华南陆块碰撞时代及拼合过程的同位素年代学和地球化学研究

中国华北、华南陆块碰撞时代的钐-钕同位素年龄证据[*]

李曙光[1]，S. R. Hart[2]，郑双根[3]，郭安林[4]，刘德良[1]，张国伟[4]

1. 中国科学技术大学地球与空间科学系，合肥 230026
2. 美国麻省理工学院地球、大气、行星科学系，美国
3. 安徽省地质矿产局 313 地质队，霍山 237200
4. 西北大学地质系，西安 710069

> **亮点介绍**：本文报道了大别山第一个 C 型榴辉岩三叠纪 Sm-Nd 同位素矿物等时线年龄，为确定华北和华南陆块碰撞时代提供了准确的年代学制约；报道了秦岭拉垃庙苏长辉长岩体古生代 Sm-Nd 同位素矿物等时线年龄，它指示在华北和华南陆块碰撞前，北秦岭有一系列蛇绿岩套仰冲事件沿华北地块南缘向外逐次发生。然而本文报道的大别山辉长岩侵入体三叠纪年龄是错的，后来锆石 U-Pb 年代学研究证明它们是早白垩世侵入体。

摘要 本文报道了秦岭-大别山造山带的 C 型榴辉岩 Sm-Nd 同位素年龄（243.9 ± 5.6 Ma），和镁铁-超镁铁岩 Sm-Nd 同位素年龄（230.6 ± 30.7 Ma 和 406.2 ± 17.4 Ma）。这些年龄表明中国华北、华南地块在三叠纪初期就沿秦岭-大别山碰撞带连接在一起了。在两地块最后碰撞以前，在古生代，有一系列蛇绿岩套仰冲事件沿华北地块南缘向外逐次发生。

关键词 华南陆块；钐-钕同位素

1 引言

构造研究表明，中国大陆是由几个古板块对接而形成的[1-7]。中国东部大陆主要是由华北地块（中朝地台）和华南地块（扬子克拉通）组成。切割中国大陆中部的秦岭-大别山脉则是这两大地块之间的碰撞带（图1）。该带广泛发育有混杂岩、蛇绿岩套及其碎块，高压低温变质带等板块缝合线所特有的特征岩石。因此，秦岭-大别山造山带历来就是国内外地质学家所注目的地区。搞清华北、华南两大地块的对接时代，及秦岭-大别山造山带的形成机制，对我们认识中国东部地壳的演化历史有重要意义。

尽管较多的人认为华北、华南地块的碰撞发生在晚古生代到早中生代（[2-4, 7-9]1)），但是这一碰撞时代问题目前仍存在着较大分歧。例如，李春昱等提出除了西秦岭在三叠纪还存有小的海盆外，华北、华南地块在晚二叠世已基本连在一起[2]。林金录等根据古地磁证据认为这两大地块直到中侏罗世尚没有完全联结起来[9]。与这些观点相反，另外一些作者则提出了一些更早的碰撞时代，如泥盆纪、晚志留世以及晚元古代等[5, 6, 10, 11]。

存在上述分歧的主要原因是，在该造山带，缺乏以解决两大地块碰撞时代为目的的可靠年代学工作。为此，我们试图用 Sm-Nd 同位素定年方法，对大别山区两大地块碰撞时形成的 C 型榴辉岩和侵入的超镁铁岩，以及分布于北秦岭并与早期古秦岭海板块俯冲事件有关的镁铁-超镁铁岩进行定年研究，以求确定华北、华南地块的碰撞时代及碰撞前可能发生的古秦岭海板块向华北地块以下俯冲的时代。

---

[*] 本文发表在：中国科学(B 辑)，1989，3：312-319
对应的英文版论文为：Li Shuguang, Hart S.R., Zheng Shuanggen, et al., 1989. Timing of collision between the North and South China Blocks: the Sm-Nd isotopc age evidence. Science in China (series B), 32: 1393-1400
1) 徐嘉炜、刘德良、李秀新，全国中新生代构造会议论文(待刊)

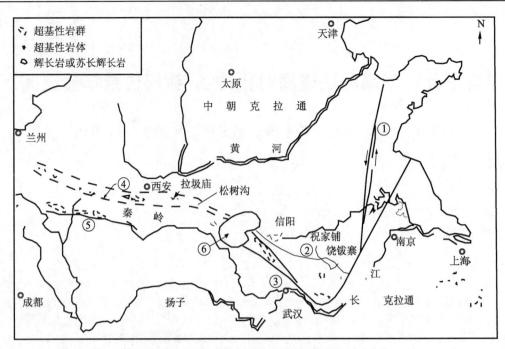

图 1 秦岭-大别山造山带镁铁-超镁铁岩的分布

在该区有两个镁铁-超镁铁岩带：在秦岭，北带沿商南-丹凤断裂④，南带沿玛沁-勉县断裂⑤延伸；在大别山，北带沿桐柏-磨子潭断裂②，南带沿广济-应山断裂③延伸；①郯城-庐江断裂，⑥南阳盆地

2 地质背景及样品

走向北西西的秦岭-大别山造山带位于中国东中部。向西，它与昆仑、祁连造山带相接；向东，它被郯城-庐江大断裂切断。南阳盆地位于该造山带中段并将其分成两部分：西部秦岭和东部大别山。在古生代期间，华北、华南地块之间是广阔的海洋，在海洋中有可能存在一些微型陆块，如北秦岭和大别山中一些古老地层[8,10]。

秦岭-大别山造山带有两条蛇绿岩或镁铁-超镁铁岩带和高压低温变质带[[2,10,12,13]1,2)]。它们平行于造山带的总构造线方向，并分别位于秦岭-大别山造山带北部和南部的重要断裂带附近（图 1）。在这里我们分别称之为北带和南带。这两个蛇绿岩和高压低温变质带表明，在秦岭-大别山造山带可能存在两个俯冲带或碰撞带。一般认为北带的镁铁-超镁铁岩和高压低温变质带在两带中发育最好。但是，由于秦岭和大别山被南阳盆地所隔，我们不能肯定秦岭的北带与大别山的北带是同一俯冲带或地缝合线。在亚洲大地构造图中，李春昱等将秦岭的南带与大别山的北带相连接作为华北、华南地块的地缝合线[2]；而秦岭的北带则被认为是古秦岭海板块向华北地块以下俯冲的俯冲带。至于其他观点可参考文献[8,10,14]。

作为第一步，我们的样品均采自北带，以求获得在北带发生碰撞或俯冲的时代。南带工作留待以后去做。

我们一共取得了三组样品并描述如下：

1. 大别山 C 型榴辉岩（R-4）

C 型榴辉岩（R-4）赋存于饶钹阿尔卑斯型超基性岩中。该超基性岩主要为斜方辉橄榄岩和纯橄岩，它以固态侵位于大别群（早元古代）混合片麻岩中（桐柏-磨子潭断裂南侧）[15]，并与围岩无明显化学交换。饶钹寨超基性岩体曾遭受了强烈的挤压作用。在该岩体边部，破碎带和片理十分发育；横跨岩体中部，沿岩体长轴还有一构造糜棱岩带[15]。平行于该糜棱岩带的榴辉岩团块和角闪岩脉，呈线状排列在距岩体边部70～100 m处。很明显，这些榴辉岩是在饶钹寨固态侵位于大陆地壳时，由高压低温变质作用形成的。此外，石榴石的化学组成（石榴石中镁铝榴石成分<30%）也证明这种产于饶钹寨超基性岩体中的榴辉岩是 C 型榴

---

1) 石铨曾等，河南地质，1983，7-14
2) 安徽饶钹寨超基性岩体的地质构造特征，南京地质矿产研究所，1977

辉岩（详情另有专文发表）。

2. 大别山祝家铺辉石岩（Dzh-1，Dzh-2）

祝家铺辉石岩体是以岩浆形式侵入到桐柏-磨子潭断裂南侧的大别群混合片麻岩中。在性质上该辉石岩属于岩浆巢堆积岩，主要含单斜辉石、斜方辉石和角闪石。该侵入体严重被围岩污染，以至于沿侵入体边部发育了辉长岩（或闪长岩）壳。无论辉石岩还是辉长岩均经历了退变质作用。其中部分辉石退变质为角闪石，从而使辉长岩退变质为闪长岩。用于测年的样品取自这种闪长岩。

3. 秦岭苏长辉长岩（QL-7）

样品采自拉圾庙基性岩体。该岩体侵入北秦岭(商南-丹凤断裂北侧)前奥陶变沉积和变火山岩系中。拉圾庙基性杂岩的主要部分为辉长岩，其他为苏长辉长岩。样品（QL-7）为苏长辉长岩，它含有两种辉石（单斜辉石和紫苏辉石）和两种不同颜色的斜长石（白色斜长石和烟色斜长石，它可能含有一些小的包体）。在该基性岩体中还发育有一些闪长岩脉（QL-8）和二长岩脉（QL-9）。尽管一些文章认为这种基性岩体是蛇绿岩套的碎片[2, 12]，但缺乏令人信服的证据。

除了 QL-7 的辉石外，所有实验测定的单矿物样品都是100%的纯净。QL-7 的辉石含有两种辉石，纯度只有95%。

3 测定方法

溶矿以前，单矿物样品首先要经过酸洗以除去矿物表层可能有的蚀变产物或吸附物。酸洗程序是根据 Shimizu 的报道[16]，并经 Richardson 修正的[17]。样品的 Sm-Nd 同位素分析是在美国麻省理工学院地球、大气及行星科学系同位素实验室进行的。分析流程已有文章描叙[18]。文中给出的 Nd 同位素比值是相对于标准样（BCR-1）——$^{143}Nd/^{144}Nd = 0.512600$。$\varepsilon^t_{Nd}$ 值计算所采用的参考值为：$(^{143}Nd/^{144}Nd)^0_{CHUR} = 0.51264$ 和 $(^{147}Sm/^{144}Nd)_{CHUR} = 0.1967$[19]。

等时线拟合采用 YORK I 回归计算，并给出 1σ 的年龄误差。年龄计算采用 $\lambda^{147}Sm = 6.54 \times 10^{-12} a^{-1}$。

4 结果与讨论

所测 Sm 和 Nd 含量及 Nd 同位素比值（附 2σ 质谱分析误差）列于表 1。等时线示于图 2-4。

对榴辉岩 (R-4)我们获得了一个非常精确的 Sm-Nd 矿物内部等时年龄 243.9 ± 5.6 Ma，其对应初始 Nd 同位素比值：$^{143}Nd/^{144}Nd = 0.51215 \pm 1$（$\varepsilon^t_{Nd} = -3.4 \pm 0.2$）（图 2）。这一年龄值大大老于饶钹寨超基性岩中角闪岩脉的角闪石 K-Ar 年龄（126 Ma）[1)]。镜下岩相研究表明，这种角闪石是由辉石、石榴子石经退变质而形成的。很明显，角闪石的 K-Ar 年龄代表的是饶钹寨超基性岩体侵位后随大别山隆起上升而经历的最后一次热事件；而榴辉岩的 Sm-Nd 年龄给出的是该超基性岩体的侵位时代。由于超基性岩的侵位可能发生在华北、华南两大陆块碰撞时或碰撞以前，因此该榴辉岩 Sm-Nd 年代可代表华北、华南地块碰撞时代的最老界线（下限）。负的初始 ε_{Nd} 值表明该榴辉岩是在饶钹寨超基性岩侵位于大陆地壳时形成的。

表1 秦岭-大别山造山带 C 型榴辉岩及镁铁-超镁铁岩 Sm，Nd 含量和 Nd 同位素比值*

样品号	岩石/矿物名称	Sm(ppm)	Nd(ppm)	$^{147}Sm/^{144}Nd$	$^{143}Nd/^{144}Nd$	ε^t_{Nd}
R-4W	C 型榴辉岩（全岩）	4.96	14.63	0.2049	0.512482±15	
R-4 Gar 1	石榴子石（1）	6.051	7.852	0.4659	0.512889±18	
R-4 Gar 2	石榴子石（2）	6.061	7.915	0.4629	0.512888±17	
R-4 Gar	石榴子石（平均）	6.056	7.884	0.4644	0.512888±12*	
R-4 Cpx	绿辉石	1.803	6.618	0.1647	0.512406±16	
Dzh-1	闪长岩	7.847	31.37	0.1512	0.511757±19	−15.9

续表

---

1) 安徽大别山基性-超基性岩及其含矿性的研究，南京地质矿产研究所，1977

样品号	岩石/矿物名称	Sm(ppm)	Nd(ppm)	$^{147}Sm/^{144}Nd$	$^{143}Nd/^{144}Nd$	ε_{Nd}^{t}
Dzh-1 Plag	斜长石	0.17	1.653	0.0623	0.511617±20	−16.0
Dzh-2	闪长岩	4.982	22.78	0.1322	0.511710±19	−16.2
QL-7	苏长辉长岩	3.904	19	0.1242	0.512476±18	+0.5
QL-7 Pyr	辉石	8.55	31.48	0.1642	0.512565±17	+1.0
QL-7 Plag·W	斜长石（白色）	1.212	8.37	0.0875	0.512344±16	−0.2
QL-7 Plag·d	斜长石（烟色）	0.363	3.748	0.0586	0.512291±16	+0.3
QL-8	闪长细晶岩	4.129	23.37	0.1068	0.512378±19	−0.5
QL-9	黑云母二长岩	6.148	33.5	0.1109	0.512432±21	+0.4

*平均误差公式：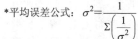

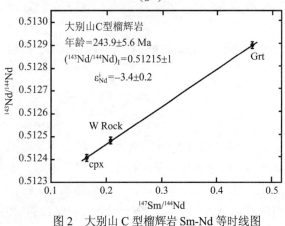

图 2 大别山 C 型榴辉岩 Sm-Nd 等时线图

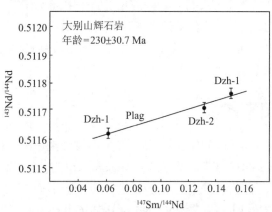

图 3 大别山辉石岩侵入体 Sm-Nd 等时线图

对祝家铺辉石岩体，我们用两个取自同一位置，斜长石含量不同的闪长岩样品（Dzh-1，Dzh-2）和一个从 Dzh-1 中分选出来的斜长石样品获得一条 Sm-Nd 等时线（图 3），并给出 230.6 ± 30.7 Ma 的年龄，它具有非常低的初始 Nd 同位素比值 $^{143}Nd/^{144}Nd = 0.51152 ± 2$ ($\varepsilon_{Nd}^{t} = -16.1 ± 0.4$)。由于样品数据点较少及可能的混染程度等原因，年龄误差较大，但它在误差范围内仍与 C 型榴辉岩的年龄一致。由于该闪长岩（原岩为辉长岩）是辉石岩浆与围岩同化混染的产物（这一点也被它们极低的初始 Nd 同位素比值所证实）。因此，这一年龄代表了祝家铺辉石岩体的侵入年龄。与 C 型榴辉岩年龄一起，它们表明在大别山北翼桐柏-磨子潭断裂南侧，超基性岩侵入和高压低温变质作用是几乎同时的。

上述地质和同位素证据均表明，祝家铺辉石岩体不是蛇绿岩套的一部分，它是以岩浆形式侵入大陆地壳的。由于它产生于地缝合线附近，因而它很可能是与华北、华南地块聚敛碰撞有关的岛弧或大陆型岩浆岩。所以，祝家铺辉石岩体的年龄有可能等于或老于真正的碰撞年龄，它给出的仍可能是华北、华南地块碰撞时代的下限。

秦岭拉圾庙苏长辉长岩样品（QL-7）在图 4 中形成一条很好的矿物内部等时线。它给出的年龄是 402.6 ± 17.4 Ma，其初始 $^{143}Nd/^{144}Nd = 0.51214 ± 1$ ($\varepsilon_{Nd}^{t} = +0.4 ± 0.2$)。此外，全岩 QL-9 样也落在 QL-7 的等时线上，而全岩 QL-8 落在该等时线下边一点。尽管这一年龄明显老于拉圾庙岩体附近的秦王山和三十里堡辉长岩体的 K-Ar 年龄（340～370 Ma）[8]，也老于最近 Mattauer 报道的北秦岭若干变质黑云母和白云母的 ^{39}Ar-^{40}Ar 年龄（328～348 Ma）[6]，但是它们都指示了北秦岭的古生代造山运动。拉圾庙苏长辉长岩的相对较高的初始 ε_{Nd} 值（+0.4 ± 0.2）说明它受地壳混染的影响不像祝家铺岩体那样严重。但是与典型的蛇绿岩套（洋壳起源的）相比，它的初始 ε_{Nd} 还是低多了，而仅与某些洋岛玄武岩和岛弧玄武岩类似（图 5）。它表明拉圾庙辉长苏长岩只可能是起源于海山的蛇绿岩碎块。403 Ma 的年龄可能代表了该苏长辉长岩的生成时代，而 328～348 Ma 的 ^{39}Ar-^{40}Ar 年龄可能指示了其仰冲事件及相伴随的在北秦岭发生的变质作用。

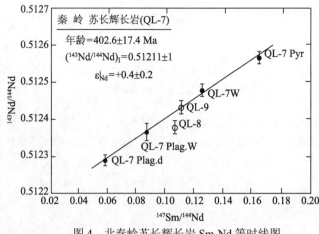

图 4　北秦岭苏长辉长岩 Sm-Nd 等时线图

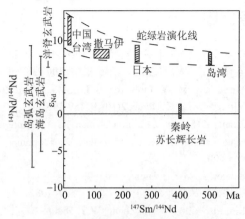

图 5　蛇绿岩套 ε^t_{Nd} 值随时间演化图

图中岛弧区范围据 White 和 Patchett[20] 及 Davidson[21]。
蛇绿岩 Nd 同位素数据取自日本 Toba[22]、阿曼、撒马伊[23]、
纽芬兰岛湾[24] 和中国台湾[23]。

5　构造意义

（1）如前所述，我们获得的大别山区与阿尔卑斯型橄榄岩侵位有关的 C 型榴辉岩 Sm-Nd 矿物等时年龄（243.6±5.6 Ma），和辉石岩侵入体 Sm-Nd 等时年龄（231±31 Ma），为确定华北、华南地块碰撞时代的下限提供了最新的年代学依据。

最近报道的南秦岭多硅白云母 ^{39}Ar-^{40}Ar 年龄为 217～236 Ma[6]。Mattauer 等将该南秦岭的 ^{39}Ar-^{40}Ar 年龄解释为碰撞后地壳内部进一步挤压，逆冲所引起的变质作用时代。这样，南秦岭的 ^{39}Ar-^{40}Ar 年龄可被视为华北、华南地块碰撞时代的上限。

综合上述 Sm-Nd 和 ^{39}Ar-^{40}Ar 年龄，我们可知中国华北、华南地块的碰撞时代应在 217～243.9 Ma 之间。华北和华南地块应在早三叠世（印支运动初期）就对接在一起了。它还表明，印支运动在中国东中部是由华北、华南板块碰撞引起的。根据构造运动的时代，大别山北翼不可能与北秦岭对比，而只能与南秦岭对比。因此，华北与华南地块之间的印支期地缝合线位于南秦岭北沿和大别山北翼的可能性是存在的[2]。

（2）拉垃庙苏长辉长岩体的 Sm-Nd 矿物内部等时年龄为 403 Ma，指示了商南-丹凤断裂以北的北秦岭古生代造山运动。根据微古生物化石证据，秦岭以西祁连山的三条蛇绿岩带也属寒武纪-奥陶纪侵位[26]。因此，在华北、华南地块最后碰撞以前的整个古生代期间内，在祁连山和北秦岭地区有若干次蛇绿岩套仰冲事件沿华北古陆块南缘向外逐次发生。

然而有一个问题尚未解决：为什么在大别山区没有古生代褶皱带或造山带？可能它至今尚未被发现，或者也许它在华北地块南缘东段确实不存在。如果后者是真的话，我们就不得不考虑这种可能性，即古秦岭海板块向华北板块以下俯冲是先在西部开始发生的，而在东部发生得较晚。

很显然，为解决上述问题和全面理解这些 Sm-Nd 年龄的构造含义，还需要在秦岭-大别山造山带的构造学、岩石学、地层学和地球化学方面做更多的工作。

致谢

在同位素测试中，Hart，Gulen，Burrhus，Salters，Kennedy，Reid 均不同程度地给予了技术上的指导和帮助；周作祯高级工程师、张绵缜、王山林工程师在大别山野外工作期间给予的大力协助；该文初稿经涂光炽教授、李春昱教授、孙贤鉥博士，Sengör 博士和张培镇博士审阅并提出建设性修改意见，作者还与许靖华教授就文中一些观点进行过有益讨论。所有这些鼓励和帮助，在此一并表示衷心的感谢。

参考文献

[1] 李春昱，中国地质科学院院报，1980，1：11-22.
[2] 李春昱、王荃、刘雪亚、汤耀庆，1：800万亚洲大地构造说明书，地图出版社，1978，l74-197.
[3] McElkinny, M. W., Embleton, B. J. J., Ma, X. and Zhang, Z., Nature, 293(1991), 212-215.
[4] KIimetz, M. P., Tectonic, 2(1983), 189-196.
[5] Zhang, Z.H. Liou, J. G. and Coleman, R. G., Geological Society of America Bulletin, 95(1984), 295-312.
[6] Mattauer, M., Matte, P., Malavielle, J., Tapponier, P., Maluski, H., Xu, Z.Q., Lu, Y.L. and Tang, Y.Q., Nature, 317(1985), 496-500.
[7] Sengör, A. M. C., Nature, 318(1995), 16-17.
[8] 王鸿祯、徐成彦、周正国，地质学报，1982，3：270-279.
[9] Lin, J.L. Fuller, M. and Zhang, W.Y. Nature, 313(1985), 444-449.
[10] 杨森楠、吴鉴、杨学忠、陆中旦，地球科学，1983，3；81-91.
[11] 谢窦克、郭坤一，中国地质科学院院报，1984，10: 167-206.
[12] 肖序常、王方国，中国地质科学院院报，1984，9: 19-30.
[13] 高延林，中国地质科学院院报，1984，10：61-76.
[14] 许志琴，地质论评，1986，1：79-89.
[15] 杨锡庸，中国地质科学院南京地质矿产研究所所刊，1983，4：81-98.
[16] Shimizu, N., Phys. Chem. Earth., 9(1975), 655-669.
[17] Richardson, S. H., Ph. D. thesis, MIT, (1984).
[18] Zindler, A., Hart, S. R., Frey, F. A. and Jakobsson, S. P., Earth Planet. Sci. Lett. 45(1079), 249-262.
[19] Jacobsen, S. B. and Wasserburg, G. J., Earth Planet. Sci. Lett. 67(1984), 137-150.
[20] White, W. M. and Patchett, J., Earth Planet. Sci. Lett. 67(1984), 167-185.
[21] Davidson, J. P., Nature, 306(1983), 253-256.
[22] Richard, P. and Allegre, C., Earth Planet. Sci. Lett. 47(1980), 65-74.
[23] McCulloch, M. T., Gregory, R. T., Wasserburg, G. J. and Taylor, H. P., Jour. of Geophys. Res., 86(1981), 2721-2735.
[24] Jacobsen, S. B. and Wasserburg, G. J., Jour. of Geophys. Res., 84(1979), 7429-7445.
[25] Jahn, B. M., Contrib. Mineral Petrol., 32(1986), 194-206.
[26] 肖序常、陈国铭、朱志直，地质学报，1987，4：281-295.

大别山北翼大别群中 C 型榴辉岩的 Sm-Nd 同位素年龄及其构造意义*

李曙光[1]，葛宁洁[1]，刘德良[1]，张宗清[2]，叶笑江[2]，郑双根[3]，彭长权[3]

1. 中国科学技术大学地球与空间科学系，合肥 230026
2. 中国地质科学院地质研究所，北京 100029
3. 安徽地矿局 313 地质队，霍山 237200

> **亮点介绍**：报道了产于大别山中酸性变质杂岩中的退变质榴辉岩 Sm-Nd 年龄。由于榴辉岩的辉石比石榴石较易退变质，故该年龄是测定来自同一标本三个不同颜色的石榴石样品，也成功获得三叠纪 Sm-Nd 同位素矿物等时线年龄；该年龄代表扬子和华北两大陆块的碰撞时代。

关键词 榴辉岩；Sm-Nd 同位素定年

秦岭-大别山造山带是华北地块与扬子地块的会聚地带。由于这两大地块的对接、碰撞，在该造山带的地缝合线附近广泛发育有各种高压低温变质岩，如蓝闪石片岩、C 型榴辉岩等，其中 C 型榴辉岩在大别山段较发育[1]。很显然，精确测定这些 C 型榴辉岩的生成年龄对探讨华北、扬子两大陆块的对接时代及造山带演化历史有重要意义。

大别山的榴辉岩按其产状可分为两类：一类产于阿尔卑斯橄榄岩体的边部，它是在岩体以固态侵位时形成的。如大别山北翼饶钹寨橄榄岩体中的 C 型榴辉岩，它的生成时代已被李曙光等用 Sm-Nd 法测定为 244 Ma，并被解释为该岩体的侵位时代[2]。由于蛇绿岩侵位一般早于陆块的对接，该超基性岩的侵位时代只能代表陆块对接时代的下限。另一类产于地层中的 C 型榴辉岩，它是由于地层中铁镁质岩石受到强大挤压力而变质形成。这种陆壳地层所受到的强大挤压力既可以在海洋板块向大陆仰冲时产生，也可以在两陆块对接碰撞时产生。因此，对该类榴辉岩定年有可能给出更接近陆块对接时代的年龄。为此，本研究对产于大别群地层中的 C 型榴辉岩进行了 Sm-Nd 同位素定年。

1 地质背景及样品

图 1 显示了大别山地区的地质、构造轮廓。东西走向的桐柏-磨子潭断裂位于大别山北翼。沿该断裂糜棱岩很发育，说明它是一深部韧性剪切带[3]。该断裂以北地层为佛子岭（或信阳）群，以南为大别群杂岩。在大别群中广泛发育有各种超镁铁岩，如阿尔卑斯型橄榄岩、辉石岩等。其中阿尔卑斯型橄榄岩被认为是以固态侵位于大别群的蛇绿岩碎块（如饶钹寨岩体）[4]。此外，该断裂南侧 C 型榴辉岩也相当发育[1]。因此，一些研究者认为桐柏-磨子潭断裂是华北陆块和扬子陆块间的一条地缝合线[5, 6]。我们认为大别地体可能为一独立微型古陆，故该断裂可视为华北陆块与大别山微古陆间的地缝合线。我们的 C 型榴辉岩样品就采自安徽大别山北翼，桐柏-磨子潭断裂南侧高坝岩附近大别群角闪斜长片麻岩中（图 1）。野外产状表明榴辉岩与该区的超基性岩体无直接关系。

镜下检查表明，该榴辉岩中的辉石全部蚀变了。因此，只有石榴石一种矿物可用于同位素分析。经仔细观察发现，高坝岩榴辉岩中石榴石的颜色不均一，颜色深者呈橙红色、浅者呈粉红色。这种颜色上的差异反映了它们化学组成上的差异，从而为我们用石榴石一种矿物获取 Sm-Nd 等时年龄提供了可能性。

---

* 本文发表在：科学通报，1989，7：522-525
对应的英文版论文为：Li Shuguang, Ge Ningjie, Liu Deliang, et al., 1989. The Sm-Nd isotopic age of C-type eclogite from the Dabie group in the Northern Dabie mountains and its tectonic implications. Chinese Science Bulletin, 19: 1625-1628

2 样品选取及分析方法

在 Sm-Nd 法中,用一种矿物获取等时年龄尚没有先例。我们用以下方法从一块标本中取得几个不同的石榴石样品。将一块标本切成两半,作为两个样品(高-1-1 和高-1-2)分别选取石榴石。其中高-1-2 号样品含有较多的橙红色石榴石。因此从高-1-2 中选取了橙红色和粉红色两个石榴石样品,从高-1-1 中只选取一个石榴石样品,共计获得三个新鲜的石榴石样品,纯度达 100%,不含任何连生体或蚀变产物。

分析前,先用 1:1 HCl,在 100℃条件下,酸洗样品 1 h,而后反复用水洗净。为避免单矿物样品不均一而引起的误差,每个样品在一次称量并全部溶解后,再定量分出一部分溶液加稀释剂以测定 Sm、Nd 含量;另一份溶液测定 Nd 同位素比值。化学分离及质谱测定在中国地质科学院地质研究所 Sm-Nd 同位素实验室完成。其 Sm、Nd 分离流程已报道[6]。由于石榴石 Nd 含量很低(见表1),尽量降低实验本底很重要。本实验过程中,空白样测定的 Nd、Sm 含量为 Nd = 34 pg,Sm = 17 pg,达到国际一流实验室水平。因此,对该石榴石样品的最低 Nd 含量来说,实验本底所引起的误差仍 $<8\times10^{-5}$。质谱分异校正是相对于 $^{146}Nd/^{144}Nd = 0.72190$。质谱测定期间,标准样 BCR-1 的 $^{143}Nd/^{144}Nd = 0.512612\pm10$。年龄计算采用 York II 双误差回归法,$\lambda_{^{147}Sm} = 6.54\times10^{-12}a^{-1}$。$^{143}Nd/^{144}Nd$ 及年龄值精度均以 2σ 误差给出。

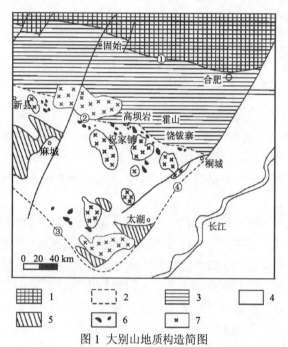

图 1 大别山地质构造简图

1. 华北陆块;2. 扬子陆块;3. 佛子岭(信阳)群为主组成的北淮阳边缘海;4. 大别群;5. 宿松群;6. 超镁铁岩体;7. 榴辉岩分布点。①合肥-明港断裂;②桐柏-磨子潭断裂;③应山-广济断裂;④郯城-庐江断裂

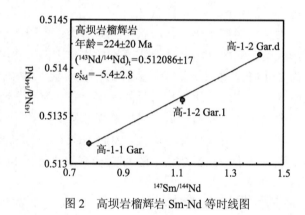

图 2 高坝岩榴辉岩 Sm-Nd 等时线图

表 1 高坝岩 C 型榴辉岩中石榴石的 Sm、Nd 含量及 Nd 同位素比值

样品号	Sm(ppm)	Nd(ppm)	$^{147}Sm/^{144}Nd$	$^{143}Nd/^{144}Nd\pm2\sigma$
高-1-1 Gar.	1.599	1.258	0.769	0.513213±27
高-1-1 Gar.1*	1.179	0.636	1.12	0.513667±36
高-1-1 Gar.d*	1.019	0.437	1.4114	0.514144±13

* Gar.1 是粉红色石榴石;Gar.d 是橙红色石榴石。

3 结果及讨论

三个石榴石样品的 Sm、Nd 含量及 $^{147}Sm/^{144}Nd$、$^{143}Nd/^{144}Nd$ 值见表1。测定结果表明,这些从同一标本中选取的三个石榴石样品具有较宽 $^{147}Sm/^{144}Nd$ 值范围,这使我们能够从这三个同种矿物样品中获得一较精确的 Sm-Nd 等时年龄。图2表明,这三个石榴石样品点在 $^{147}Sm/^{144}Nd$-$^{143}Nd/^{144}Nd$ 图上呈良好线性排列,并给出 224 ± 20 Ma 的等时年龄,其对应初始 $^{143}Nd/^{144}Nd = 0.512086\pm17$ ($\varepsilon_{Nd} = -5.4\pm2.8$)。该年龄值与该区饶钹寨超基性岩体的侵位年龄($244\pm11$ Ma)在误差范围内一致或略低一点[2]。对这两种年龄的一致性可以有不同的解释。如果将这种年龄的一致性认为是同一地质事件的结果，则高坝岩榴辉岩就可能是在一次海洋板块仰冲事件中产生的，与此同时伴随了饶钹寨阿尔卑斯型橄榄岩(蛇绿岩碎块)的侵位。另一种可能性是这种年龄的一致性反映的只是最后一次洋壳仰冲事件与陆块碰撞事件时间上的接近，高坝岩榴辉岩是陆块碰撞的产物，它的 224 Ma 年龄值代表了华北陆块与大别山微古陆的对接时代。如果将上述年龄值与秦岭-大别山造山带其他地带的年代学结果做一比较，我们将不难对上述两种可能性做出选择。

Mattauer 等报道了秦岭山阳断裂南侧，扬子陆块北缘的多硅白云母和钠闪石的 $^{39}Ar/^{40}Ar$ 年龄分别为 232 ± 5 Ma 和 216 ± 7 Ma，并将其解释为扬子陆块和华北陆块对接后，陆壳被进一步挤压而发生的陆内俯冲和缩短事件[6]。桑宝良等报道的大别山南麓宿松群由于大规模挤压剪切断裂作用而产生的动力变质作用时代也为 231 ± 48 Ma（Rb-Sr 全岩-矿物等时年龄）[8]。上述这些年龄中有的明显低于饶钹寨超基性岩侵位年龄，但它们全都与高坝岩榴辉岩的生成年龄一致。这种大范围内年龄的一致性很难用仅发生在大别山北翼的洋壳仰冲事件来说明，而应当反映的是更大规模的、华北陆块与扬子陆块的碰撞事件。高坝岩榴辉岩的年龄说明大别山作为夹在华北陆块和扬子陆块之间的一微型古陆也在三叠纪早期与华北陆块对接在一起。在对接以前，在 244 Ma 左右，在大别山微古陆北缘还发生了最后一次洋壳仰冲事件。

4 结论

(1) C 型榴辉岩中不同色调的石榴石具有不同的 Sm/Nd 值。实验表明从同一块标本中选取不同色调的石榴石以获得矿物等时年龄是可行的。

(2) 大别山北翼大别群中的 C 型榴辉岩生成年龄为 224 ± 20 Ma。这一年龄直接代表大别山微古陆与华北陆块的对接时代，也可代表扬子和华北两大陆块的对接时代。

(3) 在大别山段，陆块对接前的蛇绿岩侵位发生在 244 Ma，与陆块对接时代非常接近。这一点与北秦岭不同，在北秦岭，蛇绿岩侵位发生在泥盆-石炭纪[2]。

致谢

该项研究获得涂光炽先生热情的鼓励和支持，在此深表谢意。

参考文献

[1] 叶大年、李达周、董光复、邱秀文，华北断块的形成与演化，科学出版社，北京，1980: 120-132.
[2] 李曙光、Hart S.R.、郑双根、郭安林、刘德良、张国伟，中国科学，1989（排印中）.
[3] 刘德良、李秀新，地质科学，1984，1: 42-50.
[4] 杨锡庚，中国地质科学院南京地质矿产研究所所刊，1983，4: 81-95.
[5] 李春昱、王荃、刘雪亚、汤耀庆，1:800 万亚洲大地构造说明书，地图出版社，北京，1978: 174-187.
[6] Mattauer, M., Matte, P. H., et al., Nature, 317(1985): 496-500.
[7] 张宗清、叶笑江，中国地质科学院地质研究所所刊，1987，17.
[8] 桑宝梁、陈跃志、邵桂清，中国区域地质，1987，4: 364-370.

大别山南麓含柯石英榴辉岩的 Sm-Nd 同位素年龄*

李曙光[1]，刘德良[1]，陈移之[1]，葛宁洁[1]，张宗清[2]，叶笑江[2]

1. 中国科学技术大学地球与空间科学系，合肥 230026
2. 中国地质科学院地质研究所，北京 100037

> **亮点介绍**：首次获得南大别地区含柯石英榴辉岩的三叠纪 Sm-Nd 同位素年龄；大别山南麓榴辉岩与大别山北翼、桐柏-磨子潭断裂南侧大别群杂岩中榴辉岩一致的 Sm-Nd 同位素年龄，进一步证实整个大别造山带经历了三叠纪陆-陆碰撞的高压-超高压变质作用。

关键词 榴辉岩；Sm-Nd 同位素定年；大别山

在大别山带及其东延部分（苏北-胶南隆起）发育一条榴辉岩带，它是华北与扬子陆块在这里碰撞产生高压变质作用的有力证据[1]。最近，人们在一部分榴辉岩中发现了柯石英，它以包裹体形式存在于石榴石和绿辉石矿物晶体中[2,3] 1)。这是继挪威、西阿尔卑斯之后，世界上第三个发现柯石英的造山带，因而引起了广泛重视。含柯石英榴辉岩均分布在该造山带南翼（岳西以南）大别群的片麻岩和大理岩中。在大别山北翼的榴辉岩中尚未有发现柯石英的报道。近期来我们曾报道了两个产于大别山北翼、桐柏-磨子潭断裂南侧大别群杂岩中不同类型榴辉岩的 Sm-Nd 同位素年龄，它们均形成于印支期[4,5]。然而大别山南翼含柯石英榴辉岩的形成时代尚属未知。在大别山带是否存在多期榴辉岩，是否存在元古高压变质带仍是一个有争议的问题[2,6,7]。为此，我们对大别山南翼的含柯石英榴辉岩进行了 Sm-Nd 同位素定年研究，以确定其形成时代。

1 地质背景及样品

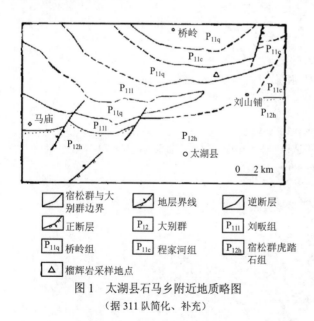

图 1 太湖县石马乡附近地质略图
（据 311 队简化、补充）

样品采自大别山南麓太湖县石马乡的大别群变质杂岩中（图 1）。Wang 等（1989）曾在该地榴辉岩中发现了柯石英[2]。该处岩性主要为黑云斜长片麻岩夹大理岩。部分黑云斜长片麻岩含有绿帘石、石榴石和少量单斜辉石，局部石榴石可富集成粉红色条带或透镜体。这说明该片麻岩经历了较高的变质作用。岩石因受挤压已强烈片理化，片理倾向正南，倾角 35°，石榴石片麻岩透镜体的长轴亦与片理方向一致。该地岩石片理产状与南面宿松群片理产状一致，说明它们均是受南北向强大挤压力而片理化的。

该处有数十个大小不同的榴辉岩块

---

* 本文发表在：科学通报，1992，37：346-349
对应的英文版论文为：Li Shuguang, Liu Deliang, Chen Yizhi, et al., 1992. The Sm-Nd isotopic age of coesite-bearing eclogite from the southern Dabie Mountains. Chinese Science Bulletin, 37: 1638-1641

1) 据张树业等（长春地质学院）和张儒媛、丛柏林（中国科学院地质研究所）在 1990 年北京高压变质带与碰撞构造讨论会上报告

以似层状、透镜状或小团块产于黑云片麻岩中。这些榴辉岩团块长度从几十厘米到几十米不等，并与围岩呈构造接触。部分榴辉岩已退变质为榴闪岩。此外，该处还发育少量蛇纹岩块体，它们局部变质为滑石片岩。上述产状特点说明该处为一构造混杂岩带，并与 Wang 等划分的第 III 类含柯石英榴辉岩产状一致[2]。在该地榴辉岩和榴闪岩中我们也发现了柯石英假象及绿辉石等高压矿物呈包裹体存在于石榴石矿物中（另有专文发表）[1]。

用于年龄测定的榴辉岩样品主要由石榴石、绿辉石和金红石组成。石榴石由于受动力挤压而破裂，并沿破裂面呈光性异常。电子探针分析表明该挤压破裂未引起成分变化。石榴石和绿辉石晶体均见有一些细小矿物包裹体。

2 实验方法

在一块岩石标本上如果只含有两种共生矿物，人们习惯于用两个单矿物样加一个全岩样的办法来获取全岩–矿物等时线。但是这种方法有缺点，因为全岩样品基本是这两种矿物的混合物，它不起独立样品的作用。尤其对榴辉岩来说，其辉石易蚀变，若辉石数据点发生偏移，则全岩数据点也随之偏移，并仍与石榴石数据点保持一条直线（混合线），从而使我们无法察觉辉石数据的问题。因此，我们将一小块标本一分为二，分别记为 ND-1 和 ND-2，并分别从中选取石榴石和绿辉石，以便获取矿物等时线。这一方法曾在测定大别山北翼榴辉岩年龄时获得成功，被证明是可行的[5]。所选单矿物纯度达 100%，无蚀变颗粒，无明显包裹物颗粒，但无法保证它们不含细小包裹体。

样品分析实验在中国地质科学院地质研究所 Sm-Nd 同位素实验室完成。样品的酸洗预处理及分析流程已有前文叙述[4,5]。实验中，空白本底为 Nd = 34 pg，Sm = 17 pg。质谱分异校正相对于 $^{146}Nd/^{144}Nd = 0.72190$。质谱测定期间，标准样 BCR-1 的 $^{143}Nd/^{144}Nd = 0.512612 \pm 10$。年龄计算采用 York II 双误差回归法，$\lambda_{^{147}Sm} = 6.54 \times 10^{-12} a^{-1}$。$^{143}Nd/^{144}Nd$ 及年龄值精度均以 2σ 误差给出。

3 结果与讨论

样品的 Sm、Nd 含量及 $^{147}Sm/^{144}Nd$、$^{143}Nd/^{144}Nd$ 的测定值见表 1。其等时线图示于图 2 测定结果表明，尽管两个绿辉石样品的 Sm/Nd 值很相似，它们的 $^{143}Nd/^{144}Nd$ 值有较明显差异。其中 ND-1 Omp 落在两个石榴石样品点的连线上，形成一条很好的等时线；而 ND-2 Omp 则落在该等时线的上方。这可能是由于 ND-2 Omp 样品中含有较多早期矿物包裹物造成的。

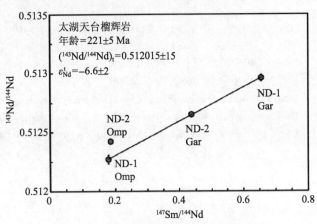

图 2 太湖天台榴辉岩的 Sm-Nd 等时线图

表 1 石马榴辉岩中单矿物的 Sm、Nd 含量及 Nd 同位素组成

样品号	Sm(ppm)	Nd(ppm)	$^{147}Sm/^{144}Nd$	$^{143}Nd/^{144}Nd \pm 2\sigma$
ND-1 Gar	2.1632	1.9970	0.65524	0.512959±14
ND-1 Omp	2.7961	9.4913	0.17820	0.512271±10
ND-2 Gar	2.1838	3.0232	0.43695	0.512650±9
ND-2 Omp	2.6435	8.6287	0.18532	0.512422±12

由两个石榴石样品盒 ND-1 Omp 所确定的等时线给出一个非常精确的矿物等时线年龄，221 ± 5 (2σ) Ma。其初始 Nd 同位素比值为 0.512015 ± 15 (2σ)，$\varepsilon_{Nd} = -6.6 \pm 2$。该年龄与桑宝梁等发表的宿松群云母钠长片麻岩的全岩–矿物 Rb-Sr 等时年龄（231 ± 48 Ma）在误差范围内一致[8]，但比它精确得多。这说明大别

山南麓含柯石英榴辉岩的形成时代与该区区域热动力变质形成时代一样。此外，该年龄也与大别山北翼高坝岩榴辉岩的 Sm-Nd 矿物等时年龄（224 ± 20 Ma）非常一致[5]。它虽然比产于超镁铁岩中的铙钹寨榴辉岩年龄（244 ± 11 Ma）略低一些，但它们均属印支期高压变质带。因此，大别山南麓含柯石英榴辉岩与大别山北翼的不含柯石英榴辉岩是在同一构造事件中形成的。

通常认为榴辉岩是消减洋壳俯冲到深部时生成的。因此榴辉岩的生成时代代表的是洋壳俯冲时代，而非陆块碰撞时代。对柯石英榴辉岩，一些作者也认为无水的榴辉岩与柯石英属同一变质阶段（stage I）[9]。据此观点，我们所测的年龄值也应代表榴辉岩原岩俯冲到最深部的时代。但是，事实上含柯石英榴辉岩中的柯石英均以包裹体形式赋存在石榴石和绿辉石矿物晶体中。这说明我们现在所见到的榴辉岩主要矿物，石榴石和绿辉石，形成于柯石英之后。如果柯石英是榴辉岩原岩俯冲到最大深度时形成的，那么包裹柯石英的石榴石和绿辉石就应当是在该榴辉岩块被推覆上升的初期再次平衡重结晶的产物。而这种推覆上升过程只能发生在陆壳俯冲过程中。因此，以石榴石、绿辉石矿物测定的含柯石英榴辉岩 Sm-Nd 同位素年龄代表的实际上是陆壳俯冲的时代。此外，如前所述，该榴辉岩年龄与该区大别群和宿松群的热动力变质作用时代一致。而发育在扬子陆块北缘的这一热动力变质作用显然与陆块碰撞和陆壳俯冲作用有关。

致谢

感谢安徽省地质矿产局 311 地质队宋勤总工程师，陆端英、刘国华、赵连吉、李立新高级工程师，在野外工作中给了很大的帮助。

参考文献

[1] Zhang, Z.M., Liou, J.G. and Coleman, R.G. Geol. Soc. Am. Bul., 95(1984), 295-312.
[2] Wang, X., Liou, J.G. and Mao, H.K., Geology, 17(1989), 1085-1088.
[3] Okay, A., Xu, S. and Sengör, A.M.C., Eur. J. Mineral., 1(1989), 595-598.
[4] 李曙光、Hart S.R.、郑双根、郭安林、刘德良、张国伟，中国科学，B 辑，1989，3: 312-319.
[5] 李曙光、葛宁洁、刘德良、张宗清、叶笑江、郑双根、彭长权，科学通报，1989，7: 522-525.
[6] 董申保，地质学报，1989，3: 273-284.
[7] Liou, J.G., Maruyama, S., Wang, X. and Graham, S., Tectonophysics, 181(1990), 97-111.
[8] 桑宝梁、陈跃志、邵桂清，中国区域地质，1987,4: 364-370.
[9] Smith, D.C., Eclogites and eclogite-facies rocks, Developments in Petrology, 12, Elsevier, Amsterdam, 1988, 1-206.

胶南榴辉岩的形成时代及成因——Sr、Nd 同位素地球化学及年代学证据*

陈移之[1]，李曙光[1]，从柏林[2]，张儒瑗[2]，张宗清[3]

1. 中国科学技术大学地球与空间科学系，合肥 230026
2. 中国科学院地质研究所，北京 100029
3. 中国地质科学院地质研究所，北京 100037

> **亮点介绍**：通过 Sm-Nd 同位素定年获得苏–鲁造山带榴辉岩变质时代与大别造山带一致，均为三叠纪，进一步确定了苏–鲁超高压变质带和大别超高压变质带是华北陆块与扬子陆块在三叠纪碰撞产生的同一个超高压变质带。榴辉岩富集的 Sr-Nd 同位素特征证实其原岩可能为扬子陆块古老基底中的镁铁质岩石。

关键词　榴辉岩；Sm-Nd 同位素定年；Sr-Nd 同位素地球化学；高压变质带

胶南隆起在构造上属秦岭–大别造山带的东延部分，只是由于郯庐断裂带将其向北错移了 500 km[1]。发育在大别山及苏北–胶南隆起的榴辉岩带是华北与扬子陆块碰撞有关的高压变质带。由于它的重要构造意义及超高压矿物——柯石英和金刚石的发现而引起广泛重视[2-4]。确定该榴辉岩带的形成时代及成因对研究中国东部两大陆块碰撞历史及榴辉岩形成、折返机制有重要意义。

李曙光等曾测定了安徽大别山三个榴辉岩的 Sm-Nd 矿物等时年龄并获得 221～244 Ma 的一致结果[5-7]。但是苏北–胶南隆起带的榴辉岩尚未报道有可靠的 Sm-Nd 同位素年龄。已报道的榴辉岩中多硅白云母 K-Ar 或 ^{39}Ar-^{40}Ar 年龄为 721～1559 Ma[8]，大大老于安徽大别山榴辉岩的 Sm-Nd 年龄。一些作者据此提出该带存在元古代榴辉岩的可能性[8,9]。在苏北–胶南隆起中是否真的存在元古代榴辉岩，有必要用更为可靠的 Sm-Nd 矿物等时线法去检查。

造山带榴辉岩的成因是多样的。它可能是消减洋壳在深部变质成因，也可能是俯冲陆壳在深部变质形成。显然这两种不同成因的榴辉岩的年龄意义及其中含发现的柯石英的构造意义均不相同。例如，如果榴辉岩的原岩是陆壳岩石，它们当中柯石英的发现就意味着由于陆壳的俯冲导致陆壳在碰撞带可叠置增厚至 100 km 以上，它的形成年龄也代表陆壳俯冲的时代；反之若榴辉岩的原岩是俯冲洋壳，则意义完全不同。因此，判断榴辉岩成因是一极为重要的研究课题。该带榴辉岩原岩性质尚不清楚。因此，本研究另一工作目的是利用 Sr、Nd 同位素示踪来判断其原岩性质及成因。

1　样品地质背景及测试方法

两个榴辉岩样品分别取自山东日照市的哑口（E87N9-1）和马家沟（E87M-1）（位置见图 1）。榴辉岩产于胶南群黑云母斜长片麻岩中。在该地的榴辉岩中已报道发现有柯石英及其假象[10]。E87N9-1 的矿物组合为石榴石+绿辉石+白云母+石英。样品片理化明显，白云母沿片理分布，形成于石榴石和绿辉石之后。E87M-1 主要由石榴石+绿辉石及少量金红石组成。考虑到全岩样品含退变质及蚀变产物，定年只选用新鲜的单矿物样品。将上述两个样品块分别切成两半，各自选出石榴石和绿辉石单矿物用于同位素分析。这样，每个样品可得两个石榴石和两个绿辉石单矿物样品。单矿物最后在双目镜下精挑，弃掉任何杂质、蚀变颗粒及明显的含包裹物颗粒。样品测试在中国地质科学院地质研究所 Sm-Nd 同位素实验室完成。实验流程前文已有描述[6,7]。

---

* 本文发表在：科学通报，1992, 37(23): 2169-2172

2 年代学结果

Sr、Nd 同位素测定结果列于表 1。样品的 Sm-Nd 矿物等时线及年龄值示于图 2。图 2（a）显示 E87N9-1 的两个绿辉石有几乎相同的 ^{143}Nd/^{144}Nd 和 ^{147}Sm/^{144}Nd 值，这表明在该样中绿辉石初始 Nd 同位素的均一性。但是两个石榴子石（gar 1 和 gar 2）不能落在与绿辉石（omp）相连的同一直线上。连接 gar 2 与 omp 1 和 omp 2 的直线给出年龄值为 221 ± 6 Ma，与大别山榴辉岩年龄相似[5-7]。然而连接 gar 1 与 omp 1 的直线给出 280 ± 14 Ma 的年龄值。由于 E87N9-1 样只有一个世代的石榴石，因此上述两条直线中只有一条是真正的等时线。基于 221 Ma 年龄值与大别山及 E87M-1 样的年龄值一致，我们认为 221 ± 6 Ma（ε_{Nd}=−13.1±0.3）代表该榴辉岩的榴辉岩相变质年龄。大别山榴辉岩 P-T-t 轨迹的研究已经证明榴辉岩中现在所见到的石榴石、绿辉石晶体是榴辉岩抬升初期退变质重结晶形成的，而早期顶峰变质阶段的榴辉岩相矿物与柯石英一样均呈包裹体残留在重结晶的石榴石和绿辉石中[1)]。因此，用这些重结晶石榴石、绿辉石矿物测定的 Sm-Nd 同位素年龄应代表榴辉岩开始抬升，早期退变质阶段时代。

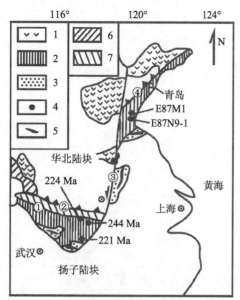

图 1 大别山和苏北-胶南隆起地质构造略图

1. 太古代；2. 大别群、胶南群变质杂岩；3. 红安群、宿松群、张八岭群和海州群；4. 已做 Sm-Nd 定年的榴辉岩采样点；5. 蓝片岩；6. 北淮阳佛子岭群、信阳群；7. 北秦岭变质杂岩
① 桐柏-磨子潭断裂；② 信阳-舒城断裂；③ 郯城-庐江断裂；④ 五莲-荣城断裂

由于榴辉岩的抬升与陆壳俯冲引发的大规模逆冲构造有关，这一年龄指示的也是陆壳俯冲时代。gar 1 的较高 ^{143}Nd/^{144}Nd 值很可能是在榴辉岩退变质重结晶时，该样品石榴子石的 Nd 同位素未在全标本范围内均一化的结果。而绿辉石可能由于 Sm、Nd 在其中扩散速度比石榴石快，封闭温度低而在标本范围内均一化了。

表 1 胶南榴辉岩的 Sm、Nd、Rb、Sr 含量及 Nd-Sr 同位素组成

样品号	矿物名称	Sm (ppm)	Nd (ppm)	^{147}Sm/^{144}Nd	^{143}Nd/^{144}Nd	$\varepsilon_{Nd(t)}$
E87N9-1-1g	石榴石	2.181	1.529	0.8629	0.513182±56	−8.9±1.1
E87N9-1-1o	绿辉石	1.253	3.471	0.2184	0.511994±36	−13.2±0.7
E87N9-1-2g	石榴石	3.553	1.756	1.224	0.513453±36	−14.2±0.7
E87N9-1-2o	绿辉石	0.9265	2.602	0.2154	0.511996±15	−13.1±0.3
E87M1-1g	石榴石	1.377	1.002	0.8313	0.512821±71	−13.4±1.4
E87M1-1o	绿辉石	0.8985	2.498	0.2176	0.511972±36	−13.6±0.6
E87M1-2g	石榴石	1.441	1.003	0.8696	0.512970±31	−11.5±0.6
E87M1-2o	绿辉石	0.6463	2.864	0.1365	0.511973±21	−11.4±0.4

样品号	矿物名称	Rb (ppm)	Sr (ppm)	^{87}Rb/^{86}Sr	^{87}Sr/^{86}Sr	$\varepsilon_{Sr(t)}$
E87N9-1-1g	石榴石	0.110			0.70585±15	20.6±2.1
E87N9-1-1o	绿辉石	0.826	77.731	0.0308	0.70597±5	22.3±0.7
E87N9-1-2g	石榴石	0.195	9.898	0.0570	0.70690±10	35.5±1.4
E87N9-1-2o	绿辉石				0.70600±6	22.7±0.8
E87M1-1g	石榴石	0.073	11.924	0.0178	0.70642±13	27.8±1.8
E87M1-1o	绿辉石	0.238	44.208	0.0156	0.70999±87	78.6±12.3
E87M1-2g	石榴石	0.073	8.494	0.0248	0.70612±4	23.4±0.6
E87M1-2o	绿辉石	0.027	104.73	0.0007	0.70571±10	18.6±1.4

注：Nd 同位素质谱分异校正相对于 ^{146}Nd/^{144}Nd = 0.72190。表中 ^{143}Nd/^{144}Nd 对于标准样 BCR-1 值等于 0.512633 给出，^{87}Sr/^{86}Sr 值相对于标准样 NBS987 值等于 0.71030。$\varepsilon_{Nd(t)}$ 为榴辉岩形成时的初始 ε_{Nd} 值。由于所有样品的 ^{87}Rb/^{86}Sr 很小，$\varepsilon_{Sr(t)}$ 未做年龄校正，可视为样品初始 ε_{Sr} 值。

---

1) 肖益林（导师李曙光），大别山榴辉岩 P-T-t 轨迹研究，中国科学技术大学硕士论文

样品 E87M-1 的情况也证明了在榴辉岩初期退变质重结晶时，在一块标本范围内 Nd 同位素未均一化的可能性（见图 2（b））。在图 2（b）中，将 gar 1 与 omp 1，和 gar 2 与 omp 2 相连可以得到两条具有相似年龄（分别为 211±18 Ma 和 208±33 Ma）和不同初始 ε_{Nd} 值（分别为 -13.5 和 -11.3）的近似平行直线。这些结果表明胶南榴辉岩也是在三叠纪变质形成的。已有研究证明榴辉岩的初期抬升速率很高，它有可能使 Sm、Nd 没有充分时间在较大尺度上扩散平衡，从而造成在一块手标本范围内同位素组成的不均一性。

3 Sr、Nd 同位素地球化学

表 1 表明该样品的初始 ε_{Nd} 值很低，达 -11～-14。它们的初始 ε_{Sr} 值绝大多数在 18.6～35.5 之间，只有一个高达 78。将这些数据投影在 ε_{Nd}-ε_{Sr} 图上（图 3），可看出它们的 Sr、Nd 同位素特征与三叠纪洋壳（MORB+OIB）不同，并且也落在洋壳与沉积物的混合趋势以下。这说明该榴辉岩的原岩不是俯冲洋壳，它们的低 ε_{Nd} 值也不可能用俯冲洋壳携带的沉积物来解释。为比较，图 3 给出了德国南部蒙柯博格古生代 MORB 成因榴辉岩的 Sr、Nd 同位素特征[11]。它们位于地幔趋势之上，与蛇绿岩类似。胶南榴辉岩的上述 Sr、Nd 同位素特征与下地壳麻粒岩及古老的上地壳玄武质岩石（如五台群细碧岩）类似（图 3）。它们非常低的 ε_{Nd} 值，要求其原岩在陆壳中有较长的存留时间。因此，它不可能是两大陆块之间的古生代消减洋壳，而是古老扬子陆块基底俯冲到深部发生榴辉岩相变质作用形成的。据此，它们的年龄（208～221 Ma）应代表扬子陆块俯冲时代，这与前述榴辉岩 P-T-t 轨迹所获的结论是一致的。如果陆块碰撞与陆撞俯冲之间没有大的时间间隔，则华北与扬子陆块的碰撞时代仍可认为是在二叠纪末或三叠纪。此外，由于该区榴辉岩中发现有柯石英[10]，本文又证明该榴辉岩原岩是扬子陆块古老基底中的镁铁质岩石，我们可以认为当时扬子陆块的俯冲深度可达 100 km 左右。此外，该区榴辉岩的 Sr、Nd 同位素特征与地幔的 EMI 富集端元相似。因此，在造山过程中因拆离作用该榴辉岩进入地幔可能是 EMI 的一种来源。

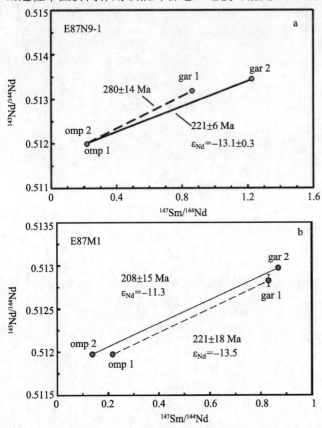

图 2 样品 E87N9-1（a）和 E87M1（b）的 Sm-Nd 矿物等时线图

年龄值用 YORK II 回归计算

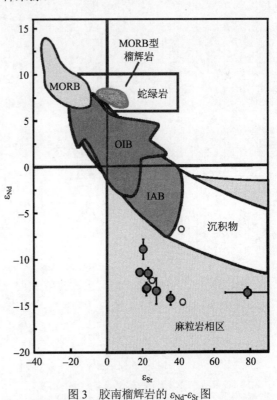

图 3 胶南榴辉岩的 ε_{Nd}-ε_{Sr} 图

红色圆圈为胶南榴辉岩；空心圆圈为五台群细碧岩；MORB 为洋中脊玄武岩；OIB 为洋岛玄武岩；IAB 为岛弧玄武岩。MORB 型榴辉岩数据据 Stosch & Lugmair (1990)[11]。五台群细碧岩数据据 Li et al. (1990)[12]，它表明新太古代浅变质玄武质岩石也可具有类似下地壳麻粒岩的 Sr、Nd 同位素组成

致谢

叶凯、仝来喜、周德进在样品制备过程中参加了部分工作，在此表示感谢。

参考文献

[1] 徐嘉炜，刘德良，李秀新，中新生代地质学术讨论会文集，地质出版社，1987，99-112.
[2] Wang, X., Liou, J.G. and Mao, H.K., Geology, 17(1989), 1085-1088.
[3] Okay, A., Xu, S. and Sengör, A.M.C., Eur. J. Mineral., 1(1989), 595-598.
[4] 徐树桐，苏文，刘贻灿，江来利，科学通报，36（1991），17，1318-1321.
[5] 李曙光, Hart S.R., 郑双根，郭安林，刘德良，张国伟，中国科学，B辑，1989，3: 312-319.
[6] 李曙光，葛宁洁，刘德良等，科学通报，34（1989），7: 522-525.
[7] 李曙光，刘德良，陈移之等，科学通报，37（1992），4: 297-300.
[8] 曹国权，王致本，张成基，山东地质，6（1989），1-14.
[9] Liou, J.G., Maruyama, S., Wang, X. and Graham, S., Tectonophysics, 181(1990), 97-111.
[10] Yang, J., Smith, D.C., 3rd international Eclogite Conference (abs), Oxford, England, 1989, 26.
[11] Stosch, H. Y., Lugmair, G.W., Earth Planet. Sci. Lett., 99（1990）230-249.
[12] Li, S.G., Hart, S.R., Wu Tieshan, Precam. Res., 47（1990）191-203.

中国中部蓝片岩的形成时代*

李曙光[1]，刘德良[1]，陈移之[1]，王松山[2]，裘冀[2]，胡世玲[2]，桑海清[2]

1. 中国科学技术大学地球与空间科学系，合肥 230026
2. 中国科学院地质研究所，北京 100029

亮点介绍： 获得中国中部大别造山带蓝片岩早三叠世 $^{40}Ar/^{39}Ar$ 变质年龄，证明其与榴辉岩带形成于同一时期；指出该带含蓝片岩地层的元古代 Rb-Sr 全岩等时线年龄指示的是它经历的第 1 次区域变质时代，不是后来发生的高压变质时代。

摘要 报道了皖中张八岭群中蓝片岩的 $^{40}Ar/^{39}Ar$ 年龄为 245.1±0.5 Ma。该年龄与扬子陆块北缘的其他高压变质岩及高压变质矿物的年龄一致，说明中国中部蓝片岩带与榴辉岩带一样形成于三叠纪。动力学分析表明，Sr 同位素在干的高压变质过程中均一化尺度很小(<1 m)，因此该带含蓝片岩地层的元古代 Rb-Sr 全岩等时线年龄指示的是它经历的第 1 次区域变质时代，不是后来发生的高压变质时代。此外，还对榴辉岩 K-Ar 年龄与 Sm-Nd 年龄的矛盾问题进行了讨论。

关键词 蓝片岩；同位素年代学；高压变质作用；秦岭-大别造山带

1 引言

在我国中部，沿扬子陆块北缘有一条长达 1700 km 的蓝片岩带，东部与北侧的大别山和苏北-胶南隆起榴辉岩带相伴，形成一巨大的高压变质带。现已查明该蓝片岩主要分布在扬子陆块北缘的中晚元古界变质火山沉积岩系中，如碧口群、耀岭河群、武当群、随县群、红安群、宿松群、张八岭群、云古群。其中所含蓝闪石类矿物为青铝闪石和镁钠闪石，共生的高压变质矿物还有多硅白云母、黑硬绿泥石等（张树业等，1989；董申保，1989）。该蓝片岩带成因与扬子陆块与华北陆块碰撞后，扬子陆块向华北陆块下俯冲引起的深部基底滑脱作用有关（许志琴等，1988；康维国等，1989）。该蓝片岩带的形成时代仍然有争议，主要有 2 种观点：

（1）该蓝片岩形成于晚元古代（725~860 Ma）。主要依据是：a. 该带蓝片岩均发育在中晚元古界变质火山岩系中，震旦系上部及下古生界中未见蓝片岩；b. 红安群含蓝片岩地层的 Rb-Sr 全岩等时线年龄为 725 Ma，与它共生的榴辉岩全岩 K-Ar 年龄为 860 Ma（张树业等，1989；董申保，1989；周高志等，1989）。

（2）该蓝片岩形成于印支期。主要依据为：a. 大别山榴辉岩的 Sm-Nd 矿物等时线年龄均为 221~224 Ma（李曙光等，1989a，1989b，1992）；b. 含蓝片岩地层的高压变质矿物（如多硅白云母、钠闪石）的 K-Ar、$^{40}Ar/^{39}Ar$ 和 Rb-Sr 矿物等时线年龄为 211~232 Ma（桑宝良等，1987；许志琴等，1988；张树业等，1989；从柏林等，1991）。如果蓝片岩和这些榴辉岩及高压变质矿物属同一高压变质作用的产物，则蓝片岩也应形成于印支期（许志琴等，1988；李曙光，1991）。

很显然，上述对不同对象使用不同方法定年所获得的年龄值是有矛盾的，并且缺乏对蓝片岩的直接有效定年是造成认识分歧的主要原因之一。对所谓蓝片岩被元古代地层控制缺乏构造分析是认识分歧的另一重要原因。对于后者，许志琴等（1988）已给予较合理的解释。本文将重点讨论年代学问题，并报道直接选自蓝片岩的多硅白云母 $^{40}Ar/^{39}Ar$ 年龄。

---

* 本文发表在：地质科学，1993，28(1)：21-27

2 张八岭群蓝片岩的 $^{40}Ar/^{39}Ar$ 年龄及其构造意义

发育于安徽滁县、嘉山一带的张八岭群是一套中元古界细碧角斑岩系。在其上部西冷组发育有蓝闪石片岩。张八岭群及其上覆震旦系和下古生界盖层之间存在一滑脱构造。在震旦系下统苏家湾组千枚岩中广泛发育 A 型褶皱，震旦系及下古生界均表现为向北西倾的同斜褶皱（荆延仁等，1989）。这说明强烈韧性剪切的震旦系下统可能为主滑脱层，上部盖层向东南方向滑脱，蓝闪石片岩分布在滑脱层以下层位。

表 1 三界多硅白云母 $^{40}Ar/^{39}Ar$ 数据（样品 Sanjie）

t(℃)	$(^{40}Ar/^{39}Ar)_m$	$(^{36}Ar/^{39}Ar)_m$	$(^{37}Ar/^{39}Ar)_m$	$^{39}Ar(10^{-14}mol)$	$^{40}Ar^{(1)}/^{39}Ar(\pm 1\sigma)$	$T \pm 1\sigma$(Ma)
1-400	23.16	0.07130	13.26	0.9477	3.248 ± 0.54	68.24 ± 11.1
2-650	16.36	0.01451	2.205	2.725	12.27 ± 0.26	245.4 ± 5.0
3-800	13.00	0.002400	0.2314	5.605	12.31 ± 0.17	245.9 ± 3.2
4-1000	12.53	0.001025	0.001561	26.58	12.22 ± 0.15	244.4 ± 2.9
5-1200	41.49	0.08326	1.6630	0.6365	17.15 ± 1.72	334.3 ± 30.6
6-1500	57.42	0.1186	1.950	0.3852	22.70 ± 3.29	430.5 ± 55.6

注：下标 m 表明为混合 Ar 比值；(1) 为放射成因 ^{40}Ar。

本工作测定的蓝闪石片岩样品采自嘉山县三界中学附近的张八岭群细碧角斑岩系中。该地的蓝片岩呈砖蓝灰色，岩石片理发育，主要由细针状镁钠闪石、多硅白云母、长石和石英组成。由于细针状镁钠闪石很难选出，故选出共生的多硅白云母进行 $^{40}Ar/^{39}Ar$ 定年，测定流程已有专文报道（王松山等，1987）。同位素分析结果见表 1，年龄谱图见图 1。

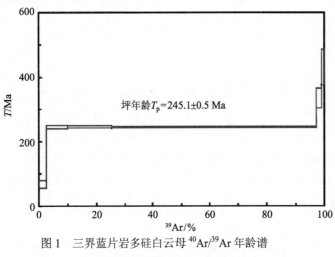

图 1 三界蓝片岩多硅白云母 $^{40}Ar/^{39}Ar$ 年龄谱

图 1 的中高温区间谱线非常平坦，视年龄变化很小，显示了一个非常好的坪年龄谱，其坪年龄值为 245.1 ± 0.5 Ma，它对应的 Ar 析出量高达 94.7%。这种在中高温阶段视年龄不随加热温度升高而升高的特征说明多硅白云母在形成之后没有受过明显的热扰动。即使曾有过热的作用，它对多硅白云母的影响不会超过多硅白云母对 Ar 的封闭温度（350 ± 50 ℃）。因此，坪年龄代表的应是多硅白云母的形成时代，也是蓝片岩的变质形成时代，而不是后期热改造事件的时代。

这个年龄值与扬子陆块北缘中晚元古界变质火山沉积岩系中非蓝片岩所含的高压变质矿物年龄一致（见表 2），说明蓝片岩及其围岩地层中的高压变质矿物都是由同一高压变质作用形成的。表 2 所列覆盖了整个蓝片岩带的蓝片岩和高压变质矿物年龄可以证明中国中部蓝片岩带形成于印支期。它们与相邻的榴辉岩带都是扬子与华北陆块碰撞的产物。

表 2 扬子陆块北缘变质火山岩系地层中蓝片岩及高压变质矿物年龄

样品号	岩石名称	采样地点层位	定年样品	T(Ma)	测试方法	数据来源
Q134	变火山岩	陕西南秦岭跃岭河群	多硅白云母	232 ± 5	$^{40}Ar/^{39}Ar$	许志琴等（1988）
Q127	变火山岩	陕西南秦岭跃岭河群	钠闪石	216 ± 7	$^{40}Ar/^{39}Ar$	许志琴等（1988）
Rd-19	云母斜长片麻岩	安徽宿松群虎踏石组	多硅白云母+斜长石+全岩	231 ± 48	Rb-Sr	桑宝良等（1987）
Rd-19	云母斜长片麻岩	安徽宿松群虎踏石组	多硅白云母	211	K-Ar	桑宝良等（1987）
Sanjie	蓝片岩	安徽嘉山三界张八岭群	多硅白云母	245.1 ± 0.5	$^{40}Ar/^{39}Ar$	本文
E87Y5-2	白云母石英片岩	江苏连云港云台山群	白云母	221.3 ± 0.7	$^{40}Ar/^{39}Ar$	从柏林等（1991）

令人感兴趣的是所报道的蓝片岩年龄在表2中是最高的,并与该带榴辉岩最高年龄值(244±11 Ma)(李曙光,1989a)相似。因此,蓝片岩的形成时代应代表扬子陆块的俯冲时代。显然在该带所有高压变质岩和高压变质矿物的年龄中,最高年龄值应当最接近陆块开始俯冲的时代,也最接近陆块碰撞时代。因此245 Ma可以作为华北与扬子陆块碰撞时代的上限。如果陆块碰撞与陆块开始俯冲之间没有大的时间间隔,则华北陆块与扬子陆块可能在二叠纪发生碰撞,在二叠纪末或三叠纪初发生陆壳俯冲。

3 变质过程中 Sr 同位素均一化尺度的动力学分析及 Rb-Sr 全岩等时线年龄的含义

张树业等(1989)曾报道了红安群Rb-Sr全岩等时线年龄(752 Ma),并将其作为蓝片岩形成于元古代的证据之一。Rb-Sr全岩等时线法是否适用于蓝片岩年龄测定是一个值得探讨的问题。用Rb-Sr全岩法测定变质时代的一个最基本的条件是在变质时,岩石的Sr同位素组成在取样所涉及的空间范围内能重新被均一化。在发生蓝片岩相高压低温变质作用时,Sr同位素均一化尺度有多大?我们作如下近似估计。

已有研究表明,扬子陆块北缘的中晚元古界变火山沉积岩系均经历了2次主要变质作用。以宿松群为例,早期变质作用为低角闪岩相,而后叠加了高压低温变质作用,张八岭群早期广泛发育的是低绿片岩相变质作用,而后叠加了高压低温变质作用(荆延仁等,1989)。在第1次变质过程中,由于温度升高,原岩石要脱水变为变质岩,因此体系中游离态水溶液(变质水)较多,它可以大大加快元素或同位素在体系内的扩散速度(Henderson,1982),使Sr同位素在较大范围内均一化。然而在第2次变质作用时,岩石由低角闪岩相或低绿片岩相向蓝闪片岩相过渡,其温度较第1次变质作用下降了或基本保持不变,只是压力增加了(康维国等,1989;荆延仁等,1989),因此在这一高压变质过程中岩石不会进一步脱水,体系内缺乏流体,它是一种干变质作用,这样Sr在该体系内的扩散速度及Sr同位素可能达到的均一化尺度将大大减小(Henderson,1982)。利用现有的扩散系数资料,在下述假设条件下对Sr同位素均一化尺度作了近似估计。

(1)第1次变质作用是"湿变质"作用,假设游离的变质水溶液湿润了所有的矿物颗粒边界。在此条件下,元素在体系内的扩散距离主要取决于它在该变质水溶液中的扩散速率。Fletcher等(1974)给出Sr在水溶液中扩散速率 $D \approx 10^{-4}\ cm^2 \cdot s^{-1}$ (Hofmann et al.,1978)。根据公式扩散距离 $x=\sqrt{DT}$,我们可获得在此条件下Sr的特征扩散距离是560 m/Ma。如果扬子陆块北缘中晚元古界变质火山沉积岩系,是在晋宁运动时期遭受了第1次变质作用,持续时间(在此时间内始终保持矿物颗粒被水溶液湿润)假定为50~100 Ma,则在该变质过程中Sr同位素均一化尺度为4~5.6 km。在此期变质作用之后,尽管该体系可能由于变质水逸出而变干了,但累积到今天的总扩散距离还要大得多,4 km实际可以作为"湿变质"过程中Sr同位素均一化尺度的下限。由于在垂直岩层走向的4 km范围内,岩石的Rb/Sr值变化幅度足以包括了该变质火山岩地层的所有Rb/Sr值,因此即使取样范围超过4 km,其Rb-Sr全岩等时线也可给出"湿变质"作用时代(Li et al.,1990)。

(2)第2次高压低温变质作用是"干变质"作用,元素的扩散完全在矿物晶体内进行。扬子陆块北缘中晚元古界变火山岩系是一套双峰式火山岩,主要岩性为石英角斑岩和细碧岩。在张八岭群、武当群、随县群、红安群中,石英角斑岩的 SiO_2 平均含量为73.17%~74.23%,与酸性火山岩相当;细碧岩的 SiO_2 平均含量为53.10%~48.44%,与玄武岩相当(周高志等,1989;梁万通等,1989)。因此,可以用玄武岩和酸性火山岩的扩散系数来估计蓝片岩相变质时Sr的扩散距离。根据Hofmann等(1978)提供的玄武岩和黑曜岩的扩散系数资料,及扩散系数补偿定律中 $\log D$ 与 $1/T$ 的线性关系(即,式中 D_0 为频率因子,Q 为活化能,R 为气体常数)(Henderson,1982),可以外推至500 ℃时(蓝片岩相最高变质温度),Sr在玄武岩玻璃和黑曜岩玻璃的扩散系数分别近似为 $3×10^{-13} cm^2 \cdot s^{-1}$ 和 $1×10^{-13} cm^2 \cdot s^{-1}$。考虑到蓝闪片岩相变质作用是在火山岩经历过低绿片岩相变质作用后进行的,以及Sr在已结晶岩石中的扩散速率低于在火山玻璃中的扩散速率,并采用该变质温度的上限,根据上述扩散系数计算的Sr在该高压变质作用时的特征扩散距离($x=\sqrt{DT}$)应是最高限。它们在石英角斑岩和细碧岩中分别是2 cm/Ma和3 cm/Ma。这一特征距离与Sr在固态地幔中的特征距离相当(Hofmann et al.,1978)。若该高压变质作用持续了100 Ma,其最大扩散距离仅有20~30 cm。即使该高压变质作用发生在晚元古代(800 Ma),而且变质温度一直保持至今,其 Sr 同位素最大均一化尺度也仅 57~84 cm。

上述分析表明，在高压低温变质过程中，该带 Sr 同位素均一化尺度<1 m，任何超过 1 m 范围的取样均不可能用 Rb-Sr 全岩等时线法测定出蓝片岩形成时代。更何况该高压变质作用可能发生在印支期，Sr 同位素均一化尺度<0.5 m。因此，采用一块标本选出单矿物做矿物等时线年龄方可保证获得蓝片岩形成时代。在<1 m 范围内，全岩 Rb/Sr 变化范围很难满足等时线要求，而较大范围取样的 Rb-Sr 全岩等时线年龄反映的应是第一次变质作用时代。

张树业等（1989）报道的 Rb-Sr 全岩等时线年龄（752 Ma）是由 4 个样品拟合的。这 4 个样品中只有一个是蓝片岩，其他 3 个样品均不含蓝闪石。显然这 4 个样品采样范围较大，至少超出了一个蓝片岩层范围，给出的年龄应是红安群经历的第 1 次变质时代。这个年龄值与朱洪山等（1979）和桑宝良等（1987）所获得的张八岭群和宿松群 Rb-Sr 全岩等时线年龄（730 ± 50 Ma 和 848 ± 73 Ma）完全一致，这 2 组年龄样品中均无蓝片岩。表明大别山古岛弧与扬子陆块在晋宁运动时期的碰撞、拼合使这一套细碧角斑岩系第 1 次变质，但并未发生高压变质作用。这可能由于在当时这套火山岩系还处在浅部盖层位置的缘故。直到印支期，它们才变为扬子陆块北缘的变质基底，从而有可能参与扬子陆块的俯冲作用。

4 榴辉岩的过剩 Ar 问题

作为该带蓝片岩形成于元古代的另一个年代学证据是若干榴辉岩全岩 K-Ar 年龄（860~875 Ma）（张树业等，1989a，1989b），及若干榴辉岩中多硅白云母、角闪石、黑云母的 $^{40}Ar/^{39}Ar$ 年龄（500~700 Ma）（Mattauer et al., 1991）。在评价这些年龄时有 2 个基本事实要给予重视：

（1）大别山和苏北-胶南隆起榴辉岩的 K-Ar 和 $^{40}Ar/^{39}Ar$ 定年结果是很分散的，如从柏林等（1991）报道的苏北东海县孟中和青龙山 2 个相邻榴辉岩体的多硅白云母 $^{40}Ar/^{39}Ar$ 年龄分别为 435.0 ± 0.7 Ma 和 874.8 ± 9.8 Ma，二者相差 1 倍。日本金泽大学石渡明测定了 8 件选自苏北-胶南榴辉岩的多硅白云母 K-Ar 年龄，其年龄值范围为 2 亿~11 亿年，主要集中在 6 亿~7 亿年（从柏林等，1991）。K-Ar 和 $^{40}Ar/^{39}Ar$ 年龄的分散性，与用 Sm-Nd 矿物等时线法测定榴辉岩年龄给出的非常一致的结果（李曙光等，1989a，1989b，1992a，1992b）形成鲜明的对照。

（2）榴辉岩的长英质围岩（片麻岩）及前述中晚元古界变火山岩系和蓝片岩中所有多硅白云母、黑云母、角闪石的 K-Ar、$^{40}Ar/^{39}Ar$ 定年均一致给出印支期结果（桑宝良等，1987；许志琴等，1988；从柏林等，1991；Mattauer et al., 1991；本文），这与榴辉岩的 Sm-Nd 定年结果相符，而与榴辉岩的 K-Ar 和 $^{40}Ar/^{39}Ar$ 定年不同。如果认为榴辉岩的 K-Ar 或 $^{40}Ar/^{39}Ar$ 年龄反映了一次晚元古代高压变质事件，那么为什么这一事件在本区长英质片麻岩、变质火山岩系和蓝片岩的 K-Ar 或 $^{40}Ar/^{39}Ar$ 年龄中得不到反映呢？

上述事实说明铁镁质的高压变质岩——榴辉岩可能存在着过剩氩问题，从而使它的 K-Ar 或 $^{40}Ar/^{39}Ar$ 年龄偏老。过剩 Ar 问题已被我们最近的工作所证实（另文详述）。因此，K-Ar 和 $^{40}Ar/^{39}Ar$ 法不适合于榴辉岩年龄测定。

参考文献

王松山，胡世玲，翟明国等，1987.应用 $^{40}Ar/^{39}Ar$ 定年技术研究清源花岗岩-绿岩地体的形成时代.岩石学报，(4): 55-62
从柏林，张儒媛，李曙光等，1991.中国苏北-胶东南高压变质带的同位素地质年代学初探.中国科学院地质研究所岩石圈构造演化开放研究实验室年报（1989—1990）.北京: 中国科学技术出版社，68-72
朱洪山，张忠奎，陈江峰等，1979.安徽张八岭群变质岩系同位素地质年龄.中国科学技术大学学报，(2): 158-165
许志琴，卢一伦，汤耀庆等，1988.东秦岭复合山链的形成: 变形、演化及板块动力学.北京: 中国环境科学出版社，101-129
李曙光，Hart S. R.，郑双根等，1989a.中国华北、华南陆块碰撞时代的 Sm-Nd 同位素年龄证据.中国科学(B 辑)，(3): 312-319
李曙光，葛宁洁，刘德良等，1989b.大别山北翼大别群中 C 型榴辉岩的 Sm-Nd 同位素年龄及其构造意义.科学通报，34(7): 522-525
李曙光，刘德良，陈移之等，1991.秦岭-大别造山带主要构造事件同位素年表及其意义.秦岭造山带学术讨论会论文集. 西安: 西北大学出版社，229-237

李曙光, 刘德良, 陈移之等, 1992a.大别山南麓含柯石英榴辉岩的 Sm-Nd 同位素年龄.科学通报, 37(4): 297-300

李曙光, 肖益林, 陈移之等, 1992b.秦岭-大别山带形成的四阶段演化模型.《七五地质科学技术研究新进展》论文集. 北京: 地质出版社（在印刷中）

周高志, 康维国, 高敬礼等, 1989.鄂北蓝片岩带基本特征.长春地质学院学报, 鄂皖蓝片岩带地质专辑: 10-17

荆延仁, 夏木林, 张良田等, 1989.皖中张八岭高压变质带基本特征.长春地质学院学报, 鄂皖蓝片岩带地质专辑: 133-151

张树业, 康维国, 1989a.扬子陆台北缘中段蓝片岩带.长春地质学院学报, 鄂皖蓝片岩带地质专辑: 1-9

张树业, 胡克, 刘晓春等, 1989b.中国中部元古代蓝片岩-白片岩-榴辉岩带-古陆内板块裂撞带三位一体特征.长春地质学院学报, 鄂皖蓝片岩带地质专辑: 152-160

董申保, 1989.中国蓝闪石片岩带的一般特征及其分布.地质学报, (3): 273-284

桑宝良, 陈跃志, 邵桂清, 1987.大别山南麓宿松群铷-锶年龄及其构造意义的探讨.中国区域地质, (4): 364-370

梁万通, 王海俊, 荆延仁等, 1989.皖中张八岭高压变质带的岩石、矿物学特征.长春地质学院学报, 鄂皖蓝片岩带地质专辑: 97-122

康维国, 张树业, 周高志等, 1989.鄂北前寒武纪高压变质带的岩石学特征及成因.长春地质学院学报, 鄂皖蓝片岩带地质专辑: 18-39

Li Shuguang, Hart S.R., Wu Tieshan,1990. Rb-Sr and Sm-Nd isotopic dating of an early Precambrian spilite-keratophyre sequence in the Wutaishan area, North China: preliminary evidence for Nd isotopic homogenization in the mafic and felsic lavas during low grade metamorphism. Precambrian Research, 47: 191-203

Henderson P., 1982. lnorganic geochemistry,Oxford, New York: Pergamon Press, 185-196

Hofmann A.W., Hart S.R. 1978. An assessment of local and regional isotopic equilibrium in the mantle. Earth Planet. Sci. Lett., 38: 44-62

Mattauer M., Matte P., Maluski H. et al., 1991. Paleozoic and Triassic plate boundary between North and South China: new structural and radiometric data on the Dabieshan (Eastern China) Paris: C. R. Acad. Sci. 312, Serie II, 1227-1233

青岛榴辉岩及胶南群片麻岩的锆石 U-Pb 年龄
——胶南群中晋宁期岩浆事件的证据*

李曙光[1]，陈移之[1]，葛宁洁[1]，刘德良[1]，张志敏[2]，张巧大[3]，赵敦敏[3]

1. 中国科学技术大学地球与空间科学系，合肥 230026
2. 山东省地质矿产局区域地质调查队，胶州 266300
3. 中国地质科学院地质研究所，北京 100037

> **亮点介绍**：首次对苏鲁榴辉岩和胶南群片麻岩进行锆石 U-Pb 定年；证明榴辉岩及其围岩片麻岩的原岩均为形成于 8 亿~9 亿年的火成岩；提出苏鲁地体在晋宁期有一次强烈的岩浆事件。

关键词 榴辉岩；胶南群；岩石 U-Pb 年龄；晋宁运动；高压变质带

已有研究表明，苏鲁地体是扬子与华北陆-陆碰撞造山带的一部分（东段）[1-4]。其主要变质岩系，胶南群，是扬子陆块北缘俯冲陆壳的基底[2-4]。在已有的文献中，人们均将胶南群划为早、中元古代（>18 亿年）地体[2,5]。组成胶南群的主要岩石是长英质片麻岩，在它中间包含有少量斜长角闪岩、大理岩、超镁铁岩和榴辉岩。大量 Sm-Nd 及 $^{40}Ar/^{39}Ar$ 同位素定年已证明了在三叠纪它们经历了与华北和扬子陆块碰撞有关的高压或超高压变质作用[6-8]，但是它们的原岩时代缺乏可靠的年龄测定。在文献中也普遍认为是早、中元古代岩石[2,5]。然而最近我们对青岛附近的退变质榴辉岩及胶南群片麻岩的结晶锆石的 U-Pb 同位素年龄测定表明该榴辉岩及其围岩（片麻岩）的原岩均为形成于 8 亿~9 亿年的火成岩。这是榴辉岩原岩时代的第一个年龄结果，并对视胶南群为单一早、中元古代岩石的传统认识提出了挑战，它启示我们必须重新认识苏鲁地体的组成及演化历史。

1 地质概况及样品

青岛东郊，崂山脚下，仰口湾海滨北侧出露一套榴辉岩相的蛇绿混杂岩（图 1），其详细地质情况已有专文介绍[2,3]。除超镁铁岩及少量硅质岩外，这套岩石主要由榴辉岩退变质形成的含石榴石斜长角闪岩组成，且其中局部保留有榴辉岩残块[3]。它们的原岩为具岛弧火山岩特征的玄武和玄武质安山岩[3]。该蛇绿混杂岩的围岩为胶南群常见的云母斜长片麻岩夹斜长角闪岩。它与榴辉岩相蛇绿混杂岩的接触带被强烈糜棱岩化。在仰口湾南侧海滨这一套胶南群片麻岩有很好出露（图 1），其所夹的斜长角闪岩局部含石榴子石。

本文在该区采取了两个大样以选取锆石单矿物样品。Y90-8 采自仰口湾北侧榴辉岩相蛇绿混杂岩中的退变质榴辉岩，灰绿色，块状，含较多石榴石。Y90-11 采自仰口湾南侧海滨采石场的胶南群云母斜长片麻岩。两个采样点相距约 3 km（图 1）。

Y90-8 的锆石主要为淡黄色，短柱状（长宽比 1∶2），四方双锥晶体，晶形完整、透明度好。Y90-11 的锆石主要为淡黄色、长柱状（长宽比 1∶3~1∶4）、四方双锥晶体，透明、晶形完整。上述锆石特征表明，它们都是岩浆结晶锆石。因此，该片麻岩原岩可能为花岗闪长质深成岩体。

我们分粒级在镜下仔细挑选其中的结晶锆石进行 U-Pb 同位素分析。分析工作在中国地质科学院地质研究所同位素实验室完成，分析流程已有前文介绍[9]。

---

* 本文发表在：科学通报，1993，38(19)：1773-1777

2 测定结果与讨论

2 个锆石样品的 U-Pb 同位素分析结果列于表 1，其一致线年龄图分别示于图 2、图 3。

表 1 及图 2 表明，片麻岩（Y90-11）的锆石在变质过程中有较多铅丢失，因而在靠近一致线下交点处呈线性分布。5 个样点拟合给出了较精确的下交点年龄 202^{+12}_{-14} Ma。这一年龄值与胶南群中榴辉岩的 Sm-Nd 矿物等时年龄[7,8]，和片麻岩中黑、白云母、角闪石的 $^{40}Ar/^{39}Ar$ 年龄[6] 一致。它说明青岛仰口湾的胶东群同样在三叠纪经历了强烈的变质作用。Y90-11 样品的 U-Pb 一致线上交点年龄为 871^{+47}_{-41} Ma，它表明该片麻岩的原岩形成时代为 8 亿~9 亿年左右。

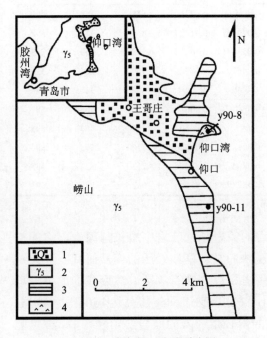

图 1 青岛仰口湾地质略图

1.第四纪；2.崂山花岗岩；3.胶南群片麻岩夹斜长角闪岩；4.仰口湾北侧榴辉岩相蛇绿混杂岩，详图见文献[2,3]。黑圆点为本文取样位置

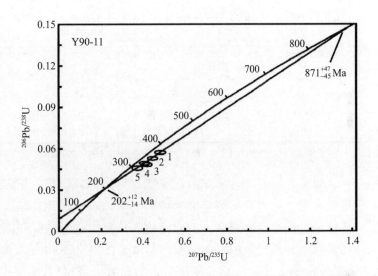

图 2 青岛胶南群片麻岩（Y90-11）锆石 U-Pb 一致线年龄图

表 1 表明榴辉岩（Y90-8）的 5 个不同粒级锆石几乎具有完全相同的 $^{206}Pb/^{238}U$ 和 $^{207}Pb/^{235}U$ 表面年龄。它们在图 3 挤在一起，因而由这 5 个点拟合而获得的一致线上、下交点年龄都具有很大误差，从而没有什么意义。在这种情况下，$^{207}Pb/^{206}Pb$ 年龄更能接近锆石的结晶时代[10]。表 1 表明该榴辉岩的 $^{207}Pb/^{206}Pb$ 年龄集中在 720 Ma 左右。锆石的逐层蒸发实验已证明锆石的外层封闭性较差，其 $^{207}Pb/^{206}Pb$ 值较低，而内层封闭性好，$^{207}Pb/^{206}Pb$ 较高，其年龄才接近锆石真实结晶年龄。然而本文给出的锆石

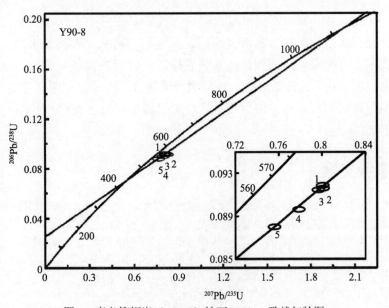

图 3 青岛榴辉岩（Y90-8）锆石 U-Pb 一致线年龄图

$^{207}Pb/^{206}Pb$ 值是锆石内外层的平均结果，因而其表面年龄较真实年龄偏低。比较 Y90-8 与 Y90-11 的 $^{207}Pb/^{206}Pb$ 表面年龄与 $^{206}Pb/^{238}U$ 和 $^{207}Pb/^{235}U$ 表面年龄的差异，我们可知 Y90-8 锆石在变质过程中 Pb 丢失要比 Y90-11 样小得多。因此，Y90-8 样的 $^{207}Pb/^{206}Pb$ 表面年龄较真实年龄的偏低程度要比 Y90-11 样小，其原岩时代很可能与其围岩（片麻岩）的原岩时代基本相同，也是 8 亿～9 亿年。这一认识与 Ames 和 Tilton（1992）的大别山榴辉岩锆石 U-Pb 年龄测定结果一致[11]。他们也认为榴辉岩原岩年龄为晚元古代[11]。

表 1 青岛胶南群片麻岩（Y90-11）和榴辉岩（Y90-8）的锆石 U-Pb 同位素分析数据 a)

样品	锆石样品量 (mg)	粒度 (μm)	放射成因 Pb 含量 (ppm)	U 含量 (ppm)	$^{206}Pb/^{204}Pb$	同位素原子比			表面年龄值（Ma）		
						$^{207}Pb/^{206}Pb$	$^{206}Pb/^{238}U$	$^{207}Pb/^{235}U$	$^{207}Pb/^{206}Pb$	$^{206}Pb/^{238}U$	$^{207}Pb/^{235}U$
Y90-11	4.43	>121	17.86	295.7	340	0.060358	0.05794	0.48219	615	363	400
	3.9	>74	21.85	423.2	4246	0.060751	0.049453	0.41423	629	311	352
	7.3	>50	19.55	346.3	6141	0.059616	0.053971	0.44364	588	339	373
	9.9	>30	18.96	393.6	4683	0.057297	0.046795	0.36969	502	295	319
	3.73	<30	18	343.2	8547	0.058941	0.05034	0.4091	564	316	348
Y90-8	9.22	>121	56.79	537.8	864	0.063265	0.09173	0.80016	716	566	597
	23.31	>50	21.7	208.7	1171	0.063673	0.09136	0.80207	730	564	598
	14.3	<50	23.02	223.1	1474	0.063391	0.091254	0.79758	720	563	595
	19.89	>97	19.78	193.3	733	0.063195	0.089422	0.77916	713	552	585
	8.66	>152	17.75	175.4	629	0.062273	0.087893	0.75466	682	543	571

a) 一致线年龄值：Y90-11 上交点年龄，下交点年龄；Y90-8 上交点年龄，下交点年龄

对榴辉岩锆石 U-Pb 年龄还可能有另外一种解释：榴辉岩的原岩是片麻岩中斜长角闪岩，它的锆石是受片麻岩原岩岩浆影响而生成的变质锆石。因此其年龄反映的是它早期经历的变质时代。如果这一解释是正确的话，那么片麻岩中的斜长角闪岩应当有和榴辉岩一样的锆石年龄。然而我们的另一项研究表明这种斜长角闪岩的锆石与榴辉岩的完全不同，它们的 $^{207}Pb/^{206}Pb$ 表面年龄高达 1504～2149 Ma（详情另有专文发表）。这说明榴辉岩的原岩较苏-鲁片麻岩中的斜长角闪岩的原岩要年轻许多，很可能是晚元古代形成的。

3 结论

胶南群片麻岩不全是早、中元古代岩石，它中间有相当一部分是 8 亿～9 亿年形成的晚元古深成岩浆岩体。青岛仰口榴辉岩的原岩可能是一套晚元古岛弧火山岩系。根据它与片麻岩围岩的构造接触关系，可能是后来的构造运动使它挤入同时代的深成岩体内。这些事实说明，晋宁运动在苏鲁地体有很强的表现。它表现为一次强烈的岩浆事件。该区晋宁期岩浆事件发生的构造背景是值得深入研究的。根据青岛仰口榴辉岩的原岩特征，我们初步认为在 8 亿～9 亿年前，苏-鲁地体可能是扬子陆块北缘的古岛弧。

参考文献

[1] 徐嘉炜，刘德良，李秀新，中新生代地质学术讨论会文集，地质出版社，1987，99-112.
[2] 曹国权，王致本，张成基，山东地质，1990，6:1-14.
[3] 李曙光，刘德良，葛宁洁等，科学通报，1991，36（13）：1161-1164.
[4] 李曙光，孙卫东，葛宁洁等，岩石学报，1992，8：351-362.
[5] 陈希道，陈允福，王桂枝，岩石学报，1992，8：40-49.
[6] 陈文寄，T. M. Harrison，M. T. Heizler 等，岩石学报，1992，8：1-17.
[7] 陈移之，李曙光，从柏林等，科学通报，1992，37（23）：2169-2172.
[8] 从柏林，张儒缓，李曙光等，中国科学院地质研究所岩石圈的构造演化开放研究实验室年报（1989—1990），中国科学

技术出版社，68-72.

[9] 伍家善，刘敦一，金龙国，地质论评，1986，32：178-184.

[10] 李曙光，刘德良，陈移之等，地质科学，1993，28：21-27.

[11] Ames, L., Tilton, G. R., EOS, *Transactions Am. Geoph. Union.*, 1992, 73：652-653.

华北与扬子陆块的碰撞时代及过程*

李曙光

中国科学技术大学地球与空间科学系,合肥 230026

> **亮点介绍**:系统总结了国家自然科学基金资助项目取得的一系列研究进展,包括首次精确地测定出大别-苏鲁超高压变质带形成于三叠纪;首次指出并证明了造山带榴辉岩中的多硅白云母含有过剩 Ar,它不适于用 $^{40}Ar/^{39}Ar$ 法对榴辉岩定年;精确测定了大别第一条榴辉岩 P-T-t 轨迹,并据此提出了榴辉岩的二阶段抬升机制;首次在大别山燕山期花岗岩中发现了超糜棱岩脉,它证明在白垩纪以后大别山仍然在抬升;提出了华北与扬子陆块拼合及秦岭-大别山形成演化过程的四阶段演化模型。该 1993 年提出的造山带演化模型除了北秦岭新元古代时期的构造归属和大别造山带碰撞后引张应该发生在早白垩世而不是晚侏罗世外,其他基本是正确的。

秦岭-大别山造山带是我国东部华北陆块与扬子陆块相互碰撞形成的。它作为我国东部天然地质界线历来受到国内外学者重视。尤其近年来在大别山及其东延部分（苏-鲁地体）发现了柯石英和金刚石,使它成为世界上最大的超高压变质带并引起广泛注意和兴趣。确定这两大陆块的碰撞时代及过程是认识秦岭-大别山造山带形成演化机制及中国东部大陆形成演化历史的关键。然而正是在这一关键问题上地质学界存在着明显的认识分歧。国家自然科学基金资助项目"华北与扬子陆块的碰撞时代"（编号:4870188）及"秦岭造山带多期蛇绿岩识别"（编号:49070165）正是试图通过对与碰撞造山运动有关的高压变质岩（榴辉岩、蓝片岩）及蛇绿岩的同位素年代学和地球化学研究以求为解决这一重要地学基础研究课题做出贡献。经过近六年的工作,在认识华北与扬子两大陆块碰撞时代和过程以及含柯石英榴辉岩的成因方面取得了重要进展。

1. 通过对该带不同类型榴辉岩系统的 Sm-Nd 矿物等时线年龄测定,首次精确地测定出大别山、苏-鲁超高压变质带形成于三叠纪（208～244 Ma）。它可以作为两大陆块碰撞的时间上限。

2. 在世界上首次指出并证明了造山带榴辉岩中的多硅白云母含有过剩 Ar,它不适于用 $^{40}Ar/^{39}Ar$ 法对榴辉岩定年,从而解决了榴辉岩 $^{40}Ar/^{39}Ar$ 年龄与 Sm-Nd 年龄矛盾这一令人困惑的问题。

3. 对 Sr 同位素在多期变质过程中的均一化尺度进行了地球化学动力学分析,指出 Rb-Sr 全岩法定年在大别山只能给出早期经历的区域变质时代,而单矿物或矿物等时线年龄才能指示后来的高压变质时代,从而解决了蓝片岩形成时代因方法不同而引起的争议并肯定了中国蓝片岩带也形成于三叠纪。

4. 精确测定了大别第一条榴辉岩 P-T-t 轨迹,并据此提出了榴辉岩的二阶段抬升机制:在 221 Ma 由陆壳俯冲引发的逆冲构造使榴辉岩从 100 km 深处快速抬升至 30～40 km 处;在 134 Ma 由于陆块碰撞后的拉伸作用,榴辉岩随山体整体抬升至地表。从而修正了国外提出的单阶段抬升模型。

5. 首次在大别山燕山期花岗岩中发现了超糜棱岩脉,它证明在白垩纪以后大别山仍然在抬升。这为大别山在陆-陆碰撞后的拉伸环境下继续隆起造山提供了又一地质证据。

6. 使用具有不同封闭温度的 Sm-Nd 和 K-Ar 定年方法,发现一些榴辉岩的围岩——片麻岩具有与榴辉岩不同的冷却历史。这为判断榴辉岩成因（原地的还是外来的）提供了重要制约条件。它表明大别山并不是整体都经历了超高压变质作用。

7. 利用 Sr、Nd 同位素示踪,证明了大别山榴辉岩不是古洋壳俯冲到深部形成的,而是由扬子陆块北缘古老基底岩石俯冲到深部形成的,从而肯定了该榴辉岩的形成时代代表扬子陆壳俯冲时代,其所含柯石英标志着陆壳已俯冲至 100 km 深处。

---

* 本文发表在:地球科学进展,1993,8(4):83-84

8. 首次发现北秦岭松树沟镁铁-超镁铁杂岩为晚元古蛇绿岩,并测定了它的侵位时代为 983 Ma。它为研究北秦岭元古代构造体制及演化提供了重要证据。

9. 用 Sm-Nd 矿物等时线法准确测定了北秦岭丹凤蛇绿岩的形成时代为 402 Ma,并通过 Sr、Nd 同位素及痕量元素示踪证明它们形成于近陆岛弧环境。这表明直到 402 Ma,华北与扬子陆块尚没有碰撞。

10. 综合已有同位素年代学结果,提出了华北与扬子陆块拼合及秦岭-大别山形成演化过程的四阶段演化模型:(1)晚元古代(10 亿年)时,华北陆块南缘裂开形成北秦岭古岛弧;(2)海西期,北秦岭古岛弧与华北大陆碰撞形成北秦岭海西造山带;(3)二叠纪末或三叠纪初,华北与扬子陆块对接。随后在整个三叠纪,扬子陆块向华北陆块下俯冲,并在俯冲基底岩石中发生高压或超高压变质作用;(4)在晚侏罗世,该造山带发生碰撞后拉伸作用,并由于重力均衡使造山带进一步抬升,同时发生强烈的花岗岩浆活动。至此,两大陆块焊合为统一大陆,碰撞过程到此结束。这一模型较完整解释了绝大部分同位素年代学、地质学、古地磁观测资料。

11. 在榴辉岩岩石学方面发现了:(1)在石榴石中呈包裹体存在的绿辉石,从而将榴辉岩变质阶段进一步划分为早期柯石英榴辉岩相顶峰变质阶段和退变质重结晶榴辉岩阶段;(2)从大别山到苏鲁,榴辉岩变质温度逐步升高以及南北大别山榴辉岩的差异,并给出合理解释;(3)在青岛附近出露有具有类似蛇绿岩岩石组合的榴辉岩相混杂岩,证明五莲-荣城断裂可能是扬子陆块北缘边界。

12. 对榴辉岩岩石学方法做出若干改进:(1)对石榴石-单斜辉石地质温度计进行了修正,提出了能充分考虑各种元素对定温影响的新的计算方法;(2)提出了同时使用石榴石镁原子数和单斜辉石四次配位铝原子数进行榴辉岩分类的新方法,从而克服了前人单纯根据石榴石或辉石成分分类不好的缺点。

13. 为更好判断蛇绿岩生成的构造环境,提出了两组新的痕量元素及 Nd 同位素判别图解,即 La-Nb-Th-Ba 判别图和 ε_{Nd}-La/Nb、Ba/Nb、Nb/Th 判别图。这些判别图克服了 Pearce 判别图判别效果不好的缺点。

14. 对大别山、苏北一带片麻岩及糜棱岩进行了较详细的显微构造研究,为认识该带宏观构造演化提供了微观证据。

以上成果已发表论文 30 篇,有关碰撞时代及榴辉岩成因的总结性论文已被欧洲地球化学协会会刊 *Chemical Geology* 录用。已发表的论文已被国内外至少 25 篇论文和专著引用(其中国际文献 7 篇)。

Collision of the North China and Yangtse Blocks and formation of coesite-bearing eclogites: timing and processes*

Shuguang Li[1], Yilin Xiao[1], Deliang Liou[1], Yizhi Chen[1], Ningjie Ge[1], Zongqing Zhang[2], Shensu Sun[3], Bolin Cong[4], Ruyuang Zhang[4], Stanley R. Hart[5] and Songshan Wang[3]

1. Department of Earth and Space Sciences, University of Science and Technology of China, Hefei 230026, Anhui, China
2. Institute of Geology, Chinese Academy of Geological Sciences, Beijing 100037, China
3. Mineral and Land Use Program, Australian Geological Survey Organization, P.O. Box 378, Canberra, A.C.T. 2601, Australia
4. Institute of Geology, Academia Sinica, Beijing 100029, China
5. Woods Hole Oceanographic Institution, Woods Hole, MA 02543, U.S.A.

亮点介绍：该文在国际刊物上最早系统报道了大别山-苏鲁地体中含柯石英榴辉岩的 Sm-Nd 矿物等时线年龄和蓝片岩多硅白云母 $^{40}Ar/^{39}Ar$ 同位素年龄，证明该造山带超高压变质作用发生在早三叠世；榴辉岩的 LREE 富集模型及 Sr、Nd 同位素组成证明其源岩不是俯冲洋壳，而是华南陆壳中的基性岩石；榴辉岩的 P-T-t 轨迹显示大别山榴辉岩经历 221 Ma 和 134 Ma 两个阶段的抬升作用；指出榴辉岩的更古老的多硅白云母 $^{40}Ar/^{39}Ar$ 年龄是过剩 Ar 引起的，无地质意义。提出华北和华南陆块在晚二叠世或早三叠世开始对接，并在三叠纪华南陆壳向北俯冲于华北陆块之下。大别山石马地区榴辉岩和它的围岩片麻岩不同的冷却历史显示部分榴辉岩及其片麻岩围岩可能是通过构造作用拼贴在一起的。

Abstract Various types of eclogite, including coesite-bearing varieties formed under high-P-high-T conditions (up to 27 kbar, 700–800 ℃), and glaucophane schist occur in the Dabie Mountains and the Su-Lu terrane, eastern China. Dating of these high-pressure rocks by the Sm-Nd mineral isochron and $^{40}Ar/^{39}Ar$ method suggests that the occurrence of the high-pressure metamorphism was during the early Triassic. LREE-enriched chondrite-normalized patterns for type-II eclogites and low initial ε_{Nd} of -14 to -3 and the $^{87}Sr/^{86}Sr$-values (0.706–0.710) for various types of eclogites suggest that their protoliths were mainly derived from Precambrian island arc or intraplate basalts in the basement of the Yangtse Block and the enriched pyroxenite layer in alpine peridotite. The P-T-t path of eclogite from the southern Dabie Mountains suggests that the uplift history of eclogite in the Dabie Mountains can be subdivided into two stages: (1) fast uplift driven by thrust during continental-continental collision and deep subduction (at 221 Ma) of the continental crust; (2) later gentle uplift with rise of the Dabie Mountains in the late Jurassic and Cretaceous (at 134 Ma). It is proposed that the collision between the North China Block and Yangtse Block began in the late Permian or early Triassic with a north dipping subduction zone. This was followed by subduction of the continental crust of the Yangtse Block under the North China Block during the Triassic. These two continental blocks were welded into a single tectonic unit in the late Jurassic or early Cretaceous. Different cooling histories for a coexisting gneiss and eclogite pair from Shima area suggest that the eclogite and their country rocks are not always coherent to each other. Some of them may have been juxtaposed through tectonic processes from different levels.

1 Introduction

Asia is well known as a composite continent formed by accretion of several continental blocks (e.g., C.Y. Li, 1980; Klimetz, 1983; Sengör, 1985). The accretionary history of the blocks can be elucidated by radiometric dating and by the P-T-t path of high-P-high-T meta morphic rocks (eclogites and glaucophane schists) occurring in the

---

* 本文发表在：Chemical Geology, 1993, 109: 89-111

suture zones.

The North China Block (NCB) and Yangtse Block (YB) are separated by the E-W-striking Qinling-Dabie orogenic belt. The Su-Lu terrane is the eastern extension of this belt. It has been displaced ~ 500 km to the north by the Tanlu fault (J. Xu et al., 1987) (Fig. 1). Abundant eclogites and glaucophane schists occur along the Dabie Mountains and Su-Lu terrane. Diamond, coesite and coesite pseudomorphs have been recently reported as inclusions in garnet, omphacite, kyanite and epidote in eclogite from the Dabie Mountains and Su-Lu terrane (Okay et al., 1989; X. Wang et al., 1989; J. Yang and Smith, 1989; R. Zhang et al., 1990; Hirajima et al., 1991; R. Zhang and Cong, 1991; S. Xu et al., 1992; Cong et al., 1993). This is the third known occurrence of coesite in deep crustal rocks besides the western Italian Alps and Norwegian Caledonides and the second occurrence of diamond of metamorphic origin besides the Kokchetav Massif, northern Kazakhstan, and it has attracted extensive interest from the geological community.

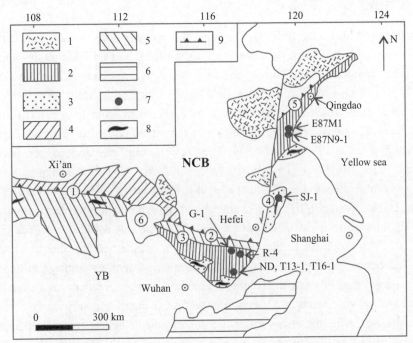

Fig. 1 Sketch map of geology and tectonics in central China and sample locations for this study. The Qinling-Dabie orogenic belt trends west to east in central China. The Nanyan Basin (⑥) is situated in the middle section of the orogenic belt dividing it into two parts: the western part is the Qinling and the eastern part is the Dabie Mountains. The Su-Lu terrane, located in eastern China, is the eastern extension of the belt and appears to be displaced to the north by the Tanlu fault (④). The suture between the North China Block (NBC) and Yangtse Block (YB) is located along the Shangnan-Danfengfault (①) in the Qinling, the Xinyang-Shucheng fault (②) in the Dabie Mountains and the Wulian-Rongchengfault (⑤) in the Su-Lu terrane. Key to other symbols:1= Archaean complexes; 2=Dabie and Jiaonan groups; 3=Hongan, Susong and Haizhou groups; 4=north Qinling terrane; 5=lower Palaeozoic and late Proterozoic strata in the south Qinling and north Dabie Mountains; 6=Banxi Group in the Yangtse Block; 7=location of the eclogite sample: 8 = glaucophane schist; 9= suture; ③ = TongboMozitan fault.

A major controversy concerning the study of this orogenic belt is the timing of collision between the NCB and YB and the age of the eclogite and glaucophane schist formation. Palaeomagnetic studies suggest that these two blocks were not fully assembled until the mid-Jurassic (McElhinny et al., 1981; Lin et al., 1985; X. Zhao and Coe, 1987; Z. Yang et al., 1991). Geologic interpretation has led to a variety of estimates of the collision age, ranging from Devonian to Jurassic (C.Y. Li, 1980; H. Wang et al., 1982; Klimetz, 1983; Zh.M. Zhang et al., 1984; Mattauer et al., 1985; Sengör, 1985; Hsu et al., 1987). We have previously reported two Sm-Nd isotopic ages of 244 ± 11 and 224 ± 20 Ma for eclogites from the northern Dabie Mountains and one Sm-Nd isotopic age of 221 ± 5 Ma for eclogite from the southern Dabie Mountains and suggested that the collision occurred in the early Triassic (S. Li et al., 1989a, b, 1992a). Recently, Ames and Tilton (1992) have also obtained a similar metamorphic age of 209 ± 2 Ma for an eclogite sample from the Dabie Mountains using U-Pb analyses on zircons. There are some remaining outstanding questions, such as whether there were multistage high-P metamorphic events in the Dabie Mountains

(Liou et al., 1990). Some eclogites and glaucophane schists in the Dabie Mountains have been considered to be Precambrian in age based on some K-Ar and ^{40}Ar/^{39}Ar dating of phengites and Rb-Sr whole-rock isochron ages (Dong et al., 1986; S. Zhang et al., 1991). Another question is related to the tectonic setting of the protoliths of the eclogites, which is important for understanding the age relationship between the time of collision and age of high-pressure metamorphism.

To further evaluate the validity of these radiometric ages and to better define the timing and processes of collision between the NCB and YB, we have made a more detailed study including Sm-Nd and Rb-Sr mineral isochron dating, rare-earth element (REE) and Sr, Nd isotope geochemistry and P-T-t path of the eclogites from the Dabie Mountains and Su-Lu terrane. In this paper, we also present a ^{40}Ar/^{39}Ar age of the glaucophane schists from this belt.

2 Geological setting

The Dabie Mountains and Su-Lu terrane (Fig. 1) represent deep parts of a collision zone between the NCB and YB. The suture line between the two blocks is situated on the north of the Dabie Mountains and the Su-Lu terrane along the Tongbo-Mozitan fault and Wulian-Rongcheng fault, respectively (Cao et al., 1990; S. Xu et al., 1992) (Fig. 1). The Dabie and Jiaonan Groups, the main rock assemblages of the Dabie Mountains and Su-Lu terrane, have similar metamorphic lithologies which are considered to be early Proterozoic in age and form parts of the YB basement. It has been generally considered that these two metamorphic complexes represent the underthrust crust (Hsu et al., 1987; Hao and Liou, 1988; Hsu, 1991). The Dabie Group was divided into two terranes by local geologists in the early 1970s (Regional Geological Survey of Anhui, unpublished report, 1992) (Fig. 2). This approach receives support from later studies (S. Xu et al., 1986; Dong et al., 1989; Okay et al., 1989; X. Wang et al., 1992). The northern Dabie terrane (NDT) is characterized by abundant migmatitic gneisses, syn- to late metamorphic granitoids, ultramafic lenses and fewer eclogites. So far, no coesite or coesite pseudomorphs have been found from the NDT. The southern Dabie terrane (SDT) is a coesite-bearing eclogite province. Fine-grained metamorphic granite, leucocratic, banded gneisses with the common mineral assemblage of quartz + plagioclase + microcline + epidote + garnet + white mica ± amphibole ± biotite ± kyanite + jadeite, and amphibolite make up the bulk of the SDT. Abundant eclogite bands and lenses, impure marble bands and very rarely ultramafic lenses occur within the leucocratic gneisses. After the discovery of coesite in Dabie eclogites, quartz pseudomorphs after coesite as garnet inclusions have been identified in marble, phengite schist and biotite gneiss by X. Wang and Liou (1991), which suggests that some felsic metamorphic rocks and marble in the SDT have been subjected to ultrahigh-pressure (UHP) metamorphism. S. Xu et al. (1986) and Dong et al. (1989) suggested that the boundary that separates these two terranes is a ductile, tectonic melange belt in a regional detachment plane that formed at great depth.

Proterozoic metasedimentary and metavolcanic rocks of the Hongan, Susong, Zhangbaling and Haizhou Groups were developed on the north margin of the YB. They are in tectonic contact with the Dabie and Jiaonan Groups at the southern edges of the Dabie Mountains and Su-Lu terrane. Petrological studies suggest that all these rocks have experienced two metamorphic events (Jing et al., 1989). Rb-Sr wholerock isochron ages of 730 ± 50 and 848 ± 73 Ma for the Zhangbaling and Susong Groups indicate that the early metamorphic event of greenschist to low amphibolite facies could be related to the late Proterozoic Yangtse orogeny (Zhu et al., 1979; Sang et al., 1987). The second metamorphic event is the high-P-high-T kinetic metamorphism of glaucophane schist facies. Glaucophane schists in this area extends discontinuously for >2000 km from Subei in the east to Shanxi in the west (Dong, 1989) (Fig. 1).

Eclogites occurring in the Dabie Mountains and Su-Lu terrane have been classified into two types according to their field occurrence and mineral assemblages. Type-I eclogites consist of garnet+diopside and occur as lenticular beds or blobs in ultramafic massifs. On the basis of their chemical compositions the protoliths of type-I eclogites are thought to be ferrogabbro (lower SiO_2 and MgO, and higher TiO_2, Al_2O_3, FeO and CaO contents) in ultramafic massifs (Table 1). Type-II eclogites consist predominantly of garnet + omphacite and occur as lenses or blocks within hornblende and biotite gneisses and marbles. Some eclogites are surrounded by amphibolite and are thus not in direct contact with the gneisses. Field relationships and chemical compositions (see Table 1) of these eclogites suggest that their protoliths are basaltic and andesitic rocks. Amphibolites in the Dabie and Jiaonan Groups may belong to this catalog.

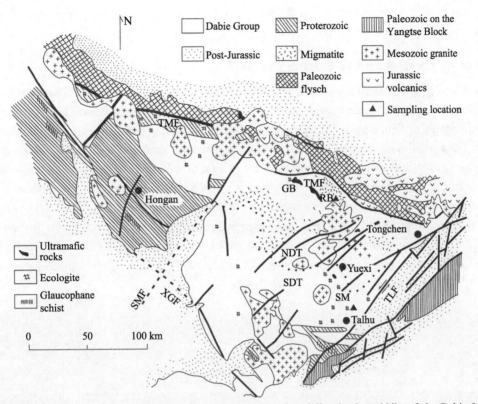

Fig. 2 Simplified geological map of the Dabie Mountains region. The dotted line in the middle of the Dabie Mountains is the boundary between the northern Dabie terrane (NDT) and the southern Dabie terrane (SDT) (after X. Wang et al., 1992). TLF= Tanlu Fault; TMF= Tongbo-Mozitan Fault; SMF= Shangcheng-Macheng Fault; XGF= Xiangfan-Guanji Fault; SM= Shima; RB= Raobazhai; GB= Gaobayan.

Table 1 Major-element composition (in wt%) of eclogites and gneiss from the Dabie Mountains

Rock type	Type-I	Type-II eclogite		Garnet-biotite gneiss
Sample	R-4	ND	T13-1	T16-1
SiO_2	39.24	44.96	59.01	67.22
TiO_2	4.66	1.1	0.88	0.84
Al_2O_3	12.14	17.89	16.1	13.98
Fe_2O_3	5.62	1.71	2.42	1.18
FeO	12.56	11.05	4.49	3.61
MnO	0.15	0.25	0.14	0.2
MgO	8.95	11.16	4.27	1.81
CaO	14.19	9.15	7.23	3.66
Na_2O	0.93	2.28	3.08	3.64
K_2O	0.19	0.09	0.94	2.66
P_2O_5	—	0.25	0.18	0.39
CO_2	0.16	0.32	0.39	0.26
H_2O^+	1.11	0.28	0.54	0.44
Total	99.93	100.49	99.77	99.89

The data were obtained at the Institute of Geophysical and Geochemical Exploratory Techniques, Chinese Academy of Geological Sciences, by X-ray fluorescence analysis except CO_2 and H_2O which were obtained by nonaqueous titration and gravimetric analysis, respectively. —= not determined.

3 Sample description

In this study, we selected 5 eclogite samples from the Dabie Mountains and Su-Lu terrane, 1 retrometamorphic eclogite and 1 garnet-biotite gneiss samples from the Dabie Mountains, and 1 glaucophane schist sample from the Zhangbaling group for age determination using the Sm-Nd and Rb-Sr mineral isochron methods or $^{40}Ar/^{39}Ar$, K-Ar dating method.

3.1 Type-I eclogites (R-4 and G-1) from the NDT

Sample R-4 was collected from the intensely deformed Raobazhai ultramafic massif (harzburgite and dunite), which was emplaced in solid state into migmatites of the Dabie Group (X. Yang, 1983). The country rocks near the contact zone are strongly mylonitizated. Foliations are well developed along the margin of the massif and a mylonite zone cross cuts the massif. The eclogite blobs, with hornblendite, are arranged in a linear fashion parallel to the trend of the mylonite zone and occur 70 – 100 m from the margin of the ultramafic massif. The mineral assemblage is garnet + diopside + amphibole + ilmenite with minor symplectite of intergrowth plagioclase + pyroxene as secondary retrograde assemblage. Chemical compositions of the garnet are similar to those of type-II eclogite with relative lower pyrope component content (Fig. 3), which suggests that the type-I eclogite is metamorphic origin (Coleman et al., 1965). Application of the garnet-clinopyroxene geothermometer (Ellis and Green, 1979) yields temperatures of 770 ℃ for eclogitic metamorphism (Ge et al., 1993). Two well-crystallized garnet and diopside separates as well as whole-rock sample were used for dating.

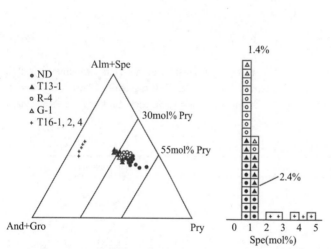

Fig. 3 Chemical composition diagram for garnet of eclogites and garnet-biotite gneiss from the Dabie Mountains. Notice the garnet compositions are distinctly different for eclogite and gneiss (see text for detailed description). The data are from Xiao and Li (1991) and Ge et al. (1993).

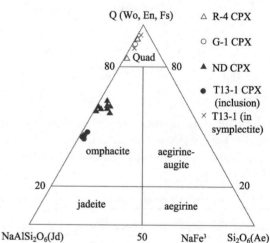

Fig. 4 Chemical composition diagram for clinopyroxenes of eclogite from the Dabie Mountains. Omphacite inclusions in garnet from sample T13-1 have higher jadeite contents than those of omphacite from sample ND. Clinopyroxenes from samples R-4 and G-1 and symplectite of sample T13-1 are Ca-Mg-Fe-pyro-xenes (diopsite or augite). The data are from Xiao and Li (1991) and Ge et al. (1993).

Sample G-1 was collected from Gaobayan village in the northern Dabie Mountains. Many eclogites and small serpentinite blocks have been emplaced into country rocks of hornblende-plagioclase gneiss in this area. Pyroxene composition of sample G-1 is diopsidic (Fig. 4), suggesting that sample G-1 is type-I eclogite, and may be relative to the neighbor serpentinites. Therefore, the eclogites and serpentinites in Gaobayan village may represent a tectonic melange. Temperature of the eclogitic metamorphism was estimated to be 750 ℃ using the garnet-clinopyroxene geothermometer (Ge et al., 1993). This sample shows strong retrogressive metamorphism. Except for garnet, almost all pyroxenes are retrograded to symplectite and amphibole. This assemblage may have been formed at epidote-amphibolite-facies conditions (P= 5–6 kbar, T = 400–530 ℃) (X. Wang et al., 1992; this paper). Since the closure temperature of Sm and Nd for garnet is likely to be >700 ℃ (Vance and O'Nions, 1990), this retrometamorphism may not have affected the Sm-Nd system in garnet. Thus only garnet was used for isotopic analysis. Colours of the garnets in this sample are heterogeneous, which allows us to obtain three different garnet fractions. We cut the specimen (G-1) into two parts named G-1-1 and G-1-2, then selected two garnet separates, the orangered one (gar.d) and the pink light one (gar.1), from sample G-1-2 and one garnet separate from sample G-1-1.

3.2 Type-H eclogites (ND and T13-1) and their country rock (T16-1) from the SDT

Samples ND and T13-1 were collected from Shima village located on the southern edge of the Dabie Mountains. Coesite and coesite pseudomorphs in eclogite have been recognized from this area (X. Wang et al., 1989). Country rocks in this area are dominated mainly by biotite gneisses with minor amounts of amphibolite and marble. In some outcrops, red garnet-biotite gneiss occurs as lenses or blobs in the biotite gneiss with successive transition in contact zone. All rocks in this area are strongly foliated. Abundant eclogites have been preserved in biotite gneiss as blocks or lenses. Some of them have undergone retrogressive amphibolization. Besides eclogites, a small amount of serpentinite, which is partly metamorphosed into talc-schist, occurs as blocks within the matrix of biotite gneiss.

Sample ND consists of garnet, omphacite and rutile. Most of the garnet and omphacite contain tiny mineral inclusions, such as quartz and amphibole. Like sample G-l, we cut a small specimen of sample ND into two parts, ND-1 and ND-2, and selected omphacite and garnet separates from each of them for isotope analyses. The mineral separates are almost 100% pure; no altered products are recognizable. However, the presence of tiny inclusions in mineral separates is possible.

Sample T13-1 is a retrograde eclogite which consists of garnet, omphacite (as inclusion in garnet), phengite, symplectite (augite + amphibole + plagioclase), epidotite, biotite, amphibole and plagioclase. Samples T13-1 and ND were collected from two separate blocks, which are 150 m apart and encased into the same biotite gneiss massif. We use samples T13-1 and ND together to define a retrograde metamorphic P-T-t path of the eclogites (see Section 6). To obtained retrometamorphic age, biotite, symplectite matrix and amphibole-rich garnet-free rock chips were selected for Rb-Sr isotopic dating.

Sample T16-1 is a garnet-biotite gneiss, which represents the country rock of eclogites in the Shima area. Chemical composition of this sample (Table 1) suggests that its protolith is likely to be dacite (Xiao and Li, 1991). The mineral assemblage of garnet, quartz, biotite and minor phengite is paragenetic, which suggests amphibolite-facies conditions (see Section 7). Colours of the garnets in this sample are also heterogeneous, but zonation of the garnet has not been observed in thin section. Three garnet separates, the red one (T16-1a), light-yellow one (T16-1b) and a mixture (T16-1M), were obtained for Sm-Nd isotopic dating. In addition, one biotite separate was obtained for K-Ar dating.

3.3 Type-H eclogites (E87N9-1 and E87M1) from the Su-Lu terrane

Sample E87N9-1 and Sample E87M1 were collected from Yakou and Majiagou villages, Rizhao county, Shandong Province. The geological setting of this area is similar to the SDT. Eclogites occurs in biotite gneisses. Coesite and coesite pseudomorphs have also been recognized in eclogites from this area (J. Yang and Smith, 1989; R. Zhang, 1992). Mineral assemblage of sample E87N9-1 consists of garnet + omphacite + phengite + quartz + rutile. This sample is clearly foliated by the development of phengite, suggesting that phengite was formed later than garnet and omphacite. Sample E87M1 consists of garnet, omphacite and minor rutile. The temperature was estimated to be ~ 750 – 780 °C for eclogites from the Su-Lu terrane by S. Zhang (1992). Following the same procedure for sample ND, four mineral separates (2 garnets + 2 omphacites) were obtained for each sample.

3.4 Glaucophane schist (SJ-1)

Sample SJ-1 was collected from Sanjie in the central Anhui Province (Fig. 1). In this area, glaucophane schists are enclosed in the Zhangbaling group, a spilite-quartz keratophyre formation, as intercalated beds. There is a regional detachment plane between the Zhangbaling group and overlying Sinian system or lower Palaeozoic erathem. According to Jing et al. (1989) and Z. Xu et al. (1991), glaucophane schists were developed only in or under the detachment plane. Sample SJ-1 consists of quartz, plagioclase, magnesian riebeckite and phengite. It is strongly foliated by the development of the magnesian riebeckite and phengite. As magnesian riebeckite occurs as very thin needles, only phengite was separated for $^{40}Ar/^{39}Ar$ isotopic dating.

4 Chronology

4.1 Analytical procedures

All mineral separates were finally obtained by careful hand-picking under binocular microscope. They were washed by 1:1 HCl at 100 ℃ for 1 h before dissolution to remove any surface contamination, and then ultrasonically cleaned in H_2O.

Sample R-4 was analyzed in 1986 at the Massachusetts Institute of Technology in the U.S.A. using techniques described by Zindler et al. (1979). Other eclogite samples (G-l, ND, E87N9-1 and E87M1) were analyzed at the isotope laboratory, Institute of Geology, Chinese Academy of Geological Sciences, Beijing, during 1988 to 1990. Experimental procedures used at Beijing are similar to those at the University de Rennes and described by Jahn et al. (1980). Samples T13-1 and T16-1 were analyzed at the isotope laboratory of the Modern Analysis Center, Nanjing University, in 1991. Blanks for the whole chemical procedures are 50–34 pg for Nd and 30–17 pg for Sm, and 1 ng for Sr. Mass fractionation was corrected against $^{146}Nd/^{144}Nd = 0.72190$. All $^{143}Nd/^{144}Nd$-values are given relative to 0.512633 for standard BCR-1, and $^{87}Sr/^{86}Sr$ values relative to 0.71030 for standard NBS987. $\varepsilon_{Nd}(t)$ and $\varepsilon_{Sr}(t)$ were calculated based on present-day reference values for CHUR (chondritic uniform reservoir): $(^{143}Nd/^{144}Nd)_{CHUR} = 0.51264$, $(^{147}Sm/^{144}Nd)_{CHUR} = 0.1967$, $(^{87}Sr/^{86}Sr)_{CHUR} = 0.7047$ and $(^{87}Rb/^{86}Sr)_{CHUR} = 0.089$. Isochron ages were calculated with York II weighted regression, and the decay constants $\lambda_{Sm} = 6.54 \times 10^{-12}$ a^{-1} and $\lambda_{Rb} = 1.42 \times 10^{-11} a^{-1}$. Ages are given with 2σ error. The Sm-Nd and partial Rb-Sr isotopic data are given in Table 2.

We would emphasize that purity of mineral separates and acid wash before sample dissolution are very critical for reliable radiometric age determination. For this reason, we only discuss our own results and the Sm/Nd ages reported by others will not be discussed in this paper due to lack of information regarding sample purity and analytical procedures.

$^{40}Ar/^{39}Ar$ dating of phengite was conducted at the Institute of Geology, Academia Sinica, following the procedure described by S. Wang et al. (1987). The following factors and constants were used: $(^{40}Ar/^{39}Ar)_K = 7.15 \times 10^{-3}$, $(^{36}Ar/^{37}Ar)_{Ca} = 2.64 \times 10^{-4}$, $(^{39}Ar/^{37}Ar)_{Ca} = 6.87 \times 10^{-4}$, $^{40}K = 5.543 \times 10^{-1} a^{-1}$ and $^{40}K/K = 1.167 \times 10^{-4}$ (atom/atom) (S. Wang et al., 1987). The Ar isotopic data are given in Table 4.

Table 2 Sm, Nd, Rb and Sr concentrations, and Nd and Sr isotope ratios for eclogites and gneiss from the Dabie Mountains and Su-Lu terrain

Sample No.	Rock or mineral	Sm (ppm)	Nd (ppm)	$^{147}Sm/^{144}Nd$	$^{143}Sm/^{144}Nd$	$\varepsilon_{Nd}(t)$ (ppm)	Rb (ppm)	Sr (ppm)	$^{87}Rb/^{86}Sr$	$^{87}Sr/^{86}Sr$	$\varepsilon_{Sr}(t)$
R-4w	eclogite	4.96	14.634	0.2049	0.512515±0.000015	-2.7					
R-4gar.1	garnet (1)	6.051	7.852	0.4659	0.512922±0.000018						
R-4gar.2	garnet (2)	6.061	7.915	0.4629	0.512921±0.000017						
R-4gar.	average	6.056	7.884	0.4644	0.512922±0.000012	-2.8					
R-4cpx.	diopside	1.803	6.618	0.1647	0.512439±0.000016	-3.0	0.1505	262.68	0.00166	0.70631±0.00002	+27.3
G-l-lgar.	garnet	1.599	1.258	0.769	0.513234±0.000027	-4.7					
G-l-2gar.1	garnet (pink)	1.179	0.636	1.120	0.513689±0.000036	-5.9					
G-1-2gar.d	garnet (red)	1.019	0.437	1.411	0.514165±0.000013	-5.0					
ND-lgar.	garnet	2.163	1.997	6552	0.512978±0.000014	-6.3					
ND-lomp.	omphacite	2.796	9.491	0.1782	0.512292±0.000010	-6.2					
Nd-2gar.	garnet	2.184	3.023	0.4370	0.512671±0.000009	-6.1					

Continued

Sample No.	Rock or mineral	Sm (ppm)	Nd (ppm)	^{147}Sm/^{144}Nd	^{143}Nd/^{144}Nd	$\varepsilon_{Nd}(t)$	Rb (ppm)	Sr (ppm)	^{87}Rb/^{86}Sr	^{87}Sr/^{86}Sr	$\varepsilon_{Sr}(t)$
ND-2omp.	omphacite	2.644	8.629	0.1853	0.512443±0.000012	−3.5					
T13-1w	eclogite						23.86	537.5	0.1262	0.706892±0.000014	+29.5
T13-1a	symplectite						7.542	54.85	0.3918	0.707107±0.000016	
T13-1-b	amphibole						48.57	151.2	0.9156	0.708828±0.000020	
T 13-l-m	symp. + amph. + plag						91.87	312.2	0.8514	0.708273±0.000018	
T13-l-bi	biotite						902.1	64.89	41.013	0.785036±0.000019	
T16-l-a	garnet (red)	2.293	1.433	0.9685	0.513451±0.000008	−6.7					
T16-l-b	garnet (yellow)	2.422	1.649	0.8885	0.513308±0.000012	−7.2					
T16-1m	garnet	2.432	1.560	0.9432	0.513381±0.000010	−7.3					
T16-1w	gneiss	7.334	34.84	0.1272	0.512176±0.000009	−7.0					
E87N9-1-1g.	garnet	2.181	1.529	0.8629	0.513182±0.000056	−8.9	0.11			0.70585±0.00015	+20.6
E87N9-1-1o	omphacite	1.253	3.471	0.2148	0.511994±0.000036	−13.2	0.826	77.731	0.0308	0.70597±0.00005	+22.3
E87Ng-1-2g	garnet	3.553	1.756	1.224	0.513453±0.000036	−14.2	0.195	9.898	0.057	0.70690±0.00010	+35.5
E87Ng-1-2o	omphacite	0.9265	2.602	0.2154	0.511996±0.000015	−13.1				0.70600±0.00006	+22.7
E87MI-lg.	garnet	1.377	1.002	0.8313	0.512821±0.000071	−13.4	0.073	11.924	0.0178	0.70641±0.00013	27.8
E87MI-lo.	omphacite	0.8985	2.498	0.2176	0.511972±0.000036	−13.6	0.238	44.208	0.0156		
E87MI-2g.	garnet	1.441	1.003	0.8696	0.512970±0.000031	−11.5	0.073	8.494	0.0248	0.70612±0.00004	+23.4
E87MI-2o.	omphacite	0.6463	2.864	0.1365	0.511973±0.000021	−11.4	0.027	104.73	0.0007	0.70571±0.00010	+18.6
QL-lw*	eclogite	7.26	34.04	0.129	0.511865±0.000011	−13.1	7.915	545.6	0.04134	0.706556±0.000011	+28.5
ZB-4w*	eclogite	7.453	31.48	0.1432	0.511652±0.000013	−17.7	9.465	658.3	0.04297	0.712901±0.000016	+118.5

*The data are from S. Li et al. (1994). Ages of samples QL-lw and ZB-4 are 226.3 ± 4.5 and 228.4 ± 6 Ma, respectivley.

4.2 Sm-Nd isotopic age of eclogites

Fig. 5 shows five Sm-Nd isochron diagrams using only mineral separate data except sampie R-4, which include a WR analysis. The nonlinearity of data points for sample R-4 in a ^{143}Nd/^{144}Nd-1/Nd diagram (inset in Fig. 5A), suggests that the straight line in Fig. 2A is a true isochron, not a mixing line. The isochron gives an age of 244 ± 11 Ma with an initial ^{143}Nd/^{144}Nd ratio of 0.51217 ± 0.00001 (ε_{Nd} = −3.0). This age is interpreted as the time of eclogite metamorphism for the Raobazhai ultramafic massif. Fig. 5B shows that a large spread in ^{147}Sm/^{144}Nd ratios of the three garnet separates from sample G-1 allows us to get a precise Sm-Nd isochron age of 224 ± 20 Ma with an initial ^{143}Nd/^{144}Nd = 0.512107 ± 0.000017 (ε_{Nd} = −4.7).

For sample ND (Fig. 5C), gar-2 data point lies on the tie-line connecting gar-1 + omph-1 of ND-1, whereas omph-2 point is far above the isochron. This situation is similar to the case reported by Thoni and Jagoutz (1992). A possible explanation is that it was caused by disequilibrium of Nd isotope in a hand specimen size (6–8 cm) during the eclogitic metamorphism. Such an isotopic disequilibrium phenomenon is more evident in sample E87M1 (see below). By excluding omph-2 three other data points in Fig. 5C yield an isochron age of 221 ± 5 Ma and initial ^{143}Nd/^{144}Nd = 0.512033 + 0.000015 (ε_{Nd} = −6.2).

Fig. 5D shows that the omphacite separates 1 and 2 from sample E87N9-1 have almost the same ^{147}Sm/^{144}Nd and ^{143}Nd/^{144}Nd ratios, whereas two garnet separates do not form a single straight line with omphacites data points. The line through garnet 2 and omphacites corresponds to an age of 221 ± 6 Ma, similar to that of sample ND-2, whereas the line through garnet 1 and the omphacites corresponds to a much older age of 280 ± 14 Ma. Petrographically, sample E87N9-1 shows only one generation of garnets, thus only one of the two "isochrons" may represent the eclogite formation age. Because the age of 221 Ma is identical with the ages of samples G-1 and ND, we prefer the age of 221 ± 6 Ma with initial ^{143}Nd/^{144}Nd = 0.511685 ± 0.000016 (ε_{Nd} = −13.1) for representing the time of high-pressure metamorphism. Like sample ND-2, the higher ^{143}Nd/^{144}Nd ratio of the garnet may also be caused by disequilibrium in the Sm-Nd isotopic system.

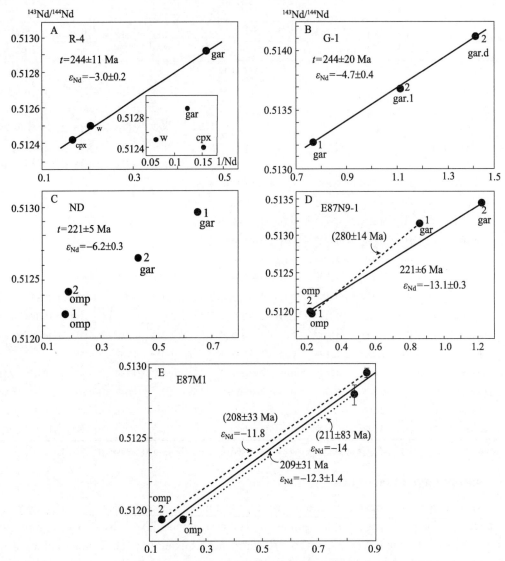

Fig. 5 Sm-Nd isochron diagrams for the eclogites from the Dabie Mountains and Su-Lu terrane. Error input in age calculation: $^{147}Sm/^{144}Nd$, ±0.01% and $^{143}Nd/^{144}Nd$, $2\sigma_m$ (see Table 2).

A striking feature of Fig. 5E is the heterogeneity of the initial Nd isotope ratios observed in sample E87M1. The lines connecting garnet 1-omphacite 1 and garnet 2-omphacite 2 are parallel and correspond to similar ages of 211 ± 83 and 208 ± 33 Ma, respectively. This is a perfect evidence for disequilibrium of Nd isotopes during the eclogitic metamorphism in a hand specimen size (6–8 cm). The best-fit line for all four data points give the same age of 209 ± 31 Ma with initial $^{143}Nd/^{144}Nd$ = 0.51203 ± 0.00007 (ε_{Nd} = −11.9).

On the basis of the above Sm-Nd isotopic ages for eclogite samples which include both type-I and -II eclogites and which were collected from a broad region, we conclude that the eclogite belt in the Dabie Mountains and Su-Lu terrane was formed during the Triassic.

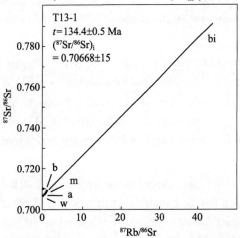

Fig. 6 Rb-Sr isochron diagram for the retrograde products of eclogite sample (T13-1) from the southern Dabie terrane.

4.3 Rb-Sr isotopic age of retrograde products in eclogite (T213-1)

As shown in Table 2 and Fig. 6, the Rb/Sr ratio of the biotite studied is much higher than other minerals analyzed. Consequently, the slope of the Rb-Sr isochron in Fig. 6 is mainly controlled by biotite. It gives an age of 134.4 ± 0.5 Ma with an initial $^{87}Sr/^{86}Sr$ value of 0.70668 ± 0.00015. This age defines the formation time of retrograde biotite and coexisting amphiboles.

4.4 Sm-Nd and K-Ar ages of garnet-biotite gneiss (T16-1)

Fig. 7 shows a Sm-Nd isochron age of 229 ± 3 Ma with an initial $^{143}Nd/^{144}Nd$ ratio of 0.51198 ± 0.00001 (ε_{Nd} = −7.1) for sample T16-1. This age is consistent within error limits with the age of 221 ± 5 Ma for eclogite sample ND hosted by T16-1 type of rock. In contrast with the Rb-Sr isotopic age of sample T13-1, biotite from sample T16-1 gives a K-Ar age of 231 ± 5 Ma (Table 3), which is identical with its Sm - Nd age. Tectonic implications for similar ages between the eclogites and their country rocks will be discussed in Section 8.

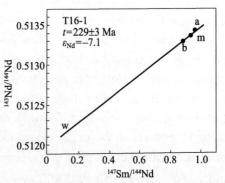

Fig.7 Sm-Nd isochron diagram for the garnet-biotite gneiss from the southern Dabie terrane.

Table 3 K-Ar isotope data of garnet-biotite gneiss (T16-1)*

Sample No.	Mineral	K(%)	$^{40}Ar(10^{-9}\ mol\ g^{-1})$	$^{40}Ar/^{40}K$	Atmospheric ^{40}Ar(%)	Age ± 2σ(Ma)
TI6-1 B	biotite	7.62	3.256	0.01432	3.08	231± 5

*The data were obtained from Institute of Geology, Nuclear Industry Ministry.

4.5 $^{40}Ar/^{39}Ar$ age of glaucophane schist (SJ-1)

Step-heating analysis of sample SJ-1, a phengite from glaucophane schist, yields a perfectly flat age spectrum over more than 94% of the total ^{39}Ar release (Table 4; Fig. 8). This suggests rapid cooling of the phengite through its closure temperature for argon. Its plateau age of 245.1 ± 0.5 Ma is very similar to the Sm-Nd mineral isochron age (244 ± 11 Ma) of sample R-4. On the basis of these data, we would suggest that both these glaucophane schist in the north margin of the YB as well as eclogites in the Dabie Mountains and Su-Lu terrane began their development in the late Permian or early Triassic through collision of the NCB and YB.

Table. 4 Analytical data for $^{40}Ar/^{39}Ar$ step experiment on Sanjie phengite (SJ-1)

Temperature(°C)	$(^{40}Ar/^{39}Ar)_m$	$(^{36}Ar/^{39}Ar)_m$	$(^{37}Ar/^{39}Ar)_m$	$^{39}Ar(10^{-14}\ mol)$	$^{40}Ar*/^{39}Ar(\pm 1\sigma)$	Apparent age ±1σ(Ma)
400	23.16	0.0713	13.26	0.9744	3.248 ± 0.54	68.24 ± 11.1
650	16.36	0.01451	2.205	2.725	12.27 ± 0.26	245.4 ± 5.0
800	13.00	0.0024	0.2314	5.605	12.31 ± 0.17	245.9 ± 3.2
1000	12.53	0.001025	0.001561	26.58	12.22 ± 0.15	244.4 ± 2.9
1200	41.49	0.08326	1.663	0.6365	17.15 ± 1.72	334.3 ± 30.6
1500	57.42	0.1186	1.95	0.3825	22.70 ± 3.29	430.5 ± 55.6

$(^{40}Ar/^{39}Ar)_m$ is the mixing $^{40}Ar/^{39}Ar$ ratio. $^{40}Ar^*$ is radiogenic ^{40}Ar.

4.6 Discussion

The geochronological information presented above would suggest that the ultrahigh and high-pressure metamorphism experienced by the Dabie Mountains and Su-Lu terrane had occurred during the early and middle Triassic (from 245 Ma to 210 Ma). In contrast to our conclusion, some much older ages have been proposed by Cao et al. (1990), Mattauer et al. (1991), and S. Zhang et al. (1991). In the following we will clarify this controversy.

Several K-Ar and $^{40}Ar/^{39}Ar$ ages which are older than 245 Ma (400−1559 Ma) have been reported for phengite separates from eclogite samples from the Dabie Mountains and Su-Lu terrane (Cao et al., 1990; Cong et al., 1991; Mattauer et al., 1991). Possible explanations include: (1) excess argon in phengite from eclogites which are enclosed by K-rich host rocks (felsic gneisses) (S. Li et al., 1994), (2) earlier eclogite formation events in this area (Mattauer et al., 1991), or (3) a combination of (1) and (2). The situation could be further complicated by Ar loss due to later heating events.

In view of the fact that Dabie Mountains region was once a late Proterozoic island arc that may have been accreted to the YB at ~ 800 Ma (Xiao and Li, 1991), consequently, the reported phengite K-Ar or $^{40}Ar/^{39}Ar$ ages of 700-800 Ma may record this collision. However, it is puzzling that phengites from gneiss and blueschist of that area do not give K/Ar and $^{40}Ar/^{39}Ar$ ages greater than 245 Ma (Mattauer et al., 1985, 1991; Sang et al., 1987; Cao et al., 1990; Eide et al. 1992; this paper). Furthermore, in our opinion there is no convincing evidence for any of the coesite-bearing eclogites in Dabie Mountains and Su-Lu terrane having been formed prior to 245 Ma. To clarify this paradoxical issue, we have performed a comprehensive isotopic dating study applying Sm-Nd, Rb-Sr and $^{40}Ar/^{39}Ar$ isotopic methods on the same eclogite samples. This approach has not been attempted by previous workers. It is encouraging to find that there is consistency between the Sm-Nd and Rb-Sr ages (220-228 Ma). In contrast, much older $^{40}Ar/^{39}Ar$ ages (876-943 Ma) were obtained for phengite separates from the same eclogite samples (S. Li et al., 1994). It is concluded that phengite from orogenic eclogites may not be suitable for $^{40}Ar/^{39}Ar$ (or K-Ar) dating because of its potential excess argon problem.

Further confusion was caused by a Rb-Sr whole-rock isochron age of 752 Ma for the Hongan Group. It have been used to support the Precambrian origin of the glaucophane schist belt (S. Zhang et al., 1991). This Rb-Sr isochron, defined by four whole-rock (schist) samples including one glaucophane schist, was interpreted as the high-pressure metamorphic age of the glaucophane schist by S. Zhang et al. (1991) and Hu et al. (1992). In our opinion, this interpretation is too simplistic, because as mentioned above the Hongan Group has experienced two metamorphic events. A kinetic analysis for Sr homogenization during the two metamorphic events is required as the Rb-Sr whole-rock isochron age only indicates the time of metamorphic event in which the scale length of Sr isotope homogenization is larger than the scale length of heterogeneity for Rb/Sr ratios in the Hongan Group (e.g., S. Li et al., 1990). Protoliths of those whole-rock samples analyzed are spilite (sample M18-4) and quartz-keratophyre (samples G12-1, G-IO and M28-2) (S. Zhang et al., 1991). The scale lengths of heterogeneity for Rb/Sr ratios of the whole rocks in the sampling area are > 20 m. When the sedimentary and volcanic rocks in the Hongan Group experienced the first metamorphism with greenschist facies, water was released to form metamorphic fluid in this system. Diffusion of Sr in this system could be fast (Henderson, 1982). Using the diffusion coefficient (D) of Sr in an aqueous fluid ($D \approx 10^{-4}$ cm^2 s^{-1}; Hofmann and Magaritz, 1976) and assuming fluid wetted all the grain boundaries, we obtain Sr characteristic transport distances (estimated using the relation $x = (Dt)^{1/2}$; Hofmann and Hart, 1978) of ~ 560 m in 1 Ma. It is therefore possible that this metamorphic event has reset the Rb-Sr whole-rock isochron age.

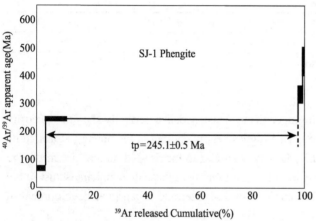

Fig. 8 $^{40}Ar/^{39}Ar$ age spectrum for the phengite of glaucophane schist from the Sanjie area.

After the first metamorphic event, metamorphic fluid could have been expelled out of this system. During the second metamorphic event, the temperature of glaucophane-facies metamorphism is similar to that of earlier greenschist-facies metamorphism (400-500 ℃) but the pressure is higher. If metamorphic rocks in the Hongan Group did not release water again during the glaucophane-facies metamorphism, diffusion of Sr in this system thus could be slow (Henderson, 1982). In view of protoliths of the glaucophane schist, we use diffusion coefficients of basalt and obsidian to estimate the characteristic transport distances of Sr in this system. Based on the Arrhenius plot (log D vs. $1/T$) for diffusion in silicates (Hofmann and Hart, 1978), diffusion coefficients of basalt and obsidian at 500 ℃ were estimated to be ~3×10^{-13} and ~1×10^{-13} cm^2 s^{-1}, respectively. Using these estimated diffusion coefficients, upper limits of the characteristic transport distances of 2-3 cm in 1 Ma were obtained for Sr in this system during the time of glaucophane-facies metamorphism. A distance of 2-3 cm is much less than the scale length of heterogeneity for the Rb/Sr ratio in the Hongan Group.

According to the above rough kinetic analysis, we conclude that Rb-Sr whole-rock isochron age of 752 Ma, as well as two other Rb-Sr whole-rock isochron ages of 730 ± 50 Ma and 844 ± 73 Ma reported by Zhu et al. (1979) and Sang et al. (1987) for the Zhangbaling and Susong Groups, respectively, probably respond to the time of the earlier greenschist metamorphism. These ages can be much older than the ages obtained from single mineral or mineral isochron which respond to the time of late glaucophane schist metamorphism. This conclusion is fully supported by our own and published data: $^{40}Ar/^{39}Ar$ ages of 245 Ma for phengite from the Sanjie glaucophane schist (this study), $^{40}Ar/^{39}Ar$ ages of 217 and 236 Ma for riebeckite and phengite, respectively, from the south Qinling (Mattauer et al., 1985), a K-Ar age of 211 Ma for phengite and a Rb-Sr whole rock-phengite isochron age of 231 ± 48 Ma from the Susong Group (Sang et al., 1987). All these ages support our contention that the glaucophane schist belt in central China was formed during the Triassic.

5 REE patterns and Sr and Nd isotopic systematics

The initial ε_{Nd}-values for all samples studied are all negative (Table 2). Type-II eclogites have initial ε_{Nd}-Values which can be as low as -17. If the protoliths of these eclogites are indeed mafic rocks of mantle origin, two interpretations for the low ε_{Nd}-values of the eclogites are possible:

(1) Protoliths of the eclogites are ancient light REE (LREE) -enriched basic rocks. They could be similar to Proterozoic amphibolites in the Dabie and Jiaonan Groups (for type-II eclogites) and enriched pyroxenite layers in alpine peridotite (for type-I eclogites).

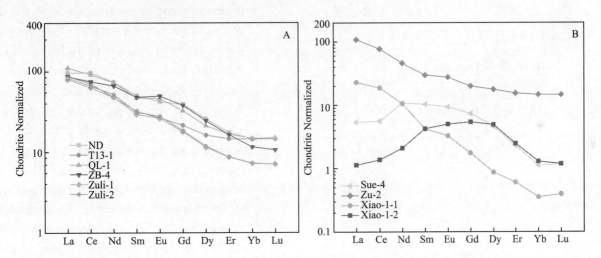

Fig. 9 Chondrite-normalized REE diagrams for eclogites, and garnet-biotite gneiss from the Dabie Mountains and Su-Lu terrane: (A) type-II eclogite; and (B) type-I eclogite.

(2) Protoliths of the eclogites were contaminated by ancient continental material before or during high-pressure metamorphism, e.g., sediment contamination of the subducted oceanic crust or felsic country rock contamination of its enclosing mafic rocks.

In order to further constrain on the origin of these eclogites, REE of eclogites and Sr isotopic compositions of some mineral separates and whole-rock samples were analyzed (see Tables 2 and 5). Because of their low $^{87}Rb/^{86}Sr$ ratios (< 0.06), all measured $^{87}Sr/^{86}Sr$ ratios of these eclogites are very close to their initial $^{87}Sr/^{86}Sr$ ratios at that time of formation ~220 Ma ago (Table 2). Coherent smooth LREE-enriched patterns of type-II eclogites (Fig. 9A) show some "island arc" or intraplate affinity for the protoliths of type-II eclogites and no relation with mid-ocean ridge basalt (MORB). If the Sm/Nd ratio of these eclogites were not significantly fractionated during eclogitic metamorphism, we can use these Sm/Nd ratios to calculate their Nd isotope model ages.

Fig. 10 shows the initial ε_{Nd}-and ε_{sr}-values of the eclogites at their metamorphic time (221 Ma). Fields of Sr and Nd isotopic compositions for various potential eclogite protoliths and MORB-type eclogites from the Munchberg Massif, southern Germany, are also shown for comparison. If the protoliths of type-II eclogite were Palaeozoic subducted oceanic crust or Palaeozoic island-arc basalt (IAB), their data points should plot in, or above the MORB-OIB-IAB-sediments trend (OIB = oceanic island basalt). This is exemplified by MORB-type eclogites from the Munchberg Massif which originated from a Palaeozoic (480 Ma) subducted oceanic crust (Stosch and Lugmair, 1990). The field for the MORB-type eclogite is very similar to those of ophiolites due to a seawater alteration effect on the protoliths. As shown in Fig. 10, all type-II eclogites from the Su-Lu terrane plot below the MORB-OIB-IAB-sediments trend falling in the granulite area. This suggests that their protoliths may have had a long period of residence time in the crust with relative lower Sm/Nd and Rb/Sr ratios. Consequently, we would suggest that their protolith are likely to be ancient LREE-enriched metabasaltic rocks, which is a part of the basement of the YB.

In contrast to type-II eclogites, type-I eclogites have a larger variation of REE contents and different chondrite-normalized patterns (Fig. 9B). For example, samples Zu-2 and Xiao-1-1 have LREE-enriched chondrite-normalized patterns with different REE contents, whereas samples Sue-4 and Xiao-l-2 have unusual convex-upward shaped chondrite-normalized patterns. This convex-upward REE pattern is a common feature in clinopyroxene of mantle xenoliths (Song and Frey, 1989) and has been observed in pyroxenite xenoliths and pyroxenite layers in alpine peridotites (McDonough and Frey, 1993). Nd and Sr isotope compositions of the sample R-4cpx of type-I eclogite from the Raobazhai massif is outstanding. It plots in the IAB field, which overlaps the granulite field (Fig. 10). In view of the fact that type-I eclogites occur within this ultramafic massif and have LREE-enriched or convex-upward chondritic normalized patterns, their protoliths are likely to be generated by cumulation of pyroxene or passage of magmas with LREE-enriched pattern through the ultramafic massif.

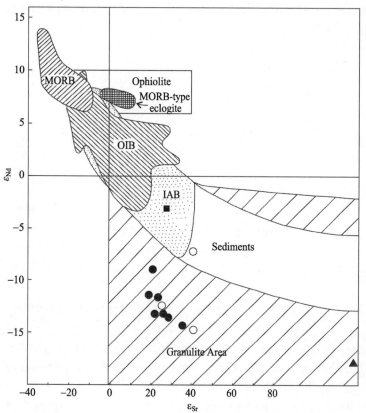

Fig. 10 Nd vs. Sr diagram for eclogites and minerals from eclogites from the Dabie Mountains and Su-Lu terrane at their metamorphic times. The fields of MORB, OIB, IAB and sediments are estimated at 221 Ma. The diagram also shows the Sr and Nd isotopic compositions of MORB-type eclogite from the Munchberg Massif, southern Germany, recalculated to 221 Ma B.P. (Stosch and Lugmair, 1990). Filled circles are minerals from type-II eclogites; filled triangles are whole-rock samples of type-II eclogites; filled square is cpx from sample R-4. Open circles are data of Late Archaean Spilite (greenschist facies) from the Wutai Group, North China, for comparison (S. Li et al., 1990).

It is worth to point out that there is a significant difference between the ε_{Nd}-Values of the eclogite and gneiss from the SDT and the eclogites from the Su-Lu terrane. Table 2 shows that eclogite and gneiss from the Shima area in the SDT have similar ε_{Nd}-values (−7.1 and −6.2), Sm/Nd ratios (0.2238 and 0.2065) (see Table 5) and Nd isotopic model ages (t_{CHUR} = 1017 and 1013 Ma), suggesting that the low ε_{Nd}-value of the eclogite was not caused by contamination of the country rock. We would suggest that protoliths for both samples are likely to be Precambrian volcanics. In contrast to samples from the SDT, eclogite samples from the Su-Lu terrane have much lower ε_{Nd}-values (−17 to −11) and older and more variable Nd isotope model ages (t_{CHUR} = 1736−2792 Ma). In view of the similar Sm/Nd ratios for the eclogites form the SDT and Su-Lu terrane (Tables 2 and 5), three explanations for the lower ε_{Nd}-Values of eclogites from the Su-Lu terrane are possible: (1) protolith ages of eclogite from the Su-Lu terrane are indeed much older than that of eclogite from the SDT, (2) protoliths of eclogite from the Su-Lu terrane have been contaminated by ancient continental material [this option applies well for sample ZB-4w with anomaly high $^{87}Sr/^{86}Sr$ ratio (0.7129) and Nd model age (2792 Ma)], or (3) a combination of options (1) and (2). Obviously, more detailed works are necessary to narrow down the options.

Table 5 REE composition (in ppm) of eclogites from the Dabie Mountains and Subei-Jaonan Rise [1]

Rock type	Type-I eclogite					Type-II eclogite					
Sample[2]	R-4	Sue-4	Zu-2	Xiao-1-1	Xiao-1-2	ND	T13-1	QL-1	ZB-4	Zuli-1	Zuli-2
La		1.28	25.67	5.44	0.27	22.85	20.86	26.3	20.5	18.92	19.71
Ce		3.41	47.15	11.51	0.84	59.36	43.38	56.06	46.02	38.71	40.69
Nd	14.634	5.03	21.42	4.91	0.98	34.58	22.25	(34.03)34.22	(31.48)31.13	22.4	24.28
Sm	4.96	1.57	4.56	0.64	0.65	7.74	4.9	(7.26)7.23	(7.453)7.35	4.52	4.78
Eu		0.54	1.6	0.19	0.29	2.44	1.54	2.68	2.89	1.52	1.61
Gd		1.51	4.11	0.36	1.11	8.01	4.41	6.68	7.79	3.66	3.8
Dy		1.17	4.5	0.22	1.24	6.57	4.11	5.38	6.21	2.92	3.01
Er		0.38	2.55	0.1	0.41	2.89	2.41	2.74	2.71	1.43	1.44
Yb		0.19	2.49	0.06	0.22	2.47	2.45	2.5	1.96	1.23	1.24
Lu		0.03	0.37	0.01	0.03	0.37	0.37	0.38	0.27	0.18	0.18

1) The REE data were obtained by inductively coupled plasma spectrometry from the analytical center of Geological Bureau, Hubei Province, P.R.C. The data in brackets were obtained by isotope dilution method (S. Li et al., 1994; this paper).

2) Sample localities: Besides sample R-4, ND and T13-1, all other eclogite samples are from the Su-Lu terrane. Sample Sue-4 is from the Suoluoshu ultramafic massif, Rizhao, Shandong Province; sample Zu-2 is from the Zhubian ultramafic massif, Junan, Shandong Province; sample Xiao-1-2 is from the Xiaopuzi ultramafic massif, Donghai, Jiangsu Province; sample QL-1 is from the Qinglongshan, Donghai, Jiangsu Province; sample ZB-4 is from the Zhubian, Junan, Shandong Province; samples Zuli-1 and Zuli-2 are from Lijiagou, Junan, Shandong Province.

6 P-T-t path and exhumation of eclogite

A detailed petrological and microprobeanalysis of two samples (ND and T13-1) from the SDT enables us to establish a P-T-t path and a plausible model for exhumation of eclogite. Petrographical study shows that these two eclogite samples have been subjected to at least four metamorphic stages (Fig. 11):

(1) Coesite eclogite-facies metamorphism (EcI). Similar to coesite, garnet and omphacite formed in this UHP stage are preserved only as inclusions in garnet. Omphacites formed during this stage contain higher jadeite contents (54–60 mole%) than those formed in the second stage (Fig. 2). Therefore, EcI may represent the peak metamorphism for the coesite-bearing eclogite. Temperatures for this EcI stage are estimated to be 809–839 ℃ using the garnet-clinopyroxene geothermometry of Ellis and Green (1979). The pressure was estimated to be > 29 kbar at T=800 ℃ based on the equilibrium curve of coesite-quartz established by Bohlen and Boettcher (1982).

(2) Recrystallized eclogite-facies metamorphism (EcII). Garnet and omphacite formed in this stage are represented by large host crystals for coesite, omphacite and garnet inclusions formed during EcI. Jadeite contents in these omphacites range from 37 to 41 mole% and temperatures were estimated to be 646–735 ℃, similar to those of type-III eclogite from the Shima area reported by X. Wang et al. (1992). Pressures are estimated to be 15–20 kbar based on the P-T relationship of eclogite facies metamorphism (Raheim and Green, 1975; Holland, 1980). The EcII stage probably represents an initial retrograde metamorphism of eclogites during their early uplift. All garnet and omphacite separates used in this study for Sm-Nd isotopic dating were formed during this stage. We would suggest that the Sm-Nd mineral isochron ages reported in this paper are likely to represent the initial uplift time of eclogites. The time of stage EcII for sample ND is indicated by its Sm-Nd mineral isochron age of 221 ± 5 Ma.

(3) Retrograde symplectite stage (Sym). Micrograined clinopyroxene (augite) and amphibole (barroisite) are intergrowth with sodium plagioclase in symplectite after omphacite. This paragenesis suggests a high-grade amphibolite-facies (600–700 ℃) condition. Application of amphibole-clinopyroxene geothermometry (Perchuk, 1969) for sample T13-1 yields a temperature of 700 ℃ for symplectite retrograde metamorphism. This temperature is similar to those estimated for the EcII stage, suggesting a very fast uplift rate at this stage. The pressure at this stage was estimated to be 10 kbar according to the jadeite contents in clinopyroxenes. This estimated P-T condition is similar to that (P=9–16 kbar, T= 600 ℃) estimated for symplectite in eclogites from the Su-Lu terrane (S. Zhang et al., 1991).

(4) Retrograde amphibolite-facies stage (Am). Coarse-grained barroisite, hornblende (richer in Al^{IV} and Al_{total}),

epidotite, biotite and plagioclase have partly replaced symplectite and phengite in this stage. This paragenesis indicates their epidote-amphibolite-facies conditions. The temperature was estimated to be ~ 400–425 ℃ based on the amphibole-plagioclase geothermometry (Perchuk, 1966 quoted from R. Zhang and Cong, 1983). The pressure estimated by Al^{IV}/Al^{VI} of amphibole is ~ 5 kbar (Rease, 1974). The time for stage Am can be defined by the Rb-Sr mineral and whole-rock isochrons. For sample T13-1, it was in the early Cretaceous (134.4 ± 0.5 Ma). In our opinion $^{40}Ar/^{39}Ar$ ages of 120–126 Ma for biotites and hornblendes from the UHP gneisses of the Dabie region reported by Eide et al. (1992) are also respondent to this retrograde stage (Am).

A retrograde metamorphic P-T-t path of eclogite from the SDT has been constructed using the information discussed above (Fig. 12). Based on this P-T-t path, the exhumation history of eclogite from the SDT can be approximately divided into two stages: (1) the early stage with a P-T path from EcI to Sym, characterized by rapid pressure decrease accompanied by relatively small temperature decrease; and (2) the second stage with a P-T path from Sym to Am, characterized by a relatively small pressure decrease accompanied by a larger temperature decrease. This P-T path suggests that there had been a rapid uplift (from 110– to 40–35 km depth) in the early stage, which may have been driven by thrust in the Triassic (~ 221 Ma) as suggested by Okay and Sengiör (1992). A steeper geothermal gradient in the second stage may be related to an extensional setting of this region in the early Cretaceous (~134 Ma). Many extensional volcanic basins and alkaline granites developed along the north side of the Dabie Mountains in the early Cretaceous support this conclusion (BGMR, 1987). This extension may be caused by decoupling of the thermal boundary layer in the subducted lithosphere (Andersen et al., 1991). Isostatic rebound of the subducted crust during the extensional time resulted in continued uplift (from 40–35 km to the surface) of the deep crust. Tectonic study of this region also suggests that the Dabie Mountains continued to rise in the Cretaceous and Tertiary (Liu et al., 1990). It leads to the final exhumation of coesite-bearing eclogites. In summary, the above two-stage tectonic model offers a reasonable explanation for the P-T-t path and exhumation history of eclogite, whereas single-stage tectonic models, such as thrusting (Rubie, 1984; Hstü, 1991; Okay and Sengör, 1992) or extension (Andersen et al., 1991), can only explain a part of the P-T-t path of eclogite.

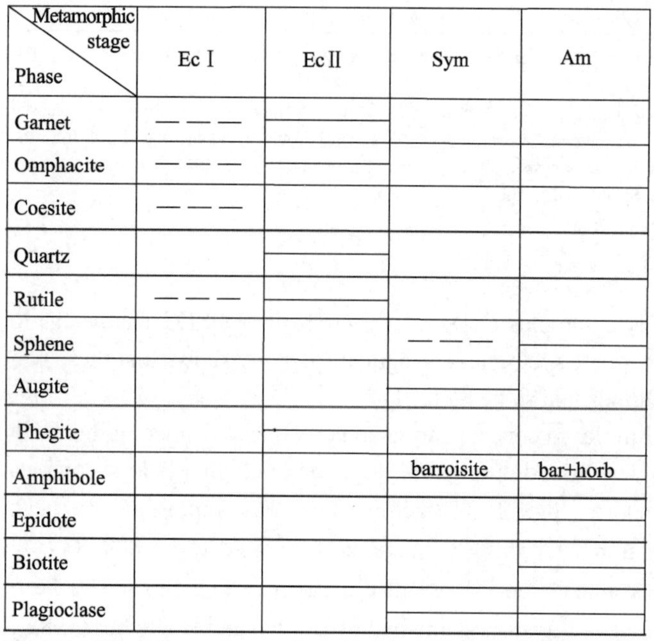

Metamorphic stage Phase	Ec I	Ec II	Sym	Am
Garnet	— — —			
Omphacite	— — —			
Coesite	— — —			
Quartz				
Rutile	— — —			
Sphene			— — —	
Augite				
Phegite				
Amphibole			barroisite	bar+horb
Epidote				
Biotite				
Plagioclase				

Fig. 11 Mineral parageneses in each metamorphic stage of eclogite.

7 Cooling rates of garnet-biotite gneiss and eclogite and implication for "foreign" or "in situ" origin

Our chronology results may give a new constraint on the controversy regarding the relationship between eclogite and their country rocks: "foreign" or "in situ"? The foreign model favours solid-state tectonic introduction of previously metamorphosed high-pressure eclogites into the low-pressure country rocks (e.g., Smith, 1980, 1988; Z. Zhao et al., 1992), whereas the "in situ" model favours cogenetic metamorphism of pre-existing mafic crustal rocks along with their adjacent gneissic country rocks at the same P-T conditions and same sub-crustal levels (e.g., Griffin and Carswell, 1985; X. Wang and Liou, 1991).

As presented above, eclogite sample ND is enclosed by T16-1 type garnet-biotite gneiss and they have very similar Sm-Nd mineral isochron ages (221 ± 5 Ma vs. 229 ± 3 Ma). However, their P-T conditions at the same time are different. Garnets from the gneiss have much lower pyrope (5.8–6.9 mole%) and higher spessartine (2.5–4.6 mole%) contents than those in eclogite (pyr: 28.7–41.7 mole%; spe: 0.6–1.4 mole%) (Fig. 2), which may imply relative low *P-T* condition for the gneiss (Green and Ringwood, 1967; Hirajima et al., 1988). Application of the garnet-biotite geothermometer (Perchuk, 1970, quoted from R. Zhang and Cong, 1983) yields a temperature of 525 ℃ for the garnet-biotite gneiss (Xiao and Li, 1991). The pressure was estimated to be 8-10 kbar at 500 ℃ according to the Si number of phengite from the gneiss (Velde, 1965). These *P-T* conditions are much lower than those of eclogite-facies metamorphism but are similar to the pressure of symplectite and temperature of retrograde amphibolite facies observed in eclogites.

One may argue that these *P-T* conditions may only reflect the amphibolite-facies retrogression of the gneiss, because high-pressure evidence in the gneiss had been completely obliterated by this retrogression. If this is true, then the age of 229 Ma should represent the time of amphibolite-facies retrogression. This age of 229 Ma is much older than the amphibolite-facies retrogression age of eclogite (134 Ma). Moreover, even though the K-Ar closure temperature for biotite is ~300 ℃ (e.g., Vance and O'Nions, 1990), which is much less than the metamorphic temperature (525 °C) of the garnet-biotite gneiss, biotite from the sample T16-1 gives a K-Ar age of 231 ± 5 Ma, which is the same as the Sm-Nd age of 229 Ma within error limit. Similar ages given by the both Sm-Nd and K-Ar systems would imply a very fast cooling rate from 525 to 300 ℃ for the garnet-biotite gneiss. In contrast, cooling from 646-735 ° to 400-425 ℃ for the eclogite sample (ND) took ~90 Ma, suggesting a much lower average cooling rate than that of the garnet-biotite gneiss. Very different cooling rates between the eclogite and garnet-biotite gneiss are consistent with the idea that they have different exhumation histories. It appears that the gneiss may never have been down to the depth where the eclogite had reached, so that temperature of the gneiss was much lower than that of eclogite at the same time. In terms of our two-stage exhumation model, the gneiss had cooled down to 300 ℃ during the first stage in the Triassic, whereas the eclogite had not cooled down to 400 ℃ before the second stage in the early Cretaceous. In this case study, a "foreign" origin for eclogites in the Shima area is favoured. Our results do not support the idea that the entire Dabie terrane experienced coesite-facies UHP metamorphism (X. Wang and Liou, 1991). We have no doubt, however, that some metapelite intercalated with eclogites, and some gneiss and marbles have also experienced UHP metamorphism.

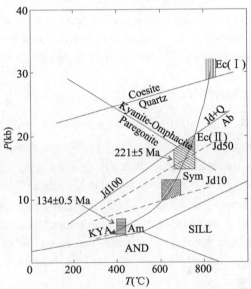

Fig. 12 *P-T* conditions for each metamorphic stage and retrograde *P-T-t* path for eclogite from the Shima in the southern Dabie terrane (see text for detailed description).

8 Collision history of the North China and Yangtse Blocks

We have concluded earlier that protoliths of eclogites and their country rocks could be derived from the basement of the north margin of the YB. On this basis, high-pressure metamorphic ages of 208-245 Ma for eclogites and glaucophane schist from the Dabie Mountains and Su-Lu terrane are likely to reflect the timing of underthrusting of continental crust of the YB. In addition, the established *P-T-t* path for eclogites suggests that the Sm-Nd mineral isochron ages reported in this paper correspond to the initial uplift time of eclogites and the uplift was driven by a thrust tectonics associated with continent collision and A-type subduction. Consequently, these ages must be regarded as upper limits of the collision time between the NCB and YB. If there was no large time interval between the collision and the initial underthrust of the YB, then 245 Ma could be close to the collision time, e.g. the NCB and YB collided in the early Triassic or late Permian. This conclusion is consistent with results of palaeomagnetic studies (McElhinny et al., 1981; Lin et al., 1985; X. Zhao and Coe, 1987; Z. Yang et al., 1991). Other lines of geological evidence also support this conclusion:

(1) The earliest post-collision continental deposits of the molasse facies are Jurassic (Hsu et al., 1987; Xue and Zhang, 1991).

(2) The earliest metamorphism and granite-forming events related to the collision in both northern and southern Qinling Mountains occurred during the Triassic (S. Li et al., 1991). Palaeozoic metamorphism and granites were only developed in northern Qinling. They may have been caused by an unrelated event of arc (northern Qinling) and continent (NCB) collision (S. Li et al., 1992b).

(3) A thick marine sequence of Permian to Triassic strata is present in west Qinling, and the thick Triassic flysch sediments were laid down before they were deformed by continental collision (C. Jiang et al., 1979). X. Zhao and Coe (1987) suggested that palaeomagnetic data from the YB and NCB, support a collision model involving rotation (X. Zhao and Coe, 1987): collision between the NCB and YB occurred initially at a point in eastern China in the late Permian or early Triassic and collision progressed westward as the YB rotated clockwise relative to the NCB. This collision model is apparently consistent with distribution of Triassic marine sediments in the Qinling-Dabie Mountains belt. Our geochronological information can also be integrated with this collision model to support that the NCB and YB were still two independent tectonic units in the Triassic (Ren et al., 1991). By the time of late Jurassic the NCB and YB palaeomagnetic poles are statistically indistinguishable (Lin et al., 1985; Z. Yang et al., 1991). Late Jurassic-early Cretaceous granites, including some alkaline rocks, and extensional basins are well developed in the Qinling-Dabie orogenic belt. Consequently, the NCB and YB were finally welded into a single tectonic unit by the time of the late Jurassic.

Acknowledgements

Shuguang Li is most grateful to L. Gulen and K. Burrhus for the technical assistance at MIT during his 1985/1986 visit and Ye Xiaojiang for her technical assistance at the Institute of Geology, CAGS, during 1987–1990. We thank Zhou Zuo-zhen, Zheng Shuanggen, Zhang Mian-zhen, Wang Shan-lin for the extensive assistance during the field trip in the north Dabie Mountains (1983) and Song Qin, Lu Duanging, Liu Guohua, Zhao Lianji and Li Lixin for the assistance during the field trip in the south Dabie Mountains (1986). Shuguang Li thanks Tu Guang-zhi for his encouragement and recommendation of this project, and K.J. Hsu, Bor-ming Jahn, J.G. Liou and R.L. Rudnick for their constructive comments and detailed reviews of an earlier draft. Shen-Su Sun publishes with the permission of the Executive Director, Australian Geological Survey Organization.

References

Ames, L. and Tilton, G.R., 1992. Timing of collision of the Sino-Korean and Yangtse Cratons, China. Eos (Trans. Am. Geophys. Union), 73(43): 652-653 (abstract).

Andersen, T.B., Jamtveit, B., Dewey, J.F. and Swensson, E., 1991. Subduction and eduction of continental crust: major mechanisms during continent-continent collision and orogenic extensional collapse, a model based on the south Norwegian Caledonides. Terra Nova, 3: 303-310.

BGMR (Bureau of Geology and Mineral Resources of Anhui Province), 1987. Regional Geology of Anhui Province. Geological Publishing House, Beijing (in Chinese, with English text).

Bohlen, S.R. and Boettcher, A.L., 1982. The quartz-coesite transformation: a pressure determination and the effects of other components. J. Geophys. Res., 87: 7073-7078.

Cao, G., Wang, Z. and Zhang C., 1990. Jiaonan Rise of Shandong Province and its boundary fault: WulianRongcheng fault— Implications for tectonics. Geol. Shandong Prov., 6:1-14 (in Chinese).

Coleman, R.G., Lee, D.E., Beatty, L.B. and Brannock, W.W., 1965. Eclogites and eclogites: theirs differences and similarities. Geol. Soc. Am. Bull., 76: 483-508.

Cong, B., Zhang, R., Li, S., Wang, S. and Chen, C.Y., 1991. Preliminary research on isotopic chronology in high pressure metamorphic zone of northern Jiangsu, eastern Shandong province, China. Acad. Sin., Inst. Geol., Lab. Lithosphere Tectonic Evol., Annu. Rep. 1989–1990. Scientific and Technological Press of China, Beijing, pp. 68-72 (in Chinese).

Cong, B., Wang, Q., Zhang, R., Zhai, M., Zhao, Z. and Li, J., 1993. Discovery of the coesite-bearing granulite in the Weihai, Shandong Province, China. Chin. Sci. Bull. (in press).

Dong, S., 1989. The general features and distributions of the glaucophane schist belts of China. Acta Geol. Sin., 35:273-284 (in Chinese).

Dong, S., Shen, Q., Sun, D. and Lu, L., 1986. Metamorphic map of China, scale 1 : 4,000,000 with an explanatory text. Geological

Publishing House, Beijing, 162 pp. (in Chinese).

Dong, S, Zhou, H., Zhang, W. and Cheng, G., 1989. A preliminary study of the movement of the Dabie Massif. Bull. Inst. Geomech., Chin. Acad., Geol. Sci., 12: 99-112.

Eide, E.A., Liou, J.G. and McWilliams, M.O., 1992. ^{40}Ar/^{39}Ar age constraints on high and ultrahigh pressure metamorphism, Hubei Province, China. Eos (Trans. Am. Geophys. Union), 73 (43): 653 (abstract).

Ellis, D.J. and Green, D.H., 1979. An experimental study of the effect of Ca upon garnet-clinopyroxene Fe-Mg exchange equilibria. Contrib. Mineral Petrol., 71 : 13-22.

Ge, N., Li, S., Peng, Z. and Liu, D., 1993. Character and origin of eclogites from the eastern section of the Dabie Mountains. Acta Geol. Sin., 67: 109-122.

Green, D.H. and Ringwood, A.E., 1967. An experimental investigation of the gabbro eclogite transformation and its petrological applications. Geochim. Cosmochim. Acta, 31: 767-833.

Griffin, W.L. and Carswell, D.A., 1985. In situ metamorphism of Norwegian eclogites: an example. In: D.G. Gee and B.A. Stuart (Editors), The Caledonide Orogem Wiley, New York, N.Y., pp. 813-822.

Hao, J. and Liou, X., 1988. A large nap-décollement structure and its evolution in the Tongbo-Dabie collision-orogeny. Sci. Geol. Sin., No. 1, pp. 1-9 (in Chinese).

Henderson, P., 1982. Inorganic Geochemistry. Pergamon, Oxford, 353 pp. (see especially pp. 185- 196).

Hirajima, T., Banno, S., Hiroi, Y. and Ohta, Y., 1988. Phase petrology of eclogites and related rocks from the Motalafjella high-pressure metamorphic complex in Spitsbergen (Arctic Ocean) and its significance. Lithos, 22: 75-97.

Hirajima, T., Ishiwatari, A., Cong, B., Zhang, R., Banno, S. and Nozaka, T., 1991. Coesite from Mengzhong eclogite at Donghai county, northeastern Jiangsu Province, China. Mineral. Mag., 54: 579-583.

Hofmann, A.W. and Hart, S.R., 1978. An assessment of local and regional isotopic equilibrium in the mantle. Earth Planet. Sci. Lett., 38: 44-62.

Hofmann, A.W. and Magaritz, M., 1976. Diffusion in silicate melts and glasses. Carnegie Inst. Washington Yearbk., 75:249-251.

Holland, J.J.B., 1980. The reaction albite =jadeite + quartz determined experimentally in the range 600-1200°C. Am. Mineral., 65: 129-134.

Hstü, K.J., 1991. Exhumation of high-pressure metamorphic rocks. Geology, 19:107-110.

Hstü, K.J., Wang, Q., Li, J., Zhou, D. and Sun, S., 1987. Tectonic evolution of Qinling Mountains, China. Eclogae Geol. Helv., 80: 735-752.

Hu, K., Zhang, S., Liu, X., Kang, W., Liu, Y. and Liang, W., 1992. Continent-continent collision zone: implications of high or ultra-high pressure metamorphic belt in Central China. Select. Pap. presented at Conf. on Science and Technological Achievements in Geology during the 7th Five-year Plan. Publishing House of Science and Technology, Beijing, pp. 264-267 (in Chinese).

Jahn, B.M., Bernard-Griffiths, J., Chariot, R., Cornichet, J. and Vidal, P., 1980. Nd and Sr isotopic compositions and REE abundances of Cretaceous MORB (Holes 417D and 418A, Legs 51, 52 and 53). Earth Planet Sci. Lett., 48: 171-184.

Jiang, C., Zhu, Z. and Kong, F., 1979. On the Liufengguan flysch. Acta Geol. Sin., 53:203-218 (in Chinese).

Jing, Y., Xia, M., Zhang, L., Liang, W. and Wang, H., 1989. Characteristics of Zhangbaling high-pressure metamorphic belt in central Anhui. J. Changchun Geol. College, Spec. Iss. on Glaucophane Schist Belt, pp. 133- 151 (in Chinese).

Klimetz, M.P., 1983. Speculations on the Mesozoic plate tectonic evolution of eastern China. Tectonics, 2:139- 166.

Li, C. Y., 1980. A preliminary study of plate tectonics of China. Chin. Acad. Geol. Sci. Bull., 2:11-22 (in Chinese).

Li, S., Hart, S.R., Zheng, S., Liou, D., Zhang, G. and Guo, A., 1989a. Timing of collision between the North and South China Blocks— Sm-Nd isotopic age evidence. Sci. China (Ser. B), 32: 1391-1400.

Li, S., Ge, N., Liou, D., Zhang, Z., Ye, X., Zheng, S. and Peng, C., 1989b. The Sm-Nd isotopic age of C-type eclogite from the Dabie Group in the Northern Dabie Mountains and its tectonic implication. Chin. Sci. Bull., 34: 1623-1628.

Li, S., Hart, S.R. and Wu, T., 1990. Rb-Sr and Sm-Nd isotopic dating of an early Precambrian spilite-keratophyre sequence in the Wutaishan area, North China: preliminary evidence for Nd-isotopic homogenization in the mafic and felsic lavas during low-grade metamorphism. Precambrian Res., 47:191-203.

Li, S., Liou, D., Chen, Y., Zhang, G. and Zhang, Z., 1991. A chronological table of the major tectonic events for Qinling-Dabie orogenic belt and its implications. Select. Pap., Conf. on Qinling Orogenic Belt. Publishing House of Northwest University, Xian, pp. 229-237 (in Chinese).

Li, S., Liu, D., Chen, Y. and Ge, N., 1992a. The Sm-Nd isotopic age of coesite-bearing eclogite from the southern Dabie Mountains. Chin. Sci. Bull., 37: 1638-1641.

Li, S., Xiao, Y., Chen, Y., Liou, D. and Ge, N., 1992b. Four stages tectonic model for evolution of the Qinling orogenic belt, China. Select. Pap. presented at Conf. on Science and Technological Achievements in Geology during 7th Five-year Plan. Publishing House of Science and Technology, Beijing, pp. 121-125 (in Chinese).

Li, S., Wang, S., Chen, Y., Zhou, H., Zhang, Z., Liu, D. and Qiu, J., 1994. Excess argon in phengite of eclogite: evidence from comparing dating of eclogite by SmNd, Rb-Sr and $^{40}Ar/^{39}Ar$ isotope methods. Chem. Geol. (Isot. Geosci. Sect.), 112 (in press).

Lin, J.L., Fuller, M. and Zhang, W.Y., 1985. Preliminary Phanerozoic polar wander paths for the North and South China Blocks. Nature (London), 313: 444-449.

Liou, J.G., Maruyama, S., Wang, X. and Graham, S., 1990. Precambrian blueschist terranes of the world. Tectonophysics, 18:97-111.

Liu, D., Ki, S., Zhu, J. and Yi, J., 1990. A discovery of ultramylonite vein on the Dabie Mountains and its geological implications. Sci. Geol. Sin., No. 2, pp. 183-186 (in Chinese).

Mattauer, M., Matte, P., Malavieille, J., Tapponnier, P., Maluski, H., Xu, Z.Q., Lu, Y.L. and Tang, Y.Q., 1985. Tectonics of the Qinling Belt: build-up and evolution of eastern Asia. Nature (London), 317: 496-500.

Mattauer, M., Matte, P., Maluski, H., Xu, Z., Zhang, Q. and Wang, Y., 1991. Paleozoic and Triassic plate boundary between North and South China—New structural and radiometric data on the Dabie-Shan (Eastern China). C.R. Acad. Sci. Paris, S6r. 2, 312: 1227-1233.

McElhinny, M.W., Embleton, B.J.J., Ma, X.H. and Zhang, Z.K., 1981. Fragmentation of Asia in the Permian. Nature (London), 293: 212-216.

Okay, A. and Sengör, A.M.C., 1992. Evidence for intracontinental thrust-related exhumation of ultrahigh-pressure rocks in China. Geology, 20:411-414.

Okay, A., Xu, S. and Sengör, A.M.C., 1989. Coesite from the Dabie Shan eclogites, central China. Eur. J. Mineral., 1: 595-598.

Perchuk, L.L., 1969. The effect of temperature and pressure on the equilibrium of natural iron-magnesium minerals. Int. Geol. Rev., 11:875-901.

Raheim, A. and Green, D.H., 1975. *P-T* paths of natural eclogites during metamorphism—a record of subduction. Lithos, 8:317-328.

Rease, P.. 1974. Al and Ti contents of hornblende, indicators of pressure and temperature of regional metamorphism. Contrib. Mineral. Petrol., 45:231-236.

Ren, J., Zhang, Z., Niu, B. and Liu, Z., 1991. On the Qinling orogenic belt—Integration of the Sino-Korean and Yangtse Blocks. Select Pap., Conf. on Qinling Orogenic Belt. Publishing House of Northwest University, Xian, pp. 99-110 (in Chinese).

Rubie, D.C., 1984. A thermal-tectonic model for highpressure metamorphism and deformation in the Sesia Zone, Western Alps. J. Geol., 92: 21-36.

Sang, B., Chen, Y. and Shao, G., 1987. The Rb-Sr ages of metamorphic series of the Susong Group at southeastern foot of the Dabie Mountains, Anhui Province, and their tectonic significance. Reg. Geol. China, No. 4, pp. 364-370 (in Chinese).

Sengör, A.M.C., 1985. East Asian tectonic collage. Nature (London), 318: 16-17.

Smith, D.C., 1980. A tectonic melange of foreign eclogites and ultramafites in the basal gneiss region, Western Norway. Nature (London), 287: 366-368.

Smith, D.C., 1988. A review of the peculiar mineralogy of the Norwegian coesite-eclogite province, with crystalchemical, petrological and geodynamical notes and extensive bibliography. In: D.C. Smith (Editor), Eclogite and Eclogite Facies Rocks. Developments in Petrology, Vol. 12. Elsevier, Amsterdam, pp. 1-206.

Song, Y. and Frey, F.A., 1989. Geochemistry of peridotite xenoliths in basalt from Hannuoba, Eastern China: implications for subcontinental mantle heterogeneity. Geochim. Cosmochim. Acta, 53:97-113.

Stosch, H.Y. and Lugmair, G.W., 1990. Geochemistry and evolution of MORB-type eclogites from the Munchberg Massif, southern Germany. Earth Planet Sci. Lett., 99: 230-249.

Thoni, M. and Jagoutz, E., 1992. Some new aspects of dating eclogites in orogenic belts: Sm-Nd, Rb-Sr, and Pb-Pb isotopic results from the Austroalpine Saualpe and Koralpe type-locality (Carinthia/Styria, southeastern Austria). Geochim. Cosmochim. Acta, 56: 347-368.

Vance, D. and O'Nions, R.K., 1990. Isotopic chronometry of zoned garnets: growth kinetics and metamorphic histories. Earth Planet Sci. Lett., 97: 227-240.

Velde, B., 1965. Experimental determination of muscovite polymorph stabilities. Am. Mineral., 50: 436-449.

Wang, H., Xu, C. and Zhou, Z., 1982. Tectonic evolutions of the continental margins on the both sides of Eastern Qinling paleo-ocean. Acta Geol. Sin. No. 3, pp. 270-279 (in Chinese).

Wang, S., Hu, S., Zhai, M., Sang, H. and Qiu, J., 1987. An application of the $^{40}Ar/^{39}Ar$ dating technique to the formation time of Qingyuan Granite-Greenstone terrain in NE China. Acta Petrol. Sin., No. 4, pp. 55-62 (in Chinese).

Wang, X. and Liou, J.G., 1991. Regional ultrahigh-pressure coesite-bearing eclogitic terrane in central China: evidence from country rock, gneiss, marble and metapelite. Geology, 19: 933-936.

Wang, X., Liou, J.G. and Mao, H.K., 1989. Coesite-bearing eclogite from the Dabie Mountains in central China. Geology, 17: 1085-1088.

Wang, X., Liou, J.G. and Maruyama, S., 1992. Coesitebearing eclogites from the Dabie Mountains, Central China: petrogenesis, *P-T* paths and implications for regional tectonics. J. Geol., 100: 231-250.

Xiao, Y. and Li, S., 1991. *P-T-t* path of eclogite from the Dabie Mountains. Master's Thesis, University of Science and Technology of China, Hefei (unpublished; in Chinese).

Xu, J., Liou, D. and Li, X., 1987. Integration between the Southern and Northern Blocks, Eastern China, in the Mesozoic. Proc. Geol. Syrup. on Mesozoic and Cenozoic. Geological Publishing House, Beijing, pp. 99-112 (in Chinese).

Xu, S., Zhou, H., Dong, S., Chen, G. and Zhang, W., 1986. Deformation and evolution of the predominant structural elements in Anhui Province, China. Sci. Geol. Sin., 2: 311-322 (in Chinese).

Xu, S., Okay, A.I., Ji, S., Sengör, A.M.C., Kiu, Y. and Jiang, L., 1992. Diamond from the Dabie Shan metamorphic rocks and its implication for tectonic setting. Science, 256: 80-92.

Xu, Z., Niu, B. and Liu, Z., 1991. Tectonic system and intracontinental plate dynamic mechanism in the Qinling-Dabie "collision-intracontinental" mountain chains. Select. Pap. Conf. on Qinling Orogenic Belt. Publishing House of Northwest University, Xian, pp. 139-147 (in Chinese).

Xue, X. and Zhang, Y., 1991. The geological age and characters of the red basins in the eastern part of the Qinling Mountains. Select. Pap., Conf. of Qinling Orogenic Belt. Publishing House of Northwest University, Xian, pp. 89-98 (in Chinese).

Yang, J. and Smith, D.C., 1989. Evidence for a former sanidine-coesite-eclogite at Lanshantou, eastern China, and the recognition of the Chinese "Su-Lu coesite-eclogite province". Int. Eclogite Conf., 3rd, Oxford. Blackwell, London, p.26 (abstract).

Yang, X., 1983. A ultramafic solid-state intrusion in the Dabie Mountains, Anhui Province. Bull. Nanjing Inst. Geol. Mineral. Res., Chin. Acad. Geol. Sci., 4: 81-95 (in Chinese).

Yang, Z., Ma, X., Besse, J., Courtillot, V., Xing, L., Xu, S. and Zhang, J., 1991. Paleomagnetic results from Triassic sections in the Ordos Basin, North China. Earth Planet. Sci. Lett., 104: 258-277.

Zhang, R., 1992. Petrogenesis of high pressure metamorphic rocks in the Su-Lu and Dianxi regions, China. Ph.D. Thesis, Kyoto University, Kyoto (unpublished).

Zhang, R. and Cong, B., 1983. Mineral Geothermometry and Mineral Geobarometry. Geological Press, Beijing, 280 pp. (in Chinese).

Zhang, R. and Cong, B., 1991. A ultrahigh-pressure metamorphism and retrograde reaction of coesite-bearing quartz eclogite from Weihai, eastern China. Eos (Trans. Am. Geophys. Union), 72: 559 (abstract).

Zhang, R., Cong, B., Hirajima, T. and Banno, S., 1990. Coesite eclogite in Su-Lu region, Eastern China. Eos (Trans. Am. Geophys. Union), 71: 1708 (abstract).

Zhang, S., Hu, K., Qiao, L. and Liu, X., 1991. The origin and rock assemblage of ancient intracontinental collapse-collision zone: a description of the high *P* metamorphic belt in central China. Select. Pap., Conf. on Qinling Orogenic Belt. Publishing House of Northwest Univ., Xian, pp. 48-56 (in Chinese).

Zhang, Z. M., Liou, J.G. and Coleman, R.G., 1984. An outline of the plate tectonics of China. Geol. Soc. Am. Bull., 95: 285-312.

Zhao, X. and Coe, R.S., 1987. Paleomagnetic constraints on the collision and rotation of North and South China. Nature (London), 327: 141-144.

Zhao, Z., Wang, Q. and Cong, B., 1992. Coesite-bearing ultrahigh pressure metamorphic rocks from Donghai, northern Jiansu Province, eastern china: "foreign" or "in situ" ? Sci. Geol. Sin., 1: 43-58.

Zhu, H., Zhang, Z., Cheng, J. and Cheng, D., 1979. Isotopic ages of metamorphic rocks from the Zhangbaling Group, Anhui Province, J. Univ. Sci. Technol. China, 9: 159-165 (in Chinese).

Zindler, A., Hart, S.R., Frey, F.A. and Jacobsen, S.P., 1979. Nd and Sr isotope rations and rare earth elements abundances in Reykjanes Peninsula basalts: evidence for mantle heterogeneity beneath Iceland. Earth Planet Sci. Lett., 45: 249-262.

北秦岭拉圾庙苏长辉长岩的痕量元素和 Sr，Nd 同位素地球化学*

李曙光[1]，陈移之[1]，张宗清[2]，叶笑江[2]，张国伟[3]，郭安林[3]，S. R. Hart[4]

1. 中国科学技术大学地球与空间科学系，合肥 230026
2. 中国地质科学院地质所，北京 100037
3. 西北大学地质系，西安 710069
4. 美国 Woods Hole 海洋研究所，马萨诸塞州

> **亮点介绍**：痕量元素和 Sr, Nd 同位素地球化学研究表明，北秦岭拉圾庙苏长辉长岩是岛弧环境下形成的蛇绿岩套的一部分。其地幔源区含有较多的陆源沉积物组分，表明北秦岭在约 400 Ma 前是华北陆块南缘的近陆岛弧。拉圾庙岩体的辉长岩和苏长辉长岩分别属于拉斑系列和钙碱系列，两者的地幔源区在痕量元素和挥发分含量、Fe^{3+}/Fe^{2+} 值以及 ε_{Nd} 值等方面存在系统性差异。

摘要 北秦岭拉圾庙苏长辉长岩体由辉长岩和苏长辉长岩组成。它们分属两个不同的岩浆系列：辉长岩属拉斑系列，TiO_2、Fe^{3+}/Fe^{2+} 和 REE 丰度较高；苏长辉长岩属钙碱系列，低 TiO_2，Fe^{3+}/Fe^{2+} 和 REE，特别是 HREE。这说明后者源区较深并受榴辉岩相控制。两种岩石均有与岛弧玄武岩类似的地球化学特征，如较高的 Ba、Pb、Sr 和较低的 Nb、Zr、Ni 含量等。拉圾庙辉长岩和苏长辉长岩的 Sr、Nd 同位素特征与蛇绿岩类似，其 ε_{Nd} 值恒定在+2 左右，ε_{Sr} 变化范围较大(-6.4~+31.2)，并与 Na_2O、H_2O^+、CO_2 含量和 Fe^{3+}/Fe^{2+} 有正相关关系。在 ε_{Nd}-Nb/Th、ε_{Nd}-La/Nb 和 ε_{Nd}-Ba/Nb 图中，清楚显示了该岩体的地幔源含较多的陆源沉积物组分。它说明古秦岭海板块在古生代俯冲时，携带有较多沉积物进入地幔。

关键词 痕量元素；Sr-Nd 同位素；苏长辉长岩；拉圾庙；北秦岭

在北秦岭南侧，商丹断裂以北，分布有许多大的苏长辉长岩体。它们顺断裂呈东西向排列，单个岩体出露面积达 10~40 km^2。许多作者认为它们是北秦岭丹凤蛇绿岩的组成部分[1-3]。在这些岩体中，除商县以南的拉圾庙苏长辉长岩体有 Sm-Nd 矿物等时线定年资料外[4]，尚未见有地球化学研究工作。这些基性岩体是否都是蛇绿岩？它们生成的构造环境如何？这些都是有待解决的重要课题。

岩石的痕量元素地球化学和 Sr、Nd 同位素地球化学可以为解决上述问题提供重要的制约。本文以拉圾庙苏长辉长岩为例，在其 Sm-Nd 同位素定年基础上进行深入的痕量元素和 Sr，Nd 同位素地球化学研究，以判断它是否具有蛇绿岩的地球化学特征，形成的构造环境和岩体中辉长岩和苏长辉长岩的成因及相互关系。

1 地质背景及样品

拉圾庙苏长辉长岩体位于商县以南约 20 km，东西分别与三十里铺辉长岩体和秦王山辉长岩体相邻，呈东西延长的豆荚状（图 1）。岩体出露面积约 19 km^2。该岩体主要为辉长岩，西部有部分苏长辉长岩，并有晚期岩脉穿插。李曙光等（1989）从一苏长辉长岩样品中获得 Sm-Nd 矿物等时线年龄为 402.6 ± 17.4 Ma[4]，

---

* 本文发表在：岩石学报，1993, 9(2): 146-157
对应的英文版论文为：Li Shuguang, Chen Yizhi, Zhang Zongqing, et al., 1994. Trace elements and Sr, Nd isotopic geochemistry of Lajimiao norite-gabbro from the North Qinling belt. Acta Geologica Sinica, 7: 137-152

它代表该岩体的结晶时代。相邻的秦王山和三十里铺岩体的角闪石 K-Ar 年龄为 340～370 Ma[1)]，它们代表了这些岩体的变质时代。岩体东侧为斜长角闪岩（变玄武岩）。张宗清等用 Rb-Sr 全岩法获得其等时年龄为 398 Ma（北秦岭变质地层同位素年代研究，附参考文献）。它的角闪石 K-Ar 年龄为 353 和 360 Ma[1)]。因此，该变玄武岩与苏长辉长岩体同时生成，同时变质，可能属于同一蛇绿岩套。340～370 Ma 的变质时代可能代表该蛇绿岩的侵位时代。

研究中共分析了 3 个辉长岩（QL-1，2，3），2 个苏长辉长岩（QL-5，6），1 个闪长玢岩（QL-9）和 1 个二长斑岩（QL-8）样品。辉长岩呈灰黑色，中粒辉长结构，主要由基性斜长石、单斜辉石、角闪石及少量磁铁矿组成。苏长辉长岩呈暗紫色，矿物粒度较辉长岩粗，主要由斜长石、紫苏辉石、单斜辉石组成，磁铁矿含量较少。两种岩石均为块状构造，无层状韵律结构，应相当岩浆房上部岩石。闪长玢岩和二长斑岩为岩体内部晚期分异岩脉，且前者早于后者。

2 样品加工及测试方法

获得可靠而准确的痕量元素及同位素数据，关键在于在样品加工过程中要防止可能的污染及选择精确的测试方法。

样品无污染粉碎在美国麻省理工学院完成。首先将样品块用棉布包裹起来并在塑料板隔离下用铁锤击碎以避免铁的污染。从中选取约 200 g 完全新鲜岩石碎块用蒸馏水洗净，烘干后，放入洗净的，有碳化钨衬里的振动盒内，粉碎至约 160 目。该组样品的稀土元素（REE）和 Rb、Sr 含量用同位素稀释法测定。其他痕量元素和主要元素用荧光光谱法测定。H_2O^+ 及 CO_2 分别用重量法及非水滴定法测定。Sr、Nd 同位素数据分别在美国麻省理工学院和中国地质科学院地质所的同位素实验室测定。实验方法已有前文描叙[4,5]。测定结果、测定单位及测定条件分别列于表 1、表 2、表 3。

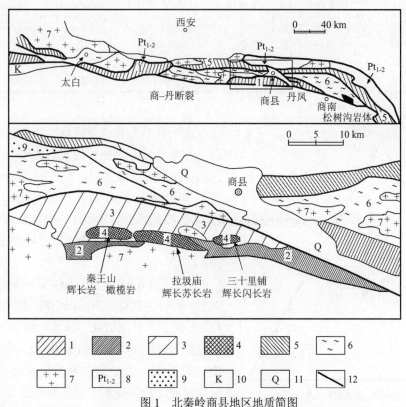

图 1 北秦岭商县地区地质简图

1.丹凤蛇绿岩；2.斜长角闪岩；3.片麻岩；4.苏长辉长岩；5.云架山和斜峪关蛇绿岩；6.秦岭群；7.花岗岩；8.宽坪群；9.上古生界；10.白垩系砾岩；11.第四系；12.断层

---

1) 西安地质矿产研究所. 陕西省秦岭-巴山地区基性-超基性岩及有关矿产研究总结报告. 1981

3 岩浆演化系列及分离结晶作用

表1表明，辉长岩与苏长辉长岩的K_2O均很低（<1%），它们都不属于碱性玄武岩浆系列。然而它们的Na_2O含量较高（>3.4%）。高Na_2O低K_2O，正是形成于海底环境的蛇绿岩特征。AMF图（图2）表明，该岩体的辉长岩和苏长辉长岩分属两个不同的岩浆系列。辉长岩属拉斑系列，而苏长辉长岩及晚期脉岩属钙碱系列。在图2中，辉长岩未落入典型拉斑系列趋势，是海水蚀变使岩石Na_2O含量增高所致[6]。钙碱系列岩石的出现，说明该苏长辉长岩体生成于岛弧环境。

表1 北秦岭拉圾庙苏长辉长岩主要元素含量（%）

样品号	QL-1	QL-2	QL-3	QL-5	QL-6	QL-8	QL-9
岩石名称	辉长岩	辉长岩	辉长岩	苏长辉长岩	苏长辉长岩	二长斑岩	闪长玢岩
SiO_2	47.14	47.32	46.46	52.81	54.49	64.91	60.37
TiO_2	1.83	1.85	1.79	0.58	0.70	0.56	0.78
Al_2O_3	16.56	16.23	17.20	18.64	18.67	14.95	15.78
Fe_2O_3	6.71	6.67	6.81	1.71	2.62	1.21	1.47
FeO	6.82	6.95	6.56	4.41	4.02	2.34	3.24
MgO	4.83	4.90	4.76	6.25	4.26	2.41	3.75
CaO	9.04	9.34	8.74	10.21	8.14	3.67	4.43
Na_2O	3.75	3.56	4.06	3.46	4.20	3.52	3.96
K_2O	0.68	0.59	0.66	0.82	0.75	3.61	3.06
P_2O_5	0.43	0.49	0.36	0.05	0.14	0.24	0.38
MnO	0.20	0.18	0.21	0.10	0.10	0.05	0.08
CO_2	0.54	0.48	0.62	0.37	0.44	1.00	0.82
H_2O^+	0.90	0.90	1.09	0.62	1.05	1.12	1.17
总计	99.43	99.46	99.32	100.03	99.58	99.59	99.29
Fe^{3+}/Fe^{2+}	0.89	0.86	0.93	0.35	0.59	0.47	0.41
CaO/Na_2O	2.41	2.62	2.17	2.95	1.94	1.04	1.12

测定单位：中国地质勘察技术研究院物化探技术研究所。

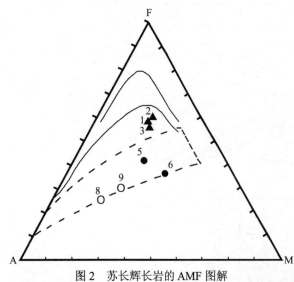

图2 苏长辉长岩的AMF图解

实线为拉斑系列演化趋势；虚线为钙碱系列演化趋势。黑三角为辉长岩样品；黑圆点为苏长辉长岩样品；空心圆为脉岩样品；图中数字为样品QL编号；该符号说明适用于本文图2、图5至图9

由于岩石元素含量随分离结晶程度不同而变化，只有对分离结晶程度相同的岩石比较其元素含量才是有意义的。据鲍温反应系列，岩石中CaO/Na_2O和MgO含量是衡量其分离结晶程度的最好标准。表1中各样品的CaO/Na_2O和MgO含量表明，除脉岩为分离结晶后期产物外，拉圾庙辉长岩和苏长辉石岩的分离结晶程度是类似的。

表1还显示辉长岩的TiO_2含量高出苏长辉长岩一倍以上，同时也具有较高的Fe^{3+}/Fe^{2+}值。这说明在辉长岩中Ti主要赋存在含高价铁的铁氧化物中。辉长岩的高Ti含量可能是含Ti磁铁矿堆晶的结果。众所周知，玄武岩浆中磁铁矿或钛铁矿的分离结晶发生在其结晶分异过程的晚期。因此，拉圾庙辉长岩不是玄武岩浆早期分离结晶形成的堆晶岩，而属于分离结晶后期形成的均质辉长岩。与辉长岩相反，苏长辉长岩的低TiO_2含量和低Fe^{3+}/Fe^{2+}值表明它不含较多的铁氧化物堆晶，因此，钙碱系列岛

弧火山岩的低Ti含量可能与源区性质有关，而与分离结晶作用无关。

辉长岩的高Fe^{3+}/Fe^{2+}值还说明该区岛弧拉斑玄武岩比钙碱系列岩浆有较高的氧逸度。考虑到拉斑玄武岩经常发育在岛弧外缘，钙碱系列火山岩发育在岛弧内部，该岩体的两种岩石可能具有不同的源区。辉长岩源区靠近岛弧外缘，较浅，由于海洋板块俯冲初期大量脱水使楔形地幔遭受了强烈的水化作用，从而导致其派生的岩浆具有高fo_2；反之苏长辉长岩的源区可能与俯冲到深部的洋壳有关。如果这一推论是正确的，则辉长岩源区应含有较多水从而使其熔融温度降低，还将导致该地幔源所产生的玄武岩浆具有较低的Ni含量[7,10]。表2表明，拉圾庙岩体的Ni含量较低，具有岛弧玄武岩浆的特征[8]，但辉长岩的Ni含量比苏长辉长岩及其晚期脉体的含量低3～7倍。这进一步证明了辉长岩源区富水的特征。

表2 北秦岭拉圾庙苏长辉长岩痕量元素含量（$\times 10^{-6}$）

样品号	QL-1	QL-2	QL-3	QL-5	QL-6	QL-8	QL-9
岩石名称	辉长岩	辉长岩	辉长岩	苏长辉长岩	苏长辉长岩	二长斑岩	闪长玢岩
Cr	20	26	16	253	67	68	121
Ni	10	12	10	77	44	34	75
V	358	361	353	131	199	80	109
Pb	4	3	5	3	5	26	17
K	5645	4898	5479	6808	6227	29970	25404
Rb	16.2	13.0	10.9	20.3	22.7	84.5	68.4
Sr	734.3	657.9	811.9	666.7	660.1	552.8	758.4
Ba	480	411	493	548	642	1168	1702
Th	3	2	1	3	3	13	8
Nb	10	10	10	5	7	14	15
Zr	56	54	59	17	41	110	151
La	21.09	21.72	24.40	12.90	16.76	—	—
Ce	40.37	47.26	51.97	23.45	33.74	—	—
Nd	22.30	23.96	32.06	11.68	12.56	23.37	33.50
Sm	4.736	4.946	7.072	2.558	2.325	4.129	6.148
Eu	1.757	1.804	2.518	1.072	0.996	—	—
Gd	4.895	4.904	7.19	2.548	1.82	—	—
Dy	4.558	4.586	6.308	2.568	1.824	—	—
Er	2.493	2.594	3.646	1.542	1.003	—	—
Yb	2.144	2.228	3.02	1.383	0.962	—	—
Lu	0.292	0.336	0.441	0.206	0.149	—	—
Y	15	16	20	13	13	18	18
P	1877	2139	1571	218	611	1048	1659
Ti	10971	11091	10731	3477	4196	3357	4676
Nb/Th	3.3	5	10	1.7	2.3	1.1	1.9
Ba/Nb	48	41.1	49.3	109.6	91.7	83.4	113.5
La/Nb	2.1	2.2	2.4	2.6	2.4	—	—

注：稀土及Rb、Sr由中国地质科学院地质研究所Sm-Nd同位素室测定。其他痕量元素由中国地质勘查技术研究院物化探研究所测定。

4 稀土及痕量元素地球化学特征

将拉圾庙辉长岩和苏长辉长岩的痕量元素用原始地幔组成标准化[11]，并按分配系数大小排序做痕量元素丰度模型图（图3）。该图显示，拉圾庙堆晶岩与海岛玄武岩（OIB）相比具有高Ba、Pb、Sr，低Nb、Zr异常。其中辉长岩具有低Th，苏长辉长岩具有低P、Ti及REE特征。高Ba、Pb、Sr和低Nb、Zr异常是岛弧玄武岩典型的地球化学特征[8,9,11,12]。如果均质辉长岩的地球化学特征可以与喷出的熔岩进行对比的话，上述特征表明拉圾庙苏长辉长岩生成于岛弧环境。已有研究表明,，岛弧玄武岩的高 Ba、Pb、Sr，低 Nb、Zr 特征与分离结晶作用无关，而是反映了岩浆源区的特征[8,9,12]。苏长辉长岩与辉长岩相比，它更贫 Nb、Zr、Ti、P 而富 Ba、Rb、K、Th。这可能受这两种岩浆源区的差异所控制，因为前已证明这两类岩石的分离结晶程度是类似的，这些痕量元素的变化与 Ca/Na 和 MgO 的变化无相关关系。

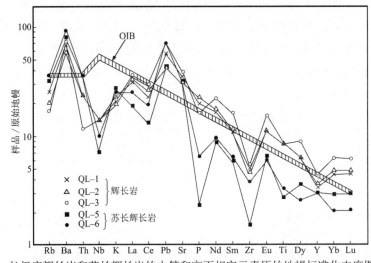

图 3　拉圾庙辉长岩和苏长辉长岩的中等和高不相容元素原始地幔标准化丰度图

为详细对比辉长岩与苏长辉长岩的 REE 分配模型，我们又单独做了它们的 REE 球粒陨石标准化丰度图（图 4）。图 4 显示拉圾庙堆晶岩均呈轻稀土富集型，其中辉长岩 REE 丰度高于苏长辉长岩，并与高稀土含量的印尼桑达（Sunda）岛弧玄武岩类似[12]。斜方辉石 REE 含量显著低于单斜辉石，这可以用来解释苏长辉长岩低 REE 丰度。但是也不排除它们源区性质的差异对岩浆 REE 丰度的影响。苏长辉长岩与辉长岩 REE 丰度模型的另一区别是它更加倾斜，其重稀土元素（HREE）含量低于一般岛弧火山岩的 HREE 含量[12]。这不可能用斜方辉石参与分离结晶过程来解释，因为根据 REE 分配系数[13]，斜方辉石比单斜辉石应更相对富 HREE。因此，苏长辉长岩的低 HREE 特征应反映源区的特征。由于岛弧钙碱系列岩浆的源较深，这里的俯冲洋壳可能已开始变为榴辉岩相，在低比例熔融时可能残留有石榴子石、金红石等矿物，从而使流体或熔体亏损 HREE、Ti、Nb 等元素。

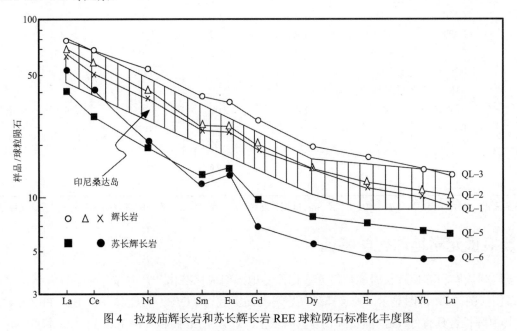

图 4　拉圾庙辉长岩和苏长辉长岩 REE 球粒陨石标准化丰度图

5 Sr-Nd 同位素地球化学

Sr、Nd 同位素研究可以为判断铁镁质杂岩是否是蛇绿岩提供可靠的依据。由于蛇绿岩作为古洋壳的一部分受海水蚀变作用的影响，其 Sr 同位素比值可在很大范围内变化，而 Nd 同位素比值保持恒定。例如洋中脊成因的蛇绿岩在 Sr-Nd 同位素图上，形成一从洋中脊区向高 Sr 同位素比值方向水平延长的特殊蛇绿岩区（图 5）[14,15]。因此，对拉圾庙苏长辉长岩进行 Sr、Nd 同位素研究可以为判断它是不是蛇绿岩提供新的证据。

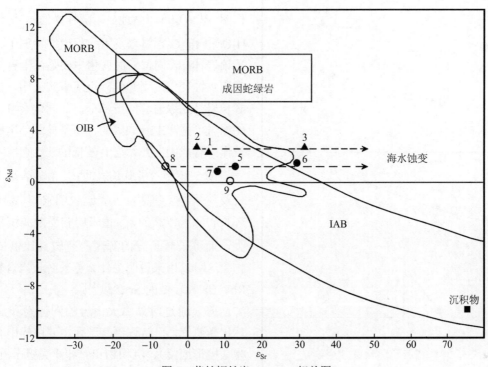

图 5 苏长辉长岩 ε_{Sr} - ε_{Nd} 相关图

MORB 洋中脊玄武岩，OIB 海岛玄武岩，IAB 岛弧玄武岩

利用该岩体的 402.6 ± 17.4 Ma Sm-Nd 矿物等时线年龄[4]，我们计算了它们的初始 ε_{Sr} 和 ε_{Nd} 值（表 3），并绘出其初始 ε_{Sr} - ε_{Nd} 相关图。

表 3 拉圾庙辉长苏长岩 Sr、Nd 同位素数据及初始 ε_{Sr} 和 ε_{Nd} 值

样品号	Rb (10^{-6})	Sr (10^{-6})	$^{87}Rb/^{86}Sr$	$^{87}Sr/^{86}Sr$	$\varepsilon_{Sr}^a(t)$	Sm (10^{-6})	Nd (10^{-6})	$^{147}Sm/^{144}Nd$	$^{143}Nd/^{144}Nd$ [b]	$\varepsilon_{Nd}^a(t)$
QL-1	16.215	734.33	0.0639	0.70496±3	+5.5±0.4	4.736	22.299	0.1285	0.512558±8	+2.3±0.2
QL-2	12.981	657.91	0.0571	0.70469±2	+2.3±0.3	4.946	23.957	0.1249	0.512569±7	+2.7±0.2
QL-3	10.856	811.93	0.0387	0.70662±3	+31.2±0.5	7.073	32.062	0.1334	0.512594±4	+2.7±0.1
QL-5	20.323	666.74	0.0883	0.70561±2	+12.8±0.3	2.558	11.678	0.1325	0.512520±10	+1.3±0.2
QL-6	22.720	660.07	0.0997	0.70683±12	+29.3±1.7	2.325	12.560	0.1120	0.512474±7	+1.5±0.2
QL-7$_{Pyr}^c$	1.812	32.194	0.1629	0.705718±32	+8.3±0.4	8.550	31.484	0.1642	0.512565±17	+0.8±0.3
QL-8 [c]	84.511	552.77	0.4427	0.70628±10	−6.4±1.4	6.148	33.503	0.1120	0.512444±21	+1.0±0.4
QL-9 [c]	68.402	758.42	0.2612	0.70649±5	+11.3±0.8	4.129	23.367	0.1068	0.512390±19	+0.1±0.3

a. $\varepsilon_{Sr}(t)$ 和 $\varepsilon_{Nd}(t)$ 为相对该辉长苏长岩形成时代 402.6 Ma 时的初始 ε_{Sr} 和 ε_{Nd} 值。误差为 2σ。

b. $^{143}Nd/^{144}Nd$ 的质谱分异相对于 $^{146}Nd/^{144}Nd$ = 0.7219 进行校正，且相应的仪器标准样（BCR-1）的测定值为 $^{143}Nd/^{144}Nd$ = 0.512612。每个比值均给出 2σ 误差。

c. QL-7$_{Pyr}$ 的 Sr、Nd 同位素数据，QL-8，QL-9 的 Nd 同位素数据是在美国麻省理工学院测定，其他数据在中国地质科学院地质所测定。

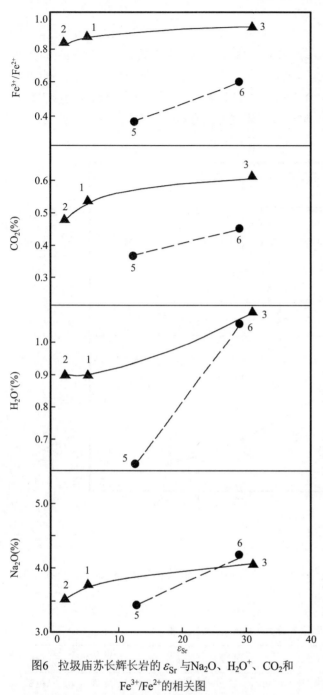

图6 拉圾庙苏长辉长岩的 ε_{Sr} 与Na_2O、H_2O^+、CO_2和 Fe^{3+}/Fe^{2+}的相关图

图5表明，该岩体的辉长岩、苏长辉长岩及其脉岩均表现有稳定的ε_{Nd}值和较大的ε_{Sr}值变化范围。它们均从岛弧玄武岩区向高ε_{Sr}方向变化，从而形成一与典型洋中脊成因蛇绿岩类似的水平相关趋势。这说明该苏长辉长岩体发育在海底环境，并遭受了海水蚀变作用或同化了海水蚀变玄武岩。将辉长岩和苏长辉长岩全岩样品的ε_{Sr}对CO_2、H_2O^+、Na_2O和Fe^{3+}/Fe^{2+}作图（图6），可清楚地显示随样品的ε_{Sr}增加，其CO_2、H_2O和Na_2O含量以及氧化程度均增加了。这也支持了该岩体Sr同位素的变化主要是由于直接的或间接的海水蚀变作用引起的。上述事实证明，拉圾庙苏长辉长岩是蛇绿岩套的一部分。其同位素特征表明，该蛇绿岩不是洋中脊成因的，而是岛弧环境成因的，它较洋中脊成因的蛇绿岩有较低的ε_{Nd}值。

图5还显示了辉长岩的ε_{Nd}值显著高于苏长辉长岩及其脉岩（参见表3），这说明它们二者源区不同。辉长岩源区较高的ε_{Nd}值表明它受陆源沉积物混染较少，而苏长辉长岩的源区含有较高比例的陆源沉积物。这可以用来说明为什么苏长辉长岩有较高的K、Rb、Ba和较低的Sr含量[11]。

前人通过对岛弧火山岩的同位素和痕量元素地球化学研究，已认识到其源区主要由洋中脊型亏损地幔（楔形地幔）、沉积物和俯冲洋壳析出流体3个端元组成[9, 14-16]。显然不同的岛弧玄武岩，其源区中3个端元的组成比例不同。但是由于在以前的研究中，未能将同位素和痕量元素研究结合起来，因而尚没能用一种图把上述3个端元同时显示出来，从而不能清楚地观察这3个端元在不同岛弧玄武岩中的地位和作用。李曙光最近提出了一组ε_{Nd}-Nb/Th，ε_{Nd}-Ba/Nb和ε_{Nd}-La/Nb图（本文略）。在这些图中，岛弧火山岩源区的3个端元组成均可同时在一张图上显示出来并呈三角形分布。根据岛弧火山岩在这些图中的位置，李曙光将它们划分成两种类型：（1）二端元型，地幔源区只含洋中脊型亏损地幔和俯冲洋壳析出流体2个端元，陆源沉积物影响较少，如阿留申和新不列颠；（2）三端元型，陆源沉积物组分在其地幔源区占有相当比例，也即其地幔源区由全部3个端元组成，如安第斯、桑达、小安德列斯。秦岭拉圾庙苏长辉长岩体的样品全部落在三端元岛弧火山岩区（图7至图9）。这说明在古生代，古秦岭海板块向北秦岭古岛弧下俯冲时携带有较多的陆源沉积物进入地幔。这可能反映了当时北秦岭是与大陆相邻的岛弧。图7至图9还同时显示了苏长辉长岩及其脉岩的源区比辉长岩的源区含有更多的陆源沉积物及俯冲洋壳析出流体的低Nb特征，从而说明苏长辉长岩的源区较深，俯冲洋壳已发生强烈的榴辉岩相变质作用和陆源沉积物的部分熔融作用。

古秦岭海俯冲板块携带有较多陆源沉积物的事实，将对北秦岭古生代和中生代的岩浆活动产生较大影响。在研究该区古、中生代花岗岩和火山岩时应当考虑这一因素。

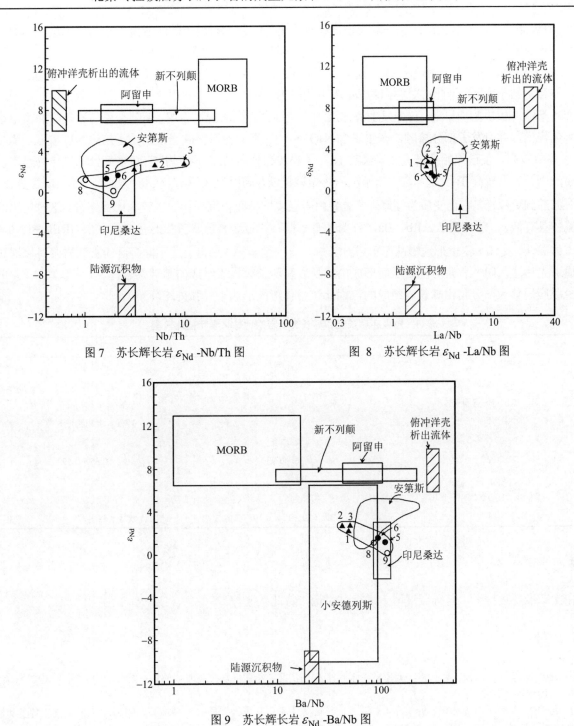

图7 苏长辉长岩 ε_{Nd}-Nb/Th 图

图8 苏长辉长岩 ε_{Nd}-La/Nb 图

图9 苏长辉长岩 ε_{Nd}-Ba/Nb 图

6 结论

1) 北秦岭拉垃庙苏长辉长岩具有类似蛇绿岩的 Sr、Nd 同位素特征，其 ε_{Nd} 值恒定在 +2 左右，ε_{Sr} 有较大变化范围（−6.4～+31.2）。它们的 ε_{Sr} 值与 Na_2O、H_2O^+、CO_2 含量和 Fe^{3+}/Fe^{2+} 值的正相关关系说明其 ε_{Sr} 值的变化与海水蚀变或同海水蚀变玄武岩有关。因此，该苏长辉长岩是蛇绿岩套的一部分。

2) 北秦岭拉垃庙苏长辉长岩所代表的蛇绿岩是岛弧成因的。主要证据为：

（1）苏长辉长岩为钙碱系列岩浆岩。

（2）痕量元素具有高 Ba、Pb、Sr，低 Nb、Zr、Ni 的典型岛弧火山岩特征。

（3）较低的初始 ε_{Nd} 值（+2 左右）。

3）拉圾庙苏长辉长岩的源区属三端元型岛弧火山岩类，它含有较多的陆源沉积物组分。这表明北秦岭在当时是华北陆块南缘的近陆岛弧，因此其海沟可以接收到大量的陆源沉积物。

4）拉圾庙岩体的辉长岩和苏长辉长岩是两个不同的岩浆系列，它们有性质不同的源区。它们之间的地球化学差异总结于表4。据此可对它们的成因提出如下模型：

在古秦岭海板块俯冲早期，首先发生海洋沉积物及蚀变洋壳的脱水作用，使其上覆岛弧外缘的楔形地幔发生水化作用，并引发了部分熔融，产生了含 H_2O 多，fo_2 高的拉斑玄武岩浆。由于熔融温度较低，橄榄石 Ni 分配系数较高，使该原始拉斑玄武岩浆贫 Ni。该岩浆在分离结晶过程晚期有较多高价铁氧化物晶出，使拉圾庙的辉长岩有较高 Ti 含量。这时陆源沉积物熔融量低，因而岩浆的 ε_{Nd} 较高，而 K、Rb、Ba 含量较低。当板块俯冲到更深部，它已变质为榴辉岩相，并由于温度增高而使沉积物发生较大比例熔融。这时析出的熔体或流体具有低 ε_{Nd} 值，富 K、Rb、Ba，并由于有石榴石和金红石等残留相的存在而亏损 HREE 和 Nb 等。这些流体混杂、交代上覆楔形地幔从而形成苏长辉长岩的地幔源。来自上述不同源区的玄武岩浆可以通过不同的通道先后注入同一岩浆房内，从而形成由辉长岩和苏长辉长岩组成的基性杂岩体。

拉圾庙岩体大部分是由辉长岩组成的，因此该岩体可能形成于岛弧外缘环境。

表 4 拉圾庙辉长岩和苏长辉长岩的地球化学特征比较表

特征	辉长岩	苏长辉长岩
岩浆演化系列	拉斑系列	钙碱系列
Nb、Zr、Ti、P 含量	较高	较低
K、Rb、Ba 含量	较低	较高（沉积物高）[16]
Sr 含量	较高	较低（沉积物低）[16]
HREE 含量	较高	较低
Fe^{3+}/Fe^{2+}	高（0.86~0.93）	低（0.35~0.59）
ε_{Nd}	较高	较低
挥发分（H_2O、CO_2）	较高	较低
Ni 含量	很低（10~12）×10^{-6}	较低（34~77）×10^{-6}

致谢

在文章准备过程中，与孙贤铁博士进行过讨论，特此致谢。

参考文献

李曙光, Hart S. R., 郑双根, 郭安林, 刘德良, 张国伟.中国华北、华南陆块碰撞时代的钐–钕同位素年龄证据.中国科学（B 辑），1989,（3）: 312–319.

李曙光, 葛宁洁, 刘德良, 张宗清, 叶笑江, 郑双根, 彭长权.大别山北翼大别群中 C 型榴辉岩的 Sm-Nd 同位素年龄及其构造意义.科学通报, 1989,（7）: 522–525.

李曙光, 张志敏, 刘德良, 陈移之, 葛宁洁, 韩宗珠, 赵广涛.青岛榴辉岩相蛇绿混杂岩——俯冲"洋壳"残片推覆体的发现及其意义.科学通报, 1991,（15）: 1161–1164.

李曙光.太古代绿岩带拉斑玄武岩与现代岛弧拉斑玄武岩 Cr、Ni 含量差异的构造环境意义.国际前寒武纪地壳演化讨论会论文集, 第 2 集, 变质岩地球化学和成矿作用, 1989: 80–94.

孙勇, 于在平, 张国伟.东秦岭蛇绿岩地球化学.秦岭造山带的形成及其演化.西安: 西北大学出版社, 1987: 65–74.

王清晨, 孙枢, 李继亮, 周达, 许靖华, 张国伟.秦岭的大地构造演化.地质科学, 1989,（2）: 129–142.

张国伟, 于在平, 孙勇, 程顺有, 薛峰, 张成立.秦岭商丹断裂边界地质体基本特征及其演化.秦岭造山带的形成及其演化.西安: 西北大学出版社, 1987: 29–47.

张宗清, 刘敦一, 傅国民. 北秦岭变质地层同位素年代研究. 北京: 地质出版社, 1994.

Davidson J P. Lesser Antilles isotopic evidence of the role of subducted sediment in island arc magma genesis. Nature, 1983, 306: 253–256.

Hart S R and Davis K E. Nickel partitioning between olivine and silicate melt. Earth Planet. Sci. Lett., 1978, 40: 203–219.

Henderson P. Inorganic Geochemistry. Oxford, Pergamon Press, 1982.

Hickey R L, Frey F A and Gerlach D C. Multiple sources for basaltic arc rocks from the southern volcanic zone of the Andes (34–41S): trace element and isotopic evidence for continental crust. Jour. Geophy. Res., 1986, 91: 5963–5983.

Perfit M R, Crust D A, Bence A E, Arculus A R and Taylor S R. Chemical characteristics of island-arc basalts: implications for mantle sources. Chem. Geol., 1989, 30: 227–256.

Ryerson F J and Watson E B. Rutile saturation in magmas: implications for Ti-Nb-Ta depletion in island-arc basalts. Earth Planet. Sci. Lett., 1987, 86: 225–239.

Sun S S and McDonough W F. Chemical and isotopic systematics of oceanic basalts: implications for mantle composition and processes, in "Magmatism in the Ocean Basins" edited by Saunders A D & Norry M J, Geol. Soc. Special Publ., 1989, (42): 313–345.

White W M and Patchett J. Hf-Nd-Sr isotopes and incompatible element abundances in island arcs: implications for magma origins and crust-mantle evolution. Earth Planet. Sci. Lett., 1984, 67: 167–185.

蛇绿岩生成构造环境的 Ba-Th-Nb-La 判别图*

李曙光

中国科学技术大学地球与空间科学系，合肥 230026

> **亮点介绍**：鉴于 Ti 对岩浆分离结晶作用很敏感，Pearce 的 Ti-Zr-Y 图判别蛇绿岩生成构造环境效果不好。该文综合前人研究数据，建立了利用四个分配系数相近的四个非常不相容元素的识别蛇绿岩形成构造环境的 Ba-Th-Nb-La 判别图解；该判别图解能很好地判断蛇绿岩生成构造环境。

摘要 Pearce 的 Ti-Zr-Y 图不能很好区分洋脊和岛弧玄武岩，Ti 对岩浆分离结晶作用很敏感，因此用这个图判别蛇绿岩生成构造环境效果不好，本文利用分配系数相近的 Ba、Th、Nb、La 四个非常不相容元素的 Ba/Nb-Ba、Nb/Th-Nb、La/Nb-La 和 Ba/Nb-Th/Nb 图解很好区分了洋脊、岛弧、洋岛玄武岩。弧后盆玄武岩同时具有洋脊和岛弧玄武岩的特征。这些元素在海水蚀变中较稳定，它们的比值在分离结晶过程中保持不变，因此这些图解有利于判别蛇绿岩(包括熔岩和均质辉长岩)生成的构造环境。

关键词 蛇绿岩；痕量元素地球化学

1 引言

如何判别蛇绿岩生成的构造环境是蛇绿岩研究中一个重要而又比较困难的问题。因为蛇绿岩在其受到变质和构造侵位后，其剖面厚度、岩石学性质都有很大变化，很难与大洋壳剖面进行严格的对比。Pearce 和 Cann（1971）及 Pearce（1975）提出用 Ti-Y-Zr 图解来判断蛇绿岩的生成环境。然而实践表明 Ti-Y-Zr 图的判别效果不好，在图上岛弧与洋中脊玄武岩有较大重叠(Coleman，1977)。这是因为该图是以玄武质熔岩资料为基础建立的，而 Ti 在分离结晶过程中非常敏感，磁铁矿和钛铁矿在岩浆演化晚期的分离结晶可使任何一种玄武岩浆中的 Ti 含量明显下降，堆晶岩中的 Ti 大大升高。也正因为如此，该图不能应用于堆晶岩。此后，Wood et al. (1979) 和 Mesehede (1986) 分别发展了不用 Ti 的 Hf-Th-Ta 和 Nb-Zr-Y 图解。前者中的 Hf、Ta 都较难准确测定，后者中洋中脊玄武岩(MORB)和岛弧玄武岩(IAB)区域又有重叠，应用效果也不理想。为此，本文根据痕量元素地球化学的最新研究成果，提出一组新的、较有效的、同时适用熔岩及均质辉长岩的判别蛇绿岩生成构造环境的地球化学图解。

2 地球化学判别图

如上所述，以前的地球化学判别图主要问题在于它们不能有效地将 MORB 和 IAB 区分开来，而 MORB 型和 IAB 型蛇绿岩恰是蛇绿岩生成构造环境的两个最主要类型(详见后面应用实例)。因此在制作新图时，首先就要选好能有效区分这两种玄武岩的指示元素。近十几年来有关岛弧火山岩的研究从不同角度证明了岛弧玄武岩的特征元素，除 Ti 以外，均与其源区有关(Ryerson and Watson, 1987; Sun and McDonough, 1989; White and Patchett, 1984; Briqueu et al., 1984; Morris and Hart, 1983)。岛弧玄武岩源区与 MORB 和板内玄武岩源区的最大差别在于它与俯冲洋壳有关。我们只要选择能直接示踪源区特征的痕量元素指标，就可以较好地将 MORB 与 IAB 分开。本文选择 Ba、Th、Nb、La 及其比值做指标，以建立一套新的能同时适用于熔岩和均质辉长岩的构造环境判别图。其理由为：

---

* 本文发表在：岩石学报，1993，9(2)：146-157

(1) 在任何岛弧玄武岩中，低 Nb 和高 Ba 异常是最突出和最稳定的地球化学特征(Perfit et al., 1980; Ryerson and Watson, 1987; Sun and McDonough, 1989; White and Patchett, 1984)。高 Nb 异常是 OIB 的典型特征(Sun and McDonough, 1989)。

(2) 它们都是非常不相容元素，其分配系数相近。因此它们的比值，尤其 Ba、Th、La 与 Nb 的比值(Nb 分配系数居中)在部分熔融和分离结晶过程中保持不变，从而可最有效地指示源区特征(Sun and McDonough, 1989)。

(3) Nb、La、Ba、Th 在海水蚀变及变质过程中是稳定或比较稳定的元素。Nb、La 的稳定性已众所周知。Hart(1971)和 Tatsumoto(1978)分别研究过 Ba 和 Th 在海水蚀变过程中的稳定性。综合他们的研究成果可以给出这些元素在海水蚀变过程中的稳定性顺序(图 1)。虽然 Ba 和 Th 在海水蚀变时的稳定性不如 Sr 和 REE，但仍比碱性元素 U、Pb 等稳定得多。蛇绿岩形成于海底环境，不可避免会受到海水蚀变。因此，Nb、La、Ba、Th 的这种稳定性对建立判别蛇绿岩生成构造环境的地球化学图解是十分重要的。

根据这一思想，本文综合了文献中 MORB、OIB、IAB 和弧后盆玄武岩(BABB)(含部分安山岩)的 Ba、Th、Nb、La 数据，并制作了 Ba/Na-Ba、Nb/Th-Nb、La/Nb-La 和 Ba/Nb-Th/Nb 图(图 2～图 5)。数据的来源是：MORB 取自 Ryerson and Watson (1987)，Langmuir et al. (1977)，Sun et al., (1979)，Price et al., (1986)，Hofmann (1988)；OIB 取自 Rhodes (1989)，Clague and Frey (1982)，Chen and Frey (1985)，Frey and Clague (1983)；IAB 取自 Ryerson and Watson (1987)，White and Patchett (1984)，Briqueu et al. (1984)，Basaltic Volcanism Study Projeet (1981)，Nye and Reid (1986)，Ewart et al. (1973)，Hickey et al. (1986)，Brown et al. (1977)，Stolz et al. (1990)，Tatsumi et al. (1991)，Carr et al. (1990)，Defan et al. (1977)；BABB 取自 Saunders and Tarney (1979)，Volpe et al. (1987)，Vallier et al. (1991)；原始地幔 (P. M.)取自 Sun and McDonough (1989)；上地壳取自 Taylor and McLennan (1985)。

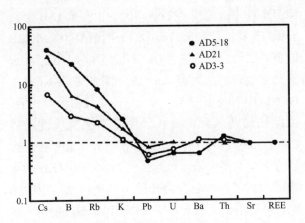

图 1 玄武岩遭受海水蚀变时元素的稳定性图（据 Hart, 1971 和 Tatsumoto, 1978 的资料综合）。AD5-18、AD21 和 AD3-3 三个样品都是大洋玄武岩标本，曲线表示这些标本边缘蚀变部分和核心新鲜部分的微量元素比值的变化

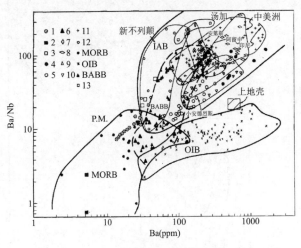

图 2 Ba/Nb-Ba 图

图例说明：1.意大利蒙特福诺蛇绿岩；2.新几内亚巴布亚蛇绿岩（玄武岩）；3.新几内亚巴布亚英云闪长岩；4.北秦岭丹凤蛇绿岩的苏长辉长岩；5.土耳其米西斯蛇绿岩；6.南斯拉夫弟爷瑞克蛇绿岩；7.安第斯 IAB；8.中美洲 IAB；9.阿留申 IAB；10.小安德烈斯 IAB；11.印尼 IAB；12.新不列颠 IAB；13.汤加 IAB；P.M.为原始地幔；其他代号见正文

这些图解表明 MORB 和 OIB 的 Ba/Nb、Nb/Th 和 La/Nb 是类似的，然而 IAB 的这些比值与 MORB 和 OIB 有很大不同。因此，在纵坐标方向上可将 IAB 与 MORB 和 OIB 分开。在横坐标方向上，MORB 与 OIB 的 Ba、Nb、La 含量有明显差异，分开程度较大。因此，这些图件可较好地将 MORB、OIB 和 IAB 三种不同构造环境形成的玄武岩分开。然而几个典型弧后盆玄武岩(BABB)的数据表明它们的分布区在所有图中均与 MORB 和 IAB 重合。这是因为弧后盆靠近岛弧一侧位于俯冲带之上，其地幔也要接受俯冲洋壳析出流体

的交代作用，因而由它产生的 BABB 具有与 IAB 相似的特征；另一方面弧后盆的扩张中心又可生成与MORB 类似的玄武岩。尽管我们不可能在任何一个痕量元素图上将 BABB 与 IAB 和 MORB 分开，但是 BABB 在上述图件中同时分布在 IAB 和 MORB 两区并形成狭长混合趋势的特征(图2、图3)可以帮助我们识别 BABB 型蛇绿岩。

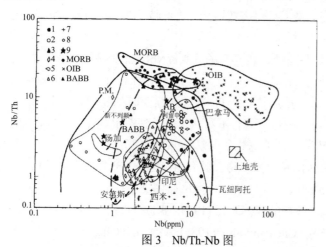

图 3 Nb/Th-Nb 图

图例说明与图 2 的相同，星号代表希腊沃瑞诺斯蛇绿岩

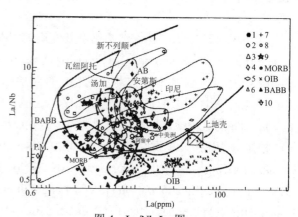

图 4 La/Nb-La 图

其他图例与图2、3 的相同；10.代表中国台湾蛇绿岩；11.代表瓦纽阿托蛇绿岩

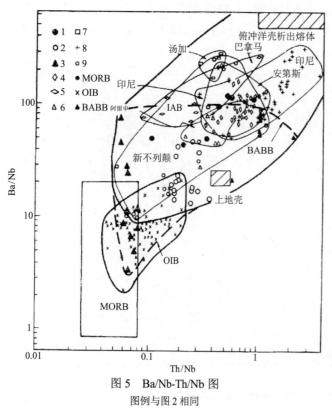

图 5 Ba/Nb-Th/Nb 图

图例与图2相同

将图2、3、4进行比较可发现 Ba/Nb-Ba 图对区分 IAB、MORB 和 OIB 效果最好，在该图中 IAB 和 MORB 只有很小部分重叠。这主要是因为 Ba、Nb 两个元素都是 IAB 最特征的元素。其次 Nb/Th-Th 图的效果也较好，因为大多数 IAB 都具有低 Th 特征。在 La/Nb-La 图中，IAB 和 MORB 的重叠区较大。这说明 Ba、Nb、Th 对上述三种玄武岩的判别功能比 La 强。因此，我们做了 Ba/Nb-Th/Nb 图(图5)。由于这两个比值分母相同，因此任何两端元混合在该图中都应呈线性趋势(Langmuir et al., 1978)。虽然图5在区分三种玄武岩方面没有多大进步，甚至使 MORB 和 OIB 的重叠区增大了，但它完全排除了分离结晶作用的影响，可适用于堆晶岩，并有助于讨论源区特征。如 OIB 在图 5 中的延长趋势是介于 MORB 和上地壳之间，它说明 OIB 的源区可能含有再循环进入地幔的陆源沉积物组分。这与 Sun and McDonough (1989)用 Ba/Nb-La/Nb 所获结论一致。大多数 IAB 在图 5 中的正相关趋势位于 OIB 趋势以上。这可能意味着 IAB 源区还加入了一种比上地壳具有更高 Ba/Nb 和 La/Nb 值的组分。它可能是俯冲洋壳在榴辉岩相条件下出熔的富硅和水的熔体和流体。因为俯冲洋壳形成的榴辉岩含有较多的金红石。在熔融程度不高时，多数金红石作为残留相保存下来(Ryerson and Watson, 1987)。它的 Nb 分配系数很高(D_{Nb}^{Rt} =30)(Green and Pearson, 1987)，因此大量 Nb 将进入残留的金红石相内并产生低 Nb 熔体。此外，在一些近大陆岛弧区，其俯冲洋壳发生部分熔融时，陆源沉积物组分也可进入熔体，从而使某些岛弧火山岩(如安第斯)在图 5 中的线性趋势不明显而必须用三端元(亏损地幔、俯冲

洋壳析出流体、陆源沉积物)混合来解释。

3 应用实例

上述图解是以玄武岩数据为基础建立起来的。使用若干已知生成构造环境有蛇绿岩资料检验这些图解判别蛇绿岩的生成构造环境的效果是必要的，因为人们可能会担心蛇绿岩所经历的海水蚀变和变质作用是否会严重影响岩石的 Ba、Th 含量。

(1) 意大利北部与瑞士接壤的蒙特福诺(Monte del Forno)蛇绿混杂岩。该蛇绿岩被认为形成于洋中脊环境，并经历区域变质和接触变质作用。在变玄武岩中的层状 Fe-Cu-Zn 矿化也与洋中脊水热蚀变玄武岩中的矿化类似(Peretti and Koppel, 1986)。根据其中辉绿岩墙和角闪岩化的枕状熔岩的 Ba 和 Nb 资料，投影到 Ba/Nb-Ba 图上(图 2)，除一个样品外，均落入 MORB 区内。这进一步证明海水蚀变和变质作用对 Ba、Nb 丰度未产生严重影响。

(2) 新几内亚的巴布亚(Papuan)蛇绿岩由超镁铁岩、辉长岩及玄武岩组成，沿该岛的所罗门海岸分布。它被认为是侏罗纪—白垩纪的洋壳，在早第三纪时，由于弧陆碰撞而仰冲侵位(Jagues and Chappell, 1980)。侵入该蛇绿岩中的云英闪长岩(始新世)与蛇绿岩无关，是岛弧安山质岩浆活动的一部分。根据 Jagues and Chappell (1980)给出的有准确 Ba、Nb 含量的 2 个玄武岩样品和 3 个云英闪长岩样品投影到 Ba/Nb-Ba 图上，它们分别落入 MORB 和 IAB 区(见图 2)，与地质判断结论一致。

(3) 中国台湾东海岸蛇绿岩具有洋中脊成因(Jahn, 1986)。根据 Jahn (1986)给出的该蛇绿岩 La 和 Nb 的数据，投影在 La/Nb-La 图上，除 2 个样品因 La/Nb<0.4 未标到图上外，全部落入 MORB 区(图 4)。显然那 2 个样品的低 La/Nb 不是 IAB 特征，而是 MORB 的特征。

(4) 拉垃庙苏长辉长岩是北秦岭丹凤蛇绿岩的一部分，相当于堆晶岩上部的均质辉长岩。它形成于岛弧环境(李曙光等，1992)，并经历了海西期角闪岩相变质作用。根据李曙光等(1992)给出的辉长岩、苏长辉长岩和脉岩的 Ba、Th、Nb、La 数据，投点在 Ba/Nb-Ba、Nb/Th-Nb、La/Nb-La 和 Ba/Nb-Th/Nb 图上均落入岛弧区(图 2~5)。这说明这些图件对蛇绿岩的均质辉长岩也有很好的判别效果。

(5) 希腊沃瑞诺斯(Vourinos)蛇绿岩是岛弧成因的(Noiret et al., 1981)。将它们 3 个玄武岩和 1 个英安岩脉样品的 La、Nb、Th 数据投影在 Nb/Th-Nb 和 La/Nb-La 图上，它们均落在岛弧区(图 3、图 4)。图 3、图 4 表明它们的地球化学特征与汤加、新不列颠等发育在洋壳之上的岛弧相近。这一点与北秦岭古岛弧不同，后者地球化学特征与大陆边缘岛弧，如小安德列斯、印尼、中美洲相似。

(6) 土耳其南部的米西斯(Misis)蛇绿岩。它是在第三纪中新特提斯闭合时发育的蛇绿混杂岩，岛弧成因(Floyd et al., 1991)。根据 Floyd et al. (1991)的数据，它们的玄武岩样品在上述一系列 Ba-Th-Nb-La 图解中均落入岛弧玄武岩区，与前人研究结论一致。

(7) 南斯拉夫第那瑞克(Dinaric)蛇绿岩。这是一套侏罗纪蛇绿混杂岩，它们侵位于一套二叠系—侏罗系的滑塌沉积物中，由杂砂岩、红色放射虫硅质岩、块状辉绿岩、枕状熔岩、辉长岩和橄榄岩碎块组成。在辉长岩及橄榄岩中有粒玄岩脉。它被鉴定为发育在弧后盆环境(Lugovic et al., 1991)。根据 Lugovic et al. (1991) 的数据，它的玄武岩和粒玄岩样品点在 Ba-Th-Nb-La 判别图中均落在 IAB 和 MORB 两个区域内。这表明它们的源区同时受俯冲洋壳及扩张脊的影响，应发育在弧后盆环境中。

上述有关洋中脊、岛弧和弧后盆成因蛇绿岩的应用实例表明，这一套 Ba-Th-Nb-La 图解对判别蛇绿岩的生成构造环境是有效的。蛇绿岩经历的海水蚀变和变质作用不能显著影响它们的 Ba、Th、Nb、La 丰度。因此，尽管洋岛(海山)成因蛇绿岩因缺乏有关痕量元素数据而未能检验，但只要它像 OIB 一样具有高 Nb 特征，这些图件就能很好地鉴别它们。

应当指出，在制作上述 Ba-Th-Nb-La 判别图时，我们没有考虑发育在大陆环境的火山岩(如大陆玄武岩、溢流玄武岩、陆内粗面安山岩等)。它们在上述图解中可能会与 IAB、OIB 有较大重叠。因此，这一套地球化学图解只适用于蛇绿岩的生成构造环境判别。对于古火山岩，如果有充分证据可以排除它的大陆成因可能性，也可应用这一套图解帮助判断其生成构造环境。

4 Ba-Th-Nb-La 判别图的制图参数

为便于对上述 Ba-Th-Nb-La 判别图的绘制和应用，我们将上述图解中 MORB、OIB 和 IAB 分布区分别用具直线边的多边形包络起来(见图6)。这些多边形的端点位置及坐标值示于图6和表1。根据这些坐标值，可以很容易地在对数坐标纸上或利用微机作图程序绘制出相应的图解。

表1 蛇绿岩生成构造环境 Ba-Th-Nb-La 判别图边界点坐标

图名	Ba/Nb-Ba		Nb/Th-Nb		La/Nb-La		Ba/Nb-Th/Nb	
坐标轴	Ba/Nb	Ba(ppm)	Nb/Th	Nb(ppm)	La/Nb-La	La(ppm)	Ba/Nb	Th/Nb
A	0.7	1.5	39	1.05	0.7	0.7	16	0.03
B	2	1.5	16	12	1.8	0.9	16	0.077
C	22	19.5	11.4	12	2.3	2.3	0.9	0.077
D	10	84	10.9	2.8	0.76	10.5	0.9	0.03
E	5.4	124	30.5	0.95	0.41	5	610	0.7
F	3	50	0.4	0.2	30	31	80	0.045
G	1.6	30.7	21	1.3	5	0.9	16	0.045
H	0.7	28	2	0.2	1.03	2	8	0.077
I	300	88	10.5	10.6	0.92	4.4	11	0.31
J	82	21.1	3.1	17.3	1.1	11	67	4
K	8.1	30.1	0.4	17.3	5	211	25	0.205
L	7.2	54	27	21	1.8	40	7	0.205
M	100	2800	11.3	130	1.06	120	2	0.053
N	22	1200	3.6	80	0.66	120	4	0.036
O	10.7	2500	7	17.3	0.5	4.4	9	0.036
P	7	2600	6.4	7				
Q	3.5	26	11.1	5				
R			24	8				

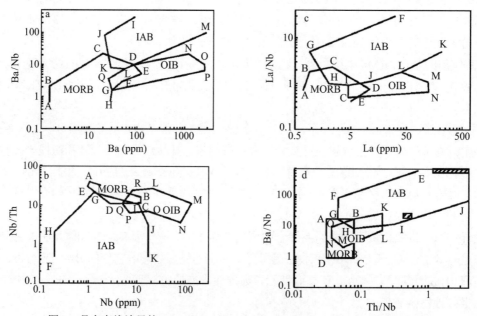

图6 具有直线边界的 Ba/Nb-Ba、Nb/Th-Nb、La/Nb-La 和 Ba/Nb-Th/Nb 图解

图中边界端点字母与表1对应

弧后盆玄武岩的分布区及边界点坐标在图6及表1中没有给出。这是因为 BABB 没有自己独立的分布。

如前所述,只要一组玄武岩样品点能同时落入 IAB 和 MORB 两区,我们就可以认为它们具有 BABB 的特征。因此,单独圈出 BABB 区没有必要。

参考文献

李曙光,陈移之,张宗清,张国伟,郭安林,1992. 北秦岭拉垃庙苏长辉长岩的痕量元素和 Sr,Nd 同位素地球化学. 地质学报(待刊).

Basaltic Volcanism Study Project, 1981. Island arc basalts, in: Basaltic volcanism on the terrestrial planets. Pergamon Press, NewYork, 193~213.

Briqueu L., Bougault H., Joron J. L., 1984. Quantification of Nb, Ta, Ti and V anomalies in magmas associated with subduction zones: petrogenetic implications. Earth Planet. Sci. Lett., 68. 297~308.

Brown G. M., Holland J. G., Sigurdsson H., Tomblin J. F., Acrulus R. J., 1977. Geochemistry of the Lesser Antilles volcanic island arc. Geochim. Cosmochim. Acta, 41: 785~801.

Carr M. J., Feifenson M. D., Bennett E. A., 1990. Incompatible element and isotopic evidence for tectonic control of source mixing and melt extraction along the Central American arc. Contrib Mineral Petrol., 105: 369~380.

Chen C. Y. and Frey F. A., 1985. Trace element and isotopic geochemistry of lavas from Haleakala volcano, east Maul, Hawaii: implications for the origin of Hawaiian basalts. Jour. Geophy. Res., 90, 8743~8768.

Clague D. A. and Frey F. A. 1982. Petrology and trace element geochemistry of the Honolulu volcanics, Oahu: implications for the oceanic mantle below Hawaii. Jour. Petrol., 23, 447~504.

Coleman R. G., 1977. Ophiolites. Springer-Verlag, New York.

Davidson J. P. 1983. Lesser Antilles isotopic evidence of the role of subducted sediment in island arc magma genesis. Nature, 306: 253~256.

Defant M. J., Clark L. F., Stewart R. H., Drummond M. S., de Boer J. Z., Maury R. C., Bellon H., Jackson T. E., Restrepo J. F., 1991. Andesite and dacite genesis via contrasting processes: the geology and geochemistry of Ei Valle volcano, Panama. Contrib. Mineral Petrol., 106: 309~324.

Ewart A., Bryan B., Gill J. B., 1973. Mineralogy and Geochemistry of the younger volcanic islands of Tonga. S. W. Pacific, Jour. Petrol., 14: 429~465.

Floyd P. A., Kelling G., Gokcen S. L., Gokcen N., 1991. Geochemistry and tectonic environment of basaltic rocks from the Misis ophiolitic melange, south Turkey. Chemical Geology, 89: 263~280.

Frey F. A., Clague D. A., 1983. Geochemistry of diverse basalt types from Loihi Seamount, Hawaii: petrogenetic implications. Earth Planet. Sci. Lett., 66: 337~355.

Green T. H., Pearson N. J., 1987. An experimental study of Nb and Ta portioning between Ti-rich Minerals and silicate liquids at high pressure and temperature. Geochim. Cosmochim. Acta, 51: 55~64.

Hart S. R., 1971. K, Rb, Cs, Sr and Ba contents and Sr isotope ratios of ocean floor basalts. Phil. Trans. Roy. Soc. Lond., A, 268: 573~587.

Henderson P. 1982. Inorganic Geochemistry. Pergamon Press, Oxford.

Hickey R. L., Frey F. A., Gerlach D. C., 1986. Multiple sources for basaltic arc rocks from the southern volcanic zone of the Andes (34o~41 oS): trace element and isotopic evidence for continental crust. Jour. Geophy., Res., 91: 5963~5983.

Hofmann A. W., 1988. Chemical differentiation of the relationship between mantle, continental crust, and oceanic crust. Earth Planet. Sci. Lett., 90: 297~314.

Jaques A. L., Chappell B.W., 1980. Petrology and trace element geochemistry of the Papuan Ultramafic Belt. Contrib. Mineral Petrol., 75. 55~70.

Jahn B. M., 1986. Mid-ocean ridge or marginal basin origin of the Eastern Taiwan ophiolite: chemical and isotopic evidence. Contrib. Mineral Petrol., 92: 194~206.

Langmuir C. H., Bender J. F., Bence A. E., Hanson G. H., Taylor S. R., 1977. Petrogenesis of basalts from the Famous area: mid-Atlantic ridge. Earth Planet. Sci. Lett., 36: 133~156.

Langmuir C. H., Vocke R. D., Hanson G. N., 1978. A general mixing equation with applications to icelandic basalts. Earth Planet. Sci. Lett., 37: 380~392.

Lugovic B., Altherr R., Raczek I., Hofmann A. W., Majer V., 1992. Geochemistry of peridotites and mafic igneous rocks from the Central Dinaric Belt, Yugoslavia. Contrib. Mineral Petrol., 106: 201~216.

Meschede M., 1986. A method of discriminate between different types of MORB and continental tholeiites with the Nb-Zr-Y diagram. Chem. Geol., 56: 207~218.

Moores E. M., 1982. Origin and emplacement of ophiolites. Rev. Geophys. Space Phys., 20: 735~759.

Morris J. D., Hart. S. R., 1983. Isotopic and incompatible element constraints on the genesis of island arc volcanics from Cold Bay and Amak island, Aleutians, and implications for mantle structure. Geochim. Cosmochim. Acta, 47: 2015~2030.

Noiret G., Montigny R., Allegre C. J., 1981. Is the Vourinos complex an island arc ophiolite? Earth Planet. Sci. Lett., 56: 375~386.

Nye C. J., Reid M. R., 1986. Geochemistry of primary and least fractionated lavas from Okmok volcano, central Aleutians: implications for arc magma genesis. Jour. Geophy. Res., 91: 10271~10287.

Palacz Z. A., Saunders A. D., 1986. Coupled trace element and isotope enrichment in the Cook-Austral-Samos islands, southwest Pacific. Earth Planet. Sci. Lett., 79; 270~280.

Pearce J. A., Cann J. R., 1971. Ophiolite origin investigated by discriminant analysis using Ti, Zr and Y. Earth Planet. Sci. Lett., 12: 339~349.

Pearce J. A., 1975. Basalt geochemistry used to investigate past tectonic environments on Cyprus. Tectonophysics, 25: 41~67.

Peretti A., Koppel V., 1986. Geochemical and lead isotope evidence for a mid-ocean ridge type mineralization within a poly-metamorphic ophiolite complex (Monte del Forno, North Italy/Switzerland). Earth Planet. Sci. Lett., 80: 252~264.

Perfit M. R., Cust D. A., Bence A. E., Arculus A. R., Taylor S. R., 1980. Chemical characteristics of island-arc basalts: implications for mantle sources. Chem. Geol., 30: 227-256.

Price R. C., Kennedy A. K., Riggs-Sneeringer M. R., Frey F. A., 1986. Geochemistry of basalts from the Indian ocean triple junction: implications for the generation and evolution of Indian ocean ridge basalts. Earth Planet. Sci. Lett., 78: 379~396.

Rhodes J. M., Wenz K. P., Neal C. A., Sparks J. W., Lockwood J. P., 1989. Geochemical evidence for invasion of Kilauea's plumbing system by Mauna Los magma. Nature, 337: 257~260.

Roden M. F., Fery F. A., Clague D. A., 1984. Geochemistry of tholeiitic and alkali lavas from the Koolau range, Oahu, Hawaii: implications for Hawaiian volcanism. Earth Planet. Sci. Lett., 69: 141~158.

Ryerson F. J., Watson E. B., 1987. Rutile saturation in magmas: implications for Ti-Nb-Ta depletion in island-arc basalts. Earth Planet. Sci. Lett., 86: 225~239.

Saunders A. D., Tarney J., 1979. The geochemistry of basalts from a back-arc spreading centre in the East Scotia Sea. Geochim. Cosmochim. Acta, 43: 555~572.

Stolz A. J., Varne R., Daries G. R., Wheller G. E., Foden J. D., 1990. Magma source components in arc-continent collision zone: the Plores-Lembata sector, Sunda arc, Indonesia. Contrib. Mineral Petorl., 105: 585-601.

Sun S. S., Nesbitt R. W., Sharaskin A. Y., 1979. Geochemical characteristics of mid-ocean ridge basalts. Earth Planet. Sci. Lett., 44: 119~138.

Sun S. S., McDonough W. F., 1989. Chemical and isotopic systematics of oceanic basalts: implications for mantle composition and processes, in "Magmatism in the ocean Basin" edited by Saunders A. D., Norry M. J., Geol. Soc. Special Publ., (42): 313~345.

Tatsumi Y., Murasaki M., Asradi E. M., Nohda S., 1991. Geochemistry of Quaternary lavas from NE Sulawesi: transfer of subduction components into the mantle wedge. Contrib. Mineral Petrol., 107: 137~149.

Tatsumoto M., 1978. Isotopic composition of lead in oceanic basalt and its implication to mantle evolution. Earth Planet. Sci. Lett., 38: 63~87.

Taylor S. R., Mclennan S. M., 1985. The continental crust: its composition and evolution, Blackwell, Oxford.

Vallier T. L., Jenner G. A., Frey F. A., Gill J. B., Davis A. S., Volpe A. M., Hawkins J. W., Morris J. D., Cawood P. A., Morton J. L., Scholl D. W., Rautenschlein M., White W. M., Williams R. W., Stevenson A. J., White L. D., 1991. Sub-alkaline andesite from Valu Fa Ridge, a back-arc spreading center in southern Lau Basin: petrogenesis, comparative chemistry, and tectonic implications. Chemical Geology, 91: 227~256.

Volpe A. M., Macdougall J. D., Hawkin J. W., 1987. Mariana Trough basalts (MTB): trace element and Sr-Nd isotopic evidence for mixing between MORB-like and arc-like melts. Earth Planet. Sci. Lett., 82: 241~254.

White W. M., Patchett J., 1984. Hf-Nd-Sr isotopes and incompatible element abundances in island arcs: implications for magma origins and crust-mantle evolution. Earth Planet. Sci. Lett., 67: 167~185.

Wood D. A., 1978. Major and trace element variations in the Tertiary lavas of eastern Iceland and their significance with respect to the Iceland geochemical anomaly. Jour. Petrol., 19: 393~436.

Wood D. A., Joron J. L., Treuil M., 1979. A re-appraisal of the use of trace elements to classify and discriminate between magma series erupted in different tectonic settings. Earth Planet. Sci. Lett., 45: 326~336.

– # ε_{Nd}-La/Nb、Ba/Nb、Nb/Th 图对地幔不均一性研究的意义
——岛弧火山岩分类及 EMII 端元的分解*

李曙光

中国科学技术大学，合肥 230026

亮点介绍：提出了一组 ε_{Nd}-La/Nb、ε_{Nd}-Ba/Nb 和 ε_{Nd}-Nb/Th 图，使岛弧火山岩源区的三个可能端元(MORB 型亏损地幔、俯冲洋壳析出流体、陆源沉积物)呈"三角形分布"显示在一张图上，并将岛弧火山岩按源区特征分为"三端元型"和"二端元型(即陆源沉积物组分很少)"两类。EMII 端元在这些图上也可以被分解成两种不同成因的富集地幔库。这是将同位素和痕量元素示踪结合起来，研究地幔不均一性及其形成机制的一个成功范例。

摘要 本文提出一组 ε_{Nd}-La/Nb、ε_{Nd}-Ba/Nb 和 ε_{Nd}-Nb/Th 图。在这些图中岛弧火山岩呈三角形分布，从而将其源区的三个主要端元(MORB 型亏损地幔、俯冲洋壳析出流体、陆源沉积物)同时显示在一张图上。据此，岛弧火山岩可根据其源区特征划分成两类：二端元型(无陆源沉积物)和三端元型。此外，EMII 型幔源岩石在这些图中也被分成两类：与岛弧火山岩类似的，具有高 La/Nb、Ba/Nb 和低 Nb/Th 的大陆溢流玄武岩和某些橄榄岩包体；具有低 La/Nb、Ba/Nb 和高 Nb/Th 的萨摩亚型海岛玄武岩。它们对应的 EMII 富集地幔端元也可分成两种：EMIIM，由俯冲洋壳析出流体交代楔形地幔而成；EMIISR，与陆源沉积物再循环进入对流上地幔有关。

关键词 钕同位素；痕量元素；岛弧火山岩；地幔不均一性；地幔交代作用

1 引言

对上地幔组成不均一性的认识主要是通过玄武岩及其他幔源岩石的同位素地球化学研究获得的。尽管人们也进行了大量的玄武岩痕量元素地球化学研究，但主要用于示踪岩浆的形成演化过程。近年来，已有一些文章利用不相容元素比值探讨地幔的不均一性及形成过程(Hofmann et al., 1986; Weaver et al., 1986; Loubet et al., 1988; Sun et al., 1989)，但这方面的研究还是初步的，尤其将同位素与痕量元素示踪结合起来研究地幔不均一性更显得不足。实际上在地幔不均一性形成的各个过程中，地幔交代作用是很重要的因素。在地幔交代流体的形成及交代过程中痕量元素可以发生明显的分异。因此，痕量元素地球化学在研究地幔不均一性及其形成机制中应发挥更重要的作用。

将同位素和痕量元素结合起来进行联合示踪是本文的主导思想。它可以将同位素和痕量元素揭示出来的特征和规律合并到一张图上去观察，从而获得更好的研究效果。

例如，同位素研究已经证明亏损的楔形地幔和俯冲洋壳携带的陆源沉积物是一些岛弧玄武岩源区的重要组成端元(Whitford et al., 1981; Davidson, 1983; White et al., 1984)。岛弧玄武岩的痕量元素地球化学研究表明俯冲洋壳析出流体是其源区的另一重要组成端元。它交代、混染上覆楔形地幔可解释岛弧玄武岩的低 Nb、Zr、Ti 和高 K、Rb 含量(Hickey et al., 1986; Ryerson et al., 1987; 李曙光, 1989)。事实上在不同岛弧玄武岩的源区中，上述三种端元组分所占比重可能是很不相同的。探讨形成这种差异的机制对了解俯冲带地球化学过程有重要意义。然而至今尚没有报道任何一个地球化学图可将上述三个可能的端元同时显示出来，以便于比较它们的作用。

---

*本文发表在：地球化学，1994，23(2)：105-114
对应的英文版论文为：Li Shuguang, 1995. Implications of ε_{Nd}-La/Nb, Ba/Nb, Nb/Th diagrams to mantle heterogeneity—Classification of island-arc basalts and decomposition of EMII component. Chinese Journal of Geochemistry, 14(2): 117-127

再如, 在 Sr-Nd 同位素图上, 以萨摩亚为代表的现代海岛玄武岩, 大陆溢流玄武岩以及一些来自金伯利岩和碱性玄武岩的橄榄岩包体均表现第二类富集地幔(EMII)的特征(Zindler et al., 1986; Wörner et al., 1986)。但是海岛玄武岩与大陆溢流玄武岩的痕量元素组成有明显差异, 前者的 Nb、Ti 含量高, 后者的 Nb、Ti 含量低(Hawkesworth et al., 1984; Sun et al., 1989)。痕量元素的这种差异是否反映了它们地幔源区性质的不同? 即所谓 EMII 端元是否不是一个单一的地幔富集端元, 而是几个具有不同成因和痕量元素组成, 但表现有相类似同位素特征的端元? 对于这一问题, 以前的研究并没有给以明确的讨论和答案。

由于上述问题都与铌在地幔过程中的行为有关, 为此本文提出钕同位素和几个与铌分配系数相近的不相容元素比值, 如 La/Nb、Ba/Nb 和 Nb/Th, 结合起来作图, 以观察它们联合示踪的效果。

2 两类岛弧火山岩及其源区特征

在本文提出的 ε_{Nd}-Ba/Nb、ε_{Nd}-La/Nb 和 ε_{Nd}-Nb/Th 图中, ε_{Nd} 值可以有效示踪岛弧火山岩的亏损地幔和陆源沉积物端元。但是由于洋壳从生成到消减的时间间隔不长, 故俯冲洋壳的钕同位素组成与亏损地幔没有大的差异, 致使 ε_{Nd} 值在示踪俯冲洋壳析出流体端元上无效。然而 Ba/Nb、La/Nb 和 Nb/Th 在俯冲洋壳因脱水及部分熔融产生流体时可以发生明显的分异(下面将详细讨论), 它们在海水蚀变和变质过程中有较好的稳定性, 并且是具有相似分配系数的高不相容元素, 其比值在部分熔融和分离结晶过程中保持不变, 可用来直接指示源区的痕量元素特征(李曙光, 1989)。因此, 这些痕量元素比值可以较好示踪俯冲洋壳析出流体端元。

图 1 总结了具有钕同位素和 Ba、Th、Nb、La 数据的岛弧火山岩资料。其中除安第斯和桑达外, 均发育在洋壳之上(White et al., 1984), 不存在陆壳混染问题。此外, 已有工作表明安第斯岛弧火山岩, 除 34°~38°S 以外, 也未受陆壳混染影响(Hickey et al., 1986)。图 1 选用的是 34°~38°S 以外的样品。图 1 显示的一个最鲜明特征是这些岛弧火山岩数据在这三个图中均呈三角形分布。该三角形的三个端元地球化学特征为: (1)高 ε_{Nd} (+7~+8.5), 高 Nb/Th、低 Ba/Nb 和 La/Nb 值; (2)低 ε_{Nd} (-3~-10), 中等 Nb/Th、Ba/Nb 和 La/Nb 值; (3)高 ε_{Nd} (+7~+8.5), 低 Nb/Th、高 La/Nb 和 Ba/Nb 值。端元(1)的特征与 MORB 相同(表 1), 它代表了亏损上地幔。端元(2)的特征与上地壳相对应(表 1), 应代表陆源沉积物, 因为本文所选用数据(除桑达外)已排除陆壳混染的可能性。端元(3)的钕同位素特征与 MORB 相同, 但它的铌含量异常低, 从而具有比陆壳还低的 Nb/Th 值和高的 La/Nb 及 Ba/Nb 值。大量研究表明, 俯冲洋壳俯冲到深部时将脱水并形成榴辉岩。在榴辉岩中金红石是一普遍存在的矿物。它在水溶液中溶解度极低。即使俯冲洋壳发生部分熔融的情况下, 熔融比例 $\leqslant 0.4$ 时, 残留相中也还有相当数量的金红石存在(Ryerson et al., 1987)。由于铌在金红石和熔体之间的分配系数 $D = 30$(Green et al., 1987), 它可以容纳大量的铌并使与之平衡的熔体贫铌。实验表明与之平衡的流体也贫铌。因此, 上述端元(3)应代表俯冲洋壳析出的流体(表 1), 该流体以及俯冲洋壳携带的陆源沉积物部分熔融产生的熔体交代, 混染上覆楔形地幔, 从而使岛弧火山岩源区由上述三个端元组成。

表 1 岛弧火山岩源区三个端元组分的钕同位素和痕量元素特征

端元名称	ε_{Nd}	Ba/Nb	La/Nb	Nb/Th
MORB 型亏损地幔	+6.5~+13	0.7~15	0.4~2.1	12~39
陆源沉积物	-8.5~-13.4	22	1.2	2.3
俯冲洋壳析出流体	+6~+8.5	350	20	0.3

数据来源: MORB 型亏损地幔(据 Zindler et al., 1986; Ryerson et al., 1987; Langmuir et al., 1977; Sun et al., 1979; Price et al., 1986; Hofmann, 1988; 李曙光[1]) 陆源沉积物(据 Ben Othman et al., 1989; O'Nions et al., 1978; Taylor et al., 1985) 俯冲洋壳析出流体(据岛弧火山岩最高 ε_{Nd}、Ba/Nb、La/Nb 值和最低 Nb/Th 值估计; White et al., 1984; Nye et alv, 1986)。

图 1 还清楚地显示了岛弧火山岩可划分为两类: (1) 源区主要由 MORB 型亏损上地幔和俯冲洋壳析出

流体组成，陆源沉积物组分很少，如阿留申和新不列颠。前人的同位素研究也指出没有任何证据表明有沉积物进入阿留申和新不列颠的岛弧火山岩中(Basaltic Volcanism Study Project, 1981；Morris et al., 1983)；(2) 源区由全部三个端元组成，如安第斯、桑达、小安德列斯。笔者建议称第(1)类为二端元型，第(2)类为三端元型。这一分类突出反映了它们源区的差异。

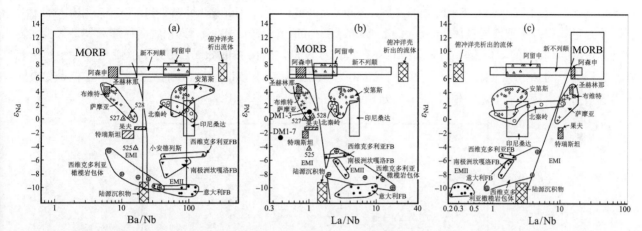

图1 岛弧火山岩 DUPAL 和萨摩亚海岛玄武岩及有代表性的大陆溢流玄武岩和橄榄岩包体(西维克多利亚、汉诺坝)的
ε_{Nd} -Ba/Nb、ε_{Nd} -La/Nb、ε_{Nd} -Nb/Th 图

(MORB. 大洋中脊玄武岩。EMI、EMII. Zindler et al.(1986)定义的地幔端元成分。) 数据来源：1. 岛弧火山岩(空心符号) (White et al., 1984；Davidson, 1983；Hickey et al., 1986；Basaltic Volcanism Study Project, 1981；Nye et al., 1986；Brown et al., 1977；Hawksworth et al., 1979)；2. DUPAL 海岛玄武岩(带斜线符号) (Weaver et al., 1986；White et al., 1982；O'Nions et al., 1977；Richardson et al., 1982；Humphris et al., 1983)，萨摩亚海岛玄武岩(Palacz et al., 1986)，橄榄岩包体(Griffin et al., 1988；Song et al., 1989)；3. 大陆溢流玄武岩(实心符号) (Rogers et al., 1985；Hergt et al., 1991)

前人也曾对岛弧火山岩进行了各种各样的分类。如 Nohda 等(1981)将岛弧划分成"大洋岛弧"和"大陆岛弧"两类。DePaolo(1988)将所有具有低 ε_{Nd} 值的岛弧火山岩都归结为是"大陆岛弧"陆壳混染的结果。用这种分类对同位素进行解释是有争议的。如 DePaolo(1988)将小安德列斯列入"大陆岛弧"并认为它的低 ε_{Nd} 值是陆壳混染的结果；但 White 等(1984)认为它是发育在洋壳之上的，Davidson(1983)的研究也证明陆壳混染不能解释其 Sr、Pb 同位素的相关趋势，它们的低 ε_{Nd} 特征是由俯冲板块携带的陆源沉积物引起的。White 等(1984)将岛弧火山岩分为同位素一致组和同位素不一致组。但这一分类只是描述性的，未指明其成因。

应当指出的是利用图1判断岛弧火山岩源区类型时，必须首先证明该岛弧火山岩未受到陆壳混染。否则即使源区是二端元型，因其岩浆上升时受到陆壳混染，该玄武岩数据会投在图1的三端元区。

图1还可以用于判断显生宙岛弧火山岩的源区特征及火山岩类型。例如，与北秦岭丹凤群岛弧火山岩有关的拉垃庙苏长-辉长岩形成于 402Ma(李曙光等,1989)，它的初始 ε_{Nd} 及有关痕量元素比值落入图内三端元区内(李曙光, 1989)。如果这一套岩石属蛇绿岩，它遭受陆壳混染的可能性不大，其源区可能为三端元型。这说明在古生代，古秦岭海板块向北秦岭古岛弧下俯冲时携带有较多的陆源沉积物进入地幔。

对于前寒武古老的岛弧火山岩系，因当时地幔和陆壳的 ε_{Nd} 值与现代不同，故对图1中各端元的 ε_{Nd} 变化范围要做相应修正(表2)。

表2 不同时代亏损地幔及大陆地壳 ε_{Nd} 值*

年龄(Ga)	亏损地幔 ε_{Nd}	大陆地壳 ε_{Nd}
0	+6.5~+13	−10~−19
0.5	+5.8~+11.6	−15~−24
1.0	+5.1~+10.1	−13~−22
2.0	+3.6~+7.3	−8~−17
2.5	+2.9~+5.3	−7~−14
3.5	+1.5~+3.0	−3~−8

*亏损地幔现代值据 MORB 统计(Zindler et al., 1986)，历史值按地幔单阶段演化公式 $\varepsilon_{Nd} = Q_{Nd}f_{Sm/Nd}T$ 计算(DePaolo, 1988)，大陆地壳 ε_{Nd} 值据 Jacobsen, 1988 估计。

3 地幔 EMII 富集端元的分解

文献中用来指示 EMII 端元的岩石有以下三种：(1) 以萨摩亚为代表的具有高 $^{87}Sr/^{86}Sr$ 值的海岛玄武岩(White et al., 1982; White, 1985; Zindler et al., 1986); (2) 富集的大陆溢流玄武岩，如意大利、南卡罗、西维克多利亚和南极洲等(Hawkesworth et al., 1984; Zindler et al., 1986; Hergt et al., 1991); (3) 交代成因的金伯利岩幔源包体和二辉橄榄岩包体(Fraser et al., 1985/1986; Griffin et al., 1988)。人们也同样注意到岛弧玄武岩也具有与 EMII 类似的同位素特征(White, 1985; Zindler et al., 1986; Sun et al., 1989)。图2显示了上述所有岩石具有类似的 Sr、Nd 同位素特征，并可用 EMII 型富集地幔或 EMII 与 DMM 的混合来解释。但是如果同时考虑痕量元素特征，上述四种岩石的一致性就不存在了。

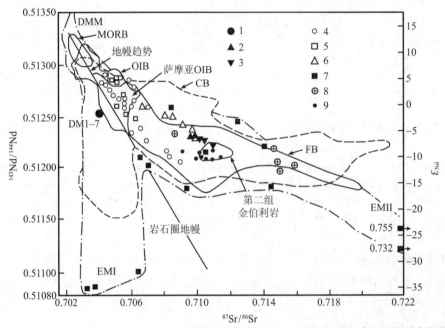

图2 海洋玄武岩(MORB+OIB)、大陆玄武岩(CB)、大陆溢流玄武岩(PB)、三端元岛弧玄武岩及各种交代成因的二辉橄榄岩包体的 Nd-Sr 同位素图

DMM、MORB、OIB、CB、FB 分布范围及金伯利岩数据据 Zindler et al., 1986。其他数据据 Davidson, 1983; Whitford et al., 1981; White et al., 1984; Griffin et al., 1988; Song et al., 1989; Rogers et al., 1985。1. 意大利 FB; 2. 西维克多利亚 FB; 3. 南极洲坎嘎洛 FB; 4. 小安德列斯; 5. 桑达; 6. 邦达; 7. 南非金伯利岩中石榴子石; 8. 澳大利亚橄榄岩包体中的辉石岩脉; 9. 汉诺坝橄榄岩包体

图1表明上地幔的两个富集端元 EMI 和 EMII 同样可以在 ε_{Nd}-La/Nb、ε_{Nd}-Nb/Th 和 ε_{Nd}-Ba/Nb 图上分开。它们基本上以 MORB 和陆源沉积物的连线为界，EMI 的代表性玄武岩，如果夫、特瑞斯坦和沃尔维斯三个样品(528、527、525)，以及汉诺坝交代二辉橄榄岩包体(DM1-3、DM1-7)均落在低 La/Nb、Ba/Nb 和高 Nb/Th 一侧；而代表 EMII 端元的意大利、南极洲坎嘎洛岛、澳大利亚西维克多利亚大陆溢流玄武岩(FB)和二辉橄榄岩中的富集的辉石岩交代脉，以及岛弧玄武岩基本都落在高 La/Nb、Ba/Nb 和低 Nb/Th 一侧。应当指出，在 ε_{Nd}-Ba/Nb 图中，相当一部分澳大利亚西维克多利亚二辉橄榄岩包体中的辉石岩交代脉样品落入了 EMI 区。这可能是由于钡相对铌有较高的活动性，从而在地幔交代过程中有较多的钡从辉石岩脉扩散进入它的围岩(橄榄岩)中去。在图1的三个图中，萨摩亚均落在 EMI 一侧。这表明尽管萨摩亚海岛玄武岩源区的 Sr、Nd、Pb 同位素特征都与 EMII 端元类似，但它的痕量元素特征与 EMII 的其他代表性岩石不同。因此，原来统称为 EMII 的富集地幔端元实际是由两种性质不同的富集地幔端元组成的，并分布在上地幔不同层圈内。

岛弧玄武岩与大陆溢流玄武岩和部分交代的二辉橄榄岩包体之所以具有相似的 La/Nb、Ba/Nb 和 Nb/Th 值，是因为它们均具有异常低的铌含量(Thompson et al., 1984; Sun et al., 1989)。如前所述，这种低铌特征与俯冲带有关。俯冲洋壳析出流体交代上覆楔形地幔所形成的富集地幔随岛弧向大陆的增生也可增生到大

陆的岩石圈地幔中去,从而成为大陆溢流玄武岩的源区。Griffin 等(1988)在分析了澳大利亚西维克多利亚的地质演化历史后,指出该区的二辉橄榄岩包体是来自一古俯冲带上部的楔形地幔。由于大陆溢流玄武岩源区的交代作用发生时代较现代岛弧火山岩老,尽管它们有相似的 La/Nb、Ba/Nb 和 Nb/Th 值,但具有更高的 $^{87}Sr/^{86}Sr$ 和更低的 $^{143}Nd/^{144}Nd$ 值。建议将这种交代成因 EMII 端元称为 EMIIM(Enriched Mantle II related to metasomatism)。

与岛弧玄武岩及 EMIIM 端元相反,萨摩亚玄武岩含有较高的铌。它们的低 La/Nb、Ba/Nb 和高 Nb/Th 值表明它们的源区与俯冲洋壳析出流体产生的地幔交代作用无关。萨摩亚玄武岩在 ε_{Nd}-La/Nb、ε_{Nd}-Nb/Th 和 ε_{Nd}-Ba/Nb 图上均表现出介于 MORB 和陆源沉积物之间的混合趋势。它支持前人根据 Sr、Nd、Pb 同位素研究所获得的结论:萨摩亚玄武岩所代表的 EMII 端元与重新进入对流上地幔的陆源沉积物有关。笔者建议称此类 EMII 端元为 EMIISR(Enriched Mantle II related to Sediment Recycling)。它分布于岩石圈以下的对流上地幔中,而 EMIIM 则分布在岩石圈地幔中。

4 DUPAL 海岛玄武岩富集地幔成因

位于 Sr-Nd 同位素图地幔趋势下面的海岛玄武岩,如阿森申、圣赫林那、布维特、果夫、特瑞斯坦、沃尔维斯脊,多集中分布在南大西洋区,从而形成海洋地幔的高 $^{87}Sr/^{86}Sr$ 异常区(DUPAL)(Hart, 1984)。这些海岛玄武岩被统称为 DUPAL OIB(Sun et al., 1989)。这些 DUPAL OIB 均表现为低 La/Nb、Ba/Nb,高 Nb/Th 特征(图1)。这说明它们不可能与俯冲带交代流体产生的地幔交代作用有关。Weaver 等(1986)根据 DUPAL OIB 在 La/Nb-Ba/Nb 图上的相关性,认为它们可以由圣赫林那代表的 OIB 源与深海沉积物混合而成。但是深海沉积物的 Sr、Nd 同位素组成位于地幔趋势右侧(Hickey et al., 1986),因此,Weaver 的沉积物混染成因说难于解释它们的 Sr、Nd 同位素特征。大量的二辉橄榄岩、第一组金伯利岩和部分煌斑岩的 Sr、Nd 同位素研究表明,交代成因的岩石圈 EMI 是确实存在的(Zindler et al., 1986)。因此,Hart 等(1986)认为 DUPAL OIB 可能是由于析离作用使交代的 EMI 岩石圈地幔物质进入对流上地幔引起的。在 ε_{Nd}-La/Nb 图中,DUPAL OIB 及河北汉诺坝交代成因的橄榄岩包体(DM1-3, DM1-7)均具有相似的低 La/Nb 值。由于 EMI 的 La/Nb、Ba/Nb、Nb/Th 均与 MORB 类似(图1),交代形成 EMI 的流体最可能来自软流圈。显然该流体较俯冲板块析出流体有较低的 $^{87}Sr/^{86}Sr$ 和较高的 $^{143}Nd/^{144}Nd$ 值。由于该交代流体自身同位素组成与亏损地幔类似,故该类型交代地幔要演化成具有较低 ε_{Nd} 值的 EMI 端元需要较长时间。它们一般应分布在古老陆壳下的古老交代岩石圈地幔中。如具有典型 EMI 特征的美国蒙大拿州煌斑岩、怀俄明州超富钾火山岩、中国汉诺坝超镁铁岩包体、南非金伯利岩均发育在古老陆块区(Fraser et al., 1985/1986;Song et al., 1989)。相反,由于俯冲洋壳析出流体本身可以具有较低的 ε_{Nd} 和较高的 $^{87}Sr/^{86}Sr$ 值,EMII 的交代形成时间可以是较年轻的。例如澳大利亚西维克多利亚的二辉橄榄岩包体是在古生代时,该区为岛弧环境时被交代富集的(Griffin et al, 1988)。

5 结论

ε_{Nd}-La/Nb、Ba/Nb、Nb/Th 图在示踪玄武岩地幔源区特征的应用中有独特的作用。

(1) 它可以使岛弧火山岩源区的三个可能端元显示在一张图上,并将岛弧火山岩按源区特征分成两种类型:二端元型和三端元型。

(2) EMII 端元在这些图上可以被分解成两种不同成因的富集地幔库:(a) EMIIM,它是俯冲板块析出流体交代楔形地幔形成的,并可随陆壳增生过程合并到大陆岩石圈地幔中去。它是大陆溢流玄武岩、某些煌斑岩、金伯利岩的源区;(b) EMIISR,它与陆源沉积物再循环进入对流上地幔有关,是萨摩亚型海岛玄武岩的源区。

(3) DUPAL OIB 与俯冲过程无关。它们的 EMI 端元具有与 MORB 类似的 La/Nb、Ba/Nb、Nb/Th 值,可能是析离的交代岩石圈地幔,其交代流体应直接来自亏损的软流圈上地幔。

参考文献

李曙光,Hart S R,郑双根等,1989.中国华北、华南陆块碰撞时代的钐–钕同位素年龄证据.中国科学（B 辑）,(3): 312–319

Basaltic Volcanism Study Project, 1981. Basaltic volcanism on the terrestrial planets. New York: Pergamon Press, 193–213

Ben Othman D, White W M, Patchett J, 1989. The geochemistry of marine sediments, island arc magma genesis, and crust-mantle recycling. Earth Planet Sci Lett, 94: 1–21

Brown G M, Holland J G, Sigurdsson H, Tomblin J F, Arculus R J, 1977. Geochemistry of the Lesser Antilles volcanic island arc. Geochim Cosmochim Acta, 41: 785–801

Cohen R S, O'Nions P K, 1982. The lead, neodymium and strontium isotopic structure of ocean ridge basalts. J Petrol, 23: 299–324

Davidson J P, 1983. Lesser Antilles isotopic evidence of the role of subducted sediment in island arc magma genesis. Nature, 306: 253–256

DePaolo D J, 1988. Neodymium Isotope Geochemistry. Berlin Heidelberg, New York, London, Paris, Tokyo: Springer-Verlag

Fraser K J, Hawkesworth C J, Erlank A J, Mitchell R H, Scott-Smith B H, 1985/1986. Sr, Nd and Pb isotope and minor element geochemistry of lamproites and kimberlites. Earth Planet Sci Lett, 76: 57–70

Griffin W L, O'Relly S Y, Stabel A, 1988. Mantle metasomatism beneath western Victoria, Australia: II. Isotopic geochemistry of Cr-diopside lherzolites and Al-augite pyroxenites. Geochim Cosmochim Acta, 52: 449–459

Green T H, Pearson N J, 1987. An experimental study of Nb and Ta partitioning between Ti-rich minerals and silicate liquids at high pressure and temperature. Geochim Cosmochim Acta, 51: 55–62

Hart S R, 1984. A large-scale isotope anomaly in the Southern Hemisphere mantle. Nature, 309: 753–757

Hawkesworth C J, O'Nions R K, Arculus R J, 1979. Nd and Sr isotope geochemistry of island arc volcanics, Grenada, Lesser Antilles. Earth Planet Sci Lett, 45: 237–248

Hawkesworth C J, Rogers N W, van Calsteren P W C, Menzies M A, 1984. Mantle enrichment processes. Nature, 317: 331–335

Hergt J M, Peate D W, Hawkesworth C J, 1991. Petrogenesis of mesozoic gondwana low-Ti flood basalts. Earth Planet Sci Lett, 105: 134–148

Hickey R L, Frey F A, Gerlach D C, 1986. Multiple sources for basaltic arc rocks from the Southern volcanic zone of the Andes, (34°–41°S): Trace element and isotopic evidence for contributions from subducted oceanic crust, mantle, and continental crust. Jour Geophys Res, 91: 5963–5983

Hofmann A W, Jochum K R, Seufert M, White W M, 1986. Nb and Pb in oceanic basalts: new constraints on mantle evolution. Earth Planet Sci Lett, 79: 33–45

Hofmann A W, 1988. Chemical differentiation of the relationship between mantle, continental crust, and oceanic crust. Earth Planet Sci Lett, 90: 297–314

Humphris S E, Thompson G, 1983. Geochemistry of rare earth elements in basalts from the Walvis Ridge: implications for its origin and evolution. Earth Planet Sci Lett, 66: 223–242

Jacobsen S B, 1988. Isotopic and chemical constraints on mantle-crust evolution. Geochim Cosmochim Acta, 52: 1341–1350

Langmuir C H, Bender J F, Bence A E, Hanson G H, Taylor S R, 1977. Petrogenesis of basalts from the famous area: Mid-Atlantic Ridge. Earth Planet Sci Lett, 36: 133–156

Loubet M, Sassi R, Donato G D, 1988. Mantle heterogeneities: a combined isotope and trace element approach and evidence for recycled continental crust materials in some OIB sources. Earth Planet Sci Lett, 89: 299–315

Morris J D, Hart S R, 1983. Isotopic and incompatible element constraints on the genesis of island arc volcanics from Cold Bay and Amak Island, Aleutians, and implications for mantle structure. Geochim Cosmochim Acta, 47: 2015–2030

Nohda S, Wasserburg G J, 1981. Nd and Sr isotopic study of volcanic rocks from Japan. Earth Planet Sci Lett, 52: 264–276

Nye C J, Reid M R, 1986. Geochemistry of primary and least fractionated lavas from Okmok volcano, central Aleutians: implications for arc magmagenesis. Jour Geophys Res, 91: 10271–10287

O'Nions R K, Hamilton P J, Evensen N M, 1977. Variations in $^{143}Nd/^{144}Nd$ and $^{87}Sr/^{86}Sr$ ratios in oceanic basalts. Earth Planet Sci Lett, 34: 13–22

O'Nions R K, Carter S R, Cohen R S, Evensen N M, Hamilton P J, 1978. Pb, Nd and Sr isotopes in oceanic ferromanganese deposits and ocean floor basalts. Nature, 273: 435–438

Palacz Z A, Saunders A W, 1986. Coupled trace element and isotope enrichment in the Cook-Austral-Samoa islands, southwest Pacific. Earth Planet Sci Lett, 79: 270–280

Richardson S H, Erlank A J, Duncan A R, Reid D L, 1982. Correlated Nd, Sr and Pb isotope variation in Walvis Ridge basalts and implications for the evolution of their mantle source. Earth Planet Sci Lett, 59: 327–342

Rogers N W, Hawkesworth C J, Parker R J, Marsh J S, 1985. The geochemistry of potassic lavas from Vulsini, Central Italy and implications for mantle enrichment processes beneath the Roman region. Contrib Mineral Petrol, 90: 244–257

Ryerson F J, Watson F B, 1987. Rutile saturation in magmas: implications for Ti-Nb-Ta depletion in island-arc basalts. Earth Planet Sci Lett, 86: 225–239

Song Y, Frey F A, 1989. Geochemistry of peridotite xenoliths in basalt from Hannuoba, eastern China: implications for subcontinental mantle heterogeneity. Geochim Cosmochim Acta, 53: 97–113

Sun S, Nesbitt R W, Sharaskin A Y, 1979. Geochemical characteristics of mid-ocean ridge basalts. Earth Planet Sci Lett, 44: 119–138

Sun S S, McDonough W F, 1989. Chemical and isotopic systematics of oceanic basalts: implications for mantle composition and processes. in: A D Saunders and M J Norry (eds.). Magmatism in Ocean Basins. Am Geol Soc Spec Publ, 42: 313–345

Taylor S R, McLeonan S M, 1985. The Continental Crusts: Its Composition and Evolution, Blackwell, Oxford

Thompson R N, Morrison M A, Hendry G L, Parry S J, 1984. An assessment of the relative roles of crust and mantle in magma genesis: an elemental approach. Philos Trans R Soc London Ser A, 310: 549–590

Weaver B L, Wood D A, Tarney J, Joron J L, 1986. Role of subducted sediment in the genesis of ocean-island basalts: geochemical evidence from South Atlantic ocean islands. Geology, 14: 275–278

White W M, 1985. Sources of oceanic basalts: radiogenic isotopic evidence. Geology, 13: 115–118

White W M, Hofmann A W, 1982. Sr and Nd isotope geochemistry of oceanic basalts and mantle evolution. Nature, 296: 821–825

White W M, Patchett J, 1984. Hf-Nd-Sr isotopes and incompatible element abundances in island arcs: implications for magma origins and crust-mantle evolution. Earth Planet Sci Lett, 67: 167–185

Whitford D J, White W M, Jezek P A, 1981. Neodymium isotopic of quaternion island arc lavas from Indonesia. Geochim Cosmochim Acta, 45: 989–995

Wörner G, Zindler A, Staudigel H, Schmincke H U, 1986. Sr, Nd and Pb isotope geochemistry of Tertiary and Quaternary alkaline volcanics from West Germany. Earth Planet Sci Lett, 79: 107–109

Zindler A, Hart S R, 1986. Chemical Geodynamics. Ann Rev Earth Planet Sci, 14: 493–571

北秦岭黑河丹凤群岛弧火山岩建造的发现及其构造意义

孙卫东[1]，李曙光[1]，肖益林[1]，孙 勇[2]，张国伟[2]

1. 中国科学技术大学地球和空间科学系，合肥 230026
2. 西北大学地质系，西安 710069

亮点介绍：发现北秦岭黑河丹凤群为一套安山岩为主的岛弧火山岩，进一步证明北秦岭古生代岛弧是近陆岛弧。

摘要 陕西省周至县黑河丹凤群有大规模安山岩浆活动的证据，地球化学研究表明这些安山岩形成于典型的岛弧环境，如轻稀土富集；在痕量元素原始地幔标准化丰度模型图上，所有样品均具有明显的 Nb、Ti 和 P 负异常；在 Ba-Nb-Th-La 图解上，所有样品均落在岛弧区等。由此可以推断黑河丹凤群有一套安山岩为主的岛弧火山岩建造。这一发现为证明北秦岭古生代岛弧构造环境提供了重要的证据，为进一步揭示秦岭造山带的构造格架和演化历史提供信息。

关键词 岛弧安山岩；地球化学；丹凤群；北秦岭

1 引言

丹凤群曾被认为是北秦岭南缘的主要蛇绿岩带，在秦岭造山带构造演化中占有重要的地位，受到较为广泛的重视(孙勇等，1988；于在平等，1988，1991；张国伟等，1988，1991；许志琴等，1991；李曙光等，1993)。但是有关丹凤群的系统的地球化学工作并不多。李曙光等（1993）曾做过其中苏长辉长岩的痕量元素和 Sr-Nd 同位素研究；孙勇等（1988）也曾做过其中一些熔岩、辉绿岩墙等的主要元素、痕量元素研究，并指出它们具有岛弧火山岩特征。从已发表的数据来看，丹凤群主要是玄武质岩石，仅有少量为安山岩，而且被认为是安山岩岩脉。北秦岭作为华北陆块南缘的古生代岛弧，在这里除了发育有岛弧型蛇绿岩以外，还理应有过大规模安山岩浆活动。北秦岭古岛弧构造环境的最终判定，需要大规模安山岩浆活动的证据。最近我们在黑河丹凤群中发现了以安山岩为主的岛弧火山岩建造。它为北秦岭古岛弧提供了关键的岩石学证据，并表明丹凤群至少不都是蛇绿岩，其中有一部分是岛弧火山岩建造。

2 地质背景及样品

丹凤群沿商丹断裂北缘出露（图1），以前认为主要由超镁铁质、镁铁质岩石组成，属北秦岭南缘蛇绿岩带（孙勇，1988）。在蛇绿岩南缘（商丹断裂附近）分布的沉积岩主要为成熟度较低的长石、石英质杂砂岩，夹薄层凝灰岩和透镜状蛇纹石化橄榄大理岩，被认为是弧前楔形体(于在平等，1988，1991)。

黑河丹凤群蛇绿岩位于周至县黑河。实验所采用的 9 个变质火山岩样品采自周至县黑河周洋公路边虎豹河口，其中 QH-3 采自丹凤群变质火山岩中的深色岩脉（野外判断为辉绿岩脉），其他样品分别采自深色和浅色两种构造岩块中（岩性见表1）。岩石片理发育，普遍遭受到绿片岩相变质，局部达低角闪岩相，有绿泥石化现象。采样跨度约 30 m。样品中主要矿物有斜长石、角闪石、石英及少量绿帘石。苏长辉长岩、辉长岩采自柳叶河小汪涧。辉长岩有蚀变，中粒辉长结构主要由斜长石、角闪石和少量绿泥石、磁铁矿组成。苏长辉长岩呈暗紫色，矿物粒度较辉长岩粗，主要由斜长石、紫苏辉石、单斜辉石和磁铁矿组成。

---

* 本文发表在：大地构造与成矿学，1995，19(3)：227-236

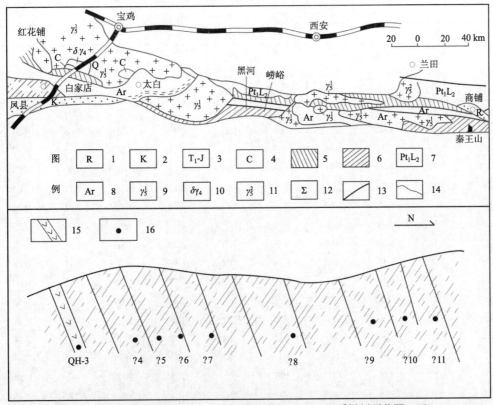

图 1 北秦岭地质略图（上）（据李曙光等，1993）及采样剖面草图（下）

1-新生界；2-白垩系；3-下三叠-侏罗系；4-石炭系；5-云架山蛇绿岩；6-丹凤蛇绿岩；7-宽坪群；8-秦岭群；9-印支期花岗岩；10-海西期花岗岩；11-加里东期花岗岩；12-松树沟超镁铁岩；13-断裂；14-地质界线；15-辉绿岩墙；16-采样点

表1 黑河丹凤群斜长角闪岩、辉长岩的主要元素（%）、痕量元素组成（$\times 10^{-6}$）

样品号	QH-3	QH-4	QH-5	QH-6	QH-7	QH-8	QH-9	QH-10	QH-11	QH-16	QH-17
岩性	深色斜长角闪岩	浅色斜长角闪岩	浅色斜长角闪岩	浅色斜长角闪岩	浅色斜长角闪岩	深色斜长角闪岩	深色斜长角闪岩	深色斜长角闪岩	深色斜长角闪岩	苏长辉长岩	辉长岩
SiO_2	50.73	65.89	63.04	60.02	62.24	58.26	59.39	58.85	51.07	47.83	47.5
TiO_2	0.85	0.42	0.53	0.48	0.53	0.59	0.67	0.88	1.23	0.88	0.21
Al_2O_3	11.75	14.75	14.90	13.7	14.63	15.7	15.1	13.25	11.72	18.77	18.72
Fe_2O_3	3.59	2.28	2.31	1.65	1.83	4.50	4.95	5.98	11.06	1.56	1.83
FeO	6.45	1.54	2.39	2.54	2.78	3.13	2.88	3.52	3.61	4.93	5.71
MgO	10.26	2.98	4.37	4.17	4.51	4.11	3.84	6.39	8.95	9.61	10.52
CaO	10.42	3.88	4.02	5.36	3.45	2.47	3.24	2.92	5.47	11.25	8.56
MnO	0.22	0.083	0.098	0.099	0.081	0.077	0.083	0.11	0.18	0.12	0.13
K_2O	0.8	0.67	0.91	0.9	1.39	4.97	3.85	0.91	1.31	0.18	0.27
Na_2O	2.18	5.81	5.08	4.86	4.36	3.81	2.55	3.74	2.44	2.59	3.16
P_2O_5	0.24	0.15	0.19	0.18	0.19	0.21	0.21	0.20	0.23	0.019	0.028
H_2O^+	1.72	1.22	1.93	2.52	2.73	2.37	2.67	2.98	2.77	1.98	3.46
CO_2	0.28	0.39	0.23	3.25	1.68	0.63	0.76	0.27	0.18	0.73	0.12
Total	99.5	100.07	100.01	99.74	100.41	100.04	100.2	100.01	100.21	99.84	100.23
La	16.67	30.78	30.47	27.3	29.87	24.85	18.79	21.49	27.73	3.16	2.76
Ce	37.6	55.09	57.28	53.02	55.69	54.63	35.44	42.63	55.7	6.748	5.425
Pr	4.73	6.04	6.34	5.93	6.51	5.67	4.58	5.75	7.31	0.9	0.72
Nd	18.56	19.62	21.85	19.78	21.74	20.74	16.96	22.18	29.02	3.92	2.906

续表

样品号	QH-3	QH-4	QH-5	QH-6	QH-7	QH-8	QH-9	QH-10	QH-11	QH-16	QH-17
岩性	深色斜长角闪岩	浅色斜长角闪岩	浅色斜长角闪岩	浅色斜长角闪岩	浅色斜长角闪岩	深色斜长角闪岩	深色斜长角闪岩	深色斜长角闪岩	深色斜长角闪岩	苏长辉长岩	辉长岩
Sm	4.23	3.55	4.03	3.72	4.08	4.01	3.67	4.76	6.01	1.05	0.66
Eu	1.2	0.9	1.02	1	1.08	1.21	0.99	1.35	1.66	0.54	0.55
Gd	4.36	2.84	3.25	3.03	3.37	3.61	3.51	4.21	5.46	1.12	0.62
Tb	0.65	0.37	0.45	0.42	0.48	0.53	0.54	0.64	0.8	0.2	0.11
Dy	3.8	2	2.39	2.18	2.51	3.07	3.11	3.59	4.57	1.05	0.52
Ho	0.76	0.4	0.48	0.44	0.5	0.63	0.66	0.72	0.91	0.22	0.12
Er	2.15	1.11	1.24	1.19	1.37	1.79	1.95	2.1	2.56	0.59	0.29
Tm	0.32	0.16	0.18	0.18	0.2	0.27	0.29	0.32	0.37	0.08	0.04
Yb	1.96	1.02	1.15	1.05	1.24	1.76	1.86	2	2.31	0.46	0.23
Lu	0.29	0.16	0.17	0.16	0.18	0.27	0.29	0.3	0.35	0.07	0.04
Y	19.56	10.7	12.29	11.23	13.11	15.71	16.62	18.09	23.24	5.13	2.48
Rb	26	22	28	40	61	122	110	31	43	2	6
Sr	444	721	696	590	572	204	221	781	386	651	740
Th	6	12	10	11	11	8	7	6	7	2	2
Cr	701	123	198	191	196	80	125	297	442	126	209
Ni	110	59	93	84	96	28	26	69	78	143	169
Ba	376	506	451	363	590	1260	1045	948	338	155	166
Zr	119	124	135	126	138	102	101	94	142	29	25
Nb	10	9	10	10	9	8	9	7	8	3	4
U	2	4	3	3	3	3	2	2	2	2	2

注：稀土元素由地矿部武汉岩矿测试中心测试；主量和痕量元素(稀土除外)由地矿部物化探研究所测试。

3 主要元素及痕量元素

在 $SiO_2\%\text{-}(K_2O+Na_2O)\%$ 变异图上（图2），变质火山岩（QH-3~QH-11）样品除采样剖面两端的 QH-3 和 QH-11 与辉长岩、苏长辉长岩一样落在玄武岩区内之外，其他样品均落在安山岩区内，其中 QH-8 落入粗安岩区内。QH-4 落入英安岩区内。这是黑河丹凤群大规模安山岩活动的重要证据。

不相容痕量元素的原始地幔标准化图（图3，原始地幔组成据 Sun S S, 1989）显示：所有变质火山岩样品都具有明显的 Nb、Ti 负异常。一般认为，Nb、Ti 负异常与俯冲作用有关，由此可以推断黑河丹凤群蛇绿岩具有与俯冲作用有关的成因特点。

变质火山岩的 REE 分配模型也与岛弧玄武岩（IAB）相似，呈现出 LREE 富集的特点，Eu

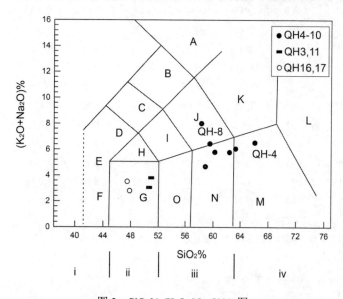

图2 $SiO_2\%\text{-}(K_2O+Na_2O)\%$ 图
i-超基性岩；ii-基性岩；iii-中性岩；iv-酸性岩。A-响岩；B-碱玄质响岩；C-响岩质碱玄岩；D-碧玄岩；E-碱玄岩；F-苦橄玄武岩；G-玄武岩；H-粗面玄武岩；I-玄武粗安岩；J-粗安岩；K-粗面岩；L-流纹岩；M-英安岩；N-安山岩；O-玄武安山岩

轻微负异常或无异常（图4）。随基性度的增大，LREE 的富集程度降低。在图4中表现为 REE 曲线彼此有不同程度的相交，显示它们可能不是简单的同源岩浆分离结晶的产物。

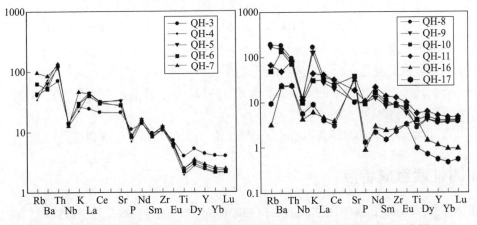

图 3 黑河丹凤群斜长角闪岩的中等和高不相容元素原始地幔标准化丰度图

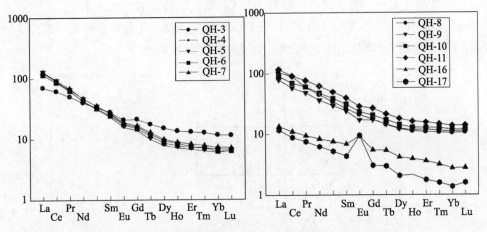

图 4 黑河丹凤群斜长角闪岩的 REE 球粒陨石标准化丰度图

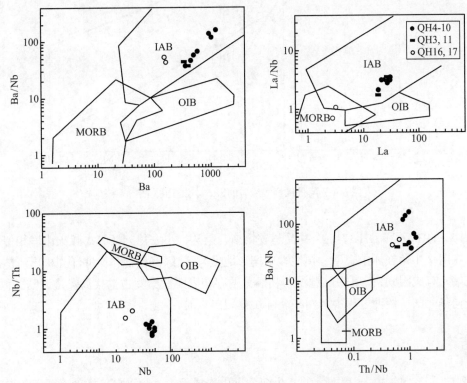

图 5 黑河丹凤群蛇绿岩的 Ba-Th-La-Nb 构造环境判别图解

辉长岩、苏长辉长岩（QH-16, 17）的 Nb、Ti 负异常不明显（图 3）。REE 含量较低，LREE 富集程度低，具有 Eu 正异常的特点（图 4）。苏长辉长岩 REE 含量高于辉长岩。Eu 正异常一般认为是在氧逸度（f_{O_2}）较低的情况下，Eu^{2+} 大量进入斜长石晶格中的结果。结合本区辉长岩、苏长辉长岩的 Fe^{3+}/Fe^{2+} 值较低的特点，作者认为其 Eu 正异常特点可能是斜长石分离结晶造成的。

在 La、Ba、Nb、Th 构造环境判别图上（据李曙光，1993），所有样品点都落在岛弧（IAB）区，也表明丹凤群变质火山岩形成于岛弧环境（图 5）。

综上所述，地球化学特点表明黑河丹凤群有一套安山岩为主的岛弧火山岩建造。

4　Sr、Nd 同位素组成特点

部分样品的同位素数据列于表 2。根据 QH-3 样的 Sm-Nd 单矿物等时线年龄（413 Ma，孙卫东等，1994）计算得到的 $\varepsilon_{Sr}(t)$ 和 $\varepsilon_{Nd}(t)$ 值同时列出。

由 $\varepsilon_{Sr}(t)$-$\varepsilon_{Nd}(t)$ 相关图（图 6），变玄武岩样品 QH-3 落在第一象限 IAB 区域的右边，显示了类似岛弧蛇绿岩的特点。辉长岩、苏长辉长岩投影在 IAB 区，没有海水蚀变影响的迹象。它们可能是深成侵入体中的早期堆晶岩，因而没有受到侵入体同化其上部受海水蚀变的熔岩的影响。

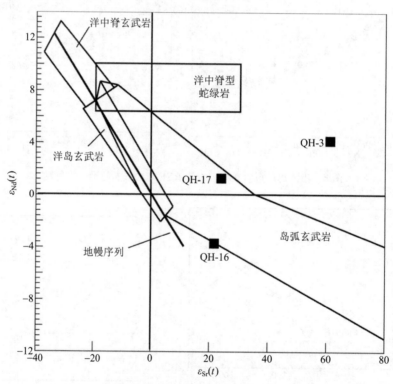

图 6　黑河丹凤群蛇绿岩的 $\varepsilon_{Sr}(t)$-$\varepsilon_{Nd}(t)$ 相关图

由图 7，QH-3，QH-16 和 QH-17 三个样品与安第斯、桑达、小安德列斯等岛弧火山岩相似，属三端元型岛弧火山岩。在图 7 中落在洋中脊型亏损地幔（楔形地幔）、陆缘沉积物和俯冲洋壳析出流体之间，表明其源区与上述三端元有成因联系。其中陆源沉积物组分的存在表明北秦岭古生代岛弧是近陆岛弧，它可能是在古秦岭洋洋壳向华北陆块俯冲过程中，由华北陆块分裂出来的。

5　讨论与结论

(1) 本文所列的采自长约 30 m 剖面上的 9 个变质火山岩样品中，除采自采样剖面两端的 QH-3（辉绿岩脉）和 QH-11（深色变质火山岩）为变玄武岩外，其他采自深色浅色构造岩片的样品均为变质安山岩。由此不难推断黑河丹凤群曾有过大规模的安山岩浆活动。

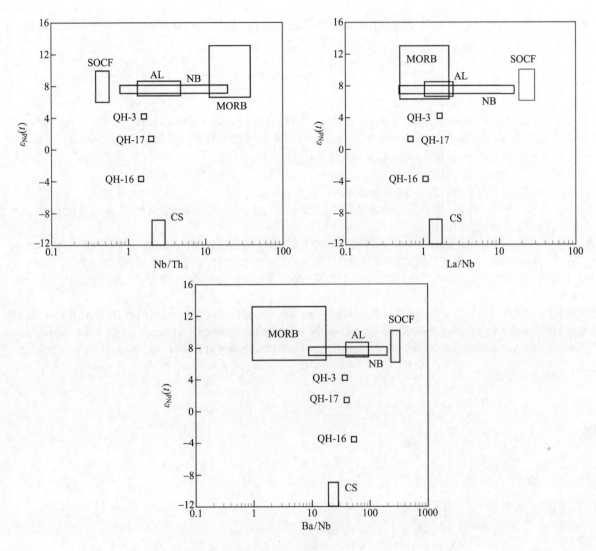

图 7 黑河丹凤群蛇绿岩的 $\varepsilon_{Nd}(t)$-Nb/Th、La/Nb、Ba/Nb 图解

图中 MORB 为洋中脊型亏损地幔；SOCF 为俯冲洋壳析出流体；CS 为陆源沉积物；AL 为阿留申岛弧火山岩；NB 代表新不列颠岛弧火山岩

(2) 黑河丹凤群变质火山岩的 REE 分配模型表现为 LREE 富集的特点，痕量元素丰度图具有明显的 Nb、Ti 负异常，在蛇绿岩生成构造环境的 Ba-Th-Nb-La 判别图上几乎所有样品点都落在岛弧玄武岩（IAB）区。上述特点表明它们形成于典型的岛弧环境。

(3) 大量地质、地球化学研究表明北秦岭是华北陆块南缘的古生代岛弧，根据板块构造理论，在这里理应有过大规模安山岩浆活动。北秦岭长期以来，没有发现大规模安山岩浆活动的证据，一些学者因此而对其古生代岛弧构造环境持谨慎的怀疑态度。

本文厘定出的黑河丹凤群安山岩为主的岛弧火山岩建造是首次在北秦岭发现的大规模安山岩浆活动的证据，它为北秦岭古岛弧构造环境的最终确定提供了重要依据，为揭示秦岭造山带构造演化格局提供了信息。同时也提醒人们：丹凤群是否需要进一步解体？

(4) 痕量元素数据及 Nd 同位素数据显示，本区变质火山岩岩浆组成介于洋中脊型亏损地幔（楔形地幔）、陆缘沉积物和俯冲洋壳析出流体三者之间（图 7），属三端元型岛弧火山岩，与拉垃庙苏长辉长岩的特征相同(李曙光等，1993)。这再一次证明北秦岭古生代岛弧是近陆岛弧，它可能是在古秦岭洋洋壳向北俯冲过程中，由华北陆块分裂出来的。

致谢

本文在导师李曙光教授的指导下完成，南京大学王银喜同志，地矿部物化探研究所周丽沂同志为样品测定提供了方便，谨此一并致谢。

参考文献

李曙光, Hart S R, 郑双根等. 1989. 中国华北、华南陆块碰撞时代的钐-钕同位素年龄证据. 中国科学(B 辑), 3: 312-319
　　李曙光, 张宗清, 张国伟等. 1993. 北秦岭拉垃庙苏长辉长岩的痕量元素地球化学.地质学报，67 (4):310-322
李曙光. 1993. 蛇绿岩生成构造环境的 Ba-Th-Nb-La 判别图.岩石学报, 9 (2)
孙勇, 于在平, 张国伟. 1988. 东秦岭蛇绿岩的地球化学.秦岭造山带的形成及其演化.西北大学出版社, 65-74
王清晨, 孙枢, 李继亮等. 1989. 秦岭的大地构造演化.地质科学, 2: 129-142
许志琴, 牛宝贵, 刘志刚等. 1991. 秦岭-大别"碰撞-陆内"型复合山链的构造体制及陆内板块动力学机制.秦岭造山带学术讨论会论文选集.西北大学出版社, 139-147
于在平, 孙勇, 张国伟. 1988. 商丹地区秦岭缝合带弧前沉积楔形体初探.秦岭造山带的形成及其演化.西北大学出版社, 48-64
于在平, 孙勇, 张国伟. 1991. 秦岭商丹沉积岩系基本地质特征.秦岭造山带学术讨论会论文选集.西北大学出版社，78-88
张国伟, 于在平, 孙勇等. 1988. 秦岭商丹断裂边界地质体基本特征及其演化.秦岭造山带的形成及其演化.西北大学出版社, 29-47
张国伟, 周鼎武, 于在平等. 1991. 秦岭岩石圈组成、结构与演化.秦岭造山带学术讨论会论文选集.西北大学出版社，121-138
Sun S S & McDonough W F. 1989. Chemical and isotopic systematics of oceanic basalts: implications for mantle composition and processes, in "Magmatism in the Ocean Basins" edited by Saunders A D & Norry M J. Geol. Soc. Special Publ., 42: 313-345

南秦岭勉略构造带黑沟峡变质火山岩的年代学和地球化学
——古生代洋盆及其闭合时代的证据*

李曙光[1]，孙卫东[1]，张国伟[2]，陈家义[3]，杨永成[3]

1. 中国科学技术大学地球和空间科学系，合肥 230026
2. 西北大学地质系，西安 710069
3. 陕西地质矿产局区调队，咸阳 712000

> **亮点介绍：** 发现南秦岭勉略构造带黑沟峡绿片岩相变质双峰式火山岩的玄武岩具有类似 MORB 的扁平的 REE 模型，和亏损的 Sr-Nd 同位素组成，Sr 和 Nd 同位素全岩等时线给出一致的三叠纪变质年龄，证明了该变质玄武岩是印支期变质的古生代洋壳残片、该勉略洋盆在三叠纪已闭合，从而为判定勉略构造带的性质及陆块拼合的时代提供了科学依据。

摘要 南秦岭勉略构造带黑沟峡变质火山岩（绿泥钠长片岩）的主要元素组成表明，它是一套双峰式火山岩系，由低 K 富 Na 拉斑玄武岩及酸性英安岩、流纹岩组成。该玄武岩具有类似 MORB 的痕量元素特征，扁平的 REE 模型，但富 Th 和 Pb。已获得该变质岩系的 Sm-Nd 全岩等时线年龄为 242±21 Ma，Rb-Sr 全岩等时线年龄为 221±13 Ma，它们在误差范围内一致，并指示该火山岩系的变质时代。变质时代的初始 ε_{Nd} 值（+6.1）指示该玄武岩源区为 MORB 型亏损地幔（ε_{Nd} 值 > +6.1）。

关键词 秦岭造山带；印支运动；双峰式火山岩；同位素定年；痕量元素地球化学

近年来，秦岭造山带构造研究的一项重要进展就是在南秦岭发现具有地缝合线性质的勉略构造带[1,2]。这一发现使得对秦岭造山带的认识由过去简单的华北与扬子两大陆块碰撞构造体制变为华北、秦岭微板块和扬子 3 个陆块碰撞的构造体制。勉略构造带能否为扬子陆块与秦岭微板块之间的地缝合线，关键是在该带陆-陆碰撞前，是否存在一个洋盆，以及该洋盆的性质及开、合历史。然而该带的蛇绿混杂岩、火山岩均缺乏较系统的岩石学、地球化学及年代学研究。针对这一关键问题，作为第一步，本文对发育在该带的变质火山岩系进行了主要元素、痕量元素及 Sr 和 Nd 同位素年代学及地球化学研究，为判定勉略构造带的性质，陆块拼合的时代提供科学依据，并就构造意义进行讨论。

1 地质与样品

勉略构造混杂岩带原称三河口群，它南以略阳断裂与前寒武碧口群相邻，北以状元碑断裂与志留系白水江群相邻。构造带内主要由强烈剪切的震旦-寒武系和泥盆-石炭系逆冲推覆岩片组成，形成自北向南的叠瓦逆冲推覆构造（图 1）。其中震旦-寒武系主要为含砾泥质岩、泥质碎屑岩、火山碎屑岩、碳酸盐岩和镁质碳酸盐岩组成；泥盆系为深水浊积岩、泥质碳酸盐岩和泥质岩；石炭系为碳酸盐岩。该带缺失奥陶、志留系岩层，与南、北两侧发育志留系而形成鲜明对照。蛇绿混杂岩和变质火山岩系亦以构造岩片形式卷入该构造带（图 1）。蛇绿混杂岩以略阳三叉子地区最为典型，它由强烈剪切变形的基性熔岩、堆晶辉长岩、超基性岩（蛇纹岩、滑石片岩）及辉绿岩墙群组成，它们周围均是韧性剪切带。在蛇绿混杂岩北侧广泛分布一套双峰式变

---

* 本文发表在：中国科学（D 辑），1996，26：223-230
对应的英文版论文为：Li Shuguang, Sun Weidong, Zhang Guowei, et al., 1996. Chronology and geochemistry of metavolcanic rocks from Heigouxia Valley in the Mian-Lue tectonic zone, South Qinling: evidence for a Paleozoic oceanic basin and its close time. Sciences in China (ser. D), 39: 300-310

质火山岩系（绿泥钠长片岩），被称为乔子沟组。它北面与泥盆系岩层相邻（图 1）。该火山岩系可能为该带古生代裂开，洋盆发育初期的产物，是本文的研究对象。

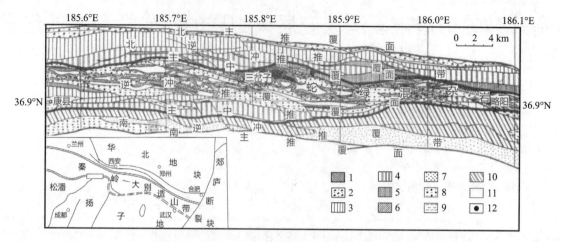

图 1 勉略构造混杂岩带康县—略阳段构造地质简图（据陕西区调队秦岭构造组）

1-蛇绿岩、超镁铁质岩岩块，2-由碎裂岩、角砾岩、各种糜棱岩组成的断裂构造混杂岩带，3-泥盆系泥质碳酸盐岩，4-泥盆系泥质岩、碎屑岩和少量泥质碳酸盐岩，5-上古生界-下中生界中基性火山岩、火山碎屑岩，6-顺层分布的碳质、碳泥质、碳硅质强剪切基质带，7-泥盆系踏坡群砾岩、砂砾岩，8-震旦-寒武系为主的含砾泥质岩、泥质岩、泥质碎屑岩、火山碎屑岩，9-太古代鱼洞子杂岩，10-石炭系略阳灰岩，11-震旦-寒武系为主的碳酸盐岩、镁质碳酸盐岩，12-采样位置

样品采自三叉子乡以东，张家坝，黑沟峡内（图 1）。该处岩石自南而北顺序为：大理岩，千枚岩夹大理岩，基性火山岩（绿泥石片岩及深色绿泥钠长片岩），向上基性度降低变为条带状绿泥钠长石片岩，并夹有少量薄层酸性浅色绿泥钠长片岩。我们在 23 m 长的剖面中采取了 9 个样品。由于强烈的剪切作用，样品片理发育并有少量的气孔杏仁构造。

2 实验方法

先将样品切成 1 cm 厚的岩片，用金刚砂纸打磨掉切片时岩片表面的铁质污染。然后用棉布包裹击碎，从中挑选 200 g 新鲜岩石碎块。将这些岩石碎块用 1:1 HCl 浸泡 12 h，再用去离子水洗净、烘干，以除去片理缝及气孔中的可能后期污染物。最后在振动盒内粉碎至 200 目。

样品的主要元素及痕量元素（稀土除外）在地质矿产部物化探研究所用 X 荧光法测定，稀土元素由地矿部武汉分析测试中心测定。Sm-Nd 及 Rb-Sr 同位素由南京大学现代分析中心同位素实验室测定，其同位素测试流程已有前文描述[3]，测试结果分别列于表 1、2。年龄及误差计算使用美国地质调查所（丹佛）K. Ludwig 编写的 Isoplot 程序（1994 年修订版）。

表 1 黑沟峡绿片岩相变质火山岩的主要元素（%）、痕量元素组成（×10⁻⁶）

样品号	QLZ-1-1	QLZ-1-2	QLZ-2-1	QLZ-2-2	QLZ-3	QLZ-4	QLZ-5	QLZ-6	QLZ-7	QLZ-8	QLZ-9
岩性	深色绿泥钠长片岩	深色绿泥钠长片岩	浅色条带状绿泥钠长片岩	浅色条带状绿泥钠长片岩	深色条带状绿泥钠长片岩	深色绿泥钠长片岩	深色绿泥钠长片岩	浅色绿泥钠长片岩	深色绿泥钠长片岩	浅色绿泥钠长片岩	深色绿泥钠长片岩
SiO_2	48.12	48.05	65.36	67.72	49.46	50.76	49.87	73.46	49.1	68.46	49.71
TiO_2	1.39	1.369	0.64	0.571	1.169	1.124	1.21	0.453	0.771	0.543	1.743
Al_2O_3	15.19	15.6	14.24	13.24	14.33	14.91	14.55	11.44	15.21	13.34	14.68
Fe_2O_3	5.09	4.07	2.95	2.48	5.18	4.22	4.8	1.81	2.03	2.64	5.73
FeO	7.48	8.4	3.04	171	5.74	7.48	6.04	1.28	6.96	2.38	7.14
MgO	7.84	7.76	2.57	2.13	5.45	5.4	5.61	1.02	7.6	1.91	6.09
CaO	6.58	5.91	3.96	3.54	9.39	6.08	8.26	3.6	5.79	3.69	8.12
MnO	0.204	0.091	0.08	0.208	0.24	0.235	0.105	0.147	0.078	0.206	0.22
K_2O	0.09	0.04	0.07	0.04	0.04	0.05	0.2	0.69	0.03	0.04	0.02

续表

样品号	QLZ-1-1	QLZ-1-2	QLZ-2-1	QLZ-2-2	QLZ-3	QLZ-4	QLZ-5	QLZ-6	QLZ-7	QLZ-8	QLZ-9
Na_2O	2.58	2.91	4.89	4.8	3.31	3.69	3.88	4.14	3.81	5.03	1.18
P_2O_5	0.143	0.133	0.12	0.102	0.103	0.104	0.086	0.053	0.83	0.105	0.19
H_2O	4.98	4.97	2.28	1.92	3.21	4.14	3.39	1.23	4.37	1.8	4.51
$CO.$	0.44	0.46	0.18	0.73	1.95	1.37	235	1.11	3.32	0.39	0.02
Total	100.3	99.98	100.44	100.11	99.6	99.88	100.44	100.46	99.27	100.46	99.53
La	4.17	4.23	10.27	9.58	5.96	5.17	3.88	14.74	2.84	9.63	8.6
Ce	1290	11.67	25.93	22	15.36	12.94	11.12	40.28	7.4	23.15	23.24
Pr	1.61	1.6	3.27	3.02	1.95	1.94	1.38	4.43	1.2	2.69	2.89
Nd	9.8	9.61	16.44	14.54	9.64	9.61	8.39	23.78	5.71	14.11	16.37
Sm	2.95	187	4.54	4.45	3.09	2.87	272	5.97	1.95	3.25	4.98
Eu	1.15	1.22	1.38	1.28	1.09	1.01	1.04	1.48	0.66	0.95	1.67
Gd	3.66	3.75	5.27	4.92	3.84	3.5	6.11	219	3.21	5.71	0.63
Tb	0.63	0.65	0.91	0.85	0.69	0.61	0.63	1.11	0.41	0.58	1.01
Dy	4.33	4.51	6.29	5.74	4.66	4.2	4.38	7.2	2.73	3.72	6.84
Ho	0.86	0.88	1.28	1.23	0.98	0.85	0.89	1.46	0.55	0.74	1.36
Er	2.67	264	3.94	3.8	2.98	2.58	2.72	2.72	4.63	1.7	2.17
Tm	0.37	0.38	0.59	0.56	0.43	0.37	0.39	0.66	0.24	0.31	0.61
Yb	2.4	2.5	4.11	3.87	2.68	2.38	2.46	4.48	1.55	1.92	3.87
Lu	0.36	0.37	0.65	0.59	0.4	0.35	0.36	0.68	0.22	0.28	0.57
Y	23.01	24.42	35.78	33.27	26.77	22.71	23.73	39.74	14.77	19.16	36.76
Rb	2	0	0	0	0	0	1	11	0	0	0
Sr	279	262	261	234	348	198	249	170	127	238	434
Th	4	6	4	3	5	4	4	1	4	2	2
Cr	126	139	18	14	214	22	232	11	363	19	21
Ni	57	62	9	7	89	19	76	4	87	8	22
Ba	37	25	42	47	28	10	44	277	23	47	29
Zr	74	74	127	124	95	89	79	78	57	121	138
Nb	5	4	6	5	7	6	5	7	5	5	9
U	1	1	2	2	1	0	1	2	2	0	1
Pb	21	36	14	10	19	27	17	123	9	77	197

表2 黑沟峡绿片岩相变质火山岩的 Sm、Nd、Rb、Sr 含量及 Sr、Nd 同位素组成

样品编号	$Sm(×10^{-6})$	$Nd(×10^{-6})$	$^{147}Sm/^{144}Nd$	$^{143}Nd/^{144}Nd±2\sigma$	$\varepsilon_{Nd}(t)$	$Rb(×10^{-6})$	$Sr(×10^{-6})$	$^{87}Rb/^{86}Sr$	$^{87}Sr/^{86}Sr±2\sigma$	$\varepsilon_{Sr}(t)$
QLZ-1-1	3.955	11.59	0.2064	0.512962±9	6.0	3.612	295.4	0.0348	0.707081±22	38.9
QLZ-1-2	3.525	9.348	0.2281	0.513001±8	6.1	2.725	278.6	0.02787	0.707001±17	38.1
QLZ-2-1	4.126	10.02	0.2492	0.513032±12	6.1	1.914	280.3	0.01946	0.706985±22	38.3
QLZ-3	3.412	10.91	0.1892	0.512932±22	6.0	1.204	369.5	0.009286	0.706949±21	38.3
QLZ-4	3.378	10.44	0.1957	0.512948±18	6.1	2.549	214.7	0.03384	0.707031±24	38.3
QLZ-5	2.997	10.05	0.1802	0.512921±16	6.0	3.657	249.3	0.04181	0.707102±18	38.9
QLZ-6	6.031	22.78	0.1602	0.512895±12	6.1	15.84	182.6	0.2472	0.707701±15	37.4
QLZ-8	2.959	11.55	0.1550	0.512882±20	6.0	4.164	225.3	0.05271	0.707112±16	38.5
OLZ-9	3.028	12.23	0.1498	0.512875±24	6.1	2.962	475.5	0.01237	0.706951±14	38.2

3 主要元素、痕量元素地球化学

图2表明，该火山岩系主要由玄武岩及少量英安岩、流纹岩组成，缺少中性岩石，表现出双模式火山岩

特征，说明它们形成于大陆裂谷环境。然而，该火山岩与一般陆内裂谷双峰式火山岩不同，它们的钾含量很低，除流纹岩（QLZ-6）外，K_2O 含量均小 0.1%，与低钾的洋中脊玄武岩或低钾岛弧拉斑玄武岩类似。该火山岩的低钾特征可由同位素稀释法测定的低 Rb 含量确证（表2）。其中英安岩的低钾特征很罕见，一个可能的解释是在变质过程中钾是活动元素，少量英安岩中的钾与周围的低钾玄武岩均一化了。应用 $FeO^*/MgO-SiO_2$ 图解判明它们的玄武质岩石属拉斑系列，仅酸性岩属钙碱系列。该火山岩富钠，除 QLZ-9 样外（其 Na_2O= 1.18%），其他样品的 Na_2O 含量为 2.58%~3.88%，大大高于新鲜的洋中脊玄武岩，略低于标准的细碧岩。玄武岩的 Ca/Na 变化范围也很大，多数在 1.51~2.84 之间，QLZ-9 样高达 6.88。Ca/Na 如此大的变化范围主要是由钠长石的绿帘石化引起的，它表明在低级变质作用时该火山岩系的 Ca 和 Na 已成为开放体系，该火山岩的原岩特征与细碧岩和石英角斑岩类似。

痕量元素的原始地幔标准化图（图3）显示该玄武岩痕量元素有如下特征：(1) Nb 与 La 含量大致相等，Nb 未显示出负异常，Ba 也未显示出正异常，这与岛弧火山岩不同；(2) 具有高 Th、Pb 异常和低 Rb、K 异常，后者排除了该玄武岩浆受陆壳混染的可能性，因此，高 Th 和 Pb 很可能反映了源区特征；考虑到 Rb 和 K 在变质过程中的活动性，应用 Ti/Zr-Ti/Y 图，同样可以证明该玄武岩来自 MORB 型地幔源并较少受陆壳混染影响，而酸性岩则源于具有陆壳特征的源区(图 5(a))；(3)除了 Th 和 Pb 外，其他痕量元素大致与 N 型 MORB 类似，而普遍低于 OIB，具有扁平的 REE 模型(图4)。应用李曙光(1993)提出的 Ba-Nb-La 判别图[4]，该玄武岩均落在 MORB 区(图 5 (b), (c))。综合上述特征，该玄武岩应属于 MORB 型，不是 OIB 和岛弧型，说明该裂谷已拉张成洋盆，洋壳已开始形成。然而该玄武岩与典型 N 型 MORB 不同之处是 Th 和 Pb 高，该特征又与一些大陆溢流玄武岩类似[5,6]，这恰好反映了该玄武岩系是由初始大陆裂谷向成熟洋盆转化阶段的产物。

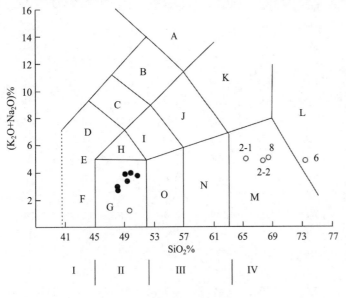

图 2 SiO_2-(K_2O+Na_2O)图

I-超基性岩，II-基性岩，III-中性岩，IV-酸性岩，A-响岩，B-碱玄质响岩，C-响岩质碱玄岩，D-碧玄岩，E-碱玄岩，F-苦橄玄武岩，G-玄武岩，H-粗面玄武岩，I-玄武粗安岩，J-粗安岩，K-粗面岩，L-流纹岩，M-英安，N-安山岩，O-玄武安山岩

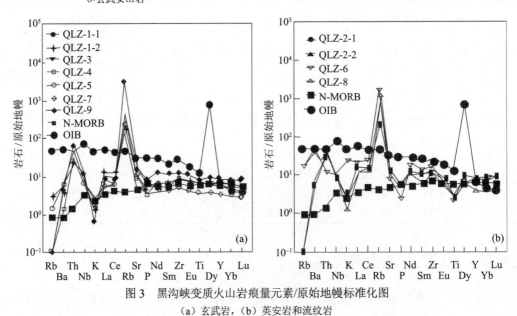

图 3 黑沟峡变质火山岩痕量元素/原始地幔标准化图
(a) 玄武岩，(b) 英安岩和流纹岩

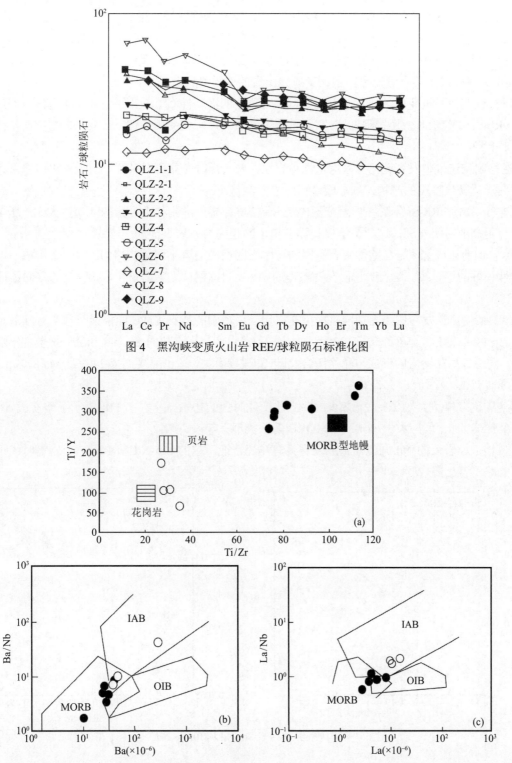

图 4 黑沟峡变质火山岩 REE/球粒陨石标准化图

图 5 显示黑沟峡变质火山岩源区特征的 Ti/Zr-Ti/Y 图[5]（a）和黑沟峡变质火山岩生成构造环境的 Ba/Nb-Ba（b）及 La/Nb-La（c）判别图[4]

●-玄武岩，○-英安岩和流纹岩

4 Sr、Nd 同位素定年及地球化学

对黑沟峡变质火山岩样品中 9 个 Sm/Nd 值差别较大的样品进行了 Sm-Nd、Rb-Sr 同位素分析(测定结果见表 2)。该组黑沟峡变质火山岩样品获得一条很好的 Sm-Nd 全岩等时线：年龄 $t = 242\pm21$ Ma，初始 $^{143}Nd/^{144}Nd$

= 0.512638±29, $\varepsilon_{Nd}(t)$ = +6.1 (图 6(a))和一条很好的 Rb-Sr 全岩等时线：年龄 t = 221±13 Ma，初始 $^{87}Sr/^{86}Sr$ = 0.706933±24, $\varepsilon_{Sr}(t)$ = +38.3 (图6(b))。由于 Rb-Sr 体系在变质过程中同位素均一化尺度很大，其变质岩全岩等时线年龄一般指示其变质时代。需要探讨的是该 Sm-Nd 全岩等时线年龄的含义。根据下列理由，242±21 Ma 的 Sm-Nd 全岩等时线年龄有可能也指示该变质火山岩的变质时代。

(1) 研究证明在低级变质条件下稀土元素可以有明显的活动性[7,8]。在研究五台群细碧-石英角斑岩系的 Sm-Nd 和 Rb-Sr 年代学时，还发现细碧-石英角斑岩系在绿片岩相变质条件下，其 Nd 同位素可被均一化并给出变质时代[9]。

(2) 在勉略构造带中发育有泥盆系深水浊积岩系，表明当时洋盆已经打开；由于该火山岩系相当于大陆初始裂谷向洋盆转化阶段的产物，其形成时代应老于泥盆纪。

(3) 该 Sm-Nd 与 Rb-Sr 年龄在误差范围内是一致的。如果考虑到稀土元素的活动性大大小于 Rb-Sr 的活动性，应有更高的封闭温度，这两个年龄实际上指示的是同一变质事件。

(4) 尽管如前述该玄武岩与酸性岩有不同的源区，因而应当有不同的初始 Nd 同位素组成，但它们仍给出一条线性很好的等时线。这表明它们在变质作用中 Nd 同位素被重新均一化了，因此该等时线应给出变质年龄。

(5) 南秦岭中广泛发育的印支期同碰撞型花岗岩已用精确的锆石 U-Pb 同位素年代学方法测定其形成时代为 206±2～220±2 Ma（陈亚东等，1995；私人通信），它们与黑沟峡的 Sm-Nd 和 Rb-Sr 全岩年龄在误差范围内一致。同碰撞型花岗岩的形成与洋盆闭合及陆-陆碰撞有关。这也间接支持 242±21 Ma 的 Sm-Nd 年龄指示的是变质时代。

考虑到大部分样品为 LREE 较微富集型，上述变质时代的初始 ε_{Nd} 值(+6.1)仅提示了该玄武岩源区 ε_{Nd} 值的下限，其原岩的 ε_{Nd} 值应大于+6.1，与 MORB 源区类似。

总之，该火山岩系的 Nd 同位素组成同样具有洋壳特征。它的 Sm-Nd 和 Rb-Sr 全岩等时线年龄，证明在三叠纪勉略洋盆已闭合。

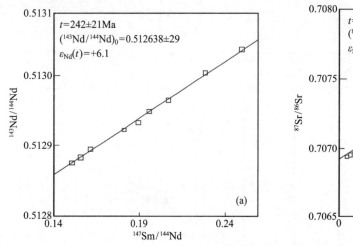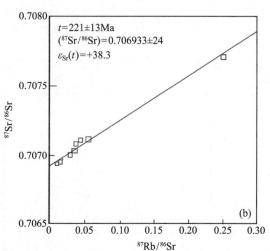

图 6 黑沟峡变质火山岩 Sm-Nd 和 Rb-Sr 等时线图

5 构造意义

华北与扬子陆块的碰撞时代一直是秦岭造山带形成与演化研究中的关键而有争议的问题。古地磁[10]、超高压变质岩的年代学[11,12]，以及一些地质研究[13,14]一致认为两大陆块的碰撞发生在二叠纪末和三叠纪初。然而，由于长期以来在东秦岭人们一直未发现有晚古生代洋壳残片，一些地质学家主张这一碰撞发生在泥盆纪[15]。勉略构造带中具有 MORB 痕量元素及 Nd 同位素特征的黑沟峡变质火山岩，是东秦岭发现的第一个印支期变质的古生代洋壳残片，为解决上述碰撞时代这一关键问题提供了非常重要的证据。它可以使上述看起来相互矛盾的各种有关陆块碰撞的事实获得统一的解释。一个新的可能的模型是：在古生代，扬子陆块北缘沿勉略构造带曾裂开形成洋盆，分出一块秦岭微陆块。该微陆块在洋盆张开的推力下北移并在泥盆纪与华北南缘对接，因此在商丹一线已不见晚古生代远洋沉积及洋壳残余。由于秦岭微陆块的质量与华北陆块相比很小，致使秦岭微陆块的动量惯性不足以使它产生强烈的陆壳俯冲和碰撞造山作用。直到三叠纪初，勉略洋盆闭合，质量巨大的华南陆块（包括扬子陆块和华夏古陆）与秦岭微陆块及华北陆块碰撞，在其巨大的动量惯性作用下，引发了陆壳俯冲及南秦岭的强烈变形及造山运动。

参考文献

[1] Sun Shu, Li J, Lin J et al. Indosinides in China and the consumption of eastern Paleotethys. Controversies in Modern Geology: Evolution of Geological Theories in Sedimentology, Earth History and Tectonics. In: Muller D W, McKenzie J A, Weissert H, eds. London: Academic Press, 1991: 363~384
[2] 张国伟, 孟庆任. 秦岭造山带的结构构造. 中国科学, B 辑, 1995, 25(9): 994~1003
[3] 王银喜, 杨杰东, 郭令智等. 浙江龙泉早元古代花岗岩的发现及基底时代的讨论. 地质论评, 1992, 38: 525~531
[4] 李曙光. 蛇绿岩生成构造环境的 Ba-Th-Nb-La 判别图. 岩石学报, 1993, 9: 146~157
[5] Hergt J M, Peate D W, Hawkesworth C J. The petrogenesis of Mesozoic Gondwana low-Ti flood basalts. Earth Planet Sci Lett, 1991, 105: 134~148
[6] Duncan A R, Erlank A J, Marsh J S. Regional Geochemistry of the Karoo Igneous province. Spec Publ Geol Soc S Afr, 1984, 13: 355~388
[7] Hellman P L, Henderson P. Are rare earth elements mobile during spilitization? Nature, 1977, 267: 38~40
[8] Hellman P L, Smith R E, Henderson P. The mobility of the rare earth elements: evidence and implication from selected terrains affected by burial metamorphism. Contrib Mineral Petrol 1979, 71:23~44
[9] Li Shuguang, Hart S R, Wu Tieshan. Rb-Sr and Sm-Nd isotopic dating of an early Precambrian spilite-keratophyre sequence in the Wutaishan area, North China: Preliminary evidence for Nd-isotopic homogenization in the mafic and felsic lavas during low-grade metamorphism. precambrian Res, 1990, 47: 191~203
[10] Lin J L, Fuller M, Zhang W Y. Preliminary Phanerazoic polar wander paths for the North and South China blocks. Nature, 1985, 313: 444~449
[11] 李曙光, Hart S R, 郑双根等. 中国华南、华北陆块碰撞时代的钐-钕同位素证据. 中国科学, B 辑, 1989, (3): 312~319
[12] Li Shuguang, Xiao Y, Liu D. Collision of the North China and Yangzi Blocks and formation of ooecite-bearing eclogites: timing and processes, Chemical Geology, 1993, 109: 89~111
[13] Hsu K J, Wang Q, Li J et al. Tectonic evolution of Qinling Mountains, China. Eclogae Geol Hev, 1987, 80: 735~752
[14] 张国伟, 周鼎武, 于在平等. 秦岭造山带岩石圈组成、结构与演化特征. 见: 秦岭造山带学术讨论会论文选集. 西安: 西北大学出版社, 1991: 121~138
[15] 任纪舜, 张正坤, 牛宝贵等. 论秦岭造山带-中朝与扬子陆块的拼合过程. 见: 秦岭造山带学术讨论会论文选集. 西安: 西北大学出版社, 1991: 99~110

A middle Silurian-early Devonian magmatic arc in the Qinling Mountains of Central China: a discussion

Shuguang Li and Weidong Sun

Department of Earth and Space Sciences, University of Science and Technology of China, Hefei 230026, Anhui, China

亮点介绍：根据已有研究成果资料，评述了 Lerch 等（1995）关于秦岭造山带北秦岭中志留—早泥盆世岩浆弧模型，肯定了该岩浆弧在早泥盆世终结的结论，但质疑将这一终结归因于华北和华南陆块古生代碰撞的结论，指出该岩浆弧的终结是南秦岭微陆块与华北陆块拼合的结果，而华南陆块与华北陆块的碰撞是三叠纪发生在南秦岭微陆块南侧的勉略构造带。

Abstract Lerch et al. (1995) documented an early-to-middle Paleozoic magmatic arc in the Qinling orogenic belt of central China by detailed field mapping, geochemistry, and geochronology in the Heihe area of North Qinling. Their U-Pb analyses of zircons from I-type granites developed in the Heihe area suggested short-lived arc activity from Middle Silurian to Early Devonian time.

This result is very interesting. The termination of calc-alkaline magmatism in the Devonian is plausible and consistent with our studies on geochemistry and geochronology of a mafic pluton developed in Lerch et al. (1995) Zone 2: we reported a Sm-Nd mineral isochron (pyr + 2 plag +WR) age of 402.6 ± 17.4 Ma for the Lajimiao norite-gabbro that intruded into the metabasaltic lava developed in the south of the Shangxian area (Li et al., 1989), then demonstrated its island arc origin using trace elements and Sr, Nd isotopes (Li et al., 1994a). Along the south margin of Zone 2, metamorphic andesite interbedded with metabasalt and intruded by mafic and granitic plutons was found recently in the Heihe area. Their major and trace element components show island arc andesite characteristics (Sun et al., 1995). The $^{40}Ar/^{39}Ar$ age of 426.0 ± 1.3 Ma (Li et al., 1996) for an amphibole from a metabasalt sample shows that these rocks underwent low-amphibolite facies metamorphism in the Middle Silurian. The Caledonian metamorphic age was regarded as the time of collision between this volcanic arc and the North China Craton (Li et al., 1992). The tectonic system of the North Qinling area between 426 and 400 Ma might be similar to that of the Andes in South America, i.e., an island arc that had already accreted to the continent while the subduction of oceanic crust continued.

However, we do not agree with Lerch et al.'s (1995) model that termination of cal-alkaline magmatism was caused by the initial contact between the North China Block (NCB) and the South China Block (SCB). This model cannot explain why continental subduction of the Yangtse Block (YB) occurred in Triassic time, as shown by the ages of ultrahigh-pressure metamorphic rocks from the Dabie Mountians (Li et al., 1989, 1993, 1994b; Amers et al., 1993; Eide et al., 1994) (Fig. 1). It also cannot explain why evidence of Caledonian metamorphism is very common in the North Qinling area but rare in the South Qining area (Li et al., 1992) and why the terrigenous clastics appeared only after the Upper Triassic in the Yangtse Block and in the whole Qinling Folded Belt (Hsu et al., 1987; Sun et al., 1991).

Instead of the model suggested by Lerch et al. (1995), we propose that the termination of magmatic activity could have been caused by contact between the Qinling microcontinent and the NCB. Besides the Northern Suture along the Shang-Dan fault in the North Qinling area, recent tectonic studies of the Qinling orogenic belt have revealed the presence of a Southern Suture Zone along the Mian-Lue tectonic zone in the South Qinling area (Sun et al., 1991; Zhang et al., 1995) (Fig. 1). The Mian-Lue tectonic zone consists of several sheared thrust slabs of

---

* 本文发表于：The Journal of Geology, 1996, 104: 501-503

Sinian-Cambrian argilic rocks and carbonatite and Devonian-Carboniferous deep-water turbidite and carbonate rock. Tectonic slabs of ophiolite and bimodal volcanic rocks were also involved in this zone. The LREE-depleted, low-K, MORB-type basalt from this zone suggests that the rift was opened to an oceanic basin (Li et al., 1996). Thus, the Qinling microcontinent may have split from the Yangtse Block before the Devonian. The basement of the Qinling microcontinent may be represented by an early Precambrian complex, such as the Fuping, Xiaomolong, or Douling complexes, developed in the South Qining area (Fig. 1). Since the Mian-Lue rift was continuously extended to be an oceanic basin, the Qinling microcontinent was moved to the north and may have connected with the NCB in Devonian time, which may have caused termination of calc-alkaline magmatism in the North Qinling area. However, due to the much smaller mass of the Qinling microcontinent than that of the NCB, momentum of the Qining microcontinent was too small to cause deep continent subduction.

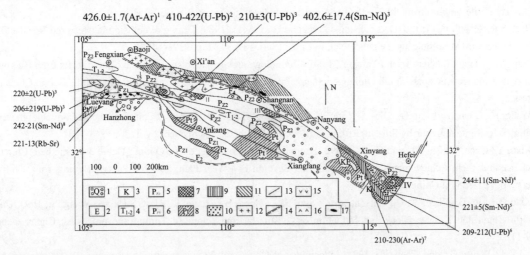

Fig.1 Geological map of the Qinling-Dabie orogenic belt. 1: Quaternary System; 2: Eogene System; 3: Cretaceous System; 4: Lower and Middle Triassic System; 5: Upper Paleozoic Erathem; 6: Lower Paleozoic Erathem; 7: Lower Proterozoic Erathem (Qinling micro-continent); 8: Middle-upper Proterozoic Erathem; 9: Lower formation of the Hong'an-Susong Group with cold eclogites; 10: Southern Dabie ultrahigh-pressure metamorphic zone; 11: Northern Qinling complex; 12: Granite; 13: Fault; 14: Suture; 15: Mian-Lue tectonic zone; 16: Heihe-Danfeng volcanic zone; 17: Mafic-ultramafic bodies. I: Fuping complex; II: Xiaomoling complex; III: Douling complex; IV: Northern Dabie terrane. F1: Shang-Dan tectonic zone (Northern suture); F2: Mian-Lue tectonic zone (Southern suture). Sources: 1: this paper; 2: Lerch et al., 1995; 3: Chen Yadong pers. Comm.; 4: Li et al., 1989; 5: Li et al., 1993: 6: Ames et al., 1993; 7: Eide et al., 1994; 8: Li et al., 1996.

The Devonian and Carboniferous turbidites in the Mian-Lue zone and Sm-Nd, Rb-Sr whole rock isochron ages of 242 ± 24 Ma and 221 ± 13 Ma for the volcanic rocks developed in this zone suggest that the basin was opened in the late Paleozoic and closed in the Triassic time. This idea is also supported by the Triassic I-type granites broadly developed to the north of the Mian-Lue tectonic zone and south of the Shang-Dan fault (Fig. 1). Precise U-Pb zircon dating has determined that their formation ages ranged from 206 to 220 Ma (Chen Yadong per. Comm.). Therefore, the South China Block, including the Yangtse and Cathaysia blocks, collided as a huge mass with the Qinling microcontinent and the NCB in the Triassic. The huge momentum of the SCB caused a deep continental subduction and formation of the ultrahigh-pressure metamorphic rocks.

This new three-microplate (i.e., NCB, Qinling microcontinent, SCB) collision model is better than the two-continent collision model broadly accepted by geologists during the last 20 years. It can reconcile the Devonian termination of calc-alkaline magmatism in the North Qinling area with the Triassic ages of ultrahigh-pressure metamorphic rocks from the Dabie Mountains.

Acknowledgments

This work is supported by the Chinese National Nature Science Foundation.

References

Ames, L.; Tilton, G.R.; and Zhou, G., 1993, Timing of collision of the Sino-Korean and Yangtse Cratons: U-Pb zircon dating of coesite-bearing eclogites: Geology, v.21, p. 339-342.

Eide, L.; McWilliams, M.O.; and Liou, J.Q., 1994, $^{40}Ar/^{39}Ar$ geochronologic constraints on the exhumation of UH-UHP metamorphic rocks in the east-central China: Geology, v. 22, p. 601-604.

Hsu, K.T.; Wang, Q.; Li, J.; Zhou, D.; and Sun, S., 1987, Tectonic evolution of Qinling Mountains, China: Eclogae Geol. Helv., v. 80, p. 735-752.

Lerch, M.F.; Xue, F.; Kroner, A.; Zhang, G.W.; and Todt, W., 1995, A Middle Silurian-Early Devonian Magmatic Arc in the Qinling Mountains of Central China: Jour. Geology, v. 103, p. 437-449.

Li, S.; Chen, Y.; Zhang, Z.; Ye, X.; Zhang, G.; Guo, A.; and Hart, S.R., 1994a, Trace elements and Sr, Nd isotope geochemistry of the Lajimiao norite-grabbro from the North Qinling belt: Acta Geol. Sinica, v. 7, p. 137-151.

Li, S.; Hart, S.R.; Zheng, S. G.; Liu, D. L.; Zhang, G. W.; and Guo, A. L., 1989, Timing of collision between the North and South China Blocks-the Sm-Nd isotopic age evidence: Sci. in China (Series B), v. 32, p. 1393-1400.

Li, S.; Sun, W.; Zhang, G.; Chen, J.; and Yang, Y., 1996, Chronology and geochemistry of metavolcanic rocks from Heigouxia valley in the Mian-Lue tectonic zone, South Qinling: evidence for a Paleozoic oceanic basin and its close time: Sci. in China (Series B), in press.

Li, S.; Wang, S.; Chen, Y.; Liou, D.; Qiu, J.; Zhou, H.; and Zhang, Z., 1994b, Excess argon in phengite from eclogite: evidence from dating of eclogite minerals by Sm-Nd, Rb-Sr, and $^{40}Ar/^{39}Ar$ methods: Chem. Geol., v. 112, p. 343-350.

Li, S.; Xiao, Y.; Chen, Y.; Liu, D.; and Ge, N., 1992, Four-stage tectonic model of the Qinling-Dabie orogenic belt, China, In Conf. on Science and Tech. Achievements in Geology during Seventh Five-Year Plan, Selected Papers: Beijing, Pub. House of Sci. and Tech., p. 121-125 (in Chinese).

Li, S.; Xiao, Y.; Liu, D.; Chen, Y.; Ge, N.; Zhang, Z.; Sun, S. S., Cong, B.; Zhang, R.; Hart, S.R.; and Wang, S., 1993, Collision of the North China and Yangtse Block and formation of coesite-bearing eclogites: Timing and processes: Chem. Geol., v. 109, p. 89-111.

Sun, S.; Li, J.; Wang, Q.; and Chen, H., 1991, Indosinides in China and the consumptions of eastern Paleotethys, in Muller, D.W.; McKenzie, J.A.; and Weissert, H., eds., Controversies in Modern Geology: Evolution of Geological Theories in Sedimentology, Earth History, and Tectonics: London, Academic Press, p. 363-384.

Sun, W.; Li, S.; Xiao, Y.; Sun, Y.; and Zhang, G., 1995, The discovery of island arc andesite from Danfeng Group Heihe, North Qinling area and its tectonic significance: Geotectonica et Metallogenia (in Chinese), v. 3, p. 227-232, 234-236.

Zhang, G.; Meng, Q.; and Lai, S., 1995, Structure and tectonics of the Qinling orogenic belt: Sci. in China (Series B), v. 38, p. 1379-1394.

北秦岭西峡二郎坪群枕状熔岩中一个岩枕的年代学和地球化学研究*

孙卫东[1]，李曙光[1]，孙　勇[2]，张国伟[2]，张宗清[3]

1. 中国科学技术大学地球和空间科学系，合肥 230026
2. 西北大学地质系，西安 710127
3. 中国地质科学院地质研究所，北京 100037

> **亮点介绍：**元素和 Sr-Nd 同位素地球化学研究表明，北秦岭西峡二郎坪群枕状熔岩具有典型的岛弧蛇绿岩的特点；从一个岩枕不同部位取样获得线性关系很好的 Rb-Sr 全岩等时线给出 401.9 ± 6.3 Ma 的年龄。由于变质作用很容易使 Sr 同位素在一个岩枕的小范围内均一化，该年龄代表了绿片岩相变质作用的时代，证明二郎坪群枕状熔岩卷入了加里东期变质作用。

摘要　河南省西峡县二郎坪群绿片岩相枕状熔岩岩枕保存完整，受构造改造轻微。在海水蚀变和后期变质作用过程中 REE，Zr，Th，Nb，Sr，Ti，P 等元素的变化较小，基本上仍能反映原岩的特点，而 Ba，U，Rb 等元素受到了明显的改造，含量变化较大。上述相对稳定的元素的地球化学特征表明，二郎坪群枕状熔岩具有典型的岛弧蛇绿岩的特点。Rb-Sr 全岩等时线年龄为 401.9 ± 6.3 Ma，代表了其绿片岩相变质的时代，表明二郎坪蛇绿岩遭受了加里东期变质作用。

关键词　痕量元素；同位素；年代学；枕状熔岩；二郎坪群；北秦岭

1 引言

二郎坪群蛇绿岩是秦岭北带蛇绿岩的重要代表，出露于商县断裂南侧，以断裂关系夹于"宽坪群"和"秦岭群"之间，在秦岭造山带演化中占有重要的地位，长期以来一直受到较为广泛的重视[1-7]，争论较大。争论的焦点集中在其形成环境、形成时代、变质时代等问题上。由于二郎坪群蛇绿岩普遍遭受海水蚀变和后期变质作用，给研究工作带来很大困难，尤其是给关于蛇绿岩原岩恢复和形成环境的研究带来不便。因此，要解决上述争论，首先要弄清海水蚀变及变质作用对蛇绿岩组成及其同位素体系的影响。

枕状熔岩岩枕的体积较小，从岩浆均一化尺度的角度来看，单个岩枕尺度内的岩浆，在喷发冷凝前基本上是均一的。而喷发后，岩枕的不同部位的结晶程度不同，受到海水蚀变及后期变质作用也略有差异，这些差异为查明海水蚀变及变质作用对熔岩痕量元素和同位素体系的影响创造了条件。

本文选用了二郎坪群枕状熔岩中受后期构造改造轻微的岩枕进行地球化学研究，分析了各元素在海水蚀变和变质作用过程中的稳定性，用稳定元素分析了其原岩的生成环境，并在此基础上，利用岩枕体积小、Rb-Sr 同位素体系易于均一的特点，获得较为可靠的变质时代。

2 地质背景及样品

河南省西峡县二郎坪群蛇绿岩位于商县断裂南侧（图 1），是秦岭北带蛇绿岩的典型代表，出露比较完整，主要由超镁铁质岩、镁铁质岩(枕状、块状)组成，上覆大量陆源碎屑岩以及碳酸盐岩，并夹有条带状

---

*本文发表在：地质评论，1996，42：144-153

放射虫硅质岩。岩石普遍遭受了绿片岩相到低角闪岩相变质,构造变形十分强烈[1,3]。其中,二郎坪群火神庙组枕状熔岩受后期构造改造较轻,经历了绿片岩相变质,岩枕保存完好,可见清晰的内粗外细的结晶环带,斑晶为斜长石,含量<5%,斑晶直径在0.5~1 mm之间。手标本可见少量的碳酸盐细脉贯入。

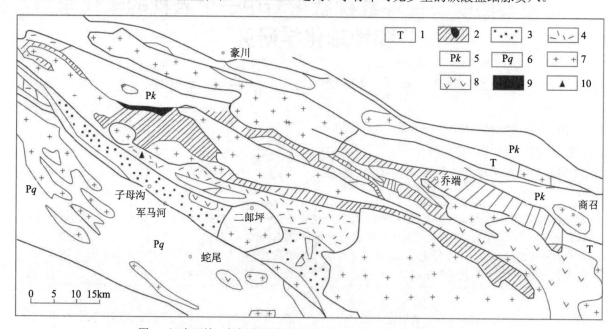

图1 河南西峡二郎坪地区地质略图(据河南省区域地质志,1989)

1.三叠系;2.二郎坪群大庙组(Pz);3.二郎坪群小寨组、子母沟组;4.二郎坪群火神庙组;5.宽坪组;6.秦岭群;7.花岗岩;8.闪长岩;9.超基性岩;10.采样点

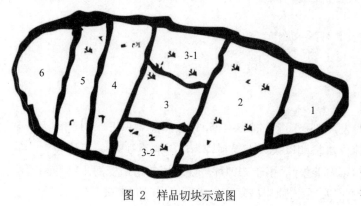

图2 样品切块示意图

本文所用枕状熔岩岩枕(QXE11)采自河南省西峡县二郎坪湾潭,在西峡至二郎坪公路674 km处路边小河的对岸。样品呈半椭球形,长约40 cm,宽约20 cm。将岩枕沿长轴方向切成6块,去掉边部风化部分,依次编号为QXE11-1—6。然后将QXE11-3分为3块,编号为QXEII-3、-31、-32,QXE11-3位于岩枕的核心部位(图2)。岩枕靠QXE11-6一端与其他岩枕相邻,在采样时边缘部分断裂,因而样品较新鲜,蚀变轻微,斑晶含量最少。

3 海水蚀变及后期变质对痕量元素的影响

可以引起岩枕成分不均一的主要地质过程有:海水蚀变、变质作用和结晶作用。对于 QXE11 岩枕,斑晶含量< 5%,主要是斜长石。如此小的结晶比例对不相容元素含量的影响可以忽略不计(< 5%);即使是作为相容元素的 Sr(D=1.3~2.9),斑晶所产生的影响也小于±10%。因此,在分析不相容元素含量变化时,我们忽略了结晶作用的影响。

二郎坪群枕状熔岩 CO_2 和 Na_2O 含量较高,含量变化也很大(表1),前者显然与后期变质过程中的碳酸盐脉的加入有关,而后者则是海水蚀变的结果。在所有样品中,QXE11-6 的 CO_2 和 Na_2O 的含量明显低于其他样品,表明该样品受海水蚀变和后期变质作用的影响最小,与手标本观察结果一致。因此,利用QXE11-6对所有样品进行均一化处理,可以比较清楚地看出海水蚀变和后期变质对各元素的影响(图3)。

由表2,表3和图3(a)可知,REE和Zr等元素的变化范围在−20%~0之间,Ni元素的变化范围在0~20%之间,表明上述元素在海水蚀变和变质作用过程中含量变化很小,基本上仍能反映原岩的特点。

表1 二郎坪群枕状熔岩的主量元素含量（%）

野外编号	岩石类型	SiO_2	TiO_2	Al_2O_3	Fe_2O_3	FeO	MgO	CaO	MnO	K_2O	Na_2O	P_2O_5	H_2O^+	CO_2	Total
QXE11-1	枕状熔岩	46.36	0.62	12.66	2.64	5.57	5.41	12.83	0.16	0.74	4.26	0.11	2.41	5.86	99.79
QXE11-2	枕状熔岩	44.54	0.61	12.43	2.99	5.47	5.70	14.03	0.16	0.58	4.10	0.12	2.58	6.22	99.67
QXE11-3	枕状熔岩	43.61	0.61	11.76	2.25	5.81	6.14	14.25	0.17	0.59	4.15	0.12	2.52	7.31	99.43
QXE11-4	枕状熔岩	45.88	0.59	12.85	2.40	5.93	6.53	11.90	0.16	0.49	4.43	0.12	3.02	5.44	99.80
QXE11-5	枕状熔岩	48.75	0.57	13.95	2.41	5.20	6.19	10.03	0.15	0.34	5.24	0.11	2.97	4.24	100.24
QXE11-6	枕状熔岩	42.30	0.82	13.44	4.44	8.45	9.95	11.64	0.19	0.36	1.88	0.13	4.74	1.95	100.30
QXE11-31	枕状熔岩	44.84	0.57	12.30	2.12	5.23	5.33	13.98	0.16	0.49	4.81	0.12	2.46	7.65	100.31
QXE11-32	枕状熔岩	45.42	0.59	2.28	2.45	6.11	6.71	12.44	0.16	0.53	4.04	0.13	3.09	5.91	99.93

分析单位：地质矿产部武汉岩矿综合测试中心。

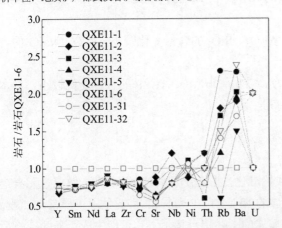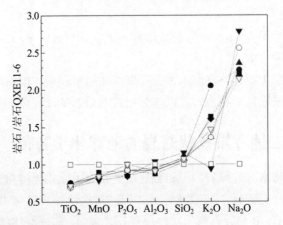

图3 二郎坪群枕状熔岩对于QXE11-6均一化图

表2 二郎坪群枕状熔岩的稀土元素含量（$\times 10^{-6}$）

野外编号	岩石类型	La	Ce	Pr	Nd	Sm	Eu	Gd	Tb	Dy	Ho	Er	Tm	Yb	Lu	Y
QXE11-1	枕状熔岩	16.61	39.45	4.70	8.10	3.64	1.04	3.17	0.44	2.71	0.54	1.52	0.23	1.47	0.22	13.92
QXE11-2	枕状熔岩	17.88	36.64	4.91	17.93	3.68	1.03	3.05	0.45	2.73	0.54	1.49	0.23	1.37	0.20	13.72
QXE11-3	枕状熔岩	18.24	40.01	5.08	18.75	3.82	1.04	3.16	0.47	2.85	0.56	1.59	0.25	1.52	0.23	14.55
QXE11-4	枕状熔岩	16.60	38.25	5.12	18.28	3.72	1.08	3.21	0.47	2.78	0.57	1.64	0.25	1.63	0.23	14.65
QXE11-5	枕状熔岩	18.85	40.40	5.17	19.40	3.91	1.08	3.39	0.50	3.06	0.60	1.71	0.26	1.63	0.24	16.01
QXE11-6	枕状熔岩	20.79	45.82	6.36	24.16	5.08	1.45	4.36	0.65	4.00	0.78	2.28	0.34	2.19	0.32	20.41
QXE11-31	枕状熔岩	17.74	38.48	5.11	18.45	3.66	1.01	3.17	0.48	2.87	0.58	1.65	0.25	1.54	0.23	15.27
QXE11-32	枕状熔岩	18.15	41.00	5.15	18.28	3.73	1.01	3.27	0.50	2.91	0.58	1.64	0.26	1.58	0.24	14.63

分析单位：地质矿产部武汉岩矿综合测试中心。

表3 二郎坪群枕状熔岩的痕量元素组成（$\times 10^{-6}$）

野外编号	岩石类型	Rb	Sr	Th	Cr	Ni	Zr	Ba	Nb	U
QXE11-1	枕状熔岩	23	169	6	646	163	64	304	5	2
QXE11-2	枕状熔岩	18	184	5	548	135	65	252	6	2
QXE11-3	枕状熔岩	17	133	3	593	170	63	268	4	4
QXE11-4	枕状熔岩	12	135	4	575	138	61	258	4	4
QXE11-5	枕状熔岩	6	123	6	531	159	60	199	4	2
QXE11-6	枕状熔岩	10	209	4	758	154	79	133	4	2
QXE11-31	枕状熔岩	14	116	4	488	163	64	225	4	4
QXE11-32	枕状熔岩	15	130	5	621	150	65	317	4	4

分析单位：地质矿产部武汉岩矿综合测试中心。

Nb，Th的变化幅度在±30%左右。出现上述现象的原因很可能是：（1）元素含量太低（只有百万分之几），测定误差较大的结果;(2)或者只是原岩不均一的反映,与后期改造无关;(3)也可能是由于海水蚀变和后期变质对 Nb,Th 的含量影响不同,其中一个过程使之增加,另一个过程使之降低,而且不同样品的海水蚀变和变质程度不同,两个过程叠加使不同样品表现出不同的特点。从总体上看,Nb,Th 受后期改造不大。

Sr 元素的变化范围在 -40%~0 之间,与前人的研究结果相比有较大差异:在海水蚀变过程中 Sr 基本上没有发生变化[8]。此外,Sr 是斜长石的相容元素,而 QXE11-6 是斑晶含量最少的样品,斑晶的影响的结果应使 QXE11-6 中的 Sr 含量减少。因此,Sr 元素的含量变化应是后期变质作用的结果。

Cr 的变化范围在 -30%~0 之间,Ni 的变化范围在 0~30% 之间,变化幅度也较小。

Rb,Ba,U 等元素的含量变化大,表明海水蚀变和后期变质作用对它们的影响较大。其中,Ba 元素的稳定性与前人的研究结果相比有较大的差异,显示了变质作用的影响。根据前人的研究结果,在海水蚀变过程中,Ba 元素含量的变化幅度很小[8],因此,二郎坪群枕状熔岩中 Ba 的含量变化应主要是后期变质作用的结果。

由图 3(b)表明主要元素中 TiO_2,MnO,P_2O_5,Al_2O_3,SiO_2 的含量变化较小,K_2O,Na_2O 的含量变化幅度较大。

综上所述,在海水蚀变和变质作用过程中,REE,Zr,Th,Nb,Sr,Ti,P 的含量变化较小,基本上仍能反映原岩的特点,可以作为判断其原岩生成环境的标准。

4 二郎坪群枕状熔岩的地球化学特征

将本区蛇绿岩中不相容的痕量元素用原始地幔组成标准化,并按分配系数大小排序,做痕量元素丰度模型图[9](图 4a)。该图显示,二郎坪群枕状熔岩具有明显的 Nb,Ti,P 负异常,Th 呈现正异常。一般认为,Nb,Ti,P 负异常与俯冲作用有关[10,11]。二郎坪群枕状熔岩的上述特点指示了它可能生成于岛弧或弧后盆地环境,是古生代沿商丹构造带发生的洋壳向北俯冲的产物。

二郎坪群枕状熔岩的 REE 分配模型与岛弧玄武岩(IAB)相似,呈现轻稀土富集的特点(图 4b)。Eu 无异常,显示在其岩浆演化过程中斜长石分离结晶作用不明显。

蛇绿岩生成构造环境判别图也表明二郎坪群枕状熔岩生成于岛弧环境。在 Nb/Th-Nb 和 La/Nb-La 图[8]上,所有样品点都落在岛弧玄武岩(IAB)区(图 5)。

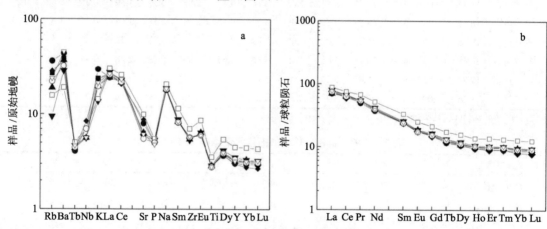

图 4 二郎坪群枕状熔岩中不相容元素原始地幔标准化和 REE 球粒陨石丰度图

由于二郎坪群枕状熔岩中 Ba 元素的含量在后期变质作用过程中受到了明显的影响,原则上讲,与 Ba 元素有关的蛇绿岩构造环境判别图已不适宜作为判断其生成的构造环境的依据。这里将 Ba/Nb-Ba 和 Ba/Nb-Th/Nb 图列出,仅供参考。在上述两图上,所有样品点同样都落在 IAB 区(图 5)。

Ti/100-Zr-Y×3 和 Ti/100-Zr-Sr/2 图解(图 6)也支持了二郎坪群枕状熔岩生成于岛弧环境的判别结果。在图 6(a)中,大部分样品落入 C 区,少数样品落在 B 区与 C 区的交界线上,排除了它们是板内玄武岩(WPB)的可能。在图 6(b)中,所有样品均落在 A 区和 B 区的交界线上,进一步排除了它们是洋中脊玄武岩(OFB)

的可能。结合主要元素组成，可以推断二郎坪群枕状熔岩属于岛弧钙碱玄武岩（CAB）。

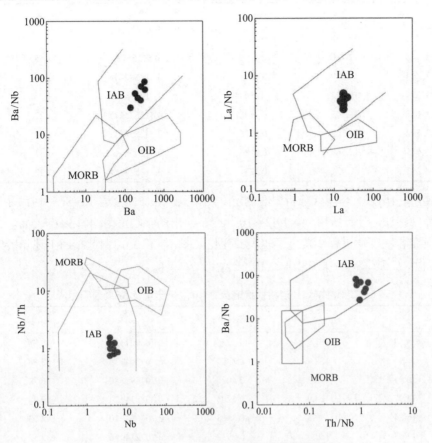

图 5 Ba-Th-Nb-La 蛇绿岩构造环境判别图

综上所述，根据痕量元素丰度模型图和蛇绿岩生成构造环境的Nb/Th-Nb，La/Nb-La图以及Ti/100-Zr-Y×3和Ti/100-Zr-Sr/2判别图解，可以推断二郎坪群枕状熔岩是形成于岛弧环境的蛇绿岩。

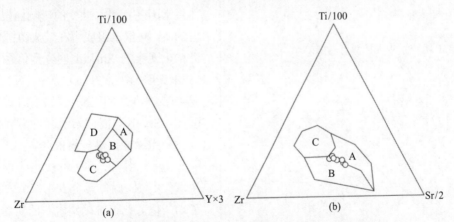

图 6 枕状熔岩的 Ti/100-Zr-Y×3(a)和 Ti/100-Zr-Sr/2(b)图解

(a)洋中脊玄武岩(OFB)投入 B 区，岛弧低钾拉斑玄武岩（LKT）投入 A 区和 B 区，岛弧钙碱玄武岩（CAB）投入 B 和 C 区，板内玄武岩（WPB）投入 D 区；(b) 岛弧低钾拉斑玄武岩（IAB）投入 A 区，钙碱玄武岩（CAB）投入 B 区，洋中脊玄武岩（OFB）投入 C 区

5 Rb-Sr 等时线年龄及 Sr、Nd 同位素地球化学

二郎坪群枕状熔岩样品的 Sm-Nd 同位素分析是在中国地质科学院地质研究所进行的，由于样品尺度太小等原因，未能形成很好的等时线。测定结果列于表4，模式年龄（T_{DM10}）在 924～1109 Ma 之间，略低于西峡长探河二郎坪群火神庙组的石英角斑岩的模式年龄，而与其中的细碧角岩的模式年龄相似[12]。

表 4 二郎坪群绿片岩相枕状熔岩的 Sm、Nd 含量及 Nd 同位素组成

样品编号	Sm ($\times 10^{-6}$)	Nd ($\times 10^{-6}$)	$^{147}Sm/^{144}Nd$	$^{143}Nd/^{144}Nd$ $\pm (2\delta)$	$\varepsilon_{Nd}(0)$	$\varepsilon_{Nd}(t)$	T_{DM10}
QXE-1	3.542	17.638	0.1215				
QXE-2	3.445	17.480	0.1192	0.512557±5	−1.58	+2.43	957
QXE-3	3.464	17.459	0.1200	0.512582±18	−1.09	+2.88	924
QXE-31	3.379	17.124	0.1194	0.512568±10	−1.36	+2.64	941
QXE-32	3.668	18.545	0.1196	0.512573±13	−1.27	+2.72	934
QXE-4	3.392	16.342	0.1256	0.512552±11	−1.68	+2.00	1034
QXE-5	3.445	17.568	0.1186	0.512458±17	−3.51	+0.53	1109
QXE-6	4.268	20.971	0.1231	0.512524±10	−2.22	+1.59	1053

Rb-Sr 同位素组成由南京大学现代分析中心测定，结果列于表 5。全岩 Rb-Sr 等时线示图 7。年龄计算用 YORK-II 双误差回归法，计算所用 $\lambda_{Rb}=1.42 \times 10^{-11}a^{-1}$，给出的年龄值为 401.9±6.3 Ma。由于二郎坪群枕状熔岩的 Rb、Sr 含量都在变质过程中发生了较大的变化（图 3，表 3），由此推断上述年龄应代表二郎坪群枕状熔岩的变质时代。$(^{87}Sr/^{86}Sr)_0 = 0.70553\pm3\times10^{-5}$。$\varepsilon_{Sr}=21.2\pm0.4$。

表 5 二郎坪群绿片岩相枕状熔岩的 Rb、Sr 含量及 Sr 同位素组成

样品编号	Rb ($\times 10^{-6}$)	Sr ($\times 10^{-6}$)	$^{87}Rb/^{86}Sr$	$^{87}Sr/^{86}Sr$ $\pm (2\delta)$	$\varepsilon_{Sr}(0)$	$\varepsilon_{Sr}(t)$
QXE-1	24.57	161.9	0.4235	0.708021±22	+49.98	+21.55
QXE-2	19.35	188.7	0.2907	0.707204±18	+38.38	+21.45
QXE-3	20.95	148.3	0.4026	0.707793±80	+46.74	+20.75
QXE-31	18.97	103.9	0.5203	0.708514±24	+56.98	+21.43
QXE-32	16.84	130.3	0.3683	0.707651±10	+44.73	+21.50
QXE-4	15.67	153.4	0.2912	0.707214±16	+38.52	+21.55
QXE-5	7.146	129.3	0.1575	0.706484±14	+28.16	+22.02
QXE-6	7.493	220.8	0.0968	0.706043±20	+21.90	+20.69

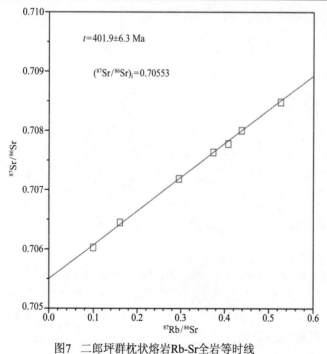

图7 二郎坪群枕状熔岩Rb-Sr全岩等时线

蛇绿岩作为古洋壳的一部分，其 Sr 同位素体系常因受到海水蚀变的影响而发生变化，蛇绿岩的 Sr 同位素组成可在较大范围内变化。而作为稀土元素的 Sm-Nd 同位素体系受后期改造的影响较小，由图 3（a），Sm、Nd 含量变化幅度在−20%～0 之间，Nd 同位素组成变化范围小。因此，Sr、Nd 同位素研究可为判断铁镁质杂岩是不是蛇绿岩提供可靠依据。

但是，由于二郎坪群枕状熔岩在加里东期变质事件中 Sr 同位素至少在整个岩枕范围内发生了均一化，已很难确定岩石的初始 Sr 同位素组成。在 ε_{Sr}-ε_{Nd} 图解上（图 8）上，$\varepsilon_{Sr}(t)$ 值稳定在+20～+22 之间，$\varepsilon_{Nd}(t)$ 保持在+0.5～+2.8 之间，多数样品仍落在岛弧玄武岩区（t 取 Rb-Sr 全岩等时年龄 401.9±6.3 Ma），少数样品落在蛇绿岩区。$\varepsilon_{Sr}(0)$ 值变化范围较大，则可能是 Sr 同位素后期演化过程中 Rb、Sr 含量不均一的结果。

6 结论

关于二郎坪群蛇绿岩的形成环境有多种意见，如洋壳、小洋盆、陆间盆地、大陆边缘盆地、弧后盆地、不成熟的岛弧等，各种观点都有其依据[1-3]。这反映了造山带蛇绿岩的复杂性：构造作用可以将形成于各种不同的构造环境的蛇绿岩混杂到一起，而样品不同必然得出不同结论。但是，如果从岛弧的发展过程考虑，自洋壳俯冲到活动大陆边缘裂开弧后盆地产生，到最后碰撞造山蛇绿岩侵位，是一个连续的过程。在其发展演化的不同阶段产生的蛇绿岩的特点各不相同，它们共同记录了岛弧发展演化的整个历史过程，因此，各种类型蛇绿岩共存正是成熟岛弧存在的证据。根据上述分析并结合地球化学证据及前人的研究成果，得出以下结论：

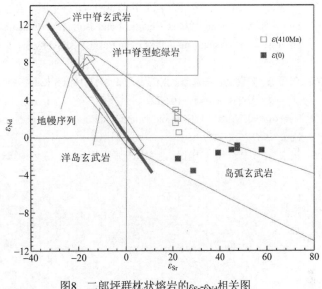

图8 二郎坪群枕状熔岩的ε_{Sr}-ε_{Nd}相关图

(1) 二郎坪群枕状熔岩的Na_2O、$\varepsilon_{Sr}(0)$及U、Rb等元素含量变化范围较大，显示了海水蚀变的影响，表明了它是二郎坪群蛇绿岩套的一部分。在变质过程中，Ba含量变化较大。

(2) 在海水蚀变和后期变质作用过程中REE、Zr、Th、Nb、Sr、Ti、P的变化较小，基本上仍能反映原岩的特点。

(3) 上述稳定元素的地球化学特点表明二郎坪群枕状熔岩形成于典型的岛弧环境。

(4) 二郎坪群枕状熔岩典型的岛弧特点与同一构造带内其他各类蛇绿岩一起表明了成熟岛弧的存在。综合考虑秦岭群两侧的地质、地球化学特点以及二郎坪群与丹凤群、宽坪群初始裂谷的关系[1-3]，认为二郎坪群形成于弧后盆地环境，是宽坪群初始裂谷进一步发展的产物。

(5) 本文选用的二郎坪群枕状熔岩岩枕长轴长仅为40 cm，绿片岩相变质可以造成Rb-Sr同位素体系在整个岩枕范围内的均一化。Rb、Sr地球化学特点也表明它们在后期变质作用过程中发生了显著的变化。因此，我们认为该Rb-Sr全岩等时线代表了枕状熔岩发生绿片岩相变质的时代。这个年龄与加里东期北秦岭地区普遍遭受的区域变质作用时间一致，反映了两者间的内在联系，表明二郎坪群枕状熔岩卷入了加里东期变质作用。

致谢

本文在导师李曙光教授的指导下完成，南京大学王银喜，地质矿产部物化探研究所周丽沂为样品测定提供了方便，谨此一并致谢。

参考文献

[1] 孙勇，于在平，张国伟. 东秦岭蛇绿岩的地球化学. 见：张国伟等著. 秦岭造山带的形成及其演化. 西安：西北大学出版社，1988: 65-74.

[2] 许志琴，牛宝贵，刘志刚，王永敏等. 秦岭-大别"碰撞-陆内"型复合山链的构造体制及陆内板块动力学机制. 见：叶连俊等主编. 秦岭造山带学术讨论会论文选集. 西安：西北大学出版社，1991: 139-147.

[3] 张国伟，周鼎武，于在平，郭安林，程顺有，李桃红，张成立，薛锋，Kroner A，Reischmann T，Altenberger U. 秦岭岩石圈组成、结构与演化. 见：叶连俊等主编. 秦岭造山带学术讨论会论文选集. 西安：西北大学出版社，1991: 121-138.

[4] 王鸿祯，徐成彦，周正国. 东秦岭古海域两侧大陆边缘区的构造发展. 地质学报，1982，56（3）：270-279.

[5] 王润三，刘文荣，车自成，刘良. 二郎坪蛇绿岩的产出环境. 秦岭-大巴山地质论文集（一），变质地质. 北京：科学技术出版社，1990: 132-141.

[6] 金守文. 二郎坪群有关问题的商榷. 河南地质，1988，6（4）21-26.

[7] 张秋生，朱永文. 东秦岭古生代蛇绿岩套. 长春地质学院院报，1984，3：1-13.

[8] 李曙光. 蛇绿岩生成构造环境的 Ba-Th-Nb-La 判别图. 岩石学报，1993，9（2）：146-157.

[9] Sun S S, McDonough W F. Chemical and isotopic systematics of oceanic basalts: implications for mantle composition and processes, in "Magmatism in the Ocean Basins" edited by Saunders A. D. & Norry M.J., Geol. Soc. Special Publ., 1989, (42): 313-345.

[10] 李曙光，陈移之，张宗清，张国伟，郭安林，Hart S R.北秦岭垃圾庙苏长辉长岩的痕量元素和 Sr, Nd 同位素地球化学. 地质学报，1993，67 (4): 310-322.

[11] Ryerson S J, Watson E B. Rutile saturation in magma: implications for Ti-Nb-Ta depletion in island-arc basalts. Earth Planet. Sci. Lett., 1987：225-239.

[12] 张宗清，刘敦一，付国民. 北秦岭变质地层秦岭、宽坪、陶湾群同位素年代学. 北京：地质出版社，1993.

Sm-Nd and Rb-Sr ages and geochemistry of volcanics from the Dingyuan Formation in Dabie Mountains, Central China: evidence to the Paleozoic magmatic arc[*]

Shuguang Li[1], Weili Han[1], Fang Huang[1], Yongfei Zheng[1], Siqing Zhang[2] and Zongheng Zhang[2]

1. Department of Earth and Space Sciences, University of Science and Technology of China, Hefei 230026, Anhui, China
2. Geological Team No.3 of the Henan Province, Xinyang 464000, Henan, China

> 亮点介绍：发现北淮阳定远组变基性火山岩具有岛弧型微量元素分配模式，并确定其喷出年龄为约445 Ma，证明定远组变质火山岩是在晚奥陶-早志留世形成于典型的岛弧环境，从而首次确定大别造山带存在古生代岩浆弧。

Abstract The meta-basic volcanic rocks from the Dingyuan formation in Beihuaiyang zone of Dabie Mountains have typical island arc geochemical feature: e.g. negative anomalies of Nb, Ti, P in trace elemental spider diagram and high LREE, Ba, Pb abundances as well as high La/Nb, Ba/Nb, Th/Nb ratios. Their Sm-Nd and Rb-Sr whole rock isochrons give the ages of 446 ± 23 Ma and 444 ± 31 Ma respectively. Based on these geochemical feature and ages as well as their tectonic setting at convergent plate boundary, we suggest that they were formed in a typical island arc environment in the late Ordovician to early Silurian. This is the first evidence to the Paleozoic magmatic arc in the Dabie Mountains.

1 Introduction

The Dabie Mountains is a collision zone between the North China block (NCB) and South China block (SCB). Various types of ultrahigh-pressure metamorphic (UHPM) rocks including coesite and diamond-bearing eclogite are exposed in the Dabie Mountains (Okay et al. 1989; Wang et al. 1989; Xu et al. 1992a). It is, in general, accepted that the Dabie Mountains can be subdivided into four tectonic subzones from the North to the South: (1) Beihuaiyang, greenschist-amphibolite facies zone; (2) North Dabie complex (terrane); (3) South Dabie ultrahigh pressure metamorphic zone and (4) Hongan-Susong high pressure metamorphic zone (Fig. 1). Many workers believe that zones 3 and 4 are subducted continental crust of the SCB, whereas the tectonic settings of zones 1 and 2 are still controversial. For example, Okay and Sengör (1993) suggest that the suture between the NCB and SCB located to the North of the zone 1; the Foziling group in zone 1 probably represents the passive continental apron deposit of the SCB. Liu and Hao (1989) and Xu et al. (1992b) argue that the boundary between zone 1 and zone 2 represents a suture between the NCB and YB; the Foziling group in zone 1 is a forearc flysch formation. Cong et al. (1994) propose that zone 2 could be an island arc near the NCB, thus the suture between the NCB and SCB should be located on the south boundary of zone 2, and the zone 1 should be a back-arc basin formation. Obviously, the tectonic setting of the zone 1 is critical to understand the tectonic evolution of the Dabie Mountains. The absence of magmatic arc in the zone 1 or zone 2 perhaps is the major difficult in understanding of their tectonic settings. This paper provides the first evidence to the Paleozoic magmatic arc in the Beihuaiyang terrane (zone 1).

---

[*] 本文发表在：Scientia Geologica Sinica, 1998, 7(4): 461-470

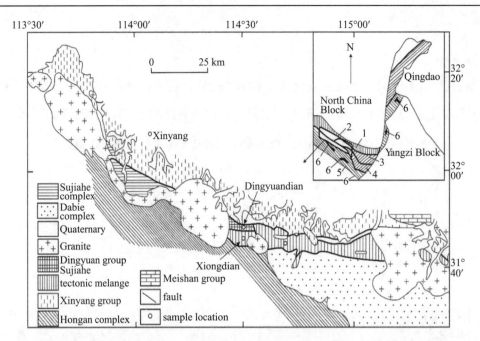

Fig. 1 Sketch geological map of the Sujiahe complex in the Dabie Mountains. 1. Baihuaiyang flyshformation; 2. Sujiahe tectonic melange; 3. North Dabie terrane; 4. South Dabie UHPM zone; 5. Hong'an-Susong HPM Zone; 6. Blueshist.

2 Geological Background

The Beihuaiyang zone is located on the northernmost side of the Dabie Mountains and bounded by a WNW-ESE trending, north dipping, sinistral strike-slip and normal fault, the Tongbai-Mozitan fault on the South (Dong et al. 1993; Okay and Sengör 1993) and the south-dipping Xinyang-Shucheng fault on the north (Xu et al. 1992b). It is characterized by greenschist-amphibolite facies and has undergone intense ductile shear deformation. It can be subdivided into two units: the lower unit of the Guishan complex of the Xinyang group in Henan Province and the Luzhenguan group in Anhui Province. It is composed of various gneiss, amphibolite, garnet-mica-schist, mica-albite-schist, marble and quartzite; their protoliths were suggested to be keratophyre, andesitic tuff and basalts. Protoliths of the K-feldspar gneiss and mica-schist could be sandstone and pelite (Xu and Hao 1988; Xu et al. 1992b). A U-Pb zircon age of 392 ± 25 Ma for acidic metavolcanics from the upper formation of the Guishan complex was reported (Ye et al. 1993). This age is given by lower intercept of a mixing line defined by magmatic and inherited zircons. Ye et al. (1993) interpret this age as formation age of the acidic volcanics. However, an amphibole sample separated from a garnet-amphibolite in the Guishan complex yields a perfect $^{40}Ar/^{39}Ar$ plateau age of 401 ± 4 Ma (Niu et al. 1994), which suggests that the lower units has undergone a metamorphism at the end of Silurian. The upper unit includes the Nanwan formation of the Xinyang group and the Foziling group, and overlies the lower unit with slightly angular unconformity (Xu and Hao 1988). It is a metaclastic flysch formation with rhythmic layers of slate, phyllite, mica-quartz-schist and quartzite and the metamorphic grade of greenschist facies. Devonian fossils in this unit (Gao et al. 1988) suggest a Devonian deposition of the Foziling group and Nanwan formation.

The Sujiahe complex is discontinuously exposed to the south of zone 1, in the western part of the Dabie Mountains, Henan Province (Fig. 1). It is bounded by the Tongbai-Mozitan fault on the north and a shear decollement zone on the south, which separates this zone from the Tongbai complex, Hongan group and a part of the Dabie complex (Ye et al. 1994; Niu et al. 1994). The Sujiahe complex, formerly named as the Sujiahe group in some Chinese literatures, was subdivided into two units, the northern metavolcano-clastic zone and south tectonic melange zone containing cold eclogites, separated by north-dipping shear fault (Fig. 1) (Ye et al. 1994).

The northern zone, named as the Dingyuan formation, is mainly composed of meta-tuff and metabasalt. The Rb-Sr whole rock isochron of tuff from the Dingyuan area yields an age of 391 ± 13 Ma with initial $^{87}Sr/^{86}Sr$ = 0.7079 ± 2 (Ye et al. 1993). This age was interpreted as the eruption age of the low-grade metavolcanics (Ye et al.

1993). However, it is also possible to indicate the timing of metamorphism. Suo et al. (1993) suggest that the Dingyuan formation (northern zone) in Sujiahe complex and Guishan complex should be merged into one unit because of their similar lithological character. The formation age and geochemistry of the metabasalt in the Dingyuan formation and Guishan complex have not been studied. This paper is focused on these issues.

The U-Pb zircon ages of 377−400 Ma for the cold eclogite from the Sujiahe tectonic melange zone suggest that the south tectonic melange zone is a Paleozoic suture between the Beihuaiyang zone and North Dabie terrane (Jian et al. 1997; Li et al. 1998).

3 Petrology

Four Metabasaltic samples (95 HN-S-1, 3, 5, 6) and 2 acidic samples (95 HN-S-2, 4) were collected from the Dingyuan formation at two stone pits in Dingyuan, Henan Province (Fig. l). The rocks in these two neighboring stone pits are almost metabasalts, but a few acidic blobs with the diameters of 10−20 cm have been observed in one stone pit. The occurrence of the metabasaltic and acidic rocks seems not to be bimodal volcanics. Two foliations can be recognized in both basaltic and acidic rocks, which suggests that the basaltic and acidic rocks have the same deformation history.

Petrography of these rocks shows typical volcanic prophyritic structures. The porphyrites in the metabasalts are clinopyroxene, augite and plagioclase. The fine grain matrix in metabasalts is composed of chlorite, amphibole, plagioclase, epidote, muscovite and minor opaque minerals. It should be indicated that the sample 95 HN-S-6 contains more abundant opaque minerals ($\sim$10%) which seriously influence its major and trace element contents. The muscovite is developed along the foliations. The pyroxene and amphibole are partly replaced by abundant chlorites or epidote. The metamorphic mineral assemblage of chlorite + epidote + muscovite suggests a high greenschist facies of the metabasalts. The porphyrite in the acidic rock is mainly plagioclase. Its fine grain matrix is composed of quartz, plagioclase and muscovite. The muscovite is developed along the foliations, but folded by the second deformation. The typical volcanic porphyritic structure of the acidic rock and its similar deformation history to the metabasalts suggest that the acidic blobs in metabasalts seem not to be the metamorphic fractional product, but more like to be boudins transformed from the acidic dyke in basalt during the first deformation.

4 Analytical

The rock was broken into the chips with diameters of 4−6 mm. 100 gram of fresh rock chips without any alteration or other secondary products, which are chosen by handpicking, were leached by 2 N HCl at about 70−80 ℃ for 1 h and then were cleaned with H_2O for three times to remove any surface contamination. The rock powder was grinded in agate ball mill.

The major elements were obtained at the Institute of geophysical and geochemical Exploratory Techniques, Chinese Academy of Geological Sciences, by X-ray fluorescence analysis except CO_2 and H_2O which were obtained by nonaqueous titration and gravimetric analysis respectively. Trace elements were analyzed by ICP-MS at the Institute of Geology, Chinese Academy. Rb-Sr and Sm-Nd isotope data were obtained at the isotope laboratory of the modern analysis center, Nanjing University, following procedures described elsewhere (Wang et al. 1992). The $^{143}Nd/^{144}Nd$ ratios were normalized relative to the value of $^{146}Nd/^{144}Nd = 0.7219$. Blanks for the whole chemical procedures are 50−60 pg for Sm and Nd 0.1 ng, for Rb and lng for Sr. The Nd and Sr isotopic standard values analyzed in this experiment are 0.511864 ± 8 for La Jolla Nd standard and 0.710241 ± 11 for NBS 987 Sr standard. Ages are calculated using "ISOPLOT" software provided by Dr. K. Ludwig and given with 2 errors. The major and trace element data are listed in Table 1 and the Sr and Nd isotopic data are listed in Table 2.

5 Major and Trace Element Geochemistry

The SiO_2-($Na_2O + K_2O$) diagram shows that the metabasalts in the Dingyuan formation are basically tholeiites except the sample 95 HN-S-1 which falls into the basaltic andesite field (Fig. 2). The AFM diagram shows that these metabasalts belong to calc-alkaline series except the sample 95 HN-S-6 which contains more abundant Fe$_2$O$_3$ than other samples (Fig. 3), because of its more abundant opaque metallic minerals than other samples.

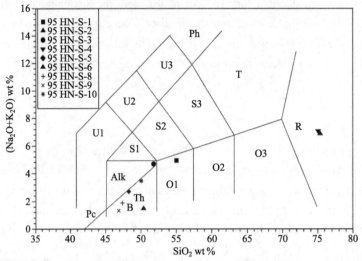

Fig. 2 SiO$_2$-(Na$_2$O + K$_2$O) classification diagram for volcanics. Pc: picritic basalt; B: basalt, alkaline basalt, subalkaline basalt; O1: basaltic andesite; O2: andesite; O3: dacite; R: rhyolite, alkali rhyolite; S1: trachybasalt, Hawaiite, potassic trachybasalt; S2: basaltic trachyandesite, olivine trachyandesite, olivine basaltic trachyte; S3: trachyandesite, andesitic trachyte; T: trachyte, alkaline trachyte, trachydecite; U1: tephrite basanite; U2: phonolitic basanite; U3: basanitic phonolite; Ph: phonolite.

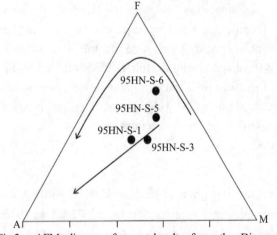

Fig.3 AFM diagram for metabasalts from the Dingyuan formation. A = K$_2$O ± Na$_2$O; F = FeO + 0.9 × Fe$_2$O$_3$; M = MgO; TH = tholeiitic series; CA = calc-alkaline series.

The striking trace elemental features of both metabasalts and acidic rocks in Dingyuan formation is high Ba, Pb, Sr contents and low Nb, Ti, P contents (Table 1). Their positive Ba, Pb, Sr anomalies and negative Nb, Ti, P anomalies are easy to be observed in the primitive mantle normalization diagram (Fig. 4A). These are the typical features of subduction zone related magmas and very similar to those of the Danfang and Erlangping metavolcanics as well as the Lajimiao norite-gabbro intrusion, which have been considered to be the Paleozoic island arc magmatism in the North Qinling (Li et al. 1994; Sun et al. 1995, 1996). In addition, high La/Nb, Ba/Nb and Th/Nb ratios of the metabasalts in Dingyuan formation are also typical island arc features (Fig. 5).

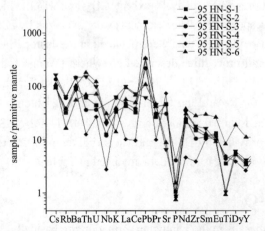
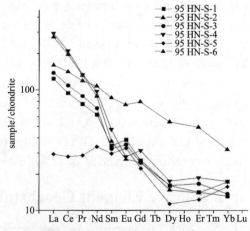

Fig. 4 A. Trace element primitive mantle normalization diagram of the metavolcanics from the Dingyuan formation; B. Rare earth element chondrite normalization diagram of the metavolcanics from the Dingyuan formation.

The REE chondrite normalization diagram (Fig. 4B) shows that all sample, except the sample 95 HN-S-6, have similar HREE flat patterns, but highly fractional LREE patterns including LREE flat pattern (95 HN-S-5) to LREE enriched (95 HN-S-1, 3) and highly enriched (acidic rocks) patterns. It reflects a heterogeneous mantle source of the metabasalts.

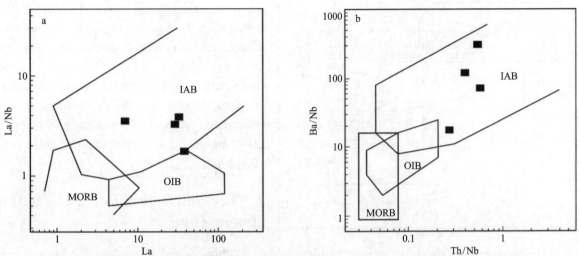

Fig. 5 Ba-Th-Nb-La classification diagram of tectonic settings for the metabasalts from the Dingyuan formation.

6 Rb-Sr and Sm-Nd Isotopic Ages and Geochemistry

The figure 6 shows that the Sm-Nd and Rb-Sr whole rock isochrons yields two consistent ages of 446 ± 23 Ma and 444 ± 31 Ma. These ages are older than the initial metamorphic ages of the Dingyuan formation and Guishan complex, such as the Rb-Sr age of 391 ± 13 Ma for metatuff in the Dingyuan formation (Ye et al. 1993) and a hornblende $^{40}Ar/^{39}Ar$ age of 401 ± 4 Ma for a garnet-amphibolite in the Guishan complex (Niu et al. 1994). These ages are also older than the formation age of 392 ± 25 Ma given by U-Pb zircon dating of acidic metavolcanics from the upper formation of the Guishan complex (Ye et al. 1993). Consequently, we interpret these ages as the formation age of the Dingyuan metabasalts rocks. It suggests that island arc magmatism in the Beihuaiyang zone was active in the later Ordovician to early Silurian. This can be compared with the island arc volcanics of the Danfang group in the North Qinling belt. The Rb-Sr whole rock isochron age of 447.8 ± 41.5 Ma for the metavolcanic rocks in the Danfang group (Sun et al. 1987) is consistent with the Rb-Sr and Sm-Nd ages reported in this paper. The Ordovician and Silurian radiolarian discovered in the chert in the Danfang group also suggests that the Danfang island arc magmatism was active in the Ordovician to Silurian (Cui et al. 1995).

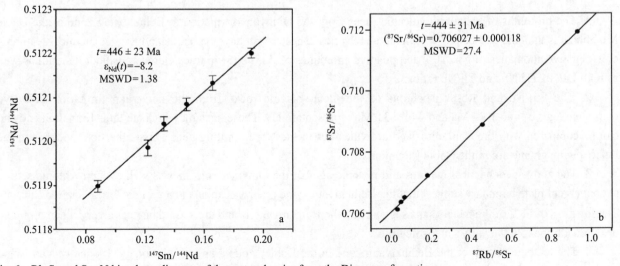

Fig. 6 Rb-Sr and Sm-Nd isochron diagram of the metavolcanics from the Dingyuan formation.

The initial $^{143}Nd/^{144}Nd$ and $^{87}Sr/^{86}Sr$ values of the metavolcanics in the Dingyuan formation are 0.511643 ± 23 ($\varepsilon_{Nd} = -8.2$) and 0.706029 ± 121 respectively (Fig. 6). This low ε_{Nd} value of -8.2 can be compared with those of the Less Antilles island arc (Davidson 1983), but is significantly lower than those of the North Qinling island arc volcanics and gabbro (Li et al. 1993; Sun et al. 1996 a, b) (Fig. 7). It suggests that the oceanic subduction under the Beihuaiyang arc could involve much more old terrestrial sediments than the North Qinling arc. Thus there must be a very old continent or microcontinent near the Beihuaiyang arc in the Ordovician. The North Dabie terrane with very high TDM values of 2.1–2.6 Ga could be more suitable candidate (Li et al. 1998).

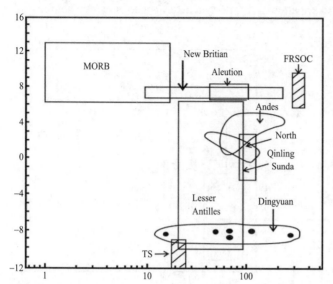

Fig. 7 ε_{Nd}-Ba/Nb diagram for comparison between the island arc basalts (after Li, 1995). MORB: mid-ocean ridge basalts; TS: terrestrial sediments; FRSOC: fluid released from subducting oceanic crust.

7 Tectonic Evolution of the Beihuaiyang Zone

As mentioned above, the Beihuaiyang zone is composed of two units. The lower unit including the Guishan complex and Dingyuan formation once was an island arc in the Ordovician to Silurian. The upper unit including the Nanwan formation and the Foziling group is a Devonian metaclastic flysch formation. The absence of volcanic rocks in the upper unit suggests the termination of the island arc magmatism in Beihuaiyang zone in the Devonian. The U-Pb zircon age of 377 Ma for the cold eclogite from the Sujiahe tectonic melange zone also suggest that the Beihuaiyang island arc may have been collided with the North Dabie microcontinent in the Devonian (Li et al. 1988). Consequently, the Beihuaiyang zone may become a remainder basin in the Devonian, which is represented by the Nanwan and Foziling metaclastic flysh formation, thus, the Nanwan formation and the Foziling group can only be compared with the Liuling group, a Devonian metaclastic flysh formation in the Qinling area, which is located to the south of the North Qinling belt.

In light of this discussion, we suggest that the tectonic setting of the Beihuaiyang zone in the Dabie Mountains is evolutionary. It once was an island arc in the Ordovician to Silurian, but became a remainder basin after collision between the Beihuaiyang island arc and North Dabie microcontinent in the Devonian.

8 Conclusions

1. The protoliths of metabasic volcanic rocks from the Dingyuan formation in Beihuaiyang zone of the Dabie Mountains is tholeiite and belong to calc-alkaline series. These rocks have typical island arc geochemical features : e. g. negative anomalies of Nb, Ti, P and positive anomalies of Ba, Pb, Sr in trace elemental spider diagram as well as high La/Nb, Ba/Nb and Tli/Nb ratios.

2. The Sm-Nd and Rb-Sr isochrons of the meta-volcanic rocks from the Dingyuan formation give two consistent ages of 446 ± 23 Ma and 444 ± 31 Ma representatively. These ages may indicate their formation age and can be compared with the Danfang island arc volcanics in the North Qinling belt. This is the first evidence to the Paleozoic magmatic arc in the Dabie Mountains.

3. The above geochemical features and isotopic ages of the Dinyuan volcanics as well as their tectonic setting at convergent plate boundary suggest that the Beihuaiyang zone once was an island arc in the Ordovician to Silurian. The lows ε_{Nd} of -8.2 of the metabasalts suggest that the Beihuaiyang island arc should be near a very older continent or microcontinent in the early Paleozoic time.

4. The tectonic setting of the Beihuaiyang zone in the Dabie Mountains is evolutionary. It once was an island arc in the Ordovician to Silurian, which is represented by the Guishan complex and Dingyuan formation. However,

after the collision between the Beihuaiyang island arc and North Dabie microcontinent in the Devonian, the Beihuaiyang zone became a remainder basin represented by the Nanwan formation and the Foziling group.

Acknowledgements

This work is supported by National Natural Science Foundation of China (No. 49573190, 49794042).

References

Cui, Z., Sun, Y., Wang, X., 1995. The discovery of Raliolarian in the Danfang ophiolite belt, North Qinling and its geological implications, China Sci. Bull., 40 (18): 1686-1688. (in Chinese).
Cong, B., Wang, Q., Zhai, M., Zhang, R., Zhao, Z. and Ye, K., 1994. Ultra-high pressure metamorphic rocks in the Dabie-Su-Lu region, China: Their formation and exhumation . The Island Arc, 3: 135 -150.
Davidson J. P., 1983. Lesser Antilles isotopic evidence of the role of subducted sediment in island genesis. Nature, 306, 253-256.
Gao, L., Liu, A., 1988, The discovery of micro-fossil from the Nawan formation of the Xinyang group and its geological implications, Geol. Review, 3 (5): 421-422 (in Chinese).
Jian, P., Yang, W., Li, Z. and Zhou, H., 1997. Isotopic chronological evidence for the Caledonian Xiongdian eclogites in the western Dabie Mountains, Acta Geol. Sinica, 71: 133-141 (in Chinese).
Li, S., Chen, Y., Zhang, Z., Ye, X., Zhang, G., Guo, A., Hart, S. R., 1994. trace element and Sr, Nd isotopic geochemistry of the Lajimiao norite-gabbro in North Qinling belt, Acta Geol. Sinica, 7(2): 137-151.
Li, S., 1995. Implications of ε_{Nd}-La/Nb, Ba/Nb, Nb/Th diagrams to mantle hetrogeneity-classification of island-arc basalts and decomposition of EMU component, Chinese J. Geochemistry, 14 (2): 117-127.
Li, S., Jagoutz, E., Zhang, Z., Lo, C. H., Chen, W. and Chen, Y., 1998. Geochemical and geochronological constrains on the tectonic outline of the Dabie Mountains, central China: A Continent-microcontinent-continent collision model, Continental Geodynamics (in press).
Li, S., Li, H., Han, W., Long, G., Zhang, Z., Zhang, S., 1998. Timing and tectonic setting of the cold eclogites from Sujiahe tectonic melange zone, Dabie Mountains : Evidences from U-Pb zircon age and geochemistry of the eclogites (Submitted to Sci. in China. Ser. D.).
Liu, X. and Hao, J., 1989. Structure and tectonic evolution of the Tongbai-Dabie range in the east Qinling collisional belt, China, Tectonics, 8(3): 637 - 645.
Niu, B., Fu, Y., Liu, Z., Ren, J., Chen, W., 1994. Main tectonothermal events and $^{40}Ar/^{39}Ar$ dating of the TongbaiDabie Mountains, Acta Geosci. Sin., (1-2): 20-34 (in Chinese).
Okay, A. T. and Sengör, A. M. C., 1989. Coesite from the Dabie Shan eclogites, central China, Eur. J. Mineral.,1,595-598.
Okay, A. T. and Sengör, A. M. C., 1993. Tectonics of an ultrahigh-pressure metamorphic terrane : the Dabie Shan/Tongbai Shan orogen, China, Tectonics, 12, 1329-1334.
Sun, Y., Yue, Z, Zhang, G., 1987. REE Geochemistry of the ophiolite in the eastern Qinling, Chinese, Sci, Bull., 32 (3) : 1654-1655 (in Chinese).
Sun, W., Li, S., Xiao, Y., Sun, Y., Zhang, G., 1995. The discovery of island arc volcanics in the Heihe area, North Qinling and its tectonic implications, Geotectonics et Metallogenia, 19 (3): 227 -236 (in Chinese).
Sun, W., Li, S., Sun, Y., Zhang, G., Zhang, Z., 1996. Chronology and geochemistry of a lava pillow in the Erlangping group at Xixia in the North Qinling Mountains, Geol. Review, 42 (2): 144-153 (in Chinese).
Suo, S., Sang, L., Han, Y., You, Z., Zhou, N., 1993. The Petrology and tectonics in Dabie precambrian metamorphic terranes, central China, Press of China Univ. of Geo Sci., Wuhan, 33-112 (in Chinese).
Wang, X., Liou, J. G. and Mao, H. K., 1989. Coesite-bearing eclogites from the Dabie Mountains in central China, Geology, 17: 1085 -1088.
Wang, Y.X., Yang, J.D., Guo, L.Z., Shi, Y.S., Hu, X.J., Wang, X. 1992. The discovery of the lower Proterozoic granite in Longquan, Zhejiang province, and the age of the basement. Geol. Rev., 38: 525-531.
Xu, G, and Hao, J., 1988. The characteristics of Foziling group and the geotectonic enviroment of its formation in the northern foot of the Dabie Mountains, Scientia Geol. Sinica, (2): 95 -109 (in Chinese).
Xu, S., Okay, A. 1., Ji, S., Sengör, A. M. C., Su, W., Liu, Y. and Jiang, Li, 1992a. Diamond from the Dabie Shan metamorphic rocks and its implication for tectonic setting, Science, 256: 80- 92.
Xu, S., Jing, L., Liu, Y. and Zhang, Y., 1992b. Tectonic framework and evolution of the Dabie Mountains in Anhui, eastern China, Acta Geol. Sinica, 66(1): 1-14 (in Chinese).
Ye, B., Jiang, P., Xu, J., Cui, F., Li, Z. and Zhang, Z., 1993. The Sujiahe terrane collage belt and its constitution and evolution along the north hill slope of the Tongbai-Dabie orogenic belt, Press of China Univ. of Geosci., Wuhan, 66-67 (in Chinese).
Ye, B., Jian, P., Xu, Z., Cui, F., Li, Z. and Zhang, Z., 1994. Timing of the Sujiahe group in the Tongbai-Dabie orogenic belt, In : " Research of isotope geochemistry", edit, by Chen, H., Zhejiang Univ. Press, Hangzhou, 175-186 (in Chinese).

Accretional history of the North and South China Blocks, and the microcontinents between them: implications for dispersion of Gondwanaland

Shuguang Li and Weidong Sun

Department of Earth and Space Sciences, University of Science and Technology of China, Hefei 230026, Anhui, China

亮点介绍：综合北秦岭、南秦岭和介于它们之间商-丹及勉-略断裂带的岩石组成，年龄及元素-同位素特点，深入探讨了华北与华南陆块及它们之间的南秦岭微陆块的古生代和三叠纪的拼贴和碰撞历史及其对冈瓦纳大陆裂解和离散的意义。

The North Qinling is a collision zone between the North China Block (NCB) and South China Block (SCB) as well as the North Qinling Microcontinent (NQM) and South Qinling Microcontinent (SQM) (Zhang et al., 1995; Yin and Huang, 1996; Li et al., 1996, 1998; Li and Sun, 1996). Except for the NCB, all of these microcontinents and SCB have been considered to be derived from the southern hemisphere or Gondwanaland. This conclusion is supported by their southern hemisphere "Dupal" characteristics of high radiogenic Pb isotopic compositions (Hart, 1984; Zhang et al., 1996; Metcalfe, 1996). Therefore, investigation of the accretional history of these blocks and microcontinents is of importance in understanding the dispersional history of the Gondwanaland.

The NQM is bound by Shang-Dan (Shangnan-Danfeng) Fault in the south and Luonan-Luanchuan Fault in the north. It consists of the Danfeng and Heihe volcanic arcs, the Qinling core complex, the Erlangping volcanic arc, and the Kuanping Group from south to north. To the south of the Shang-Dan Fault is Liuling Group (D2?), the north margin of SQM, to the north of Luonan-Luanchuan Fault is Taowan Group, the south margin of NCB. Danfeng Group lies along the northern part of Shang-Dan Fault. It consists of island-arc type metabasalts to meta-andesites of greenschist facies to low amphibolite facies, associated with low-maturity clastic rocks (e.g., Sun, Y. et al., 1988; Li et al., 1994a; Sun, W., 1994; Sun, W. et al., 1995). The Erlangping Group consists of upper greenschist facies to amphibolite facies island-arc type and mid-ocean-ridge type metabasalts suggesting a back-arc environment (Sun and Li, 1998). Cambrian-Ordovician radiolaria and Palaeozoic crinoid fossils in cherts have been reported in both Erlangping Group and Danfeng Group in the NQ (Zhang and Tang, 1983; Xiao, 1988; Cui et al., 1995; Wang et al., 1995, Niu et al., 1993). Zircon $^{207}Pb/^{206}Pb$ ages of 487 Ma to 470 Ma for plagiogranites developed in the Erlangping and Danfeng metabasalts also document their Ordovician origins (Xue et al., 1996). All these and the higher radiogenic Pb isotopic compositions ($^{206}Pb/^{204}Pb$ = 18.233–18.716; $^{207}Pb/^{204}Pb$ = 15.539–15.614; $^{208}Pb/^{204}Pb$ = 38.136–39.036; Zhang et al., 1996) which are similar to the Dupal anomaly in the South Hemisphere (Hart, 1984), suggest that the NQM was an island arc in the South Hemisphere in the early Palaeozoic time. However, we have no evidence for that the NQM was once a part of the Gondwanaland, though many Sm-Nd ages around 1000 Ma for metabasalts and ophiolites in the NQM (Li et al., 1991; Zhang et al., 1994) suggest that the NQM was once a part of the Supercontinent Rodinia.

The Rb-Sr age (405 ± 18 Ma) of the Erlangping pillow lava (this paper) and hornblende $^{40}Ar/^{39}Ar$ ages (404–435 Ma) of Erlangping Group in Tongbai Mountains (Zhai et al., 1998) suggest that the Erlangping back-arc basin was closed in Caledonian. This is supported by mineral Sm-Nd isochron age of 400 ± 16 Ma for eclogite from the Zhu-Xia Fault zone which separates the Erlangping Group and Qinling Complex (Hu et al., 1996). The Sm-Nd (414 ± 42 Ma, 413 ± 35 Ma), Rb-Sr (410.4 ± 2.7 Ma, 414 ± 35 Ma) and $^{40}Ar/^{39}Ar$ (426.0 ± 1.7 Ma) ages of the two metamorphic amphibolite samples of Danfeng Group are consistent with each other within error, which suggest that the Danfeng volcanic rocks were metamorphosed during 410–426 Ma.

Comparing the metamorphic history of the NQ with those in adjacent areas may help us to make dear whether this Caledonian metamorphic event was caused by the collision between the NCB and the SCB or between NCB and NQM. Abundant metamorphic ages of about 400 Ma have been reported for Taowan Group, the southern margin of NCB (Zhang et al., 1994). It suggests that Taowan Group was also involved in the Caledonian event of NQM (Zhang et al., 1994). In contrast, there is no reliable Caledonian metamorphic or magmatic ages reported in SQM and the northern margin of SCB. Ages of the earliest metamorphic event reported in the northern margin of SQM are $^{40}Ar/^{39}Ar$ ages of 314 Ma for biotite from Liuling Group (Mattauer et al., 1985) and 304–316 Ma for hornblende from Xinyang Group, the east equivalence of Liuling Group (Zhai et al., 1998). The oldest metamorphic age of Carboniferous period in the northern margin of SQM recorded in Liuling Group and Xinyang Group (Mattauer et al., 1985; Zhai et al., 1998) suggests that the SQM was not involved in the Caledonian event in NQM. All these suggest that the NQM had collided with NCB in later Silurian, while the collision between the NQM and the SQM along the Shang-Dan Fault (suture) did not take place at that time. This conclusion is also supported by other geochronological evidence: the I-type granites and the IAB-type mafic pluton intruded in Danfeng Group are thought to be the products of the subduction of oceanic crust (Lerch et al., 1995; Li et al., 1989,1994a; Zhang et al., 1994). Zircon U-Pb ages (422–382 Ma) and the Sm-Nd mineral isochron age (403 Ma) suggest that the oceanic subduction along Shang-Dan Suture was still active till ca. 382 Ma implying that the SQM has not been connected with NQM at that time.

The Sujiahe eclogite melange zone developed in the western Dabie Mountains has been considered to be the eastern extension of the Shang-Dan suture. The U-Pb concordant age of 377 ± 7 Ma of the metamorphic zircons from an eclogite in this zone (Li et al., 1999) is consistent with the termination time (ca. 382 Ma) of oceanic subduction along the Shang-Dan suture. It implies that the SQM could be connected to the NQM in the Devonian time.

Besides the northern suture along the Shang-Dan Fault in the North Qinling area, tectonic studies of the Qinling orogenic belt have revealed the presence of a Southern suture zone along the Mian-Lue tectonic zone which separates the SQM from the SCB (Sun et al., 1991; Zhang et al., 1995). The Mian-Lue tectonic zone consists of several sheared thrust slabs of Sinian-Cambrian gravel-bearing argillite, argillite and carbonatite and Devonian-Carboniferous deep-water turbidite and carbonate rock. In contrast to the neighbouring strata bearing the Silurian system, the Ordovician and Silurian beds are rare in the Mian-Lue tectonic zone. Tectonic slabs of ophiolite, island-arc type and bimodel volcanic rocks were also involved in this zone. All these suggest that the SCB and SQM may have been accreted together in the later Sinian-Cambrian which is consistent with the assembly time of Gondwanaland (Li and Powell, 1993) and rifted initially in the Devonian which is also consisted with the rifting time of the northern margin of Gondwana (Metcalfe, 1996). The LREE-depleted, low-K, MORB-type basalts from this zone suggests that the rift was opened to an oceanic basin (Li et al., 1996). The relative higher radiogenic Pb isotopic compositions ($^{206}Pb/^{204}Pb$ = 18.233–18.716; $^{207}Pb/^{204}Pb$ = 15.539–15.614; $^{208}Pb/^{204}Pb$ = 38.136–39.036) of basalts from this zone suggest that the Palaeozoic Mian-Lue ocean was a part of the Palaeo-Tethys system (Zhang et al., 1996). Zircons from plagiogranites developed in the Mian-Lue ophiolite yield a lower intercept age of 299 ± 67 Ma and upper intercept age of 912 ± 34 Ma. The discordant line defined by these zircon data has been interpreted as mixing line of inherited core and magmatic overgrowth of zircons. Thus the lower intercept age of 299 ± 67 Ma indicates the crystallization time of the plagiogranite, which suggests a Carboniferous ocean in the Mian-Lue zone. The Sm-Nd and Rb-Sr whole rock isochron ages of 242 ± 21 Ma and 221 ± 13 Ma for the metavolcanic rocks in this zone consistently indicate their metamorphic time suggesting the closure of the Mian-Lue oceanic basin (Li et al., 1996). This is also supported by the Triassic syncollisional granites broadly developed to the north of Mian-Lue tectonic zone and south of the Shang-Dan Fault. Detailed U-Pb zircon ages showed that their formation ages range from 206 to 220 Ma (Li and Sun, 1996). Therefore, the SCB including the Yangtze and Cathaysia blocks collided with the SQM, NQM and NCB in the Triassic which is consistent with the formation ages of ultrahigh-pressure metamorphic rocks (220–230 Ma) in the Dabie Mountains and Sulu terrane (e.g. Li et al., 1993; 1994b; Chavagnac and Jahn, 1996; Rowley et al., 1997). The similar history of tectonic evolution of the Mian-Lue tectonic zone to the Gondwanaland suggests that the SCB and SQM may have been attached to Gondwanaland during the Cambrian-Silurian and rifted from Gondwanaland in Devonian.

Acknowledgements

This work is supported by the National Nature Science Foundation of China (No. 49472144).

References

Chavagnac, V. and Jahn, B. M. (1996) Coesite-bearing eclogites from Bixiling complex, Dabie Mountains, China: Sm-Nd ages, geochemical characteristics and tectonic implications. Chem. Geol., v.133, pp.29-51.
Cui, Z.L., Sun, Y. and Wang, X.R. (1995) The discovery of Radiolaria from Danfeng ophiolite in Qinling Mountains and its geological significance. Chinese Sci. Bull., v.40, pp.1686-1688.
Hart, S.R. (1984) A large-scale isotope anomaly in the Southern Hemisphere mantle. Nature, v.309, pp.753-757.
Hu, N.G., Yang, J.X. and Zhao, D.L. (1996) Sm-Nd isochron age of eclogite from Northern Qinling Mountains. Acta Mineralogica Sinica, v. 16, pp.349-352 (in Chinese with English abstract).
Lerch, M.F., Xue, F., Kroner, A., Zhang, G.W. and Todt, W. (1995) A Middle Silurian-Early Devonian magmatic arc in the Qinling Mountains of central China. J. Geology, v.103, pp.437-449.
Li, S., Hart, S.R., Zheng, S., Liou, D., Zhang, G. and Guo, A. (1989) Timing of collision between the North and South China blocks-the Sm-Nd isotopic age evidence. Sci. Sinica (Series B), v. 32, pp.1391-1400.
Li, S., Chen, Y., Zhang, G. and Zhang Z. (1991) A 1 Ga B.P. Alpine peridotite body emplaced into the Qinling Group: evidence for the existence of the late Proterozoic plate tectonics in the North Qinling area. Geol. Rev., v.37, pp.235-242.
Li, S., Xiao, Y., Liou, D., Chen, Y., Ge, N., Zhang, S., Sun, S. S, Cong, B., Zhang, R., Hart, S.R. and Wang, S. (1993) Collision of the North China and Yangtze blocks and formation of coesite-eclogites: timing and processes. Chem. Geol., v. 109, pp.89-111.
Li, S., Chen, Y., Zhang, Z.Q., Ye, X., Zhang, G.W. and Hart, S.R. (1994a) Trace elements and Sr, Nd isotope geochemistry of the Lajimiao norite-gabbro from the North Qinling belt. Acta Geol. Sinica, v.67, pp.137-151 (in Chinese with English abstract).
Li, S., Wang, S., Chen, Y., Liou, D., Qiu, J., Zhou, H. and Zhang, Z. (1994b) Excess argon in phengite from eclogite: evidence from dating of eclogite minerals by Sm-Nd, Rb-Sr and $^{40}Ar/^{39}Ar$ methods, Chem. Geology, v. 112, pp.343-350.
Li, S. and Sun, W. (1996) A middle Silurian-early Devonian magmatic arc in the Qinling Mountains of central China: a discussion. J. Geology, v.104, pp.501-503.
Li, S., Sun, W., Zhang, G., Chen, J. and Yang, Y. (1996) Chronology and geochemistry of metavolcanic rocks from Heigouxia Valley in the Mian-Lue tectonic arc, South Qinling—Evidence for a Palaeozoic oceanic basin and its close time. Science in China, v. 39(3), pp.300-310.
Li, S., Jagoutz, E., Chen, Y., Zhang, Z., Chen, W. and Lo, C. H. (1998) Geochemical and Geochronological constrains on the tectonic outline of the Dabie Mountains, central China: a continent-microcontinent-continent collision model. Continental Dynamics, v.3, pp.14-31.
Li, S., Han, W., Long, G., Xiao, Y., Zheng, Y., Fu, B., Li, H., Zhang, Z. and Zhang, S. (1999) Timing and tectonic setting of the Sujiahe eclogite in the Dabie Mountains—U-Pb zircon age and geochemical evidences. Sci. in China (Ser. D), (in press).
Li, Z.X. and Powell, C. McA. (1993) Late Proterozoic to Early Palaeozoic, palaeomagnetism and the formation of Gondwanaland, in Gondwana 8: assembly, evolution and dispersal, edited by R.H. Findlay, R. Unrug, M.R. Banks and J.J. Veevers, A.A. Balkema, Rotterdam, pp.9-21.
Mattauer, M., Matte, P., Malavieille, J., Tapponnier, P., Maluski, H., Ku, Z., Lu, Y. and Tang, Y. (1985) Tectonics of the Qinling Belt: build-up and evolution of eastern Asia. Nature, v.317, pp.496-500.
Metcalfe, I. (1996) Report of IGCP Project 321 Fourth International Symposium-Gondwana dispersion and Asian accretion. The Australian Geologist. News Letter, No. 98, pp.23-29.
Niu, B., Liu, Z. and Ren, J. (1993) The tectonic relationship between the Qinling Mountains and Tongbai-Dabie Mountains with notes on the tectonic evolution of the Hehuai Basin. Bull. Chinese Academy Geo. Sci., v. 26, pp.1-12 (in Chinese with English abstract).
Rowley, D.B., Xue, F., Tucker, R.D., Peng, Z.X., Baker, J. and Davis, A. (1997) Age of ultrahigh pressure metamorphism and protolith of orthogneisses from the eastern Dabie Shan: U/Pb zircon geochronology. Earth Plan. Sci. Lett., v.151, pp.191-203
Sun, S., Li, J. and Lin, J. (1991) Indosinides in China and the consumption of eastern Paleo-Tethys in controversies in modern geology: evolution of geological theories in sedimentology, In: Muller, D.W., McKenzie, J.A. and Weissert, H. (Eds.) Earth history and Tectonics. London, Academic Press. 363p.
Sun, W., Li, S., Xiao, Y., Sun, Y. and Zhang, G. (1995) Discovery of Island Arc basalts in Danfeng Group and its tectonic significance. Geotectonica et Metalogenia, v.19, pp.227-236 (in Chinese with English abstract).
Sun, W. and Li, S. (1998) Pb isotopes of granitoids suggest Devonian accretion of Yangtze (South China) craton to North China craton, Comment. Geology, v.26, pp. 859-860.
Sun, Y., Yun, Z.P. and Zhang, G.W. (1988) Geochemistry of ophiolites from east Qinling area. In: Zhang, G.W. (Ed.) The formation and evolution of Qinling orogenic belt. Northwest University Press. Xi'an, pp.65-74 (in Chinese with English abstract).
Wang, X.R., Hua, H. and Sun, Y. (1995) Study on micro-fossils of Erlangping Group from Taowan area, Xixia county, Henan Province.

Acta North West Uni. (Nature Sci.), v. 25, pp.353-358 (in Chinese with English abstract).

Xiao, S.Y. (1988) North Qinling metamorphic strata. Xi'an Jiaotong Univ. Press, Xi'an, p.47 (in Chinese with English abstract).

Xue, F., Kroner, A., Reischmann, T. and Lerch, F. (1996) Paleozoic pre- and post-collision calc-alkaline magmatism in the Qinling orogenic belt, central China, as documented by zircon ages on granitoid rocks. J.Geol. Society, v.153, pp.409-417.

Yin, H. and Huang, D. (1996) Early Palaeozoic evolution of the Zhen-an-Xichuan block and the small Qinling multi-island ocean basin. Acta Geolo. Sinica, v.9, pp.1-15.

Zhai, X., Day, H.W., Hacker, B.R. and You, Z.D. (1998) Paleozoic metamorphism in the Qinling orogen, Tongbai Mountains, central China. Geology, v.26, pp.371-374.

Zhang, B., Zhang, H., Zhao, Z. and Ling, W. (1996) Geochemical subdivision and evolution of the lithosphere in East Qinling and adjacent regions—Implications for tectonics. Sci. in China (Ser. D), v. 39, pp. 245-255.

Zhang, G. W., Meng, Q. R. and Lai, S. (1995) Structure and tectonics of the Qinling orogenic belt. Science in China, Ser. B. v. 38, pp.1379-1394.

Zhang, S.C. and Tang, S.W. (1983) The discovery of early Paleozoic Radiolaria in North Qinling and plate tectonics. Shanxi Geology, v.2, pp.1-9 (in Chinese with English abstract).

Zhang, Z. Q., Liu, D. Y. and Fu, G. M. (1994) Isotopic geochronology of metamorphic rocks in the North Qinling belt. Geological Press, Beijing (in Chinese with English abstract).

大别山北部榴辉岩的地球化学特征和 Sr、Nd 同位素组成及其大地构造意义*

刘贻灿[1,2]，徐树桐[2]，李曙光[1]，江来利[2]，吴维平[2]，陈冠宝[2]，苏 文[2]

1. 中国科学技术大学地球和空间科学系，合肥 230026
2. 安徽省地质科学研究所，合肥 230001

> **亮点介绍**：本文根据岩石地球化学及 Sr、Nd 同位素组成研究，查明北大别榴辉岩的原岩主要为拉斑玄武岩和少数可能为辉长岩，并且大多数属于扬子俯冲陆壳的一部分，为扬子陆壳（下地壳）俯冲变质成因，排除了它们形成于古岛弧环境下的可能性。

摘要 大别山北部镁铁-超镁铁质岩带中榴辉岩的岩石地球化学特征及 Sr、Nd 同位素组成表明：(i) 本区榴辉岩的原岩主要为拉斑玄武岩和少数可能为辉长岩，并且大多数可能属于扬子俯冲陆壳的一部分，为扬子陆壳（下地壳）俯冲变质成因，少数可能属扬子与华北大陆板块之间的古洋壳残片；(ii) 本区榴辉岩大多具有 Nb 正异常、K 负异常等，排除了它们形成于岛弧环境下的可能性；(iii) 磨子潭-晓天断裂以南含榴辉岩和变质橄榄岩的变质镁铁-超镁铁质岩带可能代表扬子与华北两个大陆板块之间的碰撞缝合带。其中除含有扬子俯冲陆壳外，还混有古洋壳残片。

关键词 榴辉岩；Sr、Nd 同位素组成；扬子俯冲陆壳；古洋壳残片；大别山北部

大别山，因其中含柯石英[1,2]和金刚石[3,4]等超高压岩石的发现而闻名于世。但是，由于在大别山北部镁铁-超镁铁质岩带（指五河-水吼剪切带以北至磨子潭-晓天断裂以南分布区）中长期未发现有真正的榴辉岩，该带岩石组成又比较复杂[5-9]，因此对该带构造背景长期存在认识上的分歧[8-15]。此外，尽管目前人们都认为大别山属于扬子与华北两个大陆板块之间的一条碰撞型造山带，但它们之间的缝合带位置究竟在何处仍存在争议。近年来，笔者等在大别山北部地区发现了两种产状榴辉岩并作了初步报道[16]。本文主要对研究区榴辉岩的岩石地球化学特征及 Sr、Nd 同位素组成进行了研究，并就其形成构造背景及扬子与华北两个大陆板块之间的缝合带位置进行了讨论。

1 地质背景

本文研究的榴辉岩产于大别山北部镁铁-超镁铁质岩带中，大致沿磨子潭-晓天断裂以南分布(位置见图 1)。研究区镁铁-超镁铁质岩可分为变质的和未变质的两种。其中未变质的辉石岩、辉长岩（如椒子岩、沙村等）的锆石 U-Pb 年龄为 120～130 Ma[17,18]，表明为碰撞后侵入岩。

北部镁铁-超镁铁质岩带内榴辉岩的围岩大多为正片麻岩（条带状片麻岩或花岗质片麻岩），如舒城县洪庙百丈岩和华庄（样号分别为 98121 和 98122-4）、桐城市毛花岩（样号为 9801）和汪洋水库（样号为 98WY 和 98WY-1）等地；少数为面理化橄榄岩，如铙钹寨（样号为 99104-2）、黄尾河（样号为 98701 和 98702）等。其中，99104-2、9801 和 98702 样较新鲜，主要由石榴子石、绿辉石和金红石等组成，含少量石英、斜方辉石、尖晶石、斜长石及角闪石等，且 9801 样含有较多金红石；98121、98122-4 和 98WY 或 98WY-1 样退变较强，主要由石榴子石、韭闪质角闪石和斜长石等组成，含少量绿辉石（<5%）和金红石等，而残留的

---

* 本文发表在：中国科学(D 辑)，2000，30: 99-107
对应的英文版论文为：Liu Yican, Xu Shutong, Li Shuguang, et al., 2000. Eclogites from the Northern Dabie Mountains, eastern China: geochemical characteristics, Sr-Nd isotopic compositions and tectonic implications. Science in China (Ser. D), 43: 178-188

绿辉石大多呈包体形式存在于石榴子石中。最近，笔者对 98702、99104-2 和 9801 三个榴辉岩样品中用于同位素测年的绿辉石副样作砂薄片并作了电子探针分析，结果表明每个样品中所测的 20 个单颗粒均为绿辉石（Na_2O 含量分别为 3.0%～3.6%、3.5%～4.5% 和 7.5%～8.7%，硬玉端员组分分别为 20～23、25～30 和 30～53），与笔者已报道结果[16]一致，这进一步证实本文所研究的榴辉岩无疑（详细论述见刘贻灿等[①]报道）。

变质岩石学研究[16][①]表明，大别山北部榴辉岩至少可以划分出三期变质阶段，即：① 榴辉岩相变质阶段，主要矿物组合为石榴子石+绿辉石+金红石等；② 高压麻粒岩相退变质阶段，主要矿物组合为石榴子石+斜方辉石+尖晶石+单斜辉石+斜长石角闪石等，利用石榴子石-斜方辉石对[19]计算麻粒岩相阶段变质压力为 1.10～1.37 GPa；③ 角闪岩相退变质阶段，主要矿物组合为角闪石+斜长石等。本区榴辉岩中石榴子石成分属铁铝榴石-镁铝榴石系列，其中铁铝榴石、镁铝榴石和钙铝榴石端元组分分别为 25～57、18～57 和 3～30。根据石榴子石与压力相关图[20]，可分为两类石榴子石：一类形成压力≥2.3 GPa，为榴辉岩相峰期变质形成的；另一类形成压力≤1.5 GPa，为高压麻粒岩相阶段变质形成的。石榴子石成分特征及钠质单斜辉石中石英出溶等说明本区榴辉岩可能经过超高压变质阶段[20]。美国斯坦福大学 Tsai 和 Liou 等[21]虽没有在研究区发现新鲜榴辉岩但发现榴辉岩相残留体（eclogite-facies relics）并证实它们曾经过超高压变质作用。

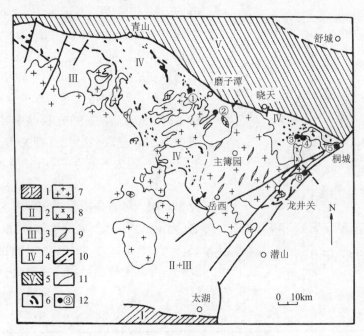

图 1　大别山（安徽部分）地质简图

1. 扬子板块俯冲盖层（宿松杂岩）；2. 超高压变质带；3. 扬子板块俯冲基底（大别杂岩）；4. 镁铁-超镁铁质岩带（可能包括部分大别杂岩）；5. 变质复理石推覆体；6. 镁铁-超镁铁质岩；7. 花岗岩；8. 辉长岩；9. 正长斑岩脉；10. 断层；11. 地质界线；12. 榴辉岩的采样点及编号：①饶钹寨、②黄尾河、③百丈岩、④华庄、⑤汪洋水库、⑥毛花岩

北部榴辉岩中已发现形成于麻粒岩相变质阶段的富 CO_2 流体包裹体以及在饶钹寨榴辉岩中发现早期榴辉岩相变质阶段富 N_2 等高盐度流体包裹体（肖益林，私人通信，2000），可能反映饶钹寨榴辉岩为洋壳成因（详见后文）以及高压麻粒岩相变质阶段存在富 CO_2 流体参与作用。

2 地球化学特征

2.1 主要化学成分及原岩性质

本区榴辉岩的 Al_2O_3、CaO 含量较高，而部分榴辉岩的 SiO_2 含量较低（表1）。其中，SiO_2、Al_2O_3、MgO、CaO 和 TiO_2 含量分别为 39.42%～47.97%、13.27%～16.81%、5.75%～9.92%、9.9%～15.01% 和 0.4%～2.15%；

---

① 刘贻灿等，2000，大别山北部榴辉岩的 Sm-Nd 年龄测定及其对麻粒岩相退变质时间的制约，待刊。

K_2O 含量较低,通常小于 0.1%(大多为 0.03%~0.06%)。对于退变较强的榴辉岩样品(如 98WY),则 K_2O 含量相对较高,表明(晚期)退变质流体可能会带进大离子亲石元素。本次分析的样品 TiO_2 含量大多数都较高,在 AFM 图(文中未附)中接近 F 端,属拉斑系列。SiO_2-Zr/TiO_2 和 Zr/TiO_2-Nb/Y 图解[22]表明,大部分样品落在亚碱性玄武岩区,而落在不易确定岩浆系列的低值区或碱性玄武岩区者为 98121 样。因此,98121 样品的 SiO_2 含量较低(39.42),且 TiO_2、Fe_2O_3+FeO、Al_2O_3 含量较高和贫 Na_2O、MgO,具有 Eu 正异常(见后文),可能反映它的原岩为拉斑系列的辉长岩;而其余样品则为拉斑玄武岩。

2.2 微量元素

Nb 含量及 Zr/Nb 值分别为 6.5~17.2 ppm 和 3.4~18.6,并且大多数具有 Nb 正异常(如样品 98702、99104-2、98121 等)(图 2),不显示与俯冲作用有关岩浆的亏损特征,且部分样品原始地幔[23]标准化后大离子亲石元素模式(图 2)类似于 E-MORB 或 N-MORB[24, 25]及南部超高压带中双河榴辉岩[26]即 K 明显负异常、Nb 正异常及 Y-Yb 正斜率。一般认为,影响岩石中 Nb 含量的矿物有金红石、韭闪质角闪石等[26, 27, 28];而且,即使在使 LREE 元素活动的过程中,Zr、Nb、Th 和 Ta 等元素仍能保持不活动[29]。本区榴辉岩 Nb 含量近于一致(多为 7~8 ppm)(图 3a,表 2),变化较小;即使退变非常强样品(如 98WY-1 样含有较多韭闪质角闪石等退变质矿物),其 Nb 含量也相似,反映后期作用(退变质作用等)对 Nb 含量影响非常小。因此,某些榴辉岩中 Nb 含量较高,如 9801 样品中 Nb 含量高达 17.2 ppm,可能与它含有较多金红石有关[26]。此外,Nb/Ta 值=3.2~21.2,但大多数为 13~16(图 3a),接近于陆壳 Nb/Ta 值(~11)[30],显示碱性玄武岩特征[27]。Green 等[28]研究也表明,尽管分析误差会影响 Nb/Ta 值,但碱性玄武岩(无论是来自大洋还是大陆的)Nb/Ta 一般都为 14~15。个别样品(如 99104-2)Nb/Ta 值较低(3.2),可能反映它与其他样品原岩分别形成于不同大地构造背景。

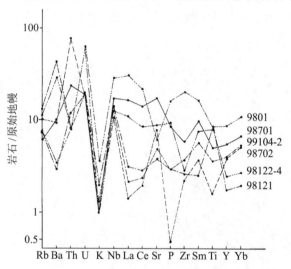

图 2 大别山北部榴辉岩的大离子亲石元素标准化图解

表 1 大别山北部榴辉岩的化学成分(%)

样号	SiO_2	Fe_2O_3	Al_2O_3	TiO_2	CaO	MgO	P_2O_5	MnO	K_2O	Na_2O	FeO	H_2O^+	烧失量	总量
98701	45.88	4.99	14.13	1.26	12.28	9.92	0.18	0.23	0.03	2.01	8.40	0.34	0.19	99.84
98702	44.08	3.65	14.23	0.40	15.01	13.85	0.01	0.21	0.04	1.29	6.61	0.92	0.20	100.5
98121	39.42	4.38	16.81	2.15	13.44	5.96	0.06	0.16	0.06	0.50	16.00	0.57	1.32	100.8
9801	47.97	2.71	13.72	2.10	9.90	5.75	0.34	0.22	0.11	2.89	13.08	0.34	1.00	100.1
98122-4	42.36	6.60	13.27	2.05	11.20	7.84	0.19	0.19	0.03	1.61	12.86	0.20	1.31	99.71
99104-2	45.39	1.08	14.59	0.90	13.08	9.74	0.06	0.22	0.06	1.61	11.69	0.13	1.16	99.71
98WY-1	46.54	3.28	16.27	1.87	10.15	5.81	0.31	0.23	0.30	2.72	10.19	0.60	1.21	99.48

注:由安徽省地质实验测试中心用化学法测定。

$(Nb/La)_N$ 值近于一致(图 3b),部分样品(如 98702)$(Nb/La)_N$ 值较高,可能是由该样品 LREE 亏损(La 含量低)造成的,而 98702 样 LREE 亏损可能与变质流体作用有关(见后文)。Huhma 等[31]和 Gruau 等[32]将镁铁-超镁铁质岩中轻稀土元素在变质过程中的活动性归结于富 CO_2 变质流体作用,Peltonen 等[29]也研究认为芬兰东北部 Jormua 蛇绿岩中变玄武岩(原岩为枕状熔岩)的轻稀土元素亏损是由富 CO_2 变质流体交代作用造成的。大别山高压-超高压或退变质期间 LREE 元素丢失现象已在南部超高压榴辉岩中被观察到[26]。$(Nb/Th)_N$ 值变化大多数可能是由 Th 含量变化造成的。Rudnick 等[33]和 Fowler[34]认为麻粒岩相变质作用期间 U、Th 行为是由副矿物相分解控制的。本区榴辉岩经历了多期变质作用（特别是麻粒岩相阶段退变质作用），不可避免地会受到变质流体、热液等改造，但这种改造可能是不均匀的，使得部分样品的 Nb/Th（如 98702）和 Nb/U（如 99104-2）值降低。由于 U、Th 含量较低，XRF（X 射线荧光光谱）的分析误差，也会引起部分岩石的 Nb/Th 和 Nb/U 等比值产生误差。

表2 大别山北部榴辉岩的微量元素含量(ppm)及有关参数

样号	Zr	Nb	Th	U	Ta	Sc	Y	Zr/Nb	Nb/Ta	Nb/U	Th/Sc	(Nb/Th)$_N$	(Nb/La)$_N$
98701	65.0	10.8	2.28	0.51	0.51	44.1	26.7	6.0	21.2	21.2	0.05	0.62	0.92
98702	24.1	6.5	7.54	0.50	0.50	57.0	18.5	3.7	13.0	13.0	0.13	0.11	6.56
98121	28.5	8.3	1.15	0.50	0.50	38.25	6.05	3.4	16.6	16.6	0.03	0.95	3.72
9801	221.4	17.2	0.78	1.70	1.31	37.95	42.35	12.9	22.0	10.1	0.02	2.89	0.78
98122-4	30.9	7.8	0.80	1.72	0.50	45.20	11.80	4.0	15.6	4.5	0.02	1.28	0.99
99104-2	40.0	7.7	0.76	0.52	2.44	39.25	18.90	5.0	3.2	14.8	0.02	1.34	5.2
98WY-1	141.8	7.6	7.61	1.44	0.54	—	—	18.6	14.0	5.3	—	0.13	0.51

注：由安徽省地质实验测试中心完成，其中，Sc、Y、La 元素用等离子光谱测定，其余元素用 X 射线荧光光谱测定。"—"表示未分析。

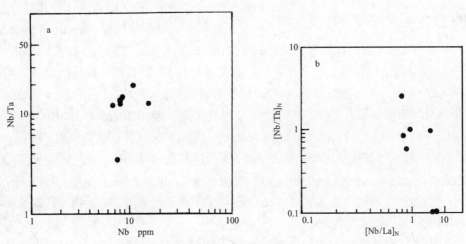

图3 Nb/Ta-Nb (a)和[Nb/Th]$_N$-[Nb/La]$_N$(b)图解

除 98702 样外，其他榴辉岩的 Th/Sc 值为 0.02～0.05（平均 0.03），类似于下地壳 Th/Sc 值[30]。Nb/U-Nb 图解[25]也表明，本区榴辉岩大多类似于陆壳玄武岩成分（图略）。在 Nb/3-TiO$_2$-Th 图解[24]上，大多数榴辉岩落在板内玄武岩区，少数在板块边缘区（大洋环境）（图略）。

2.3 稀土元素

研究区榴辉岩的稀土总量$\sum$REE=17.70～133.72 ppm，其中大多数为 20～50 ppm，见表3。图4 中 98121、98122-4、98701 和 9801 样品具有轻稀土富集的稀土配分模式，这与前面判定它们原岩大多数为大陆玄武岩的结论一致。其中 98121 样稀土含量较低并且具有正 Eu 异常，这表明它的原岩可能为辉长岩。但是，99104-2 和 98702 两个样品却具有轻稀土元素亏损的、类似于典型 N-MORB 稀土配分模式。

表3 大别山北部榴辉岩的稀土元素含量（ppm）

样号	La	Ce	Pr	Nd	Sm	Eu	Gd	Tb	Dy	Ho	Er	Tm	Yb	Lu	$\sum$REE
98701	11.70	26.8	3.4	15.4	3.8	1.1	4.5	0.75	4.7	1.0	3.1	0.46	2.9	0.48	80.09
98702	1.00	3.7	0.51	3.3	1.4	0.48	2.2	0.41	3.0	0.68	2.7	0.32	2.2	0.33	22.23
98121	2.20	5.40	0.63	3.25	0.96	0.47	1.33	0.20	1.35	0.30	0.79	0.11	0.62	0.09	17.70
9801	21.75	41.5	6.26	25.75	6.33	1.92	8.07	1.27	8.09	1.85	4.82	0.76	4.68	0.67	133.72
98122-4	7.80	15.5	2.58	10.60	2.90	0.95	3.08	0.48	2.53	0.48	1.28	0.19	1.14	0.17	49.65
99104-2	1.50	4.40	0.93	4.55	2.15	0.78	3.39	0.53	3.66	0.78	2.31	0.34	2.27	0.33	27.92

注：由安徽省地质实验测试中心用等离子光谱测定。

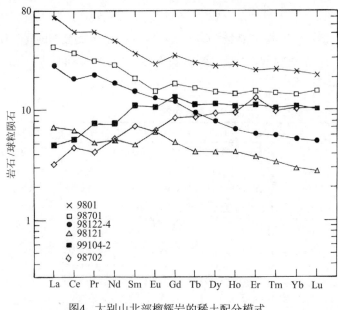

图4 大别山北部榴辉岩的稀土配分模式

3 Sr、Nd 同位素组成

同位素测试的化学处理工作在国土资源部同位素开放实验室完成,质谱分析在中国科学技术大学化学地球动力学实验室完成,具体分析方法见张宗清等[35]和刘贻灿等报道。其中,三个榴辉岩(98702、9801 和 98WY)的 Sm-Nd 等时线(全岩+石榴子石+绿辉石)年龄①表明,本区榴辉岩类似于南部超高压榴辉岩,经受过印支期超高压变质作用,且高压麻粒岩相退变质阶段冷却年龄大致为 210 Ma 左右。考虑到全岩样品中含有较多退变矿物而可能影响 Sr、Nd 同位素组成以致会影响原岩性质判断,为此,本文对六个榴辉岩样品的石榴子石 Sr、Nd 同位素成分进行了研究,见表4。结果表明,全岩钕同位素初始比 ε_{Nd}(210 Ma)大多数稍低于石榴子石(仅 98122-4 样出现异常,石榴子石出现明显正值和远大于其全岩值,将另文讨论),但基本一致反映大多数为陆壳成因,仅 99104-2 具有洋壳成分特征。本区榴辉岩的 Sr、Nd 同位素组成见表4。98702(榴辉岩)和 98407-1(橄榄岩)两个样品均采集自黄尾河镇东部上官村。它们近似的 I_{sr} 值(分别为 0.710159 和 0.710839)及钕同位素初始比 ε_{Nd}(210 Ma)= –9.7 和–10.1,均表现为陆壳岩石特征。McCulloch 等[36]研究表明,麻粒岩相变质作用对 Sm-Nd、Rb-Sr 全岩同位素体系强烈影响并造成 Sr、Nd 同位素发生再平衡;此外,由于富 CO_2 变质流体使 LREE 元素被搬运而往往造成 Sm/Nd 减小和 $\varepsilon_{Nd}(T)$ 增加。因此,98702 样的 $f_{Sm/Nd}$ = 0.5,表明轻稀土元素亏损(图4),而与负的钕同位素初始比 ε_{Nd}(210 Ma)= –9.7 不一致,可能反映近期事件即变质阶段富 CO_2 变质流体交代作用造成轻稀土元素亏损。

O'Nions 等[37]证实,洋底玄武岩的蚀变会增加 $^{87}Sr/^{86}Sr$ 值但不会明显影响 $^{143}Nd/^{144}Nd$ 值。饶钹寨榴辉岩 $^{143}Nd/^{144}Nd$ = 0.512916 和 ε_{Nd}(210 Ma)= +2.4 表现为洋壳特征,这与肖益林等(2000,私人通信)流体包裹体研究成果(即发现榴辉岩相变质阶段存在高盐度流体包裹体并认为可能代表洋壳成因)相一致;而 $^{87}Sr/^{86}Sr$ 值较高可能与海底蚀变及变质流体交代作用有关。

表 4 大别山北部榴辉岩及有关岩石的 Sr、Nd 含量及其同位素组成(T = 210Ma)

样号	Rb(ppm)	Sr(ppm)	$^{87}Rb/^{86}Sr$	$^{87}Sr/^{86}Sr$	±2σ	$I_{Sr}(T)$	$\varepsilon_{Sr}(0)$	$\varepsilon_{Sr}(T)$
98702Wr	1.22	194.81	0.0181	0.710213	21	0.710159	81.1	83.8
98702Gt	0.05	9.16	0.0145	0.710793	22	0.710750	89.3	92.2
99104-2Wr	4.24	103.19	0.1189	0.707717	20	0.707362	45.6	44.1
99104-2Gt	0.54	12.75	0.1221	0.708342	20	0.707977	54.5	52.8
9801Wr	3.92	130.30	0.0870	0.708219	20	0.707959	52.8	52.6
9801Gt	0.15	12.06	0.0372	0.708265	124	0.708154	53.4	55.3
98WYWr	7.57	190.01	0.1154	0.707490	19	0.707145	42.4	41.0
98WYGt	0.45	8.09	0.1602	0.709035	21	0.708556	64.4	61.0
98122-4Wr	0.94	188.96	0.0144	0.705388	28	0.705265	12.6	14.3
98122-4Gt	0.02	8.82	0.0069	0.705396	19	0.705375	12.7	15.9
98121Wr	1.44	77.20	0.0541	0.705597	18	0.705435	15.6	16.7
98121Gt	0.06	10.46	0.0163	0.705456	20	0.705407	13.6	16.3
98407-1Wr	1.70	168.41	0.0291	0.710926	19	0.710839	91.2	93.5

续表

样号	Sm(ppm)	Nd(ppm)	$^{147}Sm/^{144}Nd$	$^{143}Nd/^{144}Nd$	±2σ	$I_{Nd}(T)$	$\varepsilon_{Nd}(0)$	$\varepsilon_{Nd}(T)$	T_{DM}(Ga)
98702Wr	1.53	3.14	0.2947	0.512272	10	0.511867	−7.1	−9.7	−1.73
98702Gt	1.14	0.62	1.1122	0.513461	12	0.511932	+16.1	−8.4	
99104-2Wr	1.96	3.84	0.3087	0.512916	11	0.512492	+5.4	+2.4	−0.38
99104-2Gt	1.72	0.71	1.4654	0.514380	11	0.512366	+34.0	0	
9801Wr	6.24	24.40	0.1548	0.512038	11	0.511825	−11.7	−10.6	2.86
9801Gt	2.86	5.92	0.2920	0.512232	11	0.511830	−7.9	−10.5	
98WYWr	4.08	14.32	0.1732	0.512500	12	0.512263	−2.7	−2.0	2.38
98WYGt	1.19	0.43	1.6670	0.514533	12	0.512242	+37.0	−2.4	
98122-4Wr	2.78	10.54	0.1595	0.512047	12	0.511828	−11.5	−10.5	3.08
98122-4Gt	1.32	0.81	0.9858	0.514514	12	0.513159	+36.6	+15.4	
98121Wr	0.95	2.75	0.2092	0.512175	12	0.511887	−9.0	−8.7	
98121Gt	0.87	1.14	0.4596	0.512606	12	0.511974	−0.6	−7.6	
98407-1Wr	0.31	1.30	0.1442	0.512004	12	0.511806	−12.4	−10.1	2.59

注：Nd 同位素质谱分异校正相对于 $^{146}Nd/^{144}Nd=0.72190$，表中 $^{143}Nd/^{144}Nd$ 值相对于标准样 La Jolla 值等于 0.511849±9 给出；Sr 同位素采用 $^{86}Sr/^{88}Sr=0.1194$ 进行质量分馏校正，$^{87}Sr/^{86}Sr$ 值相对于标准样 NBS-987 值等于 0.710277±16 给出。98407-1Wr 为橄榄岩，其余为榴辉岩(Wr 为全岩，Gt 为石榴子石)。

本区榴辉岩的 ε_{Nd}（T=210 Ma）大多为−10 左右，个别为正值（+2.4），亏损地幔模式年龄为 2.86～3.08 Ga，见表 4。结合 ε_{Nd}-ε_{Sr} 图解（图 5），本区榴辉岩原岩大多可能来自下地壳，少数属洋壳残片。

4 大地构造意义

（1）岩石地球化学研究表明，大别山北部榴辉岩的原岩主要是拉斑玄武岩和少数可能为辉长岩。

（2）本区榴辉岩大多具有 Nb 正异常、K 负异常等，类似于南部超高压带中双河榴辉岩[26]，排除它们形成于古岛弧环境下的可能性。

（3）榴辉岩的峰期变质时代、微量元素特征及 Sr、Nd 同位素组成，说明大别山北部榴辉岩大部分可能属于扬子陆壳（下地壳）俯冲变质成因。而铙钹寨

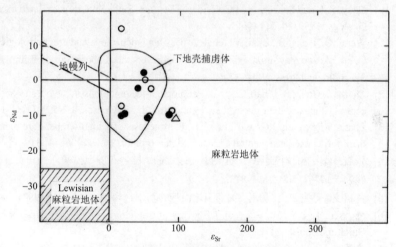

图5 大别山北部榴辉岩的 ε_{Nd}-ε_{Sr} 图解
● 榴辉岩；○ 石榴子石；△ 橄榄岩

榴辉岩（样号为 99104-2）的微量（Nb 正异常等）、稀土元素模式（轻稀土元素亏损）及 Sr、Nd 同位素组成（Nd 同位素初始比 ε_{Nd} = +2.4）等一致表明它为大洋玄武岩俯冲变质成因。笔者等[6]已在该带南部潜山县龙井关附近发现具有大洋玄武岩特征的变基性熔岩。因此，本文进一步证实研究区存在古洋壳残片。

（4）磨子潭-晓天断裂以南变质镁铁-超镁铁质岩带中除扬子俯冲陆壳外，还混有古洋壳残片，甚至可能还有来自其他构造背景下的岩石[5]（有待于进一步研究），且北部榴辉岩大致沿磨子潭-晓天断裂以南分布，以北则为华北大陆板块南部活动大陆边缘的复理石建造[8, 9, 38]。因此，大别山北部磨子潭-晓天断裂以南含榴辉岩的变质镁铁-超镁铁质岩带可能代表扬子与华北两个大陆板块之间的碰撞缝合带。

（5）本区榴辉岩在经受高压或超高压变质期间可能存在变质流体参与作用，特别是富 CO_2 变质流体参与作用，并且对部分岩石进行改造但这种改造可能是不均匀的，使得部分陆壳岩石的轻稀土元素发生亏损等。

致谢

同位素的化学分析处理工作得到中国地质科学院地质研究所张宗清研究员以及王进辉、唐索寒两位老师的具体指导和帮助；质谱测定工作得到彭子成教授的支持与指导以及张兆峰和贺剑峰两位同学的帮助。在此一并表示衷心感谢。

参考文献

[1] Okay, A.I., Xu, S. T. and Sengör, A.M.C. Coesite from the Dabie Shan eclogites, central China. Eur. J. Mineral., 1989, 1: 595-598.

[2] Wang, X., Liou, J.G. and Mao, H.K. Coesite-bearing eclogites from the Dabie Mountains in central China. Geology, 1989, 17: 1085-1088.

[3] 徐树桐，苏文，刘贻灿等. 大别山东段高压变质岩中的金刚石. 科学通报，1991, 36(17): 1318-1321.

[4] Xu, S. Okay A. I., Ji, S., et al. Diamond from the Dabie Shan metamorphic and its implication for tectonic setting. Science, 1992, 256: 80-82.

[5] 刘贻灿，徐树桐，江来利. 大别山北部斜长角闪岩类的地球化学特征及形成构造背景. 大地构造与成矿学，1998, 22（4）: 323-331.

[6] 刘贻灿，徐树桐，江来利. 大别山北部蛇绿岩的地球化学制约. 矿物岩石，1999，19（1）：68-73.

[7] 刘贻灿，徐树桐，江来利等.大别山北部中酸性片麻岩的岩石地球化学特征及其古大地构造意义. 大地构造与成矿学，1999，23（3）：222-229.

[8] 徐树桐，江来利，刘贻灿等. 大别山（安徽部分）的构造格局和演化过程. 地质学报，1992, 66(1): 1-14.

[9] 徐树桐，刘贻灿，江来利等. 大别山的构造格局和演化. 北京：科学出版社，1994. 1-175.

[10] Okay, A.I. and Sengör, A.M.C. Evidence for introcontinental thrust-related exhumation of the ultrahigh-pressure rocks in China. Geology, 1992, 20: 411-414.

[11] Wang, Q., Cong, B., Zhai, M., et al. A possible paleozoic island arc: petrological evidences from North Dabie gneiss. In: Inst. of Geol., Academia Sinia, ed. Annual Report of the Laboratory of Lithosphere Tectonic Evolution (1993-1994). Bejing: Seismological Press, 1994. 37-47.

[12] Zhai, M., Cong, B., Zhao, Z., et al. Petrological-tectonic units in the coesite-besring metamorphic terrain of the Dabie Mountains, central China and their geotectonic implication. J. SE. Geosci., 1995, 11(1): 1-13.

[13] Zhang, R.Y., Liou, J.G. and Tsai, C.H. Petrogenesis of a high-temperature metamorphic terrain: a new tectonic interpretation for the north Dabie Shan, central China. J. Metamorphic. Geol., 1996, 14: 319-333.

[14] 张旗，马宝林，刘若新等.一个消减带之上的大陆岩石圈地幔残片——安徽铙钹寨超镁铁岩的地球化学特征. 中国科学（B辑），1995, 5(8)：867-873.

[15] 董树文，孙先如，张勇等. 大别山造山带的基本结构. 科学通报，1993，38（6）：542-545.

[16] Xu, S T., Liu, Y., Su, W., et al. Discovery of the eclogite and its petrography in the Northern Dabie Mountains. Chinese Science Bulletin, 2000, 45(3): 273-278.

[17] 李曙光，洪吉安，李惠民. 大别山辉石岩-辉长岩的锆石 U-Pb 年龄及其地质意义.高校地质学报，1999, 5（3）：351-355.

[18] 葛宁结，侯振辉，李惠民等. 大别山造山带岳西沙村镁铁-超镁铁质岩体的锆石 U-Pb 年龄.科学通报，1999，44（19）：2110-2114.

[19] Wood, B.J. The solubility of alumina in orthopyroxene coexisting with garnet. Contrib. Mineral. Petrol., 1974, 46: 1-15.

[20] 刘贻灿，徐树桐，江来利等.大别山北部镁铁-超镁铁质岩带的研究新进展.安徽地质，1999，9（4）：262-267.

[21] Tsai, Chin-Ho and Liou, J.G. Eclogite-facies relics and inferred ultrahigh-pressure metamorphism in the North Dabie complex, central China. American Mineralogist, 2000, 85: 1-8.

[22] Winchester, J.A. and Floyd, P.A. Geochemical discrimination of different magma series and their differentiation products using immobile elements. Chemical Geology, 1977, 20: 325-343.

[23] Hofmann, A.W. Chemical differentiation of the Earth: relationship between mantle, continental crust and oceanic crust. EPSL, 1988, 90: 297-314.

[24] Holm, P.E. The geochemical finger prints of different tectonomagmatic environments using hydromagtophile element abundances of tholeiitic basalts and basaltic andesites. Chem. Geol., 1985, 51: 303-323.

[25] Le Roex, A.P., Dick, H.J.B and Fisher, R.L. Petrology and geochemistry of MORB from 25°E to 46°E along the southwest Indian Ridge: evidence for contrasting style of mantle enrichment. J. Petrol., 1989, 30: 947-986.

[26] Li, S., Jagoutz, E., Chen, Y., et al. Sm-Nd and Rb-Sr isotopic chronology and cooling history of ultrahigh-pressure metamorphic rocks and their country rocks at Shuanghe in the Dabie Mountains, central China. GCA, 2000, 64: 1077-1093.

[27] Byerson, F.J. and Watson, E.B. Rutile saturation in magmas: implications for Ti-Nb-Ta depletion in island-arc basalts. EPSL, 1987, 86: 225-239.

[28] Green, T.H. Significance of Nb/Ta as an indicator of geochemical processes in the crust-mantle system. Chem. Geol., 1995, 120: 347-359.

[29] Peltonen, P., Kontinen, A. and Huhma, H. Petrology and geochemistry of metabasalts from the 1.95 Ga Jormua ophiolite, Northeastern Finland. J. Petrol., 1996, 37(6): 1359-1383.

[30] Taylor, S.R. and McLennan, S.M. The continental crust: its composition and evolution. Oxford: Blackwell Scientific Publication, 1985, 1-312.

[31] Huhma, H., Cliff, R.A., Perttunen, V.,et al. Sm-Nd and Pb-Pb isotopic study of mafic rocks associated with early Proterozoic continental rifting: the Perapohja Schist belt in northern Finland Contr. Mineral. Petrol., 1990, 104: 369-379.

[32] Gruau, G., Tourpin, S., Fourcade, S., et al. Loss of isotopic (Nd, O) and chemical (REE) memory during metamorphism of komatiites: new evidence from eastern Finland. Contr. Mineral. Petrol., 1992, 112: 66-82.

[33] Rudnick, R.L., McLennan, S.M. and Taylor, S.R. Large ion lithophile elements in rocks from high-pressure granulite facies terrains. GCA, 1985, 49: 1645-1655.

[34] Fowler, M.G. Large-ion lithophile element characteristics of an amphibole facies to granulite facies transition at Gruiard Bay, North-west Scotland. J. Metamorph. Geol., 1986, 4: 345-359.

[35] 张宗清, 刘敦一, 付国民. 北秦岭变质地层同位素年代研究. 北京: 地质出版社, 1994, 5-7.

[36] McCulloch, M.T. and Black, L.P. Sm-Nd isotopic systematics of Enderby Land granulites and evidence for the redistribution of Sm and Nd during metamorphism. EPSL, 1984, 71: 46-58.

[37] O'Nions, R.K., Carter, S.R., Cohen, R.S., et al. Pb, Nd, Sr isotopes in oceanic ferromanganese deposits and ocean floor basalts. Nature, 1978, 273: 13-15.

[38] 刘贻灿, 徐树桐, 江来利等. 佛子岭群的岩石地球化学及构造环境. 安徽地质, 1996, 6(2): 1-6.

Geochemical and geochronological constraints on the suture location between the North and South China Blocks in the Dabie orogen, Central China[*]

S. -G. Li[1], F. Huang[1], Y. -H. Nie[1], W. -L. Han[1], G. Long[1], H. -M. Li[2], S. -Q. Zhang[3] and Z. -H. Zhang[3]

1. Department of Earth and Space Sciences, Laboratory for Chemical Geodynamics, University of Science and Technology of China, Hefei 230026, Anhui, China
2. Tianjin Institute of Geology and Mineral Resources, Tianjin 300170, China
3. The Third Geological Team of Henan Province, Xinyang 464000, Henan, China

> **亮点介绍**：发现大别山北缘北淮阳西段定远组变基性火山岩的原岩形成于古生代岩浆弧，其南侧浒湾构造混杂岩带形成于三叠纪，其中的低温榴辉岩的源岩是俯冲洋壳，其高压变质年龄为 301 Ma，揭示了华北板块和华南板块之间的缝合线位于浒湾构造混杂岩带。

Abstract The Dabie orogen is a collision zone between the North China Block (NCB) and South China Block (SCB), and a famous ultrahigh pressure metamorphic belt in the world. Location of the suture between the NCB and SCB in the Dabie orogen is still a controversial issue. We identify that the metabasaltic volcanic rocks from the Dingyuan Formation of the Beihuaiyang zone located in the northernmost Dabie orogen have typical geochemical feature of island arc basalt, e.g. negative Nb, Ti and P anomalies in trace elemental spidergram and high LREE, Ba and Pb abundance. Their Sm-Nd and Rb-Sr whole-rock isochron ages are 446±23 Ma and 444±31 Ma, respectively. These ages are consistent with the age of the island arc volcanics in the North Qinling belt that was the south active continental margin of the NCB in the Paleozoic. These data, together with the similarity in metamorphic history between the Dingyuan Formation and North Qinling belt, suggest that the Dingyuan formation is a Paleozoic magmatic arc on the south margin of the NCB. The Nd isotopic model ages of the Devonian metasediments from the Foziling flysch formation of the Beihuaiyang zone (T_{DM} = 1.7 to 1.9 Ga) are younger than those of the northern Paleozoic sediment cover of the SCB (T_{DM} = 1.9 to 2.1Ga). It suggests that more Phanerozoic volcanic material, which could be derived from the active continental margin of the NCB, was involved into the flysch formation. The Huwan tectonic melange zone located to the south of the Dingyuan Formation was formed in Triassic. The eclogites involved in this melange zone were formed in the Carboniferous (301 ± 13 Ma). Their protoliths could be middle Paleozoic oceanic crust or late Proterozoic to early Paleozoic island arc basalt. All these data together with the recent discovery of the Triassic eclogites in the Northern Dabie Zone place important constraints on the suture location between the NCB and SCB. It should follow the Huwan tectonic melange zone in the west and the boundary between the Northern Dabie zone and the Beihuaiyang zone in the east.

1 Introduction

The E-W trending Qinling-Dabie-Sulu orogenic belt was formed during the collision between the North China Block (NCB) and South China (or Yangtze) Block (SCB). Geologically, the belt is truncated and separated by two major geological elements — the Nanyang basin and the Tanlu fault (Fig. 1). The belt therefore can be subdivided

---

[*] 本文发表在：Physics and Chemistry of the Earth (A), 2001, 26(9-10): 655-672

into three major section — the Qinling orogen (west), the Dabie orogen (middle) and the Sulu terrane (east) (Fig. 1). It is generally believed that the Sulu terrane is the eastern extension of the Dabie orogen and was displaced ~500 km by the left-lateral movement of the Tanlu fault. Various types of ultrahigh-pressure metamorphic (UHPM) rocks including coesite- and diamond-bearing eclogite are exposed in the Dabie-Sulu orogen (Okay et al., 1989; Wang et al., 1989, 1991; Xu et al., 1992a; Zhang and Liou, 1994; Cong et al., 1995). It is generally agreed that the Dabie-Sulu UHPM belt may be the largest one on the Earth surface. Investigation of the tectonic framework and evolution of this UHPM belt is of critical importance in understanding the process of continental collision that produced the extreme metamorphic conditions. Location of the suture between the NCB and SCB in the Dabie Mountains is one of the major issues in this investigation.

It is, in general, accepted that the Dabie orogen can be subdivided into five metamorphic zones from north to south (Figs. 1 and 2): (1) Beihuaiyang greenschist-amphibolite facies zone, (2) Huwan cold eclogite melange zone which is limited in the western part of the Dabie orogen, (3) Northern Dabie complex zone, (4) Southern Dabie UHPM zone which contains coesite-bearing eclogite, (5) Hong'an Susong high-pressure metamorphic (HPM) zone which contains "cold" eclogite and blueschist. Abundant geochronological studies of the UHPM rocks and blueschist from zones 4 and 5 show that this high-pressure (HP) metamorphic belt was formed in the Triassic (Li et al., 1989, 1993, 1994a, 1999a, 2000; Ames et al., 1993, 1996; Eide et al., 1994; Hacker and Wang, 1995; Chavagnac and Jahn, 1996; Rowley et al., 1997; Hacker et al., 1998). Many of researchers believe that the zones 4 and 5 are subducted continental crust of the SCB, whereas the tectonic settings of the zones 1, 2 and 3 are still controversial. For example, Okay and Sengör (1993) suggest that the suture between the NCB and SCB is located to the North of zone 1, and zone 1 probably represents the passive continental margin of the SCB. The similar conclusion has been reached by Hacker et al. (1998) from U-Pb and Ar-Ar datings together with eclogite distribution in the Dabie orogen and by Zhou and Zheng (2000) from an investigation of low-grade metamorphic rocks in the Dabie-Sulu orogen. Liu and Hao (1989), Xu et al. (1992b) and Hacker et al. (1995) argued that the boundary between the zones 1 and 3

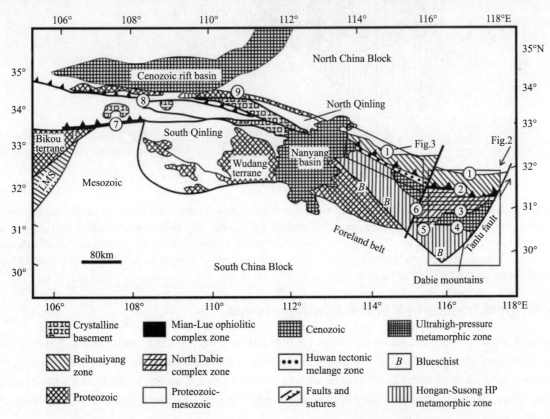

Fig. 1 Geological sketch map of the Qinling-Dabie orogenic belt. ① Xinyang-Shucheng fault; ② Tongbai-Mozitan fault; ③ Wuhe-Shuihou fault; ④ Mamiao-Taihu fault; ⑤ Xishui-Yingshan ductile shear zone; ⑥ Shang-Ma fault; ⑦ Mian-Lue fault zone; ⑧ Shang-Dan fault; ⑨ Luanchuan fault.

represents a suture between the SCB and NCB, and zone 1 is a forearc flysch formation. Cong et al. (1994) propose that zone 3 could be an island arc near the NCB, thus the suture between the NCB and SCB should be located on the south boundary of zone 3, and zone 1 should be a back-arc basin formation. Obviously, the tectonic setting of the zones 1, 2 and 3 are critical to understand the location of the suture between the NCB and SCB and tectonic evolution of the Dabie orogen. The absence of Paleozoic magmatic arc in zones 1 and 2 and a lack of Triassic eclogite in zone 3 were the major difficulties in understanding of their tectonic settings. New geochemical, isotopic and chronological data are presented in this paper, which place close constraints on the tectonic setting of the zones 1, 2 and 3.

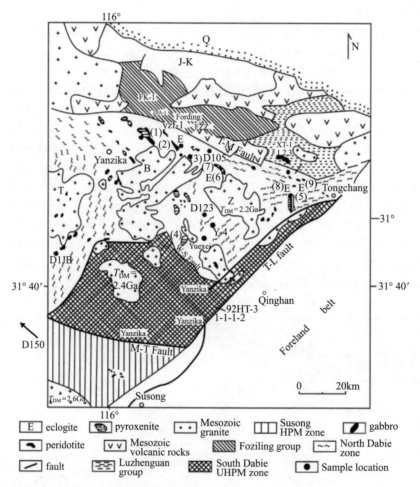

Fig. 2 Geological sketch map showing distribution of mafic-ulrtamafic rocks and eclogites in the North Dabie zone and locations of the samples in Table 3 in eastern Dabie orogen. (1) Zhujiapu pyroxenite; (2) Raobazhai peridotite, garnet pyroxenite or eclogite; (3) Renjiawan pyroxenite; (4) Zhongguan pyroxenite; (5) Jiaoyan gabbro; (6) Xiaohekou gabbro; (7) Huangweihe eclogite; (8) Baizhangyan eclogite; (9) Huazhuang eclogite. T-M Fault: Tongbai-Mozitan fault; W-S Fault: Wuhe-Shuihou fault; M-T Fault: Mamiao-Taihu fault; T-L Fault: Tan-Lu fault.

2 Geological framework of the Qinling orogen

The geologic framework and tectonic evolution of the Qinling orogen has been well investigated (e.g. Zhang et al., 1995; Meng and Zhang, 1999, 2000). In order to understand the tectonics of the Dabie orogen better, we will compare the geological units in the Dabie orogen with those in the Qinling orogen in the following section. Hence, before discussing the tectonics of the Dabie orogen, we briefly introduce the geological background of the Qinling orogen first.

The Qinling orogen is a multi-system orogenic belt with two mountain chains. It is divided into two parts, that is the North Qinling and South Qinling belts, along the Shangdan fault (Fig. 1). The North Qinling belt is regarded as a middle Paleozoic orogen (Mattauer et al., 1995) with widespread Paleozoic island-arc type magmatism represented by metavolcanic rocks in the Danfeng and Erlangping Groups (Sun et al., 1987; Li et al., 1989; Lerch et al., 1995) and metamorphism (Sun et al., 1996a, b; Zhai et al., 1998). The South Qinling belt is regarded as a Late Paleozoic to Early Mesozoic orogen with abundant Triassic granites and metamorphism (e.g. Mattauer et al., 1985; Xu et al., 1991; Li and Sun, 1996). It is, in general, accepted that the North Qinling belt evolved into an active margin of the NCB during the period from Ordovician to Silurian because of the northward subduction of the Proto-Tethyan Qinling ocean (e.g. Zhang et al., 1995; Li and Sun, 1996; Meng and Zhang, 2000). The Danfeng magmatic arc evolved along its south margin is composed of island-arc type metavolcanics (Danfeng Group) with a Rb-Sr whole rock isochron age of 447.8±41.5 Ma (Sun et al., 1987) and mafic intrusions (norite-gabbros) with a Sm-Nd mineral isochron age of 402.6±17.4 Ma (Li et al., 1989). Radiolarians in the cherts interlayered with the basalts in Danfeng Group also indicate the ages of Ordovician to Silurian (Cui et al., 1995). Geochronological studies show that this island-arc type magmatism was terminated in the Devonian (Lerch et al., 1995; Li and Sun, 1996).

Recent studies suggest that the South Qinling belt was a microcontinent between the NCB and SCB during the Middle Paleozoic time, which was separated from the SCB by Mian-Lue ophiolite complex zone (① in Fig. 1) (Zhang et al., 1995; Yin and Huang, 1995). Radiolarians in the cherts interlayered with the basalts in Mian-Lue ophiolite complex indicate the age of Carboniferous. It is argued that the initial collision between the NCB (including the North Qinling belt) and South Qinling microcontinent took place in Middle Paleozoic along the Shangdan suture(⑧ in Fig. 1) (Meng and Zhang, 2000). This idea is supported by the termination of island-arc type magmatism in the North Qinling belt in the Devonian (Lerch et al., 1995; Li and Sun, 1996). The Devonian metaclastic flysch formation (Liuling Group) developed along the Shangdan suture, which received the detritus from both the North Qinling and South Qinling belts (Yu and Meng, 1995), suggests that the ocean between the North Qinling and South Qinling belts was closed in the Devonian. A $^{40}Ar/^{39}Ar$ age of 314±6 Ma for biotite from the left-lateral strike slip fault along the Shangdan suture also indicates that this collision should be occurred before 314 Ma (Mattauer et al., 1985). On the other hand, the Sm-Nd and Rb-Sr metamorphic ages of 242±21Ma and 221±13 Ma for the metabasalts from the Mian-Lue ophiolite complex suggest that the Mian-Lue ocean could be closed in Triassic, resulting in the collision between the SCB and South Qinling microcontinent (Li et al., 1996). This idea is supported by the Triassic syncollisional granites (with U-Pb zircon ages of 206 to 220 Ma) broadly developed to the north of the Mian-Lue suture and south of the Shangdan suture (Li and Sun, 1996; Sun et al., 2000). The seismic reflection profiling in eastern Qinling shows that the South Qinling belt as a nappe overlies on the subducted continental crust of the SCB after the collision (Yuan et al., 1994).

Since the Dabie orogen is the eastern extension of the Qinling orogen, they should have the similar history of tectonic evolution. However, because the most of the upper crust over the subducted continental crust of the SCB in the Dabie orogen has been removed out by erosion, some geological records in the Qinling orogen, such as ophiolite, micro-continent, and the Triassic syncollisional granite, may not be observed in the Dabie orogen. Thus care should be taken when comparing the two orogens.

3 Beihuaiyang greenschist- to amphibolite-facies zone: Geochemical constraints

This zone is located on the northernmost part of the Dabie orogen and bounded by a WNW-ESE trending, north dipping Tongbai-Mozitan fault on the south (② in Fig. 1) (Xu et al., 1992b; Dong et al., 1993; Okay and Sengör, 1993) and the south-dipping Xinyang-Shucheng fault on the north (① in Fig. 1) (Xu et al., 1992b). It suffered metamorphism of greenschist- to amphibolite-facies and intensively ductile shear deformation with many mylonite occurrences. It may be subdivided into two units (Figs. 2 and 3). The lower unit is composed of the Guishan complex of the Xinyang Group and the Dingyuan formation in the west (see Fig. 3) and Luzhenguan Group in the east (Fig. 2). It is composed of various gneisses, amphibolite, garnet mica schist, chlorite or chlorite albite schist, marble and quartzite; their protoliths were suggested to be keratophyre, andesitic tuff and basalts. Protoliths of the K-feldspar gneiss and mica-schist could be sandstone and pelite (Xu and Hao, 1988; Xu et al., 1992b).

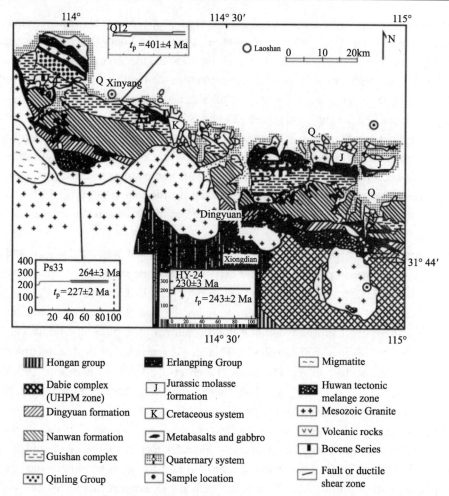

Fig. 3 Geological map showing relationship between the Guishan complex, Nanwan formation, Dingyuan formation and Huwan tectonic melange zone in the Beihuaiyang zone in western Dabie orogen. XSF: Xinyang-Shucheng fault. The $^{40}Ar/^{39}Ar$ data are from Niu et al. (1994) and Ye et al. (1993).

The upper unit includes the Nanwan formation of the Xinyang Group in the west (Fig. 3) and the Foziling Group in the east (Fig. 2). It is separated from the lower unit by ductile shear zone or fault (Fig. 3). This unit is a metaclastic flysch formation with rhythmic layers of slate, phyllite, mica quartz schist and quartzite and the metamorphic grade greenschist facies (Xu and Hao, 1988; Chen and Sang, 1995). Devonian fossils in this unit (Gao et al., 1988) suggest a Devonian deposition for the Foziling Group and Nanwan formation.

3.1 Geochronology and geochemistry of the lower unit

Ye et al. (1993) reported a U-Pb zircon age of 392±25 Ma on acidic metavolcanics from the upper formation of the Guishan complex. It is the lower intercept of disconcordia line. We interpret this age as metamorphic age of the acidic metavolcanics. More dating results support our interpretation. For example, an amphibole sample separated from a garnet-amphibolite (Q12) in the Guishan complex yields a perfect $^{40}Ar/^{39}Ar$ plateau age of 401±4 Ma (Fig. 3 of Niu et al., 1994), which indicates the metamorphic time of the lower unit. In addition, the Rb-Sr whole-rock isochron of tuff from the Dingyuan area yields an age of 391±13 Ma with an initial $^{87}Sr/^{86}Sr$ ratio of 0.7079 (Ye et al., 1993). This age might also indicate the time of metamorphism. These ages of the lower unit in the Beihuaiyang zone are consistent with the metamorphic age of the North Qinling belt, a Paleozoic island arc near the NCB (Zhang et al., 1995). For example, the hornblende and biotite from the metamorphic rocks in the North Qinling belt yield $^{40}Ar/^{39}Ar$ ages of 434 to 404 Ma and 348 to 327 Ma, respectively (Sun et al., 1996a; Zhai et al., 1998; Mattauer et al., 1985). Rb-Sr whole-rock isochron as well as Sm-Nd and Rb-Sr mineral (hornblende + plagioclase + whole rock) isochrons of amphibolites from the North Qinling belt give ages of 402 to 414 Ma (Sun et al., 1996a, b). All of these geochronological data suggest that the lower unit in the Beihuaiyang zone and the North Qinling belt experienced the same metamorphism at around 400 Ma. However, the formation age and tectonic setting of the lower unit are still unknown. In order to solve these questions, we have studied the geochemistry and geochronology of metabasalts from the Dingyuan formation.

Four metabasaltic and basaltic-andesite samples (951-N-S-1, -3, -5 and -6) and 2 acidic samples (95HN-S-2 and -4) were collected from the Dingyuan formation at one stone pit in Dingyuan (Fig. 3). The rocks in this stone pit are almost metabasalts, while a few acidic blobs with the diameters of 10 to 20 cm have been observed. Two foliations can be recognized in both basaltic and acidic rocks, suggesting that the basaltic and acidic rocks have the same deformation history. Petrography of these rocks shows typical volcanic porphyritic structure. The phenocrysts in the metabasalts are composed of augite, plagioclase and minor opaque minerals. Sample 95HN-S-6 contains more abundant opaque minerals (~10%) which severely affect its major and trace element contents. Muscovite is developed along the foliations in all basaltic samples. The matrix is mainly composed of pyroxene, plagioclase and amphibole, but most of them are partly replaced by abundant chlorites or epidote. The metamorphic mineral assemblage of chlorite + epidote + muscovite suggests a high greenschistfacies metamorphism. The phenocrysts in the acidic rock are mostly plagioclase. Its fine grain matrix is composed of quartz, plagioclase and muscovite. Muscovite developed along the foliations, but was folded by the second deformation. The typical volcanic porphyritic structure of the acidic rock and its similar deformation history to the metabasalts suggest that the acidic blobs in metabasalts are not the metamorphically fractional product, but more likely to be magmatically fractional product (deformed acidic dyke in basalt).

Major and trace elements (see Table 1) as well as Sr and Nd isotopes (see Table 2) have been analyzed for these samples. Based on their SiO_2 and $(Na_2O + K_2O)$ contents, the metabasalts in the Dingyuan formation are basically subalkaline basalts excepts the sample 95HN-S-1 which is basaltic andesite. AFM diagram (omitted) shows that these sample 95HN-S-6 that contains abundant opaque metallic metabasalts belong to calc-alkaline series excepts for the minerals, thus more Fe_2O_3 than other samples.

Table 1 Major element (%) and trace element (ppm) compositions of the metavolcanics from the Dingyuan formation[*]

elements	95 HN-S-1	95 HN-S-2	95 HN-S-3	95 HN-S-4	95 HN-S-5	95 HN-S-6
SiO_2	55.01	75.75	51.86	75.15	48.25	50.41
TiO_2	0.99	0.25	1.23	0.28	1.45	3.71
Al_2O_3	16.60	12.82	16.05	13.37	17.08	12.25
Fe_2O_3	3.90	0.87	3.69	1.12	0.38	7.22
FeO	3.40	0.32	5.52	0.30	10.36	8.34
MnO	0.107	0.011	0.133	0.007	0.153	0.155
MgO	3.96	0.45	6.30	0.58	5.59	4.56
CaO	7.82	0.90	6.03	0.70	8.95	6.84
Na_2O	4.00	5.84	2.87	5.07	1.60	1.06
K_2O	0.99	1.13	1.83	1.90	1.14	0.42
H_2O^+	2.74	0.70	3.72	1.29	4.26	4.35
CO_2	0.06	0.64	0.72	0.01	0.29	0.08
P_2O_5	0.023	0.016	0.088	0.019	0.018	0.033
Total	99.60	99.70	100.04	99.80	99.52	99.43
TFe_2O_3	7.68	1.23	9.82	1.45	11.89	16.49
Ti	4615.40	1184.40	5755.10	1351.20	6416.60	10753.00
Cr	68.52	12.21	116.02	12.37	49.98	40.45
Co	14.61	3.09	30.80	5.39	28.24	37.73
Ni	52.48	32.54	84.85	58.07	107.40	95.97
Cu	14.91	7.05	32.30	9.35	38.46	98.86
Zn	57.99	16.91	95.69	22.03	104.65	110.03
Rb	20.22	23.38	38.36	40.66	24.04	10.30
Sr	979.02	185.26	644.32	137.75	1537.80	940.06

续表

elements	95 HN-S-1	95 HN-S-2	95 HN-S-3	95 HN-S-4	95 HN-S-5	95 HN-S-6
Y	16.08	14.94	18.48	18.31	12.06	50.16
Zr	166.14	192.63	117.95	200.82	43.39	309.92
Nb	8.92	15.42	8.50	14.07	1.95	21.43
Sn	1.35	1.82	1.62	2.25	1.01	1.80
Cs	0.79	0.73	1.07	1.29	0.77	0.86
Ba	642.32	728.56	1025.10	1017.90	604.59	379.94
La	29.28	64.43	32.84	69.36	6.97	37.74
Ce	57.47	119.23	66.81	128.78	17.71	86.06
Pr	7.24	12.52	8.43	12.69	2.74	11.25
Nd	28.97	40.95	32.81	45.38	6.30	50.05
Sm	5.31	5.73	4.97	7.25	4.55	13.10
Eu	2.25	1.54	2.05	1.63	1.91	4.35
Gd	5.36	5.20	5.11	6.48	4.60	16.26
Tb	0.64	0.78	0.77	0.71	0.66	2.28
Dy	4.08	3.75	4.06	4.47	2.87	13.77
Ho	0.93	0.89	1.12	0.88	0.60	2.49
Er	2.34	2.31	2.76	3.09	2.03	8.08
Tm	0.26	0.39	0.56	0.43	0.40	0.83
Yb	2.23	2.91	2.34	2.96	2.69	5.45
Lu	0.24	0.35	0.42	0.26	0.30	0.67
Hf	3.63	7.32	4.83	7.17	1.18	12.03
Ta	1.11	1.75	1.87	2.34	0.58	2.46
Pb	113.87	7.68	15.48	4.31	23.96	23.68
Bi	0.34	0.18	0.20	0.35	0.07	0.20
Th	5.15	15.27	3.40	12.88	1.07	5.84
U	0.92	2.54	0.76	2.10	0.56	1.48
Ba/Nb	72	47.2	120.6	72.3	310	17.7
La/Nb	3.28	4.18	3.86	4.93	3.57	1.76
La/Ce	0.51	0.54	0.49	0.54	0.39	0.44
Th/Nb	0.577	0.990	0.40	0.915	0.548	0.273

* The major element contents were obtained at the Institute of Geophysical and Geochemical Exploratory Techniques, Chinese Academy of Geological Sciences, by X-ray fluorescence analysis except CO_2 and H_2O, which were obtained by nonaqueous titration and gravimetric analysis respectively. Trace element contents were analyzed by ICP-MS at the Institute of Geology, Chinese Academy. The Rb, Sr data obtained by ICP-MS are consistent with the isotope dilution data well (see Table 2). The ICP-MS Nd and Sm data for most samples are basically consistent with the isotope dilution data (Table 2). However, the ICP-MS Nd and Sm abundance of the sample 95HN-S-6 are 3 times higher than the isotope dilution abundance. Those discrepancies between the two procedures may reflect either a lower precision for the ICP-MS analysis or they may reflect the heterogeneity of the sample which contains more abundant opaque minerals. Those discrepancies are not yet resolved, but they do not affect the interpretation of the data presented here.

Table 2 Sr and Nd isotopic compositions of the metavolcanics from the Dingyuan formation

Sample	Rb(ppm)	Sr(ppm)	^{87}Rb/^{86}Sr	^{87}Sr/^{86}Sr± 2σ	Sm(ppm)	Nd(ppm)	^{147}Sm/^{144}Nd	^{143}Nd/^{144}Nd±2σ	$\varepsilon_{Nd}(t)$*
95HN-S-1	20.78	962.3	0.0628	0.706502±29	6.178	30.63	0.1220	0.511986±18	−8.5
95HN-S-2	28.95	185.6	0.4553	0.708886±17	5.369	36.98	0.0879	0.511898±13	−8.2
95HN-S-3	40.16	646.7	0.1802	0.707225±28	3.576	14.56	0.1486	0.512085±14	−8.0
95HN-S-4	45.76	143.9	0.9313	0.711910±24	6.849	31.15	0.1330	0.512040±16	−8.0
95HN-S-5	24.64	1541	0.0465	0.706351±30	3.681	13.37	0.1664	0.512134±17	−8.2
95HN-S-6	9.327	951.8	0.0285	0.706109±25	4.598	14.45	0.1925	0.512201±11	−8.3

* The Rb-Sr and Sm-Nd isotope data were obtained at the Isotope Laboratory of the Modern Analysis Center, Nanjing University, following procedures described elsewhere (Li et al., 1994a). The ^{143}Nd/^{144}Nd ratios were normalized relative to the value of ^{146}Nd/^{144}Nd = 0.7219. Blank for the whole chemical procedure are 50–60 pg for Sm an Nd, 0.1 ng for Rb and 1 ng for Sr. The Nd and Sr isotopic standard values analyzed in this experiment are 0.511864 ± 8 for La Jolla Nd Standard and 0.710241 ± 11 for NBS 987 Sr standard. Ages are calculated using "ISOPLOT" software provided by Dr. K. Ludwig and given with 2σ error. The initial $\varepsilon_{Nd}(t)$ is calculated using t = 446 Ma.

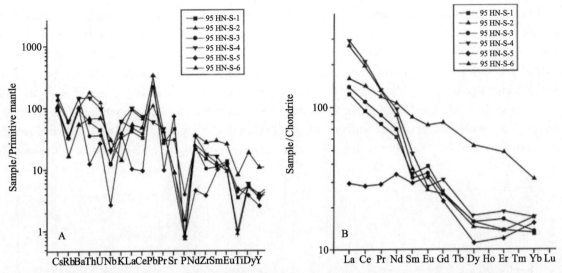

Fig. 4 Element geochemical diagrams for the Dingyuan formation. (A) Trace element primitive mantle normalized spidergram of the metavolcanics from the Dingyuan formation; (B) Chondrite normalized REE patterns of the metavolcanics from the Dingyuan formation. Chondrite values used for normalization are 1.2×(Masuda et al., 1973). Primitive mantle values used are from Sun and McDonough (1989).

The predominant trace elemental characteristics of both metabasalts and acidic rocks in the Dingyuan formation is high Ba, Pb and Sr contents but low Nb, Ti and P contents (Table 1). They display positive Ba, Pb and SR anomalies but negative Nd, Ti and P anomalies in the primitive mantle normalization diagram (Fig. 4 A). These are the typical features of subduction zone related magmas, and very similar to those of the Paleozoic island arc magmatism in the North Qinling belt (Li et al., 1994b; Sun et al., 1995, 1996b). In addition, the high La/Nb, Ba/Nb and Th/Nb ratios of the metabasalts in the Dingyuan formation are also of island arc features (Fig. 5). In addition, since the incompatible elements La, Ce and Nb are immobile during metamorphism (Hart, 1971; Li et al., 1993; Zhao, 1997), the La/Ce and La/Nb ratios of metavolcanics may reflect the ratios of their sources. Table 1 shows that both the basaltic and acidic samples, except the sample 95HN-S-6 which contains more abundant opaque minerals, have similar La/Ce and La/Nb ratios. This suggests that both the basaltic and acidic rocks may be derived from a homogeneous source. The REE chondrite normalization diagram (Fig. 4B) shows that all sample, except the sample 95HN-S-6, have similar HREE flat pattern, but highly fractional LREE patterns including LREE flat pattern(95HF-S-5) to LREE enriched (95HN-S-1, 3) and highly enriched (acidic rocks) patterns. It may suggest different degrees of partial melting of their mantle source, or high fractionation of the magma. Fig.6 Shows that the Sm-Nd and Rb-Sr whole-rock isochrons yields consistent ages of 446±23 Ma and 444±31 Ma, respectively. These

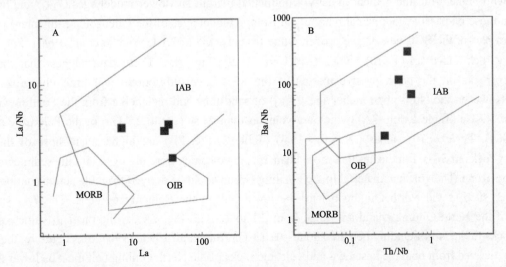

Fig. 5 La/Nb-La and Ba/Nb-Th/Nb plots showing that the metabasalts from the Dingyuan formation are characterized by high La/Nb. Ba/Nb and Th/Nb ratios, falling in the island arc basalt (IAB) field (after Li, 1993).

ages are older than the metamorphic ages of the Dingyuan formation and Guishan complex mentioned above. Therefore, we interpret these ages as the formation age of the Dingyuan metabasaltic rocks. It suggests that island arc magmatism in the Beihuaiyang zone was active in the later Ordovician to early Silurian. The island arc volcanics of the Danfeng Group in the North Qinling belt have the similar geological history. A Rb-Sr whole-rock isochron age of 447.8±41.5 Ma for the metavolcanic rocks in the Danfeng Group (Sun et al., 1987) is consistent with the Rb-Sr and Sm-Nd ages reported in this paper. Therefore, the tectonic setting of the Dingyuan formation could be similar to that of the Danfeng Froup in the North Qinling belt, i.e. it was a Paleozoic magmatic arc on the south margin of the NCB. If it is true, the suture between the NCB and SCB should lie to the south of the Dingyuan formation.

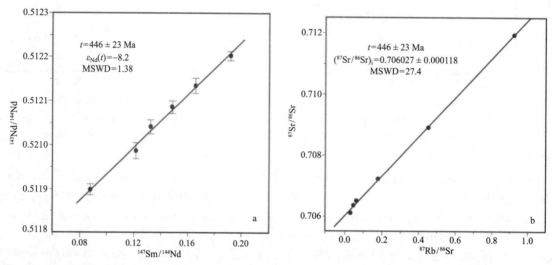

Fig. 6 Rb-Sr and Sm-Nd isochron diagram of the metavolcanics from the Dingyuan formation.

3.2 Nd isotopic compositions of flysch formation in the upper unit

The Foziling Group or the Nanwan formation has been considered as the sequences of passive continental margin of the SCB (Okay and Sengör, 1993), the forearc flysch formation of the NCB (Liu and Hao, 1989; Xu et al., 1992b), or the back arc flysch formation of the NCB (Dong et al., 1993; Cong et al., 1994), or a part of accretionary wedge scraped off from the shallow part of the Yangtze plate during the subduction of SCB beneath the NCB (Zhou and Zheng, 2000). Nd isotope model age (T_{DM}) and $\varepsilon_{Nd}(0)$ values at the present time of metasedimentary rocks from the flysch formation (upper unit) in the Beihuaiyang zone can be used to discriminate their tectonic settings. Sediment with higher volcanic content in active continental margin has lower model ages (T_{DM}) and higher $\varepsilon_{Nd}(0)$ values than those on passive continental margin. Nd isotope model ages of the Paleozoic sedimentary cover on the northeastern part of the SCB near Dabie orogen range from 1.9 Ga to 2.1 Ga (Chen et al., 1990) (Fig. 7A). These sedimentary rocks have low $\varepsilon_{Nd}(0)$ values (−14.1 to −16.9) (Fig. 7B). These data suggest that the Paleozoic sedimentary rocks on the north passive margin of the SCB have old sources and little Phanerozoic volcanic materials. However, the Nd isotope model ages (T_{DM}) of sandstone and metapelite from the Foziling Group range from 1.7 to 1.9 Ga (Table 1, Fig. 7A), which are younger than those (1.9 to 2.1 Ga) of the Paleozoic sedimentary cover in SCB. Their $\varepsilon_{Nd}(0)$ values (−9.3 to −15.3) (Table 3, Fig. 7B) are higher than those of the Paleozoic sedimentary rocks in SCB. It indicates that more Phanerozoic volcanics were involved into the sedimentary rocks of the Foziling Group. The volcanic material in the Foziling Group could be derived from the active continental margin of the NCB, such as the North Qinling belt or the lower unit in the Beihuaiyang zone. The T_{DM} values of the basement of the North Qinling island arc range from 2.1 to 1.9 Ga (Fig. 7A), while the T_{DM} values of island arc volcanic rocks in the North Qinling belt are about 1.0±0.1 Ga (Sun et al, 1995, 1996a, b). Therefore, the T_{DM} values of sediment derived from both the basement and volcanic rocks of the North Qinling belt must be lower than 1.9 Ga. In addition, the gneiss from Luzhenguan Group yields lower T_{DM} values of 1.6 to 1.8 Ga (Fig. 7A, Table 3). It indicates that the protolith of the gneiss from the Luzhenguan Group could be felsic volcanic rock or tuff that would be expected because of the island arc setting of the lower unit in the Beihuaiyang zone.

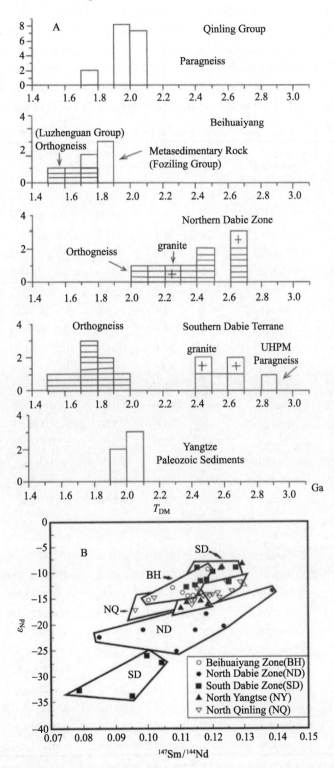

Fig. 7 Sm-Nd isotope plots of sedimentary and metamorphic rocks from various metamorphic zones in the Dabie orogen, Qinling Group and Paleozoic cover of the SCB. (A) Histograms of Nd isotopic model ages (T_{DM}); (B) $\varepsilon_{Nd}(0)$ vs $^{147}Sm/^{144}Nd$ diagram. Note: the UHP metapelite (paragneiss) have significantly higher ε_{Nd} and T_{DM} values than those of granitic gneiss (orthogneiss) in South Dabie zone. The data of granites in the Dabie orogen, data of paragneiss in Qinling Group and Paleozoic sediments in the SCB are from Xie et al., (1996), Z. Zhang et al. (1994) and Chen et al. (1990) respectively. Other data are from Table 3.

3.3 Tectonic evolution of the Beihuaiyang zone

The above geochemical and geochronological data suggest that the tectonic setting of the Beihuaiyang zone in the Dabie orogen is evolutionary. The Beihuaiyang zone is composed of two units. The lower unit of the Beihuaiyang zone including the Guishan complex and the Dingyuan formation, was an island arc or a magmatic arc on the active continental margin in the Ordovician to Silurian. This is supported by the discovery of the Paleozoic magmatic arc of the Dingyuan formation, lower T_{DM} values of the Luzhenguan Group and ~400 Ma metamorphic ages of the Guishan complex and Dingyuan formation. These features are consistent with the island arc volcanic rocks in the North Qinling belt which was the south active continental margin of the NCB. Therefore, the lower units of the Beihuaiyang zone could be the active continental margin of the NCB in the Ordovician to Silurian.

Table 3 Nd isotopic compositions and model ages of the gneisses, and metasedimentary rocks from the Dabie Mountains+

Sample	rock type*	Sample location++	Sm (ppm)	Nd (ppm)	$^{147}Sm/^{144}Nd$	$^{143}Nd/^{144}Nd \pm 2\sigma_m^8$	T_{DM}** (Ga)	data source	$\varepsilon_{Nd}(0)$
Beihuaiyang zone									
Foziling group									
92F-1	Sandstone	Foziling	2.626	14.14	0.1123	0.511897 ± 6	1.9		−14.5
PJL-1	Muc Qz Sch	Panjialing	5.769	31.52	0.1107	0.511929 ± 12	1.8		−13.8
FZL-2	Muc Qz Sch	Foziling	5.066	30.61	0.1001	0.511856 ± 12	1.7		−15.3
ZFA-1	Muc Qz Sch	Zhufoan	4.983	26.47	0.1139	0.511900 ± 12	1.9		−14.4
Luzhenguan									
XT-1-1	Gn	Xiaotan	10.41	58.57	0.1075	0.511980 ± 16	1.7		−12.8
XT-1-2	Gn	Xiaotan	7.784	39.76	0.1184	0.512160 ± 10	1.6		−9.3
XT-1-3	Gn	Xiaotan	5.788	30.45	0.1150	0.511971 ± 19	1.8		−13.0
North Dabie zone									
92 R-3	Gn	Raobazhai	18.61	98.41	0.1144	0.511557 ± 15	2.4	1	−21.1
DZh-88-6	Bi Gn	Zhujiapu	8.034	41.24	0.1178	0.511716 ± 9	2.3	1	−18.0
DZh-88-7	Gn	Zhujiapu	14.31	70.17	0.1234	0.511596 ± 7	2.6	1	−20.3
DZh-88-8	Bi-Gn-xenolith	Zhujiapu	1.474	10.51	0.0848	0.511480 ± 15	2.0	1	−22.6
92 Y-4	Hb Gn	Yuexi, Fulongzhai	5.911	25.87	0.1382	0.511947 ± 8	2.4	1	−13.5
D 123	Pl. Gn.	Yuexi, Shiguan	9.040	50.25	0.1088	0.511350 ± 6	2.6	3	−25.1
D 105	Hb. Gn.	Mozitan	6.575	40.37	0.09843	0.511560 ± 9	2.1	3	−21.0
South Dabie zone									
DX50A-1	Orthogeneiss	Bixiling	5.312	26.81	0.1198	0.512142 ± 11	1.6	1	−9.7
92 W-1	Orthogneiss	Wumiao	3.257	17.11	0.1151	0.512179 ± 14	1.5	1	−9.0
T16-1	Orthogneiss	Shima	7.334	34.89	0.1272	0.512176 ± 9	1.7	2	−9.0
Sh-1-2	Orthogneiss	Shuanghe	7.835	40.84	0.1161	0.512049 ± 32	1.7	1	−11.5
Sh-1-1	Orthogneiss	Shuanghe	6.635	33.99	0.1180	0.512027 ± 10	1.8	1	−11.4
92HT-3	Metapelite	Shuanghe	6.754	52.02	0.07854	0.510962 ± 8	2.5	1	−32.7
92H-24	Metapelite	Shuanghe	5.533	35.18	0.0951	0.510904 ± 23	2.9	1	−33.8
92HT-8	Metapelite	Shuanghe	6.890	41.84	0.09962	0.511315 ± 32	2.4	1	−25.1
921-1T-1	Metapelite	Shuanghe	11.59	71.72	0.1041	0.511210 ± 15	2.7	1	−27.3
D 118	Orthogneiss	Yingshan, Jinpu	7.467	40.36	0.1118	0.511982 ± 7	1.7	3	−12.8
D 132	Orthogneiss	Taihu, Lidu	7.137	37.70	0.1144	0.512005 ± 9	1.8	3	−12.3
D 150	Orthogneiss	Hongan	7.560	36.61	0.1249	0.512038 ± 9	1.9	3	−11.7

+ Samples Sh-1, 92HT-1 were analysed at Max-Planck Institute, Germany; Other data presented in this table were obtained at Institute of Geology, Chinese Academy of Geological Sciences, Beijing, China. Sample locations are shown in Figure 2.

* Muc-Muscovite; Qz-Quartz; Sch-Schist; Gn-Gneiss; Bi-Biotite; Hb-hornblende; Pl-plagioclase; Gr-garnet;

++ See Fig.2 for details

ε_{Nd} isotopic ratios were normalized against $^{146}Nd/^{144}Nd = 0.7219$ and adjusted to $^{143}Nd/^{144}Nd = 0.512612$ for the BCR-1 Nd standard.

** Model ages were calculated using the following equation assuming a linear Nd isotopic growth of the depleted mantle reservoir from $\varepsilon_{Nd} = 0$ at 4.56 Ga to $\varepsilon_{Nd} = d + 10$ at the present. $T_{DM} = 1/\lambda \ln\{1 + [(^{143}Nd/^{144}Nd)sample - 0.51315]/[(^{147}Sm/^{144}Nd)sample - 0.2137]\}$, $\lambda = 0.00654 Ga^{-1}$. Data source: 1. This paper; 2. S. Li et al. (1993a); 3. Xie et al. (1996).

The upper unit including the Nanwan formation and the Foziling Group is a Devonian metaclastic flysch formation (Gao et al., 1988). The relative lower T_{DM} values of the sedimentary rocks from the Foziling Group suggests that the flysch formation should be deposited near the active continental margin, such as the lower unit of the Beihuaiyang zone or North Qinling belt. However, a lack of volcanic rocks in the upper unit may suggest the termination of the island arc magmatism in the Beihuaiyang zone in the Devonian. This is supported by the Carboniferous coal series intercalated with marine shales locally overlying the Foziling Group. Extensive conglomerate beds occur along the southern limb of the Carboniferous, and contain various gravels including slate, sandstone, quartzite, black chert, granite and limestone (Ma, 1989). Some gravels (e.g. slate, sandstone) may be derived from the Foziling Group, whereas the Silurian fossils in the limestone pebbles were derived from the early Paleozoic passive continental margin of the SCB or the Gondwana continent (Ma, 1989; Xu et al., 1992b). Based on this observation, some researchers suggest the Middle Paleozoic collision between the NCB and SCB (e.g. Ma, 1989), which contradicts the Triassic age of the UHP metamorphism in the Dabie orogen. In order to reconcile this observation with the Triassic age of the UHPM rocks, a model could be proposed in that there might be a microcontinent (similar to South Qinling?) between the NCB and SCB in the Dabie area in the Middle Paleozoic time, and the ocean between the NCB and the microcontinent was closed at the end of the Devonian. The Beihuaiyang area then became a remnant basin that could receive the detritus from both the northern and southern continental margins. Thus the upper unit of the Beihuaiyang zone could correspond to the Liuling Group, which is a Devonian metaclastic flysch formation in the Qinling area and is located to the south of the North Qinling belt and on the north margin of the South Qinling belt (Zhang et al., 1995; Yu and Meng, 1995). However, the difficulty for this model is that no such a micro-continent has been identified in the Dabie orogen, though it is possible that the microcontinent overlying the subducted continental crust of the SCB in the Dabie area could be removed out by erosion. Another possible model is that the ocean between the NCB and SCB became narrow in the Dabie area in the Devonian, because Paleomagnetic results from China suggest that the collision started earlier in the east than in the west (e.g. Zhao and Coe, 1987). In any case, the upper unit of the Beihuaiyang zone would be developed near the active continental margin of the NCB in the Devonian.

4 Huwan tectonic melange zone: the suture between the NCB and SCB

The Huwan tectonic melange zone is discontinuously exposed to the south of zone 1, in the western part of the Dabie orogen (Figs. 1 and 3). It was previously named the Huwan formation of the "Sujiahe Group" in Chinese literatures. The "Sujiahe Group" was subdivided into two units, the northern metavolcano-clastic rock zone (the Dingyuan formation) and southern eclogite-bearing melange zone (the Huwan formation), which is separated by a north-dipping shear zone (Fig. 3) (Ye et al., 1994). As mentioned above, the northern zone (the Dingyuan formation) in the "Sujiahe Group" and the Guishan complex should be considered as one unit because of their similar lithological characters (Suo et al., 1993) and metamorphic history (this paper). Therefore, the southern zone (Huwan formation) of the "Sujiahe Group" is an independent tectonic melange zone. It is bounded by north dipping shear decollement zones on the north and south (Ye et al., 1994) (Fig. 3), and separates zone 1 from the Tongbai complex and Hongan terrane (Ye et al., 1994; Niu et al., 1994).

The Huwan tectonic melange zone is mainly composed of interleaved blocks and slabs, which are commonly mylonitizated. Eclogite, marble and quartzite lenses are encased in the schist and gneiss which are considered to be an argillitic matrix with metamorphism of greenschist- to amphibolite-facies. The early Paleozoic fossils found in marble block suggest that the carbonates were deposited in the Ordovician (Ye et al., 1994). Abundant geochronological data suggest that the Huwan tectonic melange was formed in the early Triassic. For example, $^{40}Ar/^{39}Ar$ age for white mica from quartz albite muscovite schist and the matrix in the melange zone are 227±2 Ma (PS33) (Niu et al., 1994) and 243±2Ma (HY-24) (Ye et al. 1993), respectively (Fig. 3). Two whole rocks (HY62 and HY89) or whole rocks + minerals (muscovite and plagioclase) Rb-Sr isochron ages for mylonites from the shear decollement zone between the melange zone and the Dingyuan formation on the north are 236±11 Ma and 225±8 Ma, respectively (Ye et al., 1993). $^{40}Ar/^{39}Ar$ ages for white micas from the shear decollement zone between the melange zone and Hongan HP zone on the south are 229±4 Ma to 235±2 Ma (Webb et al., 1999). All these ages are consistent with the peak metamorphic ages of UHPM rocks from the Dabie-Sulu orogen. However, the eclogites involved in the melange zone have older metamorphic ages than the matrix and shear decollement.

The high-pressure mineral assemblage in eclogite from the Huwan melange zone is garnet + omphacite + rutile +glaucophane or barroisite + epidote + phengite + quartz. No coesite or pseudomorph after coesite has been found in this eclogite. Compositional zonation in the garnet is evident: the core contains abundant mineral inclusions of Na-Ca-amphibole, epidote, quartz, and albite; the rim is virtually free of inclusions. This zoning pattern reflects a progressive metamorphism and chemical disequilibrium between the core and rim of a garnet. Most of the eclogites have experienced intense retrograde metamorphism. The retrograde minerals include hornblende, barroisite, epidote and albite (Liu et al., 1996).

Ye et al. (1980) and Liu et al. (1996) have analyzed mineral chemistry of the eclogites from this zone and concluded that they are cold eclogites (T = 570 to 680 ℃ and P = 10 to 15 kb). Jian et al. (1997) reported zircon U-Pb isotopic ages of the eclogites from the Xiongdian village. Four fractions of yellowish, prismatic zircon each are concordant ranging from 373 to 400 Ma within analytical uncertainties. A colorless round zircon with numerous small fluid inclusions yields a concordant age of 302±2 Ma. They interpret the oldest concordant age of 400±2 Ma as the age of HP metamorphism and the youngest concordant age of 302±2 Ma as the age of retrograde metamorphism. Recent ion microprobe (SHRIMP) data on zircon and cathodoluminescence (CL) of the zircon grains from the Xiongdian show that the inherited core in zircon grain with clear oscillatory zoning yields an age of 424±5 Ma, and overgrowth rim of zircon yield an age of 301±13 Ma (Jian et al., 2000). Zircons with unclear zoning are dated as 335±2 Ma to 408±8 Ma (Jian et al., 2000). These data clearly indicate two stages of zircon growth: the first stage at 424±5 Ma that produced oscillatory zoning and probably occurred during crystallization from a magma; the second stage at 301±13 Ma that is probably associated with HP metamorphism. The other ages ranging from 335 to 408 Ma may represent "mixed ages" due to incomplete expulsion of radiogenic lead from metamict zircon during HP metamorphism. Therefore, we suggest that the Xiongdian eclogite from the Huwan melange zone was formed at around 301 Ma instead of 424 Ma, and its protolith is the basaltic rocks of Silurian (424±5 Ma).

We have analyzed the trace element compositions of the Xiongdian eclogites (see Table 4). Though their higher SiO_2.

Table 4 Major element (%) and trace element (ppm) compositions of eclogites from the Huwan tectonic mélange zone

Elements	Hujiawan eclogite					Xiongdian eclogite	
	95HN-S-8	95HN08	95HN-S-9	95HN-S-10	95HN-S-15	95HN-S-16	95HN-S-18
SiO_2	47.40		46.89	50.05	57.07	57.22	53.02
TiO_2	0.97		0.46	0.46	0.38	0.33	0.33
Al_2O_3	17.60		12.89	14.17	15.07	13.86	13.85
Fe_2O_3	2.30		2.57	4.54	1.42	1.35	1.53
FeO	10.37		7.86	4.17	5.79	5.59	5.33
MnO	0.23		0.19	0.11	0.13	0.15	0.14
MgO	6.02		10.75	7.56	5.75	5.50	5.52
CaO	11.84		14.69	12.99	9.29	11.10	13.85
Na_2O	1.62		1.16	3.22	3.35	3.34	2.23
K_2O	0.27		0.18	0.26	1.14	0.14	0.11
H_2O	1.53		2.11	1.81	0.38	0.05	1.04
CO_2	0.17		0.30	0.08	0.68	0.96	2.76
P_2O_5	0.044		0.018	0.028	0.06	0.09	0.10
Total	100.37		100.07	99.45	100.51	99.68	99.81
TFe_2O_3	13.82		11.30	9.17	7.85	7.56	7.45
Ti	3703	2037	2028	1098	2280	1980	1980
Cr	67.21	385.7	404.0	285.0	428	704	705
Co	33.82	47.03	43.35	33.88			
Ni	77.93	159.20	145.45	138.68	88	116	110
Cu	74.56	41.71	17.10	32.43			

Continued

Elements	Hujiawan eclogite					Xiongdian eclogite	
	95HN-S-8	95HN08	95HN-S-9	95HN-S-10	95HN-S-15	95HN-S-16	95HN-S-18
Zn	81.48	68.58	64.55	72.19			
Rb	7.67	4.13	3.97	6.47	31	6	7
Sr	582.3	184.2	340.1	1416	149	345	396
Y	27.88	11.53	14.47	16.45	11.23	8.50	9.01
Zr	54.42	12.37	24.11	32.89	53	60	40
Nb	2.35	0.56	0.97	1.93	4	4	4
Sn	1.97	1.12	0.79	1.99			
Cs	0.27	0.25	0.32	0.50			
Ba	88.80	61.05	64.82	145.21	618	105	44
La	20.78	2.12	2.72	8.79	2.5	3.08	4.00
Ce	48.94	7.31	8.26	19.56	5.86	6.53	8.79
Pr	6.37	1.52	1.46	2.86	0.70	0.87	1.10
Nd	29.43	7.40	7.9	14.18	3.76	4.24	5.34
Sm	6.71	2.56	3.22	5.08	0.92	1.26	1.53
Eu	2.65	0.85	1.02	1.50	0.30	0.52	0.62
Gd	7.69	2.21	2.60	6.72	1.21	1.43	1.69
Tb	1.13	0.48	0.55	0.94	0.25	0.23	0.27
Dy	6.15	2.60	3.29	5.20	1.71	1.39	1.51
Ho	1.31	0.60	0.85	0.78	0.38	0.29	0.32
Er	3.24	2.10	2.36	1.83	1.10	0.86	0.87
Tm	0.72	0.44	0.32	0.44	0.18	0.15	0.15
Yb	3.47	1.94	2.86	3.19	1.16	0.86	0.86
Lu	0.75	0.32	0.46	0.41	0.20	0.16	0.15
Hf	2.68	1.21	1.94	1.86			
Ta	0.41	0.56	0.15	0.40			
Pb	14.76	6.80	9.53	11.98	14	25	36
Bi	0.11	0.47	<0.0086	1.10			
Th	0.99	0.20	0.22	1.01	2	3	3
U	0.30	0.14	0.19	2.14	1	1	2

* The major elements were obtained at the Institute of Geophysical and Geochemical Exploratory Techniques, Chinese Academy of Geological Sciences, by X-ray fluorescence analysis except CO_2 and H_2O which were obtained by nonaqueous titration and gravimetric analysis respectively. Trace elements of the Hujiawan eclogites were analyzed by ICP-MS at the Institute of Geology, Chinese Academy of Sciences. The Rare earth elements of the Xiongdian eclogites were analyzed at the Geological Institute of Hubei Province by ICP and their other trace elements were obtained at the Institute of Geophysical and Geochemical Exploratory Techniques by X-ray fluorescence analysis.

Table 5 Sr and Nd isotopic compositions of eclogites from Huwan tectonic melange zone[*]

Sample No.	Location	Rb (ppm)	Sr (ppm)	$^{87}Rb/^{86}Sr$	$^{87}Sr/^{86}Sr$	$(^{87}Sr/^{86}Sr)_i$	Sm (ppm)	Nd(ppm)	$^{147}Sm/^{144}Nd$	$^{143}Nd/^{144}Nd$	$\varepsilon_{Nd}(t)$
95 HN-S-8	Hujiawan	8.361	584.7	0.0416	0.704752 ± 24	0.704529	6.315	23.69	0.1613	0.512489 ± 13	−1.5
95 HN-S-9	Hujiawan	4.724	355.2	0.0387	0.705253 ± 19	0.705045	1.557	6.118	0.1539	0.512502 ± 19	−1.0
95 HN-S-10	Hujiawan	4.153	1452	0.0083	0.709235 ± 21	0.709190	3.836	11.73	0.1978	0.512580 ± 9	−1.2
Hy23-1**	Xiongdian						1.09	3.29	0.2006	0.512909 ± 2	+ 5.1

* The Sr and Nd isotope data of eclogites from Hujiawan were obtained at modem analysis center, Nanjing University, following procedures described in Table 2. The initial $(^{87}Sr/^{86}Sr)_i$ and $\varepsilon_{Nd}(t)$ are corresponding to metamorphic age of 301 Ma.
** The Nd isotopic data were quoted from Jian et al. (1997).

Table 6 U-Pb isotopic compositions of zircon from the Hujiawan eclogite (95HN-S-8)[*]

No.	Fractions Properties	Wt. (μg)	U (ppm)	Pb (ppm)	Pb com. (ng)	$^{206}Pb/^{204}Pb$	$^{208}Pb/^{206}Pb$	$^{206}Pb/^{238}U$	$^{207}Pb/^{235}U$	$^{208}Pb/^{206}Pb$	$^{206}Pb/^{238}U$	$^{207}Pb/^{235}U$	$^{207}Pb/^{206}Pb$
								Isotope atomic ratios ± 2σ			Age ± 2σ (Ma)		
1	Colorless, clear round	10	317	24	0.016	785	0.3050	0.06005 ±186	0.4484 ±196	0.05416 ±155	375.9 ±12	376.2 ±16	377.88 ±11
2	Light yellow, clear round	10	488	33	0.022	864	0.1720	0.60029 ±154	0.4509 ±161	0.05423 ±125	377.4 ±10	377.9 ±3	380.7 ±9
3	Colorless, clear short prismatic	10	191	18	0.011	944	0.1475	0.08623 ±379	0.7004 ±421	0.5891 ±223	533.2 ±23	539.0 ±32	563.9 ±22
4	Colorless, clear long prismatic	20	122	12	0.001	13025	0.09121	0.1028 ±6	0.8578 ±461	0.06053 ±355	630.7 ±3	628.9 ±34	622.4 ±31
1+2	average										376.8 ±7.3	377 ±10	379.5 ±6.7

[*] The U-Pb isotopic analysis was conducted in the Tianjin Institute of Geology and Mineral Resources. $^{208}Pb/^{235}U$ spike is used for fractions 1, 2, 3 and $^{208}Pb/^{235}U$ spike is used for fraction 4. Blanks of whole chemical procedure are: Pb = 30 pg, U = 2 pg, $^{206}Pb/^{204}Pb$ ratios has been corrected for blank and spike, other isotope atomic ratios are for radiogenic Pb. Common Pb correction and data processing were conducted using software PBDAT (1989) and ISOPLOT provided by Ludwig.

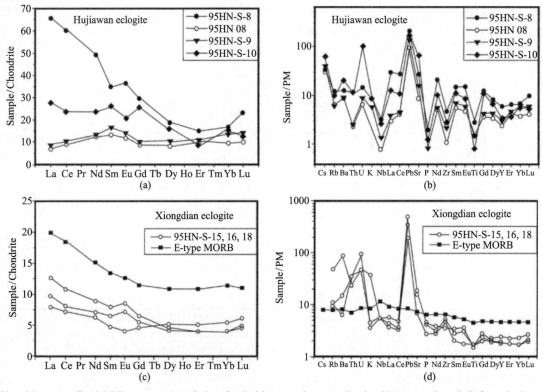

Fig. 8 Chondrite-normalized REE patterns (a and c) and primitive mantle normalized spidergrams (b and d) for eclogites from the Xiongdian and Hujiawan in Huwan tectonic melange zone. Except the mobile elements (Rb, Ba, K, Pb, Sr), REE and other trace element patterns of the Xiongdian eclogite are similar to E-MORB (c and d). Though the Hujiawan eclogites have various REE patterns including LREE depletion to LREE enriched (a), all of them are characterized by negative anomalies of Nb, P, Zr, Ti in spidergrams (b).

contents (53% to 57%) are similar to those of basaltic andesite, their high Cr (428 to 705 ppm) and Ni (88 to 116 ppm) contents are not similar to those of island-arc basalt and andesite. Their primitive mantle normalization pattern with slightly LREE enrichment and no Nb negative anomalous is similar to that of E-MORB (see Fig. 8). The Xiongdian eclogites are also plotted in the MORB field in La/Nb-La diagram (Fig. 9). The initial ε_{Nd} value at the metamorphic time (301 Ma) of the Xiongdian eclogite is +5.1 (see Table 5). If its Sm/Nd ratio has not been changed during HP metamorphism, it can be considered as ε_{Nd} value of the protolith of the Xiongdian eclogite because its whole rock

^{147}Sm/^{144}Nd ratio of 0.2006 is close to the average chondrite ^{147}Sm/^{144}Nd value of 0.1967 (Table 5). This high initial ε_{Nd} value (+5.1) is in contrast with the very low ε_{Nd} value (−5 to −17.7) of the coesite-bearing eclogites from the Southern Dabie zone (Li et al., 1993). The higher ε_{Nd} value suggests that the protolith of the Xiongdian eclogite may be a part of oceanic crust which was derived from a deplete mantle with little contamination of continental crust.

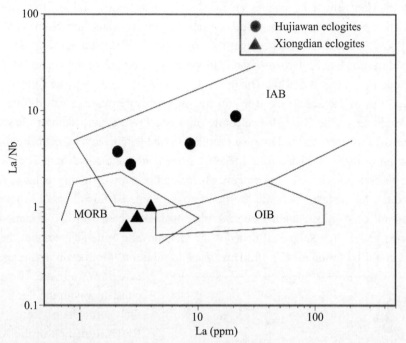

Fig. 9 La/Nb-La plot showing that the Xiongdian eclogites am characterized by lower La/Nb ratio, falling in the MORB field, while the Hujiawan eclogites are characterized by higher La/Nb ratio, falling in the IAB field (after Li et al., 1993).

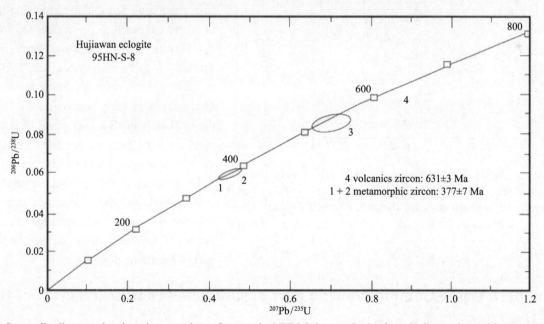

Fig. 10 Concordia diagram showing zircon analyses for sample 95HN-S-8, an eclogite from Hujiawan located in Huwan tectonic melange zone. Two fractions (1+2) of round metamorphic zircons yield a concordant age of 377±7 Ma and the prismatic zircon (fraction 4) yields a concordant age of 631±3 Ma.

However, the trace element pattern of another eclogite lense in the Hujiawan village located to the north of the Xiongdian shows the typical feature of island arc basalt with negative Nb, P, Zr and Ti anomalies in spider diagram (see Figs. 8 and 9). We also obtained several U-Pb zircon ages for the Hujiawan eclogite using conventional U-Pb zircon dating method (see Table 6 and Fig. 10). Two fractions (1 and 2) of round metamorphic zircons yield a concordant age of 377±7 Ma and the prismatic zircon (fraction 4) yields a concordant age of 631±3 Ma. If the Hujiawan eclogite and the Xiongdian eclogite experienced the HP metamorphism at the same time, the age of 377± 7 Ma may represent a "mixed age", and the real metamorphic age of the Hujiawan eclogite could be about 301 Ma. However, its concordant age of 631±3 Ma suggests that the protolith age of the Hujiawan eclogite may be older than the Xiongdian eclogite. Their initial ε_{Nd} values at 301 Ma (–1.0 to –1.5) (Table 5) are lower than that of the Xiongdian eclogite. Their Sr and Nd isotope compositions fall in the island arc basalt (IAB) field and indicate seawater alteration (Fig. 11). These data suggest that the protolith of the Hujiawan eclogite is different from that of Xiongdian eclogite and may be late Proterozoic island arc basalt erupted below the sea level.

All of the available data suggest that the Huwan tectonic melange zone may be formed in the early Triassic. The eclogites involved in this melange zone were formed in Carboniferous and had different protoliths, such as middle Paleozoic oceanic crust (the Xiongdian elcogite) and late Proterozoic island arc basalt. Since the Huwan tectonic melange zone directly contacts the Dingyuan formation (the Paleozoic magmatic arc) on the north, it should be the northmost margin of the SCB. Therefore, the Huwan melange zone could be a Triassic suture between the NCB and SCB in the western part of the Dabie orogen. The eclogites involved in this melange zone could be formed by oceanic subduction during the late Paleozoic (Carboniferous). Some island arc basalt in fore-arc region could be scraped off the island arc crust by subducting oceanic plate and subducted with the oceanic crust into the depth. However, as mentioned above, the geological data for the Beihuaiyang zone suggest that the oceanic subduction underneath the NCB could be terminated in the Devonian. In addition, the metamorphic age of 301 Ma for the eclogites from the Huwan melange zone is consistent with the $^{40}Ar/^{39}Ar$ age of 316±1 Ma for hornblende from the metamorphosed flysch formation of the Xinyang Group in the Tongbai area (Zhai et al., 1998) and $^{40}Ar/^{39}Ar$ age of 314±6 Ma of biotite from the Shangdan suture in Qinling area (Mattauer et al., 1985). Therefore, it is also possible that the Huwan tectonic melange zone may be formed by collision between the NCB and a possible microcontinent (similar to the South Qinling?) in the early Carboniferous. However, the difficulty in this interpretation is that no such a micro-continent has been identified in the Dabie orogen. More studies are needed to fully understand the tectonic setting of the Huwan melange zone, thus it is safe to say that the Huwan tectonic melange zone is the northmost boundary of the SCB in the western part of the Dabie orogen.

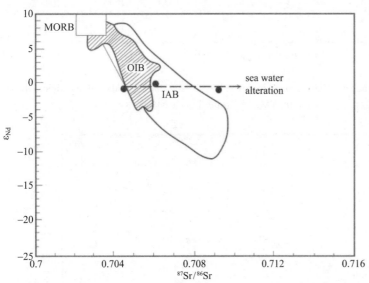

Fig. 11 ε_{Nd} vs $^{87}Sr/^{86}Sr$ diagram showing that the initial Nd and Sr isotopic compositions of eclogites from Hujiawan are similar to those of island arc basalts and characterized by sea water alteration. The fields of MORB, OIB and IAB are estimated at 300 Ma based on the data from Zindler and Hart (1996) and Davidson (1983).

5 Northern Dabie Complex zone: a part of subducted continental crust of the SCB

Northern Dabie complex zone (or Northern Dabie zone) is mostly exposed in the eastern part of the Dabie orogen (Fig. 1), and is bounded by the Shang-Ma fault on the west (⑥ in Fig. 1) and Tan-Lu fault on the east (Fig. 1). It is separated from the Beihuaiyang zone by the Tongbai-Mozitan fault (② in Fig. 1) on the north and from the Hong'an-Susong high pressure and southern Dabie UHPM zones by the Xishui-Yingshan and Wuhe-Shuihouductile shear zones on the south (⑤ and ③ in Fig. 1) (Dong et al., 1993; Suo et al., 1993; Cong et al., 1994). The Northern Dabie zone is composed mainly of felsic gneiss and migmatite with minor metapelite, marble, amphibolite, magnetite quartzite, quartzite, granulite and ultramafic rocks. Three large granitic batholiths of Cretaceous ages (120 to 125 Ma) are exposed in this zone, e.g. the Tiantangzhai (T), Baimajian (B) and Zhuboyuan (Z) granites (Fig. 2). Similar K-Ar ages for K-feldspar from migmatite with high-K leucosomes (98 to 121 Ma)(Chen et al., 1991) and Rb-Sr whole-rock isochron age for monozonitic migmatitie (111.5±4 Ma)(Jian et al., 1996) suggest that the high-K leucosomes in migmatite and the Cretaceous granite may be formed at the same time. There are two kinds of gneiss: one is deformed diorite-granodiorite, the other is banded gneiss with strong foliation. Recent U-Pb zircon datings suggest that the deformed diorite-dacite intrusions, i.e. "orthogneiss", are crystallized in Cretaceous (Xue et al., 1997; Hacker et al., 1998) while the protoliths of the banded or mylonite gneiss were formed in late Proterozoic (757±1 Ma and 707±42 Ma) (Xue et al., 1997; Xie et al., 2001). The tectonic setting of the Northern Dabie zone before or during the Triassic collision is an important issue.

There are two kinds of ultramafic rocks in the Northern Dabie zone. One is the Alpine type peridotites, consisting of harzburgite, dunite and minor amphibole pyroxenite veins. These ultramafic bodies are mainly distributed on the northern margin of the Northern Dabie zone (see Fig. 2). They are strongly deformed and thus were tectonically emplaced into mylonite gneiss. For example, the Raobazhai ultramafic massif was emplaced in solid state into migmatites of the Northern Dabie zone. Country rocks near the contact zone are strongly mylonitized. Foliations are well developed along the margin of the massif and a mylonite zone crosscuts the massif. It has been generally accepted that the Raobazhai ultramafic massif was a fragment of the lithosphere mantle (Q. Zhang et al., 1995; Li et al., 1998). The trace element compositions with Nb negative anomalous and the isotopic compositions of the garnet-pyroxenite and amphibole-pyroxenite veins in the Raobazhai massif suggest that the lithospheric mantle has been metasomatized by fluid related to oceanic subduction (Q. Zhang et al., 1995; Li et al., 1993). Its not very low initial ε_{Nd} value of –2.7 (Li et al., 1993) suggests that the related oceanic subduction could occur during the Paleozoic time. Therefore, the Raobazhai massif could be derived from the mantle wedge underneath the overthrust crust of the NCB. The Sm-Nd mineral isochron age of 244±11 Ma for the garnet-pyroxenite from the Raobazhai massif suggests that the tectonic emplacement of the ultramafic body occurred during the Triassic collision between the NCB and SCB (Li et al., 1989, 1993). The other kind of mafic-ultramafic rocks are pyroxenite-gabbro intrusions of Cretaceous ages (130 to 120 Ma), such as Zhujiapu, Renjianwan and Zhongguan pyroxenites and Jiaoyan gabbro (①, ③, ④and ⑤ in Fig. 2) (Hacker et al., 1998; Li et al, 1999b; Jahn et al., 1999). These rocks wereintruded into the banded gneiss but intruded by granitic dykes.

Since no UHP rocks has been observed in the Northern Dabie zone until very recently, various tectonic settings have been proposed for this zone, such as metaophiolite melange in the suture (Xu et al., 1992b), island arc near the NCB (Dong et al., 1993; Cong et al., 1994), a thrust plane behind the eclogite zone in the subducted continental basement of the SCB (Okay and Sengtir, 1993) or a thrust plane in the front of the eclogite zone in subducted continental crust (Maruyama et al., 1994), and the NCB hanging wall during the Triassic subduction of the SCB (Zhang et al., 1996). Recent discovery of eclogite in the Northern Dabie zone (Xu et al., 2000; Liu, 2000) and its isotopic dating (Liu et al., 2000, 2001) greatly help us to understand the tectonic setting of the Northern Dabie zone.

There are two kinds of eclogites in the Northern Dabie zone. One occurred in foliated peridotite, such as Raobazhai and Huangweihe massifs (② and ⑦ in Fig. 2) (Liu, 2000; Xiao et al., 2001). The other one is in banded gneiss, such as Baizhangyan and Huazhuang eclogites (⑧ and ⑨ in Fig. 2) (Liu, 2000). All these eclogites are located on the north margin of the Northern Dabie zone (see Fig. 2). Omphacite with high Na_2O content up to 7.6%~7.9% are preserved as inclusions in garnet. Omphacites which are outside of garnet were mostly retrograded to symplectite or diopside (Xu et al., 2000). Other scientists have observed some eclogite-facies relics, such as omphacite (Xiao et al., 2001) and microstructure of oriented quartz needles in Ca-Nd clinopyroxene (Tsai et al., 1998). So far no coesite has been found in these eclogites, thus it is unclear whether the eclogites suffered the UHP metamorphism. After the HP metamorphism, those eclogites were firstly retrograded at granulite-facies (T = 845 to 900 ℃ and P = 7 to 11 kb), and then retrograded to amphibolite (Xu et al., 2000; Xiao et al., 2001). The granulite-facies can be identified by hypersthene in symplectite (Liu, 2000; Xiao et al., 2001).

An U-Pb age for zircon from Raobazhai eclogite (230±6 Ma) (Liu et al., 2000) and Sm-Nd mineral isochron ages for Raobazhai garnet pyroxenite (garnet + diopside, 244±11 Ma) (Li et al., 1989) and Huangweihe eclogite (garnet + omphacite, 210±6 Ma and 214±6 Ma) (Liu et al., 2001) suggest that the eclogites in Northern Dabie zone were formed in the Triassic. The eclogite-bearing foliated peridotites could be scraped off the overlying lithospheric mantle by subducting continental crust and were subducted with the continental crust into depths. Hence, the northern margin of the Northern Dabie zone, where the foliated peridotites and eclogites are developed, could be close to the suture zone between the NCB and SCB. Recent U-Pb zircon dating suggests that the banded gneiss in the Northern Dabie zone, which are country rocks of the eclogite, also experienced the Triassic (226±6 Ma and 229±18 Ma) metamorphism (Liu et al., 2000; Xie et al., 2001). The different retrograde metamorphic history between the Northern Dabie eclogite and Southern Dabie eclogite as well as the older Nd model ages (2.0 to 2.6 Ga) for the orthogneiss in the Northern Dabie zone than those the Southern Dabie zone (see Table 3, Fig. 10A) suggest that the Northern and Southern Dabie zones are two different thrust planes in subducted continental crust. They have the same HP or UHP metamorphic ages, but may have different exhumation histories. The similar conclusion has been drawn by Zheng et al. (2001) from the oxygen isotope study of granulites from Northern Dabie zone. Therefore, the Wuhe-Shuihou ductile shear zone (③ in Fig.1), separating the Northern Dabie and Southern Dabie zones, is a fault in the subducted continental crust instead of the suture between the NCB and SCB. In view of the above conclusion that lower unit of the Beihuaiyang zone was the active continental margin of the NCB, the suture between the NCB and SCB could be located along the boundary between the North Dabie zone and Beihuaiyang zone in the eastern part of the Dabie orogen.

6 Conclusions

(1) The protoliths of metabasic volcanic rocks from the Dingyuan formation of the Beihuaiyang zone in the Dabie orogen is subalkaline basalt and belong to calc-alkaline series. These rocks have typical island arc geochemical features, e.g., negative Nb, Ti and P anomalies but positive Ba, Pb and Sr anomalies in trace element spider diagram as well as high La/Nb, Ba/Nb and Th/Nb ratios.

The Sm-Nd and Rb-Sr isochrons of the meta-volcanic rocks from the Dingyuan formation yield consistent ages of 446±23 Ma and 444±31 Ma, respectively. These ages may indicate that their formation age is comparable to the Danfeng island arc volcanics in the North Qinling belt. This is the first evidence for the Palezoic magmatic arc in the Dabie orogen.

The geochemical features and isotopic ages of the Dingyuan volcanics as well as their tectonic setting at convergent plate boundary suggest that the Beihuaiyang zone once could be an magmatic arc on the southern active continental margin of the NCB in the Ordovician to Silurian. Therefore, the suture between the NCB and SCB should be to the south of the Dingyuan formation in the northwestern part of the Dabie orogen.

(2) The Nd isotope model ages (T_{DM}) of metasediments from the Foziling Group range from 1.7 to 1.9 Ga, which are younger than the Nd model ages (1.9 to 2.1 Ga) of the Northern Paleozoic sedimentary cover of the SCB and those of the North Qinling basement rocks. This suggests that more Phanerozoic volcanics were involved in the sedimentary rock of the Foziling Group, indicating an active continental margin setting of the Beihuaiyang zone in the Devonian. However, if a micro-continent similar to the South Qinling belt would once exist in the Dabie area, the Beihuaiyang zone could become a remnant basin in the Devonian, which could receive detritus from both the Northern and Southern continental margins.

(3) The eclogites in the Huwan tectonic melange zone were formed in the Carboniferous (−301±13 Ma). The protolith of eclogite from Xiongdian could be Silurian (424 ± 5 Ma) oceanic crust, while the protolith of eclogite from Hujiawan is more likely to be late Proterozoic island arc basalt. However, the synkinematic minerals from the argillitic matrix and shear decollement faults of the melange zone were formed in the Triassic. Two alterative interpretations to the tectonic setting of the Huwan tectonic melange are possible: (a) it could be a Triassic suture between the NCB and SCB, and the eclogites involved in this melange zone were formed by oceanic subduction before continental collision; or (b) it could be formed by collision between the NCB and a possible micro-continent in the Carboniferous. Further studies are needed to test whether there was a micro-continent between the NCB and SCB during the Paleozoic time in the Dabie area.

(4) The discovery of the Triassic eclogite in the Northern Dabie zone suggests that the Northern Dabie zone is a part of the subducted continental crust of the SCB. The North Dabie high T/P metamorphic zone, Southern Dabie UHPM zone and Susong-Hongan HP metamorphic zone may represent three thrust planes of the subducted continental crust. In light of the conclusion (1), the suture between the NCB and SCB in the eastern part of the Dabie orogen could be located on the boundary between the Northern Dabie zone and Beihuaiyang zone.

Acknowledgements

This project is supported by Major State Basic Research Program (G1999075503) and the Natural Science Foundation of China (grant No. 49573190, 49794042). We thank Profs. J.-F. Chen, M.-G. Zhai and Y.-F. Zheng for their constructive comments which help improve the manuscript. We also thank Mr. Hou Zhenhui for his assistance in the typing work.

References

Ames, L., Tilton, G.R. and Zhou, G., 1993. Timing of collision of the Sino-Korean and Yangtze Cratons: U-Pb zircon dating of oesite-bearing eclogites. Geology, 21, 339-342.

Ames, L., Zhou, G. and Xiong, B., 1996. Geochronology and isotopic character of ultrahigh pressure metamorphism with implications for collision of the Sino-Korean and Yangtze cratons, central China. Tectonics, 15, 472-489.

Chavagnac, V., and Jahn, B.M., 1996. Coesite-bearing eclogite from the Bixiling Complex, Dabie orogen, China: Sm-Nd ages, geochemical characteristics and tectonic implication. Chem. Geol., 133, 29-51.

Chen, J.F., Zhou, T., Xing, F., Xu, X. and Foland, K.A., 1990. Provenances of low grade metamorphic and sedimentary rocks from Southern Anhui Province: evidence of Nd isotope compositions. Chinese Sci. Bull., 35, 747-750.

Chen, T.Y., Niu, B., Liu, Z., Fu, Y., and Ren, J., 1991. Geochronology of Yanshanian magmatism and metamorphism in the hinterland of the Dabie orogen and their geologic significance. Ada Geol. Sinica, 147, 329-335 (in Chinese).

Chen, Y.Z. and Sang, B., 1995. Metamorphic petrology and metamorphism of the Foziling Group in Northern Huaiyang and its age. Regional Geol. of China, (3), 280-288 (in Chinese).

Cong, B., Wang, Q., Zhai, M., Zhang, R., Zhao, Z. and Ye, K., 1994. Ultra-high pressure metamorphic rocks in the Dabie-Sulu region, China: their formation and exhumation. Island Arc, 3, 135-150.

Cong, B., Zhai, M., Carswell, D.A., Wilson, R. N., Wang, Q., Zhao, Z. and Windley, B.F., 1995. Petrogenesis of ultrahigh-pressure rocks and their country rocks at Shuanghe in Dabieshan, Central China. Eur. J. Mineral, 7, 119-138.

Cui, Z., Sun, Y., Wang, X., 1995. The discovery of raliolarian in Danfeng ophiolite belt, North Qinling and its geological implications. Chinese Sci. Bull., 40(18), 1686-1688 (in Chinese).

Dong, S. M., Sun, X., Zhang, Y., Huang, D., Wang, G., Dai, S. and Yu, B., 1993. The basic structure of the Dabieshan collision orogenic belt. Chinese Sci. Bull., 38, 542-545 (in Chinese).

Davidson, J.P., 1983. Lesser Antilles isotopic evidence of the role of subducted sediment in island arc magma genesis. Nature, 306, 253-256.

Eide, E.A., Mcwilliams, M.Q., Liou, J.G., 1994. $^{40}Ar/^{39}Ar$ geochronology and exhumation of high-pressure to ultrahigh-pressure metamorphic rocks in east-central China. Geology, 22, 601-604.

Gao, L.D., Liu, Z.G., 1988. The discovery of micro-fossil from the Nanwan formation of the Xinyang Group and its geological implications. Geol. Review, 34(5), 421-422 (in Chinese).

Hacker, B.R. and Wang, Q.C., 1995. $^{40}Ar/^{39}Ar$ geochronology of ultrahigh-pressure metamorphism in central China. Tectonics, 14(4), 994-1006.

Hacker, B.R., Ratschbacher, L., Webb, L., Dong, S., 1995. What brought them up? Exhumation of the Dabie Shan ultrahigh-pressure rocks. Geology, 23, 743-746.

Hacker, B.R., Ratschbacher, L., Webb, L., Ireland, T., Walker, D., and Dong, S., 1998. U/Pb zircon ages constrain the architecture of the ultrahigh-pressure Qinling-Dabie Orogen, China. Earth. Planet. Sci. Lett., 161, 215-230.

Hart, S. R., 1971. K, Rb, Cs, Sr and Ba contents and Sr isotope ratios of ocean floor basalts. Phil. Trans. Roy. Soc. Lond., A268, 573-587.

Jahn, B.M., Wu, F., Lo, C. H., Tsai, C. H., 1999. Crust-mantle interaction induced by deep subduction of the continental crust: geochemical and Sr-Nd isotopic evidence from post-collisional mafic-ultramafic intrusions of the northern Dabie complex, central China. Chem. Chem. Geol., 157, 119-146.

Jian, P., Ma, C. and Yang, K., 1996. Geochronological evidences of Yanshanian metamorphism-magmatism-tectonic uplifting in the eastern Dabie area, Central China. Earth Sci. 21(5), 519-523 (in Chinese).

Jian, P., Yang, W., Li, Z. and Zhou, H., 1997. Isotopic chronological evidence for the caledonian Xiongdian eclogites in the western Dabie orogen. Ada Geol. Sinica, 71, 133-141 (in Chinese).

Jian, P., Liu, D., Yang, W., Williams, I.S., 2000. Petrographical study of zircons and SHRIMP dating of the Caledonian Xiongdian eclogite, Northwestern Dabie Mountains. Acta Geol. Sinica, 74(3), 259-264.

Lerch, M.F., Xue, F., Kroner, A., Zhang, G.W. and Todt, W., 1995. A middle Silurian-Early Devonian magmatic arc in the Qinling Mountains of central China. J. Geol., 103, 437-449.

Li, S.G., Hart, S.R., Zheng, S., Liou, D., Zhang, G. and Guo, A., 1989. Timing of collision between the North and South China Blocks: Sm-Nd isotopic age evidence. Sci. China (Ser. B), 32, 1391-1400.

Li, S.G., 1993. Ba-Nb-Th-La diagrams used to identify tectonic environments of ophiolite. Acta Petro. Sinica, 9(2), 146-157 (in Chinese).

Li, S.G., Xiao, Y., Liu, D., Chen, Y., Ge, N., Zhang, Z., Sun, S.S., Cong, B., Zhang, R., Hart, S.R. and Wang, S., 1993. Collision of the North China and Yangtze Block and formation of coesite-bearing eclogites: Timing and processes. Chem. Geol., 109, 89-111.

Li, S.G., Wang, S., Chen, Y., Liou, D., Qiu, J., Zhou, H. and Zhang, Z., 1994a. Excess argon in phengite from eclogite: evidence from dating of eclogite minerals by Sm-Nd, Rb-Sr and $^{40}Ar/^{39}Ar$ methods. Chem. Geol., 112, 343-350.

Li, S.G., Chen, Y., Zhang, Z., Ye, X., Zhang, G., Guo, A., Hart, S. R., 1994b. Trace element and Sr, Nd isotopic geochemistry of the Lajimiao norite-gabbro in North Qinling belt. Acta Geol. Sinica, 7(2), 137-151.

Li, S.G. and Sun, W.D., 1996. A middle Silurian-early Devonian magmatic arc in the Qinling Mountains of Central China: a discussion. J. Geol., 104, 501-503.

Li S.G., Sun, W.D., Zhang, G.W., Chen, J. and Yang, Y., 1996. Chronology and geochemistry of metavolcanic rocks from Heigouxia Valley in the Mian-Lue tectonic zone, South Qinling: evidence for a Paleozoic oceanic basin and its close time. Sci. China (Ser. D), 39, 300-310.

Li, S.G., Nie, Y. H., Liu, D. L., Zheng, S. G., 1998. Interaction between subducted continental crust and the mantle — I. Major and trace element geochemistry of the syncollisional mafic-ultramafic intrusions in the Dabie Mountains. Sci. China (Ser. D), 41(5), 545-552.

Li, S.G., Jagoutz, E., Lo, C. H., Chen, Y., Li, Q., and Xiao, Y., 1999a. Sm/Nd, Rb/Sr and $^{40}Ar/^{39}Ar$ isotopic systematic of the ultrahigh-pressure metamorphic rocks in the Dabie-Sulu belt, central China: A retrospective view. Intern. Geol. Rev., 41, 1114-1124.

Li, S.G., Hong, J.A., Li., H.M., Jiang, L.L., 1999b. U-Pb zircon ages of the pyroxenite-gabbro intrusions in Dabie orogen an their geological implications. Geo. J. China Univ., 5(3), 351-355 (in Chinese with English abstract).

Li, S.G., Jagoutz, E., Chen, Y. and Li, Q., 2000. Sm-Nd and Rb-Sr isotopic chronology and cooling history of ultrahigh pressure metamorphic rocks and their country rocks at Shuanghe in the Dabie orogen, central China. Geochim. Cosmochim. Acta, 64, 1077-1093.

Liu, J.P., You, Z. and Zhang, Z., 1996. Eclogites from the middle and north of Dabie Mountains in southern Henan and northern Hubei, China. Sci. China (Ser. D), 39(3), 293-299.

Liu, X.H. and Hao, J., 1989. Structure and tectonic evolution of the Tongbai-Dabie range in the east Qinling collision belt, China. Tectonics, 8(3), 637-645.

Liu, Y.C., 2000. Petrology, geochemistry and isotopic chronology of the eclogites from the Northern Dabie orogen. Ph.D. Thesis, Univ. Sci. Technol. China, Hefei.

Liu, Y. C., Li, S. G., Xu, S. T., Li, H. M., Jiang, L. L., Chen, G. B., Wu, W. P., Su, W., 2000. U-Pb zircon ages of the eclogite and onalitic gneiss from the Northern Dabie orogen, China and multi-overgrowths of metamorphic zircons. Geol. J. China Univ., 6(3), 17-423 (in Chinese with English abstract).

Liu, Y.C., Li, S.G., Xu, S.T., Jahn, B. M., Zheng, Y.F., Jiang, L.L., Chen, G.B., Wu, WP, Su, W., 2001. Sm-Nd isotopic age of eclogite from the Northern Dabie zone and its geological implications. Geochimica, 30(1), 79-87 (in Chinese with English abstract).

Ma, W.P., 1989. Tectonics of the Tongbai-Dabie fold belt. J. Southeast Asian Earth Sci., 3(1-4), 77-85.

Masuda, A., Nakamura, N., Tanaka, T., 1973. Tine structures of mutually normalized rare earth patterns of chondrites. Geochim. Cosmochim. Ada, 37, 239-248.

Mattauer, M., Matte, P., Malavieille, L., Tapponnier, P., Maluski, H., Xu, Z.Q., Lu, Y.L., and Tang Y.Q., 1985. Tectonics of the Qinling Belt: build-up and evolution of eastern Asia. Nature, 293, 212-216.

Meng, Q.R., Zhang, G.W., 1999. Timing of collision of the North and South China Blocks: controversy and reconciliation. Geology, 27, 123-126.

Meng, Q. R., Zhang, G. W., 2000. Geologic framework and tectonic evolution of the Qinling orogen central China. Tectonophysics, 313, 183-196.

Niu, B.G., Fu, Y., Liu, Z., Ren, J., Chen, W., 1994. Main tectonothermal events and 40Ar/39Ar dating of the Tongbai Dabie orogen. Acta

Geosci. Sin., (1-2), 20-34 (in Chinese).

Okay, A.T., Xu, S.T. and Sengör, A.M.C., 1989. Coesite from the Dabie Shan eclogites, Central China. Eur. J. Mineral., 1, 595-598.

Okay, A.T. and Sengör, A.M.C., 1993. Tectonics of an ultrahigh-pressure metamorphic terrane: the Dabie Shan/Tongbai Shan orogen, China. Tectonics, 12, 1329-1334.

Rowley, D.B., Xue, F., Tucker, RD., Peng, Z.X., Baker, J. and Davis, A., 1997. Ages of ultrahigh pressure metamorphism and protolith orthogneisses from the eastern Dabie Shan: U-Pb zircon geochronology. Earth Planet. Sci. Lett., 152, 191-203.

Sun, S.S., McDonough, W.F., 1989. Chemical and isotopic systematics of oceanic basalts: implications for mantle composition and processes. In: Magmatism in the Oceanic Basins (eds. Saunders, A.D., Norry M.J.). Geol. Soc. Spec. Publ., 42, 313-345.

Sun, W., Li, S., Xiao, Y., Sun, Y., Zhang, G., 1995, The discovery of island arc volcanics in the Heihe area, North Qinling and its tectonic implications, Geotectonica et Metallogenica, 19 (3), 227-236 (in Chinese).

Sun, W., Li, S., Chen, W., Sun, Y. and Zhang, G., 1996a. Excess argon in metamorphic amphibole by tectonic shearing: evidence from Sm-Nd, Rb-Sr and $^{40}Ar/^{39}Ar$ dating of Plagioclase amphibolites of the Danfeng Group, North Qinling area. Geol. J. China Univ., 2(4), 382-389 (in Chinese with English abstract).

Sun, W., Li, S., Sun, Y., Zhang, G., Zhang, Z., 1996b. Chronology and geochemistry of a lava pillow in the Erlangping Group at Xixia in the North Qinling Mountains, Geol. Review, 42(2), 144-153 (in Chinese).

Sun, W., Li, S., Chen, Y., Li, Y., 2000. Zircon U-Pb dating of granitoids from South Qinling, central China and their geological significance. Geochimica, 29(3), 209-216 (in Chinese with English abstract).

Sun, Y., Yue. Z, Zhang, G., 1987. REE Geochemistry of the ophiolite in the eastern Qinling. Chinese Sci. Bull., 32 (3), 1654-1655 (in Chinese).

Suo, S.T., Sang, L., Han, Y., You Z., Zhou, N., 1993. The Petrology and Tectonics in Dabie Precambrian Metamorphic Terranes, Central China. Press of China Univ. of Geosci., Wuhan, pp. 33-112 (in Chinese).

Wang, X.M., Liou, J.G. and Mao, H.K., 1989. Coesite-bearing eclogites from the Dabie Mountains in central China. Geology, 17, 1085-1088.

Wang, X.M. and Liou, J.G., 1991. Regional ultrahigh-pressure coesite-bearing eclogitic terrane in central China: evidence from country rocks, gneiss, marble and metapelite. Geology, 19, 933-936.

Webb, L.E., Hacker, I.R., Ratschbacher, L., Mcwilliams, M.O. and Dong, S., 1999. Thermochronologic constrains on deformation and cooling history of high- and ultrahigh-pressure rocks in the Qinling-Dabie orogen, eastern China. Tectonics, 18, 621-638.

Xiao, Y.L., Hoefs, J., Van Den Kerkhof, A.M. and Li, S.G., 2001. Geochemical constrains of the eclogite and granulite facies metamorphismas recognized in the Raobazhai complex from North Dabie Shan, China. J. Metamor. Geol., 19, 3-19.

Xie, Z., Chen, J., Zhou, T., Zhang, X., 1996. Nd isotopic compositions of metamorphic and granitic rocks from Dabie orogen and their geological significance. Acta Petro. Sinica, 12(3), 401-408 (in Chinese with English abstract).

Xie, Z., Chen, J., Zhang, X., Gao, T., Dai, S., Zhou, T. and Li, H., 2001. Zircon U-Pb dating of gneiss from Shizhuhe in north Dabie and its geologic implications. Acta Petrol. Sinica, 17(1), 139-144 (in Chinese with English abstract).

Xu, G.Z. and Hao, J., 1988. The characteristics of Foziling Group and the geotectonic environment of its formation in the northern foot of the Dabie orogen. Scientia Geol. Sinica, (2), 95-109, 1988 (in Chinese).

Xu, S.T., Okay, A.I., Ji, S., Sengör, A.M.C., Su, W., Liu, Y. and Jiang, L., 1992a. Diamond from the Dabie Shan metamorphic rocks and its implication for tectonic setting. Science, 256, 80-92.

Xu, S.T., Jiang, L., Liu, Y. and Zhang, Y., 1992b. Tectonic framework and evolution of the Dabie orogen in Anhui, eastern China. Acta Geol. Sinica, 66(1), 1-14 (in Chinese with English abstract).

Xu, S.T., Liu, Y.C., Su, W., Wang, R.C., Jian, L. L., Wu, W. P., 2000. Discovery of the eclogite and its petrography in the northern Dabie Mountains. Chinese Sci. Bull., 43(19), 1651-1655.

Xu, Z.Q., Niu, B.G., Liu, Z.G. and Wang, Y.M., 1991. Tectonic system and intracontinental plate dynamic mechanism in the Qinling-Dabie "collision intracontinental" mountain chains. In: A Selection of Papers Presented at the Conference on the Qinling Orogenic Belt (eds. Ye, L.J., Qian, Zhang, G.W.). Northwest Univ. Press, Xi'an, pp. 139-147 (in Chinese).

Xue, F., Rowley, D.B., Tucker, R.D. and Peng, Z.X., 1997. U/Pb zircon ages of Granitoid rocks in the North Dabie complex, eastern Dabie Shan, China. J. Geol., 105, 744-753.

Ye, B.D., Jiang, P., Xu, J., Cui, F., Li, Z. and Zhang, Z., 1993. The Sujiahe Terrane Collage Belt and Its Constitution and Evolution along the North Hillslope of the Tongbai-Dabie Orogenic Belt. Press of China Univ. of Geosci., Wuhan, pp. 66-67 (in Chinese).

Ye, B.D., Jiang, P., Xu, Z., Cui, F., Li, Z. and Zhang, Z., 1994. Timing of the Sujiahe Group in the Tongbai-Dabie orogenic belt. In: Research of Isotope Geochemistry (ed. Che'n H.). Zhejiang Univ. Press, Hangzhou, pp. 175-186 (in Chinese).

Ye, D.N., Li, D., Dong, G. and Qiu, X., 1980. 3T type phengite and C-type eclogite from the Xinyang metamorphic belt and its tectonic implication. In: Formation and Evolution of North China Block (ed. Zhang, W. Y.). Science Press, Beijing, pp. 122-131 (in Chinese).

Yin, H. F. and Huang, D. H., 1995. The Early Paleozoic Zhenan-Xichuan block and the evolution of the small Qinling Archipelagic ocean basin. Acta Geol. Sinica, 69, 193-204 (in Chinese).

Yuan, X.C., Xu, M.C., Wang, Q.H., Tang, W.B., 1994. Eastern Qinling seismic reflection profiling. Acta Geophy. Sinica, 36(6), 749-758 (in Chinese).

Zhai, X.M., Day, H.W., Hacker, B.R. and You, Z., 1998. Palezoic metamorphism in the Qinling orogen, Tongbai Mountains, central China. Geology, 26(4), 371-374.

Zhang, G.W., Meng, Q. and Lai, S., 1995. Tectonics and structure of Qinling orogenic belt. Sci. China (Ser. B), 38(11), 1379-1394.

Zhang, Q., Ma, B., Liu, P., Zhao, D., Fan, Q. and Li. X., 1995. A fragment of lithosphere mantle above subduction zone — the geochemical features of the Raobazhai ultramafic rocks in Anhui Province. Sci. China (Ser. B), 25, 867-873 (in Chinese).

Zhang, R.Y. and Liou, J.G., 1994. Coesite-bearing eclogites in Henan Province, Central China: detailed petrography, glaucophane stability and P-T path. Eur. J. Mineral., 6, 217-233.

Zhang, R.Y., Liou, J.G. and Tsai, C.H., 1996. Petrogenesis of a high-temperature metamorphic terrane: a new tectonic interpretation for the North Dabieshan, central China. J. Metamor. Geol., 14, 319-333.

Zhang, Z.Q., Liu, D., Fu, M., 1994. Isotope Stratigraphical Chronology of the North Qinling. Geol. Press, Beijing, pp. 11-56, 177 (in Chinese).

Zhao, X. and Coe, R.S., 1987. Palaeomagnetic constraints on the collision and rotation of North and South China. Nature, 327, 141-144.

Zhao, Z.H., 1997. Principles of Trace Element Geochemistry. Science Press, Beijing. pp. 56-64 (in Chinese)

Zheng, Y. F., Fu, B., Li, Y. L., Wei, C. S., Zhou, J. B., 2001. Oxygen isotope composition of granulites from Dabieshan in eastern China and its implications for geodynamics of Yangtze plate subduction. Phys. Chem. Earth, A26, this issue.

Zhou, J.B. and Zheng, Y. F., 2000. A preliminary study of the accretionary complex of the Tangtze plate subduction. Geology of Anhui, 10(3), 173-178 (in Chinese with English abstract).

Zindler, A. and Hart, S., 1986. Chemical Geodynamics. Ann. Rev. Earth Planet. Sci., 14, 493-571.

Carboniferous and Triassic eclogites in the western Dabie Mountains, east-central China: evidence for protracted convergence of the North and South China Blocks[*]

Weidong Sun[1,2], I. S. Williams[1] and Shuguang Li[2]

1. Research School of Earth Sciences, The Australian National University, Canberra, ACT 0200, Australia
2. Department of Earth and Space Sciences, University of Science and Technology of China, Hefei 230026, Anhui, China

> 亮点介绍：根据锆石 SHRIMP U-Pb 同位素定年以及微量元素和矿物包体研究，证明西大别存在浒湾石炭纪榴辉岩和红安三叠纪榴辉岩；前者源岩为古生代俯冲洋壳，后者源岩是新元古代陆壳岩石。显示了华北与华南陆块的碰撞前，曾发生石炭洋壳俯冲，随后在三叠纪发生两个大陆的碰撞和华南陆壳的俯冲，从而给出了从洋壳俯冲过渡到陆壳俯冲的证据。

Abstract SHRIMP U-Pb dating and laser ablation ICP-MS trace element analyses of zircon from four eclogite samples from the north-western Dabie Mountains, central China, provide evidence for two eclogite facies metamorphic events. Three samples from the Huwan shear zone yield indistinguishable late Carboniferous metamorphic ages of 312±5, 307±4 and 311±17 Ma, with a mean age of 309±3 Ma. One sample from the Hong'an Group, 1 km south of the shear zone yields a late Triassic age of 232±10 Ma, similar to the age of ultra-high pressure (UHP) metamorphism in the east Qinling-Dabie orogenic belt. REE and other trace element compositions of the zircon from two of the Huwan samples indicate metamorphic zircon growth in the presence of garnet but not plagioclase, namely in the eclogite facies, an interpretation supported by the presence of garnet, omphacite and phengite inclusions. Zircon also grew during later retrogression. Zircon cores from the Huwan shear zone have Ordovician to Devonian (440–350 Ma) ages, flat to steep heavy-REE patterns, negative Eu anomalies, and in some cases plagioclase inclusions, indicative of derivation from North China Block igneous and low pressure metamorphic source rocks. Cores from Hong'an Group zircon are Neoproterozoic (780–610 Ma), consistent with derivation from the South China Block. In the western Dabie Mountains, the first stage of the collision between the North and South China Blocks took place in the Carboniferous along a suture north of the Huwan shear zone. The major Triassic continent-continent collision occurred along a suture at the southern boundary of the shear zone. The first collision produced local eclogite facies metamorphism in the Huwan shear zone. The second produced widespread eclogite facies metamorphism throughout the Dabie Mountains-Sulu terrane and a lower grade overprint in the shear zone.

Keywords eclogite; Qinling-Dabie orogen; SHRIMP U-Pb dating; trace element; zircon

1 Introduction

Despite 20 years of intensive study, the structure and history of the Qinling-Dabie orogen, which separates the North and South China Blocks, and is therefore the key to the tectonic evolution of the east Asian continent (Mattauer et al., 1985), remains a subject of active debate. Relatively sparse geochemical, isotopic and geochronological data permit a variety of tectonic models (Mattauer et al., 1985; Okay & Sengör, 1992; Xu et al., 1992; Kröner et al., 1993; Yin & Nie, 1993; Li et al., 1989, 1993, 1994, 2000; Wang et al., 1996; Hacker et al., 1998, 2000). Most researchers who have worked on the western part of the orogen (North Qinling and Tongbai Mountains) believe that the Qinling-Dabie orogenic belt formed during (Kröner et al., 1993; Xue et al., 1996a,b), or was

---

[*] 本文发表在：Journal of Metamorphic Geology, 2002, 20: 873-886

initiated in (Mattauer et al., 1985), the mid Palaeozoic. In contrast, models based on research on the ultra-high pressure (UHP) metamorphic rocks in the eastern Dabie Mountains and Su-Lu terrane mostly invoke continent-continent collision in the Mesozoic (Yin & Nie, 1993; Li et al., 1993, 1994, 2000; Wang et al., 1996; Hacker et al., 1998, 2000; Ames et al., 1996). Following discovery of the Triassic Mianlue suture in South Qinling (Zhang et al., 1995), it has been suggested that there were multiple collisions along the orogenic belt (Zhang et al., 1995; Sun, 1994; Li & Sun, 1996; Li et al., 1996; Meng & Zhang, 1999, 2000). It remains unclear, however, how the proposed sutures in the Qinling connect to those in the Dabie Mountains, and more importantly, there is disagreement over whether there are in fact two sutures in the Dabie Mountains, and if so where they might be located (e.g. Xu et al., 1994; Hacker et al., 1998, 2000).

The north-western Dabie Mountains is one area critical to deciphering the evolution of the Qinling-Dabie orogenic belt. There the junction between the western (Qinling and Tongbai Mountains) and eastern (Dabie Mountains and Su-Lu terranes) sections of the Qinling-Dabie orogenic belt is exposed, as is the transition zone between low and ultra-high pressure metamorphic rocks. Quartz eclogites from the Huwan shear zone in the north of the region are particularly important and have been closely studied petrographically (Liou et al., 1996; Liu et al., 1996) and isotopically (Ye et al., 1994; Jian et al., 1997, 2000; Xu et al., 2000). These eclogites have yielded an unusually wide range of metamorphic ages from Triassic, similar to the well-studied UHP metamorphic rocks further east (Hacker et al., 1998; Xu et al., 2000), to mid-Palaeozoic, the significance of which is the subject of much debate (Hacker et al., 1998, 2000; Ye et al., 1994; Jian et al., 1997, 2000; Xu et al., 2000; Li et al., 2001). Hacker et al. (1998, 2000), for example, interpreted a phengitic mica $^{40}Ar/^{39}Ar$ age of c. 205 Ma from eclogites developed in the Huwan Formation as evidence for Triassic HP metamorphism to the north of the North Dabie zone, the whole Dabie region thereby being an orogenic scale antiform. Conversely Jian et al. (1997) argued, based on IDTIMS zircon U-Pb dating, that the eclogites at Xiongdian were formed between about 400 and 373 Ma, with retrograde metamorphism at 301.8±0.6 Ma. This conclusion was later supported by reconnaissance SHRIMP zircon analyses (Jian et al., 2000) which were interpreted as indicating HP-UHP metamorphism of the eclogites at > 424 Ma. Phengite $^{40}Ar/^{39}Ar$ ages of the Xiongdian eclogite ranging from c. 430−350 Ma have been explained as the retrograde metamorphic age (Xu et al., 2000). In contrast, $^{40}Ar/^{39}Ar$ ages of phengite from mylonitic rocks and muscovite from weakly sheared rocks range from 269±1 to 187±1 Ma (Xu et al., 2000). These have been explained as recording a cooling event and late recrystallization, respectively (Xu et al., 2000).

Such diversity of opinion is to be expected in an area with a history as complex as that of the north-western Dabie Mountains, mainly because of the difficulties in interpreting the various isotopic ages. To help clarify the metamorphic history of the Huwan eclogites and better understand the evolution of the Qinling Dabie orogenic belt, we have studied four quartz eclogites from the north-western Dabie region: three from the Huwan shear zone (one near Hujiawan, and two from the large eclogite body near Xiongdian) and one from the adjacent Hong'an Group near Xuanhuadian. Zircon from each was dated by SHRIMP U-Pb. For the two samples from near Xiongdian, zircon trace element characteristics and mineral inclusions were also studied. These provided valuable additional constraints on the metamorphic conditions at the time of zircon growth.

2 Regional geology and samples

The north-western Dabie region (Fig. 1) is bordered by the Shangcheng- Macheng fault to the east and the Dawu fault to the west (Xu et al., 2000). It is the transition zone between the low and ultrahigh pressure metamorphic sections of the Qinling-Dabie orogenic belt—no HP eclogite assemblage has been reported west of the Dawu fault (Xu et al., 2000). Up to seven structural units have been recognized in the north-western Dabie (Xu et al., 2000) (Fig. 1). One of the most important is the Huwan shear zone, an east-west belt 5−8 km wide, consisting of mylonitic muscovite-albite-quartz schist, muscovite-feldspar gneiss, leptite, graphitic schist, marble and sheared eclogites (Ye et al., 1994; Xu et al., 2000; Hacker et al., 2000; Webb et al., 2001). To the north is the Suhe thrust sheet, which consists mainly of unmetamorphosed to greenschist facies Devonian rocks (Xu et al., 2000), including fossiliferous marble (Ye et al., 1994). To the south is the Hong'an Group, which is believed to be closely related to the UHP metamorphic terranes in the eastern Dabie Mountains (e.g. Hacker et al., 2000; Xu et al., 2000).

Eclogites occur in the Huwan shear zone as lenses or layers (Ye et al., 1994; Liu et al., 1996; Jian et al., 1997, 2000; Xu et al., 2000). They are believed to have formed under conditions similar to those in the Franciscan Complex, western USA, at a peak P-T estimated to be P_{min}=1.3–1.5 GPa and T_{min}=590–680 ℃(Ye et al., 1994). The largest eclogite body in the shear zone is exposed near Xiongdian (Fig. 1). Two samples of this body were collected from outcrops studied by previous researchers (Jian et al., 1997, 2000; Xu et al., 2000). Sample 99XD-1 was collected on the east bank of the river about 50 m east of Xiongdian village (114°28'05"E, 31°45'25"N). It was a fresh, strongly foliated quartz eclogite consisting of quartz (> 35%), garnet, omphacite and minor glaucophane, rutile and phengite. Sample 99XD-2 was collected on the river bank about 50 m south of sample 99XD-1. It is a fresh, strongly foliated banded mafic eclogite consisting of coarse (0.5–5 mm) garnet (c. 60%, > 80% in some bands), omphacite, phengite and amphibole, plus minor retrograde chlorite. The zoned garnet contained inclusions of omphacite and acicular rutile. The Xiongdian eclogites are estimated to have been metamorphosed at c. 1.3 GPa and >570 ℃ (Liu et al., 1996), similar conditions to other eclogites in the Huwan shear zone (Ye et al., 1994). A third sample of eclogite from the shear zone (95HW-S-8) was collected from an outcrop about 200 m west of Hujiawan village (114°27'46"E, 31°45'54"N). It is a metabasalt with eclogite mineralogy (Ye et al., 1994) very similar to 99XD-2; garnet, omphacite, phengite and amphibole. Negative Nb and Ti anomalies suggest that the eclogites from the Huwan shear zone had a volcanic arc protolith (Li et al., 2001).

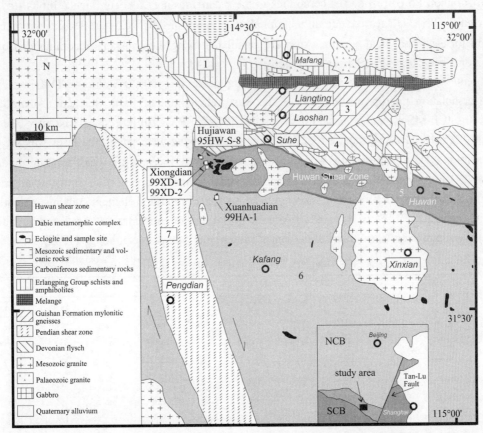

Fig. 1 Sketch geological map of the western Dabie mountains (modified after Xu et al., 2000) showing the main tectonic units, eclogite occurrences and sample locations. (1) Mafan fold belt; (2) Liangting melange belt; (3) Laoshan ductile fold belt; (4) Suhe thrust sheet; (5) Huwan shear zone; (6) Hong'an Group eclogite bearing gneiss; (7) Pengdian ductile shear zone. Inset: location of the Qinling-Dabie orogenic belt in east central China at the boundary between the North China Block (NCB) and South China Block (SCB).

A fourth sample, of Xuanhuadian eclogite (99HA-1) from the quartz eclogite belt of the Hong'an Group (Fig. 1), was collected c. 1 km south of the Huwan shear zone from an outcrop near the bridge next to the gate to the Xuanhuadian Junior High School (114°29'06"E, 31°42'26"N) (Fig. 1). It is metabasaltic (SiO_2=50%), similar to the Hujiawan eclogite, consisting of garnet, omphacite, amphibole, mica, quartz and minor rutile.

3 Analytical procedures

Zircon was separated from the four samples at the Tianjin Institute of Geology and Mineral Resources by crushing, magnetic and waterbased panning techniques designed to minimize intersample cross contamination. Following final purification by hand sorting at the Australian National University, zircon grains were mounted in epoxy together with zircon standards SL13 and AS3 and polished down to half sections. Prior to SHRIMP analysis the grains were documented by optical photomicrography, and their structures imaged by cathodoluminescence (CL) using an Hitachi S-2250 N scanning electron microscope operated at 15 kV/120 μA. The latter was essential given the multiple growth generations commonly present in individual crystals.

Zircon was analyzed for Pb-U-Th isotopes using the SHRIMP II sensitive high resolution ion microprobe at the Australian National University. Instrumental conditions and data acquisition procedures were similar to those described by Williams (1998). Pb isotopic compositions were measured directly, without correction for the small (c. 2‰/amu) mass dependent fraction. Much larger interelement fractionations were corrected by reference to AS3 (radiogenic $^{206}Pb/^{238}U=0.1859$) using a power-law relationship between Pb/U and UO/U (Claoué-Long et al., 1995). Absolute Pb, U and Th concentrations were calculated with reference to SL13 (U≈238 ppm). Each measurement consisted of 5–7 scans through the Zr, Pb, U and Th species of interest and took about 15 min Uncertainties listed in the data tables and plotted on the concordia diagrams are 1σ, and include measurement errors and, where applicable, uncertainties in common Pb correction. Uncertainties in the mean ages are 95% confidence limits (tσ, where 't' is Students t), and the $^{206}Pb/^{238}U$ ages include an additional uncertainty of 0.3%–0.5% reflecting the precision of the Pb/U calibration for each analytical session. Ages are calculated using the constants recommended by the IUGS Subcommission on Geochronology (Steiger & Jäoger, 1977).

Trace elements in zircon from the two Xiongdian eclogite samples (99XD-1 and 99XD-2) were analyzed *in situ* using a laser ablation microprobe ICP-MS at the Research School of Earth Sciences, Australian National University. The LA-ICP-MS system consisted of a Lambda Physik LPX 120I pulsed ArF excimer laser coupled to an Agilent 7500 ICP-MS. Because most zircon cores and overgrowths were small, the laser was set to a 23 μm diameter spot, with a 5-Hz repetition rate. Isotopic compositions were measured in peak-hopping mode using one point per peak. NIST 612 was used as an external standard, and a minor isotope of Si (^{29}Si) as the internal standard, assuming that the SiO_2 content of zircon is constant at 32.5%. The zircon trace element contents are listed in Tables 1 and 3. The SiO_2 content is 71.9% of NIST 612 while the trace element concentrations are listed in Table 1 (Pearce et al., 1997). REE patterns were normalized to chondrite values from Sun & McDonough (1989).

Table 1 Trace element contents of zircon from Xiongdian eclogite, sample 99XD-1

	Group 1-I				Group 1-II							
	19.1	24.1	22.1	14.2	30.1	31.1	17.1	7.1	4.1	14.1	12.1	16.1
Ti[1]	2135	2219	2517	2634	2442	2008	2396	1912	2148	2737	1946	2350
Y	458	281	306	1420	2024	730	292	134	94	212	17	84
Nb	2.4	1.7	1.4	5.4	4.1	6.1	3.3	2.2	1.8	1.3	0.67	1.3
Ba	0.2	nd	nd	0.3	nd	0.3	nd	8.0	nd	nd	nd	nd
La	nd	nd	nd	nd	nd	nd	nd	nd	nd	nd	nd	nd
Ce	4.8	2.7	3.0	0.4	0.3	0.5	0.2	0.5	0.1	0.5	nd	0.3
Pr	nd	nd	nd	nd	nd	nd	nd	nd	nd	nd	nd	nd
Nd	0.7	0.9	0.3	nd	nd	nd	nd	nd	nd	nd	nd	nd
Sm	3.4	1.9	1.8	nd	0.3	nd	nd	nd	nd	nd	nd	nd
Eu	2.0	1.1	1.0	0.3	0.3	0.2	0.3	0.3	nd	nd	nd	nd
Gd	25.1	15.1	15.6	6.4	5.6	3.5	3.3	4.1	1.3	3.3	0.4	1.8
Tb	7.4	4.9	4.8	4.1	4.4	2.2	1.9	1.4	0.7	1.7	nd	0.7
Dy	58.2	37	41.4	78.6	91.6	40.1	27.3	15.2	10.0	21.3	1.8	8.0
Ho	13.5	7.4	9.8	38.5	49.2	20.1	8.5	4.0	3.1	7.3	0.5	2.4

Continued

	Group 1- I				Group 1- II							
	19.1	24.1	22.1	14.2	30.1	31.1	17.1	7.1	4.1	14.1	12.1	16.1
Er	42.4	20.8	31.3	215.8	293.5	113.8	32.6	13.7	10.7	31.5	1.7	8.8
Tm	6.5	3.3	4.6	48.6	68.7	26.4	5.3	2.1	1.7	5.7	0.3	1.5
Yb	49.6	22.4	38.0	494.2	706.3	271.8	41.0	16.2	12.1	51.3	2.3	12.0
Lu	8.3	3.4	6.8	99.3	142.7	54.9	7.0	2.7	1.9	9.7	0.4	2.1
Hf	14280	18016	14037	15442	15133	12661	16440	13004	12419	15457	14522	12167
Ta	2.3	2.4	1.9	1.6	1.0	1.9	2.7	1.5	1.4	1.0	0.8	0.9
Th	52	35.7	38.2	6.3	2.8	6.7	0.7	2.0	0.2	1.4	0.4	1.2
U	2724	2114	1238	1147	1017	1012	817	322	203	185	137	122
Th/U	0.019	0.017	0.031	0.006	0.003	0.007	0.001	0.006	0.001	0.008	0.003	0.010
U/Nb	1138	1254	915	212	251	167	247	146	111	142	204	94
Age	326±2	328±2	373±3	323±3	323±3			324±5	322±4	310±6	312±8	290±7

	Group 1-III							Undefined			
	6.1	9.1	2.1	10.1	8.1	11.1	13.1	18.1	6.2	DL[2]	STD[3]
Ti[1]	2195	1932	2193	2327	2490	2332	2402	2400	2466	4	48.5
Y	165	62	65	63	93	26	70	2510	889	0.05	37.8
Nb	2.7	1.3	1.2	1.1	0.9	0.63	2.1	20.5	4.3	0.07	38.1
Ba	nd	nd	nd	0.5	0.5	nd	1.2	nd	nd	0.2	38.5
La	nd	nd	nd	nd	nd	nd	nd	nd	nd	0.06	35.5
Ce	0.3	0.3	0.2	0.3	0.2	0.1	0.1	2.5	49.6	0.06	38.2
Pr	nd	nd	nd	nd	nd	nd	nd	0.2	nd	0.04	37.1
Nd	nd	nd	nd	nd	nd	nd	nd	1.4	1.8	0.25	35
Sm	nd	nd	nd	nd	nd	nd	nd	1.9	3.3	0.4	36.4
Eu	nd	nd	nd	0.2	nd	nd	nd	0.9	1.4	0.14	34.6
Gd	1.5	1.1	0.8	2.4	1.3	0.6	0.8	15.2	17.6	0.34	37.1
Tb	0.8	0.5	0.4	0.8	0.7	0.2	0.5	8.0	5.3	0.22	36.3
Dy	12.9	5.8	5.4	7.3	8.3	2.4	6.7	143.9	63.9	0.3	35.7
Ho	5.0	1.8	1.8	1.7	2.6	0.7	2.3	63.3	24.4	0.06	37.9
Er	24.1	6.6	7.1	5.9	8.6	2.6	7.8	329.6	122.5	0.13	38
Tm	4.7	1.1	1.3	1.0	1.6	0.4	1.2	70.6	27.9	0.06	37.7
Yb	42.8	8.9	11.0	8.5	13.7	3.6	8.9	636.7	309.2	0.2	39.5
Lu	8.1	1.6	2.0	1.5	2.5	0.7	1.5	120.0	64.6	0.07	37.6
Hf	10819	11133	10727	11138	14062	11558	12242	16370	12702	0.12	34.8
Ta	1.5	0.8	0.7	0.7	0.5	0.5	2.5	3.7	1.9	0.07	39.8
Th	0.7	1.2	0.5	1.0	0.7	0.3	0.2	28	533	0.06	37.3
U	115	108	103	93	80	51	28	4084	645	0.08	36.9
Th/U	0.006	0.011	0.005	0.010	0.009	0.006	0.006	0.007	0.827		
U/Nb	42	86	85	83	87	82	13	200	151		
Age	295±8	308±6	322±7	298±7	322±6	307±8	293±15	337±1			

[1] All data are given as µg/g. [2] DL = detection limit determined by the background signals. [3] STD = value for NIST 612 from Pearce et al. (1997).

The accuracy of these analyses is determined mainly by two factors. Firstly, SiO_2 can vary slightly within zircon grains as a function of their Hf contents. Secondly, the ionization efficiency of Si relative to nearly all the trace elements analyzed (REE, Y, Nb, Ta, Hf, U, Th) is lower in zircon than in the NIST glass standard. This difference is a function of the laser spot size. For a spot size of 23 µm, concentrations may be underestimated by as much as 20% (S. Eggins, pers. com.). This may in part account for the lower U-values measured by laser ICP-MS compared to SHRIMP. However, the relative values determined by ICP-MS, such as the REE patterns and U/Nb and Th/U ratios are not affected by any of these factors.

Mineral inclusions in zircon were identified by optical microscope, scanning electron microscope and laser Raman spectroscopy. These provided direct information on the metamorphic conditions under which the zircon zones grew, thereby placing firmer constraints on the age of the eclogite facies metamorphism.

4 Zircon analyses

4.1 Xiongdian eclogite 99XD-1 (quartz eclogite) from the Huwan shear zone

The zircon from quartz eclogite 99XD-1 occurs as fine to medium grained (10–100 μm diameter), clear, colourless, stubby, rounded to prismatic, anhedral to euhedral crystals with little visible internal structure and very few inclusions. CL imaging shows that most grains consist of a fragment of zircon core (commonly euhedrally zoned, with weak to moderate CL) surrounded by an overgrowth (Fig. 2a–d). Most of the overgrowths have strong CL and little internal structure, but in some (< 5%) of the grains there is an inner dark overgrowth as well. About 20% of the grains are strongly luminescent throughout.

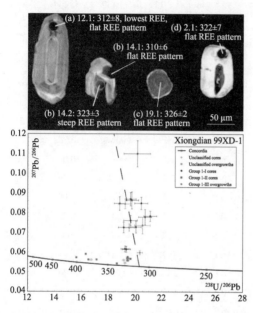

Fig. 2 Zircon U-Pb isotopic analyses, and representative CL images of zircon structures, from Xiong-dian eclogite 99XD-1 (quartz eclogite) from the Huwan shear zone. (a) Euhedrally zoned zircon with thin overgrowth; (b) Zoned zircon core with thick overgrowth; (c) CL-dark structureless zircon with very thin overgrowth; (d) Grain consisting almost entirely of overgrowth. Analytical uncertainties 1σ.

On the basis of the REE patterns and other trace element features, three main types of zircon can be recognized (Table 1, Fig. 3). The first (group 1-I) has flat heavy-REE (HREE) patterns, relatively high light- REE (LREE) contents and small negative Eu anomalies, very high U contents (1240–2720 ppm) and very high U/Nb ratios (915–1250) (Table 1, Fig. 3a). Three analyses of CL-dark cores define this group (Fig. 2c). The second (group 1-II) also is defined by analyses of CL-dark cores (Fig. 2c), but these are characterized by steep HREE patterns, high Nb contents (3–6 ppm) and moderate U/Nb ratios (170–250, Table 1, Fig. 3a). The third group (group 1-III) consists entirely of analyses of overgrowths (Fig. 2a,b&d), all of which are characterized by flat HREE patterns, low total REE contents (Fig. 3b), low U (<350 ppm), low U/Nb (<150) and very low Th/U (0.001–0.015). The REE pattern of 17.1 (CL-dark core) is superficially similar those of group 1-III, but because of its high Nb content and high U/Nb ratio, it is included in group 1-II.

These different chemical compositions reflect the various environments in which the different generations of zircon crystallized. Zircon is typically highly enriched in HREE. The relatively high LREE contents and flat HREE patterns of the group 1-I zircon therefore suggest that it was formed in an environment highly enriched in the LREE and depleted in HREE. Garnet also concentrates the HREE, so the steep HREE patterns of the group 1-II zircon suggest crystallization in the absence of garnet. In contrast, the flat HREE patterns and low total REE contents of the group 1-III zircon (the overgrowths) are consistent with crystallization in the presence of garnet (Rubatto, 2002). Those overgrowths with somewhat higher HREE and Nb are interpreted to reflect the presence of less garnet and rutile. Overall, the trace element composition of the overgrowths is indicative of zircon grown during eclogite facies metamorphism. Additional supporting evidence is provided by the presence of high pressure mineral inclusions (garnet, omphacite and phengite) only in the overgrowths (Fig. 4).

Isotopic analyses of the zircon from 99XD-1 are listed in Table 2 and plotted without common Pb correction on a Tera-Wasserburg concordia diagram in Fig. 2. The analyses are identified as representing either cores or overgrowths that appear light or dark in CL, respectively, and are further subdivided according to trace element composition where applicable. There is a wide range of U contents (40–4520 ppm), the highest U being in the dark cores and the lowest in the light overgrowths. Th/U is low to very low (0.1–0.001), even in the euhedrally zoned cores, consistent with zircon growth under high grade metamorphic conditions (Williams & Claesson, 1987).

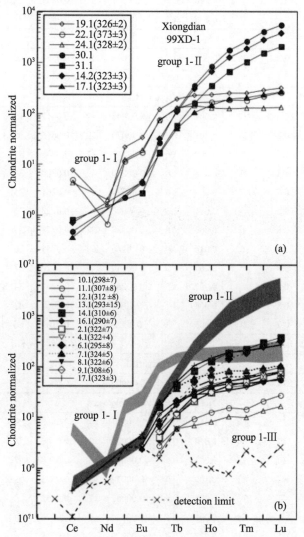

Fig. 3 Zircon chondrite-normalised REE patterns (Sun & McDonough, 1989) for Xiongdian eclogite 99XD-1 (quartz eclogite) from the Huwan shear zone measured by laser ablation ICP-MS on some of the same spots dated by SHRIMP U-Pb. Three generations of zircon growth (groups 1-I, 1-II and 1-III) can be recognized. (a) REE patterns from groups 1-I and 1-II, representing zircon cores; (b) REE patterns from group 1-III (with the ranges of groups 1-I and 1-II for comparison), representing zircon overgrowths. The composition of group 1-III is consistent with zircon growth under eclogite facies conditions.

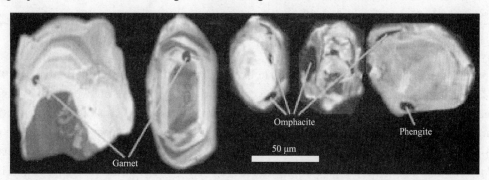

Fig. 4 CL images of zircon grains from Xiongdian eclogite 99XD-1 (quartz eclogite) from the Huwan shear zone showing inclusions of high pressure minerals (garnet, omphacite and phengite) restricted to the overgrowths, indicative of late zircon growth under eclogite facies conditions.

Table 2 SHRIMP zircon data for Xiongdian eclogite 99XD-1

Grain spot	CL domain	U ppm	Th ppm	Pb* ppm	Th/U	$^{206}Pb_c$%	$^{204}Pb/^{206}Pb$	±1σ	$^{207}Pb/^{206}Pb$	±1σ	$^{238}U/^{206}Pb$	±1σ	Inferred age(Ma)	±1σ
						Group 1-I								
19.1	Dark core	2639	92	125	0.035	0.2	0.00011	3	0.0545	6	19.2	0.1	326	2
22.1	Dark core	1585	12	86	0.008	0.3	0.00001	1	0.0561	5	16.7	0.1	373	3
24.1	Dark core	1181	41	56	0.035	0.4	0.00016	5	0.0559	6	19.1	0.1	328	2
						Group 1-II								
17.1	Dark core	1041	0.8	48	0.001	0.5	0.00038	8	0.0569	7	19.4	0.2	323	3
14.2	Dark core	1214	6.1	56	0.005	0.5	0.00054	10	0.0558	7	19.4	0.2	323	3
						Group 1-III								
7.1	Dark overgrowth	372	2.6	17	0.007	1.0	0.00068	17	0.0617	15	19.2	0.3	324	5
4.1	Dark overgrowth	266	0.2	12	0.001	1.0	0.00070	29	0.0615	15	19.3	0.3	322	4
10.1	Light overgrowth	200	1.7	9	0.008	3.6	0.00143	47	0.0848	31	20.4	0.5	298	7
14.1	Light overgrowth	159	1.3	7	0.008	3.8	0.00249	49	0.0869	34	19.6	0.4	310	6
12.1	Light overgrowth	152	0.6	7	0.004	2.6	0.00227	63	0.0761	25	19.6	0.5	312	8
9.1	Light overgrowth	151	2.2	7	0.014	2.3	0.00073	46	0.0728	30	20.0	0.4	308	6
6.1	Dark overgrowth	142	0.8	6	0.006	2.8	0.00137	53	0.0778	39	20.7	0.5	295	8
16.1	Light overgrowth	134	1.9	6	0.014	2.9	0.00186	68	0.0783	26	21.1	0.5	290	7
2.1	Light overgrowth	131	0.9	6	0.007	2.2	0.00140	40	0.0726	27	19.1	0.4	322	7
8.1	Light overgrowth	122	0.8	6	0.007	3.5	0.00288	66	0.0847	34	18.8	0.4	322	6
11.1	Light overgrowth	78	0.3	3	0.004	3.7	0.0049	10	0.0863	43	19.7	0.5	307	8
13.	Light overgrowth	37	0.2	2	0.005	6.3	0.0022	12	0.1100	66	20.1	1.0	293	15
Grain spot	CL domain	U ppm	Th ppm	Pb* ppm	Th/U	$^{206}Pb_c$%	$^{204}Pb/^{206}Pb$	±1σ	$^{207}Pb/^{206}Pb$	±1σ	$^{238}U/^{206}Pb$	±1σ	Inferred age(Ma)	±1σ
						Unclassified								
18.1	Dark core	4521	36	220	0.008	0.2	0.00009	2	0.0542	5	18.6	0.1	337	1
						Not analysed for trace elements								
23.1	Dark core	2439	25	130	0.010	0.3	0.00012	4	0.0562	5	17.1	0.1	366	3
25.1	Dark core	1399	19	64	0.014	0.5	0.00026	9	0.0566	9	19.6	0.3	319	4
20.1	Dark core	1393	10	77	0.007	0.6	0.00037	6	0.0590	6	16.3	0.1	381	3
26.1	Dark core	1392	150	83	0.108	0.3	0.00017	4	0.0573	6	15.6	0.2	398	5
28.1	Dark core	1278	7	70	0.005	0.3	0.00037	7	0.0562	7	16.5	0.1	377	3
27.1	Dark core	1167	7	54	0.006	0.5	0.00044	8	0.0570	7	19.4	0.2	322	4
21.1	Dark core	1091	35	51	0.032	0.4	0.00010	5	0.0558	9	19.4	0.2	324	3
5.2	Dark core	1061	6	49	0.006	0.6	0.00041	10	0.0575	9	19.4	0.1	322	2
5.1	Dark overgrowth	439	1.3	19	0.003	0.8	0.00072	20	0.0598	11	20.3	0.2	307	3
3.1	Light overgrowth	197	1.7	9	0.009	2.2	0.00097	41	0.0725	18	19.5	0.4	316	6
15.1	Light overgrowth	119	0.5	5	0.004	3.5	0.0019	11	0.0845	24	19.2	0.5	317	7
1.1	Light overgrowth	89	0.7	4	0.007	2.4	0.00046	37	0.0736	42	20.5	0.6	300	9

* Radiogenic.

The analyses fall into two age groups, c. 380 and c. 320 Ma. All of the higher ages come from cores. These five $^{206}Pb/^{238}U$ apparent ages are dispersed much more than expected from the analytical uncertainties (398±5 (1σ) to 366±3 (1σ) Ma) suggesting that the cores have a range of primary ages, as might be expected for protolith zircon from a metasediment. However, with U contents all in excess of 1250 ppm, there is also the very real possibility that these "ages" might be depressed by radiogenic Pb loss. The only grouping in the apparent ages is three within error at 377±9 (tσ) Ma. The younger cores also show a significant age range, but the scatter is due to just one older analysis (18.1). The other eight analyses give a weighted mean $^{206}Pb/^{238}U$ age of 324.4±3.5 (tσ) Ma. There is not a simple correlation between trace element composition and the ages of the cores. The three identified group 1-I cores have apparent ages ranging from 373±5 (1σ) Ma to 326±4 (1σ) Ma. On the other hand, both identified group 1-II cores belong to the younger group, both having an apparent age of 323±5 (1σ) Ma.

All of the overgrowth analyses cluster at c. 320 Ma, with a range of apparent ages in part explicable by the high analytical uncertainties because of the mostly low U contents, hence low radiogenic Pb. The scatter is significant, however, and is not simply due to one or two low outliers that might reflect Pb loss. The scatter is introduced by the analyses of four dark, U-rich inner overgrowths, the apparent ages of which range from 324±5 (1σ) to 295±8 (1σ) Ma. Only the two oldest are the same age within error, their weighted mean $^{206}Pb/^{238}U$ age of 323±7 (2σ) Ma being indistinguishable from the age of the younger cores. Eleven of the 12 analyses of light overgrowths give the same $^{206}Pb/^{238}U$ age within analytical uncertainty, 312±5 (tσ) Ma. The difference between this age and that of the younger cores (12.8±2.4 (1σ) Ma), the uncertainty in which is independent of the standardization error, is significant. The light overgrowths are demonstrably younger than the young cores, and probably also younger than the dark overgrowths. Both types of overgrowth have compositions consistent with growth under eclogite facies conditions, suggesting thereby that those conditions persisted from 323±7 (2σ) to 312±5 (tσ) Ma.

4.2 Xiongdian eclogite 99XD-2 (metabasalt) from the Huwan shear zone

Zircon from metabasalt sample 99XD-2 is coarser than that from 99XD-1 (mostly 100–200 μm diameter) and on average less prismatic, but otherwise of similar appearance. The grains are clear, colourless and equant, with the smooth surface and multiple dimpled crystal faces commonly observed in zircon grown at high metamorphic grade. There is no visible internal structure and inclusions are rare. CL reveals that most grains are composite, commonly consisting of a large rounded core with broad, sometimes euhedral, zoning, surrounded by a thin overgrowth rarely more than 50 μm thick (Fig. 5a–d). The CL of the cores ranges from very weak to very strong (Fig. 5a–d); CL of the overgrowths is consistently relatively strong (Fig. 5a,d&d). In a very few cases there is also a dark inner overgrowth (Fig. 5b). The striking difference relative to the zircon from 99XD-1 is the lack of cores with well developed fine euhedral zoning.

Contrasts in trace element composition between the 99XD-2 zircon growth zones are not as marked as those in the zircon from sample 99XD-1, but they nevertheless provide a guide to the conditions under which the various zones might have crystallized (Table 3, Fig. 6). Most of the cores and one CL-dark overgrowth (group 2-I) have near-parallel REE patterns enriched in HREE and with pronounced negative Eu anomalies (Fig. 6a), suggesting crystallization in the presence of plagioclase, namely not under eclogite facies conditions. This conclusion is supported by the occurrence of plagioclase (albite) and other low pressure mineral inclusions (e.g. muscovite) in some of those cores (Fig. 7). In other respects the group is heterogeneous (Table 3), with a wide range of U contents (3–2400 ppm), Nb/U (9.5–3560) and Th/U (0.01–0.3), suggesting a wide range of provenance. Two analyzed spots (group 2-II, a core and an overgrowth) have a much flatter REE pattern and no Eu anomaly (Fig. 6b), suggesting crystallization in the presence of garnet but absence of plagioclase, and therefore under eclogite facies conditions. A third group of five analyses (group 2-III), again representing both cores and overgrowths, has very steep HREE patterns and a wider range of REE contents (Fig. 6b). They differ from group 2-I in not having a

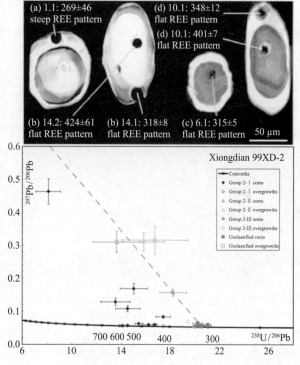

Fig. 5 Zircon U-Pb isotopic analyses, and representative CL images of zircon structures, from Xiongdian eclogite 99XD-2 (metabasalt) from the Huwan shear zone. (a) Thick overgrowth and a thin CL-dark inner layer surrounding an equant CL-bright core; (b) Thin overgrowth and a CL-dark inner layer surrounding an older, CL-bright core; (c) Young CL-dark sector-zoned core with a thin overgrowth; (d) Old sector-zoned core surrounded by young CL-bright overgrowth. Analytical uncertainties 1σ.

pronounced negative Eu anomaly. Th/U is very low (0.001–0.006), Nb high (0.4–3.7 ppm) and U/Nb low (11–140). The steep HREE and relatively high Nb of group 2-III indicate that the amounts of garnet and rutile crystallizing in competition with the zircon have decreased significantly, probably during retrograde metamorphism, possibly at amphibolite facies. The presence of phengite inclusions in the zircon overgrowths (Fig. 7) suggests that this metamorphism was also high pressure.

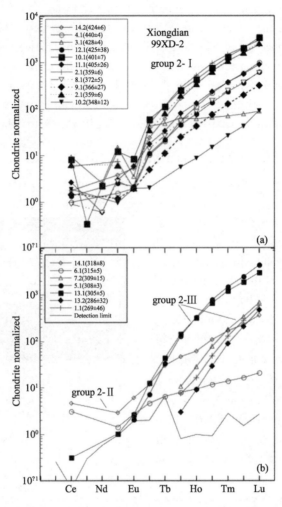

Fig. 6 Zircon chondrite-normalised REE patterns (Sun & McDonough, 1989) for Xiongdian eclogite 99XD-2 (metabasalt) from the Huwan shear zone measured by laser ablation ICP-MS on some of the same spots dated by SHRIMP U-Pb. Three generations of zircon growth (groups 2-I, 2-II and 2-III) can be recognized. (a) REE patterns from group 2-I, steep with negative Eu anomalies (plagioclase present). (b) REE patterns from groups 2-II (flat without Eu anomaly, plagioclase absent, garnet present–eclogite facies) and 2-III (steep, no garnet-retrograde, amphibolite facies)

Table 3 Trace element contents of zircons from Xiongdian eclogite, sample 99XD-2

	Group 2-I												Group 2-II		
	3.1	4.1	8.1	2.1	10.1	2.2	10.2	11.1	12.1	9.1	14.2	6.1	14.1	5.1	13.1
Ti*	1015	1079	1141	1098	1203	1110	1221	977	1063	1037	965	1218	1261	1066	985
Y	122	327	190	668	880	1056	18	270	154	84	583	21	130	730	683
Nb	1.4	0.4	0.3	2.5	1.4	2.1	0.35	0.42	0.42	0.33	0.48	0.45	0.56	3.7	2.3
Ba	nd	nd	nd	nd	nd	nd	nd	nd	nd	nd	nd	nd	nd	nd	nd
La	nd	nd	nd	nd	nd	nd	nd	nd	nd	nd	nd	nd	nd	nd	nd
Ce	5.7	0.61	0.54	3.8	5.1	3.6	1.2	1.6	0.82	0.89	1.1	1.9	2.9	0.1	0.19
Pr	0.06	nd	nd	nd	0.03	nd	nd	nd	nd	nd	nd	nd	nd	nd	nd
Nd	1.2	nd	0.28	nd	1.04	nd	nd	0.28	nd	nd	nd	nd	nd	nd	nd
Sm	2.3	0.24	0.43	0.94	1.9	1.1	nd	0.45	0.39	0.18	0.58	0.22	0.44	nd	0.16

	Group 2-I												Group 2-II		
	3.1	4.1	8.1	2.1	10.1	2.2	10.2	11.1	12.1	9.1	14.2	6.1	14.1	5.1	13.1
Eu	0.2	nd	nd	nd	0.49	nd	nd	0.34	0.13	nd	0.31	0.16	0.36	0.12	0.15
Gd	8.9	2.2	2.9	7.3	12.41	10.13	nd	3.66	2.28	1.02	4.96	0.97	2.48	1.49	2.66
Tb	1.9	0.77	1.01	3.01	4.32	4.47	0.08	1.28	0.83	0.41	2.31	0.25	1.16	1.27	1.64
Dy	16	14	14	50	65	75	1.5	20	12	6.3	44	2.1	12	33	37
Ho	3.5	6.4	5.1	20	24	30	0.5	7.6	4.4	2.4	20	0.53	3.5	19	18
Er	11	37	26	106	128	162	2.5	40	23	13	109	2.0	18	132	114
Tm	1.8	9.7	6.4	28	34	39	0.7	9.9	6.1	3.3	28	0.4	4.3	39	32
Yb	13	102	64	279	353	358	7.3	98	63	34	276	2.9	39	433	324
Lu	2.3	26	16	66	89	79	2.4	23	16	8.3	63	0.5	9.4	112	79
Hf	16670	11134	8444	14268	13158	12505	13712	7481	8937	7252	10155	12344	16962	10164	10917
Ta	1.0	0.10	0.11	2.3	0.39	1.3	nd	0.2	0.17	0.16	nd	0.21	0.33	1.03	0.93
Th	420	16	48	116	41	43	1.9	5.8	3.2	3.6	0.5	87	18	1.2	1.8
U	1727	1142	1081	614	338	291	58	25	22	17	4.5	790	91	344	324
Th/U	0.2	0.01	0.05	0.2	0.1	0.2	0.03	0.2	0.1	0.2	0.1	0.1	0.2	0.003	0.005
U/Nb	1261	2848	3565	248	247	139	164	59	54	50	9.5	1773	162	94	140
Age	428±4	440±4	372±5	359±6	401±7	383±10	348±12	405±26	425±38	366±27	424±61	315±5	318±8	308±3	305±5

	Group 2-III											
	7.2	13.2	1.1	16.1	19.1	18.2	1.2	17.2	17.1	16.2	18.1	19.2
Ti*	1058	1105	1225	1060	1259	911	1261	932	999	946	1001	1163
Y	65	22	38	95	73	221	304	95	1102	53	197	33
Nb	0.75	0.42	0.6	0.32	0.63	0.96	0.42	0.20	0.44	0.39	0.89	0.20
Ba	nd	nd	nd	nd	nd	nd	nd	nd	nd	nd	nd	nd
La	nd	nd	nd	nd	nd	nd	nd	nd	nd	nd	nd	nd
Ce	nd	nd	nd	1.5	3.7	4.0	1.2	2.6	2.4	0.72	5.4	nd
Pr	nd	nd	nd	0.03	0.03	nd	nd	nd	nd	nd	nd	nd
Nd	nd	nd	nd	0.47	0.36	0.33	0.27	0.28	0.27	0.27	nd	nd
Sm	nd	nd	nd	1.2	0.61	0.6	1.08	0.99	1.2	0.55	0.78	nd
Eu	nd	nd	nd	0.13	0.28	0.21	0.15	0.46	0.5	0.12	0.18	nd
Gd	nd	nd	nd	5.32	2.45	3.42	8.67	4.61	12.32	2.91	4.53	nd
Tb	0.13	nd	nd	1.28	0.52	1.24	2.78	1.05	5.2	0.61	1.26	nd
Dy	2.7	0.76	1.9	12	7.28	17	32.99	11	87	6.1	17	2.39
Ho	1.6	0.52	0.94	2.8	2.25	6.4	9.06	3.0	35	1.5	5.8	0.73
Er	13	4.9	8.2	9.3	10.2	33	35.21	12	177	5.2	28	3.24
Tm	4.6	2.3	3.3	1.6	2.3	8.4	6.7	2.3	42.0	1.1	6.6	0.5
Yb	59	36	47	12.3	22.5	82.3	54.9	20.8	382.6	8.3	61.2	4.3
Lu	18	12	15	2.2	6.2	19.1	11.0	4.4	83	1.7	13.2	0.7
Hf	11671	12308	17432	12095	13595	9751	14014	13558	12450	10838	10626	16041
Ta	0.42	0.39	0.71	0.17	0.24	0.32	0.15	nd	0.16	0.17	0.23	nd
Th	0.01	0.01	0.02	47	48	67	14	12	7.4	6.5	22	nd
U	30	5	14	539	239	214	137	90	84	82	35	18
Th/U	0.0003	0.002	0.001	0.1	0.2	0.3	0.1	0.1	0.1	0.1	0.6	0.002
U/Nb	39	11	24	1703	378	224	330	448	191	212	40	88
Age	309±15	286±32	269±46									

* All data are given as μg/g.

The U-Pb isotopic compositions of 26 areas on 15 zircon grains from 99XD-2 are listed in Table 4 and plotted on a concordia diagram in Fig. 5. As for 99XD-1, where applicable the analyses have been grouped according to trace element composition. There is a very wide range of U contents (3–2400 ppm), the areas with weaker CL being consistently more U rich than the areas with strong CL. The overgrowths are particularly U poor, making the corrections for laboratory common Pb very large and a precise measurement of their age very difficult. Th/U also shows wide variation, ranging from < 0.005 in the overgrowths to 0.3 in the cores. Low values are not restricted to the overgrowths, however, nor high values to the cores.

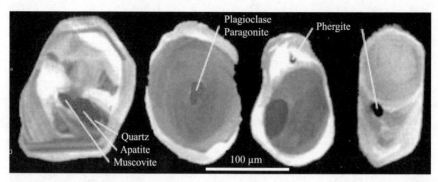

Fig. 7 CL images of zircon grains from Xiongdian eclogite 99XD-2 (metabasalt) from the Huwan shear zone showing inclusions. Plagioclase in zircon cores indicates that the cores were not formed during eclogite facies metamorphism. Phengite in the overgrowths indicates high pressure.

There is a wide range of radiogenic ^{206}Pb/^{238}U apparent ages, even larger than in 99XD-1, all due to the wide age range in the cores and one CL-dark overgrowth (440±4 (1σ) to 298±5 (1σ) Ma). These ages fall into two groups; one c. 310 Ma, which is similar to the ages of the overgrowths, and the other 350–450 Ma, all of which belong to chemical group 2-I. It is possible, if there has not been significant Pb loss, that the older zircon is predominantly of two ages, 433±9 (2σ) and 367±10 (2σ)Ma. The younger of these is the same age within error as the age of the older zircon cores in 99XD-1.

Fifteen analyses, representing groups 2-II and 2-III, and some unclassified cores and overgrowths, comprise the youngest age group. All these have the same radiogenic ^{206}Pb/^{238}U within analytical uncertainty, which gives a weighted mean age of 307.4±3.7(2σ) Ma. The HREE enrichment, negative Eu anomalies and commonly moderately high Th/U in all the older (group 2-I) zircon cores indicates that these probably did not crystallize under eclogite facies, and perhaps not even amphibolite facies, metamorphic conditions. They are best interpreted as zircon inherited from the precursor of the eclogite. In contrast, the youngest zircon has chemical features indicative of eclogite facies (group 2-II) and possibly amphibolite facies (group 2-III) growth, so the measured age of 307±4(2σ) Ma is a best estimate for the age of high grade and HP metamorphism. Although their ages cannot be resolved from the others statistically, the two group 2-II eclogite facies analyzed areas yield the oldest apparent ages (mean 316±8(2σ) Ma), and are structurally older than a CL-light zircon overgrowth (Fig. 5b,c), indicating that eclogite facies metamorphism preceded high pressure retrograde metamorphism outside the garnet stability field. The mean age of the light overgrowths (306.5±3.8 (2σ) Ma) might actually date the latter.

Table 4 SHRIMP zircon data for Xiongdian eclogite 99XD-2

Grain spot	CL domain	U ppm	Th ppm	Pb* ppm	Th/U	^{206}Pb$_c$%	^{204}Pb/^{206}Pb	±1σ	^{207}Pb/^{206}Pb	±1σ	^{238}U/^{206}Pb	±1σ	Inferred age(Ma)	±1σ
						Group 2-I								
3.1	Dark core	2400	712	163	0.297	0.2	0.00006	22	0.0569	6	14.5	0.1	428	4
4.1	Dark core	1175	22	76	0.018	0.1	0.00008	37	0.0562	10	14.1	0.1	440	4
8.1	Dark core	858	50	46	0.058	0.1	0.00009	45	0.0598	11	16.7	0.2	372	5
2.1	Dark overgrowth	733	141	40	0.192	0.2	0.00019	12	0.0553	20	17.4	0.3	359	6
10.1	Dark core	405	37	24	0.091	0.9	0.00026	11	0.0628	23	15.4	0.3	401	7
2.2	Dark core	375	54	22	0.144	0.6	0.00002	2	0.0594	19	16.2	0.4	383	10
10.2	Light core	63	2.1	3.0	0.034	3.3	0.00117	85	0.0833	39	17.4	0.6	348	12
11.1	Light core	27	5.6	1.5	0.207	5.9	0.0024	16	0.1082	95	14.5	1.0	405	26

Grain spot	CL domain	U ppm	Th ppm	Pb* ppm	Th/U	$^{206}Pb_c\%$	$^{204}Pb/^{206}Pb$	±1σ	$^{207}Pb/^{206}Pb$	±1σ	$^{238}U/^{206}Pb$	±1σ	Inferred age(Ma)	±1σ
							Group 2-I							
12.1	Light grain	26	3.2	1.6	0.121	8.1	0.0028	24	0.129	11	13.5	1.2	425	38
9.1	Light core	20	4.3	1.0	0.220	12.6	0.0108	31	0.169	16	15.0	1.1	366	27
14.2	Light core	3.1	0.2	0.2	0.074	45.0	0.024	22	0.463	38	8.1	1.0	424	61
							Group 2-II							
6.1	Dark core	900	102	42	0.113	0.4	0.00027	12	0.0561	11	19.9	0.3	315	5
14.1	Light overgrowth	118	22	5.1	0.186	1.9	0.00260	69	0.0702	27	19.4	0.5	318	8
							Group 2-III							
5.4	Dark core	1412	7.0	62	0.005	0.3	0.00002	2	0.0549	7	20.4	0.2	308	3
13.1	Dark core	496	3.0	22	0.006	0.7	0.00029	12	0.0584	23	20.5	0.4	305	5
7.2	Light grain	44	0.01	1.9	0.0003	7.1	0.0024	21	0.1175	64	18.9	0.9	309	15
13.2	Light overgrowth	8.0	0.01	0.3	0.001	28.3	0.018	11	0.310	28	15.8	1.7	286	32
1.1	Light overgrowth	4.9	0.005	0.2	0.001	29.0	0.0057	50	0.316	43	16.7	2.7	269	46
							Not analysed for trace elements							
15.3	Dark core	1581	8.6	71	0.005	0.3	0.00018	8	0.0546	10	20.0	0.2	313	3
7.1	Dark grain	1326	6.0	59	0.005	0.3	0.00018	6	0.0549	9	20.4	0.4	308	6
5.1	Dark core	1312	7.3	58	0.006	0.2	0.00011	5	0.0551	13	20.5	0.2	306	3
15.4	Dark core	412	0.7	18	0.002	0.6	0.00040	28	0.0593	15	21.0	0.3	298	5
5.2	Dark core	294	0.6	13	0.002	0.7	0.00046	21	0.0575	16	21.0	0.3	298	5
5.3	Dark core	263	0.4	12	0.002	1.3	0.00079	33	0.0639	18	20.2	0.4	307	6
15.1	Light overgrowth	21	0.01	0.9	0.001	11.5	0.0095	35	0.157	11	18.2	1.1	307	19
15.2	Light overgrowth	8.2	0.04	0.4	0.005	21.8	0.0136	81	0.310	32	13.6	1.9	332	49

* Radiogenic.

4.3 Hujiawan eclogite (95HW-S-8) from the Huwan shear zone

Zircon occurs in the Hujiawan eclogite as fine (50–100 μm), clear, colourless, anhedral, equant, dimpled grains with no visible internal structure and very few inclusions. CL imaging also shows little structure or intergranular variety of luminescence, most grains having no sign of cores, overgrowths or zoning (Fig. 8). The very few cores that are present (<5% of the crystals) have fine, weak euhedral zoning.

Twenty-three U-Pb isotopic analyses of 22 grains are listed in Table 5 and plotted on a concordia diagram in Fig. 8. All are of homogeneous zircon, none of cores. U contents are consistently low to extremely low (27–0.5 ppm), and as a consequence analytical uncertainties and common Pb corrections are relatively high. Th/U ranges widely (0.007–4.9), but the median is quite high, c. 0.7. The raw analyses, uncorrected for common Pb, plot on a mixing line between common and radiogenic end members (Fig. 8). The fit to the line is good (MSWD=1.1), implying that there has been little or no radiogenic Pb loss and that all the analyzed zircon is the same age. Each analysis must contain a mixture of common Pb components, however, c. 0.3 ppm of laboratory common Pb (the mount preparation 'blank') plus different amounts of common Pb from the zircon. The latter in this case is up to 2 ppm (analysis 1.2). In some grains this accounts for as much as 85% of the total ^{206}Pb. The intercept of the mixing line indicates that the $^{207}Pb/^{206}Pb$ of the common Pb mixture is 0.861±0.071 (1σ), similar to 400 Ma model Pb (Cumming & Richards, 1975), thus Pb of this composition was used for the common Pb corrections and calculating the inferred ages. The best estimate of the mean zircon age is given by the concordia intercept of the mixing line, 311±17 Ma, the uncertainty being the 95% confidence limit calculated from 5000 Monte Carlo trials (Ludwig, 2001). As there is no evidence for inherited zircon, this is considered the best estimate of the age of eclogite facies metamorphism at Hujiawan.

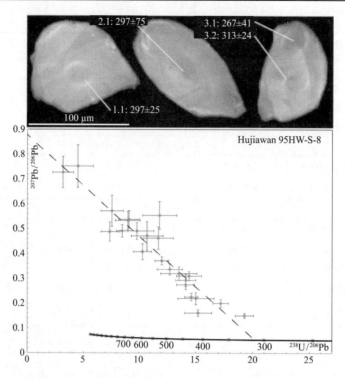

Fig. 8 Zircon U-Pb isotopic analyses, and representative CL images, from Hujiawan eclogite 95HW-S-8 (metabasalt) from the Huwan shear zone. In contrast to zircon from the other samples, these grains have very little internal structure. Their isotopic compositions are consistent with a single generation of metamorphic zircon growth and no inheritance of older zircon from the protolith. Analytical uncertainties 1σ.

4.4 Xuanhuadian eclogite (99HA-1) from the Hong'an Group

The zircon from quartz eclogite 99HA-1 occurs as fine to medium grained (50−200 μm diameter), clear, colourless, stubby, prismatic to rounded, anhedral to subhedral crystals with no visible internal structure and rare, equant, melt and mineral inclusions. CL imaging shows that most grains consist of moderately luminescent zircon with weak, broad growth banding or zoning (Fig. 9a-d). Some grains (c. 20%) are surrounded by a more luminescent, broadly or sector zoned overgrowth, and even fewer (< 5%) have a moderately luminescent discontinuous outer layer, commonly < 10 μm thick, which in rare cases forms anhedral outgrowths up to 50 μm diameter (Fig. 9b,c). About 5‰ of the grains appear to consist entirely of this outgrowth material (Fig. 9d).

U-Pb isotopic analyses of 41 areas on 38 grains are listed in Table 6 and plotted on a concordia diagram in Fig. 9. Analyses from two sessions (A and B) are identified as representing cores, centres, overgrowths or outgrowths. Cores and centres are distinguished only by the presence or absence of an overgrowth, respectively, and probably represent the same generation of zircon growth. Post-analysis CL imaging showed that several analyses accidentally overlapped a boundary between zones. These data are listed, but for clarity have not been included on the plot. All the zircon is relatively low in U (0.4−170 ppm), particularly most of the outgrowths (<10 ppm). There is a wide range of Th/U (0.005−1.9), the cores and centres being the highest (>0.8), the overgrowths mostly intermediate (0.35−0.75) and the outgrowths low (most < 0.1).

The U-Pb isotopic analyses fall into two main groups, one at c. 700 Ma and the other forming a mixing line between common Pb and radiogenic Pb at c. 230 Ma. The older group consists of analyses of cores, centres and overgrowths. The apparent ages of this group mostly range from 775±9(1σ) to 550±35(1σ) Ma, with one common Pb-rich (?discordant) area giving 345±67(1σ) Ma. Even the higher ages scatter more than expected from the analytical uncertainties, four of the analyses giving much lower ages than the rest. The remaining analyses also are scattered, but there are no obvious outliers. The weighted mean of the nine highest ages is 752±17(tσ) Ma, the large uncertainty reflecting the scatter. Because of the high likelihood of radiogenic Pb loss, this is a minimum estimate of the original age of the protolith zircon.

Table 5 SHRIMP zircon data for Hujiawan eclogite 95HW-S-8

Grain spot	U ppm	Th ppm	Pb* ppm	Th/U	$^{206}Pb_c$%	$^{204}Pb/^{206}Pb$	±1σ	$^{207}Pb/^{206}Pb$	±1σ	$^{238}U/^{206}Pb$	±1σ	Inferred age(Ma)	±1σ
11.2	27	34	1.5	1.25	12	0.0068	20	0.150	10	19.3	0.8	288	12
11.1	24	30	1.4	1.25	18	0.0088	17	0.200	15	17.2	0.7	300	14
9.1	17	21	1.2	1.23	21	0.0091	30	0.225	15	14.6	0.6	339	15
5.1	15	29	1.2	1.89	39	0.0213	40	0.371	16	12.0	0.6	318	18
18.2	9.5	47	1.2	4.93	13	0.0055	41	0.161	13	15.2	1.2	357	27
18.1	7.7	29	0.9	3.73	44	0.0303	52	0.408	32	10.3	0.8	342	36
1.1	6.7	3.1	0.3	0.46	32	0.0187	79	0.310	18	14.4	1.1	297	25
9.2	6.7	9.2	0.4	1.39	27	0.0154	47	0.275	18	14.1	0.8	324	20
4.1	5.1	5.1	0.4	1.00	54	0.0305	53	0.491	25	8.5	0.5	339	30
15.1	5.1	3.3	0.3	0.66	33	0.0268	95	0.320	27	13.5	1.6	311	38
17.1	4.3	4.3	0.2	1.00	35	0.020	12	0.337	22	12.7	1.1	320	30
3.2	3.8	2.2	0.2	0.58	30	0.0157	53	0.293	25	14.1	0.9	313	23
19.1	3.3	2.9	0.2	0.88	21	0.0183	70	0.219	24	15.0	1.6	332	36
1.2	2.6	0.9	0.1	0.33	83	0.0416	97	0.727	63	3.2	0.9	324	173
8.1	2.6	2.1	0.2	0.80	54	0.030	12	0.488	39	7.4	0.8	392	59
10.1	2.2	1.4	0.1	0.64	60	0.045	10	0.538	35	9.1	0.9	277	39
7.1	2.1	1.4	0.09	0.66	59	0.0181	62	0.534	36	9.0	1.1	283	45
6.1	2.1	1.3	0.10	0.62	52	0.0267	81	0.471	58	10.7	1.4	284	55
13.1	2.0	0.8	0.10	0.38	54	0.021	11	0.491	34	9.8	1.5	295	51
2.1	2.0	0.9	0.09	0.48	64	0.019	10	0.572	63	7.6	1.0	297	75
16.1	1.7	0.03	0.05	0.015	62	0.019	16	0.554	56	11.8	1.5	205	46
3.1	1.7	0.8	0.06	0.48	51	0.026	11	0.462	45	11.7	1.3	267	41
14.1	0.5	0.004	0.014	0.007	86	0.046	14	0.752	86	4.5	1.4	190	160

* Radiogenic.

Table 6 SHRIMP zircon data for Xuanhuadian eclogite (99HA-1)

Grain spot	CL domain	U ppm	Th ppm	Pb* ppm	Th/U	$^{206}Pb_c$%	$^{204}Pb/^{206}Pb$	±1σ	$^{207}Pb/^{206}Pb$	±1σ	$^{238}U/^{206}Pb$	±1σ	Inferred age(Ma)	±1σ
B28.1	Centre	171	321	29	1.9	2	0.00067	14	0.0771	19	8.2	0.1	732	11
B27.1	Centre	95	149	15	1.6	2	0.00141	40	0.0825	26	8.1	0.1	734	12
B25.1	Centre	51	54	7.0	1.1	5	0.00169	75	0.1029	50	8.3	0.2	696	20
B13.1	Centre	41	39	5.0	0.95	17	0.0068	14	0.198	36	8.4	0.4	609	40
B26.1	Centre	37	31	5.0	0.83	4	0.0032	12	0.0967	40	8.0	0.2	726	21
A1.1	Dark core	171	227	27	1.3	1	0.00024	9	0.0703	18	8.1	0.1	747	13
B1.1	Dark core	120	161	19	1.3	0.4	0.00019	10	0.0682	18	7.8	0.1	775	9
B24.1	Overgrowth	41	26	5.0	0.64	6	0.0026	10	0.1065	38	9.2	0.3	630	17
A1.2	Overgrowth	39	28	5.0	0.72	1	0.00071	27	0.0714	25	8.0	0.3	757	27
A2.1	Overgrowth	38	26	5.0	0.69	3	0.00202	62	0.0861	30	8.1	0.2	733	21
B19.1	Overgrowth	34	24	4.0	0.69	7	0.0016	11	0.1186	44	7.9	0.3	715	22
B2.1	Overgrowth	33	25	5.0	0.76	8	0.0033	11	0.1286	41	7.3	0.2	768	17
B1.2	Overgrowth	31	17	4.0	0.54	2	0.00107	64	0.0784	30	7.7	0.2	773	17
B21.1	Overgrowth	11	6	1.0	0.53	22	0.0101	31	0.239	20	8.8	0.5	552	35
B15.1	Overgrowth	3	2	0.3	0.60	67	0.0426	95	0.609	45	6.0	0.7	345	67
B20.1	Overgrowth	3	0.9	0.4	0.35	56	0.032	11	0.524	41	4.7	0.5	568	87
A8.1	Outgrowth	46	0.3	2.0	0.01	11	0.0064	17	0.1482	47	24.3	0.6	232	6
B22.1	Outgrowth	15	3	1.0	0.19	37	0.0214	45	0.360	22	14.9	1.1	268	22
A6.1	Outgrowth	14	0.08	0.6	0.01	10	0.0025	24	0.141	11	23.9	1.2	239	12
B16.1	Outgrowth	8	0.05	0.4	0.006	43	0.0258	68	0.411	33	19.5	1.1	185	16

Continued

Grain spot	CL domain	U ppm	Th ppm	Pb* ppm	Th/U	$^{206}Pb_c\%$	$^{204}Pb/^{206}Pb$	±1σ	$^{207}Pb/^{206}Pb$	±1σ	$^{238}U/^{206}Pb$	±1σ	Inferred age(Ma)	±1σ
B10.1	Outgrowth	7	0.06	0.4	0.008	20	0.0065	25	0.217	14	19.8	1.3	256	17
B3.1	Outgrowth	7	0.08	0.4	0.01	23	0.0140	42	0.245	24	19.5	0.9	249	15
B17.2	Dark grain	7	0.6	0.4	0.09	56	0.0280	65	0.520	42	12.4	0.9	223	30
B12.1	Outgrowth	6	0.1	0.4	0.018	72	0.0382	51	0.652	27	7.1	0.4	248	31
B17.1	Dark grain	5	2	0.3	0.44	71	0.0552	80	0.646	41	7.9	0.6	230	43
B14.1	Outgrowth	5	0.09	0.2	0.02	63	0.033	10	0.575	39	12.9	1.2	183	28
B5.1	Outgrowth	3	0.27	0.2	0.10	80	0.0429	76	0.721	44	6.9	0.6	181	51
B18.2	Outgrowth	2	0.2	0.2	0.09	73	0.064	12	0.659	45	6.6	0.7	260	58
B7.1	Outgrowth	2	0.01	0.1	0.005	66	0.0274	93	0.601	64	7.3	1.1	294	78
A3.1	Outgrowth	2	0.01	0.1	0.01	52	0.0302	97	0.521	49	12.6	2.6	244	56
B4.1	Outgrowth	1.4	0.20	0.1	0.15	69	0.041	15	0.626	64	7.1	1.1	276	78
B11.1	Outgrowth	1.1	0.09	0.05	0.08	75	0.032	11	0.677	92	10.2	1.5	156	72
A5.1	Outgrowth	1.0	0.1	0.1	0.11	60	0.067	19	0.596	72	8.8	1.8	287	81
A4.1	Outgrowth	0.8	0.01	0.04	0.01	82	0.046	13	0.80	10	5.8	0.9	198	123
B6.1	Mixture	20	10	1.0	0.50	6	0.0029	14	0.1013	62	16.4	0.7	360	16
A11.1	Mixture	17	13	2.0	0.77	4	0.00167	72	0.0938	54	9.0	0.3	653	22
A7.1	Mixture	16	2	1.0	0.15	21	0.0120	27	0.239	15	18.9	1.3	266	19
A9.1	Mixture	6	0.9	0.4	0.15	5	0.0070	30	0.102	14	19.1	1.2	311	20
B9.1	Mixture	2	0.7	0.1	0.39	31	0.0161	70	0.311	30	10.9	1.6	396	60
A10.1	Mixture	1.1	0.3	0.2	0.32	56	0.0290	96	0.563	36	4.7	0.5	574	77
B8.1	Mixture	0.4	0.2	0.03	0.46	76	0.016	12	0.682	60	4.9	1.1	315	116

* Radiogenic.

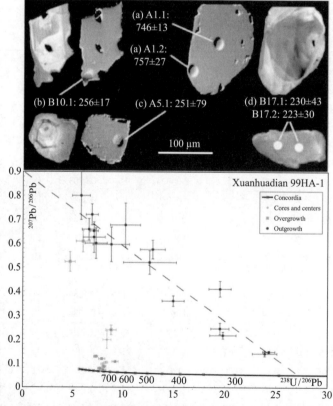

Fig. 9 Zircon U-Pb isotopic analyses, and representative CL images of zircon structures, from Xuanhuadian eclogite 99HA-1 (metabasalt) from the HP Hong'an Group. (a) Old thick overgrowth surrounding an old CL-dark core; (b,c) Examples of rare CL-dark, low-U, metamorphic zircon outgrowths. (d) Structureless grain consisting entirely of outgrowth material. Analytical uncertainties 1σ.

The mixing line defined by the outgrowth analyses, as in the case of the Hujiawan eclogite zircon, is actually a line between one radiogenic and two common Pb components, one from the sample and the other from laboratory 'blank'. Regression of the analyses yields a $^{207}Pb/^{206}Pb$ of 0.891±0.049(tσ) for the average common Pb, similar to 700 Ma model Pb (Cumming & Richards, 1975). This composition was therefore used in correcting the data. The concordia intercept of the mixing line yields an age of 232±10 Ma, the uncertainty being a 95% confidence limit calculated by the Monte Carlo technique (Ludwig, 2001). This is the best estimate of the age of the eclogite facies metamorphism of the Xuanhuadian sample.

5 Implications of the isotopic data

The zircon data from the four north-west Dabie eclogite samples examined for the present study strongly support the conclusion reached by some previous workers (Ye et al., 1994; Jian et al., 1997, 2000; Xu et al., 2000; Li et al., 2001) that eclogite facies metamorphism in the Huwan shear zone is much older than the HP-UHP metamorphism (220–230 Ma) which affected the bulk of the Qinling-Dabie orogenic belt further to the south and east (Li et al., 2000; Hacker et al., 1998, 2000). The data also show, however, that the metamorphism is probably not as old as previously thought. The ages of principal metamorphic zircon growth measured on the three Huwan shear zone samples are 312±5 (99XD-1), 307±4 (99XD-2) and 311±17Ma (95HW-S-8), respectively. These are equal within analytical uncertainty, so can be pooled to give 309±3 (tσ) Ma as the best estimate of the age of the metamorphic event. In both Xiongdian samples there is evidence that the metamorphism might have started earlier, possibly at c. 320 Ma. Nevertheless, it is obvious that the Huwan shear zone eclogites experienced a long period of metamorphism. Garnet is strongly zoned both in major and trace elements (W. Sun unpublished data). This complicates the interpretation of the mineral isochron ages (e.g. Jian et al., 1997). Other metamorphic ages of 430–350 Ma previously reported probably reflect the presence of inherited argon (Xu et al., 2000) and zircon (Jian et al., 1997, 2000), respectively.

It cannot be assumed a priori that 309±3 Ma is necessarily the age of eclogite facies metamorphism, however, as it is likely that some of the metamorphic zircon growth did not occur under peak metamorphic conditions. Both Xiongdian samples contain zircon with trace element signatures and inclusion suites consistent with growth under eclogite facies conditions, but in the metabasalt sample (99XD-2) this is less abundant than structurally younger zircon that appears to have formed at lower pressure. Although statistically indistinguishable in isotopic age from the eclogite facies zircon, most of the late zircon growth in 99XD-2 probably took place during decompression soon after the metamorphic peak.

Growth of metamorphic zircon in the Xuanhuadian eclogite from south of the Huwan shear zone is much younger, 232±10 Ma, very similar in age to the Triassic metamorphic ages (250–200 Ma) measured by Rb-Sr (Chavagnac & Jahn, 1996), Sm-Nd (Li et al., 1993, 1994, 2000), $^{40}Ar/^{39}Ar$ (Chen et al., 1992) and zircon (Hacker et al., 1998; Rowley et al., 1997) in the main part of the Qinling-Dabie orogen. There is, however, no sign of zircon growth or U-Pb isotopic disturbance at this time in any of the Huwan shear zone samples, suggesting that there were two separate eclogite facies events, one late Carboniferous and the other Triassic, and that the rocks affected by one were not affected by the other. Triassic mica $^{40}Ar/^{39}Ar$ ages reported from the Huwan shear zone (Hacker et al., 1998; Xu et al., 2000), however, indicate that that region was affected by a lower pressure Triassic event.

The inherited/protolith zircon in the Huwan shear zone samples is quite different from that in the Xuanhuadian sample, suggesting that the host rocks have different provenance. The older zircon in the Huwan shear zone samples, at 440–350 Ma, resembles the age range of Ordovician to Devonian igneous and metamorphic rocks near the southern margin of the North China Block (Kröner et al., 1993; Lerch et al., 1995; Zhai et al., 1998), implying that the rocks in the shear zone are part of that northern block. On the other hand, the older zircon in the Hong'an Group, from south of the Huwan shear zone, at 780–610 Ma matches the age of Neoproterozoic granites and gneisses that occur widely throughout the northern part of the South China Block (Kröner et al., 1993; Hacker et al., 1998; Xue et al., 1996a; Ames et al., 1996). This indicates that in the west Dabie Mountains, the Triassic suture between the North and South China Blocks lies at the southern margin of the Huwan shear zone, not at the northern contact of the

Nanwan Formation (Laoshan ductile belt) as previously suggested (Hacker et al., 1998, 2000). The presence of Carboniferous eclogites in the Huwan shear zone, however, suggests that there is probably a Carboniferous suture to the north, perhaps in fact along the Laoshan ductile belt (Fig. 1) (Hacker et al., 1998, 2000).

This being so, the metamorphic history of the Huwan shear zone and immediately adjacent Dabie metamorphic complex is a direct record of the collision of the North and South China Blocks. The first stage of the collision occurred as early as the late Carboniferous, probably with subduction of a fragment of continental crust represented by the Huwan shear zone rocks, to depths of at least 60 km (Liu et al., 1996; Ye et al., 1994). About 80 Ma later, in the late Triassic, the main collision occurred, juxtaposing the two large continental blocks, subducting the northern margin of the southern block to depths in excess of 100 km (Xu et al., 1992; Liu et al., 2001) and obducting the southern margin of the northern block, with metamorphism to at least lower amphibolite facies.

Acknowledgements

We thank Dr S.S. Sun for many constructive comments and suggestions, and Drs S. Eggins, C. Allen, D. Rubatto and L. Zhang for help with trace element analyses and the mineral inclusion identification, respectively. We also thank Drs V.C. Bennett, S. Eggins, T. Ireland, R. Armstrong, J. Hermann, D. Rubatto and J. Hong for many stimulating discussions, and Dr V. Bennett for critical comment on the manuscript. Many thanks to Drs C. Wei and B. Hacker for critical reviews. Field work and zircon separations by S. Li and W. Sun were funded by the Major State Basic Research Development Program (Grant no. 1999075503) and the Natural Science Foundation of China (49573190, 49794042). We thank Dr J. Hong for helping with the field work. WS acknowledges the support of an International Postgraduate Research Scholarship from the Australian National University.

References

Ames, L., Zhou, G. & Xiong, B., 1996. Geochronology and geochemistry of ultrahigh-pressure metamorphism with implications for collision of the Sino-Korean and Yangtze cratons, central China. Tectonics, 15, 472-489.

Chavagnac, V. & Jahn, B.M., 1996. Coesite-bearing eclogites from Bixiling complex, Dabie Mountains China: Sm-Nd ages, geochemical characteristics and tectonic implications. Chemical Geology, 133, 29-51.

Chen, W., Harrison, T. M., Heizler, M. T., Liu, R., Ma, B. & Li, J., 1992. The cooling history of melange zone in the north Jiangsu south Shandong region: evidence from multiple diffusion domain ^{40}Ar-^{39}Ar thermal geochronology. Acta Petrologica Sinica, 8, 1-17.

Claoue'-Long, J. C., Compston, W., Roberts, J. & Fanning, C. M., 1995. Two Carboniferous ages: A comparison of SHRIMP zircon dating with conventional zircon ages and ^{40}Ar/^{39}Ar analysis. In: Geochronology, Time Scales and Global Stratigraphic Correlation. Society for Sedimentary Geology Special Publication, 54 (eds Berggren, W. A., Kent, D. V., Aubrey, M.-P. & Hardenbol, J.), pp. 3-21, SEPM, Tulsa.

Cumming, G. L. & Richards, J. R., 1975. Ore lead isotope ratios in a continuously changing earth. Earth and Planetary Science Letters, 28, 155-171.

Hacker, B. R., Ratschbacher, L., Webb, L. et al., 2000. Exhumation of ultrahigh-pressure continental crust in east central China: Late Triassic-Early Jurassic tectonic unroofing. Journal of Geophysical Research, 105, 13,339-364.

Hacker, B. R., Ratschbacher, L., Webb, L., Ireland, T., Walker, D. & Dong, S., 1998. U/Pb zircon ages constrain the architecture of the ultrahigh-pressure Qinling-Dabie Orogen, China. Earth and Planetary Science Letters, 161, 215-230.

Jian, P., Liu, D., Yang, W. & Williams, I. S., 2000. Petrographical study of zircons and SHRIMP dating of the Caledonian Xiongdian eclogite, Northwestern Dabie Mountains. Acta Geologica Sinica, 74, 259-264.

Jian, P., Yang, W. & Li, Z., 1997. Isotopic geochronological evidence for the Caledonian Xiongdian eclogite in the western Dabie mountains, China. Acta Geologica Sinica, 10, 455-465.

Kröner, A., Zhang, G. W. & Sun, Y., 1993. Granulites in the Tongbai area, Qinling belt, China: geochemistry, petrology, single zircon geochronology, and implications for the tectonic evolution of eastern Asia. Tectonics, 12, 245-255.

Lerch, M. F., Xue, F., Kröner, A., Zhang, G. W. & Tod, W., 1995. A middle Silurian-early Devonian magmatic arc in the Qinling mountains of central China. Journal of Geology, 103, 437-449.

Li, S., Hart, S. R., Zheng, S., Liou, D., Zhang, G. & Guo, A., 1989. Timing of collision between the North and South China blocks—the Sm-Nd isotopic age evidence. Science Sinica (Series B), 32, 1391-1400.

Li, S., Jagoutz, E., Chen, Y. & Li, Q., 2000. Sm-Nd and Rb-Sr isotopic chronology and cooling history of ultrahigh pressure metamorphic rocks and their country rocks at Shuanghe in the Dabie Mountains, Central China. Geochimica et Cosmochimica

Acta, 64, 1077-1093.

Li, S. & Sun, W., 1996. A middle Silurian-early Devonian magmatic arc in the Qinling Mountains of central China: a discussion. Journal of Geology, 104, 501-503.

Li, S., Wang, S., Chen, Y., Liou, D., Qiu, J., Zhou, H. & Zhang, Z., 1994. Excess argon in phengite from eclogite: Evidence from dating of eclogite minerals by Sm-Nd, Rb-Sr and $^{40}Ar/^{39}Ar$ methods. Chemical Geology, 112, 343-350.

Li, S., Xiao, Y., Liou, D., Chen, Y., Ge, N., Zhang, S., Sun, S.S., Cong, B., Zhang, R., Hart, S. R. & Wang, S., 1993. Collision of the North China and Yangtze blocks and formation of coesite-eclogites: timing and processes. Chemical Geology, 109, 89-111.

Liou, J. G., Zhang, R. Y., Wang, X. M., Eide, E. A., Ernst, W. G. & Maruyama, S., 1996. Metamorphism and tectonics of high-pressure and ultrahigh-pressure belts in the Dabie-Su-Lu region, China. In: The Tectonic Evolution of Asia (eds Yin, A. & Harrison, T. M.), pp. 300-344. Cambridge University Press, Cambridge.

Liu, J. Ye, K., Maruyama, S., Cong, B. & Fan, H., 2001. Mineral inclusions in zircons from gneisses in the ultrahigh-pressure zone of the Dabie Mountains, China. Journal of Geology, 109, 523-535.

Liu, J. P., You, Z. D. & Zhong, Z. Q., 1996. Eclogites from the middle and north of Dabie mountain in southern. Henan and Northern Hubei. Science in China (Series D), 39, 293-300.

Ludwig, K. R., 2001. Isoplot/Ex, a geochronological toolkit for Microsoft Excel, Special Publication, 1a, Berkeley. Geochronology Center, Berkeley.

Mattauer, M., Matte, P. H., Malavieille, J., et al., 1985. Tectonics of the Qinling Belt: build-up and evolution of eartern Asia. Nature, 317, 496-500.

Meng, Q. R. & Zhang, G. W., 1999. Timing of the collision of the North and South China blocks: controversy and reconciliation. Geology, 27, 123-126.

Meng, Q. R. & Zhang, G. W., 2000. Geological framework and tectonic evolution of the Qinling orogen, central China. Tectonophysics, 323, 183-196.

Okay, A. I. & Sengör, A. M. C., 1992. Evidence for intracontinental thrust-related exhumation of ultrahigh-pressure rocks in China. Geology, 20, 411-414.

Pearce, N. J. G., Perkins, W. T., Westgate, J. A., et al., 1997. A Compliation of new and published major and trace element data for NIST SRM 610 and NIST SRM 612 glass reference materials. Geostandards Newsletter, 21, 115-144.

Rowley, D. B., Xue, F., Tucker, R. D., Peng, Z. X., Baker, J. & Davis, A., 1997. Age of ultrahigh pressure metamorphism and protolith of orthogneisses from the eastern Dabie Shan: U/ Pb zircon geochronology. Earth Planetary Science Letters, 151, 191-203.

Rubatto, D., 2002. Zircon trace element geochemistry: partitioning with garnet and the link between U-Pb ages and metamorphism. Chemical Geology, 184, 123-138.

Steiger, R. H. & Jäoger, E., 1977. Subcommission on geochronology; convention on the use of decay constants in geochronology and cosmochronology. Earth Planetary Science Letters, 36, 359-362.

Sun, W., 1994. Collision of Qinling orogenic belt: ophiolite and granite geochronology, Masters Thesis, University of Science and Technology of China, Hefei.

Sun, S.S. & McDonough, W. F., 1989. Chemical and isotopic systematics of oceanic basalts: implication for mantle composition and processes. Geological Society Special Publication, 42, 313-345.

Wang, Q. C., Zhai, M. G. & Cong, B. L., 1996. Regional geology. In: Ultrahigh-Pressure Metamorphic Rocks in the Dabieshan-Sulu Region of China (ed. Cong, B. L.), pp. 8-26. Science Press, Beijing.

Webb, L. E., Ratschbacher, L., Hacker, B. R. & Dong, S., 2001. Kinematics of Exhumation of High- and Ultrahigh-Pressure Rocks in the Hong'an and Tongbai Shan of the Qinling-Dabie Collisional Orogen, Eastern China, in Paleozoic and Mesozoic Tectonic Evolution of Central Asia- from Continental Assembly to Intracontinental Deformation, Special Publication 194, (eds Hendrix, M. S. & Davis, G. A.), pp. 231-245. Geological Society of America, Boulder.

Williams, I. S., 1998. U-Th-Pb geochronology by ion microprobe. In: Applications of Microanalytical Techniques to Understanding Mineralizing Processes, Review in Economic Geology, 7. (eds McKibben, M. A., Shanks, W. C. III & Ridley, W. L.), pp. 1-35, Society of Economic Geologists, Socorro.

Williams, I. S. & Claesson, S., 1987. Isotopic evidence for the Precambrian provenance and Caledonian metamorphism of high grade paragenisses from the Seve Nappes, Scandinavian Caledonides II, Ion microprobe zircon U-Th-Pb. Contributions to Mineralogy and Petrology, 97, 205-217.

Xu, B., Grove, M., Wang, C., Zhang, L. & Liu, S., 2000. $^{40}Ar/^{39}Ar$ thermochronology from the northwestern Dabie Shan: constraints on the evolution of Qinling-Dabie orogenic belt, east-central China. Tectonophysics, 322, 279-301.

Xu, S., Liu, Y. C., Jiang, L. L., Su, W. & Ji, S. Y., 1994. Tectonic framework and evolution of Dabieshan. Science Publishing, Beijing, (in Chinese with English abstract).

Xu, S., Okay, A. I., Ji, S., Sengör, A. M. C., Su, W., Liu, Y. & Jiang, L., 1992. Diamond from the Dabie Shan metamorphic rocks and its implication for tectonic setting. Science, 256, 80-82.

Xue, F., Kröner, A., Reischmann, T. & Lerch, F., 1996a. Paleozoic pre- and post-collision calc-alkaline magmatism in the Qinling orogenic belt, central China, as documented by zircon ages on granitoid rocks. Journal of the Geological Society, 153, 409-417.

Xue, F., Lerch, M. F., Kröner, A. & Reischmann, T., 1996b. Tectonic evolution of the east Qinling Mountains, China, in the Paleozoic: a review and new tectonic model. Tectonophysics, 253, 271-284.

Ye B. D., Jian, P., Xu, J. W., Cui, F., Li, Z. C. & Zhang, Z. H., 1994. Geochronological study on Huwan Group in Tongbai-Dabie orogenic belt. In: Isotopic Geochemistry Research (ed. Chen, H. S.), pp. 187-204. Zhejiang University Press, Hangzhou.

Yin, A. & Nie, S., 1993. An indentation model for the north and south China collision and the development of the Tanlu and Honam fault systems, eastern Asia. Tectonics, 12, 801-813.

Zhai, X., Day, H. W., Hacker, B. R. & You, Z. D., 1998. Paleozoic metamorphism in the Qinling orogen, Tongbai Mountains, central China. Geology, 26, 371-374.

Zhang, G. W., Meng, Q. R. & Lai, S. C., 1995. Tectonics and structures of the Qinling orogenic belt. Science in China (Series B), 38, 1379-1386.

Mid-paleozoic collision in the North Qinling: Sm-Nd, Rb-Sr and $^{40}Ar/^{39}Ar$ ages and their tectonic implications*

Weidong Sun[1,2], Shuguang Li[1], Yong Sun[3], Guowei Zhang[3] and Qiuli Li[1]

1. Department of Earth and Space Sciences, University of Science and Technology of China, Hefei 230026, Anhui, China
2. Research School of Earth Sciences, Australian National University, Canberra, ACT 0200, Australia
3. Laboratory of Continental Dynamics, Northwest University, Xi'an 710069, Shaanxi, China

> 亮点介绍：丹凤群和二郎坪群变质火山岩的 Sm-Nd、Rb-Sr 和 $^{40}Ar/^{39}Ar$ 同位素定年结果表明，北秦岭广泛存在晚志留世构造热事件，而且，该事件与北秦岭岛弧系和华北陆块南缘的碰撞密切相关。

Abstract Different isotopic geochronometers have been used to study the nature of the mid-Paleozoic tectonothermal event in the North Qinling arc system at the southern margin of the North China Block. $^{40}Ar/^{39}Ar$ age of amphibole from sample QH-3 (426±2 Ma), Rb-Sr mineral isochron age (411±5 Ma) of sample QSP-10 of the Danfeng Group and Rb-Sr whole rock isochron age (406±22 Ma) of a greenschist facies pillow lava sample of the Erlangping Group indicate a widespread Late Silurian tectonothermal event in the North Qinling area. The data also suggest a limited uplift in the North Qinling during the Late Silurian tectonothermal event. This event is restricted to the North Qinling area and the southern margin of North China Block, and not observed in the South Qinling belt and the northern margin of the South China Block. Consequently, it is related to the collision between the North China Block and a North Qinling arc system, which resulted in the closure of the North Qinling back-arc basin(s).

Key words North Qinling; geochronology; Late Silurian; collision; backarc basin

1 Introduction

Collision and subduction are significant geodynamic processes that can be observed directly. The Qinling-Dabie orogenic belt is well known for the occurrence of Triassic diamond- and coesite-bearing ultrahigh pressure metamorphic (UHPM) rocks in the Dabie Mountains and the Sulu terrane (e.g. Li et al., 1993, Li et al., 1994a, b, 2000; Jahn et al., 1995; Ames et al., 1996; Chavagnac and Jahn, 1996; Rowley et al., 1997; Hacker et al., 1995; Hacker et al., 1998). But the evolution of this 2000-km-long orogenic belt has been a controversial issue for many years. A major controversy in the tectonic evolution of this belt is the timing and processes involved in the collision between the North and the South China Blocks (NCB and SCB). Various Mesozoic continent-continent collision models have been provoked based on data collected in the eastern Dabie Mountains and Sulu terrane, the east part of the Qinling-Dabie orogenic belt (e.g. Li et al., 1989, 1993, 1994a, b, 2000; Ames et al., 1996; Okay and Sengör, 1992; Yin and Nie, 1993; Hacker et al., 1998). In contrast, some researchers, especially those who worked in the Qinling Mountains, have suggested that the collision between NCB and SCB took place during the Late Silurian (e.g. Mattauer et al., 1985; Xu et al., 1991; Ren et al., 1991; Zhang et al., 1997a). Early Mesozoic granites and metamorphic rocks in South Qinling, as well as Triassic UHPM rocks in the Dabie Mountains and Sulu terrane, are hypothesized to be the result of 'intra-continental subduction' within the South China Block (Mattauer et al., 1985; Xu et al., 1991; Zhang et al., 1997b). But this 'intra-continental subduction' model has a serious problem in explaining the formation of the Triassic UHPM rocks in the Dabie Mountains and Sulu terrane.

---

* 本文发表在：Journal of Asian Earth Sciences, 2002, 21: 69-76

One possible reason for the debate on timing of the collision is that the North Qinling is not as well studied as the Dabie Mountains in terms of geochronology. The Late Silurian tectonothermal event in the North Qinling (and Tongbai) Mountains was either dated by the $^{40}Ar/^{39}Ar$ method (Mattauer et al., 1985; Zhai et al., 1998), or bracketed with zircon ages of foliated and undeformed granites (Lerch et al., 1995; Xue et al., 1996a; Xue et al., 1996b). These ages suggest the existence of the Late Silurian tectonothermal event in the North Qinling area, represented by large scale metamorphism and igneous intrusions, nevertheless, these workers failed to give sufficient information to clearly show the nature of this event. Therefore, many geological models need to be reevaluated.

To address the nature of the supposed Late Silurian tectonothermal event in the North Qinling, metamorphic rocks from three localities of the two volcanic belts, the Danfeng and Erlangping Groups, were selected. Sm-Nd, Rb-Sr, and $^{40}Ar/^{39}Ar$ isotopic geochronometers were applied to better constrain the uplift history of the North Qinling. Based on the dating results of this paper and those of previous authors, as well as the regional geology, a back-arc closure model is proposed.

2 Geology and samples

The Qinling orogenic belt comprises two mountain chains, separated by the Shangdan Fault (e.g. Meng and Zhang, 2000). North Qinling is regarded as a mid-Paleozoic orogen (Mattauer et al., 1985) with widespread Paleozoic magmatism and metamorphism (e.g. Sun, 1994; Zhang et al., 1994; Lerch et al., 1995; Zhai et al., 1998). South Qinling is interpreted as a Late Paleozoic to Early Mesozoic orogen with abundant Triassic granite and late-Paleozoic metamorphism (Xu et al., 1991; Sun, 1994; Li et al., 1996; Li and Sun, 1996; Zhai et al., 1998). Studies also show that South Qinling consists of several micro-continents (e.g. Yin and Huang, 1995; Yin et al., 1999). Two sutures have been proposed (Zhang et al., 1995a, b, Meng and Zhang, 1999, 2000).

The North Qinling arc system is bounded by the Shangdan Fault to the south and the Luonan-Luanchuan Fault to the north. From south to north, the system consists of the Shangdan Fault zone, the Danfeng and Heihe volcanic arcs, the Qinling metamorphic core complex, the Erlangping volcanic arc, and the Kuanping Group (Fig. 1). South of the Shangdan Fault is the Liuling Group, the northern margin of the South Qinling belt. North of the Luonan-Luanchuan Fault is the Taowan Group, the southern margin of the North China Block.

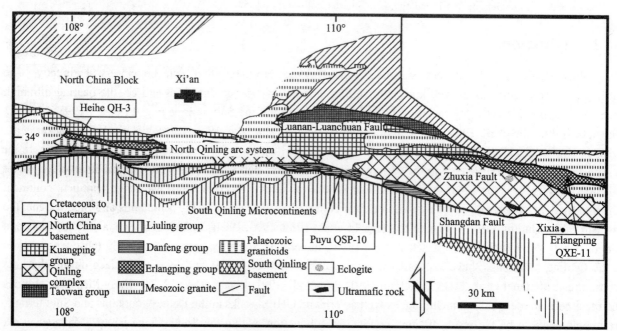

Fig. 1 Sketched geological map of the Qinling orogenic belt (modified from Xue et al., 1996a, Zhang et al., 1996). Inset delineates the major crustal Blocks in China and location of Fig. 1.

The Danfeng Group (also called Shangdan ophiolitic complex, Meng and Zhang, 2000) includes the Danfeng and Heihe volcanic arcs that lie along the northern side of the Shangdan Fault. The Danfeng Group consists of fragments of ultramafic rocks, gabbros, basalt, diabasic dikes, pillow lava and radiolarian chert ranging from a meter to more than 1 km in size. They are in fault contact (e.g. Zhang et al., 1995a, b; Meng and Zhang, 2000). The coexistence of calc-alkaline and tholeiitic basalt has been interpreted as an 'intra-oceanic island arc' setting (Zhang et al., 1995a, b). The Erlangping Group is separated from the Danfeng Group by the Proterozoic Qinling Complex and is confined between the Qinling Complex and the Kuanping Group. It consists mostly of olivine gabbros, massive and pillow basaltic lava, sheeted dikes and sills, radiolarian chert and marble (Liu et al., 1993; Meng and Zhang, 2000). Both island-arc type metabasalt/meta-andesite and mid-ocean-ridge or backarc basin type metabasalt have been found. The latter show light rare earth element depletions and Ti/V ratios ranging from 20 to 50, so, the Erlangping Group has been suggested to form in a back-arc basin (Sun et al., 1988). Cambrian-Ordovician radiolaria and Paleozoic crinoidea fossils in chert have been reported in the Erlangping and Danfeng Groups (Zhang and Tang, 1983; Xiao, 1988; Niu et al., 1993; Cui et al., 1995; Wang et al., 1995).

Plagiogranites and younger plutonic assemblages, including norite, gabbro, diorite, tonalite and granodiorite, intruded the Danfeng Group and the Qinling Complex (Li et al., 1989, 1993; Lerch et al., 1995; Xue et al., 1996a). Zircon $^{207}Pb/^{206}Pb$ evaporation ages of 487–470 Ma (Early Ordovician) have been reported for plagiogranites emplaced into the Erlangping and Danfeng metabasalt (Sun and Yu, 1991; Xue et al., 1996a). Consistent with geological relationships, the Sm-Nd isochron ages (Li et al., 1989, 1993) and zircon U-Pb ages of the younger granitoids are indeed younger and range from 422 to 383 Ma (Zhang et al., 1994; Lerch et al., 1995). Because of their island arc type trace element characteristic, Li et al. (1993) suggested that the formation of these younger intrusions were related to the subduction of oceanic crust. Alternatively, Xue et al. (1996a, b) suggested that they are post-collisional calc-alkaline plutons because they truncate earlier deformational and metamorphic features that were imprinted on the volcanic sequences and the older intrusive assemblage. The older intrusions (plagiogranites) were interpreted as pre-collisional calc-alkaline intrusions (Xue et al., 1996b).

Two amphibolite facies samples (QH-3, QSP-10) of the Danfeng Group and a greenschist facies pillow lava from the Erlangping Group (QXE-11) were selected for this work. Sample QH-3 is an amphibolite from Heihe in Zhouzhi County. LREE enrichment and Nb, Ti, and P depletions in QH-3 show characteristics of an island arc volcanic rock (Sun, 1994; Sun et al., 1995). Sample QSP-10 is a sheared amphibolite from Puyu in Shangxian County. A pillow lava (QXE-11), about 40 cm long and 20 cm wide, with clear zonal structure, was sampled from Erlangping in Xixia County. The sample was cut into six sub-samples marked QXE-11-1–6 along the long axis after the removal of weathered parts. Sample QXE-11-3 was further cut into three sections marked QXE-11-31, -32 and -33. They show IAB characteristics in Ba-Ba/Nb, La-La/Nb and Nb-Nb/Th diagrams with Nb, Ti, P negative anomalies and LREE and Th, Ba enrichments (Sun, 1994). They were selected to address the evolution of the North Qinling Arc using different isotopic geochronometers.

3 Methods

Pure mineral separates of amphibole and plagioclase from sample QSP-10 and amphibole from sample QH-3 were obtained using heavy liquid and a magnetic separator followed by handpicking under a binocular microscope. All mineral separates were washed using purified 1:1 HCl and H_2O. Pillow lava samples (QXE-11-1–6) were first cut into slices of ca. 1 cm thick, abraded and washed with purified water to avoid potential contamination and then dried. These slices were then covered with cotton cloth and broken into small pieces. About 200 g of fresh pieces were selected for each separate. They were soaked in 1:1 HCl for 1–2 h and washed with purified H_2O again. After drying, they were all crushed into powders of less than 200 mesh.

The Sm-Nd and Rb-Sr mineral isochron ages for QSP-10 and QH-3, as well as a Rb-Sr whole rock isochron age for QXE-11, were obtained in the Isotope Laboratory of the Modern Analysis Center of Nanjing University, Nanjing, People's Republic of China. The results are listed in Table 1. The analytical procedure for isotopic analysis is the same as described by Wang et al. (1992). The corresponding isochron diagrams are shown in Fig. 2, Fig. 3, Fig. 4. Plagioclase in sample QH-3 was significantly altered. Two-point isochron ages were obtained using amphibole and whole rock for reference. Sm, Nd concentrations and the Nd isotopic composition for sample QXE-11 were analyzed in the Isotope Laboratory at the Institute of Geology, Chinese Academy of Geological Sciences, Beijing,

People's Republic of China. No Sm-Nd isochron age was obtained for sample QXE-11. The Nd model ages of QXE-11 samples are similar to other volcanic rocks of the Erlangping Group (ca. 1000 Ma, Zhang et al., 1994).

Table 1 Sm, Nd, Rb, Sr contents and Nd, Sr isotope compositions of amphibolites from Puyu (QSP-10) and Heihe (QH-3) of the Danfeng Group, and pillow lava (QXE-11) of the Erlangping Group. Sm, Nd contents and Nd isotope composition for sample QXE-11 were analyzed in the Isotope Laboratory, Institute of Geology, Chinese Academy of Geological Science (Beijing). Other samples were analyzed in the Isotope Laboratory at the Modern Analysis Center of Nanjing University. The total blanks are 40 pg and 70 pg for Sm and Nd, respectively, and the $^{143}Nd/^{144}Nd$ value for the La Jolla standard is 0.511860±6 during this work

Sample no.		Rb(ppm)	Sr(ppm)	$^{87}Rb/^{86}Sr$	$^{87}Sr/^{86}Sr$	Sm (ppm)	Nd (ppm)	$^{147}Sm/^{144}Nd$	$^{143}Nd/^{144}Nd$
Amphibolite from Puyu									
QSP-10	Whole rock	18.92	229.1	0.2354	0.710578±26	1.587	5.529	0.1736	0.512809±18
	Amphibole	133.5	49.28	1.052	0.715342±18	1.219	3.591	0.2054	0.512890±14
	Plagioclase	20.54	475.5	0.1231	0.709912±22	0.3834	1.407	0.1648	0.512775±22
Amphibolite from Heihe									
QH-3	Whole rock	27.14	452.3	0.1711	0.709338±16	4.339	18.84	0.1393	0.512697±19
	Amphibole	35.47	505.4	0.2001	0.709509±13	3.878	12.28	0.1911	0.512837±14
Pillow lava basalt from the Erlangping Group									
QXE-11-1	Whole rock	24.57	161.9	0.4235	0.708021±22	3.542	17.64	0.1215	
QXE-11-2	Whole rock	19.35	188.7	0.2907	0.707204±18	3.455	17.48	0.1192	0.512557±5
QXE-11-31	Whole rock	18.97	103.9	0.5203	0.708514±24	3.379	17.12	0.1194	0.512568±10
QXE-11-32	Whole rock	16.84	130.3	0.3683	0.707651±10	3.668	18.54	0.1196	0.512573±13
QXE-11-33	Whole rock	20.95	148.3	0.4026	0.707793±8	3.464	17.46	0.12	0.512582±18
QXE-11-4	Whole rock	15.67	153.4	0.2912	0.707144±16	3.392	16.34	0.1256	0.512552±10
QXE-11-5	Whole rock	7.146	129.3	0.1575	0.706484±14	3.445	17.57	0.1186	0.512458±17
QXE-11-6	Whole rock	7.493	220.8	0.0968	0.706043±20	4.268	20.97	0.1231	0.512524±10

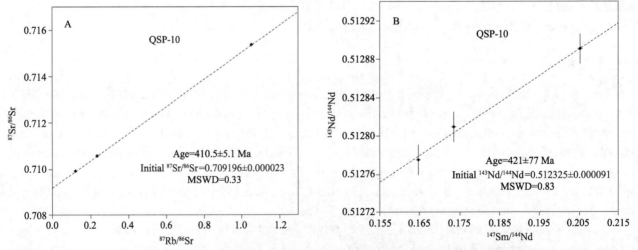

Fig. 2 Rb-Sr, Sm-Nd mineral isochron diagrams of an amphibolite sample from Puyu, QSP-10 (a, b), Danfeng Group, North Qinling Area.

A pure mineral separate of amphibole from sample QH-3 was also dated by the $^{40}Ar/^{39}Ar$ method at the Isotope Laboratory, Institutes of Geology, Chinese Academy of Geological Sciences, Beijing, People's Republic of China. Analytical results for the $^{40}Ar/^{39}Ar$ step heating experiment are listed in Table 2 while the corresponding apparent age spectrum and isochron defined by the four high temperature steps are shown in Fig. 4.

Isochron ages were calculated using Isoplot 2.0 provided by Ludwig. The input errors were 1% for ^{87}Rb/^{86}Sr, 0.3% for ^{147}Sm/^{144}Nd and 0.003% for ^{87}Sr/^{86}Sr and ^{143}Nd/^{144}Nd ratios.

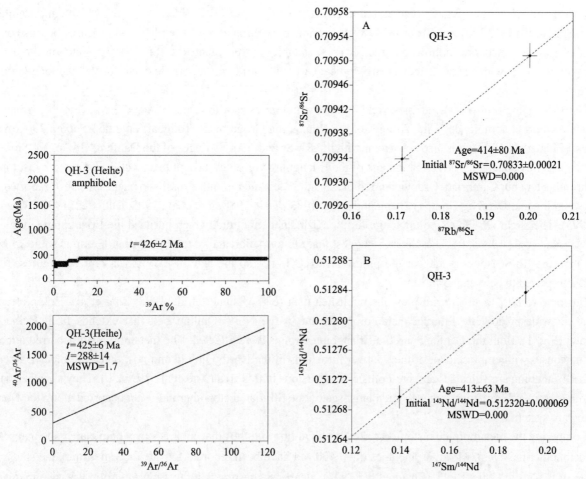

Fig. 3 ^{40}Ar/^{39}Ar age spectra for amphibole from sample QH-3 from Heihe, Danfeng Group. The four high temperature steps, containing 88.7% of the total ^{39}Ar released, define a linear array on a correlation plot of ^{40}Ar/^{36}Ar vs. ^{39}Ar/^{36}Ar. It gives an isochron age of 425±6 Ma, similar to the plateau age, with an atmospheric ^{40}Ar/^{36}Ar value of 288±14.

Fig. 4 Rb-Sr, Sm-Nd mineral isochron diagrams of an amphibolite sample from Heihe, QH-3 (a, b), the Danfeng Group North Qinling Area.

Table 2 Ar isotopic analysis by step heating experiment on amphibole of the Heihe amphibolite (weight=80 mg, J=0.01835). The correction factors for the interfering isotopes of potassium and calcium are: $(^{39}Ar/^{37}Ar)_{Ca}$=8.06×10^{-4}; $(^{40}Ar/^{37}Ar)_K$=4.78×10^{-3}; $(^{36}Ar/^{37}Ar)_{Ca}$=2.4×10^{-4}

T(℃)	$(^{40}Ar/^{39}Ar)_m$	$(^{36}Ar/^{39}Ar)_m$	$(^{37}Ar/^{39}Ar)_m$	$^{40}Ar^a/^{39}Ar$	^{39}Ar (mol) ×10^{-14}	Apparent age (Ma, 2σ)	^{39}Ar cumulative(%)
620	136.9	0.08	3.223	113.8	1.5	2034±254	0.4
780	23.29	0.0473	5.61	9.754	6.13	297±48	2.2
965	20.8	0.0356	1.423	10.38	15.43	314±30	6.7
1055	17.86	0.02	4.607	12.3	16.46	367±10	11.6
1175	18.28	0.0176	17.59	14.52	69.9	426±7	32.1
1215	16.92	0.0129	18.04	14.59	122.23	428±6	67.9
1295	18.92	0.0199	16.27	14.37	80.4	422±8	91.5
1400	21.59	0.0286	17.32	14.5	28.72	428±12	100

a Represents radiogenic ^{40}Ar.

4 Results and discussion

Sample QSP-10 yields a Rb-Sr isochron age of 411±5 Ma, a result somewhat younger than its Sm-Nd isochron age is 421±77 Ma. The large error for the latter is possibly because of the limited Sm, Nd fractionation between plagioclase and amphibole (Fig. 2). The amphibole $^{40}Ar/^{39}Ar$ age of QH-3 is 426±2 Ma (Fig. 3). The uncertainties for the Sm-Nd (413±63 Ma) and Rb-Sr (414±80 Ma) mineral isochron ages of sample QH-3 are large because only amphibole mineral separate and whole rock were available for the analyses (Fig. 4). In spite of the large uncertainties for some of the ages, they clearly indicate a Late Silurian metamorphic event for the Danfeng Group, the North Qinling.

The Erlangping pillow lava samples yield a Rb-Sr whole rock isochron age of 405±22 Ma (Fig. 5), which is consistent with the hornblende $^{40}Ar/^{39}Ar$ ages of the Erlangping Group in the Tongbai Mountains, about 300 km to the east (Zhai et al., 1998). It is also consistent with the Rb-Sr and $^{40}Ar/^{39}Ar$ ages of the Danfeng Group. No Sm-Nd isochron age was obtained. The formation age of the Erlangping volcanics has not been well constrained by isotopic geochronometers but Cambrian–Ordovician radiolaria and Paleozoic crinoidea fossils in chert have been reported in the Erlangping and Danfeng Groups (Zhang and Tang, 1983; Xiao, 1988; Niu et al., 1993; Cui et al., 1995; Wang et al., 1995). The zircon Pb–Pb evaporation age for the Xizhuanhe granitoid, which intruded the Erlangping Group, is 480 Ma (Xue et al., 1996a, b). These results suggest that the formation age of the Erlangping Group is some 75 Ma older than the Rb-Sr isochron age for the pillow lava (QXE-11), so the age of 405±22 Ma represents a resetting event, interpreted to be metamorphism.

Our new results, together with previous published data (e.g. Mattauer et al., 1985; Zhai et al., 1998), clearly indicate a widespread Late Silurian metamorphic event in the North Qinling belt that extends from Heihe to Xinyang (Fig. 1). Both the Danfeng and the Erlangping Groups were affected. The metamorphism occurred almost simultaneously with emplacement of the younger intrusions in the North Qinling and post-dated the older Danfeng and the Erlangping arc volcanics and pre-collisional intrusions in this area (Xue et al., 1996a, b). They, together with the young group of intrusions, suggest that a large scale Late Silurian tectonothermal event occurred along the North Qinling arc system.

Comparing the metamorphic histories of the North Qinling and the adjacent areas may help clarify the nature of this tectonothermal event. Metamorphic ages of ~400 Ma have been reported for the Taowan Group, the southern margin of the North China Block (Zhang et al., 1994), clearly suggesting that the Taowan Group was also involved in the Late Silurian event that affected the North Qinling belt. In contrast, no Late Silurian metamorphic or magmatic ages have been reported in the South Qinling belt and the northern margin of the South China Block, south of the Shangdan suture. The ages of the earliest metamorphic event reported in the northern margin of the South Qinling belt are $^{40}Ar/^{39}Ar$ ages of 314 Ma for biotite from the Liuling Group (Mattauer et al., 1985) and 304–316 Ma for hornblende from the Xinyang Group (Zhai et al., 1998), the eastern equivalent of the Liuling Group. They indicate that the South Qinling was not involved in the Late Silurian event. In other words, the Late Silurian tectonothermal event did not occur along the Shangdan suture (boundary between the NCB and SCB) as suggested by some authors (e.g. Ren et al., 1991; Xu et al., 1991; Zhang et al., 1991) but rather occurred between the North Qinling arc system and the North China Block.

It is also worth mentioning that no equivalent Late Silurian metamorphism has yet been confirmed for the relatively thoroughly studied east Dabie Mountains and Sulu terrane, where the UHPM rocks are exposed. This suggests that the Late Silurian tectonothermal event is likely to have been restricted to the Qinling and Tongbai (Zhai et al., 1998) areas, thereby having little to do with the collision between the NCB and SCB.

The metamorphic temperatures and pressures of these Late Silurian metamorphic rocks are low, ranging from upper greenschist to amphibole facies. This indicates that the Late Silurian tectonothermal event only led to small scale uplift. If this small scale uplift was caused by the collision between the NCB and SCB, there is a problem in explaining how the UHPM rocks were exhumed from more than 100 km depth by an intraplate collision, ca. 200 Ma after the original collision, as suggested by some authors (Mattauer et al., 1985; Xu et al., 1991; Zhang et al., 1997b).

All this evidence indicates that the Late Silurian tectonothermal event was not caused by the collision between the NCB and SCB. Instead, it is very likely to be related to the collision between the North China Block and the North Qinling arc system as a result of the closure of the small North Qinling back-arc basin(s).

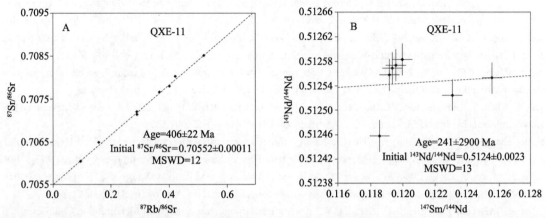

Fig. 5　Rb-Sr, Sm-Nd mineral isochron diagrams of pillow lava samples from Erlangping, QXE-11 (a, b), the Erlangping Group North Qinling Area.

5　Conclusions

Rb-Sr, Sm-Nd and $^{40}Ar/^{39}Ar$ ages of metamorphic rocks of Danfeng and Erlangping Groups indicate a widespread Late Silurian tectonothermal event in the North Qinling area. The consistence of different isotopic ages of greenschist to low-amphibole facies metamorphic rocks suggests a limited uplift in the North Qinling occurred within a short period. This event is restricted to the North Qinling arc system and the southern margin of North China Block, but not observed in the South Qinling belt and the northern margin of the South China Block. Consequently, the Late Silurian tectonothermal event does not represent the collision between the North and the South China Blocks. Instead, it is related to the collision between the North China Block and the North Qinling arc system (Fig. 1).

Acknowledgements

This is supported by the Chinese Natural Science Foundation (Grant No. 49472144). The major part of this work was done during WS's Master thesis. WS is grateful to Dr Shen-su Sun, for his constant encouragement and many constructive suggestions. Guidance and many constructive reviews from Prof. Borming Jahn and an anonymous reviewer are highly appreciated. We thank the editor and Dr C. Allen for their help that resulted in the improvement of this contribution.

References

Ames, L., Zhou, G., Xiong, B., 1996. Geochronology and geochemistry of ultrahigh-pressure metamorphism with implications for collision of the Sino-Korean and Yangtze cratons, central China. Tectonics 15, 472-489.

Chavagnac, V., Jahn, B.M., 1996. Coesite-bearing eclogites from Bixiling complex, Dabie Mountains, China: Sm-Nd ages, geochemical characteristics and tectonics implications. Chem. Geol. 133, 29-51.

Cui, Z.L., Sun, Y., Wang, X.R., 1995. The discovery of radiolaria from Danfeng ophiolite in Qingling Mountains and its geological significance. Chin. Sci. Bull. 40 (18), 1686-1688.

Hacker, B.R., Ratschbacher, L., Webb, L., Dong, S., 1995. What brought them up? Exhumation of the Dabie Shan ultrahigh-pressure rocks. Geology 23, 743-746.

Hacker, B.R., Ratschbacher, L., Webb, L., Ireland, R., Walker, D., Dong, S., 1998. U/Pb zircon ages constrain the architecture of the ultrahigh-pressure Qinling-Dabie Orogen, China. Earth Planet. Sci. Lett. 161, 215-230.

Jahn, B. M., Cornichet, J., Cong, B., Yui, T. F., 1995. Ultrahigh ε_{Nd} eclogites from an ultrahigh-pressure metamorphic terrane of China. Chem. Geol. 127, 61-79.

Lerch, M.F., Xue, F., Kroner, A., Zhang, G.W., Todt, W., 1995. A middle Silurian-early Devonian magmatic arc in the Qinling

Mountains of central China. J. Geol. 103, 437-449.

Li, S., Sun, W., 1996. A middle Silurian-early Devonian magmatic arc in the Qinling Mountains of central China: a discussion. J. Geol. 104, 501-503.

Li, S., Hart, S.R., Zheng, S., Liou, D., Zhang, G., Guo, A., 1989. Timing of collision between the North and South China Blocks—the Sm-Nd isotopic age evidence. Sci. Sin. (Ser. B) 32, 1391-1400.

Li, S., Xiao, Y., Liou, D., Chen, Y., Ge, N., Zhang, S., Sun, S. S., Cong, B., Zhang, R., Hart, S.R., Wang, S., 1993. Collision of the North China and Yangtze Blocks and formation of coesite-eclogites: timing and processes. Chem. Geol. 109, 89-111.

Li, S., Chen, Y., Zhang, Z.Q., Ye, X., Zhang, G.W., Hart, S.R., 1994a. Trace elements and Sr, Nd isotope geochemistry of the Lajimiao norite-gabbro from the North Qinling belt. Acta Geol. Sin. 67, 137-151 in Chinese with English abstract.

Li, S., Wang, S., Chen, Y., Liou, D., Qiu, J., Zhou, H., Zhang, Z., 1994b. Excess argon in phengite from eclogite: evidence from dating of eclogite minerals by Sm-Nd, Rb-Sr and ^{40}Ar-^{39}Ar methods. Chem. Geol. 112, 343-350.

Li, S., Sun, W., Zhang, G., Chen, J., Yang, Y., 1996. Chronology and geochemistry of metavolcanic rocks from Heigouxia Valley in the Mian-Lue tectonic arc, South Qinling—evidence for a Paleozoic oceanic basin and its close time. Sci. China 39 (3), 300-310.

Li, S., Jagoutz, E., Chen, Y., Li, Q., 2000. Sm-Nd and Rb-Sr isotopic chronology and cooling history of UHPM rocks and their country rocks at Shuanghe in the Dabie Mountains, central China. Geochim. Cosmochim. Acta 64 (6), 1077-1093.

Liu, G.H., Zhang, S.G., You, Z.D., Suo, S.T., Zhang, G.W., 1993. Metamorphic History of Main Metamorphic Complex in the Qinling Orogenic Belt. Geological Publishing House, Beijing p. 190; in Chinese.

Mattauer, M., Matte, P., Malavieille, J., Tapponnier, P., Maluski, H., Ku, Z., Lu, Y., Tang, Y., 1985. Tectonics of the Qinling belt: build-up and evolution of eastern Asia. Nature 317, 496-500.

Meng, Q.R., Zhang, G.W., 1999. Timing of collision of the North and South China Blocks: controversy and reconciliation. Geology 27 (2), 123-126.

Meng, Q.R., Zhang, G.W., 2000. Geologic framework and tectonic evolution of the Qinling orogen, central China. Tectonophysics 323, 183-196.

Niu, B., Liu, Z., Ren, J., 1993. The tectonic relationship between the Qinling Mountains and Tongbai-Dabie Mountains with notes on the tectonic evolution of the Hehuai Basin. Bull. Chin. Acad. Geol. Sci. 26, 1-12 in Chinese with English abstract.

Okay, A.I., Sengör, A.M.C., 1992. Evidence for intracontinental thrust-related exhumation of the ultra-high-pressure rocks in China. Geology 20, 411-414.

Ren, J.S., Zhang, Z.K., Niu, B.G., Liu, Z.G., 1991. On the Qinling orogenic belt—integration of the Sino-Korean and Yangtze Blocks. In: Ye, L.J., Qian, L.X., Zhang, G.W. (Eds.). A Selection of Papers Presented at the Conference on the Qinling Orogenic Belt. Northwest University Press, Xi'an, pp. 99-110 in Chinese with English abstract.

Rowley, D.B., Xue, F., Tucker, R.D., Peng, Z.X., Baker, J., Davis, A., 1997. Age of ultrahigh pressure metamorphism and protolith of orthogneisses from the eastern Dabie Shan: U/Pb zircon geochronology. Earth Planet. Sci. Lett. 151, 191-203.

Sun, W., 1994. The evolution history of Qinling orogenic belt: evidence from ophiolite and granites. MSc Thesis. University of Science and Technology of China, pp. 1-104 in Chinese with English abstract.

Sun, Y., Yu, Z.P., 1991. A discussion on an ancient ocean and Caledonian orogen in the East Qinling. In: Ye, L.J., Qian, L.X., Zhang, G.W. (Eds.). A Selection of Papers Presented at the Conference on the Qinling Orogenic Belt. Northwest University Press, Xi'an, pp. 167-173 in Chinese with English abstract.

Sun, Y., Yun, Z.P., Zhang, G.W., 1988. Geochemistry of ophiolites from east Qinling area. In: Zhang, G.W. (Ed.). The Formation of and Evolution of Qinling Orogenic Belt. Northwest University Press, Xi'an, pp. 65-74.

Sun, W., Li, S., Xiao, Y., Sun, Y., Zhang, G., 1995. Discovery of Island Arc basalts in Danfeng Group and its tectonic significance. Geotectonica Metalogenia 19 (3), 227-236 in Chinese with English abstract.

Wang, Y., Yang, J., Guo, L., 1992. Discovery of the early Proterozoic granite in the Longquan, Zhejiang Province and the basement time of this area. Geol. Rev., 525-530 in Chinese.

Wang, X.R., Hua, H., Sun, Y., 1995. Study on micro-fossils of Erlangping Group from Taowan area Xixia county Henan Province. Acta North West Univ. (Nature Sci. Eds) 25 (4), 353-358 in Chinese with English abstract.

Xiao, S.Y., 1988. North Qinling Metamorphic Strata. Jiaotong University Press, Xi'an p. 47; in Chinese with English abstract.

Xu, Z.Q., Niu, B.G., Liu, Z.G., Wang, Y.M., 1991. Tectonic system and intracontinental plate dynamic mechanism in the Qinling-Dabie collision-intracontinental mountain chains. In: Ye, L.J., Qian, L.X., Zhang, G.W. (Eds.). A Selection of Papers Presented at the Conference on the Qinling Orogenic Belt., pp. 139-147 in Chinese with English abstract.

Xue, F., Kroner, A., Reischmann, T., Lerch, F., 1996a. Paleozoic pre- and post-collision calc-alkaline magmatism in the Qinling orogenic belt, central China, as documented by zircon ages on granitoid rocks. J. Geol. Soc. 153, 409-417.

Xue, F., Lerch, M.F., Kroner, A., Reischmann, T., 1996b. Tectonic evolution of the east Qinling Mountains, China, in the Paleozoic: a review and new tectonic model. Tectonophysics 253, 271-284.

Yin, H.F., Huang, D.H., 1995. The Early Palaeozoic Zhen'an-Xichuan Block and the evolution of the small Qinling Archipelagic ocean basin. Acta Geol. Sin. 69 (3), 193-204 in Chinese with English abstract.

Yin, A., Nie, S., 1993. An indentation model for the north and south China collision and the development of the Tanlu and Honam Fault systems, eastern Asia. Tectonics 12, 801-813.

Yin, H.F., Wu, S.B., Du, Y.S., Peng, Y.Q., 1999. The South China is part of the Tethys multi-island ocean system. Earth Sci.-J. China Univ. Geosci. 24 (1), 1-11 in Chinese.

Zhai, X., Day, H.W., Hacker, B.R., You, Z.D., 1998. Paleozoic metamorphism in the Qinling orogen, Tongbai Mountains, central China. Geology 26, 371-374.

Zhang, S.C., Tang, S.W., 1983. The discovery of early Paleozoic radiolarian in North Qinling and plate tectonics. Shaanxi Geol. 2, 1-9 in Chinese with English abstract.

Zhang, G.W., Zhou, D.W., Yu, Z.P., Gu, A.L., Cheng, X.Y., Li, T.H., Zhang, C.L., Xue, F., 1991. Characteristics of the lithospheric composition, structure and evolution of the Qinling orogenic belt. In: Ye, L.J., Qian, L.X., Zhang, G.W. (Eds.). A Selection of Papers Presented at the Conference on the Qinling Orogenic Belt. Northwest University Press, Xi'an, pp. 121-138 in Chinese with English abstract.

Zhang, Z.Q., Liu, D.Y., Fu, G.M., 1994. Isotopic Geochronology of Metamorphic Rocks in the North Qinling Belt. Geological Press, Beijing p. 231.

Zhang, G.W., Meng, Q.R., Lai, S., 1995a. Structure and tectonics of the Qinling orogenic belt. Sci. China Ser. B 38, 1379-1394.

Zhang, Q., Han, S., Zhang, Z.Q., 1995b. Trace element and isotopic geochemistry of metabasalts from Danfeng Group (DFG) in Shangxian-Danfeng area, Shaanxi Province. Acta Petrol. Sin. 11 (1), 43-54.

Zhang, G.W., Yuan, X.C., Zhang, B.R., Chen, J.Y., Xiao, Q.H., Meng, Q.R., Sun, Y., Lu, X.X., Zhang, C.L., Yu, Z.P., Zhou, D.W., 1996. The Atlas of Qinling Orogenic Belt. Science Press, Beijing.

Zhang, H.F., Gao, S., Zhang, B.R., Luo, T.C., Lin, W.L., 1997a. Pb isotopes of granitoids suggest Devonian accretion of Yangtze (South China craton to North China craton). Geology 25, 1015-1018.

Zhang, H.F., Zhang, B.R., Ling, W.L., Gao, S., Ouyang, J.P., 1997b. Late Proterozoic crustal accretion of South Qinling: Nd isotopic study from granitic rocks. Geochimica 26, 16-24 in Chinese with English abstract.

Timing of synorogenic granitoids in the South Qinling, Central China: constraints on the evolution of the Qinling-Dabie orogenic belt *

Weidong Sun[1,2], Shuguang Li[1], Yadong Chen[3] and Yujing Li[4]

1. Department of Earth and Space Sciences, University of Science and Technology of China, Hefei 230026, Anhui, China
2. Current address: Research School of Earth Science, Australian National University, Canberra, Australian Capital Territory 0200, Australia
3. Geochronology Lab, Royal Ontario Museum, 100 Queen's Park, Toronto, Ontario M5S 206, Canada
4. Regional Geological Survey Team of Shaanxi Province, Xianyang 712009, Shaanxi, China

亮点介绍：锆石 U-Pb 同位素定年结果表明，南秦岭广泛存在 220±1–205±1 Ma 花岗岩，与华北与华南陆块之间的碰撞时代一致，属同碰撞花岗岩。

Abstract The ca. 400-km-long granitoid belt in the South Qinling is believed to be a synorogenic product of the collision between the North and South China Blocks along the Qinling-Dabie orogenic belt in central China. Single and multigrain zircon U-Pb dating of six of these granitoid bodies indicate that the granitoids were formed between 220±1 and 205±1 Ma, supporting the idea that the collision between the North and South China Blocks happened in the Triassic. The formation of these synorogenic granitoids in the South Qinling is similar in time but slightly later than the rapid exhumation of the ultrahigh-pressure metamorphic rocks and their country rocks in the eastern part of the Qinling-Dabie orogen, suggesting a close relationship between the rapid exhumation of the ultrahighpressure metamorphic rocks in the Dabie Mountains and the formation of the South Qinling granitoids. According to regional geology, the breakoff of a subducted slab, if there was any, should have occurred at a shallower depth in the South Qinling compared with that in the Dabie Mountains and Sulu terrane. This shallow breakoff is likely to have disturbed the asthenosphere seriously, leading to the melting of the overlaying lithosphere as well as the formation of the granitoid belt.

1 Introduction

The eastern Asian continent is separated into the North and the South China Blocks by the Qinling-Dabie orogenic belt. Accurate knowledge of the structure and history of the Qinling-Dabie orogenic belt is important to place firmer constraints on the evolution of this continent (e.g., Mattauer et al. 1985). A notable phenomenon in the tectonics of East Asia is the surface exposure of ultrahigh-pressure metamorphic rocks, such as coesite- and diamond-bearing eclogites in the Dabie-Sulu orogen, in the eastern part of the Qinling-Dabie orogenic belt. These ultrahigh-pressure metamorphic rocks were formed at depths of >100 km and temperatures ranging from 650 to 800 ℃ in the Triassic (e.g., Wang et al. 1989; Okay and Sengör 1992; Xu et al. 1992, 1994; Yin and Nie 1993; Yin 1994; Li et al. 1989a, 1992, 1993, 1994, 1996, 2000; Ernst and Liou 1995; Jahn et al. 1996; Hacker et al. 1998; Li 1994, 1998; Zheng et al. 1998). One of the most significant characteristics of the Qinling-Dabie orogenic belt is the contrast between the east (Dabie Mountains and Sulu terranes) and the west (Tongbai and Qinling). Triassic diamondand coesite-bearing ultrahigh-pressure metamorphic rocks occur only in the eastern part of this belt, the Sulu terrane, and Dabie Mountains. Triassic blueschist facies rocks occur in the Dabie-Sulu

---
* 本文发表在：The Journal of Geology, 2002, 110: 457-468

orogen, the Tongbai Mountains, and in the eastern part of the South Qinling (e.g., Wu 1980; Mattauer et al. 1985; Yin et al. 1991; Zhang et al. 1992; Wang et al. 1995, 1996). The western part of the South Qinling area is dominated by greenschist facies metamorphism. Except for the blueschist in the Bikou terrane, no Triassic ultrahigh-pressure metamorphic rocks have been reported. However, no synorogenic granitoids are reported in the eastern part, where the ultrahigh-pressure metamorphic rocks are exposed. In contrast, there is a ca. 400-km-long granitoid belt to the north of the Mianlue suture, roughly parallel to the Mianlue suture in the west and restricted to the north of the Daheba-Ningshan-Shanyang fault in the east Fig. 1). According to regional geological observations and detailed petrological work, this granitoid belt is a "syncollision" product (e.g., Zhang et al. 1994).

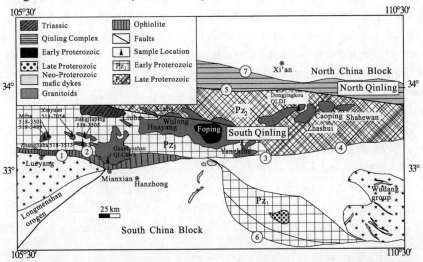

Fig. 1 Sketched geological map of the South Qinling showing the locations of the South Qinling granitoids (simplified after G. W. Zhang et al., 1996b). Also shown are major faults in the Qinling area indicated by circled numbers. 1= Mianlue suture, 2= Zhuangyuanbei fault, 3= Daheba-Ningshan fault, 4= Shayang fault, 5= Shandan suture, 6= Dabashan fault, 7= Machaoying fault.

The formation age of these South Qinling granitoid bodies is one of the most important potential constraints on the evolution of the Qinling-Dabie orogenic belt. Rb-Sr whole-rock isochron dating on three of the South Qinling granitoid intrusions suggested that they were formed from 264 to 323 Ma, in the Paleozoic (Shang and Yan 1988; B.R. Zhang et al. 1991, 1994). These Paleozoic ages for the South Qinling granites are one of the key points in the controversy on the timing of the collision between the North and the South China Blocks (e.g., B.R. Zhang et al. 1991, 1994; G. W. Zhang et al. 1991, 1995, 1996a; Li et al. 1993, 1994, 1996, 2000; Li and Sun 1996; Meng and Zhang 1999, 2000). Because of the heterogeneity of Sr isotopes in granitic intrusions caused by the effects of petrogenetic processes (such as mixing and assimilation), wholerock Rb-Sr isochron methods are usually not suitable for granite dating. Moreover, no detailed isotopic data of these three Rb-Sr whole-rock isochron ages are available to assess their reliability and geological significance. More detailed dating on the South Qinling granitoids is urgently needed to solve the controversial issues involving the evolution of the Qinling-Dabie orogenic belt.

The conventional U-Pb dating technique for zircons is a high-precision geochronological method that helps in the temporal resolution of short-lived tectonic and magmatic events in ancient orogenic belts (e.g., Schaltegger et al. 1996). This article reports the results of zircon U-Pb dating applied to six of the South Qinling granitoids. These ages will help achieve a better understanding on the evolution of the Qinling-Dabie orogenic belt.

2 Regional Geology and Samples

The Qinling orogenic belt, the western counterpart of the Dabie ultrahigh-pressure metamorphic belt (Zhang et al. 1995, 1996a; Li et al. 1996; Meng and Zhang 1999), is a multistage orogenic belt with two mountain chains and two sutures. The north suture lies mainly along the Shangdan fault, whereas the south suture, according to some authors, consists of the Mianlue ophiolite complex belt in the west and presumably the Bashan fault in the east (Zhang et al. 1996a, 1996b; Meng and Zhang 1999). North to the Shangdan fault is the North Qinling belt, a middle Paleozoic orogen (Mattauer et al. 1985) with widespread Paleozoic magmatism and metamorphism (Li et al. 1989b;

Lerch et al. 1995; Li and Sun 1996; Xue et al. 1996a, 1996b; Sun and Li 1998; Zhai et al. 1998). Between the Shangdan and Mianlue belts is the South Qinling, a later Paleozoic or Triassic orogen (Mattauer et al. 1985) with abundant Triassic granitoids and late Paleozoic metamorphism (Sun 1994; Li and Sun 1996; Li et al. 1996; Meng and Zhang 1999). South to the Mianlue belt is the South China Block (Fig. 1).

The discovery of the southern suture in the South Qinling area has been one of the most important advances in Qinling-Dabie research (Zhang et al. 1995, 1996a; Li et al. 1996; Meng and Zhang 1999) since the discovery of Mianlue tectonic melange zone. This discovery triggered a new tectonic model of the Qinling orogenic belt as a multicollision belt (e.g., Zhang et al. 1995, 1996a, 1996b; Li and Sun 1996; Li et al. 1996; Sun and Li 1998; Meng and Zhang 1999) instead of the previously accepted simple model of collision between two continents.

The Mianlue tectonic melange zone is separated from the Precambrian Bikou Group and Silurian Baishuijiang Group by the Mianlue fault in the south and the Zhuangyuanbei fault in the north, respectively (Fig. 1). It mainly consists of Sinian Cambrian and Devonian-Carboniferous sheared thrust slabs. The Sinian-Cambrian beds consist mainly of gravel-bearing argillite, argillitic and pyroclastic rocks, and carbonates. The Devonian System consists of turbidite, carbonates, argillite, quartzite, and quartzose sandstone, while the Carboniferous rocks consist mainly of carbonates with shales. In contrast, there are virtually no Ordovician and Silurian beds within the Mianlue tectonic zone (e.g., Meng et al. 1996; Meng and Zhang 1999).

Ophiolite complex and volcanic rocks appear in the Mianlue zone as tectonic slabs. The ophiolite complex mainly consists of strongly sheared metabasalts, accumulate gabbros, ultramafic rocks, and radiolarian cherts (Meng and Zhang 1999). Studies suggest that some of the metabasalts exhibit geochemical features similar to normal mid-ocean ridge basalt (Lai and Zhang 1996; Meng and Zhang 1999) and were metamorphosed in the early Triassic (242–220 Ma; Li et al. 1996).

The South Qinling granitoid belt occurs north of the ophiolite belt, roughly parallel to the Mianlue suture in the west, and goes between the Daheba-Ningshan-Shanyang fault and the Shangdan suture in the east (Fig. 1). This granitoid belt is ca. 400 km long and covers an area of ca. 6000 km^2. Granitoids within the belt have been classified into three suites: the Guangtoushan suite, the Wulong suite, and the Dongjiakou suite. Wulong suite rocks, including the Huayang, Wulong, and Yanzhiba plutons, are mostly S-type granitoids with low Na_2O, high K_2O/Na_2O, and $\delta^{18}O$ ranging from 10.4‰ to 13.0‰ (Zhang et al. 1994). Accessory minerals, such as ilmenite, monazite, and garnet, are abundant. The Dongjiangkou suite includes the Dongjiangkou, Zhashui, Caoping, and Shahewan plutons. They have been classified as I-type granitoids with high Na_2O, low K_2O/Na_2O, and $\delta^{18}O$ ranging from 4.1‰ to 9.0‰ (Zhang et al. 1994). Their initial $(^{86}Sr/^{87}Sr)_i$ ratios are lower than 0.706. Accessory minerals are mainly magnetite and sphene. The Zhashui granite of this group yields a Rb-Sr wholerock isochron age of 264 Ma (Shang and Yan 1988). The Guangtoushan suite includes Miba, Xinyuan, Jiangjiaping, Zhangjiaba, Guangtoushan, Liuba, and Xiba granitoids. Guangtoushan and Xiba granites yield Rb-Sr whole-rock ages of 323±14 Ma (Zhang et al. 1994) and 285 Ma (Shang and Yan 1988), respectively. Their accessory minerals are complicated, varying from magnetite and sphene to ilmenite, monazite, and garnet. Based on their high Na_2O, low K_2O/Na_2O, high $\delta^{18}O$ (>10‰), and low initial $(^{86}Sr/^{87}Sr)_i$ ratios (<0.706) as well as the accessory minerals, Guangtoushan suite granitoids have been classified as a transition type between I and S-types. They have been suggested to have resulted from contamination of I-type magma by continental crust at a shallow depth (Zhang et al. 1994). For the Guangtoushan and Wulong suites, three periods of intrusion have been recognized, changing from diorite, granodiorite, and quartz diorite in the first stage of intrusion to migmatitic granodiorite and adamellite in the second stage and adamellite and granite in the third stage (Yan 1985). The samples analyzed in this work are collected from the Guangtoushan suite (i.e., Zhangjiaba, Miba, Xinyuan, Guangtoushan, and Jiangjiaping granitoids) and Dongjiangkou granite (see Fig. 1).

3 Analytical Procedures

Samples collected and crushed for zircon concentrations ranged from a few to ca. 40 kg. Mineral separations were achieved using conventional methods involving Wilfley table, heavy liquids, and magnetic separation. Caution was exercised during every step to avoid cross-sample contamination. The zircon separates concentrated from the nonmagnetic fraction of each sample were carefully studied under a binocular microscope. Among a particular zircon population, only crystals that lacked visible cracks and had the fewest inclusions and other imperfections were handpicked for analysis.

Zircon U-Pb dating analyses were carried out in the Geochronology Laboratory, Royal Ontario Museum, Canada. To reduce or eliminate surface-correlated recent lead loss, all zircon fractions were abraded and polished with a pyrite medium (Davis et al. 1982; Krogh 1982) and washed after abrasion in hot 4 N HNO_3. Final cleaning in a hot HNO_3 bath in 10 mL Pyrex glass beakers on a hot plate was followed by rinsing with double-distilled H_2O and acetone. Zircon dissolution and ion-exchange techniques followed Krogh (1973). The dissolution capsules and ion-exchange columns have capacities of 0.5 cm^3 and 0.05 cm^3, respectively. A mixed ^{235}U-^{205}Pb spike was used to determine the ratios and concentrations of U and Pb (Krogh and Davis 1975; Parrish and Krogh 1987). Total U blank for zircons was <0.5 pg and Pb blank was <3 pg for single zircon and multigrain fractions. Both U and Pb were loaded together on single, degassed rhenium filaments using the silica gel technique. The isotopic data were obtained using a VG354 mass spectrometer in a single Farady cup collector or integrated Daly collector mode. A Daly to Farady conversion factor of 0.40% per atomic mass unit (AMU) was used; a mass fractionation correction of 0.10% per AMU for U and Pb was applied.

Decay constants used are those of Jaffey et al. (1971) and corrections for initial common Pb in excess of laboratory blank are calculated using Stacey and Kramers's (1975) common Pb compositions. Regression calculations to linear alignments of data points were carried out according to Davis (1982) using error estimates for individual points as shown in Fig. 2. Uncertainties for ages are quoted at the 95% confidence level.

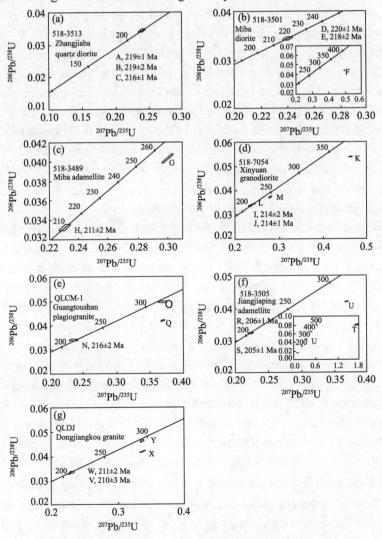

Fig. 2 U-Pb concordia diagrams for zircons from the South Qinling granitoids.

4 Results

Zircons from all samples have euhedral, short to long prismatic shapes, but they vary in abundance and size from sample to sample. No zircons show any resorption features. Results of isotopic analyses are presented in Table 1 and plotted in Fig. 2.

Zhangjiaba. The Zhangjiaba pluton is a quartz diorite intrusion (~40 km^2). About 500 zircons (40–200 μm in length) were concentrated from ~10 kg of sample 518-3513, located about 100 m south of Zhangjiaba, west of the river (lat 33° 25′ 20″ N, long 106° 19′ 40″ E). Most of the grains are colorless and high quality. Fraction A, consisting of 15 long prismatic zircons, yielded concordant U-Pb isotopic compositions with a $^{206}Pb/^{238}U$ age of 219±1 Ma. Fraction B, a short prism, gave a concordant age of 219±2 Ma. This is the likely age of the granitoid body. Fraction C, a long prismatic grain, yielded a slightly younger age of 216±1 Ma, marginally within the uncertainties of fractions A and B. This zircon may have crystallized at the same time as A and B and later lost some of its radiogenic Pb or, though less likely, represents a slightly younger event (Table 1; Fig. 2a).

Miba. The Miba pluton covers an area of ca. 200 km^2. There are three phases with coarse-grain adamellite in the center, fine-grain granodiorite in the rim, and a transition belt of adamellite to amphibole-bearing adamellite between them. Sample 518-3501 was collected from the granodioritic rim (lat 33° 30′ 30″ N, long 105° 54′ 25″ E), whereas sample 518-3489 (lat 33° 32′ 25″ N, long 105° 50′ 30″ E) was collected from the adamellitic center, both south of the Xihanshui River.

Zircons in sample 518-3501 varied in length from 50 to 200 μm. In addition to tetragonal prismatic habits, bitetragonal prisms were also common. Zircons in sample 518-3489 were similar to those in 518-3501 but generally larger (70–250 μm). In both samples, some zircons exhibited a rusty appearance.

Three analyses were undertaken for sample 518-3501. Fraction D consisted of 12 colorless, euhedral, long prismatic zircons (70–200 μm in length); a concordant age of 220±1 Ma was obtained. Fraction E contained 18 exterior tips of euhedral zircons and yielded concordant results (218±2 Ma) indistinguishable within error from that of analysis D. Since both fractions D and E consisted of multigrains, the concordant age duplicated by the two analyses must have dated an important event in the history of the rock, most likely the timing of the emplacement of the granitoid (Table 1; Fig. 2b). Fraction F contained three short prisms. They revealed an inherited component, because a $^{207}Pb/^{206}Pb$ age of 1220 Ma was obtained. An upper intercept age at ~2700 Ma was obtained when projected from 220 Ma. Since this is the age of a multigrain fraction, it does not represent the timing for any particular geological event prior to the emplacement of the granitoid.

Two fractions were analyzed from sample 518-3489 (Table 1; Fig. 2c). Fraction G, consisting of 10 short prismatic zircons, yielded discordant results with a $^{207}Pb/^{206}Pb$ age of 376±6 Ma. Fraction H, a piece of exterior tip from a short prismatic zircon, gave an age of 211±2 Ma. It represents the crystallization age of the central phase, suggesting that the magmatic event of the adamellite was later than that of the diorite (220±2 Ma). They are more likely to represent two intrusion periods than record the cooling history. When two analyses for sample 518-3489 were regressed, an upper intercept at ~820 Ma was obtained. Again, if more than one older component is represented in this multigrain fraction, it does not represent any specific geological event.

Xinyuan. The Xinyuan pluton is a granodiorite intrusion. Sample 518-7054 was collected on the east bank of the Jiugushu River at lat 33° 31′ 05″ N, long 106° 11′ 35″ E. All zircon grains (several hundred) in this sample are euhedral and colorless, and range in length from 60 to 150 μm. Tetragonal prisms are well developed; some have dipyramidal ends.

Altogether five fractions from this sample were analyzed. Two single-grain fractions (I, a long prismatic zircon, and J, a tip from a short prism) yielded concordant ages at 214±2 Ma. Fractions K, L, and M consisted of 1, 8, and 12 zircon grains, respectively. All gave discordant results with $^{207}Pb/^{206}Pb$ ages of ~682, 240, and 385 Ma. The concordant age of 214±2 Ma is likely to be the magmatic age of Xinyuan granitoid pluton. When projected from 214 Ma, the regression calculation for analyses K, L, and M yielded upper intercept ages at ~1100, ~1100, and 1300 Ma (Table 1; Fig. 2d). These three upper intercept ages are roughly consistent with each other, suggesting the involvement of basement materials (1000–1600 Ma; Zhang et al. 1995; Meng and Zhang 2000).

Table 1 U-Pb Isotopic Analyses of Zircons for Granitoids from South Qinling

Fraction, description	Weight (mg)	U (10^{-6} g)	Pb (10^{-6} g)	Pb^a (pg)	^{206}Pb/^{204}Pb^b	^{206}Pb/^{238}U	^{207}Pb/^{235}U	^{207}Pb/^{206}Pb	^{206}Pb/^{238}U (Ma;2σ)	^{207}Pb/^{235}U (Ma;2σ)	^{207}Pb/^{206}Pb (Ma;2σ)
518-3513 Zhangjiaba:											
A, 15 tips	9	133	5	3.8	723	0.03457	0.24122	0.05062	219±1	219±3	223±36
B, one short prism	3	270	9	1.6	1139	0.0346	0.24089	0.0505	219±2	219±3	218±33
C, one long prism	2	407	14	2.3	789	0.03407	0.23778	0.05062	216±1	217±3	224±29
518-3501 Miba:											
D, 12 long prisms	21	728	25	5.8	5920	0.0347	0.24189	0.05055	220±1	220±2	220±6
E, 18 tips	21	1035	41	134	381	0.03443	0.24011	0.05058	218±2	218±3	222±24
F, three short prisms	8	819	37	2.2	8563	0.04433	0.49473	0.08094	280±1	408±1	1220±5
518-3489 Miba:											
G, 10 short prisms	31	460	19	4.6	8190	0.0403	0.30073	0.05412	255±3	267±3	376±6
H, one tip	3	567	19	1.4	2559	0.03325	0.23111	0.05042	211±2	211±3	214±18
518-7054 Xinyuan:											
I, one long prism	7	437	16	3.2	2116	0.03382	0.23506	0.05042	214±2	214±2	214±16
J, one tip	2	594	19	2.5	1071	0.03374	0.23471	0.05046	214±1	214±3	216±38
K, one short prism	3	941	51	6.1	1650	0.05389	0.4624	0.06223	338±1	386±1	682±16
L, eight short prisms	11	199	7	7.4	683	0.03445	0.24215	0.05098	218±1	220±3	240±36
M, 12 long prisms	7	408	17	1.9	3579	0.0373	0.27941	0.05433	236±2	250±3	385±13
QLCM-1 Guangtoushan:											
N, one tip	2	214	8	2.9	342	0.03401	0.23681	0.0505	216±2	216±5	218±54
O, 30 prisms, UA	14	550	26	10.2	2081	0.0419	0.37011	0.06406	265±2	320±2	744±11
P, one short prism	1	268	31	2.8	729	0.11343	1.0717	0.6852	693±4	740±8	884±28
Q, one long prism	2	339	18	2.6	849	0.04956	0.37028	0.05418	312±2	320±5	379±40
518-3505 Jiangjiaping:											
R, one long prism	15	881	27	1.5	18019	0.03243	0.22466	0.05025	206±1	206±2	206±6
S, one short prism	3	358	11	3	776	0.03236	0.22438	0.05028	205±1	206±3	208±34
T, one short prism	3	415	42	1.9	3557	0.08331	1.6817	0.1464	516±2	1002±3	2304±4
U, 12 long prisms	12	205	8	3.8	1796	0.04188	0.36943	0.06398	264±1	319±2	741±10
QLDJ Dongjiangkou:											
V, seven tips	42	273	10	13.4	1856	0.03313	0.22988	0.05032	210±3	210±3	210±12
W, one tip	4	206	8	1.8	1021	0.0333	0.23131	0.05038	211±2	211±3	212±36
X, nine short prisms	40	298	13	2.7	12098	0.04169	0.33781	0.05877	263±2	296±3	559±4
Y, three long prisms	7	633	30	2.1	6272	0.04592	0.33689	5321	289±3	295±2	338±15

Note. Errors for all ages are quoted as 2σ. UA = unabraded.
^{a} Total common Pb.
^{b} Corrected for spike and fractionation only.

Guangtoushan. The Guangtoushan granite covers an area of ca. 900 km² and consists of two phases. The inner phase is mainly biotite granodiorite; the outer phase is fine-grained biotite plagiogranite (Yan 1985). Sample QLCM-1 was collected from the outer phase near Caimahe (lat 33° 14′ 45″ N, long 106° 40′ 00″ E). Zircons were scarce in this sample; only about 120 crystals were recovered from ca. 20 kg. They range in length from 40–150 μm, have a euhedral habit, and contain tiny bubble-like inclusions.

Four analyses were conducted for this sample. Fraction N consisted of a zircon tip and yielded a concordant age of 216±2 Ma. The other three analyses contained 30 (O), 1 (P), and 1 (Q) zircon grains, respectively, and all yielded discordant results with ^{207}Pb/^{206}Pb ages of ∼744, ∼884, and ∼379 Ma. When analysis N was regressed for these three analyses separately, upper intercept ages of ∼1880, ∼950, and 500 Ma were obtained. The concordant age of 216±2 Ma is likely to be the crystallization age (Table 1; Fig. 2e).

Jiangjiaping. The Jiangjiaping pluton is a finegrained adamellite intrusion. Sample 518-3505 was collected about 1 km west of Jiangjiaping village on the east bank of the river (lat 33° 30′ 15″ N, long 106° 22′ 35″ E). About 300 zircons were selected, ranging in length from 40 to 250 μm; most are less than 150 μm and some have a rusty appearance.

One multigrain and three single-grain fractions were analyzed. The first single-grain fraction (R) a large (250 μm in length; aspect ratio, 4 : 1), euhedral zircon, yielded concordant results with an age of 206±2 Ma. Single-grain analysis S produced a concordant result (205±1 Ma) indistinguishable within error from analysis R. The single-grain fraction T, a euhedral, short prismatic zircon (aspect ratio, 2 : 1), gave discordant results with a ^{207}Pb/^{206}Pb age of ca. 2304 Ma. The multigrain fraction (U) contained 12 small, euhedral zircons; its results are discordant with a ^{207}Pb/^{206}Pb age of ca. 741 Ma.

The R and S fractions with the ages of 206±1 and 205±1 Ma are likely to have grown from the granite magma without inheritance; therefore, this age represents the timing of the emplacement of this adamellitic pluton, one of the third-period intrusions of the Guangtoushan group granites. An upper intercept age of ∼2800 Ma was obtained for analysis T when a two-point regression was calculated by projecting to the concordant curve from 206 Ma. A similar calculation for analysis U yielded an upper intercept at ∼1700 Ma. Since it is a multigrain fraction, its age is likely to be an average with more than one older component. Therefore, it cannot be confidently ascribed to a specific geological event (Table 1; Fig. 2f).

Dongjiangkou. The Dongjiangkou granite covers an area of ca. 540 km². Four stages of intrusion were recognized. The Zhashui granite, part of the Dongjiangkou suite, is equivalent to the third and fourth stages of intrusion. Sample QLDJ was collected from the third-stage intrusion of the Dongjiangkou granite at lat 33° 46′ 20″ N, long 109° 10′ 00″ E. Zircons are abundant in this sample—several hundred milligrams were separated from ∼10 kg. These were all prismatic with aspect ratios varying from 2 : 1 to 4 : 1 and lengths ranging from 40 to 150 μm.

A fraction (V) composed of seven zircon tips (60–150 μm) yielded concordant results of 210±3 Ma. A concordant age (211±2 Ma) identical within error to 210±3 Ma was given by a separate, single zircon tip analysis (W). The duplicated concordant age likely represents the intrusion age of the granitoid. Fraction X, nine short prismatic zircons, gave discordant results, with a ^{207}Pb/^{206}Pb age of ca. 560 Ma. Although no core or overgrowth relationships were observed under a binocular microscope, inherited components were present in these zircons. When it was projected from ∼210 Ma, an upper intercept age at ∼1300 Ma was obtained. Fraction Y, of three long prismatic zircons, also yielded discordant results, having a ^{207}Pb/^{206}Pb age of ca. 340 Ma. Similar two-point (V, Y) regression calculation gave an upper intercept at ca. 450 Ma (Table 1; Fig. 2g).

5 Discussion

Timing of the South Qinling Granitoids and Their Geological Significance. All seven granitoid samples from the South Qinling granitoid belt yielded Triassic concordant zircon U-Pb ages, ranging from 205±1 to 220±1 Ma. Since these samples cover all types of granitoids, they suggest that there was a Triassic magmatic event in the South Qinling belt along the Mianlue suture, which lasted for about 15 Ma.

For the Guangtoushan suite, the ages of granitoids are in the order of diorite and quartz diorite > granodiorite >

adamellite, consistent with geological observations (Yan 1985). Among the three granites with Paleozoic Rb-Sr whole-rock isochron ages (264–323 Ma), the Guangtoushan pluton yields concordant zircon U-Pb age of 216±2 Ma, much younger than the previously reported Rb-Sr whole-rock isochron age of 323±14 Ma (Zhang et al. 1994). The Zhashui and Liuba granites have not been dated in this study. According to geological observation, the Zhashui granite is equivalent to the third- and fourth-stage intrusions of the Dongjiangkou granites (Shang and Yan 1988). The zircon U-Pb concordant age for the third-stage intrusion of the Dongjiangkou granite (QLDJ) is 211±2 Ma, distinctively younger than the Rb-Sr whole-rock isochron age of 264 Ma for the Zhashui granite (Shang and Yan 1988). As a member of the Guangtoushan group, according to geological observations, the Liuba quartz diorite should have formed during the first stage of intrusion for this group (Shang and Yan 1988), similar to the Zhangjiaba pluton (219±1 Ma) and the first intrusion of the Miba pluton (220±2 Ma). If so, it was probably formed in the Triassic instead of at ca. 285 Ma (Shang and Yan 1988). In other words, none of these granitic intrusions can be used to support a Paleozoic collision in the South Qinling.

The South Qinling granitoid belt has been suggested to be a syncollisional product (e.g., Zhang et al. 1994). The ages of these South Qinling granitoids are in agreement with but slightly younger than the Triassic metamorphic ages of different lithologies in the South Qinling (216±7 to 232±5 Ma [Mattauer et al. 1985], 240 Ma [Yin et al. 1991], 221±13 to 242±21 Ma [Li et al. 1996]), with an activity period of about 15 Ma. The lithologies of the South Qinling granitoids are quite complex, ranging from I-type to S-type. For these reasons, we prefer to call them synorogenic granitoids. Their age lends support to the idea that the collision between the South and the North China Blocks happened in late Triassic time along the South Qinling orogen (Li et al. 1996; Meng and Zhang 1999), similar to that in the Dabie Mountains and Sulu terranes (e.g., Li et al. 1989a, 1993, 1994, 2000; Ames et al. 1993, 1996; Chavagnac and Jahn 1996; Rowley et al. 1997; Hacker et al. 1998).

Constraints on the Origin of the South Qinling Synorogenic Granitoids and Breakoff Time. The precision of these U-Pb zircon ages for the South Qinling synorogenic granites and the metamorphic ages of Dabie ultrahigh-pressure metamorphic rocks (e.g., Li et al. 1989a, 1993, 1994, 2000; Ames et al. 1993, 1996; Chavagnac and Jahn 1996; Rowley et al. 1997; Hacker et al. 1998) are good enough to confirm that all these South Qinling synorogenic granitoids are systematically younger than the metamorphic ages of the ultrahigh-pressure metamorphic rocks and their country rocks in the Dabie and Sulu area, the eastern part of the Qinling-Dabie orogenic belt. The earliest granitoid was formed at 220±1 Ma, ca. 6–11 Ma after the peak metamorphism of the ultrahigh-pressure metamorphic rocks (226±3 Ma) and their country rocks (231±2 Ma; e.g., Li et al. 2000). It is equivalent to the cooling age of the ultrahigh-pressure metamorphic rocks at 500 °C (Li et al. 2000). The latest granitoid was formed at 205±1 Ma, ca. 8–14 Ma after the rapid exhumation of the ultrahigh-pressure metamorphic rocks and their country rocks. This age difference cannot be explained by a later collision in Qinling belt than in the Dabie-Sulu belt because the metamorphic ages of the Mianlue ophiolite (242–221 Ma; Li et al. 1996) and the blueschist (240 Ma [Yin et al. 1991], 232–216 Ma [Mattauer et al. 1985]) in the eastern Qinling are consistent with the metamorphic ages (244–221 Ma) of the high-pressure and ultrahigh-pressure metamorphic rocks in the Dabie-Sulu belt (Li et al. 1989a, 1989b, 1992, 1993, 1994, 2000; Chavagnac and Jahn 1996; Hacker et al. 1998). However, it can be better understood by a slab breakoff model (Davies and von Blanckenburg 1995). Syn- and postcollision magmatism is believed to be a result of a significant disturbance of the lithospheric root of mountain belts (Turner et al. 1992; Platt and England 1994), such as lithospheric delamination (Bird 1979), convective removal of a thickened thermal boundary layer (Houseman et al. 1981), and the breakoff of an oceanic slab from the buoyant continental lithosphere during the subduction. In their breakoff model, Davies and von Blanckenburg (1995) considered the Dabie Mountains an exception. No synorogenic magmatism has ever been identified in the Dabie Mountains and the Sulu terranes, where ultrahigh-pressure metamorphic rocks are widely distributed. "Post-orogenic magmatism" rocks were mainly formed at ca. 130 Ma (Xue et al. 1997; Hacker et al. 1998; Jahn et al. 1999; Li et al. 1999), about 100 Ma after the ultrahigh-pressure metamorphic rocks, in the Dabie Mountains. According to Davies and von Blanckenburg (1995), the breakoff in the Dabie Mountains happened at a depth of >130 km, which would lead to negligible thermal perturbation of the overriding lithosphere, so there was no major melting episode linked to the collision. In contrast to the Dabie Mountains and the Sulu terranes, there is a large

Triassic granitoid belt in the South Qinling, although no ultrahigh-pressure metamorphic rocks were ever reported there. Since slab breakoff usually happens near the boundary between continental lithosphere and oceanic lithosphere (Davies and von Blanckenburg 1995), the lack of ultrahigh-pressure metamorphic rocks indicates that the breakoff in the South Qinling, if there was any, happened at a very shallow depth, which disturbed the asthenosphere greatly and led to the formation of the synorogenic granitoids.

6 Concluding Remarks

The zircon U-Pb ages of the South Qinling synorogenic granites show that they were formed from 220±1 to 205±1 Ma. This supports a Triassic age collision between the North China and the South China Blocks, an idea previously based mainly on the ultrahigh-pressure metamorphic rocks from the Dabie Mountains and Sulu terranes (e.g., Li et al. 1989a, 1993, 1994, 2000; Ames et al. 1993, 1996; Chavagnac and Jahn 1996; Rowley et al. 1997; Hacker et al. 1998). The distribution of the Triassic South Qinling synorogenic granitoids will help us to reconstruct the Triassic suture zone all along the Qinling-Dabie orogenic belt (Fig. 3). The consistency between the formation age of the South Qinling granitoids and the rapid cooling ages of the Dabie-Sulu ultrahigh-pressure metamorphic rocks suggests that the formation of the South Qinling synorogenic granitoids may be related to the breakoff of a subducted oceanic plate. The limited distribution of synorogenical granites in the Qinling-Dabie orgenic belt can be explained by the shallower breakoff depth in the South Qinling than that of the Dabie-Sulu ultrahigh-pressure metamorphic belt.

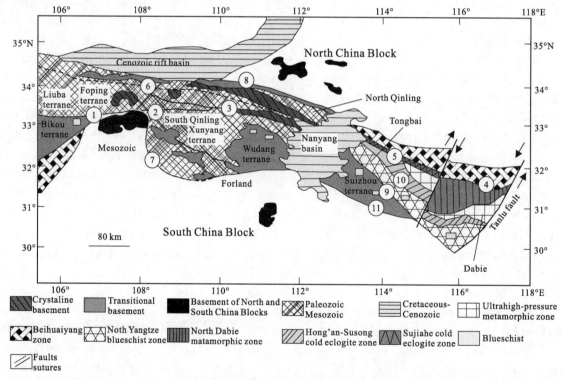

Fig. 3 Generalized geological map of the Qinling-Dabie orogenic belt, showing its geological tectonic setting, especially the relation of major faults. The west part (Qinling) is simplified after Zhang et al. (1996b) and Meng and Zhang (2000); the east part (Tongbai and Dabie) is modified after Zhang et al. (1996b) and Li et al. (2000). Numbers in circles indicate major faults: 1= Mianlue suture; 2= Daheba-Ningshan fault; 3= Shanyang fault; 4= Xiaotian-Mozitan fault; 5= tectonic boundary between Sujiahe and Hong'an groups; 6= Shangdan suture; 7= Dabashan fault; 8= Machaoying fault; 9= Dashankou fault; 10= Dawu fault; 11= Xiangfan-Guangji fault. LMS= Longmenshan orogen.

Acknowledgements

This study is part of W. Sun's master's thesis at the University of Science and Technology of China and is supported by the National Natural Science Foundation of China (grant 49472144). W. Sun is grateful to Shen-su Sun for his constant encouragement, useful suggestions, and constructive reviews. We thank Zhengxiang Li, Yongfei Zheng, and an anonymous reviewer for many constructive comments and suggestions, which resulted in improvements to the early version of this article.

References

Ames, L.; Tilton, G. R.; and Zhou, G. 1993. Timing of collision of the Sino-Korean and Yangtze cratons: U-Pb zircon dating of coesite-bearing eclogites. Geology 21:339–342.

Ames, L.; Zhou, G.; and Xiong, B. 1996. Geochronology and isotopic character of ultrahigh-pressure metamorphism with implications for collision of the SinoKorean and Yangtze cratons, central China. Tectonics 15:472–489.

Bird, P. 1979. Continental delamination and Colorado Plateau. J. Geophys. Res. 84:7561–7571.

Chavagnac, V., and Jahn, B. M. 1996. Coesite-bearing eclogites from Bixiling complex, Dabie Mountains, China: Sm-Nd ages, geochemical characteristics and tectonics implications. Chem. Geol. 133:29–51.

Davies, J. H., and von Blanckenburg, F. 1995. Slab breakoff: a model of lithosphere detachment and its test in the magmatism and deformation of collisional orogents. Earth Planet. Sci. Lett. 129:85–102.

Davis, D. W.; Blackburn, C. E.; and Krogh, E. 1982. Zircon U-Pb ages from the Wabigoon-Manitou Lakes region, Wabigoon Subprovince, northeast Ontario. Can. J. Earth Sci. 19:254–266.

Ernst, W. G., and Liou, J. G. 1995. Contrasting platetectonic styles of the Qinling-Dabie-Sulu and Franciscan metamorphic belts. Geology 23:353–356.

Hacker, B. R.; Ratschbacher, L.; Webb, L.; Ireland, T.; Walker, D.; and Dong, S. 1998. U/Pb zircon ages constrain the architecture of the ultrahigh-pressure Qinling-Dabie Orogen, China. Earth Planet. Sci. Lett. 161: 215–230.

Houseman, G. A.; McKenzie, D. P.; and Molnar, P. 1981. Convective instability of a thickened boundary layer and its relevance for the thermal evolution of continental convergent belts. J. Geophys. Res. 86: 6115–6132.

Jaffey, A. H.; Flynn, K. F.; Glendenin, L. E.; Bentley, W. C.; and Essling, A. M. 1971. Precision measurement of half-lives and specific activities of ^{235}U and ^{238}U. Phys. Rev. C4:1889–1906.

Jahn, B. M.; Cornichet, J.; Cong, B.; and Yui, T. F. 1996. Ultrahigh εNd eclogites from an ultrahigh-pressure metamorphic terrane of China. Chem. Geol. 127: 61–79.

Jahn, B. M.; Wu, F.; Lo, C. H.; and Tsai, C. H. 1999. Crustmantle interaction induced by deep subduction of the continental crust: geochemical and Sr-Nd isotopic observations from post-collisional mafic-ultramafic intrusions of the northern Dabie complex, central China. Chem. Geol. 157:119–146.

Krogh, T. E. 1973. A low contamination method for hydrothermal decomposition of zircon and extraction of U-Pb for isotopic age determinations. Geochim. Cosmochim. Acta 37:485–494.

Krogh, T. E. 1982. Improved accuracy of U-Pb zircon ages by the creation of more concordant systems using an air abrasion technique. Geochim. Cosmochim. Acta 46: 637–649.

Krogh,T. E., and Davis, G. L. 1975. The production and preparation of ^{205}Pb for use as a tracer for isotope dilution analyses. Carnegie Inst. Wash. Year Book 74: 416–417.

Lai, S. C., and Zhang, G. W. 1996. Geochemical features of ophiolite in Mianxian-Lueyang suture zone, Qingling Orogenic Belt. J. Chin. Univ. Geosci. 7:165–172.

Lerch, M. F.; Xue, F.; Kroner, A.; Zhang, G. W.; and Tod, W. 1995. A middle Silurian-early Devonian magnetic arc in the Qinling Mountains of Central China. J. Geol. 103:437–449.

Li, S.; Ge, N.; Liou, D.; Zhang, Z.; Ye, X.; Zheng, S.; and Peng, C. 1989a. The Sm-Nd isotopic age of C-type eclogite from the Dabie Group in the Northern Dabie Mountains and its tectonic implication. Chin. Sci. Bull. 34:1623–1628.

Li, S.; Hart, S. R.; Zheng, S.; Liou, D.; Zhang, G.; and Guo, A. 1989b. Timing of collision between the North and South China Blocks: Sm-Nd isotopic age evidence. Sci. China Ser. B 32:1391–1400.

Li, S.; Hong, J.; Li, H.; and Jiang, L. 1999. U-Pb ages of the pyroxenite-gabbro intrusions in Dabie Mountains and their geological implications. Geol. J. China Univ. 5:351–355.

Li, S.; Jagoutz, E.; Chen, Y.; and Li, Q. 2000. Sm-Nd and Rb-Sr isotopic chronology and cooling history of ultrahigh pressure metamorphic rocks and their country rocks at Shuanghe in the Dabie Mountains, central China. Geochim. Cosmochim. Acta 64:1077–1093.

Li, S.; Liu, D.; Chen, Y.; and Ge, N. 1992. The Sm-Nd isotopic age of coesite-bearing eclogite from the southern Dabie Mountains. Chin. Sci. Bull. 37:1638–1641.

Li, S., and Sun, W. 1996. A middle Silurian-early Devonian magmatic arc in the Qinling Mountains of central China: a discussion. J. Geol. 104:501–503.

Li, S.; Sun, W.; Zhang, G.; Chen, J.; and Yang, Y. 1996. Chronology and geochemistry of metavolcanic rocks from Heigouxia Valley in the Mianlue tectonic arc, South Qinling: observations for a Paleozoic oceanic basin and its close time. Sci. China Ser. B. 39:300–310.

Li, S.; Wang, S.; Chen, Y.; Liou, D.; Qiu, J.; Zhou, H.; and Zhang, Z. 1994. Excess argon in phengite from eclogite: observations from dating of eclogite minerals by Sm-Nd, Rb-Sr, and $^{40}Ar/^{39}Ar$ methods. Chem. Geol. 112:343–350.

Li, S.; Xiao, Y.; Liou, D.; Chen, Y.; Ge, N.; Zhang, S.; Sun, S.S.; et al. 1993. Collision of the North China and Yangtze blocks and formation of coesite-eclogites: timing and processes. Chem. Geol. 109:89–111.

Li, Z. X. 1994. Collision between the North and South China Blocks: a crustal-detachment model for suturing in the region east of the Tanlu fault. Geology 22: 739–742.

Li, Z. X. 1998. Tectonic evolution of the major East Asian lithospheric blocks since Mid-Proterozoic: a synthesis. In Flower, M.; Chung, S. L.; Lo, C. H.; and Lee, T. Y., eds. Mantle dynamics and plate interactions in East Asia. AGU geodynamics series 27. Am. Geophys. Union, Washington, D.C., p. 221–243.

Mattauer, M.; Matte, P.; Malavieille, J.; Tapponnier, P.; Maluski, H.; Xu, Z. Q.; Lu, Y. L.; and Tang, Y. Q. 1985. Tectonics of the Qinling Belt: build-up and evolution of eastern Asia. Nature 317:496–500.

Meng, Q. R., and Zhang, G. W. 1999. Timing of the collision of the North and South China Blocks: controversy and reconciliation. Geology 27:123–126.

Meng, Q. R., and Zhang, G. W. 2000. Geological framework and tectonic evolution of the Qinling orogen, central China. Tectonophysics 323:183–196.

Meng, Q. R.; Zhang, G. W.; Yu, Z. P.; and Mei, Z. C. 1996. Late Paleozoic sedimentation and tectonics of rift and limited ocean basin at southern margin of the Qinling. Sci. China Ser. D 39(suppl.):24–32.

Okay, A. I., and Sengör, A. M. C. 1992. Observations for intracontinental thrust-related exhumation of ultrahigh-pressure rocks in China. Geology 20:411–414.

Parrish, R. R., and Krogh, T. E. 1987. Synthesis and purification of ^{205}Pb for U-Pb geochronology. Chem. Geol. 63:103–110.

Platt, J. P., and England, P. C. 1994. Convective removal of lithosphere beneath mountain belts: thermal and mechanical consquences. Am. J. Sci. 294:307–336.

Rowley, D. B.; Xue, F.; Tucker, R. D.; Peng, Z. X.; Baker, J.; and Davis, A. 1997. Age of ultrahigh pressure metamorphism and protolith of orthogenisses from the eastern Dabie Shan: U/Pb zircon geochronology. Earth Planet. Sci. Lett. 151:191–203.

Schaltegger, U.; Schneider, J.L.; Maurin, J.C.; and Corfu, F. 1996. Precise U-Pb chronometry of 345–340 Ma old magmatism related to syn-convergence extension in the Southern Vosges (Central Variscan Belt). Earth Planet. Sci. Lett. 144:403–419.

Shang, R., and Yan, J. 1988. Granites in the Qinling and Bashan area. Wuhan, Chinese University of Geoscience Press, p. 69–144 (in Chinese).

Stacey, J. S., and Kramers, J. D. 1975. Approximation of terrestrial lead isotope evolution by a two-stage model. Earth Planet. Sci. Lett. 26:207–221.

Sun, W. 1994. Collision of Qinling orogenic belt: timing and processes. M.S. thesis, University of Science and Technology of China, Hefei.

Sun, W., and Li, S. 1998. Pb isotopes of granitoids suggest Devonian accretion of Yangtze (South China) craton to North China craton: comment. Geology 26: 859–860.

Turner, S.; Sandiford, M.; and Foden, J. 1992. Some geodynamic and compositional constraints on "postorogenic" magmatism. Geology 20:931–934.

Wang, Q. C.; Liu, X.; Maruyama, S.; and Cong, B. 1995. Top boundary of the Dabie UHPM rocks, central China. J. Southeast Asian Earth Sci. 11:295–300.

Wang, Q. C.; Zhai, M. G.; and Cong, B. L. 1996. Regional geology. In Cong, B. L., ed. Ultrahigh-pressure metamorphic rocks in the Dabieshan-Sulu Region of China. Beijing, Science, p. 8–26.

Wang, X.; Liou, J. H.; and Mao, H. J. 1989. Coesite-bearing eclogites from the Dabie Mountians in central China. Geology 17:1085–1088.

Wu, H. 1980. The glaucophane-schists of the eastern Qinling and northern Qilian Mountains in China. Acta Geol. Sin. 54:195–207.

Xu, S.; Liu, Y. C.; Jiang, L. L.; Su, W.; and Ji, S. Y. 1994. Tectonic framework and evolution of Dabieshan. Beijing, Science, 175 p. (in Chinese with English abstract).

Xu, S.; Okay, A. I.; Ji, S.; Sengör, A. M. C.; Su, W.; Liu, Y.; and Jiang, L. 1992. Diamond from the Dabie Shan metamorphic rocks and its implication for tectonic setting. Science 256:80–82.

Xue, F.; Kroner, A.; Reischmann, T.; and Lerch, F. 1996a. Paleozoic pre- and post-collision calc-alkaline magmatism in the Qinling orogenic belt, central China, as documented by zircon ages on granitoid rocks. J. Geol. Soc. 153:409–417.

Xue, F.; Lerch, M. F.; Kroner, A.; and Reischmann, T. 1996b. Tectonic evolution of the east Qinling Mountains, China, in the Paleozoic:

a review and new tectonic model. Tectonophysics 253:271–284.

Xue, F.; Rowley, D. B.; Tucker, R. D.; and Peng, Z. C. 1997. U/Pb ziron ages of granitoid rocks in the North Dabie complex, eastern Dabie Shan, China. J. Geol. 105:744–753.

Yan, Z. 1985. Granites in Shaanxi Province. Xi'an, Xi'an Jiaotong University Press, p. 321 (in Chinese with English abstract).

Yin, A., and Nie, S. 1993. An indentation model for the North and South China collision and the development of the Tanlu and Honam fault systems, eastern Asia. Tectonics 12:801–813.

Yin, H.F., ed. 1994. The palaeobiogeography of China. Oxford, Clarendon, 370 p.

Yin, Q.; Jagoutz, E.; and Kroner, A. 1991. Precambrian (?) blueschist/coesite bearing eclogite belt in central China. Terra Abstr. 3:85–86.

Zhai, X.; Day, H. W.; Hacker, B. R.; and You, Z. D. 1998. Paleozoic metamorphism in the Qinling orogen, Tongbai Mountains, central China. Geology 26:371–374.

Zhang, B.R.; Luo, T.; Gao, S.; Ouyang, J.; Chen, D.; Ma, Z.; Han, Y.; and Gu, X. 1994. Geochemical study of the lithosphere, tectonism and metallogenesis in the Qinling-Dabieshan region. Wuhan, Chinese University of Geoscience Press, p. 110–122 (in Chinese with English abstract).

Zhang, B.R.; Luo, T.; Gao, S.; Ouyang, J.; Gao, C.; and Li, Z. 1991. Geochemical characteristics and tectonic evolution of the earth's crust in the east Qinling and North Dabieshan mountain district. In Ye, L. J.; Qian, L. X.; and Zhang, G. W., eds. A selection of papers presented at the conference on the Qinling orogenic belt. Xi'an, Northwest University Press, p. 200–213 (in Chinese with English abstract).

Zhang, E.; Niu, D.; Huo, Y.; Li, Y.; and Zhang, L. 1992. Geological map of Qinling-Dabie mountains and adjacent region of the People's Republic of China. Beijing, Geol. Publ. House, scale 1 : 1,000,000.

Zhang, G.W.; Meng, Q. R.; and Lai, S. C. 1995. Tectonics and structures of the Qinling orogenic belt. Sci. China Ser. B 38:1379–1386.

Zhang, G.W.; Meng, Q. R.; Yu, Z. P.; Sun, Y.; Zhou, D. W.; and Guo, A. L. 1996a. Orogenic processes and dynamics of the Qinling. Sci. China Ser. D 39:225–234.

Zhang, G.W.; Zhang, B.R.; and Yuan, X.C. 1996b. Altas of orogenic processes and three-dimension lithosphere framework of Qinling orogenic belt. Beijing, Science.

Zhang, G.W.; Zhou, D.; Yu, Z.; Guo, A.; Chen, S.; Li, T.; Zhang, C.; et al. 1991. Characteristics of the lithospheric composition, structure and evolution of the Qinling orogenic belts. In Ye, L. J.; Qian, L. X.; and Zhang, G. W., eds. A selection of papers presented at the conference on the Qinling orogenic belt. Xi'an, Northwest University Press, p. 11–138 (in Chinese with English abstract).

Zheng, Y.F.; Fu, B.; Li, Y.; Xiao, Y.; and Li, S. 1998. Oxygen and hydrogen isotope geochemistry of ultrahighpressure eclogites from the Dabie Mountains and the Sulu terranes. Earth Planet. Sci. Lett. 155:113–129.

南秦岭勉略构造带三岔子古岩浆弧的地球化学特征及形成时代*

李曙光[1]，侯振辉[1]，杨永成[2]，孙卫东[1]，张国伟[3]，李秋立[1]

1. 中国科学技术大学地球和空间科学学院，合肥 230026
2. 陕西地质矿产局区域地质调查队，咸阳 712000
3. 西北大学地质学系，西安 710069

> **亮点介绍**：南秦岭勉略构造带三岔子古岩浆弧岩石微量元素以及继承锆石年龄表明其发育在南秦岭微陆块南缘的活动陆缘环境。斜长花岗岩的形成时代约为 300±61 Ma，说明勉略古洋盆此时已开始向南秦岭微陆块下消减，并可与大别山浒湾构造带进行对比，表明此时在秦岭—大别山带发生了普遍的古生代洋壳消减事件。

摘要 南秦岭勉略构造带三岔子镁铁-超镁铁杂岩可划分为两个岩块：三岔子古岩浆弧和庄科古洋壳残片（蛇绿岩）。三岔子古岩浆弧主要由岛弧型安山质熔岩、玄武及玄武安山质辉(闪)长岩、安山质岩墙、斜长花岗岩及部分超镁铁岩组成。它们具有典型的岛弧火山岩地球化学特征，如高场强元素（Nb、Ti）亏损和低 Cr、Ni 含量。该类岩石的轻稀土富集和富钾的特征及斜长花岗岩中含有 9 亿年锆石捕虏晶的特征表明它们可能发育在南秦岭微陆块南缘的活动陆缘环境。斜长花岗岩的岩浆锆石 U-Pb 年龄为 300±61 Ma，它表明勉略古洋盆在石炭纪已开始向南秦岭微陆块下消减。这一年龄和大别山浒湾构造带洋壳俯冲成因榴辉岩的形成时代（309 Ma）一致，它说明勉略洋在石炭纪可东延至大别山。三岔子古岩浆弧中类似高镁埃达克岩的存在表明这一俯冲洋壳是年轻（<25 Ma）而且较热的大洋岩石圈。

关键词 秦岭造山带；岩浆弧；锆石 U-Pb 年龄；蛇绿混杂岩

南秦岭勉略构造带是秦岭造山带扬子陆块与华北陆块及与之先期拼合的南秦岭微陆块最终碰撞的地缝合线[1,2]。该带内蛇绿混杂岩的发现，对判断其地缝合线性质起重要作用[2,3]。近年来围绕该带的蛇绿混杂岩及变质火山岩进行了一系列岩石学、地球化学及年代学研究。这些研究已证明这里既存在 MORB 型古洋壳残片[4-6]，又存在岛弧型火山岩[5,6]。黑沟峡 MORB 型变质火山岩的 Sm-Nd 和 Rb-Sr 同位素年代学研究仅给出了它们的变质时代（242~221 Ma），它可能指示了该古洋盆在三叠纪初闭合[4]。早期研究将岛弧型火山岩也归入三岔子蛇绿混杂岩，并称之为岛弧型蛇绿混杂岩[3,6]。近年来一些作者将三岔子岛弧火山岩视为独立岩块[5,7,8]，然而仍将其中的辉长岩和辉绿岩墙视为蛇绿混杂岩的一部分[8]。显然三岔子岩浆杂岩中各类岩石是否具有不同的形成构造环境，这一岛弧火山岩块究竟是代表一古岛弧，还是代表大陆边缘的古岩浆弧，仍需要更详细的工作。在年代学方面，该带缺乏能直接证明其原岩形成时代的同位素年龄证据。这些问题严重影响了我们对勉略古洋盆的形成与消减历史的研究。本文主要对该带三岔子岛弧火山岩块的主要岩石类型进行了主要元素、痕量元素及锆石 U-Pb 年代学研究，以探讨三岔子岛弧火山岩组合的地球化学特征、形成构造背景及时代。为对比，对该区相邻的代表古洋壳残片的庄科玄武岩也进行了主要元素和痕量元素地球化学研究。

1 地质与样品

勉略构造带由一系列震旦-寒武系、泥盆-石炭系和镁铁-超镁铁岩逆冲推覆岩片组成，其区域地质背景

---

* 本文发表在：中国科学(D 辑)，2003，33(12)：1163-1173
对应的英文版论文为：Li Shuguang, Hou Zhenhui, Yang Yongcheng, et al., 2004. Timing and geochronological characters of the Sanchazi magmatic arc in Mianlue tectonic zone, South Qinling. Science in China (Ser. D), 47(4): 317-328

已在较早的论文中概述[4]。该构造带三岔子乡周围集中发育了各种类型的镁铁-超镁铁岩石，并被称为三岔子"蛇绿混杂岩"（本文仍暂时沿用这一名称）[2,3]。前人工作表明三岔子"蛇绿混杂岩"可划分为两部分：三岔子乡以西以岛弧型火山岩为主；三岔子乡以东以 MORB 型玄武质火山岩为主（见图 1 (a)）[5,6]。三岔子"蛇绿混杂岩"西段由强烈剪切变形的玄武质和安山质熔岩、辉（闪）长岩、超基性岩（蛇纹岩、滑石片岩组成）及玄武或安山质岩墙和少量斜长花岗岩组成。然而这些岩石并不按典型蛇绿岩套剖面的顺序排序。超镁铁岩经常以构造关系直接与安山质火山熔岩接触，接触带强烈面理化（见图 1 (b)）。安山质熔岩和辉（闪）长岩也呈类似构造接触关系。然而也可见到细粒安山质岩墙侵入变辉（闪）长岩中（见图 1 (b) 金家河剖面）。该处斜长花岗岩（偏桥沟）及部分辉（闪）长岩（如田坝）就位于千枚岩中，然而外接触带的强烈面理化和千枚岩中未见热烘烤现象，说明它们可能为构造就位。由于有大量安山质熔岩及安山质岩墙的出现，上述岩石组合及相互关系不像是蛇绿混杂岩，而更像是一被构造强烈挤压混杂的古岩浆弧残片。三岔子"蛇绿混杂岩"东段，以强烈变形的超镁铁岩块及变质玄武岩为主，未见中酸性火山岩。变质玄武岩被一系列构造片理化带分割成若干岩片。已报道此类玄武岩具 LREE 强亏损特征[4-6]，它们代表一古洋壳残片。

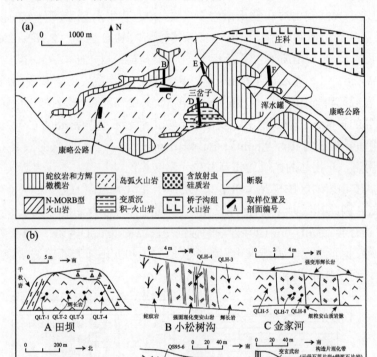

图 1　勉略构造带各种类型的镁铁-超镁铁岩石
(a) 勉略地区三岔子一带地质略图(据文献[5])；(b) 样品采样位置岩性组合示意图

本工作对三岔子地区西段各主要岩类做了较系统的采样。除了超镁铁岩因全部蛇纹石化或滑石化不能进行地球化学研究外，对变玄武质和安山质熔岩、岩墙、辉长岩和斜长花岗岩均采取了样品。对三岔子"蛇绿混杂岩"东段，因前人工作较多[4-6]，仅在庄科沟采取了若干变玄武质岩样品。取样位置及其岩石组合关系见图1。样品经镜下切片检查后，仅选取新鲜、未蚀变、无气孔及杏仁构造的样品做地球化学分析。对偏桥沟斜长花岗岩取了一个大样（100 kg），选锆石做定年研究。

2　分析方法

将全岩地球化学样品破碎成≤1cm 的岩屑，从中挑选 200 g 新鲜、无蚀变、无晚期岩脉的岩屑，用 1N HCl 浸泡，去除表面污染，然后用去离子水反复洗净，烘干，用刚玉板小型颚式破碎机细碎，最后用玛瑙球磨机磨至约200目。

样品的主要元素及痕量元素（稀土除外）在原地质矿产部物化探研究所用X荧光光谱法测定，稀土元素由原地矿部武汉分析测试中心用电感耦合等离子体发射光谱（ICP）测定，轻稀土元素（La、Ce、Pr、Nd、Sm）相对标准偏差RSD<5%，重稀土元素（Eu→Lu）相对标准偏差<10%。锆石的分选及U-Pb同位素测定由天津地质矿产研究所同位素实验室完成，其分析流程已有专文介绍[9]。数据处理及年龄计算使用美国地质调查所（丹佛）K. Ludwig编写的PBDAT和ISOPLOT软件进行。同位素质谱测定误差以2σ给出。

3 主要元素和痕量元素地球化学

3.1 主要元素组成

在SiO_2-K_2O图中（图2），庄科变玄武岩落入玄武岩和玄武安山岩区，而且因其低K_2O含量（0.01%~0.39%），属于低钾玄武岩或玄武安山岩系列（图2）。三岔子地区西段各类岩石中，仅田坝有2个细粒辉长岩样品属玄武质岩石，其他均属于玄武安山质和安山质岩石（图2）。其中沿金家河分布的被安山质岩墙穿插的强变形辉（闪）长岩为玄武安山质和安山质岩石，且因其较高的K_2O含量（2.18%~2.68%）而落入高钾玄武安山岩和安山岩区（图2）。在安山质熔岩及岩墙中，大部分样品（QLH-4、QLH-7-1、QLH-7-2、QS95-3）K_2O含量较高（2.02%~2.67%），落入高钾安山岩区。因此这一套玄武安山岩和安山质岩石，除少量样品外，均落入高钾玄武安山岩或高钾安山岩区，它们的K_2O含量均>2%。一个落入低钾安山岩区的样品（QS95-6）是文家沟与蛇纹岩相邻的样品，它较其他玄武安山岩或安山质岩石有异常高的Cr（287 ppm）、Ni（84 ppm）和H_2O（4.44%）含量和异常低的Rb（9 ppm）、Ba（216 ppm）含量，说明它的低K_2O含量是在后期变质过程中，在变质流体作用下，与相邻的蛇纹岩相互发生元素交换的结果，它不代表真实安山质岩浆的组成。另外，两个落入低钾玄武安山岩岩区的样品（QLT-3、QLT-4）是含异常高H_2O+CO_2（7.15%~9.03%）的样品，其Rb、Ba含量也较低，因此其低钾含量有可能是遭受变质流体强烈淋滤的结果。尽管钾在变质过程中是活动元素，但上述3个具有异常高H_2O+CO_2含量的强蚀变样品的低K_2O、Rb、Ba特征说明该区低级变质作用对钾主要起淋失作用。因此三岔子岛弧火山岩块中玄武安山和安山质岩石普遍富钾的特征，看来不是后期变质作用改造的结果，而反映原岩特征。

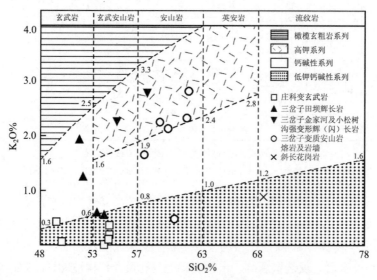

图2 岛弧火山岩的SiO_2-K_2O分类图（据文献[11]）
样品SiO_2和K_2O含量为扣除H_2O和CO_2后重新换算的重量百分比

3.2 痕量元素地球化学特征

与前人报道的结果相似，庄科变玄武岩具有与MORB类似的轻稀土亏损特征[5]（图3(a)）。然而在痕量元素原始地幔标准化图上，本文分析样品有明显的Sr、Nb正异常（图3(b)）。尽管本文样品Sr含量（107~185 ppm）

和 Nb 含量（3~5 ppm）与 N-MORB 和 E-MORB 的平均 Sr 含量（90~155 ppm）和 Nd 含量（2.33~8.30）相当[11]，但却显著高于许继锋等（1997）报道的该区 MORB 型变玄武岩的 Nb 含量（0.66~1 ppm）[5]。如果，Nb 正异常是 X 荧光光谱对低 Nb 含量（<5 ppm）样品的测量值偏高所致，但这一 Nb 含量偏高现象不影响下面对岛弧火山岩性质的判断，因为它只能削弱高场强元素负异常特征而不会造成虚假的高场强元素亏损现象。此外，本文获得的庄科变玄武岩 Th、Ba、Rb 等大离子亲石元素显著高于 MORB 平均值（见图 3 (b)）。这可能是在低级变质作用过程中这些大离子亲石元素活动的结果。

与庄科变玄武岩不同，三岔子地区西段各类火成岩（包括辉（闪）长岩、安山质熔岩及岩脉、斜长花岗岩）均表现出岛弧火山岩的地球化学特征，如 LREE 相对重稀土富集（图 3 (c)、(e)）、高场强元素（Nb、P、Ti）亏损（图 3 (d)、(f)）和大部分样品的低 Cr、Ni 含量（Cr<140 ppm；Ni<30 ppm）（见表 1）。在安山质熔岩中仅文家沟与蛇纹岩相邻的样品 QS95-6 具有较高的 MgO (4.90%) 和 Cr (287 ppm)、Ni (84 ppm) 含量。前已述，这可能是在变质过程中，它与蛇纹岩相互作用的结果。值得注意的是两个强变形辉(闪)长岩(QLH-3、QLH-5) 具有较高的 Cr（250~303 ppm）、Ni（94~110 ppm）含量，对此，我们将在下面（3.3 节）详细讨论。斜长花岗岩具有较高的 Mg$^{\#}$值（0.54）和非常低的 REE 含量，这说明它们的 Nb、P、Ti 亏损特征不是陆壳混染的结果，而反映了岛弧地幔源区特征。三岔子安山质熔岩及斜长花岗岩均有显著的 Eu 负异常（图 3 (e)），反映了斜长石作为源区残留相或分离结晶作用对岩浆组成的影响。然而斜长花岗岩的高 Mg$^{\#}$值（0.54）和它的高 SiO$_2$、Na$_2$O 及低 CaO、K$_2$O 和 REE 特征与本区辉（闪）长岩的分离结晶演化趋势很不协调。因此，斜长花岗岩不是玄武质岩浆分离结晶演化的产物。它的 Eu 负异常可能是其源区残留有斜长石所致。上述痕量元素特征说明三岔子西段的辉（闪）长岩及斜长花岗岩也发育在岛弧环境，它们与东部庄科变玄武岩无成因联系，而与西部安山质熔岩共同组成三岔子岛弧岩浆杂岩。

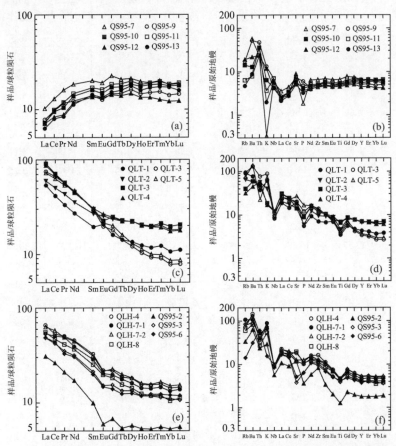

图 3 三岔子地区岩浆岩的 REE 球粒陨石标准化图及痕量元素原始地幔标准化图
(a)、(b) 庄科变玄武岩；(c)、(d) 三岔子西段辉（闪）长岩；(e)、(f) 三岔子西段安山质熔岩及岩墙

表1 三岔子地区变质岩浆岩的主要元素（%）及痕量元素（10^{-6}）组成

样品号	QLH-3	QLH-4	QLH-5	QLH-7-1	QLH-7-2	QLH-8	QLT-1	QLT-2	QLT-3	QLT-4
采样位置	小松树沟		金家河				田坝			
样品名称	强变形闪长岩	强片理化安山岩	强变形辉长岩	安山质岩脉	安山质岩脉	安山质岩脉	细粒辉长岩			
SiO_2	56.28	57.43	53.49	58.97	56.07	56.02	49.75	50.24	49.59	49.22
TiO_2	0.82	1.01	0.91	0.96	0.94	0.98	0.75	0.99	1.47	1.42
Al_2O_3	14.74	17.56	14.38	16.55	16.52	17.29	15.20	14.32	14.87	14.73
Fe_2O_3	2.07	1.72	1.99	1.57	1.75	2.32	1.96	1.86	1.62	1.05
FeO	4.75	6.02	5.61	6.40	7.15	5.49	6.07	8.27	10.64	9.37
MgO	6.15	3.21	7.10	3.26	3.39	3.91	7.37	8.39	3.88	3.63
CaO	5.92	2.76	6.57	2.15	3.11	3.85	10.40	7.75	4.58	4.46
Na_2O	3.42	4.34	3.99	3.01	3.29	5.12	2.50	3.36	5.62	6.27
K_2O	2.68	2.03	2.18	2.67	2.12	1.60	1.85	1.20	0.56	0.51
P_2O_5	0.26	0.25	0.36	0.22	0.27	0.23	0.12	0.13	0.20	0.19
MnO	0.12	0.12	0.12	0.14	0.16	0.11	0.13	0.16	0.16	0.14
H_2O	2.77	3.72	2.99	3.68	4.33	2.86	3.19	3.58	4.21	3.26
CO_2	0.20	0.49	0.17	0.92	0.98	0.15	0.87	0.08	2.94	5.77
Mg#	0.48	0.30	0.49	0.29	0.28	0.34	0.48	0.46	0.24	0.26
Total	100.18	100.66	99.86	100.50	100.08	99.93	100.16	100.33	100.34	100.02
La	17.67	15.52	17.14	12.66	14.89	13.06	12.72	14.13	21.43	20.60
Ce	40.57	32.62	38.72	30.81	35.45	28.59	25.26	30.71	43.34	42.23
Pr	5.35	4.78	5.11	4.67	4.61	3.91	3.13	4.11	5.44	5.33
Nd	21.63	21.20	21.93	18.16	20.66	17.11	12.63	16.57	21.38	20.80
Sm	4.52	4.94	4.53	4.37	4.77	3.83	2.97	4.07	4.60	4.65
Eu	1.14	1.26	1.23	1.17	1.24	1.14	1.20	1.47	1.31	1.53
Gd	3.70	4.47	4.12	4.01	4.60	3.71	3.27	4.81	5.09	5.06
Tb	0.53	0.69	0.60	0.66	0.74	0.58	0.53	0.83	0.85	0.83
Dy	3.07	4.55	3.30	4.04	4.59	3.60	3.43	5.64	5.61	5.65
Ho	0.58	0.87	0.64	0.80	0.90	0.70	0.70	1.14	1.16	1.13
Er	1.53	2.52	1.64	2.32	2.57	1.95	1.94	3.25	3.29	3.19
Tm	0.23	0.40	0.25	0.37	0.42	0.30	0.31	0.51	0.54	0.49
Yb	1.32	2.42	1.46	2.24	2.55	1.77	1.80	3.04	3.32	2.96
Lu	0.20	0.37	0.22	0.35	0.39	0.28	0.28	0.46	0.52	0.45
Y	15.32	23.52	16.68	20.89	24.24	18.74	19.37	30.30	31.44	31.77
Rb	60.6	45.0	47.7	66.8	49.9	37.9	57.2	41.4	25.4	20.2
Ba	917	1011	961	743	644	582	538	403	419	317
Th	6.6	4.4	1.9	3.3	4.1	2.3	3.2	4.8	4.0	4.9
U	2.0	2.4	1.1	0.7	0.1	0.0	0.6	0.5	0.6	1.5
Nb	6.9	6.5	6.9	6.3	6.4	6.4	5.8	8.6	8.6	8.5
Zr	152	182	147	136	139	128	78	107	127	135
Sr	424	385	565	247	263	409	406	295	299	384
Cr	250	55	303	55	57	78	136	87	10	12
Ni	94	23	110	24	28	24	25	24	8	10
V	161	171	175	167	170	183	191	206	325	266
Pb	13.5	12.4	7.9	13.9	86.2	14.8	9.2	6.3	0.9	4.0
La/Yb	13.39	6.41	11.74	5.65	5.84	7.38	7.07	4.65	6.45	6.96
Sr/Y	27.68	16.37	33.87	11.77	10.85	21.82	20.96	9.74	9.51	12.01

续表

采样位置	偏桥沟			庄科					
样品名称	斜长花岗岩	变安山岩	变安山岩	变玄武岩					
SiO_2	68.00	58.63	56.77	50.35	44.94	52.09	52.12	51.43	44.97
TiO_2	0.28	0.85	0.91	1.42	1.05	1.11	1.18	0.97	1.12
Al_2O_3	16.95	16.07	14.57	13.51	12.97	12.09	12.27	13.40	13.51
Fe_2O_3	0.04	1.10	0.21	4.95	3.93	7.12	6.27	5.04	1.74
FeO	2.29	5.99	7.24	6.45	6.15	5.39	5.94	5.66	9.63
MgO	2.79	3.09	4.90	5.85	6.64	5.42	5.68	5.08	6.65
CaO	0.32	2.07	2.90	7.77	11.22	10.13	8.89	8.96	7.49
Na_2O	7.10	5.11	5.10	3.27	2.95	2.16	2.98	3.28	3.90
K_2O	0.87	2.20	0.46	0.01	0.39	0.19	0.11	0.35	0.06
P_2O_5	0.08	0.14	0.14	0.04	0.13	0.09	0.10	0.08	0.07
MnO	0.04	0.10	0.11	0.14	0.19	0.18	0.17	0.13	0.17
H_2O	0.41	3.65	4.44	4.52	4.10	3.44	3.57	3.85	5.75
CO_2	0.39	1.19	1.72	2.39	5.59	1.16	0.72	1.85	4.63
Mg#	0.54	0.31	0.40	0.35	0.41	0.31	0.33	0.33	0.37
Total	99.56	100.19	99.47	100.67	100.25	100.57	100.00	100.08	99.69
La	7.24	11.37	11.52	2.42	1.84	1.71	1.69	1.68	1.48
Ce	15.96	25.53	26.43	7.85	5.87	5.78	6.00	5.07	4.87
Pr	1.99	3.33	3.11	1.51	1.10	0.99	1.12	0.83	0.80
Nd	7.93	14.68	14.22	8.59	5.71	6.42	6.86	5.46	5.25
Sm	1.51	3.23	3.04	3.06	2.13	2.44	2.58	2.06	2.17
Eu	0.34	0.88	0.85	1.10	0.73	0.85	0.94	0.76	0.77
Gd	1.37	3.13	2.86	4.60	2.98	3.48	3.78	2.90	3.12
Tb	0.20	0.50	0.46	0.77	0.55	0.64	0.69	0.52	0.59
Dy	1.45	3.38	3.18	5.34	4.11	4.65	5.02	3.73	4.46
Ho	0.30	0.66	0.66	1.11	0.82	0.98	1.03	0.75	0.89
Er	0.87	1.93	2.01	3.13	2.48	3.02	3.14	2.20	2.79
Tm	0.14	0.29	0.31	0.51	0.39	0.47	0.50	0.32	0.44
Yb	0.89	1.87	2.02	3.07	2.38	2.99	3.11	2.05	2.90
Lu	0.14	0.27	0.30	0.47	0.37	0.45	0.48	0.31	0.40
Y	8.31	17.78	17.09	29.97	22.89	27.74	28.87	20.18	24.33
Rb	21	54	9	0	11	9	4	12	3
Ba	439	908	216	400	363	90	55	154	61
Th	2	3	5	2	4	3	2	2	3
U	2	2	0	0	0	1	1	2	0
Nb	4	6	7	4	5	3	5	3	4
Zr	92	120	105	73	55	54	57	49	50
Sr	246	80	110	148	154	123	108	185	107
Cr	91	42	287	107	269	89	88	185	146
Ni	26	18	84	54	88	56	64	69	62
V	57	167	211	333	253	174	295	252	345
Pb	6	5	10	12	24	10	12	10	7
La/Yb	8.13	6.08	5.70	0.79	0.77	0.57	0.54	0.82	0.51
Sr/Y	29.6	4.50	6.44	4.94	6.73	4.43	3.74	9.17	4.40

3.3 高 Sr 低 Y 岩石

Defent 等 (2002)[13]指出发育在岛弧区与俯冲洋壳部分熔融有关的埃达克岩（adakite）具有 $SiO_2 \geq 56\%$、$Al_2O_3 \geq 15\%$、$Na_2O \geq 3.5\%$、$Sr > 400$ ppm、$Y \leq 18$ ppm、$Yb \leq 1.9$ ppm、$Sr/Y \geq 40$ 和 $La/Yb \geq 20$ 的特征。依据这些特征，在三岔子岛弧岩浆岩中，强变形闪长岩样品（QLH-3）具有类似的岩石化学及地球化学特征。相当于干组分，它的 $SiO_2 = 58.99\%$、$Al_2O_3 = 15.45\%$、$Na_2O = 3.58\%$、$Sr = 424$ ppm、$Y = 15.32$ ppm、$Yb = 1.32$ ppm。尽管其 $Sr/Y = 27.7$ 和 $La/Yb = 13.3$ 稍低一些，但它较该区其他岛弧型安山岩（$La/Yb < 7.37$）具有最大的轻重稀土分异模型（图 3 (c)）。此外，QLH-5 样品除 SiO_2 稍低外，其他痕量元素特征也与埃达克岩类似，并与本区其他岛弧火山岩形成鲜明反差。尽管这两个样品的某些地球化学特征（如 La/Yb、Sr/Y）还略小于典型的埃达克岩的要求，但 Xu 等 (2000)[14]报道三岔子岩块中的确存在有较典型的埃达克岩石。它们较强的 LREE 对 HREE 分异说明它们可能源于俯冲榴辉岩相洋壳的部分熔融。根据本文两个样品的较高的 MgO 含量（>6%），它们应类似于高 MgO 埃达克岩石，这意味着俯冲洋壳部分熔融产生的熔体在经过地幔时，与地幔岩石又发生了相互作用。它们较高的 Cr (250~303 ppm) 和 Ni (94~110 ppm)含量支持这一解释。Defant 和 Drummond (1990)[15]指出只有年轻（<25 Ma）的，因而较热的洋壳俯冲才能发生部分熔融。三岔子岛弧型岩浆岩中类埃达克岩的存在说明当时勉略带的俯冲洋壳是较年轻洋壳（<25 Ma）。

4 斜长花岗岩的锆石 U-Pb 定年

从斜长花岗岩中分选出来的锆石基本可分为 3 组。第 1 组为颗粒细小的无色透明柱状自形晶体锆石，晶棱锋利鲜明。它们在该样品中占大多数，其晶形及该类锆石的均一性说明它们是岩浆成因锆石。第 2 组为少量较粗粒的紫红色透明柱状晶体。第 3 组为黄色浑圆形锆石，表面带有磨圆痕迹，为典型的碎屑锆石。对无色透明锆石的 6 份样品分析（表 2），获得一条线性很好的不一致线。其上交点年龄为 913 ± 44 Ma，下交点年龄为 301 ± 85 Ma（图 4）。而我们对粗粒紫红色柱状锆石的分析，获得了一组 U-Pb 谐和年龄：$t_{206/238} = 890 \pm 17$ Ma，$t_{207/235} = 895 \pm 21$ Ma，这一谐和年龄与上述不一致线上交点年龄 913 ± 44 Ma 在误差范围内一致。这一谐和年龄说明该区低级变质作用未能导致晚元古代（~900 Ma）岩浆锆石的 Pb 丢失。因此，由第 1 组锆石所确定的不一致线不是锆石 Pb 丢失的结果，而应当是两种锆石的混合线。在 U-Pb 年龄谐和图上，由于紫红色锆石点恰好落在上述不一致线上交点处（图 4），它代表了该混合线的古老锆石端元。显然，那些无色透明锆石的不谐和性说明所有这些颗粒都是由古老的紫红色锆石和外围岩浆锆石增生边组成。因此，它的上交点应指示的是紫红色古老锆石核的形成时代，而下交点年龄应指示无色透明锆石的岩浆增生边形成时代。由于无色透明锆石是锆石样品中的主体，它们的岩浆增生边应当是在斜长花岗岩中晶出的，而紫红色柱状锆石可能是岩浆侵位过程中的捕虏晶。黄色浑圆形锆石是另一种捕虏晶，它具有更古老的 $^{207/206}Pb$ 表面年龄（1566 Ma）。将这 6 个无色透明锆石与该紫红色锆石一起拟合一条不一致线，可获得更精确的上交点年龄（913 ± 31 Ma）和下交点年龄（300 ± 61 Ma）。下交点年龄（300 ± 61 Ma）指示斜长花岗岩的形成时代，而上交点年龄或紫红色锆石谐和年龄指示的是斜长花岗岩部分同化混染物质的年龄。

上述锆石 U-Pb 年代学结果表明三岔子岛弧型斜长花岗岩的形成时代为 300 ± 61 Ma，相当于石炭纪，这与根据偏桥沟硅质岩中放射虫获得的结论是一致的[7]。它说明勉略古洋盆在石炭纪时已开始消减。无独有偶，我们对大别山西北部浠湾构造带中洋壳俯冲成因的低温榴辉岩也获得了 309 ± 3 Ma 的高压变质年龄[15]，它说明勉略构造带可以东延与浠湾构造带相连接。华北与扬子陆块碰撞前的 300 Ma 左右的古洋壳消减事件可能对秦岭-大别造山带具有普遍意义。三岔子斜长花岗岩含有的中、晚元古代锆石捕虏晶表明它侵位的地壳（千枚岩）中含有丰富的中、晚元古代陆壳物质。紫红色柱状锆石的谐和年龄 890 ± 17 Ma 与南秦岭广泛分布的耀岭河群、西乡群和三花石群火山岩年龄相当[16]，这说明三岔子岛弧火山岩块可能代表了南秦岭微陆块（在石炭纪以前已拼合到华北陆块上）南侧石炭纪活动陆缘的岩浆弧，而不是位于古勉略洋中的岛弧。

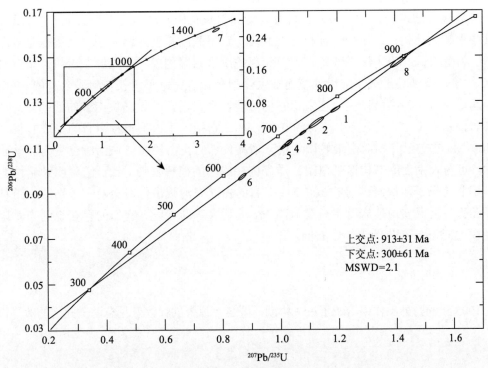

图 4 三岔子偏桥沟斜长花岗岩锆石 U-Pb 年龄谐和图

表 2 三岔子偏桥沟斜长花岗岩锆石的 U-Pb 同位素组成及年龄 a)

样品情况		浓度		样品中普通铅含量/ng	同位素原子比率					表面年龄/Ma		
点号 / 锆石特征	重量/μg	U /(μg·g⁻¹)	Pb /(μg·g⁻¹)		$^{206}Pb/^{204}Pb$	$^{208}Pb/^{206}Pb$	$^{206}Pb/^{238}U$	$^{207}Pb/^{235}U$	$^{207}Pb/^{206}Pb$	$^{206}Pb/^{238}U$	$^{207}Pb/^{235}U$	$^{207}Pb/^{206}Pb$
1 无色透明柱状自形小晶体	30	305	59	0.480	165	0.2122	0.1269(13)	1.194(14)	0.06822(34)	770±8	798±9	875±4
2 无色透明柱状自形小晶体	30	111	17	0.084	310	0.1535	0.1207(23)	1.122(27)	0.06743(93)	735±14	764±18	851±12
3 无色透明柱状自形小晶体	30	340	45	0.080	937	0.1689	0.1178(12)	1.084(13)	0.06678(32)	718±7	746±9	831±4
4 无色透明柱状自形小晶体	30	364	44	0.040	1858	0.1646	0.1113(11)	1.023(12)	0.06668(30)	680±7	716±8	828±4
5 无色透明柱状小晶体	30	278	39	0.210	289	0.1563	0.1106(13)	1.014(14)	0.06647(44)	676±8	711±10	821±5
6 无色透明柱状小晶体	30	370	40	0.055	1218	0.1676	0.0968(10)	0.869(11)	0.06512(35)	595±6	635±8	778±4
7 浅黄色浑圆形,带磨圆痕迹	15	129	58	0.330	106	0.1739	0.2565(39)	3.428(59)	0.09693(72)	1472±22	1511±26	1566±12
8 紫红色透明柱状晶体	7	402	71	0.054	489	0.1738	0.1481(28)	1.414(33)	0.06925(87)	890±17	895±21	906±11

注: ① 1.2.3.4.5.6 号点拟合不一致线, 上交点年龄为 913±44 Ma, 下交点年龄为 301±85 Ma, MSWD 值为 2.6;
② 1.2.3.4.5.6.8 号点拟合不一致线, 上交点年龄为 913±31 Ma, 下交点年龄为 300±61 Ma, MSWD 值为 2.1.

a) 表内 $^{206}Pb/^{204}Pb$ 已对实验空白(Pb=0.050 ng, U=0.002 ng)及稀释剂作了校正. 其中比率中的铅同位素均为放射成因铅同位素. 括号内的数字为 2σ 绝对误差, 例如: 0.1269(13)表示 0.1269±0.0013(2σ).

5 结论

①文献中原称的"三岔子蛇绿混杂岩"应分解成两个不同性质的岩块,三岔子乡以西的岩浆组合为古岩浆弧岩块。它包含的熔岩、辉(闪)长岩及岩墙和斜长花岗岩均为岛弧型玄武-安山质岩石。考虑到安山岩在其中占较大比例,和缺少低钾拉斑玄武岩,它不是岛弧型蛇绿混杂岩而更像是古岩浆弧;三岔子乡以东,庄科变玄武岩为古洋壳残片,它与相邻蛇纹岩共同构成一蛇绿混杂岩块。②三岔子岛弧火山岩具有高 K 性质,反映其形成于陆缘环境而非大洋型岛弧。三岔子斜长花岗岩含有的 890±17 Ma 古老岩浆锆石捕虏晶来源于南秦岭陆壳。这表明它们可能发育在南秦岭微陆块南缘的活动陆缘环境。③三岔子斜长花岗岩的锆石 U-Pb 下交点年龄为 300±61 Ma。它指示了该斜长花岗岩的形成时代。这说明勉略古洋盆在 300 Ma 左右已开始向南秦岭微陆块下消减,那时南秦岭微陆块南缘已变为活动陆缘。这一洋壳消减事件,可与大别山浒湾构造带白垩纪洋壳俯冲成因的榴辉岩带对比。它说明勉略构造带可以东延到大别山。华北与扬子陆块碰撞前的 300 Ma 左右的古洋壳消减事件对秦岭-大别造山带具有普遍意义。④三岔子岛弧型玄武安山质岩石中有类似高 Mg 埃达克岩的岩石。它们的存在说明当时该处的俯冲洋壳是较年轻(<25 Ma)和热的大洋岩石圈。

致谢

赖绍聪在 1995 年的野外工作中给予了很大协助,匿名评议人的评议意见有助于此文的改进,在此一并表示感谢。

参考文献

[1] Sun Shu, Li J, Lin J, et al. Indosinides in China and the consumption of eastern Paleotethys, In controversies in Modern Geology: Evolution of Geological Theories in: Muller D W, Mckenzie J A, Weissert H, ed. Sedimentology, Earth History and Tectonics. London: Academic Press, 1981: 363~384
[2] 张国伟,孟庆任. 秦岭造山带的结构构造. 中国科学, B 辑, 1995, 25(9): 994~1003
[3] Xu Jifeng, Zhang B R, Han Y W. Recognition of ophiolite belt and granulite in Northern Area of Mianlue, Southern Qinling, China and their implication. Journal of China Univ of Sci, 1994, 5(1): 25~27
[4] 李曙光,孙卫东,张国伟,等. 南秦岭勉略构造带黑沟峡变质火山岩的年代学和地球化学——古生代洋盆及其闭合时代的证据. 中国科学, D 辑, 1996, 26(3): 223~230
[5] 许继锋,于学元,李献华,等. 高度亏损的 N-MORB 型火山岩的发现:勉略古洋盆存在的新证据. 科学通报, 1997, 42(22): 2414~2418
[6] 赖绍聪,张国伟,杨永成,等. 南秦岭勉县-略阳结合带变质火山岩岩石地球化学特征. 岩石学报, 1997, 13(4): 564~573
[7] 冯庆来,杜远生,殷鸿福,等. 南秦岭勉略蛇绿混杂岩带中放射虫的发现及其意义. 中国科学, D 辑, 1996, 26(增刊), 78~82
[8] 张国伟,张本仁,袁学诚,等. 秦岭造山带与大陆动力学. 北京:科学出版社, 2001, 467~485
[9] Sun W D, Li S G, chen Y D, et al. Timing of synorogenic granitoids in the South Qinling, central china: constraints of on the evolution of the Qinling-Dabie orogenic belt. Journal of Geology, 2002, 110: 457~468
[10] 陆松年,李惠民. 蓟县长城系大江峪组火山岩的单颗粒锆石 U-Pb 法准确定年. 中国地质科学院院报, 1991, 22: 137~146
[11] Ewart A. The mineralogy and petrology of Tertiary-Recent orogenic volcanic rocks: with special reference to the andesite-basaltic compositional range. In: Andesites, ed. Thorpe R S, New York: John Wiley and Sons, 1982, 25~95
[12] Sun S S, McDonough S M. Chemical and isotopic systematics of oceanic basalts: implications for mantle composition and processes. In: Saunders A D & Norry M J, ed. magmatism in the Ocean Basins. Geological Society Special Publ. No. 42, 1989: 313~345.
[13] Defent M J, 许继锋, Kepezhinskes P, et al. Adakites, some variations on a theme. 岩石学报, 2002, 18(2): 127~142
[14] Xu J F, Wu Q, Yu X Y. Geochemistry of high-Mg andesites and adakitic andesite from the Sanchazi Block of the Mian-Lue ophiolitic mélange in the Qinling Mountains, central China: evidence of partial melting of the subducted Paleo-tethyan crust. Geochemical Journal, 2002, 34: 359~377

[15] Defant M J, Drummoud M S. Derivation of some modern arc magmas by melting of young subducted lithosphere. Nature, 1990, 347: 662~665
[16] Sun W D, Williams I S, Li S G. Carboniferous and Triassic eclogites in the western Dabie mountains, east-central China: evidence for protracted convergence of the North and South China Blocks. J Metamorphic Geol, 2002, 20: 873~886.
[17] 张宗清，张国伟，付国民，等. 秦岭变质地层年龄及其构造意义. 中国科学，D 辑，1996，26(3): 216~222.

大别造山带惠兰山镁铁质麻粒岩 Sm-Nd 和锆石 SHRIMP U-Pb 年代学及锆石微量元素地球化学*

侯振辉[1,2]，李曙光[1]，陈能松[3]，李秋立[1]，柳小明[2]

1. 中国科学技术大学地球和空间科学学院，合肥 230026
2. 西北大学地质学系大陆动力学重点实验室，西安 710069
3. 中国地质大学地球科学学院，武汉 430074

> **亮点介绍**：北大别罗田镁铁质麻粒岩变质作用发生在早白垩世（Sm-Nd 矿物等时线年龄为 136 ± 18 Ma），继承岩浆锆石年龄(716~780 Ma)以及微量元素特征表明罗田镁铁质麻粒岩是就位于下地壳的新元古代镁铁质岩浆岩在早白垩世大别造山带加厚下地壳根部受地幔传导热发生麻粒岩相变质作用而形成的。该麻粒岩的 Sm-Nd 变质年龄（136 ± 18 Ma）与罗田穹隆片麻岩角闪石 K-Ar 年龄（123~127 Ma）封闭温度的差异提供了罗田穹隆早白垩世快速抬升的证据。

摘要 北大别惠兰山位于罗田穹隆的核部，出露有镁铁质麻粒岩，其麻粒岩相变质矿物(石榴子石+单斜辉石+斜方辉石)Sm-Nd 等时线年龄为 136 ± 18 Ma，表明该麻粒岩的变质作用发生在早白垩世。阴极发光图像显示麻粒岩中锆石具有核-幔-边结构。锆石核具有典型岩浆锆石的韵律环带结构及稀土元素特征，其较少 Pb 丢失的锆石 SHRIMP U-Pb 年龄为 753~787 Ma，表明其原岩为新元古代镁铁质岩浆岩。幔部锆石具有切割岩浆锆石环带的蚀变结构特征，且 REE、Th、U、Y、Nb、Ta 等元素含量比岩浆锆石核低 3~10 倍，但普通 Pb 含量较高。这些特征表明幔部锆石是受热液改造的岩浆锆石，其较少 Pb 丢失的锆石 SHRIMP U-Pb 年龄(716~780 Ma)与岩浆锆石相近，指示该岩浆岩体侵位不久即经历了一次强烈的热液事件。考虑到罗田穹隆发育有强烈的早白垩世岩浆事件，因此惠兰山镁铁质麻粒岩是就位于下地壳的新元古代镁铁质岩浆岩在早白垩世大别造山带引张条件下受热发生麻粒岩相变质作用而形成的。该麻粒岩的 Sm-Nd 变质年龄（136 ± 18 Ma）与罗田穹隆片麻岩角闪石 K-Ar 年龄（123~127 Ma）的一致性，提供了罗田穹隆快速抬升的证据，这可能是大别山超高压变质岩被进一步抬升至地表的原因。

关键词：大别造山带；麻粒岩；锆石 U-Pb 年龄；Sm-Nd 同位素年龄

大别造山带麻粒岩主要出露于北大别变质杂岩中，罗田穹隆中的黄土岭长英质麻粒岩和惠兰山一带的镁铁质麻粒岩是最具有典型意义的露头(图 1)[1]。确定这些麻粒岩发生麻粒岩相变质作用的时代及成因对研究大别碰撞造山带板块俯冲前后构造环境及折返抬升历史具有重要意义。目前，长英质麻粒岩较多的地球化学和年代学[2~8][1)]工作表明，它是原岩年龄较为古老(早元古代)的副变质岩，并可能发生了多期麻粒岩相变质作用。而对于惠兰山镁铁质麻粒岩，陈能松等[2]根据麻粒岩附近的变形花岗岩的锆石 U-Pb 年龄推测其麻粒岩相减压退变质时代晚于 227 Ma。Yang 等[9]用锆石蒸发法得到了 $^{207}Pb/^{206}Pb$ 年龄为 443 ± 22 Ma，认为在加里东期曾发生了一期麻粒岩相变质作用。由此可见，该麻粒岩的年代学工作较为薄弱，并且对麻粒岩相变质作用发生的时代存在分歧，有待进一步深入工作。

本文对出露于大别造山带惠兰山的镁铁质麻粒岩进行 Sm-Nd 同位素和锆石 SHRIMP U-Pb 年代学以及锆石微区微量元素研究，以准确测定该麻粒岩的原岩及麻粒岩相变质时代，并探讨了该麻粒岩的成因及构造意义。

---

* 本文发表在：中国科学(D 辑)，2005，35(12)：1103-1111
对应的英文版论文为：Hou Zhenhui, Li Shuguang, Chen Nengsong, et al. 2005. Sm-Nd and zircon SHRIMP U-Pb dating of Huilanshan Mafic granulite in the Dabie Mountains and its zircon trace element geochemistry. Science in China (Ser. D), 48: 2081-2091.

1) 侯振辉. 大别造山带北部麻粒岩和 TTG 片麻岩的地球化学和年代学研究. 硕士学位论文，中国科学技术大学，2000

1 地质概况及样品描述

惠兰山位于罗田穹隆的核部,主要由受穹隆抬升构造作用影响而发生强烈变形的花岗岩组成,部分发育片麻理。在惠兰山顶部出露有少量麻粒岩块体。游振东等[1]对其进行了详细的岩相学研究,结果表明麻粒岩相变质作用的变质温度为682~880 ℃、压力为0.90~1.29 GPa,是在深达35 km以上的下地壳发生。

本文研究样品(01HLS-1)和游振东等[1]所研究样品采自同一露头(图1)(采样地点:30°55.18'N,115°14.66' E),为新鲜的石榴角闪二辉麻粒岩,样品手标本呈灰黑色,镜下观察麻粒岩为细粒变晶结构,主要组成矿物为:石榴子石(5%),普通角闪石(10%),紫苏辉石(10%),单斜辉石(40%),斜长石(30%),石英(3%)和不透明矿物(2%)。普通角闪石主要为麻粒岩相变质的棕色角闪石,与辉石和斜长石共生,常含有石榴石、单斜辉石、斜长石等细粒包裹体。

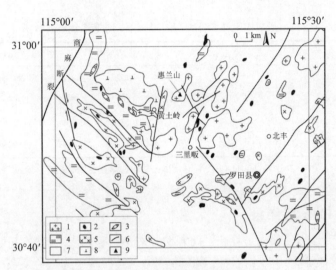

图1 大别山罗田穹隆及采样位置地质简图(据刘贻灿,2003 个人交流,改编)
1.中生代花岗岩;2.镁铁-超镁铁岩;3.大理岩;4.黑云斜长片麻岩;5.花岗闪长片麻岩;6.断层;7.英云闪长片麻岩;8.闪长岩;9.采样地点

2 分析方法

全岩及矿物的Sm-Nd同位素分析的化学处理在中国科学技术大学化学地球动力学实验室完成,分析流程见文献[10]。质谱分析工作在中国科学技术大学化学地球动力学实验室 MAT-262多接收质谱计上进行。质谱测定时同位素分馏校正正常化值 $^{146}Nd/^{144}Nd = 0.721900$,La Jalla标准的 $^{143}Nd/^{144}Nd$ 测定值为 $0.511849 ± 10(2\sigma)$,Sm、Nd化学流程的空白约 $5×10^{-10}$g。年龄计算时采用Ludwig编写的ISOPLOT/Ex (v2.06)程序[11],同位素比值采用2σ误差,分别取 $^{147}Sm/^{144}Nd = 0.2\%$ 和 $^{143}Nd/^{144}Nd = 0.005\%$。

将全岩样品碎至150目后经手工淘洗和磁选法分离出锆石,镜下观察其外形特征主要为无色透明柱状,少量为浅黄色透明柱状锆石。将上述锆石和标准锆石(Temora-417 Ma)一起制作成样品靶[12],抛光后将待测锆石进行透射光、反射光和阴极发光(CL)显微照相。阴极发光照相在中国地质科学院矿产资源研究所电子探针研究室完成。锆石的U-Th-Pb同位素分析在中国地质科学院北京离子探针中心的SHRIMPⅡ离子探针仪器上进行,每个数据点测定由 5次扫描构成,详细测定程序见有关文献[12~14]。每次计数统计过程的分析误差和单点年龄值误差均为1σ。数据处理采用Ludwig的SQUID[15]和ISOPLOT/Ex (v2.06)[11]程序。年龄计算时用实测的 ^{204}Pb 值校正普通Pb。

为更好地解释锆石SHRIMP单点年龄数据,对已完成SHRIMP U-Pb分析的锆石部位进行了微区微量元素分析,分析点位置标于相应的锆石阴极发光照片中(图2)。分析工作在西北大学地质学系大陆动力学实验室LA-ICPMS上完成,详细分析流程见有关文献[16]。实验时ArF激光束工作波长为193 nm,束斑直径为 40 μm,频率为 10 Hz,激光束能量为 140 mJ。以锆石的 SiO_2 含量作为内标,以NIST610为外标,数据处理采用GLITTER程序。西北大学LA-ICPMS对美国地质调查所(USGS)玻璃标准参考物质以及美国国家标准技术研究院（NIST）人工合成硅酸盐玻璃标准参考物质中42种元素进行了分析,分析结果表明,除La、Ce和Pr之外的REE分析的准确度和精度一般优于14%,La、Ce、Pr和其他微量元素均优于10%[16]。

3 分析结果

3.1 Sm-Nd 同位素年龄结果

麻粒岩全岩及单矿物的 Sm-Nd 同位素测定结果列于表 1。由石榴子石+单斜辉石+紫苏辉石获得一条线性很好(MSWD = 0.4)的矿物等时线(图 3),年龄为 136 ± 18 Ma。全岩数据点落在等时线下方附近,这可能是与全岩中含有少量退变质矿物有关。因此,由石榴子石、单斜辉石、紫苏辉石 3 个麻粒岩相变质矿物确定的等时线年龄 136 ± 18 Ma,指示了惠兰山麻粒岩发生麻粒岩相变质作用的时代。

表 1 惠兰山麻粒岩(01HLS-1) Sm-Nd 同位素分析数据

样品号	$Sm/\mu g \cdot g^{-1}$	$Nd/\mu g \cdot g^{-1}$	$^{147}Sm/^{144}Nd$	$^{143}Nd/^{144}Nd\ (\pm 2\sigma)$	$\varepsilon_{Nd}(t)$
全岩 Wr	8.17	32.14	0.1537	0.511901±10	−13.2
单斜辉石 Cpx	9.88	29.52	0.2022	0.511973±5	
紫苏辉石 Hy	1.47	5.60	0.1584	0.511923±14	
石榴石 Grt	4.72	6.41	0.4454	0.512183±10	

表中 $\varepsilon_{Nd}(t) = [(^{143}Nd/^{144}Nd)_{样品}(t) / (^{143}Nd/^{144}Nd)_{CHUR}(t) - 1] \times 10^4$,其中 $(^{143}Nd/^{144}Nd)_{CHUR}(0) = 0.512638$, $(^{147}Sm/^{144}Nd)_{CHUR}(0) = 0.1967$; $t = 136$ Ma。

3.2 锆石 SHRIMP U-Pb 年龄结果

样品锆石颗粒较小,大多为长柱或短柱状,CL 照片清楚地显示大多数锆石具有核-幔-边结构(图 2)。最内部为发育典型韵律环带结构的岩浆锆石核。幔部锆石在 CL 照片中为白色或灰白色,并且表现出了切割岩浆锆石核韵律环带的结构特征(图 2(a), (g)),部分白色幔部仍可见岩浆锆石韵律环带的残留(图 2(g)),它表明白色幔部锆石可能是被晚期事件改造过的岩浆锆石。锆石最外部为非常窄的灰色边(图 2(a), (b)),可能为后期变质增生锆石。

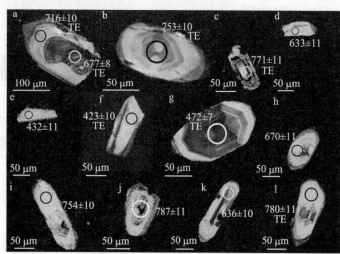

图 2 惠兰山麻粒岩(01HLS-1)锆石结构 CL 照片
图中 TE 表示锆石同时进行微量元素分析

对岩浆锆石核和白色幔部的 13 个点进行的 SHRIMP 单点年龄分析结果见表 2。由于锆石的 U 含量较低因而导致放射成因 Pb 含量很低,低的 ^{207}Pb 计数会引起较大的 ^{207}Pb 测定误差,因此文中讨论的年龄值均采用 $^{206}Pb/^{238}U$ 年龄。表 2 显示,岩浆锆石核的 Th/U 值范围为 0.7~2.1,年龄值范围为 472~787 Ma(表 2,图 4)。而白色幔部锆石的 Th/U 值范围为 0.8~1.6,年龄值范围为 423~780 Ma。同一颗锆石中(图 2)岩浆锆石核(分析点 A9, 见表 2)的年龄值为 677 ± 8 Ma,白色幔部锆石(分析点 A1, 见表 2)的年龄值为 716 ± 10 Ma,暗色的岩浆锆石核部年龄小于白色幔部年龄可能是由于核部 U 含量(182 $\mu g \cdot g^{-1}$)高于幔部 U 含量(71 $\mu g \cdot g^{-1}$),因此其蜕晶化程度较高,在后期热事件影响下导致其 Pb 丢失量高于幔部锆石所造成的。锆石边缘发育的灰色变质边因过于狭窄而未能进行离子探针年龄分析(图 3)。由于本文麻粒岩相变质矿物 Sm-Nd 年龄已证明该麻粒岩相变质作用发生在早白垩世,因此该锆石最外层灰色边可能是后期麻粒岩相变质作用时期形成的变质增生锆石部分。

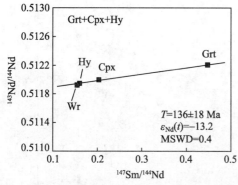
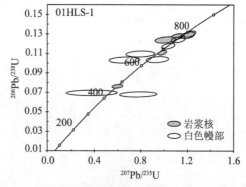

图 3 惠兰山麻粒岩(01HLS-1)矿物 Sm-Nd 等时线年龄　　　图 4 惠兰山麻粒岩(01HLS-1)锆石 U-Pb 一致曲线图

表 2 惠兰山麻粒岩(01HLS-1)锆石 SHRIMP U-Pb 年龄分析结果 a)

分析点号	CL特征	U /μg·g⁻¹	Th /μg·g⁻¹	^{232}Th $/^{238}$U	Pb* /μg·g⁻¹	Pb$_c$ /%	^{207}Pb* $/^{206}$Pb*	±%	^{207}Pb* $/^{235}$U	±%	^{206}Pb* $/^{238}$U	±%	^{206}Pb/^{238}U 年龄/Ma	^{207}Pb/^{206}Pb 年龄/Ma	^{208}Pb/^{232}Th 年龄/Ma
A1	幔	71	54	0.8	7.17	0.99	0.06360	3.2	1.030	3.5	0.1175	1.5	716±10	728±68	740±20
A3	核	140	95	0.7	15.0	1.58	0.05900	5.5	1.007	5.7	0.1238	1.5	753±10	566±120	786±32
A4	核	182	124	0.7	20.0	0.35	0.06290	2.9	1.102	3.3	0.1271	1.6	771±11	704±61	751±21
A5	幔	68	98	1.5	6.18	1.73	0.04620	13	0.6570	13	0.1031	1.8	633±11	7±310	604±26
A6	幔	43	44	1.0	2.66	9.93	0.08300	14	0.7800	14	0.06780	2.3	423±10	1269±260	538±42
A7	幔	57	54	1.0	3.63	5.50	0.03900	38	0.3800	38	0.06940	2.6	432±11	-396±1000	395±56
A8	核	153	317	2.1	10.1	2.29	0.05690	4.6	0.5960	4.8	0.07600	1.4	472±7	487±100	484.2±9.9
A9	核	182	244	1.4	17.4	0.69	0.06430	1.9	0.9820	2.3	0.1107	1.3	677±8	752±40	693±12
A10	幔	89	132	1.5	8.54	n.d.	0.05250	10	0.7930	10	0.1095	1.7	670±11	308±230	631±23
A11	幔	133	102	1.5	14.2	0.80	0.06520	3.3	1.116	3.5	0.1241	1.4	754±10	782±69	779±19
A13	核	121	156	1.3	13.5	1.27	0.06800	2.2	1.218	2.7	0.1298	1.5	787±11	870±46	811±16
A14	幔	100	160	1.6	8.93	3.82	0.06580	6.4	0.9400	6.6	0.1037	1.5	636±10	799±130	696±20
A15	幔	116	139	1.2	12.9	n.d.	0.06760	3.1	1.198	3.4	0.1285	1.5	780±11	857±64	748±25

a) 数据误差均为 1σ; Pbc 和 Pb*分别表示普通和放射成因铅; 年龄值用实测的 ^{204}Pb 值进行普通铅校正。

3.3 锆石的微区微量元素特征

为了判定上述锆石岩浆核及白色幔的成因关系, 我们对这两种已做离子探针年龄分析的锆石相应部位进行了原位激光剥蚀微量元素分析。分析结果(表 3)表明, 在微量元素球粒陨石标准化图解上(图 5(a)), 白色幔部锆石和核部岩浆锆石均表现出了正 Ce 负 Eu 异常和重稀土元素陡峭的岩浆锆石稀土元素曲线特征。二者的 Th/U 值均在 0.8~2.1 之间。但白色幔部锆石稀土元素曲线位于核部岩浆锆石的下方, 其稀土元素和 Th、U、Y、Nb、Ta 含量均比岩浆锆石核的低 3~10 倍(图 5(b))。例如同一颗锆石(图 3(a))中, 岩浆锆石核(分析点 A9)的 Y 含量为 1188 μg·g⁻¹, 稀土元素含量为 907.0 μg·g⁻¹; 而白色幔部锆石(分析点 A1)的 Y 含量为 262 μg·g⁻¹, 稀土元素含量仅为 229.2 μg·g⁻¹。这些锆石在上述微量元素特征上的差异及年龄上的接近可能是岩浆锆石从熔体中结晶后不久即受到岩浆期后热液改造的结果, 白色幔部锆石可能就是被改造过的原岩岩浆锆石(详见下文讨论部分)。

表 3 惠兰山麻粒岩(01HLS-1)锆石微量元素含量(μg·g⁻¹)

分析点	CL特征	La	Ce	Pr	Nd	Sm	Eu	Gd	Tb	Dy	Ho	Er	Tm	Yb	Lu	Y	Nb	Ta	Hf	Th	U	Th/U
A3	岩浆核	0.13	37.4	0.20	2.97	6.11	1.57	25.8	9.60	123	48.2	210	49.6	527	103	1473	4.09	0.49	8428	144	97.9	1.5
A4	岩浆核	5.24	43.8	2.52	15.4	7.68	1.10	32.6	10.4	131	47.0	208	44.8	474	79.2	1374	7.19	1.03	8027	119	213	0.6
A8	岩浆核	2.64	41.9	1.06	11.8	13.1	2.56	39.3	13.6	167	60.3	264	54.6	517	97.3	1820	2.69	0.27	6752	238	119	2.0
A9	岩浆核	2.49	71.3	0.66	4.86	5.31	1.33	23.2	8.47	102	42.0	178	39.3	365	63.1	1188	4.65	0.82	6799	183	130	1.4
A1	白色幔部	0.12	13.8	0.10	0.72	1.36	0.25	4.02	1.69	19.3	8.75	43.7	9.78	106	19.6	262	2.44	0.39	8037	52.1	71.7	0.7
A6	白色幔部	0.15	19.1	0.11	0.73	1.06	0.56	7.61	3.17	37.6	16.2	76.3	17.0	181	32.1	493	1.73	0.25	7297	41.3	39.9	1.0
A15	白色幔部	0.14	11.7	0.15	0.96	1.14	0.28	3.26	1.73	20.7	8.67	43.7	9.93	96.3	16.8	227	6.09	0.35	7392	22.7	27.4	0.8
检出限		0.05	0.05	0.05	0.25	0.35	0.08	0.05	0.06	0.22	0.04	0.06	0.12	0.05	0.05	0.11	0.10	0.06	0.32	0.16	0.16	

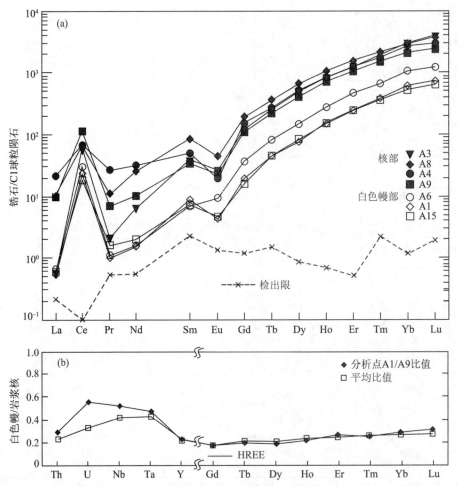

图 5 惠兰山麻粒岩锆石稀土元素球粒陨石标准化图(a)和白色幔部锆石相对岩浆锆石核的微量元素丰度(b)

C1 球粒陨石稀土元素含量取自 Sun 等[17]

4 讨论

4.1 惠兰山麻粒岩的麻粒岩相变质时代和原岩时代

惠兰山麻粒岩的麻粒岩相变质矿物 Sm-Nd 等时线年龄为 136 ± 18 Ma，这清楚表明其发生麻粒岩相变质作用的时代为早白垩世。

惠兰山麻粒岩中的锆石结构主要由岩浆锆石核和灰白色幔部锆石组成，其 SHRIMP U-Pb 单点年龄值具有如下特点：(1)变化范围较大；(2)年龄值均高于 423 Ma，岩浆锆石核部较少 Pb 丢失的单点 SHRIMP U-Pb 年龄值的高值十分接近，3 个(A3, A4, A13)岩浆核 SHRIMP 单点年龄值在 753~787 Ma 处形成较集中分布(图 4)，这表明该麻粒岩的这些具有岩浆结晶特征的锆石形成于新元古代；(3)没有发现早白垩世锆石年龄记录。那么，惠兰山麻粒岩的岩浆锆石与其原岩的关系存在两种可能性，一是其原岩为新元古代镁铁质岩浆岩，锆石在原岩形成时结晶，而麻粒岩的 Sm-Nd 变质年龄结果表明锆石在早白垩世麻粒岩相变质作用时发生少量变质增生。二是其原岩为早白垩世镁铁质岩浆岩，板底垫托就位于下地壳底部，锆石来源于镁铁质岩浆侵入过程中捕获的下地壳新元古代变质基底的锆石，而镁铁质岩浆本身由于含 Zr 量较低，没有结晶出新生变质锆石。然而，如果麻粒岩中的锆石为早白垩世镁铁质岩浆侵入过程中捕获的古老继承锆石，由于它们将受到高温玄武质岩浆（$T > 1000$ ℃）的改造作用，锆石蜕晶化部分会发生大量 Pb 丢失，将会具有如下特征：(1)存在接近 136 Ma 年龄值的锆石。但是离子探针分析结果已表明所有锆石年龄值均高于 423 Ma；(2)锆石应有蜕晶化部分和变质重结晶特征。但这些锆石均明显不具有变质重结晶锆石特征，详见下文 4.2 中讨论部分；(3)若幔部锆石是受到岩浆热改造的结果，那么由于发生较多的 Pb 丢失，其离子探针年龄值分布主体上应小于核部岩浆锆石，但实测结果明显不具有这种特征。因此本研究认为惠兰山麻粒岩样品 01HLS-1

的原岩应为新元古代镁铁质岩浆岩,和大别山超高压变质岩的原岩年龄范围(700~800 Ma)[18~22]一致,这些年龄对应了扬子陆块北缘新元古代强烈岩浆活动时期[22~24]。而样品中岩浆锆石核和白色幔部锆石的单点 SHRIMP U-Pb 年龄值变化范围较大,这种较分散的年龄值可能是由于它们在后期的麻粒岩相高级变质作用中 Pb 丢失程度不同所致。

侏罗纪末-白垩纪初,大别-苏鲁造山带从东西向构造体制转变为滨太平洋北北东向构造体制,华北板块出现区域性构造拉张[25,26],从 140 Ma 左右开始,大别山进入热隆伸展作用时期[27]。大别山增厚地壳因拉张而减薄以及深部岩石圈拆离事件[28],均可导致高温软流圈地幔上涌和大规模岩浆事件。此外,底侵的玄武岩浆也有可能提供大量的热能,使大别山的下地壳长英质岩石受热发生了部分熔融,而包裹其中的原岩时代为新元古代的镁铁-超镁铁岩包体(较难熔融)因温度升高发生麻粒岩相变质作用。随着罗田穹隆的隆升,镁铁质麻粒岩块体随长英质岩石一起迅速上升,最后剥蚀出露于地表,形成现在的惠兰山麻粒岩。如果对惠兰山镁铁质麻粒岩的这一成因认识是正确的,则该麻粒岩的变质时代(136 ± 18 Ma)就对大别山从东西向陆-陆碰撞构造体制向北北东向太平洋构造体制转变发生时代提供了重要制约,这与翟明国等[29]综合其他资料所提出的中国东部构造体制转变发生在 140 Ma 左右的见解是一致的。

4.2 热液改造锆石的微量元素特征及新元古代热液事件

虽然锆石在长期的地质过程中非常稳定,不易发生 Pb 丢失[30],但已有实验表明,在含有 F、Cl、Ca 等离子的热液作用下,即使是晶质锆石在低温下(400~600 ℃)很短时间内也很容易发生蚀变造成放射成因 Pb 丢失[31,32],而部分蜕晶质锆石则会发生大量 Pb、Hf、REE 和 U 的丢失[33]。热液携带的普通 Pb 进入锆石晶格可以造成被改造的锆石部位的普通 Pb 含量升高[34~36]。热液蚀变锆石的典型结构特征是相对原生岩浆锆石核较浅的 CL 图像和表面控制蚀变(Surface Control Alteration,SCA)环带结构[36]。

惠兰山麻粒岩锆石白色幔部具有的岩浆锆石残留韵律环带(例如图 3(c))以及和岩浆锆石类似的稀土元素标准化模式和高 Th/U 值,表明白色幔部锆石并非后期变质增生锆石。而蜕晶质锆石的重结晶作用不会导致重结晶锆石和原岩岩浆锆石 3~10 倍如此之高的微量元素差异[37],这说明白色幔部锆石并非变质重结晶锆石。岩浆锆石核部和白色幔部锆石的 SHRIMP 年龄最大值十分接近(核: 753~787 Ma; 幔: 716~780 Ma)。SHRIMP U-Pb 分析结果显示,白色幔部的普通 Pb(Pb_c%为 0.80~9.93)显著高于岩浆锆石核(Pb_c%为 0.35~2.29)(表 2)。由于变质重结晶作用也不会增加锆石的普通 Pb 含量,而热液成因锆石普遍具有较高的普通 Pb 含量[34~36]。综合考虑惠兰山麻粒岩锆石白色幔部具有与岩浆锆石韵律环带相交切的 SCA 结构特征(例如图 3(b)),普通 Pb 较岩浆锆石核要高,而微量元素比岩浆锆石低 3~10 倍的特点,3 个较少 Pb 丢失的白色幔部锆石 U-Pb 年龄集中在 716~780 Ma,它和岩浆锆石的结晶时代很相近,可以判定该锆石的改造作用为岩浆锆石形成后不久即受到的热液蚀变作用,它造成岩浆锆石大量微量元素的丢失和普通 Pb 升高。这说明该热液事件可能是同期岩浆演化晚期的热液活动,或者是该岩浆岩侵入体提供热源驱动的地下热水循环。

4.3 大别山早白垩世的快速隆升事件

Li 等[38]通过对大别山超高压变质岩的热年代学研究已经揭示现在出露于地表的大别山超高压变质岩在峰期变质 226 Ma 到 170 Ma 之间从 800℃到 300℃经历了两次快速冷却和一个等温阶段的冷却历史: 即 226 ~ 219 Ma 期间经历了第一次快速冷却阶段(40℃·Ma^{-1}),从 800℃冷却到 500℃;在 219 ~ 180 Ma 期间为一等温阶段;从 180 Ma 到 170 Ma 又经历了第二次快速冷却阶段(~ 15℃·Ma^{-1}),从 450 ℃冷却到 300 ℃[38,39]。Ratschbacher 等[25]和 Li 等[28]指出大别山早白垩世的引张和穹隆构造可导致大别山体和超高压变质岩的进一步抬升。王国灿等[40]利用穹隆不同矿物的 K-Ar 年龄大致估算了这一穹隆构造导致的山体抬升速率。本文利用获得的穹隆核部麻粒岩相岩石的 Sm-Nd 年龄可对其抬升速率给出新的制约。

罗田穹隆是东大别规模最大的穹隆,其隆升剥露影响整个大别造山带的隆升剥露格局[41]。罗田穹隆核部的斜长片麻岩中角闪石的 K-Ar 年龄为 123~127 Ma[40],对应封闭温度约 500 ℃[42],惠兰山镁铁质麻粒岩在约 136 Ma 时经历的峰期变质温度若取 800 ℃[1],则我们可以计算出罗田穹隆核部在早白垩世从 800 ℃冷却到 500 ℃的冷却速率约为 25℃·Ma^{-1}。这一快速冷却进一步证明了当时罗田穹隆的隆升速率是很快的[27,40,41]。这个冷却速率远大于王国灿等[40]用角闪石和黑云母 K-Ar 年龄计算得到的从 500℃冷却到 300℃时的速率(6.5~14.8 ℃·Ma^{-1})。这说明罗田穹隆在开始时有一快速抬升过程，是它导致了其快速的冷却过程。Kay[43]和 Li 等[28]指出，造山带岩浆作用与山体快速抬升的耦合是深部发生岩石圈拆离事件的证据。因此以罗田穹隆的快速隆升为代表的大别山第三次快速抬升过程可能是大别造山带深部发生拆离或去根事件的结果，它是导致大别山超高压变质岩在早白垩世进一步抬升出露地表的重要机制之一。

5 结论

（1）惠兰山镁铁质麻粒岩中岩浆锆石 SHRIMP U-Pb 年代学研究表明，其原岩为新元古代镁铁质岩浆岩。惠兰山镁铁质麻粒岩的麻粒岩相变质矿物 Sm-Nd 等时线年龄为 136±18 Ma，证明该麻粒岩的变质作用发生在早白垩世，与大别山大规模早白垩世岩浆事件同期。因此惠兰山镁铁质麻粒岩的成因可能是位于扬子北缘的下地壳部位新元古代镁铁质岩浆岩，在早白垩世引张构造环境下受到上涌软流圈地幔的热烘烤发生麻粒岩相变质作用。

（2）锆石 SHRIMP 年龄和微量元素特征表明，麻粒岩原岩在侵位不久即经历了一次强烈的热液事件，导致被改造锆石的普通 Pb 升高和微量元素含量降低。这部分被改造的锆石环绕在岩浆锆石核的外部，它们基本不具有韵律环带特征而和变质锆石的 CL 特征相似，但是这种锆石既不是后期变质增生锆石，也不是变质重结晶锆石，而是经历热液交代改造形成的。这种热液改造锆石的识别对解释其年代学结果具有重要的意义。

（3）将惠兰山麻粒岩的变质温度(800℃)和变质时代(136 Ma)与片麻岩围岩的角闪石 K-Ar 年龄(123~127 Ma)比较获得了罗田穹隆早白垩世从 800℃冷却到 500℃的冷却速率约为 25 ℃·Ma^{-1}，这指示了该穹隆的一次快速隆升过程。它与大别山早白垩世大规模岩浆事件的耦合关系可能暗示了造山带深部的拆离和去根过程。这可能是大别山超高压变质岩在早白垩世被进一步抬升至近地表的原因。

致谢

离子探针锆石 U-Pb 年龄分析得到了北京离子探针中心刘敦一研究员、简平研究员、宋彪研究员和陶华工程师的大力支持和帮助，锆石 CL 照相得到了中国地质科学院余静和陈振宇老师的帮助，Sm-Nd 同位素化学处理得到了陈江峰教授、钱卉工程师的大力支持和李秋立、王勇刚同学的帮助，质谱分析得到了彭子成教授和贺剑峰老师的大力支持，匿名评审人的评审意见对本文的修改有很大帮助，在此一并表示衷心的感谢。

参考文献

[1] 游振东, 陈能松, Chalokwu C I. 大别山区深部地壳的变质岩石学证迹: 罗田惠兰山一带的麻粒岩研究. 岩石学报, 1995, 11(2): 137~147
[2] 陈能松, 游振东, 索书田, 等. 大别山区中酸性麻粒岩和变形花岗岩的锆石 U-Pb 年龄. 科学通报, 1996, 41(11): 1009~1012
[3] Jian P, Yang W R, Zhang Z C. ^{207}Pb/^{206}Pb zircon dating of Huangtuling hypersthene-garnet-biotite gneiss from the Dabie mountains, Luotina county, Hubei province, China: new evidence for early Precambrian evolution. Acta Geologica Sinica, 1999, 73(1): 78~83
[4] Zhou H, Li X, Liu Y, et al. Age of granulite from Huangtuling in the Dabie Mountains: Pb-Pb dating of garnet by a stepwise dissolution technique. Chinese Sci Bull, 1999, 44: 941~944
[5] Ma C Q, Ehlers C, Xu C, et al. The roots of the Dabieshan ultrahigh-pressure metamorphic terrain: constraints from geochemistry and Nd-Sr isotope syste Matics. Precambrian Research, 2000, 102: 279~301
[6] Zheng Y F, Fu B, Li Y L, et al. Oxygen isotope composition of granulites from Dabieshan in eastern China and its implications for geodynamics of Yangtze plate subduction. Phys Chem Earth(A), 2001, 26: 673~684
[7] 吴元保, 陈道公, 夏群科, 等. 大别山黄土岭麻粒岩中锆石 LAM-ICP-MS 微区微量元素分析和 Pb-Pb 定年. 中国科学, D辑, 2003, 33(1): 20~28
[8] 吴元保, 陈道公, 夏群科, 等. 北大别黄土岭麻粒岩锆石 U-Pb 离子探针定年. 岩石学报, 2002, 18(3), 378~382

[9] Yang W R, Jian P. Geochronological study of Caledonian granulite and high-pressure gneiss in the Dabie mountains. Acta Geologica Sinica. 1998, 72(3): 264~270
[10] Foland K A, Allen J C. Mag Ma sources for Mesozoic anorogenic granites of the White Mountain Mag Ma series, New England, USA. Contrib Mineral Petrol, 1991, 109: 195~211
[11] Ludwig K R. User's Manual for Isoplot/EX, v2.06, A geochronological Toolkit for Microsoft Excel. Berkely Geochronological Center, Special Publication. 1999: 47
[12] 宋彪, 张玉海, 万渝生, 等. 锆石 SHRIMP 样品靶制作、年龄测定及有关现象讨论. 地质论评, 2002, 48(增刊): 26~30
[13] Williams I S, Claesson S. Isotopic evidence for the Precambrian provenance and Caledonian metamorphism of high grade paragneisses from the Seve Nappes, Scandinavian Caledonides: II Ion microprobe zircon U-Th-Pb. Contributions to Mineralogy and Petrology, 1987, 97: 205~217
[14] Compston W, Williams I S, Kirschvink J L, et al. Zircon U-Pb ages for the early Cambrian time-scale. J Geol Soc, 1992, 149: 171~184
[15] Ludwig K R. Squid 1.02: a user Manual. Berkeley Geochronological Center, Special Publication. 2001: 1~19
[16] Gao S, Liu X M, Yuan H L, et al. Determination of forty two Major and trace elements in USGS and NIST SRM glasses by Laser ablation-inductively coupled plas Ma- Mass spectrometry. The Journal of Geostandards and Geoanalysis, 2002, 26(2): 181~196
[17] Sun S S, McDonough W F. Chemical and isotope syste Matics of oceanic basalts: implications for Mantle composition and processes. In: Saunders A D, Norry M J, ed. Mag Matism in the Ocean Basins. Geological Society Special Publication, London, 1989, 42: 313~345
[18] Ames L, Zhou G Z, Xiong B C. Geochronology and isotopic character of ultrahigh pressure metamorphism with implications for collision of the Sino-Korean and Yangtze cratons, central China. Tectonics, 1996, 15: 472~489
[19] Rowley D B, Xue F, Tucker R D, et al. Ages of ultrahigh pressure metamorphic and protolith orthgenisses from the eastern Dabie Shan: U/Pb zircon geochronology. Earth Planet Sci Lett, 1997, 151: 191~203
[20] 徐惠芬, 杨天南, 刘福来, 等. 苏鲁高压-超高压变质带南部花岗片麻岩-花岗岩的多时代演化. 地质学报, 2001, 75 (3): 371~378
[21] Rumble D, Giorgis D, Oreland T, et al. Low δ^{18}O zircons, U-Pb dating, and the age of the Qinglongshan oxygen and hydrogen isotope ano Maly near Donghai in Jiangsu Province, China. Geochim Cosmochim Acta, 2002, 66: 2299~2306
[22] 郑永飞, 陈福坤, 龚冰, 等. 大别-苏鲁造山带超高压变质岩原岩性质: 锆石氧同位素和 U-Pb 年龄证据. 科学通报, 2003, 48(2): 110~119
[23] Hacker B R, Ratschbacher L, Webb L, et al. U/Pb zircon ages constrain the architecture of the ultrahigh-pressure Qinling-Dabie orogen, China. Earth Planet Sci Lett, 1998, 161: 215~230
[24] 李曙光, 洪吉安, 李惠民, 等. 大别山辉石-辉长岩体的锆石 U-Pb 年龄及其地质意义. 高校地质学报, 1999, 5(3): 351~355
[25] Ratschbacher L, Hacker B R, Webb L E, et al. Exhu Mation of the ultrahigh-pressure continental crust in east central China: Cretaceous and Cenozoic unroofing and the Tan-Lu fault. J Geophys Res, 2000, 105: 13303~13338
[26] 朱光, 王道轩, 刘国生, 等. 郯庐断裂带的伸展活动及其动力学背景. 地质科学, 2001, 36(3): 269~278
[27] 许长海, 周祖翼, 马昌前, 等. 大别造山带 140~85 Ma 热窿伸展作用——年代学约束. 中国科学, D 辑, 2001, 31(11): 925~937
[28] 李曙光, 黄方, 李晖. 大别-苏鲁造山带碰撞后的岩石圈拆离. 科学通报, 2001, 46(17): 1487~1491
[29] 翟明国, 朱日祥, 刘建明, 等. 华北东部中生代构造体制转折的关键时限. 中国科学, D 辑, 2003, 33(10): 913~920
[30] Lee J, Williams I, Ellis D. Pb, U and Th diffusion in nature zircon. Nature, 1997, 390: 159~162
[31] Hansen B T, Frederichsen J D. The influence of recent Pb-loss on the interpretation of disturbed U-Pb systems in zircons from igneous rocks in East Greenland. Lithos, 1989, 23: 209~223
[32] Sinha A K, Wayne D M, Hewitt D A. The hydrother Mal stability of zircon—preliminary experimental and isotopic studies. Geochim Cosmochim Acta, 1992, 56: 3551~3560
[33] Geisler T, Pidgeon R T, Kurtz R. Experimental hydrother Mal alteration of partially metamict zircon. American Mineralogist, 2003, 88: 1496~1513
[34] Watson E B, Cherniak D J, Hanchar J M, et al. The incorporation of Pb into zircon. Chemical Geology, 1997, 141: 19~31
[35] Schaltegger U, Fanning C M, Geünther D. Growth, annealing and recrystallization of zircon and preservation of monazite in high-grade metamorphism: conventional and in-situ U-Pb isotope, cathodoluminescence and microchemical evidence. Contributions to Mineralogy and Petrology, 1999, 134: 186~201
[36] Vavra G, Schmid R, Gebauer D. Internal morphology, habit and U-Th-Pb microanalysis of amphibolite-to granulite facies zircons: geochronology of the Ivrea Zone (Southern Alps). Contributions to Mineralogy and Petrology, 1999, 134: 380~404
[37] Hoskin P W O, Black L P. Metamorphic zircon for Mation by solid-state recrystallization of protolith igneous zircon. J Metamorphic Geol, 2000, 18: 423~439

[38] Li S G, Jagoutz E, Chen Y Z, et al. Sm-Nd and Rb-Sr isotopic chronology and cooling history of ultrahigh pressure metamorphic rocks and their country rocks at Shuanghe in the Dabie Mountains, central China. Geochim Cosmochim Acta, 2000, 64(6): 1077~1093

[39] Li S G, Jagoutz E, Lo C H, et al. Sm-Nd, Rb-Sr and ^{40}Ar-^{39}Ar isotopic syste Matics of the ultrahigh pressure metamorphic rocks in the Dabie-Sulu belt, central China: a retrospective view. International Geol Review, 1999, 41(12): 1114~1124

[40] 王国灿, 杨巍然. 大别山核部罗田穹隆形成的构造及年代学证据. 地球科学——中国地质大学学报, 1996, 21(5): 524~528

[41] 王国灿, 杨巍然. 大别造山带中新生代隆升作用的时空格局——构造年代学证据. 地球科学——中国地质大学学报, 1998, 23(5): 461~467

[42] Harrison T M. Diffusion of ^{40}Ar in hornblende. Contributions to Mineralogy and Petrology, 1981, 78: 324~331

[43] Kay R W, Kay S M. Delamination and delamination Mag Matism. Tectonophysics, 1993, 219: 177~189

西秦岭关子镇蛇绿混杂岩的地球化学和锆石 SHRIMP U-Pb 年龄*

李王晔[1]，李曙光[1,2]，裴先治[3]，张国伟[2]

1. 中国科学院壳幔物质与环境重点实验室，中国科学技术大学地球和空间科学学院，合肥 230026
2. 西北大学大陆动力学国家重点实验室，西安 710069
3. 长安大学地球科学与国土资源学院，西安 710054

> **亮点介绍**：微量元素地球化学研究表明，西秦岭沿天水武山断裂带出露的关子镇蛇绿混杂岩中，变中-基性深成岩浆岩包含了"古洋壳"和"古岛弧"两种类型。其中"古洋壳"成因的变辉长岩和变斜长花岗岩样品的锆石 SHRIMP U-Pb 年龄分别为 534±9 Ma 和 517±8 Ma，表明西秦岭"天水武山洋"形成于早-中寒武世；它与东秦岭的"商丹洋"共同构成了华北陆块南缘早古生代"秦岭洋"。该研究结果为东秦岭"商丹缝合带"西延与西秦岭"天水武山缝合带"相连提供了确凿的地球化学和年代学证据。

摘要 西秦岭沿天水武山断裂带产出的关子镇蛇绿混杂岩由变玄武岩、变中-基性深成岩浆岩以及少量蛇纹岩组成。变玄武岩具有平坦或轻稀土略亏损的稀土配分型式（$(La/Yb)_N = 0.63 \sim 1.24$），且不存在 Nb、Ta 和 Ti 的负异常，表明它们形成于洋中脊环境。变中-基性深成岩浆岩包含两种类型：类型 I 显示轻稀土富集（$(La/Yb)_N > 2.2$）、具有显著 Nb 负异常（$(La/Nb)_N > 2.4$）的特征，类似于岛弧火山岩；类型 II 显示轻稀土亏损到略富集（$(La/Yb)_N = 0.44 \sim 1.38$）、无显著 Nb 负异常（$(La/Nb)_N < 1.5$），这些特征与上述变玄武岩类似，表明它们也是关子镇蛇绿岩的组成部分。蛇绿岩中辉长岩和闪长岩（该闪长岩属"大洋斜长花岗岩类"）样品的锆石 SHRIMP U-Pb 年龄分别为 534±9 Ma 和 517±8 Ma，这反映关子镇蛇绿岩可能为早-中寒武世古洋壳残片。由于西秦岭的天水武山断裂带是东秦岭商丹断裂带的西延并共同被认为是华北和华南陆块之间的主缝合线，上述年龄对理解该带及古秦岭洋的演化提供了重要制约。

关键词：西秦岭；蛇绿岩；锆石 SHRIMP U-Pb 定年；大洋斜长花岗岩类

1 引言

西秦岭造山带在大地构造位置上处于古亚洲构造域和特提斯构造域交汇的特殊地段（任纪舜等，1999；张国伟等，2001），也是东秦岭与祁连和东昆仑造山带衔接的关键部位。精确测定区内出露的古岛弧岩浆岩、蛇绿岩及变质基底岩石的年龄与地球化学性质并与邻区相关岩石进行年代学对比，不仅为研究西秦岭造山带的演化，而且对查明我国中西部各主要构造单元（东秦岭、西秦岭、祁连山、东昆仑）的衔接关系提供重要制约。

张维吉等（1994）通过岩相学研究认为西秦岭天水地区出露的"李子园群"对应于东秦岭的"丹凤群"，据此提出东秦岭的商丹断裂带西延与西秦岭的天水武山断裂带相衔接。但是上述观点尚缺乏西秦岭地区相关岩石的可靠地球化学和年代学证据。

关子镇（图 1）位于天水以西 35 km，天水武山断裂东端。裴先治等（2004）已在关子镇地区识别出一套蛇绿混杂岩组合，由大套的变玄武岩、呈孤立的小岩体断续分布的变中-基性深成岩浆岩和少量蛇纹岩组成。裴先治等（2005）将变玄武岩北侧沿 NWW-SEE 方向断续分布的变辉长岩、辉长闪长岩和闪长岩体统称为"流水沟变质中基性岩浆杂岩体"，并依据三件样品的地球化学数据认为"该杂岩体属岛弧环境下形成的辉长岩-辉长闪长岩-闪长岩组合"；对流水沟口的变辉长闪长岩（未给出该样品的地球化学数据）获得了 TIMS 法锆石

---

* 本文发表在：岩石学报，2007，23(11)：2836-2844

U-Pb 年龄 507.5±3.0 Ma，间接地限定了关子镇蛇绿岩的形成时限为寒武纪。杨钊等(2006)对采自关子镇东沟的辉长岩获得了 LA-ICP-MS 法锆石 U-Pb 年龄 471±1.4 Ma，认为其代表了关子镇蛇绿岩的形成时代。

分析裴先治等(2005)三件变中-基性深成岩浆岩的地球化学数据发现它们存在主、微量元素的脱耦关系，SiO_2 与 MgO、Ni、$\sum$REE 含量不存在协同的演化关系(如变辉长闪长岩具有比变辉长岩更高的 MgO 和 Ni 含量)，这说明这三件样品不是同一岩浆分异演化的产物。杨钊等(2006)未给出定年辉长岩样品的地球化学数据，因而无法判定该辉长岩是否归属蛇绿岩组合。因此关子镇地区变中-基性深成岩浆岩产出的构造背景以及相关年龄的地质意义仍不明确，有必要对这些深成岩进行更深入的地球化学研究，区别其构造属性，寻找属于蛇绿岩组合的深成岩单元并对其进行定年，以便明确关子镇蛇绿岩的形成时代。

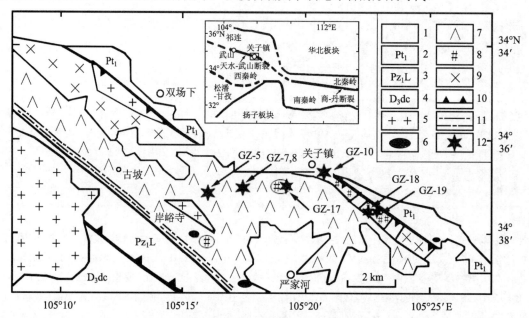

图 1 关子镇地质简图(据裴先治等，2004，修改)

1-新生界；2-古元古界秦岭群；3-下古生界李子园群；4-上泥盆统大草滩群；5-印支期花岗岩；6-蛇纹岩；7-蛇绿岩组合中变玄武岩单元；8-蛇绿岩组合中变中-基性深成岩浆岩单元；9-岛弧型变中-基性深成岩；10-逆冲断层/逆冲韧性剪切带；11-糜棱岩带；12-采样位置

本文采集了关子镇地区的三件变玄武岩样品和四件变中-基性深成岩浆岩样品进行了地球化学分析，首次在该地区识别出大洋斜长花岗岩类；选取其中两件古洋壳成因的辉长岩和闪长岩样品进行了精确的锆石 SHRIMP U-Pb 定年，直接测定关子镇蛇绿岩的形成时代，为研究东秦岭商丹断裂带的西延及古秦岭洋的形成时代提供确凿的证据。

2 地质概况

关子镇蛇绿混杂岩带夹持于南侧的早古生界"李子园群"和北侧的古元古界秦岭岩群之间，呈韧性剪切构造接触关系；该蛇绿混杂岩带向西被温泉花岗岩体吞没，向东被新生代地层覆盖后，在天水北道南与出露于李子园、木其滩、花庙河一带的变质基性火山岩带相连(裴先治等，2004)。

关子镇蛇绿岩(图 1)的主体为大套变玄武岩，其原岩为拉斑玄武岩系列(裴先治等，2004)，普遍经历了低角闪岩相变质，呈斜长角闪片岩产出。其次是变中-基性深成岩浆岩，包括变辉石岩、变辉长岩、变辉长闪长岩和变闪长岩，岩石片理或片麻理构造发育，其产状较复杂：部分呈孤立的小岩体断续分布于变玄武岩北侧，部分则以岩块的形式出露于变玄武岩中，相同岩性、不同产状的样品在岩石学特征上并无显著差别。另有少量的蛇纹岩块体零星出露于变玄武岩中。

采样位置见图 1：变玄武岩样品采自关子镇南侧的关子沟内(样品 GZ-5、GZ-7 和 GZ-8)；变中-基性深成岩浆岩样品分别采自关子沟口的变闪长岩体(样品 GZ-10)、关子沟内的变闪长岩岩块(样品 GZ-17)和变玄武岩北侧流水沟内的两个变辉长岩体(样品 GZ-18 和 GZ-19)。变玄武岩为暗绿-黑绿色，片状构造，细粒变晶结构，主要组成矿物为普通角闪石(60%~65%)、基性斜长石(35%~40%)，含少量绿帘石、阳起石等，矿物具有明显的定向排列。变辉长岩呈灰黑-黑绿色，片麻状构造，中细粒-中粒变余辉长结构，主要由普通辉石(50%~60%)、普通角闪石(10%~20%)和中基性斜长石(30%~40%)组成，含少量绿泥石、绢云母、榍石等，辉石多变为角闪石，但保留有辉石假象，普遍发育定向组构。变闪长岩呈灰色，中-细粒粒状结构，主要由普通角闪石(30%~35%)、中酸性斜长石(45%~50%)和绿泥石(5%~10%)组成，含少量石英、磁铁矿、榍石等，不含钾长石，发育弱定向组构。

3 微量元素地球化学

3.1 分析方法

样品的主要元素及微量元素的测定在西北大学大陆动力学国家重点实验室完成，主要元素由 RIX-2100 型 X 荧光光谱仪测定，分析误差 < 5%；微量元素由 ELAN6100DRC 型 ICP-MS 测定，溶样时用高纯硝酸和高纯氢氟酸在高压密闭的 Teflon 溶样器中进行，样品分析经 BHVO-1、AGV-1、BCR-2 和 G-2 国际标样监控，Ni、Rb、Y、Zr、Nb 和 LREE 等元素分析相对误差 < 5%，其他元素相对误差介于 5%~15%之间，详细的分析方法见文献(Gao et al.，1999)。分析结果见表1。

表1 关子镇变玄武岩和变中-基性深成岩浆岩样品的主量元素(wt%)及微量元素($\times 10^{-6}$)组成

样品号	GZ-5	GZ-7	GZ-8	GZ-10	GZ-17	GZ-18	GZ-19
采样地点	关子沟					流水沟	
样品名称	变玄武岩			变闪长岩		变辉长岩	
SiO_2	50.60	51.69	51.02	58.92	60.50	52.25	50.64
TiO_2	1.32	0.87	0.85	0.41	0.31	0.42	0.62
Al_2O_3	13.83	13.75	13.39	15.56	14.56	18.59	19.66
$Fe_2O_3^T$	13.49	13.04	13.25	9.33	8.72	9.32	9.74
MnO	0.35	0.18	0.18	0.13	0.12	0.14	0.16
MgO	6.76	7.07	7.23	4.00	4.31	4.61	4.48
CaO	8.74	9.51	9.59	7.67	7.49	9.12	11.21
Na_2O	3.96	3.07	3.23	2.49	2.86	3.96	0.85
K_2O	0.12	0.07	0.12	0.18	0.18	0.22	0.51
P_2O_5	0.11	0.09	0.09	0.07	0.04	0.08	0.02
烧失量	0.37	0.37	0.56	0.86	0.61	0.81	1.76
总量	99.65	99.71	99.51	99.62	99.70	99.52	99.65
V	318	438	438	275	197	240	275
Cr	135	72.4	68.1	62.1	71.6	94.8	112
Ni	73.3	82	80	27	22.4	22.6	24.1
U	0.13	0.074	0.063	0.15	0.25	0.38	0.68
Pb	12.5	1.15	5.31	1.98	1.93	2.13	2.91
Sr	228	71.2	99.8	117	115	152	188
Rb	1.29	0.99	2.66	3.78	2.53	1.94	19.8
Ba	29.3	21.7	27.0	35.3	24.6	34.1	198
Th	0.37	0.21	0.20	0.35	0.85	1.13	1.75
Nb	4.09	2.35	2.52	1.73	1.82	2.92	4.46
Ta	0.27	0.15	0.16	0.09	0.13	0.20	0.31
Zr	82.1	46.2	45.9	32.7	59.1	72.5	114
Hf	2.02	1.31	1.28	1.05	1.41	1.82	3.03
Y	30.0	26.9	26.8	20.4	13.3	24.7	28.3
La	5.17	2.52	2.39	1.25	1.40	2.58	6.07
Ce	11.2	6.27	6.14	4.38	4.59	11.0	15.9
Pr	1.76	0.95	0.94	0.69	0.64	1.01	2.03
Nd	9.45	5.28	5.20	4.24	3.53	5.01	9.48
Sm	2.95	1.90	1.87	1.61	1.06	1.67	2.62

续表

样品号	GZ-5	GZ-7	GZ-8	GZ-10	GZ-17	GZ-18	GZ-19
采样地点	关子沟					流水沟	
样品名称	变玄武岩			变闪长岩		变辉长岩	
Eu	1.24	0.63	0.65	0.41	0.32	0.57	0.71
Gd	3.74	2.40	2.35	1.84	1.37	2.39	3.17
Tb	0.74	0.52	0.52	0.39	0.28	0.51	0.62
Dy	4.78	3.60	3.59	2.65	1.90	3.53	4.13
Ho	1.06	0.85	0.85	0.62	0.45	0.84	0.95
Er	3.06	2.33	2.32	1.72	1.35	2.57	2.90
Tm	0.46	0.39	0.39	0.30	0.23	0.42	0.47
Yb	2.99	2.71	2.73	2.04	1.54	2.85	3.16
Lu	0.43	0.44	0.44	0.33	0.24	0.43	0.46
∑REE	49.03	30.78	30.38	22.47	18.91	35.39	52.61
La/Nb	1.26	1.07	0.95	0.72	0.77	0.88	1.36
$(La/Nb)_N$	1.31	1.11	0.99	0.75	0.8	0.92	1.41
$(La/Yb)_N$	1.24	0.67	0.63	0.44	0.65	0.65	1.38

3.2 微量元素地球化学特征

样品的稀土元素球粒陨石标准化图和微量元素原始地幔标准化图见图2。

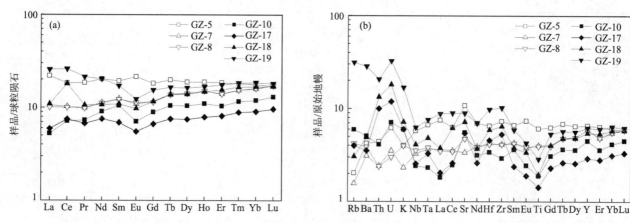

图2 样品的稀土元素球粒陨石标准化图(a)和微量元素原始地幔标准化图(b)(标准化值据Sun and McDonough, 1989)

图2显示三件变玄武岩样品(GZ-5,-7,-8)具有平坦(GZ-5)到轻稀土略亏损(GZ-7和GZ-8)的稀土配分型式，其$(La/Yb)_N$分别等于1.24和0.63~0.67，且不存在Nb、Ta和Ti的亏损，这些地球化学特征类似于洋中脊玄武岩(MORB)。

关子镇变中-基性深成岩浆岩样品(前人三件，本工作四件)在$(La/Yb)_N$-$(La/Nb)_N$图解(图3)中可大致分为两类：类型I(前人两件样品)显示轻稀土显著富集$(La/Yb)_N>2$、具有显著的Nb负异常$(La/Nb)_N>2.4$，这些特征类似于秦岭地区已识别的岛弧型中-基性岩浆岩(李曙光等，1993b；张旗等，1995；Li et al.，2004；张宗清等，2006)；类型II(前人一件样品和本工作四件样品)与上述洋中脊成因变玄武岩样品相似，显示轻稀土亏损到略富集$(La/Yb)_N<1.4$、无显著Nb负异常$(La/Nb)_N<1.5$。图2(b)中它们表现出的Ti负异常是岩浆演化晚期钛铁矿分异的结果。部分深成岩浆岩样品(GZ-17,-18,-19)表现出的Th、U的富集和样品GZ-19的大离子亲石元素(Rb、Ba、K)的富集可能与样品后期蚀变有关，因为其烧失量高达1.76%。

在判别蛇绿岩生成构造环境的La/Nb-La图解(图4)中，代表类型I的样品(前人两件)明显落在岛弧玄武岩(IAB)区域内；代表类型II的样品中，两个闪长岩GZ-10和GZ-17落在洋中脊玄武岩(MORB)区域内，其余样品(前人一件，本工作两件)则与洋中脊型变玄武岩一起落在MORB与IAB的重叠区域内或附近。

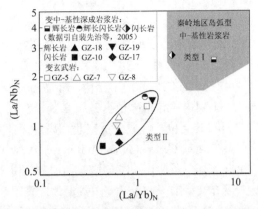
图3 关子镇蛇绿混杂岩样品的$(La/Nb)_N$-$(La/Yb)_N$图解

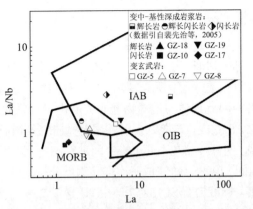
图4 关子镇蛇绿混杂岩样品的La/Nb-La图解
(据李曙光，1993a)

3.3 关子镇大洋斜长花岗岩类的识别

关子沟的两件闪长岩样品具有很低的稀土元素含量($\Sigma REE < 23 \times 10^{-6}$)和LREE略亏损的稀土配分型式，Nb相对La不亏损，在构造环境判别图解(图4)中明显落在MORB区域内，这些微量元素特征表明其属于前人定义的"大洋斜长花岗岩类(Oceanic Plagiogranite)"，是蛇绿岩岩石序列的组成部分，为缓慢扩张的洋中脊环境下基性岩浆极端分离结晶条件下的产物(Coleman and Peterman，1975；Pallister and Hopson，1981；Isma et al.，1996)。前人提出该类中酸性岩的主要鉴别标志是不含钾长石，具有非常低的全岩$K_2O(\leqslant 0.2\%)$和$Rb(\leqslant 5 \times 10^{-6})$含量(Coleman and Peterman，1975)。本工作的两件样品GZ-10和GZ-17都不含钾长石，其全岩的K_2O含量均为0.18%，Rb含量分别为3.78×10^{-6}和2.53×10^{-6}，完全符合上述指标。

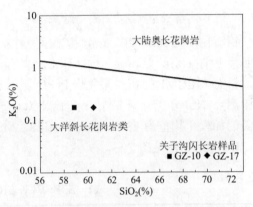
图5 区分大洋斜长花岗岩类和大陆奥长花岗岩的SiO_2-K_2O图解
(据 Coleman and Peterman，1975)

Coleman and Peterman(1975)提出了区分"大洋斜长花岗岩类"和"大陆奥长花岗岩"的SiO_2-K_2O图解(图5)，图中显示关子镇闪长岩(GZ-10和GZ-17)均落在大洋斜长花岗岩类一侧，符合前人提出的判别图解。

3.4 关子镇蛇绿混杂岩的构造背景

综合上述微量元素特征及判别图解，关子镇变玄武岩形成于洋中脊环境，这与裴先治等(2004)提出的"关子镇蛇绿岩是洋脊型蛇绿岩残片"的认识一致；变中-基性深成岩浆岩的生成环境包含了古大洋和古岛弧两种类型。流水沟辉长岩样品(GZ-18和GZ-19)尽管在判别图解中落在MORB和IAB的重叠区域内，但它们不具有显著的Nb负异常，其非常低的稀土含量($\Sigma REE < 55 \times 10^{-6}$)和钾含量($K_2O < 0.52\%$)，轻稀土亏损或略富集$(La/Yb)_N < 1.4$，富钠($K_2O/Na_2O < 1$)均与该区变玄武岩类似；关子沟闪长岩样品(GZ-10和GZ-17)在判别图解中接近变玄武岩，其地球化学特征与前人提出的"大洋斜长花岗岩类"一致。它们都是关子镇洋脊型蛇绿岩的组成部分。

变玄武岩北侧流水沟的两件辉长岩样品(GZ-18和GZ-19)显示与变玄武岩类似的微量元素特征，是洋脊型蛇绿岩的组成部分。因此，裴先治等(2005)提出的分布于变玄武岩北侧的"岛弧环境下形成的流水沟中基性岩浆杂岩体"在构造性质上应予以解体。

4 关子镇变中-基性深成岩浆岩样品的锆石 U-Pb SHRIMP 定年

为确定关子镇古洋盆的发育时代，选取洋中脊成因的流水沟辉长岩(GZ-18)和关子沟口闪长岩(GZ-10)进行锆石 SHRIMP U-Pb 定年。

4.1 分析方法

经重液和磁选法分选出锆石并在双目镜下手工挑选。每个样品选取约 200 颗锆石与数粒标准锆石 TEM 一起粘在环氧树脂靶上并抛光，具体制靶过程参见文献(宋彪等，2002)。进行 SHRIMP U-Pb 分析前，对锆石进行反射、透射光及阴极发光(CL)照相研究，以确定锆石颗粒的表面结构和内部结构。样品的锆石 CL 照相分别在中科院地质与地球物理研究所电子探针实验室(样品 GZ-10)和北京离子探针中心阴极发光实验室(样品 GZ-18)完成。制靶，反射、透射光研究及锆石的 U、Th 和 Pb 同位素组成分析在北京离子探针中心实验室完成。两件样品在不同上机时间进行测定，详细分析流程参见文献(Compston and Williams, 1984; Ireland and Gibson, 1998)。每完成三次样品测点分析，做一次标准锆石测定。应用标准锆石 TEM(417Ma)进行元素分馏校正，标准锆石 SL13 (572 Ma, U 含量 238×10^{-6})标定样品的 U、Th 和 Pb 含量。数据处理采用 Ludwig 编写的 SQUID(Ludwig, 2001)及 ISOPLOT(Ludwig, 1999)程序。普通铅分别采用 ^{204}Pb(对 GZ-18)和 ^{208}Pb 校正(对 GZ-10，该样品采用 ^{204}Pb 校正时，数据点偏离谐和线，可能与镀金不均匀有关)。

4.2 定年结果

4.2.1 流水沟辉长岩(GZ-18)定年结果

锆石粒径约 100～200 μm，主要呈短柱状(长宽比约为 2)或等粒状，晶面平直，晶棱较锐利。在阴极发光照片上(图 6(a))，锆石显示韵律环带，表明它们是岩浆结晶形成的。

锆石 SHRIMP 分析结果见表 2，所测锆石的 Th/U 为 0.35～1.47，为典型的岩浆锆石特征。22 个测点中测点 18.1 年龄显著偏小，三个测点(18.5、18.11 和 18.13)偏离谐和线，剩余 18 个测点的 ^{206}Pb/^{238}U 年龄的加权平均值为 534 ± 9 Ma ($n = 18$, MSWD = 1.1, 95%置信度)，表明流水沟辉长岩形成于早寒武世。U-Pb 谐和图见图 7。

表2 流水沟辉长岩(GZ-18)的锆石SHRIMP U-Pb年龄分析结果[a]

测点	^{206}Pb$_c$ (%)	U ($\times 10^{-6}$)	Th ($\times 10^{-6}$)	Th/U	^{206}Pb* ($\times 10^{-6}$)	^{207}Pb*/^{206}Pb* ($\pm$ %)	^{207}Pb*/^{235}U ($\pm$ %)	^{206}Pb*/^{238}U ($\pm$ %)	年龄(Ma) ^{206}Pb*/^{238}U	年龄(Ma) ^{207}Pb*/^{235}U
GZ-18.1	0.35	418	232	0.56	26.3	0.0577(3.0)	0.58(6.0)	0.0729(5.2)	454 ± 23	465 ± 22
GZ-18.2	0.11	410	197	0.48	31.0	0.0588(2.3)	0.71(4.5)	0.0879(3.9)	543 ± 20	547 ± 19
GZ-18.3	0.02	350	199	0.57	28.0	0.0582(2.6)	0.75(4.7)	0.0932(3.9)	574 ± 21	567 ± 20
GZ-18.4	0.39	447	314	0.70	34.8	0.0565(2.7)	0.70(4.7)	0.0904(3.9)	558 ± 21	541 ± 20
GZ-18.5	1.53	404	250	0.62	28.9	0.0501(8.0)	0.57(12.0)	0.0820(8.5)	508 ± 42	456 ± 43
GZ-18.6	0.16	1439	2113	1.47	100.5	0.0575(1.3)	0.64(4.1)	0.0811(3.8)	503 ± 19	504 ± 16
GZ-18.7	0.15	782	591	0.76	56.8	0.0576(1.7)	0.67(4.2)	0.0844(3.8)	522 ± 21	521 ± 17
GZ-18.8	0.77	335	195	0.58	24.0	0.0561(4.6)	0.64(6.2)	0.0829(4.2)	513 ± 21	503 ± 25
GZ-18.9	0.30	292	101	0.35	22.5	0.0575(4.2)	0.71(5.8)	0.0893(3.9)	552 ± 21	543 ± 24
GZ-18.10	0.42	541	387	0.72	40.8	0.0571(2.8)	0.69(4.8)	0.0875(3.9)	541 ± 20	532 ± 20
GZ-18.11	-0.33	316	164	0.52	25.7	0.0667(3.7)	0.87(5.4)	0.0951(3.9)	586 ± 22	638 ± 26
GZ-18.12	0.10	723	431	0.60	52.7	0.0566(2.3)	0.66(5.3)	0.0848(4.8)	424 ± 24	516 ± 22
GZ-18.13	0.49	536	426	0.79	46.2	0.0566(2.5)	0.78(4.6)	0.1000(3.9)	614 ± 23	586 ± 21
GZ-18.14	0.48	474	307	0.65	36.9	0.0554(3.3)	0.69(5.1)	0.0902(3.9)	557 ± 21	532 ± 21
GZ-18.15	0.20	438	277	0.63	31.9	0.0600(3.6)	0.70(5.3)	0.0848(4.0)	525 ± 20	540 ± 22
GZ-18.16	0.17	1288	1681	1.31	93.0	0.0572(1.6)	0.66(4.2)	0.0838(3.0)	519 ± 19	516 ± 17
GZ-18.17	0.00	933	1051	1.13	71.7	0.0619(1.5)	0.76(4.1)	0.0894(3.9)	552 ± 19	576 ± 18
GZ-18.18	0.37	428	232	0.54	33.0	0.0584(2.3)	0.72(4.5)	0.0896(3.9)	553 ± 21	552 ± 19
GZ-18.19	0.41	1101	1374	1.25	75.9	0.0566(2.5)	0.62(5.2)	0.0799(4.5)	496 ± 22	492 ± 20
GZ-18.20	0.06	666	536	0.80	47.8	0.0560(1.9)	0.65(4.4)	0.0835(3.9)	517 ± 19	505 ± 17
GZ-18.21	0.59	464	259	0.56	33.8	0.0527(2.8)	0.61(4.8)	0.0843(3.9)	522 ± 19	485 ± 18
GZ-18.22	-0.07	660	489	0.74	50.0	0.0600(1.8)	0.73(4.3)	0.0884(3.9)	546 ± 20	557 ± 18

a) 误差为 1σ；Pb$_c$ 和 Pb* 分别代表普通铅和放射成因铅；普通铅用 ^{204}Pb 校正。

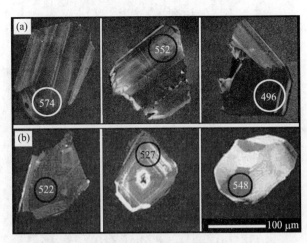

图6 流水沟辉长岩(GZ-18)(a)和关子沟口闪长岩
(GZ-10)(b)中部分锆石的CL照片,
圆圈代表测点位置,圈内数字表示该测点的年龄,单位Ma

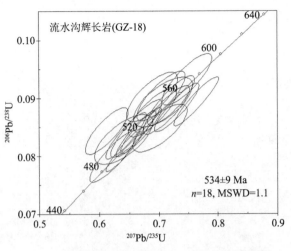

图7 流水沟辉长岩(GZ-18)锆石U-Pb年龄谐和图

4.2.2 关子沟口闪长岩(GZ-10)定年结果

锆石粒径约 100~150 μm, 主要呈短柱状(长宽比约为1.5)或等粒状, 晶面平直, 晶棱较锐利。在阴极发光照片上(图6(b)), 大多数锆石都显示韵律环带, 表明它们是岩浆结晶形成的。

锆石 SHRIMP 分析结果见表3, 所测锆石的 Th/U 为 0.15~0.37, 为典型的岩浆锆石特征。15 个测点中测点 10.3 偏离谐和线, 剩余 14 个测点的 ^{206}Pb/^{238}U 年龄的加权平均值为 517 ± 8 Ma ($n = 14$, MSWD = 0.83, 95%置信度), 表明关子沟口闪长岩形成于早–中寒武世。U-Pb 谐和图见图8。

表3 关子沟口闪长岩(GZ-10)的锆石SHRIMP U-Pb年龄分析结果[a]

测点	^{206}Pb$_c$ (%)	U (×10^{-6})	Th (×10^{-6})	Th/U	^{206}Pb* (×10^{-6})	^{207}Pb*/^{206}Pb* (± %)	^{207}Pb*/^{235}U (± %)	^{206}Pb*/^{238}U (± %)	年龄(Ma) ^{206}Pb*/^{238}U	年龄(Ma) ^{206}Pb*/^{238}U
GZ-10.1	1.01	460	121	0.26	33.0	0.0590(3.4)	0.67(4.3)	0.0827(2.7)	512 ± 14	522 ± 18
GZ-10.2	0.78	138	24	0.17	10.6	0.0571(7.9)	0.69(8.4)	0.0876(2.9)	542 ± 16	533 ± 35
GZ-10.3	2.07	39	7	0.18	3.19	0.0627(17.6)	0.77(18.0)	0.0888(3.5)	548 ± 21	578 ± 79
GZ-10.4	0.00	464	173	0.37	33.2	0.0556(1.7)	0.63(3.2)	0.0827(2.7)	512 ± 14	499 ± 13
GZ-10.5	1.43	228	67	0.29	16.7	0.0569(3.7)	0.66(4.7)	0.0837(2.8)	518 ± 15	522 ± 19
GZ-10.6	1.30	272	50	0.18	20.2	0.0611(6.2)	0.71(6.7)	0.0844(2.8)	522 ± 15	546 ± 29
GZ-10.7	0.86	428	150	0.35	31.3	0.0604(1.6)	0.70(3.1)	0.0846(2.7)	524 ± 14	542 ± 13
GZ-10.8	0.00	243	40	0.16	18.6	0.0578(3.3)	0.70(4.3)	0.0880(2.8)	544 ± 15	540 ± 18
GZ-10.9	0.36	194	59	0.30	14.6	0.0585(5.7)	0.69(6.4)	0.0851(2.8)	527 ± 16	531 ± 26
GZ-10.10	0.88	235	50	0.21	17.0	0.0546(2.4)	0.63(3.7)	0.0833(2.8)	516 ± 15	494 ± 14
GZ-10.11	0.89	297	68	0.23	21.0	0.0563(2.2)	0.63(3.5)	0.0809(2.7)	501 ± 14	494 ± 14
GZ-10.12	0.09	318	80	0.25	22.2	0.0623(3.9)	0.69(4.7)	0.0804(2.8)	499 ± 14	533 ± 20
GZ-10.13	3.35	130	24	0.18	9.30	0.0513(3.7)	0.57(4.7)	0.0808(2.9)	501 ± 15	459 ± 17
GZ-10.14	4.25	103	20	0.19	7.51	0.0612(4.0)	0.70(5.0)	0.0826(3.0)	512 ± 16	537 ± 21
GZ-10.15	0.57	170	26	0.15	12.3	0.0585(5.2)	0.67(6.0)	0.0836(2.8)	518 ± 15	524 ± 24

a) 误差为1σ; Pb$_c$和Pb*分别代表普通铅和放射成因铅; 普通铅用^{208}Pb校正。

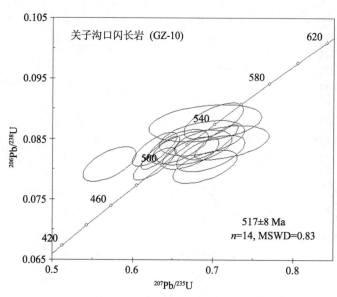

图 8 关子沟口闪长岩(GZ-10)锆石 U-Pb 年龄谐和图

5 讨论与结论

本文对关子镇地区出露的变玄武岩和变中-基性深成岩浆岩的微量元素地球化学研究表明，该区变中-基性深成岩浆岩的生成环境包含了古大洋和古岛弧两种类型。前者与关子镇变玄武岩具有一致的地球化学特征，表明它们也是洋脊型蛇绿岩组合的组成部分，对其中的辉长岩和闪长岩(该闪长岩属"大洋斜长花岗岩类")样品分别获得了锆石 SHRIMP U-Pb 年龄 534 ± 9 Ma(GZ-18)和 517 ± 8 Ma(GZ-10)，关子镇蛇绿岩深成岩单元为早-中寒武世古洋壳组成部分。这一年龄结果将华北陆块南缘的古秦岭洋的存在时代上推至 534 ± 9 Ma。

对比前人关子镇地区中-基性深成岩浆岩的定年结果发现，裴先治等(2005)获得的变辉长闪长岩年龄 507.5 ± 3.0 Ma 及杨钊等(2006)获得的辉长岩年龄 471 ± 1.4 Ma 均小于本文测定的古大洋成因的变辉长岩(534 ± 9 Ma)和变闪长岩(517 ± 8 Ma)年龄。前人并没有给出测年样品的地球化学数据，因而它们产出的构造背景仍不明确，本文研究发现关子镇中-基性深成岩浆岩的生成环境包含了古大洋和古岛弧两种类型，因此前人所得年龄结果的地质意义可能有两种解释：①测年样品形成于古大洋环境，这种情况下关子镇地区古大洋扩张作用至少持续存在了 60 Ma 以上；②测年样品形成于古岛弧环境，这种情况下关子镇地区早-中寒武世形成的洋壳在中寒武-中奥陶世发生俯冲消减，同时伴随岛弧岩浆作用。因此，今后应对关子镇蛇绿混杂岩中形成于古岛弧环境的火山岩和深成岩浆岩开展精确的年代学和地球化学研究，以明确古秦岭洋构造演化的年代学格架。

本文获得的变闪长岩年龄(517 ± 8 Ma)与东秦岭沿商丹断裂出露的富水杂岩中角闪黑云二长岩的单颗粒锆石 U-Pb 年龄(514.3 ± 1.3 Ma) (陈志宏等，2004)接近，尽管后者的成因尚存在争议。此外在天水武山断裂南侧广泛发育的泥盆系复理石建造(舒家坝群)也与商丹断裂带南侧的刘岭群一致。因此，东秦岭商丹断裂带西延与西秦岭的天水武山断裂带相连，在早古生代时它们共同构成了华北陆块南缘的活动陆缘。该古活动陆缘南侧的古秦岭洋至少在 534 ± 9 Ma 就已经存在。

致谢

锆石 CL 照相得到中科院地质与地球物理研究所徐平研究员的帮助。锆石 SHRIMP U-Pb 分析得到中国地质科学院地质所刘敦一研究员、万渝生研究员、王彦斌研究员和陶华工程师的协助。主量和微量元素分析由西北大学大陆动力学国家重点实验室柳小明、刘晔、第五春荣完成。在此一并表示衷心的感谢。

参考文献

陈志宏,陆松年,李怀坤,周红英,相振群,郭进京.2004.秦岭造山带富水中基性侵入杂岩的成岩时代：锆石U-Pb及全岩Sm、Nd同位素年代学新证据.地质通报,23(4): 322–328

李曙光.1993a.蛇绿岩生成构造环境的Ba-Th-Nb-La判别图.岩石学报,9(2): 146–157

李曙光,陈移之,张宗清,叶笑江,张国伟,郭安林,Hart SR.1993b.北秦岭拉垃庙苏长辉长岩的痕量元素和Sr,Nd同位素地球化学.地质学报,67(4): 310–322

裴先治,丁仁平,胡波,李勇,张国伟,郭军锋.2004.西秦岭天水地区关子镇蛇绿岩的厘定及其地质意义.地质通报,23(12): 1202–1208

裴先治,李勇,陆松年,陈志宏,丁仁平,胡波,李佐臣,刘会彬.2005.西秦岭天水地区关子镇中基性岩浆杂岩体锆石U-Pb年龄及其地质意义.地质通报,24(1): 23–29

任纪舜,王作勋,陈炳蔚,姜春发,牛宝贵,李锦轶,谢广联,和政军,刘志刚.1999.从全球看中国大地构造：中国及邻区大地构造图简要说明.北京：地质出版社,1–50

宋彪,张玉海,万渝生,简平.2002.锆石SHRIMP样品靶制作、年龄测定及有关现象讨论.地质论评,48(增刊): 26–30
杨钊,董云鹏,柳小明,张津海.2006.西秦岭天水地区关子镇蛇绿岩锆石LA-ICP-MS U-Pb定年.地质通报,25(11): 1321–1325
张国伟,张本仁,袁学诚,肖庆辉.2001.秦岭造山带与大陆动力学.北京：科学出版社,73–104
张旗,张宗清,孙勇,韩松.1995.陕西商县-丹凤地区丹凤群变质玄武岩的微量元素和同位素地球化学.岩石学报,11(1): 43–54
张维吉,孟宪恂,胡健民,裴先治,樊双虎.1994.祁连-北秦岭造山带接合部位构造特征与造山过程.西安：西北大学出版社,161–171
张宗清,张国伟,刘敦一,王宗起,唐索寒,王进辉.2006.秦岭造山带蛇绿岩、花岗岩和碎屑沉积岩同位素年代学和地球化学.北京：地质出版社,9–33
Coleman RG and Peterman ZE. 1975. Oceanic Plagiogranite. J. Geophys. Res., 80: 1099–1108
Compston W and Williams IS. 1984. U-Pb geochronology of zircons from lunar breccia 73217 using a sensitive high mass-resolution ion microprobe. J. Geophys. Res., 89: 525–534
Gao S, Ling WL, Qiu YM, Lian Z, Hartmann G and Simon K. 1999. Contrasting geochemical and Sm-Nd isotopic compositions of Archean metasediments from the Kongling high-grade terrain of the Yangzte craton: evidence for cratonic evolution and redistribution of REE during crustal anatexis. Geochim. Cosmochim. Acta, 63: 2071–2088
Ireland TR and Gibson CM. 1998. SHRIMP monazite and zircon geochronology of high-grade metamorphism in New Zealand. J. Metamorphic Geol., 16: 149–167
Isma A, Mathieu B and Georges C. 1996. Tectonic setting for the genesis of oceanic plagiogranites: evidence from a paleo-spreading structure in the Oman ophiolite. Earth Plant.Sci.Lett.,139: 177–194
Li SG, Hou ZH, Yang YC, Sun WD, Zhang GW and Li QL. 2004. Timing and geochemical characters of the Sanchazi magmatic arc in Mianlüe tectonic zone, South Qinling. Science in China (Series D), 47(4): 317–328
Ludwig KR. 1999. Using Isoplot/EX, version 2: a geolocronolgical Toolkit for Microsoft Excel. Berkeley Geochronological Center Special Publication 1a, 47
Ludwig KR. 2001. Squid 1.02: a user manual. Berkeley Geochronological Center Special Publication, 219
Pallister JS and Hopson CA. 1981. Samail ophiolite plutonic suite: field relations, phase variation, cryptic variation and layering, and a model of a spreading ridge magma chamber. J. Geophys. Res., 86: 2593–2644

青海东昆南构造带苦海辉长岩和德尔尼闪长岩的锆石 SHRIMP U-Pb 年龄及痕量元素地球化学
——对"祁-柴-昆"晚新元古代-早奥陶世多岛洋南界的制约

李王晔[1], 李曙光[1], 郭安林[2], 孙延贵[3], 张国伟[2]

1. 中国科学院壳幔物质与环境重点实验室, 中国科学技术大学地球和空间科学学院, 合肥 230026
2. 西北大学地质学系, 西安 710069
3. 青海省地质调查院遥感中心, 西宁 810012

> **亮点介绍**: 痕量元素地球化学研究和锆石 SHRIMP U-Pb 定年显示, 东昆南构造带苦海-阿尼玛卿地区出露的"洋岛型"苦海辉长岩的年龄为 555±9 Ma、"岛弧型"德尔尼闪长岩的年龄为 493±6 Ma。这些晚新元古代-早奥陶世中-基性岩浆岩的发育表明, 这一时期"祁-柴-昆"多岛洋的南界可达到东昆南构造带。东昆南构造带具有不同于秦岭勉略构造带的早古生代演化历史。

摘要 东昆南构造带苦海-阿尼玛卿地区出露的中-基性岩浆岩的时代存在争议。报道了苦海辉长岩和德尔尼闪长岩中岩浆锆石的 SHRIMP U-Pb 年龄分别为(555±9) Ma 和(493±6) Ma。地球化学研究表明苦海辉长岩和德尔尼闪长岩分别具有类似于洋岛玄武岩(OIB)和岛弧玄武岩(IAB)的痕量元素特征。据此, 苦海辉长岩的发育时代(555±9) Ma 及具有的 OIB 特征与北祁连的玉石沟洋壳成因蛇绿岩相似; 德尔尼闪长岩的发育时代(493±6) Ma 及具有的 IAB 特征与柴北缘的岛弧火山岩相似。苦海-阿尼玛卿地区存在晚新元古代-早奥陶世的中-基性岩浆岩, 表明这一时期"祁-柴-昆"多岛洋的南界可达到东昆南构造带。早古生代时, 东昆南构造带与秦岭造山带的勉略构造带不能对比。

关键词: 昆仑构造带; 锆石 SHRIMP U-Pb 定年; "祁-柴-昆"多岛洋; 蛇绿混杂岩

中国中西部地区存在两条主要的造山带: 北部的祁连造山带和南部的昆仑造山带, 中间发育有柴达木、中祁连等微陆块[1]。前人研究表明, "祁-柴-昆"地区存在北祁连、野马山-拉脊山、柴北缘、东昆中和东昆南等多个板块结合带[2], 其中在北祁连、柴北缘与东昆南已发现存在多个晚新元古代-早古生代岛弧火山岩带[3~5], 故这一地区在晚新元古代-早奥陶世总体呈现出"祁-柴-昆"多岛洋的构造格局[6](图1(a), (b)[(a)图据文献[7]修改, (b)图据文献[8]修改])。

前人研究已积累了一些与晚新元古代-早奥陶世"祁-柴-昆"多岛洋有关的蛇绿岩和岛弧火山岩的可靠年龄数据, 主要包括: 北祁连玉石沟蛇绿岩中辉长岩的锆石 SHRIMP U-Pb 年龄(550±17) Ma[9], 代表了洋盆开始形成的时代; 上部具 MORB 特征的基性熔岩的 Sm-Nd 等时线年龄 522~495 Ma[10], 代表了洋盆初具规模的时代。柴北缘吉绿素滩间山群火山岩中的埃达克岩的锆石 LA-ICP-MS 法 U-Pb 年龄(514.2±8.5) Ma[4], 代表了该地区洋壳俯冲作用发生的时代。此外, 对沿东昆中断裂带分布的清水泉镁铁-超镁铁质杂岩体获得了辉长岩单颗粒锆石 U-Pb 年龄(518±3) Ma[11]和(522.3±4.1) Ma[12], 但对该杂岩体产出的构造环境存在争议, 一些学者认为是蛇绿岩组合并指出其形成于大洋扩张初期或小洋盆环境[13,14]; 另一些学者则认为不是蛇绿岩组合, 而是伸展环境下的大陆板内拉斑玄武岩[15~17]。这些年代学结果表明这一"祁-柴-昆"多岛洋形成于晚新元古代-早寒武世(~550 Ma)。厘定该多岛洋的分布范围及边界位置对认识中国西北部早古生代岩石圈演化

---

* 本文发表在: 中国科学(D辑), 2007, 37: 288-294
对应的英文版论文为: Li Wangye, Li Shuguang, Guo Anlin, et al., 2007. Zircon SHRIMP U-Pb ages and trace element geochemistry of the Kuhai gabbro and the Dur'ngoi diorite in the southern east Kunlun tectonic belt, Qinghai, Western China and their geological implications. Science in China (Ser. D), 50: 331-338

历史有重要意义。

上述年代学结果反映的晚新元古代-早奥陶世"祁-柴-昆"多岛洋的南界为东昆中断裂带。在更向南的东昆南构造带东段(阿尼玛卿地区)德尔尼地区仅有石炭纪年龄数据(玄武岩全岩 $^{40}Ar/^{39}Ar$ 坪年龄(345.3±7.9) Ma[18]、锆石 SHRIMP U-Pb 年龄(308.2±4.9) Ma[19]),尚没有早古生代年龄报道。因此,一些学者认为东昆南阿尼玛卿构造带可与秦岭勉略构造带连接,同属扬子陆块西北缘的古特提斯洋体系[20]。然而在东昆南构造带西段的布青山地区发现并存有早古生代(辉长岩单颗粒锆石 U-Pb 年龄(467.2±0.9) Ma)和晚古生代两期蛇绿岩组合[7],这暗示东昆南构造带在早古生代与秦岭勉略构造带有不同的演化历史或者布青山与阿尼玛卿分属不同的构造带。为查明东昆南构造带东段是否也存在早古生代岩浆事件及这时期"祁-柴-昆"多岛洋的南部边界,我们选取东昆南构造带苦海-阿尼玛卿地区的苦海辉长岩和德尔尼闪长岩进行锆石 SHRIMP U-Pb 定年及痕量元素地球化学研究。

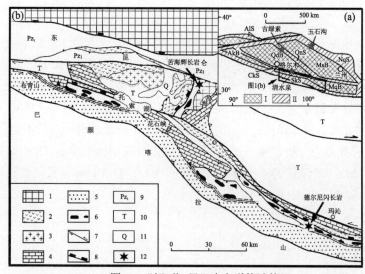

图 1 "祁-柴-昆"多岛洋构造格局

(a) "祁-柴-昆"地区构造示意图(据文献[7],修改)。I. 微陆块: MsB 为中-南祁连; QdB 为柴达木; AkB 为阿拉昆仑; MqB 为玛沁。II. 缝合带: NqS 为北祁连; AlS 为阿尔金; QnS 为柴北缘; CkS 为东昆中; SkS 为东昆南; (b) 研究区地质简图(据文献[8],修改)。1. 寒武纪变质岩; 2. 火山岩; 3. 花岗岩; 4. 晚古生代碳酸盐岩; 5. 晚古生代被动陆缘沉积; 6. 中-基性岩浆岩; 7. 断层; 8. 缝合带; 9. 早古生代被动陆缘火山-沉积岩系; 10. 三叠纪复理石沉积; 11. 第四系; 12. 本文采样点

1 地质与样品

东昆南构造带位于昆南断裂带和巴颜喀拉造山带之间,主体近东西向分布。带内的中-基性岩浆岩主要分布于西段布青山、中段玛积雪山、东段阿尼玛卿和由花石峡伸向 NE 方向的苦海-赛什塘等地区。早先认为带内岩浆事件的时代为中泥盆-早二叠世[18,19,21~23],前人据此对该带提出了"晚古生代秦-昆三向联结构造"模型[8],但布青山地区早古生代和晚古生代两期蛇绿岩并存现象的发现表明东昆南构造带有更复杂的演化历史[7],因此有必要对带内岩浆事件重新开展年代学研究。

前人指出苦海-赛什塘一带出露的蛇绿混杂岩组合呈不同规模、形态的岩块构造就位于石炭纪-早二叠世泥砂质板岩及低绿片岩相构造中[21~23]。蛇绿混杂岩主要由蛇纹石化超镁铁岩、辉长岩、辉绿岩岩墙、玄武岩及硅泥质板岩等组成,它们遭受后期构造肢解,在一条剖面上很难见到较为完整的蛇绿岩组合。前人认为这一地区蛇绿岩的时代属晚古生代,依据的同位素年龄均为 $^{40}Ar/^{39}Ar$ 高温坪年龄,即雪琼辉长岩中辉石坪年龄(368.6±1.4) Ma[22],拉龙洼基性岩墙中辉石坪年龄(393.5±3.0) Ma、斜长石坪年龄(361.4±4.2) Ma[8]。为进行锆石 U-Pb 年龄测定,我们选取了苦海辉长岩样品(KH-4),它出露于 214 国道兴海—花石峡沿线苦海北侧(358 km 里程碑附近),同时出露的还有超镁铁岩和硅质岩。这些岩类混杂在一起,其原始接触关系已无法识别。辉长岩样品新鲜,主要矿物组成为普通辉石(约 40%)、斜长石(约 40%)和角闪石(约 20%),选取样品

约 2 kg 用以分选锆石。

前人已在阿尼玛卿构造带东端、玛沁县南部的德尔尼地区识别出一套晚古生代蛇绿岩组合[18,19,24]，总体呈 NW-SE 向狭长带状分布，出露于早二叠世砂板岩与新元古代片岩、斜长角闪岩和大理岩之间。蛇绿岩由变质橄榄岩、辉石岩、辉长岩和玄武岩等组成，它们均以构造岩块或岩片的形式产出[19]。本次工作的野外观察沿德尔尼铜矿-玛沁公路进行，沿途断续出露有蛇纹石化和碳酸盐化变质橄榄岩、玄武岩和闪长岩，用于定年的德尔尼闪长岩样品(MQ-4)为上部靠近铜矿处的闪长岩(GPS 坐标：34°22′26″ N，100°08′04″ E)，主要矿物组成为角闪石(约 30%)、斜长石(约 60%)和普通辉石(约 10%)。样品经历蚀变，部分角闪石和辉石转变为绿泥石，长石表面发生绢云母化，选取样品约 2 kg 用以分选锆石。

2 锆石 SHRIMP U-Pb 定年

2.1 分析方法

经重液和磁选法分选出锆石并在双目镜下手工挑选。每个样品选取约 200 颗锆石与数粒标准锆石 TEM 一起粘在环氧树脂靶上并抛光，具体过程参见文献[25]。进行 SHRIMP U-Pb 分析前，对锆石进行反射、透射光及阴极发光(CL)照相研究，以确定锆石颗粒的表面结构和内部结构。锆石的 CL 照相在中国科学院地质与地球物理研究所电子探针实验室完成；制靶、反射、透射光研究及锆石的 U、Th 和 Pb 同位素组成分析在北京离子探针中心实验室完成。苦海辉长岩和德尔尼闪长岩在不同上机时间进行测定，详细分析流程参见文献[26, 27]。每完成 3 次样品测点分析，做一次标准锆石测定。应用标准锆石 TEM(417 Ma)进行元素分馏校正，标准锆石 SL13(572 Ma，U 含量 238 $\mu g \cdot g^{-1}$)标定样品的 U、Th 和 Pb 含量。数据处理采用 Ludwig 编写的 SQUID[28]及 Isoplot[29]程序。

2.2 定年结果

2.2.1 苦海辉长岩(KH-4)

选矿得到的多为锆石碎片，无色透明，晶面平直，晶棱较锐利。在阴极发光照片(图 2)上，锆石都显示韵律环带，表明它们是岩浆结晶形成的，多数锆石具有薄的亮边。

对该样品进行了 22 个锆石测点分析，分析结果见表 1。普通铅采用实测的 ^{204}Pb 校正，所测锆石的 Th/U 为 0.76~1.88，为典型的岩浆锆石比值。其中，3 个测点(KH-9，KH-13，KH-20)的 $^{207}Pb/^{235}U$ 年龄显著偏小，两个测点(KH-10，KH-19)的 $^{206}Pb/^{238}U$ 年龄分别偏大和偏小，剩余 17 个测点给出一致的年龄结果(图 3)，其 $^{206}Pb/^{238}U$ 年龄的加权平均值为(555±9) Ma(MSWD=1.3，置信水平 95%)，表明苦海辉长岩形成于晚震旦-早寒武世。

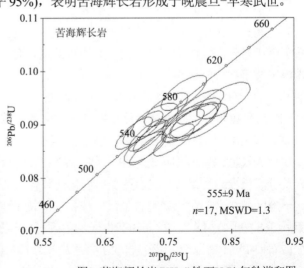

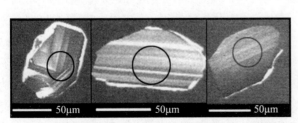

图2 苦海辉长岩中部分锆石的CL照片；圆圈代表分析点位置

图3 苦海辉长岩(KH-4)锆石U-Pb年龄谐和图

表1 苦海辉长岩(KH-4)的锆石SHRIMP U-Pb年龄分析结果 a)

测点	U/μg·g⁻¹	Th/μg·g⁻¹	Th/U	Pb*/μg·g⁻¹	普通Pb/%	$^{207}Pb^*/^{206}Pb^*$(±%)	$^{207}Pb^*/^{235}U$(±%)	$^{206}Pb^*/^{238}U$(±%)	$^{206}Pb^*/^{238}U$年龄/Ma	$^{207}Pb^*/^{235}U$年龄/Ma
KH-1	612	706	1.15	46.2	−0.04	0.0596(1.5)	0.79(3.2)	0.0878(2.8)	543±14	551±14
KH-2	580	627	1.08	42.5	0.00	0.0608(1.6)	0.79(3.2)	0.0854(2.8)	528±14	548±14
KH-3	500	685	1.37	40.0	0.21	0.0583(2.6)	0.76(3.8)	0.0928(2.8)	572±15	566±17
KH-4	266	282	1.06	20.7	−0.31	0.0628(3.0)	0.77(4.8)	0.0908(3.7)	560±20	589±21
KH-5	404	412	1.02	31.4	−0.32	0.0632(2.6)	0.71(3.9)	0.0909(2.9)	561±16	592±18
KH-6	580	852	1.47	43.4	−0.02	0.0631(2.1)	0.67(3.5)	0.0871(2.8)	538±15	573±15
KH-7	565	1035	1.83	43.5	−0.25	0.0624(2.3)	0.82(3.6)	0.0898(2.8)	555±15	581±16
KH-8	684	1080	1.58	50.3	−0.19	0.0603(2.6)	0.72(3.8)	0.0857(2.8)	530±14	546±16
KH-9	423	497	1.17	32.8	0.27	0.0536(2.8)	0.73(3.9)	0.0902(2.8)	557±15	518±16
KH-10	414	538	1.30	34.9	−0.17	0.0604(2.7)	0.62(3.8)	0.0984(2.8)	605±16	607±18
KH-11	689	1033	1.50	52.0	−0.13	0.0589(1.8)	0.71(3.3)	0.0880(2.8)	544±15	548±14
KH-12	496	486	0.98	38.6	−0.24	0.0581(3.1)	0.78(4.2)	0.0909(2.8)	561±15	556±18
KH-13	322	426	1.32	24.7	0.16	0.0503(3.1)	0.80(4.2)	0.0891(2.9)	551±15	489±16
KH-14	421	556	1.32	32.4	0.00	0.0575(2.3)	0.77(3.7)	0.0894(2.9)	552±15	544±15
KH-15	323	431	1.33	25.0	−0.61	0.0627(3.6)	0.69(4.6)	0.0907(2.8)	560±15	588±20
KH-16	490	682	1.39	39.1	−0.37	0.0622(2.6)	0.65(3.8)	0.0933(2.8)	575±16	597±17
KH-17	395	456	1.15	32.4	−0.23	0.0585(2.6)	0.55(3.8)	0.0959(2.8)	591±16	582±17
KH-18	521	646	1.24	39.6	0.19	0.0570(2.0)	0.84(3.4)	0.0884(2.8)	546±15	536±14
KH-19	1026	1931	1.88	72.1	0.05	0.0572(1.9)	0.80(3.3)	0.0818(2.7)	507±13	506±13
KH-20	364	430	1.18	29.3	0.68	0.0432(4.8)	0.79(5.6)	0.0929(2.8)	573±16	447±20
KH-21	131	85	0.76	10.6	−0.05	0.0643(3.0)	0.79(4.2)	0.0945(2.9)	559±17	596±21
KH-22	284	139	1.06	20.6	−0.38	0.0639(3.6)	0.76(4.8)	0.0905(3.1)	582±16	618±16

a) Pb*为放射成因铅, 所有误差均为 1σ, 普通铅用 ^{204}Pb 校正.

2.2.2 德尔尼闪长岩(MQ-4)

锆石无色透明, 呈短柱状(长宽比约为 2)或等粒状, 粒径约 50~150 μm。晶面平直, 晶棱锐利, 在阴极发光照片(图 4)上, 锆石具有清晰的韵律环带, 表明它们是岩浆结晶形成的, 部分锆石具有薄的亮边。

对该样品进行 11 个锆石测点分析, 分析结果见表 2。普通铅采用实测的 ^{208}Pb 校正, 所测锆石的 Th/U 为 1.43~2.05, 为典型的岩浆锆石比值。其中一个测点(MQ-1)的年龄明显偏大, 剩余 10 个测点给出 ^{206}Pb/^{238}U 年龄的加权平均值为(493±6) Ma(MSWD=6.2, 置信水平 95%) (图 5), 表明德尔尼闪长岩形成于晚寒武-早奥陶世。

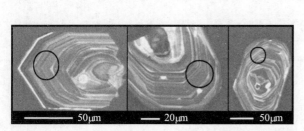

图4 德尔尼闪长岩中部分锆石的CL照片; 圆圈代表分析点位置

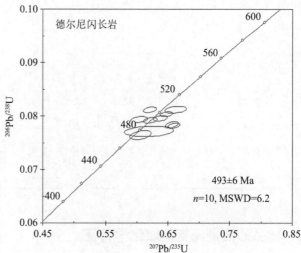

图5 德尔尼闪长岩(MQ-4)锆石U-Pb年龄谐和图

表2　德尔尼闪长岩(MQ-4)的锆石SHRIMP U-Pb年龄分析结果 b)

测点	U/μg·g⁻¹	Th/μg·g⁻¹	Th/U	Pb*/μg·g⁻¹	普通Pb/%	$^{207}Pb^*/^{206}Pb^*$(±%)	$^{207}Pb^*/^{235}U$(±%)	$^{206}Pb^*/^{238}U$(±%)	$^{206}Pb^*/^{238}U$ 年龄/Ma	$^{207}Pb^*/^{235}U$ 年龄/Ma
MQ-1	1489	1042	1.43	119	3.00	0.0599(1.4)	0.75(5.3)	0.0902(5.1)	557±32	567±23
MQ-2	457	246	1.86	30.2	0.41	0.0575(2.1)	0.61(2.3)	0.0764(0.9)	475±5	481±9
MQ-3	1380	924	1.49	91.5	0.01	0.0582(5.1)	0.62(5.2)	0.0771(1.0)	479±6	489±20
MQ-4	1523	865	1.76	105	0.34	0.0585(1.3)	0.65(1.3)	0.0803(0.4)	498±2	507±5
MQ-5	1338	736	1.82	89.8	-0.23	0.0608(1.2)	0.66(1.3)	0.0784(0.6)	486±3	513±5
MQ-6	1389	814	1.71	97.8	0.97	0.0592(2.0)	0.66(2.1)	0.0812(0.7)	503±4	516±9
MQ-7	1108	647	1.71	75.6	0.30	0.0569(1.4)	0.62(1.5)	0.0792(0.5)	491±3	491±6
MQ-8	1423	693	2.05	100.0	1.14	0.0554(1.3)	0.62(1.4)	0.0811(0.5)	503±3	490±6
MQ-9	1734	894	1.94	116	-0.11	0.0612(1.1)	0.66(1.2)	0.0782(0.5)	485±3	515±5
MQ-10	1145	665	1.72	78.4	0.13	0.0581(1.4)	0.64(1.5)	0.0796(0.5)	494±3	501±6
MQ-11	1185	646	1.83	83.4	3.08	0.0551(1.7)	0.60(1.8)	0.0793(0.6)	492±4	479±7

b) Pb*为放射成因铅，所有误差均为 1σ，普通铅用 ^{208}Pb 校正。

3 痕量元素地球化学

为明确上述年龄结果所代表的地质意义，对苦海辉长岩和德尔尼闪长岩样品进行痕量元素地球化学研究。

3.1 分析方法

样品的主要元素及痕量元素的测定在西北大学大陆动力学国家重点实验室完成。主要元素由 RIX-2100 型 X 荧光光谱仪测定，分析误差<5%；痕量元素由 ELAN6100DRC 型 ICP-MS 测定，溶样时在 200 ℃下用高纯硝酸和高纯氢氟酸在 Teflon 溶样器中进行。样品分析经 BHVO-2，AGV-1，BCR-2 和 G-2 国际标样监控，Ni、Rb、Y、Zr、Nb 和轻稀土等元素分析相对误差<5%，其他元素相对误差介于 5%~15%之间。详细的分析流程见文献[30]，分析结果见表 3。

表3　东昆南断裂带苦海辉长岩(KH-4)和德尔尼闪长岩(MQ-4)的主要元素(%)及痕量元素 (μg·g⁻¹) 组成

样品号	SiO₂	TiO₂	Al₂O₃	TFe₂O₃	MnO	MgO	CaO	Na₂O	K₂O	P₂O₅	烧失量	总量
KH-4	47.43	2.31	15.48	14.51	0.18	6.15	7.77	2.91	1.43	0.29	2.16	100.28
MQ-4	58.18	0.58	17.25	5.38	0.08	2.93	4.00	3.94	3.90	0.35	3.15	99.74

样品号	Cr	Ni	U	Pb	Sr	Rb	Ba	Th	Nb	Ta	Zr	Hf	Y	ΣREE
KH-4	96.27	101.8	0.434	7.69	431	39.43	766.7	2.06	14.83	0.99	161	3.93	28.17	108.7
MQ-4	24.38	18	11.15	97.28	801	166.2	1956	72.7	25.6	1.91	324	7.69	19.2	442.2

样品号	La	Ce	Pr	Nd	Sm	Eu	Gd	Tb	Dy	Ho	Er	Tm	Yb	Lu
KH-4	16.65	38.69	4.94	22.56	5.40	1.98	5.55	0.92	5.23	1.00	2.74	0.39	2.32	0.32
MQ-4	114.5	205	20.8	70.7	9.80	2.32	9.30	0.91	4.02	0.63	1.74	0.23	1.56	0.24

3.2 样品的痕量元素地球化学特征

样品的稀土元素球粒陨石标准化图和痕量元素原始地幔标准化图见图 6。

苦海辉长岩(KH-4)具有轻稀土富集((La/Yb)$_N$=5.15)的稀土配分型式(图 6(a)),但不存在高场强元素(Nb,Ta 和 Ti)的亏损(图 6(b)),这些特征类似于洋岛玄武岩(OIB);德尔尼闪长岩(MQ-4)具有轻稀土强烈富集((La/Yb)$_N$=52.64)的稀土配分型式(图 6(a)),并存在高场强元素(Nb,Ta 和 Ti)的显著亏损(图 6(b)),这些特征类似于岛弧玄武岩(IAB)。此外,德尔尼闪长岩较苦海辉长岩具有明显高的全岩 U、Th 和 Pb(表 3)和锆石平均 U、Th 和放射成因 Pb 含量(表 1、2),这也反映出两者在形成构造环境上存在显著差别。

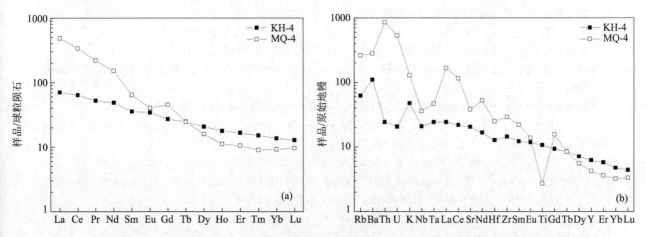

图 6 苦海辉长岩(KH-4)和德尔尼闪长岩(MQ-4)的稀土球粒陨石标准化图(a)及痕量元素原始地幔标准化图(b)
(标准化值据文献[31])

4 讨论和结论

(1) 本文测定的洋岛成因苦海辉长岩锆石 SHRIMP U-Pb 年龄为(555±9) Ma,表明其形成于晚震旦-早寒武世,它可能指示了洋壳形成时代;岛弧成因德尔尼闪长岩锆石 SHRIMP U-Pb 年龄为(493±6) Ma,表明其形成于晚寒武-早奥陶世,它指示了洋壳消减的时代。这两个年龄结果为东昆南构造带苦海-阿尼玛卿地区存在晚新元古代-早奥陶世的蛇绿岩及岛弧岩浆事件提供了新的年代学证据,它与西部的布青山带具有相同的演化历史。

(2) 前人测定的苦海地区蛇绿岩的 ^{40}Ar/^{39}Ar 年龄限于晚古生代(^{40}Ar/^{39}Ar "高温坪年龄" (368.6±1.4) Ma,(393.5±3.0) Ma 和(361.4±4.2) Ma),但它们的 ^{40}Ar/^{39}Ar 年龄谱图(具体参见文献[8, 22])均分为分别代表"高温坪"和"低温坪"的两段,显示"氩丢失谱"的特征。因此,这三个年龄存在氩丢失的可能性,由此给出小于其真正成岩年龄的年龄值。因此,苦海-赛什塘地区是否存在晚古生代蛇绿岩仍需进一步开展年代学工作。

(3) 获得的洋岛成因苦海辉长岩的形成时代(555±9) Ma 与北祁连玉石沟洋中脊成因蛇绿岩中辉长岩的形成时代(550±17) Ma[9]接近;岛弧成因的德尔尼闪长岩的形成时代(493±6) Ma 略小于柴北缘岛弧型火山岩年龄(514.2±8.5) Ma[4]。因此,东昆南的苦海-阿尼玛卿地区在晚新元古代-早奥陶世时可能也处于"祁-柴-昆"多岛洋之中。如果这一假设正确,那么晚新元古代-早奥陶世"祁-柴-昆"多岛洋的南部边界将从早先认为的"东昆中断裂带"向南推到"东昆南构造带"。

(4) 秦岭勉略构造带没有早古生代蛇绿岩,勉略洋盆在泥盆纪才打开[32,33],并在石炭纪消减[34],三叠纪闭合[35],因此在早古生代东昆南构造带与勉略构造带不是同一构造带。直到晚古生代时(泥盆纪),当"祁-柴-昆"多岛洋中所有微陆块以及南秦岭微陆块都与华北陆块碰撞拼合后,东昆南构造带和勉略构造带才形成统一的华北陆块南缘边界。

致谢

锆石 CL 照相得到中国科学院地质与地球物理研究所徐平研究员的帮助。锆石 SHRIMP U-Pb 分析得到北京离子探针中心刘敦一研究员、万渝生研究员、王彦斌研究员和陶华工程师的协助。主、微量元素测定由西北大学大陆动力学国家重点实验室柳小明、刘晔和第五春荣完成。在此一并表示衷心的感谢。

参考文献

[1] 张国伟,柳小明.关于"中央造山带"几个问题的思考.地球科学——中国地质大学学报,1998,23(5): 443–448

[2] 潘桂棠,丁俊,主编.青藏高原及邻区地质图.成都: 成都地图出版社,2004

[3] 夏林圻,夏祖春,任有祥,等.祁连、秦岭山系海相火山岩.武汉: 中国地质大学出版社,1991

[4] 史仁灯,杨经绥,吴才来,等.柴达木北缘超高压变质带中的岛弧火山岩.地质学报,2004,78(1): 52–64

[5] 边千韬,罗小泉,陈海泓,等.阿尼玛卿蛇绿岩带花岗-英云闪长岩锆石U-Pb同位素定年及大地构造意义.地质科学,1999,34(4): 420–426

[6] 殷鸿福,张克信.中央造山带的演化及其特点.地球科学——中国地质大学学报,1998,23(5): 438–442

[7] Bian Q T, Li D H, Pospelov I, et al. Age, geochemistry and tectonic setting of Buqingshan ophiolites, North Qinghai-Tibet Plateau, China. J Asian Earth Sci, 2004, 23: 577–596

[8] 孙延贵,张国伟,郭安林,等.秦昆三向联结构造及其构造过程的同位素年代学证据.中国地质,2004, 31(4): 372–378

[9] 史仁灯,杨经绥,吴才来,等.北祁连玉石沟蛇绿岩形成于晚震旦世的SHRIMP年龄证据.地质学报,2004,78(5): 649–657

[10] 夏林圻,夏祖春,徐学义.北祁连山海相火山岩岩石成因.北京: 地质出版社,1996

[11] Yang J S, Robinson P T, Jiang C F, et al. Ophiolites of the Kunlun Mountains, China and their tectonic implications. Tectonophysics, 1996, 258: 215–231

[12] 陆松年,主编.青藏高原北部前寒武纪地质初探.北京: 地质出版社,2002

[13] 高延林,吴向农,左国朝.东昆仑山清水泉蛇绿岩特征及其大地构造意义.西安地质矿产研究所所刊,1988,21: 17–28

[14] 朱云海,张克信,Pan Y M,等.东昆仑造山带不同蛇绿岩带的厘定及其构造意义.地球科学,1999,24(2): 134–138

[15] 潘裕生,周伟民,许荣华,等.昆仑山早古生代特征与演化.中国科学D辑: 地球科学,1996,26(4): 302–307

[16] 张旗,周国庆.中国蛇绿岩.北京: 科学出版社,2001

[17] 龙晓平,王立社,余能.东昆仑山清水泉镁铁质-超镁铁质岩的地球化学特征.地质通报,2004,23(7): 664–669

[18] 陈亮,孙勇,裴先治,等.德尔尼蛇绿岩 ^{40}Ar-^{39}Ar 年龄: 青藏最北端古特提斯洋盆存在和延展的证据.科学通报,2001,46(5): 424–426

[19] 杨经绥,王希斌,史仁灯,等.青藏高原北部东昆仑南缘德尔尼蛇绿岩: 一个被肢解了的古特提斯洋壳.中国地质,2004,31(3): 225–239

[20] 张国伟,孟庆仁,于在平,等.秦岭造山带的造山过程及其动力学特征.中国科学D辑: 地球科学,1996,26(3): 193–200

[21] 王秉璋,张智勇,祁生胜,等.秦昆接合部造山带非史密斯地层类型及特征的初步认识.中国区域地质,1998,17(增刊): 74–79

[22] 王秉璋,张智勇,张森琦,等.东昆仑东端苦海-赛什塘地区晚古生代蛇绿岩的地质特征.地球科学——中国地质大学学报,2000,25(6): 592–598

[23] 张智勇,殷鸿福,王秉璋,等.昆秦接合部海西期苦海-赛什塘分支洋的存在及其证据.地球科学——中国地质大学学报,2004,29(6): 691–696

[24] 陈亮,孙勇,柳小明,等.青海省德尔尼蛇绿岩的地球化学特征及其大地构造意义.岩石学报,2000,16(1): 106–110

[25] 宋彪,张玉海,万渝生,等.锆石SHRIMP样品靶制作、年龄测定及有关现象讨论.地质论评, 2002,48 (增刊): 26–30

[26] Compston W, Williams I S. U-Pb geochronology of zircons from lunar breccia 73217 using a sensitive high mass-resolution ion microprobe. J Geophys Res, 1984, 89: 525–534

[27] Ireland T R, Gibson C M. SHRIMP monazite and zircon geochronology of high-grade metamorphism in New Zealand. J Metamorph Geol, 1998, 16: 149–167

[28] Ludwig K R. Squid 1.02: A user manual. Berkeley: Berkeley Geochronological Center Special Publication, 2001. 219

[29] Ludwig K R. Using Isoplot/EX, version 2, a geolocronolgical Toolkit for Microsoft Excel. Berkeley: Berkeley Geochronological Center Special Publication, 1999. 47

[30] Gao S, Ling W L, Qiu Y M, et al. Contrasting geochemical and Sm-Nd isotopic compositions of Archean metasediments from the Kongling high-grade terrain of the Yangtze craton: evidence for cratonic evolution and redistribution of REE during crustal anatexis. Geochim Cosmochim Acta, 1999, 63: 2071–2088

[31] Sun S S, McDonough W F. Chemical and isotopic systematics of oceanic basalts: implication for mantle composition and process. In: Saunders A D, Norry M J, ed. Magmatism in the Ocean Basins. Geological Society Special Publ. No. 42. 1989. 313–345

[32] 张国伟,孟庆仁,赖绍聪.秦岭造山带的结构构造.中国科学, B 辑, 1995, 25(9): 994–1003
[33] Meng Q R, Zhang G W. Timing of collision of the North and South China blocks: controversy and reconciliation. Geology, 1999, 27: 123–126
[34] 李曙光,侯振辉,杨永成,等.南秦岭勉略构造带三岔子古岩浆弧的地球化学特征及形成时代.中国科学 D 辑: 地球科学, 2003, 33(12): 1163–1173
[35] 李曙光,孙卫东,张国伟,等.南秦岭勉略构造带黑沟峡变质火山岩的年代学和地球化学: 古生代洋盆及其闭合时代的证据.中国科学 D 辑: 地球科学, 1996, 26(3): 223–230

第三部分 超高压变质岩的冷却史、退变质 P-T-t 轨迹与多阶段差异折返机制

大别山石马地区榴辉岩 P-T-t 轨迹及其构造意义*

肖益林，李曙光

中国科技大学地球和空间科学系，合肥 230026

> **亮点介绍**：大别山南部榴辉岩的 P-T-t 轨迹研究表明：产于片麻岩中的榴辉岩变质经历了四个演化阶段，至少经历了 221 Ma（印支期）和 134 Ma（燕山期）两个阶段的抬升过程。

摘要 在大别山石马地区榴辉岩石榴石中包裹有与柯石英同期的绿辉石，据此，将这里的榴辉岩相变质作用划分为早期的柯石英榴辉岩阶段和晚期重结晶榴辉岩阶段。它们的 P-T-t 轨迹表明：产于片麻岩中的榴辉岩至少经历了两段的抬升过程：(1)在 221 Ma(印支期)时，由于陆壳俯冲所产生的大规模逆冲构造使其快速抬升；(2)在 134 Ma(燕山期) 陆壳俯冲后的拉张抬升环境下，榴辉岩随山体一起缓慢上升。

关键词 P-T-t 轨迹；榴辉岩；大别山

1 引言

大别山作为秦岭造山带的东延部分，由于发育有大量与陆-陆碰撞造山过程有关的榴辉岩，以及在部分榴辉岩中发现有柯石英和金刚石包裹体存在而引起广泛的重视。至今人们已对其进行了大量的岩石学和同位素年代学研究[1990, 1991, X.Wang et al.; 1989, 1992, 李曙光等]，这些工作对认识榴辉岩的成因及变质历史、华北陆块与扬子陆块的碰撞时代、陆-陆碰撞造山过程及构造演化均具有重大意义。然而，下列问题的解决仍有待更深入的工作。

(1)榴辉岩中发现的柯石英和金刚石等超高压矿物均是以包裹体形式存在于石榴石和绿辉石中[1989, Wang et al.; 1989, Okay et al.; 1988, Smith et al.; 1991, 徐树桐等]，这暗示着柯石英包裹体与包裹它的石榴石和绿辉石是代表两个不同阶段的变质产物。然而目前大部分文献中都将柯石英与包裹它的石榴石、绿辉石作为同一变质阶段，如[1988, Smith et al.]，并以包裹柯石英的石榴石-绿辉石矿物对计算的温度来代表柯石英形成时的温度[1989，Wang et al.; 1988, Smith et al.]，其中一种可能的解释是早已存在的石英在变质作用顶峰时变为柯石英，并被同时生长的石榴石和绿辉石所包裹[与 X.Wang 私人通信]。显然，榴辉岩相变质作用是否存在阶段性对认识榴辉岩的变质历史，以及测定其顶峰变质温度是十分重要的。

(2)榴辉岩的抬升历史和抬升机制是造山带榴辉岩研究的另一重要课题。构造学家已从动力学角度提出各种榴辉岩抬升机制[1991, Dobcratsov et al.; 1992, Okay et al.]，这些模型的一个共同特点是，榴辉岩的上升是由单一动力来源，经单阶段过程抬升的。然而榴辉岩经一次抬升过程从 100 km 深处抬升到地表是难以想象的，是否存在多阶段的抬升过程及不同动力机制是一个值得探讨的课题。

(3)大别山榴辉岩的形成时代已被用 Sm-Nd 矿物-全岩等时线法测定为 224~221 Ma [1989, 1992, 李曙光等]，然而它的构造含义取决于被测定的石榴石和绿辉石是在碰撞造山过程中哪一阶段形成的。我们已知洋壳和陆壳俯冲到一定深度都可发生榴辉岩相变质作用，那么大别山榴辉岩的 Sm-Nd 同位素年龄究竟代表的是洋壳的俯冲时代还是陆壳俯冲的时代？这是一个有待深入研究解决的问题。

显然，上述三个问题都与榴辉岩的变质演化史有关，对榴辉岩变质的 P-T-t 轨迹的研究，将为这些问题

---

*本文发表在：大地构造与成矿学，1993，17(3): 239-250

的解决提供重要依据，为此，本文拟以大别山石马地区榴辉岩为例，对该区榴辉岩的 P-T-t 轨迹的测定进行探索，并为上述问题的解决提供可能的证据。

2 地质背景与样品

大别山造山带介于华北陆块与扬子陆块之间，并出露了两陆块碰撞造山带的深部岩石。大多数地质学家认为：在印支期两陆块发生对接，并随后发生了扬子陆块的俯冲运动。

大别山东端被郯庐大断裂所截，使其东延部分苏北-胶南隆起向北错移了约五百公里；西侧的商城-麻城断裂使西侧地块南移、下降，东侧地块上升；南侧有广济-襄樊断裂，是扬子陆块陆内断裂。

大别山区广泛发育大别群变质杂岩，其主要岩性组合为：黑云斜长片麻岩、二长片麻岩、绿帘斜长片麻岩、斜长角闪岩、浅粒岩及为数不多的大理岩等。大别群与覆盖其上的宿松群呈构造接触关系[1992, Okay et al.]。大别群变质杂岩多数人认为系元古代产物。Okay 等(1989)曾根据南北大别群不同的岩性组合而将大别山地区划分为南北两个带，并认为南带是榴辉岩相而北带不是，柯石英榴辉岩均产于南带。

研究所用样品来自大别山东段南麓太湖县石马乡境内(图 1)。区内主要岩石类型有大别群斜长片麻岩、斜长角闪岩、浅粒岩及少量的大理岩、蛇纹岩和滑石片岩，此外还广泛发育有呈似层状、透镜体或小团块状产于斜长片麻岩中的榴辉岩和退变质的榴闪岩，其长度从几厘米到数十米不等。斜长片麻岩作为榴辉（闪）岩的围岩，其片理化强烈，局部可见石榴石富集成粉红色透镜体(其石榴石的含量可达 40%)以上，这些透镜体长轴在 3～20 cm 之间变化，且长轴方向与片理方向一致。

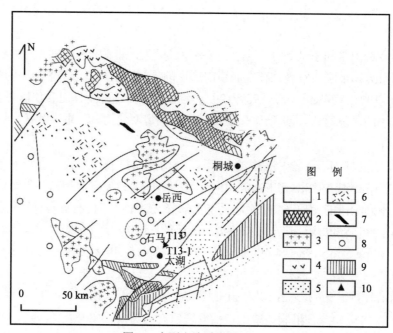

图 1 大别山地质略图

1-元古界大别群；2-古生界复理石；3-中生代花岗岩；4-侏罗纪火山岩；5-上侏罗统；6-混合岩；7-超镁铁岩；8-榴辉岩；9-蓝闪片岩；10-样品位置

在采集样品中，我们选用了一个完全新鲜的榴辉岩样品(T13')和一个退变质的榴闪岩样品(T13-1)。

T13'样品以似层状产于大别群黑云斜长片麻岩中，Wang 等 (1989)曾在同一榴辉岩体中发现柯石英；李曙光等(1992)对该样品的同一块标本进行了 Sm-Nd 同位素年龄测定；岩石标本呈翠绿玫瑰红色，中粒花岗变晶结构。

T13-1 为退变质的榴闪岩样品，亦以似层状产于大别群黑云斜长片麻岩中，岩体出露宽度约 20 m。岩石呈墨绿色，整体已强烈角闪岩化。T13'与 T13-1 样品相距约有 100 m，是本区内相距最近的两个榴辉岩体和榴闪岩体。尽管这两个样品分别来自不同的两个岩体，但基于以下理由我们认为它们具有相同的 P-T-t 演化起点，因而可以由这两块标本建立起一条榴辉岩的退变质 P-T-t 轨迹。

(1) 同位素资料证明：大别山所有产于片麻岩中的榴辉（闪）岩体的形成时代是一致的。李曙光等(1989, 1992)先后对大别山产于片麻岩中的不同榴辉（闪）岩体做的四个 Sm-Nd 同位素年龄是：224 ± 20 Ma，221 ± 5 Ma，229 ± 32 Ma，209 ± 31 Ma。因此，可以认为该区产于片麻岩中的榴辉岩均具有大致相同的榴辉岩相变质时代。

(2) 对大别山地区榴辉岩的 P-T 研究工作证明，产于片麻岩中的榴辉（闪）岩具有大致相同均顶峰变质温度：650～800 ℃ [1990, 1991, Wang et al.; 1989, Okay et al.]。

(3) 两处标本都发现过柯石英包裹体或其假象，表明二者都曾达到近 100 km 的深度。

(4) 在两处样品之间的 100 m 距离内，不存在任何晚期的断裂。

(5) Wang 等(1991)在石马地区的黑云母石榴石片麻岩(榴辉岩的围岩)的石榴石中发现柯石英假象，并认为这些片麻岩与所夹的榴辉岩块为同一高压地体。因此，高压地体中所有的榴辉岩及片麻岩、大理岩等均经历了榴辉岩相变质作用，并一起抬升至地表。本文测定的榴辉岩和榴闪岩块即发育在这一高压地体内，它们具有相同的退变质演化路径。

因此，由这两块标本所得出的榴辉岩退变质 P-T-t 轨迹应是可信的。

3 岩相学与矿物化学

新鲜榴辉岩的矿物组合为：石榴石 + 绿辉石 + 石英 + 金红石及磷灰石，其中绿辉石和石榴石呈镶嵌平衡结构，少数绿辉石呈包裹体产于石榴石中；石英含量约占 5%，金红石呈浸染状分布在石榴石和绿辉石中，含量小于 2%，磷灰石含量不超过 0.5%。

在榴闪岩中(T13-1)，绿辉石除少量以包裹体态存在于石榴石中外，其他均已强烈分解成透辉石 + 钠长石的后成合晶，继而转变为角闪石 + 钠长石；而石榴石保存较为完好，在其边缘常可见角闪石的次生反应边；有时可见多硅白云母退变为黑云母，此外，还有绿帘石、金红石、磷灰石、楣石等矿物及极少量的柯石英假象。

从上述的岩相学观察，我们把榴辉岩的变质历史划分为四个阶段，各阶段及其矿物共生组合如下：

(1)柯石英榴辉岩阶段(EcⅠ)　石榴石的核心部分及呈包裹体态存在的绿辉石和柯石英。

(2)普通榴辉岩阶段(EcⅡ)　石榴石(边缘部分) + 绿辉石 + 多硅白云母 + 石英。

(3)后成合晶阶段(Sym)　产于后成合晶中的透辉石、角闪石、钠长石。

(4)角闪岩阶段(Am)　作为石榴石次生边的角闪石 + 绿帘石 + 黑云母 + 斜长石。

在以上划分中，我们将榴辉岩相变质作用划分为柯石英榴辉岩阶段和普通榴辉岩阶段，主要是根据在大别山南部发现的柯石英[1989, Wang et al., 1989, Okay et al.]和我们这次在榴辉岩石榴石中发现的单斜辉石包裹体(照片 1-2)。据电子探针分析，石榴石中单斜辉石体的硬玉分子含量为 Jd = 50%～55%，比非包裹体态的大量绿辉石高 10% 以上，在分类图中同属绿辉石(图 2A)。Smith, D.C. (1988)在划分榴辉岩的变质阶段时，将整个榴辉岩相划分为一个阶段，但所有已发现的柯石英均以包裹体态存在于石榴石或绿辉石中[1989, Wang et al., 1989, Okay et al., 1988, Smith, D. C.]。这表明柯石英形成在前，而包裹它的主晶矿物石榴石、绿辉石形成在后。以包裹体态存在的石榴石中的绿辉石，地质产状与柯石英相同，是榴辉岩前进变质作用的产物，它和柯石英一道经受过最高峰温压的变质条件，故我们认为它与柯石英可以算同一变质阶段的矿物，由它为代表的矿物对所计算出来的温度，应更能代表(或者说更接近)榴辉岩相的顶峰变质温度。同时，这一划分也使大别山地区榴辉岩的 Sm-Nd 同位素年龄有了明确的含义：由于我们所选取的单矿物是颗粒粗大的石榴石和绿辉石，它们是在榴辉岩经变质峰期之后、开始抬升的初期结晶长大的，故所得出的年龄值代表了榴辉岩上升的初始阶段，即榴辉岩的初期退变质年龄，由于榴辉岩的初始抬升与陆壳俯冲时引发的逆冲构造有关，因此，其 Sm-Nd 同位素年龄指示的是陆壳俯冲的时代。

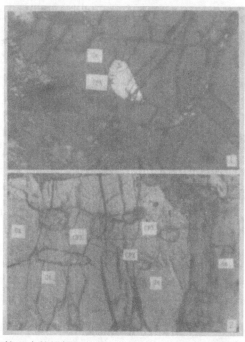

照片 1 石榴石中的绿辉石包体，4.0×10 (+)CPX: 绿辉石；Gt:石榴石
照片 2 石榴石中的绿辉石包体，4.0×10 (−)CPX: 绿辉石；Gt:石榴石

我们对榴闪岩和榴辉岩分别选取必要的单矿物进行电子探针成分分析。分析由中国地质大学完成。所用仪器主要为日本产 JEOL-733 型，能谱仪为 LINK860-2 型。我们只列出用于温度计算的绿辉石-石榴石矿物对、包裹体态绿辉石与包裹它的石榴石接触部位的分析结果和数据处理结果(见表 1)。分析结果的 Fe^{2+}/Fe^{3+} 计算按郑巧荣(1983)提出的方法。

榴辉岩(T13)中石榴石的镁铝榴石分子含量在 38%～42%之间，略高于榴闪岩(T13-1)中石榴石的镁铝榴石分子(28%～37%)，但它们在 Colemen(1965)分类图中均基本属 B 类(图 2B)；在从柏林 (1977)分类图中，所有石榴石均落在 ZC 线之上(图 2C)，显示较高的温压形成条件。

榴闪岩中产于后成合晶中的角闪石与角闪岩相阶段作为石榴石次生边的角闪石成分有显著不同，应用 Mottana (1970)分类法(图 2D): 前者属亚蓝闪石，后者为普通角闪石。

表 1 石榴石·单斜辉石电子探针分析结果（重量%）

岩性	榴辉岩 (T13')			榴闪岩 (T13-1)			榴辉岩 (T13')			榴闪岩 (T13-1)		
单矿物名称	石榴石	石榴石	石榴石	石榴石	石榴石	石榴石	单斜辉石	单斜辉石	单斜辉石	单斜辉石	单斜辉石	单斜辉石
顺序号	1	2	3	4	5	6	7	8	9	10	11	12
样品分析号	-1-GT	-2-GT	-4-GT	-1-3A	-1-3B	-1-2B	-1-GT	-2-GT	-4-GT	-1-3A	-1-3B	-1-2B
SiO_2	39.80	39.77	39.70	39.24	39.48	39.00	57.19	55.93	56.74	57.05	57.46	56.78
Al_2O_3	22.35	22.68	22.62	22.32	22.12	22.08	8.42	8.40	8.91	12.54	12.59	11.57
MgO	10.31	10.97	10.29	7.42	7.67	8.08	9.69	9.47	9.10	6.93	7.08	7.60
CaO	7.54	6.94	7.31	9.48	9.06	8.91	14.83	14.52	14.16	11.23	11.45	12.18
MnO	0.55	0.46	0.27	0.56	0.45	0.63	0.00	0.00	0.07	0.00	0.10	0.00
FeO*	18.84	18.34	19.24	20.91	20.73	20.61	3.60	4.18	3.92	3.13	2.90	3.37
(Fe_2O_3)	0.13	0.00	0.00	0.70	0.00	1.29	1.05	2.40	2.13	0.00	0.45	0.65
(FeO)	18.73	18.34	19.24	20.28	20.73	19.45	2.66	2.02	2.00	3.13	2.49	2.79
Na_2O	0.05	0.00	0.00	0.00	0.00	0.15	6.35	6.34	6.80	8.13	7.92	7.55
K_2O	0.06	0.00	0.01	0.10	0.03	0.00	0.04	0.04	0.01	0.06	0.00	0.00
TiO_2	0.16	0.02	0.00	0.00	0.13	0.00	0.12	0.16	0.16	0.15	0.08	0.00

续表

岩性	榴辉岩（T13'）			榴闪岩（T13-1）			榴辉岩（T13'）			榴闪岩（T13-1）		
单矿物名称	石榴石	石榴石	石榴石	石榴石	石榴石	石榴石	单斜辉石	单斜辉石	单斜辉石	单斜辉石	单斜辉石	单斜辉石
顺序号	1	2	3	4	5	6	7	8	9	10	11	12
样品分析号	-1-GT	-2-GT	-4-GT	-1-3A	-1-3B	-1-2B	-1-GT	-2-GT	-4-GT	-1-3A	-1-3B	-1-2B
Cr_2O_3	0.02	0.00	0.15	0.27	0.03	0.10	0.12	0.21	0.15	0.07	0.00	0.00
TOTAL	99.68	99.18	99.59	100.30	99.70	99.56	100.36	99.25	100.02	99.29	99.58	99.05
Si	2.998	2.995	2.995	2.971	3.013	2.997	2.023	2.004	2.011	2.017	2.022	2.015
Al(IV)	0.002	0.005	0.005	0.029	0.000	0.023	0.000	0.000	0.000	0.000	0.000	0.000
Al(VI)	1.983	2.008	2.007	1.962	1.989	1.964	0.351	0.355	0.372	0.523	0.522	0.484
Mg	1.158	1.231	1.157	0.837	0.837	0.920	0.511	0.506	0.481	0.365	0.371	0.408
Ca	0.609	0.560	0.591	0.769	0.741	0.729	0.562	0.558	0.538	0.425	0.432	0.463
Mn	0.035	0.029	0.017	0.029	0.029	0.041	0.000	0.000	0.002	0.000	0.003	0.000
Fe^3	0.007	0.000	0.000	0.040	0.000	0.074	0.028	0.065	0.057	0.000	0.012	0.017
Fe^2	1.180	1.155	1.214	1.284	1.323	1.242	0.079	0.061	0.059	0.093	0.085	0.083
Na	0.007	0.000	0.000	0.000	0.000	0.022	0.436	0.441	0.467	0.557	0.540	0.519
K	0.003	0.000	0.001	0.010	0.003	0.000	0.000	0.002	0.001	0.003	0.000	0.000
Ti	0.009	0.001	0.000	0.000	0.007	0.000	0.003	0.004	0.004	0.004	0.002	0.000
Cr	0.001	0.000	0.009	0.016	0.002	0.006	0.006	0.006	0.004	0.002	0.000	0.000
AND	0.241	0.000	0.000	1.370	0.000	2.524	Wo 27.77	27.80	26.74	21.28	21.95	23.240
GRO	20.175	18.823	19.839	24.974	24.983	22.340	En 25.25	25.21	23.91	18.28	18.85	20.480
ALM	39.572	38.824	40.752	43.988	44.606	42.431	Fs 3.90	3.04	2.93	4.66	4.32	4.170
SPE	1.184	0.975	0.571	0.993	0.978	1.398	Td 40.32	37.47	40.76	55.78	53.66	50.400
PYR	38.828	41.378	38.838	28.674	29.434	31.308	Ac 2.77	6.48	5.67	0.00	1.22	1.71

注：FeO^*为电子探针分析全Fe，TOTAL按FeO^*计算。

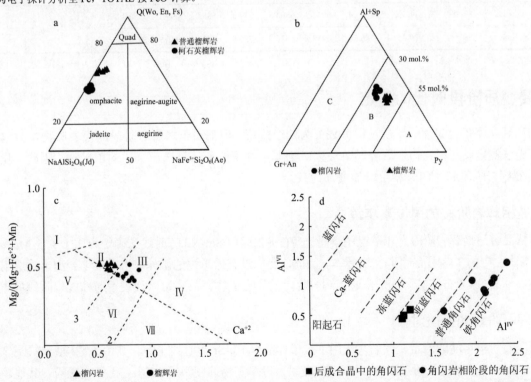

图2a 榴辉石中单斜辉石分类图；图2b 石榴石的Colemen分类图；图2c 石榴子石的从柏林分类图；图2d 榴辉(闪)岩中角闪石分类图（据A.莫塔纳，1970）

4 地质温压计的选择与温压计算

我们分别选用了 Ellis (1979)，Raheim (1974)提出的石榴石-单斜辉石地温计计算 Gt-CPX 矿物对的形成温度，从计算结果可以看出，利用包裹体态的绿辉石矿物对计算出来的温度，比利用非包裹态的绿辉石-石榴石矿物对计算的结果要高 100 ℃以上，因此，按照我们的阶段划分，柯石英榴辉岩阶段的变质温度要比普通榴辉岩阶段的温度高 100 ℃以上，即柯石英榴辉岩阶段不但有更高的压力，还具有更高的变质温度。后成合晶阶段，据 Perchuk (1969)之 CPX-Am 矿物对温度计算为 700 ℃，而其矿物组合相当于高角闪岩相，故其温度应在 600～700 ℃之间，二者是吻合的。

角闪岩相阶段之温度利用角闪石-斜长石温度计[1966, Perchuk]求得为 400～425 ℃。*

关于各阶段之压力计算，柯石英榴辉岩阶段(Ec-Ⅰ)之下限压力应用柯石英形成之下限压力，即 $P \geqslant 28$ kb ($T = 800$ ℃时)；普通榴辉岩阶段(Ec-Ⅱ)之压力据 Raheim 和 Green (1975)提出之后 K_D-P-T 图解为 15 kb$<P<$20 kb，而根据绿辉石中硬玉分子的含量计算[1980, Holland]为 17 kb$<P<$20 kb；后成合晶阶段(Sym)之压力根据 M.J.Kohn (1990)提出之 Gt-Am-Pl 矿物组合地质压力计计算为 13.8 kb，由于应用于压力计算的石榴石是前一阶段之矿物，故所得之压力应为上限压力，其下限压力系根据后成合晶中辉石之硬玉组分估算，为 10 kb；角闪石相阶段据谢窦克等(1984)的研究成果，为 4～7 kb。

各阶段温压具体数据见表 2。

表 2 大别山石马地区榴辉岩温压计算数据一览表

样品号		T13-1	T13-1	T13-1	T13-1'	T13-1'	T13-1'	T13-1	T13-1	T13-1
矿物对		CPX_1-GT	CPX_1-GT	CPX_1-GT	CPX-GT	CPX-GT	CPX-GT	AM-CPX*	AM-PL*	AM-GT*
分析号		T13-1-3A	T13-1-3B	T13-1-2B	T13-4OMP	T13'-2OMP$_2$	T13-1-2AM	T13-1-2AM	T13-1-2PL	T13-1-4GT
		T13-1-3A2	T13-1-3B2	T13-1-2B2	T13-4GT	T13-IGT	T13-1-GT	T13-1-PY	T13-1-2AM	T13-1-4AM
计算压力 (kb)		>28	>28	>28	20 15	20 15	20 15	13.80	4～7	4～7
计算温度(℃)	Elus	838.80	811.39	808.74	6934.8 6800.8	6819.3 6685.0	7499.5 7356.4	700	425	400
	Raheim	803.72	789.42	786.66	6958.6 6635.7	6780.9 6463.9	7346.7 7010.8			

-CPX_1 包裹体态绿辉石，CPX-绿辉石，GT-石榴石，PL-斜长石，AM-角闪石，T13-1-榴闪岩，T13'-榴辉岩。加 "" 者为按别尔丘克相图计算温度。

5 主要变质阶段时代(t)测定

如前所述，榴辉岩的整个变质过程可划分为四个阶段，但柯石英榴辉岩阶段的代表矿物均为包裹体存在，目前无法测定其年龄；而后成合晶阶段因矿物非常细小亦无法挑选出合适的单矿物。因此，我们只能测定普通榴辉岩阶段和角闪岩相阶段的变质时代。

5.1 普通相辉岩阶段的同位素年龄

T13'样品普通榴辉岩阶段之年龄已由李曙光等(1992)运用 Sm-Nd 等时线法确定为 221 ± 5 Ma。文中的 ND 样品与本文中的 T13'样品系同一块标本，因此，我们没有做重复的工作。由于该年龄测定时选用的是结晶粗大的石榴石-绿辉石单矿物，是普通榴辉岩阶段的代表矿物，故它可以代表普通榴辉岩阶段的年龄。

5.2 角闪石阶段的变质时代

对角闪石阶段年龄的测定，我们共选用了五个样品作 Rb-Sr 等时线年龄。其中关键样品是具有高 Rb/Sr 的，该阶段变质生成的黑云母单矿物(T13-1-Bi)。黑云母单矿物纯度为近 100%，无蚀变颗粒。其他 4 个样品分别是全岩(T13-1-全岩)，不含石榴石、以后成合晶为主的岩粒(T13-1-A)，以角闪石为主的岩粒(T13-1-B)和不含石榴石的通变质矿物混合物(T13-1-混合)。

样品测试由南京大学现代分析中心同位素室完成。实验室 Sr 全流程本底为 $2\times10^{-9}\sim5\times10^{-9}$ g，质谱测定标准样 NBS607 的 $^{87}Sr/^{86}Sr = 1.200014 \pm 12$，所有数据均符合最佳稀释法，因而具有准确度，五个样品的各项分析结果见表 3。

五个样品组成的 Rb-Sr 等时线年龄按一般线性回归计算为 $t = 134.4 \pm 0.5$ Ma (图 3)。

表 3 榴辉岩 Rb-Sr 同位素测试数据

样品号	Rb(ppm)	Sr(ppm)	Rb^{87}/Sr^{86}	Sr^{87}/Sr^{86}	误差
T13-1-全岩	23.86	537.5	0.1262	0.706892	±14
T13-1-a	7.540	54.85	0.3918	0.707107	±16
T13-1-b	48.57	151.2	0.9156	0.708828	±20
T13-1-混合	91.87	312.2	0.8514	0.708273	±18
T13-1-黑云母	902.1	64.89	41.0129	0.785036	±19
标准 NBC607	523.9	64.95		1.200014	±12

6 榴辉岩的 P-T-t 轨迹及其构造意义讨论

根据前述工作所提到的数据，我们作出大别山石马地区榴辉岩的 P-T-t 轨迹图如下(图 4)，该 P-T-t 轨迹具有如下构造意义：

(1) 在三叠纪初，华北陆块与扬子陆块发生对接，并随后发生了扬子陆块向华北陆块之下的陆壳俯冲，以及大规模的逆冲、滑脱构造如高压变质作用，这由本文所做同位素工作及李曙光等(1989, 1992)的工作得到证实，古地磁资料也证实了这一点[1987，林金录；1987，Zhao, X. et al.]。

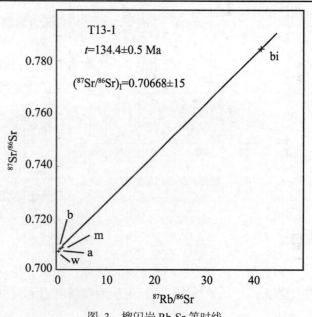

图 3 榴闪岩 Rb-Sr 等时线

(2) 大别山榴辉岩从柯石英榴辉岩阶段至后成合晶阶段，压力急剧下降，而温度下降缓慢。P 从大于 28 kb 降至 13.8 kb 以下，T 仅从 850 ℃ 降至 650 ℃ 左右，因此，该区榴辉岩在 221 Ma 左右（印支期）发生了一个近似绝热的急速抬升。这可以从我们所做榴辉岩的围岩——石榴黑云母片麻岩的同位素工作得到另一佐证。我们对片麻岩用不同方法进行了同位素定年：片麻岩中黑云母 K-Ar 年龄为 230±5 Ma，同一标本片麻岩中石榴石 Sm-Nd 等时线年龄为 229±3 Ma，两个年龄在误差范围内完全一致(详见另文)。据石榴石-黑云母地温计测定该片麻岩的变质温度为 525 ℃，其 Sm-Nd 体系在此温度下是封闭的，而黑云母的 K-Ar 封闭温度为 300 ℃，这就是说该片麻岩从 525 ℃ 因抬升下降到 300 ℃ 时所用时间是非常短的，可能只有几个百万年，用上述同位素定年方法无法区分它们，这只能解释为片麻岩在 230 Ma 左右存在一个快速抬升过程。显然，片麻岩的快速抬升与上述榴辉岩的快速抬升均发生在印支期，它们是由同一构造机制驱动的，即是扬子陆块俯冲时引起的大型逆冲构造造成了上述高压岩石的快速抬升。在这一时期内，榴辉岩被从近 100 km 处快速抬升至 40 km 以上(取 Sym 阶段上下限压力之中点)，在此过程的初期，榴辉岩开始了退变质重结晶作用，石榴石、绿辉石晶体不断生长，并包裹、交代早期的柯石英、绿辉石等矿物。

(3) 后成合晶阶段至角闪岩相阶段，榴辉岩的 P-T 轨迹较为平缓，表明在 134.4 Ma 时(燕山期)温度下降较快，此时地热增温率升高，因此这段时期可能相当于碰撞造山运动后的拉张期，大别山北侧广泛存在的侏罗纪拉张盆地及碱性火山岩活动也证明了这一点。此时引起榴辉岩抬升的动力不再是逆冲推覆构造，而是拉张条件下的重力均衡作用。碰撞造山时加厚的岩石圈由于密度大的根部物质被"拆离"而密度相对变小，从而产生向上的"浮力"。在这一时期，榴辉岩随整个山根缓慢地从 40 km 抬升至 18 km 左右(取角闪岩相阶段压力中点 5.5 kb 计算)，由于这一阶段是山体的整体抬升，故南大别山所有的变质杂岩都曾深埋于 40 km 以下，并经历了 $P = 12$ kb 左右的高压变质作用。这与 X.Wang (1991) 认为大别山为一整体高压块的观点类似，只是压力未达他认为的 27 kb 以上。

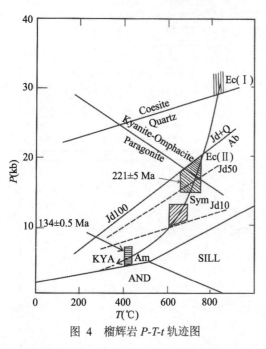

图 4 榴辉岩 P-T-t 轨迹图

7 结论

(1) 大别山南部榴辉岩的变质经历了四个演化阶段，即文中划分的柯石英榴辉岩阶段、普通榴辉岩阶段、后成合晶阶段和角闪岩相阶段。

(2) P-T-t 轨迹研究表明大别山南部榴辉岩至少经历了两个抬升阶段：a. 在 221 Ma (印支期)左右，由陆壳俯冲时形成的大规模逆冲构造驱动的快速抬升阶段；b. 在 134 Ma 左右(燕山期)，榴辉岩处于陆块碰撞后拉张环境下，由于重力均衡作用而驱动的缓慢抬升阶段。

(3) 大别山南部所有的变质杂岩都曾深埋于 40 km 以下，并经历了 $P = 12$ kb 以上的高压变质作用。

(4) 大别山南部榴辉岩中用石榴石和绿辉石定出的 Sm-Nd 同位素年龄，代表的应是榴辉岩开始抬升的退变质年龄，亦即扬子陆壳俯冲的时代。

致谢

本文样品由刘德良副教授共同采集；初稿曾与从柏林教授进行讨论，承蒙提出宝贵意见；南京大学王银喜同志，地矿部物化探研究所周丽沂同志为同位素测定及样品制备提供了方便，在此一并表示衷心的感谢。

参考文献

1965 R. G. Colemen et al., Eclogites and eclogits:their difference and similarities. Geol. Soc. Amer. Bull., 76: 483-508.
1974 A. Rahelm et al., Experimental determination of the temperature and pressure dependence of the Fe-Mg partition coefficient for coexisting garnet and clinopyroxene.Contrib.Mineral.Petrol., 48: 179-203.
1975 A. Rahelm et al., P-T paths of natural eclogites during metamorphism---a record of subduction. Lithos, 8: 413-416.
1977 从柏林, 张雯华, 榴辉岩中的石榴石. 科学通报, 8: 413-416.
1979 D.J.Ellis et al., An experimental study of the effect of Ca upon garnet-clinopyroxene Fe-Mg exchange equilibria. Contrib. Mineral. Petrol., 71: 13-22.
1983 郑巧荣, 由电子探针分析值计算 Fe^{2+} 和 Fe^{3+}. 矿物学报, 4: 55-62.
1984 谢宾克等, 大别山杂岩中的角闪石研究, 矿物学报, 3: 215-225.
1987 林金录, 华南板块的地极移动曲线及其地质意义. 地质科学, 4: 306-315.
1987 X. Zhao et al., Paleomagnetic constraints on the collision and rotation of North and South China. Nature, 327:141-144.
1988 D. C. Smith, Eclogites and eclogites-facies rocks.Elseriver, Amsterdam.
1989 A. I. Okay et al., Coesite from the Dabie Shan eclogites, central China.Eur.J.Mineral., 1: 595-598.
1989 李曙光等, 中国华北-华南陆块碰撞时代的 Sm-Nd 同位素年龄证据.中国科学(B 辑), 3: 312-319.
1989 X. Wang et al., Coesite-bearing eclogites from the Dabie Mountains in central China. Geology. 17: 1085-1088.

1990 M. J. Kohn et al., Two new geobarometers for garnet-amphibolites, with applications to southeastern Vermont.Amer, Mineralogist. 75: 89-96.

1990 X. Weng et al., Field occurrences and perology of eclogites from Dabie Mountains, Anhui, Central China. Lithos, 25: 119-131.

1991 N.L.Dobretsov et al., Blueschist and eclogites: a possible plate tectonic mechanism for their complacement from the upper mantle.Tectonophysies. 186: 253-268.

1991 X. Wang et al., Regional ultalhigh-pressure coesite-bearing eclogitic terrane in Central China: evidence from country rocks:gneiss, marble, and metapelite. Geology, 19: 933-936.

1992 A. I. Okay et al., Evidence for intracontinental thrust-related exhumation of the ultra-high-pressure rock in China. Geology, 20: 411-414.

1992 李曙光等, 大别山南麓含柯石英榴辉岩 Sm-Nd 同位素年龄. 科学通报. 4: 297-300.

P-T-t path for coesite-bearing peridotite-eclogite assciation in the Bixiling, Dabie Mountains*

Xiao Yilin[1], Li Shuguang[1], Emill Jagoutz[2] and Cheng Wen[3]

1. Department of Earth and Space Sciences, University of Science and Technology of China, Hefei 230026, China
2. Department of Cosmochemistry, Max-Planck-Institute fur Chemie, D-55020 Mainz, Germany
3. Institute of Geology, Chinese Academy of Geological Scineces, Beijing 100037, China

> 亮点介绍：研究显示碧溪岭榴辉岩经历了至少 5 个阶段的变质作用；*P-T-t* 轨迹表明其经历了 2 个阶段的抬升作用，早期退变年龄为 209 ± 6.5 Ma；围岩片麻岩与榴辉岩具有相似的变质年龄，表明它们具有相同的俯冲和抬升历史。

Abstract The peridotite/eclogite associations occur at Bixiling in the Yuexi county of Anhui province, with an outcrop area of about 0.7 square kilometer. It is one of the largest eclogite lenses in the Dabie Mountains. The main parts of this lens at this locality are eclogites and garnet peridotites, and the garnet peridotites occur as tens' belts or lenticulars (from several to ten's meters in size) within eclogites. The field evidence suggests that both eclogite and garnet peridotite have undergone the eclogite facies metamorphism. The country rocks of the eclogite/garnet peridotite associations are the mica-plagioclase gneiss of amphibolite facies. The field relations show that there is a tectonic contact between them. Besides these, some granodioritic veins relevant to the post-orogenic event were found.

So far there is little published reliable geochronological data and no *P-T-t* path study for these rocks. In the present study, we try to determine the metamorphic history and uplift machinism of the eclogites in the Bixiling.

1 Petrography, Mineral Chemistry and *P-T* Estimates

Forty-five samples including six garnet peridotites, four mica-plagioclase gneisses and thirty-three eclogites from the Bixiling were studied petrographically. Minerals from representative thin sections were analyzed by the electron probe.

Petrographical study shows that eclogite from the Bixiling domain typically contain the primary mineral assemblage garnet + clinopyroxene + phengite + kyanite + rutile + quartz ± amphibole ± epidote ± plagioclase ± biotite ± zircon. Coesite and coesite pseudomorphs have also been found in some eclogites.

Garnet in the eclogite mainly occurs as idioblastic crystals with kelyphitic rims of amphibole and/or porphyroblasts with abundant inclusions of rutile, omphacite, phengite, quartz (or coesite), etc. Based on the microprobe analysis data, the garnets in both eclogites and garnet peridotites were classified using Cong's method (Cong Bolin et al., 1977). The results show that garnet in eclogites would be formed in deep crust but garnet in peridotites would be formed in the mantle. This may indicate that the eclogites and garnet peridotites were formed around the boundary between the lower crust and mantle.

Omphacite can be divided into two kinds: the very fine-gained omphacite which is preserved only as inclusions in garnet, and the coarse-gained omphacite subjected partially to retrograde metamorphism, which coexist with coarse-gained garnets. According to the microprobe analysis data of omphacites in this study and other research (Xiao Yilin et al., 1993), the omphacites occur as inclusions in garnet generally contain higher jadeite contents (50-65 mole%) than those coarse-gained omphacites (37%-45%).

---

* 本文发表在：Chinese Science Bulletin, 1995, 40: 156-158

In all samples, coarse-grained garnet and omphacite do not show composition zoning patterns. But the omphacite which exist as inclusions in garnets sometimes show an increase of jadeite contents from the core (around 55%) to the rim (up to 65 mole%). This may imply that an increase of pressure during the prograde metamorphic event has been recorded by these inclusions.

Amphibole, if present, can be also divided into two kinds: one kind of amphibole (barroisite) is intergrowth with sodium plagioclase and micro-grained clinopyroxene (augite) in symplectite after omphacite, the other is coarse-gained hornblende which exists as the kelyphitic rims of garnet.

According to the limited conditions of P-T-t path that the chronometric and thermobarmetric information should be obtained from the exact same materials, we selected only one eclogite sample which was named DX36A for P-T estimates and geochronology. A detailed petrological study shows that this eclogite sample has been subjected at least five metamorphic stages:

Stage 1: Pre-eclogite-facies metamorphism (EcP). This stage is represented by minerals occurring as inclusions in garnet porphyroblasts. The pressure at this stage is estimated to be about 17 kbar based on the jadeite contents in the core of omphacites which occur as inclusions in garnets. Temperatures are calculated at P=17 kbar to be 719–780 ℃ using the garnet-clinopyroxene geothermometry of Ellis and Green Ellis et al. (1979) to core of omphacite inclusions and garnet in contact with these inclusions.

Stage 2: Coesite eclogite-facies metamorphism (EcC). The representative minerals of this stage are coesite + rim of the omphacite inclusions+core of garnets. The forming of coesite inclusions is characteristic of this stage. It probably represents the peak metamorphism for the coesite-bearing eclogite. Temperatures for this stage are calculated to be 841–930 ℃ using the same geothermometer as stage 1 to rim of omphacite inclusions and garnet in contact with these inclusions. The pressure is estimated to be>29 kbar based on the existence of coesite.

Stage 3: Recrystallized eclogite-facies metamorphism (EcR). The assemblage of coarse-grained garnet + omphacite + quartz + kyanite + phengite + rutile is characteristic of this stage. It probably represents an initial retrograde metamorphism of eclogites during their early uplift. Temperatures are estimated to be 770–810 ℃ using the garnet-clinopyroxene geothermometry of Ellis and Green. Pressure is estimated to be 15–20 kbar based on the P-T relationship of eclogite-facies metamorphism.

Stage 4: Retrograde symplectite stage (Sym). This stage is represented by micro-grained clinopyroxene, amphibole (barroisite) and sodic plagioclase in symplectite after omphacite. According to the mineral assemblage, this stage probably corresponds to a high-grade amphibolite-facies conditions. The temperature of 600–700 ℃ is estimated using the amphibole-clinopyroxene geothermometry. The pressure of this stage is estimated to be 9–13 kbar based on M.J.Kohn's geobarmetry Kohn et al. (1990) and the jadeite contents in clinopyroxenes.

Stage 5: Retrograde amphibolite-facies stage (Am). This stage is the last stage of metamorphism. Amphibole which occur as the kelyphitic rims of garnet, coarse-grained epidote, biotite and plagioclase are characteristic of this stage. Application of the amphibole-plagioclase geothermometry (Perchuk,1966 quoted from R.Zhang and Cong) to this stage yields the temperature of 400–450 ℃ at P = 5 kbar. The pressure estimated by geobarmetry of amphibole Rease et al . (1974) is about 5 kbar.

2 Geochronology

Information on the timing of metamorphism was obtained from minerals using Sm-Nd, Rb-Sr and Ar-Ar systematic. The Sm-Nd and Rb-Sr dating were done at the Max-Planeck-Institute in the Germany. The Ar-Ar systematic was analyzed at the Institute of Geology, Chinese Academy of Geological Sciences.

Garnet + omphacite + kyanite + rutile + amphibole + whole rock from sample DX36A yielded a Sm-Nd isochron age of 209 ± 6.5 Ma. This age is a little younger than that of eclogite in Shima (although they are indistinguishable within error limit). One possible cause for this young age is that there is a Sm-Nd isotopic disequlibrim phenomena which we can find out from the Sm-Nd concentrations of minerals in sample DASO, although this disequilibrium is slight.

Clinopyroxene and garnet have also been separated from garnet peridotite sample (93DB01) in the Bixiling.

Garnet + clinopyroxene + whole rock from 93DB01 yielded a Sm-Nd isochron age of 219 ± 24 Ma. Combination of these two ages indicates that the age of 209 ± 6.5 Ma may be considered as a minimum value of time for the eclogite beginning retrograde metamorphism.

Meanwhile, a Rb-Sr isochron age of 138 ± 8.7 Ma has been obtained for phengite + amphibole + rutile + kyanite + omphacite + whole rock from the sample DX36A. Because the sample DX36A has subjected an obvious retrograde metamorphism and thus the Rb-Sr systematic may have been reset, this age probably correspond to cooling beneath the appropriate closure temperatures, which generally estimated around 400–450 ℃ (Griffiths et al., 1991).

To determine the time of metamorphism for country rocks (gneiss), an $^{40}Ar/^{39}Ar$ plateau age of 217 ± 0.5 Ma for phengite from gneiss (sample DX53A) was obtained. Because the argon closure temperature of phengite is in the range 350 ± 25 ℃ and the gneiss subjected an amphibolite facies metamorphism, this age probably represents the time of early retrograde metamorphism of the gneiss.

3 Discussion and conclusion

A clockwise metamorphic *P-T-t* path for eclogites from Bixiling domain has been constructed using the calculated temperatures, pressures, mineral ages and textural information. Because the peak metamorphic conditions in eclogites from Bixiling domain attained > 29 kbar/841–930 ℃, the age for the peak metamorphism stage has not been obtained in this study. However, on the analogy of the age of the peak metamorphic stage for coesite-bearing from the Shima and Shuanghe (unpublished data), it may be reasonable to suppose that the age of peak metamorphic stage of the eclogites from the Bixiling is around 240 Ma.

The exhumation history of eclogite from the Bixiling can be approximately divided into two periods based on the slope of the *P-T-t* path: (1) the first period with a fast uplift rate (which is 1.42 mm/a by a rough estimate) from EcC to Sym; and (2) the second period with a slower uplift rate (about 0.21 mm/a). So the *P-T-t* path for eclogites from the Bixiling is similar to that for eclogites from the Shima(Xiao Yilin et al., 1977 ; Li Shuguang et al., 1977).

Based on the *P-T-t* path for coesite bearing eclogite from the Bixiling, following conclusions could be drawn:

1) A clockwise *P-T-t* for eclogite suggests that the formation of the eclogite in the Bixiling domain was caused by a continent-continent collisional event.

2) Eclogites from the Bixiling have been subjected to at least five stages of metamorphism during their metamorphic causes.

3) Similar to eclogites in the Shima, two-stage exhumation history is supported by the study of *P-T-t* path for eclogites in the Bixiling.

4) The initial retrograde metamorphism of eclogites probably occurred during the Triassic, and may no later than 209 ± 6.5 Ma.

5) Country rocks (gneiss) of eclogites in the Bixiling probably have the same metamorphic ages with the eclogites, although they may have been subjected much lower *P-T* conditions than the eclogites.

References

Cong Bolin et al., Chinese Science Bulletin, 1977, 8: 413-416.
Xiao Yilin et al., Geotectonic et Metallogenia, 1993, 17: 239-250.
Li Shuguang et al., Chemical Geology, 1993, 109: 89-111.
Ellis, D. J. et al., Contrib.Mineral. Petrol., 1979, 71: 13-22.
Perchuk, L. L., Int. Geol. Rev., 1969, 11: 875-901.
Kohn, M. J. et al., Amer.Mineralogist, 1990, 75: 89-96.
Rease, P., Contrib. Mineral. Petrol., 1974, 45: 231-236
Griffiths, J. B. et al., Lithos, 1991, 27: 43-57.
Jager, E.et al., Lectures in Isotope Geology, Springer, Berlin. 1979: 1-12.

Sm-Nd and Rb-Sr isotopic chronology and cooling history of ultrahigh-pressure metamorphic rocks and their country rocks at Shuanghe in the Dabie Mountains, Central China[*]

Shuguang Li[1], Emil Jagoutz[2], Yizhi Chen[1] and Qiuli Li[1]

1. Department of Earth and Space Sciences, University of Science and Technology of China, Hefei 230026, Anhui, China
2. Department for Cosmochemistry, Max-Planck-Institut für Chemie, D55020 Mainz, Germany

> 亮点介绍：系统剖析了大别山超高压变质岩石中矿物间 Sm-Nd 和 Rb-Sr 等时线年代学研究中的同位素不平衡问题，指出峰期超高压变质阶段形成的矿物之间达到了 Nd、Sr 同位素平衡，但它们与退变质矿物之间存在不平衡；通过三超高压变质矿物 Sm-Nd 等时线法给出了双河榴辉岩及其片麻岩围岩有相同的峰期变质年龄，226±3 Ma；结合低封闭温度的退变质矿物 Rb-Sr 等时线年龄首次给出榴辉岩及其围岩有相同的两阶段快速 T-t 冷却曲线。据此提出大别山超高压变质榴辉岩及其围岩组成的超高压岩片共同经历了两次快速抬升历史。

Abstract Ultrahigh-pressure metamorphic (UHPM) rocks at Shuanghe area in the Dabie Mountains occur as a UHPM block within the regional granitic gneiss. A Sm-Nd isochron defined by garnet + omphacite + rutile from coesite-bearing eclogite yield an age of 226.3±3.2 Ma. Another Sm-Nd isochron defined by garnet + 2 phengites form UHP gneiss yield a similar age of 226.5±2.3 Ma. The two consistent Sm-Nd ages defined by three UHPM minerals suggest that Nd isotopic equilibrium between UHPM minerals in these rocks during UHP metamorphism has been achieved. This may correspond to the age of peak metamorphism with an average metamorphic temperature of 800 ℃. It is well demonstrated that the retrograde metamorphism of UHPM rocks occurred in an open chemical system, while the Nd and Sr isotopic systematics of UHPM minerals may have remained closed. Nd and Sr isotopic disequilibrium between UHPM minerals and retrograde metamorphic minerals has been observed, so that the tie line of garnet or phengite and whole rock containing retrograde metamorphic minerals gives an old Sm-Nd age and a young Rb-Sr age respectively with no geological significance. However, a Rb-Sr ages of 219.0±6.6 Ma defined by phengite+garnet from a UHPM gneiss indicates that the UHPM rocks at Shuanghe cooled down to 500 ℃ at that time. It suggests that the UHPM rocks at Shuanghe experienced the first rapid cooling during 226–219 Ma. On the other hand, the Rb-Sr ages of 174±7.8–169.2±3.3 Ma defined by retrograde minerals (amphibole or biotite) with closure temperature ranging from 450–300 ℃ and intensely retrograded metamorphic rocks reflect a second rapid cooling during this time interval. This is consistent with the "rapid cooling" time of 190–170 Ma of the orthogneiss in the Dabie-Sulu UHPM belt obtained by Ar-Ar dating method. In contrast to UHPM rocks, the Nd isotopic composition of the garnet in granitic gneiss has been reset during retrograde metamorphism and is in equilibrium with those of retrograde epidote and biotite, which yields a Sm-Nd isochron age of 213 ± 5 Ma indicating the retrometamorphic time corresponding to amphibolite facies. In addition, the biotites from the granitic gneiss yield Rb-Sr ages of 171–173 Ma similar to those of the UHPM rocks. These data suggest that the country rocks (granitic gneiss) may have a similar cooling history to the UHPM rocks at Shuanghe. Two stages of rapid cooling of UHPM rocks at Shuanghe may correspond two stages of fast uplift: the initial rapid uplifting and cooling of UHPM rocks during 226–219 Ma may be caused by compression tectonics during subducting time of the continental crust; while the later rapid cooling may reflect the exhumation of the entire subducted continental crust via extension during the early-middle Jurassic.

1 Introduction

The occurrence of ultrahigh-pressure (UHP) minerals, such as coesite and diamond in crustal rocks in orogenic belts suggests that a huge amount of continental crust can be subducted to mantle depth during the

---

[*] 本文发表在：Precambrian Research, 1990, 47: 191-203

continent-continent collision (Chopin, 1984; Smith, 1984; Wang et al., 1989; Okey et al., 1989; Sobolev and Shatsky, 1990; Xue et al., 1992). This has raised an intriguing question about the mechanism how the ultrahigh-pressure metamorphic (UHPM) rocks were exhumed from the depth of over 100 km to the surface, in which process the UHP minerals are preserved during decompression and cooling rather than being destroyed? Cooling history study of UHPM rocks is the most direct means to give constraint on tectonic process of the exhumation.

The Dabie mountains and Su-Lu terrane in central China (Fig. 1) is the largest known UHPM belt on Earth. Abundant Sm-Nd mineral isochron and U-Pb zircon ages of UHPM rocks from this belt have documented the Triassic event of ultrahigh pressure metamorphism (UHPM) (e.g., Li et al., 1992, 1993, 1994, 1996, 1997; Ames et al., 1993, 1996; Chavagnac and Jahn, 1996; Rowley et al., 1997; Hacker et al. 1998). $^{40}Ar/^{39}Ar$ thermochronology of the gneiss in the Dabie Mountains and Su-Lu terrane has also been studied in order to reveal cooling history (Chen et al., 1992; Eide et al., 1994; Li et al., 1995; Chen et al., 1995; Hacker and Wang, 1995). However, presence of excess Ar in phengite and biotite from UHPM rocks cast doubt about their suitability for $^{40}Ar/^{39}Ar$ thermochronologic studies of UHP rocks (Li et al., 1994; Hacker and Wang, 1995; Arnaud and Kelley, 1995; Ruffet et al., 1995; Scailete, 1996). In fact, except for blueschist, most samples used for the $^{40}Ar/^{39}Ar$ thermochronologic studies in the Dabie-Sulu terrane are from low pressure granitic gneisses. They can only reveal cooling history of the country rocks of UHPM units in the Dabie-Sulu terrane, but can not directly reveal the cooling history of UHPM units. These studies indicate that the country rocks (granitic gneisses) of UHPM units in the Dabie-Sulu terrane have experienced a fast cooling during 190–170 Ma (Chen et al., 1992; Li Q. et al., 1995; Chen et al., 1995).

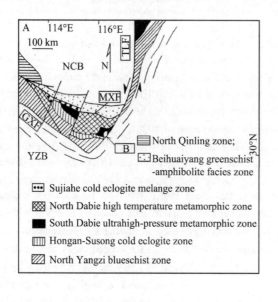
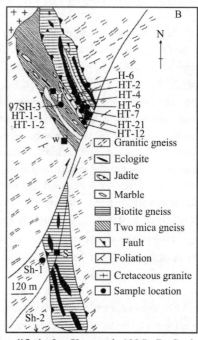

Fig. 1 A, Tectonic map of Dabie Mountains and its location in China (modified after Xue et al. 1996). B, Geological map of the Shuanghe UHPM slab and its country rocks (after Cong et al. 1995) and sample locations.

In order to better define cooling history of the UHPM rocks in the Dabie Mountains, Rb-Sr isotopic dating method in addition to Sm-Nd method has been chosen to date phengite and biotite from UHPM rocks. A high rate of initial cooling (~ 40 ℃/Ma) during 220–210 Ma for the UHPM rocks from Bixiling, Dabie Mountains has been obtained by using Sm-Nd and Rb-Sr isotopic dating methods(Chavagnac and Jahn, 1996). Similar Sm-Nd and Rb-Sr isotopic dating results for the UHPM rocks from the Qinglongshan in the Su-Lu terrane (Li et al., 1994) revealed a rapid cooling (~50 ℃/Ma) during 226–219 Ma. However, there are some important questions concerning cooling history of UHPM rocks in the Dabie-Sulu terrane that remain to be answered.

1. The reported Sm-Nd isochron ages have a large variation range from 246 Ma to 210 Ma (Li et al., 1993,1994,1996; Okey and Sengör, 1993; Chavagnac and Jahn, 1996) resulting in a large uncertainty in the peak

metamorphic age of the UHPM rocks. Most of the Sm-Nd ages were defined by two minerals (garnet + omphacite). Thus we can not judge whether Nd isotopic equilibrium between the two minerals was achieved during UHPM.

The REE data were obtained by inductively coupled plasma spectrometry at the analytical center of Geological Bureau, Hubei Province, P. R. C. The other data were obtained at the Institute of Geophysical and Geochemical Exploratory Techniques, Chinese Academy of Geological Sciences, by X-ray fluorescence analysis except CO_2 and H_2O which were obtained by nonaqueous titration and gravimetric analysis respectively.

2. Phengite Rb-Sr ages have been widely used to determine the cooling time corresponding to closure temperature of 500 ℃ (e.g., Cliff, 1985). However, the reported phengite Rb-Sr ages show considerable variation even for samples from the same UHPM unit. For example, 4 phengite samples from the Bixiling eclogite body yield Rb-Sr ages ranging 223–198 Ma (Chavagnac and Jahn, 1996). The major influence upon this age uncertainty is still unknown. Thus, the possibility of Sr isotopic disequilibrium between phengite and other minerals during cooling or retrograde metamorphism should be investigated.

3. As mentioned above, the previous studies have revealed that the initial rapid cooling of UHPM rocks during 226–210 Ma in the Dabie-Sulu terrane is much earlier than that (190–170 Ma) of their country rocks (granitic gneiss). If true, these data support the "foreign" relationship between UHPM rocks and their country rocks (e.g. Cong et al., 1995). However, if the UHPM rocks and their country rocks have "in-situ" relationships as suggested by other authors (e.g. Wang and Liu, 1991; Rowley et al., 1997; Hacker et al., 1998; Carswell et al., 1998), they should have a similar cooling history, i.e. the UHPM rocks in the Dabie-Sulu terrane should have experienced another rapid cooling during 190–170 Ma together with their country rocks and the granitic gneiss should also have experienced the earlier rapid cooling during 226–210 Ma. Try to find the evidences for these cooling events is critical to evaluate this controversy.

These problems indicate that we need to improve our understanding of the isotopic systematics of UHPM rocks during UHP or retrograde metamorphism. In this paper, we report a Rb-Sr and Sm-Nd isotopic study of an UHPM unit and its country rocks at Shuanghe, Dabie Mountains and try to answer the above questions. Based on the cooling T-t path obtained in this study, the exhumation mechanisms of the UHPM rocks in the Dabie Mountains will be discussed.

2 Geological setting and metamorphic history

The Qinling-Dabie orogenic belt is a complex collision zone of the North China Block (NCB), Yangtze Block (YB) and some microcontinents located between the NCB and YB (e.g., Zhang et al. 1995; Li et al., 1995, 1998; Li and Sun, 1996). The Dabie Mountains represent the eastern section of the belt and can be subdivided into seven metamorphic subzones from the north to the south (Fig. 1A), they are : (1) North Qinling zone; (2) Beihuaiyang greenschist-amphibolite facies zone; (3) Sujiahe cold-eclogite melange zone; (4) Northern Dabie high-temperature metamorphic zone; (5) Southern Dabie ultrahigh-pressure metamorphic zone; (6) Hongan-Susong cold eclogite zone; (7) Northern Yangtze blueschist zone. Petrological, geochemical, chronological and geological studies suggest that zone (1) is a Paleozoic island arc of the NCB; zone (2) could be a forearc complex of the NCB; zone (3) is a tectonic melange related to the collision between the NCB and Northern Dabie microcontinent; zone (4) could be a microcontinent between the NCB and YB; and the zone 5,6,7 represent a vertical section of the subducted continental crust of the YB (e.g., Li et al., 1995, 1998; Li and Sun, 1996). The ultrahigh pressure metamorphic (UHPM) rocks, including coesite or diamond-bearing eclogite associated with jadeite quartzite, garnet-epidote-micaceous paragneiss and marble with eclogite boudins are restricted within in zone (5). They are encased in amphibolite facies orthogneisses. The relationships between the UHPM units and country rocks are still a controversial issue. Cong et al.(1995) suggested a model of foreign relationship between them based on the detailed mapping and petrological studies at the Shuanghe area. While based on the more detail petrological study in the same area, Carswell et al.(1998) suggested a in situ model relationship and emphasized that it is necessary to resort to models of tectonic juxtaposition to explain the spatial association of the granitic gneiss with the undoubted UHP schists and eclogites within Dabie Mountains.

Figure 1B shows that the elongated UHPM unit with NNW-SSE trend outcrops over an area of about 1 km² at Shuanghe, and is surrounded mostly by orthogneiss. This UHPM unit is offset by a dextral strike-slip fault, and intruded by a younger granitic body to the NW. This UHPM unit is mainly composed of fine-grained and homogeneous coesite-bearing eclogite, garnet-epidote-two mica gneiss, garnet-biotite gneiss, jadeite quartzite and marble with or without eclogite nodules or boudins. The compositional layering within the UHPM unit is more evident in the northern outcrop (Fig.1B). A few tectonic fragments of actinolitite and serpentinite (several meters long) occur along the eastern boundary fault of the northern UHPM unit. Abundant eclogite lenses and boudins are interlayered or interleaved with garnet-epidote-two mica gneiss and minor garnet-biotite gneiss and marble in the eastern part of the northern unit. Whereas abundant garnet-biotite gneiss are interlayered with marble, jadeite-quartzite and minor eclogite lenses in the western part of the northern unit.

Mineral paragenesis and chemistry of the UHPM rocks and their country rocks at Shuanghe have been reported by Cong et al. (1995) and Carswell et al. (1998); and five metamorphic stages were recognized (modified after Cong et al.,1995):

(1) Pre-eclogite stage (PE) : This is recorded by phengite and barroisite inclusions within garnet porphyroblasts. this mineral assemblage may have formed prior to coesite at pressures of 6–8 kbar.

(2) Peak UHP coesite-eclogite stage (CE) . It is recorded by coesite and Jd-rich (Jd60) omphacite inclusions in garnet. The metamorphic temperatures ranging 720–880 ℃ (with an average of 800 ℃) in this stage has been estimated (Cong et al., 1995). The presence of coesite during this stage indicates pressures > 29 kb at T=800 ℃. All UHP minerals have been formed in this stage.

(3) Quartz-eclogite stage (QE). This is represented by the coexisting garnet and omphacite porphyroblasts. P-T conditions of the mineral pair were estimated to be P = 13–16 kbar, T = 630–760 ℃ (with an average of 695 ℃) (Cong et al., 1995). The garnet and omphacite could continue growth by recrystallization during this stage.

(4) Symplectite stage (Smp) : This is represented by various symplectite replacement assemblages, e.g. the omphacite and garnet have been partly replaced by symplectitic assemblages of augite + sodic plagioclase and of amphibole + sodic plagioclase respectively. The P-T conditions for this stage have been estimated to be 6–8 kbar at 470–570 ℃ (Cong et al., 1995).

(5) Post-symplectic stage (PS) : This is represented by various retrograde minerals formed after symplectite, such as actinolitic amphibole, biotite and chlorite. The P-T condition of this stage corresponds to the greenschist facies (Cong et al., 1995).

3 Sample description

Ten UHPM rock samples and two country rock samples were collected from the Shuanghe area. Sample locations are shown in Figure 1B. Their major and trace element compositions are listed in Table 1.

Table 1 Major-element (in wt%) and trace-element (in ppm) compositions of the UHPM rocks and their country rock from Shuanghe in Dabie Mountains

Sample No.	92HT-1-1	92HT-6	92HT-2	92HT-4	92HT-7	92HT-21	92HT-12	SH-1	SH-2
Rock type	UHP gneiss	UHP gneiss	retrograde quartz Eclogite	quartz eclogite	quartz eclogite	quartz eclogite	Retrograde eclogite in marble	granite gneiss	granite gneiss
SiO_2	60.71	62.99	53.37	50.97	48.82	50.78	41.13	72.75	74.65
Al_2O_3	17.42	14.34	16.39	12.58	13.85	13.36	16.73	13.87	13.74
Fe_2O_3	1.19	1.59	1.44	2.39	1.21	1.43	1.10	0.86	0.95
FeO	4.99	5.03	8.19	13.54	10.44	13.59	20.12	1.15	0.74
MgO	2.87	2.73	5.30	5.75	6.89	5.01	3.62	0.26	0.37
CaO	2.93	5.01	8.88	9.40	11.64	8.98	10.43	0.98	0.64
Na_2O	1.98	3.45	4.22	2.24	3.79	2.67	1.91	5.66	5.92
K_2O	4.40	2.34	0.42	0.01	0.05	0.01	0.27	2.96	1.86
TiO_2	0.59	0.72	0.94	2.65	1.26	2.36	2.72	0.35	0.35
P_2O_5	0.04	0.11	0.20	0.15	0.12	0.36	0.06	0.06	0.05
MnO	0.12	0.10	0.17	0.23	0.17	0.20	0.40	0.12	0.04
H_2O^+	1.75	0.88	0.06	0.07	0.04	0.05	0.85	0.57	0.22
CO_2	0.64	0.07	0.34	0.09	1.46	0.84	0.58	0.33	0.32
Total	99.63	99.36	99.92	100.07	99.74	99.64	99.92	99.93	99.88

Continued

Sample No.	92HT-1-1	92HT-6	92HT-2	92HT-4	92HT-7	92HT-21	92HT-12	SH-1	SH-2
Rock type	UHP gneiss	UHP gneiss	retrograde quartz Eclogite	quartz eclogite	quartz eclogite	quartz eclogite	Retrograde eclogite in marble	granite gneiss	granite gneiss
La	106.30	7.93	4.13	7.81	7.35	9.39	1.30	40.50	45.22
Ce	206.00	15.57	13.39	22.12	18.25	24.30	3.79	81.28	94.94
Pr	23.48	1.95	2.04	3.13	2.70	3.61	0.55	11.37	11.31
Nd	83.43	8.60	10.24	14.27	12.15	15.75	3.20	42.14	44.24
Sm	12.81	2.22	3.20	3.72	3.24	4.32	1.82	8.12	8.35
Eu	1.96	0.82	0.90	0.37	1.11	1.48	1.06	1.78	1.67
Gd	7.84	0.50	4.39	4.90	3.72	5.53	5.52	7.13	7.13
Tb	0.94	0.40	0.72	0.85	0.57	0.91	1.27	1.24	1.15
Dy	6.00	2.52	4.68	5.77	3.77	5.81	9.97	7.54	7.55
Ho	1.24	0.48	0.91	1.13	0.71	1.15	2.07	1.50	1.52
Er	3.38	1.33	2.46	3.18	1.85	3.18	6.10	4.68	4.50
Tm	0.53	0.19	0.36	0.51	0.27	0.48	0.92	0.75	0.72
Yb	3.21	1.16	1.97	2.98	1.56	2.71	5.68	4.97	4.45
Lu	0.50	0.18	0.30	0.47	0.23	0.45	0.90	0.72	0.71
Rb	153	49	8	2.2	2	0	7	56	40.2
Sr	77	171	93	80	106	184	254	107	210
Ba	1589	1091	210	20	44	21	223	1392	784
V	88	119	178	419	334	346	361		8
Cr	91	88	59	90	82	44	76	3	8
Ni	41	34	54	49	92	70	45	0	5
Y	32.90	13.41	25.25	31.31	19.12	32.19	58.35	24	35.4
Zr	162	85	140	104	86	177	245	243	325
Nb	15	8	13	11.6	9	20	19	12	11.5
Th	15	2	4	2.0	2	5	5	4	2.5
Pb	6	6	10	5.0	8	9	15		5.2
U	5	2	2	0.4	1	0	0	2	0.5
Nb/La	0.14	1.01	3.15	1.49	1.22	2.13	14.6	0.30	0.25

3.1 Petrography

Samples HT-2, HT-4, HT-7 and HT-21, were massive, fine-grained eclogites. They are mainly composed of garnet, omphacite, rutile, quartz and minor retrometamorphic minerals, such as symplectite and amphibole. Relict coesite or quartz pseudomorphs after coesite have been found in these samples as inclusion in garnet. In addition, garnet and Jd-rich omphacite inclusions have been observed in omphacite and garnet respectively. The retrometamorphism in sample HT-2 is much stronger than other samples. Garnet, omphacite and rutile from the sample HT-4 were separated for isotopic dating.

Sample HT-12, a strongly retrograded eclogite occurred as nodule in the marble (Fig. 1B). Except omphacite inclusions in garnet, all omphacites were retrograded to amphibole, but no biotite has been observed. It suggests that the retrometemorphism corresponds to the amphibolite facies. Titanite is another common retrograde mineral, and occurs as corona around ilmenite which contains relict of rutile. Recrystallized quartz and apatite coexisting with amphibole and titanite are also common in this sample. The retrograde assemblage of quartz + apatite + amphibole + titanite forms in a vein around the garnets. Garnet, amphibole, titanite and apatite were separated from this sample for isotopic dating.

Sample HT-6 is a garnet-epidote-biotite gneiss collected from a thin (1 m) gneiss layer interbedded with thick eclogite layers. It mainly consists of quartz, biotite, and epidote, together with minor garnet, phengite, rutile, titanite and thin amphibole as symplectite intergrowth with sodic plagioclase. Most of phengites are partly replaced by biotite, and some titanites occur as corona around rutiles. These observations indicate that the biotite and titanite are retrograde metamorphic products. Only biotite was separated from this sample for Rb-Sr dating.

Samples HT-1-1, HT-1-2 and 97SH-3 are garnet-epidote-two mica gneiss collected from the one outcrop in the western part of the UHPM unit (Fig. 1B). The gneiss is interbedded with marble containing eclogite nodules. It mainly consists of quartz, garnet, phengite, clinozoisite/epidote, biotite, plagioclase with minor rutile and titanite. The Sample HT-1-2 contains much less biotite than other two samples. Most phengites are developed along foliations, but a few phengite inclusions in garnet have been observed. Phengite is partly replaced by biotites and rutile by titanite, which suggest that the biotites and titanite were formed during retrograde metamorphism. Carswell et al. (1998) also indicated that clinozoisite grains in this UHP gneiss are zoned towards later more pistacite-rich epidote and sometimes cored by early allanites, which suggests retrograde origin of the epidote in these samples. Garnet, phengite, epidote and biotite were separated from these samples for isotopic dating.

Sample H-6 is a biotite gneiss from the east margin of the UHPM unit (Fig. 1B). This sample is strongly sheared and retrograded. It consist of quartz, plagioclase, biotite, and minor garnet. Biotite was separated from this sample for Rb-Sr dating.

Samples Sh-1 and Sh-2 are fine-grained granitic gneisses from the country rock of the UHPM unit. These gneiss samples consist of quartz, K-feldspar, plagioclase, biotite, garnet, epidote and clinozoisite. Contents of both phengite and garnet are lower than in the UHP gneiss while K-feldspar is conspicuously abundant. There is no obvious petrographic evidence that coesite was ever stable. Based on this mineral assemblage together with the high spessartine contents (18.6–34.9 mol%) and very low pyrope contents (1.0–3.8 mol%) of the garnet, Cong et al. (1995) suggest that the granitic gneiss has only experienced amphibolite facies metamorphism with temperature estimated to be 530 ± 20 ℃ at 4 kbar by using garnet-biotite geothermometers (Ferry and Spear, 1978). However, Carswell et al. (1998), recently, argue that the granitic gneiss also witnessed UHP conditions based on the discoveries of UHPM mineral relicts, such as rutile, phengites with Si contents as high as 3.49 and more calcic Mn-poor garnet, in the granitic gneiss at Shuanghe. Biotite and phengitic white mica (Si=3.2–3.3) appear to coexist in the late foliation but phengite grains with higher Si contents breakdown to secondary biotite and plagioclase (Carswell et al., 1998). The clinozoisite/epidote grains are often cored by allanite and show zonation from early clinozoisite towards late more Fe-rich epidote (Carswell et al., 1998). These observations suggests that the biotite and epidote in granitic gneiss are late stage products during retrograde metamorphism corresponding to 450 ℃ and 6 kbar (Carswell et al., 1998). It is also important to emphasize that the limited amount of garnet occurred in a skeletal or even atoll form around plagioclase have been observed by Carswell et al.(1998) and Xu et al.(1998), which suggests the strong influence of retrograde metamorphism upon the garnet. Garnet, biotite and epidote were separated for isotopic dating.

3.2 Major and trace element geochemistry

Chemical compositions of 4 eclogites (Table 1) with SiO_2 ranging 48.8%–53.4% and ($K_2O + Na_2O$) ranging 2.25%–4.64 % (HT-2, 4, 7, 21) suggest that their protoliths are sub-alkaline basalts. The mobility of Rb, Ba, and K during retrograde metamorphism can be observed in the primitive mantle normalized spider diagram (Fig. 2b) by comparing the strongly retrograded eclogite samples HT-2 and HT-12 with other unretrograded eclogite samples. High contents of Rb, Ba and K in sample HT-2 and HT-12 suggest that the retrograde metamorphism in eclogite occurred in a open chemical system. The retrometamorphic fluid may have introduced a lot of large iron lithophile elements into eclogite, which may potentially influence the Sr isotopic composition of retrograde metamorphic minerals and whole rock. The most of eclogite samples from Shuanghe display slight LREE enrich patterns at 10–30 x chondritic abundance without positive Eu anomaly suggesting that they are not of cumulate origin (Fig. 2a). Unlike the eclogites from the Bixiling, they do not show any negative Nb anomaly but positive anomaly in the primitive mantle-normalized spider diagrams (Fig. 2b), especially for retrograde eclogites (sample HT-2 and HT-12) with Nb/La ratios as high as 3.15–14.6 (see Table 1). The high Nb contents in some eclogites samples, e.g. 20 ppm for sample HT-21(see Table 1), could be caused by enrichment of rutile in these samples. However, the high Nb/La ratio could be also caused by loss of LREE through devolatilization during HP or UHP metamorphism and/or through retrograde metamorphic fluid. The loss of LREE during UHP and/or retrograde metamorphism can be observed more clearly from sample HT-12.

The retrograded eclogite (HT-12) in marble shows an unusual chemical composition with low MgO (3.62%) and SiO_2 (41.13%) and very high FeO (20.12%) and Zr (245 ppm) contents (Table 1). These features with relative higher Al_2O_3 and CaO contents suggest that the protolith of this sample is like to be iron-rich para-amphibolite, but not highly fractionated basalt because of its low SiO_2 content. It is interesting to note that the retrograded eclogite is strong LREE depleted (Fig. 2a) but has very low initial ε_{Nd} value (−10) at retrograde time (200 Ma) (see Fig. 5a). It suggests that the protolith of sample HT-12 should be LREE enriched and the LREE depletion is caused by a recent loss of LREE during UHP and/or retrograde metamorphism. In view of the more higher Nb/La ratios of retrograde eclogites (e.g. samples HT-2 and HT-12) than those of unretrograde eclogite (Table 1), the mobilization of LREE during retrograde metamorphism is more likely to be responsible to the loss of LREE. This may potentially influence the Nd isotopic composition of retrograde metamorphic minerals and whole rock.

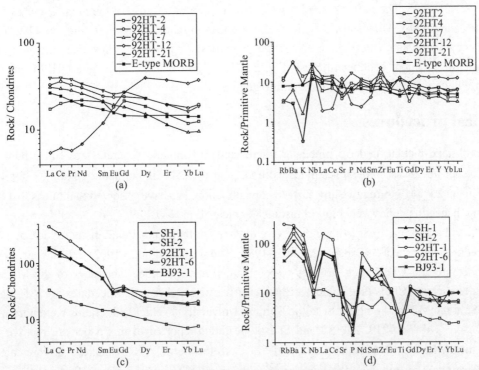

Fig. 2 REE distribution patterns and primitive mantle-normalized spider diagrams for the UHPM rocks and their country rock (granitic gneiss) from Shuanghe. (a) and (b), eclogites, the data of E-type MORB are based on Sun & Mc Donough (1989); (c) and (d), UHP gneiss and granitic gneiss, sample BT 93-1 is granitic gneiss from Bixiling for comparison (Chavagnac and Jahn, 1996).

The retrograded eclogite (HT-12) in marble shows an unusual chemical composition with low MgO (3.62%) and SiO_2 (41.13%) and very high FeO (20.12%) and Zr (245 ppm) contents (Table 1). These features with relative higher Al_2O_3 and CaO contents suggest that the protolith of this sample is like to be iron-rich para-amphibolite, but not highly fractionated basalt because of its low SiO_2 content. It is interesting to note that the retrograded eclogite is strong LREE depleted (Fig. 2a) but has very low initial ε_{Nd} value (−10) at retrograde time (200 Ma) (see Fig. 5a). It suggests that the protolith of sample HT-12 should be LREE enriched and the LREE depletion is caused by a recent loss of LREE during UHP and/or retrograde metamorphism. In view of the more higher Nb/La ratios of retrograde eclogites (e.g. samples HT-2 and HT-12) than those of unretrograde eclogite (Table 1), the mobilization of LREE during retrograde metamorphism is more likely to be responsible to the loss of LREE. This may potentially influence the Nd isotopic composition of retrograde metamorphic minerals and whole rock.

Two UHP gneiss samples, e.g. HT-1-1 and HT-6 display different geochemical characters. Sample HT-1-1 has high K_2O content with high K_2O/Na_2O (2.22) ratio and strong LREE enriched pattern with high REE abundances as well as abundant detrital zircons suggesting a sedimentary origin. However, the relative lower K_2O content with low K_2O/Na_2O (0.68) of sample HT-6 and its REE pattern similar to those of eclogites (Fig. 2d) suggest that the protolith of sample HT-6 is more like to be intermediate volcanic rocks interbedded within basaltic eclogite.

According to the petrological and geochemical features of the UHPM rocks discussed above, protoliths of the UHPM unit at Shuanghe are volcanic - sedimentary rocks deposited in a shallow marine basin. Large variation in $\delta^{18}O$ values (−2.6 to +7.0‰) of garnet and omphacite associated with a small range in δD value (−83 to −73 ‰) of phengite from eclogite at Shuanghe suggest a characteristic ^{18}O shift to lower $\delta^{18}O$ values as a result of isotopic exchange with ^{18}O-depleted meteoric water prior to the UHPM (Zheng et al., 1998). Thus, these hydrothermal altered volcanic-sedimentary rocks are expected to contain more water than deep mafic-ultramafic intrusions, such as the protoliths of the Bixiling eclogite (Chavagnac and Jahn, 1996). This difference in wet versus dry system could have an important effect on the Nd isotopic equilibrium between minerals during UHPM. It will be discussed in a latter section.

The granitic gneiss sample (Sh-1,2) are of trondhjemitic composition with high Na_2O contents (5.66%–5.92%) and Na_2O/K_2O = 1.91–3.25. Their trace element characters are similar to those of the granitic gneiss from Bixiling (Chavagnac and Jahn, 1996). They show a typical post-Archean granite REE pattern with enriched LREE, nearly flat HREE and negative Eu anomaly (Fig. 2e). Negative anomalies of Nb, P and Ti in spider diagrams are commonly observed for granitic or upper continental crust (Fig. 2e). It suggest that the Shuanghe UHPM unit and Bixiling eclogite body are both encased into the similar granitic gneiss.

4 Analytical procedures

After careful hand-picking under a binocular microscope, all mineral separates are 100% pure without any altered grain. Except for micas, mineral separates were leached by 2.5N HCl at about 70–80 ℃ for 1 hr and then for 15 minutes in cold 5% HF to remove any surface contamination. We have observed that micas lost their Rb after the acid leaching, hence the micas were only ultrasonically cleaned in H_2O.

All Sm, Nd and a part of Rb, Sr isotope data were obtained at the Max - Plank institute für Chemie, Germany, following procedures described elsewhere (Jagoutz, 1988; Thöni and Jagoutz, 1992). Rb and Sr isotope data of biotite, some phengite and whole rock samples were obtained at the isotope laboratory of the Modern Analysis Center, Nanjing University, China (see Table 3). Blanks for the whole chemical procedures are < 50 pg for Sm and Nd, 0.2 ng for Sr at the MPI and 1 ng for Sr at the Nanjing University. $^{143}Nd/^{144}Nd$ ratios were normalized against the value of $^{146}Nd/^{144}Nd$ = 0.7219. All Nd and Sr isotope ratios reported in this paper are relative to values of 0.511878 ± 14 for the La Jolla Nd standard and 0.707987 ± 28 for the E&A Sr standard. Ages are calculated using the "ISOPLOT" software of Ludwig (1994) and given with 2σ error. The uncertainty, including all sources of error, of Rb/Sr ratios is ± 2% and that of Sm/Nd ratios ± 0.5%. The Sm-Nd data and Rb-Sr data are listed in Table 2 and 3 respectively.

Table 2 Sm, Nd, concentrations and Nd isotope ratios for ultrahigh-pressure metamorphic rocks and their country rock from Shuanghe in the Dabie Mountains

Sample No.	Rock or Mineral	Sm (ppm)	Nd (ppm)	$^{147}Sm/^{144}Nd$	$^{143}Sm/^{144}Nd$ ±2σ
HT-1-1 W	UHPM gneiss	11.59	71.72	0.1041	0.511236 ± 15
HT-1-1 gar	Garnet	3.231	1.034	1.887	0.514107 ± 13
HT-1-1 phen 1	phengite	0.01484	0.1285	0.06982	0.511424 ± 60
HT-1-1 phen 2	phengite	0.007349	0.0285	0.1559	0.511539 ± 21
HT-1-1 bi	biotite	0.01828	0.04015	0.2752	0.511561 ± 3
HT-1-1 epid	epidote	17.07	32.4	0.3184	0.511525 ± 18
HT-4 W	Coesite-bearing eclogite	3.612	12.31	0.1773	0.511940 ± 17
HT-4 gar	garnet	2.529	1.263	1.21	0.513526 ± 18
HT-4 omph	omphacite	1.208	3.021	0.2418	0.512096 ± 13
HT- 4 rut	rutile	0.8092	4.569	0.107	0.511891 ± 13
HT-12 W	Retrograded ecolgite	1.448	2.289	0.3824	0.512372 ± 13
HT-12 gar	garnet	1.611	0.6713	1.451	0.514064 ± 15
HT-12 amph	amphibole	0.6028	1.056	0.345	0.512308 ± 14
HT-12 sphen	sphene	26.59	62.04	0.2591	0.512197 ± 20
HT-12 apt	apatite	202.5	761	0.1523	0.512055 ± 27
Sh-1 W	granitic gneiss	6.635	33.99	0.118	0.512040 ± 10
Sh-1 gar	garnet	3.105	2.891	0.6491	0.512778 ± 15
Sh-1 bi	biotite	0.07847	0.2574	0.1843	0.512137 ± 13
Sh-1 epid	epidotite	122.4	651.8	0.1135	0.512027 ± 9

* The Sm-Nd data were obtained at MPI.

Table 3 Rb, Sr concentrations and Sr isotope ratios for ultrahigh-pressure metamorphic rocks and their country rock from Shuanghe in Dabie Mountains

Sample No.	Rock or mineral	Rb (ppm)	Sr (ppm)	$^{87}Rb/^{86}Sr$	$^{87}Sr/^{86}Sr \pm 2\sigma$
HT-1-1 W*	UHPM gneiss	209.2	92.3	6.574	0.746840 ± 16
HT-1-1 phen*	phengite	547.5	52.51	30.23	0.790577 ± 16
HT-1-1 bi	biotite	502.1	2.811	518	2.011592 ± 26
HT-1-1 gar*	garnet	3.66	1.073	9.892	0.727364 ± 19
HT-1-2 W	UHP gneiss	207.2	91.87	6.541	0.741803 ± 26
HT-1-2 phen	phengite	471.5	61.26	22.32	0.787061 ± 24
97SH-3 W	UHP gneiss	193.7	102.4	5.548	0.735380 ± 17
97SH-3 phen	phengite	448.3	53.58	24.26	0.780487 ± 15
97SH-3 epid	epidote	14.88	1096	0.4008	0.723950 ± 17
HT-6 W*	UHP gneiss	55.31	209.2	0.7667	0.712260 ± 20
HT-6 bi	biotite	257.6	4.893	152.7	1.103687 ± 20
H-6 W	UHP gneiss	108.6	496.3	0.6346	0.715684 ± 12
H-6 bi	biotite	453.8	5.69	231.3	1.270487 ± 27
HT-12 W*	Retrograde	1.44	221.2	0.01888	0.704760 ± 23
HT-12 sphen*	eclogite	0.1653	86.77	0.005525	0.704794 ± 64
HT-12 amph*	sphene	9.548	74.49	0.3717	0.705632 ± 18
HT-12 gar*	amphibole	0.1337	7.577	0.05117	0.704811 ± 13
Sh-1 W*	garnet	52.18	90.95	1.664	0.721718 ± 11
Sh-1 bi	granitic gneiss	498.9	6.445	224.5	1.262966 ± 26

* These Samples were analyzed at MPI, and others were analyzed at the Nanjing University of China.

5 Chronological results

5.1 Sm-Nd age of eclogite HT-4

Fig. 3 shows that the data points of three UHPM minerals (garnet + omphacite + rutile) of sample HT-4 define a perfect isochron which gives an age of 226.3 ± 3.2 Ma with initial $^{143}Nd/^{144}Nd = 0.511735 \pm 11$, $\varepsilon_{Nd} = -11.9$. The whole rock lies below the mineral isochron in Fig.3, which suggests that those not unanalyzed retrograde metamorphic minerals, such as amphibole should have lower $^{143}Nd/^{144}Nd$ ratios than those of UHPM minerals. Thus, the line connecting garnet and whole rock will give an old age with no geological significance. The good linear relationship between the three HPM minerals in Fig. 3 suggests isotopic equilibrium between the UHPM minerals in eclogite from the Shuanghe area. Fine grain mineral size of the eclogite and more water contents in the precursors of the UHPM rocks at Shuanghe is likely to have facilitated isotopic equilibrium during UHP metamorphism.

5.2 Sm-Nd age of UHPM gneiss HT-1-1

Fig. 4 shows that the isochron defined by garnet and two phengites of the sample HT-1-1 gives an age of 226.5 ± 2.3 Ma with initial $^{143}Nd/^{144}Nd = 0.511310 \pm 27$, $\varepsilon_{Nd} = -20.2$. This age is consistent with the Sm-Nd age of 226.3 ± 3.2 Ma obtained from eclogite (HT-4). It can be suggested that isotopic equilibrium between the garnet and phengite has been achieved in the UHPM gneiss at Shuanghe and they remain close system during retrograde metamorphism. Retrograded biotite, epidote and whole rock plot below this isochron (Fig. 4). These results suggest that the Nd isotopic system of the whole rock sample was open during retrograde metamorphism and Nd isotopic disequilibrium between UHPM minerals and retrograde metamorphic minerals. Similar to the case in Sample HT-4, the data of Sample HT-1-1 also indicates relative lower $^{143}Nd/^{144}Nd$ ratios of the retrograde metamorphic minerals than those of UHPM minerals. Since the whole rock contain many retrograde minerals, the age of 246.0 ± 2.1 Ma given by the line connecting garnet and whole rock has no geological significance. Fig. 4 shows that the retrograde minerals (biotite and epidote) and whole rock can not fit a straight line suggesting Nd isotopic disequilibrium between the biotite and epidote in this sample. Since the much higher Nd content in epidote than biotite (see Table 2), however, the line connecting epidote and whole rock gives a reasonable retrometamorphic age of 206 ± 16 Ma which is consistent with the Sm-Nd age of 200 ± 23 Ma defined by retrograde minerals from sample HT-12 (see below).

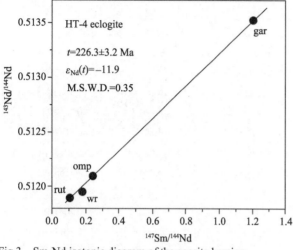

Fig.3 Sm-Nd isotopic diagram of the coesite bearing eclogite (HT-4) from Shuanghe.

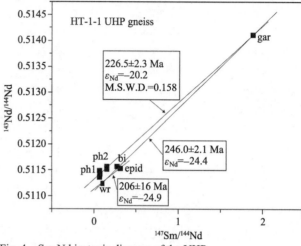

Fig. 4 Sm-Nd isotopic diagram of the UHP gneiss (HT-1-1) from Shuanghe.

5.3 Sm-Nd and Rb-Sr ages of retrograde eclogite HT-12

Sample HT-12 is an intensely retrograded eclogite. All omphacites have been replaced by amphiboles and rutiles by titanite or ilmenites in this sample. Even though samples HT-12 and HT-4 occurred in the same UHP unit, Fig. 5A shows that the line fitted by garnet + whole rock yields an age of 242.0 ± 3.6 Ma, significantly older than the age of sample HT-4. Obviously, this line is not an isochron, because the amphibole is a retrograde phase and whole rock is basically mixture of garnet and amphibole. It implies again that the retrograde fluid has a relative lower $^{143}Nd/^{144}Nd$ ratio than that of UHPM rocks at Shuanghe. Fig. 5 also shows that the line fitted by three retrograde minerals, amphibole, titanite and apatite gives a younger age of 200 ± 23 Ma with ε_{Nd} = −10.2. The good linearity (M.S.W.D. = 0.02) of this isochron suggests the Nd isotopic equilibrium between the three retrograde minerals. The age of 200 ± 23 Ma for the retrograde minerals of sample Ht-12 as well as the age of 206±16 Ma for the epidote of sample HT-1-1 roughly indicates that the recrystallization of retrometamorphic minerals with amphibolite facies in the UHP eclogite could have occurred around 200 Ma. The corresponding P-T condition of this stage has been estimated to be 450–500 ℃ at 8 kbar (Carswell et al., 1998).

Fig. 5B shows a little scatter of amphibole, whole rocks and titanite, which may be caused by the large uncertainty of $^{87}Sr/^{86}Sr$ of the titanite. Thus the Rb-Sr isochron defined by the two retrograde minerals and whole rock gives an unprecise age of 172 ± 84 Ma. While the Rb-Sr isochron defined by amphibole + whole rock gives a similar but more precise age of 174 ± 7.8 Ma (Fig. 5). Garnet falls below this isochron, which suggests that Rb-Sr system of the garnet was not reset during the retrograde metamorphism. Since the slope of the Rb-Sr isochron is mainly controlled by amphibole, this age may indicate the closure time of Rb-Sr system in amphibole. Unfortunately, we do not exactly know the closure temperature (T_c) of Rb-Sr system in amphibole. However, based on our study of North Qinling belt we have known that the Rb-Sr ages of 414 ± 1–410 ± 2 Ma defined by amphibole + plagioclase + whole rock for amphibolite are always similar to or only slightly lower than Ar-Ar ages of 426 ± 2 Ma for the amphibole (Sun et al., 1996), but significantly older than the Ar-Ar ages of 351 ± 4 Ma for the biotite (T_c =300 ℃) from the same area (Sun et al., 1996; Li et al., unpublished data). These results suggest that the T_c of Rb-Sr system in amphibole is similar or slightly lower than the T_c (500 ℃) of Ar in amphibole(Harrison, 1981). Hence, we chose 450 ± 50 ℃ as the T_c of Rb-Sr system in amphibole for the following T-t path plotting.

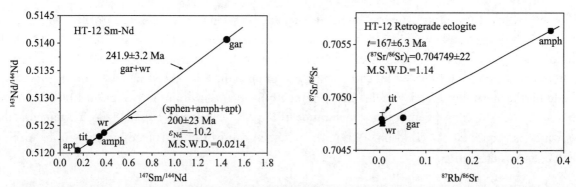

Fig. 5 Sm-Nd and Rb-Sr isotopic diagram of the retrograde eclogite (HT-12) from Shuanghe.

5.4 Rb-Sr ages of phengite and biotite from UHP gneisses

Fig. 6 shows that phengite from the samples HT-1-1, HT-1-2 and 97 SH-3 gives a large variation in Rb-Sr ages. The lines connecting phengite and garnet for sample HT-1-1 gives age of 219.0 ± 6.6 Ma, while the line connecting phengite and whole rock for these three samples give ages of 130.1 ± 4.1 Ma, 202.0 ± 5.7 Ma and 169.6 ± 4.4 Ma respectively. The age of 130.1 ± 4.1 Ma for phengite from sample HT-1-1 is even much younger than the Rb-Sr age of biotite from the same sample (see Fig. 6), though the closure temperature of phengite is 200 °C higher than that of biotite. We would suggest that the open system during retrograde metamorphism of UHPM rocks and the Sr isotopic disequilibrium between UHPM minerals and retrograde metamorphic minerals are responsible for the larger uncertainty and the reversal of phengite and biotite ages. We will discuss this later in detail.

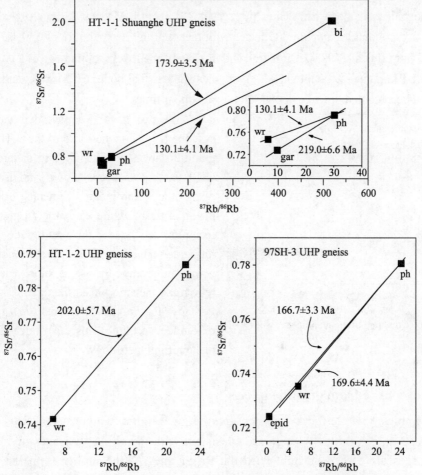

Fig. 6 Rb-Sr isotopic diagrams of the UHP gneisses (HT-1-1, HT-1-2 and 97SH-3) from Shuanghe.

Biotites + whole rocks from samples HT-1-1, HT-6 and H-6 yield ages of 173.9 ± 3.5 Ma, 181.2 ± 3.5 Ma and 169.2 ± 3.3 Ma respectively (Table 4). These ages are consistent with the Rb-Sr age of 179 ± 4 Ma defined by biotite + whole rock for the Bixiling eclogite within error limits (Chavagnac and Jahn, 1996). The age variation of about 10 Ma for biotite is much less than that of phengite from the same UHP unit. It suggests that retrograde biotite + whole rock may yields meaningful Rb-Sr age. Since biotite is high in Rb/Sr ratio, these isochron ages could be considered to indicate the time of the rock cooling through the closure temperature of biotite, which is about 300 ℃ (Dodson, 1973). The potential causes for the spread in ages will be discussed later in the discussion section.

Table 4 Rb-Sr ages defined by biotite + whole rock from the Shuanghe area.

Rock	Sample No.	Age (Ma)	Data source
UHP gneiss	HT-1-1	173.9 ± 3.5	This work
	H-6	169.2 ± 3.3	This work
	HT-6	181.2 ± 3.5	This work
granitic gneiss	Sh-1	170.8 ± 3.4	This work
	Sh-2	173.2 ± 3.4	This work
	—	181	Jahn et al. 1994

5.5 Sm-Nd age of granitic gneiss

Fig. 7 shows that garnet, biotite, epidote and whole rock from the sample Sh-1 define an isochron which gives an age of 213.3 ± 4.8 Ma with $\varepsilon_{Nd}(t) = -9.5$. The good linearity of the three minerals in Fig. 7 suggests the Nd isotopic equilibrium within the garnet, biotite and epidote. Since biotite and epidote are retrograde minerals as indicated by Carswell et al. (1998), the Nd isotopic equilibrium between garnet and retrograde minerals must be reached during retrograde metamorphism. Thus, the age of 213.3 ± 4.8 Ma may be interpreted as the retrograde metamorphic age corresponding to amphibolite facies with metamorphic temperature of 530 ± 20 ℃ (Cong et al. 1995) or 450 ℃ (Carswell et al., 1998). If it is true and in view of the Rb-Sr cooling age of 219.0 ± 6.6 Ma for the phengite from the UHP paragneiss (sample HT-1-1), our data suggest that both UHPM rocks and their country rock (granitic gneiss) had been cooled down to about 500 ℃ at similar time.

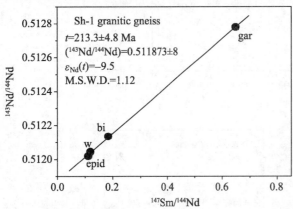

Fig. 7 Sm-Nd isotopic diagram of the granitic gneisses (Sh-1) from Shuanghe.

Now, we have observed that the Sm-Nd isotopic systematics of garnets from the UHP paragneiss and granitic gneiss at Shuanghe are different. The garnet in UHP paragneiss keeps closed and is in disequilibrium with the retrograde minerals during retrograde metamorphism, while the garnet in granitic gneiss can be reequilibrated with the retrograde minerals during retrograde metamorphism. It is still unknown that why the garnet in granitic gneiss is more easily influenced by retrograde metamorphism than garnet in UHP paragneiss. Perhaps the high-Mn garnet in granitic gneiss is retrograde origin. This is indeed a challenge to metamorphic petrology.

5.6 Rb-Sr age of biotite from granitic gneiss

Two biotite samples + whole rocks from the granitic gneiss give two consistent Rb-Sr ages of 170.8 ± 3.4 Ma and 173.2 ± 3.4 Ma (Table 4) which are consistent with their $^{40}Ar/^{39}Ar$ ages of 170–177 Ma (Li et al., 1994). These ages are also consistent with the Rb-Sr ages of the biotite from the UHPM rocks. It suggest that both UHPM rocks and their country rocks (granitic orthogneiss) were cooled down to 300 ℃ at the same time.

6 Discussion

6.1 Peak metamorphic age of the UHPM rocks in Dabie Mountains

A correct estimate of the time of peak metamorphism of the UHPM rocks is very important to the calculation of the initial cooling rate. We have reported two consistent Sm-Nd ages of 226.3 ± 3.2 Ma and 226.5 ± 2.3 Ma for eclogite and UHP paragneiss from Shuanghe respectively. Nd isotopic equilibrium between UHPM minerals during UHP metamorphism in these samples has been demonstrated by the perfect isochron defined by three UHP minerals. Another reported Sm-Nd age of 225 ± 2 Ma for eclogite (sample 92HSH-4) from the same area is consistent with our present results (Ge et al.,1995). What is the significance of these Sm-Nd ages? Do they indicate the time of UHP peak metamorphism at Shuanghe or do they only represent a cooling age? The significance of these Sm-Nd ages depends on peak metamorphic temperature of the eclogite, closure temperatures of the major HPM minerals, i.e. garnet and omphacite and diffusion of Nd isotopes during recrystallization of HPM minerals.

Closure temperatures (T_c) of the Sm-Nd system in garnet are not well defined and controversial. Based on the Sm diffusion coefficient in garnet Humphries and Cliff (1981) calculated the T_c of pyrope close to 500 °C and grossular up to 700 °C for typical metamorphic grain sized (1 mm) and orogenic cooling rate (10 °C/Ma). Mezger et al. (1992) concluded that the Tc of Sm-Nd system in garnet is 600 ± 30 °C for slow cooling rate (2–4 °C/Ma) by comparing the garnet + whole rock Sm-Nd age with other geochronologic data and temperatures estimated on the same geological unit. Other lines of evidence, however, suggest that Sm-Nd diffusion in garnet may be significantly slower than that suggested by these results. Based on the study of the Sm-Nd system in an eclogite xenolith from kimberlite, Jagoutz (1988) proposed a high T_c of 850 °C for garnet under dry condition. Hensen and Zhou (1995) also suggests that the closure temperature for the Sm-Nd system in garnet must be > 700–750 °C in an dry mafic granulite. Actually, the closure temperature for any given element in garnet is dependent on a number of factors, including grain size, availability of pore fluid, major element composition, the nature of coexisting phases, initial temperature and the cooling rate experienced by each individual sample (Dodson, 1973; Burton et al., 1995; Ganguly et al., 1998). Therefore, garnet is unlikely to possess a unique closure temperature for Nd, but may has a range from 500 °C to 850 °C depending on variable conditions. We have to consider all factors mentioned above to estimate the closure temperature for a specific sample.

A Tc range of 1040–1090 °C for clinopyroxene with radius of 1mm and cooling rates of 10–40 °C /Ma can be calculated using the Sm diffusion coefficient of diopside under high-pressure condition (Sneeringer et al., 1984). It is significantly higher than all the possible T_c (500–850 °C) of garnet. Burton et al. (1995) also observed that the diffusion rates of Sm and Nd in clinopyroxene are lower than those in garnet. Since eclogite is basically a bi-mineral rock, garnets cannot exchange Sm-Nd after the pyroxene has closed to diffusion. Since the average peak metamorphic temperatures (800 °C) of the eclogites at Shuanghe is lower than the T_c of clinopyroxene, the equilibrium of Nd isotopes between garnet and pyroxene is difficult to be achieved by diffusion. There are three possible ways to interpret the isotopic equilibration between UHP minerals:

(1) Nd isotopic equilibration may be achieved by decomposition of preexisting minerals, such as plageoclase and amphibole, under UPH condition. Decomposition of preexisting minerals and formation of garnet and omphacite could be completed during progressive eclogitic metamorphism but not necessarily at the peak UHP metamorphic stage. If the following recrystallization during whole HP and UHP metamorphic stage could not affect the Nd isotopic equilibration between garnet and omphacite because of high Tc of clinopyroxene, the Sm-Nd mineral isochron age of 226.2 ± 3.2 Ma for eclogite (HT-4) may indicate the progressive metamorphic time. However, this interpretation is incosistant with the result of the UHP gneiss sample HT-1-1, in which the absence of clinopyroxene do not affect any change of the Sm-Nd ages defined by garnet and 2 phengites. The consistence between the age of 226.3 ± 3.2 Ma for eclogite and age of 226.5 ± 2.3 Ma for UHP gneiss suggests that the high T_c of clinopyroxene is not an important factor for the isotopic equilibrium between UHPM minerals.

(2) Villa (1998) has indicated that the isotopic diffusion during fluid-assisted recrystallization is orders of magnitude faster than thermal-only volume diffusion in the absence of any recrystallization. Hence recrystallization

of garnet and omphacite could be an important factor for the isotopic equilibrium between them. As a extreme case if the Nd isotopic compositions were completely reset during recrystallization (no isotopic zoning in a mineral grain), the Sm-Nd age of 226 Ma for the eclogite and UHP gneiss at Shuanghe should indicate the time of the final stage in recrystallization process which corresponded to the quartz-eclogite stage (P = 13–16 kbar, T = 630–760 ℃) (Cong et al., 1995). However, this extreme case is unlikely because preservation of Ar isotopic zoning in phengites from Alpine and Dabie-Sulu eclogites (Scaillet, 1996; M. Cosca, personal communication 1999) suggests that isotopes is difficult to be completely equilibrated in a rapid cooling UHPM mineral grain by volume diffusion. In addition, this age interpretation is in contradiction with the U-Pb zircon ~225 Ma ages of UHPM rocks (see below).

(3) Since boundary diffusion is much fast than volume diffusion, isotopic equilibration could be easily achieved along the boundary layers of the growing minerals during recrystallization. However, when these older boundary layers were covered by new overgrowth zone, it is hard to be reequilibrated by volume diffusion. In other words, the real isotopic equilibrium between UHPM minerals is only surface equilibration which is in between the above two extreme cases. In this case, the Sm-Nd mineral isochron age of 226 ± 3 Ma may indicate the average age of recrystallization time. If the growth of garnet and omphacite were continued during the whole HP and UHP metamorphic stage, the average of recrystallization time may close to the peak metamorphic time.

The third interpretation is favoured by the authors as it is supported by the ages obtained by other independent studies using different isotopic systems. For example, the Sm-Nd isochron of the coesite - bearing eclogite (QL -1) from Qinglongshan, located in the Su-Lu terrane, defined by three UHPM minerals (garnet + omphacite + phengite) + whole rock gives an age of 226.3 ± 4.5 Ma which is almost the same to the Sm-Nd ages of the sample HT-4 and HT-1-1 from Shuanghe (Li et al., 1994). Rowley et al. (1997) reported one U-Pb zircon lower intercept age of 225.5 + 2.9 / − 6.3 Ma for eclogite (Sample MW03) from Maowu in the Dabie Mountains. This U-Pb age is consistent with our Sm-Nd ages. The SHRIMP ages of 225 ± 4 Ma on zircon from a UHPM gneiss in the Dabie Mountains is also consistent with our Sm-Nd ages (Hacker et al., 1998) . Therefore, the Sm-Nd age of 226 ± 3 Ma for eclogite from Shuanghe is more likely to be true peak metamorphic time of eclogite in the Dabie Mountains.

Chavagnac and Jahn (1996) reported 7 slight younger Sm-Nd ages ranging from 210 ± 7 Ma to 218 ± 4 Ma for the Bixiling eclogite block. Since all of these ages are given by two UHPM minerals (i.e. garnet and omphacite) isochrons, the possibility of Nd isotopic disequilibrium between UHPM minerals, which could lower the slop of tie - line between garnet and omphacite(Jagoutz, 1994), can not be rule out.

Okay and Sengör (1993) reported an older garnet+whole rock Sm-Nd age of 246 ± 8 Ma. According to their description, however, this sample is an intense retrograde eclogite in marbles as an enclave in the Wuhe area. All omphacites have been replaced by amphiboles. As mentioned above, this old age is caused by abundant retrograde minerals in the whole rock sample.

6.2 Sr-Nd isotopic systematics of UHPM rocks during retrograde metamorphism

Data presented in this paper have shown that the Nd isotopic systems of the most UHPM minerals, e. g. garnet, omphacite, rutile and phengite remain closed , while the Nd isotopic system of the whole rock containing retrograde minerals was open during retrograde metamorphism. In general, the retrograde metamorphic fluid has relative lower $^{143}Nd/^{144}Nd$ ratios than that of eclogite. Hence, the tie line of garnet and whole rock containing retrograde minerals may give a relative older Sm-Nd age than the peak metamorphic age (see Figs. 3, 4).

Similarly, Rb-Sr data of sample HT-1-1 (Fig. 6) suggest that the Sr isotopic compositions of phengite and garnet were not reset during retrograde metamorphism, whereas $^{87}Sr/^{86}Sr$ of whole rock containing epidote and biotite was modified to higher value by fluids introducing higher Rb/Sr and/or more radiogenic Sr during retrograde metamorphism. Since Sr closure temperature of clinopyroxene and garnet are similar to those of Sm-Nd in the same minerals (Sneeringer et al., 1984; Burton et al., 1995) it may be reasonable to assume that the Sr isotopic systems of garnet, omphacite and phengite remain closed. While the later generation mineral phases (e.g. biotite and epidote) grew during the retrograde metamorphism in an open chemical system. Due to the relative higher $^{87}Sr/^{86}Sr$ ration of retrograde metamorphic fluid, the later generation phases would have a higher initial $^{87}Sr/^{86}Sr$ value (Fig. 8). Thus the whole rock isotopic composition is also modified to have higher $^{87}Sr/^{86}Sr$ ratio by retrograde metamorphism (see

Fig. 8). As time went on, these two Rb-Sr isotopic systems would evolve independently (shown as the solid and dash line in Fig. 8, respectively). In this case regressions of data for these two isotopic systems separately will yield two different isochron ages that may closely represent the cooling ages after metamorphic events. However, any linear regression involving two generation phases or the line connecting phengite and retrograde whole rock (shown as the dot line in Fig. 8) should be considered as a "mixing line" or an "errochron" rather as an "isochron". The age inferred by the tie line of phengite + whole rock will be geologically meaningless, even it could be much younger than the biotite cooling age (see Fig. 8).

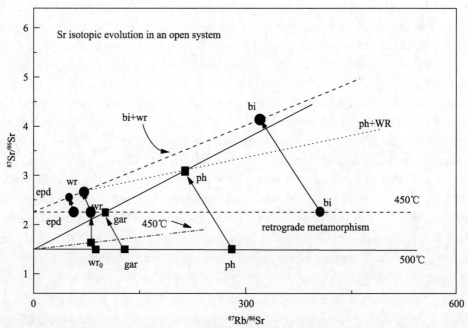

Fig. 8 Sr isotope evolution diagram of UHP minerals (phengite + garnet) and retrograde minerals (epidote + biotite) as well as whole rock in an open system (see text for explanation). wr_0 shows initial isotopic composition of whole rock before retrograde metamorphism; wr shows the whole rock isotopic composition after retrograde metamorphism.

Since the Sr closure temperature of garnet is ~750 ℃ (Burton et al., 1995), which is much higher than the Sr closure temperature (500 ℃) of phengite (Cliff, 1985), the line connecting phengite and garnet from a slow cooling rock may introduce a large error in the closure temperature age of phengite. However, since the UHPM rock cools rapidly above 500 ℃ (Li et al., 1994; Chavagnac and Jahn, 1996; Gebauer et al., 1997) and the whole rock system of some unretrograde eclogites remains closed, the phengite and garnet (or amphacite) with different closure temperatures and unretrograde whole rock are likely to fall on a straight line in a Rb-Sr isochron diagram. The slope of that line can be taken to represent the cooling age of phengite. For example, the isochrons defined by phengite + garnet (or + omphacite) + whole rock of the samples QL-1 and ZB-4 from the Sulu terrane (Li et al., 1994) and the samples BJ93-05 and BJ93-07 from Bixiling in the Dabie Mountains (Chavagnac and Jahn, 1996) yield the Rb-Sr ages of 220 ± 2 Ma, 224 ± 2 Ma, 214 ± 6 Ma and 223 ± 13 Ma respectively, which are consistent with each other within error limits. All these ages suggest that the most UHPM rocks in the Dabie-Sulu terranes may have cooled down to 500 ℃ at 219 ± 5 Ma.

In light of the above discussion, the Rb-Sr age of 219.0 ± 6.6 Ma defined by phengite + garnet for sample HT-1-1 form Shuanghe can be taken to represent the closure age of phengite at 500 ℃. This age is consistent with the above mentioned phengite Rb-Sr ages for unretrograde eclogite. Other younger Rb-Sr ages of 130.1 ± 4.1 Ma, 169.6 ± 4.4 Ma, and 202.0 ± 5.7 Ma defined by phengite + whole rock for sample HT-1-1, 97SH-3 and HT-1-2 (see Fig. 6) have no geological significance. The samples HT-1-1 and 97SH-3 contain more abundant biotite than sample HT-1-2, which could be the reason why the phengite-whole rock Rb-Sr ages of the samples HT-1-1 and 97SH-3 are much younger than that of sample HT-1-2.

It should be noted that the Rb-Sr ages defined by phengite + whole rock for the granitic gneiss in the

Dabie-Sulu terrane range 194–198 Ma (Jahn et al., 1994; Chavagnac and Jahn, 1998) which are significantly younger than the phengite Rb-Sr ages for UHPM rocks. This would not be surprised to us if we considered that the most of phengitic white mica in granitic gneiss have much lower Si contents (Si = 3.2–3.3), but are not real UHPM minerals as suggested by Carswell et al. (1998). This kind of white mica coexisting with biotite has been suggested to be formed during retrograde metamorphism at temperature 450 ℃ (Carswell et al., 1998). Therefore, the Rb-Sr ages of 194–198 Ma for white micas from the granitic gneiss can not be taken to represent the closure age at 500 ℃ but correspond to the retrometamorphic temperature of 450 ℃.

Biotite and amphibole are retrograde metamorphic minerals in UHPM rocks. Hence only on condition that the Sr isotopic system of whole rock was reset during retrograde metamorphism, the Rb-Sr age given by the line connecting biotite or amphibole and whole rock can be regarded as the best estimation of the closure age of biotite or amphibole. Intensely retrograded metamorphic rocks, such as samples HT-1-1, H-6 and HT-12, may approximately satisfy this condition. Garnet and/or phengite are the major residual UHPM minerals in intensely retrograded metamorphic rocks. Though the Sr isotopic compositions of garnet and phengite may not be reset during retrograde metamorphism, since its small proportion in this kind of rocks and low Sr content, the Sr isotopic composition of the whole rock may not be influenced significantly by these relict UHPM minerals. However, if there are more abundant residual UHPM minerals in retrograde UHPM rocks, such as sample HT-6, the line connecting biotite or amphibole and whole rock will yield an older Rb-Sr age than its closure age, because the $^{87}Sr/^{86}Sr$ of UHPM minerals, in general, may be much lower than those of retrograde metamorphic minerals. In addition to the large analytical errors for Rb/Sr this may be the major reason for the spread in biotite + whole rock Rb-Sr ages. In light of the above discussion, we would suggest that Rb-Sr ages of 169.2 ± 3.3 Ma and 173.9 ± 3.5 Ma defined by biotite + whole rock for Sample H-6 and HT-1-1 may be the best estimation of the closure age of the biotite from the Shuanghe UHPM paragneiss at 300 ℃. These ages are also consistent with the Rb-Sr ages of 170.8 ± 3.4 Ma and 173.2 ± 3.4 Ma defined by biotite + whole rock from the Shuanghe granitic gneiss. It is interesting to note that these Rb-Sr cooling ages of biotite are consistent with the Rb-Sr age (174 ± 7.8 Ma) of the amphibole from eclogite, which may suggest a rapid cooling of UHPM rocks in the Dabie Mountains during 180–170 Ma.

6.3 Cooling history of UHPM rocks

Based on the above chronology data with corresponding closure or metamorphic temperatures, a T-t path for UHPM rocks at Shuanghe has been obtained (Shown as solid line in Fig. 9). Three points in this T-t path, i.e. the peek metamorphic age of 226 ± 3 Ma, Rb-Sr cooling age of 219 ± 7 Ma for phengite and Rb-Sr cooling age of 172 ± 6 Ma for biotite, are firmly supported by other published chronology data as mentioned above. The Rb-Sr cooling age of 174 ± 7.8 Ma for amphibole is the first time such age for amphibole is reported for the Dabie UHPM rocks. Obviously, this important point on the T-t path should be confirmed by more data in the future. However, following arguments suggest that this cooling age is plausible : (1) As mentioned above, the two Sm-Nd ages of 200 ± 23 Ma and 206 ± 16 Ma for amphibole facies retrograde minerals from samples HT-12 and HT-1-1 indicate their recrystallization time. Thus their Rb-Sr cooling age should be younger than these ages. (2) A U-Pb age of around 180 Ma for rutile from the Yangkou eclogite in the Su-Lu terrane also suggests that the UHPM rocks has been cooled down to about 420 ℃ at that time range (Lu, 1998 and Li Huimin personal communication, 1999).

The T-t path in Fig. 9 shows that the cooling history of the UHPM rocks from 800 ℃ to 300 ℃ can be subdivided into three stages : two rapid cooling stages and one isothermal stage in between them. They may reflect two rapid uplift processes during the exhumation history. Even if the Rb-Sr cooling age of 174 ± 7.8 Ma for amphibole can not be substantiated by future study the T-t curve on Fig. 9 can still be divided into three stages with a bend around 200 Ma defined by two Sm-Nd ages of retrograde minerals.

The initial rapid cooling with cooling rate of about 40 ℃/Ma during 226 ± 3–219 ± 7 Ma may result from a rapid uplifting or decompression of UHPM rocks following the peak metamorphism. The T-t path shows that the temperature of UHPM rocks at Shuanghe was cooled down to 500 ℃ at 219 ± 7 Ma by the initial rapid cooling, which corresponds to the symplectite stage with pressure condition of about 6–8 kbar (Cong et al., 1995). It implies that the UHPM rocks were uplifted to the middle continental crust level by the first uplifting process. The very high initial cooling rate recorded in the Shuanghe UHPM unit is consistent with those observed on the other UHPM units, such as the Qinglongshan and Zhubian eclogites in the Sulu terrane (Li et al., 1994), Bixiling eclogite in the Dabie Mountains (Chavagnac and Jahn, 1996) and Dora Maira massif in Western Alps (Gebauer et al., 1997). Hence, it could be a common cooling pattern for UHPM rocks in collisional orogenic belt.

There was a isothermal stage between 219 ± 7 Ma to 174 ± 7.8 Ma for the UHPM rocks at Shuanghe. The P-T conditions corresponding to amphibolite facies in this stage resulted in recrystallization of the fine grained symplectite minerals. The Sm-Nd ages of 200 ± 23 Ma and 206 ± 16 Ma for retrograde minerals, e. g. amphibole, titanite and apatite from the samples HT-12 and HT-1-1 may indicate their recrystallization time (Fig. 9).

The second rapid cooling from 450 ℃ to 300 ℃ during 174 ± 7.8 to 172 ± 6 Ma may be caused by the second uplifting process of UHPM rocks. This is corresponding to the post-symplectite stage suggested by Cong et al. (1995). The UHPM rocks at Shuanghe could be uplifted to the upper crust level after the second rapid cooling.

6.4 Cooling history of the country rock

Hacker et al. (1998) reported a precise SHRIMP age of 231 ± 2 Ma for the zircons from a granitic gneiss in the South Dabie UHPM zone. This sample has identical mineralogy to granitic gneiss at Shuanghe. Thus Hacker et al. (1998) suggest that the peak

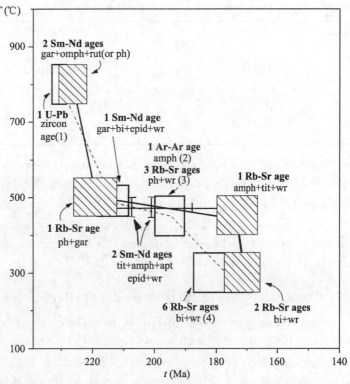

Fig. 9 T-t path of UHPM rocks and their country rocks at Shuanghe. The Shadowy squares and crosses represent UHPM rocks. The open squares represent granitic gneiss. The solid line shows the T-t path of UHPM rocks and the dash line shows the T-t path of granitic gneiss. (See text for detail explanation). Data sources: (1) Hacker et al., (1998); (2) Chen et al., (1995); (3) Jahn et al., (1994), Chavagnac and Jahn, (1998); (4) this work and Jahn et al., (1994), Chavagnac and Jahn, (1998). Other ages are presented by this paper.

metamorphic age of the granitic gneiss is indistinguishable from that of the UHPM rocks in the Dabie Mountains. In addition, Carswell et al. (1998) indicate that the peak metamorphic temperature of the granitic orthogneiss can be calculated using garnet-phengite geothermometer for the highest Si phengites, which bracket the better constrained temperatures at Pmax recorded in both the UHP paragneiss and eclogites. Therefore, the Sm-Nd cooling age of 213 ± 5 Ma corresponding to amphibolite facies (T = 530–450 ℃) for the granitic gneiss at Shuanghe suggests that the country rocks also experienced a fast cooling from 800 ℃ to 490 ± 40 ℃ during 231–213 Ma (see Fig. 9).

Because there is little white mica and no hornblende in the sample SH-1, the second rapid cooling for the granitic gneiss at Shuanghe has not been revealed in this study. However, the rapid cooling from 450 ℃ to 300 ℃ during 195–165 Ma for the country rocks, granitic gneiss, in the south Dabie zone and Su-Lu terrane has been demonstrated by several $^{39}Ar/^{40}Ar$ thermochronological studies (see Table 5). In addition, two Rb-Sr ages of 195 Ma and 188 Ma for muscovite and biotite from granitic gneiss at Hushan which is near Qinglongshan in the Su-Lu terrane have been reported (Jahn et al., 1994). As mentioned above, Carswell et al. (1998) have indicated that this muscovite is formed during retrograde metamorphism at the temperature around 450 ℃. Thus the two Rb-Sr ages also suggests a rapid cooling (~ 21 ℃/Ma) of the granitic gneiss in Su-Lu terrane during 195–188Ma. Similar

Rb-Sr ages of 194–198 Ma for muscovite and 170–184 Ma for biotite from the country rocks, tonalitic gneiss, of the Bixiling eclogites in Dabie Mountains have been recently reported by Chavagnac and Jahn (1998). Their new Rb-Sr age data suggest that the country rocks of the Bixiling eclogite have experienced a rapid cooling (8 ℃/Ma) during 196–177 Ma. We believe that the granitic gneiss at Shuanghe should have similar cooling history to other granitic gneiss in the southern Dabie zone, i. e. it undergone a rapid cooling during 190–175 Ma below 450 ℃.

Table 5 Rapid cooling events of the granitic gneiss in Dabie-Sulu belt

Author	Dating method	Rapid cooling time and rate
Chen Wenji et al. (1992)	Ar-Ar (MDD model)	172–196 Ma (10–40 ℃/Ma) 105–123 Ma (4-10 ℃/Ma)
Chen Jiangfang (1995)	Ar-Ar (Hrb. Bi)	193–180 Ma
Li Qi et al. (1995)	Ar-Ar (MDD model)	165–163 Ma(Dabie) 187–180 Ma(Sulu)

Summarized the all chronological data mentioned above, a T-t path for the regional granitic gneisses in the Dabie-Sulu terrane has been obtained (Shown as dash line in Fig. 9). It shows that the country rocks of the UHPM units in the Dabie-Sulu terrane have also experienced the two rapid cooling events during 230–213 Ma and 195–175 Ma. The similarity of the cooling histories between the UHPM units and granitic gneiss in the Dabie Mountains suggests that both the UHPM rocks and their country rocks may have exhumed together from the mantle depth to the upper crustal level.

6.5 Constraints on exhumation model of UHPM rocks

Many exhumation models of UHPM rocks in orogenic belt have been proposed (Anderson et al., 1991; Dobretsov, 1991; Hsü, 1991; Okey and Sengör, 1992; Maruyama et al., 1994; Davies and Blanekenburg, 1995; Ernst and Liou, 1995; Hacker et al., 1995; Chemenda et al., 1995,1996; Xue et al., 1996). Most of the authors suggest one initial rapid uplifting process to account for the preservation of coesite in UHPM rocks during their exhumation process, which was followed by a slow uplifting over a period of a long time. Two different exhumation mechanisms for the initial rapid uplifting, i.e. thrusting of UHPM slabs (e.g. Chemenda et al., 1995, 1996) or breakoff of subducting oceanic lithosphere (e.g. Davies and Blanekenburg, 1995; Ernst and Liou, 1995) have been emphasized by different authors. However, the two stages of rapid cooling of the UHPM rocks established above most likely correspond to two stages of fast uplift. They call for reconsideration of the exhumation model. The mechanisms of these two uplift events could be different, because they occurred separately at difference time which may correspond two different tectonic regimes in the Dabie-Sulu belt.

The Dabie-Sulu belt was in a compression regime in the early to middle Triassic, which is recorded by the UHP metamorphism in this belt. The extension tectonics in the Dabie-Sulu belt could begin at the middle Jurassic (~180–170 Ma), which is suggested by the middle Jurassic basal molasse-type sediments in the Beihuaiyang terrane which is the internal basin in the Dabie Mountains during the Jurassic and Cretaceous (Xu et al., 1994). Consequently, we suggest that the earlier rapid cooling may reflect the exhumation of the UHPM rocks including granitic gneiss from mantle depth to the middle crustal level by compression tectonics during the continental subduction (such as thrusting model suggested by Chemenda et al., 1995), while the later rapid cooling may reflect the exhumation of the entire subducted continental crust from middle crustal level to the upper crustal level via extension tectonics due to the breakoff of subducted slab as suggested by Davies and Blanckenburg (1995) and Ernst and Liou (1995). On the other hand, if both tectonic thrust and breakoff of the subducted oceanic lithosphere had occurred during the first stage of exhumation then the second stage of exhumation has to be related to another process of extension tectonics such as post-collisional delamination of the lithospheric mantle. Further research is required to evaluate such alternatives.

After the second rapid uplifting, the UHPM rocks and their country rocks may have shared the same slow cooling history (~ 4 ℃/Ma) from 300–100 ℃ during 160–110 Ma, which has been revealed by fission track ages of apatite from granitic gneiss in South Dabie terrane (Chen et al., 1995). The magmatic-structural dome formed during Cretaceous (Hacker et al., 1998) and voluminous Cretaceous alkaline volcanics and granites developed along the north side of the Dabie Mountains (Li et al., 1993) suggest the extension in the late Jurassic and Cretaceous, which may have been responsible for the slow cooling and uplift of UHPM rocks and their country rocks during this period.

7 Conclusion

(1) Neodymium isotopic equilibrium between UHPM minerals has been achieved in fine-grained eclogite at Shuanghe in Dabie Mountains during UHP metamorphism. The best estimate of the peak metamorphic time of UHPM rocks in the Dabie Mountains and Sulu terrane is 226 ± 3 Ma. Other younger Sm-Nd ages for UHPM rocks from the Dabie-Sulu terrane could be their cooling ages or resulted from Nd isotopic disequilibrium between UHPM minerals and/or analytical error.

(2) Retrograde metamorphism of the UHPM rocks in the Dabie Mountains and Sulu terrane occurred in an open chemical system. Retrograde metamorphic fluids with abundant large ion lithophile elements have influenced the isotopic composition of the retrograde metamorphic minerals. The UHPM minerals, such as garnet, omphacite, rutile and phengite have their Sr and Nd isotopic systems remained closed during the retrograde metamorphism, while Sr and Nd isotopic systematics of the retrograde metamorphic minerals, such as epidote, amphibole, biotite, growing during the retrograde metamorphism are not in equilibrium with the UHPM minerals. For this reason, tie line of garnet and UHPM whole rock containing retrograde metamorphic minerals will give a Sm-Nd "age" generally older than the peak metamorphic age with no geological significance. Similarly, tie line of phengite and whole rock from such retrograde metamorphic UHPM rocks will give a younger Rb-Sr age than the true closure age of phengite. Hence, we will not recommend to use the Rb-Sr isochron defined by phengite + whole rock in thermochronological studies.

Meaningful Rb-Sr ages of phengite from UHPM rocks can be obtained by the regression of phengite with other UHPM minerals or unretrograded metamorphic whole rock. The meaningful Rb-Sr ages of biotite or amphibole can be obtained by the regression of biotite and/or amphibole with other retrograde metamorphic minerals or strongly retrograde whole rock.

(3) Based on the Sm-Nd and Rb-Sr isochron ages with their corresponding metamorphic and closure temperatures reported in this paper, a T-t path with two rapid coolings for the UHPM rocks at Shunaghe can be obtained as shown in Fig. 9. This T-t path shows that after peak metamorphism the UHPM rocks in the Dabie-Sulu belt has experienced a rapid cooling to about 500 °C at ~219 Ma. The UHPM rocks then went through a nearly isothermal stage from ~219 Ma to ~180 Ma. Starting from ~180 Ma, the rocks cooled rapidly again to ~300 °C Considering the change of the tectonic regime in the Dabie-Sulu belt during the above cooling times : i.e. a compression regime in the early to middle Triassic and an extension regime after 180 Ma, it is plausible that the earlier rapid cooling may reflect the exhumation of the UHPM rocks from mantle depth to the middle crustal level by compression tectonics during the continental subduction, while the later rapid cooling may reflect the exhumation of the entire subducted continental crust from middle crust to the upper crustal level via extension tectonics during the early-middle Jurassic.

(4) The Nd isotopic composition of the garnet in regional granitic gneiss can be reset during retrograde metamorphism and became equilibrated with the retrograde metamorphic minerals (e.g. epidote and biotite), which is different from that of the garnet in eclogite and UHPM paragneiss. The retrograde metamorphism with amphibolite facies of the granitic gneiss at Shuanghe in the Dabie Mountains occurred at 213 ± 5 Ma, and then it was cooled down to ~300 °C at 172 ± 4 Ma. These two ages are consistent, respectively, with the Rb-Sr cooling ages of the phengite and biotite from the UHPM rocks in the Dabie-Su-lu belt. It suggests that the cooling history of the granitic gneiss may be similar to that of the UHPM rocks in the Dabie-Su-lu belt.

Acknowledgments

We would like to thank Bolin Cong, Qingchen Wang, Mingguo Zhai, Zhongyan Zhao, Huogen Dong and

Xingyun Lai for their friendly help and discussions during field trips. We are also grateful to Michael Ginther and Yohannes D. Vogel for their assistance in laboratory works at MPI and Ge Long, Weili Han for typewriting the manuscript, Shen-su Sun for spending many hours writing correspondences, offering critical comments, helping in English presentation and reviewing the draft, and an anonymous reviewer for critical comments. This study is financially supported by the National Natural Science Foundation of China (No. 49132040, 49573190, 49794042), Max-Planck-Institut für Chemie, Germany and Chinese Academy of Science (No. KZ951-A1-401).

References

Ames, L., Tilton, G.R. and Zhou, G. (1993) Timing of collision of the Sino-Korean and Yangtze cratons : U-Pb zircon dating of coesite-bearing eclogites. Geology 21, 339-342.

Ames, L., Zhou, G. and Xiong, B. (1996) Geochronology an isotopic character of ultrahigh-pressure metamorphism with implications for collision of the Sino-Korean and Yangtze cratons, cental China. Tectonics 15 (2), 472-489.

Andersen, T.B., Jamtveir, B., Dewey, J.F. and Swensson, E. (1991) Subduction and eduction of continental crust: major mechanisms during continent-continent collision an orogenic extensional collapse, a model based on the south Norwegian Caledonides. Terra Nova 3, 303-310.

Arnaud, N.O. and Kelly, S. (1995) Evidence for excess Ar during high pressure metamorphism in the Dora-Maira (western Alps, Italy), Using a Ultra-violet laser Ablation Microprobe $^{40}Ar/^{39}Ar$ technique. Contrib. Mineral. Petrol. 121, 1-11.

Burton, K.W., Kohn, M.J., Cohen, A.S. and O'Nions, R.K. (1995) The relative diffusion of Pb, Nd, Sr and O in garnet. Earth Planet. Sci. Lett. 133, 199-211.

Carswell, D.A., Wilson, R.N. and Zhai, M. (1998) The enigma of eclogite-gneiss relationship in the Ultra-high pressure terrane of Dabieshan, central China, Inter, Workshop on UHP metamorphism and Exhumation (Abstracts), Stanford Univ. A-40-44.

Chavagnac, V. and Jahn, B. M. (1996) Coesite-bearing eclogites from the Bixiling complex, Dabie Mountains, China: Sm-Nd ages, geochemical characteristics and tectonic implications. Chemical Geology 133, 29-51.

Chavagnac, V. and Jahn, B. M. (1998) Geochronological evidence for the in situ tectonic relationship in the Dabie UHP metamorphic terrane, Central China. Mineral. Magazine 62A, 312-313.

Chemenda, A.I., Mattauer, M., Malavieille, J., Bokun, A.N. (1995) A mechanism for syncollisional rock exhumation and associated normal faulting : Results from physical modeling. Earth Planet. Sci. Lett. 132, 225-232.

Chemenda, A.I., Mattauer, M., Bokun, A.N. (1996) Continental subduction and a mechanism for exhumation of high-pressure metamorphic rocks: new modeling and field data from Oman. Earth Planet. Sci. Lett. 143, 173-182.

Chen, J., Xie, Z., Liu, S., Li, X and Foland, K.A. (1995) Cooling age of Dabie orogen, China, determined by $^{40}Ar/^{39}Ar$ and fission track techniques. Sci. China (Ser. B) 38, 749-757.

Chen, W., Harrison, T., Heizerler, M.T., Liu, R., Ma, B. and Li, J. (1992) The cooling histories of the melange zone of N. Jiangsu and S. Shandong Provinces: evidence from multiple diffusion domain $^{40}Ar/^{39}Ar$ thermal geochronology. Acta Petr. Sinica 8, 1-17 (in Chinese with English abstract).

Chopin, C. (1984) Coesite and pure pyrope in high-grade blueschists of the Westen Alps: a first record and some consequences. Contrib. Mineral. Petrol. 86, 107-118.

Cliff, R.A. (1985) Isotopic dating in metamorphic belts. J. ged. Soc. London 142, 97-110.

Cong, B., Zhai, M., Carswell, D.A., Wilson, R.N., Wang, Q., Zhao, Z. and Windley, B.F. (1995) Petrogeneisis of ultrahigh-pressure rocks and their country rocks at Shuanghe in Dabieshan, Central China. Eur. J. Mineral 7, 119-138.

Davies, J.H., Blanckenburg, F.V. (1995) Slab breakoff: a model of lithosphere detachment and its test in the magmatism and deformation of collisional orogents. Earth Planet. Sci. Lett. 129, 85-102.

Dobretsov, N.L. (1991) Blueschists and eclogites : a possible plate tectonic mechanism for their emplacement from the upper mantle. Tectonophysics 186, 253-268.

Dodson, M.H. (1973) Closure temperature in a cooling geochronological and petrological system. Contrib. Mineral. Petrol. 40, 259-274.

Eide, L., McWilliams, M.O. and Liou, J.G. (1994) $^{40}Ar/^{39}Ar$ geochronologic constraints on the exhumation of HP-UHP metamorphic rocks in east-central China. Geol. 222, 601-604.

Ernst, W.G. and Liou, J.G. (1995) Contrasting plate-tectonic styles of the Qinling-Dabie-Sulu and Franciscan metamorphic belts. Geology 23, 353-356.

Ferry, J.M. and Spear, F.S. (1978) Experimental calibration of the partitioning of Fe and Mg between biotite and garnet. Contrib. Mineral. Petrol. 66, 113-117.

Ganguly, J., Tirrone, M. and Hervig, R.L. (1998) Diffusion kinetics of Samarium and Neodymium in garnet, and a method for determing cooling rates of rocks. Science 281, 805-807.

Ge, N., Li, S. and Wang, Y. (1995) Sm-Nd and Rb-Sr isotopic ages of coesite bearing eclogite in Shuanghe village, eastern Dabie Mountains, China. Chinese Sci. Bull. 40 (Supplement), 169.

Gebauer, D., Schertl, H. P., Brix, M. and Schreyer, W. (1997) 35Ma old ultrahigh-pressure metamorphism and evidence for very rapid exhumation in the Dora Maira Massif. Western Alps. Lithos 41, 5-24.

Hacker, B.R. and Wang, Q. (1995) Ar/Ar geochronology of ultrahigh-pressure metamorphism in central China. Tectonics 14, 994-1006.

Hacker, B.R., Ratschbacher, L., Webb, L. and Dong, S. (1995) What brought them up? Exhumation of the Dabie Shan ultrahigh-pressure rocks. Geology 23, 743-746.

Hacker, B.R., Ratschbacher, L., Webb, L., Ireland, T., Walker, D. and Dong, S. (1998) U/Pb zircon ages constrain the architecture of the ultrahigh-pressure Qinling-Dabie Orogen, China. Earth Planet. Sci.Lett. 161, 215-230.

Harrison, T.M. (1981) Diffusion of ^{40}Ar in hornblende. Contrib. Mineral. Petrol. 78 : 324-331.

Hensen, B.J. and Zhou, B. (1995) Retention of isotopic memory in garnets partially broken down during an overprinting granulite-facies metamorphism : implications for the Sm-Nd closure temperature. Geology 23, 225-228.

Hsü, K.J. (1991) Exhumation of high-pressure metamorphic rocks. Geology 19, 107-110.

Humphries, F.J. and Cliff, R.A. (1982) Sm-Nd dating and cooling history of Scourian granulates, Sutherland. Nature 295, 515-527.

Jagoutz, E. (1988) Nd and Sr systematics in an eclogite xenolith from Tanzania: evidence for frozen mineral equilibrium in the continental lithosphere. Geochim. Cosmochim. Acta 52, 1285-1293.

Jagoutz, E. (1994) Isotopic systematics of metamorphic rocks. Abstact of ICOG-8 at Berkeley, U.S.A. U. S. Geol. Survey Circular 1107, 156.

Jahn, B. M., Cornichet, J., Henin, O., Coz-Bouhnik, M.L. (1994) Geochemical and isotopic investigation of ultrahigh pressure (UHP) metamorphic terranes in China: Su-Lu and Dabie complexes, Abstracts of First workshop on UHP metamorphism and tectonics at Stanford, U. S. A. A, 71-74.

Liou, J.G., Zhang, R.Y. and Ernst, W.G. (1995) Occurrences of hydrous and carbonate phases in ultrahigh-pressure rocks from east-central China: implications for the role of volatiles deep in cold subduction zones. The Island Arc, 4, 362-375.

Li, Q., Chen, W., Ma, B., Wang, Q., Sun, M. (1995) Thermal evolution history after collision of North China plate with Yangtze plate. Seismology and Geol. 17, 193-203.

Li, S., Liu, D., Chen, Y. and Ge, N. (1992) The Sm-Nd isotopic age of coesite-bearing eclogite from the Southern Dabie Mountains. Chin. Sci. Bull. 37, 1638-1641.

Li, S., Xiao, Y., Liu, D., Chen, Y., Ge, N., Zhang, Z., Sun, S. S., Cong, B., Zhang, R., Hart, S.R. and Wang, S. (1993) Collision of the North China and Yangtze Blocks and formation of coesite - bearing eclogites : Timing and processes. Chem. Geol. 109, 89-111.

Li, S., Jagoutz, E., Lo, C. H., Li, H. and Chen, Y. (1994) Geochronology of ultrahigh pressure metamorphic rocks and their country rocks at Shuanghe in Dabie Mountains, central China: some new progresses. Abstact of First Workshop on UHP metamorphism and tectonics at Stanford A, 79-81.

Li, S., Wang, S., Chen, Y., Zhou, H., Zhang, Z., Liu, D. and Qiu, J. (1994) Excess argon in phengite from eclogite: evidence from dating of eclogite minerals by Sm-Nd, Rb-Sr and $^{40}Ar/^{39}Ar$ methods. Chem. Geol. 112, 343-350.

Li, S., Jagoutz, E., Zhang, Z., Chen, W. and Lo, C. H. (1995) Structure of high-pressure metamorphic belt in the Dabie Mountains and its tectonic implications. Chinese Sci. Bull. 40 (Supplement), 138-140.

Li, S. and Sun, W. (1996) A middle Silurian-early Devonian magmatic arc in the Qinling Mountains of central China: a discussion. The J. of Geol. 104, 501-503.

Li, S., Jagoutz, E., Xiao, Y., Ge, N., Chen, Y. (1996) Chronology of ultrahigh-pressure metamorphism in the Dabie mountains and Su-Lu terrane : I. Sm-Nd isotope system. Sci. China (Ser. D) 39, 597-609.

Li, S., Li, H., Chen, Y., Xiao, Y., Liu, D. (1997) Chronology of ultrahigh-pressure metamorphism in the Dabie mountains and Su-Lu terrane : II. U-Pb isotope system of zircons. Sci. China (Ser. D) 40, 200-206. (in chinese).

Li, S., Jagoutz, E., Zhang, Z., Lo, C. H., Chen, W. and Chen, Y. (1998) Geochemical and Geochronological constraints on the tectonic outline of the Dabie Mountains, central China: a continent-microcontinent-continent collision model. Continental Dynamics, 3 (No. 1-2), 14-31.

Lu, S. N, (1998) Geochronology and Sm-Nd isotopic geochemistry of Precambrian crystalline basement in eastern Shandong Province. Earth Sci. Frontiers 5, 275-283(in Chinese).

Maruyama, S., Liou, J.G. and Zhang, R. (1994) Tectonic evolution of the ultrahigh-pressure (UHP) and high-pressure (HP) metamorphic belts from Central China. The island Arc 3, 112-121.

Mezger, K., Essene, E.J. and Halliday, A.N. (1992) Closure temperatures of the Sm-Nd system in metamorphic garnet. Earth Planet Sci. Lett. 113, 397-409.

Okay, A.I., Xu, S. and Sengör A.M.C. (1989) Coesite from the Dabie Shan eclogites, central China. Eur. J. Mineral. 1, 595-598.

Okay, A.I. and Sengör, A.M.C. (1992) Evidence for intercontinental thrust-related exhumation of the ultra-high-pressure rocks in China.

Geology 20, 411-414.

Okay, A.L. and Sengör, A.M.C. (1993) Tectonics of an ultrahigh-pressure metamorphic terrane: the Dabie Shan / Tongbai Shan orogen, China. Tectonics 12, 1320-1334.

Rowley, D.B., Xue, F., Tucker, R.D., Peng, Z.X., Baker, J. and Davis, A. (1997) Ages of ultrahigh pressure metamorphism and protolith orthogneisses from the eastern Dabie Shan: U/Pb Zircon geochronology. Earth Planet. Sci. Lett. 151, 191-203.

Ruffet, G., Feraud, G., Balevre, M. and Kienast, J. R. (1995) Plateau ages and excess argon in phengites: an $^{40}Ar/^{39}Ar$ laser probe study of Alpine micas (Sesia Zone, Western Alps, northern Italy). Chem. Geol. (Isot. GeoSci. Section) 121, 327-343.

Scailete, S. (1996) Excess ^{40}Ar transport scale and mechanism in high-pressure phengites: a case study from an eclogitezed metabasite of the Dora-Maira mappe, western Alps. Geoch. Cosmoch. Acta 60, 1075-1090.

Smith, D.C. (1984) Coesite in clinopyroxene in the Caledonides and its implications for geodynamics. Nature 310, 641-644.

Sneeringer, M., Hart, S.R. and Shimizu, N. (1984) Strontium and Samarium Diffusion in diopside. Geochim. Coesmochim. Acta, 48 : 1589-1608.

Sobolev, N.V. and Shatsky, V.S. (1990) Diamond inclusions in garnets from metamorphic rocks: a new environment for diamond formation. Nature, 343, 742-746.

Sun, S.S. and McDonough, W.F. (1989) Chemical and isotopic systematics of oceanic basalts : implications for mantle composition and processes, in: Megma in the Ocean basins, A. D. Saunders & M. J. Norry eds. Geol. Soci. Special Pub. 42, 313-345.

Sun, W., Li, S., Chen, W., Sun, Y. and Zhang, G. (1996) Excess argon in metamorphic amphibole resulted from tectonic shearing: evidence from Sm-Nd, Rb-Sr and $^{40}Ar/^{39}Ar$ dating on amphibole of Danfeng group, North Qinling area. Geol. J. China Univ. 2 (4), 382-389 (in Chinese with English abstract).

Thöni, M. and Jagoutz, E. (1992) Some new aspects of dating eclogites in orogenic belts: Sm-Nd, Rb-Sr and Pb-Pb isotopic results from the Austroalpine Savalpine and Koralpe type-locality. Geochim. Cosmochim. Acta 56, 347-368.

Villa, L. M. (1998) Isotopic closure. Terra Nova 10, 42-47.

Wang, X., Liou, J.G. and Mao, H.J. (1989) Coesite-bearing eclogites from the Dabie Mountains in central China. Geology 17, 1085-1088.

Xue, F., Rowley, D.B. and Baker, J. (1996) Refolded syn-ultrahigh-pressure thrust sheets in the South Dabie complex, China: field evidence and tectonic implications. Geology 24, 455-458.

Xu, S., Okay, A. I., Ji, S., Sengör, A.M.C., Su, W., Liu, Y., Jiang, L. (1992) Diamond from the Dabie Shan metamorphic rocks and its implication for tectonic setting. Science 256, 80-82.

Xu, S., Wu, W., Su, W., Jiang, L. and Liu, Y. (1998) Meta-granitoid from the high-ultrahigh pressure metamorphic belt in the Dabie-Mountains and its tectonics significance. Acta. Petrol. Sinica, 14, 42-59 (in Chinese).

Zhang, G., Meng, Q. and Lai, S. (1995) Tectonics and structure of Qinling orogenic belt. Science in China, Ser. B 38, 1379-1394.

Zhang, R.Y., Liou, J.G. and Tsai, C.H. (1996) Petrogenesis of a high-temperature metamorphic terrane: a new tectonic interpretation for the north Dabieshan, central China. J. Metamorphic Geol. 14, 319-333.

Zheng, Y.F., Fu, B., Li, Y., Xiao, Y. and Li, S. (1998) Oxygen and hydrogen isotope geochemistry of ultrahigh-pressure eclogites from the Dabie Mountains and the Sulu terrane. Earth Planet. Sci. Lett. 155, 113-129.

大别山双河超高压变质岩及北部片麻岩的 U-Pb 同位素组成
——对超高压岩石折返机制的制约*

李曙光[1]，黄　方[1]，周红英[2]，李惠民[2]

1. 中国科学技术大学地球与空间科学系、化学地球动力学研究实验室，合肥 230026
2. 国土资源部天津地质矿产研究所，天津 300170

> **亮点介绍**：发现南大别超高压变质岩具有较低的 Pb 含量以及较高的 U/Pb 和 $^{206}Pb/^{204}Pb$ 值，而北大别超高压变质岩具有较高的 Pb 含量以及较低的 U/Pb 和 $^{206}Pb/^{204}Pb$ 值，表明南大别带出露的超高压岩石主要是深俯冲的上地壳岩石，而在北大别带具有中、下地壳的性质。因此，在大陆深俯冲过程中，俯冲陆壳拆离成若干岩片，深俯冲的超高压上地壳岩片有可能沿断层逆冲到较浅部位的超高压岩石折返机制模型。

摘要　大别山超高压变质岩及各种片麻岩的 U-Pb 同位素地球化学研究表明，出露于大别山南部的超高压变质岩具有较低的 Pb 含量（多数<4 μg/g=和较高的 U/Pb 值（多数>0.1），以及较大的 Pb 同位素变化范围和较高的放射成因 Pb（$^{206}Pb/^{204}Pb$ = 16.535~20.405）。它们表现出在俯冲过程中经历了脱水、析出流体和 Pb 丢失过程及地幔 Pb 与上地壳岩石 Pb 混合的 Pb 同位素特征。然而出露于大别山北部的片麻岩具有较高的 Pb 含量（多数>4 μg/g）和较低的 U/Pb（<0.077）以及较低的 Pb 同位素比值（$^{206}Pb/^{204}Pb$ = 15.781~16.647），并与侵入南、北大别山的中生代花岗岩的 Pb 同位素特征相同。它们表现在俯冲过程中仅有少量流体析出和 Pb 丢失以及地幔 Pb 与下地壳 Pb 混合的同位素特征。这些样品在 230 Ma 前的初始 Pb 同位素组成还表明南大别带超高压岩石的 U/Pb 值在大陆俯冲前很长一段时期内比北大别带片麻岩高。这些观测表明南大别带出露的超高压岩石主要是深俯冲的上地壳岩石，而在北大别带具有中、下地壳的性质。据此，本文提出了在大陆深俯冲过程中，俯冲陆壳拆离成若干岩片，深俯冲的超高压上地壳岩片有可能沿断层逆冲到较浅部位的超高压岩石折返机制模型。

关键词　超高压变质岩；大别山；U-Pb 同位素地球化学

碰撞造山带陆壳岩石中柯石英和金刚石的发现表明在陆-陆碰撞过程中一侧陆壳可以俯冲到 100 km 以上的深度[1~6]。在如此深度下形成的超高压变质岩是如何快速出露地表，以至于柯石英仍能保存下来已成为对大陆地球动力学的重要挑战和重要的科学问题。查清造山带的现今结构是探讨这一科学问题的重要基础，其中超高压变质带陆壳的地球化学结构研究，会对俯冲陆壳的构造岩片的划分及地壳性质提供重要制约，然而这方面研究仍十分薄弱。

大别-苏鲁碰撞造山带是世界上出露面积最大的超高压变质带[4~6]。前人仅对大别山变质岩及花岗岩的 Nd 同位素组成进行了研究，并初步观测到超高压正片麻岩的 Nd 模式年龄（t_{DM} = 1.5~1.8 Ga）比北大别正片麻岩的 Nd 同位素模式年龄（t_{DM} = 1.5~1.8 Ga）低[7~9]。然而侵入南、北大别带的燕山期花岗岩（t 约为 130 Ma）具有相同的古老的 Nd 同位素模式年龄值（t_{DM} = 2.2~2.6 Ga）[8,9]。Nd 同位素数据表明，南大别超高压带地壳具有双层结构：上部含超高压变质岩的较年轻地壳及深部较古老地壳，而北大别带已出露地表及深部地壳的 Nd 同位素组成是无差异的，均为古老地壳。

Nd 同位素模式年龄可以指示地壳的平均年龄，但不能指示地壳的性质。然而，上、下地壳的 Pb 同位

---

* 本文发表在：中国科学(D 辑)，2001，31(12)：977-984
对应的英文版论文为：Li Shuguang, Huang Fang, Zhou Hongying, et al., 2003. U-Pb isotopic compositions of the ultrahighn pressure metamorphic (UHPM) rocks from Shuanghe and gneisses from Northern Dabie zone in the Dabie Mountains, central China: constraint on the exhumation mechanism of UHPM rocks. Science in China (Ser. D), 46(3): 200-209.

素组成有显著差异。因此 Pb 同位素可以用来示踪地壳的性质。本文拟对大别山超高压变质岩及各种片麻岩进行 U-Pb 同位素研究，以查清高压和超高压变质带的 Pb 同位素结构特征及地壳性质，为探讨大别山超高压岩石折返机制提供制约。

1 地质背景

大别山造山带是华北与华南陆块的碰撞造山带，大别山东段（商-麻断裂以东）自北向南可以划分为 4 个基本构造单元：（1）北淮阳复理石带；（2）北大别高温高压变质带；（3）南大别超高压变质带；（4）宿松低温高压变质带。它们之间分别由晓天-磨子潭断裂（XMF）、五河-水吼断裂（WSF）和太湖-马庙断裂（TMF）所分割（图1）[9]。多数工作者已普遍接受北淮阳复理石带是华北陆块的南部活动陆缘；而南大别超高压变质带及宿松高压变质带属于华南陆块的俯冲陆壳。关于北大别带的构造背景过去曾存在分歧，然而最近北大别三叠纪榴辉岩的发现表明，它也是俯冲陆壳的一部分[11~13]。因此，大别山的地缝合线应该位于晓天-磨子潭断裂一线[9]。该缝合线以南的北大别、南大别及宿松带可能代表了俯冲陆壳的不同部位。

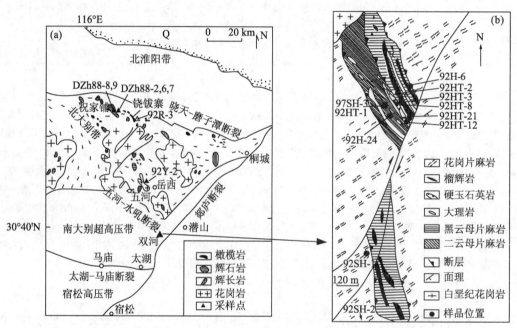

图 1 大别山东部构造地质简图及采样位置(a)和双河超高压岩石出露区地质简图及采样位置(b)

除了大量的燕山期花岗岩及变形闪长岩侵入体之外，北大别带主要由面理非常发育的条带状灰色片麻岩组成，其原岩形成时代可能在 700~800 Ma[14]。北大别带的一些超镁铁岩、榴辉岩均侵位于此类片麻岩中。本文测定的北大别带片麻岩样品均属于此类。条带状片麻岩总体上为角闪岩相变质岩，然而片麻岩中角闪石普遍交代辉石的现象说明其原始变质相要更高[15]。北大别带是大别山唯一有麻粒岩出露的地方[16]，该带发现的榴辉岩属于高温型，其峰期变质温度高达 800~850 ℃，但不含柯石英。榴辉岩的退变质作用以首先为麻粒岩相退变质作用，而后转为角闪岩相为特征[11]。这些特征表明北大别带可能是俯冲陆壳中的下地壳部分，具有较高温度，但俯冲深度不大，因而未形成柯石英。

南大别超高压变质岩带主要表现为由各种含柯石英超高压岩石，如榴辉岩、石榴橄榄岩、硬玉石英岩、含榴辉岩包体的大理岩及与上述岩石互层的超高压变泥质岩（副片麻岩）或变酸性火山岩（正片麻岩）组成的超高压岩片被包裹在大面积区域花岗岩片麻岩中。超高压岩片与围岩（花岗片麻岩）的关系曾有"外来"（foreign）和"原地"（in situ）之争。近年来，在花岗片麻岩中发现了一些高压矿物残留体，锆石中发现有柯石英或绿辉石包裹体[17,18]，说明花岗片麻岩围岩也经历了超高压变质作用。因此，南大别超高压变质带地表出露岩石均为经历了超高压变质的深俯冲陆壳岩石。本文测试的大别山超高压岩石样品主要取自双河地区，在那里上述各种超高压岩石均有出露（图 1（b））。

大别山燕山期岩浆事件非常发育，在大别山南、北带中均有壳源的中生代花岗岩侵入体出露（图 1），它们携带了深部地壳的组成信息。前人已对其长石 Pb 同位素组成做了较详细的研究[19]，已有资料显示，大别山南、北带的花岗岩长石 Pb 同位素组成没有差异，均具有较低的放射成因 Pb 同位素组成[19]，表明大别山南、北深部下地壳没有差异。

2 分析方法

岩块首先破碎成 5 mm 直径大小的岩屑，手工挑选 100 g 无次生及风化产物的新鲜岩样，用蒸馏水冲洗 3 次，烘干，而后用玛瑙球磨机将样品磨至小于 160 目粉末。

样品 U、Pb 同位素数据是在天津地质矿产研究所用 VG354 型质谱仪测得，使用同位素稀释法测得 U 和 Pb 含量。铅稀释剂为 H208，铀稀释剂为 H235。称量约 0.20 g 全岩样品（少数 Pb 含量低的样品增至 0.40 g），用 HF + HNO$_3$ 溶样。用 HBr 提取样品，并在阴离子交换树脂柱上，分别用 HBr 和 HNO$_3$ 淋洗分离 U 和 Pb。

本试验每次测定均伴有空白实验。对大多数高 Pb 含量样品（>2 μg/g），实验全流程 Pb 空白为 <2.24 ng，而对于少数低 Pb 含量（<2 μg/g）样品由于采用超纯水与酸溶液，Pb 空白<0.56 ng。因此 Pb 空白校正引入的误差将大大小于 0.1%或质谱测定误差。测定结果列于表 1。

表 1 大别山超高压变质岩及片麻岩的 U-Pb 同位素组成

样号	岩性	产地	$^{206}Pb/^{204}Pb$	$^{206}Pb/^{204}Pb$ (t = 130 Ma)	$^{207}Pb/^{204}Pb$	$^{207}Pb/^{204}Pb$ (t = 130 Ma)	$^{208}Pb/^{204}Pb$	Pb/μg·g^{-1}	U/μg·g^{-1}	U/Pb
Dzh-88-6	黑云角闪片麻岩	北大别祝家铺	15.844	15.798	15.209	15.206	36.916	8.164	0.3079	0.03772
Dzh-88-7	混合片麻岩	北大别祝家铺	16.129	16.067	15.286	15.283	37.233	8.215	0.4185	0.05095
Dzh-88-8	黑云角闪片麻岩（捕房体）	北大别祝家铺	16.384	16.322	15.367	15.364	37.418	4.531	0.2254	0.04975
Dzh-88-9	黑云角闪片麻岩（捕房体）	北大别祝家铺	16.473	16.430	15.338	15.336	37.351	3.928	0.1371	0.03492
Dzh-88-2	片麻岩	北大别祝家铺	16.707	16.647	15.531	15.528	37.867	6.894	0.3290	0.04773
92R-3	片麻岩围岩	北大别饶钹寨	15.799	15.781	15.209	15.208	36.940	6.569	0.09574	0.01457
92Y-2	混合片麻岩	北大别岳西	17.204	17.119	15.448	15.444	37.730	7.470	0.5060	0.06774
92HT-21	榴辉岩	南大别双河	17.026	16.818	15.400	15.390	37.564	3.986	0.6672	0.1674
92SH-2	花岗片麻岩	南大别双河	17.524	17.195	15.423	15.407	38.759	3.611	0.9313	0.2579
92HT-1	超高压副片麻岩	南大别双河	18.815	17.939	15.684	15.641	39.927	1.675	1.116	0.6664
92HT-2	榴辉岩	南大别双河	17.900	17.706	15.539	15.530	37.620	0.7055	0.1100	0.1559
92HT-3	超高压正片麻岩	南大别双河	18.341	17.916	15.566	15.545	38.430	3.086	1.023	0.3314
92H-6	超高压正片麻岩	南大别双河	17.343	16.524	15.278	15.238	38.368	1.342	0.8781	0.6541
92H-24	超高压正片麻岩	南大别双河	18.347	16.658	15.607	15.525	39.887	2.167	2.802	1.293
92SH-1	花岗片麻岩	南大别双河	17.712	17.590	15.392	15.386	37.794	9.45	0.9115	0.09645
92SH-3	超高压副片麻岩	南大别双河	19.029	17.876	15.588	15.532	40.274	1.553	1.355	0.8725
92HT-12	榴辉岩（大理岩包体）	南大别双河	20.687	20.405	15.881	15.867	38.057	6.299	1.339	0.2126
92HT-8	硬玉石英岩	南大别双河	20.781	19.990	15.875	15.836	42.237	2.527	1.455	0.5717

3 结果与讨论

3.1 U、Pb 含量

表 1、图 2 显示北大别片麻岩的 U、Pb 含量普遍低于下地壳平均值，但 U/Pb 值（0.015~0.068）与下地壳平均值大致相当。双河超高压岩石及花岗片麻岩的 U 含量绝大多数低于平均上地壳含量。但是，除少数榴辉岩外，它们的 U 含量高于北大别片麻岩及下地壳平均值（图 2（b））[20]。然而，超高压岩石的 Pb 含量多数低于北大别片麻岩 Pb 含量，且均大大低于上、下地壳的平均值（图 2（a））。因此，双河

超高压变质岩及其围岩的U/Pb值（0.096~1.293）显著高于北大别片麻岩的U/Pb值（0.015~0.068）及平均下地壳U/Pb值，而略高于或与上地壳平均值相当[20]。在超高压变质岩中，超高压变泥质岩的U/Pb值较其他岩石更高一些。

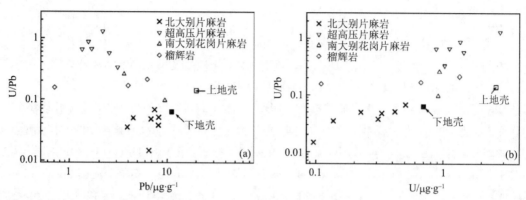

图2 大别山超高压岩石及片麻岩的U/Pb-Pb(a)和U/Pb-U(b)图。上、下地壳U和Pb平均含量及U/Pb值，据文献[20]

造成这些岩石U和Pb含量普遍偏低的主要原因可能是在俯冲过程中，俯冲陆壳岩石中的部分U和Pb会被析出的变质流体带走。斜长角闪岩的脱水实验研究证明俯冲板块在脱水过程中，Pb比U有更大的活动性，从而使脱水后的俯冲洋壳有较低的Pb含量和较高的U/Pb值[21]。上述南、北大别变质岩的Pb含量普遍低于上、下地壳平均值的事实说明华南陆壳俯冲过程中经历了脱水及Pb丢失过程。U-Pb数据还表明南大别超高压变质岩的Pb丢失量大大高于其U丢失量，也大大高于北大别片麻岩的Pb丢失量。这个事实说明南大别超高压变质岩较北大别片麻岩更强烈地经历了脱水过程。考虑到俯冲陆壳上地壳岩石较中、下地壳岩石含有更多的水，因而在俯冲过程中，脱水过程主要发生在上地壳部分；而北大别片麻岩则可能是俯冲陆壳中的中、下地壳部分。此外，图2（b）显示南大别超高压岩石的U含量仅低于平均上地壳，因此它只可能在上地壳含量水平上发生了U丢失，而北大别片麻岩可能在下陆壳含量水平上发生了U丢失。

3.2 Pb同位素

图3显示北大别片麻岩的Pb同位素组成与源于下地壳部分熔融的南、北大别带中生代花岗岩的Pb同位素组成是完全一样的，均表现出低放射成因Pb特征。在Doe等人[10]的Pb构造模型上，表现为下地壳Pb与地幔Pb的混合特征（图3）。与此相反，双河超高压岩石的Pb同位素组成具有较大的变化范围，并更富放射成因Pb。其$^{206}Pb/^{204}Pb$可高达20.781，类似于俯冲洋壳成因的HIMU端元[22,23]。在Doe等人的Pb构造模型图上，它们主要表现为地幔Pb与上地壳Pb的混合特征。为了便于和花岗岩的长石Pb进行对比，我们将样品的Pb同位素组成按花岗岩形成年龄（130 Ma）进行了年龄校正。表1和图4（a）显示，尽管超高压岩石的U/Pb值较高，校正后的同位素比值有所降低，但其总体特征没有大的变化。

为了比较南大别超高压岩石与北大别片麻岩在俯冲前没有发生U/Pb高压变质分异时Pb同位素差异，我们又将样品的Pb同位素比值按超高压变质年龄（230 Ma）[24,25]进行了年龄校正。表1和图4（b）显示，它们俯冲前的初始Pb同位素组成仍有很大差异。南大别超高压变质岩的Pb同位素变化范围要比北大别片麻岩大得多，主要是它们更富含放射成因Pb。这反映了南大别超高压变质岩的原岩在过去较长历史中有较高的U/Pb值，因而具有上地壳特征；而北大别片麻岩原岩在较长历史中具有较低的U/Pb值，因而具有中、下地壳的特征。

Zheng等人[26]报道了16个采自南大别超高压带石马到双河地区的超高压大理岩U-Pb同位素数据。这些大理石的同位素组成非常均一，其$^{206}Pb/^{204}Pb$和$^{207}Pb/^{204}Pb$的变化范围仅在18.077~18.210和15.544~15.640之间。这一同位素组成非常接近双河各种超高压变质岩的$^{206}Pb/^{204}Pb$和$^{207}Pb/^{204}Pb$平均值（18.500和15.567）（见图3、图4）。很显然，这一Pb同位素组成具有典型上地壳特征[10]。

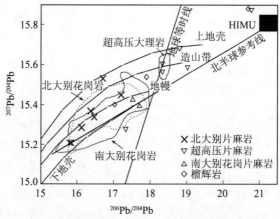

图3 大别山超高压岩石及片麻岩的 $^{207}Pb/^{204}Pb$-$^{206}Pb/^{204}Pb$ 和 $^{208}Pb/^{204}Pb$-$^{206}Pb/^{204}Pb$ 图。南、北大别花岗岩数据引自文献[19]; 大理岩数据引自文献[26]; 上、下地壳、地幔及造山带 Pb 同位素演化曲线据文献[10]; HIMU 端元据文献[22]

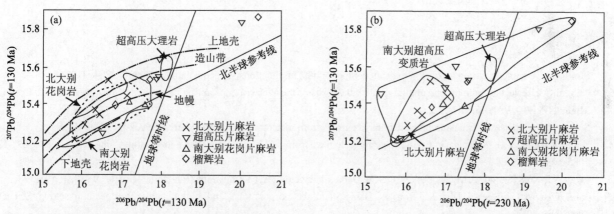

图4 大别山超高压岩石和片麻岩 t=130 Ma 时 (a)和 t=230Ma 时 (b)的 $^{207}Pb/^{204}Pb$-$^{206}Pb/^{204}Pb$ 图。(a)便于和花岗岩 (130 Ma) 的长石 Pb 进行对比; (b)显示超高压变质以前的初始 Pb 同位素组成。数据来源同图3

4 超高压岩片出露模型

根据上述 U-Pb 同位素数据，我们可以获得两个基本结论: ①南大别超高压变质带中以双河超高压岩片为代表的超高压岩石是深俯冲的上地壳岩石。双河超高压变质岩异常低的氧同位素组成 ($\delta^{18}O$) 也表明，它们在俯冲前曾在近地表经历了与大气降水的作用[27,28]; ②北大别片麻岩可能为俯冲陆壳中的中、下地壳岩石，北大别带出露的麻粒岩块体，以及该带不含柯石英的榴辉岩的高温（800~850 ℃）变质和麻粒岩相退变质特征也支持这一结论。如果这两点认识是正确的话，我们就会产生一个问题: 为什么北大别带作为俯冲中、下地壳岩石，又位于俯冲陆壳的最前沿（最靠近地缝合线）却俯冲深度不大，没有发生超高压变质作用（榴辉岩不含柯石英）; 而俯冲到 100 km 以上深度的上地壳岩石却出露在远离地缝合线的南大别带？对此，本文给出一个简单的动力学解释。陆壳俯冲是由陆块后方洋盆引张所产生的推力及先期俯冲洋壳产生的拖力联合驱动的，这些驱动力受到由俯冲陆壳与周围地幔密度差所产生的浮力的阻滞，随着俯冲陆壳体积的增大，该浮力也不断增大。由于上地壳平均密度低于下地壳，因而俯冲上地壳所受到的浮力要大于下地壳。这样，俯冲上陆壳最终随着陆壳内部产生的剪切应力的增加而断裂并产生一个逆断层，这一断层有可能发生在上地壳与中、下地壳之间，这样上地壳的俯冲推动力急剧下降，而对它的浮力依旧。正是这一浮力使深俯冲到地幔深度的上地壳岩石沿此逆冲断层上升并叠置在俯冲不太深的下地壳之上（图 5（a））。北大别带与南大别带之间的五河-水吼韧性剪切带有可能代表这一断层。以后随着北大别带中生代岩浆穿隆的发生与隆起，原来叠置在北大别带（穿隆中心）上部的超高压岩片被剥蚀掉了，使下部的俯冲不太深的中、下地壳出露地表。然而南大别带因隆起幅度低而剥蚀量较小，使得叠置在上部的深俯冲上地壳岩石得以保存下来（图 5（b））。Chemenda

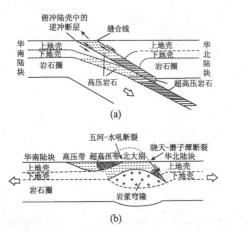

图 5 大别山超高压及高压岩石折返、出露的构造机制模型
(详见正文说明)

等人[29]做的大陆岩石圈俯冲模拟实验已证明上述过程是可以发生的。Okay 等人[30]和 Webb 等人[31]也提出过超高压岩片折返的俯冲陆壳内逆冲断层模型。本文的贡献在于通过 Pb 同位素示踪判断，区分出大别山已出露的俯冲陆壳别的不同单元的上地壳及中、下地壳性质，并指出导致超高压岩片快速上升的逆断层实际发生在上地壳与中、下地壳直接（大别山区的五河-水吼剪切带可作为这一逆断层的代表）。因此，由于逆冲作用而抬升折返的超高压岩片主要是深俯冲的上地壳岩片，而深俯冲的中、下地壳岩石还需要俯冲板块断离及其他碰撞后构造作用才有可能上升，出露于地表。

参考文献

[1] Smith D C. Coesite in clinopyroxene in the Caledonides and its implications for geodynamics. Nature, 1984, 310: 641-644.
[2] Chopin C. Coesite and pure pyrope in high-grade blueschists of the Western Alps: a first record and some consequences. Contrib. Mineral. Petrol, 1984, 86: 107-118.
[3] Sobolev N V, Shatsky V S. Diamond inclusions in garnets from metamorphic rocks. Nature, 1990, 343: 742-746.
[4] Wang X, Liou J G, Mao H G. Coesite-bearing eclogites from the Dabie Mountains in central China. Geology, 1989, 17: 1085-1088.
[5] Okay A I, Xu S, Sengör A M C. Coesite from the Dabie Shan eclogites, central China. Eur. Jour. Mineral., 1989, 1: 595-598.
[6] Xu S, Okay A I, Ji S, et al. Diamond from the Dabie Shan metamorphic rocks and its implication from tectonic setting. Science, 1992, 256: 80-82.
[7] 李曙光, 刘德良, 陈移之, 等. 扬子陆壳北缘地壳的钕同位素组成及其构造意义. 地球化学, 1994, 23(增刊): 10-17.
[8] 谢智, 陈江峰, 周泰禧, 等. 大别造山带变质岩与花岗岩的钕同位素组成及其地质意义. 岩石学报, 1996, 12(3): 401-408.
[9] Li S, Huang, F, Nie Y, et al. Geochemical and geochronological constraints on the suture location between the North and South China Blocks in the Dabie orogen, central China. Physics and Chemistry of the Earth (A), 2001, 26: 655-672.
[10] Doe B R, Zartman R E. Plumbotectonic—The model. Tectonophysics, 1981, 75: 135-162.
[11] 徐树桐, 苏文, 刘怡灿, 等. 大别山北部榴辉岩的发现及其岩相学特征. 科学通报, 1999, 44(13): 1452-1456.
[12] 刘贻灿, 李曙光, 徐树桐, 等. 大别山北部榴辉岩和英云闪长质片麻岩锆石 U-Pb 年龄及多期变质增生. 高校地质学报, 2000, 6(3): 417-423.
[13] 刘贻灿, 李曙光, 徐树桐, 等. 大别山北部榴辉岩的 Sm-Nd 年龄测定及其对麻粒岩相退变质时间的制约. 地球化学, 2001, 30(1): 79-87.
[14] Xue F, Rowley D B, Tucker R D, et al. U/Pb zircon ages of Granitois rocks in the North Dabie complex, eastern Dabie Shan, China. J of Geology, 1997, 105: 744-753.
[15] Zhang R Y, Liou J G, Tsai C H. Petrogenesis of a high-temperature metamorphic belt: a new tectonic interpretation for the northern Dabieshan, central China. J. Meta. Geol., 1996, 14: 319-333.
[16] Chen N S, Sun M, You Z D, et al. Well-preserved garnet zoning in granulite from the Dabie Mountains, central China. J. Metamorphic Geol., 1998, 16: 213-222.
[17] Carswell D A, Wilson R N, Zhai M. The enigma of eclogite-gneiss relationship in the ultra-high pressure terrane of Dabie Shan, central China. Inter. Workshop on UHP Metamorphism and Exhumation at Stanford Univ., 1998, A-40-44.
[18] Ye K, Yao Y P, Katayama I, et al. Large area extent of ultrahigh-pressure metamorphism in the Sulu ultrahighpressure terrane of East China: new implications from coesite and omphacite inclusions in zircon of granitic gneiss. Lithos., 2000, 52: 157-164.
[19] 张理刚. 东亚岩石圈块体地质. 北京: 科学出版社, 1995, 53-62.
[20] Taylor S R, McLennan S M. The continental crust: its composition and evolution. Oxford: Blackwell, 1985, 46-92.
[21] Kogiso T, Tatsumi Y, Nakono S. Trace element transport during dehydration processes in the subducted oceanic crust: 1. Experiments and implications for the origin of ocean island basalts. Earth Planet. Sci. Lett., 1997, 148: 193-205.
[22] Zindler A, Hart S. Chemical geodynamics. Ann. Rev. Earth Planet Sci., 1986, 14: 493-571.

[23] Hofmann A W. Mantle geochemistry from oceanic volcanism, Nature, 1997, 385: 219-228.
[24] 李曙光, Jagoutz E, 肖益林, 等. 大别山-苏鲁地体超高压变质年代学——Ⅰ. Sm-Nd 同位素体系. 中国科学, D 辑, 1996, 26(3): 249-257.
[25] 李曙光, 李惠民, 陈移之, 等. 大别山-苏鲁地体超高压变质年代学——Ⅱ. 锆石 U-Pb 同位素体系. 中国科学, D 辑, 1997, 27(3): 200-206.
[26] Zheng Y F, Fu B, Gong B, et al. U-Pb dating of marble associated with eclogite from the Dabie Mountains, East China. Chinese J. Geochemistry, 1997, 16(3): 193-201.
[27] Zheng Y F, Fu B, Li Y, et al. Oxygen and hydrogen isotope geochemistry of ultra high-pressure eclogites from the Dabie Mountains and the Sulu terrane. Earth Planet. Sci. Letters, 1998, 155: 113-129.
[28] Fu B, Zheng Y F, Wang Z, et al. Oxygen and hydrogen isotope geochemistry of gneisses associated with ultrahigh pressure eclogites at Shuanghe in the Dabie Mountains. Contrib. Mineral. Petrol., 1999, 134: 52-66.
[29] Chemenda A I, Mattauer M, Malavieille J, et al. A mechanism for syn-collisional rock exhumation and associated normal faulting: results from physical modeling. Earth Planet. Sci. Lett., 1995, 132: 225-232.
[30] Okay A, Sengör A M C. Evidence for intracontinental thrust-related exhumation of the ultra-high-pressure rocks in China. Geology, 1992, 20: 411-414.
[31] Webb L E, Hacker B R, Ratschacher L, et al. Thermochronologic constrains on deformation and cooling history of high- and ultrahigh-pressure rocks in the Qinling-Dabie orogen, eastern China. Tectonics, 1999, 18(4): 621-638.

Geochemical constraints of the eclogite and granulite facies metamorphism as recognized in the Raobazhai complex from North Dabie Shan, China*

Y. L. Xiao[1], J. Hoefs[1], A. M. Vau Deu Kerkhof[2] and S. G. Li[3]

1. Institute of Geochemistry, University of Gottingen, Goldschmidtstrasse 1, D-37077 Gottingen, Germany
2. Institute of Geology and Dynamics of the Lithosphere, University of Gottingen, Goldschmidtstrasse 1, D-37077 Gottingen, Germany
3. Department of Earth and Space Sciences, University of Science and Technology of China, Hefei 230026, Anhui, China

亮点介绍：饶钹寨石榴辉石岩的成分环带和矿物组成表明其先后经历了峰期榴辉岩相、重结晶榴辉岩相、麻粒岩相和角闪岩相四个阶段的变质作用，证明饶钹寨超镁铁岩块是华南陆壳俯冲过程中裹挟的华北陆块南缘岩石圈地幔碎片，并经历了陆壳深俯冲和折返过程相关的超高压和退变质作用；矿物流体包裹体成分的变化可以区分不同的变质阶段；石榴石的 O 同位素组成显示石榴石生长于封闭的流体系统，角闪岩相变质后与外来流体发生反应。

Abstract A combined study of major and trace elements, fluid inclusions and oxygen isotopes has been carried out on garnet pyroxenite from the Raobazhai complex in the North Dabie Terrane (NDT). Well-preserved compositional zoning with Na decreasing and Ca and Mg increasing from the core to rim of pyroxene in the garnet pyroxenite indicates eclogite facies metamorphism at the peak metamorphic stage and subsequent granulite facies metamorphism during uplift. A P–T path with substantial heating (from c. 750 to 900 ℃) after the maximum pressure reveals a different uplift history compared with most other eclogites in the South Dabie Terrane (SDT). Fluid inclusion data can be correlated with the metamorphic grade: the fluid regime during the peak metamorphism (eclogite facies) was dominated by N_2-bearing NaCl-rich solutions, whereas it changed into CO_2-dominated fluids during the granulite facies retrograde metamorphism. At a late retrograde metamorphic stage, probably after amphibolite facies metamorphism, some external low-salinity fluids were involved. In situ UV-laser oxygen isotope analysis was undertaken on a 7 mm garnet, and impure pyroxene, amphibole and plagioclase. The nearly homogeneous oxygen isotopic composition ($\delta^{18}O_{VSMOW}$ = c. 6.7‰) in the garnet porphyroblast indicates closed fluid system conditions during garnet growth. However, isotopic fractionations between retrograde phases (amphibole and plagioclase) and garnet show an oxygen isotopic disequilibrium, indicating retrograde fluid-rock interactions. Unusual MORB-like rare earth element (REE) patterns for whole rock of the garnet pyroxenite contrast with most ultra-high-pressure (UHP) eclogites in the Dabie-Sulu area. However, the age-corrected initial $\varepsilon_{Nd}(t)$ is −2.9, which indicates that the protolith of the garnet pyroxenite was derived from an enriched mantle rather than from a MORB source. Combined with the present data of oxygen isotopic compositions and the characteristic N_2 content in the fluid inclusions, we suggest that the protolith of the garnet pyroxenite from Raobazhai formed in an enriched mantle fragment, which has been exposed to the surface prior to the Triassic metamorphism.

Key words eclogite facies; granulite facies; metamorphic fluids; North Dabie Shan; Raobazhai complex

1 Introduction

The Dabie-Sulu ultra-high-pressure (UHP) metamorphic belt in East China is known to be the largest among recognized UHP terranes in the world. It represents the deep parts of a collision zone between the North China and

---

* 本文发表在：Journal of Metamorphic Geology, 2001, 19: 3-19

Yangtze cratons. Tectonically, the Dabie Shan is subdivided into four metamorphic units: the North Huaiyang metamorphic belt; the North Dabie Terrane (NDT); the South Dabie UHP Terrane (SDT); and the Susong metamorphic belt (Liou et al. 1995). Recent studies have demonstrated the many differences between the NDT and SDT. Jahn et al. (1999) have shown that the main rocks of the NDT, characterized by I_{Sr} of 0.709–0.710 and $\varepsilon_{Nd}(t)$ of −15 to −20, were derived by partial melting of the lower intermediate crust. Most gneisses from the SDT have higher $\varepsilon_{Nd}(t)$ values of −2 to 10. Although UHP metamorphism has had inferred based on some relic minerals (Tsai & Liou, 2000), most researchers believe that the NDT was subjected to amphibolite to granulite facies metamorphism (e.g. Wang & Liou, 1991; Li et al., 1993; Zhang et al., 1996). Another apparent difference is that eclogitic rocks have not has been undisputably identified in the NDT (Zhang et al., 1996; Jahn et al., 1999), whereas UHP eclogites are abundant in the SDT.

In this study, we concentrate on the petrology and fluid evolution of a high-pressure garnet pyroxenite from Raobazhai, in the northern part of the NDT, and aim to reconstruct its P-T-t fluid path during metamorphism. In this context, the term 'garnet pyroxenite' has been used to include samples with a pyroxene + garnet assemblage. From well-preserved chemical zoning of both garnet and clinopyroxene and variable chemical compositions of fluid inclusions in garnet, we conclude that the investigated rocks have experienced two distinguishable metamorphic events: an early eclogite facies event and a late granulite facies event. We also present data on rare earth elements (REEs), oxygen isotopes and fluid compositions in fluid inclusions, and estimate limits on the evolution of the metamorphic fluids and their origin.

2 Geological outline and samples

The NDT unit is bordered with the SDT by a major 200–300 m wide mylonitized contact zone (e.g. Hacker & Wang, 1995; Wang et al., 1995). The major lithologies of the NDT have been subdivided into three groups (Zhang et al., 1996): (i) Metamorphic complex, which consists of an abundant upper amphibolite to granulite facies rocks and migmatitic gneiss; (ii) mafic-ultramafic complex and ultramafic blocks including more than 130 mafic to mafic-ultramafic composite bodies of variable size; and (iii) Cretaceous granitic intrusive rocks containing xenoliths of gneiss, mafic and ultramafic rocks.

The Raobazhai meta-ultramafic complex is located at the southern Foziling Reservoir near Huoshan in the NDT (Fig. 1), with an outcrop area of about 3 km in length and 0.2–0.9 km in width. The complex was emplaced in a solid state into gneisses of the Dabie Group (Yang, 1983), consists mainly of Cr-spinel harzburgite, dunite and garnet pyroxenite. The country rocks near the contact zone are strongly mylonitized. Foliations are well developed along the margin of the complex.

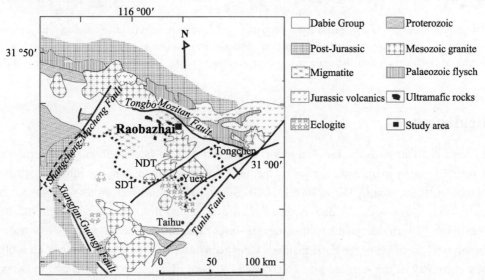

Fig. 1 Geological and tectonic sketch map of the Dabie Shan region. The dotted line in the middle of the Dabie Shan is the boundary between the North Dabie Terrane (NDT) and the South Dabie Terrane (SDT) (after Wang et al., 1992).

The studied samples were collected from a garnet pyroxenite xenolith that occurs in the intensely deformed Raobazhai ultramafic complex and is arranged in a linear fashion parallel to the trend of the mylonite zone. In the same complex, an age of 244 ± 11 Ma has had been derived-based on garnet clinopyroxene-whole-rock Sm-Nd isotope data (Li et al., 1993). The garnet pyroxenite was characterized as a type I eclogite by Li et al. (1993), although omphacite as a typical eclogite facies mineral has not been observed. Xu et al. (1994) interpreted the Raobazhai complex as a part of the an ophiolitic mélange complex, but a more detailed study suggests that it may be a mantle-derived block (Zhang et al., 1995).

The garnet pyroxenite is composed mainly of the primary phases garnet (c. 40%) and clinopyroxene (20%–30%), and secondary phases Ti-amphibole (10%–15%), plagioclase (10%–15%) and minor orthopyroxene. Rutile ± ilmenite and apatite occur as accessory minerals. Most garnet crystals are relatively large, held 2 to 8 mm. Some garnet grains inferred to be originally in contact with clinopyroxene are surrounded by thin coronae of plagioclase and amphibole. Usually, garnet contains pyroxene, amphibole, plagioclase, rutile and/or ilmenite as inclusions. Pyroxene occurs as inclusions in garnet or as a rock-forming mineral in the matrix. The pyroxene is < 1 mm in size and thus finer grained than garnet. Some pyroxene is partially retrograded into a net-like texture that is composed of clinopyroxene and plagioclase (Fig. 2a). Secondary reddish brown amphibole is in equigranular, and sometimes inclusions of early ringed minerals, such as garnet and pyroxene. Plagioclase sometimes occurs as inclusions in garnet and clinopyroxene; Most form a symplectite retrograded from clinopyroxene. Orthopyroxene occurs as an overgrowth phase of clinopyroxene, or a needle-like exsolution in clinopyroxene, indicating that orthopyroxene formed later than clinopyroxene (Fig. 2b). Rutile occurs mainly as inclusions in garnet and clinopyroxene. Some of them have retrograded into ilmenite.

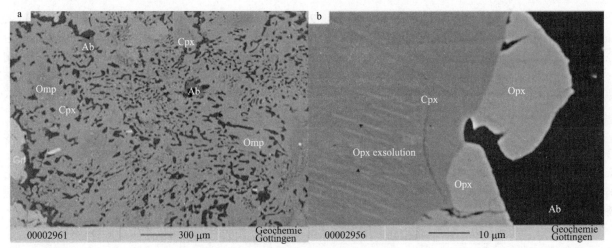

Fig. 2 Photomicrographs of the garnet pyroxenite from Raobazhai. (a) Coarse symplectite of secondary clinopyroxene and albite around primary omphacite. (b) Overgrowth of orthopyroxene (hypersthene) around clinopyroxene (see also Fig. 4a); note the needle-like exsolution of orthopyroxene at the grain margin. Cpx, clinopyroxene; Ab, albite; Grt, garnet; Omp, omphacite; Opx, orthopyroxene.

3 Analytical methods

Major and trace element analysis, fluid inclusion study and oxygen isotope laser-probe measurements were carried out on selected garnet pyroxenite samples. Chemical compositions of various minerals were obtained by electron microprobe analysis using a JXA-8900RL JEOL Superprobe. Fluid inclusions were investigated using a Linkam heating-freezing stage with a video system for ease of observation. Oxygen isotope measurement was completed by the in situ oxygen isotope analysis using ArF-laser fluorination (Fiebig et al., 1999). The experimental procedure and analytical precision of the three methods have been described in detail in Xiao et al. (2000).

Whole-rock chemistry of the samples was determined by X-ray fluorescence (XRF) and inductively coupled plasma mass spectrometry (ICPMS) techniques. XRF analysis of glass discs was undertaken using a Philips PW 1480 automated sequential spectrometer (see Hartmann, 1994). Samples (c. 100 mg) for ICPMS were digested with

1 ml of HF, 1 ml of $HClO_4$ and 4 ml of HNO_3, at 200 ℃ for about 15–24 h in Teflon beakers before evaporation to dryness and dissolution of the residue. The Measurements were performed using a VG PlasmaQuad 2 + ICPMS. Single mineral trace elements were determined using laser ablation ICPMS. The ablated particles were transported by an argon gas stream into the VG Plasma Quad2 + ICPMS. The typical parameters of analysis and the experimental procedure have been described in detail by Simon et al. (1997).

The separation was performed in two steps: cation exchange columns, with HCl chemistry, preceded a separation of Nd on Teflon columns coated with high density polyethylene (HDPE). The analyses were performed on a Finnigan MAT 262 at the Institute of Geology and Dynamics of the Lithosphere (IGDL) Gottingen, operated in static mode. The total analytical error on the Nd abundances is < 1%. The external reproducibility on La Jolla standard solution was 0.511839 ± 0.000007 (2σ, $n = 80$). The Nd isotope ratio was corrected for mass fractionation to $^{146}Nd/^{144}Nd = 0.7219$ and adjusted to recommended values for the La Jolla standard (0.511858). Blanks ranged from 100 to 200 pg, and thus no blank correction was necessary.

4 Results

4.1 Major element zoning

A careful examination for homogeneity, both within grains (zonation) and between grains (development of chemical subdomains), is critical for an evaluation of meaningful *P-T* information. The chemical compositions of representative minerals in the investigated samples are listed in Table 1. Microprobe profiles and semi-quantitative mapping indicate that both pyroxene and garnet exhibit complicated chemical zoning patterns and that these zonations are nearly symmetrical.

Table 1 Major and trace element compositions of representative minerals and whole rock from Raobazhai

	Garnet				Pyroxene					Amphibole		Plagioclase		Whole rock
	1	11	16	19	9	15	17	In-c	In-r	In	Matrix	In	Matrix	
SiO_2	40.50	40.24	39.83	40.40	55.08	51.82	53.49	53.19	53.19	41.81	41.50	61.04	60.72	44.50
TiO_2	0.06	0.06	0.08	0.05	0.12	0.13	0.06	0.29	0.15	3.43	2.56	0.01	0.05	1.42
Al_2O_3	22.14	22.14	22.25	22.60	6.94	6.43	2.33	6.43	6.70	13.00	14.87	23.40	23.41	15.70
Cr_2O_3	0.08	0.07	0.09	0.04	0.13	0.09	0.08	0.07	0.05	0.09	0.08	0.00	0.00	0.06
FeO	20.05	20.35	21.94	19.94	4.66	7.25	19.86	4.75	5.24	11.98	11.08	0.30	0.20	14.06
MnO	0.43	0.54	0.67	0.80	0.01	0.14	0.22	0.03	0.04	0.09	0.09	0.02	0.00	0.32
MgO	11.54	10.78	9.57	11.30	11.76	12.49	24.39	12.28	12.35	12.84	13.33	0.03	0.01	10.32
CaO	6.61	6.86	7.15	6.65	17.93	20.81	0.46	18.64	18.45	11.15	11.18	5.97	6.49	11.16
Na_2O	0.04	0.02	0.03	0.01	4.08	1.51	0.02	3.40	3.48	3.49	3.63	8.67	8.37	1.68
K_2O	0.00	0.01	0.01	0.00	0.02	0.00	0.01	0.00	0.01	0.05	0.03	0.02	0.03	0.05
Total	101.45	101.06	101.62	101.79	100.73	100.67	100.92	99.07	99.67	97.92	98.36	99.45	99.28	99.27
Si	2.99	3.00	2.98	2.98	1.97	1.89	1.94	1.94	1.92	6.11	5.98	2.72	2.72 FeO	12.40
Ti	0.00	0.00	0.00	0.00	0.00	0.00	0.00	0.01	0.00	0.38	0.28	0.00	0.00 Fe_2O_3	1.84
Al	1.93	1.94	1.96	1.96	0.29	0.28	0.10	0.28	0.29	2.24	2.53	1.23	1.23	
Cr	0.00	0.00	0.01	0.00	0.00	0.00	0.00	0.00	0.00	0.01	0.01	0.00	0.00	
Fe	1.24	1.27	1.37	1.23	0.14	0.22	0.60	0.14	0.16	1.46	1.33	0.01	0.01	
Mn	0.03	0.03	0.04	0.05	0.00	0.00	0.01	0.00	0.00	0.01	0.01	0.00	0.00	
Mg	1.27	1.20	1.07	1.24	0.63	0.68	1.32	0.67	0.67	2.80	2.86	0.00	0.00	
Ca	0.52	0.55	0.57	0.53	0.69	0.81	0.02	0.73	0.71	1.74	1.73	0.29	0.31 $^{143}Nd/^{144}Nd$	0.513286
Na	0.01	0.00	0.00	0.00	0.28	0.11	0.00	0.24	0.24	0.99	1.01	0.75	0.73 $\pm 2\sigma$	0.000007
K	0.00	0.00	0.00	0.00	0.00	0.00	0.00	0.00	0.00	0.01	0.01	0.00	0.00 $\varepsilon_{Nd}(0)$	12.6
Total	8.00	8.00	8.00	8.00	4.00	4.00	4.00	4.00	4.00	15.74	15.75	5.00	5.00 $\varepsilon_{Nd}(244\ Ma)$	−2.9
La (ppm)	0.002	0.016	0.013	0.098	0.067	0.118				0.002			0.058	0.187

	Garnet				Pyroxene					Amphibole		Plagioclase		Whole rock
	1	11	16	19	9	15	17	In-c	In-r	In	Matrix	In	Matrix	
Ce	0.032	0.106	0.019	0.662	0.506	0.571					0.011		0.130	0.469
Pr	0.010	0.027	0.006	0.157	0.150	0.169					0.003		0.040	0.155
Nd	0.151	0.285	0.086	1.715	1.398	1.508					0.031		0.288	0.960
Sm	0.790	0.639	0.535	1.352	0.884	0.779					0.044		0.340	1.077
Eu	0.703	0.497	0.390	0.724	0.374	0.329					0.025		0.302	0.836
Gd	3.609	2.649	2.939	3.141	0.947	1.099					0.104		1.030	2.738
Tb	1.149	0.698	0.725	0.710	0.142	0.091					0.019		0.200	0.661
Dy	11.494	6.652	7.957	6.893	0.648	0.467					0.145		1.359	4.635
Ho	3.421	2.128	2.286	2.075	0.106	0.056					0.034		0.299	1.061
Er	12.942	8.105	8.456	6.775	0.270	0.166					0.087		0.937	3.067
Tm	2.129	1.353	1.342	1.142	0.042	0.019					0.014		0.124	0.439
Yb	17.062	11.037	10.297	8.972	0.202	0.124					0.112		0.817	3.248
Lu	3.065	1.972	1.687	1.386	0.026	0.018					0.019		0.116	0.499

In, inclusion; In-c, core of inclusion; In-r, rim of inclusion; 1, 11, 16 and 19 are the analytical spot numbers on the compositional profile of garnet (Fig. 5b), whereas 9, 15 and 17 are the analytical spot numbers on the compositional profile of pyroxene (Fig. 4b). FeO*, total Fe as FeO.

4.2 Pyroxene

Most matrix pyroxene is strongly zoned. One example is to in Fig. 3 (see also Table 1, spots 9, 15 & 17). In a

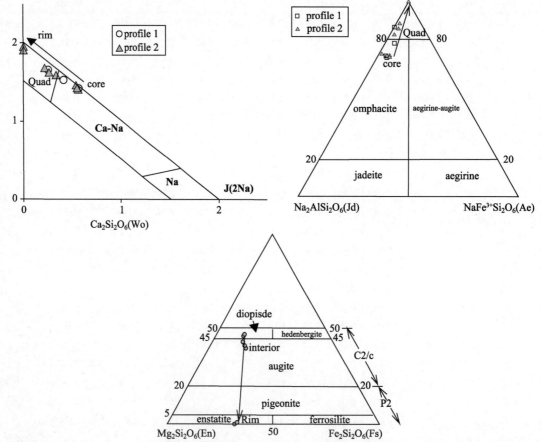

Fig. 3 Classification diagrams for pyroxene (after Morimoto et al., 1988); Quad represents the Ca-Mg-Fe pyroxene area.

Q-J classification diagram (Fig. 3a), where $Q = Ca + Mg + Fe^{2+}$ and $J = 2Na$ (Morimoto et al., 1988), profiles of pyroxene grains show that the cores belong to the Ca-Na group, whereas the margins fall into the Ca-Mg-Fe (Quad) area. In the classification diagrams of the Ca-Na and Ca-Mg-Fe groups (figure 3b, c), there are large compositional variations from omphacite through diopside to augite, which result from a decrease of jadeite content from the core to rim (from c. 30% to c. 7%). Additionally, there is the occurrence of a minor secondary phase, enstatite (hypersthene), at the very margin of the grain.

Semi-quantitative mapping of pyroxene crystals clearly demonstrates compositional zoning as shown in Fig. 4 (a, b). From core to rim, there is a decrease in Na with an increase in Ca, Mg and total Fe. It is important to note that the central portions of most pyroxene grains have a sieve-mesh texture with albite inclusions. This texture was interpreted to result from the decomposition of a jadeite-rich clinopyroxene to a jadeite-poor clinopyroxene + feldspar (e.g. Carswell, 1975). Thus, the cores of pyroxene might have had higher jadeite contents during peak metamorphic conditions than actually observed. Pyroxene inclusions in garnet are homogeneous in composition, but all of these inclusions included some exsolved rutile, which is in accordance with the high Ti content of the whole rock.

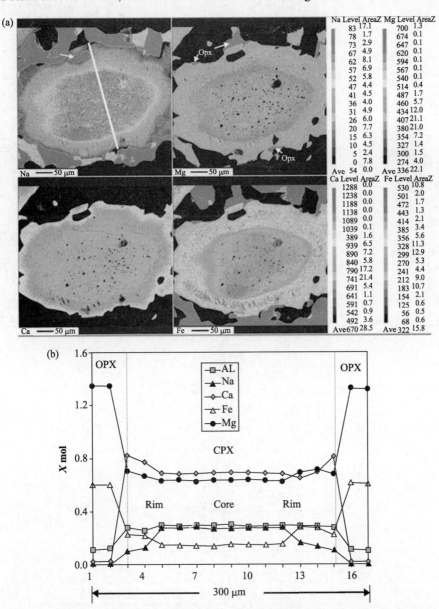

Fig. 4 (a) Semi-quantitative mapping of a pyroxene grain in the investigated samples. The map clearly shows compositional zones of Na, Mg, Ca and Fe from the core to the rim (see text for a detailed description); at the very margin, a small overgrowth of orthopyroxenes can be observed. (b) Pyroxene compositional zoning profile (rim to rim). Location of probe analysis is along the white line in (a).

4.3 Garnet

Garnet is rich in pyrope and almandine (30–43 and 39–43 mol%, respectively), with low grossular (17–25 mol%) and minor spessartine (generally < 2 mol%) contents. Some large garnet crystals (> 2 mm in size) are similar to pyroxene crystals in showing systematic, but much weaker, compositiorial zonation, whereas small garnet is generally homogeneous, having a composition equivalent to the rim of large garnet. Semi-quantitative mapping of a 3 mm wide garnet porphyroblast demonstrates zoning in Mg, Ca and Fe (Fig. 5a). A profile of this garnet (along the white line of Fig. 5a) is shown in Fig. 5(b). Two zones can be identified: an inner zone where the Mg content generally decreases and Ca increases from the core outwards, whereas Fe and Mn are almost constant; and an outer zone with an increase in Mg, Fe and Mn and a decrease in Ca.

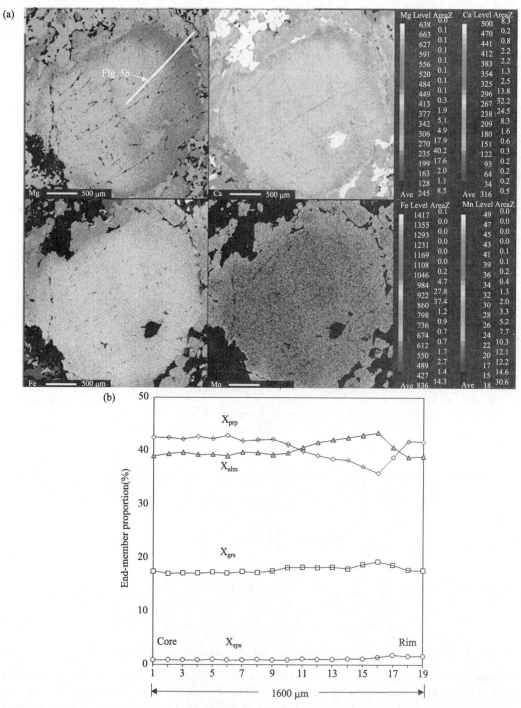

Fig. 5 (a) Semi-quantitative mapping of a garnet porphyroblast. Two-stage garnet growth marked by compositional zoning. (b) Garnet compositional zoning profile (core to rim, about 1.6 mm in length). Profile of probe analysis is along the white line in (a).

4.4 Amphibole and plagioclase

All amphibole is Ti - rich; however, the amphibole inclusions in garnet have Ti = 0.38, Al^{IV}= 1.89, Al^{VI} = 0.34 (atoms per formula unit (p.f.u.)), $Al^{VI} / (Fe^{3+} + Al^{VI})$ = 0.54 and Mg/(Mg+ Fe) = 0.52, whereas those in matrix amphibole have 0.28, 2.02, 0.51, 0.51 and 0.55, respectively (Table 1). Plagioclase inclusions in garnet and pyroxene have similar compositions, With plagioclase in symplectite containing 20–30 mol% anorthite (Table 1).

4.5 Rare earth elements

REE analyses of garnet, pyroxene and whole rock are given in Table 1, showing the well-known fact that garnet is rich in heavy rare earth elements (HREEs), whereas pyroxene is rich in middle REEs (Fig.6a, b). REE concentrations of garnet are slightly higher in the cores than in the rims. This zoning is somewhat bell shaped and roughly parallel to the bell-shaped major element zoning (Fig. 5a). The concentrations of most REEs decrease towards the garnet rims, suggesting the existence of a fractionation process. This also indicates that there was no significant fluid-rock interaction during garnet growth, which is consistent with the oxygen isotope data (see below). However, the very rim of garnet (point 1) has chondrite-normalized light rare earth element (LREE) values that are about 10 times higher than those of the next measured spot (point 2) and of the core signatures (Fig. 6a), although the heavy REE signatures are comparable. This probably resulted from fluid-rock interaction after the garnet growth. Pyroxene has convex-upward-shaped chondrite-normalized patterns (Fig. 6b), which is a common features in

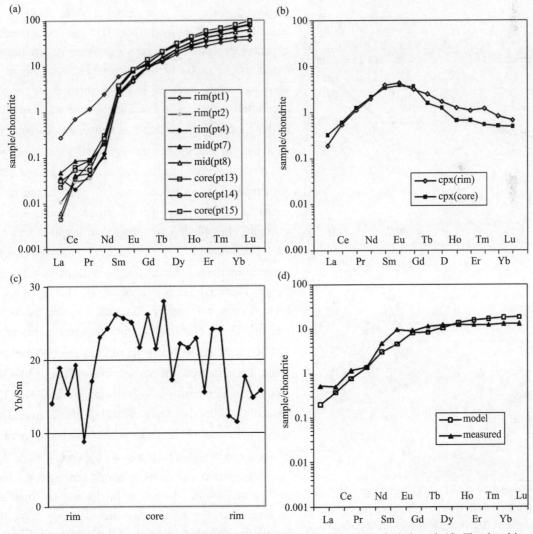

Fig. 6 Chondrite-normalized rare earth element patterns within garnet (a), pyroxene (b) and whole rock (d). The chondrite values used here are from Taylor & McLennan (1985). (c) HREE/LREE profile of the same garnet as in (a) (rim to rim). Locations of points in (a) are depicted in Fig. 9. Errors are generally smaller than the symbol size.

clinopyroxene from mantle xenoliths (Song & Frey, 1989). HREE/LREE ratios vary from 10 to 30, with the highest ratios in the cores and intermediate regions of the grain (Fig. 6c). The whole rock shows an unusual REE pattern with HREE values 10 times higher than chondrite and a severe LREE depletion (Fig. 6d); a significant Eu anomaly has not been observed. Eclogites from Wehai (Jahn et al., 1996; samples 12176 & 12177) and retrograded eclogite from Shuanghe (Li et al., 2000; sample HT-12) display roughly similar REE patterns. It is interesting to note that the rock has a negative initial $\varepsilon_{Nd}(t)$ value (−2.9) at a metamorphic age of c. 240 Ma. This suggests that the protolith of the rock should be LREE enriched, and that the severe LREE depletion could result from a loss of LREEs during its UHP or retrograde metamorphism.

4.6 Fluid inclusion studies

Fluid inclusions mainly occur in the garnet porphyr-oblasts. Pyroxene is mostly devoid of fluid inclusions, or contains only fluid inclusion relics, whose contents have been consumed by reactions with the host mineral. In the newly formed plagioclase, only a few carbonic inclusions occur. Based on their chemical compositions and textural criteria, four types of fluid inclusions were distinguished in the large garnet grains (Fig. 7).

1 High-salinity aqueous inclusions with or without N_2 are found in the centre of garnet crystals. These inclusions occur isolated or as groups (Fig. 7). Most contain halite(s) as a daughter phase. Their morphologies indicate late modification, probably during retrogression. Some of the fluid inclusions have $N_2 \pm CO_2$, which occur in variable proportions in the inclusion. During cooling, the H_2O phase of these inclusions was frozen at temperatures around −60 ℃. At about −85 ℃, sometimes very small solids, probably CO_2 and clathrate, could be observed around the gas species phase. At temperatures of about −160 ℃ and below, a bubble appeared in the gas species phase. During subsequent warming, the bubble homogenized to liquid at temperatures of −148 to −157 ℃ (Fig. 8a). This homogenization temperature range indicates that N_2 is the major component of the gas phase (van den kerkhof, 1988). Melting of solid CO_2 (if any) cannot be clearly observed. The eutectic melting of −42 to −37 ℃, and final ice melting temperatures of −15 to −24 ℃, indicate an NaCl- dominated solution for the fluid phase, but minor Ca^{2+} is possibly present (Fig. 8b). Clathrate melting temperatures of N_2 between −11 and −13 ℃ were observed. Additionally, rare clathrate melting temperatures of between 10 and 14 ℃ confirm the presence of minor CO_2. The halite melting temperatures are between 144 and 332 ℃.

2 Carbonic inclusions in garnet represent the majority of fluid inclusions, which were predominantly found in rims of large garnet grains, indicating that they were trapped during the late growth stage of the garnet (Fig. 7b). These carbonic inclusions in garnet occur as planar clusters, but they do not occur in healed fractures. Most of these

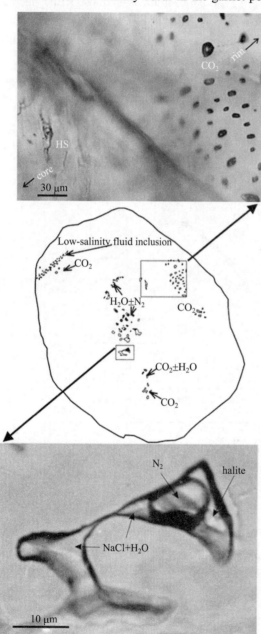

Fig. 7 Textural relations between different types of fluid inclusions in garnet. The figure combines textural relations from different grains in polished sections. HS, high-salinity aqueous inclusions.

inclusions contain fine-grained solids, indicating possible reactions between the host garnet and the fluid within fluid inclusions. They display final CO₂ melting temperatures of between −56.6 and −58.2 °C, indicative of almost pure CO_2, and homogenization temperatures to liquid between 11 and 27 °C (Fig. 8c).

3 Two-phase late secondary fluid inclusions occur in healed fractures, and show round morphologies and occur in trails. Most of these inclusions are two-phase (L + V) at room temperature with 80%–95% degrees of filling. Microthermometric results show that they have final ice melting temperatures of between 0 and −7 °C corresponding to low salinities, and homogenization temperatures of 163–332 °C (Fig. 8b).

4 H_2O-CO_2 inclusions were rarely found to coexist with high-salinity inclusions in the core of large garnet. These inclusions have been modified with filling degrees of H_2O phase that vary considerably from 20 to 80 vol.%, implying imperfect local mixing of the two fluid phases of different origins. The CO_2 phase of these inclusions shows melting temperatures from −57.5 to −56.8 °C, indicating the presence of almost pure CO_2. Their homogenization temperatures to liquid are between 21 and 25 °C.

Isochores were calculated for both the aqueous and carbonic inclusions in garnet, using the empirical equation for H_2O-NaCl and CO_2 of Brown & Lamb (1989), for a temperature in excess of 700 °C and pressure in excess of 3 kbar. The isochores for the high-salinity inclusions (± N_2) fall into the range of 10–12 kbar at 750 °C; isochores for pure carbonic inclusions indicate 3 kbar at 800 °C, whereas isochores for the low-salinity inclusions indicate 4–5 kbar at 600 °C. It should be noted that the isochores for all inclusion fluids are not conformable to peak or retrograde metamorphic pressures, indicating that most fluid inclusions were partially modified during uplift of the rock. However, due to limited fluid mobility, the compositions of the fluids in the inclusions probably still represent the primary compositions of the fluids trapped at different metamorphic conditions, which will be discussed below in more detail.

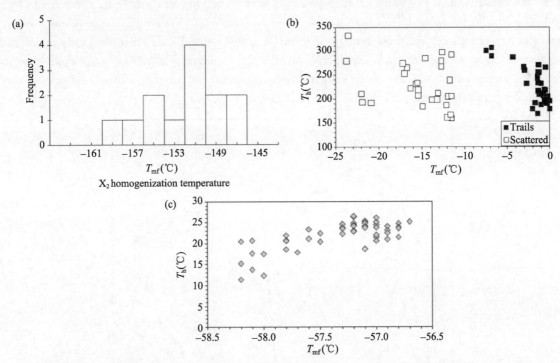

Fig. 8 (a) N_2 homogenization temperature histogram for N_2-bearing fluid inclusions. (b) Microthermometric results of aqueous inclusions in garnet. (c) Microthermometric results of carbonic inclusions in garnet.

4.7 Oxygen isotope compositions

In order to explain the compositional zoning in both garnet and pyroxene, it is important to determine whether it was largely inherited from internal cation exchanges at various P-T conditions, or from the migration of fluids; in other words, whether the zoned garnet and pyroxene grew in a closed or an open system. One approach to resolve this problem is by oxygen isotope mapping of garnet.

Using the in situ method described by Fiebig et al. (1999), $\delta^{18}O$ values of 11 spots from a garnet porphyroblast (about 7 mm in size) and 10 spots from the surrounding matrix (small garnet, pyroxene, plagioclase, amphibole) were determined. The results are listed in Table 2. Locations of analytical spots and $\delta^{18}O$ values of garnet analyses relative to V-SMOW are shown in Fig. 9.

The garnet porphyroblast, except for one analysis, is obviously homogeneous in oxygen isotopic composition, with an average $\delta^{18}O$ value of 6.6‰ ± 0.1‰. The one $\delta^{18}O$ value of 6.0‰ is from small veins of secondary amphibole close to the rim of the garnet. The garnet is almost inclusion free, and combined with the microprobe data after laser oxygen isotope measurements, indicates that no corrections for mineral inclusions are necessary. Small garnet (1–2 mm in size) is also unzoned with $\delta^{18}O$ values of 6.4‰ to 6.6‰.

It is difficult to obtain $\delta^{18}O$ values for pure pyroxene, plagioclase and amphibole because of their small crystal sizes. The pyroxene rims are usually replaced by symplectite consisting of clinopyroxene and plagioclase. Secondary amphibole occurs only as thin rims around garnet or pyroxene. Thus oxygen isotope data for pyroxene, plagioclase and amphibole represent a mixture of these minerals.

The nearly homogeneous oxygen isotopic composition of the garnet porphyroblast indicates closed system conditions during garnet growth. If any infiltrating fluids had been present during garnet growth, the garnet would become gradually zoned in ^{18}O from the centre outwards to the rim, because the isotopic front that is moving through the rock becomes less steep with increasing distance from the contact (Chamberlain & Conrad, 1993). Therefore, no external fluids infiltrated the garnet pyroxenite during garnet growth.

The plagioclase (including minor amphibole and/or secondary pyroxene), amphibole (including minor secondary pyroxene and/or plagioclase) and symplectite have $\delta^{18}O$ values of 4.8‰ to 5.7‰, 6.0‰ to 6.1‰ and 6.0‰, respectively (Table 2). According to published fractionation factors for these minerals (e.g. Zheng, 1993a, 1993b) and theoretical models of diffusional reequilibration during cooling (e.g. Eiler et al., 1992), both plagioclase and amphibole should be richer in ^{18}O than garnet. However, relative to the garnet porphyroblast, the apparent fractionations between garnet and these retrograde minerals are −1.7 to −0.3‰. These data thus indicate oxygen isotopic disequilibrium during retrograde metamorphism after the amphibolite stage. Oxygen isotopic compositions of the garnet porphyroblast and the compositional zoned pyroxene are nearly identical, indicating very high metamorphic temperatures or disequilibrium.

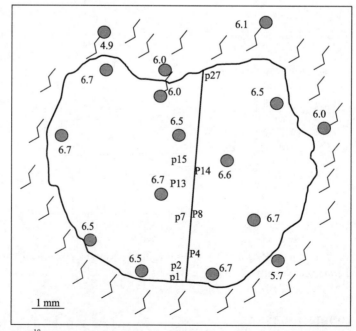

Fig. 9 Sketch of garnet showing $\delta^{18}O$ values of analysed spots. Matrix minerals are mainly plagioclase, clinopyroxene and amphibole (see Table 2 for details). Sizes of oxygen isotope laser holes are about 300 μm; not all the analysed spots of matrix are shown here. The line shows the location of a laser ICPMS profile, and numbers represent the analytical spots.

Table 2 Oxygen isotope compositions of minerals measured by UV-laser probe

Analysis No.	$\delta^{18}O$ (SMOW)	Analysis No.	$\delta^{18}O$ (SMOW)	Analysis No.	$\delta^{18}O$ (SMOW)
Coarse-grained garnet					
R-14-1	6.7	R-14-2	6.72	R-14-3	6.68
R-14-4	6.67	R-14-5	6.66	R-14-6	6.46
R-14-7	6.49	R-14-8	6.52	R-14-9	5.99
R-14-10	6.46	R-14-11	6.58		
Small garnet					
R-14-17	6.49	R-14-18	6.38	R-14-20	6.56
Pyroxene					
R-14-12	6.53	R-14-16	6.46	R-14-15	6.03 (sym)
Amphibole					
R-14-13	6.03	R-14-21	6.09		
Plagioslase					
R-14-19	5.66	R-14-14	4.85		

5 Metamorphic evolution and *P-T* conditions of the complex

It has been demonstrated that core compositions of garnet generally record information about peak or near-peak metamorphic conditions, whereas the rims record changes in the *P-T* conditions after the thermal maximum (e.g. Pearson & O'Reilly, 1991). With this approach, at least four metamorphic stages can be deduced for the metamorphic evolution of the Raobazhai complex. The *P-T* conditions of metamorphic recrystallization for the four metamorphic stages were estimated using various thermobarometers; the results are listed in Table 3 and shown in Fig. 10.

1 Peak pressure eclogite facies metamorphism (PE). The peak metamorphic stage is represented by the core compositions of pyroxene and garnet. The preservation of an omphacitic core in the zoned pyroxene indicates that the peak pressure metamorphic stage corresponds to eclogite facies metamorphism. The lower limit of the peak metamorphic pressure is estimated to be 18.5 kbar at 750 °C according to the jadeite contents in the omphacitic core of the pyroxene (c. 30 mol%) using the barometer of Holland (1980). Tsai & Liou (2000) suggested a very high metamorphic pressure for the Raobazhai complex (>25 kbar). However, as no excess SiO_2 was found in Ca-Na clinopyroxene (see Tsai & Liou, 2000) from our samples, it is assumed that the pressure of 18.5 kbar is the lower limit of the peak metamorphic pressure of the Raobazhai complex. Temperature estimates for this stage (core-core parts for the pyroxene and garnet) at 20 kbar, based on the Grt-Cpx thermometers of Aranovich & Pattison (1995), Krogh (1988) and Ellis & Green (1979), are in the range 680–860 °C (Table 3).

2 Recrystallized eclogite facies metamorphism (RE). The decrease in jadeite content in pyroxene from the core towards the rim indicates a retrograde compression. The application of Holland's (1980) barometer for the diopside or augite rims yields pressures from 14 to 17 kbar. The application of the three thermometers (Ellis & Green, 1979; Krogh, 1988; Aranovich & Pattison, 1995) to the rim-rim compositions of the coexisting garnet and pyroxene at 15 kbar gives a temperature range of 820–1000 °C (Table 3).

3 Retrograde granulite facies metamorphism (GF). The overgrowth of hypersthene at the rims of some pyroxene grains indicates that the Raobazhai complex underwent a retrograde metamorphism from recrystallized eclogite facies to granulite facies conditions. The formation of the overgrowth hypersthene can be expressed by the following decompression reaction:

$$\text{garnet} + \text{clinopyroxene} + SiO_2 = \text{orthopyroxene} + \text{plagioclase}$$

The coexistence of orthopyroxene, plagioclase and the rim of clinopyroxene (see Fig. 2b) supports this explanation. The metamorphic pressure for the retrograde granulite facies metamorphic stage is estimated to be c. 7 kbar at 900 °C using the Al content in the overgrowth hypersthene, employing the barometer of Harley (1984), whereas the application of the Fe-Mg exchange thermometer of Grt-Opx (Lee & Ganguly, 1988) yields temperatures between 900 and 950 ℃ at 7 kbar (Table 3).

4 Retrograde amphibolite facies metamorphism (AF). Secondary amphibole and plagioclase are the characteristic mineral assemblage. The metamorphic pressures are estimated to be < 5 kbar based on the relation between AlVI and Si of amphibole (Raase, 1974). The amphibole-plagioclase thermometer of Blundy & Holland (1990) yields metamorphic temperatures ranging from 800 to 850 ℃, which is in good agreement with the temperature estimates for the amphibolites and amphibolite facies gneisses from the northern Dabie Shan using the same thermometer by Zhang et al. (1996), and also corresponds to the fact that amphiboles in the studied samples usually contain high X_{Ti} (Raase, 1974).

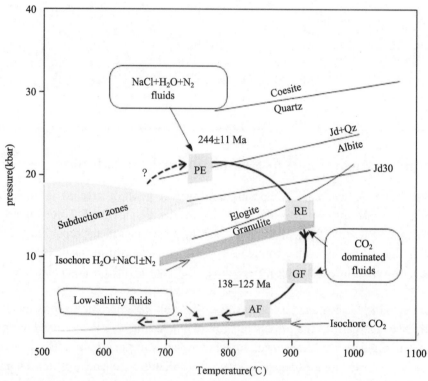

Fig. 10 *P-T-t* fluid path for the garnet pyroxenite from Raobazhai. Shaded areas are estimated *P-T* conditions for different metamorphic stages: PE, peak pressure eclogite facies metamorphism; RE, recrystallized eclogite facies metamorphism; GF, retrograde granulite facies metamorphism; AF, retrograde amphibolite facies metamorphism. Data sources: jadeite + quartz = albite (Holland, 1980); coesite = quartz (Bohlen & Boettcher, 1982); the eclogite to granulite transition is after Green & Ringwood (1967); the *P-T* conditions of subduction zones are from Spear (1993); time estimates are based on Li et al. (1993) and Hacker et al. (1998).

Table 3 *P-T* estimate of Raobazhai garnet pyroxenite

Metamorphic facies	Temperature (℃)			Pressure (kbar)
Peak eclogite	Garnet-clinopyroxene (at 20 kbar)			>18
	AP	kR	EG	
	679	729	778	
	712	769	802	
	749	800	854	
	730	782	835	
Recrystallized eclogite	Garnet-slinopyroxene (at 15 kbar)			14–17 (Holland, 1980)
	AP	kR	EG	
	827	864	900	
	848	883	916	
	900	922	959	
	824	861	898	
	961	992	1012	

Metamorphic facies	Temperature (℃)		Pressure (kbar)
Granulite	Garnet-orthopyroxene (at 7 kbar)	Lee & Ganguly (1988)	c. 7 (Harley, 1984)
	951		
	902		
	913		
	918		
Amphibolite	Plagioslase-amphibole (at 5 kbar)	Blundy & Holland (1990)	<5 (Raase, 1974)
	873		
	851		
	838		

AP, Aranovich & Pattison (1995); EG, Ellis & Green (1979); KR, Krogh (1988).

6 Discussion

6.1 *P-T-t* path and fluid evolution of the garnet pyroxenite

The fluid inclusion assemblage of a metamorphic rock is the result of a series of fluid-rock interaction processes at different stages of its *P-T* history. Different types of fluid inclusions within garnet reflect fluids trapped at different stages of garnet growth. As discussed above, the cores of garnet were formed at peak metamorphic *P-T* conditions, whereas the rims represent the retrograde stage. Thus a relative chronology for the different generations of metamorphic fluids can be deduced from the textural relations and chemical compositions of the fluid inclusions.

1 High-salinity inclusions ± N_2 in garnet cores are the recognizable earliest generation of fluid inclusions and may represent fluid compositions of the peak metamorphism (eclogite facies) of the garnet pyroxenite from the Raobazhai complex. Nitrogen is a widespread feature of eclogite facies metamorphic fluids (e.g. Andersen et al., 1989, 1993; Klemd, 1991; Klemd et al., 1993); the characteristic N_2 contents have been interpreted to be formed during subduction processes (Andersen et al., 1993; Klemd & Brocker, 1999).

In the case of Raobazhai, recent studies have demonstrated that the complex is probably of mantle origin (Zhang et al., 1995, 1996). Although the present whole rock shows a MORB-like REE pattern, the negative $\varepsilon_{Nd}(t)$ values (around −3) indicate that it cannot represent subducted oceanic crust. Thus, possible sources for nitrogen in the garnet pyroxenite include the mantle and nitrogen-rich crustal rocks, such as metasediments subducted to great depth.

Previous studies on the major composition of fluid inclusions in mantle-derived rocks have shown that carbon dioxide is a ubiquitous component in fluid inclusions in mantle xenoliths from localities through-out the world (e.g. Roedder, 1965, 1984; Murk et al., 1978; Andersen et al., 1984; Pasteris, 1987), but nitrogen is not. However, nitrogen in the mantle has been found in diamonds, and in minor amounts in phlogopite and amphibole where it may possibly substitute for potassium. The average nitrogen content of the mantle is < 2 ppm (Norris & Schaeffer, 1982), which is much lower than that in crustal rocks (c. 20 ppm; Wlotzka, 1961).

Metasedimentary rocks are one of the major lithological units of the NDT (Liou et al., 1995). Abundant UHP paragneiss, marble and jadeite quartzite occur in Shuanghe (Cong et al., 1995), where N_2 inclusions in eclogite have been described (Fu et al., 1999). There is now abundant evidence that both eclogites and their country rocks in the SDT experienced the same Triassic UHP metamorphic history (e.g. Wang & Liou, 1991; Rowley et al., 1997; Hacker et al., 1998; Rumble & Yui, 1998). Therefore, nitrogen released from such metasediments at great depths is a likely cause of the N_2 present in the eclogite facies fluids. As the N_2-bearing fluid inclusions were trapped during the peak metamorphism, the N_2-rich metamorphic fluids should have been formed during or before the peak metamorphism.

2 Carbonic inclusions in the garnet rims are assumed to have been trapped during granulite facies retrogression and thus represent the compositions of the fluid system during granulite facies metamorphism. Since the discovery of CO_2-rich fluid inclusions in granulite facies rocks, numerous studies have shown that CO_2-dominated fluid inclusions are a characteristic feature of these rocks worldwide (e.g. Touret, 1981; Lamb et al., 1987; Vry & Brown, 1991). Experimental phase equilibria investigations (Santosh, 1991) indicated that, regardless of the mechanism of granulite formation in different terranes, CO_2 plays a critical role in buffering the fluid composition and stabilizing the dry mineral assemblages in granulite. This is in accordance with the fact that some orthopyroxene overgrew the rim of the compositionally zoned pyroxene.

3 Secondary low-salinity inclusions in garnet imply that the rock interacted with an external low-salinity fluid during retrograde metamorphism. Because these inclusions occur in texturally healed fractures, they should have been trapped after garnet growth, probably after the retrograde amphibolite facies metamorphic stage.

Aranovich & Newton (1997) have demonstrated that the presence of NaCl decreases significantly the water activity (αH_2O); thus both the high-salinity aqueous inclusions (brines) and CO_2-dominated inclusions indicate low water activity during eclogite and granulite facies metamorphism. The low αH_2O, which is essential for the stability of the minor overgrown orthopyroxene around clinopyroxene grains, is explained by the lack of a free fluid phase (Touret & Huizenga, 1999).

On the other hand, compared with the published oxygen isotope compositions of UHP eclogites from the Dabie-Sulu terrane, which range from −10‰ to +10‰, indicating a definite water-rock interaction prior to subduction (e.g. Baker et al., 1997; Yui et al., 1997; Rumble & Yui, 1998; Zheng et al., 1998), the present oxygen isotope data show a very narrow spread with no obvious indications of water-rock interaction. However, as mentioned previously, the nearly homogeneous oxygen isotopic compositions in the garnet porphyroblast indicate closed system conditions during garnet growth, whereas the negative fractionations of −1.7‰ to −0.5‰ between secondary minerals (sym-plectite after pyroxene, plagioclase, amphibole) and garnet show clear isotopic disequilibrium. This disequilibrium can be best interpreted as having resulted from fluid-rock interactions during the retrograde formation of these minerals. Because different minerals have different closure temperatures to oxygen isotope exchange during exhumation, garnet being the most resistant, whereas minerals such as amphibole and plagioclase are more susceptible, differential isotopic exchange can take place between the minerals and retrograde fluids. It is unlikely that the oxygen isotope shift was caused by retrogression before or during amphibolite facies overprints, because at such high temperatures oxygen should be either in or very close to isotopic equilibrium if any interaction with an exotic fluid took place. Therefore, the oxygen isotopic compositions of the studied samples indicate that the fluid system was relatively closed during garnet growth, and that the rocks interacted with external retrograde metamorphic fluids at temperatures below amphibolite facies metamorphism, which is in accordance with the presence of the low-salinity inclusions in healed fractures.

An estimate of the metamorphic age for the Raobazhai complex comes from Sm-Nd mineral (garnet and clinopyroxene)-whole-rock isochron ages, which gave 244 ± 11 Ma on a garnet pyroxenite from the same locality (Li et al., 1993). As shown above, garnet and pyroxene are compositionally zoned; one may therefore argue that this age probably represents a mixed age between the eclogite and granulite facies metamorphism. However, Li et al. (1993) emphasized that all mineral separates were obtained by careful hand-picking under a binocular microscope in order to remove altered minerals and mineral inclusions. From the semi-quantitative mapping of the zoned pyroxene, it is obvious that the omphacitic core and diopside represent 99 vol.% of the grains, although minor overgrowth of orthopyroxene and fine-grained plagioclase occurs in the rim and core, respectively. On the other hand, the retrograde temperature of granulite facies metamorphism is up to 900 ℃ (Fig. 10), which may cause doubts as to whether or not the peak metamorphic ages have been reset. Closure temperatures of 850–900 ℃ for the Sm-Nd system in garnet were suggested by Cohen et al. (1988) and Jagoutz (1988), whereas a much lower temperature of around 600 ℃ was proposed by Bohlen et al. (1985) and Mezger et al. (1992). For the Raobazhai garnet pyroxenite, consisting only of garnet and pyroxene during eclogite and granulite facies metamorphism, diffusion of Sm-Nd in garnet is controlled mainly by clinopyroxene. In addition, the relatively large grain sizes of garnet also raise the closure temperature (e.g. Dodson, 1973). Thus, if a high closure temperature of 850–900 ℃ for the Sm-Nd system in garnet is applicable to our samples, the age of 244 ± 11 Ma is likely to represent the lower limit of the age for the peak metamorphism in Raobazhai. Because the *P-T* conditions are coincident with the inferred peak metamorphic conditions of the country rocks (Zhang et al., 1996), the amphibolite facies metamorphism in

Raobazhai may be Cretaceous, possibly in the range of 138–125 Ma (Xue et al., 1997; Hacker et al., 1998; Jahn et al., 1999).

Based on the observed textures, compositional zoning patterns in individual minerals, P-T estimates and metamorphic ages discussed in the previous sections, a possible retrograde P-T-t fluid path for the garnet pyroxenite from the Rabazhai complex, from eclogite through granulite to amphibolite facies metamorphism, is presented in Fig. 10. This P-T-t fluid path has the following characteristics.

1 During the stage of initial uplift (from peak to recrystallized eclogite facies), there is a significant temperature increase. This is quite different from the retrograde P-T paths of the UHP eclogite from the SDT, in which all the units show nearly isothermal decompression paths following the maximum pressure (e.g. Wang et al., 1992; Okay, 1193; Xiao & Li, 1993; Xiao et al., 1995). From an overview of eclogites formed in deep parts of typical continent-continent collision zones, as in the Tauern Window, Austria, the P-T paths usually show that the metamorphic temperature peak is synchronous with the maximum pressure and along a nearly isothermal decompression path (Spear, 1989). On the other hand, for eclogite formed in subduction zones, many subduction P-T paths show that the maximum pressure is often followed by a decompression path with substantial heating (Spear, 1993).In any case, the initial retrograde P-T path indicates that the Raobazhai complex must have been formed in a tectonic environment different from that of the SDT.

2 From the recrystallized eclogite to granulite facies metamorphism, the rock follows a near-isothermal decompression P-T path (from > 14 kbar at 900–1000 °C to c. 7 kbar at 900–950 °C). Bohlen (1987) and Harley (1989) divided granulite facies rocks into two types with distinct P-T paths: (i) granulites that are characterized by a period of nearly isothermal decompression (ITD); and (ii) granulites that are characterized by a period of nearly isobaric cooling (IBC). The ITD granulites most likely have formed in crust thickened by collision and their characteristic P-T paths can be generated during rapid thinning related to tectonic exhumation, whereas the IBC granulites may have formed in various settings. You et al.(1994) obtained ITD paths for the granulites from the NDT. Zhang et al. (1996) derived ITD P-T paths for the various country rocks of the Raobazhai complex. Our P-T estimates also reveal an ITD path for the granulite facies metamorphism of the Raobazhai complex, and are in accordance with the previous studies.

3 The fluid system of the Raobazhai complex changed during retrogression from eclogite to granulite facies metamorphism. The composition of the fluids evolved from N_2-bearing high-salinity solutions in the eclogite facies metamorphism towards CO_2-dominated fluids in the granulite facies metamorphism; after the amphibolite facies, the rock interacted with low-salinity, external retrograde metamorphic fluids.

6.2 Depletion of LREEs

In contrast to most Dabie-Sulu eclogites, which show LREE-rich patterns (Li et al., 1993; Jahn, 1998; Zhang & Liou, 1998; Jahn et al., 1999), the Raobazhai garnet pyroxenite displays a strong LREE depletion. The age corrected (244 Ma) $\varepsilon_{Nd}(t)$ is -2.9 (Table 1), which is in agreement with the earlier result of -3.0 obtained by Li et al. (1993) on the same rock from the Raobazhai complex, implying that the protolith was LREE-rich. This indicates that the REE pattern of the rock, especially the LREEs, has been modified, and does not represent the REE pattern of its protolith. The rock could be derived from an enriched mantle rather than any MORB-type rock (see also Zhang et al., 1995, 1996).

Coesite-bearing eclogites with high ε_{Nd} values (range, +170 to +264) from Weihai in the Sulu area show a similar REE pattern to Raobazhai. The unusual pattern was interpreted to be due to open-system behaviour in which LREEs were lost by fluid-rock interaction (Jahn et al., 1996). In the case of Raobazhai, oxygen isotope in situ analyses of coarse-grained garnet suggest negligible fluid–rock interaction during garnet growth (see 'Oxygen isotope compositions', above). This assumption is supported by REE zoning in garnet (Fig. 6a). The reconstructed primary modal mineralogy of eclogite, in conjunction with the REE abundances of pure clinopyroxene and garnet, has been used to calculate whole-rock REE patterns during peak metamorphism (e.g. Taylor & Neal, 1989; Jacob & Foley, 1999). As shown in Fig. 6(d), the calculated whole-rock REE pattern agrees well with the measured whole rock. It is thus unlikely that the retrograde fluids caused the depletion of the whole-rock LREEs, although they might be responsible for the oxygen isotope disequilibrium between garnet and secondary minerals (amphibole, plagioclase) and the slight modification of the LREEs at the very margin of garnet (Fig. 6a). Stalder et al. (1998) have experimentally shown that LREEs are always enriched in the fluid phase relative to the solid phase under P-T conditions relevant to subduction processes ($D^{(fluid/Min)}>1$). It is therefore proposed that the LREE depletion happened before peak metamorphism of the rock, i.e. probably during the dehydration or partial melting processes of the subducting slab.

7 Conclusions

1 Chemical zoning and mineral assemblages show a polymetamorphic history for the garnet pyroxenite from Raobazhai. Four metamorphic stages are recognized: (i) peak eclogite facies; (ii) recrystallized eclogite facies; (iii) granulite facies; and (iv) amphibolite facies.

2 Fluid inclusion data indicate that the earliest recognizable metamorphic fluid preserved in the Raobazhai complex is a high-salinity aqueous fluid with significant N_2. The fluid phase at the granulite facies stage was almost pure CO_2, which changed to low-salinity solutions after the amphibolite facies event. The presence of high-salinity brines and nitrogen in the metamorphic fluids thus clearly distinguishes the eclogite facies from the granulite facies metamorphism, which is characterized by an anhydrous, CO_2-dominated fluid regime.

3 Oxygen isotope mapping of a 7 mm garnet porphyroblast indicates that the fluid system was more or less closed during garnet growth. However, the rock interacted with some external low-salinity fluids after the amphibolite facies metamorphism.

Acknowledgements

Funded by grants HO 375/19-1 from the German National Science Foundation (DFG). Y. X. thanks the Max-Planck-Gesellschaft for a stipend during the period of 1997–1998. We are particularly grateful to k. Simon, G. Hartmann, A. kronz, J. Fiebig and R. Przybilla for analytical assistance. The authors thank B.-M. Jahn, D. Robinson and an anonymous reviewer for their constructive comments on an earlier draft of the manuscript. We are grateful to D. Jacob and G. Worner for discussions.

References

Andersen, T., Austrheim, H., Burke, E. & Elvevold, S., 1993. N_2 and CO_2 in deep crustal fluids: evidence from the Caledonides of Norway. Chemical Geology, 108, 113–132.

Andersen, T., Burke, E. & Austrheim, H., 1989. Nitrogen-bearing, aqueous fluid inclusions in some eclogites from the Western Gneiss Region of the Norwegian Caledonides. Contributions to Mineralogy and Petrology, 103, 153–165.

Andersen, T., O'Reilly, S. Y. & Griffin, W. L., 1984. The trapped fluid phase in upper mantle xenoliths from Victoria, Australia: implications for mantle metasomatism. Contributions to Mineralogy and Petrology, 88, 72–85.

Aranovich, L. & Newton, R. C., 1997. H_2O activity in concentrated NaCl solutions at high pressures and temperatures measured by the brucite-periclase equilibrium. Contributions to Mineralogy and Petrology, 125, 200–212.

Aranovich, L. Y. & Pattison, D. R. M., 1995. Reassessment of the garnet-clinopyroxene Fe-Mg exchange thermometer: I. Evaluation of the Pattison and Newton (l989) experiments. Contributions to Mineralogy and Petrology, 119, 16–29.

Baker, J., Matthews, A., Mattey, D., Rowley, D. & Xue, F., 1997. Fluid-rock interactions during ultra-high pressure metamorphism, Dabie Shan, China. Geochimica et Cosmochimca Acta, 61, 1685–1696.

Blundy, J. D. & Holland, T. J. B., 1990. Calcic amphibole equilibria and a new amphibole plagioclase geothermometer. Contributions to Mineralogy and Petrology, 104, 208–224.

Bohlen, S. R., 1987. Pressure-temperature-time paths and a tectonic model for the evolution of granulites. Journal of Geology, 95, 617–632.

Bohlen, S. R. & Boettcher, A. L., 1982. The quartz–coesite transformation: a pressure determination and the effects of other components. Journal of Geophysics Research, 87, 7073–7078.

Bohlen, S. R., Valley, J. W. & Essene, E. J., 1985. Metamorphism in the Adirondacks I: petrology, pressure, and temperature. Journal of Petrology, 26, 971–992.

Brown, P.E. & Lamb, W.M., 1989. P-V-T properties of fluid in the system $H_2O \pm CO_2 \pm NaCl$: new graphical presentations and implications for fluid inclusion studies. Geochimica et Cosmochimica Acta, 53, 1209–1221.

Carswell, D. A., 1975. Primary and secondary phlogopites and clinopyroxenes in garnet lherzolite xenoliths. Physics and Chemistry of the Earth, 9, 417–429.

Chamberlain, C. P. & Conrad, M. E., 1993. Oxygen-isotope zoning in garnet: a record of volatile transport. Geochimica et Cosmochimica Acta, 57, 2613–2629.

Cohen, A. S., O'Nions, R. K., Siegenthaler, R. & Griffin, W. L., 1988. Chronology of the pressure-temperature history recorded by a granulite terrane. Contributions to Mineralogy and Petrology, 98, 303–311.

Cong, B. L., Zhai, M. G, Carswell, D. A., Wilson, R. N., Wang, Q. C., Zhao, Z. Y. & Windley, B. F., 1995. Petrogenesis of ultrahigh-pressure rocks and their country rocks at Shuanghe in Dabieshan, Central China. European Journal of Mineralogy, 7, 119–138.

Dodson, M. H., 1973. Closure temperature in cooling geochro nological and petrological systems. Contributions to Mineralogy and Petrology, 40, 259–274.

Eiler, J. M., Baumgartner, L. P. & Valley, J. W., 1992. Intercrystalline stable isotope diffusion: a fast grain boundary model. Contributions to Mineralogy and Petrology, 112, 543–557.

Ellis, D. J. & Green, D. H., 1979. An experimental study of the effect of Ca upon garnet-clinopyroxene Fe-Mg exchange equilibria. Contributions to Mineralogy and Petrology, 71, 13–22.

Fiebig, J., Wiechert, U., Rumble, D. & Hoefs, J., 1999. High-precision in situ oxygen isotope analysis of quartz using an ArF laser. Geochimica et Cosmochimica Acta, 63, 687–702.

Fu, B., Touret, J.L.R.& Zheng, Y.F., 1999. Fluid inclusions in ultrahigh-pressure rocks from the Dabie Mountains, China: evidence for fluid-rock interaction and limited fluid flow during continent collision Terra Nostra (Abstracts and Program), 99 (6), 104–105.

Green, D. H. & Ringwood, A. E., 1967. An experimental investigation of the gabbro to eclogite transformation and its petrological applications. Geochimica et Cosmochimica Acta, 31, 767–833.

Hacker, B. R. & Wang, Q. C., 1995. Ar/Ar geochronology of ultra-high-pressure metamorphism in central China. Tectonics, 14, 994–1006.

Hacker, B. R., Ratschbacher, L., Webb, L., Ireland, T., Walker, D. & Dong, S., 1998. U/Pb zircon ages constrain the architecture of the ultrahigh-pressure Qinling-Dabie Orogen, China. Earth and Planetary Science Letters, 161, 215–230.

Harley, S. L., 1984. The solubility of alumina in orthopyroxene coexisting with garnet in $FeO-MgO-Al_2O_3-SiO_2$. Journal of Petrology, 25, 665–696.

Harley, S. L., 1989. The origins of granulites: a metamorphic perspective. Geological Magazine, 126, 215–247.

Hartmann, G., 1994. Late-medieval glass manufacture in the Eichsfeld Region (Thuringia, Germany). Chemie der Erde, 54, 103–128.

Holland, T. J. B., 1980. The reaction albite=jadeite+quartz determined experimentally in the range 600–1200 ℃. American Mineralogist, 65, 125–134.

Jacob, D. E. & Foley, S. F., 1999. Evidence for Archean ocean crust with low high field strength element signature from diamondiferous eclogite xenoliths. Lithos, 48, 317–336.

Jagoutz, E., 1988. Nd and Sr systematics in an eclogite xenolith from Tanzania: evidence for frozen mineral equilibria in the continental lithosphere. Geochimica et Cosmochimica Acta, 52, 1285–1293.

Jahn, B. M., 1998. Geochemical and isotope characteristics of UHP eclogites and ultramafic rocks of the Dabie Orogen: implications for continental subduction and collisional tectonics. In: When Continents Collide: Geodynamics and Geochemistry of Ultrahigh-Pressure Rocks (eds Hacker, B. R. & Liou, J. G), pp. 203–239, Kluwer Academic Publishers, Dordrecht.

Jahn, B. M., Cornichert, J., Cong, B. L. & Yui, T. F., 1996. Ultrahigh-ε_{Nd} eclogites from an ultrahigh-pressure metamorphic terrane of China. Chemical Geology, 127, 61–79.

Jahn, B. M., Wu, F., Lo, C. H. & Tsai, C. H., 1999. Crust-mantle interaction induced by deep subduction of the continental crust: geochemical and Sr-Nd isotopic evidence from post-collisional mafic-ultramafic intrusions of the northern Dabie complex, central China. Chemical Geology, 157, 119–146.

Klemd, R., 1991. Fluid inclusions in eclogite-facies metasediments from the Munchberg Gneiss Complex, NE-Bavaria (Abstract). Plinius, 5, 121–122.

Klemd, R. & Brocker, M., 1999. Fluid influence on mineral reactions in ultrahigh-pressure granulites: a case study in the Sniezknik Mts. (West Sudetes, Poland). Contributions to Mineralogy and Petrology, 136, 358–373.

Klemd, R., van den kerkhof, A. M. & Horn, E. E., 1993. High-density CO_2-N_2 inclusions in eclogite-facies metasediments of the Muenchberg gneiss complex, SE Germany. Contributions to Mineralogy and Petrology, 111, 409–419.

Krogh, E.J., 1988.The garnet-clinopyroxene Fe-Mg geotherm ometer: a reinterpretation of existing experimental data. Contributions to Mineralogy and Petrology, 99, 44–48.

Lamb, W. M., Valley, J. W. & Brown, P. E., 1987. Post-metamorphic CO_2-rich fluid inclusions in granulites. Contributions to

Mineralogy and Petrology, 96, 485–495.

Lee, H. Y. & Ganguly, J., 1988. Equilibrium compositions of coexisting garnet and orthopyroxene: experimental determina- tions in the system FeO-MgO-Al_2O_3-SiO_2, and applications. Journal of Petrology, 29, 93–113.

Li, S. G., Jagoutz, E., Chen, Y. Z. & Li, Q. L., 2000. Sm-Nd and Rb-Sr isotopic chronology and cooling history of ultrahigh-pressure metamorphic rocks and their country rocks at Shuanghe in the Dabie Mountains, Central China. Geochimica et Cosmochimica Acta, 64, 1077–1093.

Li, S. G., Xiao, Y. L., Liu, D. L., et al., 1993. Collision of the North China and Yangtse Blocks and formation of coesite- bearing eclogites: timing and processes. Chemical Geology, 109, 89–111.

Liou, J. G., Wang, Q., Zhai, M., Zhang, R. Y. & Cong, B. L., 1995. Ultrahigh-P metamorphic rocks and their associated lithologies from the Dabie Mountains, central China: a field trip guide to the 3rd international eclogite field symposium. Chinese Science Bulletin, 40, 1–41.

Mezger, K., Essene, E. J. & Holliday, A. N., 1992. Closure temperature of the Sm-Nd system in metamorphic garnets. Earth and Planetary Science Letters, 113, 397–409.

Morimoto, N., Fabries, J., Ferguson, A. K., Ginzburg, I. V., Ross, M., Seifert, F. A. & Zussman, J., 1988. Nomenclature of pyroxenes. American Mineralogist, 73, 1123–1133.

Murk, B. W., Burruss, R. C. & Hollister, L. S., 1978. Phase equilibria in fluid inclusions in ultramafic xenoliths. American Mineralogist, 63, 40–46.

Norris, T. L.& Schaeffer, O. A., 1982. Total nitrogen content of deep sea basalts. Geochimica et Cosmoschimica Acta, 46 (3), 371–379.

Okay, A. I., 1993. Petrology of a diamond and coesite-bearing metamorphic terrain: Dabie Shan, China. European Journal of Mineralogy, 5, 659–675.

Pasteris, J. D., 1987. Fluid inclusions in mantle xenoliths. In: Mantle Xenoliths (ed. Nixon, P. H.), pp. 691–707. Wiley, Chichester.

Pearson, N.J.& O'Reilly, S.Y., 1991.Thermobarometry and P-T-t paths: the granulite to eclogite transition in lower crustal xenoliths from eastern Australia. Journal of Metamorphic Geology, 9, 349–359.

Raase, P., 1974. Al and Ti contents of hornblende, indicators of pressure and temperature of regional metamorphism. Contributions to Mineralogy and Petrology, 45, 231–236.

Roedder, E., 1965. Liquid CO_2 inclusions in olivine-bearing nodules and phenocrysts from basalts. American Mineralogist, 50, 1746–1782.

Roedder, E., 1984. Fluid inclusions. Reviews in Minevalogy, 12, 503–532.

Rowley, D., Xue, F., Tucker, R., Peng, Z. X., Baker, J. & Davis, A., 1997. Ages of ultrahigh pressure metamorphism and protolith orthogneisses from the eastern Dabie Shan: U/Pb zircon geochronology. Earth and Planetary Science Letters, 151, 191–203.

Rumble, D. & Yui, T. F., 1998. The Qinglongshan oxygen and hydrogen isotope anomaly near Donghai in Jiangsu Province, China. Geochimica et Cosmochimica Acta, 62, 3307–3321.

Santosh, M., 1991. Role of CO_2 in granulite petrogenesis: evidence from fluid inclusions. Journal of Geosciences Osaka City University, 34, 1–53.

Simon, K., Wiechert, U., Hoefs, J. & Grote, B., 1997. Microanalysis of minerals by laser ablation ICPMS and SIRMS. Fresenius' Journal of Analytical Chemistry, 359, 458–461.

Song, Y. & Frey, F. A., 1989. Geochemistry of peridotite xenoliths in basalt from Hannuoba, East China: implications for subcontinental mantle heterogeneity. Geochimica et Cosmochimica Acta, 53, 97–113.

Spear, F.S., 1989. Relative thermobarometry and metamorphic P-T paths.In: Evolution of Metamorphic Belts (eds Daly, J.S., Cliff, R. A. & Yardley, B. W. D.), pp. 63–81, Geological Society, London.

Spear, F. S., 1993. Metamorphic P-T paths and tectonic evolution: examples. In: Metamorphic Phase Equilibria and Pressure-Temperature-Time Paths (ed.Spear, F. S.), pp. 737–766, Mineralogical Society of America, Washington, DC.

Stalder, R., Foley, S. F., Brey, G. P. & Horn, I., 1998. Mineral-aqueous fluid partitioning of trace elements at 900–1200 °C and 3.0–5.7 GPa: new experimental data for garnet, clinopyroxene, and rutile, and implications for mantle metasomatism. Geochimica et Cosmochimica Acta, 62, 1781–1801.

Taylor, L.A.& Neal, C.R., 1989. Eclogites with oceanic crustal and mantle signatures from the Bellsbank kimberlite, South Africa, part I: Mineralogy, petrography, and whole rock chemistry. Journal of Geology, 97, 551–567.

Taylor, S.R. & McLennan, S.M., 1985. The Continental Crust: Its Composition and Evolution. Blackwell Science, Oxford.

Touret, J. L. R., 1981. Fluid inclusions in high grade metamorphic rocks. In: Fluid Inclusions: Application to Petrology (eds Hollister, L. & Crawford, M.), pp. 182–208. Mineral Association of Canada short Course Handbook 6. Mineral Association of Canada, Toronto.

Touret, J.L. R.& Huizenga, J.M., 1999. Precambrian intraplate magmatism: high temperature, low pressure crustal granulites. Journal of African Earth Sciences, 28, 367–382.

Tsai, C.H. & Liou, J. G., 2000. Eclogite-facies relics and inferred ultrahigh-pressure metamorphism in the North Dabie complex, central-eastern China. American Mineralogist, 85, 1–8.

van den Kerkhof, A. M., 1988. The system CO_2-CH_4-N_2. In: Fluid Inclusions. Theoretical Modelling and Geological Applications. Free University Press, Amsterdam.

Vry, J.K. & Brown, P.E., 1991. Texturally-early fluid inclusions in garnets: evidence of the prograde metamorphic path? Contributions to Mineralogy and Petrology, 108, 271–282.

Wang, X. & Liou, J. G., 1991. Regional ultrahigh-pressure coesite-bearing eclogitic terrain in central China: evidence from country rocks, gneiss, marble and metapelite. Geology, 19, 933–936.

Wang, X., Liou, J. G. & Maruyama, S., 1992. Coesite-bearing eclogites from the Dabie Mountains, Central China: petrology and P-T path. Journal of Geology, 100, 231–250.

Wang, X., Zhang, R.Y. & Liou, J.G., 1995. Ultrahigh-pressure metamorphic terrane in eastern central China. In: Ultrahigh-Pressure Metamorphism (eds Coleman, R. G. & Wang, X.), pp. 356–390. Cambridge University Press, Cambridge.

Wlotzka, F., 1961. Untersuchungen zur Geochemie des Stickstoffs. Geochimica et Cosmochimica Acta, 24, 106–154.

Xiao, Y. L. & Li, S. G., 1993. *P-T-t* path and its tectonic implication for eclogite from the Shima, Dabie Mountains. Geotectonia et Metallogenia, 17, 239–250 (in Chinese with English abstract).

Xiao, Y. L., Li, S., Jagoutz, E. & Cheng, W., 1995. *P-T-t* path for coesite-bearing peridotite-eclogite association in the Bixiling, Dabie Mountains. Chinese Science Bulletin, 40, 156–158.

Xu, S. T., Liu, Y. C., Jiang, L. L., Su, W. & Ji, S. Y., 1994. Tectonic Regime and Evolution of Dabie Mountains. Science Press, Beijing (in Chinese).

Xue, F., Rowley, D. B., Tucker, R. D. & Peng, X. Z., 1997. U-Pb zircon ages of granitoid rocks in the North Dabie Complex, eastern Dabie Shan, China. Journal of Geology, 105, 744–753.

Yang, X.Y., 1983. A cold ultramafic intrusive body in Dabieshan area, Anhui Province—A reconsideration about the Raobazhai intrusive body in the Huoshan County. Bulletin of the Nanjing Institute of Geology and Mineval Resouvces, Chinese Science of Geosciences, 4, 81(in Chinese).

You, Z., Chalokwu, C. I., Chen, N. & Kuehner, S. M., 1994. Thermobarometry and geodynamics of granulites from the Dabie complex, central China. Geological Society of America, Abstracts with Programs, 26, 226.

Yui, T. F., Rumble, D., Chen, C. H. & Lo, C. H., 1997. Stable isotope characteristics of eclogites from the ultra-high-pressure metamorphic terrain, east-central China. Chemical Geology, 137, 135–147.

Zhang, Q., Ma, B. L., Liu, R. X., Zhao, D. S., Fan, Q. C., Li, Q. & Li, X. Y., 1995. A remnant of continental lithospheric mantle above subduction zone: geochemical constraints on ultramafic rock from Raobazhai area, Anhui Province. Science in China (Series B), 38, 1522–1529.

Zhang, R. Y. & Liou, J. G., 1998. Ultrahigh-pressure metamorphism of the Sulu terrain, eastern China: a prospective view. Continental Dynamics, 3, 32–53.

Zhang, R. Y., Liou, J. G. & Tsai, C. H., 1996. Petrogenesis of a high-temperature metamorphic terrane: a new tectonic interpretation for the north Dabieshan, central China. Journal of Metamorphic Geology, 3, 231–243.

Zheng, Y. F., 1993a. Calculation of oxygen isotope fractionation in anhydrous silicate minerals. Geochimica et Cosmochimica Acta, 57, 1079–1091.

Zheng, Y. F., 1993b. Calculation of oxygen isotope fractionation in hydroxyl-bearing silicates. Earth and Planetary Science Letters, 120, 247–263.

Zheng, Y. F., 2000. Fluid history of UHP metamorphism in Dabie Shan, China: a fluid inclusion and oxygen isotope study on the coesite-bearing eclogite from Bixiling. Contributions to Mineralogy and Petrology, 139, 1–16.

Zheng, Y. F., Fu, B., Li, Y. L., Xiao, Y. L. & Li, S. G., 1998. Oxygen and hydrogen isotope geochemistry of ultrahigh-pressure eclogites from the Dabie Mountains and the Sulu terrane. Earth and Planetary Science Letters, 155, 113–129.

大别山北部镁铁-超镁铁质岩带中榴辉岩的分布与变质温压条件*

刘贻灿[1,2]，徐树桐[1]，李曙光[2]，陈冠宝[1]，
江来利[1]，周存亭[1]，吴维平[1]

1. 安徽省地质科学研究所，合肥 230001
2. 中国科学技术大学地球和空间科学系，合肥 230026

亮点介绍： 本文查明了北大别榴辉岩的分布及其峰期变质 P-T 条件，重建了北大别榴辉岩的变质演化过程和 P-T 轨迹，证明该区榴辉岩经历了麻粒岩相变质叠加而区别于中大别超高压变质岩的折返历史，为大别山深俯冲陆壳的多岩片差异折返模型的建立提供了关键的岩石学证据。

摘要 大别山北部镁铁-超镁铁质岩带中的榴辉岩主要有两种产状，一是产于变形较强（面理化）的橄榄岩中，另一种是产于片麻岩中。其中，绿辉石中硬玉端元组分大多为 Jd=20 mol%～52 mol%。石榴子石的成分特征表明，产于橄榄岩和片麻岩中榴辉岩分别相当于 Coleman 的 B 型和 C 型榴辉岩。石榴子石成分特征及钠质单斜辉石中石英针状体出溶等表明，本区榴辉岩早期可能经历过超高压变质作用，且至少经历了三个变质阶段，即：①榴辉岩相峰期变质阶段，主要矿物共生组合为石榴子石+绿辉石+金红石±文石±蓝晶石±石英（或柯石英假象？），$P \geqslant 2.5$ GPa、$T=595～874$ ℃；②高压麻粒岩相变质阶段，主要矿物共生组合为石榴子石+透辉石+紫苏辉石+钛铁矿+尖晶石+斜长石等，$P=1.1～1.37$ GPa、$T=817～909$ ℃；③角闪岩相变质阶段，主要矿物共生组合为角闪石+斜长石+磁铁矿等，$P=0.5～0.6$ GPa、$T=500～600$ ℃。该区榴辉岩独特的麻粒岩相退变质阶段，表明榴辉岩在折返初期并未能上升到中上地壳，而处于下地壳，它与南部超高压变质岩有不同的 P-T 演化史即在榴辉岩相峰期变质之后经历了近等温减压（或稍升温减压）和降温减压变质过程。

关键词 榴辉岩；变质温压条件；超高压变质作用；镁铁-超镁铁质岩带；大别山北部

大别山榴辉岩，因其含有柯石英（Xu, 1987; Okay et al., 1989; Wang et al., 1989）、金刚石(Xu et al., 1992)及石英硬玉岩(徐树桐等，1991)等超高压矿物和岩石的发现而闻名于世，吸引了世界上同行的关注并纷纷来大别山进行研究。大别山地区超高压变质岩的分布说明大别山有世界上分布范围最大、出露最好、超高压矿物和岩石组合齐全的超高压变质带，除榴辉岩之外，还有石英硬玉岩、含白片岩组合的蓝晶石石英岩、片麻岩及大理岩等；同时，出露有扬子与华北大陆板块俯冲、碰撞有关的不同构造岩石单位（徐树桐等，1992，1994）（图 1），从南到北依次为：①前陆带；②扬子大陆板块俯冲盖层（宿松群和张八岭群）；③扬子大陆板块俯冲基底（大别杂岩）；④超高压变质带（榴辉岩带）；⑤镁铁-超镁铁质岩带或北部杂岩带（刘贻灿等，2000a）；⑥变质复理石等。榴辉岩相超高压变质岩主要呈面状分布于大别山南部（即大致沿潜山县龙井关—水吼—岳西县五河一线以南），向北则为镁铁-超镁铁质岩带或北部杂岩带（以下称北部）。并且，由于北部长期未发现榴辉岩以及其中岩石组合特征与南部榴辉岩带有一定差异而使不少同行将大别山造山带分成"南大别"和"北大别"（董树文等，1993；Cong et al., 1994）。

---

* 本文发表在：地质学报，2001，75(3): 385-395

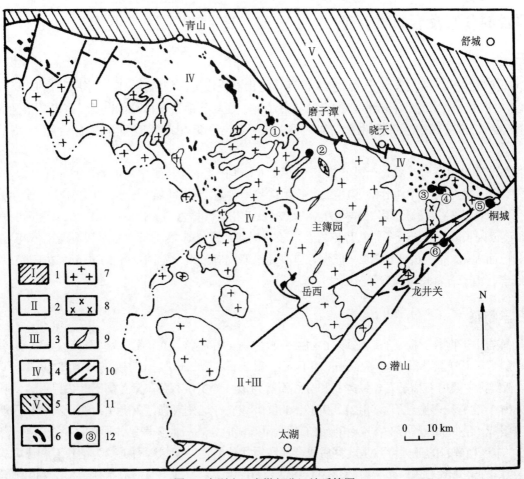

图 1 大别山（安徽部分）地质简图

1. 扬子板块俯冲盖层（宿松杂岩）；2. 超高压变质带；3. 扬子板块俯冲基底（大别杂岩）；4. 镁铁-超镁铁质岩带（可能包括部分大别杂岩）；5. 变质复理石推覆体；6. 镁铁-超镁铁质岩；7. 花岗岩；8. 辉长岩；9. 正长岩脉；10. 断层；11. 地质界线；12. 新发现的榴辉岩产地及编号：① 饶钹寨、② 黄尾河、③ 百丈岩、④ 华庄、⑤ 汪洋水库、⑥ 毛花岩

对大别山北部镁铁-超镁铁质岩带或北部杂岩带的认识之所以存在较大分歧（徐树桐等，1992，1994；Okay et al.，1992；董树文等，1993；Cong et al.，1994；Wang et al.，1993a；Li et al.，1995；Zhang et al.，1996；Xu et al.，1996；Hacker et al.，1998），一方面与该带以前未发现榴辉岩或超高压变质岩有关，另一方面可能是该带的组成比较复杂(有来自不同构造背景条件下形成的岩石)(刘贻灿等，1998，1999)，况且大多数岩石都经过多期变质、变形作用等复杂过程改造而使原来岩石已面目全非以及地球化学特征发生一定变化；此外，大别山北部带因燕山期大量花岗岩侵位及伴随混合岩化作用（徐树桐等，1992，1994；陈廷愚等，1991；牛宝贵等，1994）而使该区岩石组成及面貌复杂化。因而影响了对该带组成的正确认识，同时也影响到有关同位素年龄的解释。本区岩石类型主要有英云闪长质片麻岩、（花岗）闪长质片麻岩及二长花岗质片麻岩和少量的方辉橄榄岩、纯橄岩、石榴辉石岩、石榴二辉麻粒岩、紫苏磁铁石英岩、斜长角闪岩类、含镁橄榄石和钛斜硅镁石大理岩（刘贻灿等，2000a）等，未变质岩石类型主要有辉石岩、辉长岩、花岗岩类等。该带变质变形强烈，以角闪岩相区域变质作用为主，局部出现麻粒岩相组合。尽管有人已发现该带存在榴辉岩相峰期变质线索（徐树桐等，1994；刘贻灿等，1997）及北京大学魏春景等（1997）在该带东部官庄大麻岩发现有零星榴辉岩点（未发现新鲜榴辉岩露头），但该带西部（核部）未发现真正榴辉岩点。作者等通过进一步工作，首先在磨子潭南部黄尾河附近发现榴辉岩并首次确认它经过了麻粒岩相退变质作用（Xu et al.，1998，2000；刘贻灿等，2000a），后来又在该带其他地区发现多处新鲜榴辉岩露头点（大致沿磨子潭-晓天以南分布）（图1），这为正确认识大别山的构造格局和演化提供了重要依据。

本文主要讨论大别山北部榴辉岩的分布与变质特征及其大地构造意义。

1 榴辉岩的分布及岩石学特征

1.1 分布

大别山北部榴辉岩，主要发现于岳西县黄尾河，霍山县饶钹寨，舒城县洪庙、华庄及桐城市汪洋水库、毛花岩等地（图 1）。在该带西部地区，如金寨县燕子河、鹿吐石铺、斑竹等地也发现有退变很强的榴辉岩（表现为含石榴子石斜长角闪岩或石榴辉石岩，早期绿辉石已退变为钠长石和角闪石的合晶）（刘贻灿等，2000b）。

本区榴辉岩主要有两种产状，一是产于变形较强（面理化）的橄榄岩中，如黄尾河（样号为 98701 和 98702）、饶钹寨（样号为 99104-1 和 99104-2）等地；另一种是产于片麻岩中，如百丈岩（样号为 98121）、华庄（样号为 98122-3）、汪洋水库（样号为 98WY）及毛花岩（样号为 9801）等地。二者的远围岩均为正片麻岩（如英云闪长质片麻岩或条带状片麻岩及二长花岗质片麻岩等）。榴辉岩多呈构造透镜体产于围岩中，变形较强和已面理化，靠近核部较新鲜（但保存较少而给研究者带来一定困难），向透镜体边部退变强烈（变成榴闪岩或含石榴子石斜长角闪岩等）。

1.2 岩石学特征

本区榴辉岩中矿物有石榴子石、绿辉石、金红石、文石、蓝晶石、石英（或柯石英假象？）、单斜辉石（透辉石）、斜方辉石（紫苏辉石）、尖晶石、角闪石、斜长石和黑云母等。其中，99104-2、9801、98701 和 98702 样较新鲜，主要由石榴子石、绿辉石和金红石等组成，含少量石英、紫苏辉石、尖晶石、斜长石及角闪石等，9801 样含有较多金红石；98121、98122-3 和 98WY 样退变较强，主要由石榴子石、韭闪质角闪石和斜长石等组成，含少量绿辉石（<5%）和金红石等，而残留的绿辉石大多呈包体形式存在于石榴子石中（图版 I-1、2）。本区榴辉岩的最大特点是发育后成合晶和冠状体结构，如绿辉石退变为钠长石和透辉石的后成合晶，随压力降低，透辉石进一步分解为角闪石和钠长石的合晶；石榴子石的冠状体结构主要由细粒的角闪石+斜长石等组成的内环后成合晶以及非常细粒的透辉石+斜方辉石+斜长石等组成的外环后成合晶所构成（图版 I-2、3）。根据岩石结构和矿物之间的相互关系，大致可划分出三期变质矿物共生组合：①榴辉岩相峰期变质阶段，以石榴子石和其中的矿物包体为代表，其组合为石榴子石+绿辉石+金红石±文石±蓝晶石+石英（或柯石英假象？）（图版 I-1～4）。②高压麻粒岩相退变质阶段，以细粒的后成合晶组合为代表，主要由石榴子石与绿辉石等矿物反应形成：Gt + Omp → Di + Opx(Hy) + Pl、Omp(富硅) → Na-Cpx(Di) + Qtz、Gt → Pl(An) + Ol、Gt(Alm) + Rt → Ilm + Spl + Qtz 及 Ru → Ilm，其组合主要为石榴子石+透辉石+紫苏辉石+斜长石+钛铁矿±尖晶石±刚玉±橄榄石等（图版 I-1～5）。③以角闪石交代单斜辉石(透辉石)（图版 I-5、6）等为代表，其相应的退变质反应为：Gt + Qtz + H₂O → Hbl + Pl + Mt、Di → Hbl + Pl 及 Gt + Di + H₂O → Hbl + Pl，其组合为角闪石+斜长石+磁铁矿等（图版 I-1～4）。

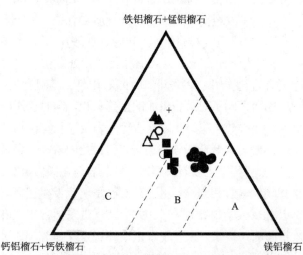

图 2 榴辉岩中石榴子石成分分类（据 Coleman, 1965）

橄榄岩中榴辉岩：●-黄尾河，■-饶钹寨；片麻岩中榴辉岩：△-百丈岩，▲-华庄，○-汪洋水库，+-毛花岩 Eclogite in the peridotite: ●-Huangweihe, ■-Naobozhai; eclogite in the gneiss: △-Baizhangyan, ▲-Huazhuang, ○-Wangyang reservoir, +-Maohuayan

2 矿物化学

2.1 石榴子石

本区榴辉岩中石榴子石成分都属于铁铝榴石-镁铝榴石系列，产于橄榄岩和片麻岩中榴辉岩分别相当于 Coleman(1965)的 B 型和 C 型榴辉岩（图 2）。不同地点或不同世代石榴子石成分仍有差异。其中铁铝榴石端元组分为 25 mol%～57 mol%、镁铝榴石端元组分为 18 mol%～57 mol%和钙铝榴石端元组分为 3 mol%～30 mol%。具代表性石榴子石的电子探针分析结果列于表 1 中。大多数石榴子石成分相近,变化不太大,但个别样品（如 99104-2）中石榴子石成分出现分带现象。对饶钹寨榴辉岩的一个石榴子石颗粒（半径为 400 μm 左右）作较密的电子探针点分析（表 2），其中心含 Ca 较高，边部相对较低，由边缘→中心：CaO 含量由 7.60%→9.61%→10.86%→9.70%→10.33%→10.22%，MnO 含量由 0.78%→0.54%→0.43%→0.23%→0.36%→0.33%，钙铝榴石端元组分由 10.6→15.5→18.9→24.7→23.7→29.8。即石榴子石核部所受到的变质压力较高，反映了变质地体在峰期变质之后的抬升降压过程（具体见后文）。它说明石榴子石的边部组成受退变质的影响，这对榴辉岩的变质温度计算（见后文）和 Sm-Nd 年龄测定（刘贻灿等，2001）都会产生影响。

2.2 单斜辉石

主要有两种：绿辉石和透辉石。绿辉石大多以包体形式产于石榴子石中（图版 I-1、2），部分以基质形式存在（图版 I-3、4）；透辉石主要以后成合晶形式与紫苏辉石、斜长石和钛铁矿等共存（图版 I-2～4），少数以基质形式存在（图版 I-5～8）。电子探针分析表明，绿辉石 Na$_2$O 为 2.51%～8.82%、Al$_2$O$_3$ 为 4.08%～10.84%、硬玉端元组分大多为 Jd=20 mol%～52 mol%（表 3，图 3）；透辉石 Na$_2$O 为 0.18%～0.79%、Al$_2$O$_3$ 为 1.04%～4.37%、硬玉端元组分大多为 Jd=1.0 mol%～3.0 mol% (表 4)。二者的区别在于前者富 Al、Na，也就是说，随着压力降低，单斜辉石中 Na$_2$O 和 Al$_2$O$_3$ 含量明显降低。

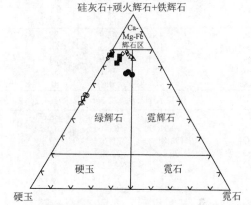

图 3 大别山北部榴辉岩中绿辉石的 WEF-JD-AE 图
■-饶钹寨、□-黄尾河 98701、◇-黄尾河 98702、△-百丈岩、●-华庄、○-汪洋水库

2.3 斜方辉石

主要产于石榴子石边部的后成合晶中，常与斜长石、尖晶石等矿物相共生（图版 I-2～5）。所有的斜方辉石都是石榴子石和绿辉石分解的产物，都是紫苏辉石。不同的样品之间，紫苏辉石的成分稍有差别（见表 4）：Na$_2$O 为 0～0.13%、硬玉端元组分为 Jd=0～1 mol%、Mg/(Mg+Fe) 值为 0.60～0.85；有些紫苏辉石的 Al$_2$O$_3$ 含量较高，最高可达 5.30%。

2.4 角闪石类

角闪石是晚期后成合晶中的主要矿物，与紫苏辉石、尖晶石、斜长石等一同形成交生结构（图版 I-1～5）。另外，广泛存在交代透辉石而形成的角闪石（图版 I-5、6）。角闪石类电子探针分析见 4。根据 Leak（1978）分类，本区榴辉岩中角闪石主要为钙质角闪石，并且大多数为韭闪石（Pargasite），部分为镁角闪石（Magnesio-Hbl）、镁钙闪石（Tschermakite）和镁–绿钠闪石（Magnesio-Hastingsite）等。这些角闪石绝大多数是在变质作用演化晚期形成的。

表 1 代表性石榴子石的电子探针分析（%）

点号	98121-gt10	98121gt-15	99104gt-10	99104gt-11	99104gt-5	99104gt-7	99104gt-9	99104gt-3	98122-3gt-1-6
SiO$_2$	37.26	38.23	39.03	38.75	38.80	38.49	37.86	38.46	37.23
TiO$_2$	0.11	0.03	0.10	0.02	0.01	0.08	0.05	0.19	0.05
Al$_2$O$_3$	22.22	23.16	22.83	23.46	23.22	22.40	23.30	23.04	22.03
FeO	19.91	22.25	17.74	17.76	18.28	17.25	18.88	17.30	24.12
Fe$_2$O$_3$	3.45	0.44	1.65	0.24	0.82	3.97	1.34	2.39	1.99
MnO	0.18	0.28	0.36	0.33	0.46	0.53	0.23	0.44	0.53
MgO	4.37	4.65	8.47	8.51	8.02	8.92	8.03	8.77	4.42
CaO	12.60	11.43	10.32	10.22	10.35	9.61	9.70	10.06	8.67
Na$_2$O	0.14	0.07	0.07	0.00	0.03	0.02	0.03	0.10	0.22
Total	100.24	100.54	100.57	99.29	99.99	101.27	99.42	100.75	99.26

续表

点号	98121-gt10	98121gt-15	99104gt-10	99104gt-11	99104gt-5	99104gt-7	99104gt-9	99104gt-3	98122-3gt-1-6
O					12				
Si	2.888	2.940	2.940	2.943	2.941	2.893	2.891	2.897	2.931
Al^{IV}	0.112	0.060	0.060	0.057	0.059	0.107	0.109	0.103	0.069
Al^{VI}	1.917	2.037	1.965	2.041	2.013	1.875	1.986	1.934	1.973
Fe3	0.201	0.025	0.094	0.014	0.047	0.224	0.077	0.135	0.118
Ti	0.006	0.002	0.005	0.001	0.001	0.005	0.003	0.011	0.003
Fe2	1.291	1.431	1.118	1.128	1.159	1.084	1.206	1.086	1.588
Mg	0.506	0.533	0.952	0.964	0.906	1.000	0.914	0.981	0.518
Mn	0.012	0.019	0.023	0.021	0.030	0.034	0.015	0.028	0.035
Ca	1.047	0.942	0.833	0.832	0.841	0.774	0.793	0.810	0.732
Na	0.021	0.011	0.010	0.000	0.004	0.003	0.005	0.014	0.034
Alm	45.2	48.9	38.2	33.2	39.5	35.7	37.3	34.6	55.3
And	10.6	1.3	4.8	0.76	2.4	12.0	4.2	7.3	6.2
Gross	26.1	30.9	23.7	29.8	26.2	15.5	24.7	21.8	19.3
Prope	17.7	18.2	32.5	35.4	30.9	35.6	33.3	35.3	18.0
Spess	0.4	0.6	0.8	0.8	1.0	1.2	0.5	1.0	1.2
样号	98122-3gt-2	98122-3gt-5	99104-2gt-2	99104-1gt	98WY-gt-19	98701-1gt	98702-2-1r	98702-2-2m	98702-2-3c
SiO_2	37.30	36.64	38.67	37.85	36.74	38.30	40.27	40.67	40.85
TiO_2	0.01	0.00	0.05	0.06	0.00	0.00	0.00	0.00	0.00
Al_2O_3	21.81	21.70	23.30	22.06	22.76	23.89	20.61	21.12	21.05
FeO	25.26	22.82	20.15	23.43	22.60	14.39	15.13	15.10	15.30
Fe_2O_3	1.56	4.52	0.00	2.05	3.63	4.48	0.06	0.77	0.91
MnO	0.28	0.32	0.62	0.56	0.67	0.51	0.55	0.39	0.38
MgO	3.88	5.18	6.87	7.27	4.84	12.97	9.66	10.22	10.14
CaO	9.24	8.77	9.78	6.32	9.46	5.90	11.90	11.65	11.75
Na_2O	0.08	0.05	0.03	0.07	0.00	0.04	0.00	0.00	0.00
Total	99.42	100.00	99.47	99.67	100.7	100.48	98.18	99.92	100.31
O					12				
Si	2.943	2.868	2.966	2.933	2.851	2.836	3.072	3.047	3.050
Al^{IV}	0.057	0.132	0.034	0.067	0.149	0.164	0.000	0.000	0.000
Al^{VI}	1.970	1.870	2.070	1.947	1.930	1.920	1.852	1.863	1.851
Fe3	0.092	0.266	0.000	0.120	0.212	0.000	0.000	0.038	0.044
Ti	0.001	0.000	0.003	0.003	0.000	0.000	0.000	0.000	0.000
Fe2	1.667	1.494	1.293	1.518	1.467	1.141	0.969	0.951	0.963
Mg	0.456	0.605	0.785	0.839	0.560	1.432	1.099	1.141	1.129
Mn	0.019	0.021	0.040	0.037	0.044	0.032	0.036	0.025	0.024
Ca	0.782	0.736	0.804	0.525	0.786	0.468	0.973	0.935	0.940
Na	0.013	0.008	0.004	0.011	0.000	0.006	0.000	0.000	0.000
Alm	57.0	52.3	44.2	52.0	49.6	37.1	31.5	31.2	31.5
And	4.7	14.0	0.0	6.2	11.5	0.0	0.0	2.0	2.3
Gross	22.0	11.8	27.5	11.8	17.0	15.1	31.6	28.6	28.5
Prope	15.6	21.2	26.9	28.7	20.3	46.5	35.7	37.4	36.9
Spess	0.6	0.7	1.4	1.3	1.6	1.1	1.2	0.8	0.8

表 2 饶钹寨榴辉岩（99104-2）具有成分分带的石榴子石的电子探针分析（%）

点号	99104-2-1	99104-2-2	99104-2-3	99104-2-4	99104-2-5	99104-2-6
SiO_2	38.52	38.49	39.46	37.86	39.03	38.75
TiO_2	0.10	0.08	0.13	0.05	0.10	0.02
Al_2O_3	22.53	22.40	19.27	23.30	22.83	23.46
FeO	18.32	17.25	17.48	18.88	17.74	17.76
Fe_2O_3	3.62	3.97	3.87	1.34	1.65	0.24
MnO	0.78	0.54	0.43	0.23	0.36	0.33
MgO	9.45	8.93	8.37	8.03	8.47	8.51
CaO	7.60	9.61	10.86	9.70	10.33	10.22
Na_2O	0.09	0.02	0.12	0.03	0.07	0.00
Total	100.92	101.29	99.99	99.42	100.58	99.29
O			12			
Si	2.902	2.893	3.021	2.891	2.940	2.943
Al^{IV}	0.098	0.107	0.000	0.109	0.060	0.057
Al^{VI}	1.901	1.875	1.738	1.986	1.965	2.041
Fe3	0.205	0.224	0.223	0.077	0.094	0.014
Ti	0.001	0.005	0.007	0.003	0.005	0.001
Fe2	1.155	1.084	1.119	1.206	1.118	1.128
Mg	1.062	1.000	0.955	0.914	0.952	0.964
Mn	0.050	0.034	0.028	0.015	0.023	0.021
Ca	0.613	0.774	0.891	0.793	0.833	0.832
Na	0.013	0.003	0.019	0.005	0.010	0.000

续表

点号	99104-2-1	99104-2-2	99104-2-3	99104-2-4	99104-2-5	99104-2-6
Alm	40.1	35.7	36.4	37.3	38.2	33.2
And	10.7	12.0	11.4	4.2	4.8	0.8
Gross	10.6	15.5	18.9	24.7	23.7	29.8
Prope	36.9	35.6	32.4	33.3	32.5	35.4
Spess	1.7	1.2	0.9	0.5	0.8	0.8

注：点号 99104-2-1~99104-2-6 表示从石榴子石边缘→中心。

表3 大别山北部榴辉岩中代表性绿辉石的电子探针分析（%）

样号 产状	98WY-1 包体	99104-1 包体	99104-2-1 包体	99104-2-2 基质	98122-3 包体	98121-1 包体	98121-2 包体	98701-1 包体	98701-2 包体	9801 基质	98702-1 包体	98702-2 基质	98WY-2 包体
SiO_2	55.29	52.23	54.45	54.86	54.57	53.06	54.29	55.49	55.89	53.50	53.45	54.44	53.57
TiO_2	0.11	0.22	0.02	0.17	0.11	0.07	0.01	0.01	0.09	0.00	0.08	0.06	0.12
Al_2O_3	10.84	6.26	7.53	5.88	4.89	4.29	4.08	10.78	10.68	9.52	4.85	4.60	9.13
FeO	6.18	6.50	2.54	4.20	4.00	4.02	7.79	1.64	6.39	7.44	3.92	3.45	7.30
Fe_2O_3	0.00	0.00	2.05	0.65	5.10	4.74	1.03	5.71	0.20	0.00	0.00	0.00	0.00
MnO	0.00	0.00	0.15	0.00	0.00	0.00	0.06	0.03	0.01	0.06	0.04	0.09	0.00
MgO	7.88	10.51	11.24	11.98	9.51	11.32	11.70	7.29	7.29	7.84	14.23	14.11	8.02
CaO	11.58	17.75	17.66	19.35	17.04	18.64	19.15	11.71	11.78	13.14	19.91	19.17	12.85
Na_2O	6.58	3.07	4.28	3.30	4.84	3.31	2.51	7.91	6.98	7.85	3.79	3.60	8.82
K_2O	0.00	0.02	0.01	0.00	0.02	0.01	0.00	0.01	0.02	0.02	0.00	0.00	0.00
Total	98.46	96.71	99.78	100.39	100.09	99.46	100.62	100.58	99.33	99.37	100.27	99.52	99.81
O						6							
Si	2.007	1.971	1.962	1.978	1.995	1.959	1.987	1.970	2.012	1.911	1.902	1.954	1.891
Al^{IV}	0.000	0.029	0.038	0.02	0.005	0.041	0.013	0.030	0.000	0.089	0.098	0.046	0.109
Al^{VI}	0.463	0.249	0.282	0.228	0.205	0.145	0.163	0.421	0.453	0.311	0.105	0.149	0.270
Fe3	0.000	0.000	0.052	0.014	0.138	0.129	0.026	0.152	0.005	0.000	0.000	0.000	0.000
Ti	0.003	0.006	0.001	0.005	0.003	0.002	0.000	0.000	0.002	0.000	0.002	0.002	0.003
Fe2	0.188	0.205	0.080	0.130	0.125	0.126	0.241	0.049	0.193	0.222	0.117	0.103	0.215
Mg	0.426	0.591	0.604	0.644	0.518	0.623	0.638	0.386	0.391	0.417	0.755	0.755	0.422
Mn	0.000	0.005	0.000	0.000	0.000	0.000	0.002	0.001	0.000	0.002	0.001	0.003	0.000
Ca	0.450	0.718	0.682	0.748	0.667	0.737	0.751	0.445	0.454	0.503	0.759	0.737	0.486
Na	0.463	0.225	0.299	0.231	0.343	0.237	0.178	0.545	0.486	0.544	0.261	0.251	0.604
K	0.000	0.001	0.000	0.000	0.001	0.001	0.000	0.000	0.001	0.001	0.000	0.000	0.000
WEF	53.5	77.2	69.6	76.7	65.6	75.9	82.1	44.7	51.6	51.3	75.7	76.1	48.2
JD	46.5	22.8	25.7	21.9	20.6	12.8	15.4	40.6	47.8	48.7	24.3	23.9	51.8
AE	0.0	0.0	4.7	1.4	13.8	11.3	2.5	14.7	0.6	0.0	0.0	0.0	0.0

表4 大别山北部榴辉岩中部分斜方辉石、单斜辉石、斜长石及角闪石电子探针分析（%）

样号	斜方辉石			单斜辉石		斜长石				角闪石				
	98122-3	98701	98702-2	98122-3	98701	99104-1	98121-9	98702	98122-3	99104	98701	98WY	98122	98702
SiO_2	53.03	49.49	53.97	52.95	50.72	58.08	48.31	52.04	51.68	39.78	40.97	45.46	42.53	41.46
TiO_2	0.00	0.06	0.09	0.16	0.60	0.00	0.00	0.00	0.05	4.45	2.57	0.32	0.04	0.43
Al_2O_3	1.28	5.30	0.00	1.26	4.37	26.10	26.07	29.86	29.77	16.00	14.79	9.83	13.89	13.53
FeO	24.40	8.72	20.31	9.11	1.25	0.55	2.14	0.00	0.27	10.78	10.97	15.00	14.21	11.92
Fe_2O_3	0.00	7.09	0.00	0.87	4.00					0.23	0.19	0.01	0.14	0.23
MnO	0.55	0.23	0.49	0.27	0.13	0.05	2.63	0.00	0.00	12.07	13.54	12.16	12.75	12.86
MgO	20.53	28.02	24.07	12.94	15.14	0.02	2.63	0.00	0.04	12.07	13.54	12.16	12.75	12.86
CaO	0.40	0.23	0.49	21.58	23.39	7.41	18.40	13.55	11.97	11.46	11.44	11.96	11.49	12.14
Na_2O	0.13	0.09	0.00	0.79	0.57	6.35	1.93	4.34	5.40	3.06	4.42	1.06	2.38	2.81
K_2O	0.02	0.00	0.01	0.02	0.02	0.07	0.01	0.03	0.00	0.00	0.04	0.19	0.02	0.12
Total	100.34	99.23	99.43	99.95	100.19	98.63	99.49	99.82	99.18	97.83	98.93	95.99	97.45	95.50
O		6			6		8					23		
Si	1.987	1.786	2.002	1.979	1.856	10.501	9.092	9.482	9.484	5.501	5.926	6.721	6.149	5.783
Al	0.057	0.226	0.000	0.054	0.188	5.558	5.779	6.407	6.435	2.755	2.519	1.711	2.364	2.718
Fe3	0.000	0.192	0.000	0.036	0.110	0.000	0.000	0.000	0.000	0.234	0.272	0.650	1.099	1.039
Ti	0.000	0.002	0.002	0.004	0.017	0.000	0.000	0.000	0.007	0.488	0.028	0.000	0.004	0.126
Fe2	0.765	0.263	0.630	0.274	0.039	0.083	0.337	0.000	0.042	1.081	1.055	1.200	0.618	0.204
Mg	1.147	1.507	1.330	0.721	0.826	0.006	0.737	0.000	0.001	2.623	2.919	2.681	2.748	3.094
Mn	0.017	0.007	0.015	0.009	0.004	0.007	0.000	0.000	0.000	0.028	0.023	0.002	0.017	0.035
Ca	0.016	0.009	0.019	0.864	0.917	1.435	3.712	2.645	2.353	1.790	1.773	1.894	1.780	1.837
Na	0.010	0.006	0.000	0.057	0.040	2.227	0.702	1.533	1.922	0.863	1.239	0.303	0.667	0.750
K	0.001	0.000	0.000	0.001	0.001	0.015	0.002	0.007	0.001	0.000	0.007	0.036	0.004	0.000
WEF	99.0	99.3	99.8	94.2	95.7									
JD	1.0	0.0	0.0	2.9	1.2									
AE	0.0	0.7	0.2	2.9	3.1									

注：斜长石、角闪石中的 FeO 为全铁。

3 变质温、压条件与 P-T 轨迹

3.1 变质温、压条件

3.1.1 超高压变质线索

尽管大别山北部榴辉岩中尚未发现有柯石英或金刚石，但下列一些矿物或矿物组合特征表明，它们也可能经历了超高压变质作用。

（1）石榴子石成分：根据石榴子石成分，作出石榴子石成分与压力相关图(Okay, 1995)，大别山北部榴辉岩中石榴子石成分大致可分出两类（图4）：一类形成压力≥2.3 GPa，为榴辉岩相峰期变质形成，相当于南部超高压带榴辉岩之柯石英/石英榴辉岩变质阶段；另一类形成压力≤1.5 GPa，主要为具有成分分带石榴子石（如99104-2样品）的边部成分和部分为强退变样品（如98122-3）中的石榴子石（见表1、2），可能为高压麻粒岩相阶段形成的。

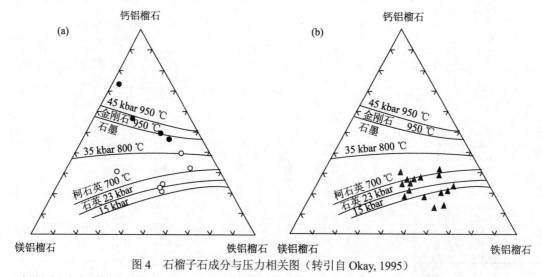

图4 石榴子石成分与压力相关图（转引自 Okay, 1995）

（a）大别山南部超高压带榴辉岩中石榴子石：●-含金刚石榴辉岩，○-柯石英/石英榴辉岩；(b) 大别山北部榴辉岩中石榴子石（▲）

（2）榴辉岩相矿物共生组合：本区榴辉岩之石榴子石中含石英、绿辉石、金红石和文石等包体以及作者在庐镇关以南与桐城之间塘湾附近大理岩中发现有含钛-斜硅镁石、镁橄榄石、金红石和文石等矿物组合（刘贻灿等，2000b）[可能类似于哈萨克斯坦不含金刚石的白云石大理岩，其形成温度、压力最小为800 ℃和2.5 GPa（Ohta et al., 1998）]。类似大理岩在大别山北部其他地方也有分布，如黄尾河等地。由于该大理岩含较多镁橄榄石和常伴有蛇纹石化，此种大理岩有可能是由超基性岩经深俯冲（至地幔深度）发生碳酸盐交代作用形成的（有待于进一步研究）；而且，其中矿物组合至少也可以划分为榴辉岩相、麻粒岩相和角闪岩相三期（详细资料，将另文发表）。

钛-斜硅镁石曾被当作一个低压矿物，但 Thompson（1992）发现它是一个地幔深度的高密度的含水硅酸盐矿物。钛-斜硅镁石的温压实验(Engi, 1980)也表明，它在大于100 km深度仍稳定。钛-斜硅镁石，以前仅发现于大别山南部超高压带中太湖毛屋的石榴辉石岩及岳西碧溪岭的石榴橄榄岩中（Okay, 1994; Zhang et al., 1995; Xu et al., 1996）。薄片内呈金黄色，呈自形-半自形的粒状，大小为0.25~1.5 μm，与石榴子石、透辉石、橄榄石及菱镁矿相接触，少数则呈包体赋存于石榴子石、透辉石或橄榄石中。太湖毛屋含钛-斜硅镁石的石榴辉石岩的温压条件为约740 ℃和>4.0 GPa (Okay, 1994)，由此表明大别山南部变质的镁铁-超镁铁岩曾俯冲到大于100 km的地幔深度。由此，推测本区榴辉岩相峰期变质压力至少应大于2.5 GPa。

（3）钠质单斜辉石中石英出溶：根据世界上几个超高压变质地体中石英针状体的出溶(Smith, 1984; Smith, 1988; Bakun-Czubarow, 1982; Gayk et al., 1995; Liou et al., 1999)及有关实验结果(Mao, 1971; Khanukhova et al., 1976; Wood and Henderson, 1978; Angel et al., 1988; Zharikov et al., 1984)表明，榴辉岩相 Ca-Na 质单斜辉石中

石英出溶是早期超高压变质的最好证据(Liou et al., 1999)。Gayk 等人(1995)认为 Münchberg Massif 地体高压麻粒岩之单斜辉石中石英出溶形成于 2.5 GPa。

本区榴辉岩中钠质单斜辉石核部常含有石英针状体出溶(大多数为 20~30 μm，少数为 120 μm，呈定向排列)并具有退变的透辉石边(图版 I-7、8)，Na_2O 含量从核部向边部递减(如从 3.19%→0.34%)，而且透辉石边无矿物包体(刘贻灿，2000)。这可能表明，核部(早期绿辉石)代表形成于超高压条件下的硅超饱和矿物，并随压力降低和硅释放而出溶石英。

以上分析表明，本区榴辉岩可能形成于 $P \geqslant 2.5$ GPa 的超高压条件下。此外，美国斯坦福大学 Tsai 和 Liou 等(2000)虽没有在研究区发现新鲜榴辉岩但发现榴辉岩相残留体(eclogite-facies relics)并认为它们曾经过超高压变质作用。实际上，在该带东延的威海地区曾发现含柯石英的麻粒岩(Wang et al., 1993b)，而该麻粒岩是由榴辉岩经过麻粒岩相退变而来的，也说明该带中榴辉岩曾经过超高压变质阶段。

3.1.2 变质温度、压力计算

以 2.5 GPa 作为计算本区榴辉岩峰期变质温度的压力值，利用石榴子石-单斜辉石地质温度计(Ellis et al., 1979; Raheim et al., 1974; Krogh, 1988)计算。三种方法对同一矿物对计算的结果十分接近(见表 5)，T=595~874 ℃，似乎可分出高温(>700 ℃，甚至 800~900 ℃)和低温(≤600 ℃)两类榴辉岩。部分榴辉岩的温度比较低(如 98121、98122)，是与它们为强退变样品有关(即受退变质影响)还是形成温度本来就比较低？还有待于进一步研究。

表 5 大别山北部榴辉岩相峰期变质温度计算结果（P=2.5 GPa）

样号	石榴子石		绿辉石		温度(℃)				K_D
	Fe^{2+}	Mg	Fe^{2+}	Mg	EG	RG	K	平均	
99104-2	1.113	0.927	0.080	0.604	739	696	705	713	9.0
98121	1.291	0.506	0.126	0.623	727	630	686	681	12.6
98122	1.673	0.462	0.125	0.518	617	599	570	595	15.0
98WY	1.467	0.560	0.188	0.426	855	796	834	828	5.9
98701	1.141	1.432	0.049	0.386	819	847	758	808	4.9
98702	0.963	1.129	0.117	0.755	908	817	897	874	5.4
9801	1.745	0.563	0.222	0.417	771	801	707	760	5.8

注：用石榴子石-绿辉石矿物对电子探针分析结果计算，其中 EG 据 Ellis and Green(1979); RG 据 Raheim and Green(1974); K 据 Krogh(1988)。

利用二辉石对(Wood and Banno, 1973)计算麻粒岩相变质温度为 817~909 ℃，利用石榴子石-斜方辉石对(Wood, 1974)计算麻粒岩相变质压力为 1.10~1.37 GPa，见表 6。

根据共存的角闪石-斜长石温、压计(Plyusnina, 1982)以及角闪石中 NaM_4-Al^{IV}(Brown, 1977)估算出角闪岩相变质温度、压力分别为 T=500~600 ℃、P=0.5~0.6 GPa，见表 6。

表 6 大别山北部榴辉岩退变质阶段温度和压力计算结果

样号	方法	矿物对	温度(℃)	压力(GPa)
98701-1	Wood and Banno (1973)	Cpx-Opx	820	
	Wood (1974)	Gt-Opx		1.22
98701-2	Wood and Banno (1973)	Cpx-Opx	845	
	Wood (1974)	Gt-Opx		1.10
98701-3	Wood and Banno (1973)	Cpx-Opx	909	
	Wood (1974)	Gt-Opx		1.37
98122-3	Wood and Banno (1973)	Cpx-Opx	817	
99104-2	Plyusnina (1982)	Amph-Pl	600	0.6
	Brown (1977)	Amph 中 NaM_4-Al^{IV}		0.6
98WY	Plyusnina (1982)	Amph-Pl	500-600	0.5-0.6

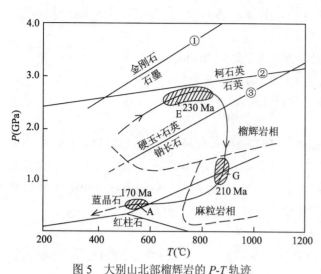

图 5 大别山北部榴辉岩的 P-T 轨迹

① Diam = Graph 转变线据 Kennedy et al.(1968); ② Cs = Q 转变线据 Chopin (1984); ③ Jd + Q = Ab 转变线据 Holland (1980)
E-榴辉岩峰期变质阶段; G-高压麻粒岩相变质阶段; A-角闪岩相变质阶段

3.2 P-T 轨迹

以上变质岩石学分析及变质温度、压力估算表明，本区榴辉岩至少经历了三个变质阶段，即：①榴辉岩相峰期变质阶段，$P \geq 2.5$ GPa、$T=595 \sim 874$ ℃；②高压麻粒岩相变质阶段，$P=1.1 \sim 1.37$ GPa、$T=817 \sim 909$ ℃；③角闪岩相变质阶段，$P=0.5 \sim 0.6$ GPa、$T=500 \sim 600$ ℃。据此，可以做出一条顺时针的 P-T 轨迹（图 5），它反映了本区榴辉岩在榴辉岩相峰期变质之后经历了近等温减压（或稍升温减压）和降温减压变质过程；它与南部超高压带中榴辉岩的 P-T 轨迹相比较，有所差异，表明它们变质作用条件的差异。该区榴辉岩的锆石 U-Pb 年龄为 230±6 Ma（刘贻灿等，2000a）及 Sm-Nd 等时线（石榴子石+绿辉石+全岩）年龄为 210 Ma（刘贻灿等，2001），表明它们均形成于印支期，且与南部超高压变质岩一样都是扬子与华北两个大陆板块相碰撞的产物。

4 结论与讨论

（1）大别山北部镁铁-超镁铁质岩带中榴辉岩主要有两种产状，一是产于变形较强（面理化）的橄榄岩中，如黄尾河、饶钹寨等地；另一种是产于片麻岩中，如洪庙、华庄及汪洋水库等地。其中，绿辉石中硬玉端元组分大多为 Jd=20 mol%～52 mol%。石榴子石的成分特征表明，产于橄榄岩和片麻岩中榴辉岩分别相当于 Coleman 的 B 型和 C 型榴辉岩。

（2）本区榴辉岩可划分出三期变质矿物共生组合：①石榴子石+绿辉石+金红石±文石±蓝晶石+石英（或柯石英假象?）；②石榴子石+透辉石+紫苏辉石+钛铁矿+尖晶石+斜长石±刚玉±橄榄石等；③角闪石+斜长石+磁铁矿等。

（3）石榴子石成分特征及钠质单斜辉石中石英针状体出溶等表明，本区榴辉岩早期可能经历过超高压变质作用，且至少经历了三个变质阶段，即：①榴辉岩相峰期变质阶段，$P \geq 2.5$ GPa、$T=595 \sim 874$ ℃；②高压麻粒岩相变质阶段，$P=1.1 \sim 1.37$ GPa、$T=817 \sim 909$ ℃；③角闪岩相变质阶段，$P=0.5 \sim 0.6$ GPa、$T=500 \sim 600$ ℃。经历了一条顺时针的 P-T 演化，且在榴辉岩相峰期变质之后经历了近等温减压（或稍升温减压）和降温减压变质过程；它与南部超高压带中榴辉岩的变质作用条件有一定差异。

（4）大别山北部榴辉岩的发现，说明北部至少其中一部分可能属于扬子俯冲陆壳的一部分。

（5）该区榴辉岩独特的麻粒岩相退变质阶段，表明榴辉岩在折返初期并未能上升到中上地壳，而处于下地壳，因而有较高的退变质温度。由于该麻粒岩相退变质阶段仅以细粒后成合晶形式出现，未重结晶，因而它可能在下地壳深度停留时间不很长，就又被进一步构造抬升至中上地壳。这与南部超高压变质岩有不同的 P-T 演化史。

（6）研究区似乎存在高温（>700 ℃，甚至 800～900 ℃）和低温（≤600 ℃）两类榴辉岩。温度较低是由退变质作用（如出现石榴子石成分分带及部分榴辉岩强退变质而引起绿辉石成分变化等）造成的，还是它们的榴辉岩相峰期变质时温度较低？以及二者之间关系如何等，还有待于进一步研究。

致谢

本文属刘贻灿博士论文的一部分，是在李曙光教授的指导下完成的；电子探针分析由南京大学王汝成教授和意大利都灵大学 Franco Rolfo 博士帮助完成；文中插图由陆益群和王晓梅女士清绘，在此一并表示感谢！

参考文献

陈廷愚, 牛宝贵, 刘志刚, 等. 1991. 大别山腹地燕山期岩浆作用和变质作用的同位素年代学研究及其地质意义. 地质学报, 65 (4): 329~336.
董树文, 孙先如, 张勇, 等. 1993. 大别山造山带的基本结构. 科学通报, 38 (6): 542~545.
刘贻灿, 徐树桐, 江来利, 等. 1997. 大别山造山带北部麻粒岩相岩石的若干特征. 安徽地质, 7 (2): 7~14.
刘贻灿, 徐树桐, 江来利, 等. 1998. 大别山北部斜长角闪岩类的地球化学特征及形成构造背景. 大地构造与成矿学, 22 (4): 323~331.
刘贻灿, 徐树桐, 江来利, 等. 1999. 大别山北部蛇绿岩的地球化学制约. 矿物岩石, 19 (1): 68~73.
刘贻灿, 李曙光, 徐树桐, 等. 2000a. 大别山北部榴辉岩和英云闪长质片麻岩锆石 U-Pb 年龄及多期变质增生. 高校地质学报, 6 (3): 417~423.
刘贻灿, 徐树桐, 李曙光, 等. 2000b. 大别山北部鹿吐石铺含石榴子石斜长角闪岩类的变质特征及 Rb-Sr 同位素年龄. 安徽地质, 10 (3): 194~198.
刘贻灿, 李曙光, 徐树桐, 等. 2001. 大别山北部榴辉岩的 Sm-Nd 年龄测定及其对麻粒岩相退变质时间的制约. 地球化学, 30 (1): 79~87.
牛宝贵, 富云莲, 刘志刚, 等. 1994. 桐柏-大别山主要构造热事件及 $^{40}Ar/^{39}Ar$ 地质定年研究. 地球学报, 15(1-2): 20~34.
魏春景, 单振刚, 张立飞, 等. 1997. 北大别榴辉岩的确定及其地质意义. 科学通报, 42 (17): 1834~1835.
徐树桐, 苏文, 刘贻灿, 等. 1991. 大别山东段高压变质岩中的金刚石. 科学通报, 36 (17): 1318~1321.
徐树桐, 江来利, 刘贻灿, 等. 1992. 大别山 (安徽部分) 的构造格局和演化过程. 地质学报, 66(1): 1~14.
徐树桐, 刘贻灿, 江来利, 等. 1994. 大别山的构造格局和演化. 北京: 科学出版社, 1~175.
Angel RJ, Gasparik T, Ross NL, et al. 1988. A silica-rich pyroxene phase with six-coordinated silicon. Nature, 335: 156~158.
Bakun-Czubarow N. 1992. Quartz pseudomorphs after coesite and quartz exsolutions in eclogitic omphacites of the Zlote Mountains in the Sudetes (SW Poland). Archiwum Mineral, 48: 3~25.
Brown EH. 1977. The crossite content of Ca-amphibole as a guide to pressure of metamorphism. J. of Petrol., 18(part 1): 53~72.
Chopin C. 1984. Coesite and pure pyrope in high-grade blueschist of the western Alps: a first record and some consequences. Contrib. Mineral. Petrol., 86: 107~118.
Coleman RG, Lee DE, Beatty LB, et al. Eclogites and eclogites: their differences and similarities. Bull. Geol. Soc. Am., 1965, 76: 483~508.
Cong B, Wang Q, Zhai M, et al. 1994. Ultrahigh-pressure metamorphic rocks in the Dabie-Su- Lu region, China: their formation and exhumation. The Island Arc, 3: 135~150.
Ellis D J, Green DH. 1979. An experimental study of the effect of Ca upon garnet-clinopyroxene Fe-Mg exchange equilibration. Contrib. Mineral.Petrol., 71: 13~22.
Engi M, Lindsley DH. 1980. Stability of Titanian clinohumite: experiments and thermodynamic analysis. Contri. Mineral. Petrol., 72: 415-424.
Gayk T, Kleinschrodt R, Langosch A, et al. 1995. Quartz exsolution in clinopyroxene of high-pressure granulite from the Munchberg Massif. Eur. J. Mineral., 7: 1217~1220.
Hacker BR, Ratschbacher L, Webb L, et al. 1998. U/Pb zircon ages contain the architecture of the ultrahigh-pressure Qinling-Dabie orogen, China. Earth and Planetary Science Letters, 161: 215~231.
Holland TJB. 1980. The reaction albite = jadeite + quartz determined experimentally in the range 600–1200 ℃. Am. Mineral., 65: 125~134.
Khanukhova LT, Zharikov VA, Ishbulatov RA, et al. 1976. Excess silica in solid solutions of—pressure clinopyroxenes as shown by experimental study of the system $CaMgSi_2O_6$-$CaAl_2SiO_2$ at 35 kilobars and 1200 ℃. Doklady Earth Sci. Sect., 229: 170~172.
Krogh EJ. 1988. The garnet-clinopyroxene Fe-Mg geothermometer—a reinterpretation of existing experimental data. Contrib. Mineral. Petrol., 99: 44~48.
Leak B E. 1978. Nomenclature of amphibole. Mineral. Mag., 42: 533~563.
Li S, Jagoutz E, Zhang Z, et al. 1995. Structure of high-pressure metamorphic belt in the Dabie Mountains and its tectonic implication. Chinese Science Bulletin, 40(supplement): 138~140.
Liou JG, Zhang RY, Ernst WG, et al. 1999. High-pressure minerals from deeply subducted metamorphic rocks. In: Hemley R.J (ed), Reviews in Mineralogy, 37: 33~96.
Mao HK. 1971. The system jadeite ($NaAlSi_2O_6$)-anorthite ($CaAl_2Si_2O_8$) at high pressures. Carnegie Inst. Year Book, 69: 163~168.
Morimoto N, Fabries J, Ferguson AK, et al. 1988. Nomenclature of pyroxenes. Schweiz. Mineral. Petrog. Mitteil., 68: 95~111.
Ohta M, Ogasawara Y, Katayama, et al. 1998. Petrology of diamond-bearing dolomite marble and diamond-free dolomitic marble from the Kokchetav massif, northern Kazakhstan. International Workshop on UHP Metamorphism and Exhumation. Stanford

University. Abstract, A96～98.

Okay AI, Sengör AMC. 1992. Evidence for intracontinental thrust-related exhumation of he ultrahigh-pressure rocks in China. Geology, 20: 411～414.

Okay AI. 1994. Saphirine and Ti-clinohumite in ultra-high pressure garnet-pyroxenite and eclogite from Dabie Shan, China. Contri. Mineral. Petrol., 116:145-155.

Okay AI. 1995. Paragonite eclogites from Dabie Shan, China: re-equilibration during exhumation? J. Metamorphic. Geol., 13: 449～460.

Okay AI, Sengör AMC. 1993. Tectonics of an ultrahigh-pressure metamorphic terrain: the Dabie Shan/Tongbai Shan orogen, China. Tectonics, 12: 1320～1334.

Okay AI, Xu Shutong, Sengor AMC. 1989. Coesite from the Dabie Shan eclogites, central China. Eur. J. Mineral., 1: 595～598.

Plyusnina LP. 1982. Geotherometry and geobarometry of plagioclase-hornblende bearing assemblage. Contrib. Mineral. Pertol., 80: 140-146.

Raheim A, Green DH. 1974. Experimental determination of the temperature and pressure dependence of the Fe-Mg partition coefficient for coexisting garnet and clinopyroxene. Contrib. Mineral. Petrol., 48: 179～203.

Smith DC. 1984. Coesite in clinopyroxene in the Caledonides and its implications for geodynamics. Nature, 310: 641～644.

Smith DC. 1988. A review of the peculiar mineralogy of the Norwegian coesite-eclogite province, with crystal-chemical, petrological, geochemical and geodynamical notes and an extensive Bibliography. In: Smith, D.C.(ed) Eclogites and Eclogite-facies Rocks. 1～206.

Thompson AS. 1992. Water in the Earth's upper mantle. Nature, 358:295～301.

Tsai Chin-Ho, Liou JG. 2000. Eclogite-facies relics and inferred ultrahigh-pressure metamorphism in the North Dabie complex, central China. American Mineralogist, 85: 1～8.

Tsai Chin-Ho, Liou JG, Ernst WG. 1998. Eclogite facies relics and retrogressed garnet peridotite in the North Dabie complex, central-eastern China, and suggested implications for regional tectonics. International Workshop on UHP Metamorphism and Exhumation. Stanford University. Abstract, A153～154.

Wang Q, Cong B, Zhai M, et al. 1993a. A possible Paleozoic island arc: petrologic evidences from North Dabie gneiss. In: Inst. Of Geol., Academia Sinia (ed), Annual Report of the Laboratory of Lithosphere Tectonic Evolution (1993–1994). Bejing: Seismological Press, 37～47.

Wang Q, Ishiwatari A, Zhao Z, et al. 1993b. Coesite-bearing granulite retrograded from eclogite in Weihai, eastern China: a preliminary study. Eur. J. Mineral., 5: 141～152.

Wang X, Liou JG, Mao HK. 1989. Coesite-bearing eclogites from the Dabie Mountains in central China. Geology, 17: 1085～1088.

Wood BJ. 1974. The solubility of alumina in orthopyroxene coexisting with garnet. Contrib. Mineral. Petrol., 46: 1～15.

Wood BJ, Banno S. 1973. Garnet-orthopyroxene and orthopyroxene-clinopyroxene relationship in simple and complex systems. Contrib. Mineral. Petrol., 42: 109～124.

Wood BJ, Henderson CMB. 1978. Compositions and unit-cell parameters of synthetic non-Stoichimetric tschermakitic clinopyroxene. Am. Mineral., 63: 66～72.

Xu S, Okay AI, Ji S, et al. 1992. Diamonds from the Dabie Shan metamorphic rocks and its implication for tectonic setting. Science, 256: 80～82.

Xu S, Jiang L, Liu Y, et al. 1996. Structural Geology and Ultrahigh Pressure Metamorphic belt of the Dabie Mountains in Anhui Province. 30th IGC Field Trip Guide T328. Beijing: Geological Publishing House, 1～40.

Xu S, Liu Y, Su W, et al. 1998. Eclogite in the northern Dabie Mountains and its tectonic implications. International Workshop on UHP Metamorphism and Exhumation. Stanford University. Abstract, A151.

Xu S, Liu Y, Su W, et al. 2000. Discovery of the eclogite and its petrography in the Northern Dabie Mountain. Chinese Science Bulletin, 45(3): 273～278.

Xu Z. 1987. Etude tectonique et microtectonique de la chaine paleozoique et triasique des Qinlings (Chine). These de Doctorat, Univ. Sci. Tech. Languedoc, Montpeuier.

Zhang R, Liou JG, Cong B. 1995. Ultrahigh-pressure metamorphosed talc-magnesite and Ti-clinohumite-bearing mafic and ultramafic complex in the Dabie Mountains, China. Journal of Petrology, 36:1011～1037.

Zhang RY, Liou JG, Tsai CH. 1996. Petrogenesis of a high-temperature metamorphic Terrain: a new tectonic interpretation for the north Dabie Shan, central China. J. Metamorphic. Geol., 14: 319～333.

Zharikov VA, Ishbulatov RA, Chudinovskikh LT. 1984. High-pressure clinopyroxene and the eclogite barrier. Sov. Geol. Geophys., 25: 53～61.

图版 I 说明

1. 汪洋水库榴辉岩，石榴子石(Gt)中含有绿辉石(Omp)包体，具有角闪石（Hbl）+斜长石（Pl）后成合晶，金红石(Ru)退变为钛铁矿(Ilm)。视域宽度为 3.3 mm，单偏光。

2. 华庄榴辉岩，石榴子石(Gt)中含有绿辉石(Omp)包体，具有斜方辉石+透辉石+斜长石组成的外环后成合晶（Sy1）和角闪石+斜长石+磁铁矿等组成的内环后成合晶（Sy2）。视域宽度为 3.3 mm，单偏光。

3. 百丈岩榴辉岩，石榴子石(Gt)中含有金红石(Ru)包体，石榴子石(Gt)与绿辉石(Omp)之间具有斜方辉石+透辉石+斜长石等组成的后成合晶（Sy1）及角闪石+斜长石组成的后成合晶（Sy2）。视域宽度为 3.3 mm，单偏光。

4. 华庄榴辉岩，石榴子石(Gt)、绿辉石(Omp)及斜方辉石+透辉石+斜长石+角闪石+斜长石组成的后成合晶（Sy）。视域宽度为 3.3 mm，单偏光。

5. 黄尾河榴辉岩（样号为 98702），斜方辉石+尖晶石+斜长石等组成的后成合晶（Sy）以及透辉石（Di）具有角闪石(Hbl)退变边。

6. 黄尾河榴辉岩（样号为 98409-2），石榴子石(Gt)及含有针状石英出溶体的透辉石(Di)具有角闪石(Hbl)退变边。视域宽度为 1.4 mm，单偏光。

7. 黄尾河榴辉岩（样号为 98702-2），石榴子石(Gt)及含有针状石英出溶体(Qtz)的钠质单斜辉石（Na-Cpx）具有无包裹体的透辉石(Di)边。视域宽度为 3.3 mm，单偏光。

8. 黄尾河榴辉岩（样号为 98702），钠质单斜辉石(Na-Cpx)核部有针状石英出溶体(Qtz)和边部退变为透辉石（Di）。视域宽度为 1.4 mm，单偏光。

图版 I

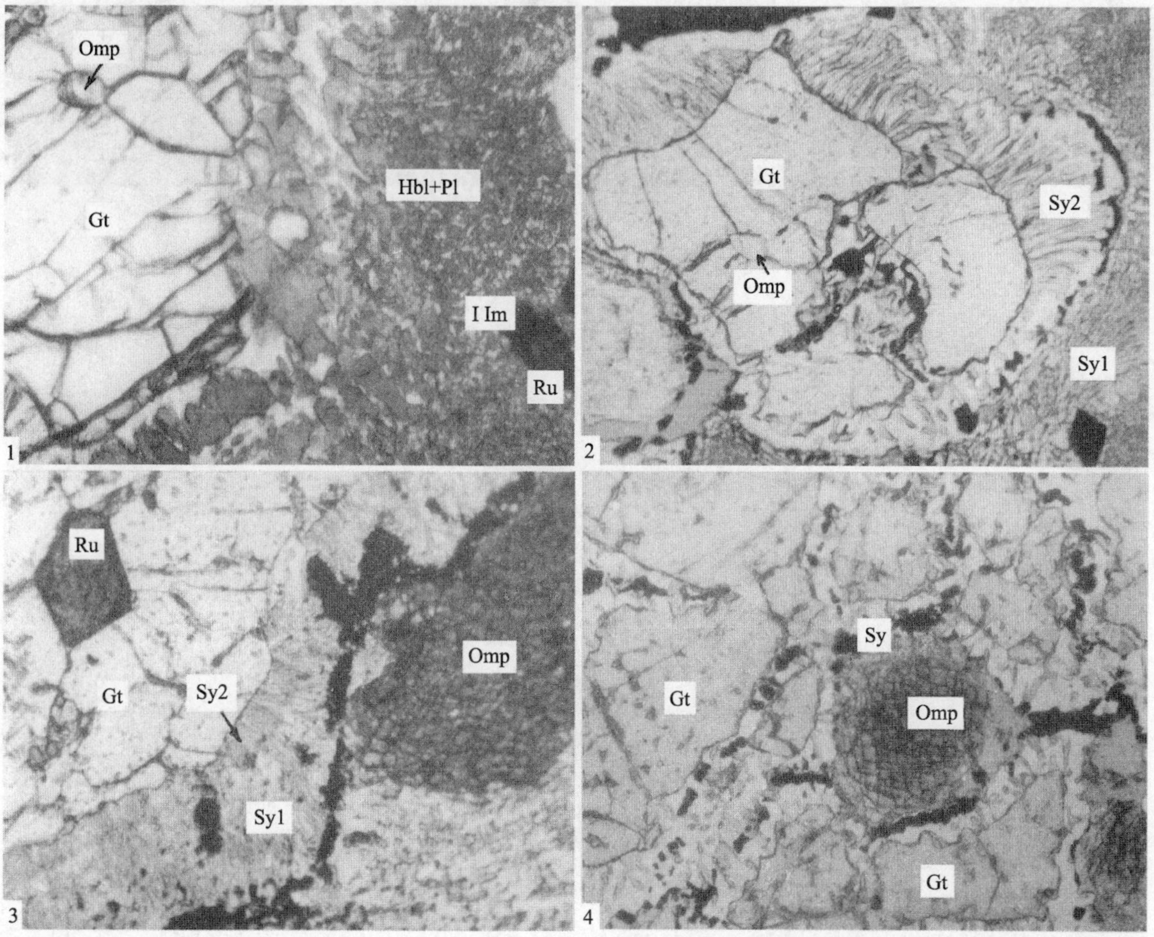

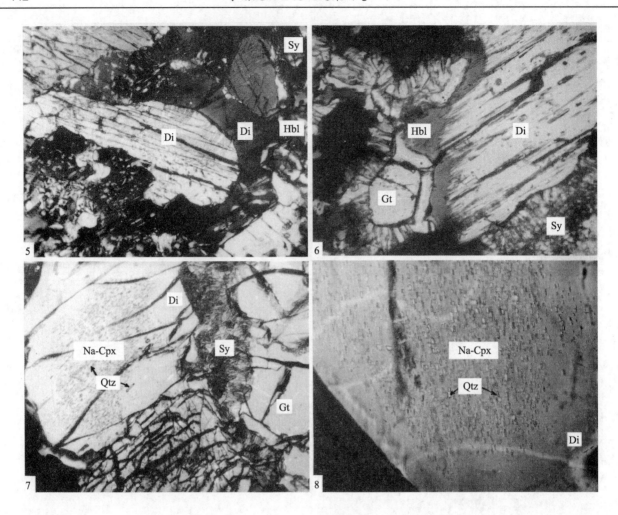

大别山北部榴辉岩的大地构造属性及冷却史*

刘贻灿[1,2]，徐树桐[2]，李曙光[1]，江来利[2]，陈冠宝[2]，吴维平[2]

1. 中国科学技术大学地球与空间科学系，安徽合肥 230026
2. 安徽省地质调查院，安徽合肥 230001

> **亮点介绍**：本文重建了北大别榴辉岩的 T-t 轨迹，揭示榴辉岩折返早期的冷却速率和抬升速率相对较慢（230 Ma → 210 Ma，近于等温减压过程），为该区榴辉岩早期折返至下地壳，减压但不降温，经历了强烈的麻粒岩相退变质作用。这为北大别深俯冲地壳岩石很少保留早期超高压变质证据提供了重要制约。

摘要　岩石地球化学及同位素年代学研究表明，大别山北部榴辉岩的大部分属印支期扬子俯冲陆壳（下地壳）的一部分；大致沿磨子潭-晓天断裂附近分布的含榴辉岩、大理岩和变质橄榄岩的镁铁-超镁铁质岩带可能代表扬子与华北 2 个大陆板块之间的变质构造混杂岩带，它应接近于扬子与华北 2 个大陆板块之间的缝合线（镁铁-超镁铁质岩带以北），而五河-水吼剪切带则可能代表扬子俯冲陆壳内部的 1 条拆离带或滑脱带。冷却史研究表明，大别山北部榴辉岩峰期变质后初期（230～210 Ma）仅抬升到下地壳水平，因而未经历降温过程并具有比大别山南部榴辉岩较低的抬升速率，峰期变质后较慢的抬升速率及相应的近于等温或局部升温阶段也许是造成大别山北部榴辉岩与南部超高压带中榴辉岩差异的重要原因之一；同时，也许是研究区榴辉岩很少见有保留早期超高压变质证据的重要原因。

关键词　榴辉岩；大地构造属性；冷却史；大别山北部

大别山，因其中含柯石英[1,2]和金刚石[3,4]等超高压岩石的发现而闻名于世。但是，在大别山北部杂岩带或镁铁-超镁铁质岩带（图1）（指五河-水吼剪切带以北至磨子潭-晓天断裂以南分布区）中长期未发现真正的榴辉岩以及该带岩石组成比较复杂[5-9]，因此对该带构造背景长期存在认识上的分歧[8-15]。近年来，笔者在大别山北部地区发现了两种产状榴辉岩并做了报道[16,17]，并根据近期研究成果[16~23]讨论了大别山北部榴辉岩的大地构造属性及冷却史，和南、北榴辉岩差异及北部榴辉岩中未发现柯石英等超高压矿物的重要原因。

1 榴辉岩的大地构造属性

人们对大别山北部杂岩带形成构造背景的认识一直存在较大分歧，最主要的原因是：①缺乏榴辉岩或榴辉岩相岩石存在的直接证据；②同位素和地球化学资料的贫乏。由此引起争论的焦点仍然是：是否有榴辉岩或榴辉岩相岩石的存在，它们的形成时代以及榴辉岩和有关岩石的地球化学性质如何？最近研究已表明，研究区存在榴辉岩[16~18]或榴辉岩相岩石[23]已无可非议[17,20,21]。

大别山北部杂岩带中的榴辉岩主要有 2 种产状，一是产于变形较强（面理化）的橄榄岩中，另一种是产于片麻岩中。它们分别相当于 Coleman 的 B 型和 C 型榴辉岩[17]。另外一种可能的产状是产于大理岩中[18]，但该种产状榴辉岩大多已强烈退变，很少保留早期绿辉石，大多数绿辉石已退变为透辉石和斜长石，并以包体形式产于石榴子石中。石榴子石成分特征及钠质单斜辉石中石英针状出溶体等表明，本区榴辉岩早期可能

---

* 本文发表在：地球科学，2003，28(1)：11-16
对应的英文版论文为：Liu Yican, Xu Shutong, Li Shuguang, et al., 2003. Tectonic affinity, T-t path and uplift trajectory of eclogites from Northern Dabie mountains, central-eastern China. Journal of China University of Geosciences, 14(1): 28-33

曾经过超高压变质作用，且至少经历了三个变质阶段[16~18]，即：①榴辉岩相峰期变质阶段，主要矿物共生组合为石榴子石+绿辉石+金红石+石英，$P \geq 2.5$ GPa、$T = 595 \sim 874$ ℃；②高压麻粒岩相退变质阶段，主要矿物共生组合为石榴子石+透辉石+紫苏辉石+钛铁矿+尖晶石+斜长石等，$P = 1.1 \sim 1.37$ GPa、$T = 817 \sim 909$ ℃；③角闪岩相变质阶段，主要矿物共生组合为角闪石+斜长石+磁铁矿等，$P = 0.5 \sim 0.6$ GPa、$T = 500 \sim 600$ ℃。北部榴辉岩以独特的麻粒岩相退变质作用及榴辉岩相峰期变质阶段无含水矿物（如多硅白云母等）形成为显著特征，并由此区别于南部榴辉岩。峰期变质矿物组合中无含水矿物形成，可能与它们的原岩来自下地壳岩石有关。

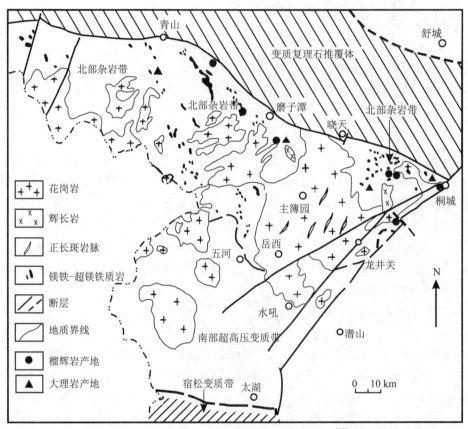

图 1 大别山（安徽部分）地质简图（据刘贻灿等[17]修改）

由于大别山北部与南部高压-超高压岩石的 Nd 模式年龄不同[24~26]，即北部一般较老而南部相对年轻。对此，部分作者解释为"北大别地体"（一般称之为镁铁-超镁铁质岩带或大别山北部）属华北地块，而"南大别地体"或"南大别超高压变质带"来源于扬子地块[27]。而另外一种解释，即"北大别地体"或其中一部分与"南大别地体"均属扬子古老变质基底[8, 9, 24]。

大别山北部杂岩带中发现有高级变质岩（榴辉岩及相关岩石），榴辉岩和英云闪长质片麻岩均经过印支期变质事件[19, 21]。目前，已证明研究区榴辉岩[17]和大理岩[28]可能曾经过超高压变质作用。而且，笔者最近研究表明，该区榴辉岩中含有斜紫苏辉石以及石榴子石由于压力降低而形成一些定向性针状矿物出溶体，如金红石、钛铁矿、单斜辉石和磷灰石等，类似于青岛仰口超高压榴辉岩中石榴子石出溶体特征（指示其峰期变质压力至少达到 5~7 GPa）[29]，由此进一步证明研究区榴辉岩经过了超高压变质作用，峰期变质压力甚至可能达到 5~7 GPa 或以上（详细资料，将另文发表）。由此说明其中至少部分岩石类似于南部超高压变质岩，也曾发生过深俯冲（至地幔深度或大于 90~100 km 的深度）。因此，它们也应该属扬子俯冲陆壳的一部分。而且，岩石地球化学及 Sr-Nd 同位素地球化学研究[20]表明，本区榴辉岩大多数为下地壳岩石变质成因。大别山中生代花岗岩及变质岩的钕模式年龄研究[25]也表明，在整个大别山造山带内广泛出露的中生代花岗岩侵入体的钕模式年龄与大别山北部变质岩的模式年龄范围一致，并且有相同的 ε_{Nd}（122 Ma）范围，表明这些花岗岩在形成时其深部源区岩石的钕同位素组成与"北大别地体"变质岩相似，也就是说，南部超高压带下伏的岩石也应为"北大别地体"变质岩。这样，"北大别地体"老的 Nd 模式年龄说明它属于扬子下地壳，而"南大别地体"年轻的 Nd 模式年龄则说明它可能为扬子上地壳。这与铅同位素研究结果[30]（研究区榴辉岩的全岩 Pb 同位素比值为 $w(^{206}Pb)/w(^{204}Pb) = 16.773\sim18.339$、$w(^{207}Pb)/w(^{204}Pb) = 15.346\sim15.516$、$w(^{208}Pb)/w(^{204}Pb) = 37.133\sim38.346$，平均值分别为 17.65、15.45 和 37.81，明显接近于下地壳成分即 $w(^{206}Pb)/w(^{204}Pb)$、$w(^{207}Pb)/w(^{204}Pb)$、$w(^{208}Pb)/w(^{204}Pb)$ 值分别为 17.62、15.35 和 38.75）相一致（详细数据资料，将另文发表）。实际上，Eide[31]也认为"北大别地体"应属于扬子俯冲-碰撞带中埋深最大的部分，只是由于折返期间递增的地热梯度使其部分熔融而成为现在的面貌，并推测其中可能有高压或超高压榴辉岩出露。

大别山北部杂岩带中除扬子俯冲陆壳外，还混有少量古洋壳残片，可能还有来自其他构造背景下的岩石[5, 6, 9, 20]，且北部榴辉岩大致沿磨子潭-晓天断裂附近分布，以北则为华北大陆板块南部活动大陆边缘的复理石建造[8, 9, 32]。周泰禧等[33]研究也表明，华北板块的南部边界应在桐柏-桐城断裂一线。

因此，大别山高压-超高压变质岩分布的北界至少应在磨子潭-晓天断裂附近，而磨子潭-晓天断裂附近含榴辉岩、大理岩和变质橄榄岩的镁铁-超镁铁质岩带可能代表扬子与华北 2 个大陆板块之间的变质构造混杂岩带，也就是说该带应接近于扬子与华北 2 个大陆板块之间的缝合线（镁铁-超镁铁质岩带以北）；五河-水吼剪切带则可能代表扬子俯冲陆壳内部的 1 条拆离带或滑脱带。

2 榴辉岩的冷却史和抬升速率

根据榴辉岩的同位素年代[18~22]（锆石 U-Pb 年龄为 230±6 Ma、石榴子石+绿辉石+全岩 Sm-Nd 年龄为 210 Ma 左右，角闪石 Rb-Sr 年龄为 172±3 Ma 和黑云母 Ar-Ar 年龄为 130 Ma）及对应的封闭温度或变质温度[17, 18]，做出榴辉岩的 T-t 轨迹（图 2）。

由 226～230 Ma（锆石 U-Pb 年龄）至 208～210 Ma（石榴子石+绿辉石+全岩 Sm-Nd 年龄）期间平均抬升速率约 4 mm/a，当然也不排除早期（220～230 Ma 左右）抬升很快而后减慢；至 210 Ma 左右时（处于下地壳，30～40 km），进入高压麻粒岩相阶段，但停留时间可能较短，然后于 170 Ma 左右时折返至中地壳（15 km 左右），此阶段平均抬升速率约为 0.5 mm/a，冷却速率平均为 10 ℃/Ma；此后以类似速率抬升至上地壳（约 10 km）（130 Ma 左右），冷却速率平均为 4 ℃/Ma。

从抬升与冷却速率轨迹看，大别山北部榴辉岩的初期平均冷却速率和平均抬升速率相对较慢，从 230 Ma 左右至 210 Ma 左右榴辉岩基本没有降温，平均抬升速率只有约 4 mm/a，这也许是研究区榴辉岩很少见有保留早期超高压变质证据的重要原因，因为慢的抬升速率及高温（≥800 ℃）条件下有可能使早期柯石英转变为石英（目前呈包体形式存在于石榴子石中）。Chavagnac 等[34]估算威海榴辉岩（对应于大别山北部榴辉岩）的早期（210～220 Ma 左右）抬升速率只有约 1 mm/a，而南部碧溪岭榴辉岩的抬升速率则平均为约 10 mm/a。因此，峰期变质后较慢的抬升速率及相应的等温阶段也许是造成大别山北部榴辉岩与南部超高压带中榴辉岩差异的重要原因之一。

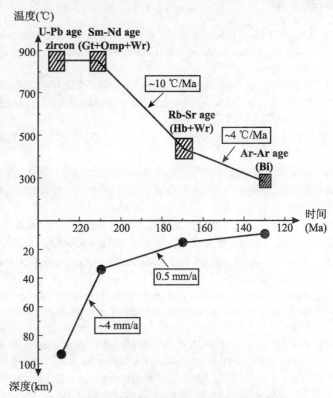

图 2 大别山北部榴辉岩的冷却史和抬升速率

由于大别山北部榴辉岩在 210 Ma 以前基本没有降温冷却,因此,在 210~170 Ma 之间(即从麻粒岩相过渡到角闪岩相)出现了较快的冷却过程。而对于大别山南部柯石英榴辉岩,由于初期 219±5 Ma 已降温至 500 ℃,在这一时期则表现为一非常缓慢的冷却过程[35]。

上述大别山南、北部榴辉岩冷却史的差异,反映了它们抬升折返路径的差异或者说具有不同的折返过程,而大别山北部榴辉岩则在第 1 次快速抬升时仅上升到下地壳水平(约 35~40 km 处)。这是造成它们冷却史差异的主要原因。

3 讨论与结论

大别山北部榴辉岩的大部分属印支期扬子俯冲陆壳(下地壳)的一部分,而南部榴辉岩大多数则属印支期扬子俯冲上地壳。磨子潭-晓天断裂附近含榴辉岩、大理岩和变质橄榄岩的镁铁-超镁铁质岩带可能代表扬子与华北 2 个大陆板块之间的变质构造混杂岩带,它应接近于扬子与华北 2 个大陆板块之间的缝合线(镁铁-超镁铁质岩带以北);而五河-水吼剪切带则可能代表扬子俯冲陆壳内部的 1 条拆离带或滑脱带。大别山北部榴辉岩峰期变质后初期(230~210 Ma)仅抬升到下地壳水平,因而未经历降温过程并具有比大别山南部榴辉岩较低的抬升速率,峰期变质后较慢的抬升速率及相应的等温阶段也许是造成大别山北部榴辉岩与南部超高压带中榴辉岩差异的重要原因之一;同时,也许是研究区榴辉岩很少见有保留早期柯石英等超高压变质证据的重要原因。

参考文献

[1] Okay A I, Xu S T, Sengör A.M.C. Coesite from the Dabie Shan eclogites, central China[J]. Eur J Mineral., 1989, 1: 595-598.
[2] Wang X., Liou J G, Mao H.K. Coesite-bearing eclogites from the Dabie Mountains in central China[J]. Geology, 1989, 17: 1085-1088.
[3] 徐树桐, 苏文, 刘贻灿, 等. 大别山东段高压变质岩中的金刚石. 科学通报, 1991, 36(17): 1318-1321.
[4] Xu S, Okay A I, Ji S, et al. Diamond from the Dabie Shan metamorphic and its implication for tectonic setting[J]. Science, 1992, 256: 80-82.
[5] 刘贻灿, 徐树桐, 江来利, 等. 大别山北部斜长角闪岩类的地球化学特征及形成构造背景[J]. 大地构造与成矿学, 1998, 22(4): 323-331.
[6] 刘贻灿, 徐树桐, 江来利, 等. 大别山北部蛇绿岩的地球化学制约[J]. 矿物岩石, 1999, 19(1): 68-73.
[7] 刘贻灿, 徐树桐, 江来利, 等. 大别山北部中酸性片麻岩的岩石地球化学特征及其古大地构造意义[J]. 大地构造与成矿学, 1999, 23(3): 222-229.
[8] 徐树桐, 江来利, 刘贻灿, 等. 大别山(安徽部分)的构造格局和演化过程[J]. 地质学报, 1992, 66(1): 1-14.
[9] 徐树桐, 刘贻灿, 江来利, 等. 大别山的构造格局和演化[M]. 北京: 科学出版社, 1994. 1-175.
[10] Okay, A.I. and Sengör, A.M.C. Evidence for introcontinental thrust-related exhumation of the ultrahigh-pressure rocks in China[J]. Geology, 1992, 20: 411-414.
[11] Wang Q, Cong B, Zhai M, et al. A possible paleozoic island arc: petrological evidences from North Dabie gneiss[A]. In: Inst Of Geol., (ed), Annual Report of theLaboratory of lithosphere tectonic evolution (1993-1994)[C]. Bejing: Seismological Press, 1994. 37-47.
[12] Zhai M, Cong B, Zhao Z, et al. Petrological-tectonic units in the coesite-besring metamorphic terrain of the Dabie Mountains, Central China and their geotectonic implication[J]. J. SE. Geosci., 1995, 11(1): 1-13.
[13] Zhang RY, Liou JG, Tsai C.H. Petrogenesis of a high-temperature metamorphic terrain: a new tectonic interpretation for the North Dabie Shan, central China. J. Metamorphic. Geol., 1996, 14: 319-333.
[14] 张旗, 马宝林, 刘若新, 等. 一个消减带之上的大陆岩石圈地幔残片——安徽饶钹寨超镁铁岩的地球化学特征[J]. 中国科学(B 辑), 1995, 5(8): 867-873.
[15] 董树文, 孙先如, 张勇, 等. 大别山造山带的基本结构[J]. 科学通报, 1993, 38(6): 542-545.
[16] Xu S T, Liu Y C, Su W, et al. Discovery of the eclogite and its petrography in the Northern Dabie Mountains[J]. Chinese Science Bulletin, 2000, 45(3): 273-278.
[17] 刘贻灿, 徐树桐, 李曙光, 等. 大别山北部镁铁-超镁铁质岩带中榴辉岩的分布与变质温压条件[J]. 地质学报, 2001, 75(3): 385-395.

[18] 刘贻灿. 大别山北部榴辉岩的岩石学、地球化学及同位素年代学研究[D]. 合肥: 中国科学技术大学, 2000.
[19] 刘贻灿, 李曙光, 徐树桐, 等. 大别山北部榴辉岩和英云闪长质片麻锆石 U-Pb 年龄及多期变质增生[J]. 高校地质学报, 2000, 6(3): 417-423.
[20] 刘贻灿, 徐树桐, 李曙光, 等. 大别山北部榴辉岩的地球化学特征和 Sr, Nd 同位素组成及其大地构造意义[J]. 中国科学(D辑), 2000, 30(增刊): 99-107.
[21] 刘贻灿, 李曙光, 徐树桐, 等. 大别山北部榴辉岩的 Sm-Nd 年龄测定及其对麻粒岩相退变质时间的制约[J]. 地球化学, 2001, 30(1): 79-87.
[22] 刘贻灿, 徐树桐, 李曙光, 等. 大别山北部鹿吐石铺含石榴子石斜长角闪岩的变质特征及 Rb-Sr 同位素年龄[J]. 安徽地质, 2000, 10(3): 194-198.
[23] Tsai C H, Liou J G. Eclogite-facies relics and inferred ultrahigh-pressure metamorphism in the North Dabie Complex, central-eastern China[J]. Am Mineral, 2000, 85: 1-8.
[24] 李曙光, 刘德良, 陈移之, 等. 扬子陆块北缘地壳的钕同位素组成及其构造意义[J]. 地球化学, 1994, 23(增刊): 10-17.
[25] 谢智, 陈江峰, 周泰禧, 等. 大别造山带变质岩和花岗岩的钕同位素组成及其地质意义[J]. 岩石学报, 1996, 12(3): 401-408.
[26] Chen J, Jahn B M. Crustal evolution of southeastern China: Nd and Sr isotopic evidence[J]. Tectonophysics, 1998, 284: 101-133.
[27] 谢智. 大别-苏鲁造山带岩石的锆石 U-Pb 年龄[D]. 合肥: 中国科学技术大学, 1998.
[28] 刘贻灿, 徐树桐, 江来利, 等. 大别山北部超高压变质大理岩及其地质意义[J]. 矿物岩石地球化学通报, 2001, 20(2): 88-92.
[29] Ye K, Cong B, Ye D L. The possible subduction of continental material to depths greater than 200 km[J]. Nature, 2000, 407: 734-736.
[30] Zartman R E, Haines S M. The plumbotectonic model for Pb isotopic systematics among major terrestrial reservoirs—a case for bi-directional transport[J]. Geochim Cosmochim Acta, 1988, 52: 1327-1339.
[31] Eide, E. A model for the tectonic history of the HP and UHPM region in east central China[A]. In: Coleman R G, Wang X, eds Ultrahigh-pressure metamorphism[C]. Cambridge: Cambridge University Press, 1995, 391-426.
[32] 刘贻灿, 徐树桐, 江来利, 等. 佛子岭群的岩石地球化学及构造环境. 安徽地质, 1996, 6(2): 1-6.
[33] 周泰禧, 陈江峰, 张巽, 等. 北淮阳花岗岩-正长岩带地球化学特征及其大地构造意义[J]. 地质论评, 1995, 41(2): 144-151. Zhou T, Chen J, Zhang X. Geochemistry of the North Huaiyang granite-syenite zone and its tectonic implication. Geological Review, 41(2): 144-151.
[34] Chavagnac V, Jahn,B M. Coesite-bearing eclogites from the Bixiling complex, Dabie Miuntains, China: Sm-Nd ages, geochemical characteristics and tectonic implications[J]. Chem Geol., 1996, 133: 29-51.
[35] Li S., Jagoutz E., Chen Y, et al. Sm-Nd and Rb-Sr isotopic chronology and cooling history of ultrahigh-pressure metamorphic rocks and their country rocks at Shuanghe in the Dabie Mountains, central China[J]. GCA, 2000, 64: 1077-1093.

Geochemistry and geochronology of eclogites from the northern Dabie Mountains, Central China*

Yican Liu[1,2], Shuguang Li[1], Shutong Xu[2], Bor-ming Jahn[3], Yong-Fei Zheng[1], Zongqing Zhang[4], Laili Jiang[2], Guanbao Chen[2], Weiping Wu[2]

1. School of Earth and Space Sciences, University of Science and Technology of China, Hefei 230026, Anhui, China
2. Anhui Institute of Geology, Hefei 230001, Anhui, China
3. Geosciences Rennes, Universite de Rennes, Rennes 35000, France
4. Institute of Geology, Chinese Academy of Geological Sciences, Beijing 100037, China

> 亮点介绍：通过元素和 Sr-Nd 同位素地球化学等方面研究，揭示了北大别榴辉岩的原岩大多数为大陆玄武质岩石，少数可能为地幔岩碎片，厘定了北大别榴辉岩相高压变质时代为 212±4 Ma。

Abstract Recent discoveries of UHP eclogites from the northern Dabie Complex (NDC) in central China provide an opportunity to improve our understanding of the tectonic setting of the NDC and exhumation mechanism of UHPM rocks. The petrology, geochemistry, Sr-Nd-O isotopic compositions and Sm-Nd isotopic chronology of some eclogites from the NDC were obtained in this study. The major elemental compositions and relatively low $\varepsilon_{Nd}(T)$ of about −10 for the eclogites suggest that the protoliths of most eclogites were ancient basaltic rocks developed in continental crust, and thus these eclogites were produced by subduction of the Yangtze continental crust. However, some of the eclogites within foliated peridotites with relatively high $\varepsilon_{Nd}(T)$ values up to +2.4 are interpreted as pieces of lithospheric mantle scraped off the overlying mantle by subducting continental crust and then subducted together into mantle depths. Therefore, the metamorphosed mafic-ultramafic belt containing eclogite and meta-peridotite blocks along the southern part of the Mozitan-Xiaotian fault zone may represent the tectonic melange produced by the subduction of the Yangtze continental crust.

The Sm-Nd age of 212 ± 4 Ma defined by garnet + omphacite is interpreted to represent a cooling age corresponding to in the retrograde stage of granulite-facies metamorphism during eclogite exhumation. The relationship between the $\delta^{18}O$ values of the eclogitic minerals such as garnet and omphacite from an eclogite sample suggests oxygen isotope equilibrium, pointing to Sm-Nd isotope equilibrium between garnet and omphacite according to comparability between O and Sm-Nd diffusivities. However, the Sm-Nd 'isochrons' defined by whole + garnet or omphacite + zonal garnet yield a large age range of 187–266 Ma, which may be produced by Nd isotopic disequilibrium between garnet and retrograded minerals in whole-rocks. Therefore, the eclogites in the NDC also underwent Triassic UHP metamorphism like the UHP eclogites in the southern Dabie Mountains. The $\delta^{18}O$ values of whole-rocks for the eclogites are 2.40‰ to 4.32‰, indicating that protolith of the eclogites in the NDC underwent hydrothermal alteration at high-temperature by surface water before subduction of the Yangtze continental crust during the Triassic period in a manner similar to protoliths of UHP rocks in the southern Dabie Mountains.

Key words eclogite; geochemistry; Sm-Nd chronology; Northern Dabie mountains; China

1 Introduction

The occurrence of ultrahigh-pressure minerals, such as coesite and diamond, in crustal rocks in orogenic belts

---

* 本文发表在：Journal of Asian Earth Sciences, 2005, 25: 431-443

suggests that a huge amount of continental crust can be subducted to mantle depth during continent-continent collision (Chopin, 1984; Smith, 1984; Okay et al., 1989; Wang et al., 1989; Sobolev and Shatsky, 1990; Xu et al., 1992). However, the tectonic framework of an orogenic belt is a fundamental factor to understanding the exhumation mechanism of ultrahigh-pressure metamorphic (UHPM) rocks.

The E-W trending Qinling-Dabie orogen is a collision zone between the North China Block and Yangtze Block. The Dabie Mountains constitute the eastern section of this orogenic belt, and are transected at its eastern end by the Tan-Lu fault. The Su-Lu orogen is believed to be the eastern extension of the Dabie Mountains orogen. The UHPM rocks are exposed in the Dabie Mountains and Su-Lu orogen, which together comprise the largest UHPM belt on the earth's surface. The UHPM rocks are composed of different kinds of metamorphic rocks such as coesite (Okay et al., 1989; Wang et al., 1989) and diamond (Xu et al., 1992) bearing eclogite, quartz jadeitite, meta-pelite, marble with eclogite nodules and white schist assemblage-bearing kyanite quartzite. It is generally accepted that the eastern Dabie Mountains can be subdivided into the four tectono-petrologic units from north to south (Liu, 2000; Li et al., 2001; Zheng et al., 2001) (Fig. 1): (1) Foziling meta-flysch (MF); (2) northern Dabie Complex (NDC); (3) southern Dabie ultrahigh-pressure metamorphic belt (SUHPMB); (4) Susong and Zhangbaling high-pressure metamorphic belt or Susong Complex (SSC). These units are separated from one other by the Xiaotian-Mozitan fault, Wuhe-Shuihou shearzone and Taihu-Mamiao fault, respectively (Fig. 1). The UHPM rocks are restricted to the SUHPMB. However, until recently, no eclogite within the NDC has been found. It has long been an unsolved problem about whether the NDC has ever been subjected to Triassic ultrahigh-pressure metamorphism. Therefore, it has been interpreted as a thermal overprinted subduction complex (Wang and Liou, 1991; Okay and Sengor, 1992), a Paleozoic Andean magmatic arc complex (Zhai et al., 1994), a metamorphic ophiolite melange (Xu et al., 1992, 1994), a Cretaceous magmatic belt (Hacker et al., 1998) or the Sino-Korean hanging wall during Triassic subduction (Zhang et al., 1996). Recently, Xu et al. (2000) and Liu et al. (2000, 2001) found several fresh eclogite outcrops in

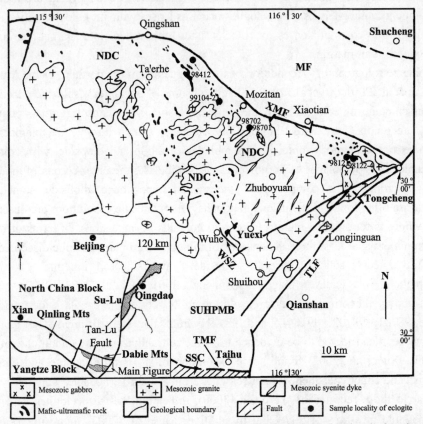

Fig. 1 Schematic geological map of the Dabie Shan orogen in Anhui Province, with inset showing the location of this area within the Triassic Qinling-Dabie-Su-Lu collision orogen in central China. The shaded area in the inset shows the UHPM belt. Sample localities with sample numbers are described in detail in the text. MF = Foziling meta-flysch, NDC = northern Dabie Complex, SUHPMB = southern Dabie UHPM belt, SSC = Susong Complex, XMF = Xiaotian-Mozitan fault, WSZ = Wuhe-Shuihou shear zone, TMF = Taihu-Mamiao fault, TLF = Tan-Lu fault.

the northern Dabie Complex (Fig. 1). Tsai and Liou (2000) and Xiao et al. (2001) also found eclogite-facies relics and inferred that they were subjected to ultrahigh-pressure metamorphism in the northern Dabie Mountains. Liu et al. (2000) reported a zircon U-Pb age of 230 ± 6 Ma from the Raobazhai eclogite in the northern Dabie Mountains. However, except for a garnet + dioside Sm-Nd isochron age of 244 ± 11 Ma for Raobazhai garnet-pyroxenite (Li et al., 1993), no metamorphic Sm-Nd age for the eclogites from the NDC has been reported. In addition, the origin of these eclogites has remained unclear. The eclogites are keys for understanding the tectonic framework and evolution of the Dabie Mountains orogen.

This paper first reports the trace element and Sr-Nd-O isotopic data as well as the Sm-Nd ages of eclogites from the northern Dabie Mountains. On the basis of these data, the suture location in the Dabie Mountains is discussed.

2 Metamorphic petrology and P-T conditions

The northern Dabie Complex is mainly composed of banded tonalitic and granitoid gneiss with minor meta-peridotite (including dunite, harzburgite and lherzolite), garnet-pyroxenite, garnet-bearing amphibolite, granulite, eclogite and marble. Most of the rocks are of amphibolite-facies; some relic mineral assemblages of eclogitic- and granulite-facies are still preserved.

The eclogites from the northern Dabie Mountains have been found at several localities (Fig. 1), for instance, Huangweihe (98701 and 98702), Raobazhai (99104-2), Baizhangyan (98121) and Huazhuang (98122-4). In addition, some retrograded eclogites or garnet-amphibolites (e.g. 98412) or garnet-pyroxenites have been found in Yanzihe, Lutoushipu and Banju (Liu, 2000). Three kinds of occurrence of the eclogites have been identified: the first occurs in the strongly deformed (foliated) peridotite such as Huangweihe and Raobazhai; the second is in gneiss, e.g. Baizhangyan and Huazhuang; the third is in marble like Yanzihe. Their country rocks are orthogneisses such as tonalitic and monzonitic granitic gneisses. The eclogites occur as lenses with strong deformation and foliation, most of them preserved in the core of lenses or blocks and retrograded into garnet-bearing amphibolite or garnet-pyroxenite towards their margin.

The metamorphic petrology and P-T condition of the eclogites from the northern Dabie Mountains have been reported in detail (Liu et al., 2001; Xu et al., 2002). Here only some of the major results are introduced.

The mineral assemblage of eclogite from the northern Dabie Mountains is garnet, omphacite and rutile; secondary minerals are quartz, corundum, diopside, hypersthene, spinel, hornblende, plagioclase and ilmenite. Omphacite is often retrograded into symplectite assemblages of albite and diopside, while diopside can further decompose into symplectite assemblages of hornblende and plagioclase; the outer corona of garnet between garnet and clinopyroxene or omphacite is composed of fine-grained hypersthene, diopside, spinel, plagioclase, etc. Omphacite and rutile are often found as inclusions in garnets. Quartz needle exsolutions are observed in the core of Na-clinopyroxene, while its retrograded margin (diopside) contains no inclusions. In terms of the texture and the relationship between minerals, at least three generations of mineral assemblages can be discerned: (1) omphacite + garnet + rutile; (2) symplectite stage of garnet + diopside + hypersthene + ilmenite + corundum + spinel + plagioclase; (3) amphibole + plagioclase + magnetite (Liu et al., 2001; Xu et al., 2002).

As mentioned above, the oriented quartz exsolution in Na-clinopyroxene is widespread in the eclogites (especially, sample 98701, 98702, 99104-2 and 98121) from the studied area. The quartz exsolution in Na-clinopyroxene has been considered to be evidence of former ultrahigh-pressure conditions (cf. Liou et al., 1999 for review). Recently, Xu et al. (2002) found some exsolution rods of rutile, apatite and clinopyroxene in the porphyroblastic garnet and clinohypersthene within eclogites from the northern Dabie Mountains. These exsolutions in garnet are similar to those reported by Ye et al. (2000), indicating that the eclogites here experienced UHP metamorphism at possible pressures of 5–7 GPa. In addition, Xu et al. (2003) found micro-diamonds from the eclogites in the northern Dabie Mountains, suggesting they also experienced UHP metamorphism at a possible pressure of >4.0 GPa. So, the eclogites from the northern Dabie Mountains formed under UHP conditions with pressures greater than 2.5 GPa or even higher than 5 GPa.

The metamorphic P-T conditions of the three metamorphic stages for the eclogites in the northern Dabie

Mountains are estimated as follows (Xu et al., 2000, 2002; Liu et al., 2001): (1) eclogite-facies stage with $P = 2.5$ GPa and $T = 808–874$ ℃, averaging 843 ℃ or even more; (2) retrogressive metamorphism of high-pressure granulite-facies stage with $P = 1.10–1.37$ GPa and $T = 817–909$ ℃, averaging 848 ℃; (3) retrogressive metamorphism of amphibolite-facies with $P = 0.5–0.6$ GPa and $T = 500–600$ ℃.

The SUHPMB is composed of the typical UHP metamorphic rocks, including eclogite, gneiss, quartz jadeitite, marble, etc. The occurrence of diamond and coesite in the rocks reveals temperature-pressure conditions of 700–850 ℃ and > 2.8 GPa for the UHP metamorphism (e.g. Wang et al., 1995; Cong, 1996). The UHP rocks sequentially experienced eclogite-, amphibolite- and green- schist-facies metamorphism (e.g. Xu et al., 1992; Wang et al., 1995; Cong, 1996). Therefore, although both the SDUHPB and NDC units experienced UHP metamorphism, they had different exhumation histories, suggesting that the two eclogite-bearing units also had separate histories (Liu et al., 2001; Zheng et al., 2001).

3 Sample descriptions

Samples 98121 and 98122-4 occur in the orthogneisses (banded or granitic gneiss). Samples 98701, 98702 and 99104-2 come from foliated peridotites. Samples 99104-2 and 98702 are slightly retrograded eclogites, which are mainly composed of garnet, omphacite and rutile with minor quartz, hypersthene, plagioclase and hornblende. Samples 98121 and 98122-4 are strongly retrograded and composed of garnet, pargasitic amphibole and plagioclase with minor omphacite (< 5%) and rutile. Garnet contents of sample 98121 are about 80% in volume. Omphacite only occurs as inclusions in garnet. So far, only part of the garnet grains from sample 99104-2 have been found to be compositionally zoned as demonstrated by microprobe analyses (Liu et al., 2001). The lower CaO and higher MnO contents of the rims relative to the cores of the garnets suggests that the rims formed during retrograde metamorphism (Liu et al., 2001). In addition, most of the amphibole, plagioclase and hypersthene minerals from these eclogite samples generally occur as symplectite.

4 Analytical procedures

4.1 Major- and trace-element analyses

Major elements were determined by wet chemical methods. Most trace elements were measured by X-ray fluorescence spectrometry (XRF), while Sc, Y and rare-earth element (REE) were obtained by inductively coupled plasma spectrometry. All these analyses were performed at the Anhui Institute of Geological Experiment. Analytical uncertainties range from ± 1 to ± 5% for major elements; around ± 5% for trace elements with concentrations ⩾ 20 ppm, and around ± 10% for those < 20 ppm. Overall analytical uncertainties are from 2 to 5% for all REE.

4.2 Isotopic analyses

Before chemical treatment, altered grains were eliminated by hand during examination under a binocular microscope. Omphacite grains from two eclogite samples (98702 and 99104-2) used for the isotopic dating were analyzed first by electron probe to confirm if they are omphacite. The electron probe analysis results demonstrate that those grains are omphacite with 17-19 and 22-28 of Jd end members, respectively (Table 1 and Fig. 2). Most of Sr-Nd chemical separation processes were performed in the Isotope Laboratory of the Chinese Academy of Geological Sciences and mass-spectrometer analyses were performed at the Laboratory for Chemical Geodynamics, University of Science and Technology of China. In order to ensure the reliability for dating, duplicate analyses of samples 98702 and 98121 for Sm-Nd were performed at Rennes. In addition, the Sm-Nd isotope data of sample 98122-4 were also analyzed at Rennes. The detailed Sr-Nd analytical procedures were described elsewhere (Chavagnac and Jahn, 1996; Li et al., 2000). Blanks for the whole chemical procedures are < 50 pg for Sm and Nd, 1 ng for Sr and 0.3 ng for Rb. $^{143}Nd/^{144}Nd$ ratios were normalized against the value of $^{146}Nd/^{144}Nd = 0.7219$ and all Nd isotope ratios are relative to the values of 0.511849 ± 9 for the La Jolla Nd standard. $^{87}Sr/^{86}Sr$ ratios were normalized against the value of $^{86}Sr/^{88}Sr = 0.1194$ and the NBS-987 Sr standard yielded $^{87}Sr/^{86}Sr = 0.710277 \pm 16$.

$\varepsilon_{Nd}(T)$ and $\varepsilon_{Sr}(T)$ were calculated based on present-day reference values for chondritic uniform reservoir (CHUR) or uniform reservoir (UR) (Faure, 1986): $(^{143}Nd/^{144}Nd)_{CHUR} = 0.512638$, $(^{147}Sm/^{144}Nd)_{CHUR} = 0.1967$, $(^{87}Sr/^{86}Sr)_{UR} = 0.7045$ and $(^{87}Rb/^{86}Sr)_{UR} = 0.0816$. Ages are calculated using the ISOPLOT software of Ludwig (1997) and given a 2σ error. Input errors used in age computations are $^{147}Sm/^{144}Nd = \pm 0.2\%$, $^{143}Nd/^{144}Nd = \pm 0.005\%$. The decay constants ($\lambda$) used in age or initial isotope ratio calculations are 0.0142 Ga^{-1} for ^{87}Rb and 0.00654 Ga^{-1} for ^{147}Sm.

Table 1 Electron probe analyses of representative omphacites used for Sm-Nd dating from eclogites (samples 98702 and 99104-2) in the northern Dabie mountains

Sample	98702-1	98702-2	98702-3	98702-4	98702-5	98702-6
SiO_2	55.38	55.37	55.38	54.67	55.26	55.10
TiO_2	0.03	0.00	0.00	0.01	0.00	0.02
Al_2O_3	4.85	4.67	5.20	5.63	4.80	5.49
FeO	0.93	2.20	1.36	0.90	0.25	1.16
Fe_2O_3	1.36	0.80	1.20	1.76	2.08	1.44
MnO	0.00	0.00	0.00	0.00	0.00	0.00
MgO	14.03	13.80	13.82	13.44	14.28	13.78
CaO	20.65	20.12	20.29	20.01	20.34	20.14
Na_2O	2.98	3.04	3.09	3.17	3.14	3.10
K_2O	0.00	0.00	0.00	0.00	0.00	0.00
Total	100.21	100.00	100.34	99.59	100.15	100.23
O	6	6	6	6	6	6
Si	1.983	1.989	1.980	1.970	1.976	1.972
Al^{IV}	0.017	0.011	0.020	0.030	0.024	0.028
Al^{VI}	0.187	0.187	0.199	0.209	0.178	0.203
Fe^{3+}	0.035	0.035	0.034	0.041	0.063	0.038
Ti	0.001	0.000	0.000	0.000	0.000	0.001
Fe^{2+}	0.029	0.053	0.039	0.034	0.001	0.036
Mg	0.749	0.739	0.737	0.722	0.761	0.735
Mn	0.000	0.000	0.000	0.000	0.000	0.000
Ca	0.792	0.774	0.777	0.773	0.779	0.772
Na	0.207	0.212	0.214	0.221	0.218	0.215
K	0.000	0.000	0.000	0.000	0.000	0.000
WEF	79.148	78.715	78.370	77.526	77.973	78.197
Jd	17.575	17.934	18.439	18.773	16.306	18.381
Ae	3.277	3.350	3.191	3.701	5.721	3.422
Sample	99104-2-1	99104-2-2	99104-2-3	99104-2-4	99104-2-5	99104-2-6
SiO_2	54.77	54.38	54.30	55.34	54.89	55.20
TiO_2	0.00	0.10	0.02	0.05	0.05	0.00
Al_2O_3	6.85	6.68	6.16	6.85	6.69	6.65
FeO	1.56	4.44	2.19	3.54	2.62	4.79
Fe_2O_3	2.95	0.40	2.08	0.80	1.92	0.08
MnO	0.00	0.00	0.00	0.00	0.00	0.00
MgO	12.01	11.90	12.23	11.96	11.75	12.09
CaO	17.82	17.41	18.53	17.43	18.18	16.78
Na_2O	4.26	3.74	3.65	4.10	4.07	3.94
K_2O	0.00	0.00	0.00	0.00	0.00	0.00
Total	100.22	99.05	99.16	100.07	100.17	99.53
O	6	6	6	6	6	6
Si	1.963	1.978	1.972	1.986	1.972	1.994

						Continued
Sample	99104-2-1	99104-2-2	99104-2-3	99104-2-4	99104-2-5	99104-2-6
Al^{IV}	0.037	0.022	0.028	0.014	0.028	0.006
Al^{VI}	0.252	0.264	0.236	0.275	0.255	0.277
Fe^{3+}	0.080	0.016	0.047	0.020	0.052	0.004
Ti	0.000	0.003	0.001	0.001	0.001	0.000
Fe^{2+}	0.046	0.130	0.076	0.108	0.079	0.143
Mg	0.642	0.645	0.662	0.640	0.629	0.651
Mn	0.000	0.000	0.000	0.000	0.000	0.000
Ca	0.684	0.678	0.721	0.670	0.700	0.649
Na	0.296	0.264	0.257	0.285	0.284	0.276
K	0.000	0.000	0.000	0.000	0.000	0.000
WEF	69.854	73.378	73.952	71.302	71.279	72.332
Jd	22.864	25.128	21.718	26.738	23.832	27.229
Ae	7.282	1.495	4.330	1.960	4.889	0.439

The compositions of omphacites were analyzed by a JEOL superprobe 733 using a wavelength dispersive system at the Analytical Center of China University of Geosciences (Wuhan). Operating conditions were 15 kV accelerating potential and 20 nA-beam current.

Oxygen isotope analysis was performed by the conventional BrF_5 method (Zheng et al., 1998). Oxygen isotope ratios were measured as CO_2 by a Finnigan Delta + mass spectrometer at the Laboratory for Chemical Geodynamics, University of Science and Technology of China. The isotopic results are reported in the conventional $\delta^{18}O$ notation of per mil relative to SMOW (standard mean-ocean water). The routine analytical error for $\delta^{18}O$ based on replicate analyses and standards is better than ± 0.2‰.

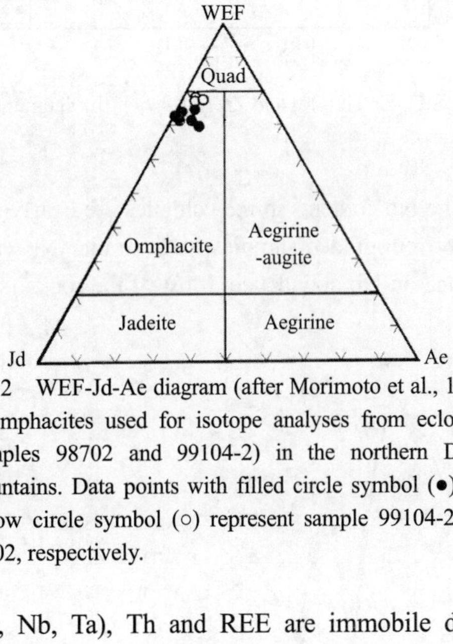

Fig. 2 WEF-Jd-Ae diagram (after Morimoto et al., 1988) of omphacites used for isotope analyses from eclogites (samples 98702 and 99104-2) in the northern Dabie Mountains. Data points with filled circle symbol (●) and hollow circle symbol (○) represent sample 99104-2 and 98702, respectively.

5 Results

5.1 Geochemical characteristics

The results of major and trace element analyses are given in Tables 2 and 3, respectively.

Because some high field strength elements (HFSE, e.g. Zr, Nb, Ta), Th and REE are immobile during metamorphic process (Peltonen et al., 1996), the concentrations of these elements are used to identify the protoliths of the eclogites.

SiO_2-Zr/TiO_2 and Zr/TiO_2-Nb/Y diagrams (Winchester and Floyd, 1977) show most samples plot in the sub-alkaline basalt field, while sample 98121 falls into the alkaline basalt field (Fig. 3). Sample 98121 has lower SiO_2 values, high TiO_2, Fe_2O_3 + FeO and Al_2O_3 values with a positive Eu anomaly (see Section 5.2), suggesting that it is of cumulate origin.

Table 2 Chemical compositions of the eclogites in the northern Dabie mountains (wt%)

Sample	SiO_2	TiO_2	Al_2O_3	FeO	Fe_2O_3	CaO	MgO	MnO	Na_2O	K_2O	P_2O_5	H_2O^C	LOI	Total
98701	45.88	1.26	14.13	8.40	4.99	12.28	9.92	0.23	2.01	0.03	0.18	0.34	0.19	99.84
98702	44.08	0.40	14.23	6.61	3.65	15.01	13.85	0.21	1.29	0.04	0.01	0.92	0.20	100.50
98121	39.42	2.15	16.81	16.00	4.38	13.44	5.96	0.16	0.50	0.06	0.06	0.57	1.32	100.83
98122-4	42.36	2.05	13.27	12.86	6.60	11.20	7.84	0.19	1.61	0.03	0.19	0.20	1.31	99.71
99104-2	45.39	0.90	14.59	11.69	1.08	13.08	9.74	0.22	1.61	0.06	0.06	0.13	1.16	99.71

Table 3 Trace element contents (ppm) and related ratios for the eclogites in the northern Dabie mountains

Sample	Zr	Nb	Th	U	Ta	Sc	Y	Zr/Nb	Nb/Ta	Nb/U	Th/Sc	Nb/Th	Nb/La
98701	65.0	10.8	2.3	0.5	0.5	44.1	26.7	6.0	21.6	21.6	0.05	4.7	0.92
98702	24.1	6.5	7.5	0.5	0.5	57.0	18.5	3.7	13.0	13.0	0.13	0.9	6.50
98121	28.5	8.3	1.2	0.5	0.5	38.3	6.1	3.4	16.6	16.6	0.03	6.9	3.77
98122-4	30.9	7.8	0.8	1.7	0.5	45.2	11.8	4.0	15.6	4.6	0.02	9.8	1.00
99104-2	40.0	7.7	0.8	0.5	2.4	39.3	18.9	5.2	3.2	15.4	0.02	9.6	5.13

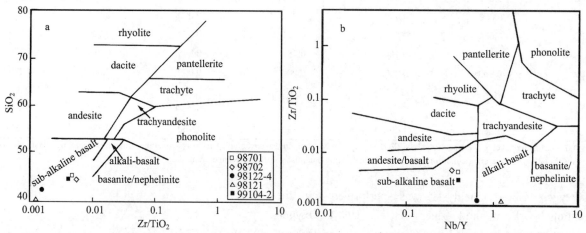

Fig. 3 SiO_2-Zr/TiO_2 (a) and Zr/TiO_2-Nb/Y (b) diagrams (after Winchester and Floyd, 1977) for the eclogites from the northern Dabie Mountains.

The Nb contents in the eclogites are nearly uniform (most are 7–8 ppm) with little variation (Table 3). Even the strongly retrograded samples have similar Nb values. Thus, late alteration and retrograde metamorphism had little influence on Nb abundance. Most of the rocks (such as samples 98702, 99104-2 and 98121) show a very distinctive positive Nb anomaly with Nb/La=3.7–6.5 (Table 3) in the primitive-mantle normalized spider diagrams (Fig. 4), which suggest that their protoliths were not subduction-related magmas. Nb/Ta ratios are 3.2–21.6 and most are 13–16 (Table 3) which are similar with Nb/Ta values of ~11 for continental crust (Taylor et al., 1985) and also alkaline basalt (Byerson and Watson, 1987). Sample 99104-2 from the Raobazhai peridotite massif has an Nb/Ta = 3.2, suggesting that its tectonic setting could be different from other samples.

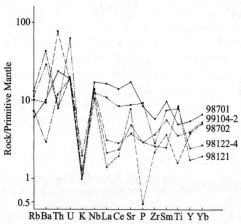

Fig. 4 Primitive-mantle-normalized spider diagrams for the eclogites from the northern Dabie Mountains. The primitive-mantle values are after Hofmann et al. (1988).

Most of the Th/Sc ratios from the eclogites are 0.02–0.05 (0.03 for mean value), in agreement with the average value of lower crust rocks (Taylor et al., 1985). The Nb/U-Nb diagram (Le Roex et al., 1989) also suggests that the protoliths for most of the eclogites were continental basalts (Fig. 5). In the Nb/3-TiO_2-Th diagram (Holm, 1985), most eclogites fall in the field of within-plate basalt and a few in the field of plate-margin-related basalt.

The eclogites have $\sum$REE = 17.70–80.09 ppm with most values between 20 and 50 ppm (Table 4, Fig. 6). Sample 98121, 98122-4 and 98701 display LREE enriched patterns consistent with a continental basalt origin. In contrast, samples 99104-2 and 98702 from foliated peridotites show typical LREE depleted patterns, suggesting that these eclogites with their host peridotites were derived from a LREE-depleted environment such as continental lithospheric mantle. It should be noted that the LREE-depleted pattern of sample 98702 is inconsistent with its low $\varepsilon_{Nd}(T)$ value (ca. −10) (see below). This suggests that sample 98702 may have experienced a recent LREE depletion event, which will be discussed in detail in Section 6.3.

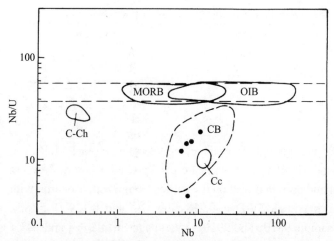

Fig. 5 Nb/U-Nb diagram (after Le Roex et al., 1989) of the eclogites from the northern Dabie Mountains. C-Ch = C-chrondrites, CC = continental crust, CB = continental basalt, MORB = middle ocean ridge basalt, OIB = ocean island basalt. Fields for MORB, OIB, C-Ch and CC from Hofmann et al. (1986); field for CB from Le Roex et al. (1989).

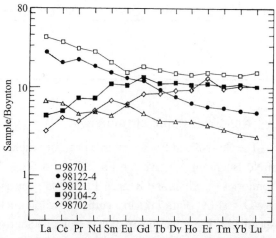

Fig. 6 REE distribution patterns for the eclogites from the northern Dabie Mountains. Chondrite values (in ppm) used for normalization are from Boynton (1984) (called as Boynton as usual): La = 0.310, Ce = 0.808, Pr = 0.122, Nd = 0.600, Sm = 0.195, Eu = 0.0735, Gd = 0.259, Tb = 0.0474, Dy = 0.322, Ho = 0.0718, Er = 0.210, Tm = 0.0324, Yb = 0.209, Lu = 0.0322.

Table 4 Rare-earth element contents (ppm) of the eclogites from the northern Dabie mountains

Sample	La	Ce	Pr	Nd	Sm	Eu	Gd	Tb	Dy	Ho	Er	Tm	Yb	Lu	∑REE
98701	11.7	26.8	3.40	15.4	3.80	1.10	4.50	0.75	4.70	1.00	3.10	0.46	2.90	0.48	80.09
98702	1.00	3.70	0.51	3.30	1.40	0.48	2.20	0.41	3.00	0.68	2.70	0.32	2.20	0.33	22.23
98121	2.20	5.40	0.63	3.25	0.96	0.47	1.33	0.20	1.35	0.30	0.79	0.11	0.62	0.09	17.70
9801	21.8	41.5	6.26	25.8	6.33	1.92	8.07	1.27	8.09	1.85	4.82	0.76	4.68	0.67	133.72
98122-4	7.80	15.5	2.58	10.6	2.90	0.95	3.08	0.45	2.53	0.48	1.28	0.19	1.14	0.17	49.65
99104-2	1.50	4.40	0.93	4.55	2.15	0.78	3.39	0.53	3.66	0.78	2.31	0.34	2.27	0.33	27.92

5.2 Oxygen isotopic compositions

The oxygen isotopic compositions of whole-rock, garnet and omphacite for samples 98702 and 99104-2 were determined and listed in Table 5. The results show that: (1) the measured oxygen isotope fractionations between omphacite and garnet are 1.77 and 0.42‰, respectively. The former corresponds to equilibrium fractionation values of 0.83‰–1.76‰ at 500–900 ℃ (Zheng, 1999) and the latter is less than the equilibrium fractionation values. This suggests that the eclogitic minerals in sample 98702 were in equilibrium with respect to oxygen isotope exchange under the conditions of eclogite-facies or granulite-facies metamorphism and the oxygen isotope systems between garnet and omphacite from sample 99104-2 were not at equilibrium. (2) After retrograde metamorphism from granulite-facies, the eclogites from sample 98702 were in a closed system and not significantly affected by exotic fluids. Otherwise the equilibrium fractionations between omphacite and garnet would have been destroyed (Zheng et al., 1998, 1999). The primary oxygen isotopic equilibrium between the garnet and omphacite in sample 99104-2 was destroyed and reequilibration by oxygen isotope exchange between them was not achieved under the conditions of retrograde metamorphism, consistent with the observed compositional zoning of garnet. Therefore, this work again suggests that the state of the equilibrium between cogenetic minerals can provide a critical test for the validity of the Sm-Nd mineral chronometer (see latter in detail). (3) The $\delta^{18}O$ values of whole-rocks for the eclogites are + 2.40‰–+ 4.32‰, indicating that the eclogites underwent hydrothermal alteration at high-temperature by surface water prior to the subduction of the Yangtze continental crust, similar to UHP eclogites and gneisses in the Dabie-Sulu orogenic belt (Yui et al., 1995; Baker et al., 1997; Zheng et al., 1998, 1999, 2001).

Table 5 Oxygen isotope compositions (δ^{18}O‰) of whole-rocks and minerals from eclogites in the northern Dabie mountains

Sample no.	Garnet	Omphacite	Whole-rock
98702	2.29	4.06	2.40
99104-2	3.80	4.22	4.32

5.3 Sm-Nd isotopic chronology

The Sm-Nd and Rb-Sr isotopic data for whole-rocks and minerals from the eclogites are given in Table 6. The Sm-Nd isochrons are given in Figs. 7 and 8. Figs. 7a and 8b show that duplicate results obtained from different laboratories (in China and Rennes, France) are consistent and correlate well on the garnet + omphacite isochron with MSWD = 1.01 (Fig. 7a), and garnet + whole-rock lines with MSWD = 0.278 (Fig. 7a) and 0.129 (Fig. 8b), respectively. The isochrons of garnet + omphacite from sample 98702 and 99104-2 yield ages of 212 ± 4 and 187 ± 5 Ma in Fig. 7a and b, respectively. However, their whole-rock compositions lie below their mineral isochrones. The line connecting garnet and whole-rock data define older ages of 221 ± 5 and 194 ± 5 Ma in Fig. 7a and b, respectively, which may reflect retrograde metamorphism in whole-rock samples (Li et al., 2000). Fig. 7a also shows that the $\varepsilon_{Nd}(T)$ value of sample 98702 is −8.7, which suggest that this eclogite was produced from subducted continental crust rather than oceanic crust. However, it should be noted that these Sm-Nd ages are less than the garnet + omphacite Sm-Nd ages of 221–228 Ma (Li et al., 1993, 1994, 1996, 2000) for coesite-bearing eclogites in the southern Dabie Mountains and Su-Lu terrain as well as the Sm-Nd age of 244 ± 11 Ma (Li et al., 1993) for the Raobazhai garnet-pyroxenite. The reasons for the age differences and their implications will be discussed in Section 6.

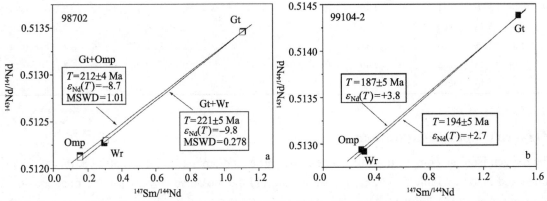

Fig. 7 Sm-Nd mineral isochrons of eclogites (samples 98702 and 99104-2) from the northern Dabie Mountains. Data points with filled square symbol (■) and hollow square symbol (□) were analyzed in China and Rennes, France, respectively.

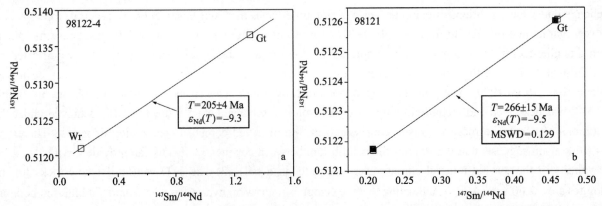

Fig. 8 Sm-Nd mineral "isochrons" of eclogites (sample 98122-4 and 98121) from the northern Dabie Mountains. Data point symbols are the same as in Fig. 7.

Figs. 7b and 8 show the Sm-Nd mineral "isochrons" of samples 99104-2, 98122-4 and 98121, which are defined only by whole-rock + garnet or by omphacite + zoned garnet. The mineral "isochron" defined by zoned garnet + omphacite from sample 99104-2 gives a younger age of 187 ± 5 Ma in Fig. 7b. This younger age may result from retrogressive zoning of garnet, which may produce lower ^{143}Nd/^{144}Nd ratios and Nd isotopic disequilibrium between zoned garnet and omphacite in sample 99104-2. The whole-rock + garnet "isochrons" yield ages of 194 ± 5, 205 ± 4 and 266 ± 15 Ma, respectively. These results show that the Sm-Nd "isochrons" defined by whole-rock + garnet, yield a large age range of 194–266 Ma, which are not completely consistent with ages determined by garnet + omphacite in Fig. 8a. This age diversity may be produced by Nd isotopic disequilibrium between garnet and retrograded minerals in whole-rocks (Li et al., 2000), because these samples are strongly retrograded eclogites and their whole-rocks contain many retrograde minerals. Hence, these ages determined by whole-rock + garnet should be used with caution.

6 Discussion

6.1 Age of the granulite-facies retrogression

As mentioned above, the garnet + Comphacite Sm-Nd age of 212 ± 4 Ma for eclogites from the northern Dabie Mountains is younger than the Sm-Nd ages of 221–228 Ma for coesite-bearing eclogites from the southern Dabie Mountains (Li et al., 1993, 2000). It has been suggested that the younger Sm-Nd isochron age for eclogites could be caused by Nd isotopic disequilibrium between garnet and omphacite (Jagoutz, 1994; Li et al., 1999; Thoni, 2002). The oxygen isotopic compositions of the mineral separates can help us determine if there is Nd isotopic equilibrium in the analyzed eclogite samples.

Generally, the oxygen diffusivities in omphacite and garnet are similar to those for Nd (Sneeringer et al., 1984; Ganguly et al., 1998; Zheng and Fu, 1998). Burton et al. (1995) suggested that the T_c for oxygen diffusion (839 °C) in garnet is even higher than that for Nd (694 °C) despite large ranges in estimated values of T_c for the latter. As demonstrated by Zheng et al. (2002) for UHP eclogite from the Su-Lu terrane, there is a direct correspondence in equilibrium or disequilibrium state between the oxygen and Nd isotope systems of eclogite minerals. The omphacite-garnet pairs with oxygen isotope equilibrium at eclogite-facies conditions yield meaningful Triassic Sm-Nd isochron ages, whereas those with oxygen isotopic dis-equilibrium give geologically meaningless non-Triassic ages (Zheng et al., 2002). Thus attainment of isotopic equilibrium in the omphacite-garnet oxygen system suggests the presence of Nd isotopic equilibrium in the same mineral pairs. As a result, if the oxygen isotope fractionation between garnet and omphacite was at equilibrium under the high temperature conditions, their Sm-Nd isotope system for the same minerals must also be at equilibrium.

As stated above, the oxygen isotope fractionations between omphacite and garnet for samples 98702 lie in the state of high temperature equilibrium. This is consistent with the observation that omphacite has δ^{18}O values higher than garnet (Table 5), consistent with the ^{18}O-rich sequence of clinopyroxene relative to garnet at thermodynamic equilibrium (Zheng, 1999). Quantitatively, the measured fractionations between omphacite and garnet for sample 98702 fall within the known range for equilibrium fractionation values at eclogite- to granulite-facies conditions. However, although omphacite has a higher δ^{18}O than garnet in sample 99104-2, its measured fractionation of 0.42‰ for omphacite-garnet pairs is significantly lower than the equilibrium fractionation values of 1.76‰–0.83‰ for omphacite-garnet pairs for temperatures of 500–900 °C (Zheng, 1999), suggesting oxygen isotopic disequilibrium between the minerals for sample 99104-2. Therefore, only the mineral Sm-Nd isotope system for sample 98702 can be inferred to be also at equilibrium and to yield meaningful isochron ages. The mineral Sm-Nd isochron age of 212 ± 4 Ma may represent the cooling age of the eclogite during exhumation. T_c of the Sm-Nd system in garnet is not well defined and controversial. Garnet may have a range of T_c from 500 to 850 °C (Li et al., 2000). The T_c for any given element in garnet depends on a number of factors such as grain size, availability of pore fluid, major element composition, nature of coexisting phases, initial temperature, and cooling rate. The peak temperatures of eclogites from the analyzed fresh eclogite samples are 808–874 °C (Liu, 2000), while the subsequent granulite-facies retrograde temperatures are 817–909 °C. In addition, some hydrous minerals such as brown hornblende and biotite formed during the granulite-facies stage, implying that there were metamorphic fluids present. The T_c of the Sm-Nd system in garnet could be 600–700 ℃ under wet conditions (Harrison and Wood, 1980; Humphries and Cliff, 1982), which is less than the temperature of granulite-facies metamorphism. Therefore, the Sm-Nd isotope system in garnet should close after retrograde metamorphism from granulite-facies, and thus, the ca. 210 Ma Sm-Nd ages of garnet + omphacite minerals represent the time of granulite-facies metamorphism. In contrast to eclogites from the northern Dabie Mountains, UHPM rocks from the southern Dabie UHPM belt did not experience granulite-facies metamorphism (Wang et al., 1989, Wang and Liou, 1991; Xu et al., 1992, 1994). Hence they quickly cooled below the T_c of Sm-Nd system in garnet and have relatively older Sm-Nd ages than eclogites from the northern Dabie Mountains. Because minerals formed during granulite-facies retrogression occur only as symplectite assemblages, the retrograde metamorphism from granulite-facies may have lasted a short time. Hence, the ca. 210 Ma age may be close to the time of eclogite-facies metamorphism. Therefore, the formation time of eclogites in the northern Dabie Mountains is similar to that of the southern Dabie Mountains (i.e. all in Triassic).

6.2 The difference of Sm-Nd ages between eclogite and garnet-pyroxenite from the NDC

Li et al. (1993) reported two Sm-Nd ages of 244 ± 11 and 224 ± 20 Ma for Raobazhai and Gaobayan (several kilometers to the west of Raobazhai) garnet-pyroxenites. The age of 244 ± 11 Ma defined by garnet + diopside + whole-rock of Raobazhai garnet-pyroxenite is significantly higher than the new age of 212 ± 4 Ma for garnet + omphacite from the Huangweihe eclogite. As mentioned above, diopside in the Raobazhai eclogite was retrograded from the omphacite and, in general, formed a retrograde rim around the relic omphacite, which is texturally similar to the retrograded rim of the garnet. Xiao et al. (2001) also observed relic omphacite in the Raobazhai garnet-pyroxenite. Therefore, the relatively old Sm-Nd age of 244 ± 11 Ma could have resulted from the lower $^{143}Nd/^{144}Nd$ ratio of diopside than that of omphacite, which results in a higher gradient for the garnet and diopside tie line. Obviously, this interpretation requires that the retrograde metamorphic fluid that caused retrograde compositional zonations in the garnet and omphacite must have had a relatively low $^{143}Nd/^{144}Nd$ ratio. Therefore, the relatively old Sm-Nd age of 244 ± 11 Ma may have been caused by Nd isotopic disequilibrium between garnet and diopside as observed by Li et al. (2000). Actually, our new Sm-Nd age of 212 ± 4 Ma for eclogite is consistent with the Sm-Nd age of 224 ± 20 Ma within error limits, which is defined by three garnet separates from the Gaobayan garnet-pyroxenite (Li et al., 1993).

6.3 Origin of eclogite protolith

In order to understand the origin of the protolith of eclogites from the northern Dabie Moutains. We also studied their Sr-Nd isotopic geochemistry (Table 6). Most eclogites have an ε_{Nd} (210 Ma) of about −10 (Table 6). These relatively low ε_{Nd} (T) values suggest that the protolith of the eclogites had LREE-enriched patterns for a long time. In addition, the trace element characteristics of the eclogites show most of them were of continental basalt origin (see Section 5.1). Therefore, the protolith of the studied eclogites were likely continental basalts instead of oceanic basalts. However, some peridotite-hosted eclogites may have been derived from enriched-mantle lithosphere. For example, samples 98702 (eclogite) and 98407-1 (peridotite) were collected from the same locality at Huangweihe town. The similar $(^{87}Sr/^{86}Sr)_I$ (at 210 Ma) values of 0.710159 and 0.710839 and ε_{Nd} (210 Ma) values of −9.7 and −10.1 for the two samples suggest similar origins. Thus, the Huangweihe eclogite that occurs within peridotite may have been derived from enriched-mantle lithosphere that was scraped off lithospheric mantle by subducting continental crust. As mentioned above, sample 98702 has a LREE depleted pattern (Fig. 6), which is inconsistent with its negative ε_{Nd} (210 Ma) value. This suggests that the protolith of the rock (sample 98702) was LREE enriched for a long period, and that the severe LREE depletion may have resulted from loss of LREE during recent geological events. Stalder et al. (1998) have experimentally shown that LREE are always enriched in the fluid phase relative to the solid phase under P-T conditions relevant to subduction processes ($D^{(fluid/Min)}$>1). If LREE preferentially partitioned into an aqueous fluid, then it is possible that fluids removed LREE from the protolith of sample 98702. However, based on oxygen isotope in situ analyses of coarse-grained garnet from the Raobazhai garnet-pyroxenite and its calculated whole-rock REE pattern, Xiao et al. (2001) considered it unlikely that retrograde fluids caused LREE depletion of the whole-rock. The REE pattern of the rock, especially the LREE, was modified and therefore does not represent the REE pattern of its protolith. It is thus proposed that LREE depletion took place before peak metamorphism of the rocks, i.e. probably during dehydration of the subducting slab or partial melting of enriched-mantle lithosphere before continental subduction.

Table 6 Whole-rock and mineral Rb-Sr and Sm-Nd isotopic data for the eclogites and related rocks from the northern Dabie mountains (T = 210 Ma)

Sample no.	Rb (ppm)	Sr (ppm)	^{87}Rb/^{86}Sr	^{87}Sr/^{86}Sr	±2σ	(^{87}Sr/^{86}Sr)$_I$
98702Wr	1.22	194.81	0.0181	0.710213	21	0.710159
98407-1Wr	1.7	168.41	0.0291	0.710926	19	0.710839

Sample no.	Sm (ppm)	Nd (ppm)	^{147}Sm/^{144}Nd	^{143}Nd/^{144}Nd	±2σ	$\varepsilon_{Nd}(T)$
98702Wr	1.53	3.14	0.2947	0.512272	10	−9.7
98702Gt	1.14	0.62	1.112	0.513461	12	
98702Omp	1.02	4.05	0.1523	0.512141	12	
*98702Wr	1.85	3.68	0.3034	0.512297	6	−9.5
*98702Gt	5.71	3.12	1.109	0.513453	13	
*98702Omp	0.83	3.19	0.1537	0.512118	6	
99104-2Wr	1.96	3.84	0.3087	0.512916	11	+2.4
99104-2Gt	1.72	0.71	1.465	0.51438	11	
99104-2Omp	1.76	3.73	0.2854	0.51294	10	
*98122-4Wr	2.71	10.88	0.1504	0.512097	4	−9.3
*98122-4Gt	1.3	0.6	1.308	0.513651	4	
98121Wr	0.95	2.75	0.2092	0.512175	12	−8.7
98121Gt	0.87	1.14	0.4596	0.512606	12	
*98121Wr	0.97	2.8	0.2097	0.512167	2	−9.7
*98121Gt	0.89	1.16	0.4626	0.512612	5	
98407-1Wr	0.32	1.3	0.1442	0.512004	12	−10.1

Sample 98407-1Wr is peridotite. Wr, Gt, Omp and Hb represent whole-rock, garnet, omphacite and hornblende, respectively. The isotope compositions for samples 98702, 98122-4 and 98121 with star symbol were measured at Rennes, France and others were obtained in China. (^{87}Sr/^{86}Sr)$_I$ is initial ratio of ^{87}Sr/^{86}Sr at T = 210 Ma

The Raobazhai eclogite (sample 99104-2) with ^{143}Nd/^{144}Nd = 0.512916 and ε_{Nd} (210 Ma) = + 2.4 seems to be of depleted mantle affinity, which is consistent with its depleted LREE pattern (Fig. 6). However, other studies reported two ε_{Nd} (T) values of −2.3 to −2.9 for garnet-pyroxenites from the Raobazhai massif (Li et al., 1993; Xiao et al., 2001). In addition, a more detailed study of the Raobazhai massif suggests that the lithospheric mantle may have been metasomatized by fluid from subducting oceanic crust (Zhang et al., 1995). Because the Huangweihe eclogite (sample 98702) with much lower ε_{Nd} (T) = −9.7 is similar in occurrence to the Raobazhai eclogite, it is suggested that the enriched lithospheric mantle represented by foliated peridotites at Raobazhai and Huangweihe is likely heterogeneous.

6.4 Constraint on the suture location in the Dabie Mountains

As mentioned above, eclogites from the northern Dabie Complex experienced Triassic UHP metamorphism. The continental crust features of their protoliths suggest that the UHP metamorphism was caused by Triassic continental subduction, similar to that of the southern Dabie UHPM belt. Recent U-Pb zircon dating suggests that the orthogneisses in the northern Dabie Mountains, which are the country rocks of the eclogite, also experienced Triassic (226 ± 6 and 229 ± 18 Ma) metamorphism (Liu et al., 2000; Xie et al., 2001). In addition, the ages for protoliths of most eclogites and orthogneisses are 700–800 Ma (e.g. Xue et al., 1997; Hacker et al., 1998; Liu et al., 2000; Xie et al., 2001; Bryant et al., 2004), similar to those in the southern Dabie Mountains. Therefore, the northern Dabie Complex should belong to the Yangtze subducted continental crust of Triassic age. Although the southern Dabie UHPM belt and northern Dabie Complex units experienced Triassic UHP metamorphism and belong to part of the Yangtze subducted continental crust, they exhibit different metamorphic processes and exhumation history, suggesting that the two eclogite-bearing units had separate histories (Liu et al., 2001; Zheng et al., 2001). There is a detachment fault between them along the Wuhe-Shuihou shear zone, which separates the northern Dabie complex and the southern Dabie UHPM belt. In addition, the eclogite-bearing foliated peridotites in the NDC may have been scraped off overlying lithospheric mantle by subduction of continental crust and were subducted with the continental crust. Hence, the northern part of the northern Dabie Complex, where foliated peridotites and eclogites are developed, could be close to the suture zone between the North China and Yangtze Blocks. This metamorphosed mafic-ultramafic belt with eclogite and meta-peridotite blocks along the southern part of the Mozitan-Xiaotian fault zone may represent tectonic melange produced during subduction of the Yangtze plate and subsequent collision between the Yangtze and North China Blocks.

7 Conclusions

(1) The eclogites in the northern Dabie Mountains formed during the Triassic period. A Sm-Nd mineral (omphacite + garnet) isochron yields an age of 212 ± 4 Ma. The relationship between the $\delta^{18}O$ values of the eclogite minerals such as garnet and omphacite suggests oxygen isotope equilibrium, pointing to the Sm-Nd isotope equilibrium between garnet and omphacite according to comparability between O and Sm-Nd diffusion coefficients. This Sm-Nd age is interpreted to represent the cooling age in the retrograde stage of granulite-facies metamorphism. The Sm-Nd "isochrons" defined by whole-rock + garnet or omphacite + zoned garnet yield a large age range of 187–266 Ma, which may be produced by Nd isotopic disequilibrium between garnet and retrograded minerals in whole-rocks.

(2) The petrologic geochemistry and Sr-Nd isotopic compositions of the eclogites suggest that the protolith for most of the eclogites is continental basalt. Most of the eclogites with lower $\varepsilon_{Nd}(T)$ values were produced from subducted continental crust and some of eclogites occurring in foliated peridotites may have been scraped off the overlying lithospheric mantle by subducted continental crust. Their positive Nb anomalies and related trace element characteristics indicate they were not the products of an island arc. The metamorphosed mafic-ultramafic belt with the Triassic eclogite and meta-peridotite blocks along the southern side of the Mozitan-Xiaotian fault zone may represent the tectonic melange produced during continental subduction and subsequent collision between the Yangtze and North China Blocks, which may be close to the suture zone between the North China and Yangtze Blocks.

(3) The $\delta^{18}O$ values of whole-rocks for the eclogites are + 2.4 to + 4.32‰, indicating that the eclogites underwent hydrothermal alteration at high-temperature by surface water prior to subduction of the Yangtze continental crust.

Acknowledgements

This study was financially supported by Chinese National Key Project for Basic Research (2003CB716500), the Natural Science Foundation of China (Nos 40172079 and 40033010), and the Major State Research Program of China (G1999075503). The first author would like to thank the staff of the Isotope Laboratory of Chinese Academy of Geological Sciences in Beijing, especially Wang Jinhui and Tang Suohan for tutoring in chemical analytical techniques. We are grateful to Prof. Peng Zicheng, Drs Zhang Zhaofeng and Huo Jianfeng for mass spectrometer analyses, and Drs Tomoharu Miyamoto, Takao Hirajima and R.Y. Zhang for their helpful suggestions and discussion. Nicole Morin and Joel Mace are highly appreciated for chemical separation and mass spectrometric analyses in Rennes. Many suggestions and constructive reviews from Professor John C. Ayers and an anonymous reviewer helped to greatly improve the manuscript. We thank the editor and Dr Jennifer Lytwyn for their help that resulted in the improvement of this contribution.

References

Baker, J., Matthews, A., Mattey, D., Rowley, D., Xue, F., 1997. Fluid-rock interactions during ultra-high pressure metamorphism, Dabie

Shan, China. Geochim. Cosmochim. Acta 61, 1685-1696.

Boynton, W.V., 1984. Geochemistry of the rare earth elements: meteorite studies, in: Henderson, P. (Ed.), Rare Earth Element Geochemistry. Elsevier, Amsterdam, pp. 63-114.

Bryant, J.L., Ayers, J.C., Gao, S., Miller, C.F., Zhang, H., 2004. Geochemical, age, and isotopic constrains on the location of the Sino-Korean/Yangtze suture and evolution of the Northern Dabie Complex, east central China. Geological Society of America Bulletin 116 in press.

Burton, K.W., Kohn, M.J., Cohen, A.S., 1995. The relative diffusion of Pb, Nd, Sr and O in garnet. Earth Planet. Sci. Lett. 133, 199-211.

Byerson, F.J., Watson, E.B., 1987. Rutile saturation in magmas: implications for Ti-Nb-Ta depletion in island-arc basalts. Earth Planet. Sci. Lett. 86, 225-239.

Chavagnac, V., Jahn, B.M., 1996. Coesite-bearing eclogites from the Bixilng complex. Dabie Mountains, China: Sm-Nd ages, geochemical characteristics and tectonic implications. Chem. Geol. 133, 29-51.

Chopin, C., 1984. Coesite and pure pyrope in high-grade blueschist of the western Alps: a first record and some consequences. Contrib. Mineral. Petrol. 86, 107-118.

Cong, B., 1996. Ultrahigh-pressure metamorphic rocks in the Dabieshan-Sulu region of China. Science Press, Beijing pp. 1-244.

Faure, G., 1986. Principles of Isotope Geology, second ed Wiley, New York pp. 200-238.

Ganguly, J., Tirone, M., Hervig, R.L., 1998. Diffusional kinetics of samarium and neodymium in garnet, and a model for determining cooling rates of rocks. Science 281, 805-807.

Hacker, B.R., Ratschbacher, L., Webb, L., Ireland, T., Walk, D., Dong, S., 1998. U/Pb zircon ages contain the architecture of the ultrahigh-pressure Qinling-Dabie orogen, China. Earth Planet. Sci. Lett. 161, 215-231.

Harrison, W.J., Wood, B.J., 1980. An experimental investigation of the partition of REE between garnet and liquid with reference to the role of defect equilibration. Contrib. Mineral. Petrol. 72, 145-155.

Hofmann, A.W., 1988. Chemical differentiation of the Earth: relationship between mantle, continental crust and oceanic crust. Earth Planet. Sci. Lett. 90, 297-314.

Hofmann, A.W., Jochum, K.P., Seufert, M., White, W.M., 1986. Nb and Pb in oceanic basalts: new constrains on mantle evolution. Earth Planet. Sci. Lett. 79, 33-45.

Holm, P.E., 1985. The geochemical finger prints of different tectonomagmatic environments using hydromagtophile element abundances of tholeiitic basalts and basaltic andesites. Chem. Geol. 51, 303-323.

Humphries, F.J., Cliff, A., 1982. Sm-Nd dating and cooling history of Scourian granulites, Sutherland, NW Scotland. Nature 295, 515-517.

Jagoutz, E., 1994. Isotopic systematics of metamorphic rocks (abstract). ICOG-8 at Berkeley Uni. U.S. Geol. Survey Circular 1107. p. 156.

Le Roex, A.P., Dick, H.J.B., Fisher, R.L., 1989. Petrology and geochemistry of MORB from 25°E to 46°E along the southwest Indian Ridge: evidence for contrasting style of mantle enrichment. J. Petrol. 30, 947-986.

Li, S., Xiao, Y., Liu, D., Chen, Y., Ge, N., Zhang, Z., Sun, S.S., Cong, B., Zhang, R., Hart, S.R., Wang, S., 1993. Collision of the North China and Yangtze blocks and formation of coesite-bearing eclogites: timing and processes. Chem. Geol. 109, 89-111.

Li, S., Wang, S., Chen, Y., Liu, D., Qiu, J., Zhou, H., Zhang, Z., 1994. Excess argon in phengite from eclogite: evidence from dating of eclogite minerals by Sm-Nd, Rb-Sr and $^{40}Ar/^{39}Ar$ method. Chem. Geol. 112, 343-350.

Li, S., Jagoutz, E., Xiao, Y., Ge, N., Chen, Y., 1996. Chronology of ultrahigh-pressure metamorphism in the Dabie mountains and Su-Lu terrane: I: Sm-Nd isotope system. Sci. Chin. (Series D) 39, 597-609.

Li, S., Jagoutz, E., Lo, C. H., Li, Q., Xiao, Y., 1999. Sm/Nd, Rb/Sr and $^{40}Ar/^{39}Ar$ isotopic systematics of the ultrahigh-pressure metamorphic rocks in the Dabie-Sulu belt, central China: a retrospective view. Int. Geol. Rev. 41, 1114-1124.

Li, S., Jagoutz, E., Chen, Y., Li, Q., 2000. Sm-Nd and Rb-Sr isotope chronology of ultrahigh-pressure metamorphic rocks and their country rocks at Shuanghe in the Dabie Mountains, central China. Geochim. Cosmochim. Acta 64, 1077-1093.

Li, S., Huang, F., Nie, Y., Han, W., Long, G., Li, H., Zhang, S., Zhang, Z., 2001. Geochemical and geochronological constraints on the suture location between the North and South China Blocks in the Dabie orogen, central China. Phys. Chem. Earth (A) 26, 655-672.

Liou, J.G., Zhang, R.Y., Ernst, W.G., Rumble, D., Maruyama, S., 1999. High-pressure minerals from deeply subducted metamorphic rocks, in: Hemley, R.J. (Ed.), Reviews in Mineralogy, vol. 37, pp. 33-96.

Liu, Y., 2000. Petrology, geochemistry and isotopic chronology of the eclogites from the northern Dabie Mountains. PhD Thesis, University of Science and Technology of China, Hefei, China. pp. 1-69.

Liu, Y., Li, S., Xu, S., Li, H., Jiang, L., Chen, G., Wu, W., Su, W., 2000. U-Pb zircon ages of the eclogite and tonalitic gneiss from the northern Dabie Mountains. China and multi-overgrowths of metamorphic zircons. Geol. J. Chin. Univ. 6, 417-423 (in Chinese with English abstract).

Liu, Y., Xu, S., Li, S., Chen, G., Jiang, L., Zhou, C., Wu, W., 2001. Distribution and metamorphic $P-T$ condition of the eclogites from the mafic-ultramafic belt in the northern part of the Dabie Mountains. Acta Geologica Sinica 75, 385-395 (in Chinese with English abstract).

Ludwig, K.R., 1997. ISOPLOT—a plotting and regression program for radiogenic-isotope data USGS Open-File Report. Version 2.92 1997 pp. 91-445.

Morimoto, N., Ferguson, A.K., Ginzburg, I.V., Ross, M., Seifert, F.A., Seifert, J., Seifert, Z., Aoki, K., Gottardi, G., 1988. Nomenclature of pyroxenes. Am. Mineral. 73, 1123-1133.

Okay, A.I., Xu, S., Sengör, A.M.C., 1989. Coesite from the Dabie Shan eclogites, central China. Eur. J. Mineral. 1, 595-598.

Okay, A.I., Sengör, A.M.C., 1992. Evidence for intracontinental thrust-related exhumation of the ultrahigh-pressure rocks in China. Geology 20, 411-414.

Peltonen, P., Kontinen, A., Huhma, H., 1996. Petrology and geochemistry of metabasalts from the 1.95 Ga Jormua ophiolite, Northeastern Finland. J. Petrol. 37, 1359-1383.

Smith, D.C., 1984. Coesite in clinopyroxene in the Caledonides and its implications for geodynamics. Nature 310, 641-644.

Sneeringer, M., Hart, S.R., Shimizu, N., 1984. Strontium and samarium diffusion in diopside. Geochim. Cosmochim. Acta 48, 1589-1608.

Sobolev, N.V., Shatsky, V.S., 1990. Diamond inclusions in garnets from metamorphic rocks: a new environment for diamond formation. Nature 343, 742-746.

Stalder, R., Foley, S.F., Brey, G.P., Horn, I., 1998. Mineral-aqueous fluid partitioning of trace elements at 900–1200 ℃ and 3.0–5.7 GPa: new experimental data for garnet, clinopyroxene, and rutile, and implications for mantle metasomatism. Geochim. Cosmochim. Acta 62, 1781-1801.

Taylor, S.R., McLennan, S.M., 1985. The Continental Crust: its Composition and Evolution. Blackwell, Oxford pp. 1-312.

Thoni, M., 2002. Sm-Nd isotope systematics in garnet from different lithologies (Eastern Alps): age results, and an evaluation of potential problems for garnet Sm Nd chronology. Chem. Geol. 185, 255-281.

Tsai, C.H., Liou, J.G., 2000. Eclogite-facies relics and inferred ultrahigh-pressure metamorphism in the North Dabie complex, central China. Am. Mineral. 85, 1-8.

Wang, X., Liou, J.G., 1991. Regional ultrahigh-pressure coesite-bearing eclogite terrane in central China: evidence from country rocks, gneiss, marble and metapelite. Geology 19, 933-936.

Wang, X., Liou, J.G., Mao, H.K., 1989. Coesite-bearing eclogites from the Dabie Mountains in central China. Geology 17, 1085-1088.

Wang, X., Zhang, R.Y., Liou, J.G., 1995. UHPM terrane in east central China, in: Coleman, R., Wang, X. (Eds.), Ultrahigh pressure metamorphism. Cambridge University Press, Cambridge, pp. 356-390.

Winchester, J.A., Floyd, P.A., 1977. Geochemical discrimination of different magma series and their differentiation products using immobile elements. Chem. Geol. 20, 325-343.

Xie, Z., Chen, J., Zhang, X., Gao, T., Dai, S., Zhou, T., Li, H., 2001. Zircon U-Pb dating of gneiss from Shizhuhe in North Dabie and its geologic implications. Acta Petrologica Sinica 17, 139-144 (in Chinese with English abstract).

Xiao, Y., Hoefs, J., van den Kerkhof, A.M., Li, S., 2001. Geochemical constraints of the eclogite and granulite facies metamorphism as recognized in the Raobazhai complex from North Dabie Shan, China. J. Metamorphic Geol. 19, 3-19.

Xu, S., Okay, A.I., Ji, S., Sengör, A.M.C., Su, W., Liu, Y., Jiang, L., 1992. Diamond from the Dabie Shan metamorphic rocks and its implication for tectonic setting. Science 256, 80-82.

Xu, S., Liu, Y., Jiang, L., Su, W., Ji, S., 1994. Tectonic regime and Evolution of Dabie Mountains. Science Press, Beijing pp. 1-175.

Xu, S., Liu, Y., Su, W., Wang, R., Jiang, L., Wu, W., 2000. Discovery of the eclogite and its petrography in the Northern Dabie Mountains. Chin. Sci. Bull. 45, 273-278.

Xu, S., Liu, Y., Jiang, L., Wu, W., Chen, G., 2002. Architecture and kinematics of the Dabie mountains orogen. University of Science and Technology of China Press, Hefei pp. 1-133.

Xu, S., Liu, Y., Chen, G., Compagnoni, R., Rolfo, F., He, M., Liu, H., 2003. New finding of micro-diamonds in eclogites from Dabie-Sulu region in central-eastern China. Chin. Sci. Bull. 48, 988-994.

Xue, F., Rowley, D.B., Tucker, R.D., Peng, Z.X., 1997. U-Pb zircon ages of granitoid rocks in the North Dabie Complex, eastern Dabie Shan, China. J. Geol. 105, 744-753.

Ye, K., Cong, B., Ye, D., 2000. The possible subduction of continental material to depths greater than 200 km. Nature 407, 734-736.

Yui, T.F., Rumble, D., Lo, C.H., 1995. Unusually low $\delta^{18}O$ ultra-high-pressure metamorphic rocks from the Sulu Terrain, eastern China. Geochim. Cosmochim. Acta 59, 2859-2864.

Zhai, M., Cong, B., Zhang, Q., Wang, Q., 1994. The northern Dabieshan terrain: a possible Andean-type arc. Intern. Geol. Rev. 36, 867-883.

Zhang, Q., Ma, B., Liu, R., Zhao, D., Fan, Q., Li, Q., Li, X., 1995. A remnant of continental lithospheric mantle above subduction zone: geochemical constrains on ultramafic rock from Raobazhai area, Anhui Province. Sci. Chin. (Series B) 38, 1522-1529.

Zhang, R.Y., Liou, J.G., Tsai, C.H., 1996. Petrogenesis of a high-temperature metamorphic terrain: a new tectonic interpretation for the north Dabie Shan, central China. J. Metamorphic Geol. 14, 319-333.

Zheng, Y. F., Fu, B., 1998. Estimation of oxygen diffusivity from anion porosity in minerals. Geochem. J. 32, 71-89.

Zheng, Y. F., Fu, B., Li, Y. L., Xiao, Y. L., Li, S. G., 1998. Oxygen and hydrogen isotope geochemistry of ultrahigh pressure eclogites from the Dabie Mountains and the Sulu terrane. Earth Planet Sci. Lett. 155, 113- 129.

Zheng, Y. F., 1999. On calculations of oxygen isotope fractionation in minerals. Episodes 22, 99-106.

Zheng, Y. F., Fu, B., Xiao, Y. L., Li, Y. L., Gong, B., 1999. Hydrogen and oxygen isotope evidence for fluid-rock interactions in the stages of pre- and post-UHP metamorphism in the Dabie Mountains. Lithos 46, 677- 693.

Zheng, Y. F., Fu, B., Li, Y.L., Wei, C.J., Zhou, J.B., 2001. Oxygen isotope composition of granulites from Dabieshan in eastern China and its implications for geodynamics of Yangtze plate subduction. Phys. Chem. Earth (A) 26, 673-684.

Zheng, Y. F., Wang, Z.R., Li, S., Li, L., 2002. Oxygen isotope equilibrium between eclogite minerals and its constrains on mineral chronometer. Geochim. Cosmochim. Acta 66, 625-634.

大别山超高压变质岩的冷却史及折返机制*

李曙光，李秋立，侯振辉，杨蔚，王莹

中国科学院壳幔物质与环境重点实验室，中国科学技术大学地球与空间科学学院，合肥 230026

> **亮点介绍**：系统总结了大别山超高压变质岩及其围岩的年代学结果，指出其经历了两次快速冷却和中间一个较长时间的等温过程；提出俯冲板片断离可能是超高压变质岩第一次快速抬升和冷却的重要机制；而已折返至中上地壳的超高压岩片在两大陆块持续汇聚和挤压背景下向北仰冲是第二次快速抬升的重要机制。通过大别山 Pb 同位素填图揭示出南大别带超高压变质岩和北大别带超高压变质岩分别源于俯冲上、下陆壳，表明在陆桥俯冲过程中上、下地壳之间可发生分离或解耦。

摘要 大别山超高压变质岩及其围岩 T-t 冷却曲线显示了超高压变质岩从 800 ℃到 300 ℃经历了三个阶段：两次快速冷却（226 ± 3 Ma 到 219 ± 7 Ma 期间从 800 ℃到 500 ℃的第一次快速冷却，180~170 Ma 期间从 450 ℃到 300 ℃的第二次快速冷却）和介于二者之间的等温过程。这一具有两次快速冷却的 T-t 曲线已被近年来获得的高精度金红石 U-Pb 年龄（218 ±1.2 Ma）(Li et al., 2003)，高压变质和退变质独居石 Th-Pb 年龄（Ayers et al., 2002)，和强面理化榴辉岩二次多硅白云母的 Rb-Sr 年龄（182.7 ± 3.6 Ma）(Li et al., 2001) 所证实。超高压变质岩的二次快速冷却事件反映了二次快速抬升过程。在东秦岭及苏鲁地体东端发育的同碰撞花岗岩 U-Pb 为 225~205 Ma，与超高压变质岩第一次快速冷却时代吻合。考虑到同碰撞花岗岩与俯冲板片断离的成因联系，这种时代耦合关系表明俯冲板片断离可能是超高压变质岩第一次快速抬升和冷却的重要机制之一。大别山 Pb 同位素填图揭示出南大别带超高压变质岩具有高反射成因 Pb 特征，因而源于俯冲的上地壳；而北大别带超高压变质岩具有低放射成因 Pb 特征，源于俯冲长英质下地壳。这表明在陆桥俯冲过程中上、下地壳之间可发生挤离（detachment）或脱耦（decoupling）。已有实验证明脱耦的上地壳在俯冲过程中可沿挤离面逆冲抬升（Chemenda et al., 1995)。同理，由于俯冲镁铁质下地壳在大别山没有出露，可以推测俯冲长英质下地壳和镁铁质下地壳之间也发生了挤离或脱耦。大陆岩石圈在不同深度存在若干低黏度带（Meissner and Mooney, 1998）是上述俯冲陆壳分层脱耦现象发生的依据。因此，俯冲上地壳及部分长英质下地壳的第一次快速抬升折返是俯冲过程中大陆地壳内部分层脱耦和俯冲板片断离的综合结果。上述过程只能使已脱耦的上地壳及部分长英质下地壳抬升折返，而未与俯冲岩石圈脱耦的下地壳在板片断离后仍可继续俯冲。俯冲板片断离后，两大陆块在晚三叠世和早-中侏罗世的继续汇聚，导致华南陆块下地壳继续俯冲，及已经脱耦并折返至中上地壳的超高压岩片向北仰冲。这一仰冲可能是导致超高压变质岩第二次快速抬升的重要机制。强面理化榴辉岩二次多硅白云母的 Rb-Sr 年龄（182.7 ± 3.6 Ma）可能记录了这一超高压岩片仰冲事件发生的时代。惠兰山基性麻粒岩年代学研究揭示了罗田穹隆在早白垩世的快速抬升，与此同时大别山发生了大规模岩浆事件。山体快速抬升与大规模岩浆事件的耦合关系指示了大别造山带早白垩世的去根作用或岩石圈拆离事件。伴随着一山体快速抬升，大别山超高压变质岩开始大面积出露地表。

关键词 超高压变质岩；冷却史；折返机制；大别造山带

1 引言

在陆-陆碰撞过程中，先期俯冲洋壳的拖曳作用，及导致陆块会聚的地幔驱动力可导致一个陆块的岩石圈向另一陆块的岩石圈下面俯冲到地幔深度。这一陆壳深俯冲作用已被陆壳岩石中所发现的柯石英和金刚石等超高压变质矿物所证实（Chopin, 1984; Smith, 1984; Wang et al., 1989; Okay et al., 1989; Soblev and Shatsky,

---

* 本文发表在：岩石学报，2005，21：1117-1124

1990; Xu et al., 1992)。这种含柯石英和金刚石包裹体的造山带变质岩我们称为超高压变质岩(它包括榴辉岩、硬玉石英岩及超高压片麻岩等),它们是陆壳岩石在俯冲到 >100 km 深度时,在高温($T \approx 700 \sim 900$ ℃),高压($P \geqslant 27$ kbar)下变质形成的。超高压变质岩形成于 100 km 以上的深度,但今天已出露于地表,且柯石英这种超高压矿物相被保留下来。这说明它们一定经历了快速的抬升折返过程。查明超高压变质岩快速折返地表的构造机制一直是大陆深俯冲研究中最重要的科学问题之一。在精确地测定超高压变质岩各种变质年龄或冷却年龄基础上构建的 T-t 冷却曲线可指示超高压变质岩的抬升冷却历史。造山带不同超高压岩片的同位素填图和精确年代学研究,同碰撞及碰撞后岩浆岩年代学和碰撞后山体抬升速率测定,可以为揭示超高压变质岩的快速抬升机制提供重要制约。近 5 年来,我们围绕这一科学问题做了较深入的化学地球动力学探讨。已有研究表明大陆岩石圈流变学性质的不均一性导致大陆岩石圈在俯冲过程中会发生与大洋岩石圈俯冲不同的构造运动。本文将对这方面的工作做一总结和讨论。

2 超高压变质岩的冷却史

应用同位素年代学原理,通过对各种超高压变质及退变质矿物进行精确定年,我们可以测定出超高压变质岩折返过程的冷却 T(温度)-t(时间)曲线,从而揭示其抬升过程。Li et al.(2000)通过对大别山超高压变质岩的热年代学研究已揭示出它们从峰期变质温度 800 ℃到 300 ℃ 经历了两次快速冷却过程:226~219 Ma 期间经历了从 800 ℃ 到 500 ℃的第一次快速冷却阶段(40 ℃/Ma);在 219~180 Ma 期间为一温度在 500 ℃到 450 ℃范围内的等温阶段,并对应一个退变质角闪岩相重结晶过程;而后在 180~170 Ma 又经历了从 450 ℃到 300 ℃的第二次快速冷却阶段(15 ℃/Ma)(图 1)。

为了验证这样一条包含两次快速冷却的 T-t 曲线,近年来我们精确测定了大别山含柯石英榴辉岩中金红石的 U-Pb 年龄(218 ± 1.2 Ma)(Li et al., 2003)。这是世界上第一个精确测定的榴辉岩金红石 U-Pb 年龄,其对应的封闭温度 T_c = 460 ℃。这一年龄支持上述第一次快速冷却事件在 219 Ma 冷却到 500 ℃的结论。此外,我们还获得了一系列强面理化

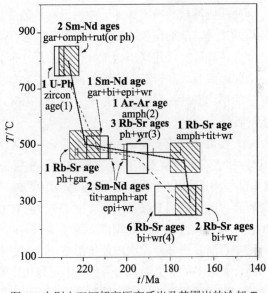

图 1 大别山双河超高压变质岩及其围岩的冷却 T-t 曲线(据 Li et al., 2000)

榴辉岩沿面理发育的二次多硅白云母 Rb-Sr 年龄(182.7 ± 3.6 Ma~187 ± 3.7 Ma)(李曙光等, 2000; 李秋立, 2003)。根据石榴石中脆性劈理与面理的关系,已有研究指出,这种面理是在角闪岩相条件下由挤压剪切构造形成的(徐树桐等, 1999)。这一沿面理发育的二次多硅白云母 Rb-Sr 年龄(平均值 185 ± 2 Ma)和对应的变质温度(450 ℃)均与上述第二次快速冷却事件相吻合,因此它是超高压变质岩第二次快速抬升的构造记录。此外,Ayers et al. (2002)报道了独居石 Th-Pb 年龄,其核部年龄(223 ± 1 Ma)指示高压变质独居石冷却到 650~700℃的时代;其退变质增生边年龄(209 ± 3 Ma)指示角闪岩相退变质重结晶时代。这些

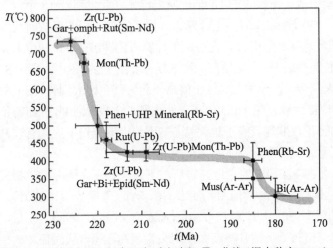

图 2 改进的大别山超高压变质岩冷却 T-t 曲线(据李秋立 2003)

结果均进一步支持了 Li et al.(2000)测定的大别山超高压变质岩二次快速冷却曲线。综合这些新的年龄数据,可获得一条更为精确的大别山超高压变质岩冷却曲线(图 2)。

3 超高压变质岩的折返机制

3.1 第一次快速抬升的机制

Davies and Blanckenburg(1995)提出俯冲板片断离（slab break off）是导致超高压岩片在浮力作用下快速抬升折返的主要机制，与此同时，由于断离俯冲板片对上地幔的扰动作用可诱发同碰撞岩浆作用。因此，查明秦岭-大别造山带同碰撞花岗岩形成时代，并与超高压变质岩二次快速冷却事件的发生时代进行对比，可以帮助我们判断俯冲板片断离是否是导致大别山超高压变质岩抬升折返的主要构造机制，以及它应对哪一次快速抬升事件负责。我们对东秦岭 6 个同碰撞花岗岩侵入体进行了系统的同位素年代学研究，获得的锆石 U-Pb 年龄为 205~220 Ma (Sun et al., 2002)。Chen et al. (2003)对苏鲁地体东部 3 个同碰撞花岗岩的锆石 U-Pb 定年也获得了 205~225 Ma 的年龄。这些年龄与大别山超高压变质岩第一次快速冷却事件的结束时代一致。这种一致性表明大别山超高压变质岩的第一次快速抬升与俯冲板片断离有关，它使俯冲陆壳失去俯冲大洋板片（榴辉岩相）的下拖力而在浮力作用下迅速反弹(Davies and Blanckenburg, 1995)。

然而，俯冲板片断离是否是导致超高压变质岩第一次快速抬升的唯一机制，以及是否所有的超高压变质岩都要在俯冲板片断离后才开始折返则是需要进一步查清的问题。此外，俯冲陆壳在俯冲板片断离后，是整体反弹抬升，从而使俯冲陆壳整体收缩回去(Wang and Liou, 1991)，大陆深俯冲作用停止，还是这种浮力仅使部分俯冲陆壳（超高压岩片）快速抬升，而大陆深部岩石圈仍继续进行俯冲，是一个令人感兴趣的问题。这个问题涉及大陆岩石圈在深俯冲过程中是像海洋板块一样保持统一刚性板块，还是由于大陆岩石圈流变学性质的不均一性而发生层间挤离（detachment）或脱耦（decoupling），从而具有与海洋板块俯冲不同的构造运动特征。

大别山超高压变质带可划分为两个一级岩片，它们是靠近地缝合线的北大别高压变质带（岩石不含柯石英），和岩石富含柯石英的南大别超高压变质带，它们以五河-水吼韧性剪切带分隔。全岩 Pb 同位素填图表明，北大别变质岩含放射成因 Pb 较低（$^{206}Pb/^{204}Pb$ = 15.844~17.204），具有下地壳特征；而南大别超高压变质岩含放射成因 Pb 较高（$^{206}Pb/^{204}Pb$ = 17.026~20.781），具有上地壳特征（李曙光等，2001；张宏飞等，2001）。然而这种全岩 Pb 同位素填图工作有两个缺点：①地表采的全岩样品的 U/Pb 和 Th/Pb 值可能因地表风化左右而发生分异，造成经年龄校正获得的初始 Pb 同位素值有较大偏差；②采自南大别超高压变质带的样品含有较多的表壳岩，而采自北大别高压变质带的样品基本都是深源的正片麻岩，二者原岩差异大，不好比较。为了克服这些缺点，检验全岩 Pb 同位素填图所获得的结论是否成立，近年来我们又开展了大别山片麻岩长石普通 Pb 填图研究。假设在超高压变质条件下，长石的 Pb 同位素组成与全岩重新平衡，则该长石普通 Pb 可代表全岩超高压变质前的初始 Pb 同位素组成。所采集的 34 个片麻岩样品（北大别带 15 个，南大别带 19 个）均是原岩为晚元古代（7 亿~8 亿年）中酸性岩浆岩的正片麻岩。由于它们原岩的形成时代和成因相同，如果它们原岩的 Pb 同位素组成没有显著差异，则南、北大别带正片麻岩俯冲前的 Pb 同位素差异是由于其原岩 U/Pb 差异造成的。这些正片麻岩长石普通 Pb 测定结果如图 3 所示。图 3 显示南大别带（S）的 $^{206}Pb/^{204}Pb$ 中心值大于北大别带（N），说明南大别带正片麻岩的 U/Pb 值高于北大别；南大别带（S）的 $^{208}Pb/^{206}Pb$ 中心值小于北大别带（N），说明南大别带正片麻岩的 Th/U 值小于北大别。这些普通 Pb 分布特征说明南大别带正片麻岩具有上地壳特征，而北大别带正片麻岩具有下地壳特征，因为上地壳较下地壳有较高的 U 含量。它验证了全岩 Pb 同位素填图所获得的基本结论是正确的。

根据 Pb 同位素填图工作，我们指出大别山华南陆块深俯冲的上地壳岩石与下地壳岩石之间应发生挤离（detachment）并形成一主逆断层，它使深俯冲的上地壳岩片向南仰冲到浅部 (图 4A)（李曙光等，2001）。这一上、下地壳之间的挤离或脱耦与上地壳下部的低黏度性质有关（Meissner and Mooney, 1998）。此后，由于早白垩世大别山的穹隆构造使北大别带（位于穹隆核心）抬升幅度大于南大别超高压带，从而使北大别带折返的俯冲上地壳岩片被剥蚀掉而出露俯冲下陆壳岩石（图 4B）。以大陆岩石圈下地壳具有低黏度性质为假设前提的构造模拟实验已证明，在陆壳俯冲过程中在下地壳低黏度带形成大的挤离面和逆断层，从而导致俯冲到地幔深度的上地壳岩片逆冲折返到浅部（Chemenda et al., 1995）。需要指出的是这一过程可以发生在俯冲板片断离以前的陆壳俯冲过程中。因此在俯冲板片发生断离以前，脱耦的俯冲上地壳已开始抬升折返，俯冲板片断离只是进一步加速了这一抬升折返过程。事实上北大别高压变质带岩石以中性的闪长质片麻岩为主，基性的榴辉岩较少，因此它仍属于下地壳上部的岩石。这部分岩石由于平均比重较小可以获得上浮推力，而俯冲的最下部地壳（lowermost crust）镁铁质榴辉岩层因其高密度而不能获得上浮力。因此可以推测，在俯冲陆壳下地壳的上部层位与下部层位之间很可能发育另一大的挤离面和逆断层，它使俯冲下地壳的上部岩片（如北大别带）逆冲上升而出露地表，而下部下地壳（包含镁铁质岩层）继续随大陆岩石圈地幔俯冲，直到陆块的汇聚碰撞过程停止。Meissner and Mooney(1998)指出下地壳内部也存在低黏度带。这为下地壳内部发生挤离或脱耦提供了条件。

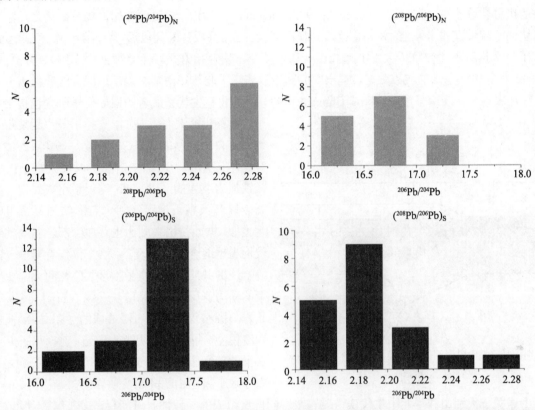

图 3　大别山北大别高压变质带和南大别高压变质带正片麻岩长石普通 Pb 同位素组成统计直方图(据李曙光等未发表数据)

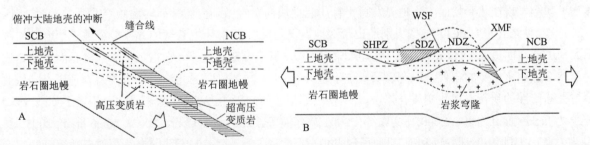

图 4　A 大陆深俯冲过程中俯冲上地壳与下地壳之间发生挤离并产生逆断层，使俯冲上地壳岩片逆冲至浅部；B 早白垩世穹隆构造使北大别带（NDZ）大幅度抬升并被强烈剥蚀从而使俯冲下地壳岩石出露地表（据 Li et al., 2003）

小结：由于大陆地壳内部流变学性质的不均一性和存在若干低黏度带，在陆壳俯冲过程中上、下地壳之间，以及下地壳内部可发生挤离（detachment）或脱耦（decoupling）。与俯冲岩石圈脱耦的上地壳及其他岩石在俯冲过程中可沿挤离面逆冲抬升。俯冲板片断离使俯冲大陆岩石圈突然失去了俯冲洋壳的下拖力而减速，它可使减速的俯冲岩石圈对已脱耦岩片的向下拖带力减弱，从而使已脱耦岩片获得的向上净浮力增加并导致其抬升过程加速。因此，大别山超高压变质岩的第一次快速抬升和冷却事件是陆壳俯冲时的壳内挤离及逆冲构造和俯冲板片断离导致的超高压变质岩片净上浮力增加的综合效果。在上述过程中，只有与俯冲岩石圈脱耦的上地壳及部分长英质下地壳折返抬升，而未与俯冲岩石圈脱耦的下部下地壳不会抬升折返。如果在板片断离后华北和华南陆块的会聚过程仍不停止，与喜马拉雅碰撞带类比，可以推测这部分未脱耦的下地壳将有可能与华南陆块岩石圈地幔一起在无俯冲洋壳牵引的情况下进行平俯冲（flat subduction），并构造板底垫托于华北岩石圈之下。

3.2 第二次快速抬升的机制

超高压变质岩的第二次快速冷却过程指示其相关构造抬升事件应发生在180 Ma左右。各种地质现象表明，侏罗纪时期大别山仍处在挤压构造体制中（周进高等，1999）。古地磁研究表明，直到晚侏罗世华北和华南陆块的极移轨迹才拼合在一起（Lin et al., 1985; Yang et al., 1991），这说明在俯冲板片断离之后华北和扬子两大陆块的汇聚过程并未停止，并一直持续到晚侏罗世。在俯冲板片断离以后，两大陆块继续汇聚过程中，究竟发生了什么构造事件导致了大别山超高压变质岩第二次快速抬升是一个重要而令人感兴趣的问题。Li Z.X.（1994）根据中国东部的航磁数据，提出郯庐断裂以东，华北与华南陆块深部地缝合线应位于南京以东一线，它较地表地缝合线南移了400 km。Chung（1999）对苏北新生代玄武岩的地球化学研究也指出苏北陆下岩石圈地幔是华北型的，支持了郯庐断裂带以东深部地缝合线南移的观点。李曙光、杨蔚（2002）通过对比中国东部华北和华南陆块以及大别山的中生代镁铁质岩浆岩的Sr-Nd-Pb同位素组成发现，在郯庐断裂以西大别山地区深部岩石圈的同位素组成与华北类似，据此得出大别山也存在深部岩石圈地缝合线较地表地缝合线南移现象。这三位作者对这一深部地缝合线的共同解释就是在侏罗纪，在两大陆块继续汇聚的动力作用下，华南陆块北缘岩石圈发生劈裂，华北岩石圈楔入华南岩石圈造成的（图5）。根据这一构造模型，华南陆块北缘的陆壳（包括已折返至地壳水平的超高压岩石）向北仰冲到华北岩石圈之上，从而导致了大别山超高压变质岩的第二次快速抬升。

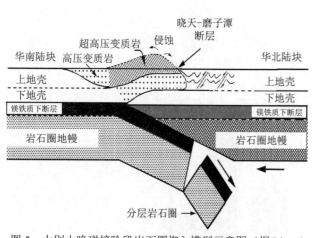

图5 大别山晚碰撞阶段岩石圈楔入模型示意图（据Li and Yang，2003）

尽管上述关于大别山和华北中生代镁铁质岩浆岩同位素组成的类似性有可能存在多种解释，但是大量其他研究均显示在秦岭-大别造山带碰撞晚期华北陆块南缘向南俯冲，而华南陆块北缘向北仰冲的事实。例如，地球物理观测发现在北秦岭的中下地壳发育平行华北地块向秦岭做陆内俯冲并向南倾斜的反射界面（张国伟等，2001）。刘福田等（2003）依据大别造山带深地震宽角反射/折射研究获得的6条二维地壳速度结构剖面显示大别山莫霍面最深（41 km）并发生错断处位于晓天-磨子潭断裂之下，它指示了深部岩石圈地缝合线的位置。然而近年来大量的年代学研究表明，位于晓天-磨子潭断裂以北的北淮阳庐镇关杂岩是华南陆块北缘未参与大陆俯冲的晚元古代浅变质岩浆岩，因此地表地缝合线应位于庐镇关杂岩以北，相当于信阳-舒城断裂一线（Hacker et al., 1998, 2000; 马文璞等，2001; 周建波等，2001; 谢智等，2002; Chen et al., 2003; 吴元保等，2004）。因此，地球物理和地表地质证据同样显示大别山仍存在深部地缝合线与地表地缝合线解耦现象，华南陆块中上地壳在北淮阳处向北仰冲覆盖于华北深部岩石圈之上。大别山北部北淮阳带直到中-晚侏罗系才出现的前陆盆地（周进高等，1999）可能就是由于华南陆壳侏罗纪向北仰冲推覆造成的。此外，Lu et al.（2004）依据对信阳地区中生代火山岩中的地幔和下地壳包体研究也提出了华北深部岩石圈楔入华南岩石圈的模型。所有这些工作都表明在侏罗纪华北、华南陆块继续汇聚的挤压条件下，华南陆块中上地壳向北仰冲作为大别山超高压变质岩第二次快速抬升的构造机制是非常可能的。大别-苏鲁超高压榴辉岩块的边部常发育强面理化榴辉岩，前已述这种面理是超高压岩石第二次快速抬升的构造记录。这些沿面理发育的二次多硅白云母平均Rb-Sr年龄（185±2 Ma）指示了榴辉岩二次快速抬升发生的时代，因而它也指示了这次仰冲构造事件发生的时代。

3.3 早白垩世岩石圈拆离和造山带去根事件及超高压变质岩的第三次快速抬升

最近人们在大别山北部中-晚侏罗系砾岩中发现了榴辉岩砾石,它证明大别山高压或超高压变质岩在中侏罗世时已折返出露于地表(Wang et al., 2002)。这说明经过两次快速抬升后,部分超高压变质岩已出露地表。然而目前出露地表尚未被剥蚀的超高压变质岩在中侏罗世时仍具有 300 ℃的温度,并未出露地表。它们出露地表还需要山体进一步抬升和剥蚀。李曙光等(2001b)指出在早白垩世大别山发生了一次岩石圈拆离事件并导致山体快速抬升。然而岩石圈拆离这种很难直接观察的深部构造事件需要更多的证据。山体快速抬升与大规模岩浆事件的耦合是指示岩石圈拆离事件发生的有力证据(Kay and Kay, 1993)。最近,我们对大别山罗田穹隆的惠兰山基型麻粒岩进行了详细的年代学研究。研究结构表明该麻粒岩是就位于中-下地壳的新元古代镁铁质岩浆岩在早白垩世大别造山带引张条件下因地幔上涌受热发生麻粒岩相变质作用而形成的。该麻粒岩的变质矿物(石榴石+紫苏辉石+单斜辉石)Sm-Nd 等时线年龄为 136 ± 18 Ma(侯振辉等,2005),变质锆石 SHRIMP U-Pb 年龄 124.0 ± 1.8 Ma(侯振辉,2003)。它们对应的麻粒岩相变质温度约为 800℃。该样品的 2 个角闪石给出一致的 Ar-Ar 年龄 119.7 ± 0.5 Ma,对应封闭温度约 500 ℃。据此,我们可获得罗田穹隆的冷却速率为约 70 ℃/Ma。其冷却速率可以与超高压变质岩的第一次快速冷却相比较。这一麻粒岩的快速冷却反映了罗田穹隆在早白垩世的一次快速抬升。与此同时,大别山在早白垩世发生了大规模岩浆事件(如 Hacker et al., 1998; Ma et al., 1998; 李曙光等, 1999; Wei et al., 2001)。这一大规模岩浆事件和山体快速抬升的时代耦合关系证明了其深部发生了岩石圈拆离或山根去根事件,它可能是大别山超高压变质岩被进一步抬升至地表的原因.

4 结论

现今出露大别山地表的超高压变质岩经历了两次快速冷却历史:第一次快速冷却发生在 226~219 Ma,从 800 ℃冷却到 500 ℃;第二次快速冷却发生在 185~170 Ma,从 450 ℃冷却到 300 ℃。这两次快速冷却反映了超高压变质岩经历的两次快速抬升事件。

由于大陆岩石圈流变学性质的不均一性,在陆壳俯冲过程中,上下陆壳之间就可发生挤离或脱耦,脱耦的俯冲上地壳可沿挤离面逆冲抬升。由于大陆壳比重较地幔轻,它俯冲进入地幔所产生的浮力随俯冲深度增大而增大。这不断增大的浮力将减缓陆壳俯冲速度,并在与前端俯冲洋壳的下拖力的联合作用下将俯冲板片拉断。俯冲板片的断离并不能使全部俯冲陆壳反弹抬升。它仅使脱耦的俯冲上陆壳在浮力作用下加速抬升折返,下陆壳在板块汇聚动力作用下可继续俯冲。继续俯冲的下陆壳内部还可发生进一步脱耦,使长英质下地壳在浮力作用下抬升折返,而榴辉岩相的镁铁质下地壳继续俯冲直至最后拆离。这是一种大陆岩石圈特有的叠瓦式俯冲过程。

俯冲板片断离后陆块的继续汇聚挤压作用还可导致缝合带两侧岩石圈的相互楔入。在此条件下,大别山华南陆块中-上地壳的向北仰冲是造成超高压变质岩第二次快速抬升的主要机制。

大别山的岩石圈拆离或山根去根事件发生在早白垩世。它导致大别山的大规模岩浆事件及山体快速抬升。与此相伴随,超高压变质岩得到进一步抬升和出露。大别-苏鲁造山带的超高压变质岩正是由于经历了多次快速抬升过程才形成了世界上最大面积出露的超高压变质带。

致谢

本工作受国家重点基础研究发展规划项目(G1999075503)和国家自然科学基金(批准号:40373009, 40173014)资助。郭敬晖审阅了此文并提供了修改建议,在此表示感谢。

参考文献

侯振辉. 2003. 大别-苏鲁造山带高级变质岩锆的地球化学、锆石微量元素特征及年代学效应. 中国科学技术大学博士论文.

侯振辉, 李曙光, 陈能松等. 2005. 大别造山带惠兰山基性麻粒岩的 Sm-Nd 和锆石 SHRIMP U-Pb 同位素年代学及锆石微量元素地球化学. 中国科学(出版中)

李秋立. 2003. 大别-苏鲁超高压的热年代学及冷却史研究. 中国科学技术大学博士论文

李曙光, 洪吉安, 李慧民等. 1999. 大别山辉石岩-辉长岩体的锆石 U-Pb 年龄及其地质意义. 高校地质学报, 5(3): 351-355

李曙光, 黄方, 周红英等. 2001a. 大别山双河超高压变质岩及北部片麻岩的同位素组成——对超高压岩石折返机制的制约. 中国科学（D 辑），31（12）：977-984

李曙光, 黄方, 李晖. 2001b. 大别-苏鲁造山带碰撞后的岩石圈拆离. 科学通报, 46(17): 1487-1491

李曙光, 孙卫东, 张宗清等. 2000. 青岛仰口榴辉岩的 Nd 同位素不平衡及二次多硅白云母 Rb-Sr 年龄. 科学通报, 45: 2223-2227

李曙光, 杨蔚. 2002. 大别造山带深部地缝合线与地表缝合线的解耦及大陆碰撞岩石圈楔入模型：中生代幔源岩浆岩 Sr-Nd-Pb 同位素证据. 科学通报, 47(24): 1898-1905

刘福田, 徐佩芬, 刘劲松等. 2003. 大陆深俯冲带的地壳速度结构—东大别造山带深地震宽角反射/折射研究. 地球物理学报, 46(3): 366-372

马文璞, 王关玉, 王果胜. 2001. 佛子岭岩群中的晋宁期深成岩带及其构造含义. 地质论评, 47(5): 476-482

吴元保, 郑永飞, 龚冰等. 2004. 北淮阳庐镇关岩浆岩锆石 U-Pb 年龄和氧同位素组成. 岩石学报, 20(5): 1007-1024

谢智, 陈江峰, 张巽等. 2002. 北淮阳新元古代基型侵入岩年代学初步研究. 地球学报, 23(6): 517-520

徐树桐, 刘贻灿, 苏文等. 1999. 大别山超高压变质带面理化榴辉岩中变形石榴石的几何学和运动学特征及其大地构造意义. 岩石学报, 15(3): 321-337

张国伟, 张本仁, 袁学诚等. 2001. 秦岭造山带与大陆动力学. 科学出版社，北京

张宏飞, 高山, 张本红等. 2001. 大别山地壳结构的 Pb 同位素地球化学示踪. 地球化学, 30: 395-401

周建波, 郑永飞, 李龙等. 2001. 扬子板块俯冲的构造加积楔. 地质学报, 57(3): 338-351

周进高, 赵宗举, 邓红婴. 1999. 合肥盆地构造演化及含油气性分析. 地质学报, 73: 15-24

Ayers J C, Dunkle S, Gao S et al. 2002. Constraints on timing of peak and retrograde metamorphism in the Dabie Shan ultrahigh-pressure metamorphic belt, east-central China, using U-Th-Pb dating of zircon and monazite. Chem Geol, 186: 315-331

Chemenda A I, Mattauer M, Malavieille J et al. 1995. A mechanism for syncollisional rock exhumation and associated normal faulting: Results from physical modeling. Earth Planet Sci Lett, 132: 225-232

Chen J F, Xie Z, Li H M et al. 2003. U-Pb zircon ages for a collision-related K-rich complex at shidao in the Sulu ultrahigh pressure terrane, China. Geochemical Journal, 37: 35-46

Chopin C. 1984. Coesite and pure pyrope in high-grade blueschists of the western Alps: a first record and some consequences. Contrib Mineral Petrol, 86: 107-118

Chung S L. 1999. Trace element and isotope characteristics of Cenogoic basalts around the Tanlu fault with implications for the Eastern Plate Boundary between North and South China. The J. of Geology, 107: 301-312

Davies J H, Vonblanckenburg F. 1995. Slab breakoff—a model of lithosphere detachment and its test in the magmatism and deformation of collisional orogens. Earth Planet Sci Lett, 129(1-4): 85-102

Hacker B R, Ratschbacher L, Webb L et al. 1998. U/Pb zircon ages constrain the architecture of the ultrahigh-pressure Qinling-Dabie Orogen, China. Earth Planet Sci Lett, 161: 215-230

Hacker B R, Ratschbacher L, Webb L et al. 2000. Exhumation of ultrahigh-pressure continental crust in east central China: Late Triassic-Early Jurassic tectonic unroofing. J. Geophysical Research, 105: 13339-13364

Hou Z H. 2003. Zirconium geochemistry, trace elemental characters of zircons and its chronological applications of high-grade metamorphic rocks in the Dabie-Sulu orogen. Ph.D thesis, USTC

Hou Z H, Li S G, Chen N S et al. 2005. Sm-Nd and zircon SHRIMP U-Pb dating of Huilanshan mafic granulite in the Dabie mountains and its zircon trace element geochemistry. Sci China (Ser. D) (in press)

Kay R W, Kay S M. 1993. Delamination and delamination magmatism. Tectonophysics, 219: 177-189

Li Q L. 2003. Thermo-chronology and cooling history of the ultrahigh-pressure metamorphic rocks in Dabie-Sulu terrane. Ph.D thesis, USTC

Li Q L, Li S G, Zheng Y F et al. 2003. A high precision U-Pb age of metamorphic rutile in coesite-bearing eclogite from the Dabie Mountains in central China: a new constraint on the cooling history. Chem Geol, 200: 255-265

Li S G, Yang W. 2003. Decoupling of surface and subsurface sutures in the Dabie orogen and a continental-collisional lithospheric-weding model: Sr-Nd-Pb isotopic evidences of Mesozoic igneous rocks in eastern China. Chinese Sci Bull, 48: 831-838

Li S G, Hong J A, Li H M, et al. 1999. U-Pb zircon ages of the pyroxenite-gabbro intrusions in Dbaie mountains and their geological implications. Geological Journal of China Universities, 5(3): 351-355(in Chinese with English abstract)

Li S G, Jagoutz E, Chen Y Z, et al. 2000. Sm-Nd and Rb-Sr isotopic chronology and cooling history of ultrahigh pressure metamorphic rocks and their country rocks at Shuanghe in the Dabie Mountains, Central China. Geochimica et Cosmochimica Acta

Li S G, Huang F, Li H. 2001a. Post-collisional delamination of the lithosphere beneath Dabie-Sulu orogenic belt. Chinese Science

Bulletin, 46(17): 1487-1490

Li S G, Sun W D, Zhang Z Q, et al. 2001b. Nd isotope disequilibrium between minerals and Rb-Sr age of the secondary phengite in eclogite from the Yangkou area, Qindao, eastern China . Chinese Sci Bull, 46(3): 252-255

Li S G, Huang F, Zhou H, et al. 2003. U-Pb isotopic compositions of the ultrahigh pressure metamorphic (UHPM) rocks from Shuanghe and gneisses from Northern Dabie zone in the Dabie mountains, central China: Constraint on the exhumation mechanism of UHPM rocks. Science in China (ser. D), 46:200-209

Li Z X. 1994. Collision between the North and South China Blocks: a crustal—detachment model for the suturing in the region east of the Tanlu fault. Geology, 22: 739-742

Lin J L, Fuller M, Zhang W Y. 1985. Preliminary Phanerogoic polar wander paths for the North and South China Blocks. Nature, 313: 444-449

Liu F T, Xu P F, Liu J S et al. 2003. The crustal velocity structure of the continental deep subduction belt: study on the eastern Dabie orogen by seismic wide-angel reflection/refraction. Chinese Journal of Geophysics, 46(3): 366-372(in Chinese with English abstract)

Lu F X, Wang C Y, Zheng J P. 2004. Lithospheric composition and structure beneath the northern margin of the Qinling orogenic belt. Science in China (D), 47(1): 13-22

Ma C Q, Li Z C, Ehlers C et al. 1998. A post-collisional magmatic plumbing system : Mesogoic granitoid plutons from the Dabieshan high-pressure and ultrahigh-pressure metamorphic zone, east-central China. Lithos, 45:431-456

Ma W P, Wang G Y, Wang G S. 2001. Jinninggian plutonic belt in the Fuziling Group and its tectonic implication. Geol. Rev., 47(5): 476-481 (in Chinese with English abstract)

Meissner R, Mooney W. 1998. Weakness of the lower continental crust: a condition for delamination, uplift and escape. Tectonophysics, 296: 47-60

Okay A I, Xu S T, S, Sengör A M C. 1989. Coesite from the Dabie Shan eclogite, Central China. Eur J Mineral, 1:595-598

Smith D C. 1984. Coesite in clinopyroxene in the Caledonides and its implications for geodynamics. Nature, 310: 641-644

Soblev N V, Shatsky V S. 1990. Diamond inclusions in garnets from metamorphic rocks - a new environment for diamond formation. Nature, 343(6260): 742-746

Sun W D, Li S G, Chen Y D et al. 2002. Timing of synorogenic granitoids in the South Qinling, central China: constraints on the evolution of the Qinling-Dabie orogenic belt. J. of Geology, 110: 457-468

Wang D X, Liu Y, Li S G et al. 2002. Lower time limit on the UHPM rock exhumation: Discovery of eclogite pebbles in the Late Jurassic conglomerates from the northern foot of the Dabie Mountains, eastern China. Chinese Sci Bull, 47(3): 231-235

Wang X M, Liou J G. 1991. Regional ultrahigh-pressure coesite-bearing eclogitic terrane in central China - evidence from country rocks, gneiss, marble, and metapelite. Geology, 19(9): 933-936

Wang X M, Liou J G, Mao H K. 1989. Coesite-bearing eclogites from the Dabie Mountains in central China. Geology, 17: 1085-1088

Wei C J, Zhang L F, Wang S G. 2001. Mesozoic high-K granitic rocks from the eastern Dabie Mountains, Central China and their geological implications. Sci in China (Ser. D), 44(6): 525-534

Wu Y B, Zheng Y F, Gong B et al. 2004. Zircon U-Pb ages and oxygen isotope compositions of the Luzhengguan magmatic complex in the Beihuaiyang zone. Acta Petrologica Sinica, 20(5): 1007-1024 (in Chinese with English abstract)

Xie Z, Chen J F, Zhang X et al. 2002. Geochronology of Neoproterozoic mafic intrusions in North Huaiyang area. Acta Geoscience Sinica, 23(6): 517-520 (in Chinese with English abstract)

Xu S T, Liu Y C, Su W et al. 1999. Geometry, kinematics and tectonic implication of the deformed garnets in the foliated eclogite from the ultra-high pressure metamorphic belt in the Dabie Mountains, eastern China. Acta Petrologica Sinica, 15(3): 321-337(in Chinese with English abstract)

Xu S, Okay A L, Ji S et al. 1992. Science, Diamond from the Dabie Shan metamorphic rocks and its implication for tectonic setting. Science, 256: 80-82

Yang Z Y, Ma X H, Besse J et al. 1991. Paleomagnetic results from triassic sections in the ordos basin, north China. Earth Planet Sci Lett, 104(2-4): 258-277

Zhang G W, Zhang B R, Yuan X C et al. 2001. Qingling orogenic belt and continental dynamics. Science Press, Beijing, China

Zhang H F, Gao S, Zhang B R et al. 2001. Pb isotopic study on crustal structure of Dabie Mountains, central China. Geochimica, 30: 395-401(in Chinese with English abstract)

Zhou J B, Zheng Y F, Li L, Xie Z. 2001. Accretionary wedge of the subduction of the Yangtze Plate. Acta Geol. Sinica, 75(3): 338-352 (in Chinese with English abstract)

Zhou J G, Zhao Z J, Deng H Y. 1999. Tectonic evolution of the Hefei basin and analysis of its petroleum potential. Acta Geol. Sinica, 73: 15-24 (in Chinese with English abstract)

大别山下地壳岩石及其深俯冲*

刘贻灿，李曙光

中国科学院壳幔物质与环境重点实验室，中国科学技术大学地球和空间科学学院，合肥 230026

> **亮点介绍**：本文根据已发表的资料，结合作者等的最新研究成果，对北大别出露的下地壳岩石的主要组成及其变质岩石学和年代学特征进行了系统总结，认为大别山下地壳岩石都参与了印支期大陆深俯冲，但具有与南部超高压岩石不同的变质演化与折返过程，提出用大陆叠瓦式深俯冲模式来解释。

摘要 本文对大别山下地壳岩石的主要组成及其变质岩石学和年代学特征进行了系统总结，并首次报道了作者等最近在罗田穹隆中发现的镁铁质下地壳岩石俯冲变质成因的榴辉岩及其有关片麻岩的 SHRIMP 锆石 U-Pb 同位素年代学的最新研究成果。大别山这种不寻常的镁铁质下地壳俯冲成因榴辉岩为研究大陆深俯冲和陆壳物质再循环以及对地幔不均一性的影响提供了重要对象和可能性。根据阴极发光图像及矿物包体组合，罗田榴辉岩中继承锆石可分为两种即具有岩浆结晶环带的岩浆锆石和含有石榴子石+紫苏辉石+斜长石等麻粒岩相变质矿物包体的变质锆石，而且这两种继承锆石常具有含石榴子石+绿辉石等榴辉岩相矿物包体的增生边。SHRIMP U-Pb 定年结果表明，继承岩浆锆石的 $^{206}Pb/^{238}U$ 年龄为 791±9 Ma、含麻粒岩相矿物包体的继承变质锆石的 $^{206}Pb/^{238}U$ 年龄为 794±10 Ma、含榴辉岩相矿物包体的锆石增生边的 $^{206}Pb/^{238}U$ 年龄值为 212±10 Ma。罗田榴辉岩的围岩——含石榴子石英云闪长质片麻岩中锆石的 CL 图像具有特征的核-幔-边结构即表现为继承岩浆锆石核-高压变质幔-退变质增生边，高压变质幔给出罗田片麻岩中印支期高压变质的精确年龄为 218±3 Ma。因此，罗田榴辉岩的原岩时代与早期麻粒岩相变质时代一致，均为晚元古代，证明华南陆块北缘晚元古代岩浆板底垫托的存在。而且，大别山下地壳岩石参与了印支期大陆深俯冲，但与南大别带具有不同的折返过程即前者经历了麻粒岩相退变质过程。此外，北大别带榴辉岩的印支期峰期变质时代(212±4 Ma)比南大别带年轻(226±3 Ma)，这可用大陆叠瓦式深俯冲模式来解释。

关键词 麻粒岩；榴辉岩；下地壳；大陆深俯冲；大别山

1 引言

大别山，因其发育有含柯石英和金刚石的超高压变质带而闻名于世。但是，从目前地表出露的岩石组成及铅同位素研究看，大别山所有折返至地表的高压-超高压变质岩主要是比重较轻的俯冲上陆壳及长英质下陆壳岩石（Liou, 1999；李曙光等, 2001；张宏飞等, 2001；Ernst, 2001），并且，现在大别造山带乃至中国东部下地壳主要是长英质的（Gao et al., 1998）。因此，人们可以推测大别山镁铁质下地壳岩石已被拆离了，然而这仍然缺乏直接证据。我们甚至不知道印支期深俯冲的华南陆壳是否有较厚的镁铁质下地壳。本文根据已发表的资料，结合作者等的最新研究成果，对大别山出露的下地壳岩石的主要组成及其变质岩石学和年代学特征进行了系统总结，并在此基础之上对有关问题进行了讨论。

2 大别山麻粒岩及下地壳岩石

众所周知，大别山从南到北，大致可划分为宿松变质带、黄镇冷榴辉岩带、南大别超高压变质带或南大别带、北大别带或北部杂岩带及北淮阳带等构造岩石单位（图1）。

---
* 本文发表在：岩石学报，2005，21(4): 1059-1066

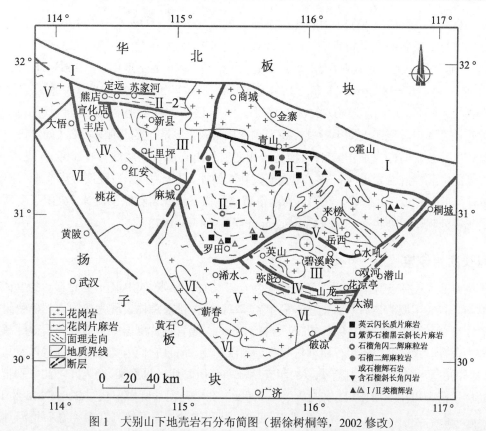

图1 大别山下地壳岩石分布简图（据徐树桐等，2002 修改）

I. 北淮阳带，II-1. 北大别带，II-2. 苏家河变质带，III. 南大别带，IV. 黄镇冷榴辉岩带，V. 大别杂岩，VI. 宿松变质带

北大别带或北部杂岩带，大致分布于磨子潭-晓天断裂以南至龙井关-水吼-五河一线以北地区，其南、北分别为超高压变质带和北淮阳带。该带变质岩的岩石类型主要有条带状片麻岩（包括英云闪长质片麻岩、花岗闪长质片麻岩及二长花岗质片麻岩）、斜长角闪岩和少量的方辉橄榄岩、纯橄岩、石榴辉石岩、石榴二辉麻粒岩、紫苏磁铁石英岩、榴辉岩、含镁橄榄石和钛斜硅镁石大理岩等，未变质岩石类型主要有辉石岩、辉长岩、花岗岩类等。该带变质变形强烈，以角闪岩相区域变质作用为主，局部出现麻粒岩相和榴辉岩相矿物组合。北部杂岩带西段即罗田及其以北地区（商城-麻城断裂以东），因以前未发现榴辉岩和局部存在古老的麻粒岩包体（如黄土岭）以及其周围的面理产状一般向外倾斜，常被称为"罗田穹隆"或"罗田片麻岩穹隆"。1998 年以前因该带未发现榴辉岩等高压变质岩，它的构造归属曾成为一个有争议的问题（如，Okay et al., 1992；董树文等，1993；徐树桐等，1994；Zhai et al., 1995；Zhang et al., 1996）。近年来，北大别带东段（安徽境内）榴辉岩（Wei et al., 1998; Xu et al., 2000）和金刚石（Xu et al., 2003）的发现以及有关变质岩石学（Tsai and Liou, 2000；刘贻灿等，2001b；Xiao et al., 2001；徐树桐等，2002；Liu et al., 2003）、岩石地球化学（Liu et al., 2000）和同位素年代学（如，刘贻灿等，2000a，2001a；谢智等，2001；葛宁洁等，2003）等方面研究，证明北大别带也经历了三叠纪高压-超高压变质作用，它属于印支期华南俯冲陆壳的一部分。最近，作者等在北大别带西段—罗田穹隆中发现镁铁质下地壳岩石俯冲变质成因榴辉岩（刘贻灿等，2005）。

铅同位素以及岩石地球化学等研究已表明，大别山南部超高压变质带岩石具有上地壳性质、北部杂岩带变质岩具有下地壳特征（Liu et al., 2000, 2003；李曙光等，2001；张宏飞等，2001；刘贻灿等，2002）。也就是说，大别山下地壳岩石主要出露于北部。但是，它主要包括哪些岩石以及主要特征如何？根据目前地表出露的岩石以及有关报道资料，结合其年代学和变质特征，现作如下分类并加以阐述。

2.1 紫苏石榴黑云斜长片麻岩

该岩石类型在大别山出露较少，主要分布于罗田穹隆核部黄土岭一带，属于古老的中酸性麻粒岩。其中，所含矿物有石榴子石、黑云母、紫苏辉石、斜长石、石英和磁铁矿等，而且，石榴子石斑晶中常含有针状金红石出熔体。吴元保等（2002）对黄土岭麻粒岩的研究表明，①该岩石中主要有三类锆石，即原岩岩浆锆石、麻粒岩相变质锆石和残留锆石；②三类锆石的 SHRIMP 年龄分别为 2723±5 Ma、2052±100 Ma 和 3443±13 Ma。前两个年龄分别与陈能松等（1996）的锆石 U-Pb 上交点年龄（2663±56 Ma）、周汉文等（Zhou et al., 1999）的石榴子石 Pb-Pb 年龄（1998±35 Ma）一致。因此，认为黄土岭麻粒岩原岩时代约为 2700 Ma、麻粒岩相变质事件的年龄为 2000 Ma 左右，并且证明大别山地区存在太古代的陆壳物质。

尽管该类麻粒岩中未发现印支期年龄及榴辉岩相变质记录，但是，其周围的岩石（如罗田一带英云闪长质片麻岩和榴辉岩等）都有印支期年龄和/或榴辉岩相变质作用记录（刘贻灿等，2005）（见后文），因此，黄土岭一带古老的中酸性麻粒岩也应参与了印支期大陆深俯冲。至于为何未发现明显印支期大陆深俯冲的岩石学记录，还有待于进一步深入研究。我们认为这很可能是该下地壳岩石缺少流体，减缓了变质反应速度的结果，类似情况在西阿尔卑斯也有发现（Compagnoni and Maffeo, 1973）。

2.2 英云闪长质片麻岩

该岩石类型是北大别带变质岩的主体。主要矿物有斜长石、石英、角闪石、磁铁矿、钛铁矿等，含少量石榴子石、单斜辉石（主要为透辉石）、金红石、紫苏辉石等。目前总体表现为角闪岩相矿物组合和局部残留麻粒岩相矿物组合（徐树桐等，1994；Zhang et al., 1996）。尽管该片麻岩中目前还未发现绿辉石，但是考虑到该片麻岩中含有石榴子石、金红石和单斜辉石等高压矿物，而且它们为榴辉岩的围岩以及与所包裹的榴辉岩经历了相同的变质演化过程和具有类似的印支期峰期变质时代（见后文），因此，北大别带中英云闪长质片麻岩也经历了榴辉岩相变质作用。刘贻灿等（2000a）在塔儿河片麻岩中获得 226±6 Ma 及 145±2 Ma 的锆石 U-Pb 谐和年龄；谢智等（2001）在石竹河片麻岩中获得 229±18 Ma 的不一致线下交点年龄；江来利等（2002）在漫水河片麻岩中获得 218±9 Ma 锆石 U-Pb 谐和年龄。这些年龄均由单颗粒锆石同位素稀释法测定。谢智等（Xie et al., 2004）采用锆石 SHRIMP 法 U-Pb 定年，在百丈岩片麻岩中获得 212±21 Ma 和 120±11 Ma 的谐和年龄。最近，作者等研究表明，罗田和塔儿河等地英云闪长质片麻岩中锆石的 CL 图像具有特征的核-幔-边结构(图 2)即表现为继承岩浆锆石核-高压变质幔-退变质增生边；通过锆石 SHRIMP 法 U-Pb 定年，获得罗田片麻岩中印支期高压变质（锆石幔部）的精确年龄为 218±3 Ma（15 个测点）以及 191±5 Ma（12 个测点）和 126±5 Ma（2 个测点）的退变质年龄（分别为锆石幔-边部过渡带和边部）（详细数据将另文发表）。另外，上述研究者都发现该类片麻岩有 700~800 Ma 的年龄记录，证明其原岩为晚元古代形成的。

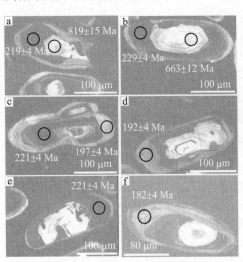

图 2　罗田穹隆中英云闪长质片麻岩中锆石 CL 图像及离子探针分析点与年龄

2.3 石榴角闪二辉麻粒岩（早白垩世）

该类岩石主要出露于惠兰山一带（游振东等，1995），属于基性麻粒岩。主要由石榴子石、角闪石、紫苏辉石、单斜辉石、斜长石、石英等组成。据侯振辉（2003）[①]研究，锆石 SHRIMP U-Pb 定年表明，继承锆石指示原岩形成时代为晚元古代；变质增生锆石的年龄为 124±2 Ma，与麻粒岩相变质矿物石榴子石+单斜辉石+紫苏辉石+全岩的 Sm-Nd 等时线年龄（136±18 Ma）在误差范围内一致，指示它发生麻粒岩相变质作用的时代为早白垩世。因此，惠兰山基性麻粒岩可能是就位于下地壳的晚元古代镁铁质岩浆岩在早白垩世大别山造山带引张条件下受热穹隆加热而发生麻粒岩相变质作用形成的。

---

① 侯振辉, 2003. 大别—苏鲁造山带高级变质岩锆的地球化学、锆石微量元素特征及年代学效应. 中国科学技术大学博士学位论文.

2.4 石榴辉石岩、石榴二辉麻粒岩、含石榴子石斜长角闪岩（具榴辉岩相印记）

该类岩石主要出露于麻城市木子店及霍山县姜河、抱儿山、杨崖、燕子河、鹿吐石铺等地。主要矿物有石榴子石、单斜辉石（主要为透辉石）、金红石、石英、斜长石、角闪石、紫苏辉石、钛铁矿、磁铁矿等，目前总体表现为麻粒岩相和/或角闪岩相矿物组合。但是，它们都曾经过了榴辉岩相变质作用以及麻粒岩相和角闪岩相退变质作用（刘贻灿等，2000b；苏文等，2000；张泽明等，2000）。它们类似于黄尾河、百丈岩等地榴辉岩的退变质特征，是由榴辉岩经强烈退变质形成的。

2.5 I 类榴辉岩（经麻粒岩相退变）

该类榴辉岩主要出露于黄尾河、百丈岩、饶钹寨等地并经过了麻粒岩相和角闪岩相退变质作用（刘贻灿等，2001b），它们不同于南大别带榴辉岩的变质过程，即后者未经历麻粒岩相退变质作用。主要矿物有石榴子石、绿辉石、透辉石、金红石、石英、斜长石、角闪石、紫苏辉石、钛铁矿、磁铁矿等。岩石学研究（Liu et al., 2003）表明该类至少经历了三个变质阶段，即：①榴辉岩相峰期变质阶段，主要矿物共生组合为石榴子石+绿辉石+金红石+石英，$P \geqslant 2.5$ GPa、$T=808 \sim 874$ ℃；②高压麻粒岩相退变质阶段，主要矿物共生组合为石榴子石+透辉石+紫苏辉石+钛铁矿+尖晶石+斜长石等，$P=1.1 \sim 1.37$ GPa、$T=817 \sim 909$ ℃；③角闪岩相变质阶段，主要矿物共生组合为角闪石+斜长石+磁铁矿等，$P=0.5 \sim 0.6$ GPa、$T=500 \sim 600$ ℃。岩石地球化学及 Pb 同位素研究证明它们为下地壳岩石俯冲变质形成（Liu et al., 2000；刘贻灿等，2002）；它们的 212 ± 4 Ma Sm-Nd 年龄（Liu et al., 2005）和 230 ± 6 Ma 锆石 U-Pb 年代（刘贻灿等，2000a）证明它们经过了印支期榴辉岩相变质作用。

2.6 II 类榴辉岩（由基性麻粒岩变质形成）

该类榴辉岩是作者等（刘贻灿等，2005）近期发现的一种特殊类型，主要出露于罗田穹隆中。主要矿物成分类似于 I 类榴辉岩。与上述 I 类榴辉岩相比较，主要差别在于该类榴辉岩中发现有榴辉岩相变质作用之前的紫苏辉石+中长石等麻粒岩相变质矿物残留，证明其俯冲前原岩是镁铁质下地壳岩石（基性麻粒岩）。根据退变质结构以及矿物之间的相互关系，该类榴辉岩经历了两个重要变质事件即麻粒岩相变质事件和榴辉岩相及其退变质事件。后者可分为 3 个变质阶段，即：①榴辉岩相变质阶段，以石榴子石及其中矿物包体为代表，主要矿物共生组合为石榴子石+绿辉石+金红石+石英；②高压麻粒岩相退变质阶段，主要矿物共生组合为石榴子石+透辉石+紫苏辉石+斜长石+钛铁矿等；③角闪岩相退变质阶段，主要矿物共生组合为角闪石+斜长石+磁铁矿等。榴辉岩中有两种继承锆石，即岩浆锆石和变质锆石核，并且二者常具有含石榴子石 + 绿辉石等矿物包体的薄边（图 3），其 SHRIMP U-Pb 定年结果表明：①具有典型岩浆结晶环带的继承岩浆锆石的 $^{206}Pb/^{238}U$ 年龄为 791 ± 9 Ma；②含石榴子石 + 紫苏辉石 + 斜长石等麻粒岩相矿物包体的变质锆石的 $^{206}Pb/^{238}U$ 年龄为 794 ± 10 Ma；③含石榴子石 + 绿辉石 ± 金红石等榴辉岩相矿物包体的锆石增生边的 $^{206}Pb/^{238}U$ 年龄值为 212 ± 10 Ma（详细数据将另文发表①）。因此，榴辉岩的原岩时代与早期麻粒岩相变质时代一致，均为晚元古代，证

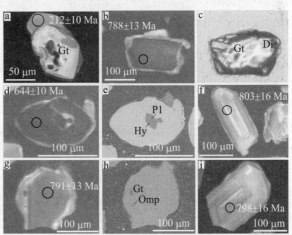

图 3 罗田榴辉岩中锆石及其所含矿物包体的显微照片以及离子探针分析点与年龄

a、b、d、f、g 和 i 为阴极发光图像；c 为显微镜下单偏光图像；e 和 h 为背散射图像。

Gt－石榴子石，Omp－绿辉石，Di－透辉石，Hy－紫苏辉石，Pl－中长石

---

① 刘贻灿，李曙光等，2005. 晚元古代华南陆块北缘镁铁质下地壳的垂向增生及印支期俯冲消减：大别山下地壳俯冲成因榴辉岩锆石 U-Pb 年代学的证据，待刊。

明华南陆块北缘晚元古代岩浆板底垫托（magmatic underplating）的存在。考虑到晚元古代华南陆块北缘存在大规模的岩浆事件(如 Rowley et al., 1997; Li et al., 2003a, b; Zheng et al., 2004)，并且，该事件与导致 Rodinia 超大陆裂解的 Plume 活动（Li et al., 1999）有关。另外，华南元古代－二叠纪沉积岩的钕同位素研究表明，在 800 Ma 左右时沉积岩的 $T_{DM}(t)$ 迅速降低和 $\varepsilon_{Nd}(t)$ 值的迅速增高，与华南晚古生代有大量新的幔源物质加入到沉积岩中有关（Li et al., 1996）。因此，这一垂向增生的镁铁质下地壳应有较大厚度。而且，华南陆块北缘镁铁质下地壳岩石参与了三叠纪大陆深俯冲作用。

3 下地壳岩石的深俯冲

大别山深俯冲下陆壳是否含有较厚的镁铁质下地壳及其去向，一直是人们关注的重要科学问题。前文已述及，北大别带下地壳岩石中的石榴辉石岩、石榴二辉麻粒岩、榴辉岩都证明经过了榴辉岩相变质作用。并且，北大别带东段榴辉岩中钠质单斜辉石的针状石英出熔体（Tsai et al., 2000；刘贻灿等，2001b）、石榴子石中针状矿物出熔体（徐树桐等，2002；Liu et al., 2003）和微粒金刚石（Xu et al., 2003）证明北大别带榴辉岩经过了超高压变质作用。尽管罗田榴辉岩中未发现可靠的超高压变质证据，但是其中有类似于东段榴辉岩的针状矿物出熔体，并且二者具有类似的变质演化过程，因此，不能排除它也经过了超高压变质作用。对于紫苏石榴黑云片麻岩、英云闪长质片麻岩，考虑它们为榴辉岩的围岩和经历了相同的变质演化过程以及具有类似的印支期峰期变质时代，而且都含有石榴子石、金红石和单斜辉石等高压矿物，因此它们也应经过了深俯冲作用。但是，北大别带的岩石组合仍是以长英质岩石为主，它仅相当于下地壳的上部长英质岩石。大别山造山带乃至华南陆块北缘现今缺乏厚层镁铁质下地壳，它们也很少出露地表，推测这些俯冲的镁铁质下地壳大多数可能已拆离再循环进入地幔。但是这一推测的假设前提是在拆离前，华南俯冲陆壳有较厚的镁铁质下地壳，本文工作已证明了这一点。这较厚的镁铁质下地壳可增加俯冲陆壳的平均比重，从而有利于大陆深俯冲的发生。

由前文可知，罗田榴辉岩及其围岩－片麻岩的锆石 SHRIMP U-Pb 年龄结果表明，含石榴子石+绿辉石等榴辉岩相变质矿物包体的变质锆石时代为 212±10 Ma（图 3）和片麻岩的峰期变质时代为 218±3 Ma，它们类似于北部杂岩带东段榴辉岩的石榴子石+绿辉石的 212±4 Ma Sm-Nd 等时线年龄（Liu et al., 2005），而明显年轻于南大别 UHP 岩石的峰期变质时代（如，226±3 Ma 的 Sm-Nd 矿物等时线年龄，Li et al., 2000）、更年轻于黄镇"冷"榴辉岩的峰期变质时代（石榴子石+绿辉石+金红石+蓝晶石的 Sm-Nd 矿物等时线年龄为 236±4 Ma，Li et al., 2004）。而且，刘福来等（Liu et al., 2004）对苏鲁超高压带中对应于南大别超高压带的超高压片麻岩锆石 SHRIMP U-Pb 年龄研究也表明，含柯石英等超高压矿物的锆石幔部年龄为 231±4 Ma，而含石英等角闪岩相退变质矿物的锆石边部年龄为 211±4 Ma。该超高压片麻岩锆石的幔、边部年龄正好分别与南大别超高压带及北大别带的峰期变质时代一致。这些表明，大别山三个主要超高压岩片的榴辉岩的峰期变质时代，由南向北逐渐变新，并且，南大别超高压带岩石的折返时代与北大别带的峰期变质时代一致。但是，它们的峰期变质温度却由南向北逐渐升高即由黄镇"冷"榴辉岩（$T<700$ ℃，一般为 570～670 ℃）(Wang et al., 1992; Okay, 1993; Zhai et al., 1995; Li et al., 2004)→南大别中温榴辉岩（T 一般为 700～800 ℃）(Okay，1993；徐树桐等，1994；Cong，1996)→北大别高温榴辉岩（$T=808～874$ ℃）(Liu et al., 2005; 刘贻灿等，2005），这种峰期变质温度的有规律变化，似乎分别与三个 HP-UHP 岩片的原岩当时所处的地壳结构有关，即南大别和北大别分别相当于上、下地壳（李曙光等，2001；张宏飞等，2001），也就是说，三个岩片原岩温度就有高、低区别。这些观察表明大别山印支期存在叠瓦式大陆深俯冲：陆壳俯冲过程中，首先在俯冲上地壳中发生层间破裂并发展成逆冲断层，该断层上部岩片（黄镇"冷"榴辉岩带）逆冲折返，下部陆壳继续俯冲；此后，在上下陆壳之间发生第二次层间破裂并发展成逆冲断层，该断层上部岩片（南大别 UHP 带）逆冲折返，下陆壳（北大别带）继续俯冲；最终，在俯冲板片断离后，三个岩片全部折返，并呈现在由南向北三个岩片峰期变质时代逐步变新、变质温度逐步升高的趋势。这种大陆壳俯冲呈叠瓦式发展的过程是大陆岩石圈内存在多个低黏滞带（Meissner and Mooney, 1998）从而导致陆壳俯冲时上、下地壳的脱

耦（decoupling），这有别于大洋俯冲板块。然而要完全证明这一俯冲模式，尚需要更多的精确定年加以验证，因为已有的罗田穹隆榴辉岩年龄误差还太大（±10 Ma），而且整个北大别带及黄镇冷榴辉岩带的精确变质年龄偏少，为此特别需要在今后的研究中加强对这两个岩片的高压-超高压变质岩进行精确定年。

4 结论

（1）大别山下地壳岩石的主要类型有紫苏石榴黑云斜长片麻岩、英云闪长质片麻岩、石榴角闪二辉麻粒岩（早白垩世）、石榴辉石岩、石榴二辉麻粒岩、含石榴子石斜长角闪岩和榴辉岩。其中，榴辉岩包括两种，一种是含有榴辉岩相变质作用前的麻粒岩相变质矿物组合，另一种不含。但二者都经过了麻粒岩相和角闪岩相退变质过程。

（2）大别山下地壳岩石都参与了印支期大陆深俯冲，但具有与南大别超高压岩石不同的变质演化与折返过程。

（3）北大别带的印支期峰期变质时代比南大别带年轻。这可用大陆叠瓦式深俯冲模式来解释。

（4）大别山镁铁质下地壳俯冲成因榴辉岩的原岩是华南晚元古代增生的镁铁质下地壳，它们参与了三叠纪大陆深俯冲。这一发现为研究大陆深俯冲与陆壳物质再循环提供了重要对象和可能性。

致谢

文章中的锆石 SHRIMP U-Pb 年龄结果是在北京离子探针中心测试的。感谢北京离子探针中心刘敦一先生、王彦斌研究员和陶华女士以及有关人员在锆石 SHRIMP U-Pb 年龄测试过程中所给予的支持和帮助！同时，感谢郭敬辉研究员和刘景波博士等在本文修改过程中提出的建设性意见！

参考文献

Compagnoni R and Maffeo B. 1973. Jadeite-bearing metagranites l. s. and related rocks in the Mount Mucrone area (Seisia-Lanzo Zne, Western Italian Alps). Schweiz. Mineral. Petrogr. Mitt., 53: 355-378.
Cong B. 1996. Ultrahigh-pressure metamorphic rocks in the Dabieshan-Sulu region of China. Beijing: Science Press.
Dong S, Sun X, Zhang Y, et al. 1993. The basic structure of the Dabieshan collision orogenic orogenic belt. Chinese Science Bulletin, 38(6): 542-545.
Ernst W G. 2001. Subduction, ultrahigh-pressure metamorphism, and regurgitation of buoyant crustal slices—implications for arcs and continental growth. Physics of the Earth and Planetary Interiors, 127: 253-275.
Gao S, Zhang B, Jin Z, et al. 1998. How mafic is the lower continental crust? Earth and Planetary Science Letters, 161: 101-117.
Ge N, Xia Q, Wu Y, et al. 2003. Zircon U-Pb ages of Yanzihe gneiss from northern Dabie, China: evidence for Jurassic metamorphism. Acta Petrologica Sinica, 19(3): 513-516 (in Chinese with English abstract).
Jiang L, Liu Y, Wu W, et al. 2002. Zircon U-Pb age and its geological implications of the gray gneiss to the northern Manshuihe in the North Dabie Mountains. Geochimica, 31(1): 66-70 (in Chinese with English abstract).
Li S, Jagoutz E, Chen Y, et al. 2000. Sm-Nd and Rb-Sr isotope chronology of ultrahigh-pressure metamorphic rocks and their country rocks at Shuanghe in the Dabie Mountains, central China. GCA, 64(6): 1077-1093.
Li S, Huang F, Zhou H, et al. 2001. U-Pb compositions of the ultrahigh pressure metamorphic (UHPM) rocks from Shuanghe and gneisses from Northern Dabie zone in the Dabie Mountains, central China: constraint on the exhumation mechanism of UHPM rocks. Chinese Science Bulletin, 31: 977-984 (in Chinese).
Li X, McCulloch M T. 1996. Secular variation in Nd isotopic composition of Neoproterozoic sediments from the southern margin of the Yangtze Block: evidence for a Proterozoic continental collision in southern China. Precambrian Research, 76: 67-76.
Li X, Li Z, Ge W, et al. 2003a. Neoproterozoic granitoids in South China: crustal melting above a mantle plume at ca. 825Ma？ Precambrian Research, 122: 45-83.
Li X P, Zheng Y, Wu Y, et al. 2004. Low-T eclogite in the Dabie terrane of China: petrological and isotopic constrains on fluid activity and radiometric dating. Contrib. Mineral. Petrol., 148: 443-470.
Li Z X, Li X H, Kinny P D, et al. 1999. The breakup of Rodinia: did it start with a mantle plume beneath South China? Earth and Planetary Science Letters, 173: 171-181.

Li Z X, Li X H, et al. 2003b. Geochronology of Neoproterozoic syn-rift magmatism in the Yangtze Craton, South China and correlations with other continents: evidence for a mantle superplume that broke up Rodinia. Precambrian Research, 122: 85-109.

Liou J G. 1999. Petrotectonic summary of less-intensively studied UHP regions. Int. Geol. Rev., 41: 571-586.

Liu F, Xu Z, Liou J G, et al. 2004. SHRIMP U-Pb ages of ultrahigh-pressure and retrograde metamorphism of gneisses, south-western Sulu terrane, eastern China. J. metamorphic Geol., 22: 315-326.

Liu Y, Xu S, Li S, et al. 2000. Eclogites from the northern Dabie Mountains, eastern China: geochemical characteristics, Sr-Nd isotopic compositions and tectonic implications. Science in China (Series D), 43(supp.): 178 - 188.

Liu Y, Li S, Xu S, et al. 2000a. U-Pb zircon ages of the eclogite and tonalitic gneiss from the northern Dabie Mountains, China and multi-overgrowths of metamorphic zircons. Geological Journal of China Universities, 6(3): 417-423 (in Chinese with English abstract).

Liu Y, Xu S, Li S, et al. 2000b. Metamorphic characteristics and Rb-Sr isotopic age of garnet-bearing amphibolite at Lutushipu in the northern Dabie Mountains. Geology of Anhui, 10(3): 194-198 (in Chinese with English abstract).

Liu Y, Xu S, Liu Y, et al. 2002. Pb isotopic characteristics of the eclogites from the northern Dabie Mountains. J. Mineral. Petrol., 22(3): 33-36 (in Chinese with English abstract).

Liu Y, Li S, Xu S, et al. 2001a. Sm-Nd dating of eclogites from northern Dabie Mountains and its constrains on the timing of granulite-facies retrogression. Geochimica, 30(1): 79~87 (in Chinese with English abstract).

Liu Y, Xu S, Li S, et al. 2001b. Distribution and metamorphic P-T condition of the eclogites from the mafic-ultramafic belt in the northern part of the Dabie Mountains. Acta Geologica Sinica, 75(3): 385-395 (in Chinese with English abstract).

Liu Y, Xu S, Li S, et al. 2003. Tectonic affinity, T-t path and uplift trajectory of eclogites from northern Dabie Mountains, central-eastern China. Journal of China University of Geosciences, 14: 28-33.

Liu Y, Li S, Xu S, et al. 2005. Geochemistry and geochronology of eclogites from the northern Dabie Mountains, central China. Journal of Asian Earth Sciences, 25: 431-443.

Liu Y, Xu S, Li S, et al. 2005. Eclogite from the subducted lower crust of the Yangtze plate within the Luotian dome and its geological implication. Earth Science—Journal of China University of Geosciences, 30(1): 71-77 (in Chinese with English abstract).

Meissner R, Mooney W. 1998. Weakness of the lower continental crust: a condition for delamination, uplift, and escape. Tectonophysics, 296: 47-60.

Okay A I and Sengör A M C. 1992. Evidence for introcontinental thrust-related exhumation of the ultrahigh-pressure rocks in China. Geology, 20: 411-414.

Okay A I. 1993. Petrology of diamond and coesite-bearing metamorphic terrane: Dabie Shan, China. Eur. J. Mineral., 5: 659-676.

Rudnick R. L. 1995. Making continental crust. Nature, 378(7): 571-578.

Su W, Xu S, Jiang L, et al. 2000. Metamorphic structure, water activities and their evolutionary features of pyrigarnite, north of Dabie Mountains, east of China. Earth Science—Journal of China University of Geosciences, 25(2): 152-158 (in Chinese with English abstract).

Tsai C H and Liou J G. 2000. Eclogite-facies relics and inferred ultrahigh-pressure metamorphism in the North Dabie Complex, central-eastern China. Am. Mineral., 85: 1-8.

Wang X, Liou J G and Marruyama S. 1992. Coesite-bearing eclogites from the Dabie Mountains, central China: petrogenesis, P-T path and implications for regional tectonics. Journal of Geology, 100: 231-250.

Wei C, Shan Z, Zhang L, et al. 1998. Determination and geological significance of the eclogites from the northern Dabie Mountains, central China. Chinese Science Bulletin, 43: 253-256.

Wu Y B, Chen D, Xia Q K, et al. 2002. SIMS U-Pb dating of zircons in granulite of Huangtuling from northern Dabieshan. Acta Petrologica Sinica, 18(3): 378-382 (in Chinese with English abstract).

Xiao Y, Hoefs J, van den Kerkhof A M, et al. 2001. Geochemical constraints of the eclogite and granulite facies metamorphism as recognized in the Raobazhai complex from North Dabie Shan, China. J. Metamorphic Geol., 19: 3 - 19.

Xie Z, Chen J, Zhang X, et al. 2001. Zircon U-Pb dating of gneiss from Shizhuhe in North Dabie and its geologic implications. Acta Petrologica Sinica, 17(1): 139-144.

Xie Z, Gao T, Chen J. 2004. Multi-stage evolution of gneiss from North Dabie: evidence from zircon U-Pb chronology. Chinese Science Bulletin, 49(18): 1963-1969.

Xu S, Liu Y, Jiang L, et al. 1994. Tectonic regime and evolution of Dabie Mountains. Beijing: Science Press, 1-175 (in Chinese with English abstract).

Xu S, Liu Y, Su W, et al. 2000. Discovery of the eclogite and its petrography in the Northern Dabie Mountains. Chinese Science Bulletin, 45(3): 273-278.

Xu S, Liu Y, Jiang L, et al. 2002. Architecture and kinematics of the Dabie Mountains orogen. Hefei: University of Science and Technology of China Press, 1-133 (in Chinese with English abstract).

Xu S, Liu Y, Chen G, et al. 2003. New finding of micro-diamonds in eclogites from Dabie-Sulu region in central-eastern China. Chinese Science Bulletin, 48: 988-994.
You Z and Chen N. 1995. The metamorphism of deeper crust in the Dabie Mountains: as evidenced by the study of granulites near Huilanshan, Luotian. Acta Petrologica Sinica, 11(2): 137-147 (in Chinese with English abstract).
Zhai M, Cong B, Zhao Z, et al. 1995. Petrologic-tectonic units in the coesite-bearing metamorphic terrain of the Dabie Mountains, central China and their geotectonic implication. J. SE. Geosci., 11(1): 1-13.
Zhang H, Gao S, Zhang B, et al. 2001. Pb isotopic study on crustal structure of Dabie Mountains, central China. Geochmica, 30(4): 395-401 (in Chinese with English abstract).
Zhang R Y, Liou J G, Tsai C H. 1996. Petrogenesis of a high-temperature metamorphic Terrain: a new tectonic interpretation for the north Dabie Shan, central China. J. Metamorphic. Geol., 14: 319-333.
Zhang Z M, Zhong Z Q, You Z D, et al. 2000. Granulite-facies retrograde metamorphism of garnet pyroxenite in Muzidian, northern Dabie Mountains. Earth Science—Journal of China University of Geosciences, 25(3): 295-301 (in Chinese with English abstract).
Zheng Y, Wu Y, Chen F, et al. 2004. Zircon U-Pb and oxygen isotope evidence for a large-scale ^{18}O depletion event in igneous rocks during the Neoproterozoic. Geochim. Cosmochim. Acta, 68: 4145-4165.
Zhou H, Liu Y, Li X, et al. 1999. Age of granulite from Huangtuling, Dabie Mountain: Pb-Pb dating of garnet by a stepwise dissolution technique. Chinese Science Bulletin, 44(10): 941-944.
董树文, 孙先如, 张勇等. 1993. 大别山造山带的基本结构. 科学通报, 38 (6): 542-545.
葛宁洁, 夏群科, 吴元保等. 2003. 北大别燕子河片麻岩的锆石 U-Pb 年龄: 印支期变质事件的确定. 岩石学报, 19 (3): 513-516.
李曙光, 黄方, 周红英等. 2001. 大别山双河超高压变质岩及北部片麻岩的 U-Pb 同位素组成——对超高压岩石折返机制的制约. 科学通报, 31: 977-984.
江来利, 刘贻灿, 吴维平等. 2002. 大别山北部漫水河灰色片麻岩的锆石 U-Pb 年龄及其地质意义. 地球化学, 31 (1): 66-70.
刘贻灿, 李曙光, 徐树桐等. 2000a. 大别山北部榴辉岩和英云闪长质片麻锆石 U-Pb 年龄及多期变质增生. 高校地质学报, 6 (3): 417-423.
刘贻灿, 徐树桐, 李曙光等. 2000b. 大别山北部鹿吐石铺含石榴子石斜长角闪岩的变质特征及 Rb-Sr 同位素年龄. 安徽地质, 10 (3): 194-198.
刘贻灿, 徐树桐, 刘颖等. 2002. 大别山北部榴辉岩的 Pb 同位素特征. 矿物岩石, 22 (3): 33-36.
刘贻灿, 李曙光, 徐树桐等. 2001a. 大别山北部榴辉岩的 Sm-Nd 年龄测定及其对麻粒岩相退变质时间的制约. 地球化学, 30 (1): 79-87.
刘贻灿, 徐树桐, 李曙光等. 2001b. 大别山北部镁铁-超镁铁质岩带中榴辉岩的分布与变质温压条件. 地质学报, 75 (3): 385-395.
刘贻灿, 徐树桐, 李曙光等. 2005. "罗田穹隆"中的下地壳俯冲成因榴辉岩及其地质意义. 地球科学——中国地质大学学报, 30(1): 71-77.
吴元保, 陈道公, 夏群科等. 2002. 北大别黄土岭麻粒岩锆石 U-Pb 离子探针定年. 岩石学报, 18 (3): 378-382.
谢智, 陈江峰, 张巽等. 2001. 大别片麻岩的锆石 U-Pb 年龄及其地质意义. 岩石学报, 17: 139-144.
徐树桐, 刘贻灿, 江来利等. 1994. 大别山的构造格局和演化. 北京: 科学出版社.
徐树桐, 刘贻灿, 江来利等. 2002. 大别山造山带的构造几何学和运动学. 合肥: 中国科学技术大学出版社. 1-133.
游振东, 陈能松. 1995. 大别山区深部地壳的变质岩石学证迹: 罗田惠兰山一带的麻粒岩研究. 岩石学报, 11 (2): 137-147.
张宏飞, 高山, 张本仁等. 2001. 大别山地壳结构的 Pb 同位素地球化学示踪. 地球化学, 30: 395-401.
张泽明, 钟增球, 游振东等. 2000. 北大别木子店石榴辉石岩的麻粒岩相退变质作用. 地球科学——中国地质大学学报, 25 (3): 295-301.

Zircon SHRIMP U-Pb dating for gneisses in northern Dabie high T/P metamorphic zone, Central China: implications for decoupling within subducted continental crust*

Yi-Can Liu[1,2], Shu-Guang Li[1] and Shu-Tong Xu[3]

1. CAS Key Laboratory of Crust-Mantle Materials and Environments, School of Earth and Space Sciences, University of Science and Technology of China, Hefei 230026, Anhui, China
2. Beijing SHRIMP Center, Chinese Academy of Geological Sciences, Beijing 100037, China
3. Anhui Institute of Geology, Hefei 230001, Anhui, China

> 亮点介绍：本文根据锆石 U-Pb 定年及锆石的矿物包体等方面研究，为北大别片麻岩经历了超高压变质作用及多阶段变质演化过程提供了矿物学和年代学制约，提出了大别山深俯冲陆壳内部的多层次拆离解耦和多岩片差异折返机制。

Abstract Zircon SHRIMP U-Pb ages and cathodoluminescence (CL) images reveal that most zircon separated from two tonalitic gneiss samples in the northern Dabie high T/P metamorphic zone (NDZ) have four different domains: (1) inherited core, with clear oscillatory zoning, low-P mineral inclusions and high Th/U ratios, indicating the Neoproterozoic age of the protolith; (2) inner-mantle, with homogeneous CL intensity, low Th/U ratios of ⩽ 0.09 and occasionally ultrahigh pressure metamorphic (UHPM) mineral inclusions such as diamond, garnet and rutile, which defined a weighted mean $^{206}Pb/^{238}U$ age of 218 ± 3 Ma, corresponding to the age of ultrahigh pressure metamorphism; (3) outer-mantle with lower Th/U ratio and retrograde mineral inclusions, which yields a retrogressive age of 191 ± 5 Ma; and (4) rim, which is black luminescence, with relatively high Th/U ratios of 0.12–0.40 and a weighted mean $^{206}Pb/^{238}U$ age of 126 ± 5 Ma, indicating an overprint of the Early Cretaceous migmatization. Consequently, four discrete and meaningful age groups have been identified. Remarkably, the U-Pb ages of both UHPM zircon and retrograded metamorphic zircon from the NDZ are significantly younger than the U-Pb ages of 238 ± 3 Ma – 230 ± 4 Ma for UHPM zircon and U-Th-Pb ages of 218 ± 1.2 Ma – 209 ± 3 Ma for rutile and monazite overgrowths (representing the cooling or retrograded metamorphic time corresponding to amphibolite-facies) from the southern Dabie UHPM zone (SDZ), respectively. Combined with reported ages for UHPM rocks from the NDZ, SDZ and Huangzhen low-T eclogite zone (HZ), we found that the metamorphic ages of these three UHPM units in the Dabie orogen gradually decrease from south to north. This age distribution suggests that the three UHPM units represent 3 exhumed crustal slices, which are decoupled from each other and have different subduction and exhumation histories. Firstly, the subducted Huangzhen crustal slice was detached from the underlying subducted continental lithosphere, and then exhumed because of the buoyancy. In the meantime, subduction of underlying continental plate continued. After the initiation of this process, lithologic separations or decoupling within subducted continental crust occurred repeatedly several times to form an imbricate megastructure in the Dabie UHPM belt.

Key words decoupling; continental subduction; zircon U-Pb dating; UHPM rocks; Dabie orogen

1 Introduction

Formation and exhumation of ultrahigh pressure metamorphic (UHPM) rocks has been a hot topic in geologic

---

* 本文发表在：Lithos, 2007, 96: 170-185

community. Subduction of continental crust into mantle depths has long been demonstrated by the discovery of UHP index minerals such as coesite and micro-diamond in supracrustal rocks in various orogenic belts (e.g., Chopin, 1984; Smith, 1984; Okay et al., 1989; Wang et al., 1989; Sobolev and Shatsky, 1990; Xu et al., 1992), the mechanism for exhumation of UHPM rocks, however, remains controversial, e.g., intracontinental thrusting and concomitant erosion (Okay and Sengör, 1992; Chemenda et al., 1996), a tectonic-wedge-extrusion model (Maruyama et al., 1994, 1996), ascent of thin tectonic slices by buoyancy-driven model (Ernst et al., 1997; Ernst, 2001), a multi-stage model involving buoyancy, wedging and thermal doming (Dong et al., 1998), coeval thrusting and normal faulting (Faure et al., 1999), a multi-stage model involving buoyancy, subvertical extrusion, normal uplift and lithospheric extension (Wang and Cong, 1999), and orogen-parallel extrusion accompanied by layer-parallel thinning (Hacker et al., 2000; Ratschbacher et al., 2000). Nonetheless, all these models for UHPM rock exhumation assumed that the whole subducted continental crust was detached from the underlying mantle lithosphere and then exhumed as a whole (e.g., Chemenda et al., 1995; Hacker et al., 2000; Massonne, 2005). Such models were based on a hypo-thesis that detachment only occurred in the lower crust, due to its more ductile behavior compared to the upper crust and the lithospheric mantle (i.e. the "jelly sandwich model"; e.g. Zuber, 1994). Modeling results of lithospheric viscosity-depth curves based on reasonable geotherms and models of lithospheric composition however, suggest that there are at least two low-viscosity zones within continental crust at different depths (Meissner and Mooney, 1998). Accordingly, several HP-UHP metamorphic crustal slices could be produced by decoupling or detachment along low-viscosity zones during subduction of continental crust, which seems more likely to be a realistic geological process for the Dabie-Su-Lu UHPM belt.

The Dabie–Su-Lu UHPM belt in east-central China can be divided into three UHPM units from north to south: the northern Dabie high T/P metamorphic zone (NDZ), the southern Dabie UHPM zone (SDZ), and the Huangzhen low-T eclogite zone (HZ). Recently, the Pb isotopic mapping on the Dabie UHPM belt has revealed the detachment between subducted upper continental crust (the southern Dabie UHPM zone) and felsic lower crust (or lower part of upper crust) (the northern Dabie high T/P metamorphic zone) (Li et al., 2003b). The intensive Sm-Nd and U-Pb geochronological studies consistently indicated the UHPM ages of 238–225 Ma for the rocks from the SDZ (e.g., Li et al., 1989, 1993; Ames et al., 1996; Rowley et al., 1997; Hacker et al., 1998; Li et al., 2000; Ayers et al., 2002). By contrast, the Sm-Nd age of 212 ± 4 Ma for eclogite from the NDZ (Liu et al., 2005a) is significantly younger than the Sm- Nd age of 226 ± 3 Ma for eclogite from the SDZ (Li et al., 2000).

In this study, we report zircon SHRIMP U-Pb age data on two samples of tonalitic gneiss from the NDZ. The purposes of this work are: (1) to precisely determine the ages of the peak and retrograde metamorphism in the NDZ; (2) to examine whether different UHP rock units in the Dabie orogen may have different peak and retrograde metamorphic ages; and (3) to present a model of within-crust lithologic decoupling and multi-slice exhumation to better interpret the age distribution pattern in the Dabie UHPM belt.

2 Geological setting and sample

The Qinling-Dabie orogen is a collision zone between the North China Block (NCB) and South China Block (SCB). The Dabie Mountains represent the eastern section of the belt, and are transected at its eastern end by the Tan-Lu fault. The Su-Lu orogen is believed to be the eastern extension of the Dabie orogen. The UHPM rocks exposed in the Dabie and Su-Lu orogen comprise the largest known UHPM belt on the Earth. The general geology of the Dabie orogenic belt has been described by many authors (e.g., Xu et al., 1994; Hacker et al., 2000; Li et al., 2001; Zheng et al., 2003; Xu et al., 2005b). It is well accepted that the Dabie orogen can be divided into five major litho-tectonic units from north to south: (1) the Beihuaiyang zone (BZ); (2) the northern Dabie high T/P metamorphic zone (NDZ); (3) the southern Dabie UHPM zone (SDZ); (4) the Huangzhen low-T eclogite zone (HZ); and (5) the Susong complex zone (SZ), separated by Xiaotian-Mozitan fault (XMF), Wuhe-Shuihou fault (WSF), Hualiangting-Mituo fault (HMF) and Taihu-Mamiao fault (TMF), respectively (see Fig. 1).

The BZ is mainly composed of the Paleozoic flysch formation (the Foziling group) (Xu et al., 1994; Li et al., 2001; Xu et al., 2005b) and unsubducted continental crust of the SCB (the Luzhenguan complex) (Okay and Sengör, 1993; Zheng et al., 2004). In addition, post-collisional volcanic-sedimentary rocks are developed in the

Jurassic-Cretaceous basins in this zone.

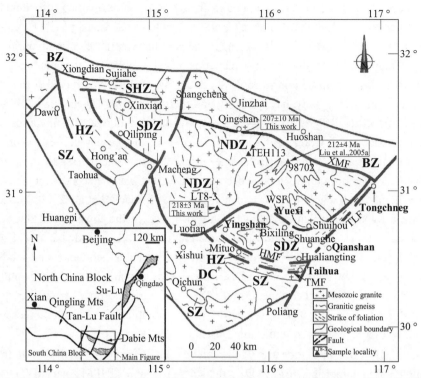

Fig. 1 Schematic geological map of the Dabie orogen, with inset showing the location of this area within the Triassic Qinling-Dabie – Su-Lu collision orogen in central China (modified after Xu et al., 2005b). The shaded area in the inset shows the UHPM belt. Sample localities with sample numbers are described in detail in the text. BZ = Beihuaiyang zone, NDZ = northern Dabie high T/P metamorphic zone, SDZ = southern Dabie UHPM zone, SZ = Susong complex zone, SHZ = Sujiahe HP metamorphic zone, DC = amphibolite-facies Dabie complex, XMF = Xiaotian-Mozitan fault, WSF = Wuhe-Shuihou fault, HMF = Hualiangting-Mituo fault, TMF = Taihu-Mamiao fault, TLF = Tan-Lu fault.

The NDZ is mainly composed of Neoproterozoic banded tonalitic and granitoid gneiss and Early Cretaceous post-collisional intrusions with minor meta-peridotite (including dunite, harzburgite and lherzolite), garnet pyroxenite, garnet-bearing amphibolite, granulite, and eclogite. The Luotian dome in the western part of the NDZ is a highly uplifted and eroded area with abundant felsic and mafic granulite lenses indicating that the NDZ could be a slice of upper part of lower crust or lower part of upper crust (Fig. 1). Although most of the rocks are of amphibolite-facies and eclogite, their relics have been identified with overprinting of granulite-facies in the NDZ (Tsai and Liou, 2000; Xu et al., 2000; Liu et al., 2001; Xiao et al., 2001; Liu et al., 2005b). Recently, the occurrence of quartz exsolution in Na-clinopyroxene (Tsai and Liou, 2000; Xu et al., 2000; Liu et al., 2001), oriented inclusions of polycrystalline rods of quartz, K-feldspar and albite in clinopyroxene (Malaspina et al., 2006), and rutile, apatite and clinopyroxene exsolution in garnet (Xu et al., 2005a) have been observed in eclogites from the NDZ. The quartz exsolution in Na-clinopyroxene has been considered to be evidence of ultrahigh-pressure conditions (cf. Liou et al., 1999 for review), while silica and K-feldspar rods are considered to be segregation products from pre-existing supersilic and K-rich clinopyroxene, respectively, that contained excess SiO_2 and K_2O at peak metamorphic conditions, stable at diamond stability field (e.g., Smith, 1984; Sobolev and Shatsky, 1990; Chopin, 2003). The rutile, apatite and clinopyroxene exsolutions in garnet are similar to those from Yangkou eclogite in the Su-Lu region reported by Ye et al. (2000), indicating that the eclogites here experienced UHP metamorphism at possible pressures of 5–7 GPa. In addition, Xu et al. (2003, 2005a) found micro-diamonds from the eclogites in the NDZ. So, the needle mineral exsolutions in garnet and clinopyroxene, and micro-diamond imply that eclogites from the NDZ subducted to mantle depths and experienced UHP metamorphism. Initial zircon U-Pb (Liu et al., 2000) and Sm-Nd (Liu et al., 2005a) geochronological studies suggest that eclogites from the NDZ formed in the Triassic, which indicates that the NDZ was involved in the Triassic continental subduction of the South China Block. The peak P-T conditions were estimated to be 808–874 ℃ or higher and > 2.5 GPa (Liu et al., 2005a) or perhaps > 5–7 GPa (Xu et al., 2005a). In addition, the overspread migmatization during 110 to 130 Ma is a typical characteristics for the NDZ (e.g., Chen et al., 1991; Hacker et al., 1998, 2000; Wang et al., 2002), which may influence some isotope systems on the gneiss and eclogite (e.g., Hacker et al., 1998; Liu et al., 2000; Xie et al., 2004).

The SDZ is composed of typical UHPM rocks, including eclogite, gneiss, quartz jadeitite, marble, etc. The occurrence of diamond and coesite in the rocks reveals temperature-pressure conditions of 700 to 850 ℃ and > 2.8 GPa or perhaps ⩾ 4 GPa for the UHP metamorphism (e.g., Xu et al., 1992; Wang et al., 1995; Cong, 1996), whereas the oriented magnetite lamellae and $FeTiO_3$ rod exsolutions in olivine from the UHPM ultramafic rocks in the SDZ suggested the peak conditions at P > 5–7 GPa or perhaps 10 GPa and T = 700–800 ℃ (e.g., Jin et al., 1998; Zhang et al., 1999). The UHP rocks from the SDZ only experienced amphibolite-and green-schist-facies retrograde metamorphism (e.g., Xu et al., 1992; Wang et al., 1995; Cong, 1996). Therefore, overprinting of granulite-facies retrograde metamorphism is unusual for eclogites from the NDZ, suggesting that the SDZ and NDZ units had different exhumation histories.

The HZ consists of mainly eclogite and gneiss. The eclogite in this zone is characterized by relatively lower metamorphic temperature, and coesite- and diamond- free eclogite, hence the Huangzhen eclogite was previously referred as HP "cold" eclogite zone by Okay (1993). Recently, Li et al. (2004) found coesite pseudomorphs from the eclogite in the Huangzhen eclogite zone and confirmed that this "cold" or lower-T eclogite zone experienced UHP metamorphism with peak conditions of 3.3 GPa and 670 ℃, and subsequent amphibolite- and green-schist-facies retrograde metamorphism.

The SZ is a green-schist–amphibolite-facies metamorphic unit with conditions of 400–500 ℃ and 0.5–0.6 GPa, which comprises mica-quartz schist, marble, amphibolite and meta-phosphorous series (e.g., Xu et al., 1994; Zhai et al., 1995; Xu et al., 2005b).

Therefore, there are three UHPM crustal slices in the Dabie orogen, i.e., the NDZ, the SDZ and the HZ. The Pb isotopic mapping in Dabie mountains reveals that UHPM rocks characterized by relative higher radiogenic Pb ($^{206}Pb/^{204}Pb$ = 17.026∼20.781) in the SDZ were derived from subducted upper continental crust, while UHPM rocks characterized by relative lower radiogenic Pb ($^{206}Pb/^{204}Pb$ = 15.844∼17.204) in the NDZ were derived from subducted lower continental crust (Li et al., 2003b). This point is supported by their different lithologic associations, hydrous mineral assemblages and metamorphic temperatures. It is well known that the NDZ is an orthogneiss unit with minor eclogite and/or granulite blocks. Recent study has demonstrated that some eclogites from the NDZ were transformed from mafic lower continental crustal rocks (Liu et al., 2005b). While the SDZ is an orthogneiss + supracrustal rocks unit, in which eclogite lens or blocks are interbedded with UHPM sediments such as paragneiss, marble and jadeite quartzite (cf., Xu et al., 1994; Zheng et al., 2003). The eclogite from the NDZ contains neither phengite nor other HP-UHP metamorphic hydrous mineral, which suggests the absence of fluid during their metamorphic process (Liu et al., 2004c, 2005a; Liu and Li, 2005). Whereas phengite-zoisite hydrous minerals are common in the eclogite from the SDZ (Cong, 1996) and lowsonite-barrositic amphibole-phengite hydrous minerals were observed in the eclogite from the HZ (Li et al., 2004), which indicates more abundant fluids in the SDZ and HZ during UHP metamorphism. Fluid inclusions in the UHPM rocks from the SDZ and NDZ are also different. The composition of metamorphic inclusions in the UHPM rocks from the SDZ is mainly aqueous fluid with various salinities, while those in the UHPM rocks from the NDZ are mainly N_2 or CO_2, suggesting more aqueous fluids in the SDZ during metamorphism (Xiao et al., 2001, 2002). In addition, the peak metamorphic temperatures of three UHP slices gradually increase from south to north, i.e. from T < 700 ℃ (generally, 570–670 ℃) for the Huangzhen eclogite in the HZ (Wang et al., 1992; Okay, 1993; Zhai et al., 1995; Li et al., 2004) to 700–800 ℃ for the UHP eclogites in the SDZ (Okay, 1993; Xu et al., 1994; Cong, 1996), and then to 808–874 ℃ for the eclogite in the NDZ (Liu et al., 2005a). The temperature variations may be related to the positions of the protoliths in subducted continental crust, indicating that protoliths come from different crustal levels (e.g., upper and lower continental crust, respectively) and originally had different temperatures. In summary, in view of more supracrustal rocks and hydrous minerals, aqueous fluid inclusions, lower peak metamorphic temperatures and higher radiogenic Pb isotopic composition of the eclogite in the HZ and SDZ, it is reasonable to propose that HZ and SDZ were from uppermost and middle parts of subducted upper continental crust in the SCB, respectively. In contrast, the NDZ is characterized by the absence of hydrous HP-UHP metamorphic mineral, N_2 or CO_2-inclusions, higher peak metamorphic temperature, minor supracrustal rocks and more abundant granulite outcrops as well as lower radiogenic Pb isotopic composition, and thus represents upper part of subducted lower continental crust or lower part of subducted upper continental crust.

Two garnet-bearing tonalitic gneiss samples (LT8-3 and TEH113) used for zircon U-Pb dating in this paper were collected from Jinjiapu and Ta'erhe, which are located in the southwest part (namely the Luotian dome) and north part of the NDZ, respectively (Fig. 1). The investigated gneisses are composed of plagioclase, quartz, amphibole, minor garnet, clinopyroxene and rutile; rare hypersthene is occasionally found (Liu and Li, 2005). In addition, micro-diamond observed in sample TEH113 suggests that the tonalitic gneiss from the NDZ experienced ultrahigh-pressure metamorphism and have similar metamorphic process to the enclosed eclogite (Liu et al., 2006).

3 Analytical method

Zircon was separated from approximately 5 kg gneiss sample by crushing and sieving, followed by magnetic and heavy liquid separation and hand-picking under binoculars. Approximately 500 zircon grains for each sample, together with a zircon U-Pb standard TEM (417 Ma), were mounted using epoxy, which was then polished until all zircon grains were approximately cut in half. The internal zoning patterns of the crystals were observed by CL image, which was analyzed at the Institute of Mineral Resources, Chinese Academy of Geological Sciences (CAGS). The representative CL images for the studied samples are presented in Figs. 2 and 3.

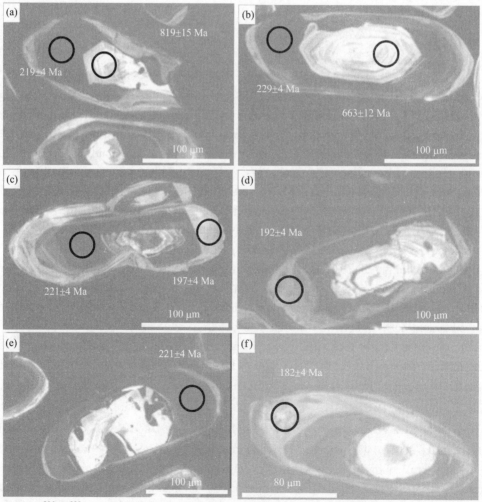

Fig. 2 CL images with $^{206}Pb/^{238}U$ ages for zircon grains from sample LT8-3 at Jinjiapu in southern part of the northern Dabie high T/P metamorphic zone.

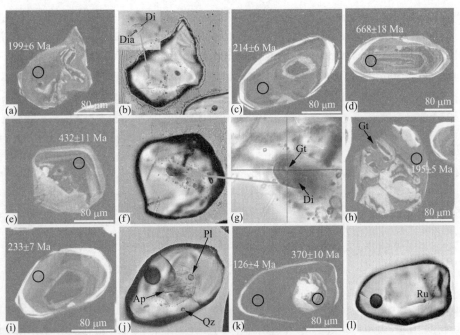

Fig. 3 Micro-photographs and the spots and ages of SHRIMP analysis of zircon grains from sample TEH113 at Ta'erhe in northern part of the northern Dabie high T/P metamorphic zone. (a), (c), (d), (f), (h), (i) and (k) are CL images with $^{206}Pb/^{238}U$ ages for zircons; (b), (f), (j) and (l) are plane-polarized light (PL) images; (g) is an enlargement part of (f) and a scanning image by Raman. Zircons (a) and (b), (e) and (f), and (i) and (j) are the same grain, respectively. Dia, Di, Pl, Ap, Qz and Ru represent diamond, diopside, plagioclase, apatite, quartz and rutile, respectively. All mineral inclusions in zircon grains were identified by Raman spectra in Fig. 4 or electron microprobe analyses (Liu et al., 2006).

Mineral inclusions in zircon have been used for diagnosis of zircon origin. Mineral inclusions were identified using Raman spectroscopy at the Analytical Centre, China University of Geosciences (Wuhan) and the Continental Dynamics Laboratory, CAGS, and/or substantiated using the electron microprobe analyzer (EMPA) at the Analytical Centre, China University of Geosciences (Wuhan). The analytical conditions on the Raman and EMPA were reported by Xu et al. (2005a) and Liu et al. (2005a). Representative Raman spectra of inclusion minerals in zircon from sample TEH113 are listed in Fig. 4.

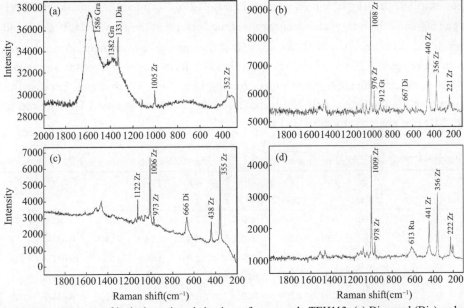

Fig. 4 Representative Raman spectra of inclusion minerals in zircon from sample TEH113. (a) Diamond (Dia) and graphite (Gra); (b) garnet (Gt) and diopside (Di); (c) diopside (Di); (d) rutile (Ru); these spectra also contain host zircon peaks at 1005–1009, 973–978, 438–441, 352–356 and 221–222 cm^{-1}.

Zircon U-Pb dating was performed by using the SHRIMP II at Beijing SHRIMP Center, CAGS. Analytical procedures are similar to those described by Compston et al. (1992) and Williams (1998). Both optical photomicrographs and CL images were taken as a guide to selection of U-Pb dating spot. The mount was vacuum-coated with a ~500 nm layer of high-purity gold. The intensity of the primary O_2- ion beam was 9 nA. Spot sizes were ~30 μm and each site was rastered for 150 s before analysis. Five scans through the mass stations were made for each age determination. Standards used were SL13, with an age of 572 Ma and U content of 238 ppm, and TEM, with an age of 417 Ma (Williams, 1998; Black et al., 2003). Data processing was carried out using the Squid and Isoplot programs (Ludwig, 2001), and measured ^{204}Pb was applied for the common lead correction. The SHRIMP results are presented in Tables 1 and 2, and in Fig. 5. The errors given in Tables 1, 2, Fig. 5 for individual analyses are quoted at the 1σ level, whereas the error for weighted mean ages given in Fig. 5, and in the text, are quoted at 2σ (95% confidence level).

4 Results of SHRIMP U-Pb dating

4.1 Sample LT8-3

Zircons separated from sample LT8-3 are 100–300 μm in length, subhedral to euhedral and long prismatic. Based on CL images four domains can be identified in most zircons (Fig. 2), i.e., bright-luminescent cores, low-luminescent inner-mantle (M1) (Fig. 2a, b, c and e), relatively bright-luminescent outer-mantle (M2) (Fig. 2d and f) and thin low-luminescent rim (Fig. 2d). Euhedral cores preserve obvious oscillatory zonation, whereas M1, M2 and rim domains are homogeneous in CL.

The cores occasionally contain some magmatic minerals such as potassium feldspar, quartz, biotite and apatite. Data from the cores yield apparent $^{206}Pb/^{238}U$ ages ranging from 394 to 819 Ma with Th/U ratios of 0.84–1.73. A concordant age among them is 819 ± 15 Ma and others are discordant age. Because oscillatory zonation and high Th/U ratios of the cores are indicative of a magmatic origin (Rubatto and Gebauer, 2000), the age of 819 Ma represents the protolith age of the gneiss. Fifteen analyses of the M1 domains record $^{206}Pb/^{238}U$ concordant ages ranging from 213 to 230 Ma, with a weighted mean age of 218 ± 3 Ma (MSWD = 2.7) and Th/U ≤ 0.02. The HP metamorphic mineral inclusions such as garnet have been observed in the M1 domains, which together with the low Th/U ratios suggest that these ages are the peak metamorphic time of the gneiss.

The M2 domains are variable in shape and in sharp contact with the M1 in CL. Twelve analyses of the M2 domains give ages from 181 to 201 Ma, with a weighted mean $^{206}Pb/^{238}U$ age of 191 ± 5 Ma (MSWD= 3.8) and Th/U ≤ 0.02. This age indicates the retrograde metamorphic time because of their lower Th/U ratios and absence of HP mineral inclusion in M2 domains. Two analyses of the M2 domains give a younger apparent age of 160 ± 4 Ma (spot 3.13) and an older age of 206 ± 3 Ma, respectively. The former is a discordant age which was probably affected by the later event, and the latter may record a mixing age of M1-M2 domains, because the M2 domain of this grain is thinner than the ion beam spot. The rim of zircon grains from sample LT8-3 are too thin (generally < 10 μm) (e.g., Fig. 2d) to be analyzed by the SHRIMP II.

Table 1 SHRIMP zircon U-Pb data for tonalitic gneiss (sample LT8-3) from the northern Dabie high T/P metamorphic zone

Spot	D	$^{206}Pb_c$ (%)	U (ppm)	Th (ppm)	Th/U	$^{206}Pb^*$ (ppm)	$^{207}Pb^*/^{206}Pb^*$	±%	$^{207}Pb^*/^{235}U$	±%	$^{206}Pb^*/^{238}U$	±%	$^{206}Pb/^{238}U$ age (Ma)	
1.1	C	0.42	179	160	0.89	16.7	0.0673	2.9	1.006	3.5	0.1084	2	663	± 12
2.1	M1	0.03	3372	36	0.01	101	0.05135	0.9	0.2474	1.9	0.03495	1.7	221	± 4
3.1	M2	0.53	350	1	0	9.59	0.0528	5.9	0.231	6.2	0.03173	1.9	201	± 4
2.2	M1	0.14	2445	32	0.01	70.6	0.05149	1.3	0.2383	2.2	0.03357	1.7	213	± 4
2.3	M1	0.18	2175	31	0.01	67.7	0.05097	1.9	0.2541	2.5	0.03616	1.7	229	± 4
2.4	M1	0.18	2351	29	0.01	66.6	0.05029	1.5	0.2282	2.3	0.0329	1.7	209	± 3
2.5	M1	0.21	3209	34	0.01	93.9	0.05128	1.5	0.2403	2.3	0.03399	1.7	215	± 4
2.6	M1	0.21	2425	37	0.01	71.1	0.05135	1.6	0.2411	2.3	0.03406	1.7	216	± 4

Spot	D	$^{206}Pb_c$ (%)	U (ppm)	Th (ppm)	Th/U	$^{206}Pb^*$ (ppm)	$^{207}Pb^*/^{206}Pb^*$	±%	$^{207}Pb^*/^{235}U$	±%	$^{206}Pb^*/^{238}U$	±%	$^{206}Pb/^{238}U$ age (Ma)
4.1	C	0.43	378	654	1.73	28.5	0.063	2.5	0.761	3.1	0.0875	1.9	541 ± 10
2.7	M1	0.18	1977	32	0.02	57	0.0518	1.9	0.2394	2.6	0.03352	1.7	213 ± 4
2.8	M1	0.12	2446	39	0.02	70.9	0.05197	1.6	0.2414	2.4	0.03368	1.7	213 ± 4
2.9	M1	0.14	1978	32	0.02	57.3	0.05203	1.5	0.2415	2.3	0.03366	1.7	213 ± 4
2.1	M1	0.11	2891	34	0.01	86.6	0.05138	1.5	0.2468	2.3	0.03484	1.7	221 ± 4
3.2	M2	2.24	424	3	0.01	11.5	0.0595	8.1	0.254	8.4	0.03101	2	197 ± 4
1.2	C	0.3	187	162	0.86	19.7	0.0674	3.2	1.136	3.7	0.1223	1.9	744 ± 14
2.11	M1	0.11	3263	39	0.01	102	0.05175	1.2	0.2594	2.1	0.03636	1.7	230 ± 4
2.12	M1	0.09	2412	22	0.01	72.5	0.05278	1.3	0.2543	2.1	0.03494	1.7	221 ± 4
2.13	M1	0.09	2566	35	0.01	77.4	0.05243	1.5	0.2537	2.3	0.03509	1.7	222 ± 4
2.14	M1	0.11	3953	35	0.01	120	0.05136	1	0.2493	2	0.03521	1.7	223 ± 4
2.15	M1	0.41	1772	27	0.01	52.8	0.0519	2	0.247	2.7	0.03453	1.7	219 ± 4
3.4	M2	0.93	250	2	0.01	6.7	0.0449	20	0.191	20	0.03091	2.2	196 ± 4
3.3	M2	0.77	643	4	0.01	16.8	0.0506	4.3	0.2107	4.6	0.03018	1.9	192 ± 3
1.3	C	0.46	174	147	0.84	20.3	0.0653	2.6	1.219	3.2	0.1354	2	819 ± 15
3.5	M2	0.16	1538	13	0.01	43.1	0.05425	1.6	0.2434	2.3	0.03255	1.7	206 ± 3
1.4	M2	1.37	539	4	0.01	14.6	0.0513	8.1	0.219	8.3	0.03101	1.9	197 ± 4
3.6	M2	1.94	389	2	0.01	9.94	0.0418	19	0.168	19	0.02917	2	185 ± 4
3.7	M2	1.49	459	1	0	11.4	0.0466	9.4	0.184	9.6	0.02855	2.1	181 ± 4
3.8	M2	1.34	694	3	0	18	0.0445	7.6	0.182	7.8	0.02977	1.9	189 ± 3
3.9	M2	1.32	515	2	0	13.6	0.0618	5.4	0.26	5.7	0.03047	1.9	193 ± 4
3.1	M2	1.93	543	5	0.01	13.7	0.0465	12	0.185	12	0.02878	2.1	183 ± 4
3.11	M2	0.7	725	9	0.01	19.8	0.0505	3.8	0.2198	4.2	0.03156	1.8	200 ± 4
3.12	M2	1.18	537	2	0	13.3	0.0556	6.8	0.219	7	0.02858	2	182 ± 3
3.13	M2	2.28	431	11	0.02	9.5	0.0599	13	0.207	13	0.02507	2.5	160 ± 4
4.2	C	0.73	596	532	0.89	32.5	0.0614	4.7	0.533	5.1	0.063	1.8	394 ± 7

Errors are 1-sigma; Pb_c and Pb^* indicate the common and radiogenic portions, respectively; common Pb corrected using measured ^{204}Pb. D-domain, C-core, M1- inner-mantle, M2- outter-mantle.

4.2 Sample TEH113

Zircon grains separated from sample TEH113 are 150–400 μm in size, and most are rounded (with length-to-width of 1:1–1:2). Similar to sample LT8-3 described above, four domains can be discerned by CL images in several zircon grains (Fig. 3c, d, i and k). Many of the cores vary in shape and are commonly embayed by metamorphic grey mantle, although clear oscillatory zoning still exists, indicating that they are of magmatic origin. Two types of mantle can be identified based on their shape and luminescent in CL: (1) most vary in shape because they embay the cores and are characterized by variable grey-luminescent (namely grey mantle or M1) (Fig. 3a, c, h and i), and (2) some with regular shape surrounding the M1 are characterized by homogeneous bright-luminescent (namely white mantle or M2) (Fig. 3c, d and i). The rim is homogenous low-luminescent (Fig. 3c, d and k). While the black overgrowth rims are very thick in some zircon grains (Fig. 3k). UHPM mineral inclusions (such as diamond) have been identified in grey mantle (see below for details), indicating that they are UHP metamorphic domains in origin. While higher Th/U ratios of 0.12–0.40 and younger ages of 126 Ma (see below for details) suggest that the black rims were magma overgrowth in origin corresponding to the regional Early Cretaceous magmatism in the NDZ.

Magmatic mineral inclusions including potassium feldspar, quartz, biotite and apatite have been identified in the cores (Fig. 3j). The analyses of 14 spots from all the cores yield apparent $^{206}Pb/^{238}U$ ages ranging from 342 to

700 Ma with Th/U ratios of 0.18–1.47. As shown in Fig. 5b, all the core data points distributed along a discordant line, which defined upper and lower intercepts at 815 ± 69 Ma and 210 ± 62 Ma (MSWD = 1.02), respectively, corresponding to the Neoproterozoic crystallization and the Triassic metamorphism. Seven analyses of the M1 domains record $^{206}Pb/^{238}U$ concordant ages ranging from 195 to 229 Ma, with a weighted mean age of 207 ± 10 Ma (MSWD = 3.3) and Th/U ⩽ 0.09. The eclogitic mineral inclusions such as rutile (Fig. 3i) were found in M1 domains. In addition, micro-diamond (Fig. 3b), garnet (Fig. 3f, g and h) and diopside (Fig. 3b, f and g) inclusions were documented by Raman spectroscopy (Fig. 4) and/or EMPA on the zircon M1 domain (Liu et al., 2006). The Raman spectrum of the diamond inclusion contains the typical peak of 1331 cm^{-1} associated with the graphite peak (1586 cm^{-1}) (Fig. 4) because diamond was partly transformed to graphite after the UHPM stage (Liu et al., 2006). The unusual paragneiss of UHPM minerals (e.g. diamond) and relatively low-P mineral (e.g. diopside) could be simply a consequence of the different reaction kinetics (Bruno et al., 2001). Similar cases had been found by some researchers, e.g., the peak eclogite-facies mineral assemblages associated with relic igneous minerals from the meta-granite in the Sesia zone (Compagnoni and Maffeo, 1973), and the preservation of gabbroic texture and relict igneous minerals in UHP rocks from Yangkou in the Su-Lu orogen (Liou and Zhang, 1996). These were most commonly attributed to a lack of fluid availability during metamorphism (e.g. Liou and Zhang, 1996; Mosenfelder et al., 2005). As mentioned above, the NDZ is a lower crust slice and absence of hydrous HP-UHP metamorphic mineral and aqueous fluid, hence, the clinopyroxene (diopside) may not transform into omphacite and, coevally, the alkalic plagioclase may not decompose into quartz + jadeite at the peak metamorphic conditions of the orthogneiss in the NDZ. More recently, the synthesis experiments using a sample with identical bulk compo- sition to the natural eclogites from the NDZ documented that the peak UHPM paragneiss consists of garnet, clinopyroxene and rutile, which was stable at $P \sim 3.5$ GPa and $T \geqslant 750–800$ ℃, whereas the clinopyroxenes are mainly diopside and augite (Malaspina et al., 2006). Therefore, the zircon M1 domains formed under UHPM conditions and the age of 207 ± 10 Ma recorded on the M1 domains with low Th/U ratios should represent the peak UHPM time of the gneiss. This age is marginally consistent with the age of 218 ± 3 Ma for sample LT8-3 within error limits. The M2 domains are thinner than the ion beam spot and contain lower U contents, thus cannot be analyzed. Most of the zircon rims were too thin to be measured, and only two thick rims can be dated and give identical age of 126 ± 4 Ma with a weighted mean $^{206}Pb/^{238}U$ age of 126 ± 5 Ma.

Table 2 SHRIMP zircon U-Pb data for tonalitic gneiss (sample TEH113) from the northern Dabie high T/P metamorphic zone

Spot	D	^{206}Pb(%)	U (ppm)	Th (ppm)	Th/U	$^{206}Pb^*$ (ppm)	$^{207}Pb^*/^{206}Pb^*$	±%	$^{207}Pb^*/^{235}U$	±%	$^{206}Pb^*/^{238}U$	±%	$^{206}Pb/^{238}U$ age (Ma)
113-2.1	R	0.00	1073	131	0.12	18.2	0.0504	3.1	0.1371	4.3	0.01975	2.9	126 ± 4
113-1.1	C	0.00	167	138	0.83	14.3	0.0662	2.5	0.911	4.5	0.0997	3.8	613 ± 22
113-2.2	C	0.51	1007	1171	1.16	60.2	0.0578	3.2	0.552	4.2	0.0693	2.7	432 ± 11
113-1.2	C	0.13	636	361	0.57	32.3	0.0617	2.1	0.502	3.6	0.0590	2.8	370 ± 10
113-1.3	C	0.00	552	362	0.66	51.7	0.0659	1.7	0.993	3.2	0.1093	2.8	668 ± 18
113-4.1	R	0.00	678	275	0.40	11.5	0.0488	3.7	0.1328	4.7	0.01975	2.9	126 ± 4
113-1.4	C	0.32	872	893	1.02	83.9	0.0639	1.8	0.984	6.6	0.1117	6.4	683 ± 42
113-1.5	M1	0.00	205	1	0.01	6.14	0.0686	4.2	0.334	5.8	0.0353	4.0	223 ± 9
113-5.1	M1	0.00	327	143	0.44	5.93	0.0666	3.7	0.1969	5.0	0.02143	3.4	137 ± 5
113-2.3	M1	0.00	468	3	0.01	13.5	0.0508	3.8	0.236	4.7	0.03371	2.9	214 ± 6
113-2.4	M1	0.00	483	6	0.01	12.6	0.0533	6.7	0.225	7.3	0.03064	2.9	195 ± 5
113-2.5	C	0.14	246	273	1.11	22.4	0.0655	2.3	0.955	3.7	0.1057	2.9	648 ± 18
113-2.6	M1	0.00	390	5	0.01	10.6	0.0578	3.0	0.253	4.2	0.03171	2.9	201 ± 6
113-2.7	C	0.10	758	135	0.18	35.5	0.0600	2.2	0.451	3.5	0.0545	2.8	342 ± 9
113-3.1	M1	0.00	317	14	0.04	10.9	0.0615	5.9	0.340	6.6	0.0401	3.0	253 ± 7
113-2.8	C	0.09	409	320	0.78	40.4	0.0647	1.9	1.022	3.4	0.1146	2.8	700 ± 19
113-2.9	C	0.18	535	785	1.47	47.0	0.0671	2.7	0.943	4.3	0.1019	3.3	626 ± 20

Continued

Spot	D	^{206}Pb$_c$(%)	U (ppm)	Th (ppm)	Th/U	^{206}Pb* (ppm)	^{207}Pb*/^{206}Pb*	±%	^{207}Pb*/^{235}U	±%	^{206}Pb*/^{238}U	±%	^{206}Pb/^{238}U age (Ma)
113-3.2	M1	0.44	737	108	0.15	30.6	0.0531	3.0	0.352	4.9	0.0481	3.9	303 ± 12
113-1.6	C	0.27	332	133	0.40	27.7	0.0639	2.2	0.854	3.6	0.0969	2.8	596 ± 16
113-2.10	M1	0.55	398	19	0.05	10.8	0.0504	7.2	0.218	7.8	0.03133	3.0	199 ± 6
113-2.11	C	0.32	985	203	0.21	48.7	0.0587	2.1	0.464	3.6	0.0573	2.9	359 ± 10
113-2.12	M1	0.00	410	7	0.02	12.2	0.0599	3.1	0.288	4.3	0.0349	2.9	221 ± 6
113-2.13	M1	0.00	719	25	0.03	19.4	0.0533	2.8	0.230	5.5	0.0314	4.8	199 ± 9
113-2.14	M1	0.00	808	120	0.15	24.7	0.0585	3.7	0.288	4.8	0.0357	3.0	226 ± 7
113-3.6	C	0.00	160	48	0.30	12.9	0.0635	2.9	0.822	4.1	0.0938	2.9	578 ± 16
113-4.2	M1	0.00	397	7	0.02	12.3	0.0503	3.1	0.251	4.3	0.0361	2.9	229 ± 7
113-2.15	M1	0.00	351	33	0.09	9.91	0.0523	3.3	0.237	4.4	0.03283	2.9	208 ± 6
113-2.16	C	0.03	1195	876	0.73	79.0	0.06058	1.3	0.642	3.0	0.0769	2.7	478 ± 13
113-3.7	M1	0.41	471	2	0.01	13.4	0.0483	5.4	0.219	6.1	0.03295	2.9	209 ± 6
113-3.8	M1	0.20	1246	4	0.00	31.1	0.0481	2.5	0.1921	3.7	0.02897	2.8	184 ± 5
113-3.9	C	0.00	206	242	1.17	13.0	0.0630	3.3	0.639	4.4	0.0736	2.9	458 ± 13
113-3.10	M1	1.02	382	21	0.05	12.2	0.0456	7.1	0.231	7.7	0.0368	2.9	233 ± 7

Errors are 1-sigma; Pb$_c$ and Pb* indicate the common and radiogenic portions, respectively; common Pb corrected using measured ^{204}Pb. D-domain, C-core, M1-inner-mantle, R-rim.

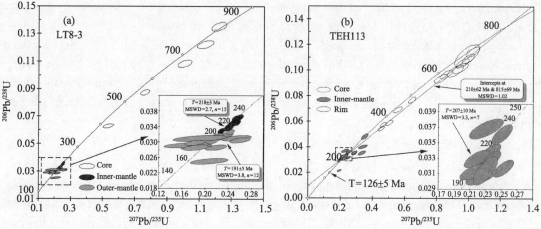

Fig. 5 Zircon SHRIMP U-Pb dating for samples LT8-3 (a) and TEH113 (b) at Jinjiapu and Ta'erhe in the northern Dabie high T/P metamorphic zone.

In summary, the zircon grains from the NDZ recorded at least four distinct age populations, i.e. Neoproterozoic (815 ± 69 Ma – 819 ± 15 Ma), 207 ± 10 Ma – 218 ± 3 Ma, 191 ± 5 Ma and 126 ± 5 Ma for core, M1, M2 and rim domains, respectively. The ages defined by M1 domains from samples TEH113 and LT8-3 are identical within error and represent peak UHP metamorphism, as indicated by diamond mineral inclusions. Because the age of 218 ± 3 Ma from sample LT8-3 is defined by fifteen data points of ^{206}Pb/^{238}U concordant ages and smaller age error, it should be the best estimate for the Triassic peak metamorphism time of the NDZ. The M2 domains contain low-P mineral inclusions such as quartz (Fig. 3j) and distributed around the M1, so 191 ± 5 Ma recorded in M2 domains suggest the retrogressive age. Because minerals formed during granulite-facies retrogression occur only as symplectite assemblages for the UHPM rocks in the NDZ, suggesting that either the granulite-facies retrograde metamorphism only lasted a short period of time (Liu et al., 2005a), or there was no fluid involved, which significantly limited mineral recrystallization or zircon overgrowth. By contrast, large-scale amphibolite-facies retrograded metamorphism occurred in the NDZ, i.e., most of the metamorphosed rocks in the NDZ show the amphibolite-facies with some relic of eclogite-and/or granulite-facies minerals. The retrograde mineral assemblage with amphibolite-facies contains abundant hydrous minerals such as amphibole and biotite, suggesting much activity of fluids in this retrograded metamorphic stage, which provided a suitable condition to the zircon overgrowth. Hence, the ca. 191 Ma may represent the retrograde time of amphibolite-facies. The age of 126 ± 5 Ma recorded on the zircon rim with higher Th/U ratios should represent the timing of migmatization in the NDZ, corresponding to a large-scale post-collisional magmatic event in the Early Cretaceous in the Dabie Mountains.

5 Discussion

5.1 Isotope chronology of three UHPM crustal slices

Due to its high closure temperature and extremely slow diffusion rates, zircon often records multiple growth stages in complex metamorphic terranes (e.g. Möller et al., 2002). The complex micro-textures of zircon grains from UHPM rocks-including irregular boundaries and various core, mantle and rim domains-revealed by CL images, suggest that many conventional zircon U-Pb ages for UHPM rocks are probably mixed ages with ambiguous geologic significance (Gebauer et al., 1997; Liu et al., 2002). The capacity of the SHRIMP ion microprobe to date small domains with high resolution makes it possible to determine different ages in a single grain, especially when accompanied by an integrated study of CL or back scatter electron (BSE) images, mineral inclusions, and geochemistry (e.g., Sun et al., 2002). These techniques thus have the potential to greatly improve our understanding of subduction and exhumation of UHPM rocks in the Dabie − Su-Lu orogen. Therefore, in order to compare the peak metamorphic ages with three different UHPM slices in the Dabie orogen, we chose the previous published data of zircon SHRIMP U-Th-Pb dating for UHPM rocks for the discussion below.

In addition, the complexity of Sm-Nd isotopic systematics in eclogite has been discussed by Thöni and Jagoutz (1992) and Li et al. (2000) in details. Most of published Sm-Nd ages are essentially isochrones given by two UHPM minerals (i.e. one garnet + one omphacite) (e.g. Li et al., 1993; Chavagnac and Jahn, 1996). Because the possibility of Nd isotopic dis-equilibrium between UHPM minerals, which could lower the slop of tie-line between garnet and omphacite (Jagoutz, 1994), and garnet + whole rock containing retrograde metamorphic minerals often give older ages (Li et al., 2000), only the Sm-Nd isochron ages given by three (or more) UHPM minerals for eclogite are chosen for the discussion below.

5.1.1 The northern Dabie high T/P metamorphic zone

Recently, a reliable Sm-Nd isochron age of 212 ± 4 Ma defined by two garnet + two omphacite from sample 98702 at Huangweihe eclogite in eastern segment of the NDZ has been reported (Liu et al., 2005a). Because eclogites from the NDZ experienced an unusual retrograde granulite-facies metamorphism, this Sm-Nd age was interpreted as a cooling age (Liu et al., 2005a). However, this Sm-Nd age is in agreement with the new reported zircon SHRIMP U-Pb age of 207 ± 10 Ma for sample TEH113 in this paper. The consistence of the above U-Pb zircon age and Sm-Nd age implies that the geological meaning of the Sm-Nd age should be reevaluated. Because the U-Pb system of zircon cannot be reset by overprinting of granulite-facies metamorphism (Mezger and Krogstad, 1997), the Sm-Nd age of 212 ± 4 Ma is not likely a cooling one but may also recorded the UHP metamorphic time. Besides the above two ages, we report a more precise U-Pb age of 218 ± 3 Ma for the UHPM overgrowth zircons from garnet-bearing tonalitic gneiss in the NDZ in this paper. Although this age is relatively higher than the U-Pb age of 207 ± 10 Ma and Sm-Nd age of 212 ± 4 Ma, they are all consistent with each other within error. Since the three samples with the above UHPM ages are distributed over south-western, middle-northern and east-northern parts of the NDZ, respectively (Fig. 1), the age of 218 ± 3 Ma can be considered to be the best estimate for the peak metamorphism time of the NDZ. As mentioned above, the NDZ sequentially experienced amphibolite-facies retrogressive overprint at 191 ± 5 Ma and tectonothermal event for the migmatization at 126 ± 5 Ma.

5.1.2 The southern Dabie UHPM zone

Li et al. (2000) dated the UHPM rocks from Shuanghe area in the SDZ, which occur as a UHPM block within the regional granitic gneiss. A Sm-Nd isochron defined by garnet + omphacite + rutile for a coesite-bearing eclogite from Shuanghe area in the SDZ yield an age of 226.3 ± 3.2 Ma (Li et al., 2000). Another Sm-Nd isochron defined by garnet + two phengites from Shuanghe UHP gneiss yield a similar age of 226.5 ± 2.3 Ma (Li et al., 2000). These two identical Sm-Nd ages defined by three UHP minerals suggest that Nd isotopic equilibrium between UHP minerals in these rocks during UHP metamorphism has been achieved. In addition, Ayers et al. (2002) obtained a concordant $^{206}Pb/^{238}U$ age of 229.6 ± 2.6 Ma with MSWD= 0.71 by twenty-five analyses of zircons for Maowu eclogite, which is in excellent agreement with the Sm-Nd ages reported by Li et al. (2000), and a weighted mean $^{206}Pb/^{238}U$ age of 238 ± 3 Ma with MSWD = 0.72 by six analyses of zircons for Shuanghe jadeite quartzite from the southern Dabie UHPM zone. Consequently, the peak UHPM age for the SDZ should be 226–238 Ma. In addition, the south-western Su-Lu terrane as a UHPM zone is comparable with the SDZ. A Sm-Nd isochron age of 226.3 ± 4.5 Ma given by garnet + omphacite + phengite (Li et al., 1994) and several zircon SHRIMP U-Pb ages ranging from 231 ± 4 Ma to 227 ± 2 Ma and 227.4 ± 3.5 Ma with UHPM mineral diagnosis of zircon origin (Liu et al., 2004a, b; Li et al., 2005a) have been reported, which strongly support the above conclusion concerning the peak UHP metamorphic age of the SDZ.

Li et al. (2000) suggested that the Rb-Sr age of 219 ± 6.6 Ma defined by phengite + garnet for UHPM gneiss from Shuanghe in the SDZ indicates the cooling time of phengite at 500 ℃. Also, the Sm-Nd age of 213.3 ± 4.8 Ma defined by garnet + biotite + epidote from granitic gneiss at Shuanghe suggests the Nd isotopic re-equilibration during retrograde amphibolite-facies metamorphism (Li et al., 2000). In addition, monazites separated from the Shuanghe jadeite quartzite display obviously core-rim zoning by CL images, and their core and rim domains yield weighted mean $^{208}Pb/^{232}Th$ age of 223 ±1 Ma and 209 ±3 Ma, respectively (Ayers et al., 2002). Because the T_{max} (700–800 ℃) of the UHPM rocks from the SDZ is higher than the T_c of 650–700 ℃ for Pb in monazite (Mezger et al., 1991; Smith and Giletti, 1997), the monazite core age should indicate the cooling time at $T = 675 ± 25$ ℃, while the rim age of 209 ± 3 Ma is corresponding to recrystallization growth of monazite occurred during retrograde amphibolite-facies metamorphism (Ayers et al., 2002). Furthermore, Li et al. (2003a) reported a precise U-Pb age of 218 ± 1.2 Ma with $T_c = 460$ ℃ for rutile from the coesite-bearing eclogite in the SDZ. Therefore, the ages of 219–209 Ma in the SDZ represent the time of amphibolite-facies retrograde metamorphism (e.g., Li et al., 2000; Ayers et al., 2002; Li et al., 2003a). The similar SHRIMP U-Pb ages of 211 ± 4 Ma and 209 ± 3 Ma for the amphibolite-facies retrograde rims in zircon from the UHP gneisses in the south-western Su-Lu terrane (Liu et al., 2004a, b) suggest that the SDZ and SW Su-Lu terrane experienced the amphibolite-facies retrograde metamorphism in the same time.

Although the UHPM and retrograde metamorphic ages of the SDZ are older than those of the NDZ, previous studies (e.g., Hacker et al., 2000; Ayers et al., 2002; Sun et al., 2002, Zheng et al., 2003, 2004; and references therein) show that most of zircon grains from the UHPM rocks in the SDZ commonly have Neoproterozoic cores, similar to those from the NDZ, suggesting the both UHPM units have similar protolith ages and belong to subducted continental crust of the SCB.

5.1.3 The Huangzhen eclogite zone

Li et al. (2004) dated the ages of the eclogite from the HZ, and gave a Sm-Nd mineral (garnet + omphacite + rutile + kyanite) isochron age of 236 ± 4 Ma and SHRIMP U-Pb zircon ages of 242 ± 3 Ma and 222 ± 4 Ma for the mantle and rim, respectively. They interpreted the UHPM metamorphic event should have occurred in the time period from 242 ± 3 Ma to 236 ± 4 Ma in the HZ, and the late retrogressive overprint occurred at 222 ± 4 Ma.

5.2 Implication for decoupling within subducted continental crust

In view of the above dating results, the peak metamorphic ages for the UHPM rocks from three different crustal slices in the Dabie orogen are different from each other. The peak metamorphic age (218 ± 3 Ma) of UHPM rocks from the NDZ is younger than the peak metamorphic age (226–238 Ma) of UHPM rocks from the SDZ, while the latter is younger than the peak metamorphic age (236–242 Ma) of UHP eclogite from the HZ. Furthermore, the amphibolite-facies retrograde metamorphic age of 191 ± 5 Ma of UHPM rocks from the NDZ is also significantly younger than the amphibolite-facies retrograde metamorphic ages of 219–209 Ma of UHPM rocks from the SDZ, while the latter is also younger than the age of 222 ± 4 Ma for retrograded metamorphic zircon from the eclogite in the HZ. All these show a decrease trend of metamorphic ages of the three UHPM crustal slices from south to north in the Dabie orogen. As described above, the three UHPM crustal slices from south to north in the Dabie orogen should be derived from subducted uppermost and middle part of upper crust and lower crust of the South China Block, respectively. Therefore, the decrease of the metamorphic ages from south to north in the Dabie orogen suggests that the detachment or a decoupling between the three UHPM crustal slices must be occurred separately from upper crust to lower crust during continental subduction. This decrease trend of the metamorphic ages of the three UHPM crustal slices from south to north in the Dabie UHPM belt can be explained by a decoupling within deeply subducted continental crust and multi-slice exhumation model (see Fig. 6).

The U-Pb ages of 244 ± 5 Ma for UHPM zircon and 221 ± 2 Ma for monazite overgrowth from a UHP granitoid gneiss cobble from the Late Jurassic Fenghuangtai Formation of the Hefei Basin (Wan et al., 2005), in the Beihuaiyang zone, are in good agreement with the zircon U-Pb ages of the Huangzhen eclogite (Li et al., 2004), suggesting the HZ exhumed at the surface by at least the Middle-Late Jurassic (Wan et al., 2005). At that time, the UHPM rocks in the SDZ and NDZ were still in upper-mid crust level (Li et al., 2000). So, the subducted Huangzhen crustal slice was firstly decoupled with the underlying subducted continental lithosphere, and then ascended driven by buoyancy forces. In the meantime, the subduction of underlying continental plate was continued (Fig. 6a).

It is interesting to note that the cooling age of 218 ± 1.2 Ma for rutile (T_c = 460 ℃) from a coesite-bearing eclogite in the SDZ (Li et al., 2003a) is identical with the peak metamorphic age of 218 ± 3 Ma for UHPM zircon from garnet-bearing tonalitic gneiss in the NDZ. It is suggested that the detachment between the SDZ and NDZ must have occurred before 218 Ma (could be at 226 Ma), which induced the uplifting of the SDZ slice along the detachment surface by thrust, while the subduction of the NDZ slice was continued (Fig. 6b).

Similarly, it can be inferred that a detachment or a decoupling within the lower continental crust could have also occurred because subducted mafic lower crust has not been exhumed in the Dabie orogen, which is supported by the existence of low-viscosity zone within lower continental crust (Meissner and Mooney, 1998). This decoupling could be occurred at 218 Ma, which caused the uplifting of upper part of felsic lower crust (i.e. the NDZ) during the time period of 218-191 Ma (Fig. 6c).

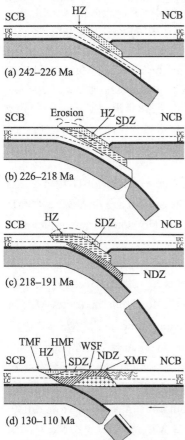

Fig. 6 A simplified tectonic model for the decoupling within deeply subducted continental crust and multi-slice exhumation in the Dabie orogen. SCB = South China Block, NCB = North China Block, UC = upper continental crust, LC = lower continental crust. The other abbreviated symbols are the same as in Fig. 1. Detailed explanations are given in the text.

Therefore, during the period of 242-191 Ma, the subduction of continental lithosphere of the South China Block was accompanied by successive uplift of decoupled crustal slices along the detachment surface by thrust. During the Cretaceous extension (130-110 Ma), doming and extensive plutonism resulted in the present structure and distribution of tectonic units (Fig. 6d), and the detachment surfaces are reworked as ductile normal faults (Faure et al., 1999; Hacker et al., 2000; Yuan et al., 2003).

6 Conclusions

(1) The zircon grains from tonalitic gneiss in the NDZ display typical zoning structures with core, mantle (including inner- and outer-mantle) and rim domains, and at least four discrete and meaningful age groups have been discerned. Residual cores show magmatic characteristics and give a concordant age of 819 ± 15 Ma, indicating that the protolith was a Neoproterozoic granitic intrusion. The inner-mantle domains with high U contents and lower Th/U ratios as well as eclogitic mineral inclusions such as garnet, diamond and rutile, give an age of 218 ± 3 Ma, representing the peak UHP metamorphic time of the eclogite and gneiss in the NDZ. The age of 191 ± 5 Ma for

outer-mantle domains recorded their amphibolite-facies retrograde metamorphic time. The age of 126 ± 5 Ma given by the rims indicates an overprinting of the Early Cretaceous magmatism and migmatization in the Dabie orogen.

(2) A decrease trend of the metamorphic ages of the three UHPM crustal slices from south to north in the Dabie orogen has been observed. This age distribution in the Dabie UHPM belt can be explained by a decoupling within deeply subducted continental crust and multi-slice exhumation model. The detachment occurred along the low-viscosity zones within subducting continental crust during the subduction process produced several decoupled crustal slices, e.g. the NDZ, SDZ and HZ in Dabie orogen. The Huangzhen crustal slice was firstly decoupled with the underlying subducted continental lithosphere, and then exhumed due to buoyancy. In the meantime, the subduction of underlying continental plate continued. After the initiation of this process, the detachment or decoupling within subducted continental crust was repeated several times to form an imbricate megastructure in the Dabie UHPM belt. The multi-slice, multi-stage exhumation processes of deeply subducted continental crust are special for continental subduction and differ from the oceanic subduction (Li et al., 2005b).

Acknowledgements

This research was funded by the NSFC (40373009 and 40572035), the Major State Research Program of China (G1999075503), and Chinese National Key Project for Basic Research (2003CB716500). We thank Dunyi Liu, Yanbin Wang, Yusheng Wan, and Hua Tao for assisting the SHRIMP analysis, Zhenyu Chen for the CL imaging, Mouchun He and Ling Yan for the Raman analysis, Guanbao Chen for the field sampling. Weidong Sun is thanked for his suggestions on the earlier versions of this manuscript, which helped clarify some ambiguities. Critical reviews by Yaoling Niu and two anonymous reviewers have helped improve the paper.

References

Ames, L., Zhou, G., Xiong, B., 1996. Geochronology and isotopic character of ultrahigh-pressure metamorphism with implications for the collision of the Sino-Korean and Yangtze cratons, central China. Tectonics 15, 472-489.

Ayers, J.C., Dunkle, S., Gao, S., Miller, C.F., 2002. Constraints on timing of peak and retrograde metamorphism in the Dabie Shan ultrahigh-pressure metamorphic belt, east-central China, using U-Th-Pb dating of zircon and monazite. Chemical Geology 186, 315-331.

Black, L.P., Kamo, S.L., Allen, C.M., Aleinikoff, J.K., Davis, D.W., Korsch, R.J., Foudoulis, C., 2003. TEMORA 1: a new zircon standard for Phanerozoic U-Pb geochronology. Chemical Geology 200, 155-170.

Bruno, M., Compagnoni, R., Rubbo, M., 2001. The ultra-high pressure coronitic and pseudomorphous reactions in a metagranodiorite from the Brossasco-Isasca Unit, Dora-Maira Massif, western Italian Alps: a petrographic study and equilibrium thermodynamic modeling. Journal of Metamorphic Geology 19, 33-43.

Chavagnac, V., Jahn, B.M., 1996. Coesite-bearing eclogites from the Bixilng complex, Dabie Mountains, China: Sm-Nd ages, geochemical characteristics and tectonic implications. Chemical Geology 133, 29-51.

Chemenda, A.I., Mattauer, M., Malavieille, J., Bokun, A.N., 1995. A mechanism for syn-collisional rock exhumation and associated normal faulting: results from physical modeling. Earth and Planetary Science Letters 132, 225-232.

Chemenda, A.I., Mattauer, M., Bokun, A.N., 1996. Continental subduction and a mechanism for exhumation of high-pressure metamorphic rocks: new modelling, field data from Oman. Earth and Planetary Science Letters 143, 173-182.

Chen, T., Niu, B., Liu, Z., 1991. Isotopic geochronology of metamorphism and Yanshanian magmatism within the Dabie Shan. Acta Geologica Sinica 65, 329-335.

Chopin, C., 1984. Coesite and pure pyrope in high grade blueschists of the western Alps: a first record and some consequences. Contributions to Mineralogy and Petrology 86, 107-118.

Chopin, C., 2003. Ultrahigh-pressure metamorphism: trace continental crust into the mantle. Earth and Planetary Science Letters 212, 1-14.

Compagnoni, R., Maffeo, B., 1973. Jadeite-bearing metagranites l. s. and related rocks in the Mount Mucrone area (Sesia-Lanzo Zone, Western Italian Alps). Schweizerische Mineralogische und Petrographische Mitteilungen 53, 355-378.

Compston, W., Williams, I.S., Kirschvink, J.L., Zhang, Z., Ma, G., 1992. Zircon U-Pb ages for the early Cambrian time-scale. Journal of the Geological Society (London) 149, 171-184.

Cong, B., 1996. Ultrahigh-Pressure Metamorphic Rocks in the Dabieshan-Sulu Region of China. Science Press, Beijing. 224 pp.

Dong, S., Chen, J., Huang, D., 1998. Differential exhumation of tectonic units and ultrahigh-pressure metamorphic rocks in the Dabie

Mountains, China. Island Arc 7, 174-183.

Ernst, W.G., 2001. Subduction, ultrahigh-pressure metamorphism, and regurgitation of buoyant crustal slices — implications for arcs and continental growth. Physics of the Earth and Planetary Interiors 127, 253-275.

Ernst, W.G., Maruyama, S., Wallis, S., 1997. Buoyancy-driven, rapid exhumation of ultrahigh-pressure metamorphosed continental crust. Proceedings of the National Academy of Sciences of the United States of America 94, 9532-9537.

Faure, M., Lin, W., Shu, L., Sun, Y., Schärer, U., 1999. Tectonics of the Dabieshan (eastern China) and possible exhumation mechanism of ultra high-pressure rocks. Terra Nova 11, 251-258.

Gebauer, D., Schertl, H. P., Brix, M., Schreyer, W., 1997. 35 Ma old ultrahigh-pressure metamorphism and evidence for very rapid exhumation in the Dora Maira Massif, Western Alps. Lithos 41, 5-24.

Hacker, B.R., Ratschbacher, L., Webb, L.E., Ireland, T., Walker, D., Dong, S., 1998. Zircon ages constrain the architecture of the ultrahigh-pressure Qinling-Dabie orogen, China. Earth and Planetary Science Letters 161, 215-230.

Hacker, B.R., Ratschbacher, L., Webb, L.E., McWilliams, M.O., Ireland, T., Calvert, A., Dong, S., Wenk, H. R., Chateigner, D., 2000. Exhumation of ultrahigh-pressure continental crust in east central China: Late Triassic-Early Jurassic tectonic unroofing. Journal of Geophysical Research 105, 13339-13364.

Jagoutz, E., 1994. Isotopic systematics of metamorphic rocks (abstract). ICOG-8 at Berkeley Uni. U.S. Geological Survey Circular 1107, 156.

Jin, Z., Jin, S., Gao, S., Zhao, W., 1998. Is the depth where UHP rocks formed in the Dabie Mountains limited to 100 ~ 150 km? Chinese Science Bulettin 43, 767-771.

Li, S., Hart, S.R., Zheng, S., Liu, D., Zhang, G., Guo, A., 1989. Timing of collision between the North and South China Blocks: Sm-Nd isotopic age evidence. Science in China. Series B, Chemistry, Life Sciences & Earth Sciences 32, 1391-1400.

Li, S., Xiao, Y., Chen, Y., Ge, N., Zhang, Z., Sun, S. S., Cong, B., Zhang, R. Y., Hart, S.R., Wang, S., 1993. Collision of the north China and Yangtse blocks and formation of coesite-bearing eclogites: timing and processes. Chemical Geology 109, 89-111.

Li, S., Wang, S., Chen, Y., Liu, D., Qiu, J., Zhou, H., Zhang, Z., 1994. Excess argon in phengite from eclogite: evidence from dating of eclogite minerals by Sm-Nd, Rb-Sr and $^{40}Ar/^{39}Ar$ methods. Chemical Geology 112, 343-350.

Li, S., Jagoutz, E., Chen, Y., Li, Q., 2000. Sm-Nd and Rb-Sr isotope chronology of ultrahigh-pressure metamorphic rocks and their country rocks at Shuanghe in the Dabie Mountains, central China. Geochimica et Cosmochimica Acta 64, 1077-1093.

Li, S., Huang, F., Nie, Y., Han, W., Long, G., Li, H., Zhang, S., Zhang, Z., 2001. Geochemical and geochronological constrains on the suture location between the North and South China Blocks in the Dabie orogen, central China. Physics and Chemistry of the Earth (A) 26, 655-672.

Li, Q., Li, S., Zheng, Y., Li, H., Massonne, H.J., Wang, Q., 2003a. A high precision U-Pb age of metamorphic rutile in coesite-bearing eclogite from the Dabie Mountains in central China: a new constraint on the cooling history. Chemical Geology 200, 255-265.

Li, S., Huang, F., Zhou, H., Li, H., 2003b. U-Pb isotopic compositions of the ultrahigh pressure metamorphic (UHPM) rocks from Shuanghe and gneisses from Northern Dabie zone in the Dabie Mountains, central China: constraint on the exhumation mechanism of UHPM rocks. Science in China (Series D) 46, 200-209.

Li, X. P., Zheng, Y., Wu, Y., Chen, F., Gong, B., Li, Y., 2004. Low-T eclogite in the Dabie terrane of China: petrological and isotopic constrains on fluid activity and radiometric dating. Contributions to Mineralogy and Petrology 148, 443-470.

Li, Q., Li, S., Hou, Z., Hong, J., Yang, W., 2005a. A combined study of SHRIMP U-Pb dating, trace element and mineral inclusions on high-pressure metamorphic overgrowth zircon in eclogite from Qinglongshan in the Sulu terrane. Chinese Science Bulletin 50 (5), 459-465.

Li, S., Li, Q., Hou, Z., Yang, W., Wang, Y., 2005b. Cooling history and exhumation mechanism of the ultrahigh-pressure metamorphic rocks in the Dabie Mountains, central China. Acta Petrologica Sinica 21, 1117-1124 (in Chinese with English abstract).

Liou, J.G., Zhang, R.Y., 1996. Occurrences of intergranular coesite in ultrahigh-P rocks from the Sulu region, eastern China: implications for lack of fluid during exhumation. American Mineralogist 81, 1217-1221.

Liou, J.G., Zhang, R.Y., Ernst, W.G., Rumble, D., Maruyama, S., 1999. High-pressure minerals from deeply subducted metamorphic rocks. In: Hemley, R.J. (Ed.), Reviews in Mineralogy, vol. 37, pp. 33-96.

Liu, Y. C., Li, S., 2005. Lower crustal rocks from the Dabie Mountains and their deep subduction. Acta Petrologica Sinica 21, 1059-1066 (in Chinese with English abstract).

Liu, Y. C., Li, S., Xu, S., Li, H., Jiang, L., Chen, G., Wu, W., Su, W., 2000. U-Pb zircon ages of the eclogite and tonalitic gneiss from the northern Dabie Mountains, China and multi-overgrowths of metamorphic zircons. Geological Journal of China Universities 6, 417-423 (in Chinese with English abstract).

Liu, Y. C., Xu, S., Li, S., Chen, G., Jiang, L., Zhou, C.W., 2001. Distribution and metamorphic P-T condition of the eclogites from the mafic-ultramafic belt in the northern part of the Dabie Mountains. Acta Geologica Sinica 75, 385-395 (in Chinese with English abstract).

Liu, F.L., Xu, Z.Q., Liou, J.G., Katayama, I., Masago, H., Maruyama, S., Yang, J., 2002. Ultrahigh-pressure mineral inclusions in zircons from gneissic core samples of the Chinese Continental Scientific Drilling Site in eastern China. European Journal of Mineralogy 14, 499-512.

Liu, F.L., Xu, Z.Q., Liou, J.G., Song, B., 2004a. SHRIMP U-Pb ages of ultrahigh-pressure and retrograde metamorphism of gneisses, south-western Sulu terrane, eastern China. Journal of Metamorphic Geology 22, 315-326.

Liu, F.L., Xu, Z.Q., Xue, H., 2004b. Tracing the protolith, UHP metamorphism, and exhumation ages of orthogneiss from the SW Sulu terrane (eastern China): SHRIMP U-Pb dating of mineral inclusion-bearing zircons. Lithos 78, 411-429.

Liu, Y. C., Li, S., Xu, S., Chen, G., 2004c. Retrogressive microstructures of the eclogites from the northern Dabie Mountains, central China: evidence for rapid exhumation. Journal of China University of Geosciences 15, 349-354.

Liu, Y. C., Li, S., Xu, S., Jahn, B. M., Zheng, Y.F., Zhang, Z., Jiang, L., Chen, G., Wu, W., 2005a. Geochemistry and geochronology of eclogites from the northern Dabie Mountains, central China. Journal of Asian Earth Sciences 25, 431-443.

Liu, Y. C., Xu, S., Li, S., Chen, G., Peng, L., 2005b. Eclogite from the subducted lower crust of the Yangtze plate within the Luotian dome and its geological implication. Earth Science-Journal of China University of Geosciences 30, 71-77 (in Chinese with English abstract).

Liu, Y. C., Li, S., Xu, S., Gu, X., 2006. Ultrahigh-pressure metamorphic evidence for gneiss from the northern Dabie complex, central China: message from zircon. Acta Petrologica Sinica 22, 1827-1832 (in Chinese with English abstract).

Ludwig, K.R., 2001. Squid 1.02: A Users Manual. Special Publication, vol. 2. Berkeley Geochronology Center. 19 pp.

Malaspina, N., Hermann, J., Scambelluri, M., Compagnoni, R., 2006. Multistage metasomatism in ultrahigh-pressure mafic rocks from the North Dabie Complex (China). Lithos 90, 19-42.

Maruyama, S., Liou, J.G., Zhang, R., 1994. Tectonic evolution of the ultrahigh-pressure (UHP) and high-pressure (HP) metamorphic belts from central China. The Island Arc 3, 112-121.

Maruyama, S., Liou, J.G., Terabayashi, M., 1996. Blueschists and eclogites of the world, and their exhumation. International Geology Review 38, 485-594.

Massonne, H.J., 2005. Involvement of crustal material in delamination of the lithosphere after continent-continent collision. International Geology Review 47, 792-804.

Meissner, R., Mooney, W., 1998. Weakness of the lower continental crust: a condition for delamination, uplift, and escape. Tectono-physics 296, 47-60.

Mezger, K., Krogstad, E.J., 1997. Interpretation of discordant U-Pb zircon ages: a evaluation. Journal of Metamorphic Geology 15, 127-140.

Mezger, K., Rawnsley, C., Bohlen, S.R., Hanson, G.N., 1991. U-Pb garnet, sphene, monazite, and rutile ages: implications for the duration of high-grade metamorphism and cooling histories Adirondack Mountains, New York. Journal of Geology 99, 415-428.

Möller, A., O'Brien, P.J., Kennedy, A., Kröner, A., 2002. Polyphase zircon in ultrahigh-temperature granulites (Rogaland, SW Norway): constraints for Pb diffusion in zircon. Journal of Metamorphic Geology 20, 727-740.

Mosenfelder, J.L., SCHERTL, H. P., Smyth, J.R., Liou, J.G., 2005. Factors in the preservation of coesite: the importance of fluid infiltration. American Mineralogist 90, 779-789.

Okay, A.I., 1993. Petrology of diamond and coesite-bearing metamorphic terrane: Dabie Shan, China. European Journal of Mineralogy 5, 659-676.

Okay, A.I., Sengör, A.M.C., 1992. Evidence for intracontinental thrust-related exhumation of the ultra-high-pressure rocks in China. Geology 20, 411-414.

Okay, A.I., Sengör, A.M.C., 1993. Tectonics of an ultrahigh-pressure metamorphic terrane: the Dabie Shan/Tongbai orogen, China. Tectonics 12, 1320-1334.

Okay, A.I., Xu, S., Sengör, A.M.C., 1989. Coesite from the Dabie Shan eclogites, central China. European Journal of Mineralogy 1, 595-598.

Ratschbacher, L., Hacker, B.R., Webb, L.E., McWilliams, M., Ireland, T., Dong, S., Calvert, A., Wenk, H. R., Chateigner, D., 2000. Exhumation of the ultrahigh-pressure continental crust in east-central China: Cretaceous and Cenozoic unroofing and the Tan-Lu fault. Journal of Geophysical Research 105, 13303-13338.

Rowley, D.B., Xue, F., Tucker, R.D., Peng, Z.X., Baker, J., Davis, A., 1997. Ages of ultrahigh pressure metamorphism and protolith orthogneisses from the eastern Dabie Shan: U/Pb zircon geochronology. Earth and Planetary Science Letters 151, 191-203.

Rubatto, D., Gebauer, D., 2000. Use of cathodoluminescence for U-Pb zircon dating by ion microprobe; some examples from the Western Alps. In: Pagel, M., Barbin, V., Blanc, P., Ohnenstetter, D. (Eds.), Cathodoluminescence in Geosciences. Springer, Berlin, pp. 373-400.

Smith, D.C., 1984. Coesite in clinopyroxene in the Caledonides and its implications for geodynamics. Nature 310, 641-644.

Smith, H.A., Giletti, B.J., 1997. Lead diffusion in monazite. Geochimica et Cosmochimica Acta 61, 1047-1055.

Sobolev, N.V., Shatsky, V.S., 1990. Diamond inclusions in garnets from metamorphic rocks: a new environment for diamond formation. Nature 343, 742-746.

Sun, W.D., Williams, I.S., Li, S., 2002. Carboniferous and Triassic eclogites in the western Dabie Mountains, east-central China: evidence for protracted convergence of the North and South China Blocks. Journal of Metamorphic Geology 20, 873-886.

Thöni, M., Jagoutz, E., 1992. Some new aspects of dating eclogites in orogenic belts: Sm-Nd, Rb-Sr and Pb-Pb isotopic results from the Austroalpine Savalpine and Koralpe-type locality. Geochimica et Cosmochimica Acta 56, 347-368.

Tsai, C. H., Liou, J.G., 2000. Eclogite-facies relics and inferred ultrahigh-pressure metamorphism in the North Dabie Complex, central-eastern China. American Mineralogist 85, 1-8.

Wan, Y., Li, R., Wilde, S.A, Liu, D., Chen, Z., Yan, L., Song, T., Yin, X., 2005. UHP metamorphism and exhumation of the Dabie Orogen, China: evidence from SHRIMP dating of zircon and monazite from a UHP granitic gneiss cobble from the Hefei Basin. Geochimica et Cosmochimica Acta 69, 4333-4348.

Wang, Q., Cong, B., 1999. Exhumation of UHP Terranes: a case study from the Dabie Mountains, eastern China. International Geology Review 41, 994-1004.

Wang, X., Liou, J.G., Mao, H.K., 1989. Coesite-bearing eclogites from the Dabie Mountains in central China. Geology 17, 1085-1088.

Wang, X., Liou, J.G., Marruyama, S., 1992. Coesite-bearing eclogites from the Dabie Mountains, central China: petrogenesis, P-T path and implications for regional tectonics. Journal of Geology 100, 231-250.

Wang, X., Zhang, R.Y., Liou, J.G., 1995. UHPM terrane in east central China. In: Coleman, R., Wang, X. (Eds.), Ultrahigh Pressure Metamorphism. Cambridge University Press, pp. 356-390.

Wang, J., Sun, M., Deng, S., 2002. Geochronological constraints on the timing of migmatization in the Dabie Shan, East-central China. European Journal of Mineralogy 14, 513-524.

Williams, I.S., 1998. U-Th-Pb geochronology by ion microprobe. Reviews in Economic Geology 7, 1-35.

Xiao, Y., Hoefs, J., van den Kerkhof, A.M., Li, S., 2001. Geochemical constraints of the eclogite and granulite facies metamorphism as recognized in the Raobazhai complex from North Dabie Shan, China. Journal of Metamorphic Geology 19, 3-19.

Xiao, Y., Hoefs, J., van den Kerkhof, A.M., Simon, K., Fiebig, J., Zheng, Y., 2002. Fluid history during HP and UHP metamorphism in Dabie Shan, China: constraints from mineral chemistry, fluid inclusions, and stable isotopes. Journal of Petrology 43, 1505-1527.

Xie, Z., Gao, T., Chen, J., 2004. Multi-stage evolution of gneiss from North Dabie: evidence from zircon U-Pb chronology. Chinese Science Bulletin 49, 1963-1969.

Xu, S., Okay, A.I., Ji, S., Sengör, A.M.C., Su, W., Liu, Y. C., Jiang, L., 1992. Diamond from the Dabie Shan metamorphic and its implication for tectonic setting. Science 256, 80-82.

Xu, S., Liu, Y. C., Jiang, L., Ji, S., Su, W., 1994. Tectonic Regime and Evolution of Dabie Mountains. Science Press, Beijing. 175 pp. (in Chinese with English abstract).

Xu, S., Liu, Y. C., Su, W., Wang, R., Jiang, L., Wu, W., 2000. Discovery of the eclogite and its petrography in the Northern Dabie Mountain. Chinese Science Bulletin 45, 273-278.

Xu, S., Liu, Y. C., Chen, G., Compagnoni, R., Rolfo, F., He, M., Liu, H., 2003. New finding of micro-diamonds in eclogites from Dabie-Sulu region in central-eastern China. Chinese Science Bulletin 48, 988-994.

Xu, S., Liu, Y. C., Chen, G., Ji, S., Ni, P., Xiao, W., 2005a. Microdiamonds, their classification and tectonic implications for the host eclogites from the Dabie and Su-Lu regions in central eastern China. Mineralogical Magazine 69, 509-520.

Xu, S., Liu, Y. C., Chen, G., Wu, W., 2005b. Architecture and kinematics of the Dabie orogen, central eastern China. Acta Geologica Sinica 79, 356-371.

Ye, K., Cong, B., Ye, D., 2000. The possible subduction of continental material to depths greater than 200 km. Nature 407, 734-736.

Yuan, X.C., Klemperer, S.L., Teng, W. B., Liu, L. X., Chetwin, E., 2003. Crustal structure and exhumation of the Dabie Shan ultrahigh-pressure orogen, eastern China, from seismic reflection profiling. Geology 31, 435-438.

Zhai, M., Cong, B., Zhao, Z., Wang, Q., Wang, G., Jiang, L., 1995. Petrologic-tectonic units in the coesite-bearing metamorphic terrain of the Dabie Mountains, central China and their geotectonic implication. Journal of Southeast Asian Earth Sciences 11, 1-13.

Zhang, R., Shu, J., Mao, H., Liou, J.G., 1999. Magnetite lamellae in olivine and clinohumite from Dabie UHP ultramafic rocks, central China. American Mineralogist 84, 564-569.

Zheng, Y. F., Fu, B., Gong, B., Li, L., 2003. Stable isotope geochemistry of ultrahigh pressure metamorphic rocks from the Dabie-Sulu orogen in China: implications for geodynamics and fluid regime. Earth-Science Reviews 62, 105-161.

Zheng, Y., Wu, Y., Chen, F., Gong, B., Li, L., Zhao, Z., 2004. Zircon U-Pb and oxygen isotope evidence for a large-scale ^{18}O depletion event in igneous rocks during the Neoproterozoic. Geochimica et Cosmochimica Acta 68, 4145-4165.

Zuber, M.T., 1994. Folding a jelly sandwich. Nature 371, 650-651.

Ultrahigh-pressure eclogite transformed from mafic granulite in the Dabie orogen, east-central China*

Y.-C. Liu[1,2], S.-G. Li[1], X.-F. Gu[1], S.-T. Xu[3] and G.-B. Chen[4]

1. CAS Key Laboratory of Crust-Mantle Materials and Environments, School of Earth and Space Sciences, University of Science and Technology of China, Hefei 230026, Anhui, China
2. Beijing SHRIMP Center, Chinese Academy of Geological Sciences, Beijing 100037, China
3. Anhui Institute of Geology, Hefei 230001, Anhui, China
4. Anhui Institute of Paleontology, Hefei 230001, Anhui, China

亮点介绍：根据变质岩石学和锆石 U-Pb 年代学等方面研究，证明北大别罗田榴辉岩为华南陆块北缘新元古代镁铁质下地壳岩石在三叠纪发生深俯冲变质成因，这为研究大陆深俯冲和陆壳物质再循环以及对地幔不均一性的影响提供了重要对象和可能性。

Abstract Although ultrahigh-pressure (UHP) metamorphic rocks are present in many collisional orogenic belts, almost all exposed UHP metamorphic rocks are subducted upper or felsic lower continental crust with minor mafic boudins. Eclogites formed by subduction of mafic lower continental crust have not been identified yet. Here an eclogite occurrence that formed during subduction of the mafic lower continental crust in the Dabie orogen, east-central China is reported. At least four generations of metamorphic mineral assemblages can be discerned: (i) hypersthene + plagioclase ± garnet; (ii) omphacite + garnet + rutile + quartz; (iii) symplectite stage of garnet + diopside + hypersthene + ilmenite + plagioclase; (iv) amphibole + plagioclase + magnetite, which correspond to four metamorphic stages: (a) an early granulite facies, (b) eclogite facies, (c) retrograde metamorphism of high-pressure granulite facies and (d) retrograde metamorphism of amphibolite facies. Mineral inclusion assemblages and cathodoluminescence images show that zircon is characterized by distinctive domains of core and a thin overgrowth rim. The zircon core domains are classified into two types: the first is igneous with clear oscillatory zonation ± apatite and quartz inclusions; and the second is metamorphic containing a granulite facies mineral assemblage of garnet, hypersthene and plagioclase (andesine). The zircon rims contain garnet, omphacite and rutile inclusions, indicating a metamorphic overgrowth at eclogite facies. The almost identical ages of the two types of core domains (magmatic = 791 ± 9 Ma and granulite facies metamorphic zircon = 794 ± 10 Ma), and the Triassic age (212 ± 10 Ma) of eclogitic facies metamorphic overgrowth zircon rim are interpreted as indicating that the protolith of the eclogite is mafic granulite that originated from underplating of mantle-derived magma onto the base of continental crust during the Neoproterozoic (c. 800 Ma) and then subducted during the Triassic, experiencing UHP eclogite facies metamorphism at mantle depths. The new finding has two-fold significance: (i) voluminous mafic lower continental crust can increase the average density of subducted continental lithosphere, thus promoting its deep subduction; (ii) because of the current absence of mafic lower continental crust in the Dabie orogen, delamination or recycling of subducted mafic lower continental crust can be inferred as the geochemical cause for the mantle heterogeneity and the unusually evolved crustal composition.

Key words Dabie orogen; eclogite; mafic lower continental crust; magma underplating

1 Introduction

The Dabie-Sulu ultrahigh pressure (UHP) metamorphic belt in east-central China is one of the largest UHP terranes in the world (Xu et al., 1994, 2005; Wang et al., 1995; Cong, 1996; Zheng et al., 2003). It was formed by

---

* 本文发表在: Journal of Metamorphic Geology, 2007, 25: 975-989

continental collision between the South China Block (SCB) and North China Block in the Triassic (e.g. Li et al., 1993, 1994, 2000; Rowley et al., 1997; Hacker et al., 1998; Ayers et al., 2002; Liu et al., 2005, 2007). It is well documented that the UHP rocks (e.g. coesite- and microdiamond-bearing eclogites) exhumed to the surface came from subducted upper and/or felsic lower continental crust with mafic boudins (e.g. Tabata et al., 1998; Liou, 1999; Carswell et al., 2000; Ernst, 2001; Li et al., 2003a; Liu et al., 2007). However, UHP eclogites formed by subduction of mafic lower continental crust have not been reported yet. Moreover, a geophysical study indicates the current absence of mafic lower continental crust in the Dabie-Sulu orogenic belt (Gao et al., 1998). Therefore, it is an interesting question as to whether voluminous mafic lower crust was involved in continental subduction in the Dabie orogen. This question is important for understanding the average density of the subducted continental lithosphere and the extent of delamination or recycling of subducted mafic lower continental crust in the Dabie orogen. Here we report petrological and geochronological observations from the UHP eclogite formed by subduction of mafic lower continental crust in the Dabie orogen, China. These eclogites provide direct windows into the formation and evolution of the subducted mafic lower continental crust of the SCB.

2 Geological setting

It is generally accepted that the Dabie orogen can be divided into five major lithotectonic units from north to south (e.g. Okay, 1993; Li et al., 2001, 2004; Xu et al., 2003, 2005; Liu et al., 2005, 2007; Zheng et al., 2005; and references therein): (1) the Beihuaiyang zone; (2) the North Dabie high-T/UHP complex zone (NDZ); (3) the Central Dabie middle-T/UHP metamorphic zone (CDZ); (4) the South Dabie low-T eclogite zone (SDZ); (5) the Susong complex zone. These five zones are separated by the respective Xiaotian-Mozitan, Wuhe-Shuihou, Hualiangting-Mituo and Taihu-Mamiao faults (see Fig. 1). Zone (1) is composed of a fore arc flysch formation of the Foziling Group (Xu et al., 1992a; Li et al., 2001) and unsubducted continental crust of the Luzhenguan complex (Okay & Sengör, 1993; Zheng et al., 2004; Jiang et al., 2005). Zones (2), (3), (4) and (5) belong to the subducted SCB.

The UHP metamorphic rocks, including coesite-bearing eclogite, UHP gneiss, quartz jadeitite and marble with eclogite nodules, are observed in the CDZ and SDZ (e.g. Xu et al., 1992b, 1994, 2005; Okay & Sengör, 1993; Wang et al., 1995; Li et al., 2004). The occurrence of diamond and coesite in the UHP rocks in the CDZ indicates the UHP metamorphism occurred at 700–850 ℃ and > 2.8 GPa (e.g. Okay et al., 1989; Wang et al., 1989; Xu et al., 1992b; Okay, 1993; Rolfo et al., 2000), whereas the peak P-T conditions on the eclogites in the SDZ were estimated at 670 ℃ and 3.3 GPa (Li et al., 2004). Wang et al. (1992) and Okay (1993) called the SDZ and CDZ units 'cold' and 'hot' eclogite terranes respectively. Both the CDZ and SDZ units sequentially experienced UHP eclogite facies and HP amphibolite facies retrograde metamorphism (e.g. Xu et al., 1992b; Wang et al., 1995; Cong, 1996; Li et al., 2004).

The NDZ is composed mainly of banded tonalitic and granitoid gneiss and post-collisional magma intrusions with minor meta-peridotite (including dunite, harzburgite and lherzolite), garnet pyroxenite, garnet-bearing amphibolite, granulite and eclogite. The oriented mineral exsolutions in garnet and clinopyroxene, and micro-diamond imply that the eclogites from the NDZ also underwent UHP metamorphism (Xu et al., 2003, 2005; Malaspina et al., 2006) at a possible pressure of > 3.5–4.0 GPa. The Triassic zircon U-Pb (Liu et al., 2000) and Sm-Nd (Liu et al., 2005) ages of the eclogites from the NDZ suggest that they also formed by subduction of the SCB in the Triassic. The metamorphic ages (Liu et al., 2000, 2007; Bryant et al., 2004; Xie et al., 2004) and UHP metamorphic evidence (Liu et al., 2007) from the banded gneisses in the NDZ suggest that the gneisses surrounding the eclogites were also involved in the Triassic deep subduction of the SCB. Following the UHP eclogite facies metamorphism, however, the eclogites from the NDZ were first subjected to granulite facies retrogression, and later to amphibolite facies overprinting (e.g. Xu et al., 2000; Liu et al., 2001, 2005). This is distinctly different from the UHP rocks in the CDZ where there is only amphibolite facies retrogression (e.g. Okay, 1993; Wang et al., 1995; Cong, 1996). Therefore, although both the CDZ and NDZ units experienced UHP metamorphism, they had different exhumation histories, suggesting that the CDZ and NDZ are two decoupled UHP slabs (Liu et al., 2001, 2003, 2005,

2007; Zheng et al., 2001; Li et al., 2005). Pb isotope investigations show that the UHP rocks from the CDZ are characterized by high radiogenic Pb, but banded gneiss from the NDZ is characterized by low radiogenic Pb (Zhang et al., 2001; Li et al., 2003a). This suggests that the exhumed UHP rocks in the CDZ were from subducted upper crust, while the UHP rocks from the NDZ were from subducted lower felsic continental crust (Li et al., 2003a). It is also supported by their different lithological associations and peak hydrous mineral assemblages and metamorphic temperatures of the three eclogite-bearing slices in the Dabie UHP metamorphic belt (cf. Liu et al., 2007; and references therein). Fluid inclusions in the UHP rocks from the CDZ and NDZ are also different. The inclusions in the UHP rocks from the CDZ are mainly aqueous fluid with various salinities, while those in the UHP rocks from the NDZ are mainly N_2 or CO_2, suggesting more aqueous fluids in the CDZ during metamorphism (Xiao et al., 2001, 2002). These further indicate that their protoliths come from different crustal levels (e.g. upper and lower continental crust, respectively), probably resulted from the variable crustal depths of decoupling during subduction (Okay & Sengör, 1993; Liu et al., 2006, 2007).

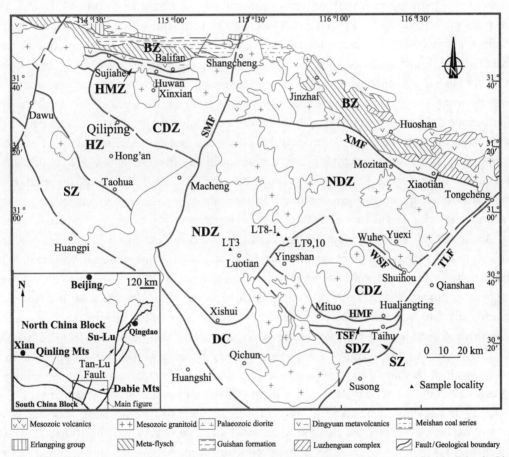

Fig. 1 Schematic geological map of the Dabie orogen, with inset showing the location of this area within the Triassic Qinling-Dabie – Su-Lu collision orogen in central China. Sample localities with sample numbers are described in detail in the text. BZ, Beihuaiyang zone; NDZ, North Dabie high-T/UHP complex zone; CDZ, Central Dabie middle-T/UHP metamorphic zone; SDZ, South Dabie low-T eclogite zone; SZ, Susong complex zone; HMZ, Huwan mélange zone; HZ, Hong'an low-T eclogite facies zone; DC, amphibolite facies Dabie complex; XMF, Xiaotian-Mozitan fault; WSF, Wuhe-Shuihou fault; HMF, Hualiangting-Mituo fault; TMF, Taihu-Mamiao fault; TLF, Tan-Lu fault.

The Luotian dome in the western segment of the NDZ is a deeply eroded area with abundant felsic and mafic granulites (Fig. 1). An unusual eclogite, with granulite facies mineral relics and later overprinting of granulite facies (see below in detail), was found as lenses in garnet-bearing banded tonalitic gneisses from the Luotian dome (see Fig. 1). Samples were collected from Jinjiapu (LT8-1, LT9 & LT10) and Sanlifan (LT3) (Fig. 1). Fresh eclogites are preserved in the core of these lenses or blocks, but there is retrogression into symplectite eclogite or garnet-bearing amphibolite from the centre towards the edge of the lens.

3 Petrography

The mineral assemblage of eclogite from the Luotian dome is garnet, omphacite, and rutile with secondary minerals of quartz, diopside, hypersthene, hornblende, plagioclase and ilmenite (Fig. 2). The compositions of representative minerals from the samples are listed in Tables 1 and 2. The most striking feature of the eclogite is the presence of two generations of hypersthene and three generations of plagioclase. The first generation of hypersthene and plagioclase occurs as inclusions in porphyroblastic garnet (Fig. 2e) or in the core domain of zircon. This plagioclase (mostly andesine with An = 33–47 mol.% and minor labradorite with An = 58 mol.%) (Table 1) also occurs as the only inclusion in the core of porphyroblastic garnet (Fig. 2f). The second generation of hypersthene occurs as a symplectite associated with diopside and plagioclase, or the retrograded rim of Na-clinopyroxene with quartz exsolution (Figs. 2c–e). The third generation of plagioclase, which is albitic (compositionally ranging from oligoclase to andesine) with An = 27–45 mol.%, occurs as symplectite associated with hornblende, that resulted from the breakdown of garnet (Figs. 2a, e), or with clinopyroxene (omphacite or diopside) (Fig. 2c). The mineral inclusions of plagioclase (andesine) ± hypersthene in the core of porphyroblastic garnet are relic minerals of the eclogite precursor whereas eclogitic facies minerals inclusions-such as omphacite and rutile-occur in the mantle of porphyroblastic garnet (Figs. 2a, b, f). Most garnet has no obvious zoning, which may be ascribed to fast exhumation and/or high-T overprinting of the eclogites. Rare garnet porphyroblasts show weak zoning from core, mantle to rim involving the variation of grossular components from 17.66, 20.02 to 18.56 mol.%, respectively (Table 2). Furthermore, plagioclase and omphacite inclusions commonly occur in the core and mantle domains of larger porphyroblastic garnet (Figs. 2a, f) respectively. This weak zoning and distribution of inclusion phases indicate the possibility that some garnet cores preserve evidence of an early, granulite facies equilibrium condition. Similarly, the mantle of this zoned garnet possibly represents the peak UHP conditions. It therefore seems very likely that the compositional zoning of garnet reflects the variation of pressure conditions relating to an evolution from early granulite to eclogite and later granulite facies overprinting as described later in detail. In addition, the hypersthene coexisting with andesine occurs as inclusions in the core domain of zircon (see below in detail). Cathodoluminescene (CL) and back-scatter electron images show that the Neoproterozoic metamorphic zircon core domains with both hypersthene and andesine inclusions of granulite facies have no fractures (Fig. 3h) and some of them contain an over-growth rim with eclogitic minerals (see latter in detail), which rules out the later or exotic origin of mineral inclusions in the zircon core. Consequently, the precursor of the eclogite was mafic granulite.

As with other diamond-bearing eclogites in the NDZ (Xu et al., 2003, 2005; Liu et al., 2005), the corona structures of the garnet from the eclogites are characterized by double symplectites. There is an inner corona composed of fine-grained amphibole and plagioclase derived from the decomposition of garnet during amphibolite facies retrogression, while the outer corona is a very narrow zone between garnet and clinopyroxene or omphacite composed of very fine-grained minerals (e.g. hypersthene, diopside and plagioclase), which was formed by retrogressive reaction between garnet and clinopyroxene under granulite facies conditions (Figs. 2a, e) (Liu et al., 2005). Ver-micular symplectites indicate retrogressive reaction or decompression breakdown during fast exhumation. In addition, the peak assemblage in these rocks is garnet + omphacite + rutile ± diamond with no hydrous minerals (e.g. phengite) (Xu et al., 2003, 2005), indicating that an anhydrous paragenesis formed at dry conditions. Therefore, the absence of fluid during their peak and initial retrograde metamorphic conditions may be related to the nature of its protolith, i.e. 'dry' mafic lower crustal rock (mafic granulite) as mentioned above.

Although no diamond or coesite has been found in the samples, the eclogites have metamorphic evolutional histories and petrographic characteristics similar to the diamond-bearing eclogites or UHP mafic rocks in the NDZ (Xu et al., 2005; Malaspina et al., 2006). For example, a spectacular microstructure of oriented quartz needles in matrix Ca-Na clinopyroxene is widespread (Figs. 2c, d) and is evidence for the prior existence of a 'supersilicic' omphacite stabilized at UHP conditions ($\geqslant 2.5$ GPa) (Tsai & Liou, 2000). Thus, it is reasonable to infer that the eclogites also underwent UHP metamorphism. On the basis of microstructural observations and mineral relationships as mentioned above, at least four generations of mineral assemblages for the eclogites can be discerned: (i) hypersthene + plagioclase ± garnet; (ii) omphacite + garnet + rutile + quartz; (iii) symplectite stage of garnet +

diopside + hypersthene + ilmenite + plagioclase; (iv) amphibole + plagioclase + magnetite. Based on the major element compositions of paragenetic minerals, metamorphic *P-T* conditions for the four metamorphic stages of the eclogite in the Luotian dome are estimated as follows (Table 3): (i) granulite facies stage at ~0.8 GPa; (ii) eclogite facies stage at ~2.5 GPa and 853–880 ℃, averaging 864 ℃; (iii) retrograde metamorphism of high-pressure granulite facies stage at 1.1–1.4 GPa and 804–857 ℃; (iv) retrograde metamorphism of amphibolite facies at 0.5–0.7 GPa and 706–777 ℃. In view of the new finding of micro- diamond from the eclogite (Xu et al., 2003, 2005) and enclosing orthogneiss (Liu et al., 2007) in the NDZ, the peak metamorphic conditions were reestimated at 900–960 ℃ (averaging 930 ℃) at 4.0 GPa, corresponding to the lowest pressure for diamond formation (Xu et al.,1992b).

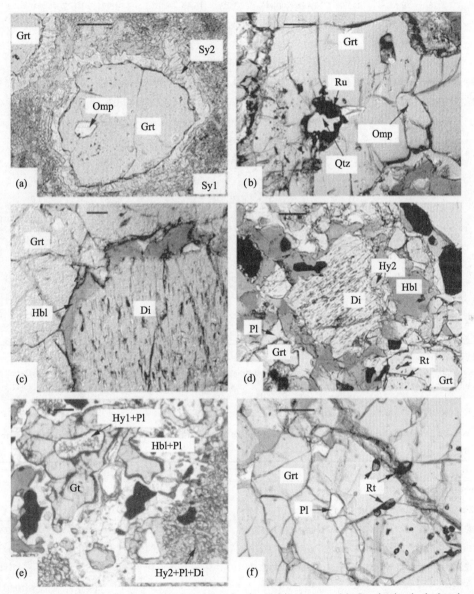

Fig. 2 Photomicrographs of eclogite from the Luotian dome in the Dabie orogen. (a) Omphacite inclusion in garnet with two generations of symplectites (Sy1 & Sy2), sample LT10; (b) Omphacite, rutile and quartz inclusions in garnet, sample LT8-1. (c) Diopside with quartz needles contains retrogressed hornblende margin, sample LT8-1. (d) Diopside with quartz needles contains retrograded hypersthene margin, sample LT8-1. (e) Hypersthene and plagioclase inclusions in garnet, sample LT3. (f) Porphyroblastic garnet contains plagioclase and rutile inclusions in the core and mantle, respectively, sample LT8-1. Hy1, first-generation hypersthene; Hy2, second-generation hypersthene; Sy1, earlier granulite facies symplectite composed of plagioclase, diopside and hypersthene; Sy2, later amphibolite facies symplectite composed of plagiosclase and hornblende. Mineral abbreviations are after Kretz (1983). The scale bar is 100 μm.

Table 1 Electron microprobe analyses of representative minerals from the eclogites in the Luotian dome

Mineral	Garnet		Omphacite		Diopside		Hypersthene			Hornblende		Plagioclase		
No.	LT8-1	LT9	LT10	LT8-1	LT8-1	LT10	LT10	LT9	LT9	LT10	LT9	LT8-1	LT8-1	LT10
Locality	m	m	i	i	m	Sy1	Sy1	i	Sy1	Sy2	Sy2	i	Sy2	Sy1
SiO_2	39.51	39.95	56.75	54.13	52.82	54.24	54.9	52.31	52.42	46.59	45.65	57.12	60.37	56.67
TiO_2	0.01	0	0	0.17	0	0.02	0	0	0	0.47	0.98	0	0	0
Al_2O_3	22.26	21.79	0.1	6.65	3.65	2.15	0.68	0.61	0.69	12.21	9.36	26.9	24.47	25.66
FeO	22.08	21.82	3.26	5.64	5.93	5.47	19.21	30.02	29.27	8.1	16.55	0.42	0.12	0.6
Cr_2O_3	0	0	0	0	0	0.1	0	0	0	0.2	0	0.03	0	0
MnO	0.31	0.23	0	0	0	0.05	0.3	0.28	0.26	0.11	0.06	0	0	0.02
MgO	8.35	5.26	9.61	12	14.41	14.43	23.72	16.15	16.7	14.97	10.27	0	0	0.92
CaO	7.47	10.74	13.76	19.28	22.04	23.28	0.52	0.61	0.64	12.1	11.31	9.17	5.75	10.26
Na_2O	0	0	6.5	2.69	1.13	0.76	0	0	0	2.04	1.63	6.35	8.68	6.43
K_2O	0	0	0	0	0	0	0	0	0	0.09	0	0	0	0
Total	99.99	99.99	99.98	100.56	99.98	99.99	99.33	99.98	99.98	96.68	95.81	99.99	99.39	100.56
O	12	12	6	6	6	6	6	6	6	23	23	8	8	8
Si	3.01	3.08	2.01	1.96	1.93	1.99	2.03	2.03	2.03	6.71	6.89	2.57	2.703	2.549
Al^{IV}	0	0	0	0.04	0.07	0.01	0	0	0	1.29	1.11	1.42	1.29	1.359
Al^{VI}	2	1.98	0.42	0.24	0.09	0.09	0.03	0.03	0.03	0.78	0.56			
Fe^{3+}	0.07	0.01	0.01	0	0.06	0	0	0	0	0.08	0.15	0	0	0
Ti	0	0	0	0.01	0	0	0	0	0	0.05	0.11	0	0	0
Fe^{2+}	1.34	1.34	0.09	0.17	0.12	0.17	0.59	0.97	0.95	0.9	1.94	0.02	0.004	0.023
Cr	0	0	0	0	0	0	0	0	0	0.02	0	0	0	0
Mg	0.95	0.6	0.51	0.65	0.79	0.79	1.31	0.93	0.96	3.21	2.31	0	0	0.062
Mn	0.02	0.02	0	0	0	0	0.01	0.01	0.01	0.01	0.09	0	0	0.001
Ca	0.61	0.9	0.52	0.75	0.86	0.9	0.02	0.02	0.03	1.87	1.83	0.44	0.276	0.495
Na	0	0	0.45	0.19	0.08	0.05	0	0	0	0.57	0.48	0.55	0.754	0.561
K	0	0	0	0	0	0	0	0	0	0.02	0	0	0	0

m, matrix; i, inclusion in garnet; sy1 and sy2 denote symplectites of granulite facies and amphibolite facies retro-metamorphism respectively. Garnet/pyroxene stoichiometries and the amount of Fe^{3+} and Fe^{2+} were estimated based on eight/four cations and the charge-balance constraint; the ferric iron content in amphibole was calculated as $Si + Al + Ti + Mg + Fe + Mn = 13$ for $O = 23$.

Table 2 Electron microprobe analyses of representative minerals in zircon and a weak zoned garnet porphyroblast from the eclogites in the Luotian dome

Mineral	Garnet					Omphacite		Hypersthene	Plagioclase	
No.	LT10	LT9	LT10	LT10	LT10	LT9	LT10	LT10	LT10Pl	LT9
Locality	Zir-C	Zir-C	M-C	M-M	M-R	Zir-R	Zir-R	Zir-C	Zir-C	Zir-C
SiO_2	40.78	38.68	41.51	40.84	41.36	54.51	56.76	53.93	56.73	55.58
TiO_2	0	0	0	0	0	0.07	0.04	0	0	0.01
Al_2O_3	22.97	21.14	22.78	22.95	23.11	9.13	9.49	0.72	27.44	28.3
FeO	18.88	24.57	17.24	17.17	17.87	7.14	1.89	21.15	0.02	0.02
Cr_2O_3	0.01	0.02	0	0	0	0	0.07	0.01	0	0.01
MnO	0.57	1.91	0.26	0.55	0.49	0.06	0.03	0.5	0	0
MgO	10.28	3.06	11.07	10.86	10.33	9.2	10.59	23.94	0	0
CaO	6.48	10.18	7.22	8.27	7.55	15.13	15.06	0.38	9.77	11.48
Na_2O	0	0.02	0	0.02	0	4.75	5.84	0.04	5.96	4.47
K_2O	0	0	0	0	0	0	0	0	0.05	0.06
Total	99.97	99.61	100.08	100.66	100.71	99.99	99.77	100.67	99.97	99.93

Continued

Mineral	Garnet					Omphacite		Hypersthene	Plagioclase	
No.	LT10	LT9	LT10	LT10	LT10	LT9	LT10	LT10	LT10Pl	LT9
Locality	Zir-C	Zir-C	M-C	M-M	M-R	Zir-R	Zir-R	Zir-C	Zir-C	Zir-C
O	12	12	12	12	12	6	6	6	8	8
Si	3.07	3.05	3.1	3.03	3.08	1.97	2.01	1.98	2.55	2.5
Al^{IV}	0	0	0	0	0	0.03	0	0.02	1.45	1.5
Al^{VI}	2.03	1.97	2	2.01	2.03	0.36	0.4	0.01		
Fe^{3+}	0.06	0.08	0.05	0.05	0.06	0	0	0.02	0	0
Ti	0	0	0	0	0	0	0	0	0	0
Fe^{2+}	1.13	1.54	1.02	1.01	1.06	0.22	0.06	0.63	0	0
Cr	0	0	0	0	0	0	0	0	0	0
Mg	1.15	0.36	1.23	1.2	1.15	0.5	0.56	1.31	0	0
Mn	0.04	0.13	0.02	0.03	0.03	0	0	0.02	0	0
Ca	0.52	0.86	0.58	0.66	0.6	0.59	0.57	0.01	0.47	0.55
Na	0	0	0	0	0	0.33	0.4	0	0.52	0.39
K	0	0	0	0	0	0	0	0	0	0
Grs	15.53	25.75	17.66	20.02	18.56					
Alm	39.73	53.27	35.9	34.8	37.26					
Prp	40.6	12.45	43.25	41.3	40.42					
Sps	1.28	4.41	0.58	1.19	1.09					
Adr	2.83	3.95	2.61	2.58	2.67					
Jd						33.88	40.31	0.1		
An									47.4	58.5

M-C, M-M and M-R denote the core, mantle and rim domains of a garnet porphyroblast in the matrix respectively. Zir-C and Zir-R represent the core and rim domains of zircon respectively. Garnet/pyroxene stoichiometries and the amount of Fe^{3+} and Fe^{2+} were estimated based on eight/four cations and the charge-balance constraint.

Table 3 *P-T* estimates of the eclogites from the Luotian dome, NDZ

Metamorphic facies	Temperature (°C)			Pressure (GPa)
Earlier granulite				Garnet-hypersthene
				0.8 (Wood, 1974)
Eclogite	Garnet-omphacite			
	K	RH	EG	
	853	860	880	2.5
	904	963	927	4
Later granulite	Diopside-hypersthene			Garnet-hypersthene
	W	WB		Wood (1974)
	804	813		1.4
	857	849		1.1
Amphibolite	Plagioclase-amphibole			Al in amphibole
	BH			BH, JR, S
	706–777			0.5–0.7

K, Krogh (1988); RH, Raheim & Green (1974); EG, Ellis & Green (1979); W, Wells (1977); WB, Wood & Banno (1973); BH, Blundy & Holland (1990); JR, Johnson & Rutherford (1988); S, Schmidt (1992). The amphibolite facies pressures are inferred from the Al content in amphibole of the symplectites after omphacite or garnet (Schmidt, 1992), though these values must be taken with caution, they are similar to those from the country rocks (granitoid gneiss) and garnet-bearing amphibolites in the NDZ (Zhang et al., 1996).

4 Zircon U-Pb geochronology

4.1 Analytical methods

Zircon was separated from ~5 kg of each sample by crushing and sieving, followed by magnetic and heavy liquid separation and hand-picking under binoculars. Approximately 500 zircon grains for each sample, together with a zircon U-Pb standard TEM (417 Ma), were mounted using epoxy, which was then polished until all zircon grains were approximately cut in half. The internal zoning patterns of the crystals were observed by CL image, which was analysed at the Institute of Mineral Resources, Chinese Academy of Geological Sciences (CAGS).

Zircon from the eclogites was dated on a SHRIMP II at Beijing SHRIMP Center. Uncertainties in ages are quoted at the 95% confidence level (2σ). Spot diameter was 30 μm for SHRIMP II. Common Pb corrections were made using measured ^{204}Pb. The SHRIMP analyses followed the procedures as described by Williams (1998). Both optical photomicrographs and CL images were taken as a guide to selection of U-Pb dating spot. Five scans through the mass stations were made for each age determination. Standards used were SL13, with an age of 572 Ma and U content of 238 ppm, and TEM, with an age of 417 Ma (Williams, 1998; Black et al., 2003). The U-Pb isotope data were treated following Compston et al. (1992) with the ISOLOT program of Ludwig (2001). Our measurement of standard zircon TEMORA 1 yielded a weighted ^{206}Pb/^{238}U age of 416.9 ± 3.5 Ma (MSWD = 2, n = 20), which is in good agreement with the recommended isotope dilution-thermal ionization mass spectrometry age of 416.75 ± 0.24 Ma (Black et al., 2003).

Mineral inclusions in zircon have been used to characterize zircon origin. Mineral inclusions were identified using Raman spectroscopy at the Analytical Centre, China University of Geosciences (Wuhan) and the Continental Dynamics Laboratory, CAGS, and/or substantiated using the electron microprobe analyser (EMPA) at the Analytical Centre, China University of Geosciences (Wuhan) and the Institute of Mineral Resources, CAGS. The analytical conditions on the Raman and EMPA were in accordance with that reported by Xu et al. (2005) and Liu et al. (2005).

The representative CL images for the studied samples are presented in Figs. 3 & 4. The compositions of representative mineral inclusions in zircon are reported in Table 2. The U-Pb data for zircon dating are listed in Tables 4 & 5.

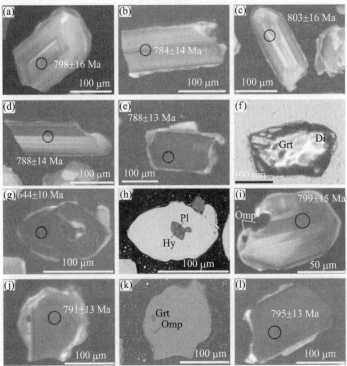

Fig. 3 Cathodoluminescene (CL) images (a–e, g, i, j and l), back-scatter electron (BSE) pictures (h and k), and plane-polarized light (PL) photo (f) for zircon from sample LT10. Zircon (e) and (f), (g) and (h), and (j) and (k) are the same grain. The open circles are analysis spots with available ^{206}Pb/^{238}U ages. Hy, hypersthene; other mineral abbreviations are the same as those in Fig. 2.

Table 4 SHRIMP zircon U-Pb data for eclogite (sample LT10) from the Luotian dome

Spot	Domain	Inclusion in domain	$^{206}Pb_c$ %	U (ppm)	Th (ppm)	Th/U	$^{206}Pb^*$ (ppm)	$^{207}Pb^*/^{206}Pb^*$	±%	$^{207}Pb^*/^{235}U$	±%	$^{206}Pb^*/^{238}U$	±%	$^{206}Pb/^{208}Pb$ age (Ma)
LT10-1.1	me, c	Gt	0.26	393	582	1.48	44.1	0.0671	1.4	1.204	2.2	0.1301	1.7	788 ± 13
LT10-1.2	me, c	No	0.7	276	520	1.88	30.1	0.0645	2.9	1.122	3.5	0.1262	1.8	766 ± 13
LT10-2.1	ma, c	No	0.57	193	277	1.43	21.6	0.0656	2.5	1.167	3.1	0.1291	1.8	783 ± 13
LT10-2.2	ma, c	No	1.28	87	103	1.18	9.16	0.0636	5.6	1.062	6	0.1211	2.1	737 ± 15
LT10-4.1	ma, c	No	0.36	240	149	0.62	21.5	0.0641	2.6	0.919	3.2	0.104	1.9	638 ± 11
LT10-3.1	ma, c	No	0.6	224	326	1.45	25.8	0.0653	2.8	1.201	3.3	0.1334	1.9	807 ± 14
LT10-3.2	me, c	No	0.75	79	85	1.07	9.34	0.067	2.8	1.256	3.5	0.136	2.2	822 ± 17
LT10-4.2	ma, c	No	0.47	195	328	1.68	21.9	0.0696	3.9	1.248	4.4	0.13	1.9	788 ± 14
LT10-3.3	me, c	No	2.81	44	40	0.91	5.25	0.0608	14	1.12	15	0.1338	3.6	810 ± 27
LT10-3.4	me, c	No	0.28	64	48	0.75	7.46	0.0733	4.2	1.37	4.8	0.1356	2.3	820 ± 18
LT10-1.4	me, c	Gt + Omp in rim	0.23	373	525	1.41	42	0.0693	1.8	1.247	2.5	0.1305	1.8	791 ± 13
LT10-4.3	me, c	Hy	0.96	375	369	0.98	17.4	0.0614	4.4	0.452	4.9	0.0534	2.1	335 ± 7
LT10-1.3	me, c	No	0.54	347	413	1.19	37.1	0.0669	2.4	1.143	3	0.1239	1.8	753 ± 13
LT10-3.5	ma, c	Omp in rim	0.41	184	155	0.84	20.9	0.0678	3.5	1.234	4	0.1319	1.9	799 ± 15
LT10-3.6	me, c	Hy, Pl	0.66	134	94	0.7	13.2	0.068	4.1	1.069	4.6	0.114	2.1	696 ± 14
LT10-3.7	me, c	No	1.71	91	61	0.67	9.52	0.0662	9.1	1.09	9.4	0.1193	2.3	726 ± 15
LT10-3.8	ma, c	No	1.07	165	202	1.22	16.8	0.0734	3.1	1.19	3.7	0.1176	2	717 ± 14
LT10-3.9	ma, c	No	0.58	70	43	0.61	6.33	0.0729	6	1.053	6.5	0.1047	2.3	642 ± 14
LT10-4.4	ma, c	No	0.56	226	182	0.8	25.2	0.0655	2.9	1.168	3.7	0.1292	2.2	784 ± 16
LT10-4.5	ma, c	No	0.96	107	152	1.42	12.3	0.0681	2.8	1.238	3.5	0.1319	2.1	798 ± 16
LT10-1.5	me, c	No	0.42	244	431	1.77	28.3	0.0667	2.5	1.239	3.1	0.1347	1.8	815 ± 14
LT10-4.7	ma, c	No	0.4	191	193	1.01	21.5	0.0684	2.2	1.231	3.1	0.1305	2.1	791 ± 16
LT10-1.6	me, c	No	0.38	326	896	2.75	36.9	0.0656	1.9	1.188	2.6	0.1313	1.8	795 ± 13
LT10-1.7	me, c	No	0.4	351	669	1.9	39.3	0.0637	3	1.14	3.5	0.1299	1.8	787 ± 14
LT10-4.6	ma, c	No	1.23	164	324	1.97	18.3	0.0702	4.1	1.242	4.5	0.1284	1.9	779 ± 14
LT10-4.8	ma, c	No	0.34	161	269	1.67	18	0.0656	2.2	1.169	2.9	0.1294	1.9	784 ± 14
LT10-4.9	ma, c	No	0	80	94	1.17	9.08	0.0711	2.4	1.3	3.2	0.1326	2.1	803 ± 16
LT10-1.8	me, c	No	0.27	464	687	1.48	51.6	0.06656	1.4	1.185	2.3	0.1292	1.8	783 ± 13
LT10-1.9	me, c	No	0.78	408	559	1.37	42.4	0.0655	2.3	1.083	3	0.1199	1.8	730 ± 12
LT10-1.10	me, c	Hy, Pl	0.3	1491	1994	1.34	135	0.06647	1	0.962	2	0.105	1.7	644 ± 10
LT10-1.11	me, c	No	0.63	264	381	1.44	30.2	0.0662	2.6	1.206	3.2	0.132	1.9	799 ± 14
LT10-1.12	me, c	No	0.15	297	700	2.36	33.3	0.0694	1.8	1.246	2.6	0.1303	1.8	790 ± 14

Pb_c and Pb^* indicate the common and radiogenic portions respectively. Pb^*: corrected for common ^{204}Pb using measured ^{204}Pb. All errors are 1σ. Mineral abbreviations are the same as those in Figs 2 & 3. me, metamorphic zircon; ma, magmatic zircon with oscillatory growth zonation; c, core.

Table 5 SHRIMP zircon U-Pb data for eclogite (sample LT9) from the Luotian dome.

Spot	Domain	Inclusion in domain	$^{206}Pb_c$ %	U (ppm)	Th (ppm)	Th/U	$^{206}Pb^*$ (ppm)	$^{207}Pb^*/^{206}Pb^*$	±%	$^{207}Pb^*/^{235}U$	±%	$^{206}Pb^*/^{238}U$	±%	$^{206}Pb/^{238}U$ age (Ma)
LT9-1-1.1	me, c	No	0.1	2446	659	0.27	107	0.058	1.6	0.4078	2.3	0.051	1.7	321 ± 6
LT9-1-2.1	ma, c	Ap	1.2	156	254	1.63	18	0.0625	4.2	1.143	4.7	0.1325	2.2	802 ± 16
LT9-1-3.1	me, c	Gt	0.96	73	48	0.66	8.73	0.0778	4.3	1.474	4.9	0.1374	2.3	830 ± 18
LT9-1-4.1	me, r	Gt in core	10.23	59	1	0.01	1.9	0.045	71	0.21	71	0.0334	4.7	212 ± 10
LT9-1-4.2	me, c	No	0.75	215	24	0.11	10	0.0669	6.9	0.499	7.1	0.0541	1.9	339 ± 6
LT9-1-4.3	me, c	No	2	148	34	0.23	11.4	0.0569	8.3	0.692	8.5	0.0882	2.1	545 ± 11

Continued

Spot	Domain	Inclusion in domain	$^{206}Pb_c$ %	U (ppm)	Th (ppm)	Th/U	$^{206}Pb^*$ (ppm)	$^{207}Pb^*/^{206}Pb^*$	±%	$^{207}Pb^*/^{235}U$	±%	$^{206}Pb^*/^{238}U$	±%	$^{206}Pb/^{238}U$ age (Ma)
LT9-1-1.2	me, c	No	0.47	924	712	0.77	103	0.06759	1.3	1.202	2.2	0.129	1.7	782 ± 13
LT9-1-4.4	me, c	No	0.22	789	357	0.45	73.8	0.06507	1.3	0.975	2.2	0.1086	1.8	665 ± 11
LT9-1-2.2	ma, c	No	0.24	227	249	1.1	23.6	0.065	2.2	1.083	3	0.1209	2.1	736 ± 14
LT9-1-1.3	me, c	No	0.1	967	1069	1.1	94.7	0.06099	1.5	0.957	2.3	0.1138	1.8	695 ± 12
LT9-1-1.4	me, c	No	0.25	1018	2266	2.22	103	0.0647	1.8	1.05	2.5	0.1177	1.8	717 ± 12
LT9-1-1.5	me, c	No	0.08	659	673	1.02	69.2	0.0644	1.9	1.084	2.6	0.1221	1.8	743 ± 13
LT9-1-1.6	me, c	No	0.21	1885	463	0.24	97.1	0.05832	1.2	0.481	2.1	0.0598	1.7	375 ± 6
LT9-1-3.3	me, r	omp	7.1	85	0	0	2.5	0.043	37	0.187	37	0.0318	3.3	202 ± 7
LT9-1-1.7	me, c	No	0.26	482	931	1.93	48.1	0.064	1.6	1.023	2.4	0.1159	1.8	707 ± 12

Pb_c and Pb^* indicate the common and radiogenic portions respectively. Pb^*: corrected for common ^{204}Pb using measured ^{204}Pb. All errors are 1σ. Mineral abbreviations are after Kretz (1983). me, metamorphic zircon; ma, magmatic zircon with oscillatory growth zonation; r, overgrowth rim; c, core.

4.2 Mineral inclusion diagnostic of zircon origin

A powerful means of linking zircon growth/recrystallization to metamorphic events is by integrated CL imaging and mineral inclusion assemblage within different domains of a zircon grain (e.g. Gebauer et al., 1997; Hermann et al., 2001; Liu et al., 2004). Zircon from samples LT10 and LT9 is granular, platy and prismatic with diameters of 100–300 μm (Figs. 3 & 4). Mineral inclusions in zircon were identified by electron microprobe (Table 2) and Raman spectroscopy (Fig. 5). Most zircon shows typical core and rim textures in CL image. Inherited zircon cores can be divided into two types. One has grey or light grey core with clear oscillatory zonation and is surrounded by a relatively strong-luminescent rim (Figs. 3a–d & 4a, b) and the second is black or deep grey, unzoned, with a weak-luminescent and surrounded by strong-luminescent rim (Figs. 3e, f, i, j, l & 4c, d, f, h). The first type usually contains apatite (Figs. 4a & 5f) and quartz minerals or no mineral inclusion and the second type contains a mineral assemblage of granulite facies, such as garnet, hypersthene and plagioclase (andesine) (Figs. 3e–h, 4f & 5a). Therefore, these two styles of zircon cores represent a magmatic and granulite facies metamorphic origin respectively. In addition, some zircon rims contain garnet and/or omphacite or rutile inclusions (Figs. 3i–k, 4e, g & 5b–d), indicating that the metamorphic overgrowth formed during eclogite facies conditions. These observations also suggest that there was an earlier metamorphic stage of granulite facies prior to the eclogite facies metamorphism in the eclogites, which is consistent with the petrographic observations described above. It is worth noting that only a few zircon grains have a thin metamorphic overgrowth rim and most grains do not. This may indicate the absence of fluid during the eclogite facies metamorphism, because the dissolution and overgrowth of zircon depend on the availability of fluids during HP metamorphism (e.g. Williams et al., 1996; Rubatto et al., 1999; Liermann et al., 2002; Zheng et al., 2004; Wu et al., 2006). The absence of fluid during the eclogite facies metamorphism is ascribed to the protolith nature of the mafic lower crust, which is in agreement with the above petrographic observations such as the absence of hydrous minerals in the peak stage.

4.3 SHRIMP zircon U-Pb dating results

A total of 32 U-Pb spot analyses were made on 30 zircon from LT10 (Table 4; Fig. 6a). The magmatic and granulite facies metamorphic zircon cores from sample LT10 record $^{206}Pb/^{238}U$ concordant ages ranging from 807 to 783 Ma and from 822 to 766 Ma (mostly 810–783 Ma), respectively, which yield weighted mean ages of 791 ± 9 Ma with MSWD of 0.44 ($n = 10$) and 794 ± 10 Ma with MSWD of 1.2 ($n = 12$) (Figs. 6a, c, d). The high Th/U ratio (0.67–2.75) of the metamorphic zircon cores (Table 4) is attributed to their granulite facies metamorphic origin, because such zircon has high Th/U ratios (e.g. Vavra et al., 1999), in contrast to UHP metamorphic zircon. It is also supported by the hypersthene + plagioclase ± garnet inclusion assemblage within the metamorphic zircon core domains mentioned above. The 15 analyses of the cores of magmatic and metamorphic zircon from sample LT9 define a discordia line with an upper intercept age of 778 ± 78 Ma and a lower intercept age of 210 ± 88 Ma (MSWD = 2.9) (Fig. 6b). This lower intercept age is consistent with a concordant age of 212 ± 10 Ma given by a metamorphic overgrowth zircon rim with a Th/U ratio of 0.01, which contains eclogitic minerals such as omphacite, garnet and rutile (Figs 3i–k, 4e, g & 5b–d), suggesting it is the result of the Triassic eclogite facies metamorphism. Consequently, the eclogites within the Luotian dome were transformed from Neoproterozoic mafic granulite by deep subduction of the continental crust of the SCB in the Triassic.

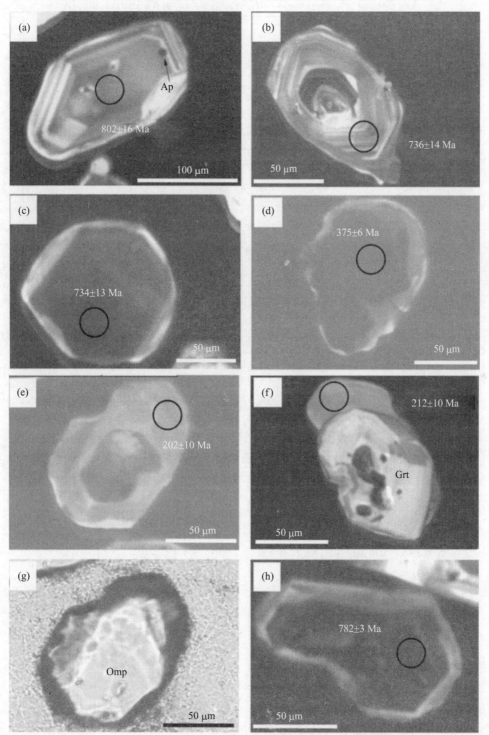

Fig. 4 Cathodoluminescene (CL) images (a–f and h) and plane-polarized light (PL) photograph (g) for zircon from sample LT9. Zircon (e) and (g) are the same grain. The analysis spot marked with open circle and its $^{206}Pb/^{238}U$ age is shown in each grain. Mineral abbreviations are after Kretz (1983).

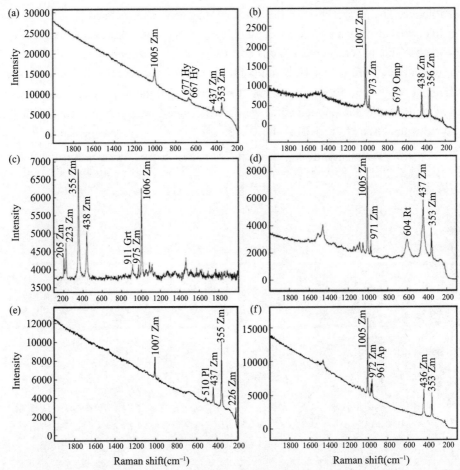

Fig. 5 Representative Raman spectra of mineral inclusions in zircon of eclogites from the Luotian dome. (a) Hypersthene. (b) Omphacite. (c) Garnet. (d) Rutile. (e) Plagioclase. (f) Apatite. These spectra also contain host zircon peaks at 205, 223–226, 353–356, 436–438, 971–975 and 1005–1007 cm^{-1}.

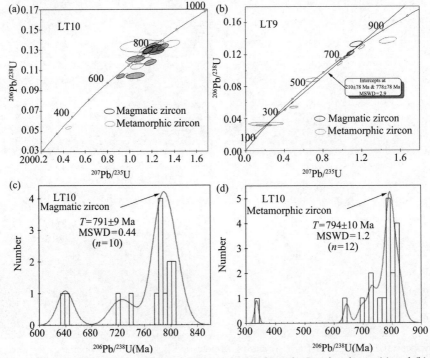

Fig. 6 Zircon SHRIMP U-Pb dating for eclogites (samples LT10 and LT9) from the Luotian dome. (a) and (b) show concordia plots for zircon from samples LT10 and LT9 respectively. (c) and (d) show histograms of apparent ^{206}Pb/^{238}U of all analyses of inherited cores

of magmatic and metamorphic zircon from sample LT10 with peaks at 791 and 794 Ma, respectively.

5 Implications for recycle of subducted continental crust

The almost identical ages of the Neoproterozoic granulite facies metamorphic and magmatic zircon from the studied eclogites (794 ± 10 and 791 ± 9 Ma respectively) imply that the precursor of the eclogite originated from Neoproterozoic basic magma under-plating on the base of continental crust in the northern margin of the SCB. This is supported by the Neoproterozoic crustal growth event in the SCB. It is well known that there was large-scale magmatic activity during the middle Neoproterozoic in the SCB, especially in the periphery of the SCB (e.g. Rowley et al., 1997; Li et al., 1999, 2003b,c; Zheng et al., 2004), which was interpreted as being related to plume activity and breakup of the supercontinent Rodinia. Furthermore, $\varepsilon_{Nd}(t)$ values and Nd model ages (T_{DM}) for Proterozoic to Permian sedimentary rocks in the SCB show a sharp change with a significant increase in $\varepsilon_{Nd}(t)$ but a decrease in T_{DM} at c. 820–750 Ma (Li & McCulloch, 1996). This Nd isotope change suggests a large increase in the proportion of juvenile mantle-derived materials from the provenance of the SCB during the Rodinia breakup, which caused the mafic lower continental crust overgrowth in the SCB formed by underplating of the Neoproterozoic mantle-derived magma. Therefore, after the Neoproterozoic, the continental lithosphere of the SCB should have a greater proportion of mafic lower continental crust and a relatively higher average density than before. In addition, early Cretaceous adakitic rocks reported in the Dabie orogen (Wang et al., 2007; Xu et al., 2007), which were produced by partial melting of the thickened mafic lower continental crust underneath the Dabie orogen, document that the thickened mafic lower continental crust remained until the collapse of the Dabie orogen in the early Cretaceous. This new occurrence of the Triassic eclogite, transformed from Neoproterozoic mafic granulite, indicates that the mafic lower continental crust formed by the Neoproterozoic basic magma underplating was involved in the Triassic continental subduction. The relatively higher average density of the subducting continental crust of the SCB may have promoted the deep subduction of the SCB in the Triassic.

Although the subducted mafic lower continental crust of the SCB could have been thick in the Triassic, geophysical investigation shows that the mafic lower continental rocks in the Dabie orogen or even in the northern margin of the SCB are scarce at present. It can be reasonably deduced that the subducted mafic lower continental crust may be delaminated and recycled into the underlying convective mantle, which could be a means of explaining the mantle heterogeneity and the evolved chemical composition of the continental crust (e.g. Kay & Kay, 1991; Rudnick, 1995; and references therein).

Acknowledgements

This research was financially supported by the National Natural Science Foundation of China (40572035 and 40634023), the Major State Research Program of China (G1999075503) and the Chinese National Key Project for Basic Research (2003 CB716500). We thank D. Liu, Y. Wang and H. Tao for their help in SHRIMP U-Pb dating on zircon, Z. Chen for the CL image, M. He and L. Yan for the Raman analysis, and H. Liu and Z. Chen for electron microprobe analysis. Y. Zheng and F. Huang are thanked for their suggestions on earlier versions of this manuscript, which helped clarify some ambiguities as well as the English presentation. D. Robinson, D. Nakamura and J. C. Ayers are thanked for the critical review of the manuscript and for many useful suggestions.

Reference

Ayers, J. C., Dunkle, S., Gao, S. & Miller, C. F., 2002. Constraints on timing of peak and retrograde metamorphism in the Dabie Shan ultrahigh-pressure metamorphic belt, east-central China, using U-Th-Pb dating of zircon and monazite. Chemical Geology, 186, 315-331.

Black, L. P., Kamo, S. L., Allen, C. M. et al., 2003. TEMORA 1: a new zircon standard for Phanerozoic U-Pb geochronology. Chemical Geology, 200, 155-170.

Blundy, J. D. & Holland, T. J. B., 1990. Calcic amphibole equilibria and a new amphibole-plagioclase geothermometer. Contributions to Mineralogy and Petrology, 104, 208-224.

Bryant, J. L., Ayers, J. C., Gao, S., Miller, C. F. & Zhang, H., 2004. Geochemical, age, and isotopic constrains on the location of the Sino-Korean/Yangtze suture and evolution of the Northern Dabie Complex, east central China. Geological Society of America Bulletin, 116, 698-717.

Carswell, D. A., Wilson, R. N. & Zhai, M., 2000. Metamorphic evolution, mineral chemistry and thermobarometry of schists and orthogneisses hosting ultra-high pressure eclogites in the Dabieshan of central China. Lithos, 52, 121-155.

Compston, W., Williams, I. S., Kirschvink, J. L., Zhang, Z. & Ma, G., 1992. Zircon U-Pb ages for the early Cambrian time-scale. Journal of the Geological Society (London), 149, 171-184.

Cong, B., 1996. Ultrahigh-Pressure Metamorphic Rocks in the Dabieshan-Sulu Region of China. Science Press, Beijing, 224 pp.

Ellis, D. J. & Green, D. H., 1979. An experimental study of the effect of Ca upon garnet-clinopyroxene Fe-Mg exchange equilibration. Contributions to Mineralogy and Petrology, 71, 13-22.

Ernst, W. G., 2001. Subduction, ultrahigh-pressure metamorphism, and regurgitation of buoyant crustal slices-implications for arcs and continental growth. Physics of the Earth and Planetary Interiors, 127, 253-275.

Gao, S., Zhang, B., Jin, Z., Kern, H., Luo, T. & Zhao, Z., 1998. How mafic is the lower continental crust? Earth and Planetary Science Letters, 161, 101-117.

Gebauer, D., Schertl, H. P., Brix, M. & Schreyer, W., 1997. 35 Ma old ultrahigh-pressure metamorphism and evidence for very rapid exhumation in the Dora Maira Massif, Western Alps. Lithos, 41, 5-24.

Hacker, B. R., Ratschbacher, L., Webb, L. E., Ireland, T. R., Walker, D. & Dong, S., 1998. U/Pb zircon ages constrain the architecture of the ultrahigh-pressure Qinling-Dabie orogen, China. Earth and Planetary Science Letters, 161, 215-230.

Hermann, J., Rubatto, D., Korsakov, A. & Shatsky, V. S., 2001. Multiple zircon growth during fast exhumation of diamondiferous, deeply subducted continental crust (Kokchetav massif, Kazakhstan). Contributions to Mineralogy and Petrology, 141, 66-82.

Jiang, L., Wolfgang, S., Chen, F., Liu, Y. C. & Chu, D., 2005. U-Pb zircon ages for the Luzhenguan Complex in northern part of the eastern Dabie orogen. Science in China (Series D), 48, 1357-1367.

Johnson, M. C. & Rutherford, M. J., 1988. Experimental calibration of an aluminum-in-hornblende geobarometer applicable to calcalkaline rocks. EOS, 69, 1511.

Kay, R. W. & Kay, S. M., 1991. Creation and destruction of lower continental crust. Geologische Rundschau, 80, 259-278.

Kretz, R., 1983. Symbols for rock-forming mineral. American Mineralogist, 68, 277-279.

Krogh, E. J., 1988. The garnet-clinopyroxene Fe-Mg geothermometer—a reinterpretation of existing experimental data. Contributions to Mineralogy and Petrology, 99, 44-48.

Li, X. & McCulloch, M. T., 1996, Secular variation in Nd isotopic composition of Neoproterozoic sediments from the southern margin of the Yangtze Block: evidence for a Proterozoic continental collision in southern China. Precambrian Research, 76, 67-76.

Li, S., Xiao, Y., Liu, D. et al., 1993. Collision of the North China and Yangtze blocks and formation of coesite-bearing eclogites: timing and processes. Chemical Geology, 109, 89-111.

Li, S., Wang, S., Chen, Y. et al., 1994. Excess argon in phengite from eclogite: evidence from dating of eclogite minerals by Sm-Nd, Rb-Sr and $^{40}Ar/^{39}Ar$ method. Chemical Geology, 112, 343-350.

Li, Z. X., Li, X. H., Kinny, P. D. & Wang, J., 1999. The breakup of Rodinia: did it start with a mantle plume beneath South China? Earth and Planetary Science Letters, 173, 171-181.

Li, S., Jagoutz, E., Chen, Y. & Li, Q., 2000. Sm-Nd and Rb-Sr isotope chronology of ultrahigh-pressure metamorphic rocks and their country rocks at Shuanghe in the Dabie Mountains, central China. Geochimica Cosmochimica Acta, 64, 1077-1093.

Li, S., Huang, F., Nie, Y. et al., 2001. Geochemical and geochronological constrains on the suture location between the North and South China Blocks in the Dabie orogen, central China. Physics and Chemistry of the Earth (A), 26, 655-672.

Li, S., Huang, F., Zhou, H. & Li, H., 2003a. U-Pb isotopic compositions of the ultrahigh pressure metamorphic (UHPM) rocks from Shuanghe and gneisses from Northern Dabie zone in the Dabie Mountains, central China: constraint on the exhumation mechanism of UHPM rocks. Science in China, Series D, 46, 200-209.

Li, X. H., Li, Z. X., Ge, W. et al., 2003b. Neoproterozoic granitoids in South China: crustal melting above a mantle plume at ca. 825 Ma? Precambrian Research, 122, 45-83.

Li, Z. X., Li, X. H., Kinny, P. D., Wang, J., Zhang, S. & Zhou, H., 2003c. Geochronology of Neoproterozoic syn-rift magmatism in the Yangtze Craton, South China and correlations with other continents: evidence for a mantle superplume that broke up Rodinia. Precambrian Research, 122, 85-109.

Li, X. P., Zheng, Y. F., Wu, Y. B., Chen, F. K., Gong, B. & Li, Y. L., 2004. Low-T eclogite in the Dabie terrane of China: petrological and isotopic constrains on fluid activity and radiometric dating. Contributions to Mineralogy and Petrology, 148, 443-470.

Li, S., Li, Q., Hou, Z., Yang, W. & Wang, Y., 2005. Cooling history and exhumation mechanism of the ultrahigh-pressure metamorphic rocks in the Dabie Mountains, central China. Acta Petrologica Sinica, 21, 1117-1124 (in Chinese with English abstract).

Liermann, H. P., Isachsen, C., Altenberger, U. & Oberhansli, R., 2002. Behavior of zircon during high-pressure, low-temperature metamorphism: case study from the Internal Unit of the Seisia zone (western Italian Alps). European Journal of Mineralogy, 14, 61-71.

Liou, J. G., 1999. Petrotectonic summary of less intensively studied UHP regions. International Geology Review, 41, 571- 586.

Liu, Y. C., Li, S., Xu, S. et al., 2000. U-Pb zircon ages of the eclogite and tonalitic gneiss from the northern Dabie Mountains, China and multi-overgrowths of metamorphic zircons. Geological Journal of China Universities, 6, 417-423 (in Chinese with English abstract).

Liu, Y. C., Xu, S., Li, S. et al., 2001. Distribution and metamorphic P-T condition of the eclogites from the mafic-ultramafic belt in the northern part of the Dabie Mountains. Acta Geologica Sinica, 75, 385-395 (in Chinese with English abstract).

Liu, Y. C., Xu, S., Li, S., Jiang, L., Chen, G. & Wu, W., 2003.Tectonic affinity, T-t path and uplift trajectory of eclogites from northern Dabie Mountains, central-eastern China. Journal of China University of Geosciences, 14, 28-33.

Liu, F. L., Xu, Z. Q., Liou, J. G. & Song, B., 2004. SHRIMP U-Pb ages of ultrahigh-pressure and retrograde metamorphism of gneisses, south-western Sulu terrane, eastern China. Journal of Metamorphic Geology, 22, 315-326.

Liu, Y. C., Li, S., Xu, S. et al., 2005. Geochemistry and geochronology of eclogites from the northern Dabie Mountains, central China. Journal of Asian Earth Sciences, 25, 431-443.

Liu, Y. C., Li, S., Gu, X. & Hou, Z., 2006. Zircon SHRIMP U-Pb dating for olivine gabbro at Wangmuguan in the Beihuaiyang zone and its geological significance. Chinese Science Bulletin, 51, 2500-2506.

Liu, Y. C, Li, S. & Xu, S., 2007. Zircon SHRIMP U-Pb dating for gneiss in northern Dabie high T/P metamorphic zone, central China: implication for decoupling within subducted continental crust. Lithos, 96, 170-185.

Ludwig, K. R., 2001. User's Manual for Isoplot/Ex (rev. 2.49): A Geochronological Toolkit for Microsoft Excel. Special Publication, No. 1a. Berkeley Geochronology Center, Berkeley, CA, 55 pp.

Malaspina, N., Hermann, J., Scambelluri, M. & Compagnoni, R., 2006. Multistage metasomatism in ultrahigh-pressure mafic rocks from the North Dabie Complex (China). Lithos, 90, 19-42.

Okay, A. I., 1993. Petrology of a diamond and coesite-bearing metamorphic terrain: Dabie Shan, China. European Journal of Mineralogy, 5, 659-675.

Okay, A. I. & Sengör, A. M. C., 1993. Tectonics of an ultrahigh- pressure metamorphic terrane: the Dabie Shan/Tongbai orogen, China. Tectonics, 12, 1320-1334.

Okay, A. I., Xu, S. & Sengör, A. M. C., 1989, Coesite from the Dabie Shan eclogites, central China. European Journal of Mineralogy, 1, 595-598.

Raheim, A. & Green, D. H., 1974. Experimental determination of the temperature and pressure dependence of the Fe-Mg partition coefficient for coexisting garnet and clinopyroxene. Contributions to Mineralogy and Petrology, 48, 179-203.

Rolfo, F., Compagnoni, R., Xu, S. & Jiang, L., 2000. First report of felsic whiteschist in the ultrahigh-pressure metamorphic belt of Dabie Shan, China. European Journal of Mineralogy, 12, 883-898.

Rowley, D. B., Xue, F., Tucker, R. D., Peng, Z. X., Baker, J. & Davis, A., 1997. Ages of ultrahigh pressure metamorphism and protolith orthogneisses from the eastern Dabie Shan: U/ Pb zircon geochronology. Earth and Planetary Science Letters, 151, 191-203.

Rubatto, D., Gebauer, D. & Compagnoni, R., 1999. Dating of eclogite facies zircons: the age of Alpine metamorphism in the Sesia-Lanzo zone (western Alps). Earth and Planetary Science Letters, 167, 141-158.

Rudnick, R. L., 1995. Making continental crust. Nature, 378, 571-578.

Schmidt, M. W., 1992. Amphibole composition in tonalite as a function of pressure: an experimental calibration of the Al-in-hornblende barometer. Contributions to Mineralogy and Petrology, 110, 304-310.

Tabata, H., Yamauchi, K., Maruyama, S. & Liou, J. G., 1998. Tracing the extent of a UHP metamorphic terrane: mineral inclusion study of zircons in gneisses from the Dabie Shan. In: When Continents Collide: Geodynamics and Geochemistry of Ultrahigh-Pressure Rocks (eds Hacker, B.R. & Liou, J.G.), pp. 261-273. Kluwer Academic Publishers, Dordrecht.

Tsai, C. H. & Liou, J.G., 2000. Eclogite facies relics and inferred ultrahigh-pressure metamorphism in the North Dabie complex, central China. American Mineralogist, 85, 1-8.

Vavra, G., Schmid, R. & Gebauer, D., 1999. Internal morphology, habit and U-Th-Pb microanalysis of amphibo-lite-to-granulite facies zircons: geochronology of the Ivrea Zone (Southern Alps). Contributions to Mineralogy and Petrology, 134, 380-404.

Wang, X., Liou, J. G. & Mao, H. K., 1989. Coesite-bearing eclogites from the Dabie Mountains in central China. Geology, 17, 1085-1088.

Wang, X., Liou, J. G. & Maruyama, S., 1992. Coesite-bearing eclogites from the Dabie Mountains, central China: petrogenesis, P-T paths, and implication for regional tectonics. The Journal of Geology, 100, 231-250.

Wang, X., Zhang, R. Y. & Liou, J. G., 1995. UHPM terrane in east central China. In: Ultrahigh Pressure Metamorphism (eds Coleman, R.G. & Wang, X.), pp. 356-390. Cambridge University Press, Cambridge.

Wang, Q., Wyman, D. A., Xu, J. et al., 2007. Early Cretaceous adakitic granites in the Northern Dabie Complex, central China:

implications for partial melting and delamination of thickened lower crust. Geochimica et Cosmochimica Acta, 71, 2609-2636.

Wells, R. A., 1977. Pyroxene thermometry in simple and complex systems. Contributions to Mineralogy and Petrology, 62, 129-139.

Williams, I. S., 1998. U-Th-Pb geochronology by ion microprobe. Reviews in Economic Geology, 7, 1-35.

Williams, I. S., Buick, I. S. & Cartwright, I., 1996. An extended episode of early Mesoproterozoic metamorphic fluid flow in the Reynolds Range, central Australia. Journal of Metamorphic Geology, 14, 29-47.

Wood, B. J., 1974. The solubility of alumina in orthopyroxene coexisting with garnet. Contributions to Mineralogy and Petrology, 46, 1-15.

Wood, B. J. & Banno, S., 1973. Garnet-orthopyroxene and orthopyroxene-clinopyroxene relationship in simple and complex systems. Contributions to Mineralogy and Petrology, 42, 109-124.

Wu, Y. B., Zheng, Y. F., Zhao, Z. F., Gong, B., Liu, X. M. & Wu, F. Y., 2006. U-Pb, Hf and O isotope evidence for two episodes of fluid-assisted zircon growth in marble-hosted eclogites from the Dabie orogen. Geochimica et Cosmochimica Acta, 70, 3743-3761.

Xiao, Y., Hoefs, J., van den Kerkhof, A. M. & Li, S., 2001. Geochemical constraints of the eclogite and granulite facies metamorphism as recognized in the Raobazhai complex from North Dabie Shan, China. Journal of Metamorphic Geology, 19, 3-19.

Xiao, Y., Hoefs, J., van den Kerkhof, A. M., Simon, K., Fiebig, J. & Zheng, Y., 2002. Fluid history during HP and UHP metamorphism in Dabie Shan, China: constraints from mineral chemistry, fluid inclusions, and stable isotopes. Journal of Petrology, 43, 1505-1527.

Xie, Z., Gao, T. & Chen, J., 2004. Multi-stage evolution of gneiss from North Dabie: evidence from zircon U-Pb chronology. Chinese Science Bulletin, 49, 1963-1969.

Xu, S., Jiang, L., Liu, Y. C. & Zhang, Y., 1992a. Tectonic framework and evolution of the Dabie Mountains in Anhui, eastern China. Acta Geologica Sinica, 5, 221-238.

Xu, S., Okay, A. I., Ji, S. et al., 1992b. Diamond from the Dabie Shan metamorphic rocks and its implication for tectonic setting. Science, 256, 80-82.

Xu, S., Liu, Y. C., Jiang, L., Su, W. & Ji, S., 1994. Tectonic Regime and Evolution of Dabie Mountains. Science Press, Beijing, 175 pp.

Xu, S., Liu, Y. C., Su, W., Wang, R., Jiang, L. & Wu, W., 2000. Discovery of the eclogite and its petrography in the Northern Dabie Mountain. Chinese Science Bulletin, 45, 273-278.

Xu, S., Liu, Y. C., Chen, G. et al., 2003. New finding of micro-diamonds in eclogites from Dabie-Sulu region in central-eastern China. Chinese Science Bulletin, 48, 988-994.

Xu, S., Liu, Y. C., Chen, G., Ji, S., Ni, P. & Xiao, W., 2005. Microdiamonds, their classification and tectonic implications for the host eclogites from the Dabie and Su-Lu regions in central eastern China. Mineralogical Magazine, 69, 509-520.

Xu, H., Ma, C. & Ye, K., 2007. Early cretaceous granitoids and their implications for the collapse of the Dabie orogen, eastern China: SHRIMP zircon U-Pb dating and geochemistry. Chemical Geology, 240, 238-259.

Zhang, R. Y., Liou, J. G. & Tsai, C. H., 1996. Petrogenesis of a high-temperature metamorphic terrane: a new tectonic interpretation for the north Dabieshan, central China. Journal of Metamorphic Geology, 14, 319-333.

Zhang, H., Gao, S., Zhang, B., Zhong, Z., Jia, W. & Wang, L., 2001. Pb isotopic study on crustal structure of Dabie Mountains, central China. Geochimica, 30, 395-401 (in Chinese with English abstract).

Zheng, Y., Fu, B., Li, Y. L., Wei, C. & Zhou, J., 2001. Oxygen isotope composition of granulites from Dabieshan in eastern China and its implications for geodynamics of Yangtze plate subduction. Physics and Chemistry of the Earth (A), 26, 673-684.

Zheng, Y., Fu, B., Gong, B. & Li, L., 2003. Stable isotope geochemistry of ultrahigh pressure metamorphic rocks from the Dabie-Sulu orogen in China: implications for geodynamics and fluid regime. Earth Science Reviews, 62, 105-161.

Zheng, Y., Wu, Y., Chen, F., Gong, B., Li, L. & Zhao, Z., 2004. Zircon U-Pb and oxygen isotope evidence for a large-scale ^{18}O depletion event in igneous rocks during the Neoproterozoic. Geochimica et Cosmochimica Acta, 68, 4145-4165.

Zheng, Y. F., Zhou, J. B., Wu, Y. B. & Zhao, Z. F., 2005. Low-grade metamorphic rocks in the Dabie-Sulu orogenic belt: a passive-margin accretionary wedge deformed during continent subduction. Internal Geology Review, 47, 851-871.

俯冲陆壳内部的拆离和超高压岩石的多板片差异折返: 以大别-苏鲁造山带为例*

刘贻灿, 李曙光

中国科学院壳幔物质与环境重点实验室, 中国科学技术大学地球和空间科学学院, 合肥 230026

> **亮点介绍**: 本文对国内外学者关于大陆碰撞过程中深俯冲陆壳的折返机制模型进行了简要评述, 并以大别-苏鲁造山带为例, 对大陆碰撞过程中俯冲陆壳内部在不同深度发生多层次地壳拆离解耦并呈多板片差异折返的关键证据进行了概括。该模型的建立, 突破了陆壳整体俯冲与整体折返的传统模式, 揭示了陆壳俯冲与洋壳俯冲的主要区别。

摘要 对国内外学者关于大陆碰撞过程中深俯冲陆壳的折返机制模型进行了简要评述, 并以大别-苏鲁造山带为例, 对大陆碰撞过程中俯冲陆壳内部在不同深度发生多层次地壳拆离解耦并呈多板片差异折返的关键证据进行概括。这些证据包括: ①大别-苏鲁超高压带北侧分布的具有华南陆壳特征, 部分经历过三叠纪变质的浅变质岩片, 它们显示了陆壳俯冲开始阶段其上部地壳与下部基底岩石的解耦; ②大别-苏鲁高压-超高压带均由若干高压-超高压岩片组成, 这些岩片由南至北变质程度逐步加深、峰期和退变质时代逐步变年轻; ③苏北大陆科学钻探揭示了超高压变质带在垂向上也是由若干岩片组成, 上部岩片具有高放射成因 Pb, 下部岩片具有低放射成因 Pb, 反映了俯冲陆壳在不同深度的解耦和折返。俯冲陆壳内部的拆离解耦和差异折返, 主要是由于大陆地壳上、下不同部位岩石组成的差异导致的力学性质差异和壳内古断层带作为流体通道而被弱化的结果。该模型的建立, 突破了陆壳整体俯冲与整体折返的传统模式, 揭示了陆壳俯冲与洋壳俯冲的主要区别。在此基础之上, 提出了有待于进一步研究的若干重要科学问题。

关键词 大陆俯冲; 超高压变质; 板片折返; 地壳拆离; 薄皮构造; 大别-苏鲁

柯石英[1~5]和金刚石[6,7]等超高压变质矿物的相继发现, 已证明巨量陆壳岩石能俯冲到地幔深度。因此近 20 年来, 超高压岩石的形成与折返机制, 一直是大陆动力学的研究热点。其中, 超高压岩石的折返机制是长期争议的焦点, 并且, 已提出多种解释模型, 如, 陆内逆冲及伴随侵蚀模式[8,9]、挤出-伸展模式[10~13]、浮力驱动模式[14]、浮力-楔入-热穹隆模式[15]、角流及浮力联合模式[16]、平行于造山带的挤出及伴随减薄模式[17,18]、连续俯冲-折返-热穹隆模式[19]等等。所有这些模式都是假定整个俯冲陆壳与下伏岩石圈地幔发生拆离解耦并整体折返(如, 文献[17, 20~22])。

与此相反, 近年来中国部分学者根据对大别-苏鲁造山带的研究成果, 提出俯冲陆壳内部曾发生多层次拆离解耦并呈多板片差异折返的模型[23~28]。已有研究表明, 俯冲陆壳不仅在深部发生拆离解耦[23~28], 而且在俯冲初期即榴辉岩相变质前在浅部不同深度也发生了地壳拆离并逆冲折返[29~34]。因此, 深俯冲陆壳是整体折返, 还是内部拆离解耦成若干岩片并相继折返已成为超高压变质岩折返机制研究的核心争议问题, 它涉及我们对大陆地壳俯冲行为与洋壳俯冲行为的差异性认识。本文以大别-苏鲁造山带为例, 总结了大陆碰撞过程中与俯冲陆壳拆离和差异折返模型相关的同位素年代学、岩石地球化学和变质岩石学等方面的证据, 并讨论了它的科学意义和有待于进一步解决的有关科学问题, 以期有助于陆壳俯冲动力学的进一步研究。

---

* 本文发表在: 科学通报, 2008, 53(16): 2153-2165

对应的英文版论文为: Liu Yican, Li Shuguang, 2008. Detachment within subducted continental crust and multi-slice successive exhumation of ultrahigh-pressure metamorphic rocks: evidence from the Dabie-Sulu orogenic belt. Chinese Science Bulletin, 53 (20): 3105-3119

1 俯冲陆壳的浅部拆离与差异折返

1.1 陆壳俯冲开始阶段其上部地壳(低级变质岩)与下部基底岩石的解耦

已有研究表明北淮阳带东段庐镇关杂岩中的花岗片麻岩和辉长岩(如山七、城冲岩体)以及苏鲁地区的五莲花岗片麻岩等的原岩形成时代为新元古代并且属于华南陆块北缘的一部分[17, 29, 32~39]。它们以低绿片岩相-绿帘角闪岩相变质作用为主。庐镇关杂岩中花岗片麻岩的角闪石Ar-Ar年龄[17]与锆石U-Pb年龄[36, 38]一致(约750 Ma),仍能记录其原岩年龄而未被后期变质作用干扰;但其中白云母的Ar-Ar年龄为241.9 ± 2.3 Ma[17],记录了与三叠纪大陆碰撞相关的变质时代。这表明这些杂岩在新元古代以后的变质反应温度都低于角闪石的Ar-Ar封闭温度(约500 ℃)而高于白云母的Ar-Ar封闭温度(约350 ℃)。因此,可以认为它们是陆壳俯冲最初阶段,最早被拆离、刮下来的"加积楔"[29, 32, 39]。

类似证据也发现于大别山西部,刘贻灿等[31]对北淮阳带西段王母观浅变质辉长岩的锆石SHRIMP U-Pb定年的研究结果表明,其形成时代为新元古代晚期(635±5 Ma)。很显然,北淮阳带西段浅变质辉长岩的形成年龄与华南陡山沱组沉积岩中普遍发育的火山凝灰岩夹层的时代(635 Ma左右)[40~42]以及华南陆块北缘(造山带南麓)湖北随州-枣阳一带发育的一期大规模基性岩墙群的时代(630~640 Ma)[1)完全一致。考虑到新元古代华南陆块北缘存在大规模的岩浆事件[17, 43~46],因而证明北淮阳带商-麻断裂以西浅变质(橄榄)辉长岩属于华南陆块北缘的一部分,它被后来的构造推覆到华北南缘并就位于华北陆块南缘的古生代岩浆弧杂岩中[31]。此外,Liu等人[19]曾报道苏家河附近一个变辉长岩的年龄为582 ± 11 Ma;陈玲等[47]也在北淮阳带西段获得柳林变辉长岩的锆石U-Pb年龄为611±13 Ma。尽管这些浅变质(橄榄)辉长岩是否属于同一期形成以及它们之间的相互关系等尚待进一步查明,但是,这些最新资料给我们一个重要启示:商-麻断裂以西的北淮阳带西段也存在从华南俯冲陆壳拆离下来的新元古代浅变质岩片。该区浅变质岩石为绿帘角闪岩相-绿片岩相[19, 31, 48, 49],并同样记录有218~258 Ma的云母K-Ar[50]或Ar-Ar[17, 19, 51]年龄。因而它们可能是印支期华南陆块发生深俯冲的初始阶段被解耦的岩片,并在后期构造作用过程中被推覆到华北南缘古生代浅变质岩系之上。因此,北淮阳带西段这些新元古代浅变质辉长岩的成因机制可能类似于周建波等[32, 34, 39]和Zheng等人[29]在研究大别-苏鲁造山带北缘浅变质岩时提出的大陆俯冲过程中被动大陆边缘变形加积楔模型。

1.2 陆壳俯冲初期的壳内解耦和低-中级变质岩片的折返

近年来,大别山东段(商城-麻城或商-麻断裂以东)及苏鲁造山带中陆续发现一些低-中级变质的新元古代岩石并具有三叠纪变质作用年龄记录,如大别山超高压带中的港河绿片岩相岩石(板岩、千枚岩)有232.2 ± 8.3 Ma Rb-Sr年龄(全岩等时线)记录和原岩时代为760~800 Ma[52]以及苏鲁地区超高压带西北侧(胶北)的粉子山群中大理岩的原岩时代为786±67 Ma和变质时代为240±44 Ma(均为锆石U-Pb年龄)[30]。由于白云母的Rb-Sr体系封闭温度为500 ℃左右[53],而变质锆石的重结晶作用至少需要600~650 ℃的温度[54],因此,大别-苏鲁造山带中这些浅变质的新元古代岩片属于华南陆块北缘的一部分并被认为参与了华南陆块的印支期俯冲,但未经历超高压变质作用。这表明这些浅变质岩系有可能是陆壳深俯冲初期在高绿片岩-角闪岩相阶段解耦的岩片[30]。Tang等人[30]提出大陆地壳俯冲过程中出现浅部地壳拆离,认为苏鲁造山带北缘粉子山群中大理岩岩片是随着大陆深俯冲的继续进行和南、北陆块的汇聚被推覆于华北陆块南缘变质的岩石单位之上的产物。

上述两种浅变质岩片的区别是前者是在陆壳俯冲初始阶段被解耦的、变质程度低,因而没发生变质锆石增生;后者被拆离解耦时,陆壳已俯冲到开始大规模脱水阶段,变质程度较高(角闪岩相)并有变质锆石增生发生,从而记录有三叠纪变质年龄。这证明俯冲陆壳在发生榴辉岩相变质前有可能在浅部不同深度发生了地壳的拆离[29~31]。

---

1) 洪吉安,马斌. 扬子北缘Marinoan期橄榄岩岩墙群及其意义. 杭州: 2005年全国岩石学与地球动力学研讨会论文摘要, 2005. 100~101

2 俯冲陆壳的深部拆离与多板片差异折返

俯冲陆壳的深部拆离与多板片差异折返,是指陆壳俯冲到较大深度(如高压或超高压榴辉岩相变质深度)发生的地壳拆离与超高压岩石的多板片差异折返。这方面证据分别来自大别-苏鲁造山带地表超高压变质岩研究和苏鲁科学钻探的地下岩芯超高压变质岩样品的研究。

2.1 大别山地表出露的超高压变质带证据

2.1.1 大别山超高压变质带由三个超高压岩片组成

大别-苏鲁造山带位于中国中部华南与华北两个大陆板块之间,是世界上最大的含柯石英[3~5]和金刚石[7]的超高压变质岩出露区。大别山超高压变质带保留有极好的岩石单位分带。从南到北,大别造山带大致可划分为宿松变质杂岩带、南大别低温榴辉岩带、中大别中温超高压变质带、北大别高温超高压杂岩带及等构造岩石单位[13, 17, 19, 28, 29, 49, 55~59](图1、表1)。它们之间依次分别被太湖-山龙、花凉亭-弥陀、水吼-五河及磨子潭-晓天等大型断裂带或韧性剪切带所分割。除了宿松变质杂岩带和北淮阳带没有发现超高压变质岩,其余3个带都发现了超高压变质岩。

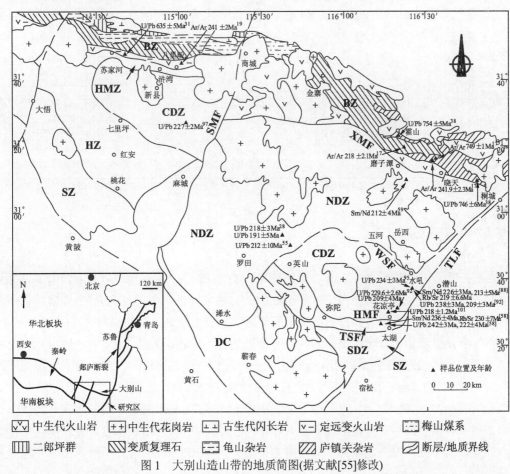

图1 大别山造山带的地质简图(据文献[55]修改)

BZ=北淮阳带,NDZ=北大别高温超高压杂岩带,CDZ=中大别中温超高压变质带,SDZ=南大别低温榴辉岩带,SZ=宿松变质杂岩带,HMZ=浒湾混杂岩带,HZ=红安低温榴辉岩带,DC=角闪岩相大别杂岩,XMF=晓天-磨子潭断裂,WSF=五河-水吼断裂,HMF=花凉亭-弥陀断裂,TSF=太湖-山龙断裂,TLF=郯庐断裂,SMF=商城-麻城断裂。图中年龄右上角编号为参考文献序号

中大别带最早发现含柯石英和金刚石的超高压岩石[4, 5, 7]并被证明是三叠纪华南板块俯冲于华北陆块之下≥120 km深度发生超高压变质形成的[60, 61]。近期研究表明,南大别低温榴辉岩中发现有柯石英假象等超高压变质证据[58]以及北大别杂岩带中发现榴辉岩[62, 63]并发现有金刚石以及其他超高压变质证据[28, 62, 64~67];同时,北大别杂岩带[28, 55, 59, 68~71]和南大别低温榴辉岩带[58]均被证明经过三叠纪超高压变质作用,因而属于

华南俯冲陆壳的一部分。因此，大别山印支期深俯冲陆壳包括3个超高压岩片即南大别低温榴辉岩带、中大别超高压变质带和北大别杂岩带[25](后文将分别简称南大别、中大别和北大别)。

表1 大别山不同岩石单位的主要特征

岩石单位	主要组成	变质特征	Pb同位素	年龄	参考文献
北淮阳带	变质复理石	绿片岩相		白云母 Ar/Ar(218±2.1)Ma	17
	庐镇关杂岩	绿片岩-绿帘角闪岩相	高放射成因Pb	U/Pb(754±5), (746±6)Ma	38
				角闪石 Ar/Ar(749±1)Ma	17
				白云母 Ar/Ar(241.9±2.3)Ma	17
	变辉长岩	绿片岩-绿帘角闪岩相		U/Pb(635±5)Ma	31
	定远变火山岩	绿片岩-绿帘角闪岩相		多硅白云母 Ar/Ar(241±2)Ma	19
北大别带	榴辉岩	超高压榴辉岩相→石英榴辉岩相→麻粒岩相→角闪岩相退变，峰期缺乏多硅白云母等含水矿物以及高的变质温度，含金刚石及石榴子石和单斜辉石中普遍发育针状出溶体，流体包裹体成分主要是N_2和CO_2	低放射成因Pb	Sm/Nd(212±3)Ma U/Pb(212±10)Ma	59 55
	花岗片麻岩			U/Pb(218±3), (191±5)Ma(r)	28
中大别带	榴辉岩	超高压榴辉岩相→石英榴辉岩相→角闪岩相，峰期含黝帘石、多硅白云母等含水矿物，流体包裹体成分主要为不同盐度的水流体	高放射成因Pb	Sm/Nd(226±3)Ma Rb/Sr(219±6.6)Ma U/Pb(234±3)Ma U/Pb(229.6±2.6), (209±4)Ma U/Pb(218±1.2)Ma	88 88 95 92 101
	石英硬玉岩			U/Pb(227±2)Ma	97
	片麻岩			U/Pb(238±3), (209±3)Ma	92
	大理岩			Sm/Nd(213±5)Ma	88
南大别带	榴辉岩	超高压榴辉岩相→石英榴辉岩相→角闪岩相，峰期富黝帘石、多硅白云母、硬柱石等含水矿物以及较低的变质温度		Sm/Nd(236±4)Ma Rb/Sr(230±7)Ma U/Pb(242±3), (222±4)Ma(r)	58 58 58
	片麻岩				
	大理岩				

注：U/Pb年龄为变质锆石或独居石的SHRIMP定年或金红石的定年结果，r表示锆石或独居石的增生边；Sm/Nd年龄为榴辉岩相或角闪岩相变质矿物3个以上分析点确定的等时线结果。

2.1.2 三个超高压岩片的差异

(1) 岩石学差异

北大别主要是一个正片麻岩单位，含有少量榴辉岩、(含石榴子石)斜长角闪岩和麻粒岩等，并且，最近研究已证明至少部分榴辉岩是由基性麻粒岩转变形成的[55]；中大别是一个正片麻岩+表壳岩单位，含有大量榴辉岩透镜体或岩块以及副片麻岩、大理岩和硬玉石英岩等；南大别则主要为榴辉岩及(含石榴子石)二云绿帘斜长片麻岩等一套副变质岩。三个超高压岩片具有不同的含水矿物组合特点(表1)。如，北大别榴辉岩中无多硅白云母或其他高压-超高压含水矿物，反映它们变质过程中缺乏流体[28,55,59]；而中大别榴辉岩中多硅白云母-黝帘石等含水矿物普遍存在，以及南大别榴辉岩中常见到低温高压变质矿物，硬柱石-冻蓝闪石，和钠云母-多硅白云母等含水矿物[49,58,72~74]，表明中大别和南大别在超高压变质作用期间富含流体以及它们在变质温度及含水矿物种类方面存在一定的差异性。另外，北大别、中大别超高压岩石中流体包裹体成分也不同，如后者主要是具有不同盐度的水流体，而前者主要是N_2和CO_2，证明超高压变质作用期间中大别比北大别富水流体[75,76]。

此外，北大别超高压变质岩以经历过独特的麻粒岩相退变质作用而区别于中大别和南大别带，显示它们具有不同的折返历史[28,55,59,63,67,77,78]。而且，三个超高压岩片的峰期变质温度由南向北逐渐升高，即由南大别低温榴辉岩($T<700$ ℃，一般为570~670 ℃)[58,79~81]→中大别中温榴辉岩(T一般为700~800 ℃)[72,73,80,82,83]→北大别高温榴辉岩($T=808\sim874$ ℃、$P=2.5$ GPa或904~963 ℃、$P=4.0$ GPa)[55,59]，这种峰期变质温度的有规律变化，可能分别与三个UHP岩片的原岩在俯冲陆壳内所处的位置有关，即南大别和北大别分别相当于上、下地壳，也就是说，三个岩片原岩温度就有高、低区别[28,55]。这也与前文所述的三个岩片的岩石组

成及其含水矿物和流体成分的特点相吻合。

因此,鉴于南大别和中大别含有较多的表壳岩、富含水矿物和流体包体以及较低的峰期变质温度并结合它们各自的岩石组合特点,推测它们分别来自俯冲上地壳的上、下部;而北大别则因峰期缺乏含水矿物、富N_2和CO_2包体、较高的峰期变质温度以及榴辉岩由镁铁质麻粒岩转变形成等而推测来自俯冲下地壳。同时考虑它们有不同的变质$P-T$演化历史,由此说明印支期华南陆壳俯冲过程中有可能在不同深度发生了壳内的解耦[27~31, 84]。

(2) 地球化学和年代学差异

由于上、下地壳的U/Pb及Pb同位素组成有明显差异,因此Pb同位素可以用来示踪地壳性质,即下部地壳相对亏损U和Th以及贫放射性成因Pb同位素组成,而上部地壳相对富集放射性成因Pb同位素组成[85, 86]。已有的Pb同位素研究表明北大别与中大别岩片的Pb同位素组成不同,中大别超高压变质岩比北大别变质岩有更高的放射成因Pb,具上地壳特征,而北大别的低$^{206}Pb/^{204}Pb$值反映了下地壳特征[23, 87],因而代表了上、下不同层位的俯冲陆壳。进一步证明大别山不同的超高压岩片分别来自不同性质的俯冲陆壳。

北大别片麻岩的锆石SHRIMP U-Pb年龄结果表明,超高压变质时代为218±3 Ma[28],而北大别榴辉岩的2石榴子石+2绿辉石的Sm-Nd等时线年龄为212±4 Ma[59]。中大别UHP岩石的峰期变质时代已被很好测定,且老于北大别。如,由三个榴辉岩相矿物确定的Sm-Nd等时线年龄为226±3 Ma[88];锆石U-Pb测定的精确年龄为225~238 Ma[43,44,89~97]。南大别低温榴辉岩的峰期变质时代最老,如石榴子石+绿辉石+金红石+蓝晶石的Sm-Nd矿物等时线年龄为236±4 Ma和锆石U-Pb年龄为242±3 Ma~243±4 Ma[58, 98]。此外,苏鲁超高压带中对应于中大别超高压带的超高压片麻岩锆石SHRIMP U-Pb年龄研究也表明,含柯石英等超高压矿物的锆石幔部年龄为231~227 Ma等[99, 100],而含石英等角闪岩相退变质矿物的锆石边部年龄为211±4 Ma。该超高压片麻岩锆石的幔、边部年龄正好分别与中大别超高压带及北大别带的峰期变质时代一致。这些表明,三个超高压岩片的峰期变质时代,由南向北逐渐变新,并且,中大别超高压带岩石的退变质年龄,如双河片麻岩退变质矿物Sm-Nd年龄为213±5 Ma[88];双河石英硬玉岩和毛屋榴辉岩中独居石的退变边年龄为209±3 Ma和209±4 Ma[92],与北大别的峰期变质时代一致。

综合已发表的年代学和岩石学资料,可以绘出大别山三个超高压岩片的冷却T-t曲线(图2)。图2反映它们明显具有不同的抬升冷却历史和演化过程,这与前文分析一致。此外,从图2还可以看出,北大别具有相对慢的冷却速率,而中大别与南大别却有快速的冷却过程。这也许是北大别榴辉岩很少见有保留早期超高压变质证据的重要原因,因为慢的抬升速率及高温(≥800 ℃)条件下有可能使超高压岩石

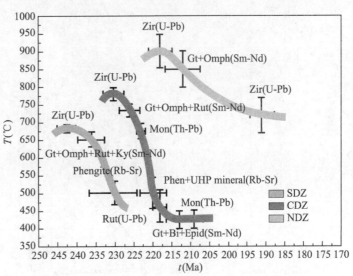

图2 大别山三个超高压岩片的冷却 T-t 曲线

SDZ-南大别低温榴辉岩带,CDZ-中大别中温超高压变质带,NDZ-北大别高温超高压杂岩带

资料来源: 文献[28, 55, 58, 59, 88, 92~95, 101]

部分或全部转变为低压矿物组合,如早期柯石英可能转变为石英(目前呈包体形式存在于石榴子石中和主晶石榴子石伴有放射状裂纹)[55,78]。

2.2 苏鲁地表出露的超高压变质带证据

苏鲁超高压变质带也可类似大别山划分出若干岩片。Xu等人[27]将苏鲁超高压变质带划分成4带:南苏鲁高压变质带(I)、中苏鲁很高压变质带(II)、北苏鲁超高压变质表壳岩带(III)和北苏鲁超高压花岗质变质岩带(IV)(图3)。根据文献[27],现将其主要特征简单介绍如下。

带I经历了 0.7~0.85 GPa、300~360 ℃的蓝片岩相变质及绿片岩相退变质作用；带II经历了 1.5~2.5 GPa、500~600 ℃的很高压变质作用及蓝片岩相退变质作用；带III主要由含柯石英超高压变质表壳岩及榴辉岩和超镁铁质岩透镜体组成，峰期变质温度和压力为 2.8~3.0 GPa、750~850 ℃；带IV的主要组分为花岗质片麻岩、少量变质表壳岩夹榴辉岩和超镁铁质岩块体，不同类型岩石都含有柯石英，证明该带经历了超高压变质作用。它们分别大致相当于大别山的宿松变质杂岩带、南大别低温榴辉岩带、中大别超高压带和北大别杂岩带。而且，南苏鲁高压变质带的峰期变质时代为 253 Ma、折返时代为 240~253 Ma 和北苏鲁超高压变质带的峰期变质时代为 220~240 Ma、折返时代为 200~220 Ma[27, 61, 95, 99, 100, 102~104]，表明从南到北，峰期变质时代和折返时代逐渐变年轻，类似于大别山。因而，在苏鲁造山带同样存在俯冲陆壳内部的多层次拆离解耦与多板片折返过程。

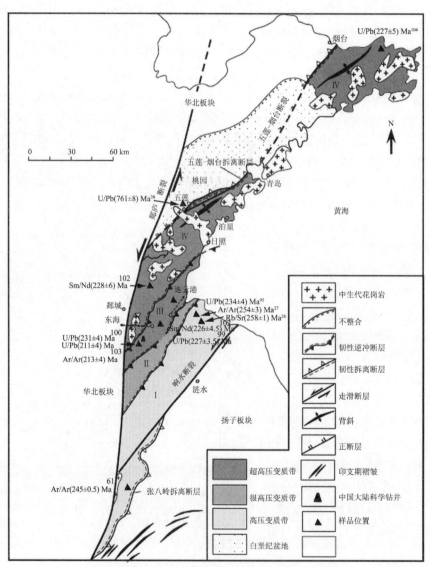

图 3 苏鲁超高压变质带构造简图(据文献[27]修改)
I-高压变质带；II-很高压变质带；III 和 IV-超高压变质带
图中年龄编号为参考文献序号

2.3 苏北大陆科学深钻岩芯揭示的超高压带垂向多岩片叠置证据

Dong 等人[105]通过反射地震剖面揭示了中大别超高压变质带在垂向上也是由若干岩片叠置而成的。在苏北东海(位于图 3 第 III 超高压变质带)实施的大陆科学钻探为直接观察这一深部岩片叠置状况提供了条件。董锋等[106]通过对苏北大陆科学钻探(CCSD)100~2000m 岩芯的普通 Pb 同位素研究揭示了苏鲁超高压带 III 地表至 2000 m 深度在垂向上超高压岩片的叠置状况(图 4)。

图 4 Pb 同位素显示该岩芯可划为 3 个岩段：100~800 m、800~1600 m 和 1600~2000 m 上、中、下 3 个岩段。上岩段具有高放射成因 Pb 特征，下岩段具有低放射成因 Pb 特征，二者有显著差别，它们分别是从俯冲大陆岩石圈解耦的上地壳和中地壳岩片。中部岩段发育有韧性剪切带，其 Pb 同位素特征介于上下岩段之间，可能是上、下岩段 Pb 同位素的混合结果，因此被解释为上、下岩片之间的拆离带。考虑到中部岩段又具有异常低 $\delta^{18}O$ 峰值[107, 108]，该拆离面可能是在地壳内部晚元古代的断裂带及冷水活动通道基础上发育起来的,该水-岩交换作用导致中部岩段原岩的低 $\delta^{18}O$ 峰值和弱的力学强度。据此，董锋等[106]提出在大陆岩石圈俯冲过程中，上部岩片同俯冲的大陆岩石圈发生拆离，率先沿着这一个拆离面逆冲折返至浅部。下岩片是随后从俯冲岩石圈解耦的中地壳岩片，并折返至浅部，并下伏在先期折返的上部岩片之下。

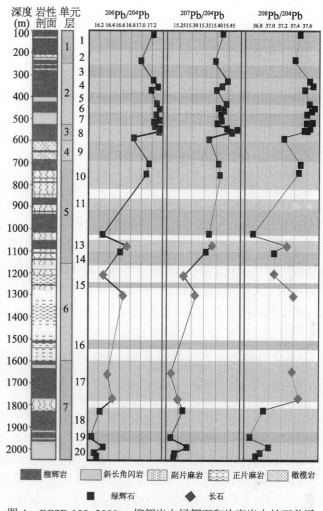

图 4 CCSD 100~2000 m 榴辉岩中绿辉石和片麻岩中长石普通 Pb 同位素比值剖面图[106]

2.4 大别山俯冲陆壳内部的多层次解耦与超高压岩石的多板片差异折返模型

上述大别山构造岩石单位的分布以及不同超高压岩片的岩石学、年代学和地球化学特征，可以用印支期华南大陆深俯冲过程中俯冲陆壳内部的多层次拆离解耦以及多板片差异折返模型[24, 25, 28]来解释：陆壳俯冲过程中，首先在俯冲上地壳中发生拆离解耦，上部岩片(如南大别低温榴辉岩带)沿拆离面逆冲折返，下部陆壳继续俯冲；此后，在上、下地壳之间发生第二次解耦，上部岩片(中大别 UHP 带) 沿拆离面逆冲折返，下地壳(北大别杂岩带)继续俯冲；最终，在俯冲板片断离后，长英质下地壳与下伏镁铁质下地壳和岩石圈地幔拆分解耦，三个岩片全部折返，并呈现出大别山(或苏鲁带)由南向北三个岩片峰期变质时代逐步变新、变质温度逐步升高的趋势。

3 俯冲陆壳内部拆离的岩石流变学证据

在陆壳俯冲过程中陆壳内部发生拆离和脱耦是陆壳内部存在力学强度薄弱带的结果。陆壳整体与岩石圈地幔脱耦并折返模型的基本假设是基于大陆岩石圈强度的果酱三明治(jelly sandwich)模型，即刚性的上地壳和地幔岩石圈夹着低黏度软弱的下地壳[22]。俯冲陆壳多层次解耦及多岩板折返模型的理论基础是大陆岩石圈强度不再是果酱三明治模型，根据合理的地热梯度和岩石圈成分所获得的岩石圈的黏度(viscosity)与深度(depth)关系的模拟结果表明，在陆壳内部至少存在两个低黏度带[109]。因此，大陆在俯冲过程中，由于大陆地壳上、下不同部位岩石组成的差异以及它们具有不同的力学性质和地球化学成分等，有可能沿不同深度的低黏度带发生壳内拆离解耦，形成几个高压-超高压变质的板片并逆冲折返。此外，对苏北大陆科学钻探(CCSD)岩芯的普通 Pb 和氧同位素研究还发现俯冲陆壳内的古断层带因其曾是流体通道及所发生的水岩作用而弱化，也能在俯冲过程中发育成壳内拆离面[106]。

4 科学意义及有待于进一步解决的问题

4.1 科学意义

板块构造理论的出现,使人们对大陆的碰撞过程有了较深入的理解。Oxburgh[110]最早根据阿尔卑斯造山带的研究,发现岩石圈曾沿下地壳与上地幔之间发生裂开(split)、上覆薄的地壳板片与下伏板片发生拆离,进而提出薄板构造(crustal flake 或 flake tectonics)概念并用来解释一些复杂的板块边缘特征。在此基础之上,许靖华[111]在研究特提斯造山带的形成过程时,发现由于在地壳的内部或底部存在拆离带(detachment horizons)以及板块运动而常常导致地壳的双重或三重叠置,提出了薄皮板块构造模式(thin-skinned plate-tectonic model),并用此模式来解释阿尔卑斯等碰撞型造山带的形成并认为碰撞型造山带主要由仰冲壳楔(overriding crustal wedge)、碰撞混杂岩(collision mélange)和俯冲壳楔(underthrusting crustal wedge)等基本构造单元所组成。目前,地壳的拆离(crustal detachment)已被认为是与板块俯冲相联系的一种重要地质过程并被理论和地震实验所证实(如,文献[112]及其所引用参考文献)。

大陆碰撞过程中的多层次地壳拆离与多板片差异折返模型的建立,是上述薄皮板块构造模式在大陆深俯冲过程中的体现和发展。它突破了传统的陆壳整体俯冲与整体折返的固定模式,同时,这也是陆壳俯冲与洋壳俯冲的主要区别。这种深俯冲陆壳呈多板片折返的过程是大陆岩石圈组成和力学性质高度不均一性,因而存在多个低黏滞带的结果,是大陆板块运动有别于大洋板块的实际例证。而且,这种俯冲陆壳内部的地壳拆离作用可能是引起壳内浅源地震的重要原因之一。因此,通过进一步研究,不仅有助于大陆动力学研究并发展板块构造理论,而且对探讨地壳浅源地震的原因和发生机制等具有重要意义。

4.2 有待于进一步解决的问题

尽管俯冲陆壳多层次解离及多岩板差异折返模型有上述证据的支持,然而仍存在两方面问题。查明这些问题尚需开展更深入的岩石学、年代学和地球化学等方面的研究工作。

(i)不同岩片提供的证据是不均衡的,部分证据的可靠性需要更深入工作加以验证。如北大别带和南大别带的精确变质年龄偏少,还难以建立如中大别带一样精确的 T-t 冷却曲线。为此,在今后的研究中加强对这两个岩片的高压-超高压变质岩进行精确定年和冷却史研究。其中,两个岩片的峰期变质时代的确定是关键。再如,前人对地表不同岩片的 Pb 同位素填图研究仅局限在中大别和北大别带,南大别带和苏鲁带都未开展工作;更为重要的是前人工作主要是做全岩 Pb,但由于地表风化等因素的影响而使全岩 U/Pb 可发生较大变化,给初始 Pb 同位素组成的年龄校正计算带来误差。因此,有必要在地表出露的各主要岩片中开展如董锋等[106]在大陆科学钻探岩芯研究中进行的超高压变质岩普通 Pb 地球化学填图。

(ii)对于陆壳俯冲过程中陆壳内部的拆离解耦及多板片差异折返的精细过程我们尚有许多问题不清楚,已提出的模型尚有一些不完善之处。

如我们并不清楚部分北淮阳浅变质岩片的变质时代是代表俯冲时代还是构造就位时代,以及新元古代晚期浅变质岩石是被何种构造运动从华南推覆并就位于华北陆块南部活动陆缘的。这是由于过去北淮阳带和造山带南麓(如随州-枣阳一带)的浅变质岩石的研究相对薄弱,如前文所述,北淮阳带西段的新元古代晚期浅变质辉长岩是否经历过三叠纪变质作用以及它们的就位机制等仍是急需解决的重要科学问题。

再如,依据上述俯冲陆壳多层次解离及多岩板差异折返模型,大别山南、中、北大别 3 个超高压岩片的峰期变质年龄差异仅代表了相应岩片拆离、折返时代的差异,因此,这 3 个岩片的俯冲深度应该不同。南大别带仅发现柯石英,尚未发现金刚石,其俯冲深度较浅。但是,已有证据表明大别山中、北大别 2 个超高压岩片的俯冲深度没有显著差别,它们均发现有金刚石[7, 28, 65, 66]。尽管石榴子石中单斜辉石+金红石+磷灰石等针状矿物出溶体显示,北大别超高压岩石的峰期变质压力可能比中大别和南大别高[66],但尚需要对北大别带开展更深入的岩石学研究,寻找更多的证据加以证实;否则,这 3 个岩片的折返过程并不都像上述模型那样简单,我们尤其需要对北大别带开展更深入的峰期变质条件、退变质岩石学和精细年代学研究,揭示它真实的折返、抬升轨迹,进一步完善大别山超高压变质带的多岩板差异折返模型。类似的问题对苏鲁Ⅲ、

IV 超高压变质带也同样存在。

致谢

感谢两位审稿人及郑永飞副主编对本文提出的建设性修改意见。

参考文献

[1] Chopin C. Coesite and pure pyrope in high grade blueschists of the western Alps: a first record and some consequences. Contrib Mineral Petrol, 1984, 86: 107-118
[2] Smith D C. Coesite in clinopyroxene in the Caledonides and its implications for geodynamics. Nature, 1984, 310: 641-644
[3] Xu Z. Etude tectonique et microtectonique de la chaine paleozoique et triasique des Qinlings (Chine). These de Doctoral, Univ Sci Tech. Languedoc, Montpeuier. 1987
[4] Okay A I, Xu S, Sengör A M C. Coesite from the Dabie Shan eclogites, central China. Eur J Mineral, 1989, 1: 595-598
[5] Wang X, Liou J G, Mao H K. Coesite –bearing eclogites from the Dabie Mountains in central China. Geology, 1989, 17: 1085-1088
[6] Sobolev N V, Shatsky V S. Diamond inclusions in garnets from metamorphic rocks: a new environment for diamond formation. Nature, 1990, 343: 742-746
[7] Xu S, Okay A I, Ji S, et al. Diamonds from the Dabie Shan metamorphic rocks and its implication for tectonic setting. Science, 1992, 256: 80-82
[8] Okay A I, Sengör A M C. Evidence for intracontinental thrust-related exhumation of the ultrahigh-pressure rocks in China. Geology, 1992, 20: 411-414
[9] Chemenda A I, Mattauer M, Bokun A N. Continental subduction and a mechanism for exhumation of high-pressure metamorphic rocks: new modelling, field data from Oman. Earth Planet Sci Lett, 1996, 143: 173-182
[10] Maruyama S, Liou J G, Zhang R. Tectonic evolution of the ultrahigh-pressure (UHP) and high-pressure (HP) metamorphic belts from central China. Island Arc, 3: 112-121
[11] Faure M, Lin W, Shu L, et al. Tectonics of the Dabieshan (eastern China) and possible exhumation mechanism of ultra high-pressure rocks. Terra Nova, 1999, 11: 251-258.
[12] 钟增球, 索书田, 游振东. 大别山高压、超高压变质期后伸展构造格局. 地球科学——中国地质大学学报, 1998, 23(3): 225-229
[13] 索书田, 钟增球, 游振东. 大别地块超高压变质期后伸展变形及超高压变质岩石折返过程. 中国科学 D 辑: 地球科学, 2000, 30(1): 9-17
[14] Ernst W G, Maruyama S, Wallis S. Buoyancy-driven, rapid exhumation of ultrahigh-pressure metamorphosed continental crust. Proce Nat Acad Sci, USA, 1997, 94: 9532-9537
[15] Dong S, Chen J, Huang D. Differential exhumation of tectonic units and ultrahigh-pressure metamorphic rocks in the Dabie Mountains, China. Island Arc, 1998, 7: 174-183
[16] Wang Q, Cong B. Exhumation of UHP Terranes: a case study from the Dabie Mountains, eastern China. Int Geol Rev, 1999, 41: 994-1004
[17] Hacker B R, Ratschbacher L, Webb L, et al. Exhumation of ultrahigh-pressure continental crust in east central China: Late Triassic-Early Jurassic tectonic unroofing. J Geophys Res, 2000, 105 (B6): 13339-13364
[18] Ratschbacher L, Hacker B R, Webb L E, et al. Exhumation of the ultrahigh-pressure continental crust ineast-central China: Cretaceous and Cenozoic unroofing and the Tan-Lu fault. J Geophys Res, 2000, 105: 13303-13338
[19] Liu X C, Jahn B M, Liu D, et al. SHRIMP U-Pb zircon dating of a metagabbro and eclogites from western Dabieshan (Hong'an Block), China, and its tectonic implications. Tectonophysics, 2004, 394: 171-192
[20] Chemenda A I, Mattauer M, Malavieille J, et al. A mechanism for syn-collisional rock exhumation and associated normal faulting: results from physical modeling. Earth Planet. Sci Lett, 1995, 132: 225-232
[21] Massonne H J. Involvement of crustal material in delamination of the lithosphere after continent-continent collision. Int Geol Rev, 2005, 47: 792-804
[22] Zuber M T. Folding a jelly sandwich. Nature, 1994, 371: 650-651
[23] 李曙光, 黄方, 周红英, 等. 大别山双河超高压变质岩及北部片麻岩的 U-Pb 同位素组成——对超高压岩石折返机制的制约. 中国科学 D 辑: 地球科学, 2001, 31: 977-984
[24] 李曙光, 李秋立, 侯振辉, 等. 大别山超高压变质岩的冷却史及折返机制. 岩石学报, 2005, 21: 1117-1124

[25] 刘贻灿, 李曙光. 大别山下地壳岩石及其深俯冲. 岩石学报, 2005, 21(4): 1059-1066
[26] 许志琴, 曾令森, 梁凤华, 等. 大陆板片多重性俯冲与折返的动力学模式——苏鲁高压-超高压变质地体的折返年龄限定. 岩石矿物学杂志, 2005, 24: 357-368
[27] Xu Z Q, Zeng L S, Liu F L, et al. Polyphase subduction and exhumation of the Sulu high-pressure–ultrahigh-pressure metamorphic terrane. Geol Soc Am Spec Paper, 2006, 403: 792-113
[28] Liu Y C, Li S, Xu S. Zircon SHRIMP U-Pb dating for gneiss in northern Dabie high T/P metamorphic zone, central China: implication for decoupling within subducted continental crust. Lithos, 2007, 96: 170-185
[29] Zheng Y F, Zhou J B, Wu Y, et al. Low-grade metamorphic rocks in the Dabie-Sulu orogenic belt: a passive-margin accretionary wedge deformed during continent subduction. Int Geol Rev, 2005, 47: 851-871
[30] Tang J, Zheng Y F, Wu Y B, et al. Zircon SHRIMP U-Pb dating, C and O isotopes for impure marbles from the Jiaobei terrane in the Sulu orogen: implication for tectonic affinity. Precambrian Res, 2006, 144: 1-18
[31] 刘贻灿, 李曙光, 古晓锋, 等. 北淮阳王母观橄榄辉长岩的锆石 SHRIMP U-Pb 年龄及其地质意义. 科学通报, 2006, 51(18): 2175-2180
[32] 周建波, 郑永飞, 李龙, 等. 扬子板块俯冲的构造加积楔. 地质学报, 2001, 75(3): 338-352
[33] 周建波, 郑永飞, 吴元保. 苏鲁造山带西北缘五莲花岗岩中锆石 U-Pb 年龄及其地质意义. 科学通报, 2002, 47(22): 1745-1750
[34] 周建波, 刘建辉, 郑常青. 苏鲁造山带浅变质岩的成因及其大地构造意义. 地质学报, 2005, 79(4): 475-486
[35] 谢智, 陈江峰, 张巽, 等. 北淮阳新元古代基性侵入岩年代学初步研究. 地球学报, 2002, 23(6): 517-520
[36] 郑永飞, 吴元保, 赵子福, 等. 大别山北麓发现新元古代低 ^{18}O 岩浆岩. 科学通报, 2004, 49(14): 1468-1470
[37] 江来利, Wolfgang Siebel, 陈福坤, 等. 大别造山带北部卢镇关杂岩的U-Pb锆石年龄. 中国科学 D 辑: 地球科学, 2005, 35: 411-419
[38] Wu Y B, Zheng Y F, Tang J, et al. Zircon U-Pb dating of water-rock interaction during Neoproterozoic rift magmatism in South China. Chem Geol, 2007, 246: 65-86
[39] Zhou J B, Wilde S A, Zhao G C, et al. SHRIMP U-Pb zircon dating of the Wulian Complex: defining the boundary between the North and South China Cratons in the Sulu Orogenic Belt, China. Precambrian Res, 2008, 162: 559-576
[40] 储雪蕾, Wolfgang T, 张启锐, 等. 南华-震旦系界线的锆石U-Pb 年龄. 科学通报, 2005, 50: 600-602
[41] Condon D, Zhu M, Bowring S, et al. U-Pb Ages from the Neoproterozoic Doushantuo Formation, China. Science, 2005, 308: 98-98
[42] Yin C, Tang F, Liu Y, et al. U-Pb zircon age from the base of the Ediacaran Doushantuo Formation in the Yangtze Gorges, South China: constraint on the age of Marinoan glaciation. Episodes, 2005, 28: 48-49
[43] Rowley D B, Xue F, Tucker R D, et al. Ages of ultrahigh pressure metamorphism and protolith orthogneisses from the eastern Dabie Shan: U/Pb zircon geochronology. Earth Planet Sci Lett, 1997, 151: 191-203
[44] Hacker B R, Ratschbacher L, Webb L E, et al. Zircon ages constrain the architecture of the ultrahigh-pressure Qinling-Dabie orogen, China. Earth Planet Sci Lett, 1998, 161: 215-230
[45] Zheng Y F, Fu B, Gong B, et al. Stable isotope geochemistry of ultrahigh pressure metamorphic rocks from the Dabie-Sulu orogen in China: implications for geodynamics and fluid regime. Earth-Sci Rev, 2003, 62: 105-161
[46] Zheng Y F, Wu Y, Chen F K, et al. Zircon U-Pb and oxygen isotope evidence for a large-scale ^{18}O depletion event in igneous rocks during the Neoproterozoic. Geochim Cosmochim Acta, 2004, 68: 4145-4165
[47] 陈玲, 马昌前, 佘振兵, 等. 大别山北淮阳构造带柳林辉长岩: 新元古代晚期裂解事件的记录. 地球科学——中国地质大学学报, 2006, 31: 578-584
[48] 叶伯丹, 简平, 许俊文, 等. 桐柏-大别造山带北坡苏家河地体拼接带及其构成和演化. 武汉: 中国地质大学出版社, 1993. 1-81
[49] 徐树桐, 刘贻灿, 江来利, 等. 大别山造山带的构造几何学和运动学. 合肥: 中国科学技术大学出版社, 2002. 1-133
[50] 陈江峰, 董树文, 邓衍尧, 等. 大别造山带钾氩年龄解释——差异抬升的地块. 地质论评, 1993, 39(1): 17-22
[51] Ratschbacher L, Franz L, Enkelmann E, et al. The Sino-Korean-Yangtze suture, the Huwan detachment, and the Paleozoic-Tertiary exhumation of (ultra) high-pressure rocks along the Tongbai-Xinxian-Dabie Mountains. In B.R. Hacker, W.C. McClelland and J. G. Liou. (Eds.), Ultrahigh-pressure metamorphic: Deep continental subduction. Geol Soc Am Spec Paper, 2006, 403: 45-77
[52] 董树文, 王小凤, 黄德志. 大别山超高压变质带内浅变质岩片的发现及意义. 科学通报, 1996, 41(9): 815-820
[53] Cliff R A. Isotopic dating in metamorphic belt. J. Ged. Soc. London, 1985, 142: 97-110
[54] Mezger K and Krogstad E J. Interpretation of discordant U-Pb zircon ages: a evaluation. J Metamorphic Geol, 1997, 15: 127-140
[55] Liu Y C, Li S, Gu X, et al. Ultrhigh-pressure eclogite transformed from mafic granulite in the Dabie orogen. J Metamorphic Geol, 2007, 25: 975-989

[56] Li S, Huang F, Nie Y, et al. Geochemical and geochronological constrains on the suture location between the North and South China Blocks in the Dabie orogen, central China. Phys Chem Earth (A), 2001, 26: 655-672

[57] 钟增球, 索书田, 张宏飞, 等. 桐柏-大别碰撞造山带的基本组成与结构. 地球科学——中国地质大学学报, 2001, 26(6): 560-567

[58] Li X P, Zheng Y F, Wu Y B, et al. Low-T eclogite in the Dabie terrane of China: petrological and isotopic constrains on fluid activity and radiometric dating. Contrib Mineral Petrol, 2004, 148: 443-470

[59] Liu Y C, Li S, Xu S, et al. Geochemistry and geochronology of eclogites from the northern Dabie Mountains, central China. J Asian Earth Sci, 2005, 25: 431-443

[60] 李曙光, 刘德良, 陈移之, 等. 大别山南麓含柯石英榴辉岩的 Sm-Nd 同位素年龄. 科学通报, 1992, 37（4）: 346–349

[61] Li S G, Xiao Y L, Liu D L, et al. Collision of the North China and Yangtze blocks and formation of coesite-bearing eclogites: timing and processes. Chem Geol, 1993, 109: 89-111

[62] 徐树桐, 苏文, 刘贻灿, 等. 大别山北部榴辉岩的发现及其岩相学特征. 科学通报, 1999, 44: 1452–1456

[63] 刘贻灿, 徐树桐, 李曙光, 等. 大别山北部镁铁-超镁铁质岩带中榴辉岩的分布与变质温压条件. 地质学报, 2001, 75 (3): 385-395

[64] Tsai C H, Liou J G. Eclogite-facies relics and inferred ultrahigh-pressure metamorphism in the North Dabie Complex, central-eastern China. Am Mineral, 2000, 85: 1-8

[65] 徐树桐, 刘贻灿, 陈冠宝, 等. 大别山、苏鲁地区榴辉岩中新发现的微粒金刚石. 科学通报, 2003, 48: 1069–1075

[66] Xu S T, Liu Y C, Chen G, et al. Microdiamonds, their classification and tectonic implications for the host eclogites from the Dabie and Su-Lu regions in central eastern China. Mineral Mag, 2005, 69: 509-520

[67] Malaspina N, Hermann J, Scambelluri M, et al. Multistage metasomatism in ultrahigh-pressure mafic rocks from the North Dabie Complex (China). Lithos, 2006, 90: 19-42

[68] 刘贻灿, 李曙光, 徐树桐, 等. 大别山北部榴辉岩和英云闪长质片麻锆石 U-Pb 年龄及多期变质增生. 高校地质学报, 2000, 6(3): 417-423

[69] 谢智, 陈江峰, 张巽, 等. 大别造山带北部石竹河片麻岩的锆石 U-Pb 年龄及其地质意义. 岩石学报, 2001, 17: 139-144

[70] 葛宁洁, 夏群科, 吴元保, 等. 北大别燕子河片麻岩的锆石 U-Pb 年龄: 印支期变质事件的确定. 岩石学报, 2003, 19(3): 513-516

[71] 薛怀民, 董树文, 刘晓春. 北大别大山坑二长花岗片麻岩的地球化学特征与锆石 U-Pb 年代学. 地球科学进展, 2003, 18(2): 192-197

[72] 徐树桐, 刘贻灿, 江来利, 等. 大别山的构造格局和演化. 北京: 科学出版社, 1994

[73] Cong B L. Ultrahigh-pressure Metamorphic Rocks in the Dabieshan-Sulu Region of China. Beijing: Science Press, 1996. 1-224

[74] Castelli D, Rolfo F, Compagnoni R, et al. Metamorphic veins with kyanite, zoisite and omphacite in the Zhu-Jia-Chong eclogite Dabie Shan, China. The Island Arc, 1998, 7: 159-173

[75] Xiao Y L, Hoefs J, van den Kerkhof A M, et al. Geochemical constraints of the eclogite and granulite facies metamorphism as recognized in the Raobazhai complex from North Dabie Shan, China. J Metamorphic Geol, 2001, 19: 3-19

[76] Xiao Y L, Hoefs J, van den Kerkhof A M, et al. Fluid History during HP and UHP Metamorphism in Dabie Shan, China: constraints from Mineral Chemistry, Fluid Inclusions, and Stable Isotopes. J Petrol, 2002, 43: 1505-1527

[77] 刘贻灿, 徐树桐, 李曙光, 等. 大别山北部榴辉岩的大地构造属性及冷却史. 地球科学, 2003, 28 (1): 11-16

[78] 刘贻灿, 徐树桐, 李曙光, 等. "罗田穹隆"中的下地壳俯冲成因榴辉岩及其地质意义. 地球科学——中国地质大学学报, 2005, 30(1): 71-77

[79] Wang X M, Liou J G and Marruyama S. Coesite-bearing eclogites from the Dabie Mountains, central China: petrogenesis, P-T path and implications for regional tectonics. J Geol, 1992, 100: 231-250.

[80] Okay A I. Petrology of diamond and coesite-bearing metamorphic terrane: Dabie Shan, China. Eur J Mineral, 1993, 5: 659-676.

[81] Zhai M G, Cong B L, Zhao Z Y, et al. Petrologic-tectonic units in the coesite-bearing metamorphic terrain of the Dabie Mountains, central China and their geotectonic implication. J SE Geosci, 1995, 11(1): 1-13

[82] Rolfo F, Compagnoni R, Xu S, et al. First report of felsic whiteschist in the ultrahigh-pressure metamorphic belt of Dabie Shan, China. Eur J Mineral, 2000, 12: 883-898

[83] Rolfo F, Compagnoni R, Wu W, et al. A coherent lithostratigraphic unit in the coesite-eclogite complex of Dabie Shan, China: geologic and petrologic evidence. Lithos, 2004, 73: 71-94

[84] Okay A I, Sengör A M C. Tectonics of an ultrahigh-pressure metamorphic terrane: the Dabie Shan/Tongbai orogen, China. Tectonics, 1993, 12: 1320-1334

[85] Zartman R E. Plumbotectonics—the model. Tectonophysics, 1981, 75: 135-162

[86] Zindler A, Hart S. Chemical geodynamics. Ann Rev Earth Planet Sci, 1986, 14: 493-571

[87] 张宏飞, 高山, 张本仁, 等. 大别山地壳结构的 Pb 同位素地球化学示踪. 地球化学, 2001, 30: 395-401

[88] Li S G, Jagoutz E, Chen Y Z, et al. Sm-Nd and Rb-Sr isotope chronology of ultrahigh-pressure metamorphic rocks and their country rocks at Shuanghe in the Dabie Mountains, central China. GCA, 2000, 64: 1077-1093

[89] Ames L, Zhou G, Xiong B. Geochronology and isotopic character of ultrahigh-pressure metamorphism with implications for the collision of the Sino-Korean and Yangtze cratons, central China. Tectonics, 1996, 15: 472-489

[90] 李曙光, 李惠民, 陈移之, 等. 大别山-苏鲁地体超高压变质年代学——II. 锆石U-Pb同位素体系. 中国科学D辑: 地球科学, 1997, 27(3): 200-206

[91] 陈道公, Isachen C, 支霞臣, 等. 安徽潜山片麻岩锆石U-Pb年龄. 科学通报, 2000, 45(2): 2110-2114

[92] Ayers J C, Dunkle S, Gao S, et al. Constraints on timing of peak and retrograde metamorphism in the Dabie Shan ultrahigh-pressure metamorphic belt, east-central China, using U-Th-Pb dating of zircon and monazite. Chem Geol, 2002, 186: 315-331

[93] Wan Y S, Li R W, Wilde S A, et al. UHP metamorphism and exhumation of the Dabie Orogen, China: evidence from SHRIMP dating of zircon and monazite from a UHP granitic gneiss cobble from the Hefei Basin. Geochim Cosmochim Acta, 2005, 69: 4333-4348

[94] Liu D Y, Jian P, Kröner A, et al. Dating of prograde metamorphic events deciphered from episodic zircon growth in rocks of the Dabie-Sulu UHP complex, China. Earth Planet Sci Lett, 2006, 250: 650-666

[95] Liu F L, Gerdes A, Liou J G, et al. SHRIMP U-Pb zircon dating from Sulu-Dabie dolomitic marble, eastern China: constraints on prograde, ultrahigh-pressure and retrograde metamorphic ages. J Metamorphic Geol, 2006, 24: 569-589

[96] Wu Y B, Zheng Y F, Zhao Z F, et al. U-Pb, Hf and O isotope evidence for two episodes of fluid-assisted zircon growth in marble-hosted eclogites from the Dabie orogen. Geochim Cosmochim Acta, 2006, 70: 3743-3761

[97] Wu Y B, Gao S, Zhang H, et al, Yuan H. Timing of UHP metamorphism in the Hong'an area, western Dabie Mountains, China: evidence from zircon U-Pb age, trace element and Hf isotope composition. Contrib Mineral Petrol, 2008, 155: 123-133

[98] Zheng Y F, Gao T S, Wu Y B, et al. Fluid flow during exhumation of deeply subducted continental crust: zircon U-Pb age and O-isotope studies of a quartz vein within ultrahigh-pressure eclogite. J Metamorphic Geol, 2007, 25: 267-283

[99] 李秋立, 李曙光, 侯振辉, 等. 青龙山榴辉岩高压变质新生锆石SHRIMP U-Pb定年、微量元素及矿物包裹体研究. 科学通报, 2004, 49 (22): 2329-2334

[100] Liu F L, Xu Z, Liou J G, et al. SHRIMP U-Pb ages of ultrahigh-pressure and retrograde metamorphism of gneisses, south-western Sulu terrane, eastern China. J Metamorphic Geol, 2004, 22: 315-326

[101] Li Q L, Li S G, Zheng Y F, et al. A high precision U-Pb age of metamorphic rutile in coesite-bearing eclogite from the Dabie Mountains in central China: a new constraint on the cooling history. Chem Geol, 2003, 200: 255-265

[102] Li S G, Wang S S, Chen Y Z, et al. Excess argon in phengite from eclogite: evidence from dating of eclogite minerals by Sm-Nd, Rb-Sr and $^{40}Ar/^{39}Ar$ methods. Chem Geol, 1994, 112: 343-350

[103] Yang J S, Wooden J L, Wu C L, et al. SHRIMP U-Pb dating of coesite-bearing zircon from the ultrahigh-pressure metamorphic rocks, Sulu terrane, east China. J Metamorphic Geol, 2003, 21: 551-560

[104] Li J Y, Yang T N, Chen W, et al. $^{40}Ar/^{39}Ar$ dating of deformation events and reconstruction of exhumation of ultrahigh-pressure metamorphic rocks in Donghai, East China: Acta Geol Sin, 2003, 77: 155-168

[105] Dong S W, Gao R, Cong B L, et al. Crustal structure of the Southern Dabie ultrahigh-pressure orogen and Yangtze foreland from deep seismic reflection profiling. Terra Nova, 2004, 16: 319-324

[106] 董峰, 李曙光, 李秋立, 等. 中国大陆科学深钻100~2000 m 超高压变质岩普通Pb同位素地球化学初步研究——俯冲陆壳内脱耦的证据. 岩石学报, 2006, 22: 1791-1798

[107] Xiao Y L, Zhang Z M, Hoefs J, et al. Ultrahigh-pressure metamorphic rocks from the Chinese Continental Scientific Drilling Project: II. Oxygen isotope and fluid inclusion distributions through vertical sections. Contrib Mineral Petrol, 2006, 152: 443-458

[108] Chen R X, Zheng Y F, Gong B, et al. Oxygen isotope geochemistry of ultrahigh-pressure metamorphic rocks from 200~4000 m core samples of the Chinese Continental Scientific Drilling. Chem Geol, 2007, 242: 51-75

[109] Meissner R, Mooney W. Weakness of the lower continental crust: a condition for delamination, uplift, and escape. Tectonophysics, 1998, 296: 47-60

[110] Oxburg E R. Flake tectonics and continental collision. Nature, 1972, 239: 202-204

[111] 许靖华. 薄壳板块构造模式与冲撞型造山运动. 中国科学A辑: 数学, 1980, 23: 1081-1089

[112] Gerya T V, Yuen D A, Maresh W V. Thermomechanical modeling of slab detachment. Earth Planet Sci Lett, 2004, 226: 101-116

Common Pb of UHP metamorphic rocks from the CCSD project (100–5000 m) suggesting decoupling between the slices within subducting continental crust and multiple thin slab exhumation*

Shuguang Li[1], Chenxiang Wang[1], Feng Dong[1], Zhenhui Hou[1], Qiuli Li[2], Yican Liu[1], Fang Huang[3] and Fukun Chen[2]

1. CAS Key Laboratory of Crust-Mantle Materials and Environments, School of Earth and Space Sciences, University of Science and Technology of China, Hefei 230026, Anhui, China
2. Institute of Geology and Geophysics, Chinese Academy of Sciences, Beijing 100029, China.
3. Department of Geology, University of Illinois at Urbana-Champaign, 245 NHB, 1301 W. Green St., IL 61801, USA

亮点介绍：通过对苏鲁超高压变质带大陆超深钻 100 m 到 5000 m 的样品中榴辉岩的绿辉石和片麻岩中的长石进行了普通 Pb 同位素系统研究；发现三个不同深度单元岩石具有不同的 Pb 同位素特征，分别显示了上/中/下俯冲陆壳的 Pb 同位素特点；在这三个岩石单元之间的韧性剪切带，是这三个超高压变质岩片折返时的拆分断裂带，其岩石具有 Pb 同位素变化大且氧同位素低特点，可能为经历过大气降水渗透的早期断层带。

Abstract In order to understand the vertical structure of the Dabie-Sulu ultrahigh-pressure metamorphic (UHPM) belt, common Pb isotopic compositions of omphacites in eclogites and feldspars in gneisses from the Chinese Continental Scientific Drilling (CCSD) project (100–5000 m) have been investigated in this study. Samples from 0 to 800 m (unit 1) in the drilling core have moderately high radiogenic Pb isotopes with small variations of $^{206}Pb/^{204}Pb$ (16.82–17.38), $^{207}Pb/^{204}Pb$ (15.37–15.49), and $^{208}Pb/^{204}Pb$ (37.21–37.72), indicating either high μ ($^{238}U/^{204}Pb$) or high initial Pb isotope ratios of their protoliths. In contrast, the samples from 1600 to 2040 m (unit 3) and most of samples from 3200 to 5000 m (unit 5) have moderately or very unradiogenic Pb (unit 3: $^{206}Pb/^{204}Pb$ from 16.05 to 16.46, $^{207}Pb/^{204}Pb$ from 15.22 to 15.29, and $^{208}Pb/^{204}Pb$ from 36.68 to 37.48; unit 5: $^{206}Pb/^{204}Pb$ from 15.52 to 15.69, $^{207}Pb/^{204}Pb$ from 15.15 to 15.27, and $^{208}Pb/^{204}Pb$ from 36.48 to 37.20), indicating either low μ or low initial Pb isotope ratios of their protoliths. Pb isotopes of samples from 800 to 1600 m (unit 2) and from 2040 to 3200 m (unit 4) in the drilling core with abundant ductile shear zones are intermediate between those of units 1 and 3 or 5 and display larger variations. Pb isotopes combined with the published oxygen isotope data of the CCSD samples reveal the original positions of the five units before the Triassic continental subduction. Units 1, 3, and 5 as three UHPM rock slabs could be derived from the subducted upper continental crust, upper-middle continental crust and lower-middle continental crust, respectively. The ductile shearing zones in units 2 and 4 could be the interfaces where the detachment and decoupling took place between the upper, upper-middle and lower-middle continental crusts. The detachment between the upper slab and subducting continental lithosphere probably occurred during continental subduction, and the upper slab (unit 1) was uplifted to a shallow depth along the detachment surface by thrusting. Units 3 and 5 may be detached later from the subducted middle and lower crust and uplifted to a shallow level underneath unit 1. The low $\delta^{18}O$ values (−4.0‰ to −7.4‰) [Xiao, Y.-L., Zhang, Z.-M., Hoefs, J., Kerkhof, A., 2006. Ultrahigh-pressure Metamorphic Rocks from the Chinese Continental Drilling Project-II Oxygen Isotope and Fluid Inclusion Distributions through Vertical Sections. Contribution Mineral Petrology 152, 443–458.; Zhang, Z.-M., Xiao, Y.-L., Zhao, X.-D., Shi, C., 2006. Fluid-rock interaction during the continental deep subduction: oxygen isotopic profile of the main hole of the CCSD

---

*本文发表在: Tectonophysics, 2009, 475: 308-317

project. Acta Petrologica Sinica 22 (7), 1941–1951.] in units 2 and 4 suggest that the detachment interfaces could be developed along an ancient fault zones which were the channels of meteoric water activity during the Neoproterozoic.

Key words Chinese Continental Scientific Drilling (CCSD) project; UHP metamorphic rocks; common Pb isotopes; Dabie-Sulu; detachment within continental crust

1 Introduction

The exhumation of ultrahigh-pressure metamorphic (UHPM) rocks from a depth of >100 km is a hot topic in the geological community. Two mechanisms of exhumation of the UHPM rocks have been proposed. Namely, the bulk subducted continental crust was detached from the underlying lithosphere mantle and then exhumed (e.g., Chemenda et al., 1995; Hacker et al., 2000; Massonne, 2005); or several HP-UHP metamorphic crustal slices are produced by decoupling between the crust slices on different depth levels within subducted continental crust and the multiple UHPM rock slices are successively exhumed (Li et al., 2003a, 2005b; Xu et al., 2006a; Liu et al., 2007).

The multi-slice exhumation model is mainly based on studies of the Dabie-Sulu UHPM belt in east-central China. The Dabie-Sulu UHPM belt was formed by continental collision between the South China Block and North China Block in the Triassic (e.g. Li et al., 1993, 1994; Rowley et al., 1997; Hacker et al., 1998; Li et al., 2000; Ayers et al., 2002; Liu et al., 2004, 2005a, b; Li et al., 2005a; Liu et al., 2007), which is the largest known UHPM belt on Earth. It is composed of the Dabie UHPM terrane in the west and Sulu UHPM terrane in the east displaced ~ 500 km to the north by the Tan-Lu fault (Fig. 1) (Li et al., 1993). Both the Dabie and Sulu UHPM terranes can be subdivided in to several HP or UHP sub-zones based on their differences in lithology, geochemistry and geochronology, which provides an opportunity to study different exhumation processes for different tectonic units. For the Dabie UHPM terrane, the Northern Dabie zone (NDZ) is different with the Southern Dabie zone (SDZ) in Pb isotopes (Zhang et al., 2001; Li et al., 2003a, 2005b), fluid inclusions (Xiao et al., 2001, 2002), metamorphic history and ages (Liu et al., 2007), supporting the multi-slice exhumation model. The multi-slice exhumation model for the Sulu UHPM terrane is also supported by the observations of the metamorphism, structure and ages for the different HP and UHP metamorphic zones (Xu et al., 2006a). However, it is not clear whether the individual UHP metamorphic zone on the surface, e.g. the SDZ in the Dabie terrane or the UHP metamorphic zone in the Sulu terrane exhumed as a single slice or multiple slices.

Uranium is a large ion lithophile element, which is more incompatible than Pb in the crustal rock-metamorphic fluid system (Kogiso et al., 1997). The lower continental crust (LCC) is depleted in U due to relatively high-grade metamorphism and thus has relatively low μ ($^{238}U/^{204}Pb$) value, while the upper continental crust (UCC) is enriched in U and has high μ value. Therefore, after long time of accumulation of radiogenic daughter isotopes (Pb), the continental crust is characterized as enrichment of radiogenic Pb in the UCC and unradiogenic Pb in the LCC (Zartman and Doe, 1981). Accordingly, Pb isotopes can be used as a tracer for the original position of rocks in the continental crust. S.-G. Li et al. (2003a) and Zhang et al. (2001) reported the whole-rock U-Pb isotopic compositions for samples from the Dabie region, showing that the Northern Dabie gneisses have the LCC-like Pb isotopes while the Southern Dabie gneisses and eclogites have the UCC-like Pb isotopes. Given the differences in Pb isotopes and lithology and metamorphic history between the NDZ and SDZ, S.-G. Li et al. (2003a, 2005b) and Liu et al. (2007) proposed that the subducted UCC was decoupled from the LCC during the continental subduction along a major thrust fault, by which the deeply subducted UCC was uplifted. Although Pb isotopes are useful means to distinguish nature of crust slices, they have not yet been investigated on the Su-Lu UHPM terrane.

Samples of previous studies on whole-rock U-Pb isotopic compositions were collected on the earth's surface (Zhang et al., 2001; Li et al., 2003a). However, because the surface samples might have experienced significant weathering and thus U/Pb fractionation, it is questionable whether the whole-rock U/Pb represents the U/Pb of the initial rocks without weathering. The measured U/Pb may not be thus suitable for accurate calculation of the initial Pb isotopic compositions of the whole-rock samples. In order to avoid this problem, one can determine the common Pb isotopes of low U/Pb minerals, which reflect the initial Pb isotopic compositions of the whole rock. Furthermore, previous studies mainly focused on surface samples that cannot provide information on the vertical spatial variation of the UHPM rocks. The Chinese Continental Scientific Drilling (CCSD) project provided an excellent opportunity to continuously sample the UHPM rocks from sub-surface. Here, we report common Pb isotopic compositions of omphacites and feldspars in eclogites and gneisses from the CCSD project. The aims of this paper are: (1) to study the vertical structure of the UHPM zone, (2) to test the decoupling hypotheses within subducting continental crust, and (3) to provide new constraints on the mechanisms of the continental crust subduction and exhumation of the UHPM rocks.

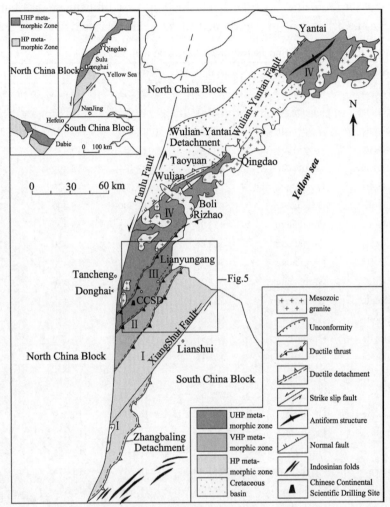

Fig. 1 Tectonic sketch map of the Sulu high-pressure and ultrahigh-pressure (HP-UHP) metamorphic belt, showing: (1) high-pressure zone I, very high-pressure (VHP) zone II, and UHP zones (III and VI) separated by ductile shear zones (modified after Xu et al., 2006a).

2 Samples and geologic background

The Sulu UHP terrane has been subdivided into 4 sub-zones from south to north, namely, the southern high-pressure zone (I), the central very high-pressure zone (II), and the northern ultrahigh-pressure (UHP) zone (III and IV) (Fig. 1) (Xu et al., 2006a). The drill site of the CCSD project is located in the UHP zone III of the southern segment of the Sulu UHP terrane, near Maobei village (N34°25′, E118°40′), about 17 km southwest of Donghai county, Jiangsu Province (Fig. 1). In this region, the Qinglongshan eclogites are well known for the first observation of significantly excess argon (Li et al., 1994) and extremely low $\delta^{18}O$ (Yui et al., 1995; Zheng et al., 1996; Rumble and Yui, 1998). The main drilling hole of the CCSD project has reached 5148 m in depth. Fifty-four samples were collected from the main drilling hole (100–5000 m in depth) of the CCSD project, which is located above the Maobei eclogite complex in the southwestern of the Maobei shearing tectonic unit. Sample names and depths are listed in Table 1.

The drilling core is mainly composed of eclogites and gneisses. The overall thickness of eclogite is ~ 1200 m (Fig. 2). Raman spectroscopy shows that coesite inclusions occur in zircons from all types of rocks in the drilling core except the ultramafic rocks (Liu et al., 2001, 2004). This indicates that most rocks in this drilling core experienced UHP metamorphism. According to the proportion of minor minerals, the eclogites can be divided into quartz eclogite (i.e. high Si eclogite), rutile eclogite (i.e. high Ti and high Ti-Fe eclogite), phengite eclogite (i.e. high Al eclogite), and common eclogite (i.e. high-Mg eclogite and normal eclogite). All eclogites have experienced variable extent of retrograde metamorphism (Zhang et al., 2004). Granitic gneiss or paragneiss is the other rock type in the main drilling hole of the CCSD project from 100 to 5000 m. Most of them are distributed between the range of 1113.14 m to 1596.22 m and below 2050 m (Fig. 2), and others are inter-layered with eclogite layers.

Table 1 Common Pb isotope compositions of omphacites from eclogite and feldspars from gneiss of the CCSD

Sample	Lithology	Mineral	Depth (m)	$^{206}Pb/^{204}Pb$	$^{207}Pb/^{204}Pb$	$^{208}Pb/^{204}Pb$	$^{208}Pb/^{206}Pb$
*B4R7P2g	Eclogite	Omphacite	108.7	17.211	15.428	37.488	2.178
B79R74P1e	Eclogite	Omphacite	235.18	16.979	15.392	37.403	2.203
*B134R166P1b	Eclogite	Omphacite	323.35	17.233	15.451	37.65	2.185
B149R127P7	Eclogite	Omphacite	347.42	17.327	15.418	37.702	2.176
B154R132P1d	Eclogite	Omphacite	354.90	17.184	15.399	37.563	2.186
*B199R176P6i	Eclogite	Omphacite	433	17.295	15.438	37.65	2.177
B200R177P1l	Eclogite	Omphacite	434.85	17.300	15.407	37.717	2.180
B211R187P2b	Eclogite	Omphacite	452.90	17.343	15.428	37.651	2.171
B218R192P1f	Eclogite	Omphacite	464.65	17.299	15.420	37.602	2.174
B247R210P2u	Eclogite	Omphacite	507.15	17.353	15.420	37.692	2.170
B250R212P1Ca	Eclogite	Omphacite	511.20	17.248	15.396	37.595	2.180
*B261R222P1a	Eclogite	Omphacite	531.50	17.259	15.448	37.663	2.182
B267R224P1f	Eclogite	Omphacite	540.00	17.378	15.486	37.561	2.161
B269R225P4m	Eclogite	Omphacite	543.55	17.363	15.472	37.556	2.163
*B287R235P1a	Eclogite	Omphacite	572.15	16.815	15.372	37.208	2.213
B355R284P1a	Eclogite	Omphacite	687.70	17.155	15.412	37.489	2.185
*B380R298P3s	Eclogite	Omphacite	728.00	17.121	15.415	37.476	2.188
*B552R399P11	Eclogite	Omphacite	1003.40	16.274	15.369	36.730	2.257
*B578R417P2a	Paragneiss	Plagioclase	1050.25	16.726	15.377	37.28	2.229
*B593R426P1aL	Eclogite	Omphacite	1074.25	16.603	15.355	37.075	2.233
*B655R462P1C	Paragneiss	Orthoclase	1183.40	16.265	15.274	37.091	2.280
*B695R483P6f	Paragneiss	Plagioclase	1262.08	16.670	15.323	37.395	2.243
*B837R572P4C	Orthogneiss	Plagioclase	1615.40	16.362	15.223	37.368	2.284
*B901R599P9b	Paragneiss	K-feldspar	1731.65	16.467	15.253	37.484	2.276
*B926R613P12n	Eclogite	Omphacite	1777.70	16.227	15.273	36.92	2.275
*B992R636P34r	Eclogite	Omphacite	1901.06	16.051	15.223	36.682	2.285
*B1018R64P33a	Eclogite	Omphacite	1939.95	16.293	15.288	37.010	2.272
*B1033R645P6a	Eclogite	Omphacite	1964.36	16.122	15.241	36.828	2.284
*B1046R647P58c	Eclogite	Omphacite	1985.97	16.132	15.237	36.780	2.280
B1100R35P19c	Orthogneiss	K-feldspar	2115.97	16.040	15.287	37.110	2.314
B1209R82P2	Orthogneiss	K-feldspar	2411.44	16.399	15.344	37.433	2.283
B1320R113P77m	Orthogneiss	K-feldspar	2641.62	16.537	15.296	37.718	2.281
B1502R151P21a	Paragneiss	K-feldspar	2958.75	16.521	15.337	37.189	2.251
B1579R14P19h	Orthogneiss	K-feldspar	3093.33	16.201	15.210	37.156	2.293
B1631R33P54u	Orthogneiss	K-feldspar	3260.12	15.812	15.165	36.588	2.314
B1692R49P7g	Orthogneiss	K-feldspar	3391.96	15.960	15.272	37.162	2.329
B1742R61P10a	Orthogneiss	K-feldspar	3481.96	16.210	15.232	37.526	2.315
B1803R79P1d	Orthogneiss	K-feldspar	3610.92	15.803	15.218	37.058	2.345
B1887R15P4c	Orthogneiss	K-feldspar	3730.64	15.519	15.149	36.484	2.351

Sample	Lithology	Mineral	Depth (m)	$^{206}Pb/^{204}Pb$	$^{206}Pb/^{204}Pb$	$^{207}Pb/^{204}Pb$	$^{208}Pb/^{204}Pb$
R1936R30P1d	Orthogneiss	K-feldspar	3830.08	15.644	15.218	36.812	2.353
R1975R41P35d	Orthogneiss	K-feldspar	3901.21	15.768	15.195	36.908	2.341
B2031R53P30r	Orthogneiss	K-feldspar	4001.83	15.777	15.248	37.042	2.348
B2084R65P12d	Orthogneiss	K-feldspar	4100.46	15.722	15.242	37.026	2.355
B2130R75P4d	Paragneiss	K-feldspar	4176.23	15.629	15.170	36.543	2.338
B2169R85P13a	Paragneiss	K-feldspar	4251.51	15.836	15.234	36.912	2.331
B2278R108P1bA	Orthogneiss	K-feldspar	4447.06	16.049	15.209	37.687	2.348
B2291R111P13a	Orthogneiss	K-feldspar	4475.29	16.189	15.207	37.764	2.333
B2388R132P1cA	Orthogneiss	K-feldspar	4644.03	15.847	15.224	37.042	2.337
B2438R142P1uA	Orthogneiss	K-feldspar	4730.31	15.886	15.237	37.119	2.337
B2477R148P23a	Orthogneiss	K-feldspar	4782.41	15.804	15.222	37.039	2.344
B2506R154P411	Orthogneiss	K-feldspar	4839.96	15.777	15.221	37.009	2.346
B2543R162P1h	Orthogneiss	K-feldspar	4898.99	15.924	15.262	37.204	2.336
B2608R179P21	Orthogneiss	K-feldspar	5028.36	15.73808	15.235	37.033	2.353

*The data are from Dong et al. (2007).

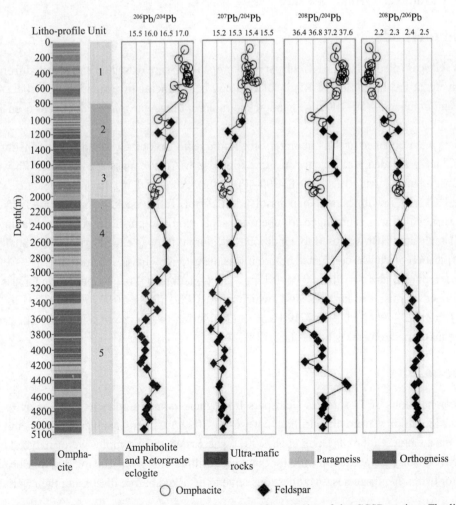

Fig. 2 Common Pb profiles of omphacite from eclogite and feldspar from gneiss of the CCSD project. The lithological profile is modified after Zhang et al. (2004).

Division of the tectonic and petrologic units in the 100–5000 m drilling core from the CCSD main hole has been studied by numerous authors (e.g., Xu et al., 2004; You et al., 2004; Xu et al., 2006b). Based on lithology, structure and Pb isotopic compositions presented in this paper, we divide the 100–5000 m drilling core into five petrologic units: (1) Unit 1 above 800 m is composed of rutile eclogites with a few layers of ultramafic rocks, gneisses, and amphibolites; (2) Unit 2 from 800 to 1600 m in the drilling core is mainly composed of paragneiss and granitic gneiss with miner ultramafic rocks and eclogites. A series of ductile shearing zone are developed in the depth of 738–1113 m, where mylonitic gneiss and mylonite occur as layers (Chen et al., 2004; Xu et al., 2004). Geophysical studies reveal a strong reflection interface at 1600 m (You et al., 2004), which is a boundary between the overlying mylonitic gneiss and underlying phengite eclogite layer (Zhang et al., 2004); (3) Unit 3 from 1600 m to 2040 m in the drilling core is mainly composed of phengite eclogite with retrograde rutile eclogite and gneiss (Xu et al., 2006b); (4) Unit 4 from 2040 m to 3200 m in the drilling core is mainly composed of paragneiss and granitic gneiss with minor eclogites. Mylonite and mylonitic gneiss occur as layers in the depth of 2650–3090 m suggesting a ductile shearing zone (Xu et al., 2006b); and (5) Unit 5 from 3200 m to 5000 m in the drilling core is mainly composed of granitic gneiss with minor eclogites (Xu et al., 2006b). All these rock units and their intervening shear zones have SE-dipping foliation and SE-plunging stretching lineation (Xu et al., 2006b). However, slight differences in striking and dipping directions between the petrologic units have been recognized. For example, systematic determination of tectonic foliations and lineations indicate different strike-dips between units 1 and 3. Rocks above 1600 m have a lineation striking of ESE (100°E) and dipping of 30°–40°, while rocks below 1600 m have a lineation striking SSE (160°E) and dipping over 50° (You et al., 2004). This indicates that units 1 and 3 are not coherent and belong to different UHPM rock slices.

3 Analytical methods

About 2 g of feldspar (plagioclase or K-feldspar) and omphacite were extracted and selected from gneisses and eclogites, respectively, at the Institution of Regional Geology and Mineral Investigation of Hebei. Alteration-free feldspar (40 mg) and omphacite (50–100 mg) handpicked under a binocular microscope were used for Pb isotope analysis.

Pb isotope data were obtained at the Laboratory for Radiogenic Isotope Geochemistry of Institute of Geology and Geophysics, Chinese Academy of Science, using a Finnigan MAT-262 mass spectrometer. Pb was purified by conventional anion-exchange method (AG1-X8, 200-400 resin). About 50 mg mineral separates were rinsed in purified water and HCl for a few times, and then dissolved in a 7 mL Teflon beaker using purified HNO_3 and HF. Pb was extracted by HBr, which was centrifuged for chemical separation and purification. After cleaning the AG1-X8 column using HCl and water, HBr was used to condition the column. Pb was eluted by HBr and finally collected by HCl. A second purification of Pb was conducted before it was measured by mass spectrometry.

The whole procedure blank for Pb is 0.05–0.1 ng. Fractionation of Pb isotopes during mass spectrometer analysis was calibrated against standard NBS981, which give $^{206}Pb/^{204}Pb$ = 16.9376 ± 0.0015 (2σ), $^{207}Pb/^{204}Pb$ = 15.4939 ± 0.0014 (2σ), and $^{208}Pb/^{204}Pb$ = 36.7219 ± 0.0033 (2σ) during the course of this study. The precision for Pb isotope data on the mass spectrometer is better than 0.1%.

4 Pb isotope results

Pb isotopic compositions of 54 samples (feldspars from gneisses and omphacite from eclogites) are listed in Table 1, in which 18 samples from the drilling core (100–2000 m) were analyzed earlier and published as preliminary results in a Chinese journal (Dong et al., 2007) whereas other 36 samples were analyzed in this study to obtained a complete Pb isotopic variation profile for the whole drilling core from 100 to 5000 m. Because Pb isotope data are obtained for two mineral species, two questions have to be asked before discussing their variations.

4.1 Do omphacites from eclogites have low U/Pb?

Feldspar has been widely used to study the common Pb of igneous or metamorphic rocks because of its low U/Pb ratio (e.g., Zhang, 1995). Recent studies show that omphacite from eclogite is also a mineral with low U/Pb. Experiment of partial melting of eclogite shows the greater partition coefficient of Pb than U between clinopyroxene and melt (Klemme et al., 2002). Q.-L. Li et al. (2003b) observed the consistency between the U-Pb mineral (rutile + omphacite) isochron age and conventional rutile U-Pb concordia age obtained by common Pb correction based on the Pb isotopic composition of omphacite in the same eclogite sample. This suggests that omphacite with low U/Pb ratio ($\mu=2.8$) can be used for common Pb correction in U-Pb dating of rutile. Therefore, omphacite can be directly used for the study of common Pb, representing the initial Pb isotopes of the whole rock. In this study, we use both feldspar and omphacite to study the initial Pb isotopic ratios and no age correction was performed.

4.2 Are the geochemical implications of the Pb isotopes of feldspar and omphacite comparable?

The feldspar and omphacite mineral separates are derived from gneiss and eclogite, respectively. The protoliths of the eclogites and gneisses from the Dabie-Sulu UHPM zones are the Neoproterozoic mafic and intermediate-felsic rocks, respectively (Zheng et al., 2003; Zhang et al., 2004; Zhao et al., 2005). Although crustal contribution of the latter is greater than the former, the whole-rock Pb isotopic compositions of the eclogites and gneisses from the southern Dabie zone are not different (Zhang et al., 2001; Li et al., 2003a). Results of this study also indicate that feldspars and omphacites from drilling core samples from 800 to 2000 m show similar trend in $^{206}Pb/^{204}Pb$ and $^{207}Pb/^{204}Pb$ variations (Fig. 2). For instance, there are large amount of gneisses in unit 2 (800–1600 m) drilling core, but two eclogites (B552R399P11 and B593R426P1aL) from the same portion of the drilling core have Pb isotopic compositions consistent with those of the gneisses, both of which show decreasing Pb isotopic ratios with increasing depth. Moreover, comparison of gneisses from unit 3 (1600–2040 m) portion (B837R572P4C and B901R599P9b) with inter-layered eclogites also shows generally similar U-Pb isotopic compositions except for different $^{208}Pb/^{204}Pb$ ratios. Notably, feldspars from 1600–2040 m drilling core have higher $^{208}Pb/^{204}Pb$ than omphacite (Fig. 1), which could be due to the protolith of the gneisses having greater contribution of the LCC. Therefore, we combine the common Pb of feldspars and omphacites to show Pb isotopic variations with increasing depth and discuss implications of the difference in Pb isotopic compositions between the three units (1, 3, and 5) of the drilling core.

The decreasing of $^{206}Pb/^{204}Pb$ and $^{207}Pb/^{204}Pb$ ratios and increasing of $^{208}Pb/^{206}Pb$ ratios with increasing of the depth are striking features of the Pb isotopic profiles of the CCSD samples (Fig. 2). As shown in Fig. 2, all samples from unit 1 (100–800 m) are omphacite-bearing, while samples from units 2 and 3 contain both omphacite and feldspar, and minerals from units 4 and 5 are feldspar. Pb isotopic composition of samples from unit 1 (100 to 800 m) shows limited variation with $^{206}Pb/^{204}Pb$ ranging from 16.82 to 17.38 with average of 17.23, $^{207}Pb/^{204}Pb$ from 15.37 to 15.49 with average of 15.42, and $^{208}Pb/^{204}Pb$ from 37.21 to 37.72 with average of 37.57. Overall they all have relatively high radiogenic Pb isotopes. Except for $^{208}Pb/^{204}Pb$ ratios, samples from unit 3 (1600 to 2040 m) also have uniform but moderately low radiogenic Pb isotopes with $^{206}Pb/^{204}Pb$ ranging from 16.05 to 16.47 with average of 16.24 and $^{207}Pb/^{204}Pb$ from 15.22 to 15.29 with average of 15.25. As mentioned above, two feldspars from 1600–2040 m drilling core have significantly higher $^{208}Pb/^{204}Pb$ than omphacites, and omphacite samples from unit 3 also show uniform low $^{208}Pb/^{204}Pb$ from 36.68 to 37.01 with average of 36.86. Except for three samples (B1742R61P10a, B2278R108P1bA, and B2291R111P13a), the other sixteen feldspar separates from unit 5 show uniform and very unradiogenic Pb isotopes with $^{206}Pb/^{204}Pb$ ranging from 15.52 to 15.96 (average: 15.78), $^{207}Pb/^{204}Pb$ from 15.15 to 15.27 (average: 15.22) and $^{208}Pb/^{204}Pb$ from 36.48 to 37.20 (average: 36.94). Although the $^{208}Pb/^{204}Pb$ ratios of unit 5 show a large variation ranging from 36.48 to 37.76, they have uniform and high $^{208}Pb/^{206}Pb$ ratios from 2.42 to 2.50, reflecting the uniform and high Th/U ratios in protolith of unit 5. In contrast, unit 1 and unit 2 have low and moderately low $^{208}Pb/^{206}Pb$ ratios, respectively, reflecting their relative lower Th/U ratios in their protoliths. The different Pb isotopic compositions of units 1, 3, and 5 clearly indicate that they were from different depths in the subducted continental crust.

Moreover, the Pb isotopic compositions of samples from unit 2 (800–1600 m) and unit 4 (2040–3200 m) of the drilling core also show transitional trends between units 1 and 3 and between units 1 and 5, respectively (Fig. 2). For unit 2, the $^{206}Pb/^{204}Pb$ varies from 16.27 to 16.73 (average: 16.51), $^{207}Pb/^{204}Pb$ from 15.27 to 15.38 (average: 15.34), and $^{208}Pb/^{204}Pb$ from 36.73 to 37.40 (average: 37.11), which are in between units 1 and 3. While for unit 4, the

^{206}Pb/^{204}Pb varies from 16.04 to 16.54 (average: 16.34), ^{207}Pb/^{204}Pb from 15.21 to 15.34 (average: 15.29), and ^{208}Pb/^{204}Pb from 37.11 to 37.72 (average: 37.32), which are higher than those of unit 3 and in between units 1 and 5.

5 Discussion

5.1 Pb isotopic constraints on original positions of the UHP units in the subducted continental crust

In the ^{207}Pb/^{204}Pb$_i$ vs. ^{206}Pb/^{204}Pb$_i$ diagram (Fig. 3A), the drilling core samples from units 1 and 5 clearly have different Pb isotope compositions with unit 1 samples having moderately high radiogenic Pb and unit 5 samples having unradiogenic Pb. Samples from units 2, 3, and 4 have intermediate ^{207}Pb/^{204}Pb and ^{206}Pb/^{204}Pb ratios. While ^{206}Pb/^{204}Pb and ^{207}Pb/^{204}Pb ratios of the drilling core samples of this study are significantly lower than the values of UHP paragneiss and eclogite nodule in marble from the Dabie orogen (Zhang et al., 2001; Li et al., 2003a) as well as the MORB and EMII (Zindler and Hart, 1986), they are generally similar to the values of eclogites and UHP orthogneiss (Zhang et al., 2001; Li et al., 2003a) as well as the post-collisional mafic-ultramafic intrusive (PCMI) rocks (Wang et al., 2005; Huang et al., 2007) from the Dabie orogen (Fig. 3A). Radiogenic Pb isotopes of the samples from unit 1 are in good agreement with omphacites from the Dabie eclogites (Fig. 3A) which were considered to be derived from subducted upper continental crust (Li et al., 2003a; Liu et al., 2007). In contrast, the samples from unit 5 have unradiogenic Pb isotopes and most of them are obviously lower than the values of the post-collisional mafic-ultramafic intrusive rocks from the Dabie orogen from a mantle source intensely effected by the recycled deeply subducted mafic lower crust from the South China Block (Huang et al., 2007), and close to the ancient "lower mafic continental crust" in the Dabie orogen (Huang et al., 2008) (Fig. 3A).

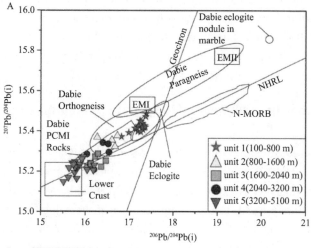

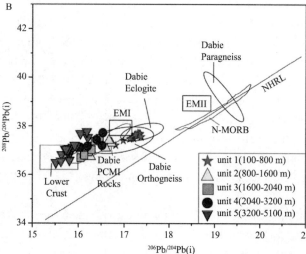

Fig. 3 ^{207}Pb/^{204}Pb-^{206}Pb/^{204}Pb and ^{208}Pb/^{204}Pb-^{206}Pb/^{204}Pb diagrams showing the comparison between common Pb of omphacite from eclogite and feldspar from gneiss of the CCSD project and initial Pb isotopic compositions (t = 230 Ma) of the rocks from the Dabie orogen. Data source: lower continental crust (LCC) are from the unradiogenic Pb endmember of the Mesozoic granites and high-Mg adakites from the Dabie orogen (Huang et al., 2008); N-MORB, EMI and EMII are from Zindler and Hart (1986); the post-collisional mafic-ultramafic intrusive (PCMI) rocks from Dabie orogen are from Huang et al. (2007) and Wang et al. (2005); eclogite, orthogneiss, paragneiss and eclogite nodule in marble from the Dabie orogen are from Zhang et al. (2001), S.-G. Li et al. (2003a) and Li et al. (unpublished data). Because of lack of Th data for some whole-rock samples, the fields of paragneiss and eclogite nodule in marble from the Dabie orogen in ^{208}Pb/^{204}Pb-^{206}Pb/^{204}Pb diagram are smaller than that in ^{207}Pb/^{204}Pb-^{206}Pb/^{204}Pb diagram or absent, respectively.

In addition, samples from the CCSD project (100–5000 m) show three trends in the ^{208}Pb/^{204}Pb vs. ^{206}Pb/^{204}Pb figure (Fig. 3B). Samples from unit 5 are arranged in left trend showing higher ^{208}Pb/^{204}Pb than other unit samples given a similar ^{206}Pb/^{204}Pb; samples from units 2, 3 and 4 are in middle trend, in good agreement with the post-collisional mafic-ultramafic intrusive rocks from the Dabie orogen; and samples from unit 1 are arranged in right trend with relative lower ^{208}Pb/^{204}Pb ratios falling out of the area of the post-collisional mafic-ultramafic intrusive rocks from the Dabie orogen, but overlapping eclogite samples from the SDZ in the

Dabie orogen (Fig. 3B). The difference in $^{208}Pb/^{204}Pb$ ratios between the rock units of the CCSD project (100–5000 m) suggests that the protolith from unit 5 is characterized by low μ and high Th/U, which is a typical feature of the lower continental crust (LCC), while the protolith of samples from unit 1 is characterized by high μ and low Th/U, a typical feature of the upper continental crust (UCC) (Zartman and Doe, 1981). The consistence in Pb isotopes between unit 1 of the CCSD drilling core and the eclogites from the SDZ also suggests that unit 1 is derived from the subducted upper continental crust, because the SDZ in the Dabie eclogites are considered to be derived from subducted upper continental crust (Li et al., 2003a; Liu et al., 2007).

The above discussions are based on the assumption that the protolith of the eclogites and granitic gneisses from the CCSD drilling core had similar initial Pb isotopic compositions. If so, the observed systematic difference in Pb isotopes should be due to the difference in the μ values and Th/U ratios of the protoliths and reflect their different depth before continental subduction. The radiogenic Pb isotopes of samples from unit 1 may reflect the high μ and low Th/U of the protolith in the UCC, while the unradiogenic Pb isotopes of samples from unit 5 may reflect the low μ and high Th/U of the protolith that originated from the LCC. However, the Pb isotopic compositions of the UHPM rocks depend on both initial Pb isotopic compositions and μ value of the protolith. If the protolith of the samples from unit 5 originally had low $^{206}Pb/^{204}Pb$ and high $^{208}Pb/^{204}Pb$, the observed difference in Pb isotopic compositions between units could be due to the Pb isotopic heterogeneity in their protoliths with similar μ values. Therefore, we need to consider oxygen isotope geochemistry to give more constraints on the reasons for the isotopic difference between the upper and lower units of the drilling core.

5.2 Oxygen isotope constraints on original positions of the UHP units in subducted continental crust

Oxygen isotopic study indicates that samples from unit 1 (0–800 m) generally have low $\delta^{18}O$ (Xiao et al., 2006) (Fig. 4), suggesting that the protolith should be at shallow crustal level in the Neoproterozoic and interacted with meteoric water (Zheng et al., 2003). Samples from unit 2 and unit 4 have very low $\delta^{18}O$ values with the negative $\delta^{18}O$ anomaly occurring at 971–1003 m ($\delta^{18}O$= −4.8–−6.5), 2552–2699 m ($\delta^{18}O$=−7.4–−4.0), and 3053–3062 m ($\delta^{18}O$= −1.3–−3.1), respectively (Fig. 4) (Xiao et al., 2006; Zhang et al., 2006). Such low $\delta^{18}O$ values indicate strong water-rock interaction in the Neoproterozoic, suggesting that the 971–1003 m, 2552–2699 m, and 3053–3062 m intervals could be meteoric channels related to ancient fault zones. As mentioned above, several ductile shear zones are developed at those intervals with negative $\delta^{18}O$ anomaly in the drilling core. On the contrary, samples from unit 3 (1600–2040 m)

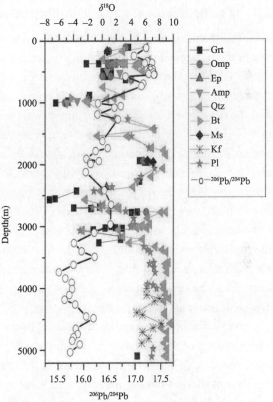

Fig. 4 Comparison between lead and oxygen isotopic compositions of the CCSD core (100–5000 m). Data source: lead isotopic data are from Table 1; oxygen isotopic data are from Zhang et al. (2006).

and unit 5 (3200–5000) have $\delta^{18}O$ of typical normal metamorphic rocks (≥5.6‰) and omphacites from eclogites having $\delta^{18}O$ of ~5.6‰ within the range of normal mantle peridotite (Fig. 4) (Xiao et al., 2006; Zhang et al., 2006). This indicates that the protoliths of units 3 and 5 could be located at relatively deep level avoiding significant water-rock interaction in the Neoproterozoic. However, the original protolith depth of unit 3 could not be too deep because of the existence of two low $\delta^{18}O$ zones at 2552–2699 m and 3053–3062 m of the drilling core (Fig. 4) (Xiao et al., 2006; Zhang et al., 2006). The repeating low $\delta^{18}O$ zones suggest that the meteoric water can enter the middle continental crust (MCC) level through a series of fault "channels". Previous studies show that meteoric water can enter the top of the MCC through a series of fault zones to react with rocks at depths of 10–15 km (Taylor, 1990). Therefore, the protolith of the samples from unit 3 could be at the upper-middle continental crust above 15 km and the protolith of the samples from unit 5 could be at a deeper level below 15 km because no low $\delta^{18}O$ zone has been observed below 3200 m in the drilling core. In summary, oxygen isotopic data suggests that the protoliths of the samples from units 1, 3, and 5 were originally at the UCC, upper-MCC and lower-MCC depths, respectively. Accordingly, the samples from unit 5 should have the lowest μ values, unit 1 has the highest μ values, and unit 3 has the intermediate. This is consistent with the implication from their Pb isotope compositions, i.e. the samples from the lower unit are more enriched in unradiogenic Pb isotopes than the samples from the upper unit (Fig. 3).

Based on the U-Pb isotopic composition of whole rocks from the Dabie orogen, S.-G. Li et al. (2003a) proposed that detachment and decoupling could occur between the UCC and LCC, and the detached upper subducted crust slice was exhumed firstly by thrust along the detached interface. In this study, we conclude that the rock units 1, 3 and 5 of the drilling core in the Sulu UHPM Zone (III) (Xu et al., 2006a) are derived from the UCC, upper-MCC and lower-MCC, respectively. This not only reinforces the suggestion by S.-G. Li et al. (2003a) but also proves the multi-slab decoupling during exhumation including decoupling between the subducted UCC and MCC as well as between the upper-MCC and the underlying lower-MCC.

5.3 The detachment and decoupling interfaces between the UHPM crust slices

The ductile shearing zones in units 2 and 4 of the drilling core from the CCSD project could be the interfaces where the detachment and decoupling between the UCC, upper-MCC and lower-MCC happened. Considering that the 738–1113 m ductile shear zone is characterized by deformation of eclogitic phases (Chen et al., 2004), such detachment within the subducted continental crust could happen during the subduction process. This study shows that the detachment interface occurring within the continental crust was a fault zone as a channel of meteoric water activity in the Neoproterozoic. The strong water-rock interaction resulted in the enrichment of H_2O in the fault zone and adjacent rocks and consequently a zone with low viscosity, which is critical for the later inner-crustal detachment and decoupling between the UCC, MCC, and LCC. Experiment models show that the UCC can be uplifted during subduction along the detachment surface (Chemenda et al., 1995). Modeling results of lithospheric viscosity-depth curves based on reasonable assumptions of geotherms and lithospheric composition suggest that there are at least two low-viscosity zones within continental crust at different depths (Meissner and Mooney, 1998). This study suggests that the realistic low-viscosity zones could be much more than the modeling results because of the existence of large fault zones in the crust. Accordingly, with increasing the depth of subduction, multiple slices in the subducting continental crust could be decoupled along the low-viscosity zones creating multi detachment surfaces within the continental crust (Meissner and Mooney, 1998).

As shown in Figs. 2 and 3, Pb isotopes of the samples from unit 2 show a transitional trend from unit 1 to unit 3. The Pb isotopic variations may indicate mixing between units 1 and 3 at the detachment interface and its adjacent area. Such big-scale ($n \times 100$ m) mixing requires the presence of fluids. As mentioned above, the extremely low $\delta^{18}O$ and enrichment of H_2O-bearing minerals in eclogites and gneisses in unit 2 suggest that it was a main channel for fluid activity in the Neoproterozoic (Xiao et al., 2006; Zhang et al., 2006). However, the detachment interface in the continental crust could be also the main channel during the subduction of the continental crust. Therefore, based on the Pb isotopic data of unit 2 only, it is not clear when the Pb isotopic mixing occurred in unit 2.

Figs. 2 and 3 also show that Pb isotopic ratios of the samples from unit 4 are slightly higher then those of unit 3, thus their Pb isotopes show a transitional trend from unit 1 to unit 5 but not from unit 3 to unit 5. This suggests that the Pb isotopic mixing could only occur due to the presence of meteoric water activity through fault zones in the Neoproterozoic, because movement of meteoric water along channels can carry Pb from the earth's surface to the depth. Metamorphic fluid activity in the detachment interface between the UHPM crust slices during the subduction of the continental crust may only cause the Pb isotopic mixing between the adjacent UHPM crust slices, i.e. the Pb isotopes of unit 4 should be produced by mixing of Pb from units 3 and 5, however, which is not observed.

In summary, the Pb isotopic mixing occurred in units 2 and 4 most probably occurred by meteoric water activity through fault zones in the Neoproterozoic. The metamorphic fluid activity in the detachment interface between the UHPM crust slices during the subduction of the continental crust may not be strong enough to significantly effect on the Pb isotopes of the ductile shearing zones.

5.4 Connection between the rock units in the drilling core and tectonic slices on the surface

Based on surface geology, Xu et al. (2006a) subdivided the UHP zone III into 4 tectonic slices from SE to NW, namely the Lianyungang (IIIa), Maobei (IIIb), Donghai (IIIc), and Shilianghe (IIId) slices, which are separated by ductile shear zones (DF6, DF7, and DF8, respectively) with abundant mylonites and mylonitic rocks (Fig. 5). These tectonic slices and their intervening shear zones have SE-dipping foliations and SE-plunging stretching lineations (Xu et al., 2006a). Because the main drilling hole (5000 m in depth) of the CCSD project is located above the Maobei tectonic slice (IIIb) (Fig. 5), and according to the three UHPM crust slices (i.e. units 1, 3 and 5) recognized from the drilling core (100–5000 m) in this study, it is reasonable to suggest that the main drilling hole of the CCSD project may pass through the Maobei (IIIb) (unit 1) and Donghai (IIIc) (unit 3) slices, and get into the Shilianghe (IIId) (unit 5) slice. Accordingly, units 2 and 4 with ductile shearing zones may connect to the DF7 and DF8 on the surface, respectively. This study also suggests that the UHP zone III in the Su-Lu terrane is an assemblage of exhumed UHPM crust slices derived from different crust levels. The interfaces between the UHPM crust slices are ancient fault zones with low viscosity which were main channels for fluid activity during neo-Proterozoic time.

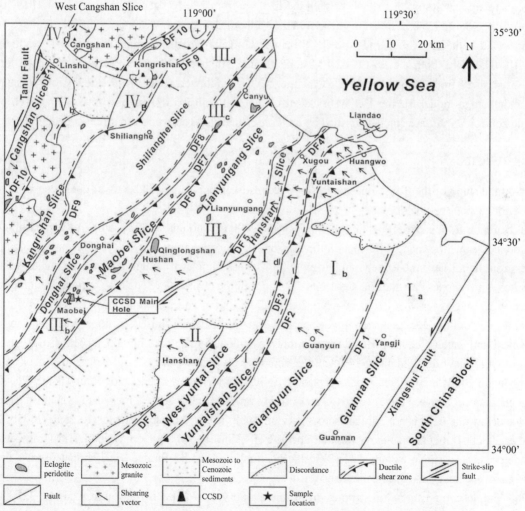

Fig. 5 Tectonic map of the Lianyungang area showing the distribution of various high-pressure and ultrahigh-pressure slices and major ductile shear zones. CCSD — Chinese Continental Scientific Drilling (modified after Xu et al., 2006a).

5.5 Comparison between the Dabie and Su-Lu UHPM zones

The Dabie UHPM belt can be subdivided in to 4 HP-UHP metamorphic zones from the south to the north, i.e. (1)

Northern Yangtze blueschist zone, (2) Hong'an-Susong cold eclogite zone, (3) Southern Dabie UHPM zone, and (4) Northern Dabie high temperature UHPM zone. It is well accepted that the Southern Dabie UHPM zone and the Su-Lu UHP zone III are comparable in petrology and geochronology, because both of them contain coesite or diamond-bearing eclogite (Xu et al., 1992; Cong et al., 1995; Zhang et al., 1995; Liu et al., 2001) and experienced the peak UHP metamorphism at 226±3 Ma or 227±4 Ma (Li et al.,1994, 2000; Li et al., 2005a; Liu et al., 2005b). The surface rocks in the Southern Dabie UHPM zone are characterized by the UCC features (Zhang et al., 2001; Li et al., 2003a). $^{206}Pb/^{204}Pb$ ratios of the UHP paragneiss and eclogite nodule in marble in the Southern Dabie UHPM zone are higher than 18.50 (Fig. 3), but no such high radiogenic Pb isotopes have been observed in the samples from the drilling core of the CCSD. $^{206}Pb/^{204}Pb$ ratios of all the samples including paragneiss from the drilling core are lower then 17.40 (Table 1 and Fig. 3). As mentioned above, all rock units in the drilling core of the CCSD project can be connected with the surface UHP sub-zones Maobei (IIIb), Donghai (IIIc), and Shilianghe (IIId) in the Su-Lu UHP zone III. If the UHPM rocks exposed on the surface in the Su-Lu UHP zone III have similar Pb isotopic compositions to the samples from the drilling core, the differences in Pb isotopic compositions of the UHPM rocks between the Southern Dabie UHPM zone and drilling core of the CCSD project suggest that the protoliths of the UHPM rocks in the Su-Lu UHP zone III were originally at a deeper level in the subducted continental crust than the surface rocks in the Southern Dabie UHPM zone. Geophysical study shows that the deep Southern Dabie UHPM zone is composed of a series of thin rock slices (Dong et al., 2004). Among these slices, there are two groups of rock slices with the first one from 7 to 15 km containing several UHPM rock slices and the second one from 15–25 km similar to gneisses in the northern Dabie zone. We envisage that the features of the first group of rocks slice in the deep Southern Dabie UHPM zone (7 to 15 km) may be similar to the rock slices in the Su-Lu UHP zone III. Therefore, both the Su-Lu UHPM zone III and Southern Dabie UHPM zone could be formed via imbricated crust slices from the subducted UCC, upper-MCC and lower-MCC. This study strongly supports the multi-slice exhumation model for the UHPM rocks in continental collision orogen and reveals that even one UHP metamorphic zone on the surface could be formed by multiple crustal slices derived from different depth levels.

6 Conclusions

Based on the study of the Pb isotopes of the omphacites in eclogites and feldspars in gneisses from the CCSD project, two conclusions can be drawn:

(1) The samples from unit 1 (100–800 m), unit 3 (1600–2000 m) and unit 5 (3200–5000 m) in the drilling core of the CCSD project show systematic difference in the Pb isotopes. Samples from unit 1 have moderately high radiogenic Pb, samples from unit 3 have moderately unradiogenic Pb whereas samples from unit 5 have very unradiogenic Pb. Pb isotopes and oxygen isotopes data indicate that units 1, 3 and 5 of the drilling core could be derived from the subducted UCC, upper-MCC and lower-MCC, respectively. The ductile shearing zones in unit 2 (800 to 1600 m) and unit 4 (2040–3200) of the drilling core could be the detachment interface between the UCC, upper-MCC and lower-MCC. The detachment and decoupling happened during continental subduction, so that the upper UHPM rock slice (unit 1) was uplifted to a shallow depth along the detachment interface by thrust. Unit 3 UHPM rock slice may be detached from the subducting middle crust latter and can also be uplifted to a shallow depth below unit 1 slice. The following detachment could occur between the lower-MCC and LCC to produce unit 5 UHPM rock slice, which was uplifted to a shallow place below unit 3. This not only reinforces previous multi-slice exhumation model (Li et al., 2005b; Xu et al., 2006a; Liu et al., 2007), but also suggests that the Su-Lu UHP zone III and Southern Dabie UHPM zone, which have been considered to be a single exhumed UHPM slab previously, could be formed via imbricated crust slices from different depth levels in the subducted continental crust.

(2) Pb isotopic compositions of samples from unit 2 (800 m to 1600 m) and unit 4 (2040–3200 m) show transitional features between units 1 and 3 or 5, respectively, which could be due to mixing processes between units 1 and 3 or 5 portions due to hydrous fluid activity in the Neoproterozoic. The ductile shear zones may be active along the detachments between units 1 and 3 or 5 UHPM crust slices and were developed on some ancient fault zones. The ancient fault zones were the main channels for meteoric water activity during the Neoproterozoic time, resulting in low-viscosity zones. Thus, the realistic low-viscosity zone numbers could be much larger than the

previous modeling results (e.g. Meissner and Mooney, 1998) because of the existence of large fault zones in the crust.

Acknowledgements

This study was financially supported by the State Key Basic Research Developed Program (2003CB716500) and the Natural Science Foundation of China (No. 40634023 and 40773013) as well as Chinese Academy of Science (kzcx2-yw-131). We thank Prof. Xu Zhiqin and all scientists and technicians working on the CCSD project for their help to collect the samples from drilling core. We also thank Prof. Xiao Yilin for a helpful discussion and references providing.

References

Ayers, J.C., Dunkle, S., Gao, S., Miller, C.F., 2002. Constraints on timing of peak and retrograde metamorphism in the Dabie Shan ultrahigh-pressure metamorphic belt, east-central China, using U-Th-Pb dating of zircon and monazite. Chemical Geology 186, 315–331.

Chemenda, A.I., Mattauer, M., Malavieille, J., Bokun, A.N., 1995. A mechanism for syn-collisional rocks exhumation and associated normal faulting: results from physical modeling. Earth and Planetary Science Letters 132, 225–232.

Chen, Y., Jin, Z.M., Ou, X.G., Jin, S.Y., Xu, H.J., 2004. Deformation features of gneiss and UHP eclogite from ductile shear zone and its relation with seismic velocity anisotropy: evidences from core samples at 680–1200 m of CCSD. Acta Petrologica Sinica 20 (1), 97–108 (in Chinese with English abstract).

Cong, B.L., Zhai, M.G., Carswell, D.A., Wilson, R.N., Wang, Q.C., Zhao, Z., 1995. Petrogeneisis of ultrahigh-pressure rocks and their country rocks at Shuanghe in Dabieshan, Central China. Eur. J. Mineral. 7, 119–138.

Dong, S.W., Gao, R., Cong, B.L., Zhao, Z.Y., Liu, X.C., Li, S.Z., Li, Q.S., Huang, D.D., 2004. Crustal structure of the southern Dabie ultrahigh-pressure orogen and Yangtze foreland from deep seismic reflection profiling. Terra Nova 16, 319–324.

Dong, F., Li, S.S., Li, Q.L., Liu, Y.C., Chen, F.K., 2007. A preliminary study of common Pb of UHP metamorphic rocks from CCSD (100–2000 m) — evidence for decoupling within subducting continental crust. Acta Petrologica Sinica 22, 1791–1798 (in Chinese with English abstract).

Hacker, B.R., Ratschbacher, L., Webb, L.E., Ireland, T.R., Walker, D., Dong, S., 1998. U/Pb zircon ages constrain the architecture of the ultrahigh-pressure Qinling-Dabie orogen, China. Earth and Planetary Science Letters 161, 215–230.

Hacker, B.R., Ratschbacher, L., Webb, L.E., McWilliams, M.O., Ireland, T., Calvert, A., Dong, S.W., Wenk, H.R., Chateigner, D., 2000. Exhumation of ultrahigh-pressure continental crust in east central China: Late Triassic–Early Jurassic tectonic unroofing. Journal of Geophysical Research 105, 13339–13364.

Huang, F., Li, S.G., Dong, F., Li, Q.L., Chen, F.K., Wang, Y., Yang, W., 2007. Recycling of deeply subducted continental crust in the Dabie Mountains, central China. Lithos 96, 151–169.

Huang, F., Li, S.G., Dong, F., He, Y.S., Chen, F.K., 2008. High-Mg adakitic rocks in the Dabie orogen, central China: implications for foundering mechanism of lower continental crust. Chemical Geology 255, 1–13.

Klemme, S., Blundy, J.D., Wood, B.J., 2002. Experimental constraints on major and trace element partitioning during partial melting of eclogite. Geochimica et Cosmochimica Acta 66, 3109–3123.

Kogiso, T., Tatsumi, Y., Nakano, S., 1997. Trace element transport during dehydration processes in the subducted oceanic crust: 1, Experiments and implications for the origin of ocean island basalts. Earth Planet. Sci. Letter 148, 193–205.

Li, S.G., Xiao, Y., Liu, D., 1993. Collision of the North China and Yangtze blocks and formation of coesite-bearing eclogites: timing and processes. Chemical Geology, 109, 89–111.

Li, S.G., Wang, S.S., Chen, Y.Z., Liu, D.L., Qiu, J., Zhou, H.X., Zhang, Z.M., 1994. Excess argon in phengite from eclogite: evidence from dating of eclogite minerals by Sm-Nd, Rb-Sr and $^{40}Ar/^{39}Ar$ methods. Chemical Geology 112, 343–350.

Li, S.G., Jagoutz, E., Chen, Y., Li, Q., 2000. Sm-Nd and Rb-Sr isotope chronology of ultrahigh-pressure metamorphic rocks and their country rocks at Shuanghe in the Dabie Mountains, central China. Geochimica et Cosmochimica Acta, 64, 1077–1093.

Li, S.G., Huang, F., Zhou, H.Y., Li, H.M., 2003a. U-Pb isotopic compositions of the ultrahigh pressure metamorphic (UHPM) rocks from Shuanghe and gneisses from Northern Dabie zone in the Dabie Mountains, central China: constraint on the exhumation mechanism of UHPM rocks. Science in China (Series D) 46, 200–209.

Li, Q.L., Li, S.G., Zheng, Y.F., Li, H.M., Massonne, H.J., Wang, Q.C., 2003b. A high precision U-Pb age of metamorphic rutile in coesite-bearing eclogite from the Dabie Mountains in central China: a new constraint on cooling history. Chemical Geology 200, 255–265.

Li, Q.L., Li, S.G., Hou, Z.H., Hong, J.A., Yang, W., 2005a. A combined study of SHRIMP U-Pb dating, trace element and mineral inclusions on high-pressure metamorphic overgrowth zircon in eclogite from Qinglongshan in the Sulu terrane. Chinese Science Bulletin 50, 459–465.

Li, S.G., Li, Q.L., Hou, Z.H., Yang, W., Wang, Y., 2005b. Cooling history and exhumation mechanism of the ultrahigh-pressure metamorphic rocks in the Dabie Mountains, central China. Acta Petrologica Sinica 21, 1117–1124 (in Chinese with English abstract).

Liu, F.L., Xu, Z.Q., Katayama, I., Yang, J.S., Maruyama, S., Liou, J.G., 2001. Mineral inclusions in zircons of pre-pilot drillhole CCSD-PP1, Chinese Continental Scientific Drilling Project. Lithos 59, 199–215.

Liu, F.L., Xu, Z.Q., Xue, H.M., 2004. Tracing the protolith, UHP metamorphism, and exhumation ages of orthogneiss from the SW Sulu terrane (eastern China): SHRIMP U-Pb dating of mineral inclusion-bearing zircons. Lithos 78, 411–429.

Liu, Y.C., Li, S., Xu, S., Jahn, B.M., Zheng, Y.F., Zhang, Z., Jiang, L., Chen, G., Wu, W., 2005a. Geochemistry and geochronology of eclogites from the northern Dabie Mountains, central China. Journal of Asian Earth Sciences 25, 431–443.

Liu, F.L., Liou, J.G., Xu, Z.Q., 2005b. U-Pb SHRIMP ages recorded in the coesite-bearing zircon domains of paragneisses in the southwestern Sulu terrane, eastern China: new interpretation. American Mineralogist 90, 790–800.

Liu, Y.C., Li, S.G., Xu, S.T., 2007. Zircon SHRIMP U-Pb dating for gneisses in northern Dabie high T/P metamorphic zone, central China: implications for decoupling within subducted continental crust. Lithos 96, 170–185.

Massonne, H.J., 2005. Involvement of crustal material in delamination of the lithosphere after continent-continent collision. International Geology Review 47, 792–804.

Meissner, R., Mooney, W., 1998. Weakness of the lower continental crust: a condition for delamination, uplift, and escape. Tectonophysics 296, 47–60.

Rowley, D.B., Xue, F., Tucker, R.D., Peng, Z.X., Baker, J., Davis, A., 1997. Ages of ultrahigh pressure metamorphism and protolith orthogneisses from the eastern Dabie Shan: U/Pb zircon geochronology. Earth and Planetary Science Letters 151, 191–203.

Rumble, D., Yui, T.F., 1998. The Qinglongshan oxygen and hydrogen isotope anomaly near Donghai in Jiangsu Province, China. Geochimica et Cosmochimica Acta 62 (20), 3307–3321.

Taylor Jr., H.P., 1990. Oxygen and hydrogen isotope constraints on the deep circulation of surface water into zones of hydrothermal metamorphism and melting. In: Norton, D.L., Bredehoeft, J.D. (Eds.) The Role of Fluids in Crustal Processes: Studies in Geophysics. National Academy Press, Washington DC, pp. 72–95.

Wang, Y.J., Fan, W.M., Peng, T.P., Zhang, H.F., Duo, F., 2005. Nature of the Mesozoic lithospheric mantle and tectonic decoupling beneath the Dabie Orogen, Central China: evidence from 40Ar/39Ar geochronology, elemental and Sr-Nd-Pb isotopic compositions of early Cretaceous mafic igneous rocks. Chemical Geology 220, 165–189.

Xiao, Y.L., Hoefs, J., van den Kerkhof, A.M., Li, S.G., 2001. Geochemical constraints of the eclogite and granulite facies metamorphism as recognized in the Raobazhai complex from North Dabie Shan, China. Journal of Metamorphic Geology 19, 3–19.

Xiao, Y.L., Hoefs, J., van den Kerkhof, A.M., Simon, K., Fiebig, J., Zheng, Y.F., 2002. Fluid history during HP and UHP metamorphism in Dabie Shan, China: constraints from mineral chemistry, fluid inclusions, and stable isotopes. Journal of Petrology 43, 1505–1527.

Xiao, Y.L., Zhang, Z.M., Hoefs, J., Kerkhof, A., 2006. Ultrahigh-pressure metamorphic rocks from the Chinese Continental Drilling Project — II Oxygen isotope and fluid inclusion distributions through vertical sections. Contribution Mineral Petrology 152, 443–458.

Xu, S.T., Okay, A.I., Ji, S., Sengör, A.M.C., Su, W., Liu, Y.C., Jiang, L.L., 1992. Diamond from the Dabie Shan metamorphic rocks and its implication for tectonic setting. Science 256, 80–82.

Xu, Z.Q., Zhang, Z.M., Liu, F.L., Yang, J.S., Tang, Z.M., Chen, S.Z., Cai, Y.C., Li, T.F., Chen, F.Y., 2004. The structure profile of 0–1000 m in the main borehole, Chinese Continental Scientific Drilling and its preliminary deformation analysis. Acta Petrologica Sinica 20 (1), 53–72 (in Chinese with English abstract).

Xu, Z.Q., Zeng, L.S., Liu, F.L., Yang, J.S., Zhang, Z.M., Williams, M.M., Liou, J.G., 2006a. Polyphase subduction and exhumation of the Sulu high-pressure-ultrahigh-pressure metamorphic terrane. Geological Society of America Special Paper 403, 93–113.

Xu, Z.Q., Wang, Q., Chen, F.Y., Liang, F.H., Tang, Z.M., 2006b. Fabric kinematics of eclogite and deep continental subduction: EBSD study of eclogite from the main hole of the Chinese Continental Scientific Drilling Project. Acta Ptrologica Sinica 22 (7), 1799–1809.

You, Z.D., Su, S.G., Liang, F.H., Zhang, Z.M., 2004. Petrography and metamorphic deformation history of the ultrahigh-pressure metamorphic rocks from the 100–2000 m core of Chinese Continental Scientific Drilling, China. Acta Petrologica Sinica 20 (1), 43–52 (in Chinese with English abstract).

Yui, T.F., Rumble, D., Luo, C.H., 1995. Unusually low $\delta^{18}O$ ultra-high-pressure metamorphic rocks from Sulu Terrain, eastern China. Geochimica et Cosmochimica Acta 59 (13), 2859–2864.

Zartman, R.E., Doe, B.R., 1981. Plumbotectonic — the model. Tectonophysics 75, 135–162.

Zhang, L.G, 1995. Block-Geology of Eastern Asia Lithosphere — Isotope Geochemistry and Dynamics of Upper Mantle, Basement and Granite. Science Press, Beijing. (in Chinese with English abstract).

Zhang, R.Y., Hirajima, T., Banno, S., et al., 1995. Petrology of ultra-high-pressure rocks from the southern Su-Lu region, eastern China. Journal of Metamorphic Geology 13, 659–675.

Zhang, H.F., Gao, S., Zhang, B.R., et al., 2001. Pb isotope study on crustal structure of Dabie mountains, central China. Geochimica 30, 395–401 (in Chinese with English abstract).

Zhang, Z.M., Xu, Z.Q., Liu, F.L., You, Z.D., Shen, Q., Yang, J.S., L i, T.F., Chen, S.Z., 2004.Petrological geochemistry of eclogites from the drilling core (100–2050 m) at main hole of the Chinese Continental Scientific Drilling (CCSD) project. Acta Petrologica Sinica 20 (1), 27–42 (in Chinese with English abstract).

Zhang, Z.M., Xiao, Y.L., Zhao, X.D., Shi, C., 2006. Fluid-rock interaction during the continental deep subduction: oxygen isotopic profile of the main hole of the CCSD project. Acta Petrologica Sinica 22 (7), 1941–1951.

Zhao, Z.F., Zheng, Y.F., Chen, B., Wu, Y.B., 2005. A geochemical study of element and Sr-Nd isotopes for eclogite and gneiss from CCSD core 734 to 933 m. Acta Petrologica Sinica 21 (2), 325–338 (in Chinese with English abstract).

Zheng, Y.F., Fu, B., Gong, B., Li, S.G, 1996. Extreme ^{18}O depletion in eclogite from the Su-Lu terrane in East China. European Journal of Mineralogy 8, 317–323.

Zheng, Y.F., Fu, B., Gong, B., Li, L., 2003. Stable isotope geochemistry of ultrahigh pressure metamorphic rocks from the Dabie-Sulu orogen in China: implications for geodynamics and fluid regime. Earth-Science Reviews 62, 105–161.

Zindler, A., Hart, S., 1986. Chemical geodynamics. Annual Reviews Earth Planet Science 14, 493–571.

A granulite record of multistage metamorphism and REE behavior in the Dabie orogen: constraints from zircon and rock-forming minerals

Shui-Jiong Wang[1,2], Shu-Guang Li[1,2], Shi-Chao An[2] and Zhen-Hui Hou[2]

1. State Key Laboratory of Geological Processes and Mineral Resources, China University of Geosciences, Beijing 100083, China
2. CAS Key Laboratory of Crust-Mantle Materials and Environments, School of Earth and Space Sciences, University of Science and Technology of China, Hefei 230026, Anhui, China

亮点介绍：通过锆石的微量元素和矿物包体识别出变质锆石的三个生长阶段；锆石 U-Pb 年龄制约了北大别在大陆俯冲和折返过程中的峰期和退变质时代；麻粒岩相和角闪岩相退变质作用分别发生在封闭和开放环境下；内部和外部流体都能改变退变质岩石中矿物的 REE 分配模式。

Abstract A combined study of mineral inclusions, U-Pb ages and trace elements was carried for zircon and coexisting minerals from granulite in the North Dabie Terrane (NDT) of the Dabie-Sulu ultrahigh-pressure metamorphic (UHP) zone, east-central China. The results provide insights into the exhumation history of NDT and into rare earth element (REE) behavior during retrogression. Besides inherited cores and one magmatic rim, zircons separated from the granulite record three episodes of metamorphism under different P-T conditions: (1) 223.8 ± 2.3 Ma for domains that contain Grt + Cpx ± Rt ± F–Ap ± Aln inclusions without plagioclase and show flat HREE patterns without negative Eu anomalies, representing peak eclogite-facies event; (2) 213.3 ± 2.1 Ma for domains that contain Pl ± Cpx ± Grt ± Qtz ± Ap inclusions and show rather flat HREE patterns with negative Eu anomalies, corresponding to granulite-facies retrogression; (3) 199.9 ± 3.3 Ma for domains that contain Amp ± Pl ± Qtz ± Ap inclusions and show high REE contents with steep HREE patterns and remarkable negative Eu anomalies, representing amphibolite-facies overprinting. Therefore, the UHP eclogite in NDT experienced decompression heating during the initial exhumation, with local hydration in the late stage of the Triassic continental collision

Garnet in the granulite is composed of a corroded core with embayed outline and spongy texture and an overgrowth rim. There is equilibrium distribution of HREE between garnet rim and granulite-facies zircon domain, confirming the geological interpretation of 213.3 ± 2.1 Ma for the granulite-facies metamorphism. There is the prograde HREE depletion in porphyroblastic garnet from core to rim and the continuous decrease of HREE from the eclogitic to granulitic zircons, suggesting that the metamorphic transformation from eclogite-facies to granulite-facies took place in a closed system. On the other hand, the amphibolitic zircons show steep HREE patterns and significantly elevated REE contents, suggesting that the amphibolite-facies metamorphism took place in an open system with introduction of external fluids containing more REE to the granulite-facies rocks. Residual clinopyroxene partially replaced by amphibole has remarkably elevated REE contents relative to euhedral primary clinopyroxene. There is equilibrium distribution of REE between the residual clinopyroxene and the amphibole, marking fluid action during the amphibolite-facies overprinting.

Key words Granulite; zircon; garnet; rare earth elements; U-Pb ages; mineral inclusions

1 Introduction

The exhumation of ultrahigh-pressure (UHP) metamorphic rocks during continental collision is commonly accompanied by multistage retrogression. Precisely and accurately dating of the retrograde metamorphism is crucial for understanding the exhumation history and metamorphic process of target rocks. Zircon, because of its highly refractory nature, high closure temperature and slow diffusion rate of Pb, can potentially preserve multiple stages of metamorphic records, and thus it is an ideal mineral for U-Pb dating of poly-metamorphic rocks (Lee et al., 1997; Katayama et al., 2001; Moller et al., 2002; Wu and Zheng, 2004). Recent advances in analytical capabilities permit in-situ investigation of complex zircon grains at high spatial resolution. In-situ analyses of trace elements and mineral inclusions in zircons also help determine metamorphic P-T conditions under which the zircons formed (e.g., Schaltegger et al., 1999; Hermann et al., 2001; Rubatto., 2002; Bingen et al., 2004; Liati, 2005; Mcclelland et al., 2006, 2009; Liu et al., 2008; Xia et al., 2009, 2010; Chen et al., 2010). However, zircons are sometimes "blind" or unresponsive to high-grade metamorphism. A careful examination using other methods such as investigation of rare earth element (REE) partition between zircon and coexisting minerals is thus required (e.g., Harley and Kelly., 2007). Such examination on mineral/mineral REE distribution is also useful to decipher the REE behavior and mineral chemistry during metamorphism (Nehring et al., 2009).

Zircon studies have contributed much to our understanding of continental subduction-zone metamorphism in the Dabie-Sulu orogenic belt (e.g., Zheng et al., 2004; Xia et al., 2009; Chen et al., 2010; Liu and Liou, 2011). Fluid effects on metamorphic growth and recrystallization of zircons are evident in UHP metamorphic rocks (e.g., Zheng, 2009; Gao et al., 2011). The Dabie orogen is composed of three major UHP metamorphic units, namely the north Dabie Terrane (NDT), the central Dabie Terrane (CDT), and the south Dabie Terrane (SDT). Three UHP units have different crustal attributes, e.g., the CDT and NDT represent the subducted upper and lower crust, respectively (Li et al., 2003; Liu et al., 2007a; Zhao et al., 2008), consistent with a multi-slice exhumation model for the UHPM rocks in the Dabie-Sulu orogen (Li et al., 2005, 2009a; Liu et al., 2007b; Liu and Li, 2008; Xu et al., 2006). The UHP rocks from CDT experienced only amphibolite-facies retrogression during exhumation. A detailed geochronological study on Shuanghe UHP rocks has defined a T-t path with two fast cooling stages, suggesting a multistage exhumation (Li et al., 2000). In contrast, the UHP rocks from NDT experienced two stages of retrograde metamorphism (granulite-facies and amphibolite-facies) during exhumation, but geochronological constraints on the exhumation of NDT are rare and debated. Only a few Sm-Nd and U-Pb isotopic ages ranging from 191 to 244 Ma have been reported for eclogites and gneisses from NDT (Li et al., 1993; Liu et al., 2005, 2007a,b, 2011a; Zhao et al., 2008; Xie et al., 2010). Although the peak eclogite-facies metamorphic time of NDT has been constrained to be 224–226 Ma (Liu et al., 2011a), one important question remains that whether other Triassic ages represent the granulite-facies or amphibolite-facies retrogression. This has bearing on the Sm-Nd closure temperature of garnet and omphacite during continental collision, and on preservation of mineral inclusions and trace elements in metamorphic zircons.

Granulite served as a precursor for UHP eclogite-facies metamorphism in Western Gneiss Region of Norway (Austraheim, 1987; Jamtveit et al., 1990), but it overprinted the UHP eclogite in the Dabie orogen (Zhang et al., 1996; Zheng et al., 2001). Despite its rare occurrence, granulite may provide an important record of metamorphic evolution during continental collision in the Dabie orogen. Zircon is one of the most important accessory minerals in metamorphic rocks because it can be used to determine both metamorphic age and facies if microbeam U-Pb dating is combined with analyses of mineral inclusions and trace elements (e.g., Liu et al., 2006; Liu and Liou, 2011). Garnet is one of the most important rock-forming minerals since it can be also used to constrain metamorphic facies if major element analysis is combined with analyses of mineral inclusions and trace elements (e.g., Xia et al., 2011; Zhou et al., 2011). In this paper, we present an integrated study of in-situ U-Pb dating, mineral inclusions and REE analyses on metamorphic zircons and coexisting minerals such as garnet, clinopyroxene, and amphibole in granulite from NDT. The results provide insights into the exhumation history of NDT and the issue about REE behavior during retrograde metamorphism.

2 Geological setting and samples

2.1 Geological background

The Dabie-Sulu orogenic belt formed by continental collision between the South China Block (SCB) and

North China Block (NCB) in Triassic (e.g., Li et al., 1993, 1994, 2000; Zheng et al., 2004). It is one of the largest UHP metamorphic zones in the world (Zheng, 2008; Liou et al., 2009). The general geology of the Dabie-Sulu orogenic belt has been described in numerous previous studies (e.g., Xu et al., 1992, 1994; Hacker et al., 1998; Li et al., 2001; Zheng et al., 2003). The Dabie orogen consists of a series of fault-bounded metamorphic units (Fig.1). They are, from north to south, (1) the Beihuaiyang low-T/LP greenschist-facies terrane (BT); (2) the North Dabie high-T/UHP granulite-facies Terrane (NDT); (3) the Central Dabie mid-T/UHP eclogite-facies Terrane (CDT); (4) the South Dabie low-T/UHP eclogite-facies Terrane (SDT); and (5) the Susong low-T/HP blueschist-facies terrane (ST).

NDT is bounded by the Wuhe-Shuihou Fault to the south and the Xiaotian-Mozitan Fault to the north. It predominantly consists of migmatitic gneisses with minor garnet-bearing amphibolites, eclogites, and granulite lenses, as well as post-collisional granitoids and mafic-ultramafic intrusions. Compared to CDT and SDT, NDT was subjected to more intensive early Cretaceous magmatism and migmatization (e.g., Wang et al., 2007; Zhao et al., 2007, 2008; Liu et al., 2010b; He et al., 2011), which may obscure the earlier metamorphic records. For this reason, whether bulk NDT was involved in the Triassic subduction to experience the UHP metamorphism or not has been the debated issue (e.g., Zhang et al., 2009; Tong et al., 2011). In the last decade, an increasing number of retrograded UHP eclogite relics and HP granulites have been reported from NDT (Tsai and Liou., 2000; Xu et al., 2000, 2003; Zhang et al., 2000; Liu et al., 2001, 2005, 2007a, b, 2011a, b; Faure et al., 2003; Malaspina et al., 2006; Lin et al., 2007). Triassic metamorphic ages and UHP metamorphic evidence (e.g., diamond inclusions from both eclogites and surrounding orthogneisses) suggest that NDT was involved in the Triassic deep subduction of SCB (Tsai and Liou, 2000; Liu et al., 2001, 2005, 2007a, 2011a, b; Malaspina et al., 2006; Xu et al., 2000, 2003, 2005).

Eclogite and gneiss in NDT are overprinted by granulite-facies metamorphism, followed by amphibolite-facies metamorphism during exhumation. Accordingly, NDT experienced three stages of metamorphism (Liu et al., 2001, 2007a; Tsai, 1998; Tsai and Liou, 2000; Xu et al., 2000): (1) eclogite-facies metamorphism at P = ~5 GPa and T = 800–870 ℃; (2) granulite-facies retrogression at P = 1.1–1.4 GPa and T = 800–910 ℃; and (3) amphibolite-facies overprinting at P = 0.5–0.6 GPa and T = 500–600 ℃. The granulite-facies overprinting on the UHP eclogite-facies rocks is exclusive in NDT. But it is absent in CDT and SDT, suggesting that NDT has a different exhumation history from CDT and SDT.

2.2 Previous dating results
2.2.1 Eclogite-facies metamorphic event

Only a few Sm-Nd and U-Pb isotopic ages related to the eclogite-facies metamorphism have been reported for NDT. Li et al. (1993) firstly reported a garnet + diopside (sample R-4) and three garnets (sample G-1) Sm-Nd isochron ages of 244 ± 11 Ma and 224 ± 20 Ma for garnet-pyroxenites in NDT, respectively. In contrast, Liu et al. (2007) utilized zircon SHRIMP U-Pb dating to obtain an age of 218 ± 3 Ma for metamorphic domains from tonalitic gneiss, and interpreted it as representing the UHP metamorphic time of NDT. Recently, Liu et al. (2011a) presented zircon U-Pb age, trace element and mineral inclusion analyses for three eclogites, and argued that the UHP metamorphic time of NDT is best estimated at 226 ± 3 Ma because of the occurrence of coesite inclusion in metamorphic zircon, whereas an age of 214 ± 3 Ma may record a later event of HP eclogite-facies recrystallization time during exhumation.

2.2.2 Granulite-facies and amphibolite-facies metamorphic events

The granulite-facies and amphibolite-facies retrograde metamorphic ages of NDT have not been directly determined yet. Some granulites in NDT, such as felsic granulite at Huangtuling and mafic granulite at Huilanshan, record their granulite-facies metamorphic times of Paleoproterozoic and Cretaceous, respectively (e.g., Chen et al., 1996, 1998; Hou et al., 2005; Wu et al., 2008). They are not related to the Triassic exhumation of NDT. For UHP rocks from NDT, researchers have given various retrograde metamorphic ages. Liu et al. (2005) reported a Sm-Nd isochron age of 212 ± 4 Ma defined by two garnet + two omphacite from eclogite and inferred it as a cooling age corresponding to the granulite-facies metamorphism, but later re-interpreted it as the HP eclogite-facies metamorphic time (Liu et al., 2011a). Liu et al. (2011a) also speculated that the granulite-facies and amphibolite-facies metamorphic ages of NDT might be ~200 Ma and 176–188 Ma, respectively. Xie et al. (2010) conducted zircon U-Pb dating for three orthogneiss samples in NDT, and got ages of 215–205 Ma for metamorphic domains. These ages were interpreted as the amphibolite-facies metamorphic time of NDT. Wang et al. (2002) reported a biotite $^{40}Ar/^{39}Ar$ plateau age of 195 ± 2 Ma from a felsic granulite at Huangtuling, and recommended it as the minimum age of granulite-facies metamorphism of NDT.

2.3 Sample description

Calc-silicate granulite samples for this study were collected from Qingtian village (31°09'06"N, 116°02'19"E) in the center of NDT (Fig. 1). The granulite occurs as boudins as large as 0.5 m within early Cretaceous migmatites (Fig. 2A). In the field, most granulite boudins are retrograded to amphibolite, while less retrograded granulites are generally preserved in the core of some boudins. Two samples are collected for this study.

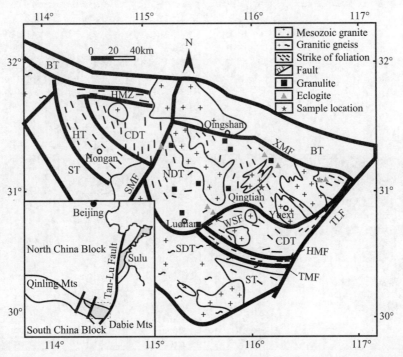

Fig. 1 Schematic geological map of the Dabie orogen, with inset showing the location of this area within the Trassic Dabie-Sulu collision orogen in central China (modified after Xu et al., 2005). The shaded area in the inset shows the UHPM belt. BT = Beihuaiyang terrane; NDT = North Dabie UHP complex terrane; CDT = Central Dabie UHPM terrane; SDT = South Dabie UHPM terrane; ST = Susong complex terrane; HMT = Huwan mélange terrane; HT = Hong'an low-T eclogite terrane; XMF = Xiaotian-Mozitan fault; WSF = Wuhe-Shuihou fault; HMF = Hualiangting-Mituo fault; TMF = Taihu-Mamiao fault; TLF = Tan-Lu fault. Locates of eclogites and high pressure granulites in NDT are also shown in this map.

Sample 0909QT-6 is a fresh granulite containing pale brown-yellow porphyroblastic garnet (~15 vol. % and one grain is up to 15 mm in size) and light green clinopyroxene (~ 50 vol. % with grain size of 0.4 to 4 mm) in the fined-grained matrix (~35 vol. %) that is composed of plagioclase, quartz, scapolite, apatite and titanite. No amphiboles have been observed in sample 0909QT-6, suggesting very weak amphibolite-facies overprinting on this sample. Large garnet porphyroblast in 0909QT-6 has a corroded core that shows ragged and embayed outline and spongy texture with calcite and clinopyroxene fillings, and a wide overgrowth rim with inclusions of clinopyroxene, quartz, ilmenite and calcite (Fig. 2B). Calc-silicate rocks with similar mineral assemblage (grandite + Cpx + Pl + Scp + Qtz) have also been reported from Yanzihe (2-34E) and Luotian (1-44A) areas in NDT (Zhang et al., 1996), recording granulite-facies metamorphic condition of $T > 800$ ℃ and $P > 0.8$ GPa. Zircons are separated from this sample for U-Pb dating.

Sample 0907QT-1 is a retrograded granulite whose margin is partially retrograded to amphibolite. It is further separated into two parts (0907QT-1A and 0907QT-1B). Sample 0907QT-1A, collected from the center of 0907QT-1, has granulite-facies assemblage that is composed of garnet (~5 vol.%), clinopyroxene (~50 vol.%), plagioclase (~30 vol.%), scapolite (~10 vol.%), quartz (~5 vol.%), as well as accessory minerals such as titanite, ilmenite, and apatite. Distinct from those in 0909QT-6, garnet in 0907QT-1A occurs as rather small grains without the corroded core instead of large porphyroblastic ones (Fig. 2C). Sample 0907QT-1B, collected from the margin of 0907QT-1, is absent of garnet and overprinted by amphibolite-facies minerals. Its mode is clinopyroxene (~40 vol.%), plagioclase (~35 vol.%), quartz (~10 vol.%), and amphibole (~15 vol.%) with other accessory minerals. Some clinopyroxene in 0907QT-1B is replaced by small amphibole (~200 μm) along cleavages or fractures (Fig. 2D). Some are free from retrogression. Large recrystallized amphibole grains (~7 mm) contain inclusions of clinopyroxene, plagioclase and other opaque minerals (Fig. 2E).

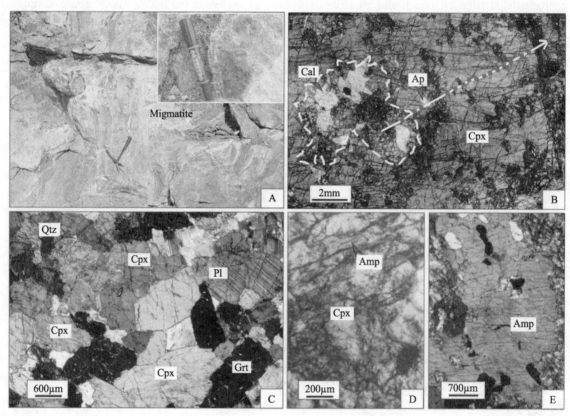

Fig. 2 Field photographs and photomicrographs showing the occurrence and textures of granulites. (A) granulites boudins within migmatites in Qingtian village; (B) porphyroblastic garnet in 0909QT-6 with a corroded core and overgrowth rim; (C) granulite-facies mineral assemblage of Grt + Cpx + Pl + Qtz; (D) clinopyroxenes are partially retrograded to amphibole in 0907QT-1B; (E) large recrystallized amphibole grain in 0907QT-1B.

3 Analytical methods

Zircons were separated from the granulite sample (0909QT-6) using standard heavy-liquid and magnetic techniques, and then handpicked under a binocular microscope. The selected crystals, together with zircon U-Pb standard Plésovice (Slama et al., 2008) and Qinghu (Li et al., 2009b), were embedded in 25 mm epoxy disks and ground to approximately half their thickness.

Cathodoluminescence (CL) image of the zircon grains were obtained at University of Science and Technology of China in Hefei, using a Quanta 400 FEG high resolution emission field environmental scanning electron microscope connected to an Oxford INCA 350 energy dispersive system and a Gatan Mono CL3+ system. The conditions during CL imaging were 10 kV and 20 nA. Mineral inclusions in the zircon grains were identified by laser Raman spectroscopy (RANISHAW RM-1000) with the 514.5 nm line of an Ar-ion laser at Key Laboratory of Continental Dynamics in Ministry of Land and Mineral Resources, Beijing. Representative Raman spectra of mineral inclusions are given in Fig. S1 (online Appendix).

U-Th-Pb analyses of zircon grains were performed using Cameca IMS-1280 at State Key Laboratory of Lithospheric Evolution in Institute of Geology and Geophysics, Chinese Academy of Sciences, Beijing. The O^{2-} primary ion beam with an intensity of ca. 10 nA was accelerated at −13 kV. The ellipsoidal spot is about 20μm × 30 μm in size. The aperture illumination mode was used with a 200 μm primary beam mass filter (PBMF) aperture to produce even sputtering over the entire analyzed area. Detail operating conditions are the same as description by Li et al. (2009b). An average of present-day crustal composition (Stacey and Kramers, 1975) is used for the common Pb assuming that the common Pb is largely surface contamination introduced during sample preparation. Zircon standard Qinghu was analyzed as unknown. The obtained weighted mean $^{206}Pb/^{238}U$ age is 159 ± 52 Ma (2σ, n = 6, Fig. S2, online Appendix), which is consistent with the SIMS age of 159.56 ± 0.69 Ma within analytical uncertainty (Li et al., 2009b). The errors for individual analyses are quoted at the 1σ level. U-Pb data were treated using ISOPLOT program of Ludwig (2001).

Trace element concentrations in minerals were analyzed using an ArF excimer laser system (GeoLas Pro, 193 nm wavelength) and quadrupole ICP-MS (PerkinElmer ElanDRCII) at CAS Key Laboratory of Crust-Mantle Materials and Environments in University of Science and Technology of China, Hefei. The analyses were carried out with pulse rate of 10 Hz and beam energy of 10 J/cm^2. The spot diameter is 32μm for zircons and 60 μm for other minerals. Detail analytical procedure was reported in Yuan et al. (2004). ^{29}Si was used as internal standards for zircon and plagioclase, and ^{43}Ca for garnet, amphibole, clinopyroxene, apatite and calcite. Certified glass reference material NIST SRM 610 was used as an external standard, which was analyzed twice for every 4 analyses. The simultaneous analysis data on NIST SRM 612 and 614 show that the accuracy and precision for most trace elements are better than 10% (Table S1, online Appendix).

Major elements in rock-forming minerals were analyzed using CAMECA SX-51 microprobe analyzer at State Kay Laboratory of Lithospheric Evolution in Institute of Geology and Geophysics, Chinese Academy of Sciences, Beijing. Analytical conditions were 15 kV accelerating voltage, 20 nA beam current and 20 s counting time. Natural and synthetic minerals were used as standards during the analyses.

4 Results

4.1 Mineral chemistry

4.1.1 Garnet

The large porphyroblastic garnet in sample 0909QT-6 consists of a corroded core and an overgrowth rim. Both are mainly grossular-andradite (grandite) solid solution ranging in composition from $And_{55}Grs_{45}$ to $And_{72}Grs_{28}$ (Table 1), with andradite decreasing and grossular increasing from core to rim (Fig. 3A).

The corroded garnet core has high REE contents of 267–285 ppm and flat HREE patterns with $(Dy/Yb)_N$ = 0.89–0.91 and negative Eu anomalies (Eu/Eu* = 0.65–0.70). In contrast, the wide garnet rims have lower REE contents of 81.7–137 ppm and more declined HREE patterns with $(Dy/Yb)_N$ = 1.00–1.51, thus making a big gap in MREE/HREE between core and rim (Fig. 3B). From the inner to outer rims, the HREE contents progressively decrease (Table 1; Fig. 3A, B). It is also notable that the inner rims (two analyzed points close to the core) have moderately negative Eu anomalies with Eu/Eu* = 0.81–0.89, while the outer rims have positive Eu anomalies with Eu/Eu* = 1.36–2.01 (Fig. 3B). Small garnets in sample 0907QT-1A also exhibit HREE contents gradually decreasing from core to rim (Fig. 4B), but have no HREE gap between core and rim. The REE patterns of garnets in 0907QT-1A are overlapped with the HREE gap between the core and rim of sample 0909QT-6, and some of them are similar to those of inner rims from 0909QT-6 that lack obviously positive Eu anomalies with $(Dy/Yb)_N$ = 0.83–1.35 and Eu/Eu* = 0.80–1.10 (Fig. 3B). In general, both garnets from 0909QT-6 and 0907QT-1A exhibit rather higher LREE and MREE contents than almandine-rich garnets from UHP eclogites in the Dabie orogen (Fig. 3B).

4.1.2 Clinopyroxene

Clinopyroxenes in sample 0909QT-6 and 0907QT-1A are mainly diopside and hedenbergite with X_{Mg} = 0.5–0.7 (Table S2 in online Appendix). No systematic variations in X_{Mg} exist between inclusions and phenocrysts. These

Table 1 Major (%) and trace element (μg·g^{-1}) data for the porphyroblastic garnet in 0909QT-6

	Grtl1 c	Grtl2 c	Grtl3 r	Grtl4 r	Grtl5 r	Grtl6 r	Grtl7 r	Grtl8 r	Grtl9 r	Grtl10 r
SiO_2	37.03	36.87	36.66	36.90	36.20	36.88	36.20	36.30	36.59	36.63
TiO_2	0.81	0.93	1.01	0.80	1.11	0.73	1.06	0.77	0.86	0.85
Al_2O_3	6.36	6.14	5.82	6.12	5.99	6.27	5.97	6.36	5.81	6.32
Cr_2O_3	0.02	0.00	0.00	0.00	0.00	0.04	0.00	0.02	0.02	0.04
FeO	21.85	21.42	21.09	21.48	21.61	20.97	21.03	20.89	20.58	20.96
MnO	0.49	0.41	0.47	0.39	0.41	0.35	0.32	0.35	0.37	0.39
MgO	0.08	0.06	0.05	0.08	0.04	0.05	0.05	0.05	0.04	0.06
CaO	33.02	33.17	32.53	32.96	32.53	33.73	33.24	33.68	33.04	33.33
NiO	0.05	0.02	0.00	0.00	0.00	0.00	0.00	0.00	0.06	0.01
Na_2O	0.00	0.00	0.02	0.00	0.00	0.01	0.00	0.00	0.00	0.00
K_2O	0.00	0.00	0.00	0.01	0.00	0.00	0.00	0.00	0.00	0.00
Total	99.71	99.03	97.66	98.73	97.89	99.04	97.87	98.43	97.36	98.59
Si	2.98	2.98	3.00	2.99	2.96	2.98	2.96	2.95	3.00	2.97
Ti	0.05	0.06	0.06	0.05	0.07	0.04	0.07	0.05	0.05	0.05
Al	0.60	0.58	0.56	0.58	0.58	0.60	0.58	0.61	0.56	0.60
Cr	0.00	0.00	0.00	0.00	0.00	0.00	0.00	0.00	0.00	0.00
Fe^{3+}	1.38	1.39	1.40	1.39	1.40	1.39	1.41	1.39	1.40	1.38
Fe^{2+}	0.09	0.06	0.05	0.06	0.08	0.03	0.03	0.03	0.01	0.05
Mn	0.03	0.03	0.03	0.03	0.03	0.02	0.02	0.02	0.03	0.03
Mg	0.01	0.01	0.01	0.01	0.00	0.01	0.01	0.01	0.00	0.01
Ca	2.84	2.87	2.85	2.86	2.85	2.92	2.91	2.94	2.90	2.90
Ura	0.07	0.01	0.00	0.00	0.00	0.12	0.00	0.08	0.06	0.14
And	69.56	70.41	71.34	70.41	70.80	69.85	70.95	69.41	71.32	69.39
Pyr	0.30	0.24	0.22	0.33	0.16	0.22	0.20	0.20	0.15	0.24
Spe	1.12	0.95	1.11	0.91	0.97	0.81	0.74	0.81	0.86	0.90
Gro	25.96	26.48	25.75	26.17	25.40	28.02	27.01	28.39	27.22	27.82
Alm	2.99	1.91	1.58	2.18	2.67	0.98	1.09	1.11	0.38	1.52
Y	293.00	287.00	75.10	68.80	57.70	47.00	39.00	28.40	26.80	34.20
Zr	189.00	154.00	129.00	119.00	183.00	158.00	169.00	130.00	135.00	159.00
La	1.93	1.49	2.04	2.64	1.83	1.87	1.77	1.76	1.78	1.82
Ce	23.70	18.30	23.20	28.40	22.70	22.30	23.20	22.90	22.50	23.20
Pr	7.49	5.82	6.18	6.86	6.79	6.54	7.35	6.80	6.61	6.79
Nd	59.20	48.70	39.50	41.50	49.10	42.80	49.50	40.80	40.00	44.80
Sm	27.40	25.30	12.00	12.20	12.60	9.96	10.90	7.70	7.98	9.61
Eu	6.67	6.88	3.54	3.26	5.97	3.81	6.21	3.34	3.08	3.79
Gd	36.40	35.60	12.20	12.30	10.50	8.81	8.22	5.86	6.02	7.32
Tb	6.15	6.17	1.75	1.79	1.46	1.26	1.12	0.76	0.77	0.99
Dy	40.30	41.20	11.20	10.80	9.13	7.12	6.39	4.68	4.33	5.64
Ho	9.38	9.56	2.53	2.28	1.95	1.52	1.37	1.02	0.91	1.18
Er	27.90	28.90	7.23	6.59	5.66	4.42	3.58	2.75	2.45	3.11
Tm	4.06	4.31	1.03	0.90	0.78	0.62	0.51	0.36	0.37	0.48
Yb	30.30	30.40	7.26	6.35	6.09	4.34	3.38	2.43	2.40	3.03
Lu	4.59	4.45	1.11	0.98	0.96	0.66	0.52	0.36	0.34	0.44
REE	285.00	267.00	131.00	137.00	135.00	116.00	124.00	102.00	99.50	112.00
$La/Sm_{(N)}$	0.05	0.04	0.11	0.14	0.09	0.12	0.11	0.15	0.14	0.12
$Dy/Yb_{(N)}$	0.89	0.91	1.03	1.14	1.00	1.10	1.27	1.29	1.21	1.24
Eu/Eu^*	0.65	0.70	0.89	0.81	1.59	1.24	2.01	1.52	1.36	1.38

Continued

	Grtl11 r	Grtl12 r	Grtl13 r	Grtl14 r	Grtl15 r	Grtl16 r	Grtl17 r
SiO_2	36.56	36.44	36.45	36.62	36.60	36.60	37.22
TiO_2	0.86	0.63	0.66	0.82	0.82	1.04	0.80
Al_2O_3	6.44	7.31	6.22	7.47	7.13	8.01	9.38
Cr_2O_3	0.01	0.00	0.00	0.00	0.00	0.00	0.03
FeO	19.79	19.56	20.89	18.84	19.31	17.29	16.48
MnO	0.37	0.27	0.34	0.29	0.33	0.31	0.15
MgO	0.07	0.05	0.08	0.09	0.06	0.12	0.08
CaO	33.68	33.75	33.25	33.89	33.56	34.31	34.51
NiO	0.00	0.02	0.01	0.00	0.00	0.00	0.01
Na_2O	0.00	0.01	0.00	0.00	0.00	0.01	0.00
K_2O	0.01	0.00	0.00	0.00	0.00	0.00	0.00
Total	97.80	98.03	97.90	98.03	97.81	97.68	98.67
Si	2.98	2.96	2.98	2.97	2.98	2.97	2.97
Ti	0.05	0.04	0.04	0.05	0.05	0.06	0.05
Al	0.62	0.70	0.60	0.71	0.68	0.77	0.88
Cr	0.00	0.00	0.00	0.00	0.00	0.00	0.00
Fe^{3+}	1.36	1.30	1.39	1.27	1.30	1.22	1.10
Fe^{2+}	0.00	0.03	0.04	0.00	0.02	0.00	0.00
Mn	0.03	0.02	0.02	0.02	0.02	0.02	0.01
Mg	0.01	0.01	0.01	0.01	0.01	0.01	0.01
Ca	2.94	2.94	2.91	2.94	2.93	2.98	2.96
Ura	0.04	0.00	0.00	0.00	0.00	0.00	0.09
And	68.68	64.94	69.88	64.08	65.47	61.36	55.39
Pyr	0.30	0.20	0.33	0.37	0.23	0.47	0.34
Spe	0.86	0.63	0.79	0.67	0.76	0.70	0.33
Gro	30.44	33.15	27.72	34.72	32.97	38.92	43.78
Alm	−0.31	1.08	1.27	0.16	0.57	−1.46	0.07
Y	31.70	24.70	25.90	18.10	20.30	26.70	30.10
Zr	158.00	147.00	185.00	153.00	143.00	178.00	170.00
La	1.59	1.33	1.56	1.42	1.36	1.19	1.11
Ce	20.10	17.50	19.50	17.70	18.60	16.50	14.80
Pr	6.25	5.52	5.99	5.47	5.73	5.36	4.98
Nd	43.30	36.80	40.50	34.60	38.30	38.60	38.10
Sm	9.97	7.77	9.23	6.65	7.98	8.86	9.46
Eu	4.21	3.56	3.68	2.68	2.78	3.95	4.58
Gd	7.54	5.89	6.78	4.64	5.48	6.66	7.54
Tb	0.96	0.77	0.83	0.63	0.72	0.85	0.96
Dy	5.17	4.30	4.46	3.34	3.88	4.72	5.11
Ho	1.11	0.87	0.86	0.69	0.70	0.90	1.03
Er	3.00	2.18	2.27	1.72	1.92	2.33	2.59
Tm	0.42	0.35	0.32	0.25	0.26	0.34	0.40
Yb	2.82	2.27	2.31	1.56	1.72	2.18	2.54
Lu	0.45	0.33	0.36	0.26	0.29	0.35	0.40
REE	107.00	89.40	98.60	81.70	89.70	92.70	93.60
$La/Sm_{(N)}$	0.10	0.11	0.11	0.14	0.11	0.09	0.08
$Dy/Yb_{(N)}$	1.23	1.27	1.29	1.43	1.51	1.45	1.35
Eu/Eu^*	1.48	1.61	1.42	1.48	1.29	1.57	1.66

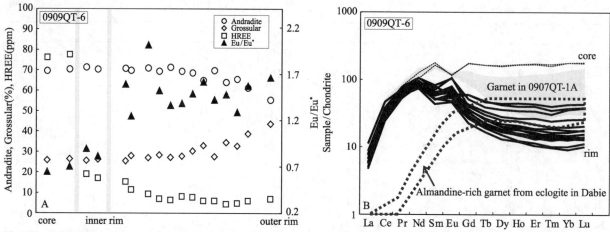

Fig. 3 A: Major and trace and trace element zoning of the porphyroblastic garnet in 0909QT-6; B: The REE patterns of the porphyroblastic garnet in 0909QT-6. Chondrite-normalization values are from Sun and McDonough (1989). Data source of almandine-rich garnet are from Tang et al. (2007).

primary clinopyroxenes are termed as Cpx I. They have concave-like REE patterns with $(La/Sm)_N$ = 2.82–12.33 and $(Dy/Yb)_N$ < 0.51 and low but variable REE contents of 0.80–7.01 ppm (Table 2; Fig. 4A, B). Clinopyroxene in sample 0907QT-B, which is partially replaced by amphibole, is diopside in composition with X_{Mg} about 0.7 (Table S2). It is distinct with Cpx I in significantly higher REE contents (27.2–46.8 ppm) and thus termed as Cpx II (Table 2; Fig. 4C). Cpx II is characterized by rather flat LREE and slightly enriched HREE patterns with $(La/Sm)_N$ = 0.81–1.00 and $(Dy/Yb)_N$ = 0.74–0.99 (Fig. 4C). Other clinopyroxene, which is free from retrogression in 0907QT-1B, has REE contents of 3.75–4.22 ppm and similar patterns with $(La/Sm)_N$ = 0.93–1.73, $(Dy/Yb)_N$ = 0.06–0.32 and Eu/Eu^* = 0.27–0.63 to Cpx I in samples 0909QT-6 and 0907QT-1A (Table 2; Fig. 4C). It is therefore termed as Cpx I as well.

Table 2 Trace element ($\mu g \cdot g^{-1}$) data for minerals in 0909QT-6, 0907QT-1A and 0907QT-1B

	0909QT-6												0907QT-1A				
	Cpx I1	Cpx I2	Cpx I3	Cpx I4	Cpx I5	Cpx I6	Cpx I7	Ap1	Ap2	Ap3	Cal1	Cal2	Grt1	Grt2	Grt3	Grt4	Grt5
Y	0.30	0.16	0.08	0.10	0.35	0.43	0.03	10.30	11.60	7.10	5.40	11.80	204.00	154.00	108.00	134.00	63.6
Zr	3.08	2.21	2.44	1.32	1.59	3.04	2.34	0.04	0.02	0.00	0.00	0.00	282.00	388.00	376.00	380.00	284.00
La	0.10	0.50	0.47	0.30	0.25	1.00	0.15	37.10	110.00	35.20	56.90	62.30	2.06	2.47	2.86	2.51	2.92
Ce	0.24	1.74	1.55	0.38	0.36	1.99	0.43	74.10	232.00	71.70	106.00	113.00	19.50	23.70	25.60	24.70	30.00
Pr	0.04	0.26	0.16	0.04	0.05	0.22	0.05	7.52	22.90	6.92	9.50	9.86	5.82	7.00	7.41	7.27	8.74
Nd	0.23	0.54	0.40	0.12	0.11	0.66	0.17	26.60	72.60	22.5	25.30	25.80	49.40	58.90	62.60	61.10	72.8
Sm	0.01	0.00	0.00	0.00	0.03	0.14	0.00	4.04	7.88	3.17	1.81	2.01	22.70	25.60	25.20	25.70	24.00
Eu	0.01	0.01	0.00	0.00	0.00	0.00	0.00	0.87	1.33	0.67	0.36	0.51	6.70	7.49	7.08	7.53	7.75
Gd	0.03	0.00	0.00	0.00	0.03	0.07	0.00	3.76	5.53	2.65	1.25	2.16	28.60	29.50	24.30	27.20	19.40
Tb	0.00	0.00	0.00	0.00	0.00	0.00	0.00	0.36	0.51	0.33	0.14	0.28	4.51	4.13	3.06	3.75	2.34
Dy	0.04	0.01	0.00	0.00	0.03	0.04	0.00	2.00	2.33	1.34	0.71	1.64	29.80	24.00	16.50	20.50	12.10
Ho	0.00	0.00	0.00	0.00	0.00	0.01	0.00	0.36	0.42	0.28	0.18	0.41	6.66	5.03	3.17	4.14	2.27
Er	0.04	0.00	0.00	0.00	0.02	0.03	0.00	0.87	0.85	0.57	0.49	1.25	20.30	13.60	8.20	11.40	5.96
Tm	0.01	0.00	0.00	0.00	0.00	0.00	0.00	0.09	0.10	0.08	0.07	0.22	3.22	2.04	1.26	1.79	0.88
Yb	0.14	0.00	0.00	0.00	0.04	0.11	0.00	0.44	0.39	0.39	0.64	1.60	23.90	14.90	9.02	12.20	6.60
Lu	0.04	0.00	0.00	0.00	0.01	0.03	0.00	0.05	0.05	0.03	0.13	0.22	3.96	2.47	1.39	2.02	1.02
REE	0.93	3.05	2.59	0.84	0.93	4.31	0.80	158.00	457.00	146.00	203.00	221.00	227.00	221.00	198.00	212.00	197.00
$La/Sm_{(N)}$	12.30	–	–	–	5.10	4.69	–	5.92	9.00	7.17	20.30	20.00	0.06	0.06	0.07	0.06	0.08
$Dy/Yb_{(N)}$	0.19	–	–	–	0.51	0.23	–	3.04	4.04	2.27	0.74	0.69	0.83	1.08	1.22	1.13	1.22
Eu/Eu^*	3.51	–	–	–	0.00	0.00	–	0.68	0.62	0.71	0.74	0.74	0.80	0.83	0.87	0.87	1.10

Continued

	0907QT-1A										0907QT-1B							
	Grt6	Cpx I1	Cpx I2	Cpx I3	Cpx I4	Cpx I5	Cpx I6	Scp	Pl1	Pl2	Cpx I1	Cpx I2	Cpx I3	Cpx II1	Cpx II2	Cpx II3	Cpx II4	Am I1
Y	61.50	0.15	0.11	0.12	0.04	0.03	0.11	0.08	0.00	0.00	0.48	0.83	1.00	8.35	11.00	7.53	6.55	23.00
Zr	357.00	12.40	6.96	11.7	7.08	2.70	9.72	4.15	0.00	0.00	3.03	4.04	4.75	17.30	12.60	14.50	13.00	35.70
La	2.47	1.04	1.00	1.20	0.66	0.22	0.91	0.44	5.38	6.79	0.35	0.45	0.36	2.76	4.22	3.32	2.43	20.40
Ce	26.00	3.10	2.84	3.31	2.31	1.03	2.84	1.35	4.70	5.13	1.34	1.59	1.04	10.20	14.60	11.00	8.40	48.80
Pr	7.65	0.42	0.39	0.48	0.27	0.15	0.39	0.22	0.26	0.26	0.23	0.26	0.16	1.78	2.55	1.72	1.40	6.59
Nd	60.30	1.44	1.48	1.72	1.26	0.56	1.30	0.71	0.46	0.58	0.84	0.94	0.88	9.16	12.50	9.30	7.40	28.50
Sm	21.80	0.24	0.14	0.17	0.05	0.04	0.03	0.07	0.00	0.07	0.14	0.17	0.25	2.20	3.06	2.16	1.56	6.54
Eu	7.05	0.00	0.01	0.04	0.03	0.00	0.03	0.01	0.39	0.34	0.02	0.01	0.04	0.37	0.47	0.39	0.27	1.46
Gd	17.70	0.00	0.09	0.07	0.02	0.00	0.06	0.04	0.00		0.22	0.17	0.12	2.11	3.02	1.51	1.68	5.70
Tb	2.06	0.00	0.00	0.00	0.01	0.01	0.00	0.01			0.02	0.02	0.03	0.26	0.38	0.27	0.22	0.71
Dy	10.40	0.01	0.01	0.00	0.00	0.00	0.01	0.02	0.00	0.00	0.03	0.10	0.18	1.73	2.03	1.43	1.27	3.74
Ho	1.75	0.00	0.00	0.00	0.00	0.00	0.01	0.00			0.02	0.02	0.04	0.31	0.40	0.30	0.25	0.82
Er	4.41	0.02	0.02	0.00	0.00	0.00	0.00	0.00	0.00	0.00	0.05	0.06	0.12	0.86	1.17	0.82	0.74	2.19
Tm	0.67	0.00	0.00	0.00	0.00	0.01	0.01	0.00	0.00	0.00	0.01	0.02	0.03	0.16	0.20	0.14	0.12	0.36
Yb	5.15	0.03	0.05	0.00	0.00	0.01	0.00	0.02	0.00	0.00	0.34	0.28	0.37	1.18	1.75	1.29	1.13	3.00
Lu	0.73	0.01	0.01	0.02	0.01	0.01	0.03	0.01	0.00	0.00	0.17	0.12	0.13	0.38	0.35	0.29	0.31	0.57
REE	168.00	6.33	6.03	7.01	4.61	2.02	5.63	2.89	11.2	13.2	3.78	4.22	3.75	33.40	46.80	33.9	27.2	129.00
La/Sm$_{(N)}$	0.07	2.82	4.66	4.68	9.33	3.70	19.70	4.17	–	67.4	1.55	1.73	0.93	0.81	0.89	0.99	1.00	2.01
Dy/Yb$_{(N)}$	1.35	0.26	0.08	–	–	0.00	–	0.44	–	–	0.06	0.23	0.32	0.99	0.78	0.74	0.76	0.83
Eu/Eu*	1.10	–	0.24	1.16	2.59	–	2.25	0.47	–	–	0.31	0.27	0.63	0.53	0.47	0.65	0.52	0.73

	0907QT-1B																
	Am I2	Am I3	Am I4	Am II1	Am II2	Am II3	Am II4	Am II5	Am II6	Am II7	Am II8	Am II9	Am II10	Am II11	Pl1	Pl2	Pl3
Y	21.80	17.00	23.30	48.00	46.30	43.30	45.40	48.10	85.90	65.20	50.10	43.40	74.60	52.90	0.00	0.07	0.02
Zr	32.10	36.10	38.10	51.30	43.40	42.20	51.00	45.50	47.10	45.60	44.70	41.80	42.90	43.00	0.00	0.00	0.06
La	17.80	17.20	21.00	20.30	20.20	17.60	21.50	27.60	34.30	31.20	25.70	28.60	33.18	29.10	5.11	4.90	4.55
Ce	46.50	37.60	51.10	69.00	68.80	60.20	72.10	91.70	141.00	111.00	84.80	90.60	134.00	100.00	4.84	5.65	4.84
Pr	6.31	4.22	6.27	11.30	11.60	10.20	11.70	14.00	24.70	19.00	13.30	13.10	23.70	14.70	0.26	0.33	0.35
Nd	31.70	17.70	27.60	58.90	60.00	52.00	59.00	63.4	127.00	98.90	64.80	59.10	128.00	68.10	0.65	0.74	0.95
Sm	6.76	3.66	5.93	14.70	14.30	13.10	14.40	12.90	30.10	24.40	15.30	12.70	31.90	14.80	0.06	0.08	0.11
Eu	1.21	0.99	1.26	2.00	2.02	1.62	2.06	2.24	3.68	3.06	2.27	2.18	3.49	2.63	0.28	0.35	0.43
Gd	5.97	2.97	4.36	12.30	12.20	11.40	12.70	9.60	25.00	18.80	12.60	10.60	25.00	11.70	0.00	0.05	0.00
Tb	0.71	0.41	0.66	1.66	1.54	1.44	1.65	1.44	3.23	2.49	1.73	1.39	3.35	1.63	0.01	0.01	0.00
Dy	4.22	2.78	3.80	8.49	8.85	8.22	8.54	8.39	17.10	13.30	9.17	7.30	17.00	8.69	0.03	0.00	0.00
Ho	0.83	0.51	0.81	1.76	1.77	1.65	1.66	1.62	3.22	2.29	1.81	1.48	3.07	1.78	0.00	0.00	0.00
Er	2.22	1.70	2.25	4.57	4.07	4.36	4.25	4.94	7.60	6.63	5.00	4.47	6.46	5.12	0.00	0.00	0.00
Tm	0.32	0.32	0.40	0.68	0.61	0.62	0.62	0.74	1.09	0.81	0.69	0.64	0.91	0.79	0.00	0.00	0.00
Yb	2.47	3.06	3.66	4.64	3.99	4.61	5.05	5.04	6.84	6.04	4.89	5.05	5.11	5.27	0.00	0.00	0.00
Lu	0.48	0.61	0.65	0.75	0.69	0.73	0.67	0.81	0.94	0.87	0.75	0.90	0.81	0.94	0.00	0.00	0.00
REE	128.00	93.70	130.00	211.00	211.00	188.00	216.00	244.00	425.00	339.00	243.00	238.00	416.00	265.00	11.20	12.10	11.20
La/Sm$_{(N)}$	1.70	3.03	2.29	0.89	0.91	0.87	0.97	1.38	0.73	0.83	1.09	1.45	0.67	1.27	56.90	37.30	25.90
Dy/Yb$_{(N)}$	1.14	0.61	0.69	1.23	1.48	1.19	1.13	1.11	1.68	1.48	1.25	0.97	2.23	1.10	–	–	–
Eu/Eu*	0.58	0.92	0.75	0.46	0.47	0.40	0.47	0.62	0.41	0.44	0.50	0.58	0.38	0.61	–	16.18	–

4.1.3 Amphibole

Two types of amphibole are recognized in sample 0907QT-1B. Type I one partially replaced clinopyroxene along the cleavage or fracture and is edenite in composition with TiO_2 between 0.49% and 0.88% (Fig. 2D; Table S3 in online Appendix). Type II one occurs as large crystals up to 7 mm in size containing clinopyroxene and plagioclase inclusions and is edenitic with TiO_2 between 1.2% and 1.8% (Fig. 2E; Table S3). Amp I has relatively low REE contents of 93.7–130 ppm and right-dipping REE patterns with $(La/Sm)_N$ =1.70–3.03 and $(Dy/Yb)_N$ = 0.61–1.14, and moderately negative Eu anomalies with Eu/Eu^* = 0.58–0.92 (Table 2; Fig.4C). Amp II is characterized by high REE contents of 188–425 ppm with low $(La/Sm)_N$ ratios of 0.67–1.45, and strong negative Eu anomalies with Eu/Eu^* = 0.38-0.62 (Table 2; Fig. 4C). There is similarity in major element compositions between Amp II and amphibole in surrounding migmatite (Table S3), suggesting that Amp II might be a recrystallized product during the early Cretaceous migmatization.

4.1.4 Other mineral phases

Plagioclase in granulites is oligoclase with An_{25-29} (Table S4 in online Appendix). It has fractionated REE patterns with very low HREE contents and strong positive Eu anomalies (Table 2; Fig.4B, C). Scapolite belongs to the meionite (Mei)-rich variety with Mei_{72-74} (Table S5 in online Appendix) and has a low REE content, flat LREE pattern and negative Eu anomaly (Table 2; Fig. 4B). Apatite is characterized by high REE contents of 146–457 ppm with $(La/Sm)_N$ = 5.92–9.00, and negative Eu anomalies with Eu/Eu^* = 0.62–0.71 (Table 2; Fig. 4A). Calcite is characterized by fractionated LREE patterns with $(La/Sm)_N$ = 20.0–20.3, and high REE contents of 203–221 ppm (Table 2; Fig. 4A).

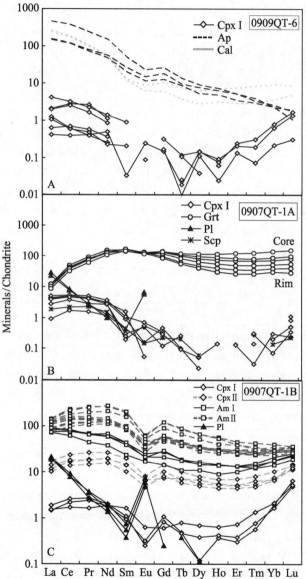

Fig. 4 Chondrite-normalized REE patterns of metamorphic mineral phases in granulitic samples. A, 0909QT-6; B, 0907QT-1A; C, 0907QT-1B. Chondrite-normalization values are from Sun and McDonough (1989).

4.2 Zircon U-Pb ages, trace elements and mineral inclusions

Zircons separated from sample 0909QT-6 are rounded and 100–400 μm in size. Most zircons (~90%) exhibit spherical to multifaceted morphology. CL image shows internal sector- to fir-tree zoning structure (Fig. 3), suggestive of metamorphic origin (e.g., Corfu et al., 2003; Wu and Zheng, 2004). For most metamorphic zircon grains, textures such as core-mantle-rim are not usually observable by the CL image. Although some zircons have homogeneous and bright-luminescent outer rims and black inner rims, they are too narrow to the U-Pb dating. Rare grains contain inherited cores (~10%) (Fig. 5).

Six of eight inherited zircon cores contain mineral inclusions of Qtz + Pl ± Ap ± Mus. They are characterized by high Th/U ratios of 0.35–0.66, extremely negative Eu anomalies with Eu^*/Eu = 0.10–0.35, and steep HREE patterns with $(Dy/Yb)_N$ = 0.12–0.27 (Table 3, Fig. 6C). U-Pb dating yields apparent $^{206}Pb/^{238}U$ ages of 244 to 474 Ma (Table 3), and defines a discordant line with an upper intercept age of 580 ± 88 Ma and a lower intercept age of 212 ± 11 Ma with MSWD=1.3 (Fig. 6A). The other two inherited cores are characterized by low Th/U ratios of 0.10 and 0.03, negligible to positive Eu anomalies with Eu/Eu^* = 0.91 and 2.11, and flat HREE patterns with $(Dy/Yb)_N$ = 0.36 and 0.16, respectively (Table 3, Fig. 6C), yielding older $^{207}Pb/^{206}Pb$ ages of 1494 Ma and 1795 Ma (Table 3).

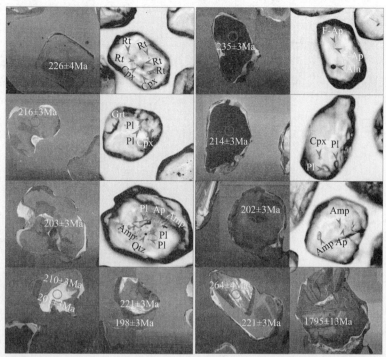

Fig. 5 Plane-polarized light images of mineral inclusions and CL images of zircon from granulites in NDT. The 30μm spots and ages of Cameca 1280 analysis are also shown.

Thirty U-Pb analyses of the metamorphic domains yield a broad age spectrum ranging from 235.4 ± 3.5 to 191.6 ± 3.0 Ma (Fig. 6B). All analyses fall near the concordia with discordance lower than 5%. In combination with the features of mineral inclusions and trace elements, nevertheless, three groups of ages can be subdivided as follows (Fig. 6B, D).

4.2.1 The first group (Zir-I)

It contains the mineral inclusion of Cpx ± Grt ± Rt without plagioclase, and/or exhibits negligible to slightly positive Eu anomalies in REE patterns. Six zircon domains contain the mineral inclusions of Rt + Cpx (three grains), Cpx + Grt (two grains), or Cpx (one grain), without plagioclase (Table 3). Other three zircon grains contain no mineral inclusions. All these zircon domains have trace elements characterized by low Th/U ratios (0.01–0.05), low REE contents (12.8–36.5 ppm), and flat HREE patterns with $(Dy/Yb)_N$ = 0.41–0.53 and negligible to slightly positive Eu anomalies with Eu/Eu^* = 0.92–1.79 (Fig. 6D). Nine U-Pb analyses yield $^{206}Pb/^{238}U$ ages of 220.9 ± 3.5 to 229.3 ± 3.5 Ma (Table 3) and define a weighted mean age of 223.8 ± 2.3 Ma with MSWD = 0.8 (Fig. 6B). One zircon grain has the oldest $^{206}Pb/^{238}U$ age of 235 ± 3.5 Ma and thus is not included in the calculation of weighted mean age for the first group, because it has the most negative Eu anomaly (Eu/Eu^*=0.74) and the lowest $(Dy/Yb)_N$ ratio of 0.39, and contains the mineral inclusions of only fluorapatite and allanite.

4.2.2 The second group (Zir-II)

These zircon domains are distinct from the first zircon group in that they either contain plagioclase inclusions or exhibit negative Eu anomalies in REE patterns. Mineral inclusions in the second zircon group are Pl + Cpx ± Grt ± Qtz ± Ap (three grains), Pl ± Qtz ± Ap (two grain) and Qtz + Grt (one grains). No rutile inclusions have been detected (Table 3). Their trace elements are characterized by low Th/U ratios (0.01–0.06), low REE contents (8.22–34.5 ppm), and flat HREE patterns with $(Dy/Yb)_N$ = 0.31–0.73. Their REE patterns are generally similar to those of the first zircon group except negative Eu anomalies (Fig. 6D). More than half of the zircons have the negative Eu anomalies with Eu/Eu^* = 0.36–0.67. Although four zircon domains exhibit negligible Eu anomaly with Eu/Eu^* = 0.94–1.17, the occurrence of plagioclase inclusions within them suggests that these zircons have also formed in the stability field of plagioclase. Ten U-Pb analyses on these zircon domains yield $^{206}Pb/^{238}U$ ages of 208.7 ± 3.2 to 217.6 ± 3.2 Ma (Table 3) and define a weighted mean age of 213.3 ± 2.1 Ma with MSWD = 0.72 (Fig. 6B).

Table 3 Results of mineral inclusions, Cameca zircon U-Pb ages, and LA-ICPMS trace element (μg·g^{-1}) compositions for the calcsilicate granulite (0909QT-6) from Qingtian village in North Dabie Terrane.

Sample (0909QT-6) Spots	Mineral inclusion	$\frac{^{207}Pb}{^{235}U}$	±σ	$\frac{^{206}Pb}{^{238}U}$	±σ	$\frac{^{208}Pb}{^{232}Th}$	±σ	ρ	$\frac{^{207}Pb}{^{206}Pb}$	±σ	$\frac{^{207}Pb}{^{235}U}$	±σ	$\frac{^{206}Pb}{^{238}U}$	±σ	Dis.%	f$_{206}$%
0909QT-6@32	—	0.15545	2.93	0.0221	1.53	0.01	5.84	0.52014	237.90	56.80	146.7	4.0	141.1	2.1	4	{0.09}
0909QT-6@11	Amp	0.21670	3.86	0.0320	1.53	0.01	6.87	0.39529	152.00	81.00	199.2	7.0	203.2	3.1	−2	{0.00}
0909QT-6@14	Amp + Ap	0.22573	2.71	0.0319	1.64	0.01	7.47	0.60414	255.60	49.00	206.7	5.1	202.4	3.3	2	{0.00}
0909QT-6@21	Pl + Qtz	0.22611	3.52	0.0312	1.51	0.01	7.99	0.42729	310.60	70.90	207.0	6.6	198.0	2.9	5	{0.00}
0909QT-6@22	Amp + Pl + Qtz + Ap	0.22352	3.19	0.0320	1.62	0.01	8.57	0.50706	224.90	62.40	204.8	5.9	203.1	3.2	1	{0.00}
0909QT-6@25	Amp + Qtz	0.22119	2.71	0.0323	1.55	0.01	9.76	0.57364	179.70	50.90	202.9	5.0	204.9	3.1	−1	{0.12}
0909QT-6@35	Amp + Qtz	0.21422	3.43	0.0314	1.60	0.01	16.64	0.46526	173.60	69.30	197.1	6.2	199.0	3.1	−1	{0.00}
0909QT-6@38	Amp + Ap	0.21502	3.06	0.0306	1.50	0.01	7.91	0.49192	241.30	60.20	197.8	5.5	194.1	2.9	2	{0.00}
0909QT-6@40	Qt + Mus	0.21419	3.06	0.0302	1.57	0.01	6.03	0.51385	263.50	59.20	197.1	5.5	191.6	3.0	3	{0.08}
0909QT-6@41	Qtz + Ap + Mus	0.21551	3.66	0.0324	2.14	0.01	15.39	0.58433	109.10	68.70	198.2	6.6	205.7	4.3	−4	{0.00}
0909QT-6@7	Pl + Qtz	0.21174	9.88	0.0318	1.52	0.01	19.09	0.15351	114.40	215.40	195.0	17.7	201.7	3.0	−3	{0.24}
0909QT-6@9	Pl + Cpx + Grt + Qtz + Ap	0.21903	4.34	0.0329	1.54	0.01	46.62	0.35437	113.20	93.10	201.1	8.0	208.7	3.2	−4	{0.00}
0909QT-6@6	—	0.23189	3.86	0.0331	1.53	0.01	8.50	0.39574	228.80	79.90	211.8	7.4	210.2	3.2	1	{0.07}
0909QT-6@27	Pl	0.22013	2.70	0.0333	1.66	0.01	12.50	0.61446	93.90	49.70	202.0	5.0	211.4	3.5	−4	{0.12}
0909QT-6@43	Pl + Qtz + Ap	0.22835	3.38	0.0335	1.51	0.01	18.78	0.44497	170.50	69.30	208.8	6.4	212.3	3.1	−2	{0.00}
0909QT-6@23	Qtz + Grt	0.23158	4.99	0.0337	1.83	0.01	10.49	0.36750	186.40	104.70	211.5	9.6	213.8	3.9	−1	{0.00}
0909QT-6@12	Pl + Cpx + Qtz	0.23760	1.95	0.0338	1.53	0.01	6.52	0.78247	243.00	27.80	216.5	3.8	214.0	3.2	1	{0.00}
0909QT-6@29	—	0.23226	3.11	0.0338	1.51	0.01	8.62	0.48545	185.40	62.20	212.1	6.0	214.5	3.2	−1	{0.00}
0909QT-6@8	—	0.23160	2.73	0.0339	1.69	0.01	7.39	0.61938	176.50	49.20	211.5	5.2	214.7	3.6	−1	{0.00}
0909QT-6@19	Pl + Cpx + Grt	0.24567	3.30	0.0342	1.61	0.01	8.87	0.48722	293.00	64.50	223.1	6.6	216.5	3.4	3	{0.00}
0909QT-6@26	—	0.24777	2.53	0.0343	1.51	0.01	10.84	0.59924	300.00	45.50	224.8	5.1	217.6	3.2	3	{0.00}
0909QT-6@37	Rt + Cpx	0.24694	2.56	0.0349	1.60	0.01	6.76	0.62454	258.20	45.20	224.1	5.2	220.9	3.5	1	{0.00}
0909QT-6@2	—	0.23695	3.26	0.0349	1.50	0.01	15.71	0.46079	157.00	66.30	215.9	6.4	221.4	3.3	−2	{0.12}

Continued

Sample (0909QT-6) Spots	Mineral inclusion	$\frac{^{207}Pb}{^{235}U}$	±σ	$\frac{^{206}Pb}{^{238}U}$	±σ	$\frac{^{208}Pb}{^{232}Th}$	±σ	ρ	$\frac{^{207}Pb}{^{206}Pb}$	±σ	$\frac{^{207}Pb}{^{235}U}$	±σ	$\frac{^{206}Pb}{^{238}U}$	±σ	Dis.%	f_{206}%
0909QT-6@20	Cpx	0.24107	3.72	0.0350	1.58	0.01	19.08	0.42508	195.5	76.4	219.3	7.4	221.5	3.4	−1	{0.20}
0909QT-6@16	Grt + Cpx	0.25232	3.09	0.0350	1.56	0.01	20.65	0.50527	296.8	59.7	228.5	6.3	221.9	3.4	3	{0.09}
0909QT-6@3	Grt + Cpx	0.25190	3.56	0.0352	1.82	0.01	31.88	0.51135	281.9	68.6	228.1	7.3	222.9	4.0	2	{0.13}
0909QT-6@31	−	0.24638	2.75	0.0352	1.59	0.01	8.83	0.57699	230.3	51.0	223.6	5.5	223.0	3.5	0	{0.00}
0909QT-6@18	Rt + Cpx	0.24405	2.78	0.0357	1.68	0.01	8.56	0.60305	176.6	50.9	221.7	5.5	226.0	3.7	−2	{0.00}
0909QT-6@34	Rt + Cpx	0.24482	3.22	0.0360	1.56	0.01	18.41	0.48574	164.4	64.4	222.4	6.4	227.9	3.5	−2	{0.15}
0909QT-6@5	−	0.25308	2.16	0.0362	1.53	0.01	24.48	0.71045	226.3	34.7	229.1	4.4	229.3	3.5	0	{0.03}
0909QT-6@4	F − Ap + Aln	0.26116	1.93	0.0372	1.50	0.01	8.31	0.77804	237.7	27.7	235.6	4.1	235.4	3.5	0	{0.00}
0909QT-6@10	−	0.28659	2.24	0.0386	1.50	0.02	3.83	0.67040	366.7	37.1	255.9	5.1	243.9	3.6	5	{0.05}
0909QT-6@1	−	0.30459	3.38	0.0419	1.50	0.02	3.95	0.44532	317.9	67.3	270.0	8.0	264.5	3.9	2	{0.15}
0909QT-6@15	Pl + Qtz + Ap	0.38594	2.87	0.0503	1.54	0.03	4.21	0.53610	440.4	53.0	331.4	8.1	316.1	4.7	5	{0.00}
0909QT-6@30	Pl + Qtz + Mus	0.39738	2.43	0.0524	1.55	0.03	4.32	0.63716	413.7	41.3	339.7	7.0	329.0	5.0	3	{0.07}
0909QT-6@42	−	0.49390	2.83	0.0650	1.51	0.03	3.67	0.53234	417.1	52.7	407.6	9.6	405.9	5.9	0	{0.19}
0909QT-6@13	Pl + Qtz	0.62549	2.05	0.0764	1.50	0.04	3.61	0.73283	582.1	30.0	493.3	8.0	474.4	6.9	4	{0.09}
0909QT-6@33	−	1.15829	1.79	0.0900	1.58	0.03	3.95	0.88065	1494.5	15.9	781.2	9.8	555.6	8.4	41	{0.14}
0909QT-6@17	−	2.12790	1.68	0.1406	1.53	0.05	3.94	0.90744	1795.0	12.8	1157.9	11.7	848.3	12.1	36	{0.04}

Continued

Spots	Ti	La	Ce	Pr	Nd	Sm	Eu	Gd	Tb	Dy	Ho	Er	Tm	Yb	Lu	Y	Th	U	Eu/Eu*	Dy/Yb$_{(N)}$	Th/U	REE	T_{Ti}(℃)
0909QT-6@32	10.20	—	—	—	—	—	—	—	—	—	—	—	—	—	—	—	—	—	—	—	—	—	742
0909QT-6@11	42.50	0.05	5.76	0.03	0.47	0.72	0.31	2.73	1.25	19.70	9.30	58.70	19.80	258.00	67.10	328.0	32.40	838	0.69	0.05	0.04	444.00	886
0909QT-6@14	15.60	0.01	8.36	0.08	1.35	2.82	1.69	12.70	3.90	38.10	11.70	46.10	8.67	78.60	15.10	342.0	83.80	397	0.86	0.32	0.21	229.00	781
0909QT-6@21	6.92	0.03	5.25	0.03	0.33	0.92	0.32	4.06	1.39	17.90	7.16	33.80	7.63	82.10	19.80	229.0	59.20	420	0.51	0.15	0.14	181.00	710
0909QT-6@22	24.90	0.02	2.42	0.01	0.00	0.45	0.25	1.67	0.67	6.17	1.65	6.19	1.14	8.04	1.70	57.0	16.90	459	0.89	0.51	0.04	30.40	828
0909QT-6@25	6.27	0.00	4.56	0.02	0.39	1.35	0.37	10.00	3.07	39.10	15.20	60.40	12.40	117.00	23.30	436.0	45.30	189	0.31	0.22	0.24	287.00	702
0909QT-6@35	18.80	0.03	10.80	0.22	3.09	5.52	1.25	32.40	10.70	122.00	45.70	202.00	39.60	342.00	67.10	1385.0	189.00	264	0.29	0.24	0.71	882.00	799
0909QT-6@38	25.10	0.02	1.47	0.00	0.00	0.07	0.07	0.87	0.34	3.13	0.90	3.56	0.72	6.28	1.34	34.5	16.60	358	0.86	0.33	0.05	18.80	829
0909QT-6@40	4.15	0.03	6.95	0.04	0.68	1.14	0.38	7.98	2.68	32.70	13.30	60.90	12.70	119.00	25.90	406.0	111.00	419	0.38	0.18	0.26	284.00	669
0909QT-6@41		0.00	3.14	0.04	0.52	1.20	0.36	7.50	2.81	33.30	11.70	50.30	10.00	84.50	17.30	342.0	19.70	202	0.37	0.26	0.10	2234.00	
0909QT-6@7	4.04	—	—	—	—	—	—	—	—	—	—	—	—	—	—	—	—	—	—	—	—	—	667
0909QT-6@9	7.93	0.00	1.23	0.00	0.04	0.39	0.20	2.72	0.80	7.86	2.07	6.57	1.14	9.51	1.83	73.0	11.40	835	0.58	0.55	0.01	34.40	721
0909QT-6@6	5.21	0.00	1.83	0.00	0.00	0.44	0.19	1.71	0.48	4.14	1.26	3.95	0.55	5.08	1.13	40.7	9.11	299	0.67	0.55	0.03	20.80	687
0909QT-6@27	7.55	0.01	1.82	0.01	0.05	0.27	0.22	1.92	0.49	4.32	1.07	4.09	0.59	4.83	0.87	42.1	10.60	362	0.94	0.60	0.03	20.50	717
0909QT-6@43	6.04	0.00	1.28	0.00	0.00	0.08	0.08	0.62	0.24	2.59	0.75	2.94	0.46	4.06	0.74	28.5	10.00	220	1.17	0.43	0.05	13.80	699
0909QT-6@23	—	0.00	1.29	0.00	0.00	0.10	0.03	0.60	0.19	2.40	0.63	2.40	0.41	3.75	0.65	25.4	10.40	220	0.42	0.43	0.05	12.40	
0909QT-6@12		0.00	1.11	0.00	0.00	0.21	0.22	2.21	0.74	7.80	1.95	6.49	1.12	10.60	2.03	65.2	10.90	709	1.00	0.49	0.02	34.50	711
0909QT-6@29	7.06	0.00	1.28	0.00	0.00	0.10	0.05	0.99	0.29	2.42	0.61	2.22	0.31	3.24	0.56	24.7	10.40	246	0.44	0.50	0.04	12.10	715
0909QT-6@8	7.38	0.00	1.42	0.00	0.00	0.16	0.04	0.70	0.31	2.21	0.65	2.80	0.39	4.75	0.72	28.0	11.90	270	0.36	0.31	0.04	14.10	729
0909QT-6@19	8.75	0.03	0.95	0.00	0.00	0.10	0.09	0.61	0.19	1.77	0.50	1.81	0.24	1.62	0.36	19.5	10.20	177	1.09	0.73	0.06	8.23	

Continued

Spots	Trace element composition (ppm)																					Eu/Eu*	Dy/Yb$_{(N)}$	Th/U	REE	T_{Ti}(°C)
	Ti	La	Ce	Pr	Nd	Sm	Eu	Gd	Tb	Dy	Ho	Er	Tm	Yb	Lu	Y	Th	U								
0909QT-6@26	–	–	–	–	–	–	–	–	–	–	–	–	–	–	–	–	–	–	–	–	–	–	–	–	–
0909QT-6@37	9.57	0.00	1.40	0.00	0.04	0.10	0.12	0.73	0.29	2.78	0.81	2.75	0.46	3.51	0.68	28.6	11.20	251	1.28	0.53	0.04	13.7	737
0909QT-6@2	12.33	0.00	1.83	0.01	0.06	0.20	0.20	1.46	0.47	4.43	1.29	4.53	0.74	5.79	1.13	45.6	15.70	347	1.12	0.51	0.05	22.1	760
0909QT-6@20	5.88	0.00	2.17	0.00	0.08	0.33	0.30	2.42	0.61	5.36	1.40	4.49	0.86	7.28	1.31	49.6	9.01	413	1.02	0.49	0.02	26.6	696
0909QT-6@16	–	–	–	–	–	–	–	–	–	–	–	–	–	–	–	–	–	–	–	–	–	–	–
0909QT-6@3	9.60	0.00	1.62	0.00	0.00	0.14	0.14	0.89	0.33	2.89	1.06	2.88	0.50	4.75	0.81	32.3	13.60	422	1.23	0.41	0.03	16.0	737
0909QT-6@31	3.28	0.00	1.33	0.00	0.00	0.10	0.24	1.70	0.79	7.00	2.16	8.69	1.44	10.80	2.24	81.0	7.42	271	1.79	0.43	0.03	36.5	652
0909QT-6@18	7.44	0.00	1.17	0.00	0.05	0.04	0.09	0.88	0.23	2.45	0.75	2.54	0.32	3.61	0.63	28.0	8.79	210	1.39	0.45	0.04	12.8	716
0909QT-6@34	7.94	0.00	2.28	0.01	0.07	0.15	0.19	1.61	0.42	4.35	1.22	5.10	0.88	6.75	1.15	42.8	20.00	514	1.18	0.43	0.04	24.2	721
0909QT-6@5	–	0.01	0.73	0.00	0.00	0.12	0.13	1.52	0.63	5.85	1.97	6.45	1.01	9.01	1.53	64.3	7.93	986	0.92	0.43	0.01	29.0	
0909QT-6@4	4.64	0.00	1.17	0.00	0.08	0.34	0.23	2.53	0.92	9.31	2.60	9.98	1.82	16.00	3.28	94.4	14.60	1113	0.74	0.39	0.01	48.2	678
0909QT-6@10	–	–	–	–	–	–	–	–	–	–	–	–	–	–	–	–	–	–	–	–	–	–	–
0909QT-6@1	10.04	0.20	28.00	0.50	8.55	13.90	1.01	74.20	22.50	246.00	88.70	373.00	72.60	645.00	132.00	2629.0	296.00	449	0.10	0.26	0.66	1707.0	741
0909QT-6@15	80.67	0.01	5.47	0.09	1.53	2.85	0.76	15.70	5.20	63.90	22.60	95.20	18.50	159.00	30.40	643.0	30.50	66	0.35	0.27	0.46	421.0	965
0909QT-6@30	–	–	–	–	–	–	–	–	–	–	–	–	–	–	–	–	–	–	–	–	–	–	–
0909QT-6@42	7.45	0.07	11.20	0.19	2.63	4.53	0.82	27.40	9.02	105.00	39.50	173.00	34.80	317.00	62.10	1236.0	183.00	360	0.23	0.22	0.51	786.0	716
0909QT-6@13	4.09	0.35	13.60	0.13	1.07	1.45	0.29	11.80	3.94	59.70	27.00	135.00	33.30	344.00	80.60	846.0	67.80	194	0.21	0.12	0.35	713.0	668
0909QT-6@33	19.56	0.11	1.17	0.03	0.46	0.66	0.51	4.42	1.37	12.80	3.76	14.40	2.51	23.60	5.26	120.0	33.40	348	0.91	0.36	0.10	71.1	804
0909QT-6@17	21.81	0.25	1.44	0.07	0.53	0.58	0.91	2.95	0.90	7.43	2.51	11.00	2.84	30.70	7.15	86.7	18.60	593	2.11	0.16	0.03	69.3	814

Dis% = $(^{207}Pb/^{235}U - ^{206}Pb/^{238}U)/(^{206}Pb/^{238}U)$

(Dy/Yb)$_N$, Chondrite normalized ratio; Eu/Eu*=Eu$_N$/(Sm$_N$×Gd$_N$)$^{0.5}$; Ti-in-zircon temperature t_{Ti}(°C)=5080/(6.01-logTi$_{zircon}$)-273 (Watson et al., 2006)

4.2.3 The third group (Zir-III)

It contains the mineral inclusions of Amp ± Pl ± Qtz ± Ap (six grains) (Table 3), and/or has high REE contents with steep HREE patterns (eight grains) (Fig. 6D), with apparent $^{206}Pb/^{238}U$ ages < 208.7 ± 3.2 Ma. Two analyses have Th/U = 0.04 and 0.05; REE = 30.4 and 18.8 ppm; Eu/Eu* = 0.89 and 0.86; and (Dy/Yb)$_N$ = 0.51 and 0.33, which are similar to the trace element composition of the second zircon group. Nevertheless, they contain the mineral inclusions of Amp + Pl + Qtz + Ap and Amp + Ap, respectively. This suggests that they would form by amphibolite-facies metamorphism (Table 3). The all other domains have REE patterns characterized by varying but high Th/U ratios (0.04-0.71), high REE contents (181–882 ppm), negative Eu anomalies (Eu/Eu* = 0.29–0.86) and steep HREE patterns with (Dy/Yb)$_N$ = 0.05–0.32 (Fig. 6D), though some of them contain mineral inclusions of only Pl + Qtz (two grains) or Qtz + Mus ± Ap (two grains). Ten U-Pb analyses on these zircon domains yield a relatively broad $^{206}Pb/^{238}U$ age ranging from 191.6 ± 3.0 to 205.7 ± 4.3 Ma, and define a weighted mean age of 199.9 ± 3.3 Ma with MSWD = 2.2 (Fig. 6B).

One zircon grain has a magmatic rim and gives a $^{206}Pb/^{238}U$ age of 141 ± 2 Ma (Table 3, Fig. 6A). This date is close to the time of migmatization in NDT (Wu et al., 2007).

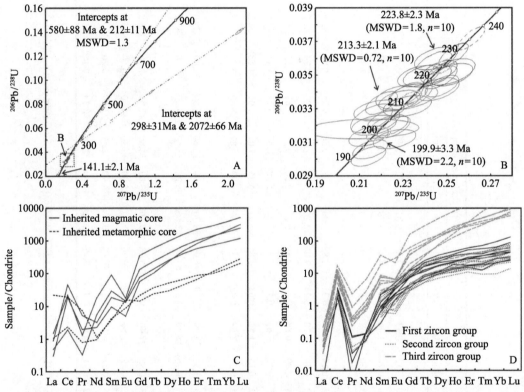

Fig.6 Concordia diagrams of zircon U-Pb dating by Cameca-1280 technique for the granulite in NDT and chondrite-normalized REE patterns of different groups of zircon domains. A: Concordia diagram for all zircon domains; B: Concordia diagram for metamorphic overgrown zircon domains; C: REE patterns of inherited metamorphic and magmatic zircon cores; D: REE patterns of the metamorphic overgrown zircon domains. Chondrite-normalized values are from Sun and McDonough (1989).

5 Geological meaning of zircon U-Pb ages

5.1 Inherited ages

Weakly oscillatory zoning for majority of the inherited cores indicates that they are igneous zircons that have been variably modified by metamorphic recrystallization (e.g., Corfu et al., 2003; Wu and Zheng, 2004). The U-Pb data from these magmatic cores suggest that they were formed during a Neoproterozoic event and later influenced by the Triassic thermal event. Other two inherited cores with flat HREE patterns may suggest that they were formed during the high-grade metamorphism in the early Proterozoic. The similar ages were also reported for UHP metamorphic rocks in the Dabie orogen (Zheng et al., 2004; Wu et al., 2008; Xia et al., 2009).

5.2 Metamorphic ages

For Triassic metamorphic zircons, mineral inclusions retained in zircons give direct information on coexisted minerals when zircons formed. However, the enclosed minerals in zircon grains may not necessarily cover the complete paragenetic mineral assemblages. Even so, the growth of minerals such as garnet (for HREE), allanite (for LREE and MREE) and plagioclase (for Eu) can influence the REE patterns of contemporarily crystallized zircons. Therefore, zircon trace elements, combined with mineral inclusions, can efficiently monitor the petrogenesis of coexisting mineral assemblages and characterize the metamorphic environment at the time when zircons formed.

(1) The first group (Zir-I): It shows low REE contents (12.8–36.5 ppm) and flat HREE patterns with $(Dy/Yb)_N$ = 0.41–0.53 and no negative Eu anomalies (Fig. 7A, B). This suggests that the zircon was synchronously precipitated with REE-enriched minerals such as allanite, apatite and garnet in the absence of plagioclase (Schaltegger et al., 1999; Rubatto, 2002; Rubatto and Hermann, 2003; Xia et al., 2010). The inferred mineral assemblages based on trace elements are coupled with mineral inclusions (Cpx ± Grt ± Rt) directly observed from zircon domains, indicating that the first group of zircon was formed under HP to UHP eclogite-facies conditions. In this regard, the Zir-I U-Pb age of 223.8 ± 2.3 Ma may represent the eclogite-facies metamorphic time. Using the Ti-in-zircon thermometer of Watson et al. (2006), Zir-I gives metamorphic temperatures of 652 to 760 ℃ with a mean of 712 ℃.

As mentioned above, the peak metamorphic ages for NDT have been a controversial issue for a long time. Li et al. (1993) first reported a Sm-Nd isochron age of 244 ± 11 Ma defined by garnet ± diopside from the Raobazhai garnet-pyroxenite. However, as discussed by Li et al. (2000) later, this age is probably meaningless as the dated diopside is a retrograded product from omphacite. On the other hand, Liu et al. (2005) obtained a Sm-Nd

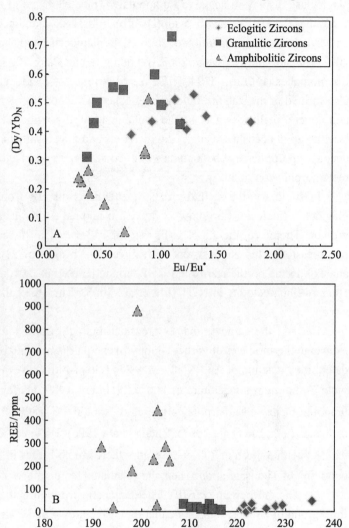

Fig. 7 Selected trace element plots of metamorphic zircons from the granulite. A: $(Dy/Yb)_N$ vs Eu/Eu*; B: REE content vs Age.

isochron age of 212 ± 4 Ma (Liu et al., 2005) from eclogite-facies metamorphic minerals (e.g., garnet, omphacite, rutile). Since NDT has undergone a unique >800 ℃ granulite-facies metamorphism during its retrogression, this age may not be responsible for the peak UHP metamorphic event in NDT, but a reworked age due to the granulite-facies retrogression. Liu et al. (2007b) reported an zircon U-Pb age of 218 ± 3 Ma for the UHP gneiss from NDT, and suggested that it is the best estimate for the peak UHP metamorphic time. However, Liu et al. (2011a) concluded that the peak UHP metamorphic age for NDT is 226 ± 3 Ma based on U-Pb dating of metamorphic zircon domains that contain coesite inclusions and show flat HREE patterns. Liu et al. (2011a) reinterpreted the zircon U-Pb age of 218 ± 3 Ma as the mean age of possible poly-metamorphic events (e.g., 224 ± 3 Ma and 214 ± 3 Ma, respectively) in zircons from their sample LT8-3. The Zir-I U-Pb age of 223.8 ± 2.3 Ma from this study is consistent within error limits with the well-constrained zircon U-Pb age of 226 ± 3 Ma by Liu et al. (2011a). Although the compositions of garnet and clinopyroxene inclusions in Zir-I are not analyzed in this study, the consistency between our U-Pb age of

223.8 ± 2.3 Ma and that recorded by coesite-bearing zircon domains from eclogite in NDT lends support to the interpretation that the first group of metamorphic zircons were formed at the UHP eclogite-facies metamorphic conditions. As a result, the first group of zircons is termed as "eclogitic zircons".

(2) The second group (Zir-II): In contrast to the eclogitic zircons, the second group of zircons contains plagioclase inclusions, indicating their growth in the stability field of plagioclase and therefore relatively lower pressure than the UHP eclogite facies. Since plagioclase is the main carrier of europium, zircon coexisting with plagioclase displays a negative Eu anomaly. Trace elements in most Zir-II are characterized by low REE contents and rather flat HREE patterns with $(Dy/Yb)_N = 0.31–0.73$ and negative Eu anomalies (Fig. 7A, B). This is consistent with synchronous precipitation of zircon with plagioclase and garnet under granulite-facies metamorphism (Rubatto, 2002; Rubatto and Hermann, 2003; Whitehouse and Platt, 2003). Although four zircon analyses exhibit negligible Eu anomalies ($Eu/Eu^* = 0.94–1.17$), the occurrence of plagioclase inclusions within them indicates that they were also formed in the stability field of plagioclase. One possibility for the apparent decoupling between trace elements and mineral inclusions may be the contamination of plagioclase inclusions during the LA-ICP-MS analysis as the laser ablation needs more analytical materials and thus results in loss of spatial resolution over SIMS. This apparent decoupling implies that both studies of zircon trace elements and mineral inclusions should be used to constrain the metamorphic condition.

In summary, the second group of zircons is distinct from the eclogitic zircons in that they either contain plagioclase inclusions or show negative Eu anomalies, indicating the important role of plagioclase during their growth. Therefore, the Zir-II U-Pb age of 213.3 ± 2.1 Ma could be the best estimate for the granulite-facies retrogression. This is consistent with zircon U-Pb ages of 210–212 Ma for granulite-facies metamorphism and anatexis in the northeastern Sulu UHP terrane (Liu et al., 2010a) and a Sm-Nd cooling age of 212 ± 4 Ma for the Huangweihua eclogite in NDT (Liu et al., 2005). Therefore, the second group of zircons is termed as "granulitic zircons".

Because the garnet is mainly grossular-andradite solid solution with low Prp + Alm components, the Fe-Mg exchange thermometer of garnet-clinopyroxene pair is not suitable for our studied sample. Based on experimentally determined reaction of An + Cal = Mei-Scp (Goldsmith & Newton, 1977), the scapolites in granulites with Mei_{72-74} yield a minimum temperature of 800 ℃ (Harley, 1989; Harley and Buick, 1992; Zhang et al., 1996). Hence, the granulite-facies metamorphic temperature should be >800℃, which is consistent with previous temperature estimates (e.g., Liu et al., 2007a; Zhang et al., 1996). However, Zrn-II yields Ti-in-zircon temperature of 667 ℃ to 729 ℃ with a mean of 706 ℃. Because there are no rutiles in Zrn-II, zircon crystallization would occur under the condition of Ti unsaturation. Then, the calculated temperatures are underestimates.

(3) The third group (Zir-III). The diagnostic criterion is the pervasive presence of amphibole inclusions in the third group of zircons, suggesting that they have grown under amphibolite-facies condition. They show steep HREE patterns with $(Dy/Yb)_N = 0.05–0.32$ and significant negative Eu anomaly with $Eu/Eu^* = 0.29–0.86$ (Fig. 7A), suggesting that the zircons grew simultaneously with plagioclase rather than garnet. The high HREE contents cannot be supplied by breakdown of garnet during the amphibolite-facies metamorphism, because abundant garnets are still present in sample 0909QT-6. Instead, an externally-derived HREE-rich fluid is appealed for the amphibolite-facies overprinting. Two analyses have low REE contents and flat HREE patterns (Fig. 6D), implying the geochemical heterogeneity in the external HREE-rich fluid. Although their REE patterns are similar to those of granulite zircons, the occurrence of amphibole inclusions within the zircons demonstrates that the zircons formed during amphibolite-facies retrogression as well. Therefore, the Zir-III U-Pb age of 199.9 ± 3.3 Ma dates the amphibolite-facies overprinting. Zrn-III shows large variation of Ti contents ranging from 4.15 to 42.49 ppm to yield Ti-in-zircon temperature of 669 ℃ to 886 ℃.

Zircon U-Pb ages of 190 to 200 Ma were also sporadically reported in previous studies. Liu et al. (2007b) reported a weight mean of 191 ± 5 Ma (MSDW=3.8) for zircon rims from a garnet-bearing gneiss and interpreted it as the amphibolite-facies retrograde metamorphic time of NDT. This age is consistent with the lowest individual age of 191.6 ± 3.0 Ma in the third group of zircons. Liu et al. (2011a) also reported a concordant U-Pb age of 199 ± 2 Ma for metamorphic zircon from eclogite in NDT, which is consistent with the presently weighted mean age for the third group of zircons. Wang et al. (2002) reported a biotite $^{40}Ar/^{39}Ar$ plateau age of 195 ± 2 Ma for felsic granulite at Huangtuling. This age might represent the lowest estimate of the amphibolite-facies metamorphism, as the closure temperature of Ar diffusion in biotite is about 300 ± 50 ℃(Harrison et al., 1985). Consequently, the Zir-III U-Pb age of 199.9 ± 3.3 Ma could be the best mean for the amphibolite-facies retrogression in NDT. The third group of zircons is thus termed as "amphibolitic zircons".

5.3 Constraints on exhumation history of NDT

The three groups of metamorphic zircons have been distinguished in the granulite from NDT on the basis of U-Pb ages, REE patterns and mineral inclusions. In this regard, a complete metamorphic evolution history can be defined for the target granulite. Obviously, it experienced UHP eclogite-facies metamorphism at 223.8 ± 2.3 Ma, followed by HP granulite-facies retrogression at 213.3 ± 2.1 Ma, and subsequent amphibolite-facies overprinting at 199.9 ± 3.3 Ma. Because the granulite outcrop is coherent with the regional orthogneiss and the three groups of zircon U-Pb ages are comparable with previous reported ages from NDT (Fig. 8), the exhumation history of granulite can be retrieved with a link to the bulk NDT.

Liu et al. (2011a) obtained the U-Pb age of 226 ± 3 Ma for the UHP eclogite-facies metamorphic event in NDT. This age may mark the termination of UHP metamorphic period in the Dabie-Sulu orogenic belt, which lasts from ~240 to ~225 Ma for the mid-T/UHP and low-T/UHP metamorphic rocks (Liu et al., 2006; Wu et al., 2006; Zheng et al., 2009; Liu and Liou, 2011). The results obtained in this study indicate that NDT experienced the UHP eclogite-facies metamorphism when CDT reached the maximum metamorphic temperature at ~226 Ma (Li et al., 2000; Zheng et al., 2002). It is known that CDT and NDT are coherent subducted units of the upper and lower crust during the Triassic continental collision (Li et al., 2003; Liu et al., 2007a, b; Zhao et al., 2008). They experienced UHP metamorphism together at mantle depths, but zircon U-Pb age records show the difference due to the fluid action on growth of metamorphic zircon (Zheng, 2009).

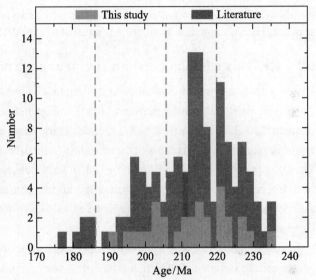

Fig. 8 Histogram plot showing the Triassic ages given in this study together with reported SHRIMP U-Pb data from eclogite and gneiss in NDT (data with error <5% and discordance <10%). Data sources: Liu et al., (2007a, b), Xie et al., (2010) and Liu et al., (2011b).

During their exhumation, CDT was decoupled with NDT and rapidly uplifted to the middle crust level with amphibolite-facies overprinting at around 219–209 Ma (e.g., Li et al., 2000; Ayers et al., 2002), but NDT only exhumed to the lower crustal level for the granulite-facies retrogression at about 215 Ma. Further collision between SCB and NCB might cause the amphibolite-facies overprinting on the NDT granulite at 199.9 ± 3.3 Ma. It is interesting that the ~200 Ma event was not recorded in CDT, suggesting the decoupling between NDT and CDT during their exhumation to the middle crustal level.

If we assume that the maximum pressure is achieved at the UHP eclogite facies whereas the maximum temperature is attained at the HP granulite facies, NDT would experience decompression heating during exhumation. The similar P-T-t path was also obtained for mid-T/UHP and low-T/UHP metamorphic rocks in the Dabie orogen (Gao et al., 2011; Zheng et al., 2011). The differences between the different UHP terranes lie in concret P-T-t path, with the highest T in the north but the lowest T in the south. While the peak UHP eclogite-facies metamorphism took place at ~240 Ma in the southern low-T zone (Li et al., 2004; Zheng et al., 2007), the peak UHP eclogite-facies metamorphism took place at ~225 Ma in the northern high-T zone. Such differences in P-T-t path between the three UHP terranes in the Dabie orogen indicate crustal detachment during the Triassic continental collision, with differential exhumation of the different slices from mantle depths.

6 REE behaviors during retrograde metamorphism

Based on the trace element data of zircon, garnet, clinopyroxene and amphibole from the granulite in NDT, we can place constraints on geochemical behaviors of REE in these minerals during the continental subduction-zone metamorphism.

6.1 Why garnet has higher LREE and MREE contents than that from eclogite?

The Ca-rich garnet (grandite) occurs in the studied calc-silicate granulite and it has higher LREE and MREE contents than those almandine-rich garnets from UHP eclogite in the Dabie orogen (Fig. 3B). This difference may be caused by the Ca-rich nature of grandite. As experimentally determined by Van Westrenen et al. (1999), HREE partition coefficients between garnets and melt, $D_{HREE(Grt/Melt)}$, are almost the same regardless of garnet composition. For instance, $D_{Yb(Grt/Melt)}$ values are 3.3 and 2.8 for $Py_{84}Gr_{16}$ and Py_9Gr_{91}, respectively; and HREE pattern of garnet becomes flatter when garnet changes to Ca-rich (e.g., $D_{Yb/Tb}$ are 3 and 1.3 for $Py_{84}Gr_{16}$ and Py_9Gr_{91}, respectively). By contrast, $D_{LREE+MREE(Grt/Melt)}$ values are very sensitive to the garnet composition and elevated to >1 when garnet becomes Ca-rich (e.g., D_{Sm} are 0.28 and 1.3 for $Py_{84}Gr_{16}$ and Py_9Gr_{91}, respectively). It is thus suggested that the apparent HREE depletion of grandite is a result of the MREE elevation due to the high Ca content.

6.2 REE distribution between zircon and garnet

If zircon and garnet crystallized simultaneously, the REE distribution between them should be in equilibrium. Therefore, REE distribution between zircon and garnet helps to understand the geological meaning of U-Pb ages (Rubatto, 2002; Whitehouse and Platt, 2003; Harley and Kelly, 2007). No REE partition coefficients are available for grossular-andradite, but it is experimentally available for almandine rich garnet (Rubatto and Hermann, 2007). Thus, the experimental HREE partition coefficients between zircon and garnet are qualitatively applicable to natural samples, because the $D_{HREE(Grt/Melt)}$ values are almost constant for different compositions of garnet (Van Westrenen et al., 1999). We can compare the calculated $D_{HREE(Zrn/Grt)}$ values with experimental results presented by Rubatto and Hermann (2007).

The apparent $D_{HREE(Zrn/Grt)}$ values, calculated using the compositions of eclogitic zircons in coupling with the garnet core in sample 0909QT-6, are far below the accepted equilibrium values (Fig. 9). This allows us to assess the two possible origins for the garnet core: (1) it was a relic from the protolith, or (2) it formed during the eclogite-facies stage but was modified for its REE by metamorphic fluid during the granulite-facies retrogression. The ragged and embayed structure for the garnet core suggests that it would have been dissolved and resorbed after its formation. REE contents of the corroded core are even higher than those of euhedral garnet in sample 0907QT-1A (Fig. 3B). Thus, the unusually high REE contents for the corroded core can be explained by its growth from REE-rich reservoir because the eclogitic garnet would preferentially uptake HREE relative to the granulitic garnet (Harley and Kelly, 2007). When comparing garnet from sample 0909QT-6 to that from 0907QT-1A, we find that the REE distribution of garnet in different granulite samples is heterogeneous. For example, garnet in 0909QT-6 occurs as large porphyroblasts with corroded cores, but that in 0907QT-1A occurs as small grains with euhedral shape. Small garnet grains in samples 0907QT-1A show characteristic REE contents that progressively decrease rimwards without a HREE gap between core and rim, suggesting that garnet was grown in a closed system. The inner rims of the porphyroblastic garnet in 0909QT-6 are very similar to garnets in 0907QT-1A in REE patterns. However, the large outer rims of garnet in 0909QT-6 show somewhat different REE patterns with positive Eu anomalies, which are analogous to those of garnet in skarns (Smith et al., 2004; Gaspar et al., 2008). Such differences may result from the effect of different components on the two samples during garnet growth. Fluid action may occur in 0909QT-6 during the later stage of granulite-facies retrogression, which could be represented by the outer rims of the porphyroblastic garnet.

Although garnet and zircon are coupled in the HREE distribution, the apparent $D_{HREE(Zrn/Grt)}$ values between the average granulitic zircons and the garnet rims fall in the range of experimental equilibrium values, suggesting that the granulitic zircons were simultaneously formed and in chemical equilibrium with the garnet rims (Fig. 9). Such coupled HREE distribution between garnet and zircon may be controlled by the high metamorphic temperature (>800 ℃) during the granulite-facies metamorphism in NDT, as the $D_{HREE(Zir/Grt)}$ values are very close to 1 at high temperatures of >850 ℃ (Rubatto and Hermann., 2007). Since sample 0909QT-6 is completely recrystallized and has not preserved the retrograde structure such as symplectite, the Zir-II U-Pb age of 213.3 ± 2.1 Ma dates the granulite-facies retrogression. The apparent $D_{HREE(Zrn/Grt)}$ values between the average amphibolitic zircons and the garnet rims is far above the accepted equilibrium values (Fig. 9), suggesting that the Zir-III has grown after the granulite-facies metamorphism, namely during the amphibolite-facies metamorphism.

6.3 REE partition between amphibole and clinopyroxene

As shown in Fig. 2, Cpx I and Cpx II are different in paragenetic relationships. While Cpx I is paragenetic with garnet and free from retrogression, Cpx II is locally replaced by Amp I. They are also distinct in trace element composition, e.g., Cpx II has high REE contents up to 15 times than Cpx I (Fig. 4). The cause for this compositional change in clinopyroxene can be evaluated by comparison of $D_{REE(Amp/Cpx)}$ values calculated using the different compositions of clinopyroxene and amphibole with accepted equilibrium $D_{REE(Amp/Cpx)}$ values. Accepted equilibrium $D_{REE(Amp/Cpx)}$ values range from 1 to 5 (Nehring et al., 2009; Pride and Muecke, 1981; Storkey et al., 2005). The apparent $D_{REE(Amp/Cpx)}$ values calculated between the average Amp I and

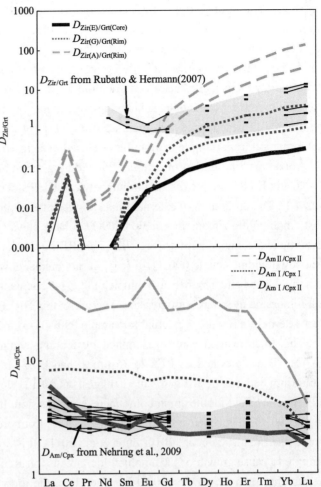

Fig. 9 REE distribution patterns between garnet and zircon, and between amphiboles and clinopyroxenes. The equilibrium $D_{REE(Zir/Grt)}$ values of Rubatto & Hermann (2007) and $D_{REE(Am/Cpx)}$ values of Nehring et al. (2009) are presented for comparison. Zir(E) = average eclogitic zircons; Zir(G) = average granulitic zircons; Zir(A) = average amphibolitic zircons; Grt(Core) = average garnet core; Grt(Rim) = garnet rims. Thick lines with the same color represent REE distribution partition between zircon and garnet rims with highest and lowest REE content, respectively.

the average Cpx II are 1.73-4.16 for Ce-Lu, and 6.00 for La, falling well in the range of equilibrium partition values (Fig.9). Combined with textural relationship, this equilibrium was likely to be established during amphibolite-facies metamorphism. Disequilibrium REE partition between Amp II and Cpx II is remarkable (Fig. 9). Textural and compositional differences between Amp II and Amp I lead us to argue that Amp II might be the recrystallized phase during later thermal event (e.g., migmatization in ~140 Ma). Disequilibrium is also prominent between Amp I and Cpx I (Fig. 9). This suggests that Cpx I represents the primary clinopyroxene that formed under the granulite-facies metamorphic conditions, without amphibolite-facies overprinting.

6.4 Closed versus open system

Mineral Sm-Nd and Rb-Sr geochronological studies indicate whether retrograde-metamorphism is operated in a closed or open system is important to the chronometric system of UHP metamorphic rocks (e.g., Li et al., 2000; Zheng et al., 2002). The garnet porphyroblast in 0909QT-6 and garnet in 0907QT-1A display systemic zoning with a decrease of REE contents from core to rim (Fig. 3A, B). Such a progressive depletion in HREE contents of garnet suggests that the garnet has grown in a closed system without supply of external REE reservoir (Rubatto, 2002;

Whitehouse and Platt, 2003). Consistently, flat to depleted HREE patterns for the eclogitic and granulitic zircons also indicate that the REE system kept closed from the eclogite-to granulite-facies retrograde metamorphism during the initial exhumation. In a closed system, the retrograde product may get the HREE supply from breakdown of garnet. It is favored by the high HREE contents of residual garnet core in 0909QT-6. The "closed system" attribution for the eclogite-facies and granulite-facies metamorphic conditions seems realistic because the scale of fluid activity during continental subduction-zone metamorphism is limited (Chen et al., 2007; Zheng et al., 2009). The mineral O isotope study of granulite from NDT by Zheng et al. (2001) also indicates that the metamorphic transformation from eclogite to granulite occurred in a closed system without infiltration of external fluids.

In contrast, infiltration of external fluids during amphibolite-facies retrogression has been documented by the Sm-Nd and Rb-Sr isotopic disequilibrium between UHP and retrograde from CDT (Li et al., 2000) and NDT (Xie et al., 2004). The infiltration of external fluids implies that amphibolite-facies metamorphism occurs in an open system. Most amphibolitic zircons in sample 0909QT-6 have elevated REE contents to about 20 times than those of both granulitic and eclogitic zircons (Fig. 7B). This REE elevation of amphibolitic zircons cannot be caused by mineral breakdown within sample 0909QT-6 (e.g., garnet breaks down to elevate the HREE contents and Dy/Yb ratios for the amphibolitic zircons), because sample 0909QT-6 has well preserved the granulite-facies porphyroblastic garnet, with minor and even rare retrogression. Therefore, this HREE elevation is more likely to be caused by introduction of an external REE source (external for sample 0909QT-6) from nearby country rocks.

This interpretation can be examined by looking at the partially retrograded sample 0907QT-1. Cpx II in 0907QT-1B has elevated its REE to 15 times than Cpx I in 0907QT-1B and 0907QT-1A (Fig. 4). The chemical equilibrium between Cpx II and Amp I suggests that REE elevation of Cpx II could be caused by the infiltration of REE-enriched fluid for the amphibolite-facies retrogression. It is possible that this REE-enriched fluid is internal for sample 0907QT-1B, because the fluid might get REE from breakdown of minerals such as garnet and allanite in the sample itself (e.g., Xia et al., 2010; Zhou et al., 2011). It is this REE-enriched fluid that could penetrate into nearby granulite to cause growth of the amphibolitic zircons, resulting in the high REE contents of amphibolitic zircons in the granulite such as sample 0909QT-6. Therefore, the growth of amphibolitic zircons in sample 0909QT-6 takes place in an open system. The field observation shows that the amphibolites-facies retrogression of granulite is more developed around the margin of boudins, while less in the core. This implies fluid flow along deformed domains with prograde infiltration into undeformed domains within UHP slices.

7 Conclusions

Polyphase zircons from calc-silicate granulite in the North Dabie Terrane record three stages of metamorphism at eclogite, granulite and amphibolite facies, respectively. NDT experienced UHP eclogite-facies metamorphism at 223.8 ± 2.3 Ma, granulite-facies retrogression at around 213.3 ± 2.1 Ma, and amphibolite-facies overprinting at 199.9 ± 3.3 Ma. It appears that the UHP metamorphic rocks in NDT terrane underwent a decompression heating during the initial exhumation, with the maximum temperature as high as granulite facies. This differs from UHP rocks in CDT and SDT where the maximum temperatures are below the granulite facies. Taken together, the three UHP slices in the Dabie orogen experienced different P-T-t paths during the Triassic continental collision, with crustal detachment during subduction and differential exhumation after peak UHP eclogite-facies metamorphism.

Metamorphic minerals in the granulite show different REE compositions with respect to the transformation from eclogite to granulite to amphibolite facies. While the eclogite-facies to granulite-facies retrogression took place in a closed system without supply of external REE source, the amphibolite-facies overprinting occurred in an open system with infiltration of external fluids. Both dehydration and hydration have taken place in NDT during the Triassic continental collision, with local transfer of aqueous fluids and trace elements between metamorphic reactants and products. The REE characteristics of precursor minerals in retrograde metamorphic rocks (e.g., garnet and clinopyroxene in the retrograde granulite or garnet) could be potentially modified by internal or external fluids during retrograde metamorphism with equilibrium and disequilibrium REE partitions between coexisting minerals.

Acknowledgments

We thank Yu Liu and Guoqiang Tang for their assistance with Cameca 1280 zircon U-Pb dating. Yican Liu, Ji Shen, Yongsheng He and Lijuan Chen are also thanked for their help on field work. Thanks are due to two anonymous reviewers for their comments that help improvement of the presentation. This study was supported by funds from the State Key Basic Research Development Program (Grant No. 2009CB825002) and the Natural Science Foundation of China (No. 40973016, 41090372 and 40921002).

References

Austrheim, H., 1987. Eclogitization of lower crustal granulites by fluid migration through shear zones. Earth and Planetary Science Letters 81, 221–232.

Ayers, J. C., Dunkle, S., Gao, S., Miller, C. F., 2002. Constraints on timing of peak and retrograde metamorphism in the Dabie Shan ultrahigh-pressure metamorphic belt, east-central China, using U-Th-Pb dating of zircon and monazite. Chemical Geology 186, 315-331.

Bingen, B., Austrheim, H., Whitehouse, M., Davis, W., 2004. Trace element signature and U-Pb geochronology of eclogite-facies zircon, Bergen Arcs, Caledonides of W Norway. Contributions to Mineralogy and Petrology 147, 671-683.

Chen, N. S., Sun, M. I. N., You, Z. D., Malpas, J., 1998. Well-preserved garnet growth zoning in granulite from the Dabie Mountains, central China. Journal of Metamorphic Geology 16, 213-222.

Chen, N. S., You, Z. D., Suo, S. T., Yang, Y., Li, H. M., 1996. U-Pb zircon ages of Dabieshan intermediate granulite and deformed granite. Chinese Science Bulletin 41, 1886-1890.

Chen, R. X., Zheng, Y. F., Gong, B., Zhao, Z. F., Gao, T. S., Chen, B., Wu, Y. B., 2007. Oxygen isotope geochemistry of ultrahigh-pressure metamorphic rocks from 200–4000 m core samples of the Chinese Continental Scientific Drilling. Chemical Geology 242, 51-75.

Chen, R.X., Zheng, Y. F., Xie, L.W., 2010. Metamorphic growth and recrystallization of zircon: Distinction by simultaneous in-situ analyses of trace elements, U-Th-Pb and Lu-Hf isotopes in zircons from eclogite-facies rocks in the Sulu orogen. Lithos 114, 132-154

Corfu, F., Hanchar, J.M., Hoskin, P. W. O., Kinny, P., 2003, Atlas of zircon textures: Reviews in Mineralogy and Geochemistry 53, 469-500.

Faure, M., Lin, W., Schärer, U., Shu, L., Sun, Y., Arnaud, N., 2003. Continental subduction and exhumation of UHP rocks. Structural and geochronological insights from the Dabieshan (East China). Lithos 70, 213-241.

Gao, X. Y., Zheng, Y. F., Chen, Y. X., 2011. U-Pb ages and trace elements in metamorphic zircon and titanite from UHP eclogite in the Dabie orogen: constraints on *P-T-t* path. Journal of Metamorphic Geology 29, 721-740.

Gaspar, M., Knaack, C., Meinert, L., Moretti, R., 2008. REE in skarn systems: a LA-ICP-MS study of garnets from the Crown Jewel gold deposit. Geochimica et Cosmochimica Acta 72, 185-205.

Goldsmith, J. R., Newton, R. C., 1977. Scapolite-plagiocalse stability relations at high pressure and temperatures in the system $NaAlSi_3O_8$-$CaAl_2Si_2O_8$-$CaCO_3$-$CaSO_4$. American Mineralogist 62, 1063-1081.

Hacker, B. R., Ratschbacher, L., Webb, L., Ireland, T., Walker, D., Shuwen, D., 1998. U/Pb zircon ages constrain the architecture of the ultrahigh-pressure Qinling-Dabie Orogen, China. Earth and Planetary Science Letters 161, 215-230.

Harley, S. L., 1989. The origins of granulites: a metamorphic perspective. Geological Magazine 126, 215-247.

Harley, S. L., Buick, I. S., 1992. Wollastonite-scapolite assemblage as indicators of granulite pressure-temperature-fluid histories: the Rauer Group, east Antarctica. Journal of Petrology 33, 693-728.

Harley, S. L., Kelly, N. M., 2007. The impact of zircon-garnet REE distribution data on the interpretation of zircon U-Pb ages in complex high-grade terrains: An example from the Rauer Islands, East Antarctica. Chemical Geology 241, 62-87.

Harrison, T., Duncan, I., McDougall, I., 1985. Diffusion of ^{40}Ar in biotite: Temperature, pressure and compositional effects. Geochimica et Cosmochimica Acta 49, 2461-2468.

He, Y. S., Li, S. G., Hoefs, J., Huang, F., Liu, S. A., Hou, Z. H., 2011. Post-collisional granitoids from the Dabie orogen: new evidence from partial melting of a thickened continental crust. Geochimica et Cosmochimica Acta 75, 3815-3838

Hermann, J., Rubatto, D., Korsakov, A., Shatsky, V. S., 2001. Multiple zircon growth during fast exhumation of diamondiferous, deeply subducted continental crust (Kokchetav Massif, Kazakhstan). Contributions to Mineralogy and Petrology 141, 66-82.

Hou, Z. H., Li, S. G., Chen, N. S., Li, Q. L., Liu, X. M., 2005. Sm-Nd and zircon SHRIMP U-Pb dating of Huilanshan mafic granulite in the Dabie Mountains and its zircon trace element geochemistry. Science in China Series D: Earth Sciences 48, 2081-2091.

Jamtveit, B., Bucber-Nurminen, K., Austrbeim, H., 1990. Fluid controlled eclogitization of granulites in deep crustal shear zones,

Bergen Arcs, Western Norway. Contributions to Mineralogy and Petrology 104, 184-193.

Katayama, I., Maruyama, S., Parkinson, C.D., Terada, K., Sano, Y., 2001. Ion micro-probe U-Pb zircon geochronology of peak and retrograde stages of ultrahigh-pressure metamorphic rocks from the Kokchetav massif, northern Kazakhstan. Earth and Planetary Science Letters 188, 185-198.

Lee, J.K.W., Williams, I.S., Ellis, D.J., 1997. Pb, U and Th diffusion in natural zircon. Nature 390, 159-162.

Li, S. G., Huang, F., Nie, Y. H., Han, W. L., Long, G., Li, H. M., Zhang, S. Q., Zhang, Z. H., 2001. Geochemical and geochronological constraints on the suture location between the North and South China blocks in the Dabie Orogen, Central China. Physics and Chemistry of the Earth, Part A: Solid Earth and Geodesy 26, 655-672.

Li, S. G., Huang, F., Zhou, H., Li, H., 2003. U-Pb isotopic compositions of the ultrahigh pressure metamorphic (UHPM) rocks from Shuanghe and gneisses from Northern Dabie zone in the Dabie Mountains, central China: constraint on the exhumation mechanism of UHPM rocks. Science in China Series D: Earth Sciences 46, 200-209.

Li, S. G., Jagoutz, E., Chen, Y. Z., Li, Q. L., 2000. Sm-Nd and Rb-Sr isotopic chronology and cooling history of ultrahigh pressure metamorphic rocks and their country rocks at Shuanghe in the Dabie Mountains, central China. Geochimica et Cosmochimica Acta 64, 1077-1093.

Li, S. G., Li, Q. L., Hou, Z. H., Yang, W., Wang, Y., 2005. Cooling history and exhumation mechanism of the ultrahigh-pressure metamorphic rocks in the Dabie Mountains, central China. Acta Petrologica Sinica 21, 1117-1124 (in Chinese with English abstract)

Li, S. G., Wang, C. X., Dong, F., Hou, Z. H., Li, Q. L., Liu, Y. C., Huang, F., Chen, F. K., 2009a. Common Pb of UHP metamorphic rocks from the CCSD project (100–5000 m) suggesting decoupling between the slices within subducting continental crust and multiple thin slab exhumation. Tectonophysics 475, 308-317.

Li, S. G., Wang, S. S., Chen, Y. Z., Liu, D. L., Qiu, J., Zhou, H. X., Zhang, Z. M., 1994. Excess argon in phengite from eclogite: evidence from dating of eclogite minerals by Sm-Nd, Rb-Sr and $^{40}Ar/^{39}Ar$ methods. Chemical Geology 112, 343-350.

Li, S. G., Xiao, Y. L., Liu, D. L., Chen, Y. Z., Ge, N. J., Zhang, Z. Q., Sun, S. S., Cong, B. L., Zhang, R. Y., Hart, S. R., 1993. Collision of the North China and Yangtse Blocks and formation of coesite-bearing eclogites: Timing and processes. Chemical Geology 109, 89-111.

Li, X. H., Liu, Y., Li, Q. L., Guo, C. H., Chamberlain, K. R., 2009b. Precise determination of Phanerozoic zircon Pb/Pb age by multicollector SIMS without external standardization. Geochemistry, Geophysics, Geosystems 10, Q04010.

Li, X. P., Zheng, Y. F., Wu, Y. B., Chen, F. K., Gong, B., Li, Y. L., 2004. Low-T eclogite in the Dabie terrane of China: Petrological and isotopic constraints on fluid activity and radiometric dating. Contributions to Mineralogy and Petrology 148, 443-470.

Liati, A., 2005. Identification of repeated Alpine (ultra) high-pressure metamorphic events by U-Pb SHRIMP geochronology and REE geochemistry of zircon: the Rhodope zone of Northern Greece. Contributions to Mineralogy and Petrology 150, 608-630.

Lin, W., Enami, M., Faure, M., Schrer, U., Arnaud, N., 2007, Survival of eclogite xenolith in a Cretaceous granite intruding the Central Dabieshan migmatite gneiss dome (Eastern China) and its tectonic implications. International Journal of Earth Sciences 96, 707-724.

Liou, J.G., Ernst, W.G., Zhang, R.Y., Tsujimori, T., Jahn, J.G., 2009. Ultrahigh-pressure minerals and metamorphic terranes—The view from China. Journal of Asian Earth Sciences 35, 199-231.

Liu, D.Y., Jian, P., Kröner, A., Xu, S.T., 2006. Dating of prograde metamorphic events deciphered from episodic zircon growth in rocks of the Dabie-Sulu UHP complex, China. Earth and Planetary Science Letters 250, 650-666.

Liu, F. L., Gerdes, A., Zeng, L. S., Xue, H. M., 2008. SHRIMP U-Pb dating, trace elements and the Lu-Hf isotope system of coesite-bearing zircon from amphibolite in the SW Sulu UHP terrane, eastern China. Geochimica et Cosmochimica Acta 72, 2973-3000.

Liu, F. L., Robinson, P. T., Gerdes, A., Xue, H. M., Liu, P. H., Liou, J. G., 2010a. Zircon U-Pb ages, REE concentrations and Hf isotope compositions of granitic leucosome and pegmatite from the north Sulu UHP terrane in China: Constraints on the timing and nature of partial melting. Lithos 117, 247-268.

Liu, F.L., Liou, J.G., 2011. Zircon as the best mineral for P-T-time history of UHP metamorphism: a review on mineral inclusions and U-Pb SHRIMP ages of zircons from the Dabie-Sulu UHP rocks. Journal of Asian Earth Sciences 40, 1-39.

Liu, S A., Teng, F. Z., He, Y. S., Ke, S., Li, S. G., 2010b. Investigation of magnesium isotope fractionation during granite differentiation: implication for Mg isotopic composition of the continental crust. Earth and Planetary Science Letters 297, 646-654.

Liu, Y. C., Li, S. G., 2008. Detachment within subducted continental crust and multi-slice successive exhumation of ultrahigh-pressure metamorphic rocks: evidence from the Dabie-Sulu orogenic belt. Chinese Science Bulletin 53, 3105-3119.

Liu, Y. C., Li, S. G., Gu, X. F., Xu, S. T., Chen, G. B., 2007a. Ultrahigh-pressure eclogite transformed from mafic granulite in the Dabie orogen, east-central China. Journal of Metamorphic Geology 25, 975-989.

Liu, Y. C., Li, S. G., Xu, S. T., 2007b. Zircon SHRIMP U-Pb dating for gneisses in northern Dabie high T/P metamorphic zone, central

China: Implications for decoupling within subducted continental crust. Lithos 96, 170-185.

Liu, Y. C., Li, S. G., Xu, S. T., Jahn, B., Zheng, Y. F., Zhang, Z. Q., Jiang, L. L., Chen, G. B., Wu, W. P., 2005. Geochemistry and geochronology of eclogites from the northern Dabie Mountains, central China. Journal of Asian Earth Sciences 25, 431-443.

Liu, Y. C., Xu, S. T., Li, S. G., Chen, G. B., Jiang, L. L., Zhou, C. T., Wu, W. P., 2001. Distribution and metamorphic p-t condition of the eclogites from the mafic-ultramafic belt in the northern part of the Dabie Mountains. Acta Geologica Sinica 75, 385-395.

Liu, Y. C., Gu, X. F., Rolfo, F., Chen, Z. Y., 2011a. Ultrahigh-pressure metamorphism and multistage exhumation of eclogite of the Luotian dome, North Dabie Complex Zone (central China): evidence from mineral inclusions and decompression textures. Journal of Asian Earth Sciences 42, 607-617

Liu, Y. C., Gu, X. F., Li, S. G., Hou, Z. H., Song, B., 2011b Multistage metamorphic events in granulitized eclogites from the North Dabie complex zone, central China: evidence from zircon U-Pb age, trace element and mineral inclusion. Lithos 122, 107-121.

Ludwig, K. R., 2001. Users manual for Isoplot/Ex (rev. 2.49): A geochronological toolkit for Microsoft Excel. Berkeley Geochronology Center Special Publication, 1, 55pp.

Malaspina, N., Hermann, J., Scambelluri, M., Compagnoni, R., 2006. Multistage metasomatism in ultrahigh-pressure mafic rocks from the North Dabie Complex (China). Lithos 90, 19-42.

Mcclelland, W., Gilotti, J., Mazdab, F., Wooden, J., 2009. Trace-element record in zircons during exhumation from UHP conditions, North-East Greenland Caledonides. European Journal of Mineralogy 21, 1135.

McClelland, W., Power, S., Gilotti, J., Mazdab, F., Wopenka, B., 2006. U-Pb SHRIMP geochronology and trace-element geochemistry of coesite-bearing zircons, North-East Greenland Caledonides. Ultrahigh-Pressure Metamorphism: Deep Continental Subduction, 23-43.

Moller, A., O'brien, P., Kennedy, A., Krner, A., 2002. Polyphase zircon in ultrahigh©\temperature granulites (Rogaland, SW Norway): constraints for Pb diffusion in zircon. Journal of Metamorphic Geology 20, 727-740.

Nehring, F., Foley, S. F., Hltt, P., 2009. Trace element partitioning in the granulite facies. Contributions to Mineralogy and Petrology 159, 493-519.

Pride, C., Muecke, G., 1981. Rare earth element distributions among coexisting granulite facies minerals, Scourian Complex, NW Scotland. Contributions to Mineralogy and Petrology 76, 463-471.

Rubatto, D., Hermann, J., 2003. Zircon formation during fluid circulation in eclogites (Monviso, Western Alps): implications for Zr and Hf budget in subduction zones. Geochimica et Cosmochimica Acta 67, 2173-2187.

Rubatto, D., Hermann, J., 2007. Experimental zircon/melt and zircon/garnet trace element partitioning and implications for the geochronology of crustal rocks. Chemical Geology 241, 38-61.

Rubatto, D., 2002. Zircon trace element geochemistry: partitioning with garnet and the link between U-Pb ages and metamorphism. Chemical Geology 184, 123-138.

Schaltegger, U., Fanning, C., G"nther, D., Maurin, J., Schulmann, K., Gebauer, D., 1999. Growth, annealing and recrystallization of zircon and preservation of monazite in high-grade metamorphism: conventional and in-situ U-Pb isotope, cathodoluminescence and microchemical evidence. Contributions to Mineralogy and Petrology 134, 186-201.

Slama, J., Kosler, J., Condon, D. J., Crowley, J. L., Gerdes, A., Hanchar, J. M., Horstwood, M. S. A., Morris, G. A., Nasdala, L., Norberg, N., 2008. Plesovice zircon—A new natural reference material for U-Pb and Hf isotopic microanalysis. Chemical Geology 249, 1-35.

Smith, M., Henderson, P., Jeffries, T., Long, J., Williams, C., 2004. The rare earth elements and uranium in garnets from the Beinn an Dubhaich aureole, Skye, Scotland, UK: constraints on processes in a dynamic hydrothermal system. Journal of Petrology 45, 457-484.

Stacey, J., Kramers, J., 1975. Approximation of terrestrial lead isotope evolution by a two-stage model. Earth and Planetary Science Letters 26, 207-221.

Storkey, A., Hermann, J., Hand, M., Buick, I., 2005. Using in situ trace-element determinations to monitor partial-melting processes in metabasites. Journal of Petrology 46, 1283.

Sun, S. S., McDonough, W., 1989. Chemical and isotopic systematics of oceanic basalts: implications for mantle composition and processes. Geological Society London Special Publications 42, 313.

Tang, H. F., Liu, C. Q., Nakai, S., Qrihashi, Y., 2007. Geochemistry of eclogites from the Dabie-Sulu terrane, eastern China: new insights into protoliths and trace element behaviour during UHP metamorphism. Lithos 95, 441-457.

Tong, L. X., Jahn, B. M., Zheng, Y. F., 2011. Diverse PT paths of the northern Dabie complex in central China and its reworking in the early Cretaceous. Journal of Asian Earth Sciences 42, 633-640

Tsai, C. H., Liou, J. G., 2000. Eclogite-facies relics and inferred ultrahigh-pressure metamorphism in the North Dabie Complex, central-eastern China. American Mineralogist 85, 1-8.

Tsai, C. H., 1998. Petrology and geochemistry of mafic-ultramafic rocks in the north of the Dabie complex, central-eastern China, PhD

Thesis, Stanford University, Stanford.
Van Westrenen, W., Blundy, J., Wood, B., 1999. Crystal-chemical controls on trace element partitioning between garnet and anhydrous silicate melt. American Mineralogist 84, 838-847.
Wang, J. H., Sun, M., Deng, S. X., 2002. Geochronological constraints on the timing of migmatization in the Dabie Shan, East-central China. Eruopean Journal of Mineralogy 14, 513-524.
Wang, Q., Wyman, D., Xu, J. F., Jian, P., Zhao, Z. H., Li, C. F., Xu, W., Ma, J. L., He, B., 2007. Early Cretaceous adakitic granites in the Northern Dabie Complex, central China: implications for partial melting and delamination of thickened lower crust. Geochimica et Cosmochimica Acta 71, 2609-2636.
Watson, E. B., Wark, D. A., Thomas, J. B., 2006. Crystallization thermometers for zircon and rutile. Contributions to Mineralogy and Petrology 151, 413-433.
Whitehouse, M., Platt, J., 2003. Dating high-grade metamorphism—constraints from rare-earth elements in zircon and garnet. Contributions to Mineralogy and Petrology 145, 61-74.
Wu, Y. B. and Zheng, Y. F., 2004. Genesis of zircon and its constraints on interpretation of U-Pb age. Chinese Science Bulletin 49, 1554-1569.
Wu, Y. B., Zheng, Y. F., Gao, S., Jiao, W.F., Liu, Y. S., 2008. Zircon U-Pb age and trace element evidence for Paleoproterozoic granulite-facies metamorphism and Archean crustal rocks in the Dabie Orogen. Lithos 101, 308-322.
Wu, Y. B., Zheng, Y. F., Zhang, S. B., Zhao, Z. F., Wu, F. Y., Liu, X.M., 2007. Zircon U-Pb ages and Hf isotope compositions of migmatite from the North Dabie terrane in China: constraints on partial melting. Journal of Metamorphic Geology 25, 991-1009.
Wu, Y. B., Zheng, Y. F., Zhao, Z. F., Gong, B., Liu, X. M., Wu, F. Y., 2006. U-Pb, Hf and O isotope evidence for two episodes of fluid-assisted zircon growth in marble-hosted eclogites from the Dabie orogen. Geochimica et Cosmochimica Acta 70, 3743-3761.
Xia, Q. X., Zheng, Y. F., Yuan, H.L., Wu, F. Y. 2009. Constrasting Lu-Hf and U-Th-Pb isotope systematics between metamorphic growth and recrystallization of zircon from eclogite-facies metagranites in the Dabie orogen, China. Lithos 112, 477-496
Xia, Q. X., Zheng, Y. F., Hu, Z.C., 2010. Trace elements in zircon and coexisting minerals from low-T/UHP metagranite in the Dabie orogen: implications for action of supercritical fluid during continental subduction-zone metamorphism. Lithos 114, 385-412.
Xia, Q. X., Zheng, Y. F., Lu, X. N., Hu, Z.C., Xu, H. J., 2011. Formation of metamorphic and metamorphosed garnets in the low-T/UHP metagranite during continental collision in the Dabie orogen. Lithos, doi: 10.1016/j.lithos.2011.10.004.
Xie, Z., Chen, J. F., Cui, Y. R., 2010. Episodic growth of zircon in UHP orthogneisses from the North Dabie Terrane of east-central China: implications for crustal architecture of a collisional orogen. Journal of Metamorphic Geology 28, 979-995.
Xie, Z., Zheng, Y. F., Jahn, B.M., Ballevre, M., Chen, J.F., Gautier, P., Gao, T.S., Gong, B., Zhou, J.B., 2004. Sm-Nd and Rb-Sr dating of pyroxene-garnetite from North Dabie in east-central China: problem of isotope disequilibrium due to retrograde metamorphism. Lithos 206, 137-158.
Xu, S.T., Liu, Y. C., Chen, G. B., Compagnoni, R., Rolfo, F., He, M. C., Liu, H. F., 2003. New finding of micro-diamonds in eclogites from Dabie-Sulu region in central-eastern China. Chinese Science Bulletin 48, 988-994.
Xu, S. T., Liu, Y. C., Chen, G. B., Ji, S. Y., Ni, P., Xiao, W. S., 2005. Microdiamonds, their classification and tectonic implications for the host eclogites from the Dabie and Su-Lu regions in central eastern China. Mineralogical magazine 69, 509-520.
Xu, S. T., Liu, Y. C., Jiang, L. L., Su, W., Ji, S. Y., 1994. Tectonic regime and evolution of Dabie Mountains. Science Press, Beijing. 175pp. (in Chinese with English abstract).
Xu, S. T., Liu, Y. C., Su, W., Wang, R. C., Jiang, L. L., Wu, W. P., 2000. Discovery of the eclogite and its petrography in the Northern Dabie Mountain. Chinese Science Bulletin 45, 273-278.
Xu, S. T., S, W., Liu, Y. C., Jiang, L. L., Ji, S. Y., Okay, A. I., Sengor, A. M. C., 1992. Diamond from the Dabie Shan metamorphic rocks and its implication for tectonic setting. Science 256, 80-82.
Xu, Z. Q., Zeng, L. S., Liu, F. L., Yang, J., Zhang, Z. M., McWilliams, M., Liou, J. G., 2006. Polyphase subduction and exhumation of the Sulu high-pressure–ultrahigh-pressure metamorphic terrane. Geological Society of America Special Papers 403, 93-113.
Yuan, H. L., Gao, S., Liu, X. M., Li, H. M., Gunther, D., Wu, F. Y., 2004. Accurate U-Pb age and trace element determinations of zircon by laser ablation-inductively coupled plasma-mass spectrometry. Geostandards and Geoanalytical Research 28, 353-370.
Zhang, R. Y., Liou, J. G., Tsai, C. H., 1996. Petrogenesis of a high-temperature metamorphic terrane: a new tectonic interpretation for the north Dabieshan, central China. Journal of Metamorphic Geology 14, 319-333.
Zhang, R. Y., Liou, J. G., Ernst, W., 2009. The Dabie-Sulu continental collision zone: a comprehensive review. Gondwana Research 16, 1-26.
Zhang, Z. M., Zhong, Z. Q., You, Z. D., Hu, K. M., 2000. Granulite-facies retrograde metamorphism of garnet pyroxenite in Muzidian, norhtern Dabie mountains. Journal of China University of Geoscience 25, 295-301. (in Chinese with English abstract)
Zhao, Z. F., Zheng, Y. F., Wei, C. S., Wu, Y. B., 2007. Post-collisional granitoids from the Dabie orogen in China: Zircon U-Pb age,

element and O isotope evidence for recycling of subducted continental crust. Lithos 93, 248-272.

Zhao, Z. F., Zheng, Y. F., Wei, C.S., Chen, F.K., Liu, X.M., Wu, F. Y., 2008. Zircon U-Pb ages, Hf and O isotopes constrain the crustal architecture of the ultrahigh-pressure Dabie orogen in China. Chemical Geology 253, 222-242.

Zheng, Y. F., 2008. A perspective view on ultrahigh-pressure metamorphism and continental collision in the Dabie-Sulu orogenic belt. Chinese Science Bulletin 53, 3081-3104.

Zheng, Y. F., 2009. Fluid regime in continental subduction zones: petrological insights from ultrahigh-pressure metamorphic rocks. Journal of the Geological Society 166, 763.

Zheng, Y. F., Chen, R. X., Zhao, Z. F., 2009. Chemical geodynamics of continental subduction-zone metamorphism: insights from studies of the Chinese Continental Scientific Drilling (CCSD) core samples. Tectonophysics 475, 327-358.

Zheng, Y. F., Fu, B., Gong, B., Li, L., 2003. Stable isotope geochemistry of ultrahigh pressure metamorphic rocks from the Dabie-Sulu orogen in China: implications for geodynamics and fluid regime. Earth-Science Reviews 62, 105-161.

Zheng, Y. F., Fu, B., Li, Y.L., Wei, C.S., Zhou, J.B., 2001. Oxygen isotope composition of granulites from Dabieshan in eastern China and its implications for geodynamics of Yangtze plate subduction. Physics and Chemistry of the Earth, Part A: Solid Earth and Geodesy 26, 673-684.

Zheng, Y. F., Gao, T. S., Wu, Y. B., Gong, B., 2007. Fluid flow during exhumation of deeply subducted continental crust: Zircon U-Pb age and O isotope studies of quartz vein in eclogite. Journal of Metamorphic Geology 25, 267-283.

Zheng, Y. F., Gao, X. Y., Chen, R. X., Gao, T.S., 2011. Zr-in-rutile thermometry of eclogite in the Dabie orogen: Constraints on rutile growth during continental subduction-zone metamorphism. Journal of Asian Earth Sciences 40, 427-451.

Zheng, Y. F., Wang, Z. R., Li, S.G., Zhao, Z. F., 2002. Oxygen isotope equilibrium between eclogite minerals and its constraint on mineral Sm-Nd chronometer. Geochimica et Cosmochimica Acta 66, 625-634.

Zheng, Y. F., Wu, Y. B., Chen, F.K., Gong, B., Li, L., Zhao, Z. F., 2004. Zircon U-Pb and oxygen isotope evidence for a large-scale ^{18}O depletion event in igneous rocks during the Neoproterozoic. Geochimica et Cosmochimica Acta 68, 4145-4165.

Zhou, L. G., Xia, Q. X., Zheng, Y. F., Chen, R. X., 2011. Multistage growth of garnet in ultrahigh-pressure eclogite during continental collision in the Dabie orogen: constrained by trace elements and U-Pb ages. Lithos 127, 101-127.

Common Pb isotope mapping of UHP metamorphic zones in Dabie orogen, Central China: implication for Pb isotopic structure of subducted continental crust[*]

Ji Shen[1], Ying Wang[1,2] and Shu-guang Li[1,3]

1. CAS key Laboratory of Crust-Mantle Materials and Environments, School of Earth and Space Sciences, University of Science and Technology of China, Hefei 230026, Anhui, China
2. Geophysical Laboratory, Carnegie Institution of Washington, 5251 Broad Branch Rd., NW, Washington, DC 20015-1305, USA
3. State Key Laboratory of Geological Processes and Mineral Resources, China University of Geosciences, Beijing 100083, China

> 亮点介绍：首次系统地研究了大别造山带正片麻岩中长石的普通 Pb 同位素组成；结合大别造山带的构造演化历史，建立了多阶段 Pb 演化模式，证实了折返的北、中和南大别带原岩来自不同的地壳深度；本工作显示 Zartman (1981) 的铅构造（Plumbotectonic）模型有助于理解大陆俯冲和超高压变质岩折返形成的造山带陆壳结构。

Abstract We report Pb isotopic compositions for feldspars separated from 57 orthogneisses and 2 paragneisses from three exhumed UHPM slices representing the North Dabie zone, the Central Dabie zone and the South Dabie zone of the Dabie orogen, central-east China. The feldspars from the gneisses were recrystallized during Triassic continental subduction and UHP metamorphism. Precursors of the orthogneisses are products of Neoproterozoic bimodal magmatic events, those in north Dabie zone emplaced into the lower crust and those in central and south Dabie zones into middle or upper crust, respectively. On a $^{207}Pb/^{204}Pb$ vs. $^{206}Pb/^{204}Pb$ diagram, almost all orthogneisses data lie to the left of the 0.23 Ga paleogeochron and plot along the model mantle evolution curve with the major portion of the data plotting below it. On a $^{208}Pb/^{204}Pb$ vs. $^{206}Pb/^{204}Pb$ diagram the most of data of north Dabie zone extend in elongate arrays along the lower crustal curve and others extend between the lower crustal curve to near the mantle evolution curve for the plumbotectonics model. This pattern demonstrates that the Pb isotopic evolution of the feldspars essentially ended at 0.23 Ga and the orthogneiss protoliths were principally dominated by reworking of ancient lower crust with some addition of juvenile mantle in the Neoproterozoic rifting tectonic zone. According to geological evolution history of the locally Dabie orogen, a four-stage Pb isotope evolution model including a long time evolution between 2.0 Ga and 0.8 Ga with a lower crust type U/Pb ratio (μ = 5–6) suggests that magmatic emplacement levels of the protoliths of the orthogneisses in the Dabie orogen at 0.8 Ga also play an important role in the Pb evolution of the exhumed UHPM slices, corresponding to their respective Pb characters at ca. 0.8–0.23 Ga. For example, north Dabie zone requires low μ values (3.4–9.6), while central and south Dabie zones require high μ values (10.9–17.2). On the other hand, Pb isotopic mixing between north and central or south Dabie zones during retro-grade metamorphism enhanced by the extensive magmatism in the Cretaceous has also been observed in the $^{207}Pb/^{204}Pb$ vs. $^{206}Pb/^{204}Pb$ and $^{208}Pb/^{204}Pb$ vs. $^{206}Pb/^{204}Pb$ diagrams. A combined study of common Pb isotopic compositions of Dabie orthogneisses and Sulu UHPM rocks from the Chinese Continental Scientific Drilling project demonstrates that a slab marked by extremely unradiogenic Pb observed in the main hole was absent in the Dabie orogen. However, occurrence of some Mesozoic granitoids with such unradiogenic character in the Dabie orogen suggests that their source may be a buried unradiogenic unit underlying below north Dabie zone. This case study clearly shows that whether the position of the Dabie data relative to the orogen curve of the plumbotectonic model is helpful in understanding the Pb isotopic structure and evolution of subducted continental crust.

[*] 本文发表在：Geochimica et Cosmochimica Acta, 2014, 143: 115-131

Key words Common Pb isotope; Continental subduction; UHPM rocks; Plumbotectonics model; Dabie-Sulu orogen

1 Introduction

Previous studies of mineral/fluid partitioning have demonstrated that U (U^{4+}, U^{5+} and U^{6+}) partition coefficients are a strong function of oxygen fugacity (fO_2). Dominant U^{6+} at high fO_2 could form the highly soluble uranyl complex during high-grade metamorphism as well as weathering (Brenan et al., 1995), which will lead to a general loss of U, Th and Pb in the trend of U > Pb > Th. Thus the U-Pb system is sensitive to crustal processes and can be significantly modified by high-grade metamorphism and weathering. This sensitivity means that the Pb isotopic system may preserve a record of crustal history (DeWolf and Mezger, 1994). The upper continental crust (UCC) has relatively high U/Pb and low Th/U, while the lower continental crust (LCC) shows large depletion in U due to granulite-facies metamorphism (Doe and Tilling, 1967; Rudnick et al., 1985), resulting in significantly lower U/Pb and higher Th/U. Consequently, the Pb isotopic composition of stabilized continental crust usually displays a strong vertical inhomogeneity that reflects this fractionation of U, Th, and Pb between the UCC and the LCC (Zartman and Doe, 1981; Taylor and McLennan, 1985). Since published, the plumbotectonics model has been widely used in assessing the sources of igneous and metamorphic rocks, and ore deposits (Cumming and Richards, 1975; Taylor et al., 1980; DeWolf and Mezger, 1994; Zhang et al., 2002). In the final version of plumbotectonics model (Zartman and Haines, 1988), all major terrestrial reservoirs were considered to be essentially open systems interacting to produce the Pb isotopic disparities between them. The variable proportions of mantle, preexisting UCC, LCC, and subcrustal lithosphere recycled by an orogen could then explain the Pb isotopic composition of the next crustal addition. However, the Pb isotopic composition of a deeply subducted continental crust, which had undergone "ultra-high pressure" UHP metamorphism and exhumation was not well understood. Whether the plumbotectonic model appropriately accounted for such deeply subducted continental crust remained unclear. For example, we don't know if Pb isotope mixing between lower and upper crust has occurred, nor if deeply subducted upper and lower crust retain their original Pb isotope signatures despite having undergone UHP metamorphism (DeWolf and Mezger, 1994).

The Dabie-Sulu orogen is a well-known UHPM belt formed by subduction of the South China Block beneath the North China Block in the Triassic (Ames et al., 1996; Li et al., 1993, 1994, 2000; Rowley et al., 1997; Hacker et al., 1998; Ayers et al., 2002), recognized after the discovery of ultrahigh-pressure (UHP) index minerals, such as coesite (Okay et al., 1989; Wang et al., 1989) and microdiamonds (Xu et al., 1992a), in eclogites and the surrounding gneisses. The orogen consists of the Dabie UHPM terrane in the west and the Sulu UHPM terrane in the east. The latter is believed to be the eastern extension of the Dabie UHPM terrane, which is displaced by ~500 km to the northeast by the Tan-Lu fault (Zhu et al., 2009). Based on distinctive features of petrology, geochemistry, geophysics, geochronology, and *T-t* cooling paths, the Dabie and Sulu HP to UHP metamorphic belts are known to be comprised of several UHP slices. They display an increasing grade of metamorphism from south to north, but a corresponding decrease in peak metamorphic age (Liu and Li, 2008). The three UHPM slices in the Dabie orogen consist of the North Dabie zone (NDZ), the Central Dabie zone (CDZ), and the South Dabie zone (SDZ). The NDZ is considered to be exhumed subducted lower continental crust, and the CDZ and SDZ are thought to be exhumed subducted middle-upper continental crustal. Therefore, the Dabie UHPM belt provides a good opportunity to test whether the plumbotectonics model is also helpful for identifying the protoliths of deeply subducted continental crusts.

In recent years, considerable attention has been devoted to the use of Pb isotope as a geochemical tool to investigate the crustal structure of the Dabie-Sulu orogenic belt (Zhang et al., 1995, 2001, 2002, 2008; Xu et al., 2002; Li et al., 2003b, 2009; Dong et al., 2006). Based on whole-rock Pb data, previous studies reported LCC-like Pb isotopic signatures for the NDZ and UCC-like Pb isotope signatures for the CDZ (Zhang et al., 2001, 2008; Li et al., 2003b). However, if the plumbotectonics can be applied to the deeply subducted continental crust, another problem of the previous studies arises as to the meaning of the whole-rock Pb isotopic compositions because such data may contain systematic biases. For example, for whole-rock samples selected from the earth's surface, weathering whole-rock samples could lead to U/Pb disturbances and thus prevent the accurate calculation of the initial Pb isotopic compositions (Li et al., 2009). Moreover, even within the same structural unit, the initial Pb isotopic composition for rocks of different lithologies may vary over a considerable range. This has been demonstrated by the observation that the Pb isotopic compositions of the paragneisses in the CDZ are significantly more radiogenic than the orthogneisses in the same zone (Li et al., 2003b; Wawrzenitz et al., 2006). Therefore, it is inappropriate to directly compare the Pb isotopic composition of different lithologies. Bearing in mind all the above considerations, in this study we analysed the Pb isotopic composition of carefully hand-picked, alteration-free minerals with low U/Pb ratios (e.g., K-feldspar and plagioclase), which ought to provide better estimates of the Pb isotope composition of the host rocks at the time of last isotopic equilibration. This time may be that of igneous crystallization or of the cessation of Pb diffusion and exchange in a rapid cooling UHPM orogen (DeWolf and Mezger, 1994). To address the second possibility, the samples were selected from metamorphic rocks with similar lithology across different HP-UHP units.

In this study, we present the Pb isotopic composition of feldspars from 57 orthogneisses and 2 paragneisses representative of the whole eastern Dabie UHPM terrane. The main purposes are: (1) to develop a common Pb isotope map of the UHPM rocks from the eastern Dabie terrane; (2) to provide constraints on the pre-metamorphic Pb evolution of the three main UHPM units from the Dabie orogen; (3) to examine the suitability of the plumbotectonics model for interpreting the UHPM belt in a collisional orogen, and (4) to discuss the relationship between the Dabie UHPM terrane and the Sulu UHPM terrane during the subduction and exhumation processes.

2 Geological setting and samples

The Dabie and Sulu terranes are located at the eastern part of the E-W trending Qinling-Dabie-Sulu orogenic belt (Fig. 1), formed during the continental collision between the South China Block and North China Block in the Triassic (Li et al., 1993, 1994, 2000; Ames et al., 1996; Hacker et al., 1998; Ayers et al., 2002). The Sulu terrane was displaced from the Dabie orogen more than 500km to the north by the sinistral NE-extended strike-slip Tan-Lu fault. UHPM rocks exposed in the Dabie and Sulu orogen comprise one of the largest known UHPM zones in the world (Li et al., 1993; Cong, 1996). The Dabie orogenic belt consists of the western Dabie orogen and the eastern Dabie orogen. The eastern Dabie orogen is bounded by the Shangcheng-Macheng fault (SMF) to the west from the western Dabie orogen and the Tan-Lu fault to the east (Li et al., 2001). It has been divided into the Beihuaiyang zone (BZ), the North Dabie high-T/UHP complex zone (NDZ), the Central Dabie middle-T/UHP metamorphic zone (CDZ), the South Dabie low-T eclogite zone (SDZ), and the Susong complex zone (SZ), which are successively separated by the Xiaotian-Mozitan fault (XMF), Wuhe-Shuihou fault (WSF), Hualiangting-Mituo fault (HMF), and Taihu-Mamiao fault (TMF) (Li et al., 2001; Liu et al., 2007b) (Fig. 1).

The NDZ is an orthogneiss unit with banded tonalitic, granitoid gneiss and minor meta-peridotite (including dunite, harzburgite and lherzolite), garnet pyroxenite, garnet-bearing amphibolite, granulite and eclogite (Liu and Li, 2005). Although coesite has not yet been discovered in the eclogite or gneiss of the NDZ, the oriented mineral exsolutions in garnet and clinopyroxene, and microdiamond in zircons from the banded gneisses imply that both the eclogites and gneisses from the NDZ underwent UHP metamorphism (Xu et al., 2003, 2005; Malaspina et al., 2006; Liu et al., 2007b). The peak P-T conditions are estimated to be 808–874 °C and >2.5 GPa (Liu et al., 2005) or possibly >5–7 GPa (Xu et al., 2005). Eclogites and gneisses in the NDZ were overprinted by granulite-facies metamorphism, followed by amphibolite-facies metamorphism during exhumation (Liu et al., 2005, 2007a; Wang et al., 2012). The entire NDZ was also subjected to intense deformation and uplift due to the intrusion of granite bodies during the Cretaceous. Three Cretaceous massive granitic batholiths including Zhubuyuan, Baimajian and Tiantangzhai intrusions are exposed in the northeastern part of the NDZ (ENDZ) due to subsequent erosion (Fig. 1). In contrast, only a few Cretaceous granitic minitype stocks and dykes are distributed on the southwestern segment of the NDZ named the Luotian dome (LT dome) (Fig. 1), suggesting a hidden granitic batholith underneath the LT dome. The UHPM rocks from the LT dome also preserve early granulite-facies mineral relics and have been overprinted by regionally extensive HP granulite-facies metamorphism, followed by penetrative amphibolite-facies retrogression during exhumation (Liu et al., 2007a).

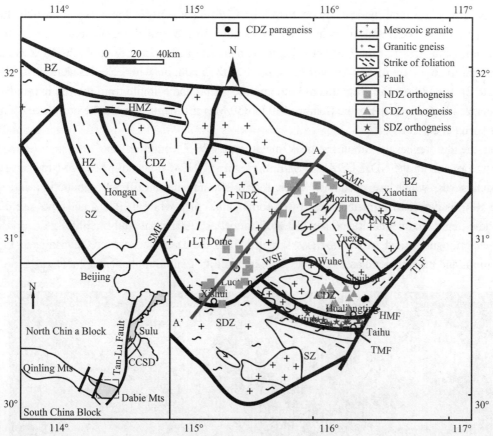

Fig. 1 Schematic geological map of the Dabie orogen (modified after Xu et al., 2005). The inset shows the location of this area within the Triassic Qinling-Dabie-Su-Lu collision orogen in central China. Sample locations with sample numbers are described in detail in the text. BZ, Beihuaiyang zone; NDZ, North Dabie high-T/UHP complex zone; CDZ, Central Dabie middle-T/UHP metamorphic zone; SDZ, South Dabie low-T eclogite zone; SZ, Susong complex zone; HMZ, Huwan melange zone; HZ, Hong'an low-T eclogite facies zone; TLF, Tan-Lu fault; TMF, Taihu-Mamiao fault; HMF, Huangliangting-Mituo fault; WSF, Wuhe-Shuihou fault; XMF, Xiaotian-Mozitan fault. Line AA' represents a cross-section, which is described in Fig. 6.

The UPHM rocks from the LT dome also preserve early granulite-facies mineral relics and have been overprinted by regionally extensive HP granulite-facies metamorphism, followed by penetrative amphibolite-facies retrogression during exhumation (Liu et al.,2007a).

The CDZ is an orthogneiss + supracrusal rocks unit, which mainly consists of orthogneisses, eclogites and meta-supracrustal rocks, including paragneisses, marbles, and jadeite quartzites. The eclogites occur as lenses or blocks interbedded with the paragneisses and marbles. The discovery of coesite and diamond in the UHPM rocks from the CDZ indicated the South China Block, subducted in the Triassic, achieved a depth of ≥140km and suffered UHP metamorphism with metamorphic conditions ranging at 700 – 850 ℃ and >2.8 GPa (Okay et al., 1989; Wang et al., 1989; Xu et al., 1992b; Li et al., 1993). UHPM rocks from the CDZ experienced only amphibolite- and greenschist-facies retrograde metamorphism (Xu et al., 1992a; Cong, 1996).

The SDZ is mainly a paragneiss unit predominantly composed of eclogites, paragneisses (mainly garnet-bearing two-mica-epidote-plagioclase gneiss), and a few orthogneisses. Coesite and diamond are absent in the eclogite of this zone. The occurrence of coesite pseudomorphs in eclogite from Huangzhen confirmed that the SDZ experienced UHP metamorphism with peak conditions of 3.3 GPa, but lower temperatures (≤700 ℃) (Okay, 1993; Li et al., 2004). UHPM rocks from the SDZ only experienced amphibolite- and greenschist-facies retrograde metamorphism.

The peak metamorphic ages of the UHP rocks from the CDZ have been well constrained by Sm-Nd UHP mineral isochrons and zircon SHRIMP U-Pb dating of coesite or diamond bearing eclogite, which gave consistent UHP metamorphic ages of 226–230 Ma (Li et al., 2000; Ayers et al., 2002; Liu et al., 2004a, b). The peak metamorphic ages determined for the SDZ eclogite provide the oldest so far obtained, e.g., a Sm-Nd UHP mineral isochron age of 236±4 Ma, and several zircon SHRIMP U-Pb ages of 242±3 Ma (Li et al., 2004). The peak metamorphic ages of the UHP rocks from the NDZ are similar to those of the CDZ, e.g., an age of 226±3 Ma defined by metamorphic zircon of the Luotian eclogite with coesite inclusion (Liu et al., 2011), and 223.8±2.3Ma defined by metamorphic zircon of eastern part of NDZ (ENDZ) granulite with eclogite-facies mineral inclusions (Wang et al., 2012). In addition, the protoliths of both granitic gneiss and eclogite from the Dabie orogen are known to have been derived from Neoproterozoic (mainly 740 to 830 Ma) rift magmatism in the north margin of the South China Block (Zheng et al., 2003).

Based on the differences in lithology, peak metamorphic P-T condition and ages as well as retrograde metamorphic history among the NDZ, CDZ and SDZ, there three UHPM units have been considered to represent 3 exhumed crustal slices, which are decoupled from each other and have different subduction and exhumation histories. The NDZ is derived from subducted lower continental crust, while the CDZ and SDZ are derived from subducted middle-upper continental crust (Li et al., 2005; Liu et al., 2007b; Liu and Li, 2008).

Fifty-seven orthogneiss and two paragneiss samples from the Dabie orogen are investigated in this study. Sampling locations and main mineral assembles are depicted in Fig. 1 and Supplemental Table 1.

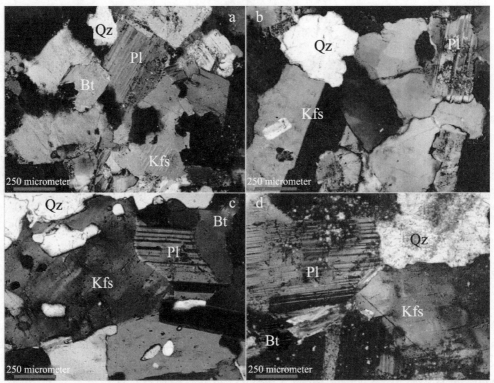

Fig. 2 Photomicrographs of coexisting plagioclase (Pl), K-feldspar (Kfs) and quartz (Qz) of the orthogneisses from the Luotian dome in the Dabie orogen. Biotites (Bt) are identified in some of them. (a) sample 10LT-2; (b) sample 10LT-6; (c) sample 10XS-1; and (d) sample 10XS-5.

The samples were collected from the eastern Dabie orogen, including the NDZ, the CDZ and SDZ, most of which are mainly granitic gneisses. Coexisting plagioclase+k-feldspar+quartz+biotite are major minerals was common in the orthogneiss samples from the whole NDZ (Fig. 2). The CDZ orthogneiss samples are granitic gneisses with a few garnet (03CP-2, 03SH-3, 02HN-4, 02HN-5), biotite or/and phengite (03CP-2, 99XF-1, 02HN-1, 2, 3, 5, 7, 15), and some other hydrous minerals (e.g. epidote, muscovite) (02HN-9-1, 02HN-10, 02HN-11, 08SH-9, 10, 08WH-1, BMJ-1). Samples from the SDZ are similar to those of the CDZ, but with more low-T/HP metamorphic or retrograde minerals such as epidote, titanite and magnetite/ilmenite. Paragenesis of these orthogneisses is general quartz+plagioclase+K-feldspar+epidote+muscovite±biotite, as well as various amounts of accessory minerals (e.g. garnet, titanite or magnetite/ilmenite). Two biotite paragneiss samples (02SH-5 and 02SH-9) were collected from Shuanghe, in the eastern part of the CDZ, occurred as compositional layers with eclogite, marble and jadeite quartzite. The general paragenesis is garnet+phengite+plagioclase+ epidote+biotite.

3 Analytical methods

About 1.5–2.0 g of feldspar (plagioclase or K-feldspar) were extracted and selected from orthogneisses at the Institution of Regional Geology and Mineral Inverstigation of Hebei province. Alteration-free feldspar (60–80 mg) carefully handpicked under a binocular microscope was used for Pb isotope analysis.

Pb chemical separation was performed at the Isotope Laboratory of University of Science and Technology of China (USTC). Twenty-four Pb isotopic analyses were performed at the Laboratory for Chemical Geodynamics, USTC, using a Finigan MAT-262 mass spectrometer. Another 35 Pb isotopic analyses were obtained with a Finigan MAT-262 mass spectrometer at the Laboratory for Radiogenic Isotope Geochemistry of the Institute of Geology and Geophysics (IGG), Chinese Academy of Science (CAS), Beijing. Pb contents were determined by the isotope dilution method. After rinsing in purified water and HCl for a few times, mineral powders were completely decomposed in a mixture of purified HNO_3-HF. Pb was separated using HBr in quartz columns with AG1-X8 of 200–400 meshes. Total procedure blanks were 50–100 pg for Pb. Fractionations of Pb isotopes during the analysis by mass-spectrometer were corrected with standard NBS981. During the course of this study, repeated analyses of standard NBS981 yielded the average values of $^{206}Pb/^{204}Pb = 16.918 \pm 0.12\%$ (2σ), $^{207}Pb/^{204}Pb = 15.468 \pm 0.13\%$ (2σ), and $^{208}Pb/^{204}Pb = 36.628 \pm 0.28\%$ (2σ) for the 24 analyses at USTC. For the rest 35 analyses at IGG, CAS, the average values of standard NBS981 were $^{206}Pb/^{204}Pb = 16.884 \pm 0.12\%$ (2σ), $^{207}Pb/^{204}Pb = 15.419 \pm 0.12\%$ (2σ), and $^{208}Pb/^{204}Pb = 36.489 \pm 0.29\%$ (2σ). Duplicate analyses gave a precision better than 99%.

4 Pb isotope results

Pb isotope compositions of 59 samples are presented in Table 1. The coexisting K-feldspar and plagioclase from 11 orthogneiss samples from the LT dome were simultaneously analysed for comparison. In previous studies, K-feldspar was considered to have an advantage over plagioclase, because the larger ionic radius of potassium allows greater substitution of Pb, while discriminating against uranium. Therefore, the ratio of U/Pb in plagioclase was supposed to be higher than coexisting K-feldspar. In our study, ten out of eleven K-feldspars samples show lower $^{206}Pb/^{204}Pb$ ratios than those of coexisting plagioclases, and only one sample (10LT-2) has similar $^{206}Pb/^{204}Pb$ ratios in both K-feldspar and plagioclase (Fig. 3). Therefore, K-feldspar rather than plagioclase was preferentially analysed for the other 48 samples, and the Pb isotopic compositions of K-feldspars were used in the subsequent discussion.

The present-day Pb isotopic compositions of orthogneiss samples from the eastern Dabie orogen are subdivided into four groups (Table 1): the Luotian dome, the ENDZ, the CDZ and the SDZ. Pb isotopic compositions of the Luotian dome orthogneisses show a large variation with $^{206}Pb/^{204}Pb$ ranging from 16.312 to 17.708, $^{207}Pb/^{204}Pb$ ranging from 15.289 to 15.473, and $^{208}Pb/^{204}Pb$ ranging from 37.152 to 38.013. The ENDZ orthogneisses also display a wide and scattered range in common Pb isotopic composition similar to the LT dome, with $^{206}Pb/^{204}Pb$ ranging from 16.213 to 17.197, $^{207}Pb/^{204}Pb$ ranging from 15.294 to 15.445, and $^{208}Pb/^{204}Pb$ ranging from 37.162 to 37.867.

Table 1 Common Pb isotopic compositions of feldspars from 57 orthogneisses and 2 paragneisses of the Dabie orogen

Sample	Lithology	Mineral	$^{206}Pb/^{204}Pb$	$^{207}Pb/^{204}Pb$	$^{208}Pb/^{204}Pb$
NDZ					
99MZT-3[a]	Orthogneiss	K-fledspar	16.918	15.445	37.668
99YZH-1[a]	Orthogneiss	K-fledspar	16.606	15.348	37.572
03JZ-4[a]	Orthogneiss	K-fledspar	16.937	15.439	37.732
03JZ-5[a]	Orthogneiss	K-fledspar	16.684	15.345	37.518
03JZ-6[a]	Orthogneiss	K-fledspar	16.797	15.390	37.685
03JZ-7[a]	Orthogneiss	K-fledspar	16.637	15.400	37.544
99MSH-1[a]	Orthogneiss	Plagioclase	16.416	15.294	37.284
03MSH-1[a]	Orthogneiss	Plagioclase	16.445	15.313	37.253

Continued

Sample	Lithology	Mineral	$^{206}Pb/^{204}Pb$	$^{207}Pb/^{204}Pb$	$^{208}Pb/^{204}Pb$
03MSH-2[a]	Orthogneiss	K-fledspar	16.213	15.334	37.162
99LN-2[a]	Orthogneiss	K-fledspar	16.315	15.373	37.260
03HS-1[a]	Orthogneiss	K-fledspar	16.385	15.331	37.281
03HS-2[a]	Orthogneiss	K-fledspar	17.105	15.393	37.424
03MZT-2[a]	Orthogneiss	K-fledspar	16.938	15.432	37.551
03SG-1[a]	Orthogneiss	K-fledspar	17.163	15.419	37.622
08MZT-4[a]	Orthogneiss	K-fledspar	17.197	15.415	37.867
Luotian dome					
97M3-1[a]	Orthogneiss	K-fledspar	17.442	15.443	37.819
10LT-1[b]	Orthogneiss	K-fledspar	17.187	15.416	37.622
		Plagioclase	17.438	15.472	37.858
10LT-2[b]	Orthogneiss	K-fledspar	17.708	15.473	37.946
		Plagioclase	17.590	15.438	37.864
10LT-3[b]	Orthogneiss	K-fledspar	16.510	15.344	37.454
		Plagioclase	17.316	15.452	37.953
10LT-4[b]	Orthogneiss	K-fledspar	16.522	15.350	37.482
		Plagioclase	16.852	15.393	37.673
10LT-5[b]	Orthogneiss	K-fledspar	17.086	15.347	37.711
		Plagioclase	17.290	15.406	37.919
10LT-6[b]	Orthogneiss	K-fledspar	17.141	15.359	37.770
		Plagioclase	17.326	15.394	37.889
10LT-7[b]	Orthogneiss	Plagioclase	17.291	15.439	37.640
10LT-8[b]	Orthogneiss	Plagioclase	17.275	15.438	37.556
10XS-1[b]	Orthogneiss	K-fledspar	16.370	15.324	37.348
		Plagioclase	17.021	15.417	37.788
10XS-2[b]	Orthogneiss	K-fledspar	16.312	15.289	37.176
		Plagioclase	16.367	15.303	37.271
10XS-4[b]	Orthogneiss	K-fledspar	17.616	15.429	38.013
		Plagioclase	17.778	15.472	38.163
10XS-5[b]	Orthogneiss	K-fledspar	16.381	15.298	37.152
		Plagioclase	16.782	15.391	37.519
10XS-6[b]	Orthogneiss	K-fledspar	16.910	15.331	37.662
		Plagioclase	17.016	15.335	37.690
00XS-1[a]	Orthogneiss	K-fledspar	16.860	15.380	37.283
CDZ					
03CP-2[a]	Orthogneiss	K-fledspar	17.291	15.441	37.583
03SH-3[a]	Orthogneiss	K-fledspar	16.894	15.377	37.334
99XF-1[a]	Orthogneiss	K-fledspar	17.336	15.459	37.743
02HN-1[a]	Orthogneiss	K-fledspar	17.114	15.438	37.403
02HN-2[a]	Orthogneiss	K-fledspar	17.344	15.457	37.493
02HN-3[a]	Orthogneiss	K-fledspar	17.209	15.440	37.368
02HN-4[a]	Orthogneiss	K-fledspar	17.542	15.466	37.598
02HN-5[a]	Orthogneiss	K-fledspar	16.796	15.374	37.078
02HN-7[a]	Orthogneiss	K-fledspar	17.383	15.459	37.625
02HN-9-1[a]	Orthogneiss	Plagioclase	17.253	15.451	37.636
02HN-10[a]	Orthogneiss	Plagioclase	17.188	15.449	37.633
02HN-11[a]	Orthogneiss	Plagioclase	17.274	15.442	37.608

Continued

Sample	Lithology	Mineral	$^{206}Pb/^{204}Pb$	$^{207}Pb/^{204}Pb$	$^{208}Pb/^{204}Pb$
02HN-15[a]	Orthogneiss	K-fledspar	17.289	15.412	37.630
08SH-10[b]	Orthogneiss	K-fledspar	17.313	15.435	37.420
08SH-9[b]	Orthogneiss	K-fledspar	17.512	15.480	37.756
08WH-1[b]	Orthogneiss	K-fledspar	17.134	15.378	37.573
BMJ-1[a]	Orthogneiss	K-fledspar	17.236	15.469	38.047
02SH-5[a]	Paragneiss	Plagioclase	19.550	15.813	38.351
02SH-9[a]	Paragneiss	Plagioclase	18.835	15.739	40.020
SDZ					
02HS-2[a]	Orthogneiss	K-fledspar	17.252	15.428	37.342
02HS-6[a]	Orthogneiss	K-fledspar	17.342	15.434	37.735
08TH-LD-1[b]	Orthogneiss	K-fledspar	17.119	15.404	37.762
08TH-LD-2[b]	Orthogneiss	K-fledspar	17.342	15.416	37.721
08TH-ZJC-1[b]	Orthogneiss	K-fledspar	16.835	15.372	37.219
08TH-HZ-2[b]	Orthogneiss	K-fledspar	16.971	15.401	37.356
08TH-FZL-1[b]	Orthogneiss	K-fledspar	17.707	15.566	38.252
08TH-LD-4[b]	Orthogneiss	K-fledspar	16.759	15.348	37.164
08TH-HZ-1[b]	Orthogneiss	K-fledspar	16.949	15.373	37.260
08TH-ZQ-1[b]	Orthogneiss	K-fledspar	16.809	15.337	36.960

[a] were measured at CAS.
[b] were performed at USTC.

The CDZ and the SDZ orthogneisses show relatively limited ranges of $^{206}Pb/^{204}Pb$ (16.413–17.543 and 16.759–17.707, respectively), $^{207}Pb/^{204}Pb$ (15.374–15.480 and 15.337–15.566, respectively) and $^{208}Pb/^{204}Pb$ (37.078–38.047 and 36.960–38.252, respectively). Moreover, the two most radiogenic samples (02SH-5 and 02SH-9, Fig. 1 and Table. 1) are the paragneisses from Shuanghe in the CDZ, which are related to the Pb isotopic feature of meta-supracrustal rocks.

5 Discussion

5.1 Common Pb isotopic characters of three UHPM units and Plumbotectonics in subducted continental crust

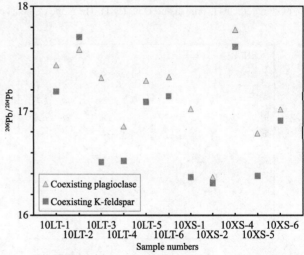

Fig. 3 $^{206}Pb/^{204}Pb$ of the coexisting K-feldspar and plagioclase of the analytical orthogneiss samples from the Luotian dome.

5.1.1 Difference between feldspar common Pb isotope and whole-rock Pb isotope

A compilation of common Pb isotopic data and published whole-rock Pb isotopic data for the Dabie UHPM eclogites and gneisses are shown in Fig. 4b. The whole-rock Pb isotopic compositions are quite distinctive from the common Pb data of the feldspar: the common Pb array occupies limited ranges of $^{206}Pb/^{204}Pb$ and $^{207}Pb/^{204}Pb$, and in general has less radiogenic Pb isotopes than the whole-rock Pb isotopes for the CDZ gneisses (Zhang et al., 2001; Li et al., 2003b). The more radiogenic whole-rock Pb were mainly from the Shuanghe UHPM paragneisses, which have comparatively high initial radiogenic Pb with the common Pb data of the feldspar from paragneisses (Fig. 4b). In addition, the U-Pb isotopic data of the whole rock may be in error caused by U/Pb fractionation during surface weathering. Differences of NDZ gneisses between whole-rock Pb and common Pb data are probably due to incomplete sampling for the whole-rock Pb studies. For example, most of the orthogneiss samples from Li et al.

(2003b) were from a small area around Zhujiapu in the northern boundary of the ENDZ, and only four orthogneisses were referenced from Zhang et al. (2001). The lack of the sampling in the Luotian dome is also responsible for the more radiogenic of the NDZ common Pb array (Fig. 5b).

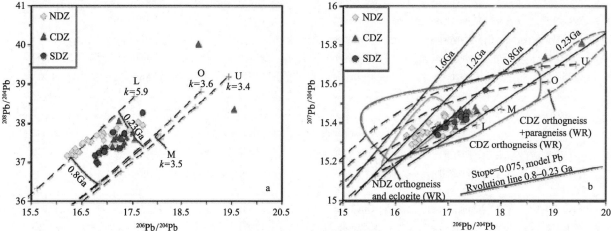

Fig. 4 Common Pb isotopic compositions of feldspars from the NDZ, the CDZ and the SDZ. On the ^{207}Pb/^{204}Pb vs. ^{206}Pb/^{204}Pb diagram (a), whole-rock Pb data from eclogites and gneisses (Zhang et al., 2001; Li et al., 2003b) in the NDZ (in blue ellipse) and the CDZ (in red solid ellipse and dash ellipse) are present for the comparison. All whole-rock Pb data have been calculated to initial values at the age t=0.23 Ga. Curves represent the evolution of idealized upper crust (U), lower crust (L) and mantle (M) Pb from the "Plumbotectonics" model (Zartman and Doe, 1981). "Paleogeochrons" are shown on the ^{207}Pb/^{204}Pb vs. ^{206}Pb/^{204}Pb diagram for 1.6, 1.2, 0.8 and 0.23 Ga, and on ^{208}Pb/^{204}Pb vs. ^{206}Pb/^{204}Pb diagram for 0.8 and 0.23 Ga. On the ^{208}Pb/^{204}Pb vs. ^{206}Pb/^{204}Pb diagram, reference ^{232}Th/^{238}U (k values) of mantle, upper crust and lower crust are shown. Cross marks along each curve indicate progressively older time in 0.4 Ga increments (Zartman and Doe, 1981). The slope of the isochron from the time interval 0.8–0.23 Ga is also shown.

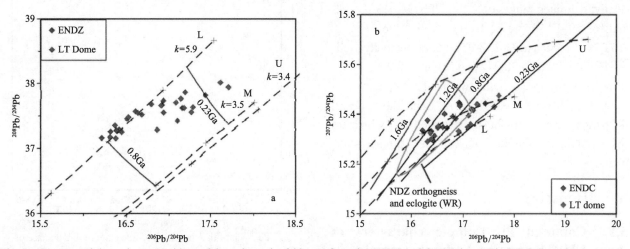

Fig. 5 Common Pb isotopic compositions of the orthogneiss feldspars from the ENDZ and the LT dome. The NDZ Growth curves and parameters, and the blue ellipse as in Fig. 4.

5.1.2 Common Pb isotopic characters and geological implications

Based on the peak metamorphic age of the UHPM rocks in the Dabie orogen, the Pb isotopic evolution of the feldspars essentially ended at 0.23 Ga as the temperature of the rocks fell below the feldspar diffusion blocking temperature. Consequently, the common Pb data of the feldspars only records the pre-metamorphic Pb evolution of the three UHPM units. On the ^{207}Pb/^{204}Pb-^{206}Pb/^{204}Pb diagram (Fig. 4b), the data arrays of the orthogneisses from the Dabie orogen all lie to the left of the 0.23 Ga paleogeochron (defined as a line joining the mantle, upper-crust, and lower-crust isotopic compositions at 0.23 Ga for the plumbotectonics model; Zartman and Doe, 1981). Therefore, their single-stage model ages are older than the expected time of isotopic closure in these feldspars immediately following the peak time of Dabie metamorphism (Li et al., 2003a, 2005). The Pb data yield single-stage ^{238}U/^{204}Pb (μ) values of 6.9–8.4, 7.5–8.3 and 7.5–8.4 for the NDZ, the CDZ and the SDZ arrays, respectively, using Canon Diabolo troilite as the starting point for Pb-evolution and 4.56 Ga for the age of the Earth.

Two paragneisses lie above the model upper crust evolution curve and display significantly more radiogenic Pb than the orthogneisses, corresponding to their sedimentary protoliths with high μ values. With respect to the Pb isotopic composition of the orthogneisses, however, only one sample (09TH-FZL-1) from the SDZ lie adjacent to the evolution curve of the plumbotectonics model orogen (Zartman and Doe, 1981). All other orthogneisses of the Dabie orogen cluster along the mantle evolution curve, with most of the data spread between the mantle and lower crustal evolution curves. Although a tight array from the CDZ lies between the mantle and orogen evolution curves, four samples (with $^{206}Pb/^{204}Pb$ = 16.796–17.289) are remarkably less radiogenic and lie between the mantle and lower crustal curves. The NDZ orthogneiss data form an array yielding a best fit line with a slope of 0.100 ± 0.014 (σ) (MSWD=13, n=30). In contrast, the CDZ and SDZ form elongate arrays with similar steeper slopes of 0.140 ± 0.026 (σ) (MSWD=4.8, n=17) and 0.139 ± 0.021 (σ) (MSWD=2.5, n=9), respectively, that are essentially collinear with each other. The slopes of all three arrays are inconsistent with a slope of 0.075, which would be given by a secondary isochron for Pb growth from 0.8 to 0.23 Ga, assuming that the protoliths of the orthogneisses in the Dabie orogen were separated from the mantle at 0.8 Ga. This suggests that the Pb isotope evolution of the orthogneisses is more complex than predicted by a two-stage evolution model.

The Pb isotopic compositions of the NDZ orthogneisses display wide variations from 1.6 Ga to 0.23 Ga in paleogeochron ages, while the CDZ and SDZ age spread is more restricted and range from 0.8 Ga to 0.23 Ga. Therefore, the NDZ array could be divided into two groups by using the 0.8 Ga paleogeochron as the division line. Group 1 with $^{206}Pb/^{204}Pb$<16.8 lie to the left of and Group 2 with $^{206}Pb/^{204}Pb$>16.8 lie to the right of the 0.8 Ga paleogeochron. For present-day Pb isotopic compositions, the Group 1 rocks that lie to the left of the 0.8 Ga paleogeochron may be best explained as having either incorporated abundant ancient lower crustal components into their protoliths, or by having had a significant reduction in the μ of their protoliths for a long time, as a result of high grade metamorphism at a lower crust level (Rudnick and Goldstein, 1990; DeWolf and Mezger, 1994). This interpretation of Group 1 rocks is also consistent with a constant single-stage k (5.7–6.1) consistent with the model lower crustal k=5.9 on the $^{208}Pb/^{204}Pb$ vs. $^{206}Pb/^{204}Pb$ diagram (Fig. 4a) that is a typical feature for ancient lower crust. In contrast, Group 2 rocks with feldspar paleogeochron ages spreading from 0.8 to 0.23 Ga on the $^{207}Pb/^{204}Pb$ vs. $^{206}Pb/^{204}Pb$ diagram nearly overlap with those of both the CDZ and SDZ arrays.

The common Pb data of the CDZ and SDZ display lower k values, higher μ values, and is more radiogenic than those characterizing the Group 1 rocks of the NDZ. This attribute could possibly be generated in two different ways: (1) by incorporation of more Paleoproterzoic lower crustal components into the protolith of the NDZ than those of the CDZ and SDZ; or (2) by a higher, middle-upper crust emplacement level of the CDZ and SDZ gneiss protoliths, leading to a subsequent higher μ and lower k evolution of their Pb. Hypothesis (1) is excluded because the zircon Hf isotopic data and model ages of all three UHPM slices are similar showing that the protoliths of the orthogneisses principally originate from reworking of the Middle Paleoproterozoic crust and juvenile early Neoproterozoic crust in an active rift zone (See details in below) (Zheng et al., 2005, 2006; Zhao et al., 2008; Xia et al., 2009). We attribute the Pb isotope differences to the various emplacement depths of the protoliths following Neoproterozoic rifting, which leads to the Pb isotopic evolution with low μ and high k for the NDZ, high μ and low k for the CDZ and SDZ, respectively. This interpretation is supported by the following quantitative modelling given below. Therefore, we propose that the emplacement depths of the protoliths of the different UHPM slices have played an important role in Pb isotopic evolution after the formation of their protoliths. The overlap between Group 2 Pb data of the NDZ and both CDZ and SDZ arrays could possibly be generated by the homogenization of Pb isotopes among the SDZ, CDZ and a part of the NDZ as a result of Pb mixing induced by the retrograde metamorphic fluid activities enhanced by the Cretaceous magmatic event (see detailed discussion in next section).

On the $^{208}Pb/^{204}Pb$-$^{206}Pb/^{204}Pb$ diagram (Fig. 4a), all Pb data for the orthogneisses plot between the mantle and lower crust curves (4 < k < 5.9), with some from the NDZ lying on the evolution curve of the model lower crust (k = 5.9). Sultan et al. (1992) attributed similar distribution pattern for Pb isotopes from the Nubian Shield in Egypt to a mixing between mantle and old crust sources. In view of global Pb data, the only environment in which k is higher than the model mantle value of 3.5, and variable, is evidenced by rocks having a long residence in the lower crust or being derived by the reworking of old lower crust (Rudnick et al., 1985; DeWolf and Mezger, 1994). The protoliths of the orthogneisses in three UHP units were derived from Neoproterozoic rift magmatism, which intruded to various depths within the continental crust. The locations of the CDZ (middle crust) and the SDZ (upper curst) are unlikely to have undergone Pb isotopic evolution with a high k ratio (> 3.5). Instead the data are best explained as arising from various degrees of mixing between model mantle and old lower crustal Pb compositions. This interpretation is consistent with the observation that the UHPM orthogneiss data of the Dabie orogen differ significantly from what would be predicted by the model orogen evolution curve (Zartman and Doe, 1981). Relative to the global average contents of an orogen, the UHPM orthogneisses of the Dabie orogen represent the part of such orogen (i.e. subducted continental crust), which was made up of the dominant pre-existing lower crustal and mantle components, with upper crustal component starved.

A recognition of the heterogeneity of an orogen was incorporated by Zartman and Haines (1988) into their revised plumbotectonics model, which identified a proximal; i.e., miogeosynclinal, and a distal; i.e., eugeosynclinal, component to the orogen. Although retaining the Pb isotope evolution curve of an average orogen, they recognized that given parts of an orogen would likely evolve along different, vertically-displaced curves on the $^{207}Pb/^{204}Pb$ vs. $^{206}Pb/^{204}Pb$ diagram. Although the Pb evolution curves of the Zartman and Doe (1981) original plumbotectonics model are retained in Fig. 4, they remain very similarly positioned in the revised model, and the orogen curve closely corresponds to the crustal growth curve of the Stacey and Kramers (1975) two-stage model. The UHPM slices of the Dabie orogen as the representative of subducted continental crust is accordingly seen as among the most upper crustal component starved part of an orogen.

5.1.3 Tectonic relationship between Luotian dome and the ENDZ within the NDZ, and implication for the Pb mixing during retrograde metamorphism

According to geologic structure of the Dabie orogen, the NDZ consist of two main parts, Luotian dome in the southwest and the ENDZ in the northeast (Fig. 1). A compilation of the common Pb data from the ENDZ and the LT Dome orthogneisses is displayed in Fig. 5, along with the model evolution curves and reference paleogeochrons. Overall, the most Pb data of these two areas overlap with each other on both $^{207}Pb/^{204}Pb$-$^{206}Pb/^{204}Pb$ and $^{208}Pb/^{204}Pb$-$^{206}Pb/^{204}Pb$ diagrams and give similar variations (Fig. 5a, b). In terms of the previous discussion, the NDZ samples could be divided into two groups by using the 0.8 Ga paleogeochron as the division line, while group 1 with $^{206}Pb/^{204}Pb<16.8$ lie to the left of and group 2 with $^{206}Pb/^{204}Pb>16.8$ lie to the right of the 0.8 Ga paleogeochron. Thirteen analysis points from the ENDZ were classified into group 1 and only four into group 2. In contrast, only five samples from the Luotian dome lie to the left of the 0.8 Ga paleogeochron (in group 1), but nine to the right (in group 2). It has been suggested that the overlap between group 2 Pb data of the NDZ and both CDZ and SDZ arrays could possibly be generated by the homogenization of Pb isotopes among the SDZ, CDZ and a part of the NDZ as a result of Pb mixing induced by the retrograde metamorphism. The more data from the Luotian dome classifying into group 2 suggest that the orthogneisses in Luotian dome could be suffered more stronger influence of Pb mixing between the NDZ and CDZ or SDZ than those of the ENDZ during retrograde metamorphism. This interpretation is consistent with their difference in surface erosion depth.

It is well known that the whole NDZ was subjected to uplifting and magma thermal events caused by the intrusions of granite batholithes during the Cretaceous (Okay, 1993; Hou et al., 2005). The extensive outcropping of the granitic batholithes in the ENDZ, rather than the Luotian dome (Fig. 1) suggests that the former seems to be a more deeply eroded area. As a consequence, the protolith of the orthogneiss samples collected from the ENDZ was located deeper than the protolith of the Luotian dome samples in the Cretaceous (see Fig. 6a). Thus, the sampling area in Luotian dome is more closer to the overlying CDZ than that of the ENDZ. The fluid activity driven by the magmatic thermal event at middle crust level may result in Pb isotopic mixing between the NDZ and CDZ slices during the Cretaceous. Obviously, the Luotian dome sampling area may suffered more stronger influence of such Pb isotopic mixing, because the Luotian dome sampling area was closer to the CDZ than that of the ENDZ (Fig. 6a and b).

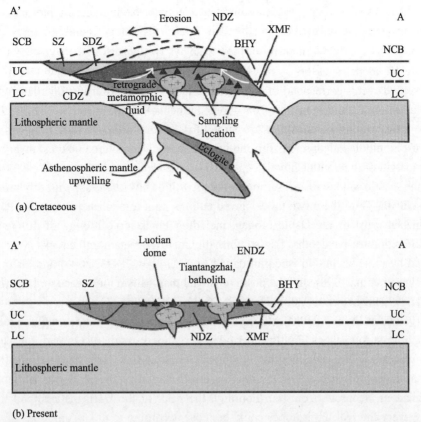

Fig. 6 Cartoons showing the positions and erosion depths of the orthogneiss sampling areas from the Luotian dome and the ENDZ within the NDZ from the Dabie orogen during the Early Cretaceous (a) and present (b) (revised after He et al., 2013). The cross-section (A-A') is according to AA' line in Fig. 1 (See text for detail explanation).

5.1.4 Geological evolution history of the Dabie orogen and Yangtze Craton

In the following, we proceed to construct a quantitative model of the Pb isotopic evolution of the Dabie UHPM slices, approach to examining the above inference, as well as revealing the relevant factors of the Pb isotope evolution within the subducted continental crust. This model would require to be consistent with the analyzed feldspar Pb isotope data set and regional geological evolution history of the Dabie orogen, as far as to that of the Yangtze Craton, both of which need to be generally assessed.

A large dataset with the isotope geochronologic systematics has been published for the Dabie UHPM rocks and ancient basement of the Yangtze Craton, which was used for the tectonic reconstruction of these two areas. The investigation of the Kongling Complex, as an oldest basement of the Yangtze Block, suggests that the initial growth of Yangtze continental crust started no later than 3500 Ma, and it suffered reworking in the Paleoproterozoic after initial growth in the Archean (Zhang et al., 2006a, b). With regard to the Dabie orogenic basement, Wu et al. (2008) and Jian et al. (2012) obtained zircon U-Pb ages for granulites and gneisses hosted by UHPM rocks from Huangtuling in the NDZ. Zircon cores with typical igneous characteristics in the granulites generated a range of $^{207}Pb/^{206}Pb$ ages from 2761±3 Ma to 3277±16 Ma (Wu et al., 2008), and 2766±8 Ma to 3089±3 Ma (Jian et al., 2012), interpreted as dating the magma emplacement of the protolith. Furthermore, some inherited zircon cores in the granulites display rather older $^{207}Pb/^{206}Pb$ ages of 2900 to 3530 Ma, suggesting that are Mesoarchean to Paleoarchean crustal remnants. These observations suggest that the growth of the subducted continental crust in Dabie orogen started at least ~2800 Ma ago. U-Pb dating of granulite-facies metamorphic zircons from the Huangtuling gneisses yield $^{207}Pb/^{206}Pb$ ages of 2029±13 Ma (Wu et al., 2008) and 2043±7 Ma (Jian et al., 2012) for a superimposed metamorphic event. These results demonstrate that the Archean crust remnants in subducted continental crust were reworked in the middle Paleoproterozoic at lower crust levels in association with granulites-facies metamorphism. In addition, combined studies of U-Pb ages and Hf isotopic composition of zircons have been carried out for the UHPM orthogneiss from the Dabie orogen (Zheng et al., 2005, 2006; Zhao et al., 2008;

Xia et al., 2009). Zircon U-Pb dating provides consistent Neoproterozoic ages for protolith crystallization, and Triassic ages for regional metamorphism. On the other hand, zircon Hf isotope analyses display a wide range of $\varepsilon_{Hf(t)}$ values between –8.5 and 14.5, while a majority of them displays two groups of $\varepsilon_{Hf(t)}$ values of (1) -8.5 to -2.9 with depleted mantle Hf model ages (T_{DM}) of 1991–2186 Ma, and (2) 2.2–12.9 with Hf model ages (T_{DM}) of 821–1259 Ma. These two events correspond to the granulite-facies metamorphism of the Huangtuling gneiss in the Dabie orogen and the assembly and breakup of the supercontinents Rodinia (Li et al., 1999b; Zheng et al., 2004), respectively. These observations demonstrate the protoliths of the orthogneisses principally originated from reworking lower crust in middle Paleoproterozoic and juvenile early Neoproterozoic crust in an active rift zone. The rifting on the north margin of the South China Block (SCB) in the Neoproterozoic would have transferred both heat and material from the mantle to the crust, resulting in the growth of juvenile crust with different intrusive levels in the continental crust depth. Thus, these two times played an important role in establishing the Pb isotope evolutions in subducted continental crust of the Dabie orogen, including Pb isotopic mixing of different components and self-evolutions in different intrusive depths. Triassic formation of UHP metamorphic rocks in the Dabie orogen were generally determined by mineral Sm-Nd isochron dating (e.g. Li et al., 1993, 2000) and zircon U-Pb dating (e.g. Ames et al., 1996; Hacker et al., 1998), and the post-collisional magmatism took place in the early Cretaceous (Jahn et al., 1999; Li et al., 1999a; Zhao et al., 2007; He et al., 2011).

Thus, the source of the Dabie UHPM orthogneisses must have undergone a complex process of Pb isotopic evolution, including at least four stages, Firstly, a period of early continental crust growth for the basement rocks of the SCB in the Archean (t_1). Secondly, the basement rocks of the SCB experimented a granulite-facies metamorphism in the Paleoproterozoic (t_2). Thirdly, the north margin of the SCB was rifting during breakup of Rodinia supercontinent in Neoproterozoic (t_3). Fourthly, the peak of the UHP metamorphism took place in the Triassic (t_4). During every interval, Pb isotopes could be those according to the assumptions of the plumbotectonic model.

Although it has been demonstrated that the orthogneisses record the presence of ancient lower crust in the Dabie orogen, the Pb isotopic data for several orthogneisses lie above the model mantle curve. A lower crustal Pb isotopic composition alone can not explain this more radiogenic kind of Pb. Thus, some amount of an old upper crustal component, characterized with lower k and higher μ, should be taken into consideration in quantitative modelling.

We provide a modestly estimation of the age of each stage to approach natural geological processes. Four time nodes are concluded below: t_1=2.8 Ga, represents the mean age of primary crustal growth of the Yangtze Craton. At this time, the juvenile upper and lower crusts were generated, and hereafter started to evolve with respective μ characters; t_2=2.0 Ga, represents the granulite-facies metamorphic age of Paleoproterzoic crust (2.1–1.8 Ga). Due to certain tectonic activity, Archean crustal component could be transferred to lower crust level. The granulite-facies metamorphism led to subsequent Pb evolution with low μ; t_3=0.8 Ga, is the age of the extensively rift magmatism during splitting of the Yangtze Block from the supercontinent Rodinia (Zheng et al., 2008). Reworking of Paleoproterzoic lower crust with the addition of Neoproterozoic mantle material generated the bimodal magmas, which intruded into the upper, middle and lower crust levels related with the protoliths of the SDZ, CDZ, and NDZ, respectively. Furthermore, Pb isotopes of three slices evolved with individual μ values until UMP metamorphism at t_4=0.23 Ga (Li et al., 1993).

5.1.5 Simulation method of a multi-stage Pb isotope evolution model

The simple modelling of Pb isotope evolution has been investigated in the following. The basic principles of presenting and interpreting the Pb isotopic data obtained in our modelling conform to the two-stage model method used by Stacey and Kramers (1975) and the three-stage model from Taylor et al. (1980). The Dabie Pb data are then plotted on diagrams and related to the upper crust, lower crust, orogen, and mantle evolution curves of Zartman and Doe (1981).

The fundamental assumptions and multi-stage method are shown in Fig. 7a and described as follows: (1) The model orogen evolution curve of Zartman and Doe (1981) is approximately equal to the second-stage; i.e., after 3.7 Ga, crustal evolution curve of Stacey and Kramers (1975) with μ = 9.74 and k = 3.78; (2) During the second stage

(t_1- t_2) the old crust separated from a mantle source region at the time (t_1) and subsequently evolved along a crustal growth curve with individual Pb isotopic character to t_2, according to its magmatic emplacement level in the crust structure (e.g. μ^L_1 for the lower crust, μ^U_1 for the upper crust). Different μ values follow the trend of $\mu^L_1 < \mu_0 < \mu^U_1$. After Pb evolution between t_1 and t_2, the Pb isotopic composition of the crust would lie along the intersections (B and D) of secondary evolution curves with different μ values and the t_1- t_2 secondary isochron (AC), according to Equations (1) and (22). The Pb composition of the mixture between mantle and crust are typified by points O and P.

$$\left(\frac{^{206}Pb}{^{204}Pb}\right)_{t_2} = \left(\frac{^{206}Pb}{^{204}Pb}\right)_{t_1} + \mu\left(e^{\lambda^{238}t_2} - e^{\lambda^{238}t_1}\right) \tag{1}$$

$$\left(\frac{^{207}Pb}{^{204}Pb}\right)_{t_2} = \left(\frac{^{207}Pb}{^{204}Pb}\right)_{t_1} + \frac{\mu}{137.8}\left(e^{\lambda^{235}t_2} - e^{\lambda^{235}t_1}\right) \tag{2}$$

(3) During the third stage (t_2- t_3) and in the case of different emplacement levels, Pb evolution of rocks crystallising from the contaminated magma, as typified by point O, would display three evolution curves (OL, OM and ON) with μ^L_2, μ^M_2 and μ^U_2, respectively. At the time (t_3), the Pb isotopic compositions of different terrestrial reservoirs lie at points L, M and N, which are collinear. The slope of line L-M-N is a function of the ages t_2 and t_3, which is parallel to t_2- t_3 tertiary isochron, according to equations (3), (4) and (5). Point P is determined in the same manner, and points L', M' and N' represent the Pb compositions of different reservoirs with μ^L_2, μ^M_2 and μ^U_2, respectively. In the case of the rocks intruding into the lower crust level at t_2, Pb compositions at t_3 would lie along line L-L'. This line is parallel to the t_1- t_2 secondary isochron (OP), following Equation (6). Pb compositions of the mixed lower crust and mantle at t_3 would fall into triangular (LL'F) space.

$$\left(\frac{^{206}Pb}{^{204}Pb}\right)^L_{t_3} = \left(\frac{^{206}Pb}{^{204}Pb}\right)^O_{t_2} + \mu^L_2\left(e^{\lambda^{238}t_3} - e^{\lambda^{238}t_2}\right) \tag{3}$$

$$\left(\frac{^{207}Pb}{^{204}Pb}\right)^L_{t_3} = \left(\frac{^{207}Pb}{^{204}Pb}\right)^O_{t_2} + \frac{\mu^L_2}{137.8}\left(e^{\lambda^{235}t_3} - e^{\lambda^{235}t_2}\right) \tag{4}$$

$$\frac{\left(\frac{^{207}Pb}{^{204}Pb}\right)^L_{t_3} - \left(\frac{^{207}Pb}{^{204}Pb}\right)^M_{t_3}}{\left(\frac{^{206}Pb}{^{204}Pb}\right)^L_{t_3} - \left(\frac{^{206}Pb}{^{204}Pb}\right)^M_{t_3}} = \frac{\left(\frac{^{207}Pb}{^{204}Pb}\right)^M_{t_3} - \left(\frac{^{207}Pb}{^{204}Pb}\right)^N_{t_3}}{\left(\frac{^{206}Pb}{^{204}Pb}\right)^M_{t_3} - \left(\frac{^{206}Pb}{^{204}Pb}\right)^N_{t_3}} = \frac{\left(\frac{^{207}Pb}{^{204}Pb}\right)^L_{t_3} - \left(\frac{^{207}Pb}{^{204}Pb}\right)^N_{t_3}}{\left(\frac{^{206}Pb}{^{204}Pb}\right)^L_{t_3} - \left(\frac{^{206}Pb}{^{204}Pb}\right)^N_{t_3}} = \frac{1}{137.8}\frac{\left(e^{\lambda^{235}t_3} - e^{\lambda^{235}t_2}\right)}{\left(e^{\lambda^{238}t_3} - e^{\lambda^{238}t_2}\right)} \tag{5}$$

$$\frac{\left(\frac{^{207}Pb}{^{204}Pb}\right)^{L'}_{t_3} - \left(\frac{^{207}Pb}{^{204}Pb}\right)^L_{t_3}}{\left(\frac{^{206}Pb}{^{204}Pb}\right)^{L'}_{t_3} - \left(\frac{^{206}Pb}{^{204}Pb}\right)^L_{t_3}} = \frac{\left(\frac{^{207}Pb}{^{204}Pb}\right)^P_{t_2} - \left(\frac{^{207}Pb}{^{204}Pb}\right)^O_{t_2}}{\left(\frac{^{206}Pb}{^{204}Pb}\right)^P_{t_2} - \left(\frac{^{206}Pb}{^{204}Pb}\right)^O_{t_2}} = \frac{1}{137.8}\frac{\left(e^{\lambda^{235}t_2} - e^{\lambda^{235}t_1}\right)}{\left(e^{\lambda^{238}t_2} - e^{\lambda^{238}t_1}\right)} \tag{6}$$

(4) During the fourth stage (t_3- t_4) rocks crystallising from the mixed magma source at t_3 evolved to the time of UHP metamorphism (t_4) shortly before the closure of Pb diffusion in feldspar. The Pb isotopes of these rocks with different μ values also plot along a fourth isochron, which is parallel to the model mantle isochron from t_3 to t_4. Due to the rocks with different initial Pb isotopic compositions, Pb isotopes of major terrestrial reservoirs would form a series of fourth-stage isochrons.

5.1.6 Modelling results of four-stage Pb evolution of the Dabie UHPM orthogneisses

Fig. 7b and c show that all common Pb data of the Dabie orthogneisses can be fitted with a series of 0.8–0.23 Ga fourth stage isochrons, with various initial Pb isotopic compositions. The different Pb isotopes of the protoliths are mainly due to the different degrees of mixing between Neoproterozoic juvenile mantle and Paleoproterzoic lower crust. For example, the data array above the model mantle evolution curve require relatively higher initial Pb isotope compositions for the fourth stage evolution, as a result of incorporation of Archean upper crustal component into Paleoproterzoic lower crust. The initial Pb isotopic compositions of crust in the fourth stage evolution could be within the scope of the gray triangle (Fig. 7b and c) confirmed by Paleoproterzoic lower crust and Neoproterozoic juvenile crust (isotopes of which approximately equal to that of Neoproterozoic mantle). In order to investigate the Pb evolution characters of different slices, two 0.8–0.23 Ga fourth stage isochrones (line 1 and 2) are chose to calculate the μ values. On the assumption that the protoliths of the Dabie orthogneisses completely derived from the reworking of 2.0 Ga lower crust with various proportions of 2.8 Ga upper and lower crusts without Neoproterozic mantle component, thus initial Pb isotopic compositions of the protoliths were displayed by the line, which has lowest $^{206}Pb/^{204}Pb$ for a given $^{207}Pb/^{204}Pb$, corresponding with line LL' in Fig. 7a. Thus, fourth stage μ values of three UHPM slices obtained would give upper limits of realistic μ values. Calculated values of the data along 0.8–0.23 Ga isochron line 1 are 3.4–8.6 for part 1 of the NDZ, 12.0–16.8 for the CDZ and 12.1–16.1 for the SDZ, while μ values from line 2 display 5.5–9.6 for the NDZ, 11.3–16.6 for the CDZ and 10.9–17.2 for the SDZ, respectively. This consequence accords with the μ characteristics of different emplacements in principle, related to theoretical μ values of 5.3–6.3, 8.3–9.3 and 11.0–13.3 for model lower crust, mantle and upper crust in plumbotectonics model (Zartman and Doe, 1981). The most radiogenic orthogneiss from the SDZ possibly requires a large amount Archean upper crustal components into its source origin.

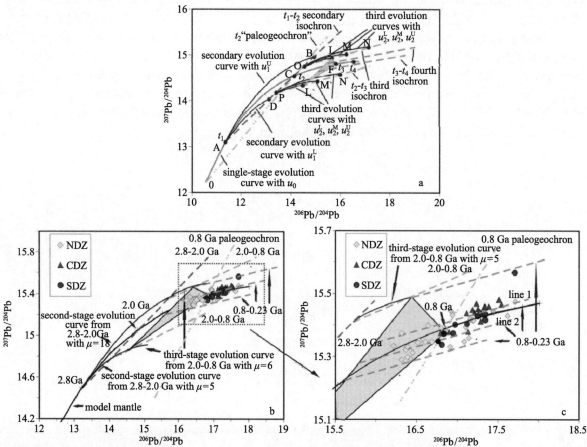

Fig. 7 Construction of four-stage model and multi-stage Pb isotopic evolution for the Dabie. (a) four-stage Pb evolution model: The specific geological processes are described in detail in the text. (b) Represents four-stage Pb isotopic evolutions modelling for the Dabie orthogneisses. (c) Represents partial enlarged view of (b). The reference μ value of upper crust curve from 2.8 to 2.0 Ga required to account for the observed spread is approximately 18, while reference μ values of lower crust curves from 2.8 to 2.0 Ga and 2.0-0.8 Ga are 5 and 6. Variation in whole-rock μ values of lower crust and upper crust are reasonable. Initial Pb isotopic compositions range of fourth stage are shown for the gray triangle. The dashed lines represent the isochrons for different time intervals, in accord with those in (a). Yellow dotted lines are 0.8 Ga paleogeochron. Other parameters are the same as in Fig. 4.

Upon the above discussion and simulation, an additional understanding of plumbotectonics model lies in exploring its suitability in the subduction zone experiencing UHP metamorphism. Although the preexisting upper continental crust accounts for a large proportion of the contributed total crust into the orogen in all versions of plumbotectonics models (Doe and Zartman, 1979; Zartman and Doe, 1981; Zartman and Haines, 1988), the UHPM orthogneisses of the Dabie orogen as among the most upper crustal component starved part of an orogen offer a new insight about the Pb isotope evolution within the subducted continental crust. The part of the subducted lower crust can apparently remain isolated from the other major reservoirs during continental subduction and be able to maintain a low μ, high k character and unradiogenic lead signature despite UHP metamorphism. On the other hand, a part of the subducted lower crust develop a higher radiogenic Pb isotope signature approach to the subducted middle and upper crust. Thus, the mixing of upper, middle and a part of lower crustal lead during exhumation of UHPM rocks and retrograde metamorphism as well as thermal events of post-collisional magmatism within the orogen tend to reduce the Pb isotopic differences, which were created by chemical fractionation and radioactive decay (Zartman and Haines, 1988).

5.2 Comparison between the Dabie orogen and the Chinese Continental Scientific Drilling project referred to the Su-Lu orogen

Similar to the Dabie terrane, the Sulu orogen can also be subdivided into four distinctive metamorphic zones from south to north, i.e. southern Sulu high-pressure zone (I), the Central Sulu very high-pressure zone (II), Northern Sulu UHP metamorphic supracrustal rocks zone (III) and the Northern Sulu UHP granitic metamorphic zone (IV) (Xu et al., 2006). From SE to NW, the Lianyungang (IIIa), Maobei (IIIb), Donghai (IIIc) and Shilianghe (IIId) slices from unit III are separated by three ductile shearing zones DF6, DF7 and DF8, respectively. Common Pb isotopic compositions of omphacites in eclogites and feldspars in gneisses from the Chinese Continental Scientific Drilling (CCSD) project (100–5000 m) have been investigated by Li et al. (2009). The main drilling hole of the CCSD project passes through the Maobei (IIIb) (unit 1 in drilling core), Donghai (IIIc) (unit 3 in drilling core) and Shilianghe (IIId) (unit 5 in drilling core) slices from the top to bottom and the two shearing units (unit 2 and 4 in drilling core) between the UHP slabs in the CCSD correspond to the DF7 and DF8 on the surface, respectively.

Comparison of the common Pb isotopic compositions between the Dabie and the CCSD displayed in Fig. 8. All of the unit 1 array in the CCSD overlap with the CDZ and SDZ ranges on both the $^{207}Pb/^{204}Pb$ vs. $^{206}Pb/^{204}Pb$ and the $^{208}Pb/^{204}Pb$ vs. $^{206}Pb/^{204}Pb$ diagrams (Fig. 8a and b), indicating a similar Pb isotope evolution with them. However, Pb data of unit 3 from the CCSD represent less radiogenic and similar k than the NDZ, which could be predictable if greater amounts of old lower crust are involved and lower emplacement level. The least

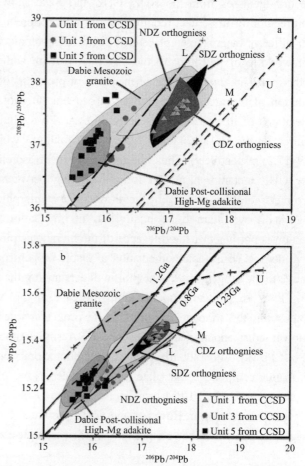

Fig. 8 Compilation of common Pb data from Dabie orthogneisses, Sulu UHPM rocks of CCSD project and Dabie Mesozoic granite for comparison. Data source: Dabie Mesozoic granite from Zhang et al. (1995); Dabie post-collisional high-Mg adakite from Huang et al. (2008); unit 1, 3, 5 of the CCSD project from Li et al. (2009). Growth curves and parameters are the same as in Fig. 4.

radiogenic and extremely high k values (5.9–7.1) occur in unit 5, which basically lies to the left of the 0.8 Ga paleoisochron and below the mantle evolution curve (Fig. 8b). These lower crust-type Pb characteristics of unit 5 are consistent with significantly more contributions of ancient lower crustal components into its magmatic source than the NDZ. Furthermore, based on the modelling above, unit 5 would display markedly lower μ than the NDZ and unit 3, indicating that the magmatic source of unit 5 intruded into a deeper crust level than them. Such a unit has not been observed in the outcrop of the Dabie orogen.

However, the Pb isotopes of three units from CCSD are almost overlapped with the range of "Dabie Mesozoic granite" including some post-collisional high-Mg adakitic rocks (Fig. 8b), which are characterized by less radiogenic Pb isotopic composition and high k. Thus, a hidden unit as the source of some post-collisional granites should exist below NDZ in the lower crust level, whose Pb isotopic compositions are consistent with unit 5 of the CCSD from the Sulu orogen. This view is in accord with the whole rock Sr-Nd isotope and Zircon U-Pb dating, Hf and O isotope of Dabie post-collisional granites (Zhao et al., 2007, 2011). In this case, the Su-Lu orogen should experience a stronger intensity of the erosion to a deeper position compared to the Dabie orogen.

6 Conclusions

Pb isotopic compositions of 59 feldspars separated from the othogneisses in the three UHPM units of the Dabie orogen provide important constraints on the pre-metamorphic evolution history of the Dabie UHPM orthogneisses. The orthogneisses from the NDZ, CDZ and SDZ show systematic differences in common Pb isotopes. The interpretation of these data requires not only the different initial Pb isotopic compositions of three units, but also their different magmatic emplacement depths. Pb isotopic features of their magma sources are basically controlled by reworking of Paleoproterozoic lower crust, with addition of Archean upper and lower crustal components. Furthermore, magmatic emplacement levels of protoliths of the orthogneisses also play an important role in the Pb evolution of the orogen, and lead to subsequent Pb evolution trends with individual μ and k features after the generations of their protoliths. The multi-stage Pb evolution model obtains the consistent results, e.g. the NDZ requires low μ (3.4–9.6) for 0.8–0.23 Ga evolution, while the CDZ and SDZ require high μ (11.3–16.8 and 10.9–17.2, respectively). Thus, the disparities in Pb isotopes between subducted lower and upper crusts do exist after UHP metamorphism. Pb isotopic mixing between the NDZ and the CDZ, SDZ occurred during the retro-metamorphism induced by the retrograde metamorphic fluid in the Cretaceous.

Finally, we return to the inquiry into the implication for plumbotectonics in subducted continental crust. The question is not whether the relevance of the plumbotectonic model is determined by how well the model's orogen curve fits the Pb isotopic composition of Dabie orogen rocks. Rather it is whether the position of the Dabie data relative to the orogen curve is helpful in understanding the makeup of the Dabie rocks. By plotting on a $^{207}Pb/^{204}Pb$ vs. $^{206}Pb/^{204}Pb$ diagram under, and in the case of the NDZ significantly under, the orogen curve of average composition, the orthogneisses of the Dabie orogen are identified as consisting of a disproportionately large amount of lower crust and/or mantle components. A similar constraint is imposed by the position of the data on the $^{208}Pb/^{204}Pb$ vs. $^{206}Pb/^{204}Pb$ diagram. The plumbotectonic model attempts to mimic the global average mix of crustal and mantle components that plate tectonic processes recycle into the formation of new crust and returned material into the mantle. The UHPM orthogneisses of the Dabie orogen do not make up the rock assemblage expected of such an average orogen. However, they have presented the opportunity to use a mathematical approach similar to that employed by plumbotectonics to trace their isotopic evolution through a complex geologic history.

Acknowledgments

This research was financially supported by the National Basic Research Program of China (2009CB825002) and the National Scientific Foundation of China (No. 41090372, 40973016). We thank Ping Xiao, Jianfeng He and Chaofeng Li for their assistance with Pb isotope analysis. Yican Liu, Shuijiong Wang, Shicao An and Chongqin Feng are also thanked for their help on field work. We highly appreciate the kind comments and suggestions from Prof. R.E. Zartman and two anonymous reviewers, which helped us to improve the manuscript. Prof. Liping Qin, Yongfei Zheng and Fukun Chen are thanked for their suggestions and discussion on the construction of this manuscript.

References

Ames, L., Zhou, G. Z., and Xiong, B. C., 1996. Geochronology and isotopic character of ultrahigh-pressure metamorphism with implications for collision of the Sino-Korean and Yangtze cratons, central China. *Tectonics* **15**, 472-489.

Ayers, J. C., Dunkle, S., Gao, S., and Miller, C. F., 2002. Constraints on timing of peak and retrograde metamorphism in the Dabie Shan Ultrahigh-Pressure Metamorphic Belt, east-central China, using U-Th-Pb dating of zircon and monazite. *Chem. Geol.* **186**, 315-331.

Brenan, J. M., Shaw, H. F., Ryerson, F. J., and Phinney, D. L., 1995. Mineral-aqueous fluid partitioning of trace elements at 900 ℃ and 2.0 GPa: Constraints on the trace element chemistry of mantle and deep crustal fluids. *Geochim. Cosmochim. Acta* **59**, 3331-3350.

Cong, B., 1996. Ultrahigh-pressure Metamorphic Rocks in the Dabieshan-Sulu region of China. Science Press.

Cumming, G. and Richards, J., 1975. Ore lead isotope ratios in a continuously changing earth. *Earth Planet .Sci. Lett* .**28**, 155-171.

DeWolf, C. P. and Mezger, K., 1994. Lead isotope analyses of leached feldspars: constraints on the early crustal history of the Grenville Orogen. *Geochim. Cosmochim. Acta* **58**, 5537-5550.

Doe, B. R. and Tilling, R. I., 1967. The distribution of lead between coexisting K-feldspar and plagioclase. *Am. Mineral.* **52**, 805–816.

Doe B. R. and Zartman R. E. 1979. Plumbotectonics , the phanerozoic. *Geochemistry of Hydrothermal Ore Deposits* 2, 22-70.

Dong, F., Li, S. G., Li, Q. L., and Chen, F. K., 2006. A preliminary study of common Pb of UHP metamorphic rocks from CCSD (100−2000 m)—evidence for decoupling within subducting continental crust. *Acta Petrol. Sin.* **22**, 1791-1798.

Hacker, B. R., Ratschbacher, L., Webb, L., Ireland, T., Walker, D., and Dong, S. W., 1998. U/Pb zircon ages constrain the architecture of the ultrahigh-pressure Qinling-Dabie Orogen, China. *Earth Planet. Sci .Lett.* **161**, 215-230.

He, Y., Li, S., Hoefs, J., Huang, F., Liu, S.-A., and Hou, Z., 2011. Post-collisional granitoids from the Dabie orogen: new evidence for partial melting of a thickened continental crust. *Geochim. Cosmochim. Acta* **75**, 3815-3838.

He Y. S., Li S. G, Hoefs J. and Kleinhanns I.C. 2013. Sr-Nd-Pb isotopic compositions of Early Cretaceous granitoids from the Dabie orogen: Constraints on the recycled lower continental crust. *Lithos* 156,204-217.

Hou, Z. H., Li, S. G., Chen, S. N., Li, Q. L., and Liu, X. M., 2005. Sm-Nd and SHRIMP U-Pb zircon geochronology and trace elements geochemistry of the Huilanshan mafic granulites in the Dabie orogen. *Science in China (Series D)* **35**, 1103-1111 (in Chinese with English abstract).

Huang, F., Li, S., Dong, F., He, Y., and Chen, F., 2008. High-Mg adakitic rocks in the Dabie orogen, central China: implications for foundering mechanism of lower continental crust. *Chem. Geol.* **255**, 1-13.

Jahn, B. M., Wu, F., Lo, C. H., and Tsai, C. H., 1999. Crust-mantle interaction induced by deep subduction of the continental crust: geochemical and Sr-Nd isotopic evidence from post-collisional mafic-ultramafic intrusions of the northern Dabie complex, central China. *Chem. Geol.* **157**, 119-146.

Jian, P., Kröner, A., and Zhou, G., 2012. SHRIMP zircon U-Pb ages and REE partition for high-grade metamorphic rocks in the North Dabie complex: insight into crustal evolution with respect to Triassic UHP metamorphism in east-central China. *Chem. Geol.* **328**, 49-69.

Li, Q., Li, S., Zheng, Y. F., Li, H., Massonne, H. J., and Wang, Q., 2003a. A high precision U-Pb age of metamorphic rutile in coesite-bearing eclogite from the Dabie Mountains in central China: a new constraint on the cooling history. *Chem. Geol.* **200**, 255-265.

Li, S., Wang, S., Chen, Y., Liu, D., Qiu, J., Zhou, H., and Zhang, Z., 1994. Excess argon in phengite from eclogite: evidence from dating of eclogite minerals by Sm-Nd, Rb-Sr and $^{40}Ar/^{39}Ar$ methods. *Chem. Geol.* **112**, 343-350.

Li, S. G., Hong, J. A., Li, H. M., and Jiang, L. L., 1999a. Zircon U/Pb dating of pyroxenites-pyroxenic plutons and its implications. *Geological Journal of China Universities* **5**, 351-355 (in Chinese with English abstract).

Li, S. G., Huang, F., Nie, Y. H., Han, W. L., Long, G., Li, H. M., Zhang, S. Q., and Zhang, Z. H., 2001. Geochemical and geochronological constraints on the suture location between the North and South China blocks in the Dabie Orogen, Central China. *Physics and Chemistry of the Earth, Part A: Solid Earth and Geodesy* **26**, 655-672.

Li, S. G., Huang, F., Zhou, H., and Li, H. M., 2003b. U-Pb isotopic compositions of the ultrahigh pressure metamorphic (UHPM) rocks from Shuanghe and gneisses from Northern Dabie zone in the Dabie Mountains, central China: constraint on the exhumation mechanism of UHPM rocks. *Sci. China Series D: Earth Sci.* **46**, 200-209.

Li, S. G., Jagoutz, E., Chen, Y., and Li, Q., 2000. Sm-Nd and Rb-Sr isotopic chronology and cooling history of ultrahigh pressure metamorphic rocks and their country rocks at Shuanghe in the Dabie Mountains, central China. *Geochim. Cosmochim. Acta* **64**, 1077−1093.

Li, S. G., Li, Q. L., Hou, Z. H., Yang, W., and Wang, Y., 2005. Cooling history and exhumation mechanism of the ultrahigh-pressure metamorphic rocks in the Dabie Mountains, central China *Acta Petrol. Sin.* **21**, 1117-1124. (in Chinese).

Li, S. G., Wang, C. X., Dong, F., Hou, Z. H., Li, Q. L., Liu, Y. C., Huang, F., and Chen, F. K., 2009. Common Pb of UHP metamorphic rocks from the CCSD project (100−5000 m) suggesting decoupling between the slices within subducting continental crust and multiple thin slab exhumation. *Tectonophysics* **475**, 308-317.

Li, S. G., Xiao, Y. L., Liou, D. L., Chen, Y. Z., Ge, N. J., Zhang, Z. Q., Sun, S. S., Cong, B. L., Zhang, R. Y., Hart, S. R., and Wang, S. S.,

1993. Collision of the North China and Yangtse Blocks and formation of coesite-bearing eclogites: Timing and processes. *Chem. Geol.* **109**, 89-111.

Li, X. P., Zheng, Y. F., Wu, Y. B., Chen, F. K., Gong, B., and Li, Y. L., 2004. Low-*T* eclogite in the Dabie terrane of China: petrological and isotopic constraints on fluid activity and radiometric dating. *Contrib Mineral Petrol* **148**, 443-470.

Li, Z. X., Li, X. H., Kinny, P. D., and Wang, J., 1999b. The breakup of Rodinia: did it start with a mantle plume beneath South China? *Earth Planet. Sci. Lett.* **173**, 171-181.

Liu, F. L., Xu, Z. Q., Liou, J. G., and Song, B., 2004a. SHRIMP U-Pb ages of ultrahigh-pressure and retrograde metamorphism of gneisses, south-western Sulu terrane, eastern China. *J. Metamorphic. Geol.* **22**, 315-326.

Liu, F. L., Xu, Z. Q., and Xue, H. M., 2004b. Tracing the protolith, UHP metamorphism, and exhumation ages of orthogneiss from the SW Sulu terrane (eastern China): SHRIMP U-Pb dating of mineral inclusion-bearing zircons. *Lithos* **78**, 411-429.

Liu, Y. C., Gu, X. F., Li, S. G., Hou, Z. H., and Song, B., 2011. Multistage metamorphic events in granulitized eclogites from the North Dabie complex zone, central China: evidence from zircon U-Pb age, trace element and mineral inclusion. *Lithos* **122**, 107-121.

Liu, Y. C. and Li, S. G., 2005. Lower crustal rocks from the Dabie Mountains and their deep subduction. *Acta Petrol. Sin.* **21**, 1059-1066 (in Chinese with English abstract).

Liu, Y. C. and Li, S. G., 2008. Detachment within subducted continental crust and multi-slice successive exhumation of ultrahigh-pressure metamorphic rocks: evidence from the Dabie-Sulu orogenic belt. *Chin. Sci. Bull.* **53**, 3105-3119.

Liu, Y. C., Li, S. G., Gu, X. F., Xu, S. T., and Chen, G. B., 2007a. Ultrahigh-pressure eclogite transformed from mafic granulite in the Dabie orogen, east-central China. *J. Metamorphic. Geol* .**25**, 975-989.

Liu, Y. C., Li, S. G., and Xu, S. T., 2007b. Zircon SHRIMP U-Pb dating for gneisses in northern Dabie high T/P metamorphic zone, central China: Implications for decoupling within subducted continental crust. *Lithos* **96**, 170-185.

Liu, Y. C., Li, S. G., Xu, S. T., Jahn, B., Zheng, Y. F., Zhang, Z., Jiang, L., Chen, G., and Wu, W., 2005. Geochemistry and geochronology of eclogites from the northern Dabie Mountains, central China. *J. Asian.Earth .Sci.* **25**, 431-443.

Malaspina, N., Hermann, J., Scambelluri, M., and Compagnoni, R., 2006. Polyphase inclusions in garnet-orthopyroxenite (Dabie Shan, China) as monitors for metasomatism and fluid-related trace element transfer in subduction zone peridotite. *Earth Planet. Sci. Lett.* **249**, 173-187.

Okay, A. I., 1993. Petrology of a diamond and coesite-bearing metamorphic terrain; Dabie Shan, China. *Eur. J. Mineral.* **5**, 659-675.

Okay, A. I., Xu, S., and Sengor, A. M. C., 1989. Coesite from the Dabie Shan eclogites, central China. *Eur. J. Mineral.* **1**, 595-598.

Rowley, D. B., Xue, F., Tucker, R. D., Peng, Z. X., Baker, J., and Davis, A., 1997. Ages of ultrahigh pressure metamorphism and protolith orthogneisses from the eastern Dabie Shan: U/Pb zircon geochronology. *Earth Planet. Sci. Lett.* **151**, 191-203.

Rudnick, R. and Goldstein, S., 1990. The Pb isotopic compositions of lower crustal xenoliths and the evolution of lower crustal Pb. *Earth Planet. Sci .Lett* .**98**, 192 - 207.

Rudnick, R., McLennan, S. M., and Taylor, S. R., 1985. Large ion lithophile elements in rocks from high-pressure granulite facies terrains. *Geochim. Cosmochim. Acta* **49**, 1645-1655.

Stacey, J. S. and Kramers, J., 1975. Approximation of terrestrial lead isotope evolution by a two-stage model. *Earth Planet. Sci. Lett.* **26**, 207-221.

Sultan, M., Bickford, M., El Kaliouby, B., and Arvidson, R., 1992. Common Pb systematics of Precambrian granitic rocks of the Nubian Shield (Egypt) and tectonic implications. *Geol Soc Am. Bull.* **104**, 456-470.

Taylor, P. N., Moorbath, S., Goodwin, R., and Petrykowski, A. C., 1980. Crustal contamination as an indicator of the extent of early Archaean continental crust: Pb isotopic evidence from the late Archaean gneisses of West Greenland. *Geochim. Cosmochim. Acta* **44**, 1437-1453.

Taylor, S. R. and McLennan, S. M., 1985. The continental crust:Its composition and evolution. *Blackwell Scientific Publications*, Oxford.

Wang, S. J., Li, S. G., An, S. C., and Hou, Z. H., 2012. A granulite record of multistage metamorphism and REE behavior in the Dabie orogen: Constraints from zircon and rock-forming minerals. *Lithos* **136**, 109-125.

Wang, X. M., Liou, J. G., and Mao, H. K., 1989. Coesite-bearing eclogite from the Dabie Mountains in Central China. *Geology* **17**, 1085-1088.

Wawrzenitz, N., Romer, R. L., Oberhansli, R., and Dong, S., 2006. Dating of subduction and differential exhumation of UHP rocks from the Central Dabie Complex (E-China): Constraints from microfabrics, Rb-Sr and U-Pb isotope systems. *Lithos* **89**, 174-201.

Wu, Y. B., Zheng, Y. F., Gao, S., Jiao, W. F., and Liu, Y. S., 2008. Zircon U-Pb age and trace element evidence for Paleoproterozoic granulite-facies metamorphism and Archean crustal rocks in the Dabie Orogen. *Lithos* **101**, 308-322.

Xia, Q. X., Zheng, Y. F., Yuan, H., and Wu, F. Y., 2009. Contrasting Lu-Hf and U-Th-Pb isotope systematics between metamorphic growth and recrystallization of zircon from eclogite-facies metagranites in the Dabie orogen, China. *Lithos* **112**, 477-496.

Xu, Q., Ouyang, J., Zhang, B., and Kuang, S., 2002. Tectonic affinity of Paleozoic stratigraphic slices along the northern margin in

Dabie Orogen, central China: evidence from Pb isotope of rocks. *Prog. Nat. Sci.* **12**, 438-444.

Xu, S. T., Liu, Y. C., Chen, G. B., Compagnoni, R., Rolfo, F., He, M. C., and Liu, H. F., 2003. New finding of micro-diamonds in eclogites from Dabie-Sulu region in central-eastern China. *Chin. Sci. Bull.* **48**, 988-994.

Xu, S. T., Liu, Y. C., Chen, G. B., Ji, S. Y., Ni, P., and Xiao, W. S., 2005. Microdiamonds, their classification and tectonic implications for the host eclogites from the Dabie and Su-Lu regions in central eastern China. *Mineral. Mag.* **69**, 509-520.

Xu, S. T., Okay, A. I., Ji, S. Y., Sengör, A. M. C., Wen, S., Liu, Y. C., and Jiang, L. L., 1992a. Diamond from the Dabie-Shan Metamorphic Rocks and Its Implication for Tectonic Setting. *Science* **256**, 80-82.

Xu, S. T., Su, W., Liu, Y. C., Jiang, L. L., Ji, S. Y., Okay, A. I., and Sengör, A. M. C., 1992b. Diamonds from High-Pressure Metamorphic Rocks in Eastern Dabie Mountain. *Chin. Sci. Bull.* **37**, 140-145.

Xu, Z. Q., Zeng, L. S., Liu, F. L., Yang, J. S., Zhang, Z. M., McWilliams, M., and Liou, J. G., 2006. Polyphase subduction and exhumation of the Sulu high-pressure-ultrahigh-pressure metamorphic terrane. *Special Papers-Geological Society of Amerca* **403**, 93.

Zartman, R. and Doe, B., 1981. Plumbotectonics—the model. *Tectonophysics* **75**, 135-162.

Zartman R. E. and Haines S. M. 1988. The plumbotectonic model for Pb isotopic systematics among major terrestrial reservoirs—a case for bi-directional transport. *Geochim. Cosmochim. Acta* 52,1327-1339.

Zhang, H. F., Gao, S., Zhang, B. R., Zhong, Z. Q., Jia, W. L., and Wang, L. S., 2001. Pb isotopic study on crustal structure of Dabie mountains, central China. *Geochimica* **30**, 395-401 (in Chinese with English abstract).

Zhang, H., Gao, S., Zhong, Z., Zhang, B., Zhang, L., and Hu, S., 2002. Geochemical and Sr-Nd-Pb isotopic compositions of Cretaceous granitoids: constraints on tectonic framework and crustal structure of the Dabieshan ultrahigh-pressure metamorphic belt, China. *Chem. Geol.* **186**, 281-299.

Zhang, L. G., Liu, J. X., and Wang, K. F., 1995. Block geology of eastern Asia Lithosphere-isotope Geochemistry and Dynamics of Upper Mantle, Basement and Granites. Science Press, Beijing. (in Chinese with English abstract).

Zhang, L., Zhong, Z. Q., Wei, L. S., and Zhang, B. R., 2008. Pb Isotope Mapping in the Tongbai-Dabie Orogenic Belt, central China. *Acta Geologica Sinica-English Edition* **82**, 126-133.

Zhang, S. B., Zheng, Y. F., Wu, Y. B., Zhao, Z. F., Gao, S., and Wu, F. Y., 2006a. Zircon U-Pb age and Hf isotope evidence for 3.8 Ga crustal remnant and episodic reworking of Archean crust in South China. *Earth Planet. Sci. Lett.* **252**, 56-71.

Zhang, S. B., Zheng, Y. F., Wu, Y. B., Zhao, Z. F., Gao, S., and Wu, F. Y., 2006b. Zircon isotope evidence for ≥3.5 Ga continental crust in the Yangtze craton of China. *Precambrian Res.* **146**, 16-34.

Zhao, Z. F., Zheng, Y. F., Wei, C. S., Chen, F. K., Liu, X., and Wu, F. Y., 2008. Zircon U-Pb ages, Hf and O isotopes constrain the crustal architecture of the ultrahigh-pressure Dabie orogen in China. *Chem. Geol.* **253**, 222-242.

Zhao, Z. F., Zheng, Y. F., Wei, C. S., and Wu, F. Y., 2011. Origin of postcollisional magmatic rocks in the Dabie orogen: implications for crust-mantle interaction and crustal architecture. *Lithos*, 99-114.

Zhao, Z. F., Zheng, Y. F., Wei, C. S., and Wu, Y. B., 2007. Post-collisional granitoids from the Dabie orogen in China: zircon U-Pb age, element and O isotope evidence for recycling of subducted continental crust. *Lithos* **93**, 248-272.

Zheng, Y. F., Fu, B., Gong, B., and Li, L., 2003. Stable isotope geochemistry of ultrahigh pressure metamorphic rocks from the Dabie-Sulu orogen in China: implications for geodynamics and fluid regime. *Earth-Sci. Rev.* **62**, 105-161.

Zheng, Y. F., Wu, R. X., Wu, Y. B., Zhang, S. B., Yuan, H., and Wu, F. Y., 2008. Rift melting of juvenile arc-derived crust: Geochemical evidence from Neoproterozoic volcanic and granitic rocks in the Jiangnan Orogen, South China. *Precambrian Res* **163**, 351-383.

Zheng, Y. F., Wu, Y. B., Chen, F. K., Gong, B., Li, L., and Zhao, Z. F., 2004. Zircon U-Pb and oxygen isotope evidence for a large-scale ^{18}O depletion event in igneous rocks during the Neoproterozoic. *Geochim. Cosmochim. Acta* **68**, 4145-4165.

Zheng, Y. F., Wu, Y. B., Zhao, Z. F., Zhang, S. B., Xu, P., and Wu, F. Y., 2005. Metamorphic effect on zircon Lu-Hf and U-Pb isotope systems in ultrahigh-pressure eclogite-facies metagranite and metabasite. *Earth Planet. Sci. Lett.* **240**, 378-400.

Zheng, Y. F., Zhao, Z. F., Wu, Y. B., Zhang, S. B., Liu, X., and Wu, F. Y., 2006. Zircon U-Pb age, Hf and O isotope constraints on protolith origin of ultrahigh-pressure eclogite and gneiss in the Dabie orogen. *Chem. Geol.* **231**, 135-158.

Zhu, G., Liu, G. S., Niu, M. L., Xie, C. L., Wang, Y. S., and Xiang, B., 2009. Syn-collisional transform faulting of the Tan-Lu fault zone, East China. *Int J. Earth. Sci.* **98**, 135-155.

Supplemental Table 1 Sampling information of 57 orthogneisses and 2 paragneisses from the Dabie orogen.

Sample	Location	Lithology	Major mineral assembles
ENDZ			
99MZT-3	Mozitan	Granitic gneiss	Quartz+K-fledspar+plagioclase+amphibole+biotite
99YZH-1	Yanzihe	Granitic gneiss	K-fledspar+Plagioclase+amphibole+quartz+biotite
03JZ-4	Jinzhai	Granitic gneiss	Plagioclase+K-feldspar+amphibole+quartz+biotite
03JZ-5	Jinzhai	Granitic gneiss	Plagioclase+K-feldspar+amphibole+quartz+biotite
03JZ-6	Jinzhai	Granitic gneiss	Plagioclase+K-feldspar+quartz+amphibole+biotite
03JZ-7	Jinzhai	Granitic gneiss	K-feldspar+Plagioclase+amphibole+quartz+biotite
99MSH-1	Manshuihe	Amphibolitic gneiss	Plagioclase+amphibole+quartz+garnet+biotite
03MSH-1	Manshuihe	Amphibolitic gneiss	Plagioclase+amphibole+quartz+garnet+biotite
03MSH-2	Manshuihe	Amphibolitic gneiss	Plagioclase+amphibole+K-feldspar+quartz+biotite
99LN-2	Lanni'ao	Granitic gneiss	Quartz+plagioclase+K-feldspar+amphibole+biotite
03HS-1	Huoshan	Granitic gneiss	Quartz+plagioclase+K-feldspar+amphibole+biotite
03HS-2	Huoshan	Granitic gneiss	Plagioclase+K-feldspar+amphibole+quartz+biotite
03MZT-2	Mozitan	Granitic gneiss	Quartz+K-fledspar+plagioclase+amphibole+biotite
03SG-1	Shiguan	Granitic gneiss	Plagioclase+K-feldspar+amphibole+quartz+biotite
08MZT-4	Mozitan	Granitic gneiss	Quartz+K-fledspar+plagioclase+amphibole+biotite
Luotian Dome			
97M3-1	Macheng	Granitic gneiss	Quartz+K-feldspar+plagioclase+biotite
10LT-1	30°46'04"N, 115°20'38"E	Granitic gneiss	Quartz+plagioclase+K-feldspar+biotite
10LT-2	30°46'04"N, 115°20'38"E	Granitic gneiss	Quartz+plagioclase+K-feldspar+biotite
10LT-3	31°07'31"N, 115°29'11"E	Granitic gneiss	Quartz+plagioclase+biotite+K-feldspar
10LT-4	31°07'31"N, 115°29'11"E	Granitic gneiss	Quartz+biotite+plagioclase+K-feldspar
10LT-5	31°06'19"N, 115°33'11"E	Granitic gneiss	Quartz+K-feldspar+plagioclase+biotite
10LT-6	31°06'19"N, 115°33'11"E	Granitic gneiss	Quartz+plagioclase+K-feldspar+biotite
10LT-7	30°47'40"N, 115°33'01"E	Granitic gneiss	Quartz+plagioclase+biotite+K-feldspar
10LT-8	30°47'40"N, 115°33'01"E	Granitic gneiss	Quartz+biotite+plagioclase+K-feldspar
10XS-1	30°30'33"N, 115°18'06"E	Granitic gneiss	Quartz+plagioclase+biotite+K-feldspar
10XS-2	30°30'33"N, 115°18'06"E	Granitic gneiss	Quartz+plagioclase+K-feldspar+biotite
10XS-4	30°37'10"N, 115°22'06"E	Granitic gneiss	Quartz+plagioclase+K-feldspar+biotite
10XS-5	30°33'09"N, 115°33'39"E	Granitic gneiss	Quartz+plagioclase+K-feldspar+biotite
10XS-6	30°39'06"N, 115°26'31"E	Granitic gneiss	Quartz+biotite+plagioclase+K-feldspar
00XS-1	Xishui	Granitic gneiss	Quartz+biotite+plagioclase+K-feldspar
CDZ			
03CP-2	Changpu	Granitic gneiss	Plagioclase+K-feldspar+quartz+amphibole+garnet+ phengite
03SH-3	Shuanghe	Granitic gneiss	Plagioclase+K-feldspar+garnet+phengite+epidote+biotite
99XF-1	Xifo	Granitic gneiss	Plagioclase+K-feldspar+amphibole+quartz+phengite
02HN-1	Qianhe	Granitic gneiss	Quartz+K-fledspar+plagioclase+amphibole+ phengite+biotite
02HN-2	Tang'an	Granitic gneiss	K-fledspar+Plagioclase+amphibole+quartz+phengite+biotite
02HN-3	Tang'an	Granitic gneiss	Plagioclase+K-feldspar+amphibole+quartz+phengite +biotite
02HN-4	Tiantou	Granitic gneiss	Plagioclase+K-feldspar+amphibole+quartz+garnet +biotite
02HN-5	Tiantou	Granitic gneiss	Plagioclase+K-feldspar+quartz+amphibole+garnet+ phengite
02HN-7	Fangbian	Granitic gneiss	K-feldspar+Plagioclase+amphibole+quartz+phengite

Continued

Sample	Location	Lithology	Major mineral assembles
CDZ			
02HN-9-1	Changputan	Granitic gneiss	Plagioclase+K-feldspar+amphibole+quartz+biotite+epidote
02HN-10	Changputan	Granitic gneiss	Plagioclase+K-feldspar+amphibole+quartz+ muscovite
02HN-11	Wumiao	Granitic gneiss	Plagioclase+K-feldspar+amphibole+quartz+ muscovite+biotite
02HN-15	Siqian	Granitic gneiss	Quartz+plagioclase+K-feldspar+amphibole+ phengite
08SH-10	Shuihou	Granitic gneiss	Quartz+plagioclase+K-feldspar+biotite+epidote
08SH-9	Shuihou	Granitic gneiss	Amphibole+quartz+biotite+K-feldspar+epidote
08WH-1	Wuhe	Granitic gneiss	Quartz+K-fledspar+plagioclase+amphibole+epidote+ biotite
BMJ-1	Baimajian	Granitic gneiss	Plagioclase+K-feldspar+amphibole+quartz+epidote+biotite
02SH-5	Shuanghe	Paragneiss	Garnet+phengite+plagioclase+epidote+biotite
02SH-9	Shuanghe	Paragneiss	Garnet+phengite+plagioclase+epidote+biotite
SDZ			
02HS-2	Tanshu'ao	Granitic gneiss	Quartz+plagioclase+K-feldspar+muscovite
02HS-6	Liufan	Granitic gneiss	Quartz+plagioclase+K-feldspar+muscovite
08TH-LD-1	97 km board along S211 road	Granitic gneiss	Quartz+plagioclase+K-feldspar+amphibole+muscovite+chlorite
08TH-LD-2	96 km board along S211 road	Granitic gneiss	Quartz+plagioclase+K-feldspar+muscovite+epidote
08TH-ZJC-1	108 km board along S211 road	Granitic gneiss	Quartz+muscovite+amphibole+K-feldspar+plagioclase +epidote
08TH-HZ-2	88 km board along S211 road	Granitic gneiss	Quartz+K-feldspar+plagioclase+muscovite+amphibole
08TH-FZL-1	109 km board along S211 road	Granitic gneiss	Quartz+amphibole+K-feldspar+plagioclase+biotite
08TH-LD-4	96 km board along S211 road	Granitic gneiss	Quartz+amphibole+biotite+K-feldspar+plagioclase
08TH-HZ-1	93 km board along S211 road	Granitic gneiss	Quartz+amphibole+biotite+K-feldspar+plagioclase
08TH-ZQ-1	108 km board along S211 road	Granitic gneiss	Quartz+muscovite+K-feldspar+plagioclase

The Dabie UHPM orthogneisses mainly occur as banded or massive blocks. The paragneisses from Shuanghe are interbeded with marble, jadeite-quartzite and banded eclogite.

第四部分 大陆俯冲过程中的流体及Nb-Ta分异

大陆俯冲过程中的流体*

李曙光，侯振辉

中国科学技术大学，合肥 230026

> **亮点介绍**：大陆地壳在俯冲过程中不同阶段，随着变质程度的升高和部分含水矿物的相继分解，流体释放的规模和活动范围均有明显的差异。角闪石中 Ar 封闭温度较高，在榴辉岩相变质过程中释放流体较少，这是造成超高压变质榴辉岩中多硅白云母含有过剩 Ar 的主要原因。

摘要 含水矿物矿物稳定性的实验研究和超高压岩石的同位素地球化学研究表明，大陆地壳在俯冲过程中，随着变质程度的升高和部分含水矿物的相继分解，会有流体释放出来。当俯冲深度接近 50 km，俯冲陆壳岩石中大量低级变质含水矿物（如绿泥石、绿帘石、阳起石）会脱水并从俯冲陆壳逸出形成流体流。这一流体流可溶解带走俯冲陆壳内已从云母类矿物逸出的放射成因 Ar 及部分 U、Pb，并导致 $w(U)/w(Pb)$ 升高。这一阶段逸出的流体有可能交代、水化仰冲壳楔，为其发生部分熔融形成同碰撞花岗岩或加速山根下地壳的榴辉岩化创造条件。在俯冲深度为 50~100 km，变镁铁质岩石中的角闪石相继分解并释放出 H_2O。由于变镁铁质岩石在陆壳中所占比例较少，这一阶段释放的水不能形成大规模的流体流，因而不能使体系内的过剩 Ar 大量散失，但足以形成局部循环，加速变镁铁质岩石及其互层或邻近围岩的榴辉岩化变质反应。在俯冲深度 >100 km 的超高压变质阶段，仅有少量的含水矿物分解，而多硅白云母仍保持稳定。这时俯冲陆壳内只可能有少量粒间水存在，从而导致俯冲陆壳与周围软流圈地幔不能发生充分的相互作用。

关键词 大陆俯冲；流体；超高压变质作用

碰撞造山带陆壳岩石中柯石英及金刚石的发现证明，在碰撞造山过程中一侧陆壳可以俯冲到 100 km 以上的深度[1~5]。俯冲陆壳随着温度、压力及变质温度的升高能否释放大量流体以及所产生的地质、地球化学效应是一个非常重要和有意思的科学问题，并引起广泛关注[6~8]。对俯冲洋壳的深入研究，人们已了解海水蚀变的俯冲洋壳及其所携带的沉积物在俯冲过程中，随着变质程度的加深，可释放大量的水，并形成流体流（fluid flux）。该流体流进入并交代上覆楔形地幔，从而引发地幔部分熔融，形成岛弧型玄武 - 安山质岩浆作用[9]。然而俯冲陆壳因其组成及俯冲环境与俯冲洋壳不同，在俯冲过程中流体析出的深度、析出量及地质地球化学效应一定会有其特殊性。人们主要关心以下几个问题：①陆壳在俯冲过程中能否脱水产生流体流？它产生的深度及影响的范围如何？尤其是它能否像俯冲洋壳那样在地幔深度产生流体流从而交代影响仰冲盘的楔形地幔？②在深俯冲的超高压变质条件下，俯冲陆壳中有无游离水流体存在？如果有，它的地球化学效应是什么，以及有多大影响范围？针对这些问题，人们主要从以下 3 个途径进行研究：①实验岩石学及变质含水矿物的稳定性研究；②超高压矿物流体包裹体研究；③超高压变质岩的稳定同位素及放射性同位素地球化学研究。由于流体包裹体及稳定同位素研究已有郑永飞专文论述[10]，本文主要对实验岩石学及放射性同位素示踪领域的研究进展进行讨论和评述。

1 变质含水矿物稳定性的实验研究及俯冲陆壳析出流体的可能性

俯冲陆壳除了在较浅部（<(25±15) km）可通过压实及去挥发分作用产生流体外，在较深部（>25 km），

---

* 本文发表在：地学前缘，2001，8(3)：123-129

只有通过含水矿物的脱水变质反应才能产生游离态水[11]。因此，这方面工作试图通过高温、高压实验方法，研究各种含水变质矿物相（如云母类、闪石类、帘石类、绿泥石类）的 T (温)-p (压)稳定区间，建立它们的 p-T 相图，从而帮助我们推测俯冲陆壳在到达何等深度或 T-p 条件下，哪些含水矿物变得不稳定而脱水，从而使俯冲陆壳有析出流体的可能性。Ernst（2000）对这一领域工作有很好的评述[11]。

除了碳酸盐岩以外，俯冲陆壳岩石主要有 3 种岩石组合类型：①变玄武岩＋变辉长岩；②蛇纹岩化超镁铁岩；③长英质岩石。它包括变质硬砂岩＋变泥质岩＋片麻岩及变质花岗岩。它们在俯冲过程中脱水情况是不一样的。

(1) 对于变玄武岩及变辉长岩来说，它的含水矿物主要是绿泥石、绿帘石及角闪石。在温度低于 600~650 ℃、压力<1.2~1.5 GPa 时，绿泥石和绿帘石与阳起石可共生存在（图1）。当温度>650 ℃、压力升高时，绿泥石、绿帘石及阳起石类闪石就会消失，角闪石主要以蓝闪石和非闪石形式存在，而且随着压力升高，其数量急剧减少。在到达超高压条件下（T=750~900 ℃，p>27 GPa），所有的 Ca-Na 闪石都要分解。因此，变玄武岩及变辉长岩在俯冲过程中随着温度、压力的升高伴随有显著的脱水过程并释放出相当数量的水。然而应当指出的是，如果变玄武岩含有较多的 K，则在压力升高以后会出现多硅白云母这一含水矿物，而且它的温度、压力稳定区高达 900~1000 ℃ 和 9 GPa。因此在高压及超高压条件下，多硅白云母的出现可以吸收一部分水。图 1 又显示，角闪石大约在 2.2 GPa 时消失，这相当于角闪榴辉岩向绿帘石榴辉岩转变阶段，因此，变玄武岩和变辉长岩主要在此以前有较多的水析出。此后只有绿帘石、黝帘石及蓝闪石等少量含水矿物在超高压条件下分解，因而只能有很少量游离态水产生。超高压矿物中流体包裹体的研究已证明超高压阶段游离态水的存在[14,15]。由于俯冲洋壳基本上是由变玄武岩和变辉长岩组成，因此，当它俯冲到 70~80 km 时（p=2.2~2.4 GPa），将会有大量的闪石矿物脱水并产生流体流。如果脱水反应速度较慢，这一过程还可持续到更深的深度。

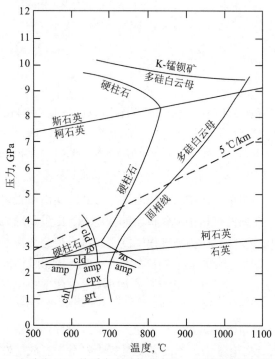 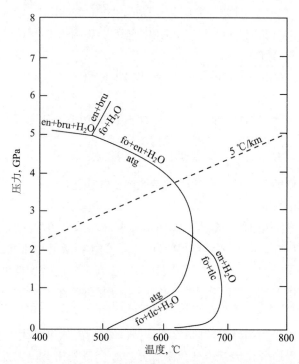

图 1 玄武岩-H_2O 体系含水矿物稳定区 p-T 相图（据文献[11]，[12]）俯冲带的地热梯度为 5 ℃/km（据文献[12]）

图 2 MgO-SiO_2-H_2O 体系叶蛇纹石稳定区相图（据文献[11]，[13]）俯冲带的地热梯度为 5 ℃/km（据文献[12]）

Amp-角闪石；Chl-绿泥石；Cld-硬绿泥石；Cpx-单斜辉石；Grt-石榴石；Zo-黝帘石

Atg-蛇纹石；Bru-水镁石；En-顽火辉石；Fo-镁橄榄石；Tlc-滑石

(2) 对蛇绿纹石化的超镁铁岩来说，图 2 显示，在角闪岩相变质条件下（T=550~620 ℃，p=0.5~1.0 GPa），叶蛇纹石会分解成橄榄石＋滑石，并释放出大量的水，这类岩石的脱水过程对俯冲大洋板块有较重要的意义。

因为大洋岩石圈顶部可能有大量水化（蛇纹石化）的超镁铁岩。但对俯冲陆壳来说，蛇纹石化橄榄岩含量很少，它们的脱水量仅对自身及其围岩的榴辉岩化过程产生影响。

(3) 在长英质岩石中，含水矿物主要是绿泥石及黑云母和白云母，实验工作表明，除绿泥石在550~600 ℃以上要脱水并消失外（图1），白云母+黑云母组合可以在很高的温度下保持稳定[16~21]。例如，在俯冲带地热梯度条件下，多硅白云母至少可在3.5~4.0 GPa时仍保持稳定[19]。由于黑云母和白云母是陆壳长英质岩石中最重要的含水矿物，而长英质岩石在俯冲陆壳中又占有最大体积，因此，Ernst(2000)认为陆壳岩石在俯冲到100 km以上的超高压条件下，将不可能有显著的水卷入超高压变质作用。由于缺少游离态水，长英质岩石仍可在经历超高压变质时保持长石+石英+云母组合的亚稳状态而不能形成超高压矿物组合。然而应当指出的是，由于绿泥石的脱水，长英质陆壳岩石仍可能在俯冲到50 km（p=1.5 GPa）以前释放出较多水。此外，在俯冲陆壳中仍可能有相当数量的各种变质的古老基性、中基性火山岩。它们的脱水过程可否在俯冲陆壳内形成流体流及其影响范围尚有待通过其他途径进行研究。

2 陆壳深俯冲过程中流体活动的放射性同位素示踪

流体的存在可提高元素的活动性并加快同位素的平衡，流体流的活动则可使被溶解元素快速带出，因而可导致放射性同位素母体/子体的分异。因此，超高压岩石的放射性同位素及母体/子体值可以用来示踪流体活动。目前已对过剩Ar及U-Pb同位素进行了研究并对陆壳深俯冲过程中的流体活动给出了重要的制约。

(1) 自从Li 等(1994)发现超高压榴辉岩中的多硅白云母含有大量过剩Ar以来[22]，人们对其过剩Ar的来源及成因进行了大量的研究[23~29]。最新的Ar-Ar激光分析研究成果表明，多硅白云母的Ar是从原岩自身继承来的，而非外来流体带入的[26,27]。Li 等(1999)对大别－苏鲁超高压变质带各种岩石中云母的表面年龄的统计表明，榴辉岩中的多硅白云母含过剩Ar最多，同样经历过高压变质的花岗片麻岩中的云母基本不含过剩Ar，而与榴辉岩互层的片麻岩中的云母含少量过剩Ar，介于二者之间（图3）[30]。对于这种现象的一个可能的解释就是，长英质花岗片麻岩原岩（花岗岩）中的主要含K矿物为钾长石、黑云母和白云母，它们的Ar封闭温度很低（<350 ℃），因此，在陆壳俯冲初期，这些含K矿物中积累的放射成因Ar就释放出来了。花岗片麻岩中的云母

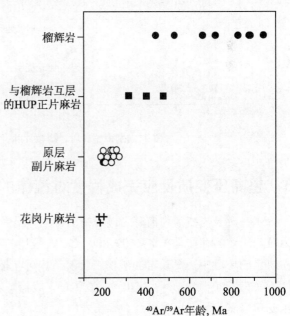

图3 大别－苏鲁造山带超高压变质岩中云母的 $^{40}Ar/^{39}Ar$ 表面年龄（横坐标）与其岩性和产状（纵坐标）的关系（Li et al, 1999）（据文献[30]）

不含过剩Ar的事实说明，这些释放出来的放射成因Ar并没有保存在体系内，而是伴随俯冲陆壳逸出的流体而散失了。因此在高压及超高压条件下，花岗片麻岩中^{40}Ar分压很低，从而使此时形成的多硅白云母不含过剩Ar。它说明长英质陆壳岩石在俯冲初期（榴辉岩相变质以前）也有大量流体逸出。然而榴辉岩的原岩是基性斜长角闪岩类，其主要含K矿物为角闪石，Ar封闭温度接近500 ℃，因此它积累的放射成因Ar要到变质温度高于500~600 ℃时才会完全释放出来，这时俯冲陆壳已经历榴辉岩相变质。多硅白云母过剩Ar的继承成因说明，此时角闪石释放出的放射成因Ar并没有大量散失到岩石体系以外，因而榴辉岩体系内仍保持较高的^{40}Ar分压，从而使这时形成的多硅白云母含较多过剩Ar。在榴辉岩相变质时，体系内放射成因Ar没有大量丢失，说明这一阶段尽管角闪石会释放出水，但并未形成大规模的流体流，这时游离态水的活动范围还是有限的。

(2) 前人研究已经发现俯冲洋壳析出流体可使俯冲洋壳的U-Pb体系发生分异。实验研究表明，俯冲板

块在脱水过程中，Pb 比 U 有更大的活动性。从而使脱水后的俯冲洋壳有较低的 Pb 含量和较高的 $w(U)/w(Pb)$ 值[31]。因此，观测超高压岩石的 $w(U)/w(Pb)$ 值变化可以帮助我们判断俯冲陆壳是否发生过大量的脱水及流体逸出过程。我们对大别山双河超高压岩石的 U、Pb 含量及 $w(U)/w(Pb)$ 值用同位素稀释法进行了精确的分析[32]。双河超高压变质岩为榴辉岩与大理岩、变泥质岩互层，它们保持了良好的火山沉积岩系的层状结构及地球化学特征[33,34]。氧同位素研究表明，它们在俯冲前也曾与天水进行过同位素交换[35]。因此双河超高压变质岩的原岩应是接近地表的上地壳岩石。测定结果表明，大别山超高压变质岩的 Pb 含量普遍低于上、下地壳的平均组成；绝大多数样品的 U 含量低于上地壳平均组成，但高于下地壳平均组成，因此样品的 $w(U)/w(Pb)$ 值均高于上地壳的平均值（图 4）[32]。这表明双河超高压变质岩遭受了 U、Pb 的丢失，其中 Pb 的丢失量远高于 U 的丢失量。这完全符合俯冲板块析出流体所形成的 U、Pb 分异特征。这一事实说明，双河超高压变质岩在俯冲过程中经历了一次脱水和流体逸出过程。结合 Ar 同位素制约，这一流体逸出过程可能发生在榴辉岩相变质以前。

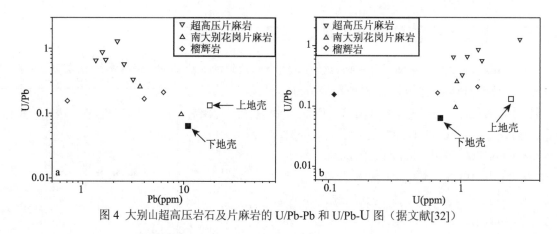

图 4 大别山超高压岩石及片麻岩的 U/Pb-Pb 和 U/Pb-U 图（据文献[32]）

3 超高压变质反应完成程度对流体的制约

水在辉长岩变质为榴辉岩过程中的作用早已受到许多研究者的注意[36,37]。Ahrens 和 Shubert(1975)[36] 认为粒间流体的存在是在 $T<600\sim800\ ℃$、在板块俯冲的时间范围内完成辉长岩向榴辉岩转变的必要条件，而在缺水的环境中，辉长岩的矿物组合及结构可在榴辉岩变质条件下残留下来[35]。事实上，在超高压变质条件下（$T=700\ ℃$, $p>2.7\ GPa$），流体对完成榴辉岩化变质反应同样是重要的。例如，在我国青岛仰口超高压变质岩露头区就可观察到在辉长岩中心部位因缺少流体而保留下斜长石、角闪石、辉石等矿物组合及火成岩结构，而仅在岩体边部的强烈变形地带（因而流体易于活动）辉长岩才变为榴辉岩，而且有粒间柯石英形成[38,39]。在西阿尔卑斯的 Dora Maira 超高压变质带甚至可见到经历过超高压变质的花岗岩中有残留的黑云母存在[40]。这些事实说明，即使在超高压变质条件下，如果缺少流体，其变质反应速率仍然是很慢的，以致岩石在陆壳俯冲所允许的时间范围内仍未能完成其变质反应。通常辉长岩和花岗岩侵入体因缺乏角闪石类含水矿物，在岩体内部可形成缺乏流体的环境，但在岩体边部，因构造变形强，可使来自围岩的流体渗入，从而加快榴辉岩化反应。但是对于变玄武岩来说，它们因含有大量角闪石，在高压及超高压变质时不缺乏流体，因此很少出现高压变质反应不完全的情况。双河超高压变质岩的岩石学研究表明，那些与榴辉岩互层的长英质变泥质岩均可形成高压矿物组合，而不含榴辉岩的花岗片麻岩仍基本保持角闪岩相[39]。这说明变玄武岩释放出来的流体不仅以粒间流体形式存在，还可在与它互层的岩石中形成局部流体循环，但并未影响到花岗片麻岩的内部。

4 结论

上述研究工作表明，俯冲陆壳尽管主要由长英质岩石组成，它在俯冲深度到达 50 km 以前，由于绿泥石、绿帘石、阳起石等矿物的脱水，可以有大量的流体析出。这些析出的流体可以带走陆壳中云母释放出来的放射成因 Ar，并导致俯冲陆壳岩石的 U、Pb 丢失和 $w(U)/w(Pb)$ 升高。这一阶段析出的流体由于深度较浅，不能对仰冲盘的岩石圈地幔楔产生显著影响，但可以水化、交代仰冲壳楔，为它们以后发生部分熔融形成同碰撞花岗岩，或加速山根下地壳的榴辉岩化创造条件。在俯冲陆壳俯冲到 50~100 km 之间时，俯冲陆壳中的变玄武岩及变辉长岩的含 Na 角闪石会逐步分解，释放出游离态水。然而，这一阶段由于变玄武岩和变辉长岩在陆壳中所占比例较低，因而释放的水量尚不足以形成大规模流体流。它可以在变玄武岩和变辉长岩内部，以及与它们互层或近距离围岩中形成局部的流体循环，以加速促进这些岩石的榴辉岩化过程，但不能使该体系内角闪石释放的放射成因 Ar 大量逃逸。在陆壳俯冲到 100 km 以上深度并发生超高压变质时，仅有少量的残留含 Na 角闪石被分解。超高压矿物中高盐度流体包裹体的存在[14,15]证明这一阶段俯冲陆壳中仍有少量粒间流体存在，但这些流体的活动范围很小。因此，它不能导致俯冲陆壳与周围的软流圈发生充分的化学相互作用，但足以保证超高压变质反应得以完成。

参考文献

[1] Smith D C. Coesite in clinopyroxene in the Caledonides and its implications for geodynamics. Nature, 1984, 310: 641-644
[2] Chopin C. Coesite and pure pyrope in high-grade blueschists of the western Alps: a first record and some consequences. Contrib Mineral Petrol, 1984, 86: 107-118
[3] Wang X, Liou J G, Mao H J. Coesite-bearing eclogites from the Dabie Mountains in central China. Geology, 1989, 17: 1085-1088
[4] Sobolev N V, Shatsky V S. Diamond inclusions in garnets from metamorphic rocks. Nature, 1990, 343: 742-746
[5] Xu S, Okay A L, Ji S, et al. Diamond from the Dabie Shan Metamorphic rocks and its implication for tectonic setting. Science, 1992, 256: 80-82
[6] 李曙光. 大陆俯冲化学地球动力学. 地学前缘，1998，5(4)：211-234
[7] 李曙光，聂永红，Hart S R，等. 俯冲陆壳与上地幔的相互作用——II. 大别山同碰撞镁铁-超镁铁岩的 Sr、Nd 同位素地球化学. 中国科学(D 辑)，1998，28(1)：18-22
[8] Liou J G, Zhang R Y, Ernst W G. Occurrences of hydrous and carbonate phases in ultrahigh-pressure rocks from east-central China: implications for the role of volatiles deep in cold subduction zones. The Island Arc, 1995, 4: 362-375
[9] Tatsumi Y, Eggins S M. Subduction zone magmatism. Cambridge Blackwall, 1995, 211
[10] 郑永飞，徐宝龙，周根陶. 矿物稳定同位素地球化学研究[J]. 地学前缘，2000，7(2)：299-320
[11] Ernst W G. H_2O and ultrahigh-pressure subsolidus phase relations for mafic and ultramafic systems. In: Ernst W G and Liou J G ed. Ultra-high pressure metamorphism and geodynamics in collision-type orogenic belts, Bellwether Publishing Ltd, 2000, 121-129
[12] Schmidt M W, Poli S. The stability of lawsonite and zoisite at high pressure: Experiments in CASH to 92 kbar and implications for the presence of hydrousphases in subducted lithosphere. Earth Plan Sci Lett, 1994, 124: 105-118
[13] Wunder B, Schreyer W. Antigorite: High-pressure stability in the system $MgO-SiO_2-H_2O$ (MSH). Lithos, 1997, 41: 213-227
[14] Xiao Y L, Hoefs J, Van den kerkhof A M, et al. Fluid history of UHP metamorphism in Dabie Shan, China: a fluid inclusion and oxygen isotope study on the coesite-bearing eclogite from Bixiling. Contrib Mineral Petrol, 2000, 139: 1-16
[15] Xiao Y L, Hoefs J, Van Den Kerkhof A M, et al. 2001. Geochemical constrains of the eclogite and granulite facies metamorphism as recognized in the Raobazhai complex from North Dabie Shan, China. J. Metamorphic Geol., 19, 3-19
[16] Gardien V, Thompson A B, Frujic D, et al. Experimental melting of biotite+plagioclase+quartz±muscovite assemblages and implications for crustal melting. Jour Geophys Res, 1995, 100: 15581-15591
[17] Le Breton N, Thompson A B. Fluid-absent(dehydration) metling of biotite in the early stages of crustal anatexis. Conrtib Mineral Petrol, 1988, 99: 226-237
[18] Luth R W. Experimental studies of the system phlogopite-diopside from 3.5 to 17 GPa. Amer Mineral, 1997, 82: 1198-1209
[19] Skjerlie K P, Johnston A D. Vapor-absent melting from 10 to 20 kbar of crustal rocks that contain multiple hydrous phases: Implications for anatexis in the deep to very deep continental crust and active continental margins. Jour Petrol, 1996, 37: 661-691
[20] Massone H J, Szpurka Z. Thermodynamic properties of white micas on the basis of high-pressure experiments in the systems $K_2O-MgO-Al_2O_3-SiO_2-H_2O$ and $K_2O-FeO-Al_2O_3-SiO_2-H_2O$. Lithos, 1997, 41: 229-250
[21] Vielzeuf D, Holloway J R. Experimental determination of the fluid-absent melting relations in the pelitic system. Consequences for crustal differentiation. Contrib Mineral Petrol, 1988, 98: 257-276
[22] Li S G, Wang S S, Chen Y Z, et al. Excess argon in phengite from eclogite: evidence from dating of eclogite minerals by Sm-Nd,

Rb-Sr and $^{40}Ar/^{39}Ar$ methods. Chemical Geology, 1994, 112: 343-350

[23] Arnaud N O, Kelley S P. Evidence for excess Ar during high pressure metamorphism in the Dora Maira (western Alps, Italy), using a ultraviolet laser ablation microprobe $^{40}Ar/^{39}Ar$ technique. Contrib Mineral Petrol, 1995, 121: 1-11.

[24] Scaillet S. Excess ^{40}Ar transport scale and mechanism in high-pressure phengites: case study from an eclogitized metabasite of the Dora Maira nappe, Western Alps. Geochim Cosmochim Acta, 60: 1075-1090

[25] Reddy S M, Kelley S P, Wheeler J, et al. A $^{40}Ar/^{39}Ar$ laser probe study of micas from the Sesia Zone, Italian Alps: implications for metamorphic and deformation histories. J Metamorph Geol, 1996, 14: 493-508.

[26] Boundy T M, Hall C M, Li G. Fine-scale isotopic heterogeneities and fluids in the deep crust: a $^{40}Ar/^{39}Ar$ laser ablation and TEM study of muscovites from a granulite-eclogite transition zone, Earth Plan Sci Lett, 1997, 148: 223-242.

[27] Ruffet G, Gruau G, G. Féraud M. Rb-Sr and ^{40}Ar-^{39}Ar laser probe dating of high-pressure phengites from the Sesia Zone (Western Alps): underscoring of excess argon and new age contraints on the high-pressure metamorphism. Chem Geol, 1997,141: 1-18.

[28] Scaillet S. K-Ar ($^{40}Ar/^{39}Ar$) geochronology of ultrahigh pressure rocks. In: Hacker B R, Liou J G (eds.). When Continents Collide: Geodynamics and Geochemistry of Ultrahigh-Pressure Rocks, Kluwer Academic Publishers, Dordrecht, 1998, 161-201.

[29] Giorgis D, Cosca M, Li, S G. Distribution and significance of extraneous argon in UHP eclogite (Sulu terrain, China): insight from in situ $^{40}Ar/^{39}Ar$ UV-laser ablation analysis. Earth Plan Sci Lett, 2000, 181: 605-615

[30] Li S G, Jagoutz E, Lo C H, et al. Sm/Nd, Rb/Sr, and $^{40}Ar/^{39}Ar$ isotopic systematics of the ultrahigh pressure metamorphic rocks in the Dabie-Sulu belt, central China: a retrospective view. International Geol Review, 1999, 41(12): 1114-1124

[31] Kogiso T, Tatsumi Y, Nakono S. Trace element transport during dehydration processes in the subducted oceanic crust: 1. Experiments and implications for the origin of ocean island basalts. Earth Plan Sci Lett, 1997, 148: 193-205

[32] 李曙光，黄方，周红英，等. 大别山双河超高压变质岩及北部片麻岩的 U-Pb 同位素组成——对超高压岩石折返机制的制约[J]. 中国科学(D 辑：地球科学)，2001(12)：977-984.

[33] Li S G, Jogout E, Chen Y Z, et al. Sm-Nd and Rb-Sr isotopic chronology and cooling history of ultrahigh pressure metamorphic rocks and their country rocks at Shuanghe in the Dabie Mountains, central China. Geochim. Cosmochim. Acta, 2000, 64(6): 1077-1093

[34] Cong B, M Zhai, D A Carswell, et al. Petrogenesis of ultrahigh-pressure rocks and their country rocks at Shuanghe in Dabieshan, central China. Eur J Mineral, 1995, 7: 119-138

[35] Zheng Y F, Fu B, Li Y, et al. Oxygen and hydrogen isotope geochemistry of ultrahigh-pressure eclogites from the Dabie Mountains and the Sulu terrane. Earth Plan Sci Lett, 1998, 155: 113-129

[36] Ahrens T J, Schubert G. Gabbro-eclogite reaction rate and its geophysical significance. Review of Geophysics and Space Physics, 1975, 13: 383-440

[37] Austrheim H R, Griffin W L. Shear deformation and eclogite formation within granulite-facies auorthosites of the Bergen Arcs, western Norway. Chemical Geol., 1985, 50: 267-281

[38] Zhang R Y, Liou J G. Partial transformation of gabbro to coesite-being eclogite from Yangkou, the Sulu terrane, eastern China. J Metamorphic Geol, 1997, 15: 183-202

[39] Ye K, Hirajima T, Ishiwatari A. Significance of interstitial coesite in eclogite from Yangkou, Qingdao City, eastern Shandong Province. Chinese Sci Bull, 1996, 41(5): 1047-1050(in Chinese)

[40] Tilton G R, Ames L, Schertl H P, et al. Reconnaissance isotopic investigations on rocks of an undeformed granuite contact with in the coesite-bearing unit of the Dora Maira Massif. Lithos, 1997, 41: 25-36

Making continental crust through slab melting: constraints from niobium-tantalum fractionation in UHP metamorphic rutile[*]

Yilin Xiao[1,2], Weidong Sun[3,4], Jochen Hoefs[1], Klaus Simon[1], Zeming Zhang[5], Shuguang Li[2] and Albrecht W. Hofmann[4]

1. Geowissenschaftliches Zentrum der Universität Göttingen, Goldschmittstrasse 1, Göttingen D-37077, Germany
2. CAS Key Laboratory of Crust-Mantle and Environments, School of Earth and Space Sciences, University of Science and Technology of China, Hefei 230026, Anhui, China
3. Key Laboratory of Isotope Geochronology and Geochemistry, Guangzhou Institute of Geochemistry, Chinese Academy of Sciences, Guangzhou 510640, Guangdong, China
4. Max-Planck Institut f. Chemie, Postfach 3060, Mainz D-55020, Germany
5. Institute of Geology, Chinese Academy of Geological Sciences, Beijing 100037, China

亮点介绍：利用激光-ICPMS 原位分析技术，第一次揭示了金红石高度的 Nb/Ta 分异现象；在此基础上，详细探讨了俯冲板片的热结构和变质脱水及熔融过程与 Nb/Ta 的分馏关系及形成机制；并提出了一个初步的模型，合理地解释了目前处于争论焦点的关于大陆地壳和亏损地幔为什么都具有低于球粒陨石的 Nb/Ta 值的悖论。

Abstract The formation of the continental crust (CC) is one of the most important processes in the evolution of the silicate Earth. Exactly how the CC formed is the subject of ongoing debate that focuses on its subchondritic Nb/Ta ratio. Nb and Ta are "geochemical identical twins," so they usually do not fractionate from each other. Here, we show that rutile grains from hydrous rutile-bearing eclogitic layers recovered from drillcores in the Dabie-Sulu ultrahigh pressure terrain have highly variable Nb/Ta values (ranging from 5.4 to 29.1, with an average of 9.8 ± 0.6), indicating major fractionation of Nb and Ta most likely occurred during blueschist to amphibole-eclogite transformation in the absence of rutile. It is suggested that the released fluids with subchondritic Nb/Ta were transported to, and retained by, hydrous rutile-bearing eclogite in colder regions, resulting in suprachondritic Nb/Ta ratios for drier eclogite in hotter regions. Further dehydration of hydrous rutile-bearing eclogites cannot transfer the fractionated Nb/Ta values to the CC due to the low solubility of Nb and Ta in fluids in the presence of rutile, while dehydration-melting results in a major component of the CC, the tonalite-trondhjemite-granodiorite (TTG) component, which is responsible for the low Nb/Ta of the CC. Consequently, residual eclogites have variable but overall suprachondritic Nb/Ta.

1 Introduction

The formation of the Earth's continental crust (CC) is of critical importance to understand the evolution of the Earth. Currently, there are several competing models with respect to the dominant process that has shaped the major chemical feature of the continental crust. Most geologists believe melting of subducted slabs in the early history of the Earth is the major way that shaped the chemical characteristics of the CC (McDonough, 1991; Rapp and Watson, 1995; Martin, 1999; Rudnick et al., 2000; Foley et al., 2002; Rapp et al., 2003), followed by modifications (Rudnick, 1995; Gao et al., 2004), and post Archean accretions of arc-magmatism and oceanic plateaus (Abbott et al., 1997; Reynaud et al., 1999) likely through complicated processes (Collins, 2002). By contrast, Kamber et al. (2002) suggested that Archean TTG was formed in the same way as modern arc rather than slab melting as adakite.

---

[*]本文发表在：Geochimica et Cosmochimica Acta, 2006, 70 (18): 4770-4782

Alternatively, others proposed that melting at the bottom of the thickened oceanic crust (Smithies, 2000) was more important. How subducted slabs were melted and how the distinct chemical features of the CC were produced is, however, still a controversial topic. The debate is mainly focused on the unique subchondritic Nb/Ta value of the CC, and more specifically, how Nb and Ta fractionated from each other and consequently how, in detail, the CC was built (Rudnick et al., 2000; Foley et al., 2002; Rapp et al., 2003; Xiong et al., 2005).

Nb and Ta share the same valence state (+5) and have similar atomic radii for octahedral coordination (Shannon, 1976), thus behaving identically in most geochemical fractionation processes linked to the evolution of the mantle (Jochum et al., 1986, 1989; Sun and McDonough, 1989; Rudnick et al., 2000). As a result, the Nb/Ta has long been regarded as constant and chondritic (17.5) in major silicate reservoirs of the Earth (Jochum et al., 1986, 1989; Sun and McDonough, 1989), and MORB, OIB, and BABB, (e.g., Sun and McDonough, 1989; Kamber and Collerson, 2000; Sun et al., 2003), although some depleted samples may have Nb/Ta ratios considerably lower than the chondritic value (Niu and Batiza, 1997; Sun et al., 2003). In contrast, the Nb/Ta of the CC is 12–13 (Barth et al., 2000; Rudnick and Fountain, 1995), much lower than the other major silicate reservoirs of the Earth. Therefore, the negative Nb and Ta anomalies and subchondritic Nb/Ta values are among the most distinct features of the CC (Taylor and McLennan, 1985; Rudnick and Fountain, 1995; Plank and Langmuir, 1998; Barth et al., 2000), and are of critical importance for understanding the formation of the CC (McDonough, 1991; Foley et al., 2002; Rudnick et al., 2000; Rapp et al., 2003).

In the last decade, analytical techniques for Nb and Ta have been improved considerably, represented by laser ablation-ICP-MS, which solves the problems of incomplete dissolution of accessory minerals that host Nb and Ta and of memory-effects in solution ICP-MS particularly for samples with high Ta concentration. The precision of Nb/Ta for laser ablation-ICP-MS data can be as good as 2%, whereas the accuracy depends heavily on the external standards used. Nonetheless, the Nb/Ta of NIST610 glass is reasonably well confined (Sun, 2003). Isotope dilution method has also been applied to Ta analyses, which have revealed large variations in Nb/Ta (Munker et al., 2003; Weyer et al., 2003). Based on data obtained using isotope dilution method, a new chondritic Nb/Ta of 19.9 ± 0.6 has been suggested; meanwhile, Nb/Ta of 14.0 ± 0.3 has been suggested for the primitive mantle, using a correlation between Zr/Hf and Nb/Ta, assuming a chondritic Zr/Hf value in the primitive mantle (Munker et al., 2003). Considering all factors together, in this paper we take Nb/Ta = 17.5 as the chondritic value and as the value for the primitive mantle (McDonough and Sun, 1995).

Given that rutile is the dominant carrier of Nb and Ta (Foley et al., 2000; Schmidt et al., 2004; Klemme et al., 2005; Xiong et al., 2005) and a common minor phase in high-grade metamorphic rocks (Chopin et al., 1991; Carswell et al., 1996; Gao et al., 1999; Liou et al., 1998; Tsujimori, 2002; Zack et al., 2002; Spandler et al., 2003), rutile has long been regarded as the controlling factor that resulted in Nb, Ta depletion in the continental crust, while the subchondritic Nb/Ta of the CC has been attributed to the melting of subducted slabs in the presence of rutile (Rudnick et al., 2000; McDonough, 1991), probably in garnet amphibolite or eclogite facies (Rapp and Watson, 1995; Rapp et al., 2003). This model has been challenged by partitioning data, which indicate that rutile favors Ta over Nb (Foley et al., 2002; Schmidt et al., 2004; Klemme et al., 2005; Xiong et al., 2005), such that partial melting of rutile-bearing eclogites with chondritic Nb/Ta results in suprachondritic Nb/Ta in the melts (Foley et al., 2002). Considering that Nb is more compatible than Ta in amphibole, a low degree of partial melting of subducted oceanic crust in the form of amphibolite was proposed (Foley et al., 2002), which, however, was questioned by Rapp et al. (2003) due to the difficulty to plausibly explain other major and trace element features of the CC. Therefore, the subchondritic Nb/Ta ratio of the CC was interpreted as the result of melting of hydrous rutile-bearing eclogitic sources with initially low Nb/Ta ratios (Rapp et al., 2003; Xiong et al., 2005), e.g., arc crust and special basalts from seamounts (Rapp et al., 2003). Nevertheless, the arc crust is not likely to be a major source of TTG, while basalts with low Nb/Ta ratios are rare and usually very depleted in incompatible elements (Niu and Batiza, 1997). Therefore, none of these models can account for all the Nb and Ta features of the continental crust. Here, we report highly variable but overall subchondritic Nb/Ta ratios in eclogitic rutile, indicating major Nb, Ta fractionation in subducted slabs that can conceivably reconcile different lines of observations.

2 Geological background and samples

As the largest ultrahigh-pressure (UHP) metamorphic terrain so far found worldwide, the Dabie-Sulu orogenic belt, which represents a Triassic collision zone between the North China Block and Yangtze Block, has attracted extensive interest from the geoscience community (Li et al., 1993, 1994, 2000; Liou and Zhang, 1995; Zhang et al., 1995, 1998, 2003, 2005; Carswell et al., 2000; Liou et al., 2000; Yang and Jahn, 2000; Sun et al., 2002). Research activities over the past decade have documented a number of characteristic features of this area, including rapid subduction to depths of greater than 100 km followed by rapid initial uplift (Xu et al., 1992; Li et al., 1993), the abundance of hydroxyl-bearing UHP mineral phases (Okay, 1995; Zhang et al., 1995), interaction between meteoric water and the rocks (Yui et al., 1995; Zheng et al., 1998), and variable fluid phases during UHP metamorphism (Xiao et al., 2000; Fu et al., 2001; Xiao et al., 2002).

The Chinese Continental Scientific Drilling Program (CCSD) in the southern limb of the Su-Lu ultrahigh-pressure metamorphic terrain in eastern China aims to reconstruct the formation and exhumation mechanisms of UHP terrains. The project consists of three pilot holes and a 5000 m main hole (Xu, 2004). ZK 703 is one of the pilot holes and about 70 m away from the main hole, with a depth of 558 m and a core-recovery >85%. The penetrated rocks consist mainly of a rutile-rich eclogitic layer with a composite thickness of about 300 m, which is interlayered by gneiss, phengite-quartz schist, jadeite quartzite and kyanite quartzite (Fig. 1). Geological survey indicates that this layer covers an area of about 3500 m×200 m (Xu et al., 1998). Rutile occurs in all eclogites with volume percents of 1%–5%. Coesite relics as inclusions in garnet and omphacite suggest that the eclogitic layer has been subducted to depths >100 km and subjected to peak metamorphic temperatures of 700–880 ℃ (Zhang et al., 2004).

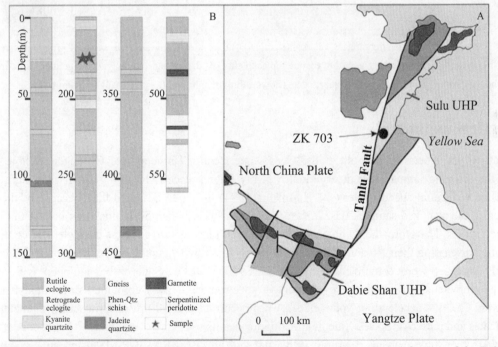

Fig.1 (A) Simplified geological map of the Dabie-Sulu orogenic belt showing major geological features of the Dabie-Sulu UHP metamorphic belt and the location of the CCSD drill site. (B) Simplified lithological profile of ZK703.

Investigated samples were collected from the rutile-rich eclogitic layer of ZK 703, including an eclogitic sample and a nearby quartz vein sample. The eclogitic sample (703-27a) consists mainly of garnet, omphacite, albite + amphibole symplectite, and coarse-grained rutile (~5 vol%, some grains are up to 6 mm in size). Petrological observations indicate that the assemblage of coarsegrained garnet + omphacite + rutile represents the peak of an early eclogite-facies metamorphic stage. Most garnet grains are relatively large, ranging from 1 to 2 mm. As shown in Table 1, garnet is compositionally homogeneous, and is rich in almandine and grossular (~49 mol % and ~30 mol %, respectively), with low pyrope (<20 mol %) and minor spessartine (around 1 mol %). Omphacite occurs either in the matrix or as inclusions in garnet, with jadeite contents up to 58 mol %. During retrograde metamorphism from eclogite- to amphibolite- and/or greenschist-facies, the eclogite-facies minerals were partially or totally replaced by symplectitic mineral coronas. Retrograde reactions include the formation of amphibole coronas around garnet, replacement of omphacite by symplectite of amphibole + plagioclase + clinopyroxene. The nearby quartz vein sample (703-27b) is composed of quartz and coarse-grained rutile (up to 8 mm in size). Quartz-garnet and quartz-rutile oxygen isotopic thermometers suggest peak metamorphic temperatures of 830 and 620 ℃ for the eclogite and the vein samples, respectively (Xiao, unpublished data).

Rutile in both samples occurs as euhedual to subhedral grains, up to 2–8 mm in diameter. The five rutile grains selected for analyses are 6–8 mm in size. Thin titanite replacement (10–20 μm) surrounding coarse-grained rutile in the eclogitic sample indicates only very limited retrogression for rutile (Fig. 2A). Coarse-grained rutiles in the quartz vein show even less retrogression compared to those in the eclogitic sample, with only a thin ilmenite rim (of a few μm) at the margin (Fig. 2B).

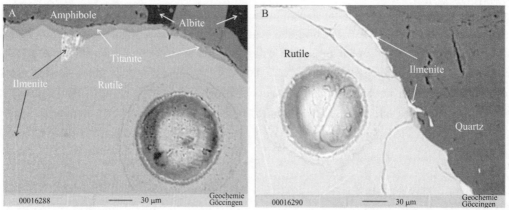

Fig.2 Back-scattered electron scanning images of coarse-grained rutiles in the investigated samples. The holes represent laser ablation-ICP-MS spots. Note there is only very thin titanite replacement (10–20 μm) surrounding coarse-grained rutile in the eclogitic sample (A), whereas coarse-grained rutile in the quartz vein shows a thin ilmenite rim (a few μm) at the margin (B).

3　Analytical method

All analyses have been carried out at the Geoscience Centre, University of Göttingen. Whole rock major elements of the eclogitic sample were determined by X-ray fluorescence (XRF) on glass discs using a Philips PW 1480 automated sequential spectrometer (see Hartmann, 1994). It has to be noted that, because of the scarcity of sample material, the analyzed sample 703-27a (see Table 1) with less than 50 g, does not include the investigated coarse-grained rutiles. Therefore, the "whole rock" value is not representative for the trace element content of rutile, especially when discussing their Nb and Ta characteristics as these two elements are mainly controlled by rutile.

Minerals analyses were completed using a JXA-8900RL JEOL Superprobe equipped with wavelength-dispersive spectrometer (WDS) and energy dispersive spectrometer (EDS) combined micro-analyzer. Operating conditions were 15.0 kV accelerating voltage, 15 nA beam current and 10 μm electron beam diameter. Standards included silicates and pure oxides (see Xiao et al., 2005 for details).

Nb, Ta, and other trace elements of rutile were analyzed by laser (LA) ICP-MS using an ArF-Excimer laser source "Compex 110" (λ= 193 nm; repetition rate of 10 Hz) coupled to a Perkin-Elmer DRC II ICP-MS. All samples have been prepared as polished thin sections (ca. 200 μm thick). ^{49}Ti was used as an internal standard; NIST610 was used as a calibration standard. Reproducibility and accuracy of trace element concentrations were assessed to be better than 10%. Ablation spots were around 80μm, with the laser energy of ~200 mJ. The ablated aerosol was carried to the ICP source with Ar gas.

To check for systematic shifts of measured Nb and Ta concentrations and Nb/Ta ratios, we measured a few MPI-DING glasses (Atho-g1, KL2-g, StHs6/80) (Jochum et al., 2000) and a pure rutile crystal (R10 from T. Zack, Heidelberg) with Ti-content ranging from 0.147 to 60 wt%. Obtained ratios show little systematic dependence on increasing Ti-content (Table 2 and Fig. 4): a positive deviation of <5% for MPI-DING glasses (mostly <3%) when compared to well calibrated values of (Jochum et al., 2000), and ~ 9% for pure rutile (R10) when compared to Zack's data (unpublished) that have been measured at the Mineralogisches Institut in Heidelberg. Our LA-ICP-MS measurements gave a mean Nb/Ta value of 6.2 ± 0.1 (2σ = 1.4%) calculated from three repeated measurements for R10, close to the average value of 5.6 ± 0.3 given by Zack (unpublished data).

Table 1 Representative mineral and whole rock compositions of the investigated eclogite

		Grt(rim)	Grt(core)	Cpx(rim)	Cpx(core)	Cpx(in)[a]	Pl	Am(sym)[a]	Am(grt)[a]	Am(rut)[a]		Wholerock[b]
SiO_2		38.9	38.7	56.6	56.8	57	66.9	40.2	41.8	42.4		43.2
TiO_2		0.062	0.091	0.068	0.074	0.076	0.016	0.533	0.137	0.609		4.49
Al_2O_3		21.5	21.4	13.5	13.5	13.6	20.4	14.5	16.8	15.5		14.9
Cr_2O_3		0.017	0.015	0.028	0.017	0.019	0.012	0	0	0.018		0.01
FeO^c		23.4	23.5	4	4.01	4.31	0.25	14.9	16.1	14.4		16.5
MnO		0.563	0.669	0	0.017	0.06	0.017	0.155	0.107	0.238		0.34
MgO		5.21	5.07	6.19	6.27	6.14	0.02	11.1	8.09	9.71		5.83
CaO		10.7	10.7	9.92	9.87	9.93	0.69	10	8.09	8.87		9.69
Na_2O		0.025	0.026	8.62	8.72	8.84	11.5	3.88	5.04	4.45		2.93
K_2O		bd	bd	bd	bd	bd	0.09	1.19	0.32	0.85		0.01
											P_2O_5	0.873
Total		100	100	98.9	99.2	100	99.8	96.4	96.5	97		98.8
Si		3	2.99	2.01	2.01	2.01	2.93T-site				V(ppm)	161
Ti		0.004	0.005	0.002	0.002	0.002	0.001Si	6.02	6.22	6.28	Co	19.2
Al		1.96	1.96	0.57	0.56	0.56	1.05Al(4)	1.98	1.78	1.72	Ni	14.9
Cr		0.001	0.001	0.001	0.0005	0.001	0.0004M(123)				Y	71.5
Fe^{3+}		0.029	0.048	0	0.008	0.022	0.009Ti	0.06	0.015	0.068	Zr	244
Fe^{2+}		1.48	1.47	0.119	0.111	0.105	Al(6)	0.576	1.17	0.977	Nb	9.2
Mn		0.037	0.044	0	0.001	0.002	0.001Cr	0	0	0.002	La	9.33
Mg		0.6	0.585	0.328	0.331	0.322	$0.001Fe^{3+}$	0.705	0.479	0.352	Ce	18.7
Ca		0.886	0.89	0.378	0.375	0.374	0.032Mg	2.48	1.8	2.14	Nd	12.9
Na		0.004	0.004	0.594	0.599	0.603	0.972Mn	0.02	0.013	0.03	Sm	5.04
K		bd	bd	bd	bd	bd	$0.005Fe^{2+}$	1.16	1.53	1.43	Lu	0.966
Total		8	8	4	4	4	5M(4)				Hf	5.17
							Ca	1.61	1.29	1.41	Ta	0.644
	Prp	0.2	0.196Jd	0.582	0.577	0.578Ab	0.963Na(M4)	0.39	0.709	0.592	W	11.1
	Grs	0.295	0.297Ac	0	0.008	0.022An	0.032A-site				Pb	4.94
	Alm	0.493	0.492Di	0.285	0.288	0.289Or	0.005Na(A)	0.738	0.746	0.686	Th	0.746
	Spe	0.012	0.015Hed	0.103	0.096	0.094K	K	0.227	0.06	0.161	U	0.237

[a] Cpx (in), omphacite inclusion in garnet; Am (Sym), amphibole occurring in symplectite; Am (grt) and Am (rut), amphiboles around garnet and rutile, resepctively.
[b] Major elements of whole-rock (703-27a) were determined by XRF, whereas trace elements were analyzed using laser-ICP-MS on the same XRF glass disc.
[c] Total Fe as FeO.

Table 2 Comparison between measured and reference values of various standards

	Ti	V	Zr	Nb	Hf	Ta	W	Pb	Th	U	Nb/Ta
atho-g1	1470	3.5	675	70	18.3	4.19	12.3	5.59	10.5	3.35	16.7
Ref.value[a]	1470	4.4	524	61.9	13.6	3.81		5.7	7.48	2.35	16.2
KL2-g-1	15600	420	215	16.9	5.7	1.08	1.07	2.63	1.46	1.1	15.6
KL2-g-2	15600	416	218	16.9	5.69	1.07	0.599	2.39	1.47	0.851	15.8
Ref.value[a]	15600	370	159	15.8	4.14	0.97	0.4	2.2	1.03	0.55	16.3
StHs6/80-G	4100	115	176	7.63	4.46	0.461	0.085	9.67	3.22	0.87	16.6

Continued

	Ti	V	Zr	Nb	Hf	Ta	W	Pb	Th	U	Nb/Ta
Ref.value[a]	4100	96	120	7.1	3.16	0.418		10.2	2.22	1.03	17
ML3B-G-1	12500	373	174	9.45	4.47	0.608	0.533	1.77	0.776	0.654	15.5
ML3B-G-2	12500	374	175	9.52	4.6	0.595	0.518	1.76	0.82	0.659	16
Ref.value[a]	12500		126	9	3.32	0.55		1.45	0.54	0.44	16.4
R10.1(core)	600000	1350	1100	2970	53.9	476	124	0.575	0.007	63.4	6.24
R10.2(middle)	600000	1361	1060	3070	51.7	502	128	0.041	0.022	64.7	6.12
Ref.value[b]	600000		788	2730	38.9	504					5.42
R10.3(rim)	600000	1700	943	2870	45.3	460	95.1	0.352	0.097	51.5	6.24
Ref.value[b]	600000		769	2590	37.4	456					5.68

[a] Published values by Jochum et al. (2000).
[b] Values from Thomas Zack (unpublished data).

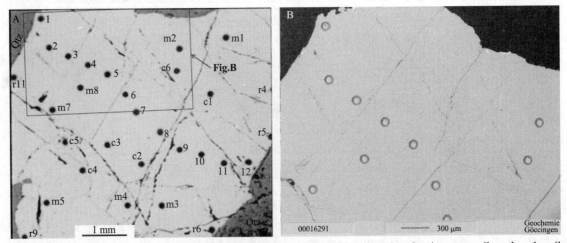

Fig. 3 Photomicrographs of an investigated rutile (grain 27b3), showing that all analyzed points are rutile rather than ilmenite or titanite. Continuous numbers (1, 2, 3...) indicate rim-core-rim analyses; whereas analytical spots starting with c, m, and r represent the geographic core, middle zone, and rim of the grain, respectively.

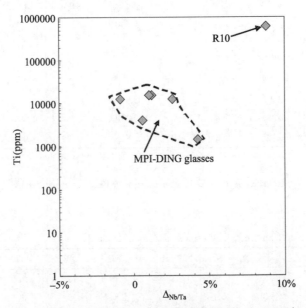

Fig. 4 Ti vs. $\Delta_{Nb/Ta}$ plot for reference samples (MPI-DING glasses and rutile R10). $\Delta_{Nb/Ta}$ = (measured Nb/Ta-published reference value)/published reference value. The ratio shows little systematic dependence on increasing Ti-content: a positive deviation of <5% (mostly <3%) for MPIDING glasses, and ~9% for pure rutile (R10) when compared to Zack's data (unpublished), which are based on an average of two measurements.

After "in situ" LA-ICP-MS measurements, thin sections have been re-polished and analyzed by electron microprobe. All analyzed spots have been qualitatively checked by electron microprobe in order to ensure that LA-ICP-MS measurements were done on rutile rather than on retrograde ilmenite or titanite. Furthermore, Nb, Ta and other trace element concentrations of 49 selected LA-ICP-MS spots were conducted with the JXA-8900RL JEOL Superprobe, with the purpose to rule out any Nb and Ta fractionations due to laser ablation and electron bombardment (Table 3). Analyses were performed at 25 kV acceleration voltage, 80 nA beam current and 8 μm probe diameter. Peak counting and background counting were, respectively, set at 15 and 10 s for Ti, 150 and 100 s for Si, Al, and Fe, 200 and 200 s for V, Cr, Sn, and 300 and 300 s for Nb, Ta, and Zr, taking altogether ca.21 min for each analysis. Every 10 analyses were bracketed by a synthetic rutile standard for zero-concentration countrates on the peaks and to exclude any machine drift (Zack et al., 2004). Detection limits were calculated to be 30 ppm for Cr and Al, 40 ppm for Fe, V and Zr, 60 ppm for Ta and Sn, 70 ppm for Nb and Si, and 1400 ppm for Ti (2σ).

Table 3 Nb, Ta, and other trace element concentrations in rutile grains from drilling hole in eastern China

	Ti^a	V	Zr	Nb	Hf	Ta	W	Th	U	Nb/Ta	V^b	Zr^b	Nb^b	Ta^b
Grain 27a-1														
-1	599000	1380	228	727	9.01	130	22.1	0.023	1.47	5.59	936	141	512	n.d.
-2	600000	1200	196	991	7.6	57.9	20.2	n.d.	1.09	17.1	969	138	997	n.d.
-3	601000	1160	194	1250	6.93	127	17.5	0.004	1.18	9.9	963	155	1070	87.6
-4	599000	1160	196	1020	6.87	84.2	18.3	0.2	2.07	12.1	941	155	884	50.8
-5	599000	1420	227	1330	9.44	125	24.7	0.71	1.3	10.7	932	122	877	80.3
-6	601000	1190	184	1220	6.96	145	17.4	0.083	1.51	8.38	922	126	1050	116
-7	597000	1230	179	1510	7.82	143	24.2	0.02	1.49	10.6	1060	144	1220	112
-8	596000	1180	171	1320	6.51	155	18.4	0.005	0.86	8.5	1230	120	1260	133
-9	596000	1180	180	1410	6.71	155	31.7	0.037	1.22	9.08	1130	149	1100	148
-10	591000	1160	176	1360	6.78	149	24.7	0.006	0.79	9.12	1130	146	1000	55.7
-11	600000	1250	174	1320	5.78	138	18	0.016	1.11	9.58	990	100	1090	172
-12	597000	1180	188	1520	7.61	157	20.3	0.014	1.26	9.68	997	127	1170	197
-13	586000	1210	171	1330	6.85	162	23.6	n.d.	0.85	8.23	1180	134	905	179
-14	593000	1110	188	429	7.68	24	15.8	0.004	0.64	17.8	1110	144	317	n.d.
-15	594000	1130	178	531	7.11	27.6	23.6	n.d.	0.64	19.2	1080	170	449	n.d.
-16	601000	1120	185	678	6.96	46.5	17.5	0.017	1.17	14.6	971	130	505	n.d.
Grain 27a-2														
-1	600000	1380	218	1000	8.87	125	20.1	0.081	1.48	8.03				
-2	600000	1230	187	2200	9.08	248	36.3	0.005	1.08	8.88				
-3	600000	1090	164	1180	7.19	161	16.2	0.034	1.02	7.35				
-4	600000	1070	171	909	7.72	62.1	14.7	0.045	1.18	14.6				
-5	600000	1130	190	1900	7.81	179	26.8	n.d.	3.38	10.6				
-6	599000	1120	195	2060	8.41	182	41.4	n.d.	3.15	11.3		175	1860	155
-7	600000	1100	198	1450	7.76	180	29	n.d.	3.17	8.04				
-8	597000	1090	189	1390	6.93	144	22.8	0.008	2.22	9.64		195	1470	161
-9	600000	1120	193	1740	8.99	169	47.4	0.022	3	10.3				
-10	600000	1130	201	1680	9.45	160	40.9	0.008	3.69	10.5				
-11	600000	1100	215	1550	8.62	181	42.8	0.022	3.87	8.53				
-12	600000	1070	200	1100	8.33	69.8	46.2	n.d.	2.84	15.7				
-13	600000	1210	189	1440	8.58	170	45.7	0.009	3.53	8.44				
-14	596000	1050	176	590	8.33	40.1	33.8	0.009	1.29	14.7	1110	147	452	60.6
c1	599000	1080	193	1370	8.27	132	27.5	0.007	5.59	10.3			1110	128
c2	600000	1080	196	1220	9.14	128	17.9	0.008	4.93	9.5				

	Ti[a]	V	Zr	Nb	Hf	Ta	W	Th	U	Nb/Ta	V[b]	Zr[b]	Nb[b]	Ta[b]
Grain 27a-2														
c3	598000	1110	182	1590	7.64	191	24.4	n.d.	2.78	8.34			1370	170
c4	600000	1080	185	1960	7.48	172	42.8	0.006	2.4	11.4				
c5	600000	1090	180	1450	6.5	116	29.7	0.008	1.56	12.5				
m1	598000	1110	197	1960	8.35	172	46.4	0.007	4.74	10.8			1470	86.8
m2	600000	1060	168	774	7.96	70.2	15.4	n.d.	2.42	11				
m3	600000	1150	181	2800	8.47	240	63.5	n.d.	1.43	11.7				
m4	600000	1030	174	861	8.44	103	16.7	0.022	1.28	8.34				
m5	600000	1040	174	951	6.58	172	18.8	n.d.	1.36	5.52				
m6	600000	1070	173	739	7.25	58.1	13.7	n.d.	2.18	12.7				
m7	600000	1080	172	1220	6.89	137	18.9	n.d.	3.59	8.93			1060	114
m8	600000	1120	174	1580	6.95	152	34.1	0.029	2.15	10.4				
m9	600000	1100	182	1510	8.17	153	35.2	0.005	2.33	9.88				
r1	600000	1060	164	604	6.16	42.5	20.8	n.d.	1.66	14.2				
r2	599000	1070	171	596	7.18	42.7	21.5	n.d.	4.31	14	951	137	514	n.d.
r3	600000	1090	154	995	5.78	149.8	19.1	0.009	1.86	6.64				
r4	600000	1090	165	685	7.1	127.3	13.5	n.d.	1.22	5.38				
r5	600000	1020	159	460	6.53	35	15.4	0.01	0.98	13.1				
r6	600000	1050	154	750	5.38	102	13.5	0.009	0.878	7.39				
r7	597000	1090	151	1540	7.96	175	38.2	n.d.	1.12	8.81	978	151	1190	188
r8	597000	1110	180	1560	7.86	181	29.2	n.d.	1.37	8.63	959	140	1170	58.1
r9	600000	1030	175	366	7.79	29.6	12.7	0.019	1.13	12.4				
r10	600000	1040	159	391	6.37	30.1	15.5	0.018	1.52	13				
r11	600000	1030	164	417	7.19	35.4	22.5	0.348	2.1	11.8				
r12	600000	1050	159	338	6.78	23.8	13.2	n.d.	1.5	14.2				
r13	600000	1030	138	409	5.72	31.5	13.5	0.008	1.06	13				
r14	600000	1030	164	362	5.98	23.9	13.3	0.008	1.3	15.1				
r15	600000	1060	170	929	6.31	118	32.7	n.d.	1.15	7.85				
Grain 27b-1														
-1	600000	1090	325	1180	11.7	150	32.4	0.651	4.52	7.89				
2	600000	1110	477	1480	14.8	135	39.3	1.19	6.57	10.9				
3	600000	1160	431	1500	15.4	134	44.1	0.898	6.8	11.2				
4	600000	1440	311	1510	86.3	129	65.8	9.38	12.1	11.7				
5	600000	1160	198	1520	9.88	155	45.3	0.009	9.12	9.81				
6	600000	1150	194	1500	8.02	157	48.7	0.041	9.31	9.53				
7	600000	1110	199	1490	8.12	156	48.2	0.023	9.51	9.58				
8	600000	1190	209	1590	9	158	49.2	0.01	8.6	10.1				
9	600000	1190	209	1590	10.1	161	45.9	0.012	9.53	9.89				
10	600000	1110	197	1570	8.49	147	51.2	0.01	10.9	10.6				
11	600000	1160	209	1710	9.28	193	55.4	n.d.	8.23	8.87				
12	600000	1160	205	1740	9.6	201	65.6	0.061	9.58	8.63				
14	600000	1150	195	1700	8.64	184	58.9	n.d.	6.56	9.21				
15	600000	1120	201	1560	9.82	124	49.1	0.01	5.71	12.6				
16	600000	1160	199	1600	8.46	137	50.5	0.028	5.36	11.7				
17	600000	1140	194	1260	8.49	129	42.4	0.054	5.51	9.78				
18	600000	1050	193	660	6.52	34.4	31.8	n.d.	4.74	19.2				
rim1	600000	1480	253	1800	9.77	208	53.9	0.023	5.88	8.66				

Continued

	Ti[a]	V	Zr	Nb	Hf	Ta	W	Th	U	Nb/Ta	V[b]	Zr[b]	Nb[b]	Ta[b]
Grain 27b-1														
rim2	600000	1090	186	830	9.32	28.6	22.9	0.012	4.74	29.1				
Grain 27b-2														
-1	600000	1070	189	1080	7.06	147	34.1	n.d.	3.88	7.38	1010	148	883	90.1
-2	599000	1120	199	1570	9.79	143	40.3	n.d.	5.57	11	1010	170	1430	60.6
-3	598000	1170	208	1590	9.52	144	45.7	0.006	6.38	11	1020	177	1430	102
-4	595000	1210	212	1650	8.61	145	48.7	0.007	6.79	11.4	1020	161	1380	148
-5	598000	1130	210	1570	9.66	150	46.6	0.016	6.45	10.5	976	176	1440	128
-6	597000	1120	209	1560	9.17	148	45.3	0.007	8.26	10.5	1010	157	1370	148
-7	593000	1150	214	1580	8.8	149	44.5	0.015	7.43	10.6	1110	189	1420	181
-8	598000	1130	214	1570	8.55	150	41.6	n.d.	7.43	10.4	950	178	1360	164
-9	598000	1120	215	1590	8.78	150	43.1	0.046	7.63	10.6	1040	145	1380	91.7
-10	598000	1110	218	1580	8.91	156	43	n.d.	8.03	10.1	1050	165	1430	98.3
-11	600000	1120	197	1290	6.79	148	32.3	n.d.	4.38	8.73	989	153	935	154
Grain 27b-3														
-1	598000	1360	212	1630	9.98	184	45	0.01	8.93	8.85	941	150	827	72.1
-2	600000	1350	220	1780	9.98	167	46.5	n.d.	6.32	10.6	932	168	1600	138
-3	599000	1350	228	1800	10.7	187	50.5	0.009	6.75	9.61	1010	162	1590	143
-4	600000	1350	229	1750	10.2	185	53.1	n.d.	8.27	9.43	969	167	1530	155
-5	594000	1350	237	1730	11.5	180	52.8	n.d.	8.38	9.61	982	180	1530	179
-6	597000	1340	236	1690	11	169	49.8	0.008	8.88	10	963	193	1490	175
-7	600000	1350	237	1770	11.5	170	54.1	n.d.	9.56	10.4	978	184	1410	102
-8	599000	1340	240	1860	9.94	213	62.4	0.011	8.32	8.73	1000	183	1380	115
-9	598000	1360	226	1840	10.9	235	69	n.d.	8.45	7.84	978	146	1460	176
-10	599000	1330	224	1850	8.82	225	69	0.008	7.44	8.2	971	160	1430	159
-11	597000	1350	221	1880	10.4	223	67.6	0.03	6.5	8.43	990	173	1410	80.3
-12	600000	1290	207	1150	8.91	114	43.4	n.d.	4.61	10	995	181	1000	86.8
c1	600000	1400	227	1920	10	176	55.3	0.008	7.55	10.9				
c2	600000	1390	240	2080	10.1	219	64.4	n.d.	7.27	9.5				
c3	600000	1340	230	1800	9.74	169	54.6	n.d.	8.21	10.7				
c4	600000	1370	241	1830	10.3	173	52.4	n.d.	9.6	10.6				
c5	600000	1410	213	1840	9.35	165	50.5	n.d.	5.76	11.2				
c6	600000	1350	231	1880	10.1	168	51.5	n.d.	8.09	11.2				
m1	600000	1370	233	1880	10.7	200	54.2	n.d.	5.89	9.42				
m2	600000	1380	224	1850	10.7	157	55.5	n.d.	5.1	11.8				
m3	600000	1370	229	1810	10	168	50.3	0.008	8.15	10.8				
m4	600000	1340	232	1850	8.88	173	52.2	0.01	6.53	10.7				
m5	600000	1340	220	1770	9.46	183	52.1	0.026	5.34	9.66				
m6	600000	1370	212	1740	8.53	148	43.5	0.019	7.37	11.7				
m7	600000	1360	229	1810	9.47	161	47.1	n.d.	8.34	11.2				
m8	600000	1350	231	1880	9.32	172	46.4	0.01	4.98	10.9				
r1	600000	1310	195	1210	9.34	149	35.7	0.059	4.86	8.08				
r2	600000	1310	206	1480	9.85	174	43.7	0.046	4.39	8.53				
r3	600000	1310	182	1300	7.66	172	26.7	n.d.	3.43	7.53				
r4	600000	1320	221	1200	9.1	80.5	57.3	n.d.	6.21	14.9				
r5	600000	1260	216	745	7.3	44.1	24.1	0.01	5.21	16.9				
r6	600000	1340	208	1110	9.05	149	32.5	0.062	4.5	7.41				

Continued

	Ti[a]	V	Zr	Nb	Hf	Ta	W	Th	U	Nb/Ta	V[b]	Zr[b]	Nb[b]	Ta[b]
Grain 27b-3														
r7	600000	1350	212	1100	8.87	178	34.8	0.1	3.69	6.17				
r8	600000	1320	206	1420	8.06	168	38.6	n.d.	4.83	8.46				
r9	600000	1360	211	1480	8.79	170	36.6	0.01	4.97	8.75				
r10	600000	1300	243	1510	7.94	178	40.3	0.1	4.84	8.5				
r11	600000	1330	213	1340	8.41	151	39.3	n.d.	4.27	8.95				
r12	600000	1330	218	1400	9.61	169	30.8	n.d.	4.47	8.3				
r13	600000	1360	221	1840	8.46	202	50.6	n.d.	3.44	9.12				
r14	600000	1300	211	1220	9.74	120	44.5	0.022	3.33	10.2				

Five rutile grains were analyzed: 27a-1 and 27a-2 are from the eclogitic sample, whereas 27b-1, 27b-2, and 27b-3 are from the nearby quartz vein. Continuous numbers (1, 2, 3. . .) indicate rim-core-rim profile analyses; whereas analytical spots starting with c, m, r represent the geographic core, the middle zone, and the rim of the grain, respectively. n.d. not detected.

[a] A minor isotope of Ti (^{49}Ti) was used as an internal standard, assuming that Ti content in rutile is 60 wt% when not measured.
[b] Values measured by electron microprobe.

4 Fractionated Nb/Ta in rutile

Rim-core-rim analyses were carried out on five rutile grains ranging from 5 to 8 mm in size. To reveal two-dimensional inhomogeneities of Nb and Ta in rutile, more analyses on the core (c), the mantle (m) and the rim (r) of two rutile grains were carried out besides the profile analysis (Fig. 3). The Nb and Ta contents range from 338 to 2800, 23.8–248 ppm, respectively, with Nb/Ta values ranging from 5.40 to 29.1 (Table 3). Compared to the eclogitic sample, rutile grains in the quartz vein usually have higher Nb and Ta contents. Remarkably, although concentrations of Nb and Ta measured by electron microprobe are 20%–30% lower than those from the LA-ICP-MS, both methods indicate that rutile grains have subchondritic Nb/Ta ratios with high Nb, Ta contents in the cores, which change sharply to suprachondritic Nb/Ta ratios with lower Nb, Ta contents near the rims, forming "Nb/Ta spikes". In a few cases, the Nb/Ta decreases and the Nb, Ta contents increase further to the rims (Fig. 5). Such abrupt variations cannot be explained by a closed system; instead, the results suggest that rutile has collected Nb and Ta from at least two sources—first a subchondritic source, then a suprachondritic one, and in some cases, finally back to the subchondritic source. Consistently, the Nb/Ta spikes are also separated in Nb/Ta versus Nb and U diagrams, with systematically lower U and Nb (Fig. 6). All these features indicate that both Nb and Ta were mobile. Nonetheless, the average Nb/Ta of all the analyses weighted by Nb, Ta contents is 9.8 ± 0.6, which is much lower than the chondritic value (17.5) (McDonough and Sun, 1995), and also lower than that of Dabie eclogites sampled from the surface (Sun, unpublished data) and their protoliths, and Neoproterozoic mafic dykes in south China (Li et al., 2002), indicating significant Nb,Ta fractionation. Remarkably, the average Nb/Ta ratio of ~10 is even lower than the Nb/Ta values of the CC (12–13) (Barth et al., 2000).

It is worth to consider whether the high Nb/Ta spikes observed in the present study are just

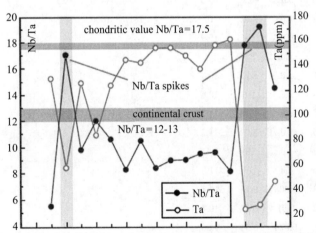

Fig. 5 Nb/Ta variations within a rutile grain (27a-1) from eclogite. Also shown are the Nb/Ta ratio for chondrites (McDonough and Sun, 1995) and for the CC (Barth et al., 2000). The cores of rutile grains have subchondritic Nb/Ta lower than that of the CC with high Nb, Ta contents. The Nb/Ta usually increases sharply to suprachondritic coupled with lower Nb, Ta contents near the rims, forming "Nb/Ta spikes". The Nb/Ta then drops quickly to subchondritic with high Nb and Ta. Thus, the Nb/Ta ratios can be separated into three groups: the cores and the rims with Nb/Ta spikes in between.

mineral-scale features, reflecting prograde mineral breakdown reactions, or alternatively, reflecting the characteristics of "zone refining" dehydrations. So far, to our knowledge, no mechanism is known that can fractionate Nb/Ta by a factor of 2 on the mineral-scale of rutile. Therefore, we consider that the high Nb/Ta spikes can be better explained by a "zone refining" dehydration model (see below).

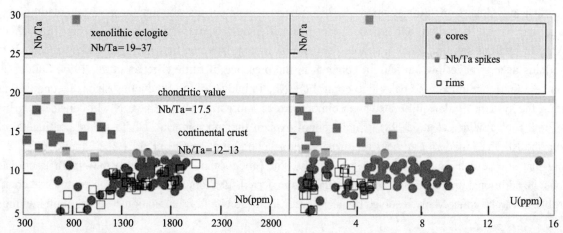

Fig. 6 Nb/Ta versus Nb and U for investigated rutile grains. Also shown are the Nb/Ta for chondrites (McDonough and Sun, 1995), the CC (Barth et al., 2000) and xenolithic eclogites (Rudnick et al., 2000). The cores and the rims are similar to each other, forming positive correlations, whereas the Nb/Ta spikes fall off the lines, with lower Nb and U contents, indicating that the Nb, Ta fractionation was controlled by at least two different processes (see text).

5 Discussion and conclusion

5.1 Mobility and fractionation of Nb and Ta

The highly variable Nb/Ta ratios in the studied rutile indicate major fractionation between the two elements in the subducted slab. Moreover, the presence of thick rutile-rich layers suggests that Ti is mobile during subduction. These observations are remarkable because Nb, Ta and Ti are all regarded as conservative elements that are not mobile during subduction as indicated by arc magmas (Pearce and Peate, 1995).

It has been experimentally shown that Nb and Ta are highly soluble in fluids in the absence of rutile (Stalder et al., 1998), and that Nb and Ta are immobile in the presence of rutile (e.g., Brenan et al., 1994). Low Nb/Ta ratios found by Henry et al. (1996) in a rutile bearing eclogitic vein compared to its wall rocks was taken as an indicator of Nb and Ta mobility in the presence of rutile. However, in our view, these low ratios can be better explained by the hypothesis that Nb and Ta were mobile before rutile appears, and then were retained by newly formed rutile under suitable pressures and temperatures. We propose that the Nb/Ta fractionation and the mobility of Ti are probably due to dehydration processes that occurred at shallower depths before rutile appears, or when rutile is unstable. These facts also imply that other Ti-minerals, e.g., titanite and ilmenite, have not been involved during the dehydration processes that mobilized Ti, Nb, Ta, and fractionated Nb/Ta, because of the high Ti, Nb, Ta concentrations in these minerals (Barth et al., 2002; Prowatke and Klemme, 2005).

Ti-minerals (mostly rutile, titanite, and sometime also ilmenite) have been reported from blueschists and eclogites (Chopin et al., 1991; Carswell et al., 1996; Liou et al., 1998; Gao et al., 1999; Tsujimori, 2002). Experiments with a tholeiitic basalt at temperatures of 600–950 °C show that titanite at low temperatures and ilmenite at high temperatures occur essentially at pressures <1.5 GPa (Liou et al., 1998), whereas rutile appears at pressures >1.5 GPa (Liou et al., 1998; Xiong et al., 2005). Interestingly, titanite and ilmenite in our samples are retrograde products of rutile, occurring as thin rims surrounding rutile grains or exsolutions (Fig. 2). No titanite inclusion in rutile grains has been identified. In fact, in blueschists and eclogites, titanite often appears as retrograde product of rutile (Carswell et al., 1996; Gao et al., 1999; Tsujimori, 2002). These facts indicate that rutile is likely to be the first Ti-mineral in many blueschists and eclogites, possibly because the subducting slab was not heated up to temperatures suitable for titanite formation before pressures reached 1.5 GPa. It has not been investigated experimentally yet at what temperature titanite starts to appear. Assuming the lowest temperature (600 ℃) for the long-run-duration experiments (Liou et al., 1998) is the onset temperature of titanite formation, the dehydration that fractionated Nb/Ta should have occurred at pressures <1.5 GPa and temperatures <600 ℃.

Dehydration has been proposed as the main mechanism that fractionates Nb from Ta (Ionov and Hofmann, 1995; Kamber and Collerson, 2000; Rapp et al., 2003) because element distributions in most magmatic processes are charge and radius controlled, and as such, the most efficient way of fractionating Nb from Ta is chemical complexing in hydrous solution. Such a simple dehydration model, however, has not been widely accepted: First, hydrous fluids usually have low Nb and Ta contents in the presence of rutile (Brenan et al., 1994; Stalder et al., 1998), and are therefore unlikely to be able to control the Nb, Ta budget of the subducting slabs. Second, if there is no rutile in the system, fluids equilibrated with clinopyroxene and garnet should have Nb/Ta considerably higher than the bulk rock (Stalder et al., 1998), which cannot explain the subchondritic Nb/Ta of the continental crust. Moreover, the Nb/Ta in modern arc rocks are usually not subchondritic (Eggins et al., 1997; Munker et al., 2004), indicating that simple dehydration is unlikely to be the sole process that contributed to the Nb/Ta fractionation. There must be additional processes that have selectively transferred subchondritic Nb/Ta to the continental crust. All these problems can be conceivably resolved by dehydration-induced Nb/Ta fractionation ± dehydration melting.

5.2 "Zone refining" dehydration model

Subducting slabs are not homogenous in terms of temperature and water contents at the early stage of subduction. In general, the temperature increases while the water content varies from the surface of the slab to the mantle lithosphere before subduction (Hacker et al., 2003). During subduction, the slab is heated up from both the bottom and the top. So, at the early stage of subduction, the subducting slab may have a "sandwich" structure, with colder regions bounded by hotter layers both on the slab surface and inside the slab. In the hotter parts of the slab, major dehydration occurs during blueschist to amphibole-eclogite transformation (BAT) that takes place in the absence of rutile (<1.5 GPa) (Fig. 7). In this case, Nb and Ta budgets are controlled by the partitioning between fluids and amphibole.

In a Ti-mineral-free system, amphibole has higher Nb, Ta compared to other minerals in rocks ranging from mantle peridotite (Witt-Eickschen and Harte, 1994; Ionov and Hofmann, 1995; Chazot et al., 1996) to syenites and granites (Marks et al., 2004). There are no partitioning data of amphibole versus fluid for Nb and Ta so far. Nonetheless, it has been demonstrated that amphibole prefers Nb over Ta compared to melt and clinopyroxene (Ionov and Hofmann, 1995; Foley et al., 2002). The lower the mg number in amphibole, the more it prefers Nb (Foley et al., 2002). The $D_{amphibole/CPX}$ for Nb is about 2.5 times higher than that for Ta, estimated for high mg amphibole from veins in mantle xenoliths (Ionov and Hofmann, 1995), and can be as high as 8 times for quartz syenite (Marks et al., 2004). The $D_{amphibole/CPX}$ ratio of Nb versus Ta in metabasaltic rocks should lie between 2.5 and 8. Fluids also prefer Nb to Ta compared with CPX, with $D_{fluids/CPX}$ for Nb slightly less than 2 times higher than that for Ta (Stalder et al., 1998). Therefore, amphibole prefers Nb over Ta compared with fluids. In other words, fluids in equilibrium with amphibole should have low Nb/Ta. This is supported by the fact that amphibole favors Nb over Ta in systems that have been affected by fluids (Ionov and Hofmann, 1995; Marks et al., 2004). Therefore, the BAT fractionates Nb from Ta, resulting in subchondritic and suprachondritic Nb/Ta in the released fluids and residual amphibole eclogites, respectively. Ti is also mobile before the appearance of rutile, whereas hydrous minerals, such as amphibole and lawsonite, are stable in cold regions. Therefore, hydrous fluids, together with trace and major elements (Nb, Ta, Ti, Si, etc.) released during the BAT in the hotter regions, are gradually transferred towards colder regions inside the subducting slab and subsequently retained by hydrous minerals. This is consistent with the observation that the investigated rutile-bearing hydrous eclogite layers are situated at depths between 40 and 500 m of the drillcores ZK703 and 300–600 m of the CCSD main hole (Zhang et al., 2004).

The mobility of Ti during early dehydration of the hotter regions leads to Ti depletion, which can result in rutilefree residual amphibole-eclogites even at high pressures in the dehydrated regions. In case there is no rutile, dehydration of these residual amphibole eclogites with suprachondritic Nb/Ta will further elevate the Nb/Ta in fluids, because both clinopyroxene and garnet favour Ta over Nb (Stalder et al., 1998). This can plausibly explain the

Nb/Ta spikes of the rutile grains (Fig. 5). The low Nb/Ta ratios at the very margin were probably formed during retrograde metamorphism.

Such a "zone refining" dehydration process results in subchondritic Nb/Ta values in the colder and wetter regions, with complementary suprachondritic Nb/Ta in the hotter and dryer regions of the slabs. The dehydration induced Nb/Ta fractionation was probably even more efficient and more substantial in the early history of the Earth, as the oceanic crust was thicker with larger geothermal variations within subducting slabs (Rudnick et al., 2000).

In addition to dehydration of basaltic portions of the slab, dehydration of serpentinites is likely to be an important source of fluids (Hattori and Guillot, 2003). These fluids should have passed though the basaltic layer after being released, which might also fractionate Nb and Ta in case P-T conditions are favourable.

5.3 The fate of the fractionated Nb/Ta signature and the formation of the CC

For cold geotherms such as in modern subduction zones, amphibole breaks at pressures of 2.5 GPa (Schmidt and Poli, 1998; Xiong et al., 2005), long after rutile appears (Liou et al., 1998; Xiong et al., 2005). No major slab melting occurs, except for the subduction of young hot oceanic crust, which produces adakite. Consequently, Ti, Nb, and Ta are all retained in rutile, and thus are not mobile during amphibole-eclogite to eclogite transformation. Therefore, the fractionated Nb/Ta features are preserved and descended to greater depths. As a result, the fractionated Nb/Ta signature cannot be systematically transferred to modern arc magmas. This explains the lack of systematic subchondritic Nb/Ta in many modern arc volcanic rocks (Munker et al., 2004).

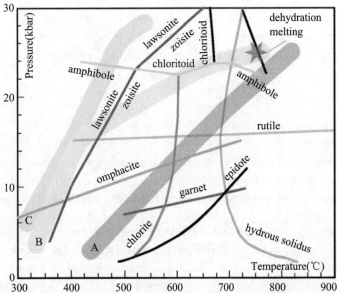

Fig. 7 Phase relations for subducting mid-ocean ridge basalts (modified after Schmidt and Poli, 1998 and Xiong et al., 2005) and different P-T regions (A–C) illustrating Nb/Ta fractionation in different portions of a subducting slab. Hot regions (A) inside a subducting slab are dehydrated at low pressures before rutile appears, releasing fluids with subchondritic Nb/Ta ratios. This portion of the slab will not melt during subduction because of its low water contents. In contrast, cold regions (B and C) in a subducting slab will not experience major dehydration before rutile appears and act as a sink of Nb, Ta and Ti in fluids released from the BAT in hot regions, which subsequently gathers fractionated Nb/Ta in rutile. In the Archean when the geotherm was high (B), hydrous eclogites were heated up relatively fast during further subduction, and were melted through dehydration-melting indicated by the star (Rapp et al., 2003; Rapp and Watson, 1995). For cold geotherm such as in modern convergent margins (C), hydrous eclogites will not be melted, but will be further dehydrated in the presence of rutile, without additional Nb/Ta fractionation.

The fractionated Nb/Ta ratios of the subducting slabs can be transferred to the CC through slab melting. As subduction continues, the temperature of the subducting slab increases and becomes more homogenous. At this stage melting of the subducting slab is mainly controlled by H_2O contents. In the Archean, when the geotherm was hot, large proportions of the colder and wetter regions would be melted through dehydration-melting when they were heated up with increasing depths (Fig. 7). This resulted in large quantities of TTG magmas with subchondritic Nb/Ta values, as well as depletions of Nb, Ta and heavy rare earth elements because of residual rutile and garnet, respectively. The residues of the dehydration-melting have even lower Nb/Ta values. In contrast, the dry regions (early dehydrated inner portions) of the slab with suprachondritic Nb/Ta ratios cannot be melted. Therefore, refractory residual eclogites have highly variable Nb/Ta from subchondritic to suprachondritic, with an overall suprachondritic average Nb/Ta as observed in xenolithic eclogites from kimberlite pipes (Rudnick et al., 2000), which is complementary to the continental crust.

Acknowledgment

YLX's research was supported by the German National Science Foundation (DFG, HO 375/22) and partly by the National Science Foundation of China (40472036). WDS was supported by the Natural Science Foundation of China (40525010), the Chinese Academy of Science and the Alexander von Humboldt Foundation. We dedicate this paper to Dr. Shen-su Sun, a great scientist, who initiated this study with stimulating discussions and constructive suggestions, and helped to shape the findings reported here before he passed away. Valuable comments by Alan Brandon, Balz Kamber, Carsten Münker, and an anonymous reviewer materially improved the original manuscript and are greatly appreciated.

References

Abbott, D.H., Drury, R., Mooney, W.D., 1997. Continents as lithological icebergs: the importance of buoyant lithospheric roots. Earth Planet. Sci. Lett. 149 (1–4), 15–27.

Barth, M.G., McDonough, W.F., Rudnick, R.L., 2000. Tracking the budget of Nb and Ta in the continental crust. Chem. Geol. 165 (3–4), 197–213.

Barth, M.G., Rudnick, R.L., Horn, I., McDonough, W.F., Spicuzza, M.J., Valley, J.W., Haggerty, S.E., 2002. Geochemistry of xenolithic eclogites from West Africa, part 2: origins of the high MgO eclogites. Geochim. Cosmochim. Acta 66 (24), 4325–4345.

Brenan, J.M., Shaw, H.F., Phinney, D.L., Ryerson, F.J., 1994. Rutileaqueous fluid partitioning of Nb,Ta,Hf,Zr,U and Th—implications for high-field strength element depletions in island-arc basalts. Earth Planet. Sci. Lett. 128 (3–4), 327–339.

Carswell, D.A., Wilson, R.N., Zhai, M., 1996. Ultra-high pressure aluminous titanites in carbonate-bearing eclogites at Shuanghe in Dabieshan, central China. Mineral. Mag. 60 (400), 461–471.

Carswell, D.A., Wilson, R.N., Zhai, M., 2000. Metamorphic evolution, mineral chemistry and thermobarometry of schists and orthogneisses hosting ultra-high pressure eclogites in the Dabieshan of central China. Lithos 52 (1–4), 121–155.

Chazot, G., Menzies, M.A., Harte, B., 1996. Determination of partition coefficients between apatite, clinopyroxene, amphibole, and melt in natural spinel lherzolites from Yemen: implications for wet melting of the lithospheric mantle. Geochim. Cosmochim. Acta 60(3), 423–437.

Chopin, C., Henry, C., Michard, A., 1991. Geology and petrology of the coesite-bearing terrain, Dora-Maira Massif, Western Alps. Eur. J. Miner. 3 (2), 263–291.

Collins, W.J., 2002. Hot orogens, tectonic switching, and creation of continental crust. Geology 30 (6), 535–538.

Eggins, S.M., Woodhead, J.D., Kinsley, L.P.J., Mortimer, G.E.,Sylvester,P.,McCulloch,M.T.,Hergt,J.M., Handler, M.R., 1997. A simple method for the precise determination of ≥40 trace elements in geological samples by ICPMS using enriched isotope internal standardisation. Chem. Geol. 134 (4), 311–326.

Foley, S., Tiepolo, M., Vannucci, R., 2002. Growth of early continental crust controlled by melting of amphibolite in subduction zones. Nature 417 (6891), 837–840.

Foley, S.F., Barth, M.G., Jenner, G.A., 2000. Rutile/melt partition coefficients for trace elements and an assessment of the influence of rutile on the trace element characteristics of subduction zone magmas. Geochim. Cosmochim. Acta 64 (5), 933–938.

Fu, B., Touret, J.L.R., Zheng, Y.F., 2001. Fluid inclusions in coesitebearing eclogites and jadeite quartzite at Shuanghe, Dabie Shan (China). J. Metamorph. Geol. 19 (5), 529–545.

Gao, J., Klemd, R., Zhang, L., Wang, Z., Xiao, X., 1999. P-T path of high-pressure/low-temperature rocks and tectonic implications in the western Tianshan Mountains, NW China. J. Metamorph. Geol. 17 (6), 621–636.

Gao, S., Rudnick, R.L., Yuan, H.L., Liu, X.M., Liu, Y.S., Xu, W.L., Ling, W.L., Ayers, J., Wang, X.C., Wang, Q.H., 2004. Recycling lower continental crust in the North China craton. Nature 432 (7019), 892–897.

Hacker, B.R., Abers, G.A., Peacock, S.M., 2003. Subduction factory—1. Theoretical mineralogy, densities, seismic wave speeds, and H_2O contents. J. Geophys. Res. Solid Earth 108 (B1). doi:10.1029/ 2001JB00112.

Hartmann, G., 1994. Late-medieval glass manufacture in the Eichsfeld Region (Thuringia, Germany). Chem. der Erde 54, 103–128.

Hattori, K.H., Guillot, S., 2003. Volcanic fronts form as a consequence of serpentinite dehydration in the forearc mantle wedge. Geology 31 (6), 525–528.

Henry, C., Burkhard, M., Goffe, B., 1996. Evolution of synmetamorphic veins and their wallrocks through a western Alps transect: no evidence for large-scale fluid flow. Stable isotope, major- and trace-element systematics. Chem. Geol. 127 (1–3), 81–109.

Ionov, D.A., Hofmann, A.W., 1995. Nb-Ta-rich mantle amphiboles and micas—implications for subduction-related metasomatic trace-element fractionations. Earth Planet. Sci. Lett. 131 (3–4), 341–356.

Jochum, K.P., Dingwell, D.B., Rocholl, A., Stoll, B., Hofmann, A.W., Becker, S., Besmehn, A., Bessette, D., Dietze, H.J., Dulski, P.,

Erzinger, J., Hellebrand, E., Hoppe, P., Horn, I., Janssens, K., Jenner, G.A., Klein, M., McDonough, W.F., Maetz, M., Mezger, K., Munker, C., Nikogosian, I.K., Pickhardt, C., Raczek, I., Rhede, D., Seufert, H.M., Simakin, S.G, Sobolev, A.V., Spettel, B., Straub, S., Vincze, L., Wallianos, A., Weckwerth, G., Weyer, S., Wolf, D., Zimmer, M., 2000. The preparation and preliminary characterisation of eight geological MPI-DING reference glasses for in-site microanalysis. Geostandards Newsl. J. Geostandards Geoanal. 24 (1), 87–133.

Jochum, K.P., McDonough, W.F., Palme, H., Spettel, B., 1989. Compositional constraints on the continental lithospheric mantle from trace elements in spinel peridotite xenoliths. Nature 340, 548.

Jochum, K.P., Seufert, H.M., Spettel, B., Palme, H., 1986. The solar system abundances of Nb, Ta and Y and the relative abundances of refractory lithophile elements in differentiated planetary bodies. Geochim. Cosmochim. Acta 50, 1173–1183.

Kamber, B.S., Collerson, K.D., 2000. Role of 'hidden' deeply subducted slabs in mantle depletion. Chem. Geol. 166 (3–4), 241–254.

Kamber, B.S., Ewart, A., Collerson, K.D., Bruce, M.C., McDonald, G.D., 2002. Fluid-mobile trace element constraints on the role of slab melting and implications for Archaean crustal growth models. Contrib. Mineral. Petrol. 144 (1), 38–56.

Klemme, S., Prowatke, S., Hametner, K., Gunther, D., 2005. Partitioning of trace elements between rutile and silicate melts: implications for subduction zones. Geochim. Cosmochim. Acta 69 (9), 2361–2371.

Li, S., Jagoutz, E., Chen, Y., Li, Q., 2000. Sm-Nd and Rb-Sr isotopic chronology and cooling history of ultrahigh pressure metamorphic rocks and their country rocks at Shuanghe in the Dabie Mountains, central China. Geochim. Cosmochim. Acta 64 (6), 1077–1093.

Li, S., Wang, S., Chen, Y., Liu, D., Ji, Q., Zhou, H., Zhang, Z., 1994. Excess argon in phengite from eclogite: evidence from dating of eclogite minerals by Sm-Nd, Rb-Sr and $^{40}Ar/^{39}Ar$ methods. Chem. Geol. 112 (3–4), 343–350.

Li, S., Xiao, Y., Liou, D., Chen, Y., Ge, N., Zhang, Z., Sun, S.S., Cong, B., Zhang, R., Hart, S.R., Wang, S., 1993. Collision of the North China and Yangtse blocks and formation of coesite-bearing eclogites: timing and processes. Chem. Geol. 109 (1–4), 89–111.

Li, X.H., Li, Z.X., Zhou, H.W., Liu, Y., Kinny, P.D., 2002. U-Pb zircon geochronology, geochemistry and Nd isotopic study of Neoproterozoic bimodal volcanic rocks in the Kangdian Rift of South China: implications for the initial rifting of Rodinia. Precambrian Res. 113 (1–2), 135–154.

Liou, J.G, Zhang, R.Y., 1995. Significance of ultrahigh-P talc-bearing eclogitic assemblages. Mineral. Mag. 59 (394), 93–102.

Liou, J.G, Zhang, R.Y., Ernst, W.G., Liu, J., McLimans, R., 1998. Mineral parageneses in the Piampaludo eclogitic body, Gruppo di Voltri, Western Ligurian Alps. Schweiz. Mineral. Petrograph. Mitteilungen 78 (2), 317–335.

Liou, J.G, Zhang, R.Y., Jahn, B.M., 2000. Petrological and geochemical characteristics of ultrahigh-pressure metamorphic rocks from the Dabie-Sulu terrane, east-central China. Int. Geol. Rev. 42 (4), 328–352.

Marks, M., Halama, R., Wenzel, T., Markl, G., 2004. Trace element variations in clinopyroxene and amphibole from alkaline to peralkaline syenites and granites: implications for mineral-melt trace-element partitioning. Chem. Geol. 211 (3–4), 185–215.

Martin, H., 1999. Adakitic magmas: modern analogues of Archaean granitoids. Lithos 46 (3), 411–429.

McDonough, W.F., 1991. Partial melting of subducted oceanic-crust and isolation of its residual eclogitic lithology. Philos. Trans. R. Soc. Lond. Math. Phys. Eng. Sci. 335 (1638), 407–418.

McDonough, W.F., Sun, S.S., 1995. The composition of the Earth. Chem. Geol. 120 (3–4), 223–253.

Munker, C., Pfander, J.A., Weyer, S., Buchl, A., Kleine, T., Mezger, K., 2003. Evolution of planetary cores and the earth-moon system from Nb/Ta systematics. Science 301 (5629), 84–87.

Munker, C., Worner, G., Yogodzinski, G., Churikova, T., 2004. Behaviour of high field strength elements in subduction zones: constraints from Kamchatka-Aleutian arc lavas. Earth Planet. Sci. Lett. 224 (3–4), 275–293.

Niu, Y.L., Batiza, R., 1997. Trace element evidence from seamounts for recycled oceanic crust in the eastern Pacific mantle. Earth Planet. Sci. Lett. 148 (3–4), 471–483.

Okay, A.I., 1995. Paragonite eclogites from Dabie-Shan, China—reequilibration during exhumation. J. Metamorph. Geol. 13 (4), 449–460.

Pearce, J.A., Peate, D.W., 1995. Tectonic implications of the composition of volcanic arc magmas. Annu. Rev. Earth Planet. Sci. 23, 251–285.

Plank, T., Langmuir, C.H., 1998. The chemical composition of subducting sediment and its consequences for the crust and mantle. Chem. Geol. 145 (3–4), 325–394.

Prowatke, S., Klemme, S., 2005. Effect of melt composition on the partitioning of trace elements between titanite and silicate melt. Geochim. Cosmochim. Acta 69 (3), 695–709.

Rapp, R.P., Shimizu, N., Norman, M.D., 2003. Growth of early continental crust by partial melting of eclogite. Nature 425 (6958), 605–609.

Rapp, R.P., Watson, E.B., 1995. Dehydration melting of metabasalt at 8–32 kbar—implications for continental growth and crust-mantle recycling. J. Petrol. 36 (4), 891–931.

Reynaud, C., Jaillard, E., Lapierre, H., Mamberti, M., Mascle, G.H., 1999. Oceanic plateau and island arcs of southwestern Ecuador:

their place in the geodynamic evolution of northwestern South America. Tectonophysics 307 (3–4), 235–254.
Rudnick, R.L., 1995. Making continental-crust. Nature 378 (6557), 571–578.
Rudnick, R.L., Barth, M., Horn, I., McDonough, W.F., 2000. Rutilebearing refractory eclogites: missing link between continents and depleted mantle. Science 287 (5451), 278–281.
Rudnick, R.L., Fountain, D.M., 1995. Nature and composition of the continental-crust—a lower crustal perspective. Rev. Geophys. 33(3), 267–309.
Schmidt, M.W., Dardon, A., Chazot, G., Vannucci, R., 2004. The dependence of Nb and Ta rutile-melt partitioning on melt composition and Nb/Ta fractionation during subduction processes. Earth Planet. Sci. Lett. 226 (3–4), 415–432.
Schmidt, M.W., Poli, S., 1998. Experimentally based water budgets for dehydrating slabs and consequences for arc magma generation. Earth Planet. Sci. Lett. 163, 361–379.
Shannon, R.D., 1976. Revised effective ionic-radii and systematic studies of interatomic distances in halides and chalcogenides. Acta Crystallographica Section A 32(SEP1), 751–767.
Smithies, R.H., 2000. The Archaean tonalite–trondhjemite–granodiorite (TTG) series is not an analogue of Cenozoic adakite. Earth Planet. Sci. Lett. 182 (1), 115–125.
Spandler, C., Hermann, J., Arculus, R., Mavrogenes, J., 2003. Redistribution of trace elements during prograde metamorphism from lawsonite blueschist to eclogite facies, implications for deep subduction-zone processes. Contrib. Mineral. Petrol. 146 (2), 205–222.
Stalder, R., Foley, S.F., Brey, G.P., Horn, I., 1998. Mineral aqueous fluid partitioning of trace elements at 900–1200 degrees C and 3.0–5.7 GPa: new experimental data for garnet, clinopyroxene, and rutile, and implications for mantle metasomatism. Geochim. Cosmochim. Acta 62 (10), 1781–1801.
Sun, S.S., McDonough, W.F., 1989. Chemical and isotopic systematics of oceanic basalts; implications for mantle composition and processes. In: Saunders, A.D., Norry, M.J. (Eds.), Magmatism in the Ocean Basins, vol. 42. Geological Society of London, pp. 313–345.
Sun, W., Bennett, V.C., Eggins, S.M., Arculus, R.J., Perfit, M.R., 2003. Rhenium systematics in submarine MORB and back-arc basin glasses: laser ablation ICP-MS results. Chem. Geol. 196 (1–4), 259–281.
Sun, W.D., 2003. The Subduction Factory, a Perspective from Rhenium and Trace Element Geochemistry of Oceanic Basalts and Eclogites. Ph.D. Thesis, Australian National University.
Sun, W.D., Williams, I.S., Li, S.G., 2002. Carboniferous and Triassic eclogites in the western Dabie Mountians, east-central China: evidence for protracted convergence of the North and South China Blocks. J. Metamorph. Geol. 20, 873–886.
Taylor, S.R., McLennan, S.M., 1985. The Continental Crust: Its Composition and Evolution. Blackwell, Oxford.
Tsujimori, T., 2002. Prograde and retrograde P-T paths of the late Paleozoic glaucophane eclogite from the Renge metamorphic belt, Hida Mountains, southwestern Japan. Int. Geol. Rev. 44 (9), 797–818.
Weyer, S., Munker, C., Mezger, K., 2003. Nb/Ta, Zr/Hf and REE in the depleted mantle: implications for the differentiation history of the crust-mantle system. Earth Planet. Sci. Lett. 205 (3–4), 309–324.
Witt-Eickschen, G., Harte, B., 1994. Distribution of trace-elements between amphibole and clinopyroxene from mantle peridotites of the Eifel (Western Germany)—an ion-microprobe study. Chem. Geol. 117 (1–4), 235–250.
Xiao, Y.L., Hoefs, J., Kronz, A., 2005. Compositionally zoned Cl-rich amphiboles from North Dabie Shan, China: monitor of high-pressure metamorphic fluid/rock interaction processes. Lithos 81 (1–4), 279–295.
Xiao, Y.L., Hoefs, J., van den Kerkhof, A.M., Fiebig, J., Zheng, Y., 2000. Fluid history of UHP metamorphism in Dabie Shan, China; a fluid inclusion and oxygen isotope study on the coesite-bearing eclogite from Bixiling. Contrib. Miner. Petrol. 139 (1), 1–16.
Xiao, Y.L., Hoefs, J., van den Kerkhof, A.M., Simon, K., Fiebig, J., Zheng, Y.F., 2002. Fluid evolution during HP and UHP metamorphism in Dabie Shan, China: constraints from mineral chemistry, fluid inclusions and stable isotopes. J. Petrol. 43 (8), 1505–1527.
Xiong, X.L., Adam, J., Green, T.H., 2005. Rutile stability and rutile/melt HFSE partitioning during partial melting of hydrous basalt: implication for TTG genesis. Chem. Geol. 218 (3–4), 339–359.
Xu, S., Okay, A.I., Ji, S., Sengör, A.M.C., Su, W., Liu, Y., Jiang, L., 1992. Diamond from the Dabie Shan metamorphic rocks and its implication for tectonic setting. Science 256 (5053), 80–82.
Xu, Z.Q., 2004. The scientific goals and investigation progresses of the Chinese Continental Scientific Drilling Project. Acta Petrol. Sin. 20 (1), 1–8.
Xu, Z.Q., Yang, W.C., Zhang, Z.M., Yang, T.N., 1998. Scientific significance and site-selection researches of the first Chinese Continental Scientific Deep Drillhole. Cont. Dyn. 3, 1–13.
Yang, J.J., Jahn, B.M., 2000. Deep subduction of mantle-derived garnet peridotites from the Su-Lu UHP metamorphic terrane in China. J. Metamorph. Geol. 18 (2), 167–180.

Yui, T.F., Rumble, D., Lo, C.H., 1995. Unusually low delta-O-18 ultrahigh-pressure metamorphic rocks from the Sulu Terrain, Eastern China. Geochim. Cosmochim. Acta 59 (13), 2859–2864.

Zack, T., Kronz, A., Foley, S.F., Rivers, T., 2002. Trace element abundances in rutiles from eclogites and associated garnet mica schists. Chem. Geol. 184 (1–2), 97–122.

Zack, T., von Eynatten, H., Kronz, A., 2004. Rutile geochemistry and its potential use in quantitative provenance studies. Sediment. Geol. 171 (1–4), 37–58.

Zhang, R.Y., Hirajima, T., Banno, S., Cong, B., Liou, J.G., 1995. Petrology of ultrahigh-pressure rocks from the southern Su-Lu region, eastern China. J. Metamorph. Geol. 13 (6), 659–675.

Zhang, Z.M., Xu, Z.Q., Liu, F.L., You, Z.D., Shen, K., Yang, J.S., Li, T.F., Chen, C.Z., 2004. Geochemistry of eclogites from the main hole (100 to 2050 m) of the Chinese Continental Scientific Drilling Project. Acta Petrol. Sin. 20 (1), 27–42.

Zhao, R.X., Liou, J.G., Zhang, R.Y., Wooden, J.L., 2005. SHRIMP U-Pb dating of zircon from the Xugou UHP eclogite, Sulu terrane, eastern China. Int. Geol. Rev. 47 (8), 805–814.

Zheng, Y.F., Fu, B., Gong, B., Li, L., 2003. Stable isotope geochemistry of ultrahigh pressure metamorphic rocks from the Dabie–Sulu orogen in China: implications for geodynamics and fluid regime. Earth Sci. Rev. 62 (1–2), 105–161.

Zheng, Y.F., Fu, B., Li, Y.L., Xiao, Y.L., Li, S.G, 1998. Oxygen and hydrogen isotope geochemistry of ultrahigh-pressure eclogites from the Dabie Mountains and the Sulu terrane. Earth Planet. Sci. Lett. 155 (1–2), 113–129.

致　　谢

　　本论文集《李曙光院士论文选集（卷一）》经过三个多月的整理、归纳，最终汇编成册。论文集编撰中得到许多老师和同学的支持，大家解决了诸多问题，如概括论文的亮点介绍、重绘年代久远论文的图件、逐一对照修订原文并统一排版格式等。在此，我们衷心感谢研究生马海波、王沛杰、王泽宁、王萃平、王寅文、王晰、王照雪、卢卓、朱政、刘文然、刘宇晨、刘纯韬、刘春阳、孙媛滢、孙瑶、严倩倩、杜云峰、李丹妮、李东旭、李佳美、李欣、李孟伦、李璐瑶、杨如意、杨诗湘、杨锡铭、汪洋、宋柯馨、张子达、张法礼、张寅楚、陈缘、范明峰、罗柏川、周文军、秦政、原成帅、徐勇、高雪、盛思彰、常慧、董栩含、舒梓坦、蔡荣华、潘寒以及中国科学技术大学相关老师和同学的帮助和支持。千言万语汇成一句话：谢谢你们。